ASTRONOMY AND ASTROPHYSICS ABSTRACTS

A Publication of the Astronomisches Rechen-Institut Heidelberg
Member of the Abstracting Board of the International
Council of Scientific Unions

Volume 13
Literature 1975, Part 1

Edited by S. Böhme U. Esser W. Fricke
U. Güntzel-Lingner I. Heinrich F. Henn D. Krahn
L. Schmadel H. Scholl G. Zech

Published for
Astronomisches Rechen-Institut
by
Springer-Verlag Berlin Heidelberg GmbH 1975

Astronomisches Rechen-Institut
Heidelberg
Director: Prof. Dr. W. Fricke

Astronomy and Astrophysics Abstracts
Editor-in-Chief: F. Henn

Astronomy and Astrophysics Abstracts
is prepared under the auspices
of the International Astronomical Union

ISBN 978-3-662-12300-3   ISBN 978-3-662-12298-3 (eBook)
DOI 10.1007/978-3-662-12298-3

Preface

Astronomy and Astrophysics Abstracts, which has appeared in semi-annual volumes since 1969, is devoted to the recording, summarizing and indexing of astronomical publications throughout the world. It is prepared under the auspices of the International Astronomical Union (according to a resolution adopted at the 14th General Assembly in 1970).

Astronomy and Astrophysics Abstracts aims to present a comprehensive documentation of literature in all fields of astronomy and astrophysics. Every effort will be made to ensure that the average time interval between the date of receipt of the original literature and publication of the abstracts will not exceed eight months. This time interval is near to that achieved by monthly abstracting journals, compared to which our system of accumulating abstracts for about six months offers the advantage of greater convenience for the user.

Volume 13 contains literature published in 1975 and received before August 15, 1975; some older literature which was received late and which is not recorded in earlier volumes is also included.

We acknowledge with thanks contributions to this volume by Dr. J. Bouška, who surveyed journals and publications in the Czech language and supplied us with abstracts in English, and by the Commonwealth Scientific and Industrial Research Organization (C.S.I.R.O.), Sydney, for providing titles and abstracts of papers on radio astronomy. We want to acknowledge valuable contributions to this volume by Zentralstelle für Atomkernenergie-Dokumentation, Leopoldshafen, which supported our abstracting service by sending us retrospective literature searches.

It is a pleasure to express our warmest thanks again to Miss Helga Ballmann, Mrs Monika Betz, Mrs Karola Gudé, Miss Lore Kiefert, and Mrs Ingrid Wolf, who typed the text of this volume on IBM 72 Composers and compiled the pages from abstract slips in a perfect form for offset reproduction, to Mrs Elisabeth Feigenbutz for punching material for the author index.

Heidelberg, September 1975

Siegfried Böhme
Ute Esser
Walter Fricke
Ulrich Güntzel-Lingner
Inge Heinrich

Frieda Henn
Dietlinde Krahn
Lutz Schmadel
Hans Scholl
Gert Zech

Contents

Positional Astronomy, Celestial Mechanics

Space Research

Theoretical Astrophysics

Sun

Earth

Planetary System

Stars

Introduction

Astronomical bibliographies

Astronomy and Astrophysics Abstracts begins documentation and abstracting as from the year 1969. For information on astronomical literature before this date consultation of one of the following bibliographies is suggested:

(1) J. J. de Lalande, Bibliographie Astronomique, Paris 1803 (this work covers the time from 480 B. C. to the year 1803, VIII + 966 pages).

(2) J. C. Houzeau, A. Lancaster, Bibliographie générale de l'astronomie, Volume I (in two parts), Bruxelles 1882, 1887, Volume II, Bruxelles 1889. The complete title of Volume II is "Bibliographie générale de l'astronomie ou catalogue méthodique des ouvrages, des mémoires et des observations astronomiques, publiés depuis l'origine de l'imprimerie jusqu'en 1880". A new edition of these volumes was prepared by D. W. Dewhirst in 1964.

(3) Bibliography of Astronomy, 1881 - 1898. The literature of this period was recorded on standard slips by the Observatoire Royal de Belgique. From the material (some 52,000 items) a microfilm version was produced by University Microfilms Limited, Tylers Green, High Wycombe, Buckinghamshire, England, in 1970.

(4) Astronomischer Jahresbericht, 1899 gegründet von Walter Wislicenus, herausgegeben vom Astronomischen Rechen-Institut in Heidelberg (formerly in Berlin), Verlag W. de Gruyter, Berlin. For the period from 1899 to 1968 sixty-eight volumes were published, each of which, in general, covers the literature of one year.

(5) Bulletin Signalétique – Section 120: Astronomie, Physique Spatiale, Géophysique. Published by Centre de Documentation du Centre National de la Recherche Scientifique, Paris. This publication is a continuation of "Bibliographie Mensuelle de l'Astronomie" founded in 1933 by the Société Astronomique de France. The publication is continued.

(6) Referativnyj Zhurnal. Founded in 1953 and published by Vsesoyuznyj Institut Nauchnoj i Tekhnicheskoj Informatsii, Akademiya Nauk, Moskva. The publication is continued.

Concept of Astronomy and Astrophysics Abstracts

This abstracting service aims to present a comprehensive documentation of the literature in all fields of astronomy and astrophysics. It appears in semi-annual volumes, two of which cover the literature of a calendar year. The half-yearly period of issue is regarded as an optimal period of time for summarizing papers into subject categories and for the presentation of abstracts as quickly as possible after the publication of the original literature. The time limits at which the documentation begins and ends for a volume are not sharply defined, except in the sense that all literature will be covered which was received by the editors within these limits.

Vol. 13 is devoted to the recording, summarizing and indexing of astronomical publications of the year 1975 received from January 1, 1975 to August 15, 1975; it also records a number of papers issued before 1975 but received within the period of time.

The main characteristics of the concept of Astronomy and Astrophysics Abstracts may be summarized briefly.

(1) Titles of papers are given in the language of their authors whenever possible. If they are not in English but supplied with English translations they will be given in English. Abstracts are presented in English, French or German. Titles of papers in Russian are given in English.

(2) Authors' abstracts are used whenever possible. As a rule, popular articles were not abstracted; however their titles are usually given with the notation "Popular article".

(3) As a rule, each paper has been classified into one of 108 numbered subject categories and allocated a serial number within the category. In this way each item is numbered by six figures, the first three of which indicate the number of the category. Three further figures indicate the serial number within the category, which was allocated in the order of the receipt of the abstract. Reference to an abstract in Volume 1 is indicated by "01" before the number of the category; for example, 01.074.028, denotes Volume 1, category 0.74, abstract 028, Vol. 2 is indicated by "02", etc., Vol. 13 by "13".

A paper may have been classified into more than one category. Then its abstract has been allocated a number in one of the categories involved, and in the other category (or categories) the paper has been indicated by the title and a reference to the abstract number.

Papers whose authors are not named were treated like those with authors' names, with one exception: reports from correspondents of journals whose names were unknown were not numbered.

(4) There are categories which suggest the presentation of the material in subject groups. For instance, a subject group may be formed by all information received on the same solar eclipse, comet, nova, etc. The unsorted presentation of such material in a subject category would be inconvenient for the user, even if the individual comet, etc. were included in the subject index.

The following subject categories are subdivided into subject groups:

008 Observatories, Institutes. The publications of observatories and astronomical institutes are listed in alphabetical order of the towns of the institutions, each town forming a numbered subject group.

010 Societies, Associations, Organizations. The publications of each one form a subject group. The groups are presented in alphabetical order.

079 Solar eclipses. All publications related to one solar eclipse form a subject group.

103 Comets: Listed Objects. All publications related to the same comet form a numbered group.

124 Novae. All publications related to one nova form a subject group.

125 Supernovae. All publications related to one supernova form a subject group.

(5) Border fields of astronomy and astrophysics have been taken into account by presenting titles of papers occasionally without abstracts. The selection of papers for inclusion has been made according to the degree of relevance to astronomical research.

Transliteration of the Russian alphabet

The transliteration of the Russian alphabet in use in Astronomy and Astrophysics Abstracts is presented here.

А	а	a	Р	р	r
Б	б	b	С	с	s
В	в	v	Т	т	t
Г	г	g	У	у	u
Д	д	d	Ф	ф	f
Е	е	e	Х	х	kh
Ё	ё	e	Ц	ц	ts
Ж	ж	zh	Ч	ч	ch
З	з	z	Ш	ш	sh
И	и	i	Щ	щ	shch
Й	й	j	Ъ	ъ	″
К	к	k	Ы	ы	y
Л	л	l	Ь	ь	′
М	м	m	Э	э	eh
Н	н	n	Ю	ю	yu
О	о	o	Я	я	ya
П	п	p			

This transliteration was recommended by the Abstracting Board of the International Council of Scientific Unions in 1969. It is essentially the same as the transliteration proposed by the Academy of Sciences, Moscow, and used by the Referativnyj Zhurnal. It may be noted that the letters can be read and printed by usual data processing machines.

If the names of Russian authors in the literature are transliterated very different from this scheme we present the names in the form in which they are given in the references cited and in addition in round brackets according to our transliteration table.

Sources of information

The majority of sources of information for this volume are given in section **001 Periodicals** and in section **008 Observatories, Institutes**. The term "periodical" has been used in its widest sense for publications in a sequence of undetermined duration, even if the intervals of appearance are not regular. Section 001 records 294 periodicals with their full titles and with abbreviations which are in use in Astronomy and Astrophysics Abstracts. It may be noted that the titles of the periodicals are given in their original languages, and that Russian titles have been transliterated applying the transliteration given above. Section 008 records 125 periodicals; these are publication series of observatories and astronomical institutes which have not been included in section 001. The abbreviations of the titles of the periodicals have been given so that in most cases they permit recognition of the full title without recourse to the key in section 001. The steadily growing number of periodicals makes it necessary to use more extensive abbreviations and to abandon the use of very condensed ones.

Other abstracting journals have been consulted in order to examine the degree of completeness of our service. Occasionally, in particular in Physics Abstracts, Referativnyj Zhurnal, and Bulletin Signalétique abstracts of papers were found which had not come to our attention. In such cases Astronomy and Astrophysics Abstracts cites these paper and gives in general reference to the abstracting service which acted as the source.

Classification into a scheme of subject categories

The subdivision of astronomy and its border fields into sub-ject categories is facilitated by the fact that the astronomical objects appear to be particularly well suited for the formation of categories. Sun, moon, earth, planets, comets, and meteorites, the various kinds of stars, galaxies, radio sources, quasars, and pulsars etc. suggest natural subdivisions. It may be assumed that such subdivisions can be maintained for long periods of time. Experience shows, however, that progress in research may imply changes in the classification scheme, in particular, in fields where the expansion of knowledge is explosive.

A few explanatory remarks may be in order on some of the subject categories. In section 003 books on astronomy and astrophysics and its border fields are listed which came to our notice from January 1975 to August 1975. References to book reviews are given if the review appeared quickly.

For completeness of documentation, personal notes (section 006) and obituaries (section 007) are listed. In section 012 (Proceedings of Colloquia, Congresses, Meetings, and Symposia) the proceedings etc. are listed with titles and editors. The individual papers are classified into their corresponding subject categories, but mostly not included in the subject index. The main subjects of the symposia are cited in the index under section 012.

Errata to papers communicated by the authors are listed at the end of the corresponding subject categories.

Author index and subject index

The subject category and the serial number forming six figures for each abstract have been used as a means of reference in the author index and the subject index. These references are more precise than page references. They offer considerable advantages in indexing by means of data processing machines, and they are more convenient for the user.

The author index of this volume contains 7887 names. A complete reference comprises six figures, three for the subject category and three for the serial number within the category. In the case of more than one reference to abstracts in one category, the number of the category is given only once and not repeated in the immediately following references. The total number of papers (some do not give names of authors) recorded in this volume is about 6450.

We consider the subject index as an approximation to an optimal index covering all fields of astronomy and astrophysics and their border fields. The assigning of one or more key words to a paper is undoubtedly a difficult task. Some journals have started giving key words together with the titles of papers. These key words are chosen by the authors themselves and are in many cases identical with our designations of subject categories with no additional specification. In fact, in some cases it may be more useful to refer to a subject category as a whole than to an item number, in particular, if the total number of abstracts in a category is very small, and if more specific key words do not provide a proper description of the paper.

While each volume is scheduled to contain an author index and a subject index, the magnetic tapes containing the index information will be used to produce separate index volumes (authors and subjects) at intervals of a few years.

The text of the publication was typed on IBM 72 Composers in the editorial office, and it was given to the printer in a form ready for offset reproduction. For the preparation of the indexes a new sorting program has been developed which is based on the IBM SORT-Program. The introduction of small and capital letters in the layout caused some difficulties. Special programs had to code the capital letters into small ones. For the layout a TN chain for a 1403 IBM high speed printer was used. All the programs are written in PL/I. The computations are carried out on an IBM 370/168.

Abbreviations

AAS	American Astronomical Society		Geogr.	Geography, etc.
AAVSO	American Association of Variable Star Observers		Geophys.	Geophysics, etc.
			Ges.	Gesellschaft
Abh.	Abhandlungen		Glav.	Glavnyj (Main)
Abstr.	Abstract		Gos.	Gosudarstvennyj (State)
Abt.	Abteilung		HRD	Herzsprung-Russell diagram
Acad.	Academy, etc.		Hydrogr.	Hydrography, etc.
Accad.	Accademia		IAF	International Astronautical Federation
Adv.	Advances		IAU	International Astronomical Union
AG	Astronomische Gesellschaft		ICSU	International Council of Scientific Unions
AIAA	American Institute of Aeronautics and Astronautics		IEEE	Institute of Electrical and Electronics Engineers
AJB	Astronomischer Jahresbericht		Industr.	Industry, etc.
Akad.	Akademie		Inform.	Information
An.	Anales, etc.		Inst.	Institute, etc.
Ann.	Annals, etc.		Instn.	Institution
Arch.	Archiv, etc.		Ionosph.	Ionosphere, etc.
Ark.	Arkiv		Issled.	Issledovaniya (Research)
ASA	Astronomical Society of Australia		Ist.	Istituto
Asoc.	Asociación		Izv.	Izvestiya (News)
ASP	Astronomical Society of the Pacific		Jb.	Jahrbuch
Ass.	Association		JO	Journal des Observateurs
ASSA	Astronomical Society of Southern Africa		Journ.	Journal
Astrofis.	Astrofisica, etc.		Kl.	Klasse
Astrofiz.	Astrofizika, etc.		Lab.	Laboratory
Astron.	Astronomy, etc.		Mag.	Magazine
Astronaut.	Astronautics, etc.		Mat.	Matematica, etc.
Astrophys.	Astrophysics, etc.		Math.	Mathematics, etc.
ASV	Astronomical Society of Victoria		Mech.	Mechanics, etc.
ASWA	Astronomical Society of Western Australia		Med.	Mededelingen
Atmosph.	Atmosphere, etc.		Medd.	Meddelande, Meddelser
BA	Bulletin Astronomique		Mekhan.	Mekhanika, etc.
BAA	British Astronomical Association		Mém.	Mémoires
BAN	Bulletin of the Astronomical Institutes of the Netherlands		Mem.	Memoirs, Memorandum, etc.
			Meteorol.	Meteorology, etc.
Ber.	Berichte		MIT	Massachusetts Institute of Technology
BIH	Bureau International de l'Heure (Paris)		Mitt.	Mitteilungen
Bol.	Boletin		MVS Sonneberg	Mitteilungen über Veränderliche Sterne, Sonneberg
Boll.	Bolletino			
Bull.	Bulletin		Nachr.	Nachrichten
Byull.	Byulleten' (Bulletin)		NASA	National Aeronautics and Space Administration
Circ.	Circular			
Cl.	Classe		Nat.	Naturwissenschaftlich, etc.
Coll.	Collection		Naut.	Nautics, etc.
Commun.	Communication		NBS	National Bureau of Standards
Comun.	Comunicazioni		NRAO	National Radio Astronomy Observatory (Green Bank)
Contr.	Contributions, etc.			
COSPAR	Committee on Space Research		NRL	Naval Research Laboratory (Washington)
C.S.I.R.O.	Commonwealth Scientific Industrial Research Organization		Obs.	Observatory, etc.
			OSA	Optical Society of America
Dep.	Department		Oss.	Osservatorio, Osservazioni, etc.
Diss.	Dissertation		Ped.	Pedagogika, etc. (Pedagogics)
Div.	Division		Phil.	Philosophical
Dokl.	Doklady (Reports)		Phys.	Physics, etc.
ESO	European Southern Observatory		Planet.	Planetary
ESRO	European Space Research Organization		Priklad.	Prikladnoj (Applied)
Fis.	Fisica, etc.		Proc.	Proceedings
Fiz.	Fizika, etc.		Progr.	Progress, etc.
Fys.	Fysica, etc.		Pubbl.	Pubblicazioni
Géod.	Géodésie, etc.		Publ.	Publications
Geod.	Geodesy, etc.		Rap.	Raportoj
Geofis.	Geofisica, etc.		RAS	Royal Astronomical Society
Geofiz.	Geofizika, etc.		RAS Canada	Royal Astronomical Society of Canada
Geofys.	Geofysik, etc.		Rech.	Recherches
Geol.	Geology, etc.		Rend.	Rendiconti

Abbreviations

Rep.	Report	Techn.	Technics, etc.
Repr.	Reprint	Tekhn.	Tekhnika, etc.
Res.	Research	Teor.	Teoreticheskij
Rev.	Review, etc.	Terr.	Terrestrial, etc.
Ric.	Ricerche	TH	Technische Hochschule
Roy.	Royal, etc.	Theor.	Theoretical
SAF	Société Astronomique de France	Tidssk.	Tidsskrift
SAI	Società Astronomica Italiana	Trans.	Transactions
SAO	Smithsonian Astrophysical Observatory	Trudy	Trudy (Publications)
SAS	Société Astronomique de Suisse	Tsentr.	Tsentral'nyj (Central)
Sci.	Science, etc.	Tsirk.	Tsirkulyar (Circular)
Sect.	Section	TU	Technical University
Ser.	Series, etc.	Uch. Zap.	Uchenye Zapiski (Treatise)
S. I. R.	Service International Rapide des Latitudes	Univ.	University, etc.
Sitz.-Ber.	Sitzungsberichte	URSI	Union Radio Scientifique Internationale
Soc.	Society	Verh.	Verhandlungen
Soobshch.	Soobshcheniya (Communications)	Veröff.	Veröffentlichungen
Sternw.	Sternwarte	Wet.	Wetenschappen
Stud. Cerc.	Studii şi Cercetari	Wiss.	Wissenschaften, etc.
Supl.	Suplemento	Zeitschr.	Zeitschrift
Suppl.	Supplement	ZfA	Zeitschrift für Astrophysik
SuW	Sterne und Weltraum	Zhurn.	Zhurnal (Journal)

Periodicals, Proceedings, Books, Activities

001 Periodicals

AAS Photo-Bull.
AAS (American Astronomical Society) Photo-Bulletin. Published by the Working Group on Photographic Materials. Produced by Eastman Kodak Co., Rochester, N. Y.

Acad. Roy. Belgique, Bull. Cl. Sci.
Académie Royale de Belgique, Bulletin de la Classe des Sciences (Koninklijke Academie van België, Mededelingen van de Klasse der Wetenschappen). 5ᵉ Série. Palais des Académies, Bruxelles.

Acta Astron.
Acta Astronomica. Publisher: Komitet Astronomii, Polskiej Akademii Nauk, Warszawa - Kraków.

Acta Astronaut.
Acta Astronautica. Journal of the International Academy of Astronautics. Publisher: Pergamon Press Inc., Elmsford, New York, U.S.A.; Pergamon Press Ltd., Oxford, England.

Acta Astron. Sinica
Acta Astronomica Sinica. Published by Purple Mountain Observatory, Academia Sinica, Nanking, China.

Acta Cosmologica
Acta Cosmologica. Published by Obserwatorium Astronomiczne Universytetu Jagiellońskiego, Kraków, Poland.

Acta Geophys. Sinica
Acta Geophysica Sinica. Chinese Academy of Sciences, Department of Geophysical Research. Published by Science Press, Peking, Peoples Republic of China.

Acta Phys. Austriaca
Acta Physica Austriaca. Publisher: Springer-Verlag, Wien.

Actas Acad. Nacional Cienc. Lima
Actas de la Academia Nacional de Ciencias Exactas, Fisicas y Naturales de Lima. Lima - Peru.

Acta Univ. Carolinae Math. Phys.
Acta Universitatis Carolinae, Mathematica et Physica. Administrace: Matematicko-fyzikálni fakulta University Karlovy, Praha.

Adv. Astron. Astrophys.
Advances in Astronomy and Astrophysics. Publisher: Academic Press, New York – London.

AIAA Journ.
AIAA Journal. A Publication of the American Institute of Aeronautics and Astronautics devoted to Aerospace Research and Development. Published by the American Institute of Aeronautics and Astronautics, New York, N.Y

American Scient.
American Scientist. Society of Sigma Xi, New Haven, Conn.

Ann. Françaises Chronométrie Micromécanique
Annales Françaises de Chronométrie et de Micromécanique, publication annuelle de l'Observatoire de Besançon, du Centre Technique de l'Industrie Horlogère et de la Société Française de Chronométrie et de Micromécanique. Rédaction et administration: Observatoire de Besançon. Publiées avec le concours du Centre National de la Recherche Scientifique et des organismes corporatifs.

Ann. Géophys.
Annales de Géophysique. Revue Internationale trimestrielle, publiée par le Centre National de la Recherche Scientifique, Paris.

Ann. Obs. Astron. Météorol. Toulouse
Annales de l'Observatoire Astronomique et Météorologique de Toulouse. Publisher: Gauthier-Villars, Paris.

Ann. Physics
Annals of Physics. Publisher: Academic Press Inc., New York, N.Y.

Ann. Physik
Annalen der Physik. 7. Folge. Publisher: Johann Ambrosius Barth, Leipzig.

Ann. Physique
Annales de Physique. Publisher: Masson et Cie., Paris.

Ann. Soc. Sci. Bruxelles
Annales de la Société Scientifique de Bruxelles. Série I: Sciences Mathématiques, Astronomiques et Physiques. Published by Institut de Physique, Heverlé-Louvain.

Annual Rep. Astron. Inst. Greece
Annual Reports of the Astronomical Institutes of Greece. Published by the Greek National Committee for Astronomy. Academy of Athens, Research Center for Astronomy and Applied Mathematics.

Annual Rev. Astron. Astrophys.
Annual Review of Astronomy and Astrophysics. Publisher: Annual Reviews Inc., Palo Alto, California.

Ann. Univ.-Sternw. Wien
Annalen der Universitäts-Sternwarte Wien. In Kommission bei Ferd. Dümmlers Verlag, Bonn.

Anzeiger. Österreich. Akad. Wiss. Math.-Nat. Kl.
Anzeiger. Österreichische Akademie der Wissenschaften. Mathematisch-Naturwissenschaftliche Klasse. Publisher: Springer-Verlag, Wien.

Applied Optics
Applied Optics. A monthly publication of the Optical Society of America. Published for the Optical Society of America by the American Institute of Physics, New York, N. Y.

Arch. Sci. Genève
Archives des Sciences, éditées par la Société de Physique et d'Histoire Naturelle de Genève. Publisher: Imprimerie Kundig, Genève. Subscription address: Librairie Payot, Genève.

Ark. Astron.
Arkiv för Astronomi. Utgivet av Kungliga Svenska Vetenskapsakademien, Stockholm. Printed by Almqvist & Wiksell, Stockholm.

Ark. Geofys.
Arkiv för Geofysik. Kungliga Svenska Vetenskapsakademien, Stockholm. Printed by Almqvist & Wiksell, Stockholm.

Artificial Satellites
Artificial Satellites. Publication of Polish Scientific Institutions. Polish Academy of Sciences, National Committee of Geophysics and Geodesy, National Committee for Space Research, Warsaw. Publishing Office: Palac Kultury i Nauki, Warszawa.

Asoc. Argentina Astron. Bol.
Asociación Argentina de Astronomía. Boletin. Editor: Instituto Argentino de Radioastronomía, Provincia de Buenos Aires, Argentina. Printer: Talleres Gráficos "Renovación", La Plata, República Argentina.

Astrofizika
Astrofizika. Izdatel'stvo Akademii Nauk Armyanskoj SSR, Erevan. [An English translation is published in "Astrophysics".]

Astrofiz. Issled. Izv. Spets. Astrofiz. Obs.
Astrofizicheskie Issledovaniya. Izvestiya Spetsial'noj Astrofizicheskoj Observatorii. Akademiya Nauk SSSR. Publishers: Izdatel'stvo "Nauka", Leningradskoe Otdelenie, Leningrad.

Astron. Astrophys.
Astronomy and Astrophysics. A European Journal. Published by Springer-Verlag, Berlin – Heidelberg – New York.

Astron. Astrophys. Suppl. Ser.
Astronomy and Astrophysics. Supplement Series. A European Journal. Published by the Astronomical Institute Lausanne and Geneva Observatory, Switzerland, on behalf of the Board of Directors.

Astronautik
Astronautik. Organ der Hermann-Oberth-Gesellschaft e.V. Astronautik-Verlag, Druckerei H. Brandt, Delmenhorst (Germany).

Astron. in der Schule
Astronomie in der Schule. Zeitschrift für die Hand des Astronomielehrers. Herausgegeben vom Verlag Volk und Wissen, Berlin. Redaktion: Sternwarte Bautzen.

Astron. Journ.
The Astronomical Journal. Published for the American Astronomical Society by the American Institute of Physics, New York, N.Y. Editorial Office: Department of Astronomy, Columbia University, New York, N.Y.

Astron. Nachr.
Astronomische Nachrichten. Publisher: Akademie-Verlag, Berlin.

Astron. Tidssk.
Astronomisk Tidsskrift. Edited by Astronomisk Selskab, København; Norsk Astronomisk Selskap, Oslo; Svenska Astronomiska Sällskapet, Stockholm. Printed by John Griegs Boktrykkeri, Bergen.

Astron. Tsirk.
Astronomicheskij Tsirkulyar, izdavaemyj Byuro Astronomicheskikh Soobshchenij Akademii Nauk SSSR. Tipografiya Astrosoveta AN SSSR, Moskva.

Astron. Vestn.
Astronomicheskij Vestnik. Publishers: Izdatel'stvo "Nauka", Moskva.

Astron. Zhurn. Akad. Nauk SSSR
Astronomicheskij Zhurnal. Akademiya Nauk SSSR. Publishers: Izdatel'stvo "Nauka", Moskva. [An English translation is published in "Soviet Astronomy AJ"].

Astrophysics
Astrophysics. A cover-to-cover translation of Astrofizika (USSR). Consultants Bureau, New York, N.Y.

Astrophys. Journ.
The Astrophysical Journal. Published for The American Astronomical Society by The University of Chicago Press, Chicago, Illinois.

Astrophys. Journ., (Letters)
The Astrophysical Journal. Letters to the Editors. Published for The American Astronomical Society by The University of Chicago Press, Chicago, Illinois.

Astrophys. Journ., Suppl. Ser.
The Astrophysical Journal. Supplement Series. Published for The American Astronomical Society by The University of Chicago Press, Chicago, Illinois.

Astrophys. Letters
Astrophysical Letters. An International EXPRESS Journal. Published monthly by Gordon and Breach Science Publishers Ltd., New York – London – Paris.

Astrophys. Space Sci.
Astrophysics and Space Science. An International Journal of Cosmic Physics. Published by D. Reidel Publishing Company, Dordrecht – Holland.

Atti Accad. Nazionale Lincei. Mem.
Atti della Accademia Nazionale dei Lincei. Serie Ottava. Memorie. Classe di Scienze fisiche, matematiche e naturali. Sezione I: Matematica, Meccanica, Astronomia, Geodesia e Geofisica. Published by Accademia Nazionale dei Lincei, Roma.

Atti Accad. Nazionale Lincei. Rend.
Atti della Accademia Nazionale dei Lincei. Serie Ottava. Rendiconti. Classe di Scienze fisiche, matematiche e naturali. Published by Accademia Nazionale dei Lincei, Roma.

Atti Soc. Astron. Italiana
Atti della Società Astronomica Italiana. Publisher: Tipografia Baccini & Chiappi, Firenze (Italy).

Australian Journ. Phys.
Australian Journal of Physics. Published by the Commonwealth Scientific and Industrial Research Organization, East Melbourne, Victoria.

Australian Journ. Phys. Astrophys. Suppl.
Australian Journal of Physics, Astrophysical Supplement. Published by Commonwealth Scientific and Industrial Research Organization, East Melbourne, Victoria.

BAV Rundbrief
BAV Rundbrief. Mitteilungsblatt der Berliner Arbeitsgemeinschaft für Veränderliche Sterne. Editor: BAV Berliner Arbeitsgemeinschaft für Veränderliche Sterne eV., Berlin.

BBSAG Bull.
Bedeckungsveränderlichen Beobachter der Schweizerischen Astronomischen Gesellschaft, [Swiss Astronomical Society's Eclipsing Variable Observers], Bulletin. To be obtained from R. Diethelm, Winterthur, Switzerland.

Bild der Wiss.
Bild der Wissenschaft. Zeitschrift über die Naturwissenschaften und die Technik in unserer Zeit. Publisher: Deutsche Verlagsanstalt, Stuttgart.

Bol. Inst. Mat., Astron., Fis. Univ. Nacional Córdoba
Boletin del Instituto de Matematica, Astronomia y Fisica, Universidad Nacional de Córdoba (R. A.).Dirección General de Publicaciones, Córdoba (Argentina).

Bol. Liga Latinoamericana Astron.
Boletin de la Liga Latinoamericana de Astronomia. Publicado por la Asociacion Argentina Amigos de la Astronomia, Buenos Aires, Argentina.

Boll. Geod. Sci. Affini
Bolletino di Geodesia e Scienze Affini. Pubblicazione dell'Istituto Geografico Militare, Firenze.

British Astron. Ass. Circ.
British Astronomical Association, Circular. Editorial Office: 97 Hawkswood Drive, Hailsham, Sussex.

Bull. American Astron. Soc.
Bulletin of the American Astronomical Society. Published for the American Astronomical Society by the American Institute of Physics Inc., New York, N. Y.

Bull. Astron. Inst. Czechoslovakia
Bulletin of the Astronomical Institutes of Czechoslovakia. Published under the auspices of the Czechoslovak Academy of Sciences by Academia, Praha. Editor: Astronomical Institutes of the Czechoslovak Academy of Sciences, Praha.

Bull. Astron. Soc. India
Bulletin of the Astronomical Society of India. Edited and published by M. S. Vardya, Tata Institute of Fundamental Research, Bombay on behalf of the Astronomical Society of India, Osmania University, Hyderabad.

Bull. Géod.
Bulletin Géodésique, being the Journal of the International Association of Geodesy. Nouvelle Série. Publié par le Bureau Central de l'Association Internationale de Géodésie, Paris.

Bull. Geograph. Survey Inst.
Bulletin of the Geographical Survey Institute. Published by the Geographical Survey Institute, Ministry of Construction, Tokyo, Japan.

Bull. Obs. Astron. Beograd
Bulletin de l'Observatoire Astronomique de Béograd.

Editor: Observatoire Astronomique de Béograd. Printed by Naucna delo, Béograd.

Bull. Signal.
Bulletin Signalétique. Section 120: Astronomie, Physique spatiale, Géophysique. Centre de Documentation du Centre Nationale de la Recherche Scientifique, Paris.

Bull. Signal.
Bulletin Signalétique. Bibliographie des Sciences de la Terre. Section 220, Cahier A: Minéralogie, Géochimie, Géologie extraterrestre. Centre de Documentation du C.N.R.S., Paris; Département Documentation du B.R. G.M., Orléans.

Bull. Soc. Roy. Sci. Liège
Bulletin de la Société Royale des Sciences de Liège. L'Université, Liège.

Byull. Abastuman. Astrofiz. Obs.
Abastumanskaya Astrofizicheskaya Observatoriya, Gora Kanobili. Byulleten'. Akademiya Nauk Gruzinskoj SSR. Publishers: Izdatel'stvo "Metsniereba", Tbilisi.

Byull. Stantsij Optichesk. Nablyud. Iskusstv. Sputnikov Zemli
Byulleten' Stantsij Opticheskogo Nablyudeniya Iskusstvennykh Sputnikov Zemli. Published by Astronomicheskij Sovet Akademii Nauk SSSR, Moskva.
Beginning with number 60 (1971) the title of the publication changed in Nablyudeniya Iskusstvennykh Nebesnykh Tel.

Canadian Journ. Phys.
Canadian Journal of Physics. Published by the National Research Council of Canada, Ottawa. Printed in Canada by the University of Toronto Press, Toronto, Ont.

Celestial Mechanics
Celestial Mechanics. An International Journal of Space Dynamics. Publishers: D. Reidel Publishing Company, Dordrecht—Holland.

Ciel et Terre
Ciel et Terre. Bulletin de la Société Belge d'Astronomie, de Météorologie et de Physique du Globe. Administration: Avenue Circulaire, 3, Bruxelles. Printed by Imprimerie R. Louis, Bruxelles.

Circ. d'Information
Circulaire d'Information. Union Astronomique Internationale. Commission des Etoiles Doubles. Address: Observatoire de Meudon, Meudon, France.

Coelum
Coelum. Periodico bimestrale per la Divulgazione dell' Astronomia. Editor: Osservatorio Astronomico Universitario di Bologna.

Comments Astrophys. Space Phys.
Comments on Astrophysics and Space Physics. A Journal of Critical Discussion of the Current Literature. Comments on Modern Physics: Part C. Publishers: Gordon and Breach Science Publishers, Inc., New York — London

Comptes Rendus Acad. Bulg. Sci.
Comptes Rendus de l'Académie bulgare des Sciences. (Doklady Bolgarskoj Akademii Nauk). Sofia.

Comptes Rendus Acad. Sci. Paris
Comptes Rendus hebdomadaires des Séances de l'Académie des Sciences, publié avec le concours du Centre

National de la Recherche Scientifique. Imprimerie: Gauthier-Villars, Paris.

Contr. Atmosph. Phys.
Contributions to Atmospheric Physics – Beiträge zur Physik der Atmosphäre. Publisher: Friedrich Vieweg & Sohn, Braunschweig.

COSPAR Inform. Bull.
COSPAR. Information Bulletin. Address: COSPAR Secretariat, Paris.

Deutsche Geod. Kommission Bayer. Akad. Wiss.
Deutsche Geodätische Kommission bei der Bayerischen Akademie der Wissenschaften. Reihe A: Höhere Geodäsie; Reihe B: Angewandte Geodäsie; Reihe C: Dissertationen; Reihe D: Tafelwerke; Reihe E: Geschichte und Entwicklung der Geodäsie. Published by Verlag der Bayerischen Akademie der Wissenschaften, München.

Dokl. Akad. Nauk
Doklady Akademii Nauk SSSR. Seriya Matematika, Fizika. Publishers: Izdatel'stvo "Nauka", Moskva.

Dunsink Obs. Publ.
Dunsink Observatory Publications. The Observatory of the School of Cosmic Physics, Dublin Institute for Advanced Studies, Dublin.

Earth Extraterr. Sci.
Earth and Extraterrestrial Sciences. Published by Gordon and Breach Science Publishers, London.

Earth Planet. Sci. Letters
Earth and Planetary Science Letters. A Letter Journal devoted to the Development in Time of the Earth and Planetary System. Publisher: North-Holland Publishing Company, Amsterdam.

Electronics (*USA*)
Electronics. Published by McGraw-Hill Publishing Company, New York, N.Y., U.S.A.

Electronics Letters (*GB*)
Electronics Letters. Published by Institution of Electrical Engineers, London, England.

El Universo
El Universo. Organo de la Sociedad Astronomica de Mexico, Mexico, D. F.

Endeavour
Endeavour. A review of the progress of science, published in four languages by Imperial Chemical Industries Limited, London.

ESO Bull.
European Southern Observatory, Bulletin. Edited by European Southern Observatory. Office of the Director: Hamburg.

ESO Techn. Rep.
European Southern Observatory, (ESO), Technical Report. Published by the European Southern Observatory Telescope Project Division, CERN, Geneva, Switzerland.

Feinwerktechn. & Messtechn.
F & M. Feinwerktechnik und Messtechnik. Fusion of "Feinwerktechnik" and "Messtechnik" (formerly Zeitschrift für Instrumentenkunde) beginning with Jahrgang

82, No. 5 (1974). Publishers: Karl Hanser Verlag, München, Germany.

Gaz. Astron. Mém.
Gazette Astronomique. Mémoires van het Sterrenkundig Genootschap van Antwerpen, (de la Société d'Astronomie d'Anvers), Antwerpen. Printer: «De Voorzorg», A. Van Leuvenhaege, Antwerpen.

General Relativ. Gravitation
General Relativity and Gravitation. Published under the auspices of the International Committee on General Relativity and Gravitation GRG. Publishing Office: Plenum Publishing Company Limited, London.

Geochim. Cosmochim. Acta
Geochimica et Cosmochimica Acta. Journal of the Geochemical Society. Publishing House: Pergamon Press, Ltd., Oxford.

Geodezja Kartografia
Geodezja i Kartografia. Komitet Geodezji Polskiej Akademii Nauk. Publisher: Państwowe Wydawnictwo Naukowe, Warszawa.

Geomagn. Aeronom.
Geomagnetizm i Aehronomiya. Akademiya Nauk SSSR. Izdatel'stvo "Nauka", Moskva [An English translation is published in "Geomagnetism and Aeronomy", American Geophysical Union, Washington, D.C.].

Geophys. Journ.
The Geophysical Journal of the Royal Astronomical Society. Published for the Royal Astronomical Society by Blackwell Scientific Publications, Oxford – Edinburgh.

Geophys. Res. Letters
Geophysical Research Letters. Published monthly by the American Geophysical Union, Washington, D.C., U.S.A.

Gerlands Beiträge Geophys.
Gerlands Beiträge zur Geophysik. Publisher: Akademische Verlagsgesellschaft Geest & Portig K.-G., Leipzig.

Glasnik Mat.
Glasnik Matematicki. Published by the Society of Mathematicians and Physicists of the S. R. of Croatia. Publisher: Drustvo Matematicara i Fizicara S. R. Hrvatske, Zagreb.

Helvetica Phys. Acta
Helvetica Physica Acta. Schweizerische Physikalische Gesellschaft. Publisher: E. Birkhäuser, Basel.

IAU Circ.
International Astronomical Union, Circular. Central Bureau for Astronomical Telegrams, Smithsonian Astrophysical Observatory, Cambridge, Mass.

Icarus
Icarus. International Journal of Solar System Studies. Publisher: Academic Press, New York – London.

ICSU Bull.
ICSU Bulletin. International Council of Scientific Unions. Secretariat: 51, Bd de Montmorency, Paris, France.

IEEE Spectrum
IEEE Spectrum. Published monthly by the Institute of Electrical and Electronics Engineers, Inc., New York, N. Y.

IEEE Trans. Aerospace Electron. Systems
IEEE Transactions on Aerospace and Electronic Systems. Published by the Institute of Electrical and Electronics Engineers, New York, N.Y., U.S.A.

IEEE Trans. Antennas Propagation
IEEE Transactions on Antennas and Propagation. Published by the Institute of Electrical and Electronics Engineers, New York, N.Y., U.S.A.

IEEE Trans. Electron Devices
IEEE Transactions on Electron Devices. Published by the Institute of Electrical and Electronics Engineers, New York, N.Y., U.S.A.

IEEE Trans. Microwave Theory Techn.
IEEE Transactions on Microwave Theory and Techniques. Published by the Institute of Electrical and Electronics Engineers, New York, N.Y., U.S.A.

Indian Journ. Pure and Applied Phys.
Indian Journal of Pure and Applied Phys. Council of Scientific and Industrial Research, New Delhi, India.

Inform. Bull. Southern Hemisphere
Information Bulletin for the Southern Hemisphere. Editorial Office: Observatorio Astronómico, La Plata, Argentina.

Inform. Bull. Variable Stars
Commission 27 of the I.A.U. Information Bulletin on Variable Stars. Konkoly Observatory, Budapest.

Infrared Physics
An International Research Journal. Publisher: Pergamon Press Ltd., Oxford – London – New York.

International Journ. Theor. Phys.
International Journal of Theoretical Physics. Publisher: Plenum Publishing Company, Donington House, London.

Irish Astron. Journ.
The Irish Astronomical Journal. A Quarterly Publication under the auspices of the Observatories of Armagh and Dunsink. Subscription address: Managing Editor, Irish Astronomical Journal, Armagh Observatory, Northern Ireland.

Izv. Akad. Nauk Armyan. SSR
Izvestiya Akademii Nauk Armyanskoj SSR. Fizika. Publisher: Izdatel'stvo AN Armyanskoj SSR, Erevan.

Izv. Glav. Astron. Obs. Pulkovo
Izvestiya Glavnoj Astronomicheskoj Observatorii v Pulkove. Akademiya Nauk SSSR. Izdanie Glavnoj astronomicheskoj observatorii v Pulkove, Leningrad.

Izv. Komissii Fiz. Planet
Izvestiya Komissii po Fizike Planet. Akademiya Nauk SSSR. Astronomicheskij Sovet. Moskva.

Izv. Krymskoj Astrofiz. Obs.
Izvestiya Krymskoj Astrofizicheskoj Observatorii. Akademiya Nauk SSR. Publishers: Izdatel'stvo "Nauka", Moskva.

Jenaer Rundschau (Jena Review)
Jenaer Rundschau (Jena Review). Publisher: VEB Verlag Technik, Berlin.

JETP Letters
JETP Letters. A translation of JETP Pis'ma v Redaktsiyu of the Academy of Sciences in the USSR. Published semimonthly by the American Institute of Physics, Lancaster, Pennsylvania.

Journ. Astronaut. Sci.
The Journal of the Astronautical Sciences. Published by the American Astronautical Society Inc., Alexandria, VA.

Journ. Astron. Soc. Victoria
The Journal of the Astronomical Society of Victoria. Printed by D. Buscombe Printers, Glen Waverley, Victoria.

Journ. Astron. Soc. Western Australia
The Journal of the Astronomical Society of Western Australia. Edited by the Astronomical Society of Western Australia, Perth, W. A.

Journ. Atmosph. Sci.
Journal of the Atmospheric Sciences. Published by the American Meteorological Society, Boston, Mass.

Journ. Atmosph. Terr. Phys.
Journal of Atmospheric and Terrestrial Physics. Publishers: Pergamon Press, Oxford – London – New York.

Journ. British Astron. Ass.
Journal of the British Astronomical Association. Subscription address: British Astronomical Association, Burlington House, Piccadilly, London.

Journ. British Interplanet. Soc.
Journal of the British Interplanetary Society. Printed in Great Britain by Unwin Brothers Ltd., The Gresham Press, Old Woking, Surrey, and published by The British Interplanetary Society, London.

Journ. Fluid Mechanics
Journal of Fluid Mechanics. Published by Cambridge University Press, London – New York.

Journ. Geophys.
Journal of Geophysics / Zeitschrift für Geophysik. Publisher: Springer-Verlag, Berlin–Heidelberg–New York

Journ. Geophys. Res.
Journal of Geophysical Research. An International Scientific Publication. Published three times a month by the American Geophysical Union, Washington, D. C. First section: Space Physics; Second section: Physics and chemistry of the solid earth, planetology, geodesy; Third section: Oceans and atmospheres.

Journ. History Astron.
Journal for the History of Astronomy. Publisher: Science History Publications Ltd., Chalfont St Giles, Buckinghamshire, England. American Representative: Neale Watson Academic Publications, Inc., New York City, U.S.A.

Journ. Navigation
The Journal of Navigation. Published quarterly by The Royal Institute of Navigation at the Royal Geographical Society, London.

Journ. Optical Soc. America
Journal of the Optical Society of America. Publisher: American Institute of Physics, New York.

Journ. Phys. A (Math., Nuclear, General)
Journal of Physics A, (Mathematical, Nuclear and
General). Europhysics Journal. Published by the Insti-
tute of Physics and Physical Society, London, England,
in association with the American Institute of Physics,
New York.

Journ. Physique
Journal de Physique. Publication de la Société Française
de Physique, Paris.

Journ. Plasma Phys.
Journal of Plasma Physics. Publishers: Cambridge
University Press, London.

Journ. Proc. Roy. Soc. New South Wales
Journal and Proceedings of the Royal Society of New
South Wales. Publisher: Science House, Sydney, N.S.W.
(Australia).

Journ. Quant. Spectrosc. Radiat. Transfer
Journal of Quantitative Spectroscopy & Radiative Transfer. Publisher: Pergamon Press, Oxford — New York.

Journ. Roy. Astron. Soc. Canada
The Journal of the Royal Astronomical Society of Canada, devoted to the advancement of astronomy and allied
sciences. Printed by the University of Toronto Press,
Toronto, Ontario, Canada.

Kometn. Tsirk. *Kiev*
Kometnyj Tsirkulyar. Gruppa po Issledovaniyu Komet
Astrosoveta i Mezhduvedomstvennyj Geofizicheskij
Komitet Akademii Nauk SSSR. Kievskij Universitet im.
T. G. Shevchenko.

Komety i Meteory
Komety i Meteory. Akademiya Nauk Tadzhikskoj SSR.
Astronomicheskij Sovet Akademii Nauk SSSR. Publish-
ers: Izdatel'stvo "Donish", Dushanbe.

Kosmich. Issled.
Kosmicheskie Issledovaniya. Akademiya Nauk SSSR.
Publishers: Izdatel'stvo "Nauka", Moskva [An English
translation is published as "Cosmic Research", Consultants Bureau, New York, N.Y.].

Kozmos
Kozmos. Popular Astronomical Journal of the Slovak
Central Observatory in Hurbanovo. Publisher: Slovenská
ústredná hvezdáren v Hurbanove.

L'Astronomie
L'Astronomie et Bulletin de la Société Astronomique de
France. Revue mensuelle. Rédaction: Société Astrono-
mique de France, Paris.

L'Universo
L'Universo. Rivista dell'Instituto Geografico Militare.
Direzione, Redazione e Amministrazione: Istituto Geo-
grafico Militare, Firenze.

Magnitnye Polya Solnech. Pyaten
Magnitnye Polya Solnechnykh Pyaten. (Supplements to
Solnechnye Dannye. Byulleten' (*Solar Data*)). Publishers:
Izdatel'stvo "Nauka", Leningrad.

Math. Rev.
Mathematical Reviews. Published by the American
Mathematical Society, Providence, R. I.

Mem. Fac. Sci. Kyoto Univ.
Memoirs of the Faculty of Science, Kyoto University.
Series of Physics, Astrophysics, Geophysics, and Chemis-
try. Printed by Yamashiro Printing Publishing Co. Ltd.,
Kamigyo, Kyoto.

Mem. Roy. Astron. Soc.
Memoirs of the Royal Astronomical Society. Published
for the Royal Astronomical Society by Blackwell Scientific Publications, Oxford — Edinburgh.

Mem. Soc. Astron. Italiana
Memorie della Società Astronomica Italiana. Nuova Serie. Pubblicate sotto gli auspici del Consiglio Nazionale
dell Ricerche. Publisher: Tipografia Baccini & Chiappi,
Firenze.

Mercury
Mercury. The Journal of the Astronomical Society of the
Pacific. Published by the Astronomical Society of the
Pacific, San Francisco, California.

Meteoritics
Meteoritics. The Journal of the Meteoritical Society. Pub-
lished quarterly by The Meteoritical Society and Arizona
State University Bureau of Publications. Editorial address:
Center for Meteorite Studies, The Arizona State Universi-
ty, Tempe, Arizona.

Meteoritika
Akademiya Nauk SSSR. Komitet po Meteoritam. Publishers: Izdatel'stvo "Nauka", Moskva.

Microwave Journ. (*USA*)
Microwave Journal. To be obtained from 610 Washington
Street, Dedham Plaza, Dedham, Massachusetts, U.S.A.

Microwaves (*USA*)
Microwaves. Published by Hayden Microwaves Corporation, New York, N.Y., U.S.A.

Minor Planet. Bull.
The Minor Planet Bulletin. Bulletin of the Minor Planets
Section of the Association of Lunar and Planetary
Observers. Editorial Office: R. G. Hodgson, Dordt
College, Sioux Center, Iowa, U.S.A.

Mitt. Astron. Ges.
Mitteilungen der Astronomischen Gesellschaft, Hamburg.
Printed by G. Braun, GmbH, Karlsruhe.

Monthly Notes Astron. Soc. Southern Africa
Monthly Notes of the Royal Astronomical Society of
Southern Africa. Published by the Astronomical Society
of Southern Africa, Royal Observatory, Cape Province,
South Africa.

Monthly Notes International Polar Motion Service
Monthly Notes of the International Polar Motion Service.
Published by the Central Bureau, International Latitude
Observatory of Mizusawa, Mizusawa-shi, Iwate-ken, Japan.

Monthly Notices Roy. Astron. Soc.
Monthly Notices of the Royal Astronomical Society.
Published for the Royal Astronomical Society by Black-
well Scientific Publications, Oxford — Edinburgh.

Moon
The Moon. An International Journal of Lunar Studies.
Publisher: D. Reidel Publishing Company, Dordrecht —
Holland.

MVS Sonneberg
Mitteilungen über Veränderliche Sterne. Edited by Sternwarte Sonneberg (Zentralinstitut für Astrophysik, Bereich Sternphysik) der Deutschen Akademie der Wissenschaften zu Berlin.

Nablyud. Iskusstv. Nebesn. Tel
Nablyudeniya Iskusstvennykh Nebesnykh Tel. Published by Astronomicheskij Sovet Akademii Nauk SSSR, Moskva.

Nachr. Akad. Wiss. Göttingen
Nachrichten der Akademie der Wissenschaften in Göttingen. II. Mathematisch-Physikalische Klasse. Vandenhoeck & Ruprecht, Göttingen.

Nachr. Karten-, Vermessungswesen
Nachrichten aus dem Karten- und Vermessungswesen. Editor: Institut für Angewandte Geodäsie (Abt. II des Deutschen Geodätischen Forschungsinstituts). Published by Verlag des Instituts für Angewandte Geodäsie, Frankfurt a. M.

Nature
Nature. Editorial and Publishing Offices: Macmillan Journals Limited, 4 Little Essex Street, London; 711 National Press Building, Washington, D. C.

Naturwissenschaften
Die Naturwissenschaften. Publisher: Springer-Verlag, Berlin – Heidelberg – New York.

Nauchn. Informatsii
Nauchnye Informatsii. Astronomicheskij Sovet Akademii Nauk SSSR, Moskva.

Nuovo Cimento
Il Nuovo Cimento. Rivista Internazionale e Organo della Società Italiana di Fisica, Series A, B. Publisher: Nicola Zanichelli, Editore, Bologna.

Nuovo Cimento Lettere
Lettere al Nuovo Cimento, a Cura della Società Italiana di Fisica. Editrice Compositori, Bologna.

Nuovo Cimento Suppl.
Supplemento al Nuovo Cimento. Publisher: Nicola Zanichelli, Editore, Bologna.

Observations Artificial Earth Satellites
Observations of Artificial Satellites of the Earth (Nablyudeniya Iskusstvennykh Sputnikov Zemli). Magyar Tudományos Akadémia Csillagvizsgáló Intézete, Budapest.

Observatory
The Observatory. A Review of Astronomy. Publishers: The Editors of "The Observatory", Royal Greenwich Observatory, Herstmonceaux Castle, Hailsham, Sussex, England.

Optik
Optik. Zeitschrift für das gesamte Gebiet der Licht- und Elektronenoptik. Publishers: Wissenschaftliche Verlagsgesellschaft mbH., Stuttgart.

Origins of Life
Origins of Life (Formerly Space Life Sciences). An International Journal. Publisher: D. Reidel Publishing Company, Dordrecht–Holland.

Orion
Orion. Zeitschrift der Schweizerischen Astronomischen Gesellschaft (SAG). Bulletin de la Société Astronomique de Suisse (SAS). Printed by A. Schudel & Co. AG, Riehen, Suisse.

Österreich. Zeitschr. Vermessungswesen
Österreichische Zeitschrift für Vermessungswesen und Photogrammetrie. Editor and Publisher: Österreichischer Verein für Vermessungswesen und Photogrammetrie, Wien, Austria.

Peremennye Zvezdy, Byull.
Peremennye Zvezdy, Byulleten', izdavaemyj Astronomicheskim Sovetom Akademii Nauk SSSR. Published by Astronomicheskij Sovet Akademii Nauk SSSR, Moskva.

Peremennye Zvezdy, Prilozhenie
Peremennye Zvezdy, Prilozhenie (The Variable Stars, Supplement). Astronomicheskij Sovet Akademii Nauk SSSR, Moskva.

Phil. Mag.
The Philosophical Magazine. A Journal of Theoretical, Experimental and Applied Physics. Eighth Series. Publisher: Taylor & Francis, Ltd., London.

Phil. Trans. Roy. Soc. London
Philosophical Transactions of the Royal Society of London. Series A, Mathematical and Physical Sciences. Published by the Royal Society, London.

Phys. Abstr.
Physics Abstracts. Science Abstracts, Series A. An INSPEC Publication, published by The Institution of Electrical Engineers, London.

Phys. Ber.
Physikalische Berichte. Herausgegeben von der Deutschen Physikalischen Gesellschaft e.V. und von der Deutschen Akademie der Wissenschaften zu Berlin. Physik-Verlag, Weinheim, West Germany.

Phys. Blätter
Physikalische Blätter. Physik-Verlag, Mosbach/Baden.

Phys. Bull.
Physics Bulletin. Published by the Institute of Physics and the Physical Society, London, England.

Phys. Earth Planet. Interiors
Physics of the Earth and Planetary Interiors. A journal devoted to observational and experimental studies of the Earth and Planetary interiors and their theoretical interpretation by the physical sciences. Publisher: North-Holland Publishing Company, Amsterdam, Netherlands.

Phys. Fluids
The Physics of Fluids. Published by the American Institute of Physics, New York, N.Y.

Phys. Letters
Physics Letters. Volumes A and B. Publisher: North-Holland Publishing Company, Amsterdam.

Phys. Rev. A
Physical Review A, General Physics. Published for the American Physical Society by the American Institute of Physics, Lancaster, Pa., and New York, N.Y.

Phys. Rev. B
Physical Review B, Solid State. Published for the American Physical Society by the American Institute of Physics, Lancaster, Pa., and New York, N. Y.

Phys. Rev. C
Physical Review C, Nuclear Physics. Published for the American Physical Society by the American Institute of Physics, Lancaster, Pa., and New York, N.Y.

Phys. Rev. D
Physical Review D, Particles and Fields. Published for the American Physical Society by the American Institute of Physics, Lancaster, Pa., and New York, N.Y.

Phys. Rev. Letters
Physical Review Letters. Published weekly by The American Physical Society, New York, N. Y.

Phys. Today
Physics Today. Published by the American Institute of Physics, New York, N.Y.

Physica
Physica. Publishers: North-Holland Publishing Company, Amsterdam, The Netherlands, on request of the Foundation "Physica", Utrecht.

Physica Scripta
Physica Scripta. (Formerly Arkiv för Fysik). Published by the Royal Swedish Academy of Sciences, Stockholm.

Pis'ma v Astron. Zhurn.
Pis'ma v Astronomicheskij Zhurnal. Akademiya Nauk SSSR. Publishers: Izdatel'stvo 'Nauka', Moskva.

Planet. Space Sci.
Planetary and Space Science. Pergamon Press, Oxford – London – New York.

Plasma Physics
Plasma Physics. Publisher: Pergamon Press, Oxford, England.

Pokroky
Pokroky matematiky, fyziky a astronomie. Editor: Jednota čs. matematiků a fyziků. Publisher: Academia, Praha.

Postępy Astron.
Postępy Astronomii. Czasopismo Poświecone Upowszechnianiu Wiedzy Astronomicznej. Polskie Towarzystwo Astronomiczne, Warszawa. Printed in Poland by Pánstwowe Wydawnictwo Naukowe, Lódź.

Priroda
Priroda. Publishers: Izdatel'stvo "Nauka", Moskva.

Proc. Astron. Soc. Australia
Proceedings of the Astronomical Society of Australia. Published for the Society by Sydney University Press, Sydney.

Proc. Cambridge Phil. Soc.
Proceedings of the Cambridge Philosophical Society (Mathematical and Physical Sciences). Publishers: Cambridge University Press, London.

Proc. IEEE
Proceedings of the IEEE. Published monthly by the Institute of Electrical and Electronics Engineers, Inc., New York, N. Y.

Proc. IREE Australia
Proceedings of the Institution of Radio and Electronics Engineers, Australia. Science House, Sydney, N.S.W., Australia.

Proc. Koninkl. Nederl. Akad. Wet.
Koninklijke Nederlandse Akademie van Wetenschappen. Proceedings. Series B, Physical Sciences. Publishers: North-Holland Publishing Company, Amsterdam.

Proc. National Acad. Sci. U.S.A.
Proceedings of the National Academy of Sciences of the United States of America. Published monthly by the National Academy of Sciences, Washington, D.C.

Proc. Roy. Soc. London
Proceedings of the Royal Society of London. Series A: Mathematical and Physical Sciences. Published by the Royal Society, London.

Progr. Theor. Phys. Japan
Progress of Theoretical Physics. Published for the Research Institute for Fundamental Physics and the Physical Society of Japan. Publication Office: Progress of Theoretical Physics, Yukawa Hall, Kyoto University, Kyoto, Japan.

Progr. Theor. Phys. Suppl.
Supplement of the Progress of Theoretical Physics. Published for the Research Institute for Fundamental Physics and The Physical Society of Japan. Publication Office: Progress of Theoretical Physics, Yukawa Hall, Kyoto University, Kyoto, Japan.

PTB Mitt.
PTB Mitteilungen. Amts- und Mitteilungsblatt der Physikalisch-Technischen Bundesanstalt, Braunschweig – Berlin.

Publ. Astron. Soc. Japan
Publications of the Astronomical Society of Japan. Published by the Astronomical Society of Japan. Office of the Society: Tokyo Astronomical Observatory, Mitaka, Tokyo. Agent: Maruzen Co. Ltd. (Export Department), Nihonbashi, Tokyo, Japan.

Publ. Astron. Soc. Pacific
Publications of the Astronomical Society of the Pacific. Published in Provo, Utah, by the Astronomical Society of the Pacific, San Francisco, California. Printed by Brigham Young University Press, Provo, Utah.

Publ. Roy. Obs. Edinburgh
Publications of the Royal Observatory, Edinburgh. Published by The Royal Observatory, Edinburgh, Scotland.

Publ. Tartu Astrofiz. Obs.
W. Struve nimelise Tartu Astrofüüsika Observatooriumi, Publikatsioonid. Eesti NSV Teaduste Akadeemia, Tartu.

Quarterly Journ. Roy. Astron. Soc.
Quarterly Journal of the Royal Astronomical Society. Published for the Royal Astronomical Society by Blackwell Scientific Publications, Oxford.

Radio Sci.
Radio Science. Published by the American Geophysical Union, Richmond, Virginia.

Referativ. Zhurn. 51. Astron.
Referativnyj Zhurnal. 51. Astronomiya. Vsesoyuznyj Institut Nachnoj i Tekhnicheskoj Informatsii. Moskva.

Referativ. Zhurn. 52. Geod. i Aehros"emka
Referativnyj Zhurnal. 52. Geodeziya i Aehros"emka. Vsesoyuznyj Institut Nauchnoj i Tekhnicheskoj Informatsii. Moskva.

Referativ. Zhurn. 62. Issled. kosm. prostranstva
Referativnyj Zhurnal. 62. Issledovanie Kosmicheskogo Prostranstva. Vsesoyuznyj Institut Nauchnoj i Tekhnicheskoj Informatsii. Moskva.

Rep. Progr. Phys.
Reports on Progress in Physics. Published by The Institute of Physics and the Physical Society, London.

Rev. Geophys. Space Phys.
Reviews of Geophysics and Space Physics (formerly Reviews of Geophysics). Published by the American Geophysical Union, Richmond, Virginia.

Revista Astron.
Revista Astronomica. Organo de la Asociación Argentina Amigos de la Astronomia, Buenos Aires.

Rev. Mexicana Astron. Astrofis.
Revista Mexicana de Astronomia y Astrofisica. Dirección: Instituto de Astronomia, Universidad Nacional Autónoma de México, México, D.F.

Rev. Modern Phys.
Reviews of Modern Physics. Published for The American Physical Society by the American Institute of Physics, Lancaster, Pa., and New York, N.Y.

Rev. Sci. Instruments
Reviews of Scientific Instruments. Published by the American Institute of Physics, Lancaster, Pa., and New York, N.Y.

Rezul'taty Nablyud. Iskusstv. Sputnikov Zemli
Rezul'taty Nablyudenij Iskusstvennykh Sputnikov Zemli. Published by Astronomicheskij Sovet Akademii Nauk SSSR, Ryazanskij Gosudarstvennyj Pedagogicheskij Institut, Ryazan'.

Rezul'taty Nablyud. Sovet. Iskusstv. Sputnikov Zemli
Rezul'taty Nablyudenij Sovetskikh Iskusstvennykh Sputnikov Zemli. Published by Astronomicheskij Sovet Akademii Nauk SSSR, Moskva. Replaced after No. 140 by Rezul'taty Nablyudenij Iskusstvennykh Sputnikov Zemli.

Říše hvězd
Říše hvězd. Czechoslovak popular astronomical journal. Publisher: Orbis, Praha.

Roy Astron. Soc. New Zealand Publ.
Publications of Variable Star Section, Royal Astronomical Society of New Zealand. Publication Office: Greerton, Tauranga, New Zealand.

Rumanian Sci. Abstr.
Rumanian Scientific Abstracts. Natural Sciences. Publishers: The Scientific Documentation Centre of the Academy of the Socialist Republic of Romania, Bucureşti.

Sci. American
Scientific American. Published monthly by Scientific American, Inc., New York, N.Y.

Science
Science. American Association for the Advancement of Science, Washington, D.C.

Scient. Sinica
Scientia Sinica. Edited by Editorial Committee of Scientia Sinica, Peking. Published by Science Press, Peking, China.

Sci. Progr. Découverte
Science Progrès Découverte (formerly Science Progrès, La Nature). Revue publiée avec la participation du Palais de la Découverte. Published by Dunod, Editeur, Paris. Imprimerie Bayeusaine, Bayeux.

Sci. Rep. Tôhoku Univ.
The Science Reports of the Tôhoku University. First Series (Physics, Chemistry, Astronomy). Published by the Faculty of Science, Tôhoku University, Sendai, Japan.

Sitzungsber. Akad. Wiss. Berlin
Sitzungsberichte der Akademie der Wissenschaften der DDR. Klasse für Mathematik, Physik und Technik. Publisher: Akademie-Verlag, Berlin.

Sitzungsber. Bayer. Akad. Wiss.
Bayerische Akademie der Wissenschaften. Mathematisch-Naturwissenschaftliche Klasse. Sitzungsberichte. Publisher: Verlag der Bayerischen Akademie der Wissenschaften, München.

Sitzungsber. Heidelberger Akad. Wiss.
Sitzungsberichte der Heidelberger Akademie der Wissenschaften. Mathematisch-Naturwissenschaftliche Klasse. Publisher: Springer-Verlag, Heidelberg.

Sitzungsber. Österreich. Akad. Wiss.
Sitzungsberichte. Österreichische Akademie der Wissenschaften. Mathematisch-Naturwissenschaftliche Klasse. Abteilung II: Mathematik, Astronomie, Meteorologie und Technik. Publisher: Springer-Verlag, Wien.

Sky Telescope
Sky and Telescope. Published by Sky Publishing Corporation, Cambridge, Mass.

Smithsonian Contr. Astrophys.
Smithsonian Contributions to Astrophysics. Smithsonian Institution Astrophysical Observatory, Cambridge, Mass. Printed by Smithsonian Institution Press, City of Washington. For sale by the Superintendent of Documents, U. S. Government Printing Office, Washington, D. C.

Smithsonian Year
Smithsonian Year. Annual Report of the Smithsonian Institution, including the financial report of the Executive Committee of the Boards of Regents. Published by the Smithsonian Institution, Washington, D.C.

Solar Physics
Solar Physics. A Journal for Solar Research and the Study of Solar Terrestrial Physics. Publishers: D. Reidel Publishing Company, Dordrecht–Holland.

Solnechnye Dannye Byull.
Solnechnye Dannye. Byulleten'. *(Solar Data).* Publishers: Izdatel'stvo "Nauka", Leningradskoe Otdelenie, Leningrad.

Soobshch. Byurakan. Obs.
Soobshcheniya Byurakanskoj Observatorii. Akademiya Nauk Armyanskoj SSR, Erevan.

Soobshch. Gos. Astron. Inst. Shternberg
Soobshcheniya Gosudarstvennogo Astronomicheskogo Instituta im P.K. Shternberga. Publishers: Izdatel'stvo Moskovskogo Universiteta, Moskva.

Southern Stars
Southern Stars. The Journal of the Royal Astronomical Society of New Zealand (Inc.). Address of the Society: P.O. Box 3181, Wellington C1, New Zealand.

Soviet Astron.
Soviet Astronomy. A translation of Astronomicheskij Zhurnal (Astronomical Journal). Published by the American Institute of Physics, New York, N.Y.

Spaceflight
Spaceflight. A Publication of the British Interplanetary Society. Printed by Eyre & Spottiswoode Limited at Grosvenor Press, Portsmouth, and published by the British Interplanetary Society, London.

Space Sci. Instrum.
Space Science Instrumentation. An International Journal of Scientific Instruments for Aircraft, Balloons, Sounding Rockets, and Spacecraft. Published by D. Reidel Publishing Company, Dordrecht–Holland.

Space Sci. Rev.
Space Science Reviews. Publishers: D. Reidel Publishing Company, Dordrecht – Holland.

Sterne
Die Sterne. Zeitschrift für alle Gebiete der Himmelskunde Johann Ambrosius Barth, Leipzig.

Sternenbote
Sternenbote. Monatsschrift für Österreichs Amateurastronomen. Publisher: Astronomisches Büro, Hermann Mucke, Wien.

Stockholms Obs. Ann.
Stockholms Observatoriums Annaler. Printed by Almquist & Wiksell, Stockholm.

Strolling Astronomer
The Strolling Astronomer. The Journal of The Association of Lunar and Planetary Observers, Publication Office: The Strolling Astronomer, Box 3 AZ, University Park, New Mexico.

Stud. Cerc. Astron.
Studii şi Cercetǎri de Astronomie. Editura Academiei Republicii Socialiste România. Editorial Office: Observatorul Astronomic, Bucureşti.

Stud. Geophys. Geod.
Studia geophysica et geodaetica. Published for the Geophysical Institute of the Czechoslovak Academy of Sciences by Academia, Praha.

Stud. Soc. Sci. Torunensis
Studia Societatis Scientiarum Torunensis, Toruń – Polonia. Sectio F (Astronomia).

Stud. Univ. Babeş-Bolyai
Studia Universitatis Babeş-Bolyai. Series Mathematica-Physica. Publishers: Intreprinderea Poligrafica, Cluj.

SuW
Sterne und Weltraum. Astronomische Monatsschrift. Publisher: Verlag Sterne und Weltraum Dr. Vehrenberg, Düsseldorf, Germany.

Tellus
Tellus, a bi-monthly Journal of Geophysics. Svenska Geofysiska Foreningen. Printed in Sweden by Almqvist & Wiksells Boktryckeri AB, Uppsala.

Trans. Astron. Obs. Yale Univ.
Transactions of the Astronomical Observatory of Yale University. Published by the Observatory, New Haven.

Trans. IAU
Transactions of the International Astronomical Union. Published and distributed for the IAU (UAI) by D. Reidel Publishing Company, Dordrecht – Holland/Boston – U.S.A.

Trans. Roy. Soc. Canada
Transactions of the Royal Society of Canada. Published by the Royal Society of Canada, National Research Building, Ottawa.

Trudy Astrofiz. Inst. Alma-Ata
Trudy Astrofizicheskogo Instituta, Alma-Ata. Akademiya Nauk Kazakhskoj SSR. Publishers: Izdatel'stvo "Nauka" Kazakhskoj SSR, Alma-Ata.

Trudy Glav. Astron. Obs. Pulkovo
Trudy Glavnoj Astronomicheskoj Observatorii v Pulkove. Akademiya Nauk SSSR. Izdanie Glavnoj astronomicheskoj observatorii v Pulkove, Leningrad.

Trudy Inst. Teor. Astron., *Leningrad*
Trudy Instituta Teoreticheskoj Astronomii. Akademiya Nauk SSSR. Publishers: Izdatel'stvo "Nauka", Leningrad.

Trudy Tashkent. Astron. Obs.
Trudy Tashkentskoj Astronomicheskoj Observatorii. Akademiya Nauk Uzbekskoj SSR. Publishers: Izdatel'stvo "FAN" Uzbekskoj SSR, Tashkent.

Tsirk. Astron. Inst. Tashkent
Tsirkulyar Astronomicheskogo Instituta. Akademiya Nauk Uzbekskoj SSR. Izdatel'stvo "FAN" Uzbekskoj SSR, Tashkent.

Tsirk. Astron. Obs. L'vov
Tsirkulyar. Astronomicheskaya Observatoriya. L'vovskij Ordena Lenina Gosudarstvennyj Universitet imeni Ivana Franko. Publisher: Izdatel'stvo L'vovskogo Universiteta, L'vov.

Umschau
Umschau in Wissenschaft und Technik, vereinigt mit Weltraumfahrt – Raketentechnik. Umschau Verlag Breidenstein KG, Frankfurt am Main.

Urania Barcelona
Urania. Revista de Astronomia y Ciencias Afines. Organo de la Sociedad Astronómica de España y América, Barcelona; Unión Nacional de Astronomia y Ciencias Afines, Madrid.

Urania Kraków
Urania. Miesiecznik Polskiego Towarzystwa Miłośników

Astronomii, Kraków. Publisher: Krakowska Drukarnia Prasowa, Kraków.

Vasiona
Vasiona. Revue d'Astronomie et d'Astronautique. Bulletin de la Société Astronomique "R. Bosković", Beograd.

Veröff. Astron. Rechen-Inst. Heidelberg
Veröffentlichungen des Astronomischen Rechen-Instituts Heidelberg. Verlag G. Braun, Karlsruhe.

Veröff. Sternw. Sonneberg
Deutsche Akademie der Wissenschaften zu Berlin. Institut für Sternphysik. Veröffentlichungen der Sternwarte in Sonneberg. Publisher: Akademie-Verlag, Berlin.

Vesmír
Vesmír. Přírodovědecky časopis Čs. akadmie věd. Publisher: Academia, Praha.

Vestn. Khar'kov. Univ.
Vestnik Khar'kovskogo Universiteta. Seriya Astronomicheskaya. Publishers: Izdatel'stvo Khar'kovskogo Universiteta, Khar'kov.

Vestn. Kiev. Univ.
Vestnik Kievskogo Universiteta. Seriya Astronomii. Publishers: Izdatel'stvo Kievskogo Universiteta, Kiev.

VJS Naturforsch. Ges. Zürich
Vierteljahresschrift der Naturforschenden Gesellschaft in Zürich. Printer and Publisher: Leeman AG, Zürich.

Wiss. Zeitschr. Friedrich-Schiller Univ. Jena
Wissenschaftliche Zeitschrift der Friedrich-Schiller-Universität. Jena. Mathematisch-Naturwissenschaftliche Reihe. Edited by the Rektor der Friedrich-Schiller-Universität Jena.

Wiss. Zeitschr. Humboldt-Univ. Berlin
Wissenschaftliche Zeitschrift der Humboldt-Universität zu Berlin. Mathematisch-Naturwissenschaftliche Reihe. Edited by the Rektor der Humboldt-Universität, Berlin.

Zeitschr. Naturforschung
Zeitschrift für Naturforschung. Europhysics Journal. Teil a: Astrophysik, Physik, Physikalische Chemie. Published by Verlag der Zeitschrift für Naturforschung, Tübingen, Germany.

Zeitschr. Physik
Zeitschrift für Physik. Publisher: Springer-Verlag, Berlin–Heidelberg–New York.

Zemlya i Vselennaya
Zemlya i Vselennaya. Astronomiya, Geofizika, Issledovaniya Kosmicheskogo Prostranstva. Nauchno-Populyarnyj Zhurnal Akademii Nauk SSSR. Publishers: Izdatel'stvo "Nauka", Moskva.

Zenit
Populair wetenschappeliik maandblad over sterrenkunde/weerkunde/ruimtevaart/ruimte-onderzoek/aanverwante wetenschappen en technieken. Bureau: Stichting De Koepel, Utrecht.

Zentralblatt Math. Grenzgebiete – Math. Abstr.
Zentralblatt für Mathematik und ihre Grenzgebiete Mathematics Abstracts. Publisher: Springer-Verlag, Berlin–Heidelberg–New York.

Zvaigžņota Debess
Latvijas PSR Zinātņu Akadémijas Radioastrofizikas Observatorijas Populārzinatnisks Gadalaiku Izdevums. Izdevnieciba "Zinātne", Riga.

002 Bibliographical Publications

002.001 **Les manuscrits de la bibliothèque de l'Observatoire de Paris.** G. Feuillebois.
Journ. History Astron., Vol. 6, 72 - 74 (1975).

002.002 **Bibliography of interstellar travel and communication – 2.** E. F. Mallove, R. L. Forward.
Journ. British Interplanet. Soc., Vol. 28, 191 - 219 (1975).

002.003 **The Astrophysical Journal: 1974 author and subject index** to volumes 187 - 194, parts 1 and 2 and to the Supplement Series, volumes 27 and 28.
Compiled by L. Newman.
Astrophys. Journ., Vol. 194, No. 3, Part 3, 100 pp. Price $ 5.00 (1975).

002.004 **Bibliography of interstellar travel and communication – 3.** E. F. Mallove, R. L. Forward.
Journ. British Interplanet. Soc., Vol. 28, 405 - 434 (1975).

002.005 **Annotations on the papers on geomagnetism and aeronomy published in "News of Universities.**
Radiophysics", 1973, Vol. 16, Nos. 4, 6 - 12; 1974, Vol. 17, No. 1.
Geomagn. Aeronom., Vol. 15, 584 - 588 (1975). In Russian.

002.006 **Catálogo de las obras y publicaciones periódicas que existen en esta Biblioteca y que corresponden a los siglos XV, XVI, XVII y XVIII.**
Inst. y Obs. de Marina, San Fernando (Cádiz), España, Biblioteca No. 1, 113 pp. (1974/75).

002.007 **Centre de Données Stellaires. Information Bulletin No. 8.** J. Jung (Editor).
Compiled at Observatoire de Strasbourg, 27 pp. (1975).
The individual contributions are included in their corresponding subject categories – see abstracts 041.036, 041.039, 112.013, 113.063 - 113.065, 114.074, 114.079, 114.080.

002.008 **Rassegna delle riviste e notizie brevi.** P. Maffei.
Coelum, Vol. 43, 30 - 35, 78 - 80, 116 - 129 (1975).

002.009 **Astronomical notebook.** J. S. Griffith.
Journ. British Interplanet. Soc., Vol. 28, 51 - 56, 134 - 139, 278 - 283, 347 - 355, 435 - 436 (1975).

002.010 **Astronomy and Astrophysics Abstracts. Vol. 12, Literature 1974, Part II.**
S. Böhme, U. Esser, W. Fricke, U. Güntzel-Lingner, F. Henn, D. Krahn, H. Scholl, G. Zech (Editors).
Published for Astronomisches Rechen-Institut, Heidelberg by Springer-Verlag, Berlin – Heidelberg – New York. 10 + 699 pp. Price DM 86.00; (US $ 37.00) [Subscription price DM 68.80 (US $ 29.60)] (1975).

002.011 **Bibliography.** Z. Kopal, M. Moutsoulas, J. W. Salisbury, F. B. Waranius (Editors).
The Moon, Vol. 12, 113 - 124 (1975).

002.012 **Black holes, 1970–1974.**
Inspec Bibliography Series No. 1, with an introduction by B. Carter.
Institution of Electrical Engineers, London. 83 pp. Price £ 2.85, $ 7.50 respectively (1974). – Reviews in Monthly Notes Astron. Soc. Southern Africa, Vol. 34, 17; 1975 (*G. F. R. Ellis*); Sky Telescope, Vol. 49, 395 (1975).

Bibliography of various optical testing methods.
See Abstr. 031.024.

The materials for the study of the polar motion before 1891, collected in the library of the Tokyo Astronomical Observatory. See Abstr. 045.011.

003 Books (Astronomy and Astrophysics)

003.001 **Problems in stellar atmospheres and envelopes.**
B. Baschek, W. H. Kegel, G. Traving (Editors).
Springer-Verlag, Berlin–Heidelberg–New York. 19 + 375 pp.
Price DM 48.00 (1975). – This volume is dedicated to
Albrecht Unsöld on the occasion of his seventieth birthday.
The individual contributions are included in their correspond-
ing subject categories – see abstracts 063.013, 064.025,
064.028 - 064.030, 080.019, 114.027, 114.028, 126.009,
131.049, 132.021.

003.002 **Astrophysics. Part A: Optical and infrared.**
N. Carleton (Editor).
Methods of experimental physics: Vol. 12. Academic Press
New York – London, 19 + 587 pp. Price $43.50, DM 159.00
respectively (1974). – The individual contributions are included in their corresponding subject categories – see abstracts
031.021 - 031.022, 031.216, 034.024 - 034.033, 036.003,
061.022.

003.003 **Astrometriya i Astrofizika, Vypusk 24.**
Ya. S. Yatskiv (Editor).
Respublikanskij Mezhvedomstvennyj Sbornik. Akademiya
Nauk Ukrainskoj SSR, Glavnaya Astronomicheskaya Observatoriya. Izdatel'stvo "Naukova dumka", Kiev. 136 pp. Price
1 Rbl. 20 Kop. (1974). In Russian. – The papers included are
abstracted in their corresponding subject categories – see
abstracts 011.003, 034.037, 041.015 - 041.018, 044.006,
044.007, 064.039, 066.050, 072.032, 082.049, 102.022,
103.110, 113.024, 114.039, 122.042 - 122.044.

003.004 **Novae and supernovae.** Yu. P. Pskovskij.
Edited by Eh. R. Mustel'.
Akademiya Nauk SSSR; Seriya "Problemy nauki i tekhni-
cheskogo progressa". Izdatel'stvo "Nauka", Moskva. 208 pp.
Price 70 Kop. (1974). In Russian. – Reviews in Referativ.
Zhurn. 51. Astron., 3.51.59; Zemlya i Vselennaya, 1975, p.
93 (1975).

003.005 **Cosmological hypothesis. Attempt of a historical-
methodological study.**
O. S. Gevorkyan. Edited by B. M. Kedrov.
Akademiya Nauk SSSR, Institut Istorii Estestvoznaniya i
Tekhniki. Izdatel'stvo "Nauka", Moskva. 144 pp. Price 50 Kop.
(1974). In Russian. – Review in Referativ. Zhurn. 51. Astron.,
2.51.1 (1975).

003.006 **Planet earth. (Its past, present and future).**
F. Yu. Zigel'. Edited by Yu. G. Reshetov.
Izdatel'stvo "Mycl'", Moskva. 223 pp. Price 70 Kop. (1974).
In Russian.

003.007 **Plasma on the earth and in space.**
V. N. Oraevskij.
Izdatel'stvo "Naukova dumka", Kiev. 171 pp. Price 32 Kop.
(1974). In Russian. – Review in Referativ. Zhurn. 51. Astron.,
5.51.42 (1975).

003.008 **Nova-like stars and novae.** V. G. Gorbatskij.
Izdatel'stvo "Nauka", Glavnaya Redaktsiya Fiziko-
Matematicheskoj Literatury, Moskva. 184 pp. Price 77 Kop.
(1974). In Russian.

003.009 **Profiles of selected Fraunhofer lines for different
positions center-limb on the solar disk.**
Eh. A. Gurtovenko, R. I. Kostyk, T. V. Orlova, V. I. Troyan,
G. L. Fedorchenko. Edited by N. N. Morozhenko.
Akademiya Nauk Ukrainskoj SSR, Glavnaya Astronomiches-

kaya Observatoriya. Izdatel'stvo "Naukova dumka", Kiev. 224
pp. Price 1 Rbl. 45 Kop (1975). In Russian.

003.010 **Stars, their birth, life and death.**
I. S. Shklovskij.
Izdatel'stvo "Nauka", Glavnaya Redaktsiya Fiziko-Mate-
maticheskoj Literatury, Moskva. 368 pp. Price 88 Kop. (1975).
In Russian.

003.011 **Annual Review of Earth and Planetary Sciences,
Vol. 3.**
F. A. Donath, F. G. Stehli, G. W. Wetherill (Editors).
Annual Reviews Inc., Palo Alto, California. 9 + 463 pp.
Price $ 12.50 (1975). – For the individual contributions
within the subject scope of Astronomy and Astrophysics
Abstracts – see abstracts 094.224, 015.012.

003.012 **Instationary stars and methods of their investigation.
Phenomena of instationarity and stellar evolution.**
A. A. Boyarchuk, Yu. N. Efremov (Editors).
Izdatel'stvo "Nauka"; Glavnaya Redaktsiya Fiziko-Matema-
tischeskoj Literatury, Moskva. 376 pp. Price 2 Rbl. 30 Kop.
(1974). In Russian. – The individual papers are included in
their corresponding subject categories – see abstracts 114.054,
114.055, 117.027, 122.057, 122.058, 125.039, 141.076,
142.079.

003.013 **Problems of cosmic physics. Vypusk 9.**
S. K. Vsekhsvyatskij (Editor).
Mezhvedomstvennyj Nauchnyj Sbornik. Izdatel'skoe Obedinenie
"Vishcha Shkola", Izdatel'stvo pri Kievskom Gosudarstvennom
Universitete, Kiev. 183 pp. Price 1 Rbl. 14 Kop. (1974). In
Russian. – The papers included are abstracted in their corresponding subject categories – see abstracts 031.406, 062.037,
062.038, 072.046, 072.047, 077.051, 078.017, 082.057,
084.247, 102.026 - 102.028, 103.109, 103.113, 103.114,
103.115, 104.022 - 104.024, 105.104, 107.006, 131.111,
132.030.

003.014 **Taking into account the influence of astroclimate
on determination of the exact time.**
Uchenye Zapiski Latvijskogo Gosudarstvennogo Universiteta
im. P. Stuchki, Vol. 220, Astron., vyp. (No.) 11, 92 pp.
(1975). In Russian. – The papers included are abstracted in
their corresponding subject categories – see abstracts 031.237,
031.238, 032.027, 041.024, 044.010.

003.015 **Astrometriya i Astrofizika, Vypusk 25.**
M. Ya. Orlov (Editor).
Respublikanskij Mezhvedomstvennyj Sbornik. Akademiya
Nauk Ukrainskoj SSR, Glavnaya Astronomicheskaya Observatoriya. Izdatel'stvo "Naukova dumka", Kiev. 120 pp. Price
1 Rbl. (1975). In Russian. – The papers included are abstracted in their corresponding subject categories – see
abstracts 041.025, 041.026, 044.011, 064.059, 064.060,
065.062, 072.049, 082.068, 082.069, 094.233, 094.568,
122.060.

003.016 **Morphological calalogue of galaxies. V. Catalogue
of 1637 galaxies from −33° to −45° declination.**
B. A. Vorontsov-Vel'yaminov, V. P. Arkhipova.
Moskovskij universitet, Moskva. 67 pp. Price 82 Kop. = Trudy
Gos. Astron. Inst. Shternberga, Vol. 46, vyp. (No.) 1 (1974).
In Russian. – Abstr. in Referativ. Zhurn. 51. Astron., 5.51.716
(1975).

003.017 **Physical foundations of aeronautics. Physics of**

space. Textbook for students of aviation at universities.   V. P. Budrakov, F. Yu. Zigel'.
Atomizdat, Moskva, 232 pp. Price 75 Kop. (1975). In Russian. Review in Referativ. Zhurn. 62. Issled. kosmich. prostranstva, 5.62.63 (1975).

003.018   **Magnetohydrodynamics.** Swiss Society of Astronomy and Astrophysics, Fourth Advanced Course, Saas Fee, 1974.   L. Mestel, N. O. Weiss.
Edited by Astronomical Institute, University of Basel, Binningen/Switzerland. Published and sold by: Geneva Observatory, Sauverny/Switzerland. 8 + 248 pp. Price sfr 30.00 (1974).

003.019   **Rotational motions in fine solar structures.**   B. Rompolt.
Acta Univ. Wratislav. No. 252, Mat., Fiz., Astron. XVII, 139 pp. = Contr. Wrocław Astron. Obs. No. 18 (1975). − Published by Państwowe Wydawnictwo Naukowe, Warszawa−Wrocław. Price zł 30.00 (1975). − Contents: 1. Rotational mass motions and twisted magnetic fields in solar phenomena; 2. Spectral evidence for spiral motions in prominences; 3. Spiral structures in prominences deduced from Hα filtergrams; 4. Evidences for rotation in solar filaments − deduced from the narrow-passband filtergrams; 5. Spectral features to be expected from rotational and expansional motions in fine solar structures; 6. Profiles of Hα line resulting from rotating optically thin filaments of prominences.

003.020   **The heritage of Copernicus: theories "pleasing to the mind".** The Copernican volume of the National Academy of Sciences.   J. Neyman (Editor).
The MIT Press, Cambridge, Mass. − London. 10 + 542 pp. Price $ 25.00 (1974). − Review in Science, Vol. 188, 838 - 839; 1975 (C. H. Waddington). − The individual contributions within the subject scope of Astronomy and Astrophysics Abstracts are included in their corresponding categories − see abstracts 004.078 - 004.081, 051.025, 066.101, 162.078.

003.021   **The redshift controversy.**   G. B. Field, H. Arp, J. N. Bahcall, with an introduction by G. B. Field.
W. A. Benjamin Inc., Advanced Book Program, Reading, Mass. 15 + 324 pp. Price $ 11.00 paper or $ 19.50 cloth (1973). Review in Observatory, Vol. 95, 67 - 69; 1975 (M. Rowan-Robinson). − The volume contains besides the papers − see abstracts 141.102, 158.135 − reprints of papers selected by J. N. Bahcall, p. 131 - 225 and reprints of papers selected by H. Arp, p. 227 - 314.

003.022   **Tables K₁₁: Two-star fix without use of altitude difference method. Vol. III N (Lat. 20°−29°30′ N).**   S. Kotlarić.
Edited by Hidrografskog Instituta Jugoslavenske Ratne Mornarice, Split. 25 + 367 pp. (1974). In Serbo-Croatian.

003.023   **J. C. Poggendorff: Biographisch-literarisches Handwörterbuch der exakten Naturwissenschaften.**
Edited by Sächsischen Akademie der Wissenschaften zu Leipzig under the redaction of H. Salié.
Akademie-Verlag, Berlin Band VIIb - Teil 5, 3. Lieferung, 160 pp. Price DM 24.00 (1975).

003.024   **Solar-terrestrial physics. Part 1.**   S.-I. Akasofu, S. Chapman.
Translated from the English edition. "Mir", Moskva. 384 pp. Price 2 Rbl. 71 Kop. (1974). In Russian. − Review in Referativ. Zhurn. 51. Astron., 3.51.510 (1975).

003.025   **Waves and satellites in the near-earth plasma. (Studies in Soviet space).**

Ya. L. Al'pert. Translated from Russian by J. B. Barbour. Consultants Bureau, New York − London. 9 + 196 pp. Price $ 42.00. Plenum Publishing Corporation, New York − London. 196 pp. Price $ 35.00 (1974). − (From Nature, Vol. 255, No. 5506, p. V (1975)).

003.026   **The mariner's astrolabe.** An exhibition at the Royal Scottish Museum.   R. G. W. Anderson.
Royal Scottish Museum, Edinburgh. 36 pp. Price 25p. (1972). Review in Journ. History Astron., Vol. 6, 71 (1975).

003.027   **Introduction to general relativity.**   H. A. Atwater.
The Pergamon Groups of Companies, Oxford − New York. International Series of Monographs in Natural Philosophy, Vol. 63. 8 + 224 pp. Price $ 9.00 (1974). − (From Science, Vol. 187, 188 (1975)).

003.028   **Radio meteor investigations of the upper atmosphere circulation.**   P. B. Babadzhanov, B. L. Kashcheev, V. A. Nechitajlenko, V. V. Fedynskij.
Donish, Dushanbe. 172 pp. Price 1 Rbl. 8 Kop. (1974). In Russian. − Review in Referativ. Zhurn. 51. Astron., 2.51. 417 (1975).

003.029   **Materie und Raum.**   E. Baier, W. W. Weiss.
Verlagsgesellschaft Schulfernsehen. 212 pp. Price DM 26.80 (1974). − (From Sky Telescope, Vol. 49, 253 (1975)).

003.030   **The planetarium of Giovanni de' Dondi, citizen of Padua.**   G. H. Baillie, H. A. Lloyd, F. A. B. Ward.
Antiquarian Horological Society, Monograph No. 9, London. 8 + 156 pp. Price £ 7.50 (1974). − Essay Review in Journ. History Astron., Vol. 6, 126 - 131; 1975 (A. J. Turner).

003.031   **Astro filters for observation and astrophotography.**   R. Barbera, C. Capen, G. Carvalho, Sr., R. Steeg.
The Optica b/c Company, 4100 MacArthur, Oakland, California. 103 pp. Price $ 4.95 (1973). − Review in Strolling Astronomer, Vol. 25, 118; 1975 (J. R. Smith).

003.032   **The astronomical telescope.**   B. V. Barlow.
The Wykeham Science Series. Wykeham Publications Ltd., London − Winchester. 8 + 213 pp. Price cloth £ 3.25; paper £ 2.50 (1975). − Review in Nature, Vol. 255, 432; 1975 (D. W. Hughes).

003.033   **Astronomie Indienne.** Investigation des textes Sanskrit et des données numériques.   R. Billard.
Publ. Vol. 83, École Française d'Extrême-Orient, Paris. 6 + 204 pp. (1971). − Essay review in Journ. History Astron., Vol. 6, 65 - 66; 1975 (B. Chatterjee).

003.034   **Weltraumbilder.** − Die dritte Entdeckung der Erde.   J. Bodechtel, H.-G. Gierloff-Emden.
Paul List Verlag KG, Schulbuch- und Lehrmittelverlag, München. 200 pp. Price DM 58.00 (1974). − Review in Sternenbote, 18. Jahrgang, p. 111 - 112 (1975).

003.035   **The Milky Way.**   B. J. Bok, P. F. Bok.
Fourth edition, revised and enlarged. Harvard University Press, Cambridge, Mass. 6 + 273 pp. Price $ 15.00 (1974). − Review in Strolling Astronomer, Vol. 25, 163 - 164; 1975 (R. Doel). − Contents: (1) Presenting the Milky Way; (2) The data of observation; (3) The sun's nearest neighbors; stellar populations; (4) Moving clusters and open clusters; (5) Pulsating stars and globular clusters; (6) The whirling Galaxy; (7) The nucleus of our Galaxy; (8) The interstellar gas; (9) Dark nebulae and cosmic grains; (10) The spiral structure of the Galaxy; (11) Our changing Galaxy.

003.036 **A concise history of science in India.**
D. M. Bose, S. N. Sen, B. V. Subbarayappa
(Editors).
Indian National Science Academy, New Delhi. 18 + 690 pp.
Price £ 6.00 (1971). – Essay review in Journ. History
Astron., Vol. 6, 135 - 137; 1975 (*W. A. Blanpied*).

003.037 **Instability phenomena and stellar evolution.**
A. A. Boyarchuk, Yu. N. Efremov (Editors).
Nauka, Moskva. 375 pp. Price 2 Rbl. 30 Kop. (1974).
In Russian. – Review in Referativ. Zhurn. 51. Astron.,
2.51.60 (1975).

003.038 **The galactic club: intelligent life in outer space.**
R. N. Bracewell.
W. H. Freeman, San Francisco, Calif., – Reading – New York.
141 pp. Price hardcover $ 6.95, £ 3.70 respectively; paperback
$ 3.95, £ 1.90 respectively (1975). – Reviews in Nature, Vol.
255, 513; 1975 (*P. C. W. Davies*); Sky Telescope, Vol. 49, 395
(1975).

003.039 **The end of the world.** F. M. Branley.
Thomas Y. Crowell Company, New York, N.Y.
42 pp. Price $ 3.50 (1974). – Review in Sky Telescope, Vol.
49, 113 (1975).

003.040 **Scottish scientific instrument-makers 1600–1900.**
D. J. Bryden.
Royal Scottish Museum Information Series, Edinburgh. 59 pp.
(1972). – Review in Journ. History Astron., Vol. 6, 71 (1975).

003.041 **Dynamics of the Venus atmosphere.**
N. I. Burangulov, S. S. Zilitinkevich, V. V. Ker-
zhanovich, A. S. Monin, M. K. Rozhdestvenskij, A. S.
Safraj, V. G. Turikov, D. V. Chalikov.
Nauka, Leningrad. 184 pp. Price 1 Rbl. 11 Kop. (1974).
In Russian. – Review in Referativ. Zhurn. 51. Astron., 4.51.
269 (1975).

003.042 **Ewige Kalender** (Wieczne kalendarze).
A. W. Butkewitsch (*A. V. Butkevich*), M. S. Selikson.
(Kleine Naturwissenschaftl. Bibliothek, Physik, t. 23).
BSB B. G. Teubner Verlagsgesellschaft, Leipzig. 124 pp.
Price M 5.90 (1974). – Reviews in Astron. in der Schule, 12.
Jahrgang, p. 69; 1975 (*W. König*); Urania Kraków, Vol. 46,
26; 1975 (*L. Zajdler*).

003.043 **The geometry of the stars.** J. P. Calk.
Exposition, Hicksville, N.Y. 104 pp. Price $ 5.00
(1975). – (From Science, Vol. 188, 1230 (1975)).

003.044 **Planets, comets, meteors, and reference guide.**
C. Capen.
Twenty super color slides. Hansen Planetarium, Salt Lake
City. Price $ 6.00 (1975). – Review in Strolling Astronomer,
Vol. 25, 165; 1975 (*J. R. Smith*).

003.045 **The natural philosophy of Galileo: essays of the**
origins and formation of classical mechanics.
M. Clavelin.
Translated from the French edition by A. J. Pomerans.
MIT Press, Cambridge, Mass. – London. 26 + 498 pp. Price
$ 25.00 (1974). – Reviews in Nature, Vol. 253, 145;
1975 (*J. D. North*); Science, Vol. 187, 944 - 945; 1975
(*M. S. Mahoney*).

003.046 **The Viking mission to Mars.** W. R. Corliss.
National Aeronautics and Space Administration,
Goddard Space Flight Center, Greenbelt. NASA SP-334,
available from Superintendent of Documents, U.S. Government Printing Office, Washington, D.C. 77 pp. Price $ 1.20
(1974). – Review in Sky Telescope, Vol. 49, 113 (1975).

003.047 **The science of astronomy.** H. Crull, W. Kaufmann.
Harper & Row, Publishers Inc., New York, 466 pp.
Price $ 8.50 (1974). – Review in Phys. Today, Vol. 28, No. 1,
p. 95 - 97; 1975 (*R. Berendzen*).

003.048 **The earth in the looking glass.** L. Darden.
Doubleday & Company, Inc., New York, N.Y.
324 pp. Price $ 7.95 (1974). – (From Sky Telescope, Vol.
49, 321 (1975)).

003.049 **The physics of time asymmetry.** P. C. W. Davies.
Leighton Buzzard; Surrey University Press. University of California Press, Berkeley. 18 + 214 pp. Price £ 6.50,
$ 15.75 respectively (1974). – Reviews in Nature, Vol. 253,
485; 1975 (*W. H. McCrea*); Science, Vol. 188, 841 - 842;
1975 (*R. Schlegel*).

003.050 **Atlas of the planets.** V. de Callataÿ, A. Dollfus,
translated from the French edition by M. Collon.
University of Toronto Press. Heinemann Educational Books
Ltd., London. 8 + 152 pp. Price $ 15.00, £ 5.00 respectively
(1974). – Reviews in Journ. British. Astron. Ass., Vol. 85,
293; 1975 (*J. H. Rogers*); Journ. History Astron., Vol. 6,
71 (1975).

003.051 **On the works of the academician V. G. Fesenkov**
on atmospheric optics and the structure of the
terrestrial atmosphere. N. B. Divari.
Sbornik "Atmosfernaya optika". Nauka. 5 pp. (1974). In
Russian. – Review in Referativ. Zhurn. 51. Astron., 2.51.24
(1975).

003.052 **Dynamic Astronomy.** R. T. Dixon.
Second edition. Prentice-Hall, Inc., Englewood
Cliffs, N. J. 8 + 440 + 12 pp. Price $ 10.50 (1975). – Reviews
in Sky Telescope, Vol. 49, 395 (1975); Strolling Astronomer,
Vol. 25, 167; 1975 (*J. R. Smith*).

003.053 **Die Sonne – Stern des Lebens.** G. Doebel.
Reihe Astrophysik für Jedermann, Astrokosmos,
"Kosmos", Gesellschaft der Naturfreunde, Franck'sche
Verlagshandlung, Stuttgart. 184 pp. Price DM 19.80 (1975).

003.054 **Cosmic rays: variations and space explorations.**
L. I. Dorman.
Translated from the second Russian edition. North-Holland
Publishing Company, Amsterdam – Oxford; American
Elsevier Publishing Company, Inc., New York. 15 + 675 pp.
Price Dfl. 215, $ 82.75 respectively (1974). – Review in
Nature, Vol. 255, 661; 1975 (*J. J. Quenby*).

003.055 **Astronomie heute - Geschichte einer alten**
Wissenschaft.
J. Dorschner, C. Friedemann, S. Marx, W. Pfau.
Edition Leipzig, Verlag für Kunst und Wissenschaft, Leipzig.
208 pp. Price M 56.00 (1974). – Review in Astron. in der
Schule, 12. Jahrgang, p. 69 - 70; 1975 (*M. Schukowski*).

003.056 **Astronomie vom Altertum bis heute.**
J. Dorschner C. Friedemann, S. Marx, W. Pfau.
Umschau-Verlag, Frankfurt a.M.; Pinguin-Verlag, Innsbruck.
208 pp. Price DM 45.00 (1975).

003.057 **Materie und Antimaterie.** M. Duquesne.
Deutsche Verlags-Anstalt, Stuttgart. 114 pp.
Price DM 12.80 (1974). – (From SuW, Vol. 14, 33 (1975)).

003.058 **Les molécules interstellaires.**

P. Encrenaz.
Delachaux et Niestlé, Neuchâtel, Switzerland. 142 pp.
Price Sw.Fr. 75.00 (1974). – (From Science, Vol. 188, 746 (1975)).

003.059 **The planet-girded suns.** S. L. Engdahl.
Antheneum Publishers, New York. 201 pp. Price $ 7.50 (1974). – Review in Strolling Astronomer, Vol. 25, 166 - 167; 1975 (*B. M. Frank*).

003.060 **Raumzeitdenken – Zwangsvorstellung Unendlichkeit.** H. J. Fahr.
"Texte und Thesen", Vol. 39. Verlag A. Fromm, Osnabrück. 82 pp. Price DM 6.00 (1973). – Review in SuW, Vol. 14, 138 - 139; 1975 (*G. D. Roth*)

003.061 **L'Orologio Notturno.** M. Felli.
Edizioni Arnaud e Istituto e Museo di Storia della Scienza, Florence. 16 pp. (1974). – (From Journ. History Astron., Vol. 6, 142 (1975)).

003.062 **Volcanoes of the earth, moon and Mars.**
G. Fielder, L. Wilson.
Paul Flek Ltd., London, 200 pp. Price £ 6.00 (1974). – (From SuW, Vol. 14, 106 (1975)).

003.063 **Pioneer Odyssey: encounter with a giant.**
R. O. Fimmel, W.Swindell, E. Burgess.
National Aeronautics and Space Administration, Washington, D.C. 171 pp. Price $ 5.50 (1974). – Review in Strolling Astronomer, Vol. 25, 165 - 166; 1975 (*V. W. Capen*).

003.064 **Naturwissenschaftliche Weltsicht und christlicher Glaube.** E. Föhr.
Verlag Herder KG., Freiburg i. Br. – Basel. 304 pp. Price DM 34.00 (1974). – Review in SuW, Vol. 14, 140; 1975 (*A. Oberstatter*).

003.065 **Greenwich Observatory. The Royal Observatory at Greenwich and Herstmonceux, 1675 - 1975.**
Volume 1: Origins and early history (1675 - 1835), E. G. Forbes. Volume 2: Recent history (1836 - 1975), A. J. Meadows. Volume 3: The buildings and instruments, D. Howse.
Taylor and Francis, London. Vol. 1: 15 + 204 pp.; Vol. 2: 11 + 135 pp.; Vol. 3: 19 + 178 pp. Price £ 25.00 the set (1975).

003.066 **Die Sache mit der Schöpfung, eine Geschichte der Kosmologie von der Mythologie zur Astrophysik.**
W. Frese.
BLV Verlagsgesellschaft mbH, München – Bern – Wien. 272 pp. Price DM 25.00 (1973). – Review in Sternenbote, 18. Jahrgang, p. 19 (1975).

003.067 **Space puzzles – curious questions and answers about the solar system.** M. Gardner.
G. Bell & Sons, Ltd., London. 96 pp. Price £ 1.95 (1974). Review in Journ. British. Astron. Ass., Vol. 85, 296 - 297; 1975 (*H. Miles*).

003.068 **Einstein collection.** V. A. Ginzburg (Editor). Compiled by U. I. Frankfurt.
Nauka, Moskva. 422 pp. Price 1 Rbl. 87 Kop. (1974). In Russian. – Review in Priroda, No. 2.75, p. 118 (1975).

003.069 **The nebular variables.** J. S. Glasby.
Pergamon Press, Oxford – New York. International series of monographs in natural philosophy, Vol. 69. 9 + 208 pp. Price $ 22.50 (1974). – Contents: (1) General information on nebular variables; (Part I) RW Aurigae variables;

(Part II) T Orionis variables; (Part III) T Tauri variables; (Part IV) Peculiar nebular variables. – Review in Sky Telescope, Vol. 49, 113 (1975).

003.070 **A preface to astronomy.** M. Goran.
Technomic Publishing Co., Westport, Conn. 112 pp. Price $ 12.00 (1975). – Review in Sky Telescope, Vol. 49, 395 (1975).

003.071 **Dialektischer Materialismus und Naturwissenschaften in der UdSSR.** Erster Teil: Quantenmechanik, Relativitätstheorie, Ursprung und Aufbau des Weltalls.
L. R. Graham.
S. Fischer Verlag, Frankfurt/Main, 255 pp. Price DM 19.80 (1974). – (From SuW, Vol. 14, 106 (1975)).

003.072 **Starfall.** B. Grissom, H. Still.
Thomas Y. Crowell Company, New York. 276 pp. Price $ 7.95 (1974). – Review in Sky Telescope, Vol. 49, 393 - 394; 1975 (*R. N. Watts, Jr.*).

003.073 **Very old secrets of the sky.** A. A. Gurshtejn.
Prosveshchenie, Moskva. 256 pp. (1973). In Russian. Review in Priroda, No. 1.72, p. 121 - 122; 1975 (*V. A. Krat*).

003.074 **M. Kopernik. Obehy nebeskych sfér.**
Slovak translation of 'De revolutionibus', with an introduction and essay on the reception of Copernicanism in Slovakia by V. Guth, J. Tibenský, a commentary and notes by Z. Horský.
Slovak Academy of Sciences, Bratislava. 535 pp. (1974). (From Journ. History Astron., Vol. 6, 142 (1975)).

003.075 **The intensity interferometer: its application to astronomy.** R. Hanbury Brown.
Taylor and Francis Ltd., London; Halsted (Wiley), New York. 16 + 184 pp. Price £ 6.00, $ 18.75 respectively (1974). Review in Journ. History Astron., Vol. 6, 71 (1975). – Contents: (1) The story of how and why the stellar intensity interferometer at Narrabri came to be built; (2) A simple explanation of how an intensity interferometer works; (3) The theory of coherent light; (4) The principles of three types of interferometer; (5) The theory of a practical stellar intensity interferometer; (6) Laboratory tests; (7) Two early intensity interferometers; (8) The Narrabri stellar interferometer; (9) How the observations at Narrabri were made; (10) Analyzing the data; (11) Results; (12) Future possibilities.

003.076 **Quantum mechanics in chemistry.** M. W. Hanna.
W. A. Benjamin, Inc., Menlo Park, Calif. 2nd edition. 15 + 262 pp. Price $ 12.50 (1974). – Review in Space Sci. Rev., Vol. 17, 161; 1975 (*A. Schadee*).

003.077 **The new Mars: the discoveries of Mariner 9.**
W. K. Hartmann, O. Raper.
National Aeronautics and Space Administration, NASA SP-337 (available from Superintendent of Documents, U.S. Government Printing Office, Washington, D.C. 179 pp. Price $ 8.75 (1974). – (From Sky Telescope, Vol. 49, 181 (1975)).

003.078 **Plasma instabilities and nonlinear effects.**
A. Hasegawa.
Springer-Verlag, Berlin – Heidelberg – New York. Physics and Chemistry in Space, Vol. 8, 11 + 217 pp. Price DM 73.00 (1975). – Contents: (1) Introduction to plasma instabilities; (2) Microinstabilities – instabilities due to velocity space nonequilibrium; (3) Macroinstabilities – instabilities due to coordinate space nonequilibrium; (4) Nonlinear effects associated with plasma instabilities.

003.079 **Cosmic ray physics. Part 2. Astrophysical aspect.**

S. Hayakawa.
Translated from the English edition. "Mir", Moskva. 342 pp.
Price 1 Rbl. 90 Kop. (1974). In Russian. – (From Referativ.
Zhurn. 51. Astron., 3.51.710 (1975)).

003.080 **Astrobiologie.** J. Herrmann.
Franckh'sche Verlagshandlung/Kosmos-Verlag,
Stuttgart. 176 pp. Price DM 18.80 (1974). – Review in
Sternenbote, 18. Jahrgang, p. 18 (1975).

003.081 **Zwischen Sonne und Jupiter.** Weltraumbilder der
Planeten. H. Heuseler.
Deutsche Verlags-Anstalt, Stuttgart. 141 pp. Price DM 69.00
(1975).

003.082 **Das Radiouniversum.** Einführung in die Radioastro-
nomie.
J. S. Hey, translated from the English edition by H. Scheffler.
Verlag Chemie GmbH, Weinheim, Bergstr. 8 + 280 pp. Price
DM 32.80 (1974). – Reviews in Sternenbote, 18. Jahrgang,
p. 17 - 18 (1975); SuW, Vol. 14, 176 - 177; 1975 (*K. Rohlfs*).

003.083 **Action at a distance in physics and cosmology.**
F. Hoyle, J. V. Narlikar.
W. H. Freeman & Co. Ltd., San Francisco – Reading, 12 +
266 pp. Price $15.00, £7.80 respectively (1974). – Reviews
in Journ. British. Astron. Ass., Vol. 85, 179; 1975 (*H. Bondi*);
Nature, Vol. 254, 223; 1975 (*G. J. Suggett*); Science, Vol.
187, 943; 1975 (*D. J. Raine*).

003.084 **Foundations of satellite geodesy.** A. A. Izotov,
V. I. Zubinskij, N. L. Makarenko, A. M. Mikisha.
Nedra, Moskva. 317 pp. Price 2 Rbl. 10 Kop. (1974). In
Russian. – Review in Referativ. Zhurn. 62. Issled. kosmich.
prostranstva, 2.62.327 (1975).

003.085 **Die spezielle Relativitätstheorie.** J. Kahra.
Aulis Verlag Deubner & Co. KG, Köln. 120 pp.
Price DM 12.80 (1973). – (From SuW, Vol. 14, 33 (1975)).

003.086 **Astronomy – Activities and experiments.**
L. J. Kelsey, D. B. Hoff, J. S. Neff.
Kendall/Hunt Publishing Company, Dubuque, Iowa. 183 pp.
Price $6.95 (1974). – Review in Sky Telescope, Vol. 49,
395 (1975).

003.087 **Elementare Plasmaphysik.**
R. Kippenhahn, C. Möllenhoff.
Bibliographisches Institut, Mannheim - Wien - Zürich.
(B.-I. Wissenschaftsverlag). 297 pp. Price DM 54.00 (1975).

003.088 **'Himmelskunde für jedermann'** .
A. Krause, C. Fischer.
Franckh'sche Verlagshandlung/Kosmos-Verlag, Stuttgart.
7th edition. 303 pp. Price DM 29.50 (1974).

003.089 **Bürgels Himmelskunde – Entdeckungsreisen zu
fernen Welten.** E. Krug (Editor)
New revised edition. Bertelsmann Lexikon - Verlag, Gütersloh.
304 pp. Price DM 26.00 (1975). – Review in Orion, 33. Jahrgang, p. 91; 1975 (*E. Wiedemann*).

003.090 **Textbook on spherical astronomy.** K. A. Kulikov.
3rd enlarged and revised edition. Textbook for
students of astronomy. Nauka, Moskva. 232 pp. Price 47
Kop. (1974). In Russian. – (From Referativ. Zhurn.,
51. Astron., 6.51.40 (1975)).

003.091 **Physics of planet Venus.**
A. D. Kuz'min, M. Ya. Marov.
Nauka, Moskva. 408 pp. Price 2 Rbl. 41 Kop. (1974). In

Russian. – Review in Referativ. Zhurn. 51. Astron., 5.51.
252; 62. Issled. kosmich. prostranstva, 5.62.196 (1975).

003.092 **Star and planet spotting.** P. Lancaster Brown.
Blandford Press, London, 155 pp. Price £2.15
(1974). – Review in Journ. British. Astron. Ass., Vol. 85,
291 (1975).

003.093 **The voyages of Apollo.** The exploration of the
moon. R. S. Lewis.
Quadrangle (New York Times), New York. 12 + 308 pp.
Price $12.50 (1974). – (From Science, Vol. 188, 872 (1975)).

003.094 **Man and the stars. Contact and communication
with other intelligence.** D. Lunan.
Souvenir Press, Ltd., London, 324 pp. Price $7.05 (1974). –
Review in Southern Stars, Vol. 26, 18; 1975 (*G. A. Eiby*).

003.095 **Royal Greenwich Observatory.** W. H. McCrea.
Her Majesty's Stationery Office, London. 80 pp.
Price £1.00 (1975). – Review in Journ. British. Astron. Ass.,
Vol. 85, 370; 1975 (*E. A. Beet*).

003.096 **Positional Astronomy.** D. McNally.
Frederick Muller, London. 13 + 375 pp. Price
cloth £8.50, paper £4.25 (1975). – Review in Nature, Vol.
255, 660; 1975 (*H. A. Couper*).

003.097 **Earth – space – moon.**
S. N. Minchin, A. T. Ulubekov.
National Aeronautics and Space Administration. NASA TT F-
800. (To be ordered from the National Technical Informa-
tion Service, Springfield, Va.) 298 pp. $7.25 (1974) - (From
Sky Telescope, Vol. 49, 181 (1975)).

003.098 **The concise atlas of the universe.** P. Moore.
Rand McNally & Company, Chicago, IL; Mitchell
Beazley Limited, London. 190 pp. Price $19.95 (1974). –
(From Science, Vol. 188, 252 (1975)).

003.099 **Guide to the stars.** P. Moore.
Lutterworth Press, Guildford – London. 251 pp.
Price £4.95 (1974).

003.100 **1975 yearbook of astronomy.**
P. Moore (Editor).
Sidgwick and Jackson Ltd., London; W. W. Norton and Co.,
Inc., New York, N. Y. 221 pp. Price $8.95 (1974). – Review
in Strolling Astronomer, Vol. 25, 117; 1975 (*R. W. Gordon*).

003.101 **Black holes in space.** P. Moore, I. Nicolson.
Orbach & Chambers Ltd., London, 126 pp. Price
£2.25 (1974). – Review in Journ. British. Astron. Ass., Vol.
85, 292 - 293; 1975 (*V. Barocas*).

003.102 **Mouvement d'un satellite artificiel de la terre.**
B. Morando.
Gordon and Breach Science Publishers Ltd., London. 225 pp.
Price £12.80, £4.80 respectively (1974). – Reviews in Bull.
Géod., Nouvelle Sér., Année 1975, No. 116, p. 210 (*J. Ville-
crose*); Orion, 33. Jahrgang, p. 92; 1975 (*E. Antonini*).

003.103 **New data on Mars. Collection of articles.**
Translated from the English edition. Edited by and
with a preface of V. I. Moroz.
Mir, Moskva. 196 pp. Price 1 Rbl. 47 Kop. (1974). In Russian. – Review in Referativ. Zhurn. 51. Astron., 5.51.259;
62. Issled. kosmich. prostranstva, 5.62.200 (1975).

003.104 **Stable and random motions in dynamical systems.**
J. Moser.

Princeton University Press, Princeton, N. J. 198 pp. Price $7.50 (1973). – Review in Phys. Today, Vol. 28, No. 3, p. 47 - 48; 1975 (*D. Saari*).

003.105 **Sonne, Satelliten, Kometen und Blitze.** Forscher am Wendelstein. R. Müller.
Rosenheimer Verlagshaus A. Förg, Rosenheim, BRD. 92 pp. Price DM 16.80 (1974). – Reviews in Sternebote, 18. Jahrgang, p. 79 (1975); SuW, Vol. 14, 104; 1975 (*H. J. Staude*).

003.106 **Stars and outer space made easy.** C. S. Mundt.
Naturegraph, Healdsburg, Calif. 2nd edition. 96 pp. Price $2.95 (1974).(From Science, Vol. 187, 1004 (1975)).

003.107 **Carbonaceous meteorites.** B. Nagy.
Developments in solar system and space science, Vol. 1. Elsevier Scientific Publishing Company, Amsterdam – Oxford – New York. 14 + 748 pp. Price Dfl. 210, $80.95 respectively (1975). – Review in Science, Vol. 188, 1007; 1975 (*B. Mason*).

003.108 **Astrophotography – from film to infinity.** J. Newton, D. R. Hankin.
Astronomical Endeavors Publishing Co., Buffalo, N.Y. 42 pp. Price $3.00 (1974). – Review in Strolling Astronomer, Vol. 25, 162 - 163; 1975 (*J. R. Smith*).

003.109 **The asteroids.** A. E. Nourse.
Franklin Watts, Inc., New York. 59 pp. Price $3.90 (1975). – (From Sky Telescope, Vol. 49, 252 (1975)).

003.110 **The theory of fifth dimensional space.** J. O'Callaghan.
Exposition Press, Hicksville, N. Y. 60 pp. Price $4.00 (1974). (From Science, Vol. 187, 342 (1975)).

003.111 **Mariner 9 navigation.** W. J. O'Neil et al.
Jet Propulsion Laboratory, Technical Report 32 - 1586 [Available from ESRO/ELDO Space Documentation Service]. 270 pp. Price $17.50 (1973). – Review in Journ. British Interplanet. Soc., Vol. 28, 294 - 295; 1975 (*W. I. McLaughlin*).

003.112 **Astrophysics of gaseous nebulae.** D. E. Osterbrock.
W. H. Freeman & Co., San Francisco – Reading. 14 + 251 pp. Price £ 8.90, $ 17.00 respectively (1974). – Reviews in Nature, Vol. 254, 222; 1975 (*F. D. Kahn*); Sky Telescope, Vol. 49, 113 (1975). – Contents: (1) General introduction; (2) Photoionization equilibrium; (3) Thermal equilibrium; (4) Calculation of emitted spectrum; (5) Comparison of theory with observations; (6) Internal dynamics of gaseous nebulae; (7) Interstellar dust; (8) H II regions in the galactic context; (9) Planetary nebulae.

003.113 **The rings of Saturn.** F. D. Palluconi, G. H. Pettengill (Editors).
National Aeronautics and Space Administration, Washington, D.C. NASA SP-343 (available from Superintendent of Documents, U.S. Government Printing Office, Washington, D.C.). 222 pp. Price $ 3.25 (1974). – Review in Sky Telescope, Vol. 49, 395 (1975).

003.114 **Die dritte Entdeckung der Erde.** (Das neue Ziel der Raumfahrt). G. Paul.
Econ Verlag GmbH, Düsseldorf – Wien. 267 pp. – (From SuW, Vol. 14, 33 (1975)).

003.115 **Early physics and astronomy.** O. Pedersen, M. Pihl.

Elsevier Scientific Publishing Company, Amsterdam – London-New York. 413 pp. Price $35.50 (1974). – Review in Sky Telescope, Vol. 49, 113 (1975).

003.116 **Chemical composition of the earth.** A. I. Perel'man.
'Znanie', seriya 'Nauka o Zemle', No. 1. Moskva. 64 pp. Price 12 Kop. (1975). In Russian. – Review in Priroda, No. 5.75, p. 123 (1975).

003.117 **Novae and supernovae.** Yu. P. Pskovskij.
Nauka, Moskva. 208 pp. Price 70 Kop. (1974). In Russian. – Review in Referativ. Zhurn. 51. Astron., 3.51.59 (1975).

003.118 **The physics of stellar interiors.** V. C. Reddish.
Crane, Russak & Company, Inc., New York. 107 pp. Price $8.75 (1974). – (From Phys. Today, Vol. 28, No. 4, p. 93 (1975)).

003.119 **Black holes, gravitational waves and cosmology: an introduction to current research.** M. Rees. R. Ruffini, J. A. Wheeler.
Topics in astrophysics and space physics, Vol. 10. Gordon and Breach Science Publihsers Ltd., New York – London. 410 pp. Price $29.50 (1974). – Review in Nature, Vol. 255, 661; 1975 (*J. Gribbin*).

003.120 **Vita di Galileo.** M. L. Righini Bonelli.
Editore Nardini, Firenze. 240 pp. Price Lire 3.800 (1974). – Reviews in Ciel et Terre, Vol. 91, 252; 1975 (*H. Michel*); Coelum, Vol. 43, 131; 1975 (*L. Rosino*).

003.121 **Atlas of Surveyor 5 television data.** R. M. Batson, R. Jordan, K. B. Larson.
National Aeronautics and Space Administration, Washington, D.C. NASA SP-341 (available from the Superintendent of Documents, U.S. Government Printing Office, Washington, D.C.). 597 pp. Price $ 14.40 (1974). – (From Sky Telescope, Vol. 49, 252 (1975)).

003.122 **Greenwich Observatory: 300 years of astronomy.** C. A. Ronan (Editor).
Times Books, London. 63 pp. Price 75 p. (1975). – Review in Journ. British Astron. Ass., Vol. 85, 370; 1975 (*H. Hatfield*).

003.123 **Astronomy: a handbook.** G. D. Roth (Editor).
Translated from the second German edition by A. Beer.
Springer-Verlag, Berlin – Heidelberg – New York. 610 pp. Price DM 44.90, $ 19.40 respectively (1975).

003.124 **Homogeneous relativistic cosmologies.** M. P. Ryan, Jr., L. C. Shepley.
Princeton series in physics. Princeton University Press, Princeton, N. J. 15 + 321 pp. Price cloth $ 15.00, £ 7.90 respectively; paper $ 7.50, £ 3.95 respectively. – (From Phys. Today, Vol. 28, No. 4, p. 93 (1975)).

003.125 **Nachbarn im Kosmos.** Leben und Lebensmöglichkeiten im Universum.
C. Sagan, J. Agel, with a preface by H. von Ditfurth.
Translated from the American edition by C. Francke.
Kindler Verlag GmbH, München. 214 pp. Price DM 34.00 (1975).

003.126 **Other worlds.** C. Sagan.
Bantam Books, Inc., New York. 160 pp. Price $ 1.95 (1975). – (From Sky Telescope, Vol. 49, 253 (1975)).

003.127 **The astrolabe.** H. N. Saunders.

Micro Instruments Ltd., Oxford. 35 pp. Price £ 3.30.
Review in Journ. History Astron., Vol. 6, 138 - 140; 1975
(*J. Henderson*).

003.128 **Abenteuer Weltall.** C. Schurbohm.
Franz Schneider Verlag, München — Wien. 136 pp.
Price DM 7.80 (1974). — (From SuW, Vol. 14, 106 (1975)).

003.129 **"It is I, sea gull".** M. R. Sharpe.
Thomas Y. Crowell Company, New York. 214 pp.
Price $ 5.95 (1975). — (From Sky Telescope, Vol. 49, 321 (1975)).

003.130 **Plasma scattering of electromagnetic radiation.**
J. Sheffield.
Academic Press, Inc., New York. 12 + 306 pp. Price $ 29.00 (1975). — (From Science, Vol. 188, 746 (1975)).

003.131 **Prywatne życie Mikołaja Kopernika.**
J. Sikorski.
Wydawn. Pojezierze—Olsztyn. 275 pp. Price 40 zł (1973).
Review in Urania Kraków, Vol. 46, 91 - 92; 1975 (*H. Korpikiewicz*).

003.132 **Radio astronomy.** F. G. Smith.
A Pelican book. Penguin Books, Ltd., Harmondsworth, Middlesex — Baltimore. 4th revised edition. 270 + 22 pp. Price 80 p., $ 3.50 respectively (1974). — Review in Sky Telescope, Vol. 49, 395 (1975).

003.133 **Milestones of science.** R. A. Sparrow.
Buffalo Museum of Science, Buffalo, New York.
308 pp. $ 25.00 (1970). — Review in Journ. History Astron., Vol. 6, 71; 1975 (*O. Gingerich*).

003.134 **The awakening interest in science during the first century of printing, 1450 — 1550.**
M. B. Stillwell.
The Bibliographical Society of America, New York. 29 + 401 pp. Price $ 25.00 (1970). — Review in Journ. History Astron., Vol. 6, 70; 1975 (*O. Gingerich*).

003.135 **Everyman's astronomy.** R. H. Stoy (Editor).
J. M. Dent & Sons Ltd., London; St. Martin's Press, Inc., New York. 493 pp. Price £ 4.50, $ 10.00 respectively (1974). — Reviews in Astron. Tidssk., Årg. 8, p. 94 (1975); Monthly Notes Astron. Soc. Southern Africa, Vol. 34, 78 - 79; 1975 (*A. H. Jarrett*); Nature, Vol. 254, 223; 1975 (*J. Gribbin*); Sky Telescope, Vol. 49, 395 (1975).

003.136 **New science in the solar system.**
P. Stubbs (Editor).
New Science Publications, London. 64 pp. Price £ 1.00 (1975). — Review in Journ. British Astron. Ass., Vol. 85, 292 (1975).

003.137 **Lunar science — a post-Apollo view.** Scientific results and insights from the lunar samples.
S. R. Taylor.
Pergamon Press Inc., New York — Toronto — Oxford — Sydney — Braunschweig. 19 + 372 pp. Price $ 16.50 (1975). — Contents: (1) Introduction; (2) Lunar geology; (3) The surface of the moon; (4) The maria; (5) The highlands; (6) The interior of the moon; (7) The origin and evolution of the moon; Epilogue: On the usefulness of manned space flight.

003.138 **Stonehenge, Carnac, Brogar and Islay.**
A. Thom and family.
Science History Publications, Chalfont St. Giles, Bucks. 113 pp. Price £ 3.00 (1974). — Review in Journ. British Astron. Ass., Vol. 85, 183; 1975 (*C. Ronan*).

003.139 **Relativity, thermodynamics and cosmology.**
P. Tolman.
Translated from the English edition. Nauka, Moskva. 520 pp. Price 2 Rbl. 52 Kop. (1974). In Russian. — Review in Referativ. Zhurn. 51. Astron., 4.51.849 (1975).

003.140 **Elementare Kosmologie. Die Weltmodelle der klassischen und der relativistischen Gravitationstheorie.** H.-J. Treder.
Akademie-Verlag, Berlin. Wissenschaftliche Taschenbücher, Vol. 154. 153 pp. (1975).

003.141 **Über Prinzipien der Dynamik von Einstein, Hertz, Mach und Poincaré (Zur Geometrisierung der Relativität der Beschleunigungen und der Trägheit).**
H.-J. Treder (Editor).
Akademie-Verlag, Berlin. Veröff. Bereich Kosmische Physik, Heft 4. 75 pp. (1974). — Review in Bull. Astron. Inst. Czechoslovakia, Vol. 26, 126; 1975 (*Z. Horák*).

003.142 **Annual review of fluid mechanics, Vol. 6.**
M. Van Dyke, W. G. Vincenti (Editors).
Annual Reviews, Palo Alto, California. 371 pp. Price $ 15.00 (1974). — Review in Icarus, Vol. 24, 139 - 140; 1975 (*A. P. Ingersoll*).

003.143 **Die Quasare.** Probleme und Deutungsversuche der Forschung. P. Véron, R. Engel.
Deutsche Verlags-Anstalt, Stuttgart. 210 pp. Price DM 19.80 (1975). — Review in Sternenbote, 18. Jahrgang, p. 95; 1975 (*J. Pfleiderer*).

003.144 **Outline of astronomy.** Vols. 1 and 2.
H. H. Voigt. Translated from the German edition by L. Plaut, N. Houk.
Noordhoff International Publishing, Academic Book Services, Groningen — Leiden, Netherlands. 556 pp. Price Dfl. 60.00 (1974). — Review in Sky Telescope, Vol. 49, 113 (1975); Vol. 49, 318 - 320; 1975 (*G. S. Mumford*).

003.145 **Abriß der Astronomie.** H. H. Voigt.
Bibliographisches Institut, Mannheim — Wien — Zürich. (B.I.-Wissenschaftsverlag). Second revised edition. 9 + 540 pp. Price DM 38.00 (1975).

003.146 **Informational relations in biological, solar and geophysical phenomena and elements of their forecast.**
K. S. Vojchishin, Ya. P. Dragan, V. I. Kuksenko, V. N. Mikhajlovskij.
Nauk. dumka, Kiev. 207 pp. Price 1 Rbl. 22 Kop. (1974). In Russian. — Review in Referativ. Zhurn. 51. Astron. 4.51.543 (1975).

003.147 **Children of the universe. The tale of our existence.**
H. von Ditfurth.
Atheneum, New York. 301 pp. Price $ 10.95 (1974). — Review in Journ. Roy. Astron. Soc. Canada, Vol. 69, 50 - 51; 1975 (*K. Hujer*).

003.148 **The astrolabe.** R. S. Webster.
Paul MacAlister and Associates, Lake Bluff, Ill.
24 pp. Price $ 18.00, $ 12.00 respectively (1974). — Reviews in Journ. History Astron., Vol. 6, 138 - 140; 1975 (*J. Henderson*); Sky Telescope, Vol. 49, 246 - 248; 1975 (*R. N. Mayall*).

003.149 **Aberrations of the symmetrical optical system.**
W. T. Welford.
Academic Press, Inc., New York. 240 pp. Price $ 20.00 (1974). (From Phys. Today, Vol. 28, No. 4, p. 89 (1975)).

003.150 **The monument builders.**

R. Wernick and the editors of Time–Life Books. Time–Life Books, New York. 160 pp. Price $ 7.95 (1973). Review in Journ. History Astron., Vol. 6, 71 (1975).

003.151 **The mathematical papers of Isaac Newton.**
D. T. Whiteside (Editor).
Cambridge University Press, New York. (Vol. 5): 24 + 628 pp. Price $ 65.00 (1972). (Vol. 6): 36 + 614 pp. Price $ 72.50 (1975). – Review in Science, Vol. 188, 826 - 827; 1975 (*M. S. Mahoney*).

003.152 **Astronomy and geodesy. No. 4.**
Trudy Tomsk. un-t, 251. Tomsk, Tomsk. un-t. 93 pp. Price 53 Kop. (1973). In Russian. – (From Referativ. Zhurn. 51. Astron., 11.51.51 (1974)).

003.153 **The collected contributions of Fred L. Whipple.**
Smithsonian Astrophysical Observatory, Cambridge, Mass., 15 + 2004 pp. (1972). – Review in Irish Astron. Journ., Vol. 11, 203 - 205; 1974 (*E. Öpik*).

003.154 **Einstein symposium. 1973.**
Otdelenie yader. fiz. AN SSSR. Nauka, Moskva. 421 pp. Price 1 Rbl. 87 Kop. (1974). In Russian. – Review in Referativ. Zhurn. 51. Astron., 6.51.60 (1975).

003.155 **Johannes Hevelius and his catalog of stars.**
The J. Reuben Clark, Jr., Library, Brigham Young University Press, Provo, Utah. 90 pp. Price $ 8.00 (1971). Review in Journ. History Astron., Vol. 6, 69; 1975 (*O. Gingerich*).

003.156 **Mars as viewed by Mariner 9.**
National Aeronautics and Space Administration, Washington, D.C. NASA SP-329 (available from Superintendent of Documents, U.S. Government Printing Office, Washington, D.C.). 225 pp. Price $ 8.15 (1974). – Review in Sky Telescope, Vol. 49, 395 (1975).

003.157 **Meteorites of the Caucasus and meteorite showers.**
Nauka, Moskva. 183 pp. Price 1 Rbl. 20 Kop. (1974). In Russian. – Review in Referativ. Zhurn. 51. Astron., 4.51.427 (1975).

003.158 **New science in the solar system.**
A New Scientist special review.
New Scientist, London. 64 pp. Price £ 1.00 (1975). – (From Phys. Today, Vol. 28, No. 5, p. 62 (1975)).

003.159 **Problems of modern physics. Collection of papers in memoriam of academician B. P. Konstantinov.**
Nauka, Leningrad. 363 pp. Price 2 Rbl. 11 Kop. (1974). In Russian. – Review in Referativ. Zhurn. 51. Astron., 2.51.54 (1975).

003.160 **Radio telescopes. Submillimeter and X-ray telescopes.**
Trudy fiz. in-ta. AN SSSR, 77. Nauka, Moskva. 214 pp. Price 1 Rbl. 56 Kop. (1974). In Russian. – (From Referativ. Zhurn., 51. Astron., 6.51.52 (1975)).

003.161 **The precise alignment survey of a 5-kilometre radio telescope aerial array for the Cavendish Laboratory, Cambridge University.** Ordnance Survey, Professional Papers, New Series, No. 27.
Ordnance Survey, Southampton. 46 pp. Price £ 2.50 (1974). Review in Observatory, Vol. 95, 56 - 57; 1975 (*R. F. Griffin*).

003.162 **Erde und Weltall, Daten und Fakten zum Nachschlagen.**
Lexikon-Institut Bertelsmann.
Verlag Bertelsmann, Gütersloh. 324 pp. Price DM 24.00. (From SuW, Vol. 14, 106 (1975)).

004 History of Astronomy, Chronology

004.001 **Tobias Mayer's claim for the longitude prize. A study in 18th century Anglo-German relations.**
E. G. Forbes.
Journ. Navigation, Vol. 28, 77 - 90 (1975).
This paper describes the historical background to the Göttingen Professor Tobias Mayer's bid under the terms of the Act 12 Queen Anne, cap. XX (1714) to obtain a reward for having constructed lunar tables of unprecedented accuracy (to about ± $\frac{1}{2}'$) which promised to make both useful and practicable the lunar-distance method of finding longitude at sea.

004.002 **Ancient observatories.** A. J. Meadows.
Nature, Vol. 253, 395 (1975).

004.003 **Survey of three megalithic sites in Argyllshire.**
M. E. Bailey, J. A. Cooke, R. W. Few, J. G. Morgan, C. L. N. Ruggles.
Nature, Vol. 253, 431 - 433 (1975).
Three sites showing simple alignments of standing stones in Argyllshire were tested for possible astronomical significance by members of the Cambridge University Astronomical Society during August and September 1973.

004.004 **The eminent scientist and encyclopaedist Biruni from Central Asia.** B. M. Kedrov.
Vopr. istorii estestvozn. i tekhn. Vyp. (No.) 2 - 3 (47 - 48). Moskva, Nauka, 1974, p. 60 - 68. In Russian. – Abstr. in Referativ. Zhurn. 51. Astron., 2.51.14 (1975).

004.005 **An important find concerning the history of mathematics, astronomy and optics.**
B. A. Rozenfel'd.
Vopr. istorii estestvozn. i tekhn. Vyp. (No.) 2 - 3 (47 - 48). Moskva, Nauka, 1974, p. 123 - 124. In Russian. – Abstr. in Referativ. Zhurn. 51. Astron., 2.51.16 (1975).

004.006 **N. Copernic and the Polish culture.** I. F. Behlza.
Vopr. istorii estestvozn. i tekhn. Vyp. (No.) 2 - 3 (47 - 48). Moskva, Nauka, 1974, p. 77 - 81. In Russian. Abstr. in Referativ. Zhurn. 51. Astron., 2.51.17 (1975).

004.007 **Stonehenge as a possible lunar observatory.**
A. Thom, Archibald S. Thom, Alexander S. Thom.
Journ. History Astron., Vol. 6, 19 - 30 (1975).

004.008 **Levi ben Gerson's analysis of precession.**
B. R. Goldstein.
Journ. History Astron., Vol. 6, 31 - 41 (1975).

004.009 **Megalithic astronomy – a prehistorian's comments.**
R. J. C. Atkinson.
Journ. History Astron., Vol. 6, 42 - 52 (1975).

004.010 **A refined computation of the perigee angle in Ptolemy's Mercury model.** D. T. Whiteside.
Journ. Histroy Astron., Vol. 6, 57 (1975).

004.011 **Ancient astronomy in Mexico and Central America.**
L. F. Rodriguez.
Mercury, Vol. 4, No. 1, p. 24 - 27 (1975).

004.012 **Kalendersystem och astronomi hos mayafolket, II.**
C. Schalén.
Astron. Tidssk., Årg. 8, p. 1 - 15 (1975).

004.013 **Truth and heresy over earth and sky.**

D. G. King-Hele.
Observatory, Vol. 95, 1 - 12 (1975). – This paper is the written version of the Halley Lecture for 1974, delivered in Oxford on 1974 May 7.

004.014 **Die Kosmogonie Immanuel Kants. II.**
H. Lambrecht.
Sterne, Vol. 51, 29 - 39 (1975).

004.015 **Die astronomischen Grundlagen des französischen Revolutionskalenders.** P. Aufgebauer.
Sterne, Vol. 51, 40 - 48 (1975).

004.016 **Let's be useless.** N. Goodman.
Journ. British. Astron. Ass., Vol. 85, 93 - 114 (1975). – Presidential address.

004.017 **The myth of a supernova in 1664.** J. Ashbrook.
Sky Telescope, Vol. 49, 154, 166 (1975).

004.018 **The struggles between the Confucian and the Legalist Schools viewed from Tsu Chung-chih's calender reform.**
Dep. Astron., Peking Teachers Univ.
Acta Astron. Sinica, Vol. 15, 101 - 103 (1974). In Chinese.

004.019 **The struggles between the Confucian and the Legalist Schools and the development of ancient astronomy in China during Chin and Han dynasties.** K. Shi.
Acta Astron. Sinica, Vol. 15, 104 - 112 (1974). In Chinese.

004.020 **On the anniversary of Copernicus.**
A. Opolski.
Postępy Astron., Vol. 23, 61 - 64 (1975). In Polish.

004.021 **The origins of the Greenwich Observatory.**
E. G. Forbes.
Journ. British. Astron. Ass., Vol. 85, 213 - 216 (1975).

004.022 **The old Dominion Observatory.**
A. E. Covington.
Journ. Roy. Astron. Soc. Canada, Vol. 69, 86 - 88 (1975).

004.023 **The tercentenary of the Royal Observatory at Greenwich.** D. H. Sadler.
Journ. Navigation, Vol. 28, 121 - 128 (1975).

004.024 **Charles Messier: 'amateur uit de achttiende eeuw'. I. II.** M. Drummen.
Zenit, Vol. 2, 12 - 15, 39, 55 - 60 (1975).

004.025 **Astronomy in ancient literate societies: Introduction to some basic astronomical concepts.**
R. R. Newton.
Phil. Trans. Roy. Soc. London, Ser. A, Vol. 276, (see 012. 004), 5 - 20 (1974).
The goal of astronomy among ancient peoples was probably to develop methods of dealing with certain appearances in a coordinate system based upon the observer's horizon. These appearances include eclipses, the positions at which celestial objects rise and set, and methods of telling both the time of day and the time of the year by observations of risings and settings. This paper is concerned with discussions of these appearances. It deals with theories of motion only to the extent needed in discussing the appearances.

004.026 **Scientific astronomy in antiquity.** A. Aaboe.

Phil. Trans. Roy. Soc. London, Ser. A, Vol. 276, (see 012.004), 21 - 42 (1974).

The character and content of Babylonian scientific or mathematical astronomy, as we know it from texts of the last half millennium B.C., are sketched. This late-Babylonian astronomy is set in contrast to earlier Babylonian astronomy as well as to the kinds of astronomy found in other ancient cultures, and an attempt is made at a very broad classification of such pre-scientific astronomies.

004.027 **Babylonian observational astronomy.**
  A. Sachs.
Phil. Trans. Roy. Soc. London, Ser. A, Vol. 276, (see 012. 004), 43 - 50 (1974).

The cuneiform texts from ancient Assyria and Babylonia that are preserved offer direct evidence for systematic astronomical observation in two widely separated periods.

004.028 **Ancient Egyptian astronomy.**   R. A. Parker.
  Phil. Trans. Roy. Soc. London, Ser. A, Vol. 276, (see 012.004), 51 - 65 (1974).

The authors have shown that Egyptian astronomy, in a quantitative sense, was almost non-existent. To it one may award the determination of the length of the year, the division of the day into 24 hours and the decan names in the zodiac. Overshadowing all these is the pictorial element of which the Egyptian was a master, and the portrayals of astronomical figures on ceilings and other monuments continue to interest and charm us.

004.029 **Astronomy in ancient and medieval China.**
  J. Needham.
Phil. Trans. Roy. Soc. London, Ser. A. Vol. 276, (see 012. 004), 67 - 82 (1974).

Chinese astronomy differed from that of the Western world in two important respects: (i) it was polar and equatorial rather than planetary and ecliptic, (ii) it was an activity of the bureaucratic state rather than of priests or independent scholars. In cosmology, China developed three doctrines: (i) the Kai Thien universe, a domical geocentrism not unlike early Babylonian ideas, (ii) the Hun Thien universe, essentially the recognition of the primary celestial spherical coordinates, and (iii) the Hsüan Yeh system, which accepted the Hun Thien as methodologically necessary but viewed the heavenly bodies as lights of unknown nature floating in infinite empty space. Instrumentation developed early, armillary rings being in use by the end of the −2nd century and the complete armillary sphere by the end of the +1st.

004.030 **Maya astronomy.**   J. E. S. Thompson.
  Phil. Trans. Roy. Soc. London, Ser. A, Vol. 276, (see 012.004), 83 - 98 (1974).

No people in history has shown such interest in time as the Maya. They had three concurrent counts: 365-day years; 360-day 'years' (*tuns*) for calculations; and a 260-day sacred almanac. Solar eclipse and Venus synodical revolutions are tabulated. The Maya successfully predicted eclipses, but were unaware of which would be visible to them. Means were astronomical; ends, astrological.

004.031 **Two uses of ancient astronomy.**   R. R. Newton.
  Phil. Trans. Roy. Soc. London, Ser. A, Vol. 276, (see 012.004), 99 - 116 (1974).

Observations from ancient astronomy are useful in studying the non-gravitational accelerations of the earth and moon, and recent developments in this study are reviewed. Such a study necessarily involves astronomical chronology and simultaneously shows some limitations in its use. Limitations include lack of veracity in many records, errors in dating events, and uncertainty in calculating the circumstances of ancient eclipses of the sun. These limitations are studied both quantitatively and by example.

004.032 **Contributions to the discussion on astronomy in ancient literate societies.**
A. Digby, F. R. Stephenson.
Phil. Trans. Roy. Soc. London, Ser. A, Vol. 276, (see 012. 004), 117 - 121 (1974).

004.033 **Neolithic science and technology.**
  R. J. C. Atkinson.
Phil. Trans. Roy. Soc. London, Ser. A, Vol. 276, (see 012. 004), 123 - 131 (1974).

The purpose of this paper is to furnish a background of information about the state of neolithic technology, against which the evidence for the beginnings of astronomy in prehistoric Europe can be examined in greater detail.

004.034 **Voyaging stars: aspects of Polynesian and Micronesian astronomy.**   D. Lewis.
Phil. Trans. Roy. Soc. London, Ser. A, Vol. 276, (see 012. 004), 133 - 148 (1974).

In Polynesia and Micronesia, where concepts are virtually identical, astronomy and navigation form one inseparable science. Sky 'domes', stellar zones, the seasons, yam growth and the Pleiades, as well as time spans of settlement, are mentioned in the paper, and attention is drawn to apparent maritime technological parallels elsewhere.

004.035 **Astronomical significance of prehistoric monuments in Western Europe.**   A. Thom.
Phil. Trans. Roy. Soc. London, Ser. A, Vol. 276, (see 012. 004), 149 - 156 (1974).

The accuracy of megalithic man's linear measurements is shown by recent work in Orkney and Brittany. Everywhere his geometry is based on integral right-angled triangles. The sun was observed at the solstices and equinoxes. The evidence that the moon was observed in its extreme positions is extensive but the most interesting sites are those which show the small perturbation of the inclination of the lunar orbit.

004.036 **Astronomical alinements in Britain, Egypt and Peru.**
  G. S. Hawkins.
Phil. Trans. Roy. Soc. London, Ser. A. Vol. 276, (see 012. 004), 157 - 167 (1974).

Photogrammetric air surveys have been made of Stonehenge, the Great Temple of Amon-Re, Karnak, and the desert lines near Nasca, Peru. New astronomical alinements have been found for the trilithons and station stones at Stonehenge. The Amon-Re temple alined precisely, within the limits of present-day measurements, with the rising of the sun at midwinter at the time of rebuilding of the structure by Tuthmosis III. No significant astronomical alinements were found for the desert lines.

004.037 **Archaeological tests on supposed prehistoric astronomical sites in Scotland.**
E. W. MacKie, with an appendix: 'Petrofabric analysis' by J. S. Bibby.
Phil. Trans. Roy. Soc. London, Ser. A, Vol. 276, (see 012. 004), 169 - 194 (1974).

The Thom theories; Two solstice sites; Dating the standing stone sites; Astronomer priests in iron age Britain.

004.038 **Climate, vegetation and forest limits in early civilized times.**   H. H. Lamb.
Phil. Trans. Roy. Soc. London, Ser. A, Vol. 276, (see 012. 004), 195 - 230 (1974).

004.039 **Hunting quanta.**   D. G. Kendall.
  Phil. Trans. Roy. Soc. London, Ser. A, Vol. 276, (see 012.004), 231 - 266 (1974). − In loving memory of

Noel Bryan Slater (1912–1973).

Thom considers that many of the dimensions of megalithic sites can be expressed in terms of a quantum of 5.44 ft (1.66 m). The problem of detecting a quantum when its size is not known in advance has been studied in pioneer papers by Broadbent (1955, 1956). With the greatly increased volume of evidence now available, a renewed attack on this problem seems called for, and a new approach, based on a Fourier analysis, will be outlined here.

004.040 **Contributions to the discussion on ancient astronomy: the unwritten evidence.**
A. H. A. Hogg, L. E. Maistrov, A. Penny, J. E. Wood, H. L. Porteous, W. S. Reith.
Phil. Trans. Roy. Soc. London, Ser. A, Vol. 276, (see 012. 004), 267 - 271 (1974).

004.041 **History of formation of the scientific ideas of Copernicus.** G. S. Razdimakha.
Filos. probl. suchasn. prirodozn. Mizhvid. nauk. zb., 1974, vip. (No.) 37, p. 48 - 52. In Ukrainian. – Abstr. in Referativ. Zhurn. 51. Astron., 3.51.3 (1975).

004.042 **Quand commence et quand finit la canicule?**
J. Denoyelle, L. Dufour.
Ciel et Terre, Vol. 91, 137 - 142 (1975).

004.043 **Spektroskopie, al meer dan 170 jaar.**
W. van Rensbergen.
Zenit, Vol. 2, 136 - 137, 158 (1975).

004.044 **Keplers droom van de maan.** W. Kastelein.
Zenit, Vol. 2, 138 - 139 (1975).

004.045 **Leonardo da Vinci (1452–1519): over de hemel èn de tijd.** T. de Vries.
Zenit, Vol. 2, 155 - 156 (1975).

004.046 **Megalithic astronomy: a prehistorian's view.**
A. Fleming.
Nature, Vol. 255, 575 (1975).

004.047 **Three hundred years of Greenwich.** G. Smith.
Nature, Vol. 225, 581 - 583 (1975).

004.048 **English astronomy before 1675.** J. L. Russell.
Nature, Vol. 255, 583 - 587 (1975).

004.049 **The Greenwich Observatory: origins and early development, 1675–1835.** E. G. Forbes.
Nature, Vol. 255, 587 - 592 (1975).

004.050 **Airy and after.** A. J. Meadows.
Nature, Vol. 255, 592 - 595 (1975).

004.051 **Greenwich Observatory in the twentieth century.**
A. Hunter.
Nature, Vol. 255, 596 - 599 (1975).

The author traces the history and future of observational astrophysics at the RGO in the twentieth century and the continuing role of the RGO in positional astronomy.

004.052 **Sidelights on the Royal Observatory.**
P. S. Laurie.
Nature, Vol. 255, 599 - 602 (1975).

004.053 **Astronomy three hundred years ago.**
O. Gingerich.
Nature, Vol. 255, 602 - 606 (1975).

004.054 **Thābit Ibn Qurra between Ptolemy and Copernicus:** an analysis of Thābit's solar theory.
K. P. Moesgaard.
Arch. History Exact Sci., Vol. 12, 199 - 216 (1974).

004.055 **Copernicus iaponicus.** S. Yajima.
Japan. Studies Hist. Sci., Vol. 12, 1 - 3 (1974).

004.056 **Ptolemy, Azarquiel, Ibn al-Shātir, and Copernicus on Mercury. A study of parameters.** W. Hartner.
Arch. Internat. Histoire Sci., Vol. 24, 5 - 25 (1974).

The author compares the 'models of Azarquiel, Ibn al-Shātir and Copernicus with the one displayed in Ptolemy's "Almagest". Formulae for converting one into the other are given.

004.057 **Nicholas Copernicus and his traditions in Toruń and in Poland.** R. S. Ingarden.
Rep. Math. Phys., Vol. 5, 1 - 5 (1974).

004.058 **Levi ben Gerson's preliminary lunar model.**
B. R. Goldstein.
Centaurus, Vol. 18, 275 - 288 (1974).

004.059 **On the moon's elliptic inequality, evection, and variation and Horrox's 'new theory of the moon'.**
N. T. Jørgensen.
Centaurus, Vol. 18, 316 - 318 (1974).

004.060 **Contribution à l'étude de la diffusion du De Revolutionibus de Copernic. Inventaire des exemplaires des deux premières éditions conservés en France.**
R. Taton, M. Cazenave.
Revue Histoire Sci., Vol. 27, 307 - 328 (1974/75).

As part of an international search, the results are given of the inquiry made in 1973–1974 by the French Committee on Nicolas Copernicus in French libraries and private collections concerning first editions of Copernicus' "De Revolutionibus" (1543, 1566) and "De Lateribus" (1542) and Rheticus' "Narratio prima" (1540, 1541).

004.061 **Copernic au Canada français: l'interdit, l'hypothèse et la thèse.** C. Galarneau.
Revue Histoire Sci., Vol. 27, 329 - 333 (1974/75).

004.062 **Ideas on the structure of the earth among the inhabitants of ancient Armenia.** V. M. Manoyan.
Trudy IV Zakavkaz. konf. po istorii nauki, posvyashch. 50-letiyu obrazovaniya SSSR. Sovet po istorii estestvozn. i tekhn. AN ArmSSR, 1972. Erevan, 1974, p. 345 - 354. In Russian. – Abstr. in Referativ. Zhurn. 51. Astron., 5.51.7 (1975).

004.063 **Criticism of Chaldean astrology by Anania Shirakazi.**
Eh. L. Danielyan.
Trudy IV Zakavkaz. konf. po istorii nauki, posvyashch. 50-letiyu obrazovaniya SSSR. Sovet po istorii estestvozn. i tekhn. AN ArmSSR, 1972. Erevan, 1974, p. 220 - 224. In Russian. – Abstr. in Referativ. Zhurn. 51. Astron., 5.51.9 (1975).

004.064 **Dissemination of Copernicus' doctrine in Lithuania till the end of the 18th century.**
Ya. A. Matviishin.
Funkts. i differents.-raznostn. uravneniya, Kiev, 1974, p. 156 - 176. In Russian. – Abstr. in Referativ. Zhurn. 51. Astron., 5.51.19 (1975).

004.065 **The Caracol tower at Chichen Itza: an ancient astronomical observatory?**
A. F. Aveni, S. L. Gibbs, H. Hartung.
Science, Vol. 188, 977 - 985 (1975).

The authors have examined the possible astronomical aspects of the principal architectural elements of the Caracol, treating each part in the chronological order in which it was constructed.

004.066 **E. C. Pickering in the history of variable star astronomy.** D. Hoffleit.
Journ. American Ass. Variable Star Observers, Vol. 1, 3 - 8 (1972).

004.067 **On the "Iwafune" or stone-vessel of Masuda; was it the oldest astronomical observatory founded 675 A.D. in Japan?** K. Saito.
Tokyo Astron. Obs., Report No. 65, Vol. 17, 213 - 260 (1975). In Japanese.

004.068 **The planet Venus in Mayan astronomy.** L. Zajdler.
Urania Kraków, Vol. 46, 167 - 175 (1975). In Polish.

004.069 **Archimedes's measurement of the sun's apparent diameter.** A. E. Shapiro.
Journ. History Astron., Vol. 6, 75 - 83 (1975).

004.070 **Tychonian observations, perfect numbers, and the date of creation: Longomontanus's solar and precessional theories.** K. P. Moesgaard.
Journ. History Astron., Vol. 6, 84 - 99 (1975).

004.071 **Further work on the Brogar lunar observatory.** A. Thom, A. S. Thom.
Journ. History Astron., Vol. 6, 100 - 114 (1975).

004.072 **The simultaneous 'discovery' of internal motions in spiral nebulae.** N. S. Hetherington.
Journ. History Astron., Vol. 6, 115 - 125 (1975).

004.073 **Astronomical archives in Canada.** R. A. Jarrell.
Journ. History Astron., Vol. 6, 143 - 147 (1975).
In the course of researching a history of Canadian astronomy, the author has visited or contacted all Canadian libraries, observatories, and archives containing astronomical materials of importance. Only Canadian sources are described here.

004.074 **Simon Newcomb and nineteenth century positional astronomy.** A. L. Norberg.
Diss. University of Wisconsin, 1974.

004.075 **The educational and intellectual background of American astronomers, 1825–1875.**
M. Rothenberg.
Diss. Bryn Mawr College, Pennsylvania, 1974.

004.076 **Newton and Clairaut on the motion of the lunar apse.** P. P. Chandler II.
Diss. University of California, San Diego, 1975.

004.077 **The hissing phenomenon.** R. A. Ham.
Journ. British Astron. Ass., Vol. 85, 317 - 323 (1975).

004.078 **Introduction: Nicholas Copernicus (Mikolaj Kopernik): an intellectual revolutionary.**
J. Neyman.
The heritage of Copernicus, (see 003.020), p. 1 - 22 (1974).

004.079 **Harlow Shapley and the discovery of the center of our Galaxy.** B. J. Bok.
The heritage of Copernicus, (see 003.020), p. 26 - 62 (1974).

004.080 **Edwin Hubble and the universe outside our Galaxy.** D. W. Goldsmith.
The heritage of Copernicus, (see 003.020), p. 63 - 94 (1974).

004.081 **Explosive events in the universe.** W. Zonn.
The heritage of Copernicus, (see 003.020), p. 95 - 115 (1974).

004.082 **Another look at Eratosthenes' and Posidonius' determinations of the earth's circumference.**
I. Fischer.
Quarterly Journ. Roy. Astron. Soc., Vol. 16, 152 - 167 (1975).

004.083 **Der Stern der Magier.** K. Ferrari d'Occhieppo.
Anzeiger Österreich. Akad. Wiss., Phil.-hist. Kl., Vol. 111, 319 - 345 = Astron. Mitt. Wien, Suppl. (1974/75).
I. The star of the Magi in the view of the late Babylonian astronomy; II. The names of the Magi: Bithisarea, Melichior, Gathaspar.

004.084 **Kepler's second law in England.**
V. E. Thoren, O. Gingerich, B. Welther.
British Journ. History Sci., (*GB*), No. 27, Vol. 7, 243 - 258 (1974).
Discusses the authors' interpretation of the work of Russell and Whiteside that Kepler's law was not being used, i.e. not accepted or not known, between 1650 and 1670.

004.085 **Mathematical treatment of astronomical observations (a historical essay).** O. B. Sheynin.
Arch. History Exact. Sci., Vol. 11, 97 - 126 (1973).

004.086 **The new science of motion: a study of Galileo's 'De motu locali'.** W. L. Wisan.
Arch. History Exact Sci., Vol. 13, 103 - 306 (1974).

004.087 **Hipparchus on the distances of the sun and moon.** G. J. Toomer.
Arch. History Exact Sci., Vol. 14, 126 - 142 (1974).

Les manuscrits de la bibliothèque de l'Observatoire de Paris. See Abstr. 002.001.

The planetarium of Giovanni de' Dondi, citizen of Padua. See Abstr. 003.030.

Astronomie Indienne. See Abstr. 003.033.

A concise history of science in India.
See Abstr. 003.036.

The natural philosophy of Galileo: essays of the origins and formation of classical mechanics.
See Abstr. 003.045.

Astronomie heute – Geschichte einer alten Wissenschaft. See Abstr. 003.055.

Astronomie vom Altertum bis heute.
See Abstr. 003.056.

On N. Copernic's life and research work (materials for discussions of pupils). See Abstr. 014.001.

Errata

004.901 **Erratum: "Copernicus and modern astronomy"** [Zemlya i Vselennaya, 1973, No. 4, p. 11 - 18].
V. A. Ambartsumyan.
Zemlya i Vselennaya, 1975, No. 2, p. 32. In Russian.

005 Biography

005.001 **Friedrich Wilhelm Argelander − zum hundertsten Todestag.** L. Brandt.
SuW, Vol. 14, 40 - 42 (1975).

005.002 **Life and work of Biruni.** P. G. Bulgakov.
Vopr. istorii estestvozn. i tekhn. Vyp. (No.) 2 - 3 (47 - 48). Moskva, Nauka, 1974, p. 68 - 75. In Russian. Abstr. in Referativ. Zhurn. 51. Astron., 2.51.15 (1975).

005.003 **Bruno H. Bürgel. Ein Bericht zur 100. Wiederkehr seines Geburtstags am 14. November 1975.**
E. Krug.
Orion, 33. Jahrgang, p. 9 - 12 (1975).

005.004 **Lobachevskij − astronomer.** N. I. Idel'son.
Zemlya i Vselennaya, 1975, No. 1, p. 38 - 45. In Russian.

005.005 **Petersburg astrophysicist of the XVIIIth century.** A. I. Eremeeva.
Zemlya i Vselennaya, 1975, No. 1, p. 62 - 66. In Russian. Franz Ulrich Theodor Epinus.

005.006 **Camille Flammarion et la Roumanie.** I. M. Stefan.
L'Astronomie, 89e année, p. 165 - 168 (1975).

005.007 **J. S. Bailly − 1736−1793.** R. G. Daniels.
Journ. British. Astron. Ass., Vol. 85, 224 - 237 (1975).

005.008 **Der einsame Weg zur Höhe. Zur 100. Wiederkehr des Geburtstages von Bruno H. Bürgel.**
E. Krug, with the preface to an unpublished biography of Bruno H. Bürgel by C. Hoffmeister.
SuW, Vol. 14, 148 - 149 (1975).

005.009 **Newton − a man of his times.** G. C. Dyer.
Journ. Navigation, Vol. 28, 209 - 217 (1975).

005.010 **A Titan of the High Renaissance (to the 500th anniversary of the birthday of Nicolaus Copernicus).**
O. V. Golubeva.
Sb. tr. Mosk. obl. ped. in-t, 1974, vyp. (No.) 3, p. 5 - 11. In Russian. − Abstr. in Referativ. Zhurn. 51. Astron., 4.51.24 (1975).

005.011 **Nouvelles recherches sur la biographie de Nicolas Copernic. Fondements méthodiques et résultats.**
M. Biskup.
Revue Histoire Sci., Vol. 27, 289 - 306 (1974/75).
Current state of research in Poland and in the two Republics of Germany into the life of Copernicus, his public activity and the role of certain social milieus in Royal Prussia and Warmia in Copernicus' scientific development and in his cultural environment; an examination of various bibliographic sources on these subjects. New elements concerning certain aspects of the life of Copernicus, principally during the Warmian epoch, which shed new light on the personal life of this scientist.

005.012 **Yrjö Väisälä.** T. J. Kukkamäki.
Proc. Finnish Acad. Sci. Letters 1972, p. 59 - 66 (1974). − Memorial address given January 14, 1972.

005.013 **Veikko Aleksanteri Heiskanen.** R. A. Hirvonen.
Proc. Finnish Acad. Sci. Letters 1972, p. 75 - 83 (1974). − Memorial address given May 12, 1972.

005.014 **Friedrich Wilhelm August Argelander, 22.3.1799 − 17.2.1875.** K.-B. Menzel.
BAV Rundbrief, 24. Jahrgang, p. 21 - 26 (1975).

005.015 **Notice sur la vie et l'œuvre de Pierre Tardi (1897 - 1972).** H. Lacombe.
Comptes Rendus Acad. Sci. Paris, Vie Académique, Vol. 280, 173 - 183 (1975).

005.016 **A. D. Thackeray.** G. G. Cillié.
Monthly Notes Astron. Soc. Southern Africa, Vol. 34, 34 - 38 (1975).

005.017 **Argelander: Een ten onrechte vergeten sterrenkundige.** G. W. E. Beekman.
Zenit, Vol. 2, 209 - 212 (1975).

005.018 **Vincent Ferraro as a pioneer of hydromagnetics.** T. G. Cowling.
Quarterly Journ. Roy. Astron. Soc., Vol. 16, 136 - 144 (1975). − Ferraro Memorial lecture, 1975 January 17.

005.019 **Encyclopaedist from Chwarezm (to the 1000th anniversary of the birthday of Abu'l Raihan Biruni).**
L. S. Khrenov.
Geod. raboty na Urale. Sverdlovsk, 1974, p. 193 - 217. In Russian. − Abstr. in Referativ. Zhurn. 51. Astron., 6.51.3 (1975).

005.020 **Harlow Shapley in variable comments.** D. Hoffleit.
Journ. American Ass. Variable Star Observers, Vol. 2, 1 - 6 (1973).

005.021 **Revolutionist in the role of a canon.** B. Popović.
Vasiona, Vol. 22, 81 - 86 (1974). In Serbo-Croatian.

005.022 **Albert Einstein.** H. Lambrecht.
Sterne, Vol. 51, 65 - 68 (1975).

006 Personal Notes

E. Anders received the Leonard Medal of the Meteoritical Society.
Meteoritics, Vol. 9, 425 - 426 (1974).

W. Fricke received the Prix Janssen.
L'Astronomie, 89ᵉ année, p. 37 (1975).

J. L. Greenstein received the Gold Medal of the Royal Astronomical Society.
Quarterly Journ. Roy. Astron. Soc., Vol. 16, 220 (1975).

J. L. Greenstein received the Gold Medal of the Royal Astronomical Society.
Sky Telescope, Vol. 49, 300 (1975).

V. Guth, 70th birthday.
Říše hvězd, Vol. 56, 32 (1975). In Czech.

S. W. Hawking received the RAS Eddington Medal and the Pius XI Medal.
Phys. Today, Vol. 28, No. 6, p. 75 (1975).

S. Hawking received the Eddington Medal.
Quarterly Journ. Roy. Astron. Soc., Vol. 16, 220 (1975).

G. H. Herbig, 1975 Henry Norris Russell Lecturer.
Phys. Today, Vol. 28, No. 5, p. 66 (1975).

A. Hewish received the Nobel prize in Physics for 1974.   V. K. Kapahi.
Bull. Astron. Soc. India, Vol. 2, 76 (1974).

A. Hewish wins the Nobel prize of Physics 1974.
G. Miley, F. Israel.
Zenit, Vol. 2, 124 - 128 (1975).

C. Jaschek, director of Centre de Données Stellaires, Strasbourg.
Centre de Données Stellaires, Inform.Bull. No. 8, (see 002.007), p. 1 (1975).

E. J. Öpik received the Gold Medal of the Royal

Astronomical Society.
Quarterly Journ. Roy. Astron. Soc., Vol. 16, 220 (1975).

R. Penrose received the RAS Eddington Medal.
Phys. Today, Vol. 28, No. 6, p. 75 (1975).

R. Penrose received the Eddington Medal.
Quarterly Journ. Roy. Astron. Soc., Vol. 16, 220 (1975).

V. C. Reddish, Astronomer Royal for Scotland and Regius Professor of Astronomy in the University of Edinburgh.
Monthly Notes Astron. Soc. Southern Africa, Vol. 34, 4 (1975).

V. C. Reddish, Director of the Royal Observatory, Edinburgh.
Observatory, Vol. 95, 36 (1975).

M. Ryle received the Nobel prize in Physics for 1974.   V. K. Kapahi.
Bull. Astron. Soc. India, Vol. 2, 76 (1974).

M. Ryle wins the Nobel prize of Physics 1974.
G. Miley, F. Israel.
Zenit, Vol. 2, 124 - 128 (1975).

F. G. Smith, director of the Royal Greenwich Observatory, Herstmonceux Castle, Sussex.
Monthly Notes Astron. Soc. Southern Africa, Vol. 34, 4 - 5 (1975).

A. Unsöld, 70th birthday.   K. Kromphardt.
Phys. Blätter, 31. Jahrg., p. 175 - 177 (1975).

A. Unsöld, 70th birthday.   H.-H. Voigt.
SuW, Vol. 14, 112 - 113 (1975).

L. Woltjer, director general of the European Southern Observatory.
Sky Telescope, Vol. 49, 291 (1975).

007 Obituaries

C. G. Abbot, 1872–1973.   D. Chalonge.
L'Astronomie, 89ᵉ année, p. 118 - 119 (1975).

M. Bidault de l'Isle, 1920 November 27 - 1975
January 23.   L. Tartois.
L'Astronomie, 89ᵉ année, p. 217 - 219 (1975).

A. Blagonravov died 1975 February 4.
Spaceflight, Vol. 17, 128 (1975).

J. Bosler, 1878 - 1973 September 25.
C. Fehrenbach.
L'Astronomie, 89ᵉ année, p. 220 - 221 (1975).

P.-E.-E. Bourgeois, 1898 February 13 - 1974 May 11.
S. Arend.
Ciel et Terre, Vol. 91, 1 - 4 (1975).

J. A. Carroll, 1899 January 8 - 1974 May 2.
D. H. Sadler, F. M. Sadler.
Quarterly Journ. Roy. Astron. Soc., Vol. 16, 100 - 103
(1975).

G. M. Clemence, 1908 August 16 - 1974 November
22.   R. L. Duncombe.
Phys. Today, Vol. 28, No. 3, p. 59, 61 (1975).

G. M. Clemence, 1908 August 16 - 1974 November
22.   D. H. Sadler.
Quarterly Journ. Roy. Astron. Soc., Vol. 16, 210 - 214 (1975).

G. M. Clemence died 1974 November 22.
Sky Telescope, Vol. 49, 93 (1975).

G. M. Clemence died 1974 November 22.
P. Herget.
Sky Telescope, Vol. 49, 215 - 216 (1975).

L. Detre died 1974 October 5.
BAV Rundbrief, 24. Jahrgang, p. 40 (1975).

L. Detre, 1906 - 1974.   L. Rosino.
Coelum, Vol. 43, 113 (1975).

L. Detre died 1974 October 5.
SuW, Vol. 14, 83 (1975).

T. E. de Vries died 1975 January 12.
T. de Groot.
Zenit, Vol. 2, 3 (1975).

S. Einarsson, 1879 - 1974.   J. Phillips.
Mercury, Vol. 4, No. 1, p. 7, 23 (1975).

P. W. Gast, 1930 - 1973.
Geochim. Cosmochim. Acta, Vol. 39, 93 - 94 (1975).

A. Gougenheim, 1902 January 31 – 1975 March 21.
H. Lacombe.
Comptes Rendus, Acad. Sci. Paris, Vie Académique, Vol. 280,
118 - 122 (1975).

B. J. Harris died 1974 December 23.
Journ. Astron. Soc. Victoria, Vol. 28, 10 (1975).

R. Jonckheere, 1888 July 25 – 1974.
C. Fehrenbach.

L'Astronomie, 89ᵉ année, p. 35 - 37 (1975).

H. Kienle, 1895 October 22 - 1975 February 15.
D. Labs.
Phys. Blätter, 31. Jahrgang, p. 222 - 223 (1975).

H. Kienle, 1895 - 1975 February 15.
Sky Telescope, Vol. 49, 368 (1975).

H. Kienle died 1975 February 15.
SuW, Vol. 14, 83 (1975).

H. Kienle, 1895 - 1975 February 15.
H. Haffner, A. Kizilirmak.
SuW, Vol. 14, 184 - 186 (1975).

G. Kuiper (1905–1973).   J.-C. Pecker.
L'Astronomie, 89ᵉ année, p. 84 - 85 (1975).

K.-O. Kiepenheuer, 1910 - 1975 May 23.
SuW, Vol. 14, 188 (1975).

E. M. Lindsay, 1907 - 1974.   P. Moore.
Journ. British Astron. Ass., Vol. 85, 358 - 359 (1975).

E. M. Lindsay, 1907 - 1974 July 27.
P. A. Wayman.
Quarterly Journ. Roy. Astron. Soc., Vol. 16, 215 - 217 (1975).

L. M. Milne-Thomson, 1891 May 1 - 1974 August
21.   D. H. Sadler.
Quarterly Journ. Roy. Astron. Soc., Vol. 16, 218 - 219 (1975).

R. A. Naef died 1975 March 13.
L. Baldinelli.
Giorn. Ass. Astrofili Bolognesi, No. 38, p. 8 (1975).

R. A. Naef, 1907 - 1975 March 13.   F. Egger.
Orion, 33. Jahrgang, p. 84 - 85 (1975).

R. A. Naef died 1975 March 13.
SuW, Vol. 14, 113 (1975).

R. A. Naef, 1907 - 1975 March 13.
W. Sandner.
SuW, Vol. 14, 209 (1975).

J. Pokrzywnicki, 1892 – 1974 December 28.
Meteoritics, Vol. 10, 99 (1975).

J. Pokrzywnicki, 1892 - 1974.   B. Lang.
Urania Kraków, Vol. 46, 98 - 100 (1975). In Polish.

J. Procházka died 1975 January 5.
B. Maleček.
Říše hvězd, Vol. 56, 101 (1975). In Czech.

G. T. Railton, 1901 - 1974.   L. P. Lee.
Southern Stars, Vol. 26, 15 (1975).

R. O. Redman died 1975 March 6.
Monthly Notes Astron. Soc. Southern Africa, Vol. 34, 16
(1975).

R. O. Redman died.
Nature, Vol. 254, 371 (1975).

R. O. Redman died 1975 March 6.
Observatory, Vol. 95, 115 (1975).

S. V. Rublev, 1930 February 26–1974 November 10.
Astrofiz. Issled., Izv. Spets. Astrofiz. Obs., Vol. 7, p. 247 (1975). In Russian.

J. Schaedler-Amstein died 1975 May 7.
E. Wiedemann.
Orion. 33. Jahrgang, p. 85 (1975).

F. P. Scott died 1974 November 1.
Sky Telescope, Vol. 49, 18 (1975).

K. W. Steinmetz, 1920 - 1974 August 29.
Strolling Astronomer, Vol. 25, 126 (1975).

W. Studer died 1975 May 17.   E. Wiedemann.
Orion, 33. Jahrgang, p. 93 (1975).

A. Thomson died 1974 October 17.
J. F. Heard.
Journ. Roy. Astron. Soc. Canada, Vol. 69, 32 - 34 (1975).

G. A. van Biesbroeck, 1880 January 21 - 1974
February 23.   S. Arend, J. Dommanget.

Quarterly Journ. Roy. Astron. Soc., Vol. 16, 104 - 105 (1975).

H. Witkowski, 1920 - 1974.   A. Woszczyk.
Urania Kraków, Vol. 46, 90 - 91 (1975). In Polish.

K. Wurm, 1899 - 1975.   L. Rosino.
Coelum, Vol. 43, 114 - 115 (1975).

K. Wurm died 1975 February 16.
Orion, 33. Jahrgang, p. 82 (1975).

K. Wurm died 1975 February 16.
SuW, Vol. 14, 113 (1975).

W. Zonn died 1975 February 28.
Observatory, Vol. 95, 115 (1975).

W. Zonn, 1905 July 14 - 1975 February 28.
J. Smak.
Urania Kraków, Vol. 46, 130 - 132 (1975). In Polish.

F. Zwicky, 1898 - 1974 February 8.
A. Wilson.
Quarterly Journ. Roy. Astron. Soc., Vol. 16, 106 - 108 (1975).

008 Observatories, Institutes

Reports, communications and publications of observatories and astronomical institutes are recorded in this section; included are numbered series of reprints. Whenever possible, the numbers of the abstracts referring to the publications are given. Observatories and institutes are listed in alphabetical order of their towns. In some cases observatory publications do not give the name of the town; the following list which gives names and towns of some institutions may serve as an aid in such cases.

Aarne Karjalainen Observatory	**Oulu**, Finland
Algonquin Radio Observatory	**Lake Traverse**, Ontario, Canada
Allegheny Observatory	**Pittsburgh**, Pennsylvania
Archenhold-Sternwarte	**Berlin-Treptow**, Germany
Arthur J. Dyer Observatory	**Nashville**, Tennessee
Astronomical Latitude Station, Polish Academy of Sciences	**Borowiec**, Poland
Bosscha Observatory	**Lembang**, Indonesia
Boyden Observatory	**Bloemfontein**, South Africa
Bureau International de l'Heure	**Paris**, France
Cajigal Observatory	**Caracas**, Venezuela
California Institute of Technology	**Pasadena**, California
Cape of Good Hope	**Cape Town**, South Africa
Carter Observatory	**Wellington**, New Zealand
Catalina Station	**Tucson**, Arizona
Cavendish Laboratory	**Cambridge**, England
Ceskoslovenská Akademie Ved Astronomický Ustav	**Praha**, Czechoslovakia
Chamberlin Observatory, University of Denver	**Denver**, Colorado
Commonwealth Observatory	**Canberra**, Australia
Corralitos Observatory	**Las Cruces**, New Mexico
David Dunlap Observatory, University of Toronto	**Richmond Hill**, Ontario
Dearborn Observatory	**Evanston**, Illinois
Department of Astronomy and Observatory, Univ. California	**Los Angeles**, California
Department of Astronomy, University of Texas	**Austin**, Texas
Division Radiophysics, C.S.I.R.O. University Grounds	**Sydney**, N.S.W., Australia
Dominion Astrophysical Observatory	**Victoria**, British Columbia
Dominion Observatory	**Ottawa**, Ontario
Dominion Radio Astrophysical Observatory	**Penticton**, British Columbia
Dudley Observatory	**Albany**, New York
Dunsink Observatory	**Dublin**, Ireland
Engelhardt Observatory	**Kazan**, U.S.S.R.
European Southern Observatory	**Hamburg**, Federal German Republic
Five College Observatories	**Amherst**, Massachusetts
Florida State University Radio Observatory	**Tallahassee**, Florida
Flower and Cook Observatories, University of Pennsylvania	**Philadelphia**, Pennsylvania
Fraunhofer Institut	**Freiburg**, Federal German Republic
Georgetown Observatory	**Washington**, D.C.
Goddard Space Flight Center	**Greenbelt**, Maryland
Goethe Link Observatory, University of Indiana	**Bloomington**, Indiana
Hale Observatories	**Pasadena**, California
Harvard College Observatory	**Cambridge**, Massachusetts

Harvard Radio Astronomy Station	**Cambridge, Massachusetts**
Haystack Observatory	**Westford, Massachusetts**
Heinrich-Hertz-Institut	**Berlin-Adlershof**, Germany
High Altitude Observatory, University of Colorado	**Boulder, Colorado**
Institute for Astronomy, University of Hawaii	**Honolulu, Hawaii**
Institute for Theoretical Astronomy (Institut Teoreticheskoj Astronomii)	**Leningrad**, U.S.S.R.
Institute of Theoretical Astrophysics, Blindern	**Oslo**, Norway
Inter-American Observatory	**La Serena**, (Cerro-Tololo), Chile
International Latitude Observatory	**Mizusawa**, Japan
Joint Institute for Laboratory Astrophysics (JILA)	**Boulder, Colorado**
Kandilli Observatory	**Istanbul, Turkey**
Kansas University Observatory	**Lawrence, Kansas**
Kapteyn Astronomical Laboratory	**Groningen**, Netherlands
Karl-Schwarzschild-Observatorium	**Tautenburg**, German Democratic Republic
Kenneth Mees Observatory	**Rochester**, New York
Kwasan Observatory	**Kyoto**, Japan
Lamont-Hussey Observatory	**Bloemfontein**, South Africa
Leander McCormick Observatory University of Virginia	**Charlottesville**, Virginia
Lee Observatory	**Beirut**, Lebanon
Leopold-Figl-Observatorium	**Wien**, Austria
Leuschner Observatory	**Berkeley**, California
Lick Observatory	**Santa Cruz**, (Mount Hamilton), California
Lindheimer Astronomical Research Center	**Evanston**, Illinois
Lockheed Solar Observatory	**Saugus**, California
Lohrmann-Observatorium für Geodätische Astronomie	**Dresden**, German Democratic Republic
Louisiana State University Observatory	**Baton Rouge**, Louisiana
Lowell Observatory	**Flagstaff**, Arizona
Lunar and Planetary Laboratory	**Tucson**, Arizona
Max-Planck-Institut für Astronomie	**Heidelberg**, Federal German Republic
Max-Planck-Institut für Physik und Astrophysik	**München**, Federal German Republic
Max-Planck-Institut für Radioastronomie	**Bonn**, Federal German Republic
McDonald Observatory	**Fort Davis**, Texas
McMath Hulbert Observatory	**Pontiac**, Michigan
Michigan State University Observatory	**East Lansing**, Michigan
Molonglo Radio Observatory, University of Sydney	**Sydney**, New South Wales
Mount Cuba Observatory	**Wilmington**, Delaware
Mount John Observatory	**Lake Tekapo**, New Zealand
Mount Palomar Observatory	**Pasadena**, California
Mount Wilson Observatory	**Pasadena**, California
Mullard Radio Astronomy Observatory	**Cambridge**, England
Narrabri Observatory, University of Sydney	**Sydney**, New South Wales

National Bureau of Standards **Washington, D. C.**
~ .ional Observatory,USA **Kitt Peak, Arizona**
National Radio Astronomy **Charlottesville, Virginia**
  Observatory **Green Bank, West Virginia**
    Tucson, Arizona
New Mexico State
  University Observatory **Las Cruces, New Mexico**
Nizamiah Observatory **Hyderabad, India**
Nuffield Radio Astronomy
  Laboratories, Jodrell Bank
  University of Manchester **Manchester, England**
Observatoire Royal de Belgique **Uccle, Belgium**
Observatorio de Cartuja **Granada, Spain**
Observatorio del Ebro **Tortosa, Spain**
Observatorio Fabra **Barcelona, Spain**
Observatory, University of
  Michigan **Ann Arbor, Michigan**
Ohio State University
  Radio Observatory **Columbus, Ohio**
Ole Roemer-Observatoriet **Aarhus, Denmark**
Onsala Space Observatory **Gothenburg, Sweden**
Owens Valley Radio **Big Pine, California**
  Observatory
Perkins Observatory, Ohio State
  and Wesleyan Universities **Delaware, Ohio**
Purple Mountain Observatory **Nanking, China**
Radcliffe Observatory **Pretoria, South Africa**
Remeis-Sternwarte **Bamberg,**
    Federal German Republic
Republic Observatory **Johannesburg, South Africa**
Rosemary Hill Observatory **Gainesville, Florida**
Royal Radar Establishment,
  Radio Astronomy Division **Malvern, England**

Sagamore Hill Radio Observatory **Bedford, Massachusetts**
Saint-Michel, l'Observatoire **Haute Provence, France**
San Fernando Observatory **El Segundo, California**
Smithsonian Astrophysical
  Observatory **Cambridge, Massachusetts**
Specola Astronomica Vaticana **Castel Gandolfo, Italy**
Specola di Padova **Asiago, Italy**
Sproul Observatory **Swarthmore, Pennsylvania**
Sternberg Observatory **Moscow, U.S.S.R.**
Steward Observatory,
  University of Arizona **Tucson, Arizona**
United States Naval Observatory **Washington, D.C.**
University of Florida,
  Radio Observatory **Gainesville, Florida**
University of Illinois Observatory **Urbana, Illinois**
University of Michigan
  Observatories **Ann Arbor, Michigan**
University of South Florida
  Observatory **Tampa, Florida**
Uttar Pradesh State Observatory **Naini Tal, India**
Van Vleck Observatory **Middletown, Connecticut**
Wallace Observatory **Cambridge,.Massachusetts**
Warner and Swasey Observatory **Cleveland, Ohio**
Washburn Observatory **Madison, Wisconsin**
West Melton Observatory **Christchurch, New Zealand**
Wilhelm-Förster Sternwarte **Berlin**
Yale University Observatory **New Haven, Connecticut**
Yerkes Observatory **Williams Bay, Wisconsin**
Zentralinstitut für Astrophysik,
  Sternwarte Babelsberg, (Fach-
  bereich Kosmische Physik) **Potsdam-Babelsberg, German**
    Democratic Republic

008.001 Albany

Dudley Observatory and State University of New York at Albany (SUNYA), Albany, New York. – Observatory reports. J. W. Erkes, J. L. Weinberg. Bull. American Astron. Soc., Vol. 7, 31 - 39 (1975).

This report covers activities for Dudley Observatory and for the SUNYA Department of Astronomy and Space Science from October 1971 to September 1974, and for the SUNYA Space Astronomy Laboratory from October 1973 to September 1974.

Dudley Observatory, *Albany, New York,* **Reports,** No. 9 (A. G. D. Philip, D. S. Hayes, 13.012.008).

Dudley Observatory, *Albany, New York,* **Reprint** Nos. B39 (D. S. Hallgren, C. L. Hemenway, V. A. Mohnen, C. D. Tackett, 12.082.065), B41 (D. S. Hallgren, D. C. Schmalberger, C. L. Hemenway, 12.082.066), B42 (C. L. Hemenway, J. W. Erkes, J. M. Greenberg, D. S. Hallgren, D. C. Schmalberger, 12.082.067), B45 (C. L. Hemenway, 13.105.137), B46 (C. L. Hemenway, 13.082.087), B47 (C. L. Hemenway, 13.106.045), B48 (W. F. Tozer, D. E. Beeson, 12.082.095), B49 (C. L. Hemenway, D. S. Hallgren, C. D. Tackett, 13.105.138).

Dudley Observatory, *Albany, New York.* **Reprint** Nos. C40 (A. G. D. Philip, 10.113.071), C41 (A. G. D. Philip, 10.131.133), C47 (A. G. D. Philip, J. W. Sulentic, 09.160.014), C50 (A. G. D. Philip, K. D. Philip, 09.113.011), C51 (A. G. D. Philip, 08.113.045), C52 (R. D. Mercer, L. Dunkelman, C. L. Ross, A. Worden, 12.031.083), C53 (J. L. Weinberg, M. S. Hanner, H. M. Mann, P. B. Hutchison, R. Fimmel, 12.106.047), C54 (A. G. D. Philip, 09.153.013), C56 (A. G. D. Philip, 09.154.014), C57 (D. W. Schuerman, 09.117.037), C59 (Y. K. Minn, J. M. Greenberg, 09.131.008), C61 (H. E. Bond,

A. G. D. Philip, 09.113.056), C62 (Y. K. Minn, J. M. Greenberg, 09.131.010), C64 (M. S. Hanner, J. L. Weinberg, 09.106.003), C65 (J. M. Greenberg, R. T. Wang, 09.131.091), C67 (K. G. Henize, J. L. Weinberg, 09.054.007), C68 (M. S. Hanner, J. L. Weinberg, 12.106.074), C70 (R. H. Giese, M. S. Hanner, C. Leinert, 10.106.034), C72 (D. Clarke, 11.031.037), C73 (D. Clarke, 11.031.040), C75 (J. M. Greenberg, 10.131.146), C78 (J. M. Greenberg, A. J. Yencha, 10.131.182), C81 (R. D. Mercer, L. Dunkelman, R. E. Evans, 13.106.046; R. M. MacQueen, C. L. Ross, R. E. Evans, 13.074.095), C82 (G. E. Mavko, D. S. Hayes, J. M. Greenberg, W. A. Hiltner, 11.131.042), C84 (P. M. Millman, A. F. Cook, C. L. Hemenway, 13.104.035), C85 (A. F. Cook, C. L. Hemenway, P. M. Millman, A. Swider, 13.104.036), C86 (J. M. Greenberg, 13.131.140), C87 (H. Patashnick, G. Rupprecht, D. W. Schuerman, 12.102.001).

008.002 Alger

Université d'Alger. Annales de l'Observatoire Astronomique d'Alger, Tome 4, Fasc. 1 (A. Ghezloun, M. Benhocine, J. Pham-Van, 13.041.023; L. A. Ourassine, 13.158.087; D. A. Andrienko, L. A. Ourassine, J.-C. Pham-Van, M. A. Svetchnikov, 13.103.100; D. A. Andrienko, L. A. Ourassine, 13.099.230; D. A. Andrienko, 13.102.029).

008.003 Ames, Iowa

Erwin W. Fick Observatory, Iowa State University, Ames, Iowa. – Observatory report. W. I. Beavers. Bull. American Astron. Soc., Vol. 7, 50 - 52 (1975).

008.004 Amherst

Five College Astronomy Department, Amherst
College, Amherst, Massachusetts; Hampshire College, Amherst,
Massachusetts; Mount Holyoke College, South Hadley, Massachusetts; Smith College, Northampton, Massachusetts; University of Massachusetts, Amherst, Massachusetts. – Observatory report.   W. M. Irvine.
Bull. American Astron. Soc., Vol. 7, 52 - 55 (1975).

008.005 Ann Arbor

University of Michigan, Ann Arbor, Michigan. – Observatory report.   W. A. Hiltner.
Bull. American Astron. Soc., Vol. 7, 144 - 147 (1975).

008.006 Arecibo

Arecibo Observatory, Arecibo, Puerto Rico. – Observatory report.   H. D. Craft.
Bull. American Astron. Soc., Vol. 7, 10 - 13 (1975).

008.007 Armagh

Contributions from the Armagh Observatory,
Nos. 81 (E. J. Öpik, 11.097.019), 82 (D. J. Mullan, 11.065.
024), 84 (E. M. Lindsay, 11.123.001), 85 (E. J. Öpik, 12.
097.004), 86 (E. M. Lindsay, 12.123.017).

Armagh Observatory, Leaflet, Nos. 122 (E. J. Öpik,
12.015.011), 123 (E. J. Öpik, 12.015.012), 124 (E. Öpik, 12.
002.043).

008.008 Athens

Astronomical Institute, National Observatory of
Athens. – Annual report 1973.   D. Kotsakis.
Annual Rep. Astron. Inst. Greece 1973, p. 3 - 6 (1974).

Astronomical Institute, National Observatory of
Athens. – Annual report 1974.   E. Mariolopoulos.
Annual Rep. Astron. Inst. Greece 1974, p. 3 - 6 (1975).

Department of Astronomy, University of Athens.
Annual report 1973.   D. Kotsakis.
Annual Rep. Astron. Inst. Greece 1973, p. 7 - 10 (1974).

Department of Astronomy, University of Athens.
Annual report 1974.   M. Moutsoulas.
Annual Rep. Astron. Inst. Greece 1974, p. 7 - 9 (1975).

Department of Astronomy, Technical University
of Athens. – Annual report 1973.   J. Argyrakos.
Annual Rep. Astron. Inst. Greece 1973, p. 16 (1974).

Department of Astronomy, Technical University
of Athens. – Annual report 1974. G. Argyrakos.
Annual Rep. Astron. Inst. Greece 1974, p. 15 (1975).

Research Center for Astronomy and Applied
Mathematics, Academy of Athens. – Annual report 1973.

C. Macris.
Annual Rep. Astron. Inst. Greece 1973, p. 17 - 19 (1974).

Research Center for Astronomy and Applied
Mathematics, Academy of Athens. – Annual report 1974.
C. Macris.
Annual Rep. Astron. Inst. Greece 1974, p. 16 - 18 (1975).

Research Center for Astronomy and Applied
Mathematics, Academy of Athens, Contributions Series I
(Astronomy), Nos. 36 (C. Poulakos, 13.072.057), 37 (C. J.
Macris, 13.073.084), 38 (J. Xanthakis, C. Poulakos, B.
Tritakis, 13.085.019).

University of Athens, New Chair of Astrophysics.
S. Svolopoulos.
Annual Rep. Astron. Inst. Greece 1973, p. 27 (1974).

Chair of Astrophysics, University of Athens.
Annual report 1974.   S. Svolopoulos.
Annual Rep. Astron. Inst. Greece 1974, p. 26 (1975).

008.009 Atlanta, Georgia

Fernbank Observatory, Fernbank Science Center,
Atlanta, Georgia. – Observatory report.   R. R. Hayward.
Bull. American Astron. Soc., Vol. 7, 49 - 50 (1975).

008.010 Austin

McDonald Observatory, University of Texas at
Austin, Austin, Texas. – Observatory report.
H. J. Smith, D. S. Evans.
Bull. American Astron. Soc., Vol. 7, 318 - 335 (1975).

008.011 Baton Rouge

Louisiana State University Observatory, Baton
Rouge, Louisiana. – Observatory report.   A. U. Landolt.
Bull. American Astron. Soc., Vol. 7, 127 - 129 (1975).

008.012 Bedford

Sagamore Hill Radio Observatory, Air Force
Cambridge Research Laboratories, Bedford, Massachusetts.
Observatory report.   D. A. Guidice.
Bull. American Astron. Soc., Vol. 7, 199 - 201 (1975).

008.013 Beograd

Publications de l'Observatoire Astronomique de
Beograd, No. 19 (G. M. Popović, 13.118.022).

008.014 Berkeley

University of California, Berkeley, Los Angeles, San
Diego, and Lick Observatories. I. Berkeley Campus. – Observatory report.   L. V. Kuhi.
Bull. American Astron. Soc., Vol. 7, 281 - 293 (1975).

008.015   Berlin

Veröffentlichungen der Wilhelm-Foerster-Stern-
warte Berlin, No. 37 (W. Nehls, 13.014.017).

008.016   Berlin-Adlershof

Zentralinstitut für solar-terrestrische Physik
(Heinrich-Hertz-Institut).
Akad. Wiss. DDR, Jahrbuch 1974, p. 257 - 263 (1975).

Heinrich-Hertz-Institut, Solare Beobachtungsergebnisse. Akademie der Wissenschaften der DDR, Zentralinstitut
für Solar-Terrestrische Physik (Heinrich-Hertz-Institut),
Berlin-Adlershof. HHI Solar Data, Vol. 25, 1974 November –
December; Vol. 26, 1975 January – March (E. A. Lauter,
C.-U. Wagner, A. Böhme, F. Fürstenberg, D. Scholz, S. Böhm,
13.075.007).

008.017   Berlin-Treptow

Blick in das Weltall, Archenhold-Sternwarte Berlin-
Treptow. Astronomische Veranstaltungen und Mitteilungen
für Sternfreunde, 23. Jahrgang, Nos. 1 - 6 (1975).

008.018   Big Pine, California

Owens Valley Radio Observatory, California Institute of Technology, Big Pine, California. – Observatory report.   G. J. Stanley.
Bull. American Astron. Soc., Vol. 7, 186 - 189 (1975).

008.019   Bloemfontein

University of the Orange Free State, Boyden Observatory, Department of Astronomy.   A. H. Jarrett.
Monthly Notes Astron. Soc. Southern Africa, Vol. 34, 23 - 24
(1975). – Report for 1974.

American exodus from Boyden Observatory.
A. H. Jarrett.
Monthly Notes Astron. Soc. Southern Africa, Vol. 34, 32 - 33
(1975).

008.020   Bloomington

Goethe Link Observatory, Indiana University,
Bloomington, Indiana. – Observatory report.
F. K. Edmondson.
Bull. American Astron. Soc., Vol. 7, 122 - 126 (1975).

008.021   Bologna

Pubblicazioni dell'Osservatorio Astronomico,
Universitario di Bologna, Vol. 10, Nos. 20 (P. Battistini, A.
Bonifazi, A. Guarnieri, 09.121.070), 21 (V. Castellani, P.
Giannone, A. Renzini, 08.065.055), 22 (V. Castellani,
P. Giannone, A. Renzini, 10.122.069), 23 (C. Cacciari, 11.
122.093), 24 (P. Battistini, A. Bonifazi, A. Guarnieri, 12.121.

018), 25 (F. Fusi-Pecci, A. Renzini, 13.064.046).

008.022   Bombay

Radio astronomy centre, Ootacamund.
G. Swarup.
Bull. Astron. Soc. India, Vol. 2, 26 - 27 (1974).
This report describes work being carried out in the field
of radio astronomy at the Tata Institute of Fundamental
Research using a large radio telescope located at Ootacamund.

008.023   Bonn

Mitteilungen der Astronomischen Institute Bonn,
Nos. 142 (O. Hachenberg, W. Priester, H. Schmidt, 10.008.
019), 143 (O. Hachenberg, W. Priester, H. Schmidt, 12.008.
015).

Veröffentlichungen der Astronomischen Institute
Bonn, No. 88 (W. Wiemer, 13.113.060).

Max-Planck-Gesellschaft zur Förderung der Wissen-
schaften, M. P. I. f. Radioastronomie, Bonn, Sonderdruck,
Ser. A, Nos. 1 (W. M. Goss, L. E. B. Johansson, J. Elldér,
B. Höglund, Nguyen-Q-Rieu, A. Winnberg, 10.131.071), 2
(K. H. Hesse, W. Sieber, R. Wielebinski, 10.141.517), 3
(N. J. Keen, W. E. Wilson, C. G. T. Haslam, D. A. Graham,
P. Thomasson, 10.125.018), 4 (W. Sieber, 10.141.524), 5
(E. Fürst, O. Hachenberg, W. Zinz, W. Hirth, 10.077.061), 6
(H. J. Wendker, J. W. M. Baars, W. J. Altenhoff, 10.141.055),
7 (J. Barsuhn, 13.022.071), 8 (W. T. Sullivan, III, D. Downes,
10.131.230), 9 (H. E. Matthews, W. M. Goss, A. Winnberg,
H. J. Habing, 10.131.145), 10 (C. G. T. Haslam, W. E. Wilson,
D. A. Graham, G. C. Hunt, 11.157.004), 11 (W. M. Goss,
Nguyen-Quang-Rieu, A. Winnberg, 10.131.231), 12 (M.
Grewing, C. M. Walmsley, 11.131.006), 13 (W. M. Goss,
A. Winnberg, H. J. Habing, 11.114.018), 14 (U. Mebold,
O. Hachenberg, C. A. Laury-Micoulaut, 11.131.506), 15
(P. G. Mezger, 11.131.106), 16 (C. Andersson, L. E. B.
Johansson, W. M. Goss, A. Winnberg, Nguyen-Quang-Rieu,
11.131.019), 17 (T. L. Wilson, 11.131.513), 18 (R. E. Hills,
J. N. Bahcall, 13.141.096), 19 (L. E. B. Johansson, B. Höglund,
A. Winnberg, Nguyen-Q-Rieu, W. M. Goss, 11.131.092), 20
(K. H. Hesse, R. Wielebinski, 11.141.314), 21 (A. Greve,
C. D. McKeith, N. E. McKeith, 11.076.010), 22 (P. Stumpff,
J. Schraml, 11.033.043), 23 (P. G. Mezger, E. B. Churchwell,
T. A. Pauls, 11.155.063), 24 (O. Hachenberg, U. Mebold,
11.131.546), 25 (A. Greve, 11.073.051), 26 (L. F. Smith, 12.
131.221), 27 (I. I. K. Pauliny-Toth, K. I. Kellermann, 12.141.
053), 28 (P. G. Mezger, L. F. Smith, E. Churchwell, 11.131.
538), 29 (E. Churchwell, P. G. Mezger, W. Huchtmeier, 11.
131.539), 30 (P. G. Mezger, 11.155.072), 31 (J. Barsuhn, 12.
131.026), 32 (A. Greve, C. D. McKeith, 12.076.001), 33 (E.
Preuß, I. Pauliny-Toth, A. Witzel, 12.033.001), 34 (E. Church-
well, 12.131.538), 35 (A. J. Green, 12.155.030), 36 (I. I. K.
Pauliny-Toth, A. Witzel, S. Gorgolewski, 12.099.201), 37 (D.
Downes, T. L. Wilson, 12.125.003), 38 (T. Pauls, P. G. Mezger,
E. Churchwell, 12.131.504), 39 (E. M. Berkhuijsen, R.
Wielebinski, 12.158.031), 40 (H. J. Habing, W. M. Goss,
H. E. Matthews, A. Winnberg, 12.131.522), 41 (D. Hoang-
Binh, C. M. Walmsley, 12.131.523), 42 (R. Sancisi, W. M. Goss,
C. Anderson, L. E. B. Johansson, A. Winnberg, 12.131.075),
43 (A. Greve, 12.114.105), 44 (W. K. Huchtmeier, E. Church-
well, 12.159.005), 45 (I. I. K. Pauliny-Toth, A. Witzel, E.
Preuss, 12.141.055), 46 (E. M. Berkhuijsen, 12.125.024), 47
(D. Downes, T. L. Wilson, 12.131.015), 48 (F. F. Gardner,

T. L. Wilson, P. Thomasson, 13.131.502), 49 (E. Fürst, O. Hachenberg, W. Hirth, 12.080.044), 50 (A. Greve, 13.031. 035), 51 (W. H. Hocking, G. Winnewisser, 13.022.072), 52 (J. R. Baker, E. Preuss, J. B. Whiteoak, P. Zimmermann, 12.158.138), 53 (V. Pankonin, A. Parrish, Y. Terzian, 13.131. 505), 54 (A. Winnberg, Nguyen-Q-Rieu, L. E. B. Johansson, W. M. Goss, 13.131.023), 55 (A. Greve, 13.071.011), 56 (D. A. Graham, U. Mebold, K. H. Hesse, D. L. Hills, R. Wielebinski, 13.131.504), 57 (R. Schröder, H. J. Wendker, P. Stumpff, 13.103.100), 58 (L. F. Smith, 13.131.519), 59 (G. Winnewisser, 13.131.128), 60 (G. Winnewisser, 13.131.112).

008.024 Borowiec

Polish Academy of Sciences, Astronomical Latitude Station, Borowiec — Poland, Circular Nos. 132 - 133 (13.044. 018).

008.025 Boulder

Joint Institute for Laboratory Astrophysics of the National Bureau of Standards and the University of Colorado, Boulder, Colorado. — Observatory report.　　P. S. Conti. Bull. American Astron. Soc., Vol. 7, 76 - 85 (1975).

National Oceanic and Atmospheric Administration, Space Environment Services Center, Boulder, Colorado. R. B. Doeker. Bull. American Astron. Soc., Vol. 7, 169 (175).

008.026 Cambridge, England

University of Cambridge, Institute of Astronomy. D. Lynden-Bell. Quarterly Journ. Roy. Astron. Soc., Vol. 16, 170 - 183 (1975). — Report for the year ending 1974 August 31.

008.027 Cambridge, Massachusetts

Center for Astrophysics, Harvard College Observatory and Smithsonian Astrophysical Observatory, Cambridge, Massachusetts. — Observatory reports.　　G. B. Field. Bull. American Astron. Soc., Vol. 7, 201 - 206 (1975).

Smithsonian Contributions to Astrophysics. Smithsonian Institution Astrophysical Observatory, (United States Government Printing Office, Washington), Nos. 16 (C. H. Payne-Gaposchkin, 13.122.112), 17 (C. H. Payne-Gaposchkin, 13.122.113).

Smithsonian Institution. Astrophysical Observatory. Research in Space Science. SAO Special Reports, Nos. 362 (R. L. Kurucz, E. Peytremann, 13.022.086), 363 (M. R. Williamson, E. M. Gaposchkin, 13.081.031), 364 (H. Kinoshita, 13.043.011), 365 (B. A. Romanowicz, 13.052.064).

008.028 Cape Town

University of Cape Town: Department of Astronomy; Department of Applied Mathematics. B. Warner, G. F. R. Ellis.

Monthly Notes Astron. Soc. Southern Africa, Vol. 34, 19 - 23 (1975). — Report for 1974.

008.029 Cardiff, Wales

University College, Cardiff, Astronomical Communications Nos. 6 (N. C. Wickramasinghe, 13.131.003), 7 (M. G. Edmunds, N. C. Wickramasinghe, 12.065.030), 8 (M. G. Edmunds, 13.155.018), 9 (N. C. Wickramasinghe, T. Lukes, M. J. Dempsey, 12.063.038), 10 (T. L. John, D. J. Morgan, A. R. Williams, 13.022.082), 11 (N. C. Wickramasinghe, 12.131.099), 12 (T. L. John, 13.022.003), 13 (T. L. John, D. J. Morgan, 13.022.002).

008.030 Carmel Valley, California

Monterey Institute for Research in Astronomy, Carmel Valley, California. — Observatory report. W. B. Weaver. Bull. American Astron. Soc., Vol. 7, 152 - 153 (1975).

008.031 Castel Gandolfo

Specola Vaticana. Annual report 1974: Report of the Astronomical Observatory; Report of the Astrophysical Laboratory.　　P. J. Treanor, J. Junkes. Printed in Vatican City, 15 pp. (1975).

Ricerche Astronomiche, Specola Vaticana, Città del Vaticano, Vol. 8, Nos. 24 (G. V. Coyne, 13.121.084), 25 (H. H. Swope, 13.121.085), 26 (D. J. K. O'Connell, 13.121. 086), 27 (G. V. Coyne, 13.126.019).

Ricerche Spettroscopiche, Laboratorio Astrofisico della Specola Vaticana, Vol. 3, No. 8 (J. Junkes, 13.031.029).

008.032 Catania

Report from the Catania Astrophysical Observatory (1973). List of the papers sent to the printer during 1973. G. Godoli. Oss. Astrofis. Catania, Pubbl. Nuova Ser. No. 151, 10 pp. (1974).

Report from the Catania Astrophysical Observatory (1974). List of the papers sent to the publisher during 1974. G. Godoli. Oss. Astrofis. Catania, Pubbl. Nuova Ser. No. 154, 10 pp. (1975).

Osservatorio Astrofisico di Catania, Pubblicazione Nuova Serie Nos. 151 (G. Godoli, 13.008.032), 153 (G. Godoli, 13.080.048), 154 (G. Godoli, 13.008.032), 155 (G. Godoli, 13.075.008).

008.033 Charlottesville

Leander McCormick Observatory, University of Virginia, Charlottesville, Virginia. — Observatory report. L. W. Fredrick. Bull. American Astron. Soc., Vol. 7, 137 - 140 (1975).

National Radio Astronomy Observatory, Charlottesville, Virginia; Green Bank, West Virginia; and Tucson, Arizona. — Observatory report.　　D. S. Heeschen, D. E. Hoog. Bull. American Astron. Soc., Vol. 7, 170 - 178 (1975).

008.034　Chicago

The University of Chicago, Department of Astronomy and Astrophysics, Chicago, Illinois. The Yerkes Observatory, Williams Bay, Wisconsin. — Observatory reports. E. N. Parker, L. M. Hobbs. Bull. American Astron. Soc., Vol. 7, 25 - 30 (1975).

008.035　Cincinnati

Minor Planet Circulars (MPC), Nos. 3779 - 3828 (P. Herget, 13.098.067).

008.036　Cleveland, Ohio

Warner and Swasey Observatory, Case Western Reserve University, Cleveland, Ohio. — Observatory report. W. P. Bidelman. Bull. American Astron. Soc., Vol. 7, 218 - 222 (1975).

008.037　College Park, Maryland

University of Maryland, College Park, Maryland. Observatory report.　　F. J. Kerr. Bull. American Astron. Soc., Vol. 7, 132 - 137 (1975).

008.038　Columbus

The Observatories of the Ohio State and Ohio Wesleyan Universities, Columbus and Delaware, Ohio. — Observatory reports.　　A. Slettebak. Bull. American Astron. Soc., Vol. 7, 181 - 185 (1975).

Ohio State University Radio Observatory, Columbus, Ohio. Observatory report.　　J. Kraus. Bull. American Astron. Soc., Vol. 7, 185 - 186 (1975).

008.039　Copenhagen

Copenhagen University Observatory Reprint Nos. 241 (B. G. Jørgensen, B. Reipurth, 09.103.005), 242 (J. O. Petersen, 10.122.008), 243 (L. Hansen, 10.115.007), 244 (B. Strömgren, 10.114.140), 245 (J. V. Clausen, 10.114.139), 246 (J. Andersen, 11.118.004), 247 (J. Andersen, A. H. Batten, R. W. Hilditch, 11.121.014), 248 (B. Grønbech, 11. 123.017), 249 (B. Gustafsson, P. Kjærgaard, S. Andersen, 12. 114.011), 250 (J. O. Petersen, 12.122.005), 252 (J. Andersen, J. V. Clausen, 12.032.008), 254 (H. E. Jørgensen, J. O. Petersen, 12.065.058), 255 (A. Sundman, L. O. Lodén, B. Nordström, 12.114.020), 258 (H. E. Jørgensen, H. U. Nørgaard-Nielsen, 13.122.013), 260 (B. Grønbech, 13.121.006).

008.040　Delaware

The Observatories of the Ohio State and Ohio Wesleyan Universities, Columbus and Delaware, Ohio. — Observatory reports.　　A. Slettebak. Bull. American Astron. Soc., Vol. 7, 181 - 185 (1975).

Contributions from the Perkins Observatory, Ohio State — Ohio Wesleyan Universities. Series I, No. 156 (R. F. Wing, 12.113.032).

Contributions from the Perkins Observatory, Ohio State — Ohio Wesleyan Universities. Series II, Nos. 39 (T. Roark, B. Roark, G. W. Collins II, 11.132.024), 40 (J. H. Baumert, 11.115.012), 41 (P. C. Keenan, R. F. Garrison, A. J. Deutsch, 12.122.060), 42 (G. W. Collins II, 12.114.015), 43 (E. R. Craine, 12.114.013), 44 (D. Weistrop, 12.155.026), 45 (N. D. Delamater, G. W. Collins II, G. H. Newsom, 12.022. 056), 46 (G. O. Boeshaar, S. J. Czyzak, L. H. Aller, 12.133. 027), 47 (G. H. Newsom, 12.022.077), 48 (L. H. Aller, S. J. Czyzak, C. D. Keyes, G. Boeshaar, 13.159.019), 49 (P. Lee, L. H. Aller, J. B. Kaler, S. J. Czyzak, 12.133.004).

008.041　Dresden

Mitteilungen des Lohrmann-Observatoriums der Technischen Universität Dresden, Nos. 27 (J. Dittrich, S. Wächter, 13.031.261), 28 (W. Albrecht, 13.031.262), 29 (H. Potthoff, 12.031.180).

008.042　Dublin

Dunsink Observatory 1968 – 1973. P. A. Wayman. Irish Astron. Journ., Vol. 11, 173 - 179 (1974).

Communications of the Dublin Institute for Advanced Studies, Series C. Dunsink Observatory Publications, Vol. 1, No. 7 (C. J. Butler, P. A. Wayman, 13.123.037).

Contributions from the Dunsink Observatory, Nos. 11 (T. Kiang, 09.103.103), 12 (S. H. Plagemann, 10.141.048).

Dunsink Observatory Reprints Nos. 69 (P. A. Wayman, 07.004.016), 70 (P. A. Wayman, 08.004.031), 71 (E. M. Lindsay, P. A. Wayman, 08.122.088), 72 (P. A. Wayman, 08.005.027), 73 (P. A. Wayman, 08.113.028), 74 (T. Kiang, P. A. Wayman, 09.103.103), 75 (C. J. Butler, P. B. Byrne, 09.142.109), 76 (P. B. Byrne, C. J. Butler, 09.142.121), 77 (T. Kiang, 13.098.020), 78 (T. Kiang, 13.098.026), 79 (T. Kiang, 13.103.104), 80 (M. V. Norris, 10.122.060), 81 (C. J. Butler, 10.031.010), 82 (P. A. Wayman, 10.035.003), 84 (M. J. Stift, 12.115.001), 85 (M. J. Stift, 12.065.110), 86 (W. M. Dumpleton, I. E. Elliott, 12.031.015).

008.043　East Lansing, Michigan

Michigan State University, Department of Astronomy and Astrophysics and the Observatory, East Lansing, Michigan. — Observatory report.　　P. D. Noerdlinger. Bull. American Astron. Soc., Vol. 7, 147 - 149 (1975).

008.044 Edinburgh

Royal Observatory, Edinburgh. H. A. Brück. Quarterly Journ. Roy. Astron. Soc., Vol. 16, 197 - 207 (1975). – Report for the year ending 1974 December 31.

Communications from the Royal Observatory, Edinburgh, Nos. 172 (B. McInnes, M. F. Walker, 12.082.033), 178 (D. H. Morgan, K. Nandy, 12.131.159), 179 (B. N. G. Guthrie, 12.160.002), 182 (W. McD. Napier, R. J. Dodd, 12.064.027), 183 (K. Nandy, D. H. Morgan, V. C. Reddish, 12.161.003).

008.045 El Segundo, California

The Aerospace Corporation, El Segundo, California. – Observatory reports. G. A. Paulikas. Bull. American Astron. Soc., Vol. 7, 1 - 7 (1975). This report reviews the astronomy research carried out at The Aerospace Corporation during 1974. The report describes the activities of the San Fernando Observatory, the research in millimeter-wave radio astronomy as well as the space astronomy research.

008.046 Evanston

Lindheimer Astronomical Research Center and Dearborn Observatory, Evanston, Illinois; Corralitos Observatory, Las Cruces, New Mexico. – Observatory reports. J. A. Hynek. Bull. American Astron. Soc., Vol. 7, 120 - 122 (1975).

008.047 Flagstaff

Lowell Observatory, Flagstaff, Arizona. – Observatory report. J. S. Hall. Bull. American Astron. Soc., Vol. 7, 129 - 132 (1975).

Lowell Observatory Bulletin, *Flagstaff, Arizona,* No. 162, Vol. 8, No. 2 (H. L. Giclas, R. Burnham, Jr., N. G. Thomas, 13.112.008).

008.048 Gainesville

University of Florida Observatories, Gainesville, Florida. – Observatory reports. F. B. Wood, A. G. Smith. Bull. American Astron. Soc., Vol. 7, 55 - 59 (1975).

008.049 Genève

Publications de l'Observatoire de Genève, Série A, Fasc. 81 (P. Steiger, 13.113.061; P. Rigaud, O. Steiger, D. Huguenin, 13.082.088).

008.050 Gothenburg

Research Laboratory of Electronics and Onsala Space Observatory, Chalmers University of Technology, Gothenburg, Sweden. **Research Report,** Nos. 120 (O. E. H.

Rydbeck, E. Kollberg, Å. Hjalmarson, A. Sume, J. Elldér, W. M. Irvine, 13.131.121), 123 (B. T. Cato, B. O. Rönnäng, P. T. Lewin, O. E. H. Rydbeck, K. S. Yngvesson, A. G. Cardiasmenos, J. F. Shanley, 13.131.529), 124 (Å. Hjalmarson, A. Sume, J. Elldér, O. E. H. Rydbeck, E. Moore, R. Huguenin, A. Sandqvist, P. O. Lindblad, P. Lindroos, 13.131.122).

008.051 Green Bank

National Radio Astronomy Observatory, Charlottesville, Virginia; Green Bank, West Virginia; and Tucson, Arizona. – Observatory report. D. S. Heeschen, D. E. Hoog. Bull. American Astron. Soc., Vol. 7, 170 - 178 (1975).

National Radio Astronomy Observatory, *Green Bank,* Reprints, Series A, Nos. 355 (L. E. Snyder, D. Buhl, 11.131.060), 356 (K. I. Kellermann, B. G. Clark, D. B. Shaffer, M. H. Cohen, D. L. Jauncey, J. J. Broderick, A. E. Niell, 11.141.060), 357 (F. N. Owen, 11.160.016), 358 (J. M. Rankin, D. B. Campbell, D. C. Backer, 11.141.315), 359 (D. S. De Young, M. S. Roberts, 11.160.013), 360 (F. H. Briggs, 11.100.010), 361 (P. M. Harvey, K. B. Bechis, W. J. Wilson, J. A. Ball, 11.114.132), 362 (B. L.'Ulich, E. K. Conklin, 11.103.101), 363 (D. S. Heeschen, 11.008.032, 11.008.052, 11.008.133), 364 (J. F. C. Wardle, R. A. Sramek, 11.158.088), 365 (R. N. Manchester, J. H. Taylor, Y. Y. Van, 11.141.330), 366 (B. E. Turner, 11.103.101), 367 (R. H. Sanders, 11.141.049), 368 (H. S. Murdoch, 11.141.025), 369 (G. R. Knapp, 11.131.542), 370 (G. R. Knapp, 11.131. 543), 371 (B. Zuckerman, J. A. Ball, 11.131.544), 372 (W. C. Saslaw, M. J. Valtonen, S. J. Aarseth, 11.151.050), 373 (K. Y. Lo, K. P. Bechis, 11.114.161), 374 (G. R. Knapp, F. J. Kerr, 11.158.113), 375 (P. O. Lindblad, 12.157.003), 376 (S. D. Peterson, G. S. Shostak, 12.158.001), 377 (T. Velusamy, M. R. Kundu, 11.125.024), 378 (R. N. Manchester, J. H. Taylor, 12.141.306), 379 (D. C. Backer, 11.141.343), 380 (R. H. Sanders, G. T. Wrixon, 11.155.054), 381 (H. S. Liszt, R. W. Wilson, A. A. Penzias, K. B. Jefferts, P. G. Wannier, P. M. Solomon, 11.132.034), 382 (F. O. Clark, D. Buhl, L. E. Snyder, 11.131.134), 383 (L. E. Snyder, D. Buhl, P. R. Schwartz, F. O. Clark, D. R. Johnson, F. J. Lovas, P. T. Giguere, 12.131.016), 384 (R. A. Sramek, H. M. Tovmassian, 12.158.013), 385 (P. E. Clegg, P. A. R. Ade, M. Rowan-Robinson, 11.141.073), 386 (G. R. Knapp, F. J. Kerr, 12.125.002), 387 (W. B. Burton, T. M. Bania, 12.155. 004), 388 (Aa. Sandqvist, 12.157.001), 389 (R. A. Sramek, H. M. Tovmassian, 12.158.034), 390 (B. Margon, S. Bowyer, R. Cruddace, C. Heiles, M. Lampton, T. Troland, 12.142.028), 391 (G. W. Brandie, A. H. Bridle, 12.141.043), 392 (W. B. Burton, T. M. Bania, 12.155.005), 393 (J. W. Findlay, J. M. Payne, 13.033.015), 394 (G. T. Wrixon, M. V. Schneider, 12.077.004), 395 (W. J. Wilson, P. R. Schwartz, E. E. Epstein, W. A. Johnson, R. D. Etcheverry, T. T. Mori, G. G. Berry, H. B. Dyson, 12.131.011), 396 (M. Morris, P. Palmer, B. E. Turner, B. Zuckerman, 12.141.603), 397 (K. H. Johnson, 12.141.064), 398 (B. Zuckerman, D. Buhl, P. Palmer, L. E. Snyder, 11.131. 062), 399 (A. H. Bridle, E. B. Fomalont, 12.141.063), 400 (J. J. Broderick, R. L. Brown, 12.141.035), 401 (K. J. Gordon, C. P. Gordon, F. J. Lockman, 12.131.515), 402 (B. Balick, R. H. Sanders, 12.155.023), 403 (R. L. Brown, R. H. Gammon, G. R. Knapp, B. Balick, 12.131.089), 404 (M. Morris, B. Zuckerman, B. E. Turner, P. Palmer, 12.132.010), 405 (D. Buhl, L. E. Snyder, F. J. Lovas, D. R. Johnson, 12.132. 027), 406 (K. R. Lang, 12.077.031), 407 (N. Kaifu, M. Morimoto, K. Nagane, K. Akabane, T. Iguchi, K. Takagi, 12.131. 030), 408 (B. R. Hermann, J. R. Dickel, 10.125.037), 409 (G. R. Knapp, F. J. Kerr, 12.155.038), 410 (J. J. Condon, M. J. Yerbury, D. L. Jauncey, 12.100.024), 411 (R. L. Brown, 12.141.622), 412 (R. M. Hjellming, R. L. Brown, L. C.

Blankenship, 12.141.084), 413 (B. E. Turner, 12.131.100),
414 (K. D. Tucker, M. L. Kutner, P. Thaddeus, 12.131.116),
415 (J. J. Condon, L. L. Dressel, 12.141.077), 416 (E. R.
Seaquist, P. C. Gregory, R. A. Perley, R. H. Becker, J. B.
Carlson, M. R. Kundu, R. C. Bignell, J. R. Dickel, 12.142.036),
417 (P. F. Bowers, R. H. Cornett, 12.122.068), 418 (B. Balick,
R. H. Gammon, R. M. Hjellming, 12.131.549), 419 (H. M.
Tovmassian, Y. Terzian, 12.158.153), 420 (D. M. Gibson,
R. M. Hjellming, 12.117.033), 421 (A. S. Milman, 12.131.102),
422 (R. C. Bignell, 12.133.023), 423 (J. J. Condon, D. L.
Jauncey, 12.141.093), 424 (J. F. C. Wardle, P. P. Kronberg,
12.141.101), 425 (B. Balick, R. L. Brown, 12.155.077), 426
(B. E. Turner, B. Balick, D. D. Cudaback, C. Heiles, R. J. Boyle,
12.131.545), 427 (B. E. Turner, C. E. Heiles, 12.131.167),
428 (G. W. Brandie, A. H. Bridle, M. J. L. Kesteven, 12.141.
071), 429 (K. I. Kellermann, 12.141.123), 430 (C. C. Counsel-
man III, S. M. Kent, C. A. Knight, I. I. Shapiro, T. A. Clark,
H. F. Hinteregger, A. E. E. Rogers, A. R. Whitney, 12.066.089),
431 (F. J. Kerr, P. F. Bowers, 12.141.625), 432 (E. B.
Fomalont, A. H. Bridle, M. M. Davis, 12.141.095), 433
(R. B. Loren, W. L. Peters, P. A. Vanden Bout, 13.131.004),
434 (N. Kaifu, D. Buhl, L. E. Snyder, 13.131.009), 435
(R. L. Brown, G. R. Knapp, T. B. H. Kuiper, E. N. Rodriguez
Kuiper, 13.132.001), 436 (F. N. Owen, 13.160.003), 437
(F. W. Peterson, C. King III, 13.141.013), 438 (W. C. Saslaw,
13.151.004), 439 (G. R. Knapp, R. L. Brown, T. B. H. Kuiper,
13.132.011), 440 (A. S. Milman, G. R. Knapp, F. J. Kerr,
S. L. Knapp, W. J. Wilson, 13.131.027), 441 (A. S. Milman,
G. R. Knapp, S. L. Knapp, W. J. Wilson. 13.131.028), 442
(D. S. Heeschen, D. E. Hogg, 13.008.051).

National Radio Astronomy Observatory, *Green
Bank,* **Reprints,** Series B, Nos. 448 (M. R. Kundu, R. H. Becker,
T. Velusamy, 11.073.005), 449 (B. E. Turner, 11.131.049),
450 (F. H. Briggs, 11.100.207), 451 (C. M. Wade, 11.041.054),
452 (D. Wills, W. D. Cotton, 11.141.080), 453 (B. E. Turner,
12.131.001), 454 (W. C. Saslaw, 12.158.084), 455 (W. B.
Burton, 13.033.028), 456 (I. I. K. Pauliny-Toth, K. I. Kellermann, 12.141.053), 457 (W. F. Huebner, L. E. Snyder, D.
Buhl, 12.103.104).

008.052 Greenbelt

Goddard Space Flight Center, Greenbelt, Maryland,
GSFC Documents X-660-74-368 (R. Ramaty, B. Kozlovski,
R. E. Lingenfelter, 13.076.033), X-660-75-74 (H. T. Wang,
R. Ramaty, 13.076.034), X-660-75-79 (R. Ramaty, 13.076.
035), X-660-75-96 (R. Ramaty, C. J. Crannell, 13.076.036),
X-661-75-39 (S. H. Pravdo, E. A. Boldt, 13.143.057), X-661-
75-114 (S. S. Holt, 13.142.095), X-661-75-115 (S. S. Holt,
13.034.069), X-663-75-44 (J. H. Trainor, F. B. McDonald,
D. E. Stilwell, B. J. Teegarden, W. R. Webber, 13.099.081),
X-920-74-334 (C. A. Wagner, 13.052.055), X-920-75-62
(P. Musen, 13.081.020), X-921-74-145 (F. J. Lerch, C. A.
Wagner, J. A. Richardson, J. E. Brownd, 13.081.021),
X-921-74-242 (J. G. Marsh, B. C. Douglas, D. M. Walls, 13.
046.022), X-921-74-250 (J. G. Marsh, B. C. Douglas, D. M.
Walls, 13.046.023), X-921-74-275 (M. A. Khan, 13.081.022),
X-921-75-56 (D. P. Rubincam, 13.052.056).

Goddard Space Flight Center, Greenbelt, Maryland,
Separate prints (J. G. Marsh, S. Vincent, 13.081.032), (V. K.
Balasubrahmanyan, A. T. Serlemitsos, 12.078.022), (T. L.
Cline, U. D. Desai, 13.061.011), (D. P. Rubincam, 13.094.
004), (F. C. Jones, T. J. Birmingham, 13.062.054), (T. J.
Birmingham, F. C. Jones, 13.011.037).

008.053 Greenwich

Royal Greenwich Observatory.
E. M. Burbidge, A. Hunter.
Quarterly Journ. Roy. Astron. Soc., Vol. 16, 70 - 99 (1975).
Report for the year ending 1973 December 31.

Tercentenary of the Royal Greenwich Observatory.
H. Hatfield.
Journ. British Astron. Ass., Vol. 85, 315 - 316 (1975).
RGO tercentenary exhibition.

The origins of the Greenwich Observatory.
See Abstr. 004.021.

**The tercentenary of the Royal Observatory at
Greenwich.** See Abstr. 004.023.

Three hundred years of Greenwich.
See Abstr. 004.047.

**The Greenwich Observatory: origins and early
development, 1675–1835.** See Abstr. 004.049.

Airy and after. See Abstr. 004.050.

Greenwich Observatory in the twentieth century.
See Abstr. 004.051.

Sidelights on the Royal Observatory.
See Abstr. 004.052.

**H. M. Nautical Almanac Office, Royal Greenwich
Observatory, Library Reprint,** Nos. 311 (A. T. Sinclair, 12.
100.214), 312 (P. J. N. Davison, J. L. Culhane, L. V. Morrison,
13.134.002), 313 (A. T. Sinclair, 13.042.026).

008.054 Groningen

Nederlandse Vereniging voor Weer- en Sterrenkunde.
Observations of Variable Stars. Report (Kapteyn Astronomical
Laboratory, Groningen–Netherlands), No. 27 (L. Plaut, H.
Feijth, 13.123.007).

008.055 Hamburg

Deutsches Hydrographisches Institut, Hamburg.
**Astronomische Zeit- und Breitenbestimmungen, Empfangs-
zeiten von Zeitsignalen,** 1974 July – September (13.044.019).

**European Southern Observatory. Annual Report
1974.** L. Woltjer.
Printed in the Federal Republic of Germany by Lütcke &
Wulff, Hamburg. 56 pp. (1975). – Contents: Research
activities; The ESO 3.6 metre telescope project; Danish
national 1.5 metre telescope project; Other telescopes and
auxiliary instruments; Buildings and grounds; Administrative
matters; Council, committees, working groups.

Some aspects of ESO. A. Blaauw.
Monthly Notes Astron. Soc. Southern Africa, Vol. 34, 53 - 54
(1975).

The Messenger — El Mensajero, No. 3.
Edited by European Southern Observatory (ESO), Hamburg,
8 pp. (1975).

008.056 Heidelberg

Astronomy and Astrophysics Abstracts, Vol. 12
(S. Böhme, U. Esser, W. Fricke, U. Güntzel-Lingner, F. Henn,
D. Krahn, H. Scholl, G. Zech, 13.002.010).

Astronomisches Rechen-Institut in Heidelberg,
Mitteilungen, Serie A, No. 87 (H. Scholl, R. Giffen, 12.098.
032).

Astronomisches Rechen-Institut in Heidelberg,
Mitteilungen Serie B, Nos. 44 (S. Isobe, 12.155.074), 45
(S. Isobe, 12.155.075), 46 (S. J. Aarseth, M. Hénon, R. Wielen,
12.151.044).

008.057 Holmdel, New Jersey

Bell Telephone Laboratories, Crawford Hill Labora-
tory, Holmdel, New Jersey. — Observatory report.
A. A. Penzias.
Bull. American Astron. Soc., Vol. 7, 24 - 25 (1975).

008.058 Honolulu

University of Hawaii, Institute for Astronomy,
Honolulu, Hawaii. — Observatory report.   J. T. Jefferies.
Bull. American Astron. Soc., Vol. 7, 62 - 69 (1975).

008.059 Houston

Rice University Department of Space Physics and
Astronomy, Houston, Texas. — Observatory report.
F. C. Michel.
Bull. American Astron. Soc., Vol. 7, 194 - 196 (1975).

008.060 Hyderabad

Observing programmes of Nizamiah and Rangapur
observatories.   N. B. Sanwal.
Bull. Astron. Soc. India, Vol. 2, 31 (1974). — Abstract.

Nizamiah & Rangapur Observatories and Department
of Astronomy, Osmania University, Hyderabad, India, Reprint
Nos. 50 (K. Desikachary, M. Parthasarathy, N. Kameswara Rao,
06.122.159), 51 (K. S. Sastry, 07.151.087), 52 (M. Parthasa-
rathy, 08.121.070), 53 (N. B. Sanwal, M. Parthasarathy,
K. D. Abhyankar, 09.121.004), 54 (N. B. Sanwal, M. B. K.
Sarma, M. Parthasarathy, K. D. Abhyankar, 11.121.006), 55
(N. B. Sanwal, M. Parthasarathy, 11.122.002), 56 (K. D.
Abhyankar, M. Parthasarathy, N. B. Sanwal, M. B. K. Sarma,
11.121.007), 57 (G. M. Ballabh, 10.151.028), 58 (A. Potdar,
G. M. Ballabh, 11.151.008), 59 (S. M. Alladin, K. S. Sastry,
G. M. Ballabh, 11.158.079), 60 (K. D. Abhyankar, 12.155.
088).

008.061 Ioannina

Department of Astronomy, University of Ioannina.
Annual report 1973.   D. Miliotis.
Annual Rep. Astron. Inst. Greece 1973, p. 24 (1974).

Department of Astronomy, University of Ioannina.
Annual report 1974.   D. Miliotis.
Annual Rep. Astron. Inst. Greece 1974, p. 23 (1975).

008.062 Iowa City

University of Iowa, Iowa City, Iowa. — Observatory
report.   J. S. Neff.
Bull. American Astron. Soc., Vol. 7, 311 - 315 (1975). — This
report describes the astronomical activities of the Department
of Physics and Astronomy for the period 1 October 1973
through 30 September 1974.

008.063 Izmir

Scientific Reports of the Faculty of Science, Ege
University, Izmir, No. 218 — Astron. No. 16 (N. Güdür, 13.
121.087).

008.064 Kharkov

Vestnik Khar'kovskogo Universiteta. Izdatel'skoe
Obedinenie 'Vishcha Shkola', Izdatel'stvo pri Khar'kovskom
Gosudarstvennom Universitete, Khar'kov, 1974, No. 117
[Ser.] Astronomiya, Vypusk (No.) 9 (Yu. V. Aleksandrov, D.
F. Lupishko, T. A. Lupishko, 13.097.093; Yu. V. Aleksandrov,
D. F. Lupishko, V. P. Tishkovets, 13.097.094; V. N. Dudinov,
V. S. Tsvetkova, V. A. Krishtal', N. A. Khovanskij, 13.031.
279; N. N. Evsyukov, 13.094.259; M. F. Khodyachikh, 13.099.
096; K. N. Kuz'menko, N. S. Olifer, L. S. Pavlenko, V. Kh.
Pluzhnikov, 13.041.078; V. I. Turenko, 13.044.032; K. N.
Derkach, N. G. Zuev, V. M. Kirpatovskij, K. N. Kuz'menko,
13.032.038; P. P. Pavlenko, L. S. Pavlenko, 13.034.113; K. N.
Derkach, 13.041.079).

008.065 Kiev

Astrometriya i Astrofizika, Kiev, Vyp. (Nos.)
24 (Ya. S. Yatskiv, 13.003.003), 25 (M. Ya. Orlov, 13.003.
015).

Kievskij ordena Lenina gosudarstvennyj universitet
im. T. G. Shevshenko, Astronomicheskaya Observatoriya
(Astron. Obs.), Preprint Nos. 2 (P. R. Romanchuk, V. N.
Krivodubskij, 13.073.009), 3 (P. R. Romanchuk, 13.072.011).

008.066 Kitt Peak

Das Sonnen-Teleskop von Kitt Peak. (Robert R.
McMath solar telescope).   See Abstr. 032.012.

008.067 **Kodaikanal**

Indian Institute of Astrophysics.
M. K. V. Bappu.
Quarterly Journ. Roy. Astron. Soc., Vol. 16, 184 - 188 (1975).
Report for the year 1973 April 1 to 1974 March 31.

The Indian Institute of Astrophysics.
K. R. Sivaraman.
Bull. Astron. Soc. India, Vol. 1, 4, 12 (1973).

008.068 **Krim**

The Crimean Astrophysical Observatory.
V. M. Mozhzherin.
Zemlya i Vselennaya, 1975, No. 1, p. 46 - 55. In Russian.

008.069 **Kyoto**

Contributions from the Kwasan and Hida Observatories, University of Kyoto, Nos. 208 (J. Kubota, H. Kurokawa, T. Kureizumi, 13.073.073), 209 (K. Maeda, N. Oda, K. Nakayama, 13.073.074), 210 (S. Ebisawa, 13.097.065), 211 (S. Miyamoto, 13.097.066), 212 (T. Tamenaga, T. Kureizumi, J. Kubota, 10.073.076), 213 (J. Kubota, T. Tamenaga, I. Kawaguchi, 10.073.077), 214 (H. Kurokawa, K. Nakayama, T. Tsubaki, M. Kanno, 11.073.050), 215 (J. Kubota, T. Kureizumi, S. Koyama, 12.073.045), 216 (M. Kanno, T. Tsubaki, H. Kurokawa, 13.079.103), 217 (S. Miyamoto, 13.097.067), 218 (S. Miyamoto, 13.097.068), 219 (S. Miyamoto, 13.097.069), 220 (S. Miyamoto, 13.092.011), 221 (A. Hattori, T. Akabane, 13.097.070), 222 (J. Kubota, T. Tamenaga, I. Kawaguchi, R. Kitai, 12.072.032), 223 (S. Ebisawa, 13.097.071), 224 (S. Ebisawa, 13.097.072).

008.070 **La Plata**

Observatorio Astronómico de la Universidad Nacional de La Plata. Serie Astronómica, Vol. 38 (C. Jaschek, E. Hernandez, A. Sierra, A. Gerhardt, 13.113.023).

Separata Astronómica, Observatório Astronómico — La Plata — Argentina, Nos. 131 (A. Feinstein, H. G. Marraco, 11.114.011), 132 (R. P. Platzeck, J. M. Simon, 13.031.039), 133 (A. Feinstein, H. G. Marraco, J. C. Muzzio, 10.153.013), 134 (M. Jaschek, E. Brandi, 11.114.041), 135 (H. Levato, S. Malaroda, 12.153.006), 136 (A. Feinstein, J. C. Forte, 11.153.033), 137 (A. Feinstein, 12.113.038), 138 (J. C. Muzzio, H. G. Marraco, A. Feinstein, 12.155.033), 139 (H. Levato, 12.116.019).

008.071 **Las Cruces**

Lindheimer Astronomical Research Center and Dearborn Observatory, Evanston, Illinois; Corralitos Observatory, Las Cruces, New Mexico. — Observatory reports.
J. A. Hynek.
Bull. American Astron. Soc., Vol. 7, 120 - 122 (1975).

New Mexico State University, Department of Astronomy, Las Cruces, New Mexico. — Observatory report.

H. A. Beebe.
Bull. American Astron. Soc., Vol. 7, 178 - 180 (1975).

008.072 **La Serena, Chile**

Kitt Peak National Observatory, Tucson, Arizona; Cerro Tololo Inter-American Observatory, La Serena, Chile. Observatory reports.   B. T. Lynds.
Bull. American Astron. Soc., Vol., 7, 87 - 120 (1975).

008.073 **Lawrence, Kansas**

Kansas University Observatory, Lawrence, Kansas. — Observatory report.   S. J. Shawl.
Bull. American Astron. Soc., Vol. 7, 86 - 87 (1975).

008.074 **Leiden**

Leidse Sterrewacht, wel en wee.
H. W. Verheyen.
Zenit, Vol. 2, 84 - 88 (1975).

008.075 **Livermore, California**

Lawrence Livermore Laboratory, University of California, Livermore, California. — Observatory report.
C. B. Tarter.
Bull. Amercian Astron. Soc., Vol. 7, 316 - 318 (1975).

008.076 **London, Canada**

The Observatories of the University of Western Ontario, London, Canada. — Observatory reports.
W. H. Wehlau.
Bull. American Astron. Soc., Vol. 7, 230 - 231 (1975).

008.077 **Los Angeles**

University of California, Berkeley, Los Angeles, San Diego, and Lick Observatories. II. Los Angeles Campus.
G. Abell.
Bull. American Astron. Soc., Vol. 7, 293 - 297 (1975).

008.078 **Madison, Wisconsin**

Washburn Observatory, University of Wisconsin, Madison, Wisconsin. — Observatory report. R. C. Bless.
Bull. American Astron. Soc., Vol. 7, 223 - 227 (1975).

008.079 **Manchester**

University of Manchester, Nuffield Radio Astronomy Laboratories, Jodrell Bank.   B. Lovell.
Quarterly Journ. Roy. Astron. Soc., Vol. 16, 62 - 69 (1975).
Report for the year ending 1974 August 31.

008.080 Mauna Kea, Hawaii

Das Mauna Kea-Observatorium auf Hawaii.
M. Lammerer, H. Treutner.
Orion, 33. Jahrgang, p. 3 - 8 (1975).

008.081 Middletown, Connecticut

Van Vleck Observatory, Wesleyan University,
Middletown, Connecticut. – Observatory report.
A. R. Upgren.
Bull. American Astron. Soc., Vol. 7, 215 - 216 (1975).

008.082 Minneapolis

University of Minnesota, Minneapolis, Minnesota.
Observatory report.   E. P. Ney.
Bull. American Astron. Soc., Vol. 7, 150 -152 (1975).

University of Minnesota, Minneapolis, Minnesota,
Separate prints (W. J. Luyten, 13.112.010; 13.115.013;
13.112.011), (W. J. Luyten, 13.112.012).

008.083 Mizusawa

Annual Report of the International Polar Motion
Service 1972, (S. Yumi, 13.045.015).

Bulletins, Time Service of the Mizusawa Observa-
tory, Vol. 17, No. 1 - 12, 1972 (S. Takagi, M. Aihara, K.
Yokoyama, T. Hara, G. Murakami, T. Okuda, 13.044.012).

Monthly Notes of the International Polar Motion
Service, 1974 Nos. 11, 12; 1975 Nos. 1 - 5 (13.045.025).

Proceedings of the International Latitude Observa-
tory of Mizusawa, No. 14 ((1) I. Okamoto, T. Sasao, 13.045.
010; (2) N. Sekiguchi, 13.045.011; (3) C. Kakuta, 13.011.022;
(4) T. Gotō, N. Kikuchi, 13.082.078; (5) H. Ōkawa, Y. Gotō,
13.045.012; (6) T. Hara, 13.035.006; (10) T. Okuda, C.
Sugawa, K. Hosoyama, T. Suzuki, T. Sato, 13.081.019; (12)
H. Ishii, 13.045.013; (13) K. Yokoyama, 13.045.014; (14)
C. Sugawa, N. Kikuchi, 13.082.079).

Publications of the International Latitude Observato-
ry of Mizusawa, Vol. 9, No. 1 (K. Yokoyama, 13.045.006;
S. Takagi, 13.045.007; I. Naito, 13.082.077; M. Ooe, 13.045.
008; M. Ooe, 13.045.009).

Annual report of geophysical observations made at
the International Latitude Observatory of Mizusawa for the
year 1972.   T. Okuda.
Published by the International Latitude Observatory of
Mizusawa, Japan. 44 pp. (1974).

Annual report of the meteorological observations
made at the International Latitude Observatory of Mizusawa
for the year 1972.   T. Okuda.
Published by the International Latitude Observatory of
Mizusawa, Japan. 2 + 30 pp. (1973).

008.084 Mons

Report for the years 1973–1974. L. Houziaux.
Centre Univ. Etat Mons, Fac. Sci., Dép. Astrophys., Commun.
No. 44, 13 pp. (1975).

Communications du Département d'Astrophysique
de la Faculté des Sciences de Mons. Mons Astrophysical
Papers, Nos. 40 (A. Delcroix, 12.065.171), 41 (E. Peytremann,
11.064.049), 42 (E. Peytremann, 12.064.006), 43 (Y.
Andrillat, C. Fehrenbach, L. Houziaux, 12.124.102), 44
(L. Houziaux, 13.008.084), 45 (E. Peytremann, 13.064.032),
46 (S. Volonté, 13.062.057).

008.085 Montevideo, Uruguay

Departamento de Astronomía y Física, Facultad de
Humanidades y Ciencias, Universidad de la República, Monte-
video, Uruguay. **Publicación,** No. 13 (R. E. Caligaris, J. C.
Grangel, 13.022.089).

008.086 Moskva

Soobshcheniya Gosudarstvennogo Astronomiches-
kogo Instituta im. P. K. Shternberga. Izdatel'stvo Moskovs-
kogo Universiteta, Nos. 187 (M. U. Sagitov, 13.022.075;
M. U. Sagitov, 13.043.009; V. L. Panteleev, 13.081.025;
L. G. Gugel', 13.043.010; Kh. G. Tadzhidinov, 13.022.076),
188 (E. B. Kostyakova, 13.133.030; N. M. Artyukhina,
P. N. Kholopov, 13.153.024; N. M. Artyukhina, P. N.
Kholopov, 13.153.025; T. A. Uranova, G. S. Tsarevskij,
13.152.003; D. K. Karimova, E. D. Pavlovskaya, M. S.
Toropova, 13.112.005; G. A. Starikova, 13.118.017), 189
(B. A. Vorontsov-Vel'yaminov, G. Ivanišević, 13.158.112;
N. N. Yakimova, 13.115.011), 190 (Yu. A. Shokin, 13.031.
246; N. S. Blinov, E. N. Fedoseev, 13.044.014; L. V. Rykhlova.
13.045.018; N. S. Blinov, K. G. Belelin, I. N. Makarov,
13.044.015; A. A. Volchkov, 13.031.247), 191 (A. A. Orlov,
13.042.068; V. S. Zenkov, I. I. Kalinnikov, V. M. Popov,
M. U. Sagitov, 13.022.077), 197 - 198 (I. N. Glushneva,
A. V. Kharitonov, I. B. Voloshina, E. A. Glushkova, V. T.
Doroshenko, E. A. Kolotilov, M. F. Novikova, I. G. Petrovskaya, V. T. Rebristyj, V. M. Tereshchenko, T. S. Fetisova,
L. D. Frishberg, 13.114.081).

Trudy Gosudarstvennogo Astronomicheskogo
Instituta im. P. K. Shternberga. Izdatel'stvo Moskovskogo
Universiteta, Vol. 45 (M. S. Zverev, A. G. Oborneva,
13.041.030; L. M. Khommik, V. A. Korobova, O. A. Kozina,
13.041.031; D. K. Karimova, E. D. Pavlovskaya, M. S.
Toropova, 13.112.004; N. B. Frolova, 13.041.032; D. N.
Ponomarev, N. B. Frolova, 13.041.033; A. A. Orlov, N. A.
Solovaya, 13.042.065; N. A. Solovaya, 13.042.066; G. I.
Shirmin, 13.042.067; M. S. Yarov-Yarovoj, 13.021.004),
Vol. 46, vyp. (No.) 1 (B. A. Vorontsov-Vel'yaminov, V. P.
Arkhipova, 13.003.016).

008.087 München

Max-Planck-Institut für Physik und Astrophysik,
Institut für Extraterrestrische Physik, Garching bei München,
MPI – PAE/Extraterr. 107 (K.-U. Dettmar, L. Haser, 13.031.
267).

Max-Planck-Institut für Physik und Astrophysik,
Institut für Extraterrestrische Physik, Garching bei München,
Separate prints (S. Drapatz, K. W. Michel, 12.131.151; G.
Kanbach, C. Reppin, V. Schönfelder, 13.084.414).

008.088   Naini Tal

Uttar Pradesh State Observatory.   S. D. Sinvhal.
Bull. Astron. Soc. India, Vol. 1, 23 - 24 (1973).

008.089   Nashville

Dyer Observatory, Vanderbilt University, Nashville,
Tennessee. – Observatory report.   A. M. Heiser.
Bull. American Astron. Soc., Vol. 7, 47 - 49 (1975).

008.090   Neuchâtel

Rapport d'activité pour l'exercice 1974.
J. Bonanomi, G. Fischer.
Observatoire Cantonal de Neuchâtel, 23 pp. (1975).

Observatoire de Neuchâtel, Bulletin. Série B, 1974
September – 1975 March (13.044.021); Série D, 1974 September – December (13.044.022).

008.091   Oslo

Institute of Theoretical Astrophysics, Blindern–
Oslo. Report, Nos. 40 (O. Engvold, T. Kozar, B. Rompolt,
13.073.110), 41 (R. Brahde, 13.031.268), 42 (F. Albregtsen,
T. Hansen, 13.034.075).

Institutt for Teoretisk Astrofysikk, Blindern–Oslo.
Småtrykk, Nos. 82 (T. S. Ringnes, 13.094.571), 83 (E.
Jensen, 13.102.038), 84 (Ø. Hauge, 12.092.017).

008.092   Ottawa

The old Dominion Observatory.
See Abstr. 004.022.

Publications of the Earth Physics Branch, Department of Energy, Mines and Resources, Ottawa, Canada, Vol.
46, No. 1 (J. Hruska, 13.084.269).

008.093   Palo Alto

Lockheed Palo Alto Research Laboratory, Palo Alto,
California. – Observatory report.   B. M. McCormac.
Bull. American Astron. Soc., Vol. 7, 126 - 127 (1975).

008.094   Paris

Bureau International de l'Heure. Annual report for
1974.   R. Michard (Editor).
Printing Office: Observatoire de Paris. 5 + A15 + B45 + C19

pp. (1975). – Contents: Methods of computation of the universal time, the coordinates of the pole, the atomic time;
detailed results and residuals; time comparisons and coordination of time; information on time signals.

Bureau International de l'Heure, (B.I.H.), Circu-
laires D98 - D104 (13.044.024).

008.095   Pasadena

Hale Observatories, operated by Carnegie Institution
of Washington and California Institute of Technology, Pasa-
dena, California. Annual report of the director, 1973–1974.
H. W. Babcock, J. B. Oke.
Reprinted from Carnegie Institution, Washington, Year Book,
Vol. 73, 119 - 187 (1974).

008.096   Patras

Department of Astronomy, University of Patras.
Annual report 1973.   B. Barbanis.
Annual Rep. Astron. Inst. Greece 1973, p. 25 - 26 (1974).

Department of Astronomy, University of Patras.
B. Barbanis.
Annual Rep. Astron. Inst. Greece 1974, p. 24 - 25 (1975).

008.097   Philadelphia

Flower and Cook Observatory, University of Penn-
sylvania, Philadelphia, Pennsylvania. – Observatory report.
B. S. P. Shen.
Bull. American Astron. Soc., Vol. 7, 60 - 62 (1975).

008.098   Pittsburgh

Allegheny Observatory, University of Pittsburgh,
Pittsburgh, Pennsylvania. – Observatory report.
J. H. Kiewiet de Jonge.
Bull. American Astron. Soc., Vol. 7, 7 - 10 (1975).

008.099   Potsdam

Zentralinstitut für Astrophysik.
Akad. Wiss. DDR, Jahrbuch 1974, p. 11, 243 - 248 (1975).

Mitteilungen des Astrophysikalischen Observatori-
ums Potsdam, Nos. 154 (K.-H. Rädler, 10.062.047), 155
(K.-H. Rädler, 11.062.029), 156 (K.-H. Rädler, 11.062.030),
157 (F. Krause, G. Rüdiger, 11.062.031), 163 (A. S. Nikolow,
W. Schöneich, 11.031.025), 164 (F. Krause, G. Rüdiger, 11.
062.054).

Publikationen des Astrophysikalischen Observatori-
ums zu Potsdam, No. 108 = Vol. 32, Heft 1 (G. Dautcourt,
13.066.109; G. Dautcourt, 13.160.034; G. Rüdiger, 13.
062.058).

008.100 Praha

Académie Tchécoslovaque des Sciences, Institut Astronomique, **Station de l'Heure à Prague**, Série 7, No. 1 (L. Webrová, V. Ptáček, 13.044.025).

008.101 Pretoria

Communications from the Radcliffe Observatory, Pretoria, Nos. 136 (T. Lloyd Evans, 11.154.013), 141 (A. D. Thackeray, 12.114.003), 142 (W. L. Martin, 12.158.002), 143 (A. D. Thackeray, B. L. Webster, 12.124.100), 144 (A. J. Wesselink, 12.154.001), 145 (A. D. Thackeray, G. Hill, 12.119.001).

Radcliffe Observatory, *Pretoria*, Reprints, Nos. 139 (P. J. Andrews, M. W. Feast, T. Lloyd Evans, A. D. Thackeray, J. W. Menzies, 11.122.134), 140 (P. J. Andrews, I. S. Glass, T. G. Hawarden, 12.158.301), 141 (A. D. Thackeray, 12.124.020), 142 (T. Lloyd Evans, 12.112.008), 143 (W. L. Martin, 12.121.016), 144 (M. W. Feast, 12.122.102), 145 (A. D. Thackeray, 12.008.084).

008.102 Princeton, New Jersey

Princeton University Observatory, Princeton, New Jersey. – Observatory report.   L. Spitzer, Jr. Bull. American Astron. Soc., Vol. 7, 189 - 194 (1975).

008.103 Pulkovo

The future of solar physics at the Pulkovo Observatory.   V. A. Krat. Uspekhi fiz. nauk, Vol. 114, 159 (1974). In Russian.

008.104 Pulsnitz

Veröffentlichungen der Sternwarte Pulsnitz (Sachsen) No. 10 (J. Classen, 12.094.104, 12.094.168).

008.105 Richmond Hill

David Dunlap Observatory, University of Toronto, Richmond Hill, Ontario, Canada. – Observatory report. D. A. MacRae. Bull. American Astron. Soc., Vol. 7, 39 - 47 (1975).

David Dunlap Observatory, University of Toronto. D. A. MacRae. Quarterly Journ. Roy. Astron. Soc., Vol. 16, 39 - 61 (1975). Report for the period 1973 July 1 to 1974 June 30.

Communications from the David Dunlap Observatory, University of Toronto, Richmond Hill, Ontario, Canada, Nos. 418 (S. van den Bergh, 12.158.012), 419 (E. R. Seaquist, P. C. Gregory, T. R. Clarke, 12.141.044), 420 (W. Herbst, 12.153.013), 421 (B. F. Madore, S. van den Bergh, D. H. Rogstad, 12.158.020), 422 (V. A. Hughes, A. Woodsworth, P. C. Gregory, E. R. Seaquist, 12.142.031), 423 (F. J. Ahern, C. Pritchet, 12.031.120), 424 (J. P. Vallée, 12.158.103), 425 (W. E. Harris, 12.154.012), 426 (C. T. Bolton, 12.114.039),

427 (S. van den Bergh, 12.158.076), 428 (R. Racine, 12.153.014), 429 (C. T. Bolton, R. F. Garrison, D. Salmon, N. Geffken, 12.125.101), 430 (S. van den Bergh, 12.125.041), 431 (P. G. Martin, J. R. P. Angel, 12.131.103).

008.106 Riga

Uchenye Zapiski Latvijskogo Gosudarstvennogo Universiteta im. P. Stuchki, Vol. 220, Astron., vyp. (No.) 11, (13.003.014).

008.107 Rochester

C. E. Kenneth Mees Observatory, University of Rochester, Rochester, New York. – Observatory report. J. G. Duthie. Bull. American Astron. Soc., Vol. 7, 140 - 144 (1975).

C. E. Kenneth Mees Observatory, University of Rochester, Rochester, N. Y., **Reprint**, Nos. 52 (J. M. Sorvari, 12.115.017), 54 (S. Sofia, H. M. Van Horn, 12.061.079), 56 (J. G. Duthie, B. Renaud, M. P. Savedoff, 12.122.106).

008.108 Roma

Monthly Bulletin. Osservatorio Astronomico di Roma, Nos. 200 - 202 (M. Cimino, M. Torelli, F. Casamassima, V. Croce, 13.075.010).

Photographic Journal of the Sun, Osservatorio Astronomico di Roma, Nos. 87 - 92 (M. Cimino, 13.075.009).

008.109 San Diego

University of California, Berkeley, Los Angeles, San Diego, and Lick Observatories. III. San Diego Campus. Bull. American Astron. Soc., Vol. 7, 297 - 301 (1975).

008.110 San Fernando

Memoria de las actividades en 1974. Inst. y Obs. de Marina, San Fernando (Cádiz), España, 14 pp. (1975).

Catálogo de las obras y publicaciones periódicas que existen en esta Biblioteca y que corresponden a los siglos XV, XVI, XVII y XVIII.   See Abstr. 002.006.

008.111 Santa Cruz

University of California, Berkeley, Los Angeles, San Diego, and Lick Observatories. IV. Lick Observatory, D. E. Osterbrock; **V. Board of Studies in Astronomy and Astrophysics,** W. G. Mathews. – Observatory reports. Bull. American Astron. Soc., Vol. 7, 301 - 311 (1975).

University of California. **Lick Observatory Bulletin,** Nos. 629 (B. F. Jones, 09.112.002), 630 (K. H. Nordsieck, 08.158.052), 631 (A. R. Klemola, E. A. Harlan, 08.101.020),

632 (P. P. Eggleton, J. Faulkner, B. P. Flannery, 09.065.
029), 633 (A. R. Klemola, 09.099.027), 634 (W. L. Burke,
13.031.248), 635 (S. Vasilevskis, B. J. McNamara, 10.043.
003), 636 (K. M. Cudworth, 10.118.007), 637 (J. P. Simpson,
10.132.038), 638 (R. W. O'Connell, 10.114.241), 639
(P. W. Baker, 11.115.005), 640 (M. F. Walker, 10.082.108),
641 (G. H. Herbig, J. Lorre, 11.141.606), 642 (G. H. Herbig,
11.141.607), 643 (G. H. Herbig, 11.141.604), 644 (J. S.
Miller, 11.132.015), 645 (W. G. Mathews, 11.141.058), 646
(P. Bodenheimer, J. P. Ostriker, 12.141.304), 647 (J. Bregman, D. Butler, E. Kemper, A. Koski, R. P. Kraft, R. P. S.
Stone, 13.113.052), 648 (G. R. Blumenthal, 11.022.011),
649 (R. L. Milton, 11.126.017), 650 (R. W. O'Connell, J. D.
Scargle, W. L. W. Sargent, 12.158.010), 651 (D. A. Keeley,
12.131.088), 652 (E. L. Robinson, R. E. Nather, A. Kip-
linger, 12.122.033), 653 (G. E. Langer, R. P. Kraft, K. S.
Anderson, 11.114.106), 654 (J. D. Scargle, L. J. Caroff,
C. B. Tarter, 11.141.062), 655 (E. A. Harlan, 11.114.152),
656 (K. H. Nordsieck, 11.034.082), 657 (B. McInnes, M. F.
Walker, 12.082.033), 658 (G. H. Herbig, 13.132.036), 659
(W. R. Alschuler, 13.114.012), 660 (R. P. S. Stone, 12.114.
086), 661 (E. L. Robinson, R. P. Kraft, 11.153.030), 662
(E. L. Robinson, 12.124.016), 663 (J. A. Baldwin, E. M.
Burbidge, G. R. Burbidge, C. Hazard, L. B. Robinson, E. J.
Wampler, 12.141.081), 664 (D. Burstein, L. H. McDonald,
13.154.002), 666 (K. M. Cudworth, 12.112.010), 667
(D. E. Osterbrock, 12.133.031), 668 (A. R. Klemola, L. B.
Robinson, S. Vasilevskis, 12.031.155), 669 (G. H. Herbig,
13.131.026), 670 (R. H. Durisen, 13.065.007), 671 (K. M.
Cudworth, 12.133.038), 672 (G. H. Herbig, 12.131.173),
673 (R. D. Schwartz, 13.132.005), 674 (M. F. Walker, 12.
158.195), 675 (C. D. Pike, 12.122.109), 676 (W. L. Burke,
13.162.024), 677 (M. F. Walker, C. D. Pike, J. D. McGee,
12.158.184), 678 (L. R. Cathey, 12.154.019), 679 (M. F.
Walker, C. D. Pike, J. D. McGee, 12.141.133), 681 (L. V.
Mirzoyan, J. S. Miller, D. E. Osterbrock, 13.160.009), 682
(J. A. Baldwin, E. M. Burbidge, L. B. Robinson, E. J. Wampler,
13.158.301), 683 (R. D. Schwartz, 13.064.020), 687 (J. D.
Bregman, D. M. Rank, 13.133.005), 690 (J. A. Baldwin, 13.
141.034), 693 (D. Butler, 13.122.091).

The University of California. **Contributions from
the Lick Observatory, Santa Cruz, California,** Nos. 361
(M. F. Walker, 07.114.122), 362 (J. Faulkner, 07.154.019),
363 (M. A. Smith, 08.114.071), 364 (R. W. O'Connell, R. P.
Kraft, 08.158.007), 365 (M. A. Smith, 08.152.002), 366
(P. Goldreich, D. A. Keeley, J. Y. Kwan, 09.131.013), 367
(J. Faulkner, B. P. Flannery, B. Warner, 08.126.002), 368
(R. P. Kraft, 09.142.061), 369 (W. L. Burke, 08.066.041),
370 (J. S. Miller, 09.134.004), 371 (M. F. Walker, 08.159.005),
372 (J. D. Scargle, 09.064.011), 373 (J. S. Miller, 10.133.018),
374 (M. A. Smith, 09.114.093), 375 (I. Iben, Jr., R. S. Tuggle,
08.065.098), 376 (I. Iben, Jr., 08.065.097), 377 (M. J. Blanc-
Vaziaga, G. Cayrel, R. Cayrel, 09.114.060), 378 (S. M. Faber,
09.151.007), 379 (S. V. M. Clube, 09.155.042), 380 (J. S.
Miller, 10.133.029), 381 (P. Bodenheimer, J. P. Ostriker,
09.065.022), 382 (J. P. Ostriker, P. Bodenheimer, 09.065.
023), 383 (A. Barnes, J. D. Scargle, 10.062.009), 385 (J.
Bregman, D. Butler, E. Kemper, A. Koski, R. P. Kraft, R. P. S.
Stone, 10.142.072), 386 (P. Goldreich, D. A. Keeley, J. Y.
Kwan, 09.131.162), 387 (D. Butler, R. P. Kraft, J. S. Miller,
L. B. Robinson, 09.122.008), 388 (L. B. Robinson, E. J.
Wampler, 09.158.016), 389 (J. S. Miller, L. B. Robinson,
E. J. Wampler, 09.158.017), 390 (L. B. Robinson, E. J.
Wampler, 09.160.007), 391 (K. H. Nordsieck, 10.151.011),
392 (K. H. Nordsieck, 10.158.044), 393 (B. P. Flannery,
G. H. Herbig, 10.133.004), 394 (P. A. Strittmatter, R. F.
Carswell, E. M. Burbidge, C. Hazard, J. A. Baldwin, L. Robin-
son, E. J. Wampler, 10.141.023), 395 (J. A. Baldwin, E. M.
Burbidge, C. Hazard, H. S. Murdoch, L. B. Robinson, E. J.

Wampler, 10.141.075), 396 (J. P. Ostriker, P. Bodenheimer,
10.065.020), 397 (E. J. Wampler, L. B. Robinson, J. A.
Baldwin, E. M. Burbidge, 09.141.090), 399 (G. R. Blumenthal,
W. H. Tucker, 12.142.057), 400 (J. S. Miller, 12.133.016),
401 (J. Faulkner, 12.124.009), 402 (P. Bodenheimer, 12.099.
085).

008.112 Santiago de Chile

**Departamento de Astronomía, Universidad de
Chile,** Facultad de Ciencias Fisicas y Matematicas, Observa-
torio Astronomico Nacional, Cerro Calán, Santiago de Chile,
Publicaciones, Vol. 2, No. 4 (H. Wroblewski, L. Panaiotov,
S. Vásquez, 13.098.064; H. Moreno, 13.113.062; H. Moreno,
J. Maza, 13.082.095), No. 5 (P. R. Loyola, V. N. Shishkina,
13.041.038).

Universidad de Chile, Departamento de Astronomía,
Santiago. Separata (F. Noël, 08.044.004), (A. Gutiérrez-
Moreno, H. Moreno, 10.114.142), (H. Moreno, 10.114.143),
(C. Anguita, 11.041.044), (G. Carrasco, 11.041.045), (F.
Noël, K. Czuia, P. Guerra, 12.041.009).

008.113 Seattle, Washington

University of Washington Astronomy Department,
Seattle, Washington. – Observatory report. G. Wallerstein.
Bull. American Astron. Soc., Vol. 7, 227 - 229 (1975).

008.114 Sendai

Sendai Astronomiaj Raportoj, Nos. 156 (Y. Sabano,
M. Tosa, 13.131.035) 157 (T. Aikawa, Z. Hitotuyanagi,
13.158.065), 158 (M. Kubo, 13.162.019).

008.115 Siding Spring

**Inauguration of Anglo-Australian telescope, Siding
Spring, New South Wales, October 16, 1974.**
Monthly Notes Astron. Soc. Southern Africa, Vol. 34, 5 - 6
(1975).

008.116 Skalnaté Pleso

**Contributions of the Astronomical Observatory
Skalnaté Pleso,** Vol. 5 (J. Sýkora, 13.074.089; M. Antal,
13.103.008; M. Antal, 13.098.055; A. Hajduk, 13.104.031;
J. Tremko, 13.034.072).

008.117 Sonneberg

**Zentralinstitut für Astrophysik. Mitteilungen über
Veränderliche Sterne,** *Sonneberg,* Vol. 6, No. 8 (R. Hudec,
13.114.071; P. Ahnert, 13.121.069; P. Ahnert, 13.121.070;
P. Ahnert, 13.121.071; P. Ahnert, 13.121.072; E. Splittgerber,
13.123.029; W. Wenzel, H. Geßner, 13.142.098; W. Wenzel,
13.124.104; I. Meinunger, 13.123.030; 13.123.031), Vol. 7,
No. 1 (L. Meinunger, 13.123.032).

008.118 Stanford, California

Stanford Radio Astronomy Institute, Stanford, California. – Observatory report. R. N. Bracewell. Bull. American Astron. Soc., Vol. 7, 207 - 208 (1975).

008.119 Stony Brook, New York

State University of New York at Stony Brook, Department of Earth and Space Sciences, Stony Brook, New York. – Observatory report. R. F. Knacke. Bull. American Astron. Soc., Vol. 7, 208 - 210 (1975).

008.120 Strasbourg

Publications de l'Observatoire de Strasbourg, Vol. 3, Fasc. 1 (A. Valbousquet, 13.041.040), 2 (P. Lacroute, A. Valbousquet, 12.041.004), Vol. 4, Fasc. 1 (P. Bacchus, P. Lacroute, 11.041.059), 2 (P. Lacroute, 11.041.056).

008.121 Swarthmore

Sproul Observatory, Swarthmore College, Swarthmore, Pennsylvania. – Observatory report. S. L. Lippincott, W. D. Heintz. Bull. American Astron. Soc., Vol. 7, 106 (1975).

008.122 Sydney

Sydney Observatory. W. H. Robertson. Quarterly Journ. Roy. Astron. Soc., Vol. 16, 208 (1975). Report for the year ending 1974 December 31.

Sydney Observatory Papers, Nos. 71 (K. P. Sims, 13.112.009), 72 (H. Wood, 11.041.053).

Division of Radiophysics, C.S.I.R.O., Sydney, Radiophysics Publication RPP 1659 (D. K. Milne, 09.133.025), 1684 (G. A. Dulk, 10.074.106), 1688 (S. Suzuki, 12.077.028), 1697 (N. Fourikis, M. W. Sinclair, R. D. Brown, J. G. Crofts, P. D. Godfrey, 12.131.120),1701 (D. B. Melrose, 12.078.019), 1706 (O. B. Slee, J. G. Ables, R. A. Batchelor, S. Krishna-Mohan, V. R. Venugopal, G. Swarup, 11.141.308), 1709 (K. V. Sheridan, D. J. McLean, S. F. Smerd, 12.071.008), 1710 (S. F. Smerd, K. V. Sheridan, R. T. Stewart, 13.077.004), 1712 (N. R. Labrum, R. A. Duncan, 12.077.042), 1715 (K. Kai, K. V. Sheridan, 11.074.044), 1716 (S. F. Smerd, K. V. Sheridan, R. T. Stewart, 12.077.058), 1718 (M. M. Komesaroff, J. G. Ables, D. J. Cooke, P. A. Hamilton, P. M. McCulloch, 12.141.333), 1723 (W. K. Huchtmeier, A. E. Wright, 12.158.119), 1725 (J. B. Whiteoak, F. F. Gardner, 12.158. 120), 1727 (D. J. McLean, 12.077.052), 1731 (R. T. Stewart, M. K. McCabe, M. J. Koomen, R. T. Hansen, G. A. Dulk, 11. 074.063) 1734 (B. A. Peterson, J. G. Bolton, A. J. Shimmins, 12.141.013), 1745 (G. A. Dulk, K. V. Sheridan, 11.074.062), 1747 (G. J. Nelson, K. V. Sheridan, 12.077.054), 1750 (P. A. Hamilton, P. M. McCulloch, J. G. Ables, M. M. Komesaroff, D. J. Cooke, 12.141.332), 1754 (R. T. Stewart, R. A. Howard, F. Hansen, T. Gergely, M. Kundu, 11.074.064), 1759 (N. Fourikis, B. MacA. Thomas, 13.033.029), 1761 (T. Takakura, S. Yousef, 11.077.066), 1765 (T. W. Cole, J. G. Ables, 12.034. 003), 1767 (J. P. Wild, 12.077.033), 1774 (N. Fourikis, M. W.

Sinclair, 12.131.204), 1777 (D. S. Mathewson, M. N. Cleary, J. D. Murray, 11.159.004), 1789 (N. Fourikis, K. Takagi, M. Morimoto, 12.131.031), 1792 (J. B. Whiteoak, 12.131.202), 1797 (F. F. Gardner, B. J. Robinson, 12.131.205), 1794 (J. W. Brooks, M. W. Sinclair, G. A. Manefield, 12.131.203), 1798 (R. D. Robinson, 12.077.085), 1799 (R. T. Stewart, 12.077.070), 1803 (R. A. Duncan, 12.077.084), 1806 (T. W. Cole, 12.033. 068), 1809 (J. B. Whiteoak, F. F. Gardner, 13.158.011), 1818 (J. B. Whiteoak, D. H. Rogstad, I. A. Lockhart, 12.131.111), 1819 (T. W. Cole, 13.031.040).

Division of Radiophysics, C. S. I. R. O., Sydney, Australia. Separate prints (O. B. Slee, P. Y. Lee, 12.083.111; B. J. Robinson, J. W. Brooks, P. D. Godfrey, R. D. Brown, 12.131.225; D. B. Melrose, 13.062.056; J. G. Bolton, A. J. Shimmins, J. V. Wall, 13.141.084; J. G. Bolton, P. W. Butler, 13.141.085; J. V. Wall, A. J. Shimmins, J. G. Bolton, 13.141. 086; A. J. Shimmins, J. G. Bolton, J. V. Wall, 13.141.087; R. X. McGee, L. M. Newton, R. A. Batchelor, 13.132.032; D. K. Milne, J. R. Dickel, 13.141.060; F. F. Gardner, J. B. Whiteoak, D. Morris, 13.141.103; O. B. Slee, C. S. Higgins, 13.141.104).

008.123 Tartu

Tartu Astronoomia Observatoorium, Preprints, Nos. 5 (= Tartu Astron. Obs. Teated, No. 52), 6 (= Tartu Astron. Obs. Teated, No. 53), 7 (R. E. Gershberg, L. Luud, 13.114.077).

Tartu Astronoomia Observatoorium, Teated Nos. 50 (H. Eelsalu, 13.155.011), 51 (H. Eelsalu, 13.155.012), 52 (J. Pelt, 13.031.263), 53 (T. Kipper, J. Sitska, 13.031.264).

008.124 Thessaloniki

Astronomical Department, University of Thessaloniki. – Annual report 1973. G. Contopoulos. Annual Rep. Astron. Inst. Greece 1973, p. 11 - 15 (1974).

Astronomy Department, University of Thessaloniki. Annual report 1974. G. Contopoulos. Annual Rep. Astron. Inst. Greece 1974, p. 10 - 14 (1975).

Department of Geodetic Astronomy, University of Thessaloniki. – Annual report 1973. L. N. Mavridis. Annual Rep. Astron. Inst. Greece 1973, p. 20 - 23 (1974).

Department of Geodetic Astronomy, University of Thessaloniki. – Annual report 1974. L. N. Mavridis. Annual Rep. Astron. Inst. Greece 1974, p. 19 - 22 (1975).

008.125 Tokyo

Annals of the Tokyo Astronomical Observatory, University of Tokyo, Second Series, Vol. 14, Nos. 3 (S. Isobe, 13.131.132), 4 (S. Isobe, 13.132.037; S. Isobe, 13.131.133; S. Isobe, 13.131.134; S. Isobe, 13.131.135; M. Crézé, S. Isobe, 13.131.136); Vol. 15, Nos. 1 (Y. Yamashita, 13.124. 100; Y. Yamashita, 13.114.072), 2 (H. Yasuda, R. Fukaya, H. Hara, T. Ina, 13.041.034; H. Yasuda, N. Miyauchi, 13. 041.035).

Tokyo Astronomical Bulletin, Tokyo Astronomical Observatory, Second Series, Nos. 236 (H. Yasuda, R. Fukaya,

H. Hara, S. Isobe, H. Ishii, Y. Adachi, T. Ina, 13.041.041), 237 (M. Saitō, H. Sato, E. Watanabe, K. Okida, H. Ogata, C. Hukusaku, H. Sugai, 13.121.104).

Tokyo Astronomical Observatory, Reprints Nos. 457 (S. Manabe, M. Miyamoto, 13.155.019), 458 (K. Akabane, Y. Chikada, 13.103.100), 459 (S. Kikuchi, A. Okazaki, 13. 103.100), 460 (K. Ogura, K. Ishida, 13.132.013), 463 (T. Hirayama, 11.073.030), 464 (K. Nariai, 12.124.024), 465 (T. Daishido, N. Kawano, N. Kawajiri, M. Inoue, M. Konno, 12.142.020), 470 (Y. Uchida, 12.074.107).

Time and Latitude Bulletins, Tokyo Astronomical Observatory, Vol. 48, Nos. 7 - 12 (K. Osawa, 13.044.026).

University of Tokyo, Tokyo Astronomical Observatory, Report (No. 65), Vol. 17, No. 2 (K. Saito, 13.004.067; K. Fujiwara, T. Kato, S. Hokugo, K. Miyazaki, 13.035.007; M. Saitō, 13.031.250; S. Hamana, S. Yajima, 13.034.074; H. Shibasaki, T. Yamaguchi, N. Oshima, M. Noguchi, 13.032. 029; S. Iijima, 13.034.089; T. Yamaguchi, H. Nishimura, 13.032.031);(No. 66), Vol. 17, No. 3 (I. Shimizu, T. Hirayama, Y. Ohki, M. Fukatsu, T. Oe, Y. Shimizu, E. Hiei, 13.034.090; R. Fukaya, 13.082.085; Y. Iizuka, 13.035.009; N. Kobayashi, N. Oshima, M. Noguchi, 13.035.010; T. Miyaji, 13.033.026; F. Moriyama, E. Hiei, A. Tokuya, H. Miyazaki, 13.079.100; H. Sekiguchi, S. Aiba, H. Nakajima, 13.033.027).

Data Report of Hydrographic Observations. Series of Astronomy and Geodesy, Maritime Safety Agency, Tokyo, Japan, No. 9 (T. Mori, Y. Ganeko, Y. Harada, M. Sasaki, M. Yamaguti, 13.096.011; 13.046.026).

008.126 Toledo

Ritter Astrophysical Research Center of the University of Toledo, Toledo, Ohio. – Observatory report. A. N. Witt. Bull. American Astron. Soc., Vol. 7, 196 - 198 (1975).

008.127 Tonantzintla

Boletin del Instituto de Tonantzintla, Vol. 1, No. 3 (A. D. Andrews, 13.113.053; G. Haro, E. Chavira, 13.122. 097; D. Malacara, A. Cornejo, 13.031.036; L. R. Berriel, 13. 031.249).

008.128 Torino

Attività dell'Osservatorio. See Abstr. 047.020, p. 27 - 35.

Contributi dell'Osservatorio Astronomico di Torino (Pino Torinese), Nos 69 (V. Zappalà, 10.103.108), 78 (M. G. Fracastoro, 11.118.017), 79 (H. Debehogne, S. Vaghi, V. Zappalà, 11.103.124), 80 (F. Scaltriti, M. A. Vogliotti, V. Zappalà, 13.103.100), 81 (M. G. Fracastoro, 13.092.025), 82 (M. A. Vogliotti, V. Zappalà, 13.098.074), 85 (F. Scaltriti, V. Zappalà, 13.098.003).

Osservatorio Astronomico di Torino, Pino Torinese. Time Service, Bulletin No. 7 (C. Moranzino, 13.044.030).

Pubblicazioni Varie Fuori Serie dell'Osservatorio Astronomico di Torino (Pino Torinese), No. 59 (13.047.020).

008.129 Trieste

Pubblicazione Osservatorio Astronomico di Trieste, Nos. 435 (B. Cester, 07.122.059), 436 (G. Sedmak, 07.031. 011), 437 (R. Stalio, 08.064.051), 438 (R. Stalio, 08.114.151), 439 (C. Aydin, 08.064.003), 440 (C. Aydin, 08.114.085), 442 (P. L. Selvelli, 08.114.036), 443 (A. Abrami, 07.077.053), 444 (P. L. Selvelli, 09.114.037), 445 (M. Hack, 08.116.013), 446 (B. Cester, M. Pucillo, 08.121.012), 447 (R. Faraggiana, 09. 114.009), 448 (13.075.004), 449 (13.075.004), 450 (P. Zlobec, 13.077.058), 451 (13.075.004), 452 (B. Cester, M. Pucillo, 08.121.103), 453 (13.075.004), 454 (M. Hack, 09.121.017), 455 (G. Sedmak, 09.034.025), 456 (P. Santin, 09.077.067), 457 (13.075.004), 458 (13.075.004), 459 (G. Sedmak, 09.034.056), 460 (G. Bisiacchi, U. Flora, M. Hack, 11.121.008), 461 (13.075.004), 462 (13.075.004), 463 (G. A. De Biase, G. Sedmak, 11.034.058), 464 (G. Sedmak, 11.031.001), 465 (B. Cester, M. Pucillo, 10.121.020), 466 (A. Abrami, 13.077.059), 467 (P. Zlobec, 13.077.060), 468 (P. Santin, P. Zlobec, 13.077.061), 469 (B. Cester, M. Pucillo, 11.121.018), 470 (P. L. Selvelli, 11.114.025), 471 (R. Stalio, 11.114.022), 472 (S. Engin, 11.114.068), 473 (S. Engin, 12.114.012), 474 (S. Engin, 12.114.004), 475 (13.075.004), 476 (13.075.004), 477 (M. Hack, J. B. Hutchings, Y. Kondo, G. E. McCluskey, M. Plavec, R. S. Polidan, 11.121.059), 478 (U. Flora, M. Hack, 13.121.001), 479 (M. Hack, 12.121.052), 480 (A. M. Boesgaard, F. Praderie, D. S. Leckrone, R. Faraggiana, M. Hack, 12.114.151), 481 (R. Stalio, 12.114.117), 482 (A. Abrami, 13.077.062), 483 (G. Sedmak, 13.034.086).

008.130 Tucson

Kitt Peak National Observatory, Tucson, Arizona; Cerro Tololo Inter-American Observatory, La Serena, Chile. Observatory reports.   B. T. Lynds. Bull. American Astron. Soc., Vol. 7, 87 - 120 (1975).

National Radio Astronomy Observatory, Charlottesville, Virginia; Green Bank, West Virginia; and Tucson, Arizona. – Observatory report.   D. S. Heeschen, D. E. Hoog. Bull. American Astron. Soc., Vol. 7, 170 - 178 (1975).

University of Arizona, Department of Planetary Sciences and Lunar and Planetary Laboratory, Tucson, Arizona. – Observatory report.   C. P. Sonett. Bull. American Astron. Soc., Vol. 7, 14 - 24 (1975).

008.131 Uccle

Bulletin Astronomique. (Astronomisch Bulletin). **Observatoire Royal de Belgique** (Koninklijke Sterrenwacht van België), Vol. 8, No. 3 (H. Debehogne, 13.098.065; J. Dommanget, E. Van Dessel, 13.096.016; C. Gonze, R. Gonze, 13.075.011; J. Dommanget, 13.041.042; J. Dommanget, 13. 118.023; R. Van de Wiele, 13.118.024).

Observatoire Royal de Belgique (Koninklijke Sterrenwacht van Belgie), **Communications** (Mededelingen), Série A, Nos. 23 (P. Cugnon, 06.131.035), 24 (E. W. Elst, 09.122.017), 25 (A. G. Velghe, 13.114.085), 26 (H. Debehogne, 12.041.005), 27 (H. Debehogne, 07.031.001), 28 (H. Debehogne, 09.031.028), 29 (N. Grevesse, A. J. Sauval, 10.071. 007).

Observatoire Royal de Belgique (Koninklijke Sterrenwacht van Belgie), **Communications** (Mededelingen), Série B, Nos. 87 (P. Cugnon, 10.131.159), 88 (C. De Jager, L. Neven, 08.071.037), 90 (P. Melchior, 13.081.037).

008.132 **Uppsala**

Uppsala Astronomiska Observatoriums Annaler, Band 5, No. 8 (Å. Wallenquist, 13.153.032).

Uppsala Astronomical Observatory, Report No. 6 (C.-I. Lagerkvist, 13.098.066).

008.133 **Utrecht**

Utrechtse Sterrekundige Overdrukken, Sterrewacht "Sonnenborgh", Utrecht, Nos. 264 (C. Chiuderi, R. Giachetti, H. Rosenberg, 10.077.079), 265 (E. L. van Dessel, J. Houtgast, D. Koelbloed, 10.071.083), 266 (C. de Jager, R. Hoekstra, K. A. van der Hucht, T. M. Kamperman, H. J. Lamers, A. Hammerschlag, W. Werner, J. G. Emming, 11.034.001), 267 (J. Buurman, 10.072.077), 268 (G. A. Stevens, 11.061. 013), 269 (M. Kuperus, M. A. Raadu, 11.073.010), 270 (J. Buurman, 11.072.008), 271 (C. de Jager, 11.071.007), 272 (P. Hoyng, G. A. Stevens, 11.061.017), 273 (H. C. Spruit, 11. 080.015), 274 (A. C. Brinkman, D. R. Parsignault, E. Schreier, H. Gursky, E. M. Kellogg, H. Tananbaum, R. Giacconi, 11. 142.042), 275 (H. F. van Beek, L. D. de Feiter, C. de Jager, 11.076.018), 276 (A. C. Brinkman, J. Heise, C. de Jager, 12. 142.141), 277 (H. C. Spruit, C. Zwaan, 13.080.053), 278 (J. Kuijpers, 11.077.039), 279 (Solar Radio Group Utrecht, 11. 077.051), 280 (L. D. de Feiter, 11.073.058), 281 (K. A. van der Hucht, H. J. Lamers, 11.114.130), 282 (R. Hoekstra, K. A. van der Hucht, T. Kamperman, H. J. Lamers, 13.034.104), 283 (H. van de Stadt, T. de Graauw, J. C. Shelton, C. Veth, 13.034.105), 284 (C. Zwaan, 12.072.002), 285 (E. Gurtovenko, V. Ratnikova, C. de Jager, 12.071.004), 286 (W. J. Weber, T. de Graaf, 13.061.065), 287 (M. Kuperus, 12.080.040), 288 (W. J. Weber, H. Rosenberg, 12.074.029), 289 (H. F. van Beek, L. D. de Feiter, C. de Jager, 12.076.049), 290 (L. D. de Feiter, 12.073.124), 291 (M. Burger, K. A. van der Hucht, 12.114.164), 292 (H. J. Lamers, K. A. van der Hucht, R. Hoekstra, R. Faraggiana, M. Hack, 12.114.163), 293 (C. Zwaan, 12.072.038), 294 (K. A. van der Hucht, H. J. Lamers, 12.114.162), 295 (M. Kuperus, 13.080.054).

008.134 **Victoria**

Dominion Astrophysical Observatory, Victoria, B.C. J. B. Hutchings. Journ. Roy. Astron. Soc. Canada, Vol. 69, 97 (1975).

Publications of the Dominion Astrophysical Observatory, Victoria, B. C., Vol. 14, Nos. 11 (C. L. Morbey, J. M. Fletcher, 13.031.409), 12 (D. Crampton, W. A. Fisher, 13.114.086), 13 (R. E. Stencel, C. R. Cowley, 13.114.087).

008.135 **Villanova**

Villanova University, Department of Astronomy, Villanova, Pennsylvania. – Observatory report. E. F. Jenkins. Bull. American Astron. Soc., Vol. 7, 217 - 218 (1975).

008.136 **Warsaw**

Warsaw University Observatory and Astronomical Institute, Polish Academy of Sciences, Reprint Nos.358 (W. Dziembowski, M. Kozłowski, 12.122.049), 359 (J. S. Stodółkiewicz, 12.151.031), 360 (J. S. Stodółkiewicz, 12.151.045), 361 (J. Madej, 12.063.046), 362 (S. L. Piotrowski, S. M. Ruciński, I. Semeniuk, 12.121.062), 363 (S. L. Piotrowski, 13.117.014), 364 (A. Kruszewski, I. Semeniuk, 13.160.012), 365 (G. Sitarski, 13.103.116), 366 (S. Grzędzielski, G. Sitarski, 13.106.035).

Politechnika Warszawska, Obserwatorium Astronomiczno-Geodezyjne w Józefosławiu, (Warsaw Technical University, Astronomic-Geodetical Observatory at Józefosław), **Latitude Circular,** Nos. 48 (B. Kołaczek, M. Dukwicz-Łatka, 13.045.027), 51, 52 (L. Pieczyński, 13.045.028).

008.137 **Washington**

U. S. Naval Observatory, Washington, D. C. (1 July 1973 – 30 June 1974). – Observatory report. K. A. Strand. Bull. American Astron. Soc., Vol. 7, 211 - 215 (1975).

Astronomical Papers prepared for the use of the American Ephemeris and Nautical Almanac, Vol. 22, Part II (E. S. Jackson, 13.041.043).

Publications of the United States Naval Observatory, *Washington,* Second Series, Vol. 22, Part 6 (F. J. Josties, C. C. Dahn, V. V. Kallarakal, M. Miranian, G. G. Douglass, J. W. Christy, A. L. Behall, R. S. Harrington, 13.118.026).

U.S. Naval Observatory, Washington, D. C. Time Service Publications, Series 4, Nos. 413 - 438; Series 7, Nos. 366 - 391; Series 14, No. 18 (13.044.027 – 13.044.029).

United States Naval Observatory, *Washington, D.C.,* **Circular,** Nos. 149 (P. M. Janiczek, G. H. Kaplan, 13.021.007), 150 (Index to United States Naval Observatory Circulars Nos. 101 - 150).

National Bureau of Standards, Washington, D. C. L. Hagan. Bull. American Astron. Soc., Vol. 7, 165 - 169 (1975).

National Aeronautics and Space Administration, Washington, D. C. N. W. Boggess. Bull. American Astron. Soc., Vol. 7, 153 - 165 (1975).

008.138 **Westford, Massachusetts**

Haystack Observatory, Northeast Radio Observatory Corporation (EROC), Westford, Massachusetts. – Observatory report. P. B. Sebring. Bull. American Astron. Soc., Vol. 7, 69 - 73 (1975).

008.139 **Wien**

Astronomische Mitteilungen Wien, Nos. 15 (M. G. Firneis, F. J. Firneis, 13.079.100), 16 (K. Ferrari d'Occhieppo,

13.118.025), Suppl. (K. Ferrari d'Occhieppo, 13.004.083).

008.140 Williams Bay, Wisconsin

The University of Chicago, Department of Astronomy and Astrophysics, Chicago, Illinois. The Yerkes Observatory, Williams Bay, Wisconsin. – Observatory reports. E. N. Parker, L. M. Hobbs. Bull. American Astron. Soc., Vol. 7, 25 - 30 (1975).

008.141 Williamstown, Massachusetts

Hopkins Observatory, Williams College, Williamstown, Massachusetts. – Observatory report.   J. M. Pasachoff. Bull. American Astron. Soc., Vol. 7, 73 - 75 (1975).

008.142 Wrocław

Contributions from the Wrocław Astronomical Observatory No. 18 (B. Rompolt, 13.003.019).

Wrocław Astronomical Observatory, Reprint Nos.93 (B. Szczodrowska-Kozar, 13.042.084), 94 (B. M. Sylwester, 12.074.058), 95 (J. A. Sylwester, 12.074.059).

008.143 Yorktown Heights, New York

IBM Thomas J. Watson Research Center, Yorktown Heights, New York.   G. Lasher. Bull. American Astron. Soc., Vol. 7, 75 - 76 (1975).

008.144 Zelenchukskaya

Chronicle. Astrofiz. Issled., Izv. Spets. Astrofiz. Obs., Vol. 7, p. 245 - 246 (1975). In Russian.

Astrofizicheskie Issledovaniya. Izvestiya Spetsial'noj Astrofizicheskoj Observatorii, Vol. 7 (R. N. Kumajgorodskaya, N. M. Chunakova, 13.114.322; N. F. Vojkhanskaya, G. N. Alekseev, 13.122.028; Yu. P. Korovyakovskij, Yu. V. Sukha-

rev, 13.117.017; Yu. P. Korovyakovskij, 13.117.018; A. A. Korovyakovskaya, 13.061.020; A. I. Shapovalova, 13.158.051; B. P. Artamonov, F. Börngen, P. Notni, 13.158.052; L. G. Antropova, B. P. Artamonov, F. Börngen, 15.158.053; N. V. Bystrova, I. A. Rakhimov, 13.157.005; N. V. Bystrova, I. A. Rakhimov, 13.157.006; I. V. Gosachinskij, 13.155.023; T. B. Pyatunina, 13.131.517; V. M. Bogod, A. N. Korzhavin, 13. 077.028; N. G. Peterova, 13.077.029; G. N. Alekseev, V. G. Shtol', 13.031.212; A. A. Korovyakovskaya, Yu. P. Korovyakovskij, E. L. Chentsov, 13.031.213; G. M. Beskin, A. M. Bogudlov, Eh. A. Vitrichenko, O. A. Evseev, 13.031.015; Eh. A. Vitrichenko, F. K. Katagarov, V. G. Lipovetskaya, 13.031.016; G. M. Beskin, A. M. Bogudlov, Eh. A. Vitrichenko, O. A. Evseev, S. M. Soldatov, 13.031.017; G. M. Beskin, A. M. Bogudlov, Eh. A. Vitrichenko, O. A. Evseev, 13.031.018; G. M. Beskin, Eh. A. Vitrichenko, A. I. Samojlov, A. G. Shcherbakov, 13.031.019; A. F. Dravskikh, 13.033.008; N. L. Kajdanovskij, 13.033.009; N. L. Kajdanovskij, 13.033.010; M. N. Kajdanovskij, A. A. Stotskij, 13.033.011; G. S. Golubchin, 13.033.012; N. N. Mikhel'son, 13.032.017; V. S. Rylov, V. G. Shtol', 13. 031.214).

Soobshcheniya Spetsial'noj Astrofizicheskoj Observatorii, Akademiya Nauk SSSR, vyp. (No.) 10 (K. V. Bychkov, 13.132.010; R. N. Kumajgorodskaya, N. M. Chunakova, 13. 114.307; T. A. Uranova, G. S. Tsarevskij, 13.113.009), 12 (G. N. Alekseev, 13.031.204; V. I. Abramenko, A. V. Ipatov, N. M. Lipovka, A. K. Obolenskij, 13.157.001; N. S. Soboleva, O. N. Shivris, 13.033.004; R. M. Kirakosyan, K. S. Mosoyan, O. B. Petrosyan, V. G. Gevorkyan, 13.033.005).

008.145 Zürich

Tätigkeitsbericht der Eidgenössischen Sternwarte Zürich für das Jahr 1974.   M. Waldmeier. Zürich, 8 pp. (1975).

Astronomische Mitteilungen der Eidgenössischen Sternwarte Zürich, Nos. 331 (M. Waldmeier, 13.074.107), 332 (M. Waldmeier, 13.079.107), 333 (M. Waldmeier, 13.075. 012), 334 (S. Cortesi, 13.082.093), 335 (J. Dürst, 13.074. 108), 336 (A. Zelenka, 13.071.001), 337 (M. Waldmeier, 13. 074.018), 338 (M. Waldmeier, 12.074.084), 341 (M. Waldmeier, 13.074.109).

Quarterly Bulletin on Solar Activity (Zürich), Nos. 185 - 186 (M. Waldmeier, R. Howard, G. Olivieri, M. Bernot, H. Tanaka, 13.075.013).

009 Notes on Observatories, Planetaria, and Exhibitions

009.001 **Mt. Hood Community College solar observatory.**
J. Del Wiseman, Jr.
Sky Telescope, Vol. 49, 82 - 86 (1975).

009.002 **The planetarium as an analogue computer.**
H. Mucke.
Journ. History Astron., Vol. 6, 53 - 57 (1975).

009.003 **Armagh astronomy centre.** T. Murtagh.
Spaceflight, Vol. 17, 66 - 68 (1975).

009.004 **A major planetarium for Tucson, Arizona.**
O. R. Norton.
Sky Telescope, Vol. 49, 143 - 146 (1975).

009.005 **Five amateur observatories.**
E. R. Matthew, J. M. Cope, A. Keeley, J. C. Baize,
H. Anderson.
Sky Telescope, Vol. 49, 208 - 212 (1975).

009.006 **Armagh Planetarium expands.** T. Murtagh.
Sky Telescope, Vol. 49, 224 - 225 (1975).

009.007 **Potchefstroom University for C.H.E., Department
of Physics.** P. H. Stoker.
Monthly Notes Astron. Soc. Southern Africa, Vol. 34, 24 - 25
(1975). – Report for 1974.

009.008 **Rhodes University, Department of Physics.**
E. E. Baart.
Monthly Notes Astron. Soc. Southern Africa, Vol. 34, 26
(1975). – Report for 1974.

009.009 **University of South Africa, Department of Mathematics and Astronomy.** J. Wolterbeek.
Monthly Notes Astron. Soc. Southern Africa, Vol. 34, 27
(1975). – Report for 1974.

009.010 **University of the Witwatersrand, Departments of
Applied Mathematics and Physics.**
B. L. Fanaroff.
Monthly Notes Astron. Soc. Southern Africa, Vol. 34, 27
(1975). – Report for 1974.

009.011 **Twenty years of the Silesian planetarium.**
M. Pańków, K. Rudnicki.
Postępy Astron., Vol. 23, 73 - 74 (1975). In Polish.

009.012 **Sterrenwachten in Californië en Arizona bedreigd
door stadslicht.** G. W. E. Beekman.
Zenit, Vol. 2, 47 - 49 (1975).

009.013 **A new comet observatory on South Baldy.**
J. C. Brandt, S. A. Colgate, R. W. Hobbs, W. Hume,
S. P. Maran, E. P. Moore, R. G. Roosen.
Mercury, (Journ. Astron. Soc. Pacific), Vol. 4, No. 2, p. 12 -
13 (1975).

009.014 **Observatory for the youth.** S. Ya. Grechko.
Zemlya i Vselennaya, 1975, No. 2, p. 86 - 87.
In Russian.

009.015 **The Stanford solar observatory.**
P. H. Scherrer, J. M. Wilcox, L. Svalgaard, P. H.
Dittmer, T. L. Duvall.
Bull. American Astron. Soc., Vol. 7, 350 - 351 (1975).
Abstr. AAS.

009.016 **Eine neue Feriensternwarte in Cuxhaven.**
M. Koch.
Orion, 33. Jahrgang, p. 89 - 90 (1975).

009.017 **Vom Atomkern bis in den Weltraum. Ein Bericht
über die Aufgaben und Forschungsziele des Max-
Planck-Instituts für Kernphysik in Heidelberg.**
J. Kiko, U. Schmidt-Rohr.
Umschau, 75. Jahrgang, p. 359 - 367 (1975).

009.018 **Aufgaben und Ergebnisse der Forschungsarbeiten
im Zentralinstitut für Physik der Erde der Akademie der Wissenschaften der DDR.** H. Kautzleben.
Akad. Wiss. DDR, Zentralinst. Phys. Erde, Veröff. No. 29,
(see 013.012), p. 5 - 20 (1974).

009.019 **Das Planetarium Wernigerode.** J. Schön.
Astron. in der Schule, 12. Jahrgang, p. 64 - 65
(1975).

009.020 **Universitäts-Sternwarte Graz 1972.**
H. Haupt.
Sternenbote, 18. Jahrgang, p. 33 - 36 (1975).

009.021 **Department of Natural Philosophy, University of
Aberdeen.** R. V. Jones.
Quarterly Journ. Roy. Astron. Soc., Vol. 16, 168 - 169 (1975).
Report for the year ending 1974 September 30.

009.022 **Department of Physics, Queen Mary College, Uni-
versity of London.** J. A. Bastin.
Quarterly Journ. Roy. Astron. Soc., Vol. 16, 189 (1975).
Report for the year ending 1970 September 30.

009.023 **Royal Aircraft Establishment, Farnborough. Geo-
physical studies in space department.**
D. G. King-Hele.
Quarterly Journ. Roy. Astron. Soc., Vol. 16, 190 - 196 (1975).
Report for the year ending 1974 December 31.

009.024 **Public observatory in Prague and its stations in
Dáblice and Kleť.** O. Hlad.
Říše hvězd, Vol. 56, 117 - 119, 141 - 142 (1975). In Czech.

009.025 **Public observatory and planetarium in Brno.**
O. Oburka.
Říše hvězd, Vol. 56, 36 - 37 (1975). In Czech.

009.026 **Public observatory in Valašské Meziříčí.**
B. Maleček.
Říše hvězd, Vol. 56, 77 - 78, 102 - 103 (1975). In Czech.

Astronomical site prospects in New South Wales.
See Abstr. 082.136.

010 Societies, Associations, Organizations

010.001 **American Association of Variable Star Observers (AAVSO)**

Annual report of the director 30 September 1972.
M. W. Mayall.
Journ. American Ass. Variable Star Observers, Vol. 1, 80 - 83 (1972).

Annual report of the director, 30 September 1973.
M. W. Mayall.
Journ. American Ass. Variable Star Observers, Vol. 2, 92 - 95 (1973).

Annual report of the director, 30 September 1974.
J. A. Mattei.
Journ. American Ass. Variable Star Observers, Vol. 3, 82 - 91 (1974).

Minutes of the general meeting of the A.A.V.S.O. held at Steward Observatory, Tucson, Arizona, May 27, 1972.
C. B. Ford.
Journ. American Ass. Variable Star Observers, Vol. 1, 22 - 26 (1972). – Included are the treasurer's report and the report of the ad hoc Committee on Data Processing.

Minutes of the general meeting of the AAVSO held at the Coffin School, Nantucket, Mass., October 14, 1972.
C. B. Ford.
Journ. American Ass. Variable Star Observers, Vol. 1, 71 - 79 (1972). – Included are Committee reports and AAVSO Treasurer's report.

Minutes of the general meeting of the AAVSO held at the Conference Center, University of Virginia, Charlottesville, Virginia, May 19, 1973. C. B. Ford.
Journ. American Ass. Variable Star Observers, Vol. 2, 43 - 48 (1973). – Included are committee reports and treasurer's report.

Minutes of the general meeting of the AAVSO held at Blantyre Castle, Lenox, Mass. October 20, 1973.
C. B. Ford.
Journ. American Ass. Variable Star Observers, Vol. 2, 84 - 91 (1973). – Included are committee reports and treasurer's report.

Minutes of the general meeting of the AAVSO held at the University of Manitoba, Winnipeg, Manitoba, Canada on Saturday, June 28, 1974. C. B. Ford.
Journ. American Ass. Variable Star Observers, Vol. 3, 36 - 42 (1974). – Included are committee reports and treasurer's report.

Minutes of the general meeting of the AAVSO held at the Treadway Williams Inn, Williamstown, Massachusetts, on October 19, 1974. C. B. Ford, R. N. Mayall.
Journ. American Ass. Variable Star Observers, Vol. 3, 72 - 81 (1974). – Included are Committee reports and the AAVSO treasurer's report for the year ending September 30, 1974.

Our directors – past and present. C. B. Ford.
Journ. American Ass. Variable Star Observers, Vol. 2, 50 - 51 (1973).

010.002 **American Astronomical Society (AAS)**

Meetings of the Society – see Bull. American Astron. Soc., Vol. 7, 233 - 272, 336 - 391 (1975).

010.003 **Association of Lunar and Planetary Observers (ALPO)**

Announcements.
Strolling Astronomer, Vol. 26, 170 - 171 (1975).

Minor Planet Section news and other Section news.
Minor Planet Bull., Vol. 2, 24, 30, 41, 44 (1975).

List of materials and services supplied by A.L.P.O. recorders. J. R. Smith.
Strolling Astronomer, Vol. 25, 160 - 161 (1975).

010.004 **Astronomical Society of Australia (ASA)**

No publication received.

010.005 **Astronomical Society of Czechoslovakia**

No publication received.

010.006 **Astronomical Society of the Pacific (ASP)**

No publication received.

010.007 **Astronomical Society of Southern Africa (ASSA)**

Notices.
Monthly Notes Astron. Soc. Southern Africa, Vol. 34, 1, 31, 77 (1975).

010.008 **Astronomical Society of Victoria (ASV)**

Annual report. A. E. Coombs, R. J. C. Lawrence.
Journ. Astron. Soc. Victoria, Vol. 28, 2 - 5 (1975).

Treasurer's report. D. H. Walker.
Journ. Astron. Soc. Victoria, Vol. 28, 7 (1975).

Statement of receipts and expenditure for the year ended 1974 December 31. J. H. White.
Journ. Astron. Soc. Victoria, Vol. 28, 8 - 9 (1975).

Astrophotographic Section. A. Stern.
Journ. Astron. Soc. Victoria, Vol. 28, 5 (1975).

Auroral Section. T. B. Tregaskis.
Journ. Astron. Soc. Victoria, Vol. 28, 7 (1975).

Current Phenomena Section. J. B. Trainor.
Journ. Astron. Soc. Victoria, Vol. 28, 12 - 13 (1975).

Historical Section. P. Simon.
Journ. Astron. Soc. Victoria, Vol. 28, 13 - 14 (1975).

Instrument Making Section. C. N. Chatfield.
Journ. Astron. Soc. Victoria, Vol. 28, 14 (1975).

Lunar and Planetary Section. B. S. Adcock.
Journ. Astron. Soc. Victoria, Vol. 28, 5 (1975).

Nova Search Section. B. S. Adcock.
Journ. Astron. Soc. Victoria, Vol. 28, 15 (1975).

Variable Stars Section. T. B. Tregaskis.
Journ. Astron. Soc. Victoria, Vol. 28, 11 - 12 (1975).

010.009 **Astronomical Society of Western Australia (ASWA)**

Report on proceedings of ordinary meetings.
Journ. Astron. Soc. Western Australia, Vol. 26 - 28, January - April (1975).

010.010 **Astronomische Gesellschaft (AG)**

No publication received.

010.011 **Astronomisk Selskab København**

No publication received.

010.012 **British Astronomical Association (BAA)**

Notices.
Journ. British Astron. Ass., Vol. 85, 82 - 84, 194 - 198 , 306 - 307 (1975).

The annual general meeting of the Association held on 1974 October 30.
N. J. Goodman, G. E. Stone, H. R. Hatfield.
Journ. British. Astron. Ass., Vol. 85, 85 - 88 (1975).

Meetings of the Association.
Journ. British Astron. Ass., Vol. 85, 89 - 92, 199 - 207, 308 - 314 (1975).

New members elected.
Journ. British Astron. Ass., Vol. 85, 184 - 189, 298 - 300, 375 - 380 (1975).

Lunar Section. P. Moore.
Journ. British Astron. Ass., Vol. 85, 139 - 144, 265 - 267 (1975).

Mars Section. E. H. Collinson.
Journ. British Astron. Ass., Vol. 85, 336 - 341 (1975).

Mercury and Venus Section. J. H. Robinson.
Journ. British Astron. Ass., Vol. 85, 145 - 149, 268 - 270 (1975).

Meteor Section.

British Astron. Ass. Circ. No. 562 (1975).

Meteor Section. K. B. Hindley.
Journ. British Astron. Ass., Vol. 85, 150 - 155, 342 - 345 (1975).

Solar Section. V. Barocas, M. R. Whippey.
Journ. British Astron. Ass., Vol. 85, 136 - 138, 255 - 264, 332 - 335 (1975).

Variable Star Section. J. E. Isles.
British Astron. Ass. Circ. Nos. 561, 562 (1975).

Variable Star Section. J. E. Isles.
Journ. British Astron. Ass., Vol. 85, 156 - 160, 271 - 279, 346 - 354 (1975).

Nova search programme. Report for 1973 – 74 session. C. V. Borzelli.
Journ. British. Astron. Ass., Vol. 85, 278 - 279 (1975). – Report of Variable Star Section.

010.013 **British Interplanetary Society**

Society news.
Journ. British Interplanet. Soc., Vol. 28, 364 - 366 (1975).

The report of the council for the year ended 31 December 1974. K. W. Gatland.
Spaceflight, Vol. 17, 156 - 157 (1975).

010.014 **Committee on Space Research (COSPAR)**

No publication received.

010.015 **European Space Research Organization (ESRO)**

No publication received.

010.016 **International Astronautical Federation (IAF)**

Astronautics in Amsterdam. The 25th I.A.F. congress. L. Parks.
Spaceflight, Vol. 17, 42 - 50 (1975).

010.017 **International Astronomical Union (IAU)**

67th symposium of the International Astronomical Union. J. M. Kreiner.
Postępy Astron., Vol. 23, 75 - 76 (1975). In Polish.

International Astronomical Union, Information Bulletin, Nos, 33, 34. 43 + 15 pp. (1975).
G. Contopoulos.
Contents: The sixteenth general assembly; Executive committee; Commissions; IAU symposia and colloquia; Other scientific meetings; IAU publications; Other publications; International organizations; Membership.

010.018 Meteoritical Society

Abstracts of papers presented at the 37th annual meeting of the Meteoritical Society, August 7–9, 1974. University of California at Los Angeles, Los Angeles, California. Meteoritics, Vol. 9, 313 - 423 (1974). – See Abstr. 091.007; 091.008; 094.119 - 094.128; 094.420 - 094.430; 097.009; 098.004; 105.016 - 105.078.

010.019 Nederlandse Vereniging voor Weer- en Sterrenkunde

Jaarvergadering N.V.W.S.
T. de Groot, S. J. van Leverink, G. Comello, E. J. A. Meurs. Zenit, Vol. 2, 95 - 96 (1975).

010.020 Polskie Towarzystwo Astronomiczne (PTA)

No publication received.

010.021 Polskie Towarzystwo Miłośników Astronomii (PTMA)

PTMA chronicle.
Urania Kraków, Vol. 46, 23 - 25, 54 - 59, 78 - 91, 152 - 157 (1975). In Polish.

010.022 Royal Astronomical Society (RAS)

Meetings of the Society.
Observatory, Vol. 95, 37 - 44, 73 - 78 (1975).

Meetings of the Society.
Quarterly Journ. Roy. Astron. Soc., Vol. 16, 110 - 115, 220 - 232 (1975).

010.023 Royal Astronomical Society of Canada (RAS Canada)

No publication received.

010.024 Royal Astronomical Society of New Zealand (RAS New Zealand)

The Royal Astronomical Society of New Zealand (Inc.). 52nd annual report of council for the year ended 1974 September 30.   N. J. Rumsey, P. D. Cain. Southern Stars, Vol. 26, 19 - 23 (1975).

Variable Star Section.   F. M. Bateson. Southern Stars, Vol. 26, 24 - 30 (1975). – Report for the year ended 1974 September 30.

010.025 Schweizerische Astronomische Gesellschaft (SAG)

L'assemblea generale della SAS a Locarno. R. Roggero.

Orion, 33. Jahrgang, p. 86 - 87 (1975).

Jahresbericht des SAG-Zentralpräsidenten anlässlich der Generalversammlung vom 3. Mai 1975 in Locarno. W. Studer. Orion, 33. Jahrgang, p. 87 (1975).

010.026 Sociedad Astronómica de México

No publication received.

010.027 Società Astronomica Italiana (SAI)

No publication received.

010.028 Société Astronomique de France (SAF)

Les séances de la Société.   B. Clouet. L'Astronomie, 89e année, p. 169 - 171, 200, 216 (1975).

Commission des Surfaces Planétaires.
J. Dragesco, A. Poirier. L'Astronomie, 89e année, p. 23 - 34, 93 - 108 (1975).

Commission du Soleil.   M.-J. Martres. L'Astronomie, 89e année, p. 52 - 54, 208 (1975).

Commission des Cadrans Solaires.   R. Sagot. L'Astronomie, 89e année, p. 55 - 58 (1975).

Commission des Instruments.   J. Funel. L'Astronomie, 89e année, p. 109 - 114 (1975).

Comité Camille Flammarion.
L'Astronomie, 89e année, p. 4 - 5 (1975).

Activité de la Cinémathèque S.A.F. en 1973–1974. J. A. Leclerc. L'Astronomie, 89e année, p. 6 (1975).

Une réunion commune à toutes les commisions de la Société Astronomique de France.   J.-C. Pecker. L'Astronomie, 89e année, p. 50 - 51 (1975).

010.029 Société Astronomique "R. Boškovic"

See Abstr. 010.045.

010.030 Société Chronométrique de France

No publication received.

010.031 Société Belge d'Astronomie, de Météorologie et de Physique du Globe

Séance mensuelle.   M. Ducuroir. Ciel et Terre, Vol. 91, 247 - 250 (1975).

010.032   Svenska Astronomiska Sällskapet

Svenska Astronomiska Sällskapet; Astronomiska
Sällskapet Tycho Brahe; Göteborgs Astronomiska Klubb.
Styrelsens berättelse för år 1974.
T. Elvius, G. Darsenius, P.-Å. Björklund, J. O. Stenflo, H.
Wilhelmsson, L. Ekelund.
Separate print Svenska Astronomiska Sällskapet, Stockholm.
8 pp. (1975).

010.033   VAGO (Astronomical-Geodetical Society of the
          USSR)

No publication received.

010.034   Vereniging voor Sterrenkunde, België

No publication received.

010.035   Fifth meeting of the Canadian Astronomical
          Society at the University of Manitoba, Winnipeg,
November 7 - 9, 1974.   P. M. Millman.
Journ. Roy. Astron. Soc. Canada, Vol. 69, 38 - 43 (1975).

010.036   4. Frühjahrstagung des VdS in Würzburg am 5.
          April 1975.   E. Wiedemann.
Orion, 33. Jahrgang, p. 80 - 81 (1975).

010.037   Société Royale d'Astronomie d'Anvers, [Koninklijk
          Sterrenkundig Genootschap van Antwerpen].
Cinquante-cinquième rapport 1974.   J. Storms.
Imprimerie: «La Prévoyance», Antwerpen. 11 + 13 pp. (1975).
In French and Flemish.

010.038   Censimento statistico delle associazioni e gruppi di
          astrofili Italiani.
Compiled and published by Ass. Astrofili Bolognesi, 44 pp.
(1974). — 41 associations and groups of "Astrofili Italiani"
are filed. Their addresses, observatories, astronomical instruments as well as the program of their scientific work are listed.

010.039   The Nantucket Maria Mitchell Association. Seventy-
          third annual report for the year ending December
31, 1974.
Edited by The Nantucket Maria Mitchell Ass., Nantucket,
Mass., 64 pp. (1975). — Included is the annual report of the
director of Maria Mitchell Observatories by D. Hoffleit.

010.040   The Irish Astronomical Society. Belfast Centre
          progress 1960–1973.   D. E. Beesley.
Irish Astron. Journ., Vol. 11, 187 - 192 (1974).

010.041   Comité Belge des Astronomes Amateurs.
          R. Charles.
Ciel et Terre, Vol. 91, 251 (1975).

010.042   Constitution of the Astronomical Society of India.
          Bull. Astron. Soc. India, Vol. 1, 5 - 9 (1973).
This constitution consists of two parts, viz: A: Memorandum
and B: Bye-laws.

010.043   La vie de l'Association.   É. Schweitzer.
          A.F.O.E.V. Bull., Vol. 8, 7 - 8, 44 - 45, 93 (1974);
Vol. 9, 26 (1975). — Concerning Association Française des
Observateurs d'Étoiles Variables.

010.044   Asociación Peruana de Astronomía.
          Bol. As. Peruana Astron. (APA), Vol. 7, No. 155
(1974). — 1974 June - September.

010.045   Osnovana podružnica Astronomskog društva
          'Ruđer Bošković' in Novom Sadu — 'ADNOS'
(New branch of the Astronomical Society 'Ruđer Bošković'
in Novi Sad — 'ADNOS').   J. Francisti.
Vasiona, Vol. 22, 121 (1974).

010.046   XXII redovna godišnja skupština društva
          (XXIIth annual meeting of the Society).   A. Tomić.
Vasiona, Vol. 22, 121 - 122 (1974).

010.047   I.U.A.A. International Union of Amateur Astro-
          nomers. Proceedings of the second general assembly,
(Malmö 1972, July 31st - August 4th).   A. Leani (Editor).
I.U.A.A. Publ. PADUS, Cremona (Italy), 47 pp. (1974).

010.048   Nachrichten der Vereinigung der Sternfreunde e.V.
          SuW, Vol. 14, 28 - 30, 65 - 67, 99 - 100, 135 - 137,
170 - 174, 208 - 212 (1975).

011 Reports on Colloquia, Congresses, Meetings, Symposia, and Expeditions

011.001 **Structure and dynamics of spiral galaxies.**
V. Icke, J. Pringle.
Nature, Vol. 253, 312 - 313 (1975).
Informal discussion day at the Institute of Astronomy, Cambridge, 1975 January 9.

011.002 **Chronicle.** Yu. Z. Mavashev, K. A. Shafeeva.
Geliotekhnika, 1974, No. 4, p. 75 - 76. In Russian.
Abstr. in Referativ. Zhurn. 62. Issled. kosmich. prostranstva, 2.62.8 (1975). − Concerning the 24th congress of IAF.

011.003 **Orientation of coordinate systems in space and investigation of systematic errors of star catalogues.**
E. P. Fedorov.
Astrometriya i Astrofizika, *Kiev*, vyp. (No.) 24, (see 003.003), p. 125 - 130 (1974). In Russian.
Report on the All-Union conference of scientists working in the field of astrometry, Kiev, Oct. 15 - 17 (1973).

011.004 **Atlanta planetarium conference.** G. Lovi.
Sky Telescope, Vol. 49, 26 - 27 (1975).

011.005 **Solar influences on the weather.** R. H. Olson.
Nature, Vol. 253, 686 (1975). − Symposium on solar−weather relationships sponsored jointly by the American Meteorological Society and the Solar Physics Branch of the American Astronomical Society, Denver 1975 January 23.

011.006 **International CETI review meeting 1974.**
A. T. Lawton.
Journ. British Interplanet. Soc., Vol. 28, 220 - 222 (1975).
Amsterdam, 1974 Oct. 4.

011.007 **Soviet-American conference on the cosmochemistry of the moon and planets.** E. L. Ruskol.
Zemlya i Vselennaya, 1975, No. 1, p. 58 - 62. In Russian.

011.008 **4th conference on ultraviolet and X-ray spectroscopy of astrophysical and laboratory plasmas, Harvard University, 9 - 11 September 1974.** D. J. Lovell.
Applied Optics, Vol. 14, 7 - 8 (1975).

011.009 **Report from Gainesville.** J. B. Irwin.
Sky Telescope, Vol. 49, 164 - 166 (1975). − Meeting of the AAS 1974 December 10 - 13.

011.010 **Between earth and sun.** L. J. C. Woolliscroft.
Nature, Vol. 254, 562 (1975).
A MIST (Magnetosphere, Ionosphere and Solar-Terrestrial relations) meeting, arranged by the Royal Astronomical Society, was held at the University of Exeter on March 25 and 26.

011.011 **Status report on the inner planets.** J. Gribbin.
Nature, Vol. 254, 657 - 658 (1975).
On April 11 the Royal Astronomical Society, the Royal Meteorological Society and the Geological Society held a joint meeting. Though the meeting was entitled 'Mariner 10 results on Venus and Mercury', it included results on other planets and data obtained from other sources.

011.012 **The third solar wind conference: a summary.**
C. T. Russell.
Space Sci. Rev., Vol. 17, 435 - 447 = Inst. Geophys. Planet. Phys., Univ. Calif., *Los Angeles*, Publ. No. 1354-51 (1975).

Review paper − see 012.002.
The third solar wind conference was convened from March 25 to 29, 1974 at the Asilomar Conference Grounds, Pacific Grove, California. The conference consisted of nine sessions dealing with solar abundances; the history and evolution of the solar wind; the structure and dynamics of the solar wind; the structure and dynamics of the solar corona; macroscopic and microscopic properties of the solar wind; cosmic rays as a probe of the solar wind; spatial gradients; stellar winds; and interactions with objects in the solar wind. This paper summarizes the invited and contributed talks presented at the conference.

011.013 **Compte rendu de la Commission Gravimétrique Internationale.** S. Coron.
Bull. Géod., Nouvelle Sér., No. 115, p. 83 - 97 (1975). − Paris, 2 - 6 Septembre 1974.

011.014 **A report on the conference on Time Service and Latitude Service.**
Acta Astron. Sinica, Vol. 15, 246 (1974). In Chinese.
Brief note.

011.015 **4th Summer School on cosmology in Opole.**
K. Rudnicki.
Postępy Astron., Vol. 23, 76 - 77 (1975). In Polish.

011.016 **Radiopulsare, Neutrinoströme und kosmisches Urgewinsel. Die VII. Texaskonferenz für relativistische Astrophysik in Dallas, Texas, vom 16. bis 20. Dezember 1974.** R. Breuer.
Phys. Blätter, 31. Jahrgang, p. 224 - 228 (1975).

011.017 **Scientific session of the Department of General Physics and of Astronomy of the USSR Academy of Sciences, Leningrad, April 3 - 6, 1974.**
Uspekhi fiz. nauk, Vol. 114, 153 - 160 (1974). In Russian.

011.018 **Symposium international sur les mesures de distances terrestres par procédé électromagnétique et sur la réfraction dans les mesures angulaires. (Stockholm 19−24 Août 1974).** P. L. Baetslé, M. Louis.
Bull. Géod., Nouvelle Sér., Année 1975, No. 116, p. 199 - 204 (1975).

011.019 **Cosmochemistry of the moon and planets. Conference in Moscow.** Yu. I. Belyaev.
Vestn. AN SSSR, 1974, No. 12, p. 88 - 91. In Russian.
Abstr. in Referativ. Zhurn. 51. Astron., 5.51.23 (1975).

011.020 **First All-Union astronomical and geodetical conference.** L. S. Khrenov.
Geod. raboty na Urale. Sverdlovsk, 1974, p. 23 - 25. In Russian. − Abstr. in Referativ. Zhurn. 51. Astron., 5.51.25 (1975).

011.021 **Intercosmos symposium on results of satellite observations.** G. Gerstbach.
Österreich. Zeitschr. Vermessungswesen, 62. Jahrgang, p. 177 - 178 (1975). In German. − Short report on the symposium which took place in Budapest, 1974 October 21 - 24.

011.022 **The abstracts of the meetings of the Research Group of Geodynamics.** C. Kakuta.
Proc. International Latitude Obs. Mizusawa, No. 14, p. 24 -

29 (1974). In English and Japanese.

011.023 "Stellar wounds" on the earth and their diagnostics by geophysical methods.
A. I. Dabizha, V. V. Fedynskij.
Zemlya i Vselennaya, 1975, No. 3, p. 56 - 64. In Russian.
Conference in Moscow, 1975, Jan. 6 - 9.

011.024 Summaries of papers presented at R.A.S. specialist discussion on "The universal background radiation",
held on 1974 October 11.
Observatory, Vol. 95, 79 - 83 (1975). – For the individual contributions – see 061.054, 061.055, 066.096 - 066.098.

011.025 Joint meeting of the Royal Meteorological Society and the Royal Astronomical Society on "The atmosphere of Jupiter", held on 1974 November 8.
G. E. Hunt.
Observatory, Vol. 95, 83 - 85 (1975).

011.026 Royal Astronomical Society specialist discussion on results obtained from the UK 48-inch Schmidt telescope plates.
Observatory, Vol. 95, 85 - 90 (1975). – 1974 December 13.

011.027 Seminar on Nicholaus Copernicus and astronomy.
S. N. Sen.
Bull. Astron. Soc. India, Vol. 1, 10 (1973). – New Delhi, 1973 February 19 - 20.

011.028 Brief report on the 13th International Cosmic Ray Conference, Denver, Colorado, August 17 - 30, 1973.
J. V. Narlikar.
Bull. Astron. Soc. India, Vol. 1, 22 (1973).

011.029 Paris symposium on gravitational waves.
P. C. Vaidya.
Bull. Astron. Soc. India, Vol. 1, 45 (1973). – 1973 June 18 - 22.

011.030 13th International Cosmic Ray Conference.
M. V. K. Appa Rao.
Bull. Astron. Soc. India, Vol. 1, 46 (1973).

011.031 Report on the Fifth Lunar Conference.
N. Bhandari.
Bull. Astron. Soc. India, Vol. 2, 25, 43 (1974).

011.032 Eighth ESLAB symposium on H II regions and the galactic centre. S. N. Tandon.
Bull. Astron. Soc. India, Vol. 2, 58 (1974).

011.033 Report on the IAU symposium No. 67.
S. D. Sinvhal, S. C. Joshi.
Bull. Astron. Soc. India, Vol. 2, 70 - 72 (1974). – Concerning the symposium on "Variables in relation to the evolution of stars and stellar systems", Moscow, 1974 July 29 - August 3.

011.034 Report on the I.A.U. symposium No. 69.
S. M. Alladin.
Bull. Astron. Soc. India, Vol. 2, 72 - 73 (1974). – Concerning

the symposium on "Dynamics of stellar systems", Besançon, 1974 September 9 - 13.

011.035 Symposium on "Some aspects of astrophysics".
R. K. Varma.
Bull. Astron. Soc. India, Vol. 2, 73, 75 (1974). – Ahmedabad, 1974 August 19 - 24.

011.036 Fifth seminar on general relativity and gravitation.
J. V. Narlikar.
Bull. Astron. Soc. India, Vol. 3, 18 (1975). – Powai, Bombay, 1974 December 27 - 29.

011.037 Cosmic ray diffusion – report of the Workshop in Cosmic Ray Diffusion Theory.
T. J. Birmingham, F. C. Jones.
National Aeronautics and Space Administration, Washington, D.C., NASA Technical Note, NASA TN D-7873, 7 + 18 pp. = Goddard Space Flight Center, Greenbelt, Maryland, Separate print (1975).
A workshop in Cosmic Ray Diffusion Theory was held at Goddard Space Flight Center on May 16–17, 1974. Topics discussed were: 1) Cosmic ray measurements as related to diffusion theory; 2) quasi-linear theory, nonlinear theory, and computer simulation of cosmic ray pitch-angle diffusion; and 3) magnetic field fluctuation measurements as related to diffusion theory. This report summarizes the proceedings.

011.038 The XIXth Herstmonceux conference.
Monthly Notes Astron. Soc. Southern Africa, Vol. 34, 78 (1975).

011.039 The All-Union symposium on submillimeter and millimeter wave propagation in the atmospheres of the earth and planets. K. A. Agabekyan.
Izv. vyssh. ucheb. zavedenij. Radiofizika, Vol. 17, 1742 - 1748 (1974). In Russian. – Abstr. in Referativ. Zhurn. 51. Astron., 6.51.137; 62. Issled. kosmich. prostranstva, 6.62.263 (1975). Gorkij, 1974 January 28 - 30.

011.040 International seminar concerning the theme "Particle acceleration and nuclear reactions in space".
G. E. Kocharov.
Vestn. AN SSSR, 1975, No. 1, p. 100 - 103. In Russian.
Abstr. in Referativ. Zhurn. 62. Issled. kosmich. prostranstva, 6.62.4 (1975). – Leningrad, 1974, August 19 - 21.

011.041 Protokoll der 120. Sitzung der Schweiz. Geodätischen Kommission vom 22. Juni 1974 in der Universität Bern.
Société Helvétique des Sciences Naturelles, (Schweiz. Naturforschende Gesellschaft). Spross + Co., Kloten. 14 pp. (1975).

011.042 Second European astronomical meeting.
V. Bahýl, D. Chochol.
Kozmos, Vol. 6, 23 (1975). In Slovak.

011.043 Lunar petrology conference.
W. I. Ridley, A. M. Reid, P. R. Brett.
EOS Trans. American Geophys. Union, Vol. 55, 4 - 15 (1974).

012 Proceedings of Colloquia, Congresses, Meetings, and Symposia

012.001 **New problems of astrophysics.**
Publications of the Astrophysical Winter School, Arkhyz, Jan. 21 - 30, 1972. Izdatel'stvo Moskovskogo Universiteta. 166 pp. Price 40 Kop. (1974). In Russian. − The individual contributions are included in their corresponding subject categories − see abstracts 066.011, 066.012, 141.312 - 141.315, 141.602, 151.005, 158.019, 162.011, 162.012.

012.002 **International solar-terrestrial physics symposium,**
Sao Paulo, Brazil, 17−24 June, 1974, with an introduction by S. A. Bowhill.
Space Sci. Rev., Vol. 17, 171 - 614 (1975). − The individual contributions are included in their corresponding subject categories − see abstracts 011.012, 073.005, 073.025, 074.029, 074.030, 083.037, 084.011, 084.012, 084.234 - 084.237, 084.403, 084.404, 106.021, 106.022.

012.003 **Proceedings of the Fifth Lunar Science Conference,**
Houston, Texas, March 18−22, 1974, sponsored by the NASA Johnson Space Center and the Lunar Science Institute. In memoriam P. W. Gast.
W. A. Gose (Managing editor).
Vol. 1: Mineralogy and petrology, (with lunar orbital data maps, 8 pp. and a glossary of lunar terms, 11 pp.); Vol. 2: Chemical and isotope analyses, organic chemistry; Vol. 3: Physical properties.
Pergamon Press, New York−Oxford−Toronto−Sydney− Braunschweig. Geochim. Cosmochim. Acta, Suppl. 5, 24 + 12 + 11 + 3134 + 32 + 20 + 20 pp. Price $ 100.00 (1974). − The individual papers are included in their corresponding subject categories − see abstracts 074.040, 094.005, 094.006, 094.138 - 094.219, 094.432 - 094.552, 143.027.

012.004 **The place of astronomy in the ancient world.**
A joint symposium of the Royal Society and the British Academy.
Organized by D. G. Kendall, S. Piggott, D. G. King-Hele, I. E. S. Edwards. F. R. Hodson (Editor).
Published for the British Academy by Oxford University Press, London, 276 pp. Price £ 13.00 = Phil. Trans. Roy. Soc. London, Ser. A, Vol. 276, No. 1257 (1974). − The individual contributions are included in their corresponding subject categories − see abstracts 004.025 - 004.040.

012.005 **Asteroids, comets, meteoric matter.** International Astronomical Union, 22nd Colloquium, held in Nice, France, April 4−6, 1972.
C. Cristescu, W. J. Klepczynski, B. Milet (Editors).
Editura Academiei Republicii Socialiste România, 333 pp. (1974). − The individual contributions are included in their corresponding subject categories − see abstracts 031.219 - 031.222, 041.014, 042.033 - 042.035, 098.016 - 098.028, 099.035, 099.036, 102.012 - 102.020, 103.104, 103.105, 103.106, 103.107, 103.108, 103.109, 104.015 - 104.020.

012.006 **Materials on the All-Union conference on cosmic ray physics, Kharkov, 25 - 28 September, 1973.**
Izv. AN SSSR. Ser. fiz., Vol. 38, 1785 - 2008 (1974). In Russian.

012.007 **Satellite Dynamics.** COSPAR−IAU−IUTAM Symposium, São Paulo, Brazil, June 19 - 21, 1974.
G.E.O. Giacaglia (Editor).
Springer-Verlag, Berlin−Heidelberg−New York. 8 + 376 pp. Price DM 64.00 (1975). − The individual contributions within

the subject scope of Astronomy and Astrophysics Abstracts are included in their corresponding categories − see abstracts 042.045, 052.020 - 052.035, 054.013.

012.008 **Multicolor photometry and the theoretical HR diagram.** Proceedings of a conference held at the State University of New York at Albany, October, 1974.
A. G. D. Philip, D. S. Hayes (Editors), with a summary by D. C. Crawford.
Dudley Obs. Rep. No. 9, 8 + 523 pp. Price $ 10.00 (1975). The individual papers are included in their corresponding subject categories − see abstracts 034.050, 034.051, 034.055, 064.061 - 064.063, 065.063, 111.002, 113.028 - 113.047, 114.056 - 114.062, 114.347, 114.348, 115.008, 115.009, 121.041, 122.061 - 122.063, 154.012, 154.013, 155.038, 158.089.

012.009 **Proceedings of the workshop on coherent detection in astronomy,** held at Rhenen, The Netherlands, 25 and 26 April 1974.
H. G. Van Bueren (Editor), with a preface by H. G. Van Bueren, T. De Graauw, H. Van de Stadt.
Space Sci. Rev., Vol. 17, (No. 5), 617 - 736 (1975). − The individual contributions are included in their corresponding subject categories − see abstracts 022.068, 022.069, 031.240 - 031.243, 034.057 - 034.063, 061.048, 071.044, 097.063.

012.010 **Transactions of the fifth All-Union conference on the problem "Astrophysical phenomena and radiocarbon", Tbilisi, October 4 - 6, 1973.**
Tbilis. un-t, fiz.-tekhn. in-t AN SSSR. Tbilisi, Tbilis. un-t. 370 pp. Price 2 Rbl. 10 Kop. (1974). In Russian.

012.011 **A discussion on the origin of the cosmic radiation,** held 20 and 21 February 1974.
With introductory remarks by G. D. Rochester, A. W. Wolfendale.
Phil. Trans. Roy. Soc. London, Ser. A, Vol. 277, (No. 1270), p. 317 - 501 (1975). − The individual contributions are included in their corresponding subject categories − see abstracts 061.002, 141.352, 141.353, 143.045 - 143.053.

012.012 **Origin of cosmic rays.** Proceedings of the NATO Advanced Study Institute held in Durham, England, August 26 - September 6, 1974.
J. L. Osborne, A. W. Wolfendale (Editors).
D. Reidel Publishing Company, Dordrecht−Holland/Boston− U.S.A. 10 + 466 pp. Price Dfl 105.00, $ 39.50 respectively (1975). − The individual contributions are included in their corresponding subject categories − see abstracts 061.051, 061.052, 125.051 - 125.053, 143.062 - 143.073.

012.013 **Conference on optical observing programs on galactic structure and dynamics.** Proceedings.
T. Schmidt-Kaler (Editor).
Published by Astron. Inst. Ruhr-Univ., Bochum. 8 + 261 pp. (1975). − The individual contributions are included in their corresponding subject categories − see abstracts 031.253, 034.087, 061.053, 065.082, 065.083, 114.073, 131.138, 142.101, 151.035 - 151.037, 153.026, 155.044 - 155.049, 157.011, 158.124.

012.014 **Thackeray symposium. S.A.A.O.-January 1975.**
Monthly Notes Astron. Soc. Southern Africa, Vol. 34, 33 - 76 (1975). − For the individual contributions − see abstracts 005.016, 008.000, 112.006, 113.054, 113.055, 114.075, 122.098 - 122.102, 124.006, 142.102, 153.027,

155.050, 159.016, 159.017.

012.015 **Ferraro Memorial Meeting,** with a preface by
P. C. Kendall, H. Rishbeth.
Geophys. Journ. Roy. Astron. Soc., Vol. 41, (No. 3), p. 307 -
446 (1975). – For the individual contributions within the
subject scope of Astronomy and Astrophysics Abstracts –
see abstracts 062.053, 083.071 - 083.073, 084.266 - 084.268,
097.084, 099.086 - 099.088.

012.016 **Atti del convegno internazionale sulla "Rotazione
della terra e osservazioni di satelliti artificiali".**
Proceedings of the international meeting on "Earth's rotation
by satellite observations". Cagliari, 16–18 April 1973.
E. Proverbio (Editor).
Rend. Seminario Fac. Sci. Univ. Cagliari, Suppl. Vol. 44,
[Graficoop Soc. Tipografica Editoriale, Bologna], 2 + 120 pp.
(1974). – The individual contributions are included in their
corresponding subject categories – see abstracts 031.256,
031.257, 044.016, 044.017, 045.020 - 045.024, 054.017,
081.028 - 081.030.

012.017 **Solar gamma-, X-, and EUV radiation.** International
Astronomical Union, Symposium No. 68, held in
Buenos Aires, Argentina, 11 - 14 June 1974. Organized by the
IAU in cooperation with COSPAR.
S. R. Kane (Editor).
D. Reidel Publishing Company, Dordrecht–Holland/ Boston–
U.S.A. 12 + 439 pp. Price Dfl. 145.00 (1975). – The individual
contributions are included in their corresponding subject
categories – see abstracts 073.085 - 073.097, 074.096 - 074.
099, 076.038 - 076.059.

012.018 **High energy astrophysics and its relation to
elementary particle physics.** Lectures of the NATO
Advanced Study Institute, held in Erice, Sicily, Italy, June
16 - July 6, 1972. K. Brecher, G. Setti (Editors).
The MIT Press, Cambridge, Mass. – London. 6 + 591 pp.
Price $ 9.95 (1974). – The individual contributions are
included in their corresponding subject categories – see
abstracts 022.084, 022.085, 061.058, 066.100, 141.100,
141.101, 158.132, 158.133, 158.134, 162.074 - 162.077.

012.019 **Modern problems of positional astrometry.**
(19th Astrometrical Conference of the USSR,
Moscow, 27 - 30 June 1972). V. V. Podobed (Editor).
Moskovskij Gosudarstvennyj Universitet im.M. V. Lomono-
sova, Gosudarstvennyj Astronomicheskij Institut im. P. K.
Shternberga. Izdatel'stvo Moskovskogo Universiteta. 327 pp.
Price 1 Rbl. 96 Kop. (1975). In Russian. – The individual
papers are included in their corresponding subject categories
see abstracts 031.271 - 031.275, 032.032 - 032.037, 033.033,
034.107, 034.108, 041.044 - 041.077, 044.031, 046.028,
082.094, 094.013, 094.014, 094.250 - 094.257, 094.268 -
094.273, 096.017, 112.014 - 112.016.

012.020 **Solar wind three.** Proceedings of a conference
sponsored jointly by the Universities of California
and Arizona and by the National Aeronautics and Space
Administration, at the Asilomar Conference Grounds,
Pacific Grove, California. C. T. Russell (Editor).
Published by Institute of Geophysics and Planetary Physics,
University of California, Los Angeles, 6 + 487 pp. Price
$ 10.00 (1974). – The individual contributions are included
in their corresponding subject categories – see abstracts
062.060, 063.036, 064.079, 073.118, 074.119 - 074.150,
077.068, 078.031, 078.032, 084.270, 084.271, 084.272,

093.038, 093.039, 094.260, 099.097, 099.098, 102.040,
103.100, 106.048 - 106.053, 107.013, 117.041, 143.095,
143.096. – Review in Planet. Space Sci., Vol. 23, 1015,
1975 (*P. D. Hudson*).

012.021 **Instrumentation in astronomy II.** Conference held
at Tucson, Arizona, 1974 March 4 - 6. Society of
Photo-optical Instrumentation Engineers, Palos Verdes,
California. 8 + 252 pp. (1974). – The individual contributions
are included in their corresponding subject categories – see
abstracts 031.041, 031.042, 031.281, 031.414 - 031.417,
032.039 - 032.044, 034.114 - 034.127, 036.007.

012.022 **Optique spatiale (space optics).**
A. Maréchal, G. Courtès (Editors).
Gordon and Breach Science Publishers Ltd., London – New
York – Paris. 6 + 389 pp. Price $ 26.50 (1974). – Reviews in
Journ. Optical Soc. America, Vol. 65, 746 - 747; 1975 (*F. W.
Paul*); Strolling Astronomer, Vol. 25, 164 - 165; 1975 (*R. E.
Cox*).

012.023 **Proceedings of symposium on earth's gravitational
field and secular variations in position.**
R. S. Mather, P. V. Angus-Leppan (Editors).
Published by the School of Surveying – The University of New
South Wales, Sydney. 726 pp. Price Aust. $ 20.00. – Review in
Bull. Géod., Nouvelle Sér., Année 1975, No. 116, p. 209; 1975
(*J. Villecrose*).

012.024 **The physics of mesospheric (noctilucent) clouds.**
Proceedings of a conference, Riga, Latvia, Nov. 1968.
J. Ikaunieks (Editor).
Israel Program for Scientific Translation, Jerusalem, (U.S.
distributor, International Scholarly Book Services, Portland,
Ore.), 8 + 156 pp. Price $ 16.00 (1973). – (From Science,
Vol. 188, 48 (1975).

012.025 **Transactions of the plenary session of the Commission of the Astronomical Council of the Soviet
Academy of Sciences for study of the earth's rotation, Poltava,
October 24, 1972.**
AN USSR. Poltav. gravimetr. observ. Kiev, "Nauk. dumka".
74 pp. Price 25 Kop. (1974). In Russian. – Review in Referativ. Zhurn. 51. Astron., 4.51.171 (1975).

012.026 **Structure and dynamics of the upper atmosphere.**
(Proceedings of the Second Course of the Interna-
tional School of Atmospheric Physics).
F. Verniani (Editor).
Developments in Atmospheric Science, 1. Elsevier Scientific
Publishing Company, Amsterdam – London – New York. 13 +
535 pp. Price Dfl. 160, $ 61.50 respectively (1974). – Review
in Nature, Vol. 254, 89; 1975 (*T. Beer*).

012.027 **Evolution and physical properties of meteoroids.**
Proceedings of the I.A.U. Colloquium No. 13, Al-
bany, N.Y., June 14–17, 1971.
C. L. Hemenway, P. M. Millman, A. F. Cook (Editors).
Scientific and Technical Information Office, NASA, Washing-
ton, D.C., NASA SP-319, 6 + 378 pp. Price $ 4.30 (1973).
Review in Irish Astron. Journ., Vol. 11, 206; 1974 (*E. Öpik*).

012.028 **Large space telescope – a new tool for science.**
Proceedings of a conference held in Washington
1974.
American Institute of Aeronautics and Astronautics, New
York, N.Y. 124 pp. Price $ 26.00 (1974). – (From Sky Tele-
scope, Vol. 49, 321 (1975)).

013 Reports on Astronomy in Various Countries and Particular Fields, International Cooperation

013.001 **Astrophysical researches in the A. F. Ioffe Physical-Technical Institute of the USSR Academy of Sciences.** M. M. Bredov.
Probl. Sovrem. fiz. Leningrad, Nauka, 1974, p. 141 - 146. In Russian. − Abstr. in Referativ. Zhurn. 51. Astron., 2.51.34 (1975).

013.002 **Von der Astronomie in Indien.** H. Letsch.
Jenaer Rundschau (Jena Review), 20. Jahrgang, p. 10 - 12 (1975).

013.003 **Astronomy in New Zealand.** N. J. Rumsey.
Southern Stars, Vol. 25, 171 - 207 (1974). − The Hudson lecture to the Wellington Branch of the Royal Society of New Zealand, 1974 July 31.

013.004 **X-ray astronomy in India.** H. P. Mama.
Spaceflight, Vol. 17, 98 - 101 (1975).

013.005 **Extraterrestrial radioastronomical observations.** R. Schreiber.
Postępy Astron., Vol. 23, 11 - 19 (1975). In Polish.
    The review is an attempt to describe the achievements of extraterrestrial radio astronomy during the last 10 years. Substantial progress, especially in the observations of the solar corona and the interplanetary space, is reported. Observations of planets and the Galaxy are also mentioned. Some remarks concerning future experiments are briefly outlined.

013.006 **Origins of Canadian government astronomy.** R. A. Jarrell.
Journ. Roy. Astron. Soc. Canada, Vol. 69, 77 - 85 (1975).

013.007 **A foundation for research.** W. A. Fowler.
Science, Vol. 188, 414 - 420 (1975).
    The return on investment in scientific research during the first 25 years of support by the National Science Foundation is discussed. Over this period there has been a veritable explosion in astronomical science, and here NSF has played an important and in many way the important role.

013.008 **De nouveaux horizons dans l'hémisphère austral.** R. S. Stobie.
Orion, 33. Jahrgang, p. 71 - 74 (1975).

013.009 **Development of astronomy.** O. V. Dobrovol'skij.
Nauka sov. Tadzhikistana. Dushanbe, "Donish", 1974, p. 28 - 38. In Russian. − Abstr. in Referativ. Zhurn. 51. Astron., 4.51.32 (1975).

013.010 **Investigation of the cosmos and scientific-technical progress.** B. N. Petrov.
Vopr. filosofii, 1974, No. 10, p. 36 - 45. In Russian.

013.011 **Thirty years from the liberation of Czechoslovakia.**
Bull. Astron. Inst. Czechoslovakia, Vol. 26, 129 - 132 (1975). − Editorial note on research at astronomical institutes of Czechoslovakia.

013.012 **Aufgaben und Ergebnisse der Forschungsarbeiten des Zentralinstituts für Physik der Erde.**
Edited on the occasion of the 25th anniversary of the foundation of the German Democratic Republic by H. Kautzleben. Akad. Wiss. DDR, Forschungsber. Kosm. Phys., Zentralinst. Phys. Erde, Veröff. No. 29, 173 pp. (1974). − For the individual contributions within the subject scope of Astronomy and Astrophysics Abstracts see abstracts 009.018, 045.019, 046.024, 046.025, 081.026.

013.013 **Astronomy and astrophysics at PRL, past, present and future.** R. Pratap.
Bull. Astron. Soc. India, Vol. 2, 47 - 50 (1974).

013.014 **Astronomical work in the Ural district.** K. A. Barkhatova, Z. N. Shukstova.
Geod. raboty na Urale. Sverdlovsk, 1974, p. 10 - 17. In Russian. − Abstr. in Referativ. Zhurn. 51. Astron., 6.51.22 (1975).

013.015 **Astronomy.**
Nauka i chelovechestvo. 1975. Moskva, Znanie, 1974, p. 335 - 370. In Russian. − Abstr. in Referativ. Zhurn. 51. Astron., 6.51.75 (1975).

013.016 **Advances in astronomy in the year 1974.** J. Grygar.
Říše hvězd, Vol. 56, 47 - 55, 70 - 74, 92 - 100, 106 - 113, 127 - 137 (1975). In Czech.

013.017 **Czechoslovak astronomy in the years 1945 - 1975.** L. Křivský.
Říše hvězd, Vol. 56, 41 - 47 (1975). In Czech.

013.018 **Popularization of astronomy in Czechoslovakia after the year 1945.** O. Hlad.
Říše hvězd, Vol. 56, 65 - 70 (1975). In Czech.

013.019 **Scientific work on Czechoslovak public observatories in the years 1945 - 1975.** O. Oburka.
Říše hvězd, Vol. 56, 81 - 86 (1975). In Czech.

014 Teaching in Astronomy

014.001 On N. Copernic's life and research work (materials for discussions of pupils). A. I. Kritskij.
Sbornik st. po metodike prepodav. mat. i fiz. Minsk, Vishehjsh. shkola, 1974, p. 104 - 113. In Russian. — Abstr. in Referativ. Zhurn. 51. Astron., 2.51.53 (1975).

014.002 Talcott Mountain Science Center.
A. Schneider, R. Judd.
Sky Telescope, Vol. 49, 4 - 7 (1975).

014.003 Course in astronomy at the school of the future.
E. P. Levitan.
Zemlya i Vselennaya, 1975, No. 1, p. 80 - 84. In Russian.

014.004 Photographing the moon with a school telescope.
Yu. K. Lyubimov.
Zemlya i Vselennaya, 1975, No. 1, p. 89 - 91. In Russian.

014.005 Über eine Möglichkeit für die Einbeziehung von wissenschaftlichen Arbeitsmethoden der Astronomie in den Astronomieunterricht. M. Schukowski.
Sterne, Vol. 51, 51 - 58 (1975).

014.006 Some national projects in astronomy education.
A. Fraknoi.
Mercury, (Journ. Astron. Soc. Pacific), Vol. 4, No. 2, p. 3, 23 (1975).

014.007 Astronomical equipment for schools.
Eh. F. Brazhnikova.
Tsirkulyar Vses. astron.-geod. o-va, 1974, No. 26, p. 81 - 86. In Russian. — Abstr. in Referativ. Zhurn. 51. Astron., 3.51.43 (1975).

014.008 Observation of the moon and planets in lessons of astronomy. A. G. Konyaeva.
Vopr. metodiki prepodav. fiz. Vyp. (No.) 2. Tula, 1973, p. 21 - 24. In Russian. — Abstr. in Referativ. Zhurn. 51. Astron., 3.51.44 (1975).

014.009 Zum Begriffssystem für den Astronomieunterricht.
M. Schukowski.
Astron. in der Schule, 12. Jahrgang, p. 36 - 40 (1975).

014.010 L'insegnamento dell'astronomia nella scuola. Testi della tavola rotonda internazionale.
M. Rigutti, V. Barocas, K. Chilton, V. Deasy, A. Leani, R. A. Naef, O. Oburka, K. Ziolkowski (participants).
Published by Ass. Astrofili Bolognesi, 19 pp. (1975).

014.011 Some questions concerning the teaching of astronomy at pedagogical colleges by correspondence.
O. N. Chajko.

014.012 Problems of the universe in optional courses in 8-class high schools. I. S. Chumak.
XII nauch. konf. Novokuz. gos. ped. in-t. Vyp. (No.) 3. Sekts. fiz.-mat. i obshchetekhn. distsiplin. Tezisy dokl. Novokuznetsk, 1974, p. 54 - 57. In Russian. — Abstr. in Referativ. Zhurn. 51. Astron., 4.51.43 (1975).

014.013 A teachers course in astronomy. P. A. Wayman.
Irish Astron. Journ., Vol. 11, 180 - 186 (1974).

014.014 Results of the olympiad in Kiev and astronomical education. M. L. Divinskij.
Zemlya i Vselennaya, 1975, No. 3, p. 86 - 87. In Russian.

014.015 Teaching of astronomy. D. G. Wenzel.
Bull. Astron. Soc. India, Vol. 1, 40 - 41 (1973).

014.016 Zum zielgerichteten methodischen Einsatz der zum Stoffkomplex HRD vorhandenen Unterrichtsmittel.
H. Risse.
Astron. in der Schule, 12. Jahrgang, p. 56 - 59 (1975).

014.017 Einfache parallaktische Montierung in Holzbauweise. W. Nehls.
Veröff. Wilhelm-Foerster-Sternw., Berlin, No. 37, 30 pp. (1975).

014.018 The universe in the classroom: A laboratory exercise on the inverse square law of light intensity.
H. Kruglak.
Mercury, (Journ. Astron. Soc. Pacific), Vol. 4, No. 3, p. 9 (1975).

014.019 Some remarks on teaching astronomy in Bulgaria.
R. Radkov.
Mat. i fizika (NRB), Vol. 17, No. 6, p. 45 - 50 (1974). In Bulgarian.

014.020 Extramural work in astronomy. B. Iliev.
Mat. i fizika (NRB), Vol. 17, No. 6, p. 51 - 52 (1974). In Bulgarian.

014.021 Astronomy at school to-day and to-morrow.
M. Nikolić.
Vasiona, Vol. 22, 98 - 101 (1974). In Serbo-Croatian.

014.022 Importance of astronomy to general education.
V. Vaný́sek, J. Svatoš.
Říše hvězd, Vol. 56, 1 - 3 (1975). In Czech.

015 Miscellanea

015.001 On the problem of the real existence of objects in space appearing to be envelopes from orbital rings surrounding a star. G. I. Pokrovskij.
Trudy sed'mykh chtenij posvyashch. razrab. nauch. naslediya i razvitiyu idej K. Eh. Tsiolkovskogo, Kaluga, 1972. Sekts. Probl. raket. i kosmich. tekhn. Moskva, 1974, p. 73 - 77. In Russian. – Abstr. in Referativ. Zhurn. 62. Issled. kosmich. prostranstva, 2.62.524 (1975).

015.002 More night-sky paradox.
B. W. Shore, A. D. Allen, E. R. Harrison.
Phys. Today, Vol. 28, No. 2, p. 15, 69, 71 (1975).

015.003 Notes of a biology-watcher. The world's biggest membrane. L. Thomas.
Southern Stars, Vol. 25, 220 - 223 (1974).

015.004 On communications with extra-terrestrial or alien intelligences. R. P. Haviland.
Journ. British Interplanet. Soc., Vol. 28, 161 - 167 (1975).

015.005 The sky of Capella. J. C. Holmes.
Sky Telescope, Vol. 49, 147 - 149 (1975).

015.006 Time markers in interstellar communication.
G. W. Pace, J. C. G. Walker.
Nature, Vol. 254, 400 - 401 (1975). – Letter.

015.007 K. Eh. Tsiolkovskij and "Cosmic philosophy".
A. D. Ursul, Yu. A. Shkolenko.
Trudy vos'mykh chtenij, posvyashch. razrabotke nauch. naslediya i razvitiyu idej K. Eh. Tsiolkovskogo, Kaluga, 1973, Sekts. "Issled. nauch. tvorchestva K. Eh. Tsiolkovskogo". Moskva, 1974, p. 3 - 15. In Russian. – Abstr. in Referativ. Zhurn. 51. Astron., 3.51.1 (1975).

015.008 Did life come from outer space? A. I. Oparin.
Spaceflight, Vol. 17, 214 (1975).

015.009 Interstellar messengers. R. Bracewell.
Mercury, (Journ. Astron. Soc. Pacific), Vol. 4, No. 2, p. 4 - 11 (1975).

015.010 Some of K. Eh. Tsiolkovskij's ideas about other civilizations in cosmic space. I. A. Kol'chenko.
Trudy vos'mykh chtenij, posvyashch. razrabotke nauch. naslediya i razvitiyu idej K. Eh. Tsiolkovskogo, Kaluga, 1973, Sekts. "Issled. nauch. tvorchestva K. Eh. Tsiolkovskogo". Moskva, 1974, p. 55 - 62. In Russian. – Abstr. in Referativ. Zhurn. 51. Astron., 3.51.2 (1975).

015.011 On the scientific content of K. Eh. Tsiolkovskij's work "Dreams about the earth and sky and effects of the universal gravitation". R. S. Shikhranov.
Trudy vos'mykh chtenij, posvyashch. razrab. nauch. naslediya i razvitiyu idej K. Eh. Tsiolkovskogo, Kaluga, 1973. Sekts. "Issled. nauch. tvorchestva K. Eh. Tsiolkovskogo", Moskva, 1974, p. 83 - 98. In Russian. – Abstr. in Referativ. Zhurn. 62. Issled. kosmich. prostranstva, 3.62.8 (1975).

015.012 Advances in the geochemistry of amino acids.
K. A. Kvenvolden.
Annual Rev. Earth Planet. Sci., Vol. 3, (see 003.011), 183 - 212 (1975). – Parts of this paper are concerned with amino acids in meteorites and lunar samples.

015.013 Come avvicinarsi all'astronomia. Testi della tavola rotonda.
W. Magri, P. Andrenelli, L. Baldinelli, A. Betti, A. Leani (participants).
Published by Ass. Astrofili Bolognesi, 12 pp. (1975).

015.014 Le souvenir du calendrier Julien dans les traditions populaires. F. Biraud.
L'Astronomie, 89e année, p. 242 - 244 (1975).

015.015 Pravila za pisanje članaka prema uputstvu Medunarodne Astronomske Unije (Rules for correct spelling according to the instructions of the International Astronomical Union).
Vasiona, Vol. 23, 29 - 31 (1975).

015.016 Einige biologische Aspekte außerirdischen Lebens.
F. Jungnickel.
Sterne, Vol. 51, 91 - 101 (1975).

015.017 An explanation for the absence of extraterrestrials on earth. M. H. Hart.
Quarterly Journ. Roy. Astron. Soc., Vol. 16, 128 - 135 (1975).
No intelligent beings from outer space are now present on earth. It is suggested that this fact can best be explained by the hypothesis that there are no other advanced civilizations in our Galaxy. Reasons are given for rejecting all alternative explanations of the absence of extraterrestrials from earth.

015.018 Cosmic perspective: in search of the galactic library.
E. Duckworth.
Mercury, (Journ. Astron. Soc. Pacific), Vol. 4, No. 3, p. 10 - 13 (1975).

015.019 Proximity of galactic civilizations. B. M. Oliver.
Icarus, Vol. 25, 360 - 367 (1975).
Approximate expressions are derived for the number of civilizations within a few tens of light years of each other since intelligent life first evolved in the Galaxy. The number is proportional to the square of the usual selectivity factors and to the first power of the longevity. Arguments are presented for expecting intelligent life in certain multiple star systems, and the number of coexistent civilizations in such systems is estimated.

015.020 Interstellar archaeology and the prevalence of intelligence. J. Freeman, M. Lampton.
Icarus, Vol. 25, 368 - 369 (1975).
Simple calculations indicate that the spatial density in the Galaxy of extinct civilizations, and of planets inhabited by intelligent creatures who do not have technical civilizations, may be quite large.

015.021 Universal music? S. von Hoerner.
Psychology of Music, Vol. 2, No. 2, p. 18 - 28 (1974).

015.022 Zur Verwendung von Schriftsprachensendungen bei CETI (Communication with extra-terrestrial intelligence). J. Lehmann.
Sterne, Vol. 51, 102 - 109 (1975).

CETI signal to Messier 13.
Spaceflight, Vol. 17, 56, 76 (1975).

Planetary systems and extraterrestrial life.
See Abstr. 117.039.

Applied Mathematics, Physics

021 Mathematics, Computing

021.001 **A note on an attempt at more efficient Poisson series evaluation.**
P. J. Shelus, W. H. Jefferys III.
Celestial Mechanics, Vol. 11, 75 - 78 (1975).
Eliminating many of the trigonometric function calls by a suitable series transformation has resulted in a substantial reduction of on-line computing time while very long Poisson series are being evaluated. A further reduction has been realized by applying a short SNOBOL processor to the FORTRAN coding of the transformed series which eliminates many of the multiplication operations during the course of series evaluation.

021.002 **Numerical solutions in remote sensing.**
J. Y. Wang, R. Goulard.
Applied Optics, Vol. 14, 862 - 871 (1975).

021.003 **Das Wesen der Ausgleichung nach der Methode der kleinsten Quadrate, wenn Ausgangsgrößen vorhanden sind.** W. K. Hristov.
Österreich. Zeitschr. Vermessungswesen, 62. Jahrgang, p. 97 - 110 (1974).

021.004 **On the application of more accurate numerical integration methods to celestial mechanics.**
M. S. Yarov-Yarovoj.
Trudy Gos. Astron. Inst. Shternberga, Vol. 45, 179 - 200 (1974). In Russian.
The Runge-Kutta method including the first four derivatives is recommended.

021.005 **The compression of data resulting from photon counters.** J. M. Beckers.
Space Sci. Instrum., Vol. 1, 153 - 159 (1975).
A method is presented to compress in an efficient way the counts which result from random event detectors such as photon counters.

021.006 **Concerning a criterion for the validity of the first order smoothing approximation.** I. Lerche.
Journ. Math. Phys., *New York*, Vol. 15, 1967 - 1971 (1974).
The author uses the dynamo equations in simple form to illustrate criteria for establishing the validity of the first order smoothing approximation. A general principle is then established. Variational methods are employed throughout.

021.007 **Fortran automatic typesetting system.**
P. M. Janiczek, G. H. Kaplan, with a preface by R. L. Duncombe.
United States Naval Obs., *Washington,* Circ. No. 149, 71 pp (1974).
(I) General discussion; (II) Subroutine descriptions; (III) Appendices: (A) Examples, (B) FATS use from other programming languages, (C) Program debugging, (D) Grid diagrams, (E) FATS control section and entry names, (F) FATS subroutine summary.

Methods of investigation of astronomical optics. II. Hartmann method. See Abstr. 031.016.

Methods for search of variable star periods. See Abstr. 031.263.

Stray light in solar observations. A method of numerical integration. See Abstr. 031.268.

The theoretical instrumental profile of the combination telescope-pinhole photometer and its effect upon observations of umbra intensities. See Abstr. 034.075.

One of the solutions for the Laplace equation and its physical interpretation. See Abstr. 042.015.

Das Dreikörperproblem am Computer-Bildschirm. See Abstr. 042.076.

Literal algebra for satellite dynamics. See Abstr. 052.030.

Satellite perturbations with automated Poisson series manipulations. See Abstr. 052.038.

General relativity and satellite orbits. See Abstr. 052.056.

On a problem of uniqueness regarding H-function calculations. See Abstr. 063.016.

A computer program for the analysis of the 1972 chromospheric spectrum of 31 Cygni. See Abstr. 121.018.

022 Physical Papers Related to Astronomy and Astrophysics

022.001 The motion of charged particles in a strong electro-magnetic field and curvature radiation.
Yu. V. Chugunov, V. Ja. Eidman (*V. Ya. Ehjdman*),
E. V. Suvorov.
Astrophys. Space Sci., Vol. 32, L7 - L10 (1975).
The features of the relativistic charge particle motion and emission due to the radiative damping in the strong electromagnetic fields are investigated. It is shown that the radiative force responsible for curvature radiation is associated with the particle drift in an inhomogeneous magnetic field. The adiabatic trajectory is obtained for the relativistic particle, moving in a strong static electromagnetic field, particle energy being determined by the balance of the work of the electric field and the energy losses through curvature radiation.

022.002 The free–free transitions of the negative chlorine ion. T. L. John, D. J. Morgan.
Monthly Notices Roy. Astron. Soc., Vol. 170, 1 - 4 (1975).
The free–free continuous absorption coefficient of Cl^- is calculated by multichannel techniques using close coupling continuum wave functions determined neglecting exchange.

022.003 The free–free transitions of atomic and molecular negative ions in the infrared. T. L. John.
Monthly Notices Roy. Astron. Soc., Vol. 170, 5 - 6 (1975).
The free–free absorption coefficient of the negative ions He^-, Ne^-, Ar^-, Kr^-, Xe^-, Li^-, Na^-, Cs^-, Hg^-, N^-, O^-, H_2^-, N_2^-, O_2^-, CO^-, CO_2^- and H_2O^- is given in the temperature range $50-25000$ K. Calculations are based on theoretical and experimental momentum loss cross-section data.

022.004 Classification of spectra of K, Ca, Sc, Ti, V and Cr and extrapolation to Fe solar flare lines.
B. C. Fawcett, R. W. Hayes.
Monthly Notices Roy. Astron. Soc., Vol. 170, 185 - 197 (1975).
Potassium, calcium, titanium, vanadium and chromium emission lines are classified in the same isoelectronic sequences as Fe XIX to Fe XXIII. Most of the lines identified belong to the $2p^n-2p^{n-1}3d$ or $2s^2 2p^k-2s 2p^{k+1}$ transition arrays. The data presented permit the calculation of the wavelengths of Fe XIX to Fe XXIII emission lines which are emitted from solar flares.

022.005 Newtonian relative mechanics. H.-J. Treder.
Astron. Nachr., Vol. 296, 1 - 7 (1975). In German.
The author proves that a closed system of N Newtonian particles with the same atomic masses m (according to the hypothesis of Hertz) can be described by a Huygensian relative dynamics in the rest system of the center of masses. In this reference system the Hamiltonian, the Lagrangian, the virial, the angular momentum etc. of the N particle system are depending on the difference quantities, only.

022.006 Absolute integrated intensity and individual line parameters for the $6.2\,\mu$ band of NO_2.
A. Goldman, F. S. Bonomo, W. J. Williams, D. G. Murcray, D. E. Snider.
Journ. Quant. Spectrosc. Radiat. Transfer, Vol. 15, 107 - 112 (1975).
The absolute integrated intensity of the $6.2\,\mu$ band of NO_2 at 40°C was determined from quantitative spectra at ~ 10 cm^{-1} resolution by the spectral band model technique. A value of 1430 ± 300 cm^{-2} atm^{-1} was obtained. Individual line parameters, positions, intensities and ground state energies were derived, and line-by-line calculations were compared with the band model results and with the quantitative spectra ob-tained at ~ 0.5 cm^{-1} resolution.

022.007 Measurement of the HeI 4471 Å profile at an electron density of 10^{15} cm^{-3}.
A. J. Barnard, D. C. Stevenson.
Journ. Quant. Spectrosc. Radiat. Transfer, Vol. 15, 123 - 125 (1975).
The profile of the HeI 4471 Å line and its forbidden component was measured using a pulsed arc plasma (electron density 10^{15} cm^{-3}, temperature 1.5 eV) as the light source. The profile is in good agreement with recent calculations in which the ion motion has been taken into account.

022.008 Measurements of the electronic transition moments of C_2-band systems.
D. M. Cooper, R. W. Nicholls.
Journ. Quant. Spectrosc. Radiat. Transfer, Vol. 15, 139 - 150 (1975).

022.009 Of fundamental electrodynamics and astrophysics.
J. C. Byrne, R. R. Burman.
Nature, Vol. 253, 27 (1975). – Letter.

022.010 The determination of the electron to proton inertial mass ratio via molecular transitions.
R. I. Thompson.
Astrophys. Letters, Vol. 16, 3 - 4 (1975).
It is demonstrated that the wavelengths of molecular transitions are sensitive to the ratio of electron to proton inertial mass. Observation of molecular transitions can therefore provide checks on the invariance of this ratio in distant objects. If confirmed, the recent observation of H_2 absorption lines in QSO spectra would allow a determination of m_e/m_p for these objects.

022.011 Comment on "A note on ionization equilibrium".
C. A. Rouse.
Astrophys. Journ., Vol. 195, 585 (1975).
The recent derivation by Lafferty of an expression for the time rate of change of the average charge of multiply charged ions is considered. After simplifying a more general expression derived here, the condition for the validity of Lafferty's equation is demonstrated. It is shown that, contrary to his conclusion, detailed balance is still required to obtain the Elwert relation. Concerning 11.022.001 [Astrophys. Journ., Vol. 187, 209 - 210 (1974)].

022.012 The blackbody as a standard light source.
H. Tüg.
Astron. Astrophys., Vol. 37, 249 - 255 (1974). In German.
A new blackbody light source radiating at the melting point of platinum is described for measuring the absolute spectral energy distribution of stars in the wavelength range $3000-10000$ Å.

022.013 Absolute transition probabilities for some lines of neutral calcium. G. Smith, J. A. O'Neill.
Astron. Astrophys., Vol. 38, 1 - 4 (1975).
Accurate new measurements of transition probabilities are presented for two line multiplets of the neutral calcium spectrum which are often used in solar abundance analyses. A critical comparison is made with the results of other workers and the relationship between the new absolute values and those used in recent solar abundance analyses is discussed.

022.014 On the broadening of iron lines by neutral hydro-gen. M. G. Edmunds.

Astron. Astrophys., Vol. 38, 137 - 139, with a correction, Vol. 41, 119 (1975).

A simple formula for the effective van der Waals broadening of iron spectral lines by neutral hydrogen is deduced from published line broadening calculations, and justified by consideration of solar iron lines.

022.015 On OH formation in collision between H and O⁻.
B. Huron, F. Tran Minh.
Astron. Astrophys., Vol. 38, 165 - 167 (1975). In French.

Results show that the reaction $O^- + H \rightarrow OH + e^-$ must be considered again for the OH formation in interstellar matter.

022.016 The photoionization and dissociative photoionization of H_2, HD, and D_2.
A. L. Ford, K. K. Docken, A. Dalgarno.
Astrophys. Journ., Vol. 195, 819 - 824 (1975).

The cross sections for photoionization of H_2 to the bound and continuum vibrational levels of the $1s\sigma_g$ state of H_2^+ have been calculated in the range 804–425 Å ionizing radiation. The dipole matrix elements have been computed as a function of ejected electron energy and internuclear distance using a simple Coulomb approximation to the final-state scattering wave function. The relative population of the v' levels of H_2^+ at 584 Å, the ratio of production of H^+ to H_2^+ as a function of photon energy, and the total photoionization cross section are in good agreement with experimental measurements.

022.017 Single-domain grain size limits for metallic iron.
R. F. Butler, S. K. Banerjee.
Journ. Geophys. Res., Vol. 80, 252 - 259 (1975).

Theoretical examination of possible nonuniform spin configurations in metallic iron indicates that circular spin (CS) is the lowest-energy nonuniform arrangement. The upper grain size limit (d_0) to single-domain (SD) behavior is thus defined by the SD to CS transition. The size and shape criteria for the stable SD behavior of metallic iron help to explain (1) the low SD content of lunar samples, (2) the widespread occurrence of superparamagnetic behavior and viscous magnetization in lunar soils and low metamorphic grade breccias, (3) the changes in the magnetic properties of breccias during annealing and (4) the increased SD content of shocked breccias. The narrow grain size limits for SD behavior also suggest that magnetostatic interaction between metal grains in the solar nebula is not a viable mechanism for iron-silicate fractionation.

022.018 Of atoms, mountains, and stars: a study in qualitative physics. V. F. Weisskopf.
Science, Vol. 187, 605 - 612 (1975).

022.019 The broadening of calcium II H and K lines by helium. G. L. Hammond.
Astrophys. Journ., Vol. 196, 291 - 305 (1975).

Laboratory measurements of widths and shifts of the Ca II H and K lines have been made under conditions simulating those in the atmospheres of metallic-line white dwarfs, namely, densities of 10^{21} helium atoms cm^{-3} and temperatures of 4500° to 6000°K. There are significant differences in the shifts and damping constants of the two fine-structure components. The impact theory is evidently applicable to the Ca II/He system under these conditions, and the observed damping constants are fortuitously close to the predictions of conventional Lindholm-Foley/van der Waals theory. The widths and shifts have also been analyzed in terms of Lennard-Jones potential theory.

022.020 Radiative lifetimes for the $A^1\Pi$ and $B^1\Delta$ states of the CH^+ molecule with application to the CH^+ abundance in Zeta Ophiuchi. N. H. Brooks, W. H. Smith.

Astrophys. Journ., Vol. 196, 307 - 310 (1975).

Using an electron-beam phase-shift apparatus, the authors have measured radiative lifetimes of 270 ± 75 ns and 290 ± 75 ns, respectively, for the $v' = 1$ and 2 levels of the $A^1\Pi$ state of the CH^+ molecule, and lifetimes of 180 ± 50 ns and 165 ± 50 ns for the $v' = 0$ and 1 levels, respectively, of the $B^1\Delta$ state of CH^+.

022.021 Collision strengths for [N II], [O III], [Ne II] and [Ne III]. M. J. Seaton.
Monthly Notices Roy. Astron. Soc., Vol. 170, 475 - 486 (1975).

New calculations give collision strengths for [N II], [O III], [Ne II] and [Ne III] correct to within 10 per cent. The results are tabulated in a form convenient for the interpretation of observations.

022.022 The excitation of several iron and calcium lines in the visible spectrum of the solar corona.
H. E. Mason.
Monthly Notices Roy. Astron. Soc., Vol. 170, 651 - 689 (1975).

New atomic data have been obtained for the coronal ions Fe X, Fe XI, Fe XIV, Ca XII, Ca XIII and Ca XV, using a computer package developed at University College, London. Energy levels and radiative transition probabilities have been computed allowing for configuration interaction and relativistic effects. The electron scattering problem has been solved using the 'distorted wave' approach. The atomic data, level populations and emission rates for the ions studied, are tabulated for various physical conditions appropriate to the solar corona.

022.023 Line intensities of CO_2 in the 2.7 micron region.
H. D. Downing, L. R. Brown, R. H. Hunt.
Journ. Quant. Spectrosc. Radiat. Transfer, Vol. 15, 205 - 210 (1975).

022.024 Wavelengths and transition probabilities for the isoelectronic sequence of oxygen.
U. I. Safronova.
Journ. Quant. Spectrosc. Radiat. Transfer, Vol. 15, 223 - 229 (1975). In Russian.

Perturbation theory was used for calculations of transition probabilities for the transitions $1s^2 2s^2 2p^4 - 1s^2 2s 2p^5 - 1s^2 2p^6$ of the isoelectronic sequence of oxygen. The results may be useful in interpreting soft X-ray line radiation from these transitions, as observed in the solar corona and in high-temperature laboratory-generated plasmas.

022.025 Collision-broadened linewidths of tetrahedral molecules—II. Computations for CH_4 lines broadened by N_2, O_2, He, Ne and Ar.
G. D. T. Tejwani, P. Varanasi, K. Fox.
Journ. Quant. Spectrosc. Radiat. Transfer, Vol. 15, 243 - 254 (1975).

022.026 Intensity and half-width measurements in the 1.525 μm band of acetylene.
P. Varanasi, B. R. P. Bangaru.
Journ. Quant. Spectrosc. Radiat. Transfer, Vol. 15, 267 - 273 (1975).

022.027 Observation of hydrogen and helium satellites in a turbulent plasma.
C. C. Gallagher, M. A. Levine.
Journ. Quant. Spectrosc. Radiat. Transfer, Vol. 15, 275 - 279 (1975).

022.028 Term analysis of Fe VI. J. O. Ekberg.
Physica Scripta, Vol. 11, 23 - 30 (1975).

The spectrum of Fe VI has been observed. From the complete set of $3d^3$ levels all forbidden lines of Fe VI from 1387 to 10000 Å have been predicted. All but two lines in the spectrum of the star RR Telescopii previously identified as forbidden Fe VI lines have been confirmed and 6 of the otherwise unexplained lines have been identified as [Fe VI].

022.029 Transitions $2s^2 2p^k - 2s 2p^{k+1}$ of the N I and C I iso-electronic sequences.
U. Feldman, G. A. Doschek, R. D. Cowan, L. Cohen.
Astrophys. Journ., Vol. 196, 613 - 616 (1975).

Transitions of the type $2s^2 2p^k - 2s 2p^{k+1}$ have been identified for the elements from titanium through iron for ions of the nitrogen isoelectronic sequence and for the elements titanium through nickel for ions of the carbon isoelectronic sequence. Wavelengths, intensity estimates and energies are given. The lines were identified from EUV spectra obtained from laser-produced plasmas. The energy differences of levels of the ground configuration for the C I isoelectronic sequence are compared with extrapolations based on semiempirical equations derived by Edlén. Wavelengths of forbidden Fe XXI lines that can be seen in low-density plasmas such as solar flare plasmas are predicted. The laser plasma spectrum is briefly compared with available solar flare spectra.

022.030 Valence excited states of CH. III. Radiative lifetimes.
J. Hinze, G. C. Lie, B. Liu.
Astrophys. Journ., Vol. 196, 621 - 631 (1975).

Theoretical calculations on transition probabilities between low-lying states of CH, using accurate ab initio electronic wave functions, have yielded radiative lifetimes for the $v = 0$ vibrational levels of the $A^2\Delta$, $B^2\Sigma^-$, and $C^2\Sigma^+$ states of 388, 260, and 89 ns, respectively. These results are expected to be accurate within ±15 percent. Based on these theoretical results, an effort has been made to assess the validity of available experimental data.

022.031 Nonthermal rotational distribution of CO($A^1\Pi$) fragments produced by dissociative excitation of CO_2 by electron impact. M. J. Mumma, E. J. Stone, E. C. Zipf.
Journ. Geophys. Res., Vol. 80, 161 - 167 (1975).

Bands of the CO($A^1\Pi - X^1\Sigma^+$) fourth-positive group (4PG) were excited by electron impact dissociative excitation of CO_2, and their rotational profiles were measured. The results are discussed in terms of their applicability to the Mariner observations of the CO 4PG in the dayglow of Mars.

022.032 Reduced absorption of the nonthermal CO($A^1\Pi$-$X^1\Sigma^+$) fourth-positive group by thermal CO and implications for the Mars upper atmosphere. M. J. Mumma, H. D. Morgan, J. E. Mentall.
Journ. Geophys. Res., Vol. 80, 168 - 172 (1975).

The CO($A^1\Pi$) was produced with a nonthermal rotational distribution by impact of 30-eV electrons on CO_2. The rotationally hot fourth-positive bands, CO($A^1\Pi$-$X^1\Sigma^+$), were spectrally dispersed, and the reduced absorption of the (v',0) bands by thermal CO (300°K) was measured (v'= 0,1). The nonthermal bands are absorbed less strongly by CO (300°K) than are thermal bands. The fractional transmission of the hot bands was measured over a wide range of CO absorber pressures (0 – 1000 millitorrs). The results lead to more accurate values for the slant column density and local number density of CO in the Mars upper atmosphere. The new values are 5.7×10^{15} cm^{-2} and 1.5×10^8 cm^{-3} at 150 km and 2.5×10^{15} cm^{-2} and 6.8×10^7 cm^{-3} at 170 km, respectively. The thermal and nonthermal sources of CO($A^1\Pi$) are shown to be of approximately equal strength at 170 km.

022.033 Statistical-band-model analysis and integrated in-

tensity for the 21.8 μm bands of HNO_3 vapor.
A. Goldman, F. S. Bonomo, W. J. Williams, D. G. Murcray.
Journ. Optical Soc. America, Vol. 65, 10 - 12 (1975).

022.034 Spectrum of Al VII in the vuv (extreme vacuum ultraviolet). F. P. J. Valero.
Journ. Optical Soc. America, Vol. 65, 197 - 198 (1975).

022.035 Observations of the fundamental rotation-vibration band of CN. R. R. Treffers.
Astrophys. Journ., Vol. 196, 883 - 886 (1975).

Three lines of the fundamental rotation-vibration band of CN have been observed in the laboratory using a King furnace. The transition moment squared has been found to be $R_e^2 = 7.5 \pm 3.1 \times 10^{-4}$ atomic units. In addition, the transition moment of the 0–0 band of the red system of CN has been measured to be $R_e^2 = 0.48 \pm 0.28$ atomic units ($f_{00} = 3.3 \pm 1.9 \times 10^{-3}$).

022.036 Microwave absorption spectrum of the CO^+ ion.
T. A. Dixon, R. C. Woods.
Phys. Rev. Letters, Vol. 34, 61 - 63 (1975).

The $K = 0 \to 1$ pure rotational spectrum of CO^+ ($X^2\Sigma$, $v = 0$) has been detected in a direct absorption experiment, in which microwave radiation is passed through a $3\frac{1}{2}$-m-long dc glow discharge in carbon monoxide or a helium-carbon-monoxide mixture.

022.037 Experimental oscillator strengths for some Ni I and Ni II multiplets in the spectral range 2250 Å –
2550 Å. A. Goly, J. Moity, S. Weniger.
Astron. Astrophys., Vol. 38, 259 - 262 (1975). In French.

Relative oscillator strengths of eleven Ni I lines and thirty-eight Ni II lines in the wavelength range from 2250 to 2550 Å have been measured.

022.038 Precision measurement of relative oscillator strengths–II. Fe I transitions from levels $a^5 D_4$ (0.00 eV) and $a^5 D_3$ (0.05 eV).
D. E. Blackwell, P. A. Ibbetson, A. D. Petford.
Monthly Notices Roy. Astron. Soc., Vol. 171, 195 - 208 (1975).

Relative oscillator strengths have been measured for 16 low excitation potential lines of Fe I, and a photometric scale covering nearly six decades set up. The accumulated probable error over these six decades is about 5 per cent. A critical comparison is made with the results of other methods of measurement and with the results of calculation.

022.039 The continuous absorption coefficient of the negative oxygen ion in the infrared.
T. L. John, R. J. Williams.
Monthly Notices Roy. Astron. Soc., Vol. 171, 7P - 9P (1975).

The free-free continuous absorption coefficient of O^- is determined with accurate configuration interaction wave functions. The authors' results show that O^- is probably below the threshold of detectability in infrared stellar spectra.

022.040 Intensity measurements in the 4.5μ fundamental of $CH_3 D$ at low temperatures.
S. Sarangi, P. Varanasi.
Journ. Quant. Spectrosc. Radiat. Transfer, Vol. 15, 291 - 294 (1975).

The integrated intensities of the J-multiplets $R(4)$ through $P(12)$, including the Q-branch, of the 4.5μ fundamental of $CH_3 D$ have been measured at 100°K, 150°K, 200°K, and 298°K. Comparison of the measured line strengths with values calculated using symmetric-top formulae suggests strong intensity anomalies.

022.041 Mean lifetimes of excited levels of Ar(II)–II.

Theoretical considerations. F. Bely-Dubau, C. Camhy-Val, A. M. Dumont.
Journ. Quant. Spectrosc. Radiat. Transfer, Vol. 15, 375 - 377 (1975).

022.042 Stark broadening tables for He I λ 4922Å.
A. J. Barnard, J. Cooper, E. W. Smith.
Journ. Quant. Spectrosc. Radiat. Transfer, Vol. 15, 429 - 437 (1975).

The Stark broadening of He I λ 4922Å and its forbidden components by both ions and electrons is calculated using a theory that includes ion dynamic effects. Tables are presented for temperatures from 5000°K to 40,000°K covering the density range 10^{13} cm^{-3} to 10^{16} cm^{-3} for both helium and hydrogen ionic perturbers.

022.043 A critical survey of atomic transition probabilities for Cu I. A. Bielski.
Journ. Quant. Spectrosc. Radiat. Transfer, Vol. 15, 463 - 472 (1975).

A critical survey of experimentally-determined atomic transition probabilities for 102 spectral lines of CuI has been carried out. Many of the values have been brought up to date and renormalized on the basis of recent measurements.

022.044 Mean lifetimes of excited levels of Ar II. I. Time correlation measurements. C. Camhy-Val, A. M. Dumont, M. Dreux, L. Perret, C. Vanderriest.
Journ. Quant. Spectrosc. Radiat. Transfer, Vol. 15, 527 - 530 (1975).

022.045 The microwave frequencies, line parameters, and spectral constants for $^{14}NH_3$.
R. L. Poynter, R. K. Kakar.
Astrophys. Journ., Suppl. Ser., No. 277, Vol. 29, 87 - 96 (1975).

By combining the authors' precision measurements of 119 microwave inversion lines for $^{14}NH_3$ with the reported millimeter and far-infrared spectral measurements, they have obtained improved values for the rotational and centrifugal distortion constants in the ground vibrational state. Fits of the microwave inversion lines were compared for 10, 15, and 21-term polynomial and exponential models. Computed values are also given for the line widths, absorption coefficients, and energy levels.

022.046 Thorium comparison spectrum.
J. B. Breckinridge, A. K. Pierce, C. P. Stoll.
Astrophys. Journ., Suppl. Ser., No. 278, Vol. 29, 97 - 112 (1975).

Wavelengths for approximately 1850 thorium emission lines between 2720 and 8805 Å are given. The accuracy of these measures appears to be better than ±0.001 Å. These measures, made from photographic plates taken with the 13.7-m focal length grating spectrograph at the McMath Solar Telescope at Kitt Peak, are part of work done for the solar wavelength atlas program (Pierce and Breckinridge, Kitt Peak Contributions 559, 1973). The authors suggest they be used for general stellar and solar comparison wavelengths.

022.047 Frequency dependence of collisional depolarization of resonance radiation.
F. Schuller, W. Behmenburg.
Zeitschr. Naturforschung, Vol. 30a, 442 - 444 (1975).

The authors have studied collisional depolarization of resonance radiation with special attention given to the line wings. It is shown that in the one particle quasistatic limit the fluorescence radiation is completely depolarized, independently of frequency and perturber density. By comparison with the integrated polarization degree it is concluded, that the polarization state of the radiation must exhibit a frequency dependence.

022.048 Resonance broadening and oscillator strength of the mercury absorption line 1850 Å.
D. Gebhard, W. Behmenburg.
Zeitschr. Naturforschung, Vol. 30a, 445 - 450 (1975).

A new method for the determination of f-values of resonance transitions from wing-measurements of selfbroadened resonance absorption lines is described. The method is applied to the mercury resonance transition $6^1 S_0 \rightarrow 6^1 P_1$, λ 1850 Å. The resulting f-value of 1.08±0.05 agrees well with those obtained from other methods.

022.049 Intensity measurements on seven band systems of C_2 between 2000 and 12,000 Å.
D. M. Cooper, R. W. Nicholls.
Journ. Roy. Astron. Soc. Canada, Vol. 69, 39 (1975). − Abstr. Canadian Astron. Soc.

022.050 On free-free absorption by Cl$^-$. M. S. Vardya.
Observatory, Vol. 95, 50 - 51 (1975). − Letter.

022.051 Experimentally determined absolute oscillator strengths of Ti I, Ti II, and Ti III.
J. R. Roberts, P. A. Voigt, A. Czernichowski.
Astrophys. Journ., Vol. 197, 791 - 798 (1975).

Absolute oscillator strengths of Ti I, Ti II, and Ti III transitions in the wavelength region from 2440 Å to 3500 Å have been experimentally determined using a wall-stabilized arc.

022.052 Oscillator strengths and excited-state lifetimes in the sodium sequence.
C. Laughlin, M. N. Lewis, Z. J. Horak.
Astrophys. Journ., Vol. 197, 799 - 804 (1975).

A first-order Z-expansion method is used to calculate oscillator strengths and radiative lifetimes in the sodium sequence. The computed $3s-3p$ and $3p-3d$ oscillator strengths for the more highly ionized members of the sequence are generally in satisfactory agreement with the experimental values. Lifetime results for $4p$ and $4d$ states of sodium-like ions are presented graphically in a manner which demonstrates their regularity as a function of Z^{-1} and allows interpolation to be carried out.

022.053 Systematic trends of Hartree-Fock oscillator strengths along the sodium isoelectronic sequence.
E. Biemont.
Journ. Quant. Spectrosc. Radiat. Transfer, Vol. 15, 531 - 542 (1975).

022.054 Production de béryllium-7 dans le fer et le silicium par des protons de 0,6 et 24 GeV.
S. Regnier, P. Paillard, G. Simonoff.
Comptes Rendus Acad. Sci. Paris, Sér. B, Vol. 280, 513 - 514 (1975).

022.055 High resolution measurements of atomic and molecular lifetimes using the high frequency deflection technique. P. Erman.
Phys. Scripta, Vol. 11, 65 - 78 (1975).

022.056 Extended identifications of odd energy levels of Si I: Lu−Fano graphical analysis.
C. M. Brown, S. G. Tilford, M. L. Ginter.
Journ. Optical Soc. America, Vol. 65, 385 - 388 (1975).

The Lu−Fano graphical technique has been applied to the analysis of the strongly interacting $J = 2$ and $J = 3$ Rydberg levels of Si I which arise from the $3pnd$ configurations. Approximately 20 new $J = 3$ and 40 new $J = 2$ energy levels are reported. A periodic intensity variation in the $^1D_2-3pnd$

J = 3 transitions is discussed. Twenty-three new $3pnd\ ^1P^{\circ}{}_1$ levels are reported.

022.057 Transition probabilities in the spectra of Ne I.
R. A. Lilly.
Journ. Optical Soc. America, Vol. 65, 389 - 391 (1975).

022.058 Transitions $2s^2\ 2p-2s\ 2p^2$ in the B I isoelectronic sequence.
G. A. Doschek, U. Feldman, L. Cohen.
Journ. Optical Soc. America, Vol. 65, 463 - 464 (1975).

022.059 Extension of the analysis of triply ionized aluminum (Al IV). M.-C. Artru, V. Kaufman.
Journ. Optical Soc. America, Vol. 65, 594 - 599 (1975).

A total of 225 new lines of Al IV have been observed in the wavelength range 400–4700 Å, leading to the determination of all of the levels of the $2p^5 4p$, $4d$, $4f$, $5s$, $5f$, and $5g$ configurations. Results of parametric calculations for these configurations are presented. An ionization energy of $967\,804 \pm 15\ \mathrm{cm}^{-1}$ has been derived.

022.060 Mean life of the $5p\ ^2P_{3/2}$ resonance level in Ag I.
J. Z. Klose.
Astrophys. Journ., Vol. 198, 229 - 233 (1975).

022.061 Dielectronic recombination of Ne IX-, F VIII- and O VII-ions. A. Pospieszczyk.
Astron. Astrophys., Vol. 39, 357 - 370 (1975).

022.062 Lifetime measurements of some levels belonging to the $3d^8\ 4s(a^2F)\ 4p$ configuration of Ni I.
J. Heldt, H. Figger, K. Siomos, H. Walther.
Astron. Astrophys., Vol. 39, 371 - 375 (1975).

022.063 Excitation of forbidden Fe VI lines.
R. H. Garstang, W. D. Robb.
Bull. American Astron. Soc., Vol. 7, 268 (1975). – Abstr. AAS.

022.064 Ground state centrifugal distortion constants of vinyl isocyanide, CH_2-CH-NC, from the microwave and millimeter wave rotational spectra.
K. Yamada, M. Winnewisser.
Zeitschr. Naturforschung, Vol. 30a, 672 - 689 (1975).

022.065 Light as a fundamental particle. S. Weinberg.
Phys. Today, Vol. 28, No. 6, p. 32 - 37 (1975).
Prepared from an invited paper at the Washington meeting of the Optical Society of America, Spring 1974.

022.066 Le cinquantième anniversaire de la mécanique ondulatoire. J.-C. Pecker.
L'Astronomie, 89e année, p. 227 - 230 (1975).

022.067 La mécanique ondulatoire et son rôle dans l'évolution des idées. B. d'Espagnat.
L'Astronomie, 89e année, p. 231 - 241 (1975).

022.068 Lasers as local oscillators. E. Wiesendanger.
Space Sci. Rev., Vol. 17, 721 - 730 (1975). – Presented at the workshop on coherent detection in astronomy, held at Rhenen, The Netherlands, 25 and 26 April 1974 – see 012.009.

022.069 Tunable lasers. H. G. Häfele.
Space Sci. Rev., Vol. 17, 731 - 736 (1975). – Presented at the workshop on coherent detection in astronomy, held at Rhenen, The Netherlands, 25 and 26 April 1974 – see 012.009.

022.070 Détermination expérimentale de l'élargissement et du déplacement Stark de raies d'atomes de silicium ionisés. A. Lesage, M. Miller.
Comptes Rendus Acad. Sci. Paris, Sér. B, Vol. 280, 645 - 647 (1975).

Les profils des raies correspondant aux transitions $(^2D-{}^2P^0)$; $(^2D-{}^2F^0)$; $(^2P^0-{}^2D)$; $(^2P^0-{}^2S)$ et $(^2S-{}^2P^0)$ de l'atome de silicium ionisé ont été mesurés dans l'onde de choc réfléchie d'un tube à choc classique. La densité électronique du plasma a été déterminée à partir de la raie Hβ observée dans le plasma. Les résultats des mesures sont comparés avec ceux d'autres expérimentateurs et avec les prévisions des théoriciens.

022.071 Nonempirical calculations on the electronic spectrum of the molecular ion C_2^-. J. Barsuhn.
Journ. Phys. B, Atomic Molecular Phys., Vol. 7, 155 - 162 = Max-Planck-Inst. Radioastron., Bonn, Sonderdruck, Ser. A, No. 7 (1974).

022.072 Determination of the structure of monothioformic acid by microwave spectroscopy.
W. H. Hocking, G. Winnewisser.
Journ. Chem. Soc., *London,* Chem. Commun., 1975, p. 63 - 64 = Max-Planck-Inst. Radioastron., Bonn, Sonderdruck, Ser. A, No. 51 (1975).

022.073 Ultraviolet absorption lines arising on metastable states. P. D. Noerdlinger, S. E. Dynan.
Astrophys. Journ., Suppl. Ser., No. 283, Vol. 29, 185 - 191 (1975).

The authors present a catalog and finding list of stronger ultraviolet (900–4000 Å) absorption lines arising on relatively low-lying metastable states of astronomically common ions. Applications to QSO absorption regions and extended stellar atmospheres are suggested.

022.074 Kako nastaje radiozračenje (Generation of radio radiation). S. M. Dimitrijević.
Vasiona, Vol. 23, 4 - 9 (1975).

022.075 On the torque of mutual gravitation of test masses.
M. U. Sagitov.
Soobshch. Gos. Astron. Inst. Shternberga, No. 187, p. 3 - 12 (1973). In Russian.

022.076 Special cases of expansion of the torque of mutual gravitation of test masses. Kh. G. Tadzhidinov.
Soobshch. Gos. Astron. Inst. Shternberga, No. 187, p. 46 - 49 (1973). In Russian.

022.077 The influence of a light stream on a high-quality pendulum in a vacuum.
V. S. Zenkov, I. I. Kalinnikov, V. M. Popov, M. U. Sagitov.
Soobshch. Gos. Astron. Inst. Shternberga, No. 191, p. 35 - 38 (1974). In Russian.

022.078 Calculations and measurements of the dynamic Stark effect in hydrogen.
W. R. Rutgers, H. W. Kalfsbeek.
Zeitschr. Naturforschung, Bd. 30a, 739 - 749 (1975).

022.079 Measurement of Stark broadening and shift of the helium I line λ = 4471.5 Å and its forbidden component λ = 4470 Å for electron densities $7 \times 10^{14} \lesssim N_e$ $[\mathrm{cm}^{-3}] \lesssim 3 \times 10^{16}$.
C. S. Diatta, A. Czernichowski, J. Chapelle.
Zeitschr. Naturforschung, Bd. 30a, 900 - 903 (1975).

022.080 Determination of the effective mass of the localized electron in dense helium gas from space-charge–limited currents. J. Howell, M. Silver.
Phys. Rev. Letters, Vol. 34, 921 - 924 (1975).

022.081 ¹D autoionization series in helium.
D. Burch, J. Bolger, C. F. Moore.
Phys. Rev. Letters, Vol. 34, 1067 - 1070 (1975).

022.082 **The free-free transitions of Li⁻ by a 'multi-channel asymptotic method'.**
T. L. John, D. J. Morgan, A. R. Williams.
Journ. Phys. B, Atomic Molecular Phys., Vol. 7, 1990 - 1993 =
Univ. College, Cardiff, Astron. Commun. No. 10 (1974).

022.083 **Far infrared measurements of selected optical materials at 1.6 K.**   J. A. Alvarez, R. E. Jennings, A. F. M. Moorwood.
Infrared Physics, Vol. 15, 45 - 49 (1975).
Measurements at 1.6 K of the transmittance and refractive index of quartz, polyethylene, polytetrafluoroethylene and polychlorotrifluoroethylene have been made, using a Michelson interferometer operating in the phase modulated mode.

022.084 **Statistical thermodynamics of strong interactions.**
R. Hagedorn, with notes by R. K. P. Zia, R. Bland.
High energy astrophysics and its relation to elementary particle physics, (see 012.018), p. 255 - 296 (1974).

022.085 **Strong interactions, gravitation and cosmology.**
A. Salam, with notes by J. K. Lawrence.
High energy astrophysics and its relation to elementary particle physics, (see 012.018), p. 441 - 452 (1974).

022.086 **A table of semiempirical gf values. Part 1: Wavelengths: 5.2682 nm to 272.3380 nm. Part 2: Wavelengths: 272.3395 nm to 599.3892 nm. Part 3: Wavelengths: 599.4004 nm to 9997.2746 nm.**
R. L. Kurucz, E. Peytremann.
Smithsonian Astrophys. Obs., *Cambridge, Mass.,* Special Report 362, Part 1–3, 17 + 1219 pp. (1975).
The authors tabulate gf values for 265,587 atomic lines selected from the line data used by Kurucz, Peytremann, and Avrett (1974) to calculate line-blanketed model atmospheres. These data are especially useful for line identification and spectral synthesis in solar and stellar spectra. Except for 10,000 lines taken from the literature, the gf values have been calculated semiempirically by using scaled Thomas-Fermi-Dirac radial wavefunctions and eigenvectors found through least-squares fits to observed energy levels. Included in the calculation were the first five or six stages of ionization for sequences up through nickel. Published gf values have been included for elements heavier than nickel. The tabulation is restricted to lines with wavelengths less than 10 μm. The data are also available on magnetic tape in a slightly different form.

022.087 **Theory of first rotational lines in transitions of diatomic molecules.**   A. Schadee.
Astron. Astrophys., Vol. 41, 203 - 212 (1975).
Correct expressions for the energies of the first rotational levels (those with $J < \Lambda + S$) and the Hönl-London factors for transitions involving these levels are derived for doublet and triplet states and transitions in diatomic molecular spectra.

022.088 **Hönl-London factors for ³Π – ³Σ and ³Σ – ³Π transitions with intermediate coupling.**
A. Schadee.
Astron. Astrophys., Vol. 41, 213 - 215 (1975).
Hönl-London factors of rotational lines for ³Π – ³Σ and ³Σ – ³Π transitions of diatomic molecules are calculated, allowing for non-zero values of both the splitting constant λ of the Σ state and the coupling constant Y of the Π state.

022.089 **Semiempirical helium intermolecular pair potential. Third virial coefficients at intermediate temperatures.**
R. E. Caligaris, J. C. Grangel.
Chem. Phys., [North-Holland Publ. Company, Amsterdam], Vol. 2, 249 - 252 = Univ. Republica, Fac. Humanidades y Ciencias, Montevideo, Dep. Astron. Fis., Publ. No. 43 (1973).

Die Lichtgeschwindigkeit im Gravitationsfeld der Sonne nach der Relativitätstheorie.   See Abstr. 066.037.

Forbidden-line excitation data for certain coronal lines.   See Abstr. 074.011.

Population théorique des sous niveaux Zeeman relatifs à la raie 5303 Å de Fe XIV.
See Abstr. 074.048.

Photoionization of barium clouds via the ³D metastable levels.   See Abstr. 082.007.

Weak interactions and cosmology.
See Abstr. 162.077.

Errata

022.901 **Erratum: 'Empirical computation of collisional ionization rates of atoms and ions by electrons'**
[Rev. Mexicana Astron. Astrofis., Vol. 1, 5 - 9 (1974)].
J. Cantó, E. Daltabuit.
Rev. Mexicana Astron. Astrofis., Vol. 1, 269 (1974).

022.902 **Erratum: 'The phase diagram of graphite grains'**
[Publ. Astron. Soc. Japan, Vol. 26, 197 - 206 (1974)].   M. Takada, M. Hirai.
Publ. Astron. Soc. Japan, Vol. 27, 195 (1975).

Astronomical Instruments and Techniques

031 Astronomical Optics, Methods of Observation and Reduction, Data Processing, Automation

Astronomical Optics

031.001 **Über den durch Blenden begrenzten geometrischen Gesichtsfelddurchmesser.** K. J. Stepputat.
SuW, Vol. 14, 21 (1975).

031.002 **A Cooke photovisual lens in a compensated cell.**
D. W. Dewhirst.
Sky Telescope, Vol. 49, 24 - 25 (1975).

031.003 **Umwandlung eines Schmidt-Spiegels in ein licht-starkes Teleskop.** E. Wiedemann.
Orion, 33. Jahrgang, p. 20 (1975).

031.004 **Interference fringes obtained on Vega with two optical telescopes.** A. Labeyrie.
Astrophys. Journ., (*Letters*), Vol. 196, L71 - L75 (1975).
Two small telescopes spaced by 12 m have been operated as a Michelson stellar interferometer. The system may be expanded progressively into a telescope array. Plans are made for starting a multinational array with a pair of 1.5-m telescopes.

031.005 **Evaluation of large aberrations using a lateral-shear interferometer having variable shear.**
M. P. Rimmer, J. C. Wyant.
Applied Optics, Vol. 14, 142 - 150 (1975).
A variable shear lateral shearing interferometer consisting of two holographically produced crossed diffraction gratings is used to test nonrotationally symmetric wavefronts having aberrations greater than 100 wavelengths and slope variations of more than 400 wavelengths/diameter. Comparisons are made with results of Twyman-Green interferometric tests for wavefront aberrations of up to thirty wavelengths. The results indicate that small wavefront aberrations can be measured as accurately with the lateral-shear interferometer as with the Twyman-Green interferometer and that aberrations that cannot be measured at all with a Twyman-Green interferometer can be measured to about 1% accuracy or better.

031.006 **Null Ronchi test for aspherical surfaces: comment.**
P. B. Fellgett, A. E. Gee.
Applied Optics, Vol. 14, 279, with a reply by D. Malacara, A. Cornejo, p. 279 (1975). – Concerning the paper by Malacara and Cornejo (Applied Optics, Vol. 13, 1778 - 1780 (1974) – see Abstr. 12.031.076).

031.007 **Irradiance distribution, resolution, and size estimates in diffraction limited imagery of extended circular targets.** Yu. Mekler, J. Otterman, Z. Ravnoy.
Applied Optics, Vol. 14, 503 - 508 (1975).

031.008 **Holographic simulation of parabolic mirrors.**
G. D. Mintz, D. K. Morland, W. M. Boerner.
Applied Optics, Vol. 14, 564 - 565 (1975).

031.009 **Effect of temperature gradients on the wave aberration in athermal optical glasses.**
F. Reitmayer, H. Schroeder.
Applied Optics, Vol. 14, 716 - 720 (1975).
Temperature gradients that are caused by partial heating in optical elements may result in wave aberrations. The athermal glasses developed within the past two years exhibit considerably reduced thermal wave aberrations compared to the classical optical glasses. Evidence is given that the total wave aberration is due not only to the change of optical path ΔW_T computed from α, n, and dn/dT, but also to that, upon the occurrence of thermal stresses, an additional wave aberration ΔW_S must be taken into consideration.

031.010 **Michelson interferometer with frustrated-total-internal-reflection beam splitter.**
M. Daehler, P. A. R. Ade.
Journ. Optical Soc. America, Vol. 65, 124 - 130 (1975).
An efficient beam splitter for a Michelson interferometer can be made from a pair of prisms, using the process of frustrated total internal reflection. Although the phases and irradiances of the beams reflected and transmitted by the beam splitter depend on the polarization, the phase difference between the two interferometer beams vanishes for all polarizations and the transmittance can be made polarization insensitive by suitable design. The interferometer transmits radiant power within a wavelength band approximately two octaves wide, and rejects all other radiant power. The design, laboratory tests, and astronomical applications of an interferometer for the millimeter and submillimeter regions are discussed. An observation of the day-sky spectrum is presented.

031.011 **Anastigmatic catadioptric telescopes.**
R. J. Lurie.
Journ. Optical Soc. America, Vol. 65, 261 - 266 (1975).
A conic mirror and an afocal achromatic refractive corrector located at the proximate geometrical focus of the conic define a family of anastigmatic telescopes of which the Schmidt telescope is a special member. In general, the corrector must consist of at least two elements whose surfaces can be spherical. Equations are given for the third-order design of thin two-element correctors. A solution that satisfies the condition of minimum higher-order aberrations is given for a corrector for a paraboloidal mirror. An example is given of a trigonometrically corrected paraboloidal-mirror anastigmat, and its aberrations are discussed. Conic-mirror anastigmats with mirrors of eccentricity $e \sim 1$ are a potentially valuable class of telescopes that merit consideration by the astronomical community.

031.012 **Das Protuberanzenfernrohr. Eine Bau- und Gebrauchsanleitung.** H. Treutner.
Orion, 33. Jahrgang, p. 51 - 55 (1975).

031.013 **To determine the focal length of specialized mirrors.**
D. C. Salter.
Journ. British. Astron. Ass., Vol. 85, 124 - 125 (1975).

031.014 **Parabolizing a 610 mm zero expansion mirror.**
S. R. Dunlop.

Journ. British. Astron. Ass., Vol. 85, 126 - 128 (1975).

031.015 **Methods of investigation of astronomical optics. I.**
An analysis of a complex of methods.
G. M. Beskin, A. M. Bogudlov, Eh. A. Vitrichenko, O. A. Evseev.
Astrofiz. Issled., Izv. Spets. Astrofiz. Obs., Vol. 7, p. 163 - 166 (1975). In Russian.

An analysis of a complex of methods for investigation of a parabolic astronomical mirror is given. The complex includes five methods: two types of Hartmanns' method; two television photometric methods; and the method of photoelectric measurement of the image. The complex of methods considered here allows to get quantitative, complete and objective data on an astronomical mirror.

031.016 **Methods of investigation of astronomical optics. II.**
Hartmann method.
Eh. A. Vitrichenko, F. K. Katagarov, V. G. Lipovetskaya.
Astrofiz. Issled., Izv. Spets. Astrofiz. Obs., Vol. 7, p. 167 - 181 (1975). In Russian.

Computer assisted methods of investigation of parabolic mirrors by Hartmanns' method in an illuminating beam coming from the center of curvature is developed. A program of automatic processing of Hartmann pictures using an «M-222» computer is written in «ALGOL-60». The program allows to treat both extrafocal and intrafocal Hartmann pictures.

031.017 **Methods of investigation of astronomical optics. III.**
An improved Foucault-Philbert method.
G. M. Beskin, A. M. Bogudlov, Eh. A. Vitrichenko, O. A. Evseev, S. M. Soldatov.
Astrofiz. Issled., Izv. Spets. Astrofiz. Obs., Vol. 7, p. 182 - 187 (1975). In Russian.

An improved optico-mechanical Foucault–Philbert system for control of astronomical optics is described. The use of a rotation unit for the television camera of the device makes it possible to construct accurate charts of the optics under investigation. A semi-transparent plate used in the system allows to make measurements on the optical axis of the device increasing hereby its accuracy and making it more convenient in use.

031.018 **Methods of investigation of astronomical optics. IV.**
Analysis of the image of a point (artificial star) by
television techniques.   G. M. Beskin, A. M. Bogudlov,
Eh. A. Vitrichenko, O. A. Evseev.
Astrofiz. Issled., Izv. Spets. Astrofiz. Obs., Vol. 7, p. 188 - 193 (1975). In Russian.

A new quantitative method of image investigation is described. The method permits to construct the energy distribution in the scattering circle and to perform its isophotometry.

031.019 **Methods of investigation of astronomical optics. V.**
A ring-diaphragm photometer.   G. M. Beskin,
Eh. A. Vitrichenko, A. I. Samojlov, A. G. Shcherbakov.
Astrofiz. Issled., Izv. Spets. Astrofiz. Obs., Vol. 7, p. 194 - 198 (1975). In Russian.

A technique used for measuring energy distribution in the image of a point given by an astronomical mirror is considered. It allows the determination of the most important value: the resolution of an astronomical mirror. Using this method one obtains the highest accuracy compared with the other methods of investigation discussed in these series.

031.020 **Progress on the CFHT 3.6 m mirror at the D.A.O.**
K. O. Wright.
Journ. Roy. Astron. Soc. Canada, Vol. 69, 46 - 48 (1975).

031.021 **Reshaping and stabilization of astronomical images.**
D. M. Hunten.
Astrophysics. Part A, (see 003.002), p. 193 - 220 (1974).

Reshaping of images: definitions, throughputs of spectroscopes, image slicers, Bowen image slicer, installation, Richardson image slicer, image transformer, telescope slicer of Fastie, fiber optics; Stabilization of images: automatic guiding, dynamics of a servo mirror, other drivers, image trackers, tracking devices, pulse-counting image dissector.

031.022 **Aplanatic two-mirror telescopes; a systematic study.**
3: The Schwarzschild-Couder configuration.
C. L. Wyman, D. Korsch.
Applied Optics, Vol. 14, 992 - 995 (1975).

A systematic performance analysis of aplanatic Schwarzschild-Couder type telescopes has been carried out by means of a ray trace program. The classic Schwarzschild and Couder designs were studied in detail. It was found that those types of telescopes, in contrast to the Gregorian and the Cassegrainian configuration, suffered severely from higher order aberration.

031.023 **Simple laser interferometer with variable shear and**
tilt.   P. Hariharan.
Applied Optics , Vol. 14, 1056 - 1057 (1975).

031.024 **Bibliography of various optical testing methods.**
D. Malacara, A. Cornejo, M. V. R. K. Murty.
Applied Optics, Vol. 14, 1065 - 1080 (1975).

A bibliography of various methods of optical testing has been compiled, with the papers grouped by subject, e.g., Fizeau interferometer. Only interferometer papers—or those on similar devices—that have direct relevance to shop testing are included. This reasonably complete compendium should be of use to workers in optical fabrication and testing.

031.025 **Bragg-angle blazing of diffraction gratings.**
A. Hessel, J. Schmoys, D. Y. Tseng.
Journ. Optical Soc. America, Vol. 65, 380 - 384 (1975).

The Bragg condition $\lambda = 2d \sin \theta$ is presented as a necessary condition for perfect (100 %) blazing of infinite, perfectly conducting diffraction gratings that produce only a single diffracted order, $n = -1$. As an illustration, the rectangular-profile grating is analyzed. Design blaze curves, showing the required depth as a function of angle of incidence, for each polarization, are obtained. Intersections of corresponding blaze curves for the two polarizations yield parameters of gratings perfectly blazed simultaneously in both polarizations. The angle of incidence at which this is achieved depends on the shape of the grating. Gratings so designed have a theoretical 100 % efficiency at the design frequency, and above 90 % over a bandwidth of more than 10 %.

031.026 **Ray tracing in gradient-index media.**   D. T. Moore.
Journ. Optical Soc. America, Vol. 65, 451 - 455 (1975).

A new procedure for the tracing of rays in inhomogeneous media is described. The method is based on a polynomial solution of the differential equation for ray paths. The index of refraction is expressed in a polynomial in the optical-direction coordinate and in the coordinate in the direction orthogonal to the optical axis. The technique is incorporated into a lens-design program to provide optimization of gradient indices. The results of the methods have been verified in two other ways.

031.027 **Photon noise and atmospheric noise in active optical systems.**   F. J. Dyson.
Journ. Optical Soc. America, Vol. 65, 551 - 558 (1975).

A general theoretical analysis is made of the performance of optical systems that are designed to give diffraction-limited images of astronomical objects by compensating effects of atmospheric seeing in real time. The heart of any such system is a feedback algorithm, which expresses the controlled displacements of mirror surfaces as functions of the output of optical sensors. The statistical behavior of the system is calculated, assuming the feedback algorithm to be linear but otherwise unrestricted. Applications of the general theory to particular optical systems are briefly discussed. In principle, systems optimized in this way should be able to give images of arbitrarily high resolution of astronomical objects brighter than about magnitude 14.

031.028 **Die Sichtbarkeit von Sternen und Planeten bei Tageslicht, in der Dämmerung und bei Beobachtungen aus tiefen Schächten.** J. Feitzinger.
SuW, Vol. 14, 156 - 159 (1975).

031.029 **Über das geometrische Spektrum gekreuzter Halbprismenpaare. II. Geometrisch und optisch gleichwertige Halbprismen in symmetrischer Kreuzung nach Art des Herschel-Risley-Doppelprismas bei einfachem Strahlendurchgang.** J. Junkes.
Ric. Spettrosc., Lab. Astrofis. Specola Vaticana, *Castel Gandolfo*, Vol. 3, (No. 8), 381 - 529 (1974).
The present part II of the researches on the geometric spectrum of crossed half-prisms deals with the geometric spectrum and its dispersion for symmetrically crossed Herschel-Risley double prisms at a single passing of the rays.

031.030 **Infrared reflectance of silicon oxide and magnesium fluoride protected aluminum mirrors at various angles of incidence from 8 μm to 12 μm.**
J. T. Cox, G. Hass, W. R. Hunter.
Applied Optics, Vol. 14, 1247 - 1250 (1975). − Rapid communication.

031.031 **Multilayer antireflection coatings: theoretical model and design parameters.** · K. Rabinovitch, A. Pagis.
Applied Optics, Vol. 14, 1326 - 1334 (1975).

031.032 **Optical materials and processes. Part 1: mirror materials.** G. R. Nankivell.
Southern Stars, Vol. 26, 8 - 10 (1975).

031.033 **Optical systems of satellite-borne submillimeter telescopes.** A. S. Khajkin.
Trudy Fiz. in-ta. AN SSSR, Vol. 77, 56 - 79 (1974). In Russian. − Abstr. in Referativ. Zhurn. 62. Issled. kosmich. prostranstva, 5.62.128; 51. Astron., 6.51.264 (1975).

031.034 **Reflectors for X-ray telescopes.**
I. L. Bejgman, L. A. Vajnshtejn, Yu. P. Vojnov, V. P. Shevel'ko.
Trudy Fiz. in-ta. AN SSSR, Vol. 77, 14 - 32 (1974). In Russian. − Abstr. in Referativ. Zhurn. 62. Issled. kosmich. prostranstva, 5.62.135; 51. Astron., 6.51.266 (1975).

031.035 **Achromatic Schmidt corrector plates.**
A. Greve.
Optik, Vol. 42, 87 - 95 = Max-Planck-Inst. Radioastron., Bonn, Sonderdruck, Ser. A, No. 50 (1975).
The mathematical expressions for two types of achromatic Schmidt corrector plates are derived. Both types of achromatic corrector give for an extended wavelength region an on-axis image whose Strehl number is $\geqslant 0.8$.

031.036 **The Talbot effect in the Ronchi test.**
D. Malacara, A. Cornejo.
Bol. Inst. Tonantzintla, Vol. 1, 193 - 196 (1974).

031.037 **Focusing conditions for the spherical concave grating. II. Rowland surfaces.**
A. Danielsson, P. Lindblom.
Optik, Vol. 41, 465 - 478 (1975).

031.038 **Wave tracing for optical systems.**
E. Byckling, A. Friberg.
Optik, Vol. 42, 147 - 154 (1975).
The authors derive a recursion relation which is analogous to ray tracing equations and can be used to calculate the propagation of wave components through an optical system. The method is applicable, in particular, to diffraction limited systems and to the evaluation of diffraction effects.

031.039 **The method of the caustic for measuring optical surfaces.** R. P. Platzeck, J. M. Simon.
Optica Acta, Vol. 21, 267 - 276 = Separata Astron., Obs. Astron., La Plata, Argentina, No. 132 (1974).
The method of the caustic, as used in testing optical systems or surfaces, has been examined for two measuring techniques, namely bisecting slit images with a wire and photographic recording of slit images for later measurement. Diffraction effects have been considered and the sensitivity of the methods estimated, the better results being obtained with the photographic method.

031.040 **An improved acousto-optical light modulator.**
T. W. Cole.
Optics Commun., (*Netherlands*), Vol. 13, 192 - 193 = Div. Radiophys., C.S.I.R.O., Sydney, Radiophys. Publ. RPP 1819 (1975).
A simplification of the conventional acousto-optical light modulator and deflector arrangement is presented. The collimating lens, modulator, and focusing lens are combined into one unit − a modulator whose faces are curved rather than plane.

031.041 **Optical synthesis of different wavefronts by kinoform principle and fabrication of a Schmidt plate.**
C. Abitbol, J. J. Clair, L. H. Torres.
Instrumentation in astronomy II, (see 012.021), p. 223 - 226 (1974).

031.042 **Use of screen rotation in testing large mirrors.**
I. Ghozeil.
Instrumentation in astronomy II, (see 012.021), p. 247 - 252 (1974).

031.043 **An introduction to integrated optics.** H. Kogelnik.
IEEE Trans. Microwave Theory Techn., Vol. MTT - 23, 2 - 16 (1975).
Deals briefly with dielectric waveguides, coupled waves and periodic structures, fabrication techniques and includes over 100 references. − *OJC*

Aberrations of the symmetrical optical system.
See Abstr. 003.149.

Erste Erfahrungen mit der prismatischen Schmidt-Platte des Tautenburger 2-m-Spiegelteleskops.
See Abstr. 032.008.

Versatile nebular insect-eye Fabry-Perot spectrograph. See Abstr. 034.014.

Calculation of arbitrary-order diffraction efficiencies of thick gratings with arbitrary grating shape.
See Abstr. 034.016.

The theoretical instrumental profile of the combination telescope-pinhole photometer and its effect upon observations of umbra intensities. See Abstr. 034.075.

Focal predictions for spectrographs employing a Maksutov camera. See Abstr. 034.093.

Methods of Observation and Reduction

031.201 Infrared heterodyne spectroscopy of astronomical and laboratory sources at 8.5 μm.   M. Mumma, T. Kostiuk, S. Cohen, D. Buhl, P. C. von Thuna.
Nature, Vol. 253, 514 - 516 (1975).

The first successful infrared heterodyne spectrometer featuring semi-tuneable semiconductor diode lasers was constructed and used near 8.5 μm to make laboratory measurements of line profiles in N_2O and to detect thermal emission from Mars and from the moon.

031.202 Aperture synthesis in X-ray astronomy.   A. M. Cruise.
Monthly Notices Roy. Astron. Soc., Vol. 170, 305 - 312 (1975).

A new method of X-ray imaging is described. By combining a number of collimators it is possible to synthesize a useful aperture to record structure in cosmic X-ray sources. Results from computer simulations show this method to be a practical solution to imaging at many photon energies and in many fields of physics.

031.203 Ionospheric direct measurement techniques.   S. J. Bauer, A. F. Nagy.
Proc. IEEE, Vol. 63, 230 - 249 (1975).

The most important physical parameters of the ionosphere which have been studied extensively over the years are: 1) the temperature, density, chemical composition, and directed motion (wind) of the ionized and neutral gas particles; and 2) the electric and magnetic fields. This review will discuss direct in situ techniques used on sounding rockets and satellites to measure these physical parameters.

031.204 On methods of high-accuracy electrophotometric observations of variable stars.   G. N. Alekseev.
Soobshch. Spets. Astrofiz. Obs., *Zelenchukskaya*, vyp. (No.) 12, p. 5 - 42 (1974). In Russian.

Factors which influence the accuracy of electrophotometry and ways of error reduction for prolonged measurements of small light variations are considered. Drafts are presented for the optico-mechanical part and recording devices of the photometer which would allow to make measurements of light with an error of 0.1%.

031.205 Precise measurements of radial velocity using a Lirepho microphotometer.
P. Heinzel, P. Hadrava.
Bull. Astron. Inst. Czechoslovakia, Vol. 26, 90 - 91 (1975).

A simple method suitable for precise radial-velocity measurements by means of the Zeiss-Lirepho microphotometer is briefly described. The testing showed that the precision of the new method is quite comparable to the precision reached with a classical Abbé comparator. In addition, the method allows for measuring velocities of broad or asymmetric lines and of the weak lines invisible in the comparator.

031.206 Curvature in Hβ transformations.   D. Kilkenny.
Astron. Journ., Vol. 80, 134 - 136 (1975).

Observations of Hβ standard stars at two sites have resulted in transformations to the standard system which are slightly curved. From filter simulations it is probable that the nonlinearity is caused by differences between standard and instrumental intermediate band filters. An estimate is made of the maximum deviations from the standard filters which still produce linear transformations.

031.207 An attempted explanation for the discrepancy between internal and external errors in statistical adjustments.   M. G. Firneis, F. J. Firneis.
Astron. Nachr., Vol. 296, 95 - 100 (1975).

It is shown that underparameterization of models used in statistical adjustments is at least in part responsible for the discrepancies between so-called internal and external rms errors (standard deviations), frequently encountered in derived sets of statistical parameters.

031.208 Removal of instrument signature from Mariner 9 television images of Mars.   W. B. Green, P. L. Jepsen, J. E. Kreznar, R. M. Ruiz, A. A. Schwartz, J. B. Seidman.
Applied Optics, Vol. 14, 105 - 114 (1975).

The Mariner 9 spacecraft was inserted into orbit around Mars in November 1971. The two vidicon camera systems returned over 7300 digital images during orbital operations. The high volume of returned data and the scientific objectives of the television experiment made development of automated digital techniques for the removal of camera system-induced distortions from each returned image necessary. This paper describes the algorithms used to remove geometric and photometric distortions from the returned imagery. Enhancement processing of the final photographic products is also described.

031.209 Submillimeter-wave atmospheric and astrophysical spectroscopy.   J. E. Beckman, J. E. Harries.
Applied Optics, Vol. 14, 470 - 485 (1975).

031.210 Image information by means of speckle-pattern processing.   P. H. Deitz.
Journ. Optical Soc. America, Vol. 65, 279 - 285 (1975).

031.211 Spektroskopie der Sonne, Planeten und Sterne. Eine Anleitung für Anfänger.   R. Schneider.
Orion, 33. Jahrgang, p. 55 - 58 (1975).

031.212 Observations of rapidly fluctuating objects. I. Techniques and instrumentation.
G. N. Alekseev, V. G. Shtol'.
Astrofiz. Issled., Izv. Spets. Astrofiz. Obs., Vol. 7, p. 148 - 155 (1975). In Russian.

Techniques and an apparatus applied at the Special Astrophysical Observatory of the USSR Academy of Sciences for observing rapidly fluctuating objects are described. The instrumentation used with an on-line M-222 computer allows to register brightness variability of astronomical objects in a time interval from 0^{s}002 to 1000^s.

031.213 Automation of processing of high-dispersion spectrograms of O−A-type stars with the help of a computer.
A. A. Korovyakovskaya, Yu. P. Korovyakovskij, E. L. Chentsov.
Astrofiz. Issled., Izv. Spets. Astrofiz. Obs., Vol. 7, p. 156 - 162 (1975). In Russian.

Methods for automatic determination of photometric parameters and wavelengths of isolated spectral lines with the help of a digital microphotometer and a computer are described. The methods allow processing of high-dispersion spectrograms of stars of earlier spectral classes avoiding losses of the accuracy given by the plate.

031.214 Convectometer, an apparatus for measuring optical irregularities.   V. S. Rylov, V. G. Shtol'.
Astrofiz. Issled., Izv. Spets. Astrofiz. Obs., Vol. 7, p. 240 - 244 (1975). In Russian.

A convectometer manufactured at the Special Astrophysical Observatory of the USSR Academy of Sciences is designed for measuring optical irregularities in local volumes (observatory domes, large spectrographs, telescope tubes) with fluctuation amplitudes within 1.5°C to 0.004°C. A description of the apparatus and results of testing are presented.

031.215 A method to identify the reference stars with an

electronic computer.   L. Wu.
Acta Astron. Sinica, Vol. 15, 241 - 245 (1974). In Chinese.
A method to identify the reference stars is given, when the equatorial coordinates of the center of a plate are only roughly known.

031.216   Observational technique and data reduction.
         A. T. Young.
Astrophysics. Part A, (see 003.002), p. 123 - 192 (1974).
Atmospheric extinction: basic error analyses, random and systematic errors in photometry, errors in reduction methods, actual error laws, how much time to spend on extinction, concluding remarks on extinction; Transformation to a standard system: transformations for blackbodies, transformations in general, matching response functions, mathematical models.

031.217   Astronoom heeft diverse soorten 'meetlatten'
         voorhanden.   G. W. E. Beekman.
Zenit, Vol. 2, 107 - 111 (1975).

031.218   Notes on observing methods and programs for new
         observers.   M. Taylor.
Journ. American Ass. Variable Star Observers, Vol. 3, 57 - 58 (1974).

031.219   Méthode des moindres carrés (erreurs).
         H. Debehogne.
IAU Colloquium No. 22, (see 012.005), p. 51 - 59 (1974).

031.220   Contribution au problème de l'aberration différ-
         entielle.   H. Debehogne, E. van Hemelrijck.
IAU Colloquium No. 22, (see 012.005), p. 61 - 67 (1974).

031.221   New determination of the plate constants of the
         Bordeaux zone of the Astrographic Catalogue.
P. Herget.
IAU Colloquium No. 22, (see 012.005), p. 69 - 70 (1974).

031.222   Les erreurs sur les étoiles de référence dans la
         méthode des dépendances et leurs conséquences.
H. Debehogne, R. R. de Freitas Mourao.
IAU Colloquium No. 22, (see 012.005), p. 77 - 82 (1974).

031.223   Continuum source and a focusing technique for the
         80–500-Å spectral range: improvements.
A. M. Cantù, G. Tondello.
Applied Optics, Vol. 14, 996 - 998 (1975).

031.224   Controlled generation and absolute calibration of
         radiant incidence in low-background test facilities
for infrared space sensors.   R. H. Meier.
Applied Optics, Vol. 14, 1021 - 1028 (1975).
Meaningful prelaunch calibrations of ir sensors designed to be operated from space platforms require low-background test facilities that permit both the controlled generation and the absolute calibration of radiant incidence over wide dynamic ranges. Such a test facility is presently under development. It comprises a source assembly, a collimator, and a spectroradiometer. The source can emit either broadband or monochromatic radiation. In conjunction with a reference blackbody, the spectroradiometer monitors both the radiant incidence and the spectral distribution of the flux that enters the sensor under test.

031.225   Optimum conditions for heterodyne detection of
         light.   L. Mandel, E. Wolf.
Journ. Optical Soc. America, Vol. 65, 413 - 420 (1975).

031.226   Determination of areas and temperatures from
         composite blackbody spectra.

C. S. Kelley, J. C. Bremer.
Journ. Optical Soc. America, Vol. 65, 559 - 564 (1975).

031.227   Time-resolved Fourier spectroscopy.
         R. E. Murphy, F. H. Cook, H. Sakai.
Journ. Optical Soc. America, Vol. 65, 600 - 604 (1975).

031.228   On a weight system for reduction of observations
         of pairs for screw evaluation.
I. V. Dzhun', I. I. Maksimova, E. I. Obrezkova.
Vrashchenie i prilivn. deformatsii Zemli. Vyp. (No.) 6. Kiev, "Nauk. dumka", 1974, p. 120 - 125. In Russian. – Abstr. in Referativ. Zhurn. 51. Astron., 3.51.136 (1975).

031.229   Automatic solar image motion measurements.
         S. A. Colgate, E. P. Moore.
Solar Physics, Vol. 41, 487 - 498 (1975).
The solar seeing image motion has been monitored electronically and absolutely with a 25 cm telescope at three sites along the ridge at the southern end of the Magdalena Mountains west of Socorro, New Mexico. The uncorrelated component of the variations of the optical flux from two points at opposite limbs of the solar disk was continually monitored in 3 frequencies centred at 0.3, 3 and 30 Hz. The frequency band of maximum signal centred at 3 Hz showed the average absolute value of image motion to be somewhat less than 2″.

031.230   Absolute radiometric calibration of detectors
         between 200–600 Å.
E. B. Saloman, D. L. Ederer.
Applied Optics, Vol. 14, 1029 - 1034 (1975).

031.231   Problems of oversampling with SEC vidicon televi-
         sion systems.
E. J. Devinney, D. Fischel, D. A. Klinglesmith.
Bull. American Astron. Soc., Vol. 7, 270 (1975). – Abstr. AAS.

031.232   Identification of natural groups from direct photo-
         graphs and from very low dispersion spectra.
M. F. McCarthy.
Bull. American Astron. Soc., Vol. 7, 270 (1975). – Abstr. AAS.

031.233   On the accuracy of measuring magnetic fields using
         Fourier transform techniques.   T. D. Tarbell.
Bull. American Astron. Soc., Vol. 7, 351 (1975). – Abstr. AAS.

031.234   Error in bandwidth-limited measurement of optical
         flux.   P. E. Tallant.
Bull. American Astron. Soc., Vol. 7, 351 (1975). – Abstr. AAS.

031.235   Ultra-high resolution infrared spectroscopy of
         laboratory sources, the moon, and Mars using he-
terodyne techniques.   M. Mumma, T. Kostiuk, S. Cohen,
D. Buhl, P. C. Von Thuna.
Bull. American Astron. Soc., Vol. 7, 390 (1975). – Abstr. AAS.

031.236   Comparative analysis of different methods for
         estimating the parameters in the problem of V. G.
Kurt.   L. S. Gurin, N. P. Ivanova, V. S. Mokrov, K. A. Tsoj.
AN SSSR. In-t kosmich. issled. Pr-186. Moskva, 1974. 35 pp.
In Russian. – Abstr. in Referativ. Zhurn. 62. Issled. kosmich.
prostranstva, 4.62.118 (1975).

031.237   Photoelectric device for determination of the mean
         transit moment of stars.   A. V. Ivanov.
Uch. Zap. Latv. Gos. Univ., Vol. 220, Astron., vyp. (No.) 11, (see 003.014), p. 55 - 79 (1975). In Russian.
The paper contains the basic principles for the design of a device specially meant to determine the mean moment of transit of a star, accounting for accidental ejections. The

necessary coefficient of amplifications of the amplifier is estimated. Integral microschemes have been used for operational amplifier circuits.

031.238 Device for direct registration of mean transit moments of stars with transfer of results upon a digital printer. M. Ogriņš.
Uch. Zap. Latv. Gos. Univ., Vol. 220, Astron., vyp. (No.) 11, (see 003.014), p. 80 - 90 (1975). In Russian.

The paper presents a description of a device for registering moments of transits of stars. A reversible impulse counter with two inputs is used, permitting direct registration of the mean moment.

031.239 Sampling in imaging systems.
W. D. Montgomery.
Journ. Optical Soc. America, Vol. 65, 700 - 706 (1975).

031.240 Coherent detection applied to optical and infrared astronomy: an introduction. H. G. Van Bueren.
Space Sci. Rev., Vol. 17, 621 - 628 (1975). – Presented at the workshop on coherent detection in astronomy, held at Rhenen, The Netherlands, 25 and 26 April 1974 – see 012.009.

031.241 Some filter techniques in the far infrared.
J. J. Wijnbergen.
Space Sci. Rev., Vol. 17, 657 - 658 (1975). – Presented at the workshop on coherent detection in astronomy, held at Rhenen, The Netherlands, 25 and 26 April 1974 – see 012.009.

031.242 Heterodyne spectroscopy of astronomical and laboratory sources at 8.5 μm using diode laser local oscillators.
M. Mumma, T. Kostiuk, S. Cohen, D. Bühl, P. C. von Thuna.
Space Sci. Rev., Vol. 17, 661 - 667 (1975). – Presented at the workshop on coherent detection in astronomy, held at Rhenen, The Netherlands, 25 and 26 April 1974 –see 012.009.

031.243 Conclusions from the discussions on observational results of heterodyne spectroscopy in astronomy.
M. Mumma.
Space Sci. Rev., Vol. 17, 669 - 670 (1975). – Presented at the workshop on coherent detection in astronomy, held at Rhenen, The Netherlands, 25 and 26 April 1974 – see 012.009.

031.244 Phase-contrast detection of telescope seeing errors and their correction. R. H. Dicke.
Astrophys. Journ., Vol. 198, 605 - 615 (1975).

The technique of phase-contrast microscopy applied to a telescope permits the detection and correction of phase errors in real time. These corrections can be made separately for two or more levels of the atmosphere. It may be possible to correct the seeing errors in a circular patch up to $2'$ in diameter about a reference star $m_v \sim 11$ or brighter.

031.245 Calcul de la période d'un phénomène cyclique.
B. Rutily.
Comptes Rendus Acad. Sci. Paris, Sér. B, Vol. 280, 689 - 690 (1975).

Une nouvelle méthode de calcul est proposée pour la recherche des périodes dans les phénomènes vibratoires. L'application de cette méthode à l'astrophysique nous renseigne en particulier sur la notion de «périodicité» dans les étoiles variables de type δ Scuti.

031.246 On the reduction of measured coordinates to standard coordinates with terms depending on color and magnitude of stars. Yu. A. Shokin.
Soobshch. Gos. Astron. Inst. Shternberga, No. 190, p. 3 - 19 (1974). In Russian.

Five reduction models including linear terms of coordi-nates and color, first and second power terms of image diameter in different combinations are investigated.

031.247 On the calculation of the curvature of the parallel in observations with the PZT. A. A. Volchkov.
Soobshch. Gos. Astron. Inst. Shternberga, No. 190, p. 37 - 39 (1974). In Russian.

031.248 Gravitational-wave antenna design to detect random gravitational waves. W. L. Burke.
Phys. Rev. D, Particles and Fields, Vol. 8, 1030 - 1035 = Lick Obs. Bull. No. 634 (1973).

A design is proposed for an antenna system capable of detecting random gravitational waves and separating their effects from random fluctuations in the antenna. The strongest known signal should be the random signal from all of the binary stars in the Galaxy. This design may allow one to penetrate the background noise of the earth itself and recover this signal.

031.249 Restauración de imagenes con conocimiento apriori. L. R. Berriel.
Bol. Inst. Tonantzintla, Vol. 1, 197 - 201 (1974).

031.250 Deformation of absorption line profiles due to secular change of the grating. M. Saitō.
Tokyo Astron. Obs., Report No. 65, Vol. 17, 284 - 291 (1975). In Japanese.

031.251 A simple method to extract information on anisotropy of particle fluxes from spin-modulated counting rates of cosmic ray telescopes.
K. C. Hsieh, Y. C. Lin, J. D. Sullivan.
Journ. Geophys. Res., Vol. 80, 2305 - 2308 (1975).

A simple method to extract information on anisotropy of particle fluxes from data collected by cosmic ray telescopes on spinning spacecraft but without sectored accumulators is presented. Application of this method to specific satellite data demonstrates that it requires no prior assumption on the form of angular distribution of the fluxes; furthermore, self-consistency ensures the validity of the results thus obtained. The examples show perfect agreement with the corresponding magnetic field directions.

031.252 Influence of probe surface properties on plasma measurements in the ionosphere and magnetosphere.
A. Pedersen.
Space Sci. Instrum., Vol. 1, 111 - 123 (1975).

031.253 Extended speckle interferometry.
G. P. Weigelt.
Conference on optical observing programs on galactic structure and dynamics, (see 012.013), p. 173 - 176 (1975). In German.

031.254 A stepping motor drive for small telescopes.
D. G. McCartan, A. de Sa.
Journ. British Astron. Ass., Vol. 85, 324 - 328 (1975).

031.255 A new method of image restoration.
C. R. Subrahmanya.
Bull. Astron. Soc. India, Vol. 3, 13 - 14 (1975).

031.256 Caratteristiche spettrali del metodo di regolarizzazione di Whittaker. E. Proverbio, S. Uras.
Rend. Seminario Fac. Sci. Univ. Cagliari, Suppl. Vol. 44, (see 012.016), 75 - 103 (1974).

031.257 Sulla eliminazione delle osservazioni considerevolmente diverse da tutte le altre. G. Cresci.
Rend. Seminario Fac. Sci. Univ. Cagliari, Suppl. Vol. 44, (see

012.016), 111 - 115 (1974).

031.258 Photographische Astrometrie ohne Messapparat.
W. Dienes.
Sternenbote, 18. Jahrgang, p. 3 - 11 (1975).

031.259 Come e che cosa osservare (5ª parte).
L. Baldinelli.
Giorn. Ass. Astrofili Bolognesi, No. 38, p. 3 - 6 (1975).

031.260 Discovery of Herbig-Haro objects by [S II] interference photography. S. van den Bergh.
Publ. Astron. Soc. Pacific, Vol. 87, 405 - 406 (1975).
It is shown that [S II] interference filter photography is a powerful technique for the discovery of Herbig-Haro objects.

031.261 Der Einfluß der Saalrefraktion im Meridianhaus des Lohrmann-Observatoriums auf die Breitenbeobachtungen nach Horrebow-Talcott.
J. Dittrich, S. Wächter.
Vermessungstechnik, 20. Jahrgang, p. 393 - 395 = Mitt. Lohrmann-Obs. Techn. Univ. Dresden, No. 27 (1972).

031.262 Ergebnisverbesserungen beim Reduktionsverfahren nach Turner-Vick zum Anschluß von Einzelobjekten an Nachbarsterne. W. Albrecht.
Vermessungstechnik, 21. Jahrgang, p. 212 - 215 = Mitt. Lohrmann-Obs. Techn. Univ. Dresden, No. 28 (1973).

031.263 Methods for search of variable star periods.
J. Pelt.
Tartu Astron. Obs., Teated, No. 52, 24 pp. = Preprint No. 5 (1975). In Russian.
A general form of period-finding algorithms is presented. The well-known methods by Lafler-Kinman (1965) and Jurkevich (1971) are considered as particular cases of this algorithm. New realizations are also described and specific algorithms are presented in the form of ALGOL-60 procedures. By the author's experience the so-called method of bounded phase differences is one of the most effective realizations of this scheme.

031.264 On digital reduction of stellar spectrograms.
T. Kipper, J. Sitska.
Tartu Astron. Obs., Teated, No. 53, 21 pp. = Preprint No. 6 (1975).
Observed stellar spectra are always distorted by instrumental broadening effects and contain random fluctuations of intensity due to inevitable noise of any recording system. The authors use the fast Fourier transform procedure to perform filtering and deconvolution in order to reduce the smearing effects of the apparatus function in photographically recorded stellar spectra.

031.265 Considerazioni sull' isogonismo o non delle principali proiezioni cartografiche e su alcuni procedimenti operativi. L. Turturici, A. Vassallo.
Boll. Geod. Sci. Affini, Anno 33, p. 407 - 422 = Ist. Geod. Idrografia Nautica, *Napoli,* Pubbl. Nuova Ser., No. 2 (1974).

031.266 On a graphical method of determination of the azimuth of Polaris. N. A. Uvarov.
Geod. i kartografiya, 1974, No. 12, p. 26 - 29. In Russian.
Abstr. in Referativ. Zhurn. 51. Astron., 6.51.209; 52. Geodeziya i Aehrosemka, 6.52.102 (1975).

031.267 Heterodyn-Detektion als Mittel zur hochauflösenden Spektralanalyse schwacher Linienstrahlung im fernen Infrarot. K.-U. Dettmar, L. Haser.
Max-Planck-Inst. Phys. Astrophys., Inst. Extraterr. Phys., München, MPI– PAE/Extraterr.107, 2 + 38 pp. (1974).

031.268 Stray light in solar observations. A method of numerical integration. R. Brahde.
Inst. Teor. Astrophys., Blindern—Oslo, Rep. No. 41, 2 + 49 pp. (1974).
The correction of solar observations for stray light requires some empirical stray light function, with parameters to be determined from observations of the solar limb and the aureole. A formula $\Psi_s(\rho) = A/(B^{2q} + \rho^{2q})$ where q is a new parameter, has been tried out on 52 scans of the solar and lunar limbs obtained during the partial eclipse of February 25, 1971 in 7 wavelengths ranging from $\lambda 3870$ to $\lambda 22000$. It is shown that this formula can be used with success in all of the cases. The corrections for stray light in a sunspot is shown for two cases. Finally the Fortran programs are presented with comments.

031.269 The search for new variables with a blink microscope.
B. A. Capron.
Journ. American Ass. Variable Star Observers, Vol. 2, 63 - 66 (1973).

031.270 Visual techniques for faint observations with small telescopes. E. H. Mayer.
Journ. American Ass. Variable Star Observers, Vol. 3, 19 (1974).

031.271 The role of image vibration in astrometry and measures for excluding its influence on observations.
I. G. Kolchinskij.
19th Astrometrical Conference 1972, (see 012.019), p. 14 - 25 (1975). In Russian.

031.272 Photoelectric observations of meridian passages of stars according to a new method.
H. Potthoff.
19th Astrometrical Conference 1972, (see 012.019), p. 142 - 144 (1975). In Russian.

031.273 The influence of moving gears of a micrometer on the results of astrometrical observations.
I. Rusu, K. N. Tavastsherna.
19th Astrometrical Conference 1972, (see 012.019), p. 159 - 161 (1975). In Russian.

031.274 On personal errors in observing disc-shaped objects.
E. M. Nenakhova, A. S. Kharin.
19th Astrometrical Conference 1972, (see 012.019), p. 168 - 170 (1975). In Russian.

031.275 On perspectives of increasing the accuracy of daytime observations. A. S. Kharin.
19th Astrometrical Conference 1972, (see 012.019), p. 170 - 173 (1975). In Russian.

031.276 Optimal scan of the sky from a rocket (computer aided). R. A. Nidey.
AIAA Journ., Vol. 12, 1162 - 1164 (1974).

031.277 The moon as source for G/T measurements.
K. G. Johannsen, A. Koury.
IEEE Trans. Aerospace Electron. Systems, Vol. AES-10, 718 - 727 (1974).
The moon provides a power flux density higher by at least one order of magnitude compared to the strongest radio star, and the resulting y factors are usable. G/T ratios determined from moon measurements agree well with expected values.

031.278 Apodization of telescopes working in a turbulent medium. M. De, L. N. Hazra, P. Sengupta.
Optica Acta, (*GB*), Vol. 22, 125 - 140 (1975).

The authors present a complete analytical solution of the problem of far-field diffraction by apertures of a partially space-coherent wave-field. The effect of atmospheric turbulence on the far-field diffraction pattern has been studied as regards the intensity distribution, the factor of encircled energy distribution and the effective central illumination on the same. Numerical results are presented for a few representative apertures. Indications are given regarding the convenient amenability of the approach to the problem of optimization.

031.279 **Reduction of astronomical images with methods of coherent optics.** V. N. Dudinov, V. S. Tsvetkova, V. A. Krishtal', N. A. Khovanskij.
Vestn. Khar'kov. Univ., No. 117, (Ser. Astron., vyp. (No.) 9), p. 19 - 26 (1974). In Russian.

031.280 **The effect of errors in composed zone-plates.** J. V. Feitzinger.
Optik, Vol. 41, 180 - 190 (1974). In German.
To strengthen the light gathering power of zone-plates their diameters were increased. By this procedure the correct position of the zones is disturbed. Calculations show, that if concentricity of the outer zones and a displacing smaller than the half width of the last open zone is preserved the image quality is enhanced.

031.281 **Some astronomical applications for precision high-speed computer interfaced microdensitometry.** J. D. Wray, G. F. Benedict
Instrumentation in astronomy II, (see 012.021), p. 137 - 146 (1974).

031.282 **Tips für die Astropraxis.** SuW, Vol. 14, 58 - 60, 131 - 134 (1975).

031.283 **On the probability distribution of line-of-sight fluctuations of optical signals.** J. W. Strohbehn, T.-I. Wang, J. P. Speck.
Radio Sci., (*USA*), Vol. 10, 59 - 70 (1975).

The compression of data resulting from photon counters. See Abstr. 021.005.

The Santiago Grant machine and Zeeman coudé data reductions. See Abstr. 031.411.

Space reflectors for radar and astronomy. See Abstr. 032.009.

Polarization techniques. See Abstr. 034.029.

The instrumentation and techniques of infrared photometry. See Abstr. 034.030.

The ESO spot sensitometer. See Abstr. 034.097.

A clock pulse generator for the Lunar Laser Ranging System. See Abstr. 035.009.

A radioastronomical inertial coordinate system based on measurement of arcs between radio sources. See Abstr. 041.029.

Cylinders as gravitational radiation telescopes. See Abstr. 066.009.

Photoelectric speckle interferometry of the solar granulation. See Abstr. 071.008.

Far-infrared solar brightness measured with a balloon-borne lamellar-grating interferometer. See Abstr. 071.012.

X-ray spectra of solar active regions. See Abstr. 076.041.

Influence of exposure time on spectral properties of turbulence-degraded astronomical images. See Abstr. 082.070.

Infrared heterodyne spectroscopy of CO_2 on Mars. See Abstr. 097.063.

Why image Uranus? See Abstr. 101.006.

Observations of comet Kohoutek (1973f) at the satellite station 1147 Ondřejov 2. See Abstr. 103.100.

Metal abundances of RR Lyrae stars established from low resolution scanner spectrophotometry. See Abstr. 122.061.

Data Processing, Automation

031.401 **The analysis of data from rotation modulation collimators.** A. M. Cruise, A. P. Willmore.
Monthly Notices Roy. Astron. Soc., Vol. 170, 165 - 175 (1975).
Three methods of analysing data from rotation modulation collimators are discussed and only one is found to be at all satisfactory. A new algorithm has been developed which decreases the time required to perform this cross-correlation analysis considerably. Corrections to the image intensity have been derived for two special cases of spin axis motion and these are used to evaluate the results of two X-ray observations from sounding rockets.

031.402 **Algorithm for automatic identification of plate stars with a catalogue.**

L. A. Vorontsova, G. P. Chejdo.
Avtometriya, 1974, No. 4, p. 103 - 111. In Russian. – Abstr. in Referativ. Zhurn. 51. Astron., 2.51.711 (1975).

031.403 **Programming system OPAL for analytical theories of celestial mechanics.** A. L. Kutuzov.
Pis'ma v Astron. Zhurn., Vol. 1, No. 1, p. 43 - 46 (1975). In Russian.
A programming system in computer M–20 language has been constructed which permits to perform all basic operations with literal Poisson series.

031.404 **Digital and electronic-analogue methods of isophotometric processing of astronomical photographic pictures.** M. F. Shabanov.
Astron. Zhurn. Akad. Nauk SSSR, Vol. 52, 404 - 414 (1975). In Russian. English translation in Soviet Astron., Vol. 19, No. 2.
The use of digital computers and electronic devices for

isophotometric processing of photographic pictures is considered. The algorithms of isophote construction using a computer are worked out. Examples of isophotometric processing of lunar and solar corona photographs by the electronic–analogue method are given. The effect of photographic noise on the accuracy of isophotometry of extended objects and stars is investigated.

031.405   Digital image processing.   B. R. Hunt.
Proc. IEEE, Vol. 63, 693 - 708 (1975).
A review of the field of digital image processing is presented, with concentration upon image formation and recording processes, digital sampling and digital image display, and with in-depth coverage of image coding and image restoration. New results in image restoration are also presented, covering restoration by use of an eye-model constraint and nonlinear restoration by maximization of the posterior density function.

031.406   Coordinate-meter for semi-automatic measuring of registergrams.   Yu. S. Romanov.
Problems of cosmic physics. Vyp. (No.) 9, (see 003.013), p. 159 - 162 (1974). In Russian.

031.407   On the use of algorithms of optimum filtration in systems of automatic tracking of moving objects.
A. E. Sazonov, V. A. Orlov.
VI Vses. soveshch. po probl. upr., 1974. Ref. dokl. Ch. 3. Moskva, Nauka, 1974, p. 295 - 298. In Russian. – Abstr. in Referativ. Zhurn. 51. Astron., 4.51.104 (1975).

031.408   The ESO telescope control systems.
J. van der Lans, S. Lorensen.
ESO, Technical Rep. No. 6, 17 pp. (1975).
A telescope control system is described providing programmable control of the tracking and setting velocities as well as accurate presetting to objects stored in a prepared catalog. This system is a subset of that developed for the ESO 3.6 m telescope and has been operational on the ESO 1 m photometric telescope on La Silla since 1973, and on the ESO Schmidt since 1974.

031.409   A simultaneous two-channel system for lunar occultation observations.
C. L. Morbey, J. M. Fletcher.
Publ. Dominion Astrophys. Obs., Victoria, Vol. 14, (No. 11), 271 - 281 = NRC No. 14374 (1974).
A data acquisition system is described that is used in the observations of disappearances or reappearances of stars occulted by the moon. Analyses of several representative occultation traces are presented.

031.410   The Grant machine.
J. Rickard, W. Nees, F. Middelburg.
European Southern Obs., Bull. No. 12, p. 5 - 19 (1975).
ESO-Chile has had in operation an automated Grant comparator-microdensitometer for about five years. The machine has two principal functions: (a) the accurate measurement of the positions of stellar lines for radial velocity (RV) determinations, (b) microdensitometry (MD) of stellar spectrograms.

031.411   The Santiago Grant machine and Zeeman coudé data reductions.   H. J. Wood.
European Southern Obs., Bull. No. 12, p. 21 - 24 (1975).
The ESO Chile long-screw Grant measuring machine, coupled to a small computer for data acquisition, has been used in a programme of line identification, radial velocity determination and the determination of the magnetic fields of stars. It is the purpose of this paper to briefly describe the operation of this system.

031.412   Programmes for the reduction of radial velocity

measurements.   A. Ardeberg, E. Maurice.
European Southern Obs., Bull. No. 12, p. 25 - 28 (1975).

031.413   A data acquisition programme for photometric measurements.   A. Ardeberg, F. Middelburg.
European Southern Obs., Bull. No. 12, p. 29 - 44 (1975).
The authors present the programme used for data acquisition with the HP 2114B computer in connection with the ESO photometer (de Vries, 1966). They include a discussion concerning the on-line reduction programmes for UBV photometry.

031.414   An automated photometric telescope.
A. P. Linnell, S. J. Hill.
Instrumentation in astronomy II, (see 012.021), p. 63 - 70 (1974).

031.415   A quadrant photosil autoguider for the 17/24 inch Schmidt telescope at the Institute of Astronomy Cambridge.   J. V. Jelley, A. N. Argue, J. P. Choisser.
Instrumentation in astronomy II, (see 012.021), p. 89 - 93 (1974).

031.416   Enhancement of ground based astronomical photographs by digital image processing.   P. H. Richter.
Instrumentation in astronomy II, (see 012.021), p. 215 - 221 (1974).

031.417   Infrared stellar image profiles by digital superposition.   D. R. Curott, B. Atwood.
Instrumentation in astronomy II, (see 012.021), p. 227 - 230 (1974).

031.418   Electro-optical processing of phased-array antenna data.   D. Casasent, F. Casasayas.
IEEE Trans. Aerospace Electron. Systems, Vol. AES - 11, 65 - 75 (1975).

031.419   A statistical technique for processing radio interferometer data.   G. D. Papadopoulos.
IEEE Trans. Antennas Propagation, Vol. AP - 23, 45 - 53 (1975).
This paper describes a maximum-likelihood processing technique which results in a synthesized beam that is more uniform, has lower sidelobes, and higher resolution than the conventional methods of data analysis. – MB

031.420   The coherent-optical processing of radio signals.
T. W. Cole.
Australian Physicist, Vol. 12, No. 3, p. 45 - 46 (1975).
A brief summary of the 100 to 200 MHz electro-optic spectrograph developed at CSIRO radiophysics. – JBS

031.421   Electro-optical processing in radio astronomy.
T. W. Cole.
Optica Acta, (GB), Vol. 22, 83 - 92 (1975).
Electro-optical techniques are beginning to find application in radio astronomy. A brief survey of radio-astronomy instruments and the processing of the radio signals is given. It is shown how a comparatively simple electro-optical processor can not only perform the functions of current, electronic devices but can also allow processing not practical by other techniques. As an illustration an electro-optical spectrograph is described and some observational results are shown. Possible future applications of such instruments are given.

An automated system for photoelectric photometry.
See Abstr. 034.070.

Advantages and disadvantages of various spectrum scanner systems.   See Abstr. 034.087.

Algorithms of celestial mechanics. (Materials for mathematical providing of computers). See Abstr. 042.082.

Use of computer literal programs to solve the main problem in satellite theory. See Abstr. 052.019.

Acquisition and description of Mariner 10 television science data at Mercury. See Abstr. 092.015.

IPL processing of the Mariner 10 images of Mercury. See Abstr. 092.016.

032 Astronomical Instruments

032.001 **3.6-m-Teleskop der Europäischen Süd-Sternwarte in Chile.**
SuW, Vol. 14, 1, 20 (1975).

032.002 **Schliff des Spiegels einer Schmidt-Kamera.** H. O. von Seggern.
SuW, Vol. 14, 58 - 60 (1975).

032.003 **A new three-mirror unobstructed reflector.** A. Kutter.
Sky Telescope, Vol. 49, 46 - 48 (1975).

032.004 **Notes on four great reflectors.**
Sky Telescope, Vol. 49, 79 - 81 (1975).

032.005 **More about the tri-schiefspiegler.** A. Kutter.
Sky Telescope, Vol. 49, 115 - 120 (1975).

032.006 **1-m-Spiegelteskope mit Ritchey-Chrétien- und Coudé-System aus Jena in Indien.** H. Artus.
Jenaer Rundschau (Jena Review), 20. Jahrgang, p. 3 - 9 (1975).

032.007 **Acht Jahre Arbeitserfahrungen mit dem 2-m-Spiegelteleskop aus Jena.** I. A. Aslanov.
Jenaer Rundschau (Jena Review), 20. Jahrgang, p. 13 - 17 (1975).

032.008 **Erste Erfahrungen mit der prismatischen Schmidt-Platte des Tautenburger 2-m-Spiegelteleskops.**
F. Börngen, N. Richter, P. Notni, H. Oleak, W. Wenzel.
Jenaer Rundschau (Jena Review), 20. Jahrgang, p. 18 - 25 (1975).

032.009 **Space reflectors for radar and astronomy.** J. C. Yater.
Applied Optics, Vol. 14, 526 - 536 (1975).
A new concept to utilize large flat optical reflecting surfaces in space to increase by several orders of magnitude the sensitivity and resolution of earth laser radar and astronomy measurements is described. The physical principles on which simple structures can maintain the optical reflectance gratings in space are derived, and the data processing requirements of the measurements are discussed. Space and ground system designs are given for a high resolution earth resources laser radar sensor, a synchronous earth and planetary science laser radar system, and an astronomy observation system including a variable very long compound grating interferometer system.

032.010 **Scanning Kirkpatrick-Baez X-ray telescope to maximize effective area and eliminate spurious images; design.** J. W. Kast.
Applied Optics, Vol. 14, 537 - 545 (1975).
The author considers the design of a Kirkpatrick-Baez grazing-incidence X-ray telescope to be used in a scan of the sky and analyzes the distribution of both properly reflected rays and spurious images over the field of view. To obtain maximum effective area over the field of view, it is necessary to increase the spacing between plates for a scanning telescope as compared to a pointing telescope. Spurious images are necessarily present in this type of lens, but they can be eliminated from the field of view by adding properly located baffles or collimators. Results of a computer design are presented.

032.011 **White light coronagraph in OSO-7.**
M. J. Koomen, C. R. Detwiler, G. E. Brueckner, H. W. Cooper, R. Tousey.
Applied Optics, Vol. 14, 743 - 751 (1975).

032.012 **Das Sonnen-Teleskop von Kitt Peak. (Robert R. McMath solar telescope).** E. Wiedemann.
Orion, 33. Jahrgang, p. 35 - 38 (1975).

032.013 **Opening the far frontier.**
K. W. Gatland, A. T. Lawton.
Spaceflight, Vol. 17, 51 - 54 (1975).
A review of the giant astronomical instruments being prepared in the Soviet Union and the United States for a new era of optical and radio observations.

032.014 **Southern hemisphere astronomy.**
Spaceflight, Vol. 17, 55 - 56 (1975). − Concerning the Anglo-Australian 3.9 m telescope.

032.015 **Penn State's new 60-inch telescope.**
F. R. Zabriskie.
Sky Telescope, Vol. 49, 219 - 221 (1975).

032.016 **A folded refractor exclusively for the sun.**
J. Dragesco.
Sky Telescope, Vol. 49, 323 - 326 (1975).

032.017 **Compensation for the field rotation of an altazimuth telescope.** N. N. Mikhel'son.
Astrofiz. Issled., Izv. Spets. Astrofiz. Obs., Vol. 7, p. 237 - 239 (1975). In Russian.
The necessity of using two photoelectric guides (two-coordinate and one-coordinate) to compensate for the field rotation in an altazimuth telescope is grounded.

032.018 **A miniaturized spectrohelioscope.** F. N. Veio.
Journ. British. Astron. Ass., Vol. 85, 242 - 244 (1975).

032.019 **20 cm Celestron Schmidt-Cassegrain telescoop.**
M. A. M. van Venrooy.
Zenit, Vol. 2, 89 - 91 (1975).

032.020 **Le télescope de six mètres du Caucase.**
J. Heidmann.
L'Astronomie, 89ᵉ année, p. 178 - 180 (1975).

032.021 **Sur un support de miroirs de télescopes à roulements à billes.** A. Hamon.
L'Astronomie, 89ᵉ année, p. 201 - 207 (1975).

032.022 **Interferometric recording of the deflections of towers and telescopes.** R. H. Hammerschlag.
Applied Optics, Vol. 14, 885 - 889 (1975).
Wind generates vibrations in towers and in telescopes placed in the open air. Interferometers for measuring these vibrations were developed. Applications to scale models of a telescope led to improvements of the telescope design. An optoelectronic system using three photocells for scanning the interferometer fringes allows the application of an inexpensive laser with several longitudinal modes as a light source for the interferometers.

032.023 **On some instrumental errors of Danjon's astrolabe.**
N. I. Panchenko.
Vrashchenie i prilivn. deformatsii Zemli. Vyp. (No.) 6. Kiev, "Nauk. dumka", 1974, p. 118 - 120. In Russian. – Abstr. in Referativ. Zhurn. 51. Astron., 3.51.148 (1975).

032.024 **Progress with CTIO's 4-meter telescope.**
V. M. Blanco.
Bull. American Astron. Soc., Vol. 7, 238 (1975). – Abstr. AAS.

032.025 **A three-mirror glancing incidence X-ray telescope for solar X-ray astronomy.** R. B. Hoover.
Bull. American Astron. Soc., Vol. 7, 351 (1975). – Abstr. AAS.

032.026 **Statistical study of the levels of a Zeiss zenith telescope.** V. K. Budz'ko.
Vrashchenie i prilivn. deformatsii Zemli. Vyp. (No.) 6. Kiev, "Nauk. dumka", 1974, p. 100 - 105. In Russian. – Abstr. in Referativ. Zhurn. 51. Astron., 4.51.201 (1975).

032.027 **Introduction of automatic CS for the transit instrument of the Astronomical Observatory at the Latvian State University.** K. Šteins.
Uch. Zap. Latv. Gos. Univ., Vol. 220, Astron., vyp. (No.) 11, (see 003.014), p. 3 - 16 (1975). In Russian.
It is shown that the correction value determined from astronomical observations by means of the photoelectric transit instrument is only slightly dependent on conditions of observation (within 4 ± 2 ms). This systematic error is not connected with wind direction effects on the correction value. A description is given of a grating with a moving window which permits use of the instrument for registration of mean transit moments.

032.028 **Mirror X-ray telescope for an astrophysical orbital station.**
I. L. Bejgman, L. A. Vajnshtejn, Yu. P. Vojnov, D. A. Goganov, N. I. Komyak, S. L. Mandel'shtam, I. P. Tindo, N. A. Shatskij, A. I. Shurygin.
Trudy Fiz. in-ta. AN SSSR, Vol. 77, 3 - 13 (1974). In Russian. Abstr. in Referativ. Zhurn. 62. Issled. kosmich. prostranstva, 5.62.134; 51. Astron., 6.51.265 (1975).

032.029 **Utilization and maintenance of 91 cm reflecting telescope at the Dōdaira Station.**
H. Shibasaki, T. Yamaguchi, N. Oshima, M. Noguchi.
Tokyo Astron. Obs., Report No. 65, Vol. 17, 298 - 315 (1975).

In Japanese.

032.030 **A kinematic mounting.** E. Høg.
Astron. Astrophys., Vol. 41, 107 - 109 (1975).
Kinematic mounting of a mirror on a face-plate by means of three cemented thin springs instead of the well-known "hole, slot and plane" system is described and tested.

032.031 **An improvement of the AFU camera at the Dodaira Station.** T. Yamaguchi, H. Nishimura.
Tokyo Astron. Obs., Report No. 65, Vol. 17, 325 - 328 (1975). In Japanese.

032.032 **Preliminary investigation of the variation of the instrumental parameters of the Charkov meridian circle in declination.**
K. N. Derkach, N. G. Zuev, K. N. Kuz'menko.
19th Astrometrical Conference 1972, (see 012.019), p. 134 - 137 (1975). In Russian.

032.033 **Investigation of the meridian circle of the Babelsberg Observatory.** J. Liebert, H. Sandig.
19th Astrometrical Conference 1972, (see 012.019), p. 145 - 148 (1975). In Russian.

032.034 **Some results of investigation of the eccentricity of the Toepfer meridian circle.** K. G. Gnevysheva.
19th Astrometrical Conference 1972, (see 012.019), p. 148 - 154 (1975). In Russian.

032.035 **Investigation of the circle division errors of the Pulkovo photographic vertical circle with the method of turning one circle relative to the other.**
V. A. Naumov.
19th Astrometrical Conference 1972, (see 012.019), p. 154 - 156 (1975). In Russian.

032.036 **Investigation of the flexure of the Pulkovo photographic vertical circle (PVC).**
A. A. Naumova, V. A. Naumov.
19th Astrometrical Conference 1972, (see 012.019), p. 156 - 158 (1975). In Russian.

032.037 **On some systematic errors of flexure determination with horizontal collimators at day-time.**
M. Miyatov.
19th Astrometrical Conference 1972, (see 012.019), p. 173 - 175 (1975). In Russian.

032.038 **Investigation of the pivots of the Kharkov meridian circle.** K. N. Derkach, N. G. Zuev, V. M. Kirpatovskij, K. N. Kuz'menko.
Vestn. Khar'kov, Univ., No. 117, (Ser. Astron., vyp. (No.) 9), p. 57 - 60 (1974). In Russian.

032.039 **Large Space Telescope (LST) observational capability of the solar system.** L. A. Klein.
Instrumentation in astronomy II, (see 012.021), p. 3 - 13 (1974).

032.040 **X-ray telescopes.** W. D. Antrim, Jr. R. L. Hall.
Instrumentation in astronomy II, (see 012.021), p. 15 - 34 (1974).

032.041 **A stabilized large-aperture far-infrared telescope gondola.** N. L. Hazen.
Instrumentation in astronomy II, (see 012.021), p. 41 - 47 (1974).

032.042 **Airborne infrared astronomy telescope.**
R. M. Cameron.

Instrumentation in astronomy II, (see 012.021), p. 49 - 55 (1974).

032.043 Current status of the multiple mirror telescope.
W. F. Hoffmann.
Instrumentation in astronomy II, (see 012.021), p. 57 - 62 (1974).

032.044 The NASA 48″ telescope.   R. F. Delgado.
Instrumentation in astronomy II, (see 012.021), p. 71 - 79 (1974).

Anglo-Australian telescope (AAT).

Journ. British Interplanet. Soc., Vol. 28, 59 - 63 (1975).

The astronomical telescope.   See Abstr. 003.032.

Anastigmatic catadioptric telescopes.
See Abstr. 031.011.

Reshaping and stabilization of astronomical images.
See Abstr. 031.021.

Personal and instrumental error jumps in the photo-electric time service device of the Astronomical Observatory of the Latvian State University in the years 1970 and 1973.
See Abstr. 044.010.

033   Radio Telescopes and Equipment

033.001 Automatic wind radar. I. Problems and principles of designing.
B. L. Kashcheev, V. A. Nechitajlenko.
Radiotekhnika. Resp. mezhved. temat. nauch.-tekhn. sb., 1974, vyp. (No.) 31, p. 38 - 45. In Russian. − Abstr. in Referativ. Zhurn. 51. Astron., 2.51.100 (1975).

033.002 Automatic wind radar. II. Preliminary data processing.   V. A. Nechitajlenko.
Radiotekhnika. Resp. mezhved. temat. nauch.-tekhn. sb., 1974, vyp. (No.) 31, p. 45 - 51. In Russian. − Abstr. in Referativ. Zhurn. 51. Astron., 2.51.101 (1975).

033.003 Radio telescope for spectral investigations in the wavelength range of 1.1 - 1.7 mm.
Yu. Yu. Kulikov, A. A. Shvetsov.
Astron. Zhurn. Akad. Nauk SSSR, Vol. 52, 199 - 201 (1975). In Russian. English translation in Soviet Astron., Vol. 19, No. 1.
    The design and operation modes of the 3-channel radio telescope are capable of spectrum measurements in the 1.1 - 1.7 mm region. Test results are given with preliminary data of atmospheric transparency measurements made in the vicinity of the telluric CO line.

033.004 On the possibility of variable profile antenna observations of sources at different altitudes with a fixed secondary mirror.   N. S. Soboleva, O. N. Shivris.
Soobshch. Spets. Astrofiz. Obs., *Zelenchukskaya,* vyp. (No.) 12, p. 51 - 64 (1974).
    A possibility to use a variable profile antenna at different altitudes without rearrangement of the secondary mirror is considered. Diagrams are presented to show the limits within which such observations are possible. This method allows to improve considerably the quality of relative coordinate measurement of sources.

033.005 A broad-band Dicke radiometer with a tuned radio-frequency receiver at 13 cm wavelength.
R. M. Kirakosyan, K. S. Mosoyan, O. B. Petrosyan, V. G. Gevorkyan.
Soobshch. Spets. Astrofiz. Obs., *Zelenchukskaya,* vyp. (No.) 12, p. 65 - 72 (1974). In Russian.
    A tuned radio-frequency radiometer with an input two-

stage parametric amplifier at 13-cm wavelength intended for RATAN-600 is designed and experimentally investigated. The sensitivity is about $0.01°K$ for $\tau = 2$ sec.

033.006 A new high-speed solar spectrograph for meter and decameter wavelengths.
S. R. Mosier, J. Fainberg.
Solar Physics, Vol. 40, 501 - 509 (1975).
    A new high-speed digital solar radio spectrograph has been designed and is being operated at the Clark Lake Radio Observatory in California. The spectrograph design attempts to optimize sensitivity, dynamic range, and frequency-time resolution while utilizing modern high-speed computer data-handling techniques. The system is described and initial data observations are presented.

033.007 Arecibo's giant radio telescope upgraded.
    Sky Telescope, Vol. 49, 140 - 142, 146 (1975).

033.008 The influence of the tracking error and antenna deformations on the phase stability of a radio interferometer.   A. F. Dravskikh.
Astrofiz. Issled., Izv. Spets. Astrofiz. Obs., Vol. 7, p. 199 - 206 (1975). In Russian.
    The definition of the interferometer baseline is discussed. It is proposed for the antennas with intersecting axes of rotation to use as a base the vector connecting the points of intersection of the axes. The formula which describes the influence of the tracking error of the phase of the interference signal is given.

033.009 Formation of the multibeam directivity diagrams of variable profile antennas (VPA).
N. L. Kajdanovskij.
Astrofiz. Issled., Izv. Spets. Astrofiz. Obs., Vol. 7, p. 207 - 213 (1975). In Russian.
    A possibility of formation of multibeam directivity diagrams of the variable profile antenna (VPA) is considered. The diagrams are to be used for simultaneous measurements of both coordinates of celestial bodies and also of brightness distribution of extended sources.

033.010 Automatic radio source tracking in observations

with the variable profile antenna.
N. L. Kajdanovskij.
Astrofiz. Issled., Izv. Spets. Astrofiz. Obs., Vol. 7, p. 214 - 222 (1975). In Russian.

A method of automatic source tracking in observations with the variable profile antenna (VPA) is considered. The gain of temperature sensitivity using different methods of tracking has been estimated.

033.011 Code converter for automation of radio astronomical observations.
M. N. Kajdanovskij, A. A. Stotskij.
Astrofiz. Issled., Izv. Spets. Astrofiz. Obs., Vol. 7, p. 223 - 225 (1975). In Russian.

A description is given of a code converter allowing to record results of radio-astronomical observations obtained with the help of a digital voltmeter EhTsV-3 on a standard punched tape by means of a perforator for direct introduction into a «Minsk-22» computer.

033.012 Automatic control system for flat and circular mirrors of a variable profile antenna.
G. S. Golubchin.
Astrofiz. Issled., Izv. Spets. Astrofiz. Obs., Vol. 7, p. 226 - 236 (1975). In Russian.

A control principle for circular and flat elements of a variable profile antenna upon the pattern of the radio telescope RATAN-600 is stated. The choice of a mode of control and translation of the angle of rotation of the shaft into a digital code are grounded. The control structure of circular and flat reflectors is described, the expected error of fixing the coordinates is determined.

033.013 The very large array. D. S. Heeschen.
Sky Telescope, Vol. 49, 344 - 351 (1975).

033.014 On some constructions of multi-frequency feed for solar radio telescopes.
V. M. Vyatkina, V. G. Ioganson.
Radioizluchenie Solntsa. Vyp. (No.) 3. Leningrad, Leningr. un-t, 1974, p. 109 - 118. In Russian. — Abstr. in Referativ. Zhurn. 51. Astron., 4.51.101 (1975).

033.015 An instrument for measuring deformations in large structures. J. W. Findlay, J. M. Payne.
IEEE Trans. Instrument. Measurement, Vol. IM-23, 221 - 226 = National Radio Astron. Obs., *Green Bank*, Repr. Ser. A, No. 393 (1974).

An instrument that has been developed for measuring the deformations in shape that result from the movement of a radio telescope reflector is described. A radar technique is used to measure distances from near the focal point of the reflector to selected points on the reflector surface. The short term accuracy of the instrument is ±0.003 in) and when used on the 140-ft telescope in Green Bank, W. Va., good agreement was found between calculated and measured deformations.

033.016 Conformity of reflector deformations of steerable radio telescopes.
P. D. Kalachev, A. N. Kozlov, V. B. Tarasov, V. N. Titov.
Trudy Fiz. in-ta AN SSSR, Vol. 77, 128 - 136 (1974). In Russian. — Abstr. in Referativ. Zhurn. 51. Astron., 5.51.86 (1975).

033.017 On extremal sizes of a steerable paraboloid reflector of a radio telescope. P. D. Kalachev.
Trudy Fiz. in-ta AN SSSR, Vol. 77, 137 - 146 (1974). In Russian. — Abstr. in Referativ. Zhurn. 51. Astron., 5.51.87 (1975).

033.018 Experimental investigation of constructions of

aerodynamic compensators with references to
paraboloids. V. E. D'yachkov, S. L. Myslivets, V. P. Nazarov.
Trudy Fiz. in-ta AN SSSR, Vol. 77, 147 - 156 (1974). In Russian. — Abstr. in Referativ. Zhurn. 51. Astron., 5.51.88 (1975).

033.019 Paraboloid antenna of a radio telescope with a radially balanced main reflector.
P. D. Kalachev, V. P. Nazarov, I. A. Emel'yanov, V. L. Shubeko, V. B. Khavaev.
Trudy Fiz. in-ta AN SSSR, Vol. 77, 157 - 162 (1974). In Russian. — Abstr. in Referativ. Zhurn. 51. Astron., 5.51.89 (1975).

033.020 Study of elastic properties of a steerable paraboloid antenna of a radio telescope.
P. D. Kalachev, V. E. D'yachkov.
Trudy Fiz. in-ta AN SSSR, Vol. 77, 163 - 178 (1974). In Russian. — Abstr. in Referativ. Zhurn. 51. Astron., 5.51.90 (1975).

033.021 System of automatic data reduction for radio astronomy. M. V. Konyukov, V. Yu. Bunakov.
Trudy Fiz. in-ta AN SSSR, Vol. 77, 179 - 186 (1974). In Russian. — Abstr. in Referativ. Zhurn. 51. Astron., 5.51.91 (1975).

033.022 The synchronous tracking drive system of the RTI-7.5/250 MVTU radio telescope.
A. A. Parshchikov, I. A. Emel'yanov.
Trudy Fiz. in-ta AN SSSR, Vol. 77, 187 - 192 (1974). In Russian. — Abstr. in Referativ. Zhurn. 51. Astron., 5.51.92 (1975).

033.023 The reflector-type radio telescope RTI-7.5/250 with a steerable paraboloid antenna.
P. D. Kalachev, V. P. Nazarov, A. A. Parshchikov, B. A. Rozanov.
Trudy Fiz. in-ta AN SSSR, Vol. 77, 193 - 210 (1974). In Russian. — Abstr. in Referativ. Zhurn. 51. Astron., 5.51.93 (1975).

033.024 Radio telescopes of large resolving power.
M. Ryle.
Science, Vol. 188, 1071 - 1079 (1975). — This article is the lecture which was delivered in Stockholm, Sweden, on 12 December 1974 when the author received the Nobel Prize in Physics.

033.025 An intercontinental array — a next-generation radio telescope. G. W. Swenson, Jr., K. I. Kellermann.
Science, Vol. 188, 1263 - 1268 (1975).

New techniques permit construction of a radio telescope with extreme angular resolution.

033.026 Reflectivity measurement of the 6-meter millimeter wave telescope. T. Miyaji.
Tokyo Astron. Obs., Report No. 66, Vol. 17, 377 - 388 (1975). In Japanese.

033.027 Wide-band square-law detector with excellent characteristics.
H. Sekiguchi, S. Aiba, H. Nakajima.
Tokyo Astron. Obs., Report No. 66, Vol. 17, 417 - 424 (1975). In Japanese.

033.028 Remarks on the Very Large Array (VLA) radio telescope project. W. B. Burton.
Journ. Franklin Inst., Vol. 298, 299 - 306 = National Radio Astron. Obs., *Green Bank*, Repr. Ser. B, No. 455 (1974).

The Very Large Array (VLA) radio telescope which is

presently under construction in central New Mexico is considered the highest priority new astronomical facility in the U.S.A. during the next decade. The instrument will consist of 27 separate parabolic antennas each of which will be movable along the arms of a Y-configuration.

033.029 Beam-separation constraints in a beam-switched radio telescope. N. Fourikis, B. Mac A. Thomas.
Proc. Instn. Radio Electron. Engineers Australia, Vol. 35, 199 - 201 = Div. Radiophys. C.S.I.R.O., Sydney, Radiophys. Publ. RPP 1759 (1974).

When a two-beam switching technique is used to minimise those fluctuations at the output of a radio telescope which are caused by tropospheric effects, the separation between the two beams of the radio telescope should be minimised. The authors examine the constraints imposed by the geometry of certain feeds, the electrical isolation between them and the distortion of the radiation patterns when such minimisation is sought.

033.030 Works in the field of construction of narrow pencil-beam radio telescopes at the Leningrad Polytechnical Institute.
B. V. Braude, N. A. Esepkina, V. Yu. Petrun'kin.
Trudy Leningr. politekhn. in-ta, 1974, No. 339, p. 130 - 143. In Russian. – Abstr. in Referativ. Zhurn. 51. Astron., 6.51.114 (1975).

033.031 The registration system of a very-long-baseline interferometer. L. R. Kogan, L. I. Matveenko.
AN SSSR. In-t kosmich. issled. Moskva, 1974, 34 pp. In Russian. – Abstr. in Referativ. Zhurn. 51. Astron., 6.51.122 (1975).

033.032 A high resolution decameter multichannel radio spectrograph. A. Lecacheux, C. Rosolen.
Astron. Astrophys., Vol. 41, 223 - 227 (1975).

The authors describe a decameter multichannel spectrograph which allows to have simultaneously a 20 kHz frequency resolution and a 0.1 ms time resolution. Its characteristics have been defined for the study of narrow band solar bursts ("split pairs", for instance) and of the fine structure of the Jovian emission ("modulation lanes" and "millisecond bursts"). They give samples of solar and Jovian storms recently observed with this receiver, in Nançay Observatory.

033.033 A radio interferometer as geodetical instrument.
B. A. Dubinskij.
19th Astrometrical Conference 1972, (see 012.019), p. 257 - 259 (1975). In Russian.

033.034 Radio astronomy at Bell Laboratories.
L. C. Tillotson.
Bell Lab. Record (*USA*), Vol. 52, 310 - 317 (1974).

Describes the way in which sensitive radio-receiving equipment designed at Bell Labs has turned out to be useful for radio astronomy, and has been responsible for several fundamental discoveries concerning the nature of the universe.

033.035 Spaceborne very-long-baseline radio astronomy interferometry. T. A. Heppenheimer.
Journ. Spacecraft and Rockets, (*USA*), Vol. 11, 268 - 270 (1974).

This paper describes a space-based VLBI synthetic aperture system. It is suggested that space operations offer simpler data processing, improved astronomical image information, and capacity for system growth.

033.036 An investigation of new primary feeds in the Effelsberg radio telescope.
G. F. Koch, O. Lochner, H. Scheffer, R. Wohlleben.
Nachrichtentechn. Zeitschr., (*Germany*), Vol. 28, No. 2, p. 41 -

46 (1975). In German.

In order to achieve a high aperture efficiency with low spillover, primary feeds are needed which have a sector-shaped beam for large reflector angular apertures. In addition to the waveguide feed with an offset choke structure, two types of coaxial feeds were installed in the radiotelescope in 1973, and the values obtained for aperture efficiency, radiation pattern and spillover were measured. The relatively great differences between the measured aperture efficiency values and those calculated for an ideal reflector are discussed.

033.037 Zur Positions-Kalibrierung des Effelsberger Radioteleskops. P. Brosche.
Allgemeine Vermessungs-Nachr. (AVN), Vol. 82, 35 - 37 (1975).

The read-off values of azimuth and elevation are corrected for instrumental errors. Taking into account further systematic effects of unknown origin, the r.m.s. errors in both coordinates can be reduced to $3''.4$ and $4''.4$ respectively (at a wavelength $\lambda = 6$ cm).

033.038 Sidelobe levels of a large radio telescope employing a combination of physical and resistive tapering.
D. Wynne, F. S. Chute, C. R. James.
IEEE Trans. Antennas Propagation, Vol. AP-23, 278 - 283 (1975).

The main beam sidelobe levels, and effective collecting area of a proposed radio telescope array, operating at 12 MHz, are considered.

033.039 A gain-stabilizing detector for use in radio astronomy. M. J. Yerbury.
Rev. Sci. Instruments, Vol. 42, 169 - 179 (1975).

The effect of excess noise on radio astronomy measurements and its origin are discussed. A new type of radiometer was designed. A theoretical analysis gives the basis of a practical gain stabilizing detector using an optimized noise-adding principle. The effective system temperature was derived.

033.040 Approximate formulae for microstrip transmission lines. L. W. Cahill.
Proc. IREE Australia, Vol. 35, (No.10), 317 - 321 (1974).

033.041 Submillimeter detection and mixing using Schottky diodes. H. R. Fetterman, B. J. Clifton, P. E. Tannenwald, C. D. Parker.
Applied Phys. Letters, Vol. 24, No. 2, p. 70 - 72 (1974).

Schottky diodes have been used as harmonic mixers in the 0.1 - 1.0 mm wavelength region. – *MWS*

033.042 The moon as source for G/T measurements.
K. G. Johannsen, A. Koury.
IEEE Trans. Aerospace Electron. Systems, Vol. AES – 10, 718 - 727 (1974).

A technique, making use of the moon as a calibrated noise source, for determining the receiving gain to system temperature (G/T) ratio for small aperture antennas. – *MWS*

033.043 Coupling between crossed-dipole feeds.
J. B. Andersen, H. Schjaer-Jacobsen, H. A. Lessow.
IEEE Trans. Antennas Propagation, Vol. AP–22, 641 - 646 (1974).

The antennas are used as feeds for a parabolic reflector, and the effect of coupling on the secondary fields is analysed. Polarization loss is discussed. – *ACM*

033.044 Rain-induced cross-polarization at centimeter and millimeter wavelengths. T. S. Chu.
Bell Syst. Techn. Journ., Vol. 53, 1557 - 1579 (1974).

033.045 The design of a bandpass filter with inductive strip –

planar circuit mounted in waveguide.
Y. Konishi, K. Uenakada.
IEEE Trans. Microwave Theory Techn., Vol. MTT–22, 869 - 873 (1974).
 The equivalent circuit of an inductive strip inserted in the middle of a waveguide parallel to the E-plane is analysed. A design theory for the bandpass filter of this type is derived from the equivalent circuit. – *JWB*

033.046 A proposed multiple-beam microwave antenna for earth stations and satellites. E. A. Ohm.
Bell Syst. Techn. Journ., Vol. 53, 1657 - 1665 (1974).

033.047 Radioheliograph of the Royal Observatory of Belgium. R. J. Wislez, R. Gonze.
Electr. Commun., Vol. 49, 218 - 226 (1974).
 Description of a Mills-cross type instrument operating at 408 MHz, and used mainly as a radioheliograph. – *JWB*

033.048 Pulsed Gunn-diode oscillator: 40 W at 16 GHz.
R. Stevens, D. Tarrant, F. A. Myers.
Electronics Letters, Vol. 10, 531 - 533 (1974).

033.049 Millimeter-wave integrated circuits cost less using dielectric waveguides. M. M. Chrepta, H. Jacobs.
Microwave Journ., Vol. 17, No. 11, p. 45 - 47 (1974).

033.050 A high speed serial FFT processor.
L. J. De Lorenzo, P. Gottlieb.
Proc. IREE Australia, Vol. 35, 353 - 359 (1974).

033.051 Thick film techniques for hybrid integrated microwave circuits. W. Funk, W. Schilz.
Radio and Electronic Engineer, (GB), Vol. 44, 504 - 508 (1974).
 Fabrication techniques of thick-film integrated microwave circuits for frequencies up to 10 GHz are discussed. – *SS*

033.052 Large lateral feed displacements in a parabolic reflector. W. A. Imbriale, P. G. Ingerson, W. C. Wong.
IEEE Trans. Antennas Propagation, Vol. AP - 22, 742 - 745 (1974).
 Radiation patterns computed from both scalar and vector theories are compared with experimental results and the range of validity of the approximate analysis is indicated. – *ACM*

033.053 A new multimode rectangular horn antenna generating a circularly polarized elliptical beam.
C. C. Han, A. N. Wickert.
IEEE Trans. Antennas Propagation, Vol. AP -22, 746 - 751 (1974).

033.054 Dual-bandwidth loop speeds phase lock.
A. T. Anderson, D. E. Sanders, R. S. Gordy.
Electronics, Vol. 48, 116 - 117 (1975).
 A technique for extending the acquisition on locking bandwidth of a phase locked loop, by using a dual loop filter system with 'soft' switching between the two. – *MWS*

033.055 Take the trouble out of diode mounting.
R. L. Eisenhart.
Microwaves, Vol. 13, No. 11, p. 78, 80 - 81 (1974).
 Rapid technique for determining the equivalent circuit elements for the typical coaxial diode mount. Curves are developed for line impedances ranging from 10 to 70 ohms. – *MWS*

033.056 The dual-mode filter – a realization.
R. V. Snyder.
Microwave Journ., Vol. 17, No. 12, p. 31 - 33, 63 (1974).

Concerned primarily with tunable band-pass/-stop filters for short cm wavelengths. – *KJW*

033.057 New broadband balun. H. Bex.
Electronics Letters, Vol. 11, No. 2, p. 47 - 48 (1975).
 A modified rat-race with an open-circuit at the uncoupled port. – *TWC*

033.058 A corrugated horn antenna using V-shape corrugations. C. A. Mentzer, L. Peters, Jr., F. B. Beck.
IEEE Trans. Antennas Propagation, Vol. AP - 23, 93 - 97 (1975).
 This paper investigates the properties of corrugated horns with V-shaped corrugations. It is suggested that for millimetre antennas other corrugation shapes than rectangular are desirable.

033.059 Harmonically pumped stripline down-converter.
M. V. Schneider, W. W. Snell, Jr.
IEEE Trans. Microwave Theory Techn., Vol. MTT - 23, 271 - 275 (1975).

033.060 Distortion in variable-capacitance diodes.
R. G. Meyer, M. L. Stephens.
IEEE Journ. Solid-State Circuits, Vol. SC-10, 47 - 54 (1975).
 Distortion in variable capacitance diodes is analyzed and methods for its reduction given. – *DJC*

033.061 Radiation pattern of a corrugated conical horn in terms of Laguerre-Gaussian functions.
C. Aubry, D. Bitter.
Electronics Letters, Vol. 11, 154 - 156 (1975).

033.062 Cross-polarization measurements using a composite feed. S. I. Ghobrial, A. R. Obeid, F. Nakhla.
Electronics Letters, Vol. 11, 85 - 86 (1975).
 A report of measurements made on a deep (f/d = 0.25) paraboloid using a five dipole feed to minimize cross-polarization. A 9 dB improvement over a single dipole was achieved. – *DNC*

033.063 Millimeter-wave microstrip oscillators.
B. S. Glance, M. V. Schneider.
IEEE Trans. Microwave Theory Techn., Vol. MTT-22, 1281 - 1283 (1974).

033.064 Millimeter-wave receivers and their applications in radio astronomy. T. G. Phillips, K. B. Jefferts.
IEEE Trans. Microwave Theory Techn., Vol. MTT-22, 1290 - 1292 (1974).
 Discussions from the point of view of applications to line radio astronomy and descriptions of Kitt Peak 'heterodyne-bolometer receivers', a recent version of which has a double-sideband noise temperature of 300 K at around 230 GHz. – *SS*

033.065 A low-noise room-temperature 12-GHz parametric amplifier. S. D. Lacey, B. T. Hughes, J. C. Vokes.
IEEE Trans. Microwave Theory Techn., Vol. MTT-22, 1329 - 1331 (1974).

033.066 Antenna investigation of a statistically inhomogeneous atmosphere.
N. A. Armand, A. N. Lomakin, V. A. Sarkisyanz.
Radio Sci., (USA), Vol. 10, 87 - 95 (1975).

033.067 Very wideband corrugated horns. Z. Frank.
Electronics Letters, Vol. 11, 131 - 133 (1975).
 An expression for the input impedance of a short-circuited, linearly-tapered transmission line is derived. Use of tapered

slots in a corrugated horn may lead to bandwidth greater than 3 to 1. Experimental results are given for a horn operating over 7.5–18 GHz. – *ACM*

033.068 **Operational amplifier integrators for the measurement of the delay times of microwave transistors.**
D. D. Cohen, R. A. Zakarevicius.
IEEE Journ. Solid-State Circuits, Vol. SC-10, 19 - 27 (1975).

A simple low-frequency technique for the measurement of subnanosecond delay times with an operational amplifier connected as a summing integrator is described. – *DJC*

033.069 **Wideband negative-current mirror.** B. Gilbert.
Electronics Letters, Vol. 11, 126 - 127 (1975).

A wideband current mirror circuit employing only NPN transistors is claimed to have several advantages over well known circuits employing both NPN and PNP types. – *DJC*

Opening the far frontier. See Abstr. 032.013.

Radio astronomy and astrometry. See Abstr. 041.045

Long-range communications with Pioneer 10 at Jupiter. See Abstr. 053.007.

Observations of Jovian S-bursts with an electro-optical radio spectrograph. See Abstr. 099.017.

034 Astronomical Accessories (Spectrometers, Photometers, etc.)

034.001 The Lockheed diode array magnetograph.
R. C. Smithson.
Solar Physics, Vol. 40, 241 - 246 (1975).
A new magnetograph using a solid state monolithic linear silicon diode array has been constructed at Lockheed Solar Observatory. This magnetograph uses a digital image processor, and makes data available both in digital and analog form. The diode array detector is capable of a signal-to-noise ratio of 2000:1 or better when cooled to a temperature of −40 deg centigrade. Thus, intensity differences of the order of one part in a thousand may easily be detected without signal averaging. This instrument may be considered a prototype for an instrument using a two-dimensional array. The magnetograph is now fully operational, and is being used to produce data for statistical studies of solar magnetic field diffusion.

034.002 Protection of solar magnetograph photomultipliers from overloading. N. I. Kobanov.
Issled. po geomagnetizmu, aehron. i fiz. Solntsa. Vyp. (No.) 31. Moskva, Nauka, 1974, p. 104 - 106. In Russian. − Abstr. in Referativ. Zhurn. 51. Astron., 2.51.96 (1975).

034.003 Device for light polarization control by an electro-optical switch. N. I. Kobanov.
Issled. po geomagnetizmu, aehron. i fiz. Solntsa. Vyp. (No.) 31. Moskva, Nauka, 1974, p. 107 - 109. In Russian. − Abstr. in Referativ. Zhurn. 51. Astron., 2.51.97 (1975).

034.004 Drift in interference filters. J. Walker.
Nature, Vol. 253, 592 - 593 (1975).

034.005 Infrared detectors in remote sensing.
H. Levinstein, J. Mudar.
Proc. IEEE, Vol. 63, 6 - 14 (1975).
The history of the development of infrared detectors is briefly reviewed. The parameters describing the performance of detectors are described and the measurement procedures outlined. Recent developments in intrinsic HgCdTe and PbSnTe, extrinsic silicon, and pyroelectric detectors are reviewed. Several applications of detector technology to specific remote sensing problems are discussed.

034.006 Infrared detector arrays by new technologies.
C. Corsi.
Proc. IEEE, Vol. 63, 14 - 26 (1975).

034.007 High-performance $8-14$-μm $Pb_{1-x}Sn_xTe$ photodiodes. C. A. Kennedy, K. J. Linden, D. A. Soderman.
Proc. IEEE, Vol. 63, 27 - 32 (1975).

034.008 Heterojunction III − V alloy photodetectors for high-sensitivity 1.06-μm optical receivers.
R. C. Eden.
Proc. IEEE, Vol. 63, 32 - 37 (1975).

034.009 Erfahrungen mit dem Universal-Astro-Gitterspektrographen aus Jena. E. H. Geyer.
Jenaer Rundschau (Jena Review), 20. Jahrgang, p. 26 - 30 (1975).

034.010 A cross-dispersed echelette spectrograph and a study of the spectrum of the QSO 1331 + 170.
R. F. Carswell, R. L. Hilliard, P. A. Strittmatter, D. J. Taylor, R. J. Weymann.
Astrophys. Journ., Vol. 196, 351 - 361 (1975).
A simple modification of the Cassegrain spectrograph of the Steward Observatory 90-inch (2.3 m) telescope is described. The system utilizes a grating operating in the 5−13th orders and a quartz prism to cross-disperse these orders in an echelle format. Complete spectral coverage from 3100 Å to about 9000 Å on a 40-mm image tube is possible with a dispersion of 44 Å mm^{-1} at 4200 Å. With this equipment, an image-tube spectrogram of the absorption-rich quasi-stellar object 1331 + 170 has been obtained. A large number of previously undetected absorption lines are shown to be present in this object.

034.011 Diffraction grating ruling engine with piezoelectric drive. I. R. Bartlett, P. C. Wildy.
Applied Optics, Vol. 14, 1 - 3 (1975).

034.012 Improvement of birefringent filters. 3: Effect of errors on wide field elements. A. M. Title.
Applied Optics, Vol. 14, 445 - 449 (1975).
The properties of nontunable and tunable Lyot wide field elements are examined when the components of the elements deviate from their proper values. Special emphasis is put on determining what variations cause light to be transmitted at the transmission minima. The analysis shows that the nine- and ten-element plastic waveplates described in Paper 2 of this series can be used to make a Lyot filter that is tunable from 3500 Å to 10,000 Å.

034.013 Johnson noise limited operation of photovoltaic InSb detectors.
D. N. B. Hall, R. S. Aikens, R. Joyce, T. W. McCurnin.
Applied Optics, Vol. 14, 450 - 453 (1975).
Photovoltaic indium antimonide detectors have been operated at temperatures $\lesssim 77$ K with sufficiently low background radiation levels that Johnson noise limited performance is realized. Under such conditions the noise equivalent power (NEP) is completely determined by the detector operating temperature, resistance, and quantum efficiency. Optimization of these parameters in the manufacture of commercially available detectors has led to 5-μm NEP's as low as 10^{-15} W. The particular preamplifier is critical to the achievement of Johnson noise limited operation and is described in detail.

034.014 Versatile nebular insect-eye Fabry-Perot spectrograph. J. Meaburn.
Applied Optics, Vol. 14, 465 - 469 (1975).
The design and performance of an insect-eye F.P. spectrograph used on the 249-cm Isaac Newton telescope, which can also be converted into a nebular filter camera, is presented. This device has several novel features, including a pressure-controlled optically contacted etalon and an image tube as a detector.

034.015 New stigmatic, coma-free, concave-grating spectrograph. J.-D. F. Bartoe, G. E. Brueckner.
Journ. Optical Soc. America, Vol. 65, 13 - 21, with a correction, p. 617 (1975).
Tandem use of two classical concave gratings makes it possible to design a double-dispersion spectrograph that is essentially free of astigmatism and coma over a large wavelength range near normal incidence. The first grating is used as a Wadsworth collimator. Light of different wavelengths is dispersed by the Wadsworth collimator so that it illuminates different portions of the second grating. The second grating is placed so that it acts as a Wadsworth camera, in which the light bundle of a certain wavelength illuminating a particular section of the second grating is diffracted along the local normal of that section. In this way, the Wadsworth condition

for stigmatic and coma-free imaging is almost fulfilled for all wavelengths. Only two reflecting surfaces are needed. The instrument is a double-dispersion spectrograph with additive dispersion. It does not use an intermediate slit, but has the stray-light-suppression characteristics of such a mount. A comparison of its imaging capabilities with other stigmatic concave-grating spectrographs is presented.

034.016 **Calculation of arbitrary-order diffraction efficiencies of thick gratings with arbitrary grating shape.**
S. F. Su, T. K. Gaylord.
Journ. Optical Soc. America, Vol. 65, 59 - 64 (1975).

034.017 **Infrared multi-color photometry and polarimetry.**
H. Okuda, T. Maihara, S. Sato.
Mem. Fac. Sci.,Kyoto Univ., Ser. Phys., Astrophys., Geophys., Chem., Vol. 34, 229 - 241 (1974).

A multi-color photometer has been built for ground based infrared astronomical observations in near infrared i.e. between 1 and 4 microns. A liquid nitrogen cooled lead sulphide photoconductor is used for the detector. It can be used also for a polarimeter by inserting a rotating polarizer. The signal from the detector is recorded on a magnetic tape simultaneously with the mark indicating the rotation angle. The data are digitized by an A-D converter and analysed by computer processing.

034.018 **A balloon borne liquid nitrogen cooled infrared radiometer.** T. Maihara, H. Okuda, T. Sugiyama.
Mem. Fac. Sci.,Kyoto Univ., Ser. Phys., Astrophys., Geophys., Chem., Vol. 34, 341 - 351 (1974).

An infrared radiometer has been built to observe the diffuse galactic light in the infrared region. The whole telescope is cooled by liquid nitrogen to reduce the intense emission from the telescope. It was launched by a balloon on Oct. 9, 1971. Small size irregularities were found to be present in the OH airglow emission. This is a great obstacle for the observation of the galactic light. The observation of the OH airglow emission would give a useful probe for the studies of dynamic properties of the thermosphere.

034.019 **A Fourier transform spectrometer for observations of stars in the intermediate infrared.**
R. R. Treffers.
Astron. Astrophys., Vol. 38, 345 - 350 (1975).

A description is given of a rapid scanning Fourier transform spectrometer used to obtain spectra of stars in the 10 and 20 micron atmospheric windows, with resolutions of as high as 1 cm^{-1} The performance of this instrument and the problems of high background radiation are discussed.

034.020 **A double beam photoelectric photometer for astronomical applications.**
E. H. Geyer, M. Hoffmann.
Astron. Astrophys., Vol. 38, 359 - 362 (1975).

The double beam photoelectric photometer allows two stars within the field of a Cassegrain telescope to be monitored. The measured brightness differences of the two stars are largely independent of atmospheric transparency changes and thin cirrus clouds. The photometer is especially suited for observations of the light curves of rapid variables. The achieved time resolution of 5−10 s with a 1% photometric accuracy is typical for 12 th magnitude stars at a 1 m telescope.

034.021 **Fraunhofer selective diffraction produced by structure elements in sodium vapour.**
Y. Öhman, L. Lindberg.
Journ. Quant. Spectrosc. Radiat. Transfer, Vol. 15, 283 - 289 (1975).

Observations are reported of selective diffraction produced by structure elements in sodium vapour separated from intermediary empty spaces by thin sheets of iron. The phenomena observed can be explained by the phase shifts in sodium vapour due to anomalous dispersion.

034.022 **Characteristics of the silicon diode vidicon.**
P. Crane, M. Davis.
Publ. Astron. Soc. Pacific, Vol. 87, 207 - 216 (1975).

A series of laboratory tests of the silicon diode vidicon has yielded valuable information of particular importance to astronomers. The results confirm that these devices are extremely good sensors for low-contrast high-optical flux problems in astronomy.

034.023 **A stable precision astronomical electrometer amplifier.** J. P. Oliver.
Publ. Astron. Soc. Pacific, Vol. 87, 217 - 220 = Rosemary Hill Obs., Univ. Florida, Gainesville, Contr. No. 57 (1975).

A highly reliable DC electrometer amplifier is described. The five-magnitude range of fine gain steps, built-in calibration and sky bucking source, and easily adjustable time constant all make this device particularly useful and easy to work with at the telescope.

034.024 **Photomultipliers: their cause and cure.**
A. T. Young.
Astrophysics. Part A, (see 003.002), p. 1- 94 (1974).

An idealized photomultiplier; Basic physics of photomultipliers: photoemission, secondary emission; Real photomultipliers: materials and construction, undesirable properties of photomultipliers; Photomultipliers and system components: pulse-height distributions, detection: the signal/ noise ratio.

034.025 **Other components in photometric systems.**
A. T. Young.
Astrophysics. Part A, (see 003.002), p. 95 - 122 (1974).

Optical systems: the telescope and atmosphere, filters and spectrographs; Calibration problems and standard sources: light sources, electronic systems; Principles of photometer design.

034.026 **Two-dimensional electronic recording. I. Phosphor output image tubes.** E. J. Wampler.
Astrophysics. Part. A, (see 003.002), p. 237 - 251 (1974).

Principles of operation; astronomical applications; future developments.

034.027 **Two-dimensional electronic recording. II. Electrographic tubes.** G. E. Kron.
Astrophysics. Part A, (see 003.002), p. 252 - 276 (1974).

Electronic focusing; magnetic shielding, the Lallemand electronic camera; the US Navy electronic camera; the McGee spectracon; other developments; emulsions and development; characteristics of the electronic camera.

034.028 **Two-dimensional electronic recording. III. Television systems for astronomical applications.**
J. L. Lowrance, P. Zucchino.
Astrophysics. Part A, (see 003.002), p. 277 - 313 (1974).

Review of TV sensor types; system design considerations; integrating TV system operation; television data processing; observational results with integrating television.

034.029 **Polarization techniques.** K. Serkowski.
Astrophysics. Part A, (see 003.002), p. 361 - 414 (1974).

Analyzers for linearly polarized light; Retarders; Depolarizers; Optimum design of an astronomical polarimeter; Instrumental corrections; Astronomical polarimetry in the future: television and image tube techniques.

034.030 The instrumentation and techniques of infrared photometry. F. J. Low, G. H. Rieke.
Astrophysics. Part A, (see 003.002), p. 415 - 462 (1974).
Detectors; Associated apparatus; Telescope design; Modulation and space filtering techniques; Atmospheric limitations; The infrared photometric system; Observing procedure.

034.031 Diffraction grating instruments. D. J. Schroeder.
Astrophysics. Part A, (see 003.002), p. 463 - 489 (1974).
General spectrometer considerations; diffraction gratings; grating spectrometers; echelle spectrometers.

034.032 Fourier spectrometers.
H. W. Schnopper, R. I. Thompson.
Astrophysics. Part A, (see 003.002), p. 491 - 529 (1974).
Historical background; Theory of Fourier transform spectroscopy: the Michelson interferometer, mathematics of Fourier transform spectroscopy, resolution and apodization, sampling, off-center interferograms, phase correction, noise; Fourier spectroscopy in practice: the interferometer, data systems, optics, limiting magnitude.

034.033 Fabry-Perot instruments of astronomy.
F. L. Roesler.
Astrophysics. Part A, (see 003.002), p. 531 - 569 (1974).
The ideal Fabry-Perot interferometer; Application of the Fabry-Perot interferometer as a spectrometer; Multiple Fabry-Perot spectrometers; Observation of astronomical sources with Fabry-Perot spectrometers: Examples of basic Fabry-Perot spectrometer design for astronomical observations; Adjustment and calibration of Fabry-Perot spectrometers; Comparison with other instruments.

034.034 Eenvoudige belichtingsmeter.
G. P. Pols, L. Delvoye.
Zenit, Vol. 2, 97 - 100 (1975).

034.035 Cryogenic infrared grating spectrometer.
D. P. McNutt, K. Shivanandan, M. Daehler, P. D. Feldman.
Applied Optics, Vol. 14, 1116 - 1119 (1975).
A liquid-helium-cooled Ebert-Fastie grating spectrometer for use in a sounding rocket is described. Twelve detectors and associated filters separate the $5-70$-μm spectral range into twelve intervals, each of which is scanned as the grating is rotated. The instrument was launched into an aurora from Fort Churchill, Canada, but a cryogenic failure occurred early in the flight, and only a small amount of data was obtained.

034.036 Self-scanned Digicon: a digital image tube for astronomical spectroscopy.
R. G. Tull, J. P. Choisser, E. H. Snow.
Applied Optics, Vol. 14, 1182 - 1189 (1975).
The authors have successfully fabricated, tested, and operated a digital image tube consisting of a magnetically focused image intensifier tube in which a self-scanned linear array of 1024 silicon photodiodes operating in the EBS mode serves as the photoelectron image detector, amplifier, and intermediate image storage device. Integral on-chip MOS shift registers driven by an external clock sequentially interrogate the photodiodes through MOS FET multiplex switch arrays. Each output frame is a sequence of 256 analog pulses on each of four video lines. Laboratory and observing tests show that the output signal is photoelectron shot noise limited over at least a range of $1-10^5$ detected photoelectrons per picture element, indicating that single photon detection is achieved. High-resolution astronomical spectroscopy has been carried out in the coudé spectrograph of the 2.7-m telescope at McDonald Observatory. Examples are shown.

034.037 Additional noise in multistage image converters generated by a signal.
K. L. Mench, M. G. Sosonkin.
Astrometriya i Astrofizika, *Kiev*, vyp. (No.) 24, (see 003.003), p. 120 - 124 (1974). In Russian.

034.038 A photometer for calibration of star imitators.
D. I. Stepanov, E. S. Kupriyanov, N. N. Raimov.
Optiko-mekh. prom-st'. Nauch.-tekhn. zh., 1974, No. 8, p. 29 - 33. In Russian. – Abstr. in Referativ. Zhurn. 51. Astron., 3.51.193 (1975).

034.039 Digital video system for rapid spectral and spatial mapping of planets.
B. J. Duncan, T. D. Faÿ, W. Wamsteker.
Bull. American Astron. Soc., Vol. 7, 237 (1975). – Abstr. AAS.

034.040 A quantum limited near infrared vidicon camera system. M. T. Sandford II, J. P. Jekowski.
Bull. American Astron. Soc., Vol. 7, 269 - 270 (1975). Abstr. AAS.

034.041 Initial results from a scanning Stokes polarimeter.
L. L. House, T. G. Baur, H. K. Hull.
Bull. American Astron. Soc., Vol. 7, 349 (1975). – Abstr. AAS.

034.042 A photometer for measuring the brightness of features near the extreme solar limb.
G. A. Chapman.
Bull. American Astron. Soc., Vol. 7, 351 (1975). – Abstr. AAS.

034.043 A spectrum scanning Stokes polarimeter.
T. G. Baur, L. L. House, H. K. Hull.
Bull. American Astron. Soc., Vol. 7, 351 (1975). – Abstr. AAS.

034.044 Superheterodyne receiver for spectroscopy at 10 microns.
D. Buhl, M. Mumma, T. Kostiuk, T. A. Clark.
Bull. American Astron. Soc., Vol. 7, 390 (1975). – Abstr. AAS.

034.045 De spektroheliograaf en zijn mogelijkheden.
G. Jager.
Zenit, Vol. 2, 145 - 149 (1975).

034.046 Twee F-24 camera's geautomatiseerd.
H. Betlem.
Zenit, Vol. 2, 178 - 180 (1975).

034.047 Photoelectric device for photometry of astronomical objects in the atmospheric transparency window at $1.7 - 2.6\,\mu$m. V. M. Kolesnikov, A. G. Shcherbakov.
Vestn. Belorus. un-ta, 1974, ser. 1, No. 3, p. 77 - 79. In Russian. – Abstr. in Referativ. Zhurn. 51. Astron., 4.51.253 (1975).

034.048 Two-channel cooled receiver for board telescopes of the submillimeter region.
A. A. Kobzev, V. I. Lapshin, S. V. Solomonov, A. S. Khajkin.
Trudy Fiz. in-ta. AN SSSR, 1974, p. 77, 80 - 84. In Russian. Abstr. in Referativ. Zhurn. 62. Issled. kosmich. prostranstva, 4.62.122 (1975).

034.049 Determination of the value of a revolution of a micrometer from observations of large arcs in declination. V. K. Budz'ko.
Vrashchenie i prilivn. deformatsii Zemli. Vyp. (No.) 6. Kiev, "Nauk. dumka", 1974, p. 125 - 126. In Russian. – Abstr. in Referativ. Zhurn. 52. Geodeziya i Aehrosemka, 4.52.123 (1975).

034.050 Advances in instrumentation for stellar photometry. D. W. Latham.
Dudley Obs. Rep. No. 9, (see 012.008), p. 111 - 119 (1975).

Over the past several years enormous progress has been made in the development of better light detectors and data-handling techniques which promise huge gains over older methods when applied to stellar photometry. Although a few stellar photometrists reported that they were changing over to the new photomultipliers with III—V cathodes or to the new type IIIa—J and 127 plates from Kodak, these new detectors have not yet come into general use. This paper summarizes the advantages and a few potential problems of these new detectors.

034.051 UBVRI photometry with a single photomultiplier. J. D. Fernie.
Dudley Obs. Rep. No. 9, (see 012.008), p. 399 (1975). Abstract.

034.052 SIT vidicon with magnetic intensifier for astronomical use.
S. A. Colgate, E. P. Moore, J. Colburn.
Applied Optics, Vol. 14, 1429 - 1436 (1975).

The authors have coupled a magnetic intensifier to an intensified silicon target vidicon. The output and controls are digital and designed to optimize the computer analysis of the astronomical data. A coarse sweep 128×128 lines gives a large ($\simeq 10:1$) pixel to pixel modulation. Photoelectron noise is comparable to readout noise. Tube preparation is by a single light flash, ten beam erases, and a shift in cathode potential before integrate. The real time numerical picture analysis programs find stars, integrate their intensity, determine background, calculate variations, etc. and, in general, are far more useful than visual displays. Star intensities are linear with integrate time and reproducible to 3%.

034.053 Silicon diode array vidicons at the telescope: observational experience.
T. B. McCord, M. F. Frankston.
Applied Optics, Vol. 14, 1437 - 1446 (1975).

The authors have used silicon vidicon tubes in a two-dimensional integrating imaging device at the telescope to obtain photometric images of astronomical objects. In this article the authors report on the performance of the imaging system they have experienced under actual observing conditions. The imaging system and the procedures they have devised to use it are described. Exposures reproduce to better than 0.05% of full scale intensity for each picture element. The response of the entire imaging system is linear with $\gamma = 1$ for at least nearly 5 orders of magnitude to within at least 0.44% of full scale intensity and probably much better. Images made through the telescope of extended sources agree photometrically with photomultiplier measurements to about 1%. Photometry of stars reproduces to better than 0.5%.

034.054 Spatially multiplexed infrared camera. D. W. Davies.
Journ. Optical Soc. America, Vol. 65, 707 - 711 (1975).

A spatially multiplexed infrared camera is described. The camera records an image by measuring the coefficients of an orthogonal series expansion of the two-dimensional radiance distribution. By choice of appropriate functions, significant gains in the picture signal-to-noise ratio can be realized. The author has constructed a camera utilizing this multiplexing principle to record images of astronomical objects at infrared wavelengths, where conventional scanning systems perform badly because of low signal levels. Test results verify that the camera exhibits the increase of efficiency expected of a multiplexing system, allowing exposure times two order of magnitude shorter than necessary for comparable scanning systems.

034.055 A proposed successor to the Narrabri stellar intensity interferometer. J. Davis.
Dudley Obs. Rep. No. 9, (see 012.008), p. 199 - 214 (1975).

The Australian Federal Government announced in its budget for 1974—75 that it had approved in principle a grant to the University of Sydney to construct a large stellar interferometer. It has included in its budget an initial grant for a detailed design study of the instrument. The background to this announcement, and the preliminary design and potential of a new instrument, are described.

034.056 Extending the interferometer. A. L. Fymat.
Phys. Today, Vol. 28, No. 6, p. 15, 80 - 81, 83 (1975). — Letter.

034.057 Comparison between stellar interferometers. H. Van de Stadt.
Space Sci. Rev., Vol. 17, 673 - 676 (1975). — Presented at the workshop on coherent detection in astronomy, held at Rhenen, The Netherlands, 25 and 26 April 1974 — see 012.009.

034.058 Berkeley heterodyne interferometer. A. Betz.
Space Sci. Rev., Vol. 17, 677 - 680 (1975). — Presented at the workshop on coherent detection in astronomy, held at Rhenen, The Netherlands, 25 and 26 April 1974 — see 012.009.

034.059 Technological requirements associated with a shuttle-borne infrared interferometer. J. B. Farrow.
Space Sci. Rev., Vol. 17, 681 - 686 (1975). — Presented at the workshop on coherent detection in astronomy, held at Rhenen, The Netherlands, 25 and 26 April 1974 — see 012.009.

034.060 An infrared astrometric interferometer. J. Gay, A. Journet.
Space Sci. Rev., Vol. 17, 687 - 688 (1975). — Presented at the workshop on coherent detection in astronomy, held at Rhenen, The Netherlands, 25 and 26 April 1974 — see 012.009.

034.061 Signal-to-noise ratio and other characteristics of heterodyne radiation receivers. T. G. Blaney.
Space Sci. Rev., Vol. 17, 691 - 702 (1975). — Presented at the workshop on coherent detection in astronomy, held at Rhenen, The Netherlands, 25 and 26 April 1974 — see 012.009.

034.062 Investigation of a superheterodyne mixing system with an 'open structure' mixer at 337 μm wavelength. W. Reinert.
Space Sci. Rev., Vol. 17, 703 - 707 (1975). — Presented at the workshop on coherent detection in astronomy, held at Rhenen, The Netherlands, 25 and 26 April 1974 — see 012.009.

034.063 Detectors for infrared heterodyne mixing and detection. T. De Graauw.
Space Sci. Rev., Vol. 17, 709 - 719 (1975). — Presented at the workshop on coherent detection in astronomy, held at Rhenen, The Netherlands, 25 and 26 April 1974 — see 012.009.

034.064 Substantial errors of a positional contact micrometer. A. V. Gozhij.
Vrashchenie i prilivn. deformatsii Zemli, Vyp. (No.) 6. Kiev. Nauk. dumka, 1974, p. 127 - 137. In Russian. — Abstr. in Referativ. Zhurn. 51. Astron., 5.51.197 (1975).

034.065 Polarization devices of the submillimeter region. V. I. Lapshin.
Trudy Fiz. in-ta AN SSSR, Vol. 77, 117 - 127 (1974). In Russian. — Abstr. in Referativ. Zhurn. 51. Astron., 5.51.233; 62. Issled. kosmich. prostranstva, 5.62.131 (1975).

034.066 Apparatus for measurement of cosmic rays aboard

sounding balloons.
V. A. Vorob'ev, N. A. Mikirova, V. Yu. Prichesnyaev.
Trudy In-t prikl. geofiz. Gl. upr. gidrometeorol. sluzhby pri
Sov. Min. SSSR, 1974, vyp. (No.) 29, p. 45 - 47. In Russian.
Abstr. in Referativ. Zhurn. 62. Issled. kosmich. prostranstva,
5.62.137 (1975).

034.067 **P1 instrument for measurement of the parameters
of plasma near an artificial earth satellite.**
S. K. Chapkynov, T. N. Ivanova, M. Kh. Petrunova.
Nauch. pribory, No. 5, Moskva, 1974, p. 39 - 42. In Russian.
Abstr. in Referativ. Zhurn. 62. Issled. kosmich prostranstva,
5.62.142 (1975).

034.068 **Automatic pointing device of a horizontal solar
telescope.** M. Klvaňa.
Bull. Astron. Inst. Czechoslovakia, Vol. 26, 186 - 189 (1975).
    The paper gives a description of the principle of a point-
ing system of a solar telescope using stepped motors, operating
under ground-level conditions with an accuracy of 1−2
seconds of arc under the assumption of an image of good
quality and a cloudless atmosphere. The properties of the
photoelectric sensor used are described and the whole pointing
system is characterized according to the experimental data
observed.

034.069 **Si (Li) X-ray astronomical spectroscopy.**
S. S. Holt.
Goddard Space Flight Center, Greenbelt, Maryland, GSFC
Document X-661-75-115, 27 pp. (1975). − Paper presented
at the Symposium on the techniques of solar and cosmic X-ray
spectroscopy, Mullard Space Science Laboratory.
    The general considerations involved in the choice of Si
(Li) as a non-dispersive spectrometer for X-ray astronomy are
discussed. In particular, its adaptation to HEAO-B is described
as an example of the space-borne application of Si (Li) technology.

034.070 **An automated system for photoelectric photometry.**
A. P. Linnell, S. J. Hill, E. F. Brandt.
Publ. Astron. Soc. Pacific, Vol. 87, 273 - 283 (1975).
    A system for automated photoelectric photometry in-
cludes a 24-inch reflector, an automated sequential photometer,
and a minicomputer. The minicomputer performs all data-
gathering and control functions.

034.071 **The use of electronographic tubes for spectrophotometric observations.** E. A. Mallia.
Astrophys. Space Sci., Vol. 35, L21 - L25 (1975).
    It is shown that the performance of spectracons is fairly
close to the limit set by the photocathode. Image drift appears
to be an as yet unresolved problem, at least in the particular
tubes available to us. The emulsion characteristics make grain
counting difficult to carry out, but for stellar spectra it
proved possible to combine low densities with unwidened
spectra. In this way one can shorten exposures appreciably
compared to more conventional use of spectracons. For
spectrophotometry G5 and not L4 is probably the best
general purpose emulsion. At low densities the presence of
dirt particles in emulsions limits their usefulness.

034.072 **The photoelectric photometry with a 60 cm reflector
at the Skalnaté Pleso Observatory.** J. Tremko.
Contr. Astron. Obs. Skalnaté Pleso, Vol. 5, 159 - 180
(1973/75).
    A study of the colour and magnitude systems of a photo-
electric photometer attached to the 60-cm reflector at the
Skalnaté Pleso Observatory showed the colour system to
depend on ambient temperature, while measurements of the
atmospheric extinction revealed the presence of the azimuth
effect which attained several hundredths of magnitude. An

interpretation is given of the photoelectric observations of
three RR Lyrae type variables obtained recently at the
Skalnaté Pleso Observatory. New elements of light variations
were derived and a secular change of the period of W CVn was
found. On the basis of the photoelectric observations and using
Fernie's method, the distance modulus of the investigated
variables was derived.

034.073 **A FET (*Field effect transistor*) operational amplifier
circuit for photoelectric photometry.**
A. J. Stokes.
Journ. American Ass. Variable Star Observers, Vol. 1, 60 - 61
(1972).

034.074 **Improvements on the monochromatic heliograph.**
S. Hamana, S. Yajima.
Tokyo Astron. Obs., Report No. 65, Vol. 17, 292 - 297 (1975).
In Japanese.

034.075 **The theoretical instrumental profile of the combination telescope-pinhole photometer and its effect
upon observations of umbra intensities.**
F. Albregtsen, T. Hansen.
Inst. Teor. Astrophys., Blindern−Oslo, Rep. No. 42, 19 pp.
(1974).
    The convolution of the theoretical diffraction pattern of
a circular telescope aperture with a circular scanning aperture
has been performed. A computer program is given. This
instrumental profile is compared to a straylight function at
the wavelength 3.8 μm. The influence of the instrumental
profile upon the umbra of a circular sunspot is discussed. The
influence turns out to be considerable in the infrared wavelength region.

034.076 **Modification of a dispersion interferometer for
investigation of nonstationary processes.**
I. P. Stakhanov.
Kosmich. Issled., Vol. 13, 375 - 380 (1975). In Russian.

034.077 **A high resolution spectroheliograph operating at
soft X-ray wavelengths.**
W. M. Glencross, D. H. Brabban.
Space Sci. Instrum., Vol. 1, 5 - 15 (1975).
    A spectroheliograph suitable for locating sources emitting
soft X-radiation within any narrow waveband is described. The
spectral resolution is achieved using a Bragg plane crystal
spectrometer, while the element providing spatial information
is a rotation collimator.

034.078 **A sounding rocket spectroheliometer for photo-
metric studies at extreme ultraviolet wavelengths.**
J. G. Timothy, R. M. Chambers, A. M. D'Entremont, N. W.
Lanham, E. M. Reeves.
Space Sci. Instrum., Vol. 1, 23 - 49 (1975).

034.079 **An image-stabilized telescope-ten channel ultraviolet
spectrometer for sounding rocket observations.**
J. W. Giles, W. R. McKinney, C. S. Freer, H. W. Moos.
Space Sci. Instrum., Vol. 1, 51 - 59 (1975).
    A sounding rocket borne pointing telescope-ultraviolet
spectrometer is described. The secondary mirror of the Cassegrain telescope is servo-controlled to obtain one arc sec pointing accuracy. A LiF prism Czerny-Turner polymonochromator
analyzes radiation from 1150 - 1950 Å. Ten photomultiplier
detectors in the focal plane with pulse counting electronics
increase the sensitivity over previous instruments. A programmable entrance slit mechanism with two apertures permits
spatial resolution of planetary targets and reduction of the
terrestrial airglow signal on stellar targets.

034.080 **A correlation technique for magnetometer zero**

level determination. P. C. Hedgecock.
Space Sci. Instrum., Vol. 1, 83 - 90 (1975).

034.081 **A rocket borne absorption cell for high resolution spectroscopy of the He I line at 584 Å.**
J. P. Delaboudinière, C. Carabetian.
Space Sci. Instrum., Vol. 1, 91 - 109 (1975).

034.082 **A focussing X-ray collector and its response in flight.** R. C. Catura, D. T. Roethig.
Space Sci. Instrum., Vol. 1, 141 - 151 (1975).

A single focussing X-ray reflector system designed for rocket observation of cosmic X-ray sources is described. The reflector has an angular resolution of one minute of arc in X-rays. Focussing properties of this reflector are discussed and the response observed as the flight instrument was scanned over the galactic X-ray source, Sco X-1, is presented.

034.083 **Airborne infrared polarimeter.**
D. L. Coffeen, J. Hämeen-Anttila, R. H. Toubhans.
Space Sci. Instrum., Vol. 1, 161 - 175 (1975).

034.084 **The Pioneer XI high field fluxgate magnetometer.**
M. H. Acuña, N. F. Ness.
Space Sci. Instrum., Vol. 1, 177 - 188 (1975).

034.085 **The Viking Mars lander camera.**
F. O. Huck, H. F. McCall, W. R. Patterson, G. R. Taylor.
Space Sci. Instrum., Vol. 1, 189 - 241 (1975).

The two facsimile cameras feature an array of 12 silicon photodiodes, including six spectral bands for color and near-infrared imaging with an angular resolution of 0.12° and four focus steps for broadband imaging with an improved angular resolution of 0.04°. The scanning rates are synchronized to the lander data transmission rates of 16000 bits s^{-1} to the Viking Orbiters as relay stations and 250 bits s^{-1} directly to earth. Image data can also be stored on a lander tape recorder. About 10^7 bits of image data will be transmitted during the mission.

034.086 **The multifunction photoelectric photometer of Torino Observatory.** G. Sedmak.
Reprinted from Astrometric Conference, Torino, May 1974, 5 pp. = Oss. Astron. Trieste Pubbl. No. 483 (1975).

034.087 **Advantages and disadvantages of various spectrum scanner systems.**
T. Schmidt-Kaler, W. Schlosser, R. Rudolph.
Conference on optical observing programs on galactic structure and dynamics, (see 012.013), p. 177 - 190 (1975). In German.

034.088 **A concave grating infra-red spectrophotometer using Hulthen-Lind mounting.**
A. V. Datar, A. D. Tillu.
Bull. Astron. Soc. India, Vol. 2, 34 - 35 (1974). – Abstract.

034.089 **A couple of frequency synthesizers as an offsettable sidereal convertor.** S. Iijima.
Tokyo Astron. Obs., Report No. 65, Vol. 17, 316 - 324 (1975). In Japanese.

034.090 **Microphotometer controlled by "mini-computer".**
I. Shimizu, T. Hirayama, Y. Ohki, M. Fukatsu, T. Oe, Y. Shimizu, E. Hiei.
Tokyo Astron. Obs. Report No. 66, Vol. 17, 329 - 356 (1975). In Japanese.

034.091 **Photomètre à cellule photomultiplicatrice.**
É. Schweitzer.
A.F.O.E.V. Bull., Vol. 8, 38 - 40 (1974).

034.092 **L'intensificateur d'images R.C.A. 4550.**
M. Duruy.
A.F.O.E.V. Bull., Vol. 8, 83 - 89 (1974); Vol. 9, 11 (1975).

034.093 **Focal predictions for spectrographs employing a Maksutov camera.**
D. J. Taylor, E. G. Schmidt.
Publ. Astron. Soc. Pacific, Vol. 87, 357 - 365 (1975).

An analysis has been made of a spectrograph with a Maksutov camera. Because of chromatic aberration in the corrector lens, the focus and plate tilt are sensitive to dispersion and wavelength. Simple approximate formulae for calculating collimator focus and plate tilt are found and numerical results, applicable to a particular commercial spectrograph, are presented. The choice of field flattener lens is discussed.

034.094 *RI* **photometry with a gallium-arsenide photocathode.** D. Weistrop.
Publ. Astron. Soc. Pacific, Vol. 87, 367 - 368 (1975).

An *RI* photoelectric system using a Ga-As photocathode and glass filters is described. The system transforms satisfactorily to the Kron *RI* system.

034.095 **A rapid filter change photometer.** J. M. Sorvari.
Publ. Astron. Soc. Pacific, Vol. 87, 443 - 447 (1975).

A photometer designed to change filters on a short time scale is described. This device is simple and compact and has proved to be very reliable. Evaluation of the photometer indicates that photometric data can be obtained on many nights ordinarily considered unsuitable for photometry.

034.096 **The coudé spectrograph of the 1.52 m telescope.**
H. J. Wood, B. Wolf, E. Maurice.
European Southern Obs., Bull. No. 11, p. 5 - 19 (1975).

034.097 **The ESO spot sensitometer.** H. J. Wood.
European Southern Obs., Bull. No. 11, p. 21 - 31 (1975).

034.098 **The Cassegrain spectrograph RV Cass.**
E. Maurice.
European Southern Obs., Bull. No. 11, p. 33 - 42 (1975).

This article gives the necessary data for obtaining a "theoretical" knowledge of the Cassegrain spectrograph RV Cass and for preparing an observing programme.

034.099 **Fabry-Perot emission line scanner.** J. Rickard.
European Southern Obs., Bull. No. 11, p. 43 - 47 (1975).

034.100 **Image tube camera.** J. Rickard.
European Southern Obs., Bull. No. 11, p. 49 - 58 (1975).

034.101 **Note on maximizing signal from photoconductive detectors.** C. M. Penchina.
Infrared Physics, Vol. 15, 9 - 11 (1975).

It is demonstrated that the use of constant bias current, rather than matched load resistor, optimizes the low-frequency signal - to - noise ratio of photoconductive detectors.

034.102 **Proportional seal counters of soft X-rays for space experiments.** S. V. Viktorov, G. E. Kocharov, G. A. Matveev, V. I. Chesnokov.
Izv. AN SSSR Ser. fiz., Vol. 39, 435 - 444 (1975). In Russian. Abstr. in Referativ. Zhurn. 62. Issled. kosmich. prostranstva, 6.62.225 (1975).

034.103 **A new driving device for astrolabe micrometers.**
G. Billaud, H. Llop.

Astron. Astrophys., Vol. 41, 237 - 238 (1975). In French.

The astrolabe micrometer is driven by an old fashioned electric motor. A new type of motor designed by Sopelem and Paris Observatory remedies to the drawbacks and improves the quality of observations. The new motor is tachometrically controlled and a feeding back generator replaced the old mechanical differential. The whole is quite steady and perfectly silent. The first results show that, as a mean, the weights are increased of 25 %.

034.104 The Utrecht orbiting stellar spectrophotometer S 59 R. Hoekstra, K. A. van der Hucht, T. Kamperman, H. J. Lamers.
Space Optics, National Acad. Sci. Washington, p. 825 - 834 = Utrechtse Sterrekundige Overdrukken No. 282 (1974).

034.105 Near-infrared heterodyne interferometer for the measurement of stellar diameters.
H. van de Stadt, T. de Graauw, J. C. Shelton, C. Veth.
Space Optics, National Acad. Sci. Washington, p. 441 - 457 = Utrechtse Sterrekundige Overdrukken No. 283 (1974).

034.106 The AMAS — the Astrometric Multiplexing Area Scanner. A new technique for measuring stellar positions and magnitudes. C. D. Smith.
Journ. American Ass. Variable Star Observers, Vol. 2, 29 - 34 (1973).

034.107 Application of pendulum levels in astrometry. B. K. Bagil'dinskij.
19th Astrometrical Conference 1972, (see 012.019), p. 161 - 164 (1975). In Russian.

034.108 Results of an investigation of the automatic coordinate measuring machine in measurements of photographic binaries. L. M. Zatsiorskij, A. A. Kiselev.
19th Astrometrical Conference 1972, (see 012.019), p. 221 - 226 (1975). In Russian.

034.109 A device for collecting cosmic dust at altitudes between 30 and 60 km. R. Wlochowicz.
Canadian Aeronaut. and Space Journ., Vol. 20, 341 - 348 (1974).

034.110 The optical sensors of the Netherlands astronomical satellite (ANS). I. The sun sensors. A. J. Smets.
Philips Techn. Rev., (*Netherlands*), Vol. 34, 208 - 212 (1974).

Describes the construction and operation of the sun sensors to direct the +Z-axis of the ANS towards the sun: six coarse sensors give the total field of view covering the whole 4π of the celestial sphere; two intermediate sensors measure angular displacements of the satellite about the X- and Y-axis, up to $\pm 36°$; two fine sensors do the same jobs as the intermediate sensors.

034.111 The optical sensors of the Netherlands astronomical satellite (ANS). II The horizon sensor.
P. van Dijk.
Philips Techn. Rev., (*Netherlands*), Vol. 34, 213 - 217 (1974).

Describes the characteristics and the operation of the ANS horizon sensor which is an infrared telescope that scans a plane normal to the line between the satellite and the sun by means of a brushless d.c. motor drive rotating mirror.

034.112 The optical sensors of the Netherlands astronomical satellite (ANS). III. The star sensor.
W. J. Christis.
Philips Techn. Rev., (*Netherlands*), Vol. 34, 218 - 224 (1974).

The attitude control system holds the astronomical instruments pointed at the stellar objects under observation by means of data from the star sensor. This sensor comprises a fast response camera tube, detection and tracking circuits, data processing logic and power supplies.

034.113 On guide bars and scales of the KIM-3 instrument. P. P. Pavlenko, L. S. Pavlenko.
Vestn. Khar'kov. Univ., No. 117, (Ser. Astron., vyp. (No.) 9), p. 60 - 64 (1974). In Russian.

034.114 The narrow-angle telescope for the visual imaging subsystem of the Mariner-Venus/Mercury (1973) spacecraft. L. Larks.
Instrumentation in astronomy II, (see 012.021), p. 35 - 40 (1974).

034.115 A self-scanned digital image tube. J. P. Choisser, R. E. Nather, R. G. Tull.
Instrumentation in astronomy II, (see 012.021), p. 83 - 87 (1974).

034.116 Television system optimization for astronomy and special applications. F. L. Schaff.
Instrumentation in astronomy II, (see 012.021), p. 95 - 101 (1974).

034.117 The Ranicon-A resistive anode image converter for the soft X-ray and vacuum ultraviolet.
M. Lampton, F. Paresce, S. Bowyer.
Instrumentation in astronomy II, (see 012.021), p. 103 - 107 (1974).

034.118 A 512-channel photodiode array for solar observations. R. B. Dunn, D. M. Rust, G. E. Spence.
Instrumentation in astronomy II, (see 012.021), p. 109 - 115 (1974).

034.119 Thermal image projector/recorder. C. M. Redman.
Instrumentation in astronomy II, (see 012.021), p. 117 - 126 (1974).

034.120 Airborne far infrared solar spectroscopy. W. G. Mankin, J. A. Eddy, R. H. Lee, R. M. MacQueen.
Instrumentation in astronomy II, (see 012.021), p. 133 - 136 (1974).

034.121 The rapid scanning Michelson interferometer as an infrared spectrometer in astronomy.
T. H. Morgan, R. C. Baldwin, A. E. Potter.
Instrumentation in astronomy II, (see 012.021), p. 147 - 150 (1974).

034.122 The extreme ultraviolet spectrograph. J.-D. F. Bartoe, G. E. Brueckner, J. D. Purcell, R. Tousey.
Instrumentation in astronomy II, (see 012.021), p. 153 - 158 (1974).

034.123 The photoelectric spectroheliometer on ATM. E. M. Reeves, J. G. Timothy, M. C. E. Huber.
Instrumentation in astronomy II, (see 012.021), p. 159 - 173 (1974).

034.124 Design characteristics of a Skylab soft X-ray telescope. E. J. Walsh, T. I. Sokolowski, G. M. Miller, K. L. Cofield, J. D. Douglas, B. J. Lewter, H. O. Burke, A. J. Davis.
Instrumentation in astronomy II, (see 012.021), p. 175 - 184 (1974).

034.125 The X-ray spectrographic telescope.

G. S. Vaiana, A. S. Krieger, R. Petrasso, J. K. Silk, A. F. Timothy.
Instrumentation in astronomy II, (see 012.021), p. 185 - 205 (1974).

034.126 The High Altitude Observatory white light corona- graph. R. M. MacQueen, J. T. Gosling, E. Hildner, R. H. Munro, A. I. Poland, C. L. Ross.
Instrumentation in astronomy II, (see 012.021), p. 207 - 212 (1974).

034.127 The Mariner '69 and '71 high-resolution camera and recent additions to its lens design program. G. W. Wilkerson.
Instrumentation in astronomy II, (see 012.021), p. 237 - 245 (1974).

034.128 Two high resolution velocity vector analyzers for cosmic dust particles. S. Auer.
Rev. Sci. Instruments, Vol. 42, 127 - 135 (1975).

Two new position sensitive detectors are described to build an instrument with an angular resolution of about $1°$ and a velocity resolution of about 1 %. The measurement accuracy is discussed. Composition analysis was combined with the velocity vector analyzer and results reported in comparison with the HEOS 2 detector.

034.129 The induction-coil magnetometer experiment for the Helios solar probe. G. Dehmel.
Internationale Elektron. Rundschau, (Germany), Vol. 29, No. 1 - 2, p. 1 - 4 (1975). In German.

A detailed description is given of the functions, design and construction of the 'Helios A' solar probe, to be launched from the Kennedy Space Center in December 1975, including a block diagram of the electronics for the induction-coil magnetometer.

034.130 A flat-panel TV display system in monochrome and color. Y. Amano.
IEEE Trans. Electron Devices, Vol. ED-22, 1 - 7 (1975).

034.131 Contribution à la détermination des positions lunaires par l'observation d'occultations au moyen d'un micromètre à double image. C. Meyer.
Thesis Sci. Phys., Univ. Paris VI. AO-CNRS-9856, 84 pp. (1974).

On décrit un nouveau micromètre à double image construit spécifiquement en vue de l'observation, au bord éclairé, des occultations d'étoiles par la lune. La réduction des observations a conduit à comparer les profils lunaires de Watts et Weimer; on montre qu'il est possible de trouver une surface de référence commune par rapport à laquelle les écarts locaux entre les profils ne présentent plus un caractère systématique. On obtient finalement une précision de $\pm0.''13$ sur la longitude et de $\pm0.''21$ sur la latitude.

Far infrared measurements of selected optical materials at 1.6 K. See Abstr. 022.083.

Interference fringes obtained on Vega with two optical telescopes. See Abstr. 031.004.

Michelson interferometer with frustrated-total-internal-reflection beam splitter. See Abstr. 031.010.

On methods of high-accuracy electrophotometric observations of variable stars. See Abstr. 031.204.

Heterodyn-Detektion als Mittel zur hochauflösenden Spektralanalyse schwacher Linienstrahlung im fernen Infrarot. See Abstr. 031.267.

An evolution of X-ray astronomy instrumentation. See Abstr. 051.006.

Magnetometer experiments in the European Space Research Organisation's HEOS satellites. See Abstr. 054.016.

L'étude de la couronne blanche à bord de Concorde 001 au cours de l'éclipse totale de soleil du 30 juin 1973. See Abstr. 074.032.

Rocket investigations of corpuscular radiation at different geomagnetic latitudes. I. Complex of the apparatus for investigations of corpuscular radiation aboard MR-12 meteorological rockets. See Abstr. 082.083.

Search for faint red stars. See Abstr. 113.046.

Wavelength dependence of interstellar polarization and ratio of total to selective extinction. See Abstr. 131.031.

Temporal X-ray astronomy with a pinhole camera. See Abstr. 142.095.

Absorption-line redshifts of galaxies in remote clusters obtained with a sky-subtraction spectrograph using an SIT television detector. See Abstr. 160.018.

035 Clocks and Frequency Standards

035.001 **Zeitzeichenempfänger.** J. Biel.
SuW, Vol. 14, 22 (1975).

035.002 **Atomic frequency standards: a survey.**
H. W. Hellwig.
Proc. IEEE, Vol. 63, 212 - 229 (1975).
This survey reviews the more recent historical background of atomic frequency standards leading to the present developments. A discussion of the underlying physical and engineering principles is given. Modern atomic frequency standards, including their performance, are compared quantitatively, and projections are attempted at likely future developments and performance characteristics. New developments include passive hydrogen devices, saturated absorption stabilized lasers, ion storage devices, and atomic beams in the far infrared and infrared region, as well as new techniques to evaluate frequency biases such as those encountered in cesium and hydrogen standards.

035.003 **Neuartige Sonnenuhr-Konstruktionen.**
W. Brunner-Bosshard.
Orion, 33. Jahrgang, p. 44 - 47 (1975).

035.004 **A sundial on an office ceiling.** W. W. Shrader.
Sky Telescope, Vol. 49, 217 - 218 (1975).

035.005 **Zeitbewahrung mit Riefler-Uhren.** K. Erbrich.
SuW, Vol. 14, 197 - 203 (1975).

035.006 **A relativistic effect on an atomic clock.**
T. Hara.
Proc. International Latitude Obs. Mizusawa, No. 14, p. 52 - 53 (1974). In Japanese.
The effect of the relativistic redshift on an atomic clock on the earth's surface is discussed. Gravities of the earth, planets, Galaxy and the local group of galaxies are taken into consideration. The redshift is constant on the geoid of the earth.

035.007 **Precise clock comparison by means of TV signals.**
(Between Tokyo Astronomical Obs. and Kanozan Geodetic Obs.).
K. Fujiwara, T. Kato, S. Hokugo, K. Miyazaki.
Tokyo Astron. Obs., Report No. 65, Vol. 17, 261 - 283 (1975). In Japanese.

035.008 **Electronic registration of time.** R. Fangor.
Urania Kraków, Vol. 46, 136 - 143, 176 - 181 (1975). In Polish.

035.009 **A clock pulse generator for the Lunar Laser Ranging**
System. Y. Iizuka.
Tokyo Astron. Obs., Report No. 66, Vol. 17, 365 - 371 (1975). In Japanese.

035.010 **LORAN comparator at the Dodaira Station.**
N. Kobayashi, N. Oshima, M. Noguchi.
Tokyo Astron. Obs., Report No. 66, Vol. 17, 372 - 376 (1975). In Japanese.

Estimate of the accuracy of quartz clocks of the united Time Service of the Sternberg Astronomical Institute and the General Research Institute of Geodesy and Cartography. See Abstr. 044.015.

036 Photographic Auxiliaries

036.001 **New analytic expressions of photographic characteristic curves.**   A. E. S. Green, R. D. McPeters.
Applied Optics, Vol. 14, 271 - 272 (1975). − Letter.

036.002 **Investigation of the Kodak IIIa−J plates characterized by large signal-to-noise ratio.**
I. I. Brejdo, O. M. Mikhajlova.
Astron. Zhurn. Akad. Nauk SSSR, Vol. 52, 415 - 419 (1975).
In Russian. English translation in Soviet Astron., Vol. 19, No. 2.

The investigation of the Kodak IIIa−J plates has shown that these plates are characterized by high resolution, high contrast, small granularity and large signal-to-noise ratio. With the use of an astrosensitometer the plates are shown to be able to registrate essentially fainter stars than the Kodak 103a-O, ORWO ZU-2 and other plates at the same background brightness. The gain in stellar magnitudes can reach 2^m.

036.003 **Detective performance of photographic plates.**
D. W. Latham.
Astrophysics. Part A, (see 003.002), p. 221 - 235 (1974).

Photographic photometry; Signal-to-noise and detective quantum efficiency; Detective performance of Kodak Spectroscopic Plates, types IIa-O, 103a-O, and IIIa-J.

036.004 **Indirekte Astrofarbenfotografie nach dem modifizierten Dreifarbenverfahren.**
E. Alt, J. Rusche.
Orion, 33. Jahrgang, p. 67 - 71 (1975).

036.005 **Einsatzmöglichkeiten des „Photomicrography Monochrome Films SO-410" von Kodak für die Mond- und Planetenphotographie.**
K. Schiefer, U. Schiefer.
SuW, Vol. 14, 204 - 205 (1975).

036.006 **Amateur photographic astronomy.**   M. B. Ward.
Publ. Variable Star Section, Roy. Astron. Soc. New Zealand, No. 2 (C 74), p. 13 - 20 (1974). − A personal view.

036.007 **Parasite images. General considerations and design precepts [astronomical photography].**
J. E. Simmons.
Instrumentation in astronomy II, (see 012.021), p. 231 - 235 (1974).

The ESO spot sensitometer.   See Abstr. 034.097.

Astronomical photography. Part B: Solar corona photography.   See Abstr. 074.095.

Astronomical photography. Part A: Zodiacal light photography.   See Abstr. 106.046.

Positional Astronomy. Celestial Mechanics

041 Positional Astronomy, Astrometry, Star Catalogues and Atlases

041.001 Barnards Pfeilstern—auch für Amateure ein interessantes Objekt.   B. Wedel.
SuW, Vol. 14, 93 - 95 (1975).

041.002 Algorithm for computing equatorial coordinates of stars and of the sun.
E. V. Yakshevich, L. L. Vagushchenko, V. A. Sinyaev.
Trudy TsNII mor. flota, 1974, vyp. (No.) 190, p. 3 - 8. In Russian. – Abstr. in Referativ. Zhurn. 51. Astron., 2.51.135 (1975).

041.003 On the determination of the errors $\Delta\alpha_a$ and $\Delta\mu_a$ for stars of the Boss GC catalogue.
E. V. Vityazeva.
Fiz. poluprovodn. soedin. slozhn. sostava. Ehlista, Kalmytsk. un-t, 1974, p. 172 - 174. In Russian. – Abstr. in Referativ. Zhurn. 51. Astron., 2.51.137 (1975).

041.004 Reduction of astronomical observations made with mean solar time chronometers.   V. I. Vajtsekyan.
Geod. i kartografiya, 1974, No. 10, p. 20 - 23. In Russian. Abstr. in Referativ. Zhurn. 51. Astron., 2.51.145 (1975).

041.005 The relativity effect of deflection of light in astrometry.   L. Ya. Arifov, R. K. Kadyev.
Astron. Zhurn. Akad. Nauk SSSR, Vol. 52, 164 - 170 (1975). In Russian. English translation in Soviet Astron., Vol. 19, No. 1.
The influence of the effect of deflection of light by the gravity field of the sun on the results of astrometric measurements is discussed. The relativity corrections to coordinates of stars and change of latitude are found. A principally new method of measurement of absolute star parallaxes and a new method of research of Einstein's effect by using a physical clock are suggested.

041.006 Compression of ephemerides.
A. Deprit, W. Poplarchek, A. Deprit-Bartholomé.
Celestial Mechanics, Vol. 11, 53 - 58 (1975).
Approximations in the norm L_1 by Chebyshev polynomials are generated to represent astronomical ephemerides over large intervals of time.

041.007 Observations of Saturn obtained with the astrolabe at Paris Observatory during the winter of 1972– 1973.   S. Débarbat, N. Capitaine, F. Chollet, H. Choplin, M. Feissel, A. Journet, S. K. Lam, D. Proust, F. Meyer.
Astron. Astrophys., Suppl. Ser., Vol. 19, 389 - 393 (1975). In French.
This paper contains the results of the second programme of observations at the Paris Observatory; results of the first one have been published previously (Chollet et al. 1973). During the 1972–1973 winter 19 east transits and 17 west transits were observed. The programme will be pursued in accordance with a resolution of IAU Colloquium No. 20 and IAU Commission 8 (15th General Assembly): It is recommended that programmes of observations of fundamental stars, with meridian circles and astrolabes, should include members of the solar system.

041.008 L'astrométrie: hier, aujourd'hui, demain.

S. Débarbat, J. Lévy.
L'Astronomie, 89e année, p. 7 - 22 (1975).

041.009 Raumastrometrie.   E. Høg.
SuW, Vol. 14, 123 - 124 (1975).

041.010 Sur les observations du soleil faites avec l'astrolabe de Danjon.   F. Laclare.
Comptes Rendus Acad. Sci. Paris, Sér. B, Vol. 280, 13 - 15 (1975).
On décrit la méthode d'observation et de réduction des hauteurs correspondantes du soleil. La précision interne sur les premiers résultats semble être du même ordre que celle obtenue par des observations méridiennes.

041.011 The limiting magnitude of the ESO (B) Survey.
H.-E. Schuster, R. M. West.
Astron. Astrophys., Vol. 38, 161 - 164 (1975).
The limiting magnitude of the ESO (B) Survey depends on the image size (the seeing disc) and is found to be in the interval $m_B(\text{lim}) = 21\overset{m}{.}0 - 21\overset{m}{.}5$. It is the same for the on-glass and on-film atlas copies as for the original plates.

041.012 Systematic differences astrolabe—FK 4 of type Δa_a and $\Delta\delta_a$ obtained at Santiago, Chile.
F. Noël.
Astron. Astrophys., Vol. 38, 197 - 202 (1975).
From the results given in the "First astrolabe catalogue of Santiago" an investigation was made to disclose the existence of systematic differences: Astrolabe—FK 4 and FK 4 Sup. of Δa_a and $\Delta\delta_a$ type. Some of the results obtained, compared with similar data deduced from observations made with the astrolabe of Quito and with a small transit instrument at Santiago, as well as a remarkable consistence in the group corrections obtained with quite different methods of computation allow the assumption that these differences could be explained by systematic errors of the same type in the FK 4 and FK 4 Sup.

041.013 Improvement of the system of a fundamental catalogue by observations of minor planets.
V. N. Bojko.
Astron. Zhurn. Akad. Nauk SSSR, Vol. 52, 431 - 440 (1975). In Russian. English translation in Soviet Astron., Vol. 19, No. 2.
The precision of determination of the corrections to the elements of the orbits of the earth and minor planets as well as the corrections of equinox and equator and systematic errors of the fundamental system is investigated on the basis of hypothetical optical and radar observations of minor planets. Mutual correlation of the unknowns is studied. Conclusions concerning the conditions allowing to increase the precision of determination of the correction of the equinox are drawn.

041.014 On the improvement of star positions from minor planet observations.   V. I. Orelskaya.
IAU Colloquium No. 22, (see 012.005), p. 39 - 50 (1974).

041.015 On the determination of the directions of star catalogue axes.   E. P. Fedorov.
Astrometriya i Astrofizika, *Kiev*, vyp. (No.) 24, (see 003.003),

p. 3 - 14 (1974). In Russian.

The orientation of the coordinate system of a star catalogue may be determined by the angles through which the axes of the system should be turned to make them coincide with the axes of an ideal equatorial system. Two methods enable these angles to be determined: 1. Comparison of the coordinates of the sun derived from concurrent fundamental and differential meridian observations. 2. Comparison of the positions of planets observed on the system of the catalogue with their ephemeris positions.

041.016 On the relation between the conceptions "fundamental" and "inertial" systems in astrometry.
V. V. Podobed.
Astrometriya i Astrofizika, Kiev, vyp. (No.) 24, (see 003.003), p. 15 - 17 (1974). In Russian.

The paper deals with the relation between the geometric coordinate system determined by a fundamental catalogue and the frame to which planetary theories are referred in celestial mechanics. Methods of constructing the inertial coordinate system are discussed.

041.017 On the compilation of the Preliminary General Catalogue of Fundamental Faint Stars (PFKSZ₂).
A. N. Kur'yanova, D. D. Polozhentsev, Ya. S. Yatskiv.
Astrometriya i Astrofizika, Kiev, vyp. (No.) 24, (see 003.003), p. 18 - 21 (1974). In Russian.

The paper deals with the problem of compilation of the PFKSZ₂. A method of comparison of compiled catalogues is given.

041.018 Catalogue of differences of proper motions in declination of 34 Washington zenith stars.
E. I. Obrezkova.
Astrometriya i Astrofizika, Kiev, vyp. (No.) 24, (see 003.003), p. 22 - 25 (1974). In Russian.

For the purpose of deriving differences of proper motions of the Washington zenith stars the observation of scale pairs was undertaken. The results obtained are given.

041.019 Astrometric positions of the planet Pluto in the years 1971–1974.
C. Barbieri, M. Capaccioli, G. Pinto.
Astron. Journ., Vol. 80, 412 - 414 (1975).

This paper contains 41 astrometric positions of the planet Pluto from 1971 to 1974. Comparison between these positions supplemented by those for 1970 taken from Paper I (Barbieri et al. 1972), and the ephemeris published by Kaplan et al. (1972), shows excellent agreement with differences seldom exceeding 1 arcsec in both coordinates. Nevertheless, the systematic behavior of the residuals requires further refinements of the orbit and observations for several years to come.

041.020 Determination of a new fundamental reference coordinate system (FK5) for astronomy.
W. Fricke.
Bull. American Astron. Soc., Vol. 7, 339 - 340 (1975). Abstr. AAS.

041.021 The new JPL planetary ephemeris.
E. M. Standish, Jr.
Bull. American Astron. Soc.,Vol. 7, 343 (1975). − Abstr. AAS.

041.022 An evaluation of the tables of systematic differences FK4 minus GC and FK4 minus N30 for the epoch of 1970.0. F. A. Fajemirokun.
Bull. Géod., Nouvelle Sér., Année 1975, No. 116, p. 149 - 166 (1975).

The tables of systematic differences between the FK4 catalogue and the GC and N30 catalogues, published by Brosche et al. (1964) are evaluated for the epoch of 1970.0.

Mean star positions in the three catalogue systems are updated to the epoch of 1970.0 and systematic corrections are applied to the GC and N30 positions. Residuals are obtained before and after systematic corrections, and these residuals are compared. Results show that most of the systematic differences between the three fundamental catalogue systems are eliminated if tabulated corrections are applied to the GC and N30 mean positions of stars.

041.023 Résultats des observations faites à Alger avec l'astrolabe impersonnel A. Danjon OPL 8. Temps et latitude 1973. A. Ghezloun, M. Benhocine, J. Pham-Van.
Ann. Obs. Astron. Alger, Vol. 4, Fasc. 1, p. 1 - 13 (1974).

041.024 Catalogue for automatic photoelectric observations of stars from the Time Service Catalogue.
K. Šteins, A. V. Ivanov.
Uch. Zap. Latv. Gos. Univ., Vol. 220, Astron., vyp. (No.) 11, (see 003.014), p. 29 - 54 (1975). In Russian.

041.025 Determination of positions in a "stellar" coordinate system. V. K. Drofa.
Astrometriya i Astrofizika, Kiev, vyp. (No.) 25, (see 003. 015), p. 66 - 76 (1975). In Russian.

A method is considered for determining coordinates in the system directly from the observations; the axes of the system being fixed on two selected (reference) stars. Instrumental errors, order of observations and their processing are analyzed.

041.026 Methods and results of determination of zero points and periodical errors in star catalogues.
D. P. Duma.
Astrometriya i Astrofizika, Kiev, vyp. (No.) 25, (see 003.015), p. 77 - 87 (1975). In Russian.

More than 80 determinations of corrections to the equinox ΔA and to the equator $\Delta\delta_0$ of the FK3, FK4 and other catalogues are analyzed.

041.027 Herleitung und Erprobung eines erweiterten Verfahrens zur Bestimmung eines Instrumentalsystems von Örtern und Eigenbewegungen von Sternen. H. Schwan.
Diss. Nat. Gesamtfakultät, Ruprecht-Karl-Univ., Heidelberg. 4 + 127 pp. (1975).

A method has been developed to determine the systematic differences Δ in a statistically reliable manner. The possible dependence of Δ simultaneously on the position (α, δ) and the magnitude m is taken into account by developing Δ in a series of products of spherical harmonics with Hermite polynomials. Which of the functions enter in the final development is tested statistically. The instrumental system of the Washington 6″-meridian circle (catalogues W_{25}, $W1_{50}$, $W2_{50}$, $W3_{50}$, $W4_{50}$) has been derived by determining its systematic deviation in position and proper motion against the FK4.

041.028 On the determination of absolute declinations of equatorial stars. A. A. Mikhajlov.
Pis'ma v Astron. Zhurn., Vol. 1, No. 5, p. 39 - 42 (1975). In Russian.

A list of approximate coordinates of 100 equatorial stars has been compiled. It is requested that observatories regularly observe the declinations of these stars as this will lead to a general improvement of the declination system.

041.029 A radioastronomical inertial coordinate system based on measurement of arcs between radio sources.
A. F. Dravskikh, G. A. Krasinskij, A. M. Finkel'shtejn.
Pis'ma v Astron. Zhurn., Vol. 1, No. 5, p. 43 - 48 (1975).

The possibility to construct a radioastronomical coordinate system based on measurement of angular distances between quasars with the accuracy ~0".001 using the VLBI method is shown.

041.030 **Catalogue d'étoiles géodesiques de Moscou.**
M. S. Zverev, A. G. Oborneva.
Trudy Gos. Astron. Inst. Shternberga, Vol. 45, 3 - 48 (1974). In Russian.

Publication d'un catalogue d'étoiles géodesiques composé d'après les observations de Zverev (α, δ) et de Tolgsky (δ) faites à Moscou dans les années 1932–1939 à l'aide du cercle méridien de Repsold. Le catalogue contient 1612 étoiles dont 1331 on été prises au catalogue Pu$_{25}$ et 268 étoiles des catalogues de 1642 (1631) étoiles qui on servi d'étoiles de repère. Le catalogue a été traité dans le systeme des catalogues de 1642 (1631) étoiles d'après la methode de Zimmermann. Les coordonnées sont données pour l'époque des observations et l'équinoxe 1950.0. L'erreur quadratique moyenne d'une observation est $\epsilon_a \times \cos \delta = \pm 0^s.023$, $\epsilon_\delta = \pm 0".40$.

041.031 **Catalogue d'ascensions droites et de déclinaisons des étoiles KSZ (zone +45° – +60°) obtenu au cercle méridien Repsold de Moscou.**
L. M. Khommik, V. A. Korobova, O. A. Kozina.
Trudy Gos. Astron. Inst. Shternberga, Vol. 45, 49 - 86 (1974). In Russian.

Le catalogue d'ascensions droites et de déclinaisons des étoiles KSZ (zone +45° – +60°) est composé d'après les observations faites à l'aide du cercle méridien Repsold de l'observatoire de Moscou au cours des années 1957–1964. Le catalogue est rapporté au système d'ensemble PFKSZ et FK3R. Une description de la méthode d'observation est donnée. L'erreur quadratique moyenne d'une observation d'ascension droite est égale à $\pm 0^s.015$ sec δ, telle de déclinaison à $\pm 0".44$.

041.032 **Photographic catalogue of the coordinates for 323 Moscow zenith stars.** N. B. Frolova.
Trudy Gos. Astron. Inst. Shternberga, Vol. 45, 95 - 105 (1974). In Russian.

041.033 **Comparison of the catalogue of the Moscow zenith zone with the PZT catalogue and with five other catalogues.** D. N. Ponomarev, N. B. Frolova.
Trudy Gos. Astron. Inst. Shternberga, Vol. 45, 106 - 118 (1974). In Russian.

The new Moscow photographic zenith zone catalogue (MZK) containing 323 stars of the +55°20' – +56°10' zone in the AGK3R system is compared with the catalogues of right ascensions and declinations of Moscow PZT stars, the Yale zone catalogue, AGK2, AGK3, AGK3R and the catalogue of declinations of the latitude programme stars KSZ, Belgrad. Systematic differences $\Delta a_a \cos \delta$, $\Delta \delta_a$, $\Delta a \cos \delta_m$, $\Delta \delta_m$, $\Delta a \cos \delta_{sp}$, $\Delta \delta_{sp}$ are considered.

041.034 **The catalog of 743 bright stars between the declinations, –10° and –30°.**
H. Yasuda, R. Fukaya, H. Hara, T. Ina.
Ann. Tokyo Astron. Obs., Second Ser., Vol. 15, (No. 2), 61 - 107 (1975).

The present catalog contains the results of the observations of bright stars made with the Tokyo meridian circle between March 1963 and September 1969. It also contains the comparison of the results with the FK4 Sup. and the GC. FK4 star observations made in connection with this catalog are compared with the FK4.

041.035 **Magnitude effect in the observations of southern reference stars and bright stars made with Tokyo meridian circle.** H. Yasuda, N. Miyauchi.
Ann. Tokyo Astron. Obs., Second Ser., Vol. 15, (No. 2), 108 - 116 (1975).

The positions of southern reference stars and bright stars observed with the Tokyo meridian circle were examined for residual-magnitude effects. The FK4 star observations made in connection with those stars were also examined for the northern and southern celestial spheres separately. In observations of right ascensions, there exists no evidence of a significant magnitude effect, whereas in declination a significant magnitude effect was found for the stars in the southern celestial hemisphere.

041.036 **The Catalogue of Stellar Identifications, progress report III.** F. Ochsenbein, M. Bischoff.
Centre de Données Stellaires, Inform. Bull. No. 8, (see 002. 007), p. 2 - 7 (1975).

The present state of the catalogue of stellar identifications and a statistical description of the different types of observational data available are given. Two tables deal with the overlap between spectroscopic data and other types of data, specially HD and MK classifications.

041.037 **AGK3. Star catalogue of positions and proper motions north of –2°5 declination, derived from photographic plates taken at Bergedorf and Bonn in the years 1928–1932 and 1956–1963.** Planned by O. Heckmann, produced and edited by W. Dieckvoss, in collaboration with H. Kox, A. Günther, E. Brosterhus, with a historical review by O. Heckmann. Vol. 1: +90°. . . +65° decl., 41 + 283 pp.; Vol. 2: +65°. . . +50° decl., 2 + 389 pp.; Vol. 3: +50°. . . +40° decl., 2 + 352 pp.; Vol. 4: +40°. . . +30° decl., 2 + 380 pp.; Vol. 5: +30°. . . +20° decl., 2 + 418 pp.; Vol. 6: +20°. . . +12° decl., 2 + 322 pp.; Vol. 7: +12°. . . +5° decl., 2 + 354 pp.; Vol. 8: +5°. . . –2° decl.,2 + 342 pp. Hamburger Sternwarte, Hamburg-Bergedorf. Price DM 193.85 (1975).

041.038 **Third Santiago-Pulkovo Fundamental Stars Catalogue (SPF 3). A catalogue in R.A. of 671 fundamental bright stars of the zone +40° to –90°.**
P. R. Loyola, V. N. Shishkina.
Dep. Astron., Univ. Chile, Obs. Astron. Nacional, Cerro Calán, Santiago de Chile, Publ. Vol. 2, (No. 5), 159 - 180 (1974).

A method and a program were prepared to determine right ascensions, which included 671 fundamental bright stars (FK4: 621 stars and N30: 654 stars). In order to do this work a Zeiss transit instrument (d = 100 mm, f = 1000 mm) was prepared at Pulkovo and was brought to Chile by the end of 1962. About 12000 observations were made with this instrument during 1963 and 1964, observing each star 8 times as a minimum. Some preliminary results have already been published.

041.039 **Star catalogs and files available at the Stellar Data Center.**
C. Jaschek, F. Ochsenbein, M. Guilbaut.
Centre de Données Stellaires, Inform. Bull. No. 8, (see 002. 007), p. 22 - 25 (1975).

041.040 **Étude de l'incertitude à craindre dans la résolution de catalogues photographiques utilisant le recouvrement de clichés.** A. Valbousquet.
Thesis Univ. Louis Pasteur Strasbourg, 99 pp. = Publ. Obs. Astron. Strasbourg, Vol. 3, Fasc. 1 (1974).

041.041 **Meridian observation of major planets and some minor planets, 1968 – 1973.**
H. Yasuda, R. Fukaya, H. Hara, S. Isobe, H. Ishii, Y. Adachi, T. Ina.
Tokyo Astron. Bull., Second Ser., No. 236, p. 2743 - 2771 (1975).

This bulletin contains positions of major planets (Venus, Mars, Jupiter, Saturn, Uranus, and Neptune) and of some minor planets (Ceres, Pallas, Juno, and Vesta) observed with the Tokyo meridian circle during the period 1968 – 1973.

During this period, the meridian circle was also used to fulfill the Tokyo Astronomical Observatory commitment to the international southern reference star program. The Bright Stars between the declinations −10° and −30° and northern PZT stars were also observed.

041.042 La représentation des positions observées d'un mobile animé d'un mouvement linéaire uniforme. Application aux cas des couples stellaires optiques et des passages des planètes inférieures sur le soleil.
J. Dommanget.
Bull. Astron. Obs. Roy. Belgique, Vol. 8, 161 - 163 (1974).

Lorsque les mesures des positions successives d'un mobile se font de manière indépendante dans chacune des coordonnées x et y, le calcul par la méthode des moindres carrés des mouvements en x et en y séparément conduit à la même solution que si l'on détermine le mouvement du mobile par la même méthode, sur la base de toutes les mesures considérées globalement.

041.043 A discussion of the observations of Neptune 1846−1970. E. S. Jackson.
Astron. Papers, U.S. Naval Obs., *Washington, D.C.,* Vol. 22, (Part II), 95 - 204 (1974). [For sale by the Superintendent of Documents, U.S. Government Printing Office, Washington, D.C., Price $ 3.10].

041.044 The situation in meridian astrometry.
A. P. Gulyaev.
19th Astrometrical Conference 1972, (see 012.019), p. 5 - 13 (1975). In Russian.

041.045 Radio astronomy and astrometry.
Yu. N. Parijskij, A. F. Dravskikh.
19th Astrometrical Conference 1972, (see 012.019), p. 25 - 42 (1975). In Russian.

041.046 Comparison of different methods for investigation of the differences of positions and proper motions of stars. Ya. S. Yatskiv, A. N. Kur'yanova, A. A. Molotaj.
19th Astrometrical Conference 1972, (see 012.019), p. 43 - 59 (1975). In Russian.

041.047 The problem of fitting relative coordinates to a fundamental system. L. M. Khommik.
19th Astrometrical Conference 1972, (see 012.019), p. 81 - 87 (1975). In Russian.

041.048 The PFKSZ' catalogue and its comparison with AGK3R with the method of expansion of the differences of coordinates by spherical functions.
D. D. Polozhentsev, A. N. Kur'yanova.
19th Astrometrical Conference 1972, (see 012.019), p. 87 - 91 (1975). In Russian.

041.049 Approaches to increase the accuracy of meridian observations demonstrated by the work of the Pulkovo horizontal meridian circle.
G. I. Pinigin, L. A. Sukharev, G. M. Timashkova.
19th Astrometrical Conference 1972, (see 012.019), p. 92 - 94 (1975). In Russian.

041.050 The system of right ascensions of the horizontal meridian circle of the Pulkovo Observatory.
G. I. Pinigin.
19th Astrometrical Conference 1972, (see 012.019), p. 95 - 100 (1975). In Russian.

041.051 On determination of the absolute right ascensions of stars during the polar night. G. M. Petrov.
19th Astrometrical Conference 1972, (see 012.019), p. 100 -

102 (1975). In Russian.

041.052 Method for determining absolute right ascensions.
E. Bem.
19th Astrometrical Conference 1972, (see 012.019), p. 102 - 104 (1975). In Russian.

041.053 Dependence of the accuracy of determination of the equinox correction on the distance earth − planet. D. P. Duma.
19th Astrometrical Conference 1972, (see 012.019), p. 105 - 107 (1975). In Russian.

041.054 Determination of the zero points of the FK4 catalogue from meridian observations of the moon.
V. A. Fomin.
19th Astrometrical Conference 1972, (see 012.019), p. 107 - 110 (1975). In Russian.

041.055 List of additional SRS stars on areas with galaxies of the southern sky. G. D. Baturina.
19th Astrometrical Conference 1972, (see 012.019), p. 116 - 127 (1975). In Russian.

041.056 Preliminary results of absolute determinations of right ascensions with the large Pulkovo transit instrument (LPTI) in Chile. D. D. Polozhentsev, T. A. Polozhentseva, R. Tajbo, G. M. Timashkova.
19th Astrometrical Conference 1972, (see 012.019), p. 128 - 134 (1975). In Russian.

041.057 On the possibility of analyzing the $\Delta\alpha_\delta$ system of a fundamental catalogue from differential observations. E. V. Khrutskaya.
19th Astrometrical Conference 1972, (see 012.019), p. 137 - 139 (1975). In Russian.

041.058 On the zenith discontinuity.
V. N. Pyshnenko, R. T. Fedorova.
19th Astrometrical Conference 1972, (see 012.019), p. 139 - 141 (1975). In Russian.

041.059 Determination of the declinations of the sun, Mercury and Venus from observations made with the Pulkovo vertical circle taking into account day-time flexure. G. S. Kosin.
19th Astrometrical Conference 1972, (see 012.019), p. 166 - 168 (1975). In Russian.

041.060 On photographic observations of the sun at equal heights with stars. A. G. Zhestkov.
19th Astrometrical Conference 1972, (see 012.019), p. 175 - 177 (1975). In Russian.

041.061 Observations of major planets with the normal astrograph in Pulkovo. N. M. Bronnikova, L. V. Zhukov, L. S. Koroleva, G. V. Panova.
19th Astrometrical Conference 1972, (see 012.019), p. 178 - 183 (1975). In Russian.

041.062 Photographic positional observations of the Galilean satellites of Jupiter and Saturn satellites in Pulkovo.
T. P. Kiseleva, L. S. Koroleva, G. V. Panova.
19th Astrometrical Conference 1972, (see 012.019), p. 183 - 185 (1975). In Russian.

041.063 Photographic positional observations of major planets with the AKD in Pulkovo.
T. P. Kiseleva, N. A. Shakht.
19th Astrometrical Conference 1972, (see 012.019), p. 185 -

186 (1975). In Russian.

041.064 **Observations of Venus and Mars with the wide-angle astrograph of the Sternberg Astronomical Institute.**
B. S. Vozdvizhenskij.
19th Astrometrical Conference 1972, (see 012.019), p. 187 - 190 (1975). In Russian.

041.065 **Photographic observations of Venus at the Main Astronomical Observatory of the Academy of Sciences of the Ukrainian SSR in 1967 and 1969.**
A. B. Onegina, E. M. Sereda.
19th Astrometrical Conference 1972, (see 012.019), p. 190 - 193 (1975). In Russian.

041.066 **Photographic observations of Venus in Pulkovo with the 26″ refractor.**
N. M. Bronnikova, A. A. Kiselev.
19th Astrometrical Conference 1972, (see 012.019), p. 193 - 194 (1975). In Russian.

041.067 **On the conversion of photographic observations of selected minor planets into a homogeneous co-ordinate system.** V. I. Orel'skaya.
19th Astrometrical Conference 1972, (see 012.019), p. 195 - 200 (1975). In Russian.

041.068 **Results of photographing galaxies and first results of determination of absolute proper motions of stars.** L. P. Panteleeva.
19th Astrometrical Conference 1972, (see 012.019), p. 217 - 219 (1975). In Russian.

041.069 **Application of the multi-factor dispersion analysis in photographic astrometry.** S. G. Valeev.
19th Astrometrical Conference 1972, (see 012.019), p. 219 - 221 (1975). In Russian.

041.070 **Catalogue of declinations of the stars of the list of latitude programs from observations with the zenith telescope of the Moscow Observatory.**
I. M. Kalinina.
19th Astrometrical Conference 1972, (see 012.019), p. 226 - 230 (1975). In Russian.

041.071 **Advantages and possibilities of the zenith and zonal programs of observations with a zenith telescope.**
S. V. Drozdov.
19th Astrometrical Conference 1972, (see 012.019), p. 230 - 232 (1975). In Russian.

041.072 **New result of a determination of declinations of stars from observations with a zenith telescope.**
L. S. Bratolyubova.
19th Astrometrical Conference 1972, (see 012.019), p. 232 - 233 (1975). In Russian.

041.073 **Analysis of results of observations of Talcott pairs compiled from stars of a list of the Catalogue of Faint Stars.** N. I. Karchevskaya.
19th Astrometrical Conference 1972, (see 012.019), p. 233 - 236 (1975). In Russian.

041.074 **Construction of an absolute system of declinations of stars with the help of astrolabes.**
E. I. Krejnin, S. A. Murri.
19th Astrometrical Conference 1972, (see 012.019), p. 236 - 242 (1975). In Russian.

041.075 **On the possibility of using observations of artificial earth satellites in fundamental astrometry.**

D. P. Duma.
19th Astrometrical Conference 1972, (see 012.019), p. 242 - 246 (1975). In Russian.

041.076 **On objectifying observations according to the method of equal heights.** K. Steinert.
19th Astrometrical Conference 1972, (see 012.019), p. 246 - 250 (1975). In Russian.

041.077 **The influence of random and systematic errors of catalogues on observational results according to Zinger's method.** V. I. Sergienko.
19th Astrometrical Conference 1972, (see 012.019), p. 250 - 257 (1975). In Russian.

041.078 **Catalogue of declinations of FKSZ stars in the FK4 system compiled from observations of V. A. Mikhajlov.**
K. N. Kuz'menko, N. S. Olifer, L. S. Pavlenko, V. Kh. Pluzhnikov.
Vestn. Khar'kov. Univ., No. 117, (Ser. Astron., vyp. (No.) 9), p. 37 - 50 (1974). In Russian.

041.079 **Results of observations made with the meridian circle of the Kharkov Observatory in 1968 - 1973.**
K. N. Derkach.
Vestn. Khar'kov. Univ., No. 117, (Ser. Astron., vyp. (No.) 9), p. 64 - 72 (1974). In Russian.

Positional Astronomy.   See Abstr. 003.096.

Orientation of coordinate systems in space and investigation of systematic errors of star catalogues. See Abstr. 011.003.

A method to identify the reference stars with an electronic computer.   See Abstr. 031.215.

New determination of the plate constants of the Bordeaux zone of the Astrographic Catalogue. See Abstr. 031.221.

Les erreurs sur les étoiles de référence dans la méthode des dépendances et leurs conséquences. See Abstr. 031.222.

On the reduction of measured coordinates to stand-ard coordinates with terms depending on color and magnitude of stars.   See Abstr. 031.246.

Algorithm for automatic identification of plate stars with a catalogue.   See Abstr. 031.402.

An infrared astrometric interferometer. See Abstr. 034.060.

A new driving device for astrolabe micrometers. See Abstr. 034.103.

A dynamical determination of the general preces-sion in longitude.   See Abstr. 043.006.

The effect of systematic differences between refer-ence positions on the second harmonic of the earth's gravita-tional potential.   See Abstr. 046.007.

Catalogue of stars observed photoelectrically. See Abstr. 113.023.

Catalogue of southern stars embedded in nebulosity. See Abstr. 132.017.

042 Celestial Mechanics, Figure of Celestial Bodies

042.001 **A grid search for three-dimensional motions and three new types of such motions.**
P. G. Kazantzis, C. L. Goudas.
Astrophys. Space Sci., Vol. 32, 95 - 113 (1975).

A grid search method aimed at locating 'all' doubly symmetric orbits of the three-dimensional restricted problems of one, two, etc. revolutions is developed and applied numerically. Three new types of orbits have thus been located and a second order 'predictor-corrector' method is applied in order to determine a certain number of members of the families of which the 'located' orbits are members. The stability of these members is also discussed.

042.002 **New types of motion in Störmer's problem.**
A. Mavraganis, C. L. Goudas.
Astrophys. Space Sci., Vol. 32, 115 - 138 (1975).

Application of boundary value techniques in the case of an electron performing relativistic motions within a magnetic dipole such as that of the earth supplemented by a scanning process by means of which the entire phase space of the problem can be investigated, six new types of periodic motion have been discovered and computed. The stability of these motions is investigated and their direct bearing on formation and shape of the Van Allen zones of the earth is discussed.

042.003 **A generalization as to periodic solutions of disturbed elliptic restricted three-body problem.**
V. Matas.
Bull. Astron. Inst. Czechoslovakia (BAC), Vol. 26, 30 - 33 (1975).

Provided that perturbations due to the gravitational field and the radiation pressure from a disturbing body and a presence of resisting medium are considered to effect on the elliptic restricted three-body problem and in addition provided that a commensurability of motions of the three primaries (including the disturbing one) occurs, some special periodic motions of the infinitesimal body are shown to exist instead of the five equilibrium solutions (the libration points). No restrictions are made as regards the orbital elements of the three primaries.

042.004 **A contribution to a stability of the libration points of the elliptic restricted three-body problem.**
V. Matas.
Bull. Astron. Inst. Czechoslovakia (BAC), Vol. 26, 34 - 38 (1975).

On the basis of a formerly published (Matas, 1973) transformation and separation of the equations of variation corresponding to the libration points of the elliptic restricted three-body problem into Hill's equations, regions of stability and non-stability of the libration points have been derived here. A comparison is made to confront the transition curves obtained in the present paper and results of other authors based on different approaches to the problem.

042.005 **Tidal friction and generalized Cassini's laws in the solar system.** W. R. Ward.
Astron. Journ., Vol. 80, 64 - 70 = Calif. Inst. Techn., Div. Geol. Plan. Sci., *Pasadena*, Contr. No. 2282 (1975).

The tidal drift toward a generalized Cassini state of rotation of the spin axis of a planet or satellite in a precessing orbit is described. Generalized Cassini's laws are applied to several solar system objects and the location of their spin axes estimated. Of those considered only the moon definitely occupies state 2 with the spin axis near to the normal of the invariable plane. Most objects appear to occupy state 1 with the spin axis near to the orbit normal. Iapetus could occupy either

state depending on its oblateness. In addition, the resonant rotation of Mercury is found to have little effect on the tidal drift of its spin axis toward state 1.

042.006 **Families of symmetric periodic orbits of the restricted three body problem, when the perturbing mass is small. II.** P. Guillaume.
Astron. Astrophys., Vol. 37, 209 - 218 (1974).

This paper completes the work of an earlier article (Guillaume, 1969) concerning the symmetric periodic solutions of the restricted problem for small values of the mass ratio of P_1 and P_2. 1. The symmetry equations are expanded to first order for the case of the symmetric periodic solutions of the second kind, and are shown to be valid even if the basic solutions (relative to the case $\mu=0$) collide with P_1. 2. The equations of the symmetric periodic characteristics of the first and second kinds are obtained in global finite form.

042.007 **Das Dreikörperproblem am Computer.**
K. Steiner.
SuW, Vol. 14, 96 - 98 (1975).

042.008 **Analytical developments of the inverse of the distance.** N. Abu-El-Ata, J. Chapront.
Astron. Astrophys., Vol. 38, 57 - 66 (1975). In French.

The aim of the paper is the construction of an analytic expression for inverse powers of the distance between two planets. This is done by means of (1) Laplace coefficients and (2) Legendre polynomials.

042.009 **The solution of the integral equations of the theory of a figure for a quadratic law of density distribution.** V. N. Zharkov, A. V. Kozenko.
Astron. Zhurn. Akad. Nauk SSSR, Vol. 52, 125 - 138 (1975). In Russian. English translation in Soviet Astron., Vol. 19, No. 1.

An analytic solution of the equation of the figure in the form of power series with recurrent relations is found. The parameters e, k, h are found to be of third order. It is shown that the functional form of e, k, h is conserved in the third approximation.

042.010 **Particular solutions in the limited problem of the motion of three spheroids.** V. V. Vidyakin.
Astron. Zhurn. Akad. Nauk SSSR, Vol. 52, 152 - 158 (1975). In Russian. English translation in Soviet Astron., Vol. 19, No. 1.

The problem of the existence of libration points in the limited problem of the translational-rotational motion of a homogeneous spheroid satellite under gravitation of two spheroids is considered.

042.011 **On the equilibrium of "daily" planet satellites taking into account the planet's triaxiality.**
I. S. Kozlov.
Astron. Zhurn. Akad. Nauk SSSR, Vol. 52, 159 - 163 (1975). In Russian. English translation in Soviet Astron., Vol. 19, No. 1.

Investigation on the problem of the stability of equilibrium of a "daily" satellite of a rotating triaxial planet the potential of which is approximated by the force function of four fixed centers. It is shown that two of the four positions of relative equilibrium found are unstable, the other two stable in first approximation with regard to the position and velocity vector.

042.012 **Periodic orbits in the planar general three-body problem.** R. Broucke, D. Boggs.
Celestial Mechanics, Vol. 11, 13 - 38 (1975).

The article contains a numerical study of periodic solutions of the planar general three-body problem. Several new periodic solutions have been discovered and are described. In particular, there is a continuous family with variable masses, extending all the way from the elliptic restricted problem to the general problem with three equal masses. All examples have special symmetry properties which are described in detail. Finally the authors suggest some important applications to the natural satellites of the solar system.

042.013   A short derivation of the Sperling-Burdet equations.
       M. Silver.
Celestial Mechanics, Vol. 11, 39 - 41 (1975).

A short derivation is given of the regularized equations of motion for the perturbed two-body problem. This method is then applied to the slightly modified time transformation $dt/ds=r/\omega$.

042.014   Existence of periodic orbits of the second kind in the elliptic restricted problem of three bodies.
A. Sergysels-Lamy.
Celestial Mechanics, Vol. 11, 43 - 51 (1975).

Hale's method is used to show the existence of symmetric periodic orbits of the second kind for the particular case of the elliptic restricted problem of three bodies. In this treatment the author also obtains a new proof of the existence of periodic orbits of the first and second kinds in the circular restricted problem.

042.015   One of the solutions for the Laplace equation and its physical interpretation.   E. I. Burstein.
Celestial Mechanics, Vol. 11, 79 - 94 (1975).

In this paper it is confirmed once more that there exists the general solution of Laplace's equation in ellipsoidal coordinates which satisfies the Stäckel theorem and which was derived earlier by M. Jarov-Jarovoi and S. J. Madden. The author interprets physically the general solution in real space as potentials of layers of charge and double layers in which the distribution of densities is defined by Green's formula.

042.016   Asymptotic expansions in the perturbed two-body problem with application to systems with variable mass.   F. Verhulst.
Celestial Mechanics, Vol. 11, 95 - 129 (1975).

The main theorems of the theory of averaging are formulated for slowly varying standard systems and the author shows that it is possible to extend the class of perturbation problems where averaging might be used. The application of the averaging method to the perturbed two-body problem is possible but involves many technical difficulties which in the case of the two-body problem with variable mass are avoided by deriving new and more suitable equations for these perturbation problems.

042.017   Secular perturbations in general planetary theory.
       V. A. Brumberg, L. S. Evdokimova, V. I. Skripnichenko.
Celestial Mechanics, Vol. 11, 131 - 138 (1975).

Based on a general planetary theory, the secular perturbations in the motion of the eight major planets (excluding Pluto) have been derived in polynomial form. The linear terms of second order with respect to the planetary masses and the nonlinear terms of first order up to the fifth (and partly seventh) degree with respect to eccentricities and inclinations were taken into account in the right-hand members of the secular system.

042.018   On the force centre in the generalized three-body problem.   M. Zelený.
Bull. Astron. Inst. Czechoslovakia, Vol. 26, 114 - 116 (1975).

In the paper it is proved that in a non-collinear system of three mass points which interact among each other with quite arbitrary central forces which act along the lines joining the mass points and satisfying the principle of action and reaction, the directions of the resultant forces, acting on the individual bodies, intersect at a single point.

042.019   Close encounters of small bodies and planets.
       S. J. Weidenschilling.
Astron. Journ., Vol. 80, 145 - 153 = Mass. Inst. Techn., Planet. Astron. Lab., *Cambridge, Mass.,* Contr. No. 111 (1975).

A small body in an orbit which crosses that of a planet may be eliminated by collision or ejected from the solar system. A method is developed for computing the probabilities of these fates. With some simplifying assumptions, the probabilities can be expressed as functions of the relative velocity only, rather than the orbital elements. The ejection probability does not vary greatly with velocity. The results are in agreement with those of Bandermann and Wolstencroft, but not with those of Öpik; a reason for this disagreement is suggested. This method can be adapted to compute probabilities of generalized orbital changes. Some applications are discussed.

042.020   Dynamical history of coplanar two-satellite systems.   E. L. Ruskol, E. V. Nikolajeva (*Nikolaeva*), A. S. Syzdykov.
The Moon, Vol. 12, 3 - 10, 11 - 18 (1975). In Russian and English.

One of the possible early states of the earth-moon system was a system of several large satellites around the earth. The dynamical evolution of coplanar three-body systems is studied; a planet (earth) and two massive satellites (proto-moons) with geocentric orbits of slightly different radii. Such configurations may arise in multiple satellite systems receding from a planet due to tidal friction. The lifetime of a system of several proto-moons is mainly determined by their tidal interactions with the earth. For conditions which the authors have considered, the most probable result of the evolution was coalescence of satellites as the consequence of the collisions.

042.021   Numerisk beregning over bevaegelser i planetsystemet.   O. Møller.
Astron. Tidsskr., Årg. 8, p. 16 - 38 (1975).

042.022   Sur l'équivalence de la théorie de Hori-Mersman et de la théorie de Deprit en théorie des perturbations.
J. Meffroy.
Comptes Rendus Acad. Sci. Paris, Sér. A, Vol. 280, 21 - 23 (1975).

On indique une expression analytique générale liant entre eux, dans une théorie des perturbations, les opérateurs D de Hori-Mersman et les opérateurs L de Deprit lorsque les deux théories sont équivalentes.

042.023   Linear analysis of one type of second species solutions.   P. Guillaume.
Celestial Mechanics, Vol. 11, 213 - 254 (1975).

From Breakwell-Perko's matching theory, the author deduces input–output equations which make it possible to give a linear description of an arc of a second species solution starting far from the moon and also finishing far from the moon after a close approach to it. This allows a global analysis of the second species solutions. Special attention is drawn to the periodic symmetric second species solutions.

042.024   The representation of the force functions of the earth's and moon's gravitation in ellipsoidal coordinates.   E. I. Burstein.
Celestial Mechanics, Vol. 11, 255 - 264 (1975).

This paper derives asymptotic expansions of ellipsoidal coordinates in Cartesian coordinates and an expansion in spherical harmonics of the dominant term for the solution of

Laplace's equation corresponding to the gravitational force function for a two-dimensional finite body. On comparing the expansion of the dominant term derived here with known expansions of the force functions of the earth's and moon's gravitation the author obtains values for the semimajor axes and eccentricities of the singular ellipses of these bodies in terms of the second degree harmonic coefficients c_{20} and c_{22}.

042.025 Group method of calculating the perturbations of minor planets.
E. Z. Hotimskaya, M. S. Petrovskaya.
Celestial Mechanics, Vol. 11, 265 - 274 (1975).

An expansion of the disturbing function R is suggested for the special case when the orbits of a group of minor planets have approximately equal major semi-axes a_i. R is presented as a Taylor series in the deviations of a_i from some mean value a_0. The expansion has an analytical form with respect to $\alpha = a_0/a$, a being the major semi-axis of the disturbing planet. The poles corresponding to $\alpha = 1$ are taken out of the series, which contributes to the convergence of the expansion. The domain of convergence is given.

042.026 The orbital resonance amongst the Galilean satellites of Jupiter. A. T. Sinclair.
Monthly Notices Roy. Astron. Soc., Vol. 171, 59 - 72 (1975).

The resonance amongst three of the Galilean satellites is described in a way intended to demonstrate the similarities it has to the normal two-satellite resonance. The hypothesis that the resonance was formed by the action of tidal forces is discussed. The problem is too complicated to reach any firm conclusions, but the tidal hypothesis does not seem to be a satisfactory explanation. The inner three satellites of Uranus have an approximate relationship amongst their mean motions similar to that of the Galilean satellites. The type of libration possible in this system is briefly discussed.

042.027 On a Newton-Moser type method. O. H. Hald.
Numerische Math., Vol. 23, 411 - 426 (1975).

The author proves that a variant of Moser's iterative method for solving nonlinear equations is quadratically convergent and gives error bounds. He estimates the amount of arithmetic for the method and compares it to Newton's method. Finally the method is used to solve a problem with small divisors.

042.028 Point-connected rigid bodies in a topological tree.
P. W. Likins.
Celestial Mechanics, Vol. 11, 301 - 317 (1975).

Identical equations of motion are shown to emerge for a system of $n + 1$ rigid bodies all interconnected by n points, each of which is common to two bodies, by means of each of several derivation procedures, all of which employ a kinematical identity developed by Hooker and Margulies. Thus the previously published Hooker-Margulies/Hooker equations are shown to be the natural result of several derivation procedures other than the Newton-Euler method originally used.

042.029 Problème restreint des trois corps appliqué à un gyrostat. M. Pascal.
Celestial Mechanics, Vol. 11, 319 - 336 (1975).

This paper is concerned with an extension of the classical restricted problem of three bodies when the smallest body is not considered to be a point mass. It is assumed that the smallest body consists of a solid hub and symmetric rotors rotating at constant relative angular velocities. The mass center of the gyrostat satellite is presumed to occupy one of the five librations points $L_1, ..., L_5$ of the classical restricted problem of three bodies. Assuming that the gyrostatic moment can have arbitrary constant values, the author finds the set of positions of relative equilibrium of the gyrostat satellite. He then proceeds to define the domains of stability and instability.

042.030 Equations of motion for interconnected rigid and elastic bodies: a derivation independent of angular momentum. W. W. Hooker.
Celestial Mechanics, Vol. 11, 337 - 359 (1975).

Equations of motion are derived for systems of rotationally interconnected bodies in which the terminal bodies may be flexible and the remaining bodies are rigid. The model for a flexible body assumes that the elastic deformation is representable as a time-varying linear combination of given mode shapes. The paper also derives the appropriate form for gravitational terms, so that the equations can be used for flexible satellites. Also included are expressions for kinetic energy and angular momentum so that in case these are theoretically constant, they can be used to monitor the accuracy of the numerical integration. It is shown how interbody constraint forces and torques (which do not appear in the equations of motion) can be recovered from quantities available in this formulation, and also how to treat state variables which are prescribed functions of time.

042.031 Un formulaire pour le calcul des perturbations d'ordres élevés dans les problèmes planétaires.
J. Chapront, P. Bretagnon, M. Mehl.
Celestial Mechanics, Vol. 11, 379 - 399 (1975).

The authors present formulas in compact form for constructing high order planetary perturbations with respect to the disturbing masses. They have been built by an iterative process and give the variations of osculating elements. Singularities due to vanishing eccentricities and inclinations are not present in the differential equations. All elementary operations are manipulations of Fourier series with numerical coefficients, and great care has been taken to economize algebraic operations. Results are presented in vectorial, complex and scalar forms. Two illustrations are given (Jupiter and Saturn, Venus and Earth).

042.032 Quasi-periodic intermediate orbits for the major planets and resonances of zero order.
V. A. Brumberg, L. S. Evdokimova, V. I. Skripnichenko.
Astron. Zhurn. Akad. Nauk SSSR, Vol. 52, 420 - 430 (1975). In Russian. English translation in Soviet Astron., Vol. 19, No. 2.

Particular quasi-periodic solutions of the equations of motion of the major planets in rectangular heliocentric coordinates have been constructed by means of successive iterations with respect to the planetary masses. These solutions are presented as exponential series in multiples of the mean longitudes of the planets, the sum of all exponential indices in every term being equal to zero. The corresponding inequalities in the planetary motion are of zero order in the eccentricities and inclinations of the planetary orbits. The resonance terms due to close commensurabilities between three or more mean motions of the planets have been revealed. These resonance terms are of zero order in the eccentricities and inclinations. With respect to the planetary masses they are analytically at least of the second order but numerically they are comparable with the terms of the first order.

042.033 On the generalized restricted problem of three bodies. G. N. Duboshin.
IAU Colloquium No. 22, (see 012.005), p. 85 (1974). – Abstract.

042.034 Sur l'application de la méthode de Laplace-Danjon à la détermination des orbites.
V. Ionescu Vlăsceanu.
IAU Colloquium No. 22, (see 012.005), p. 111 - 119 (1974).

042.035 Development of the theory of planetary perturbations in non-classical cases.
M. S. Petrovskaya, T. V. Ivanova, N. I. Lobkova.
IAU Colloquium No. 22, (see 012.005), p. 121 - 131 (1974).

042.036 **Doubly-symmetric motions in the elliptic problem.**
G. Macris, G. A. Katsiaris, C. L. Goudas.
Astrophys. Space Sci., Vol. 33, 333 - 340 (1975).

Poincaré's continuation method is applied to the elliptic restricted problem for the computation of four families of doubly symmetric three-dimensional periodic orbits emanating from similar orbits corresponding to zero eccentricity. The results are given in four tables and the orbits are characterized in regard to their stability.

042.037 **Restricted problem stability and the solar system.**
V. V. Markellos, C. L. Goudas, G. A. Katsiaris.
Astrophys. Space Sci., Vol. 33, 341 - 346 (1975).

Reference periodic orbits are determined accurately for the planets of the solar system, using a restricted problem model with the sun and Jupiter as the two primaries. The prediction is verified that stability of the planetary orbits should imply stability of those reference orbits that are simple-periodic.

042.038 **Highly perturbed periodic oscillations around a small primary.**
V. V. Markellos, P. E. Moran, W. Black.
Astrophys. Space Sci., Vol. 33, 385 - 419 (1975).

A global picture of the families of simple periodic orbits in terms of their characteristics is given in the part of the plane (γ, ξ_1) of representation near the singularity corresponding to the small primary, for the case of the restricted problem with a small value of the mass parameter μ. The value used for μ is smaller than the critical value of Routh. The picture is found to be qualitatively different from the one corresponding to μ larger than the critical value. By means of the stability parameters four new families, consisting of asymmetric periodic orbits, are shown to exist as bifurcations of families of symmetric periodic orbits.

042.039 **On the evolution of families of periodic orbits approaching a small primary.**
V. Markellos, P. E. Moran, W. Black.
Astrophys. Space Sci., Vol. 33, L29 - L32 (1975).

A network of families of periodic orbits is obtained approximately for the case $\mu = 0.1$ of the restricted problem using a direct grid search. Only such orbits of the third body are considered that cross the synodical line of the primaries 'outside' the smaller of the two, perpendicularly, and in the direction of rotation of the system.

042.040 **The present state of the problem of the motion of the major planets.**
V. A. Brumberg, G. A. Krasinskij.
Uspekhi fiz. nauk, Vol. 114, 377 (1974). In Russian. – Abstr. in Referativ. Zhurn. 51. Astron., 3.51.109 (1975).

042.041 **Stability of Riemann ellipsoids.** N. B. Sotina.
Vestn. Mosk. un-ta. Mat., mekh., 1974, No. 5, p. 98 - 105. In Russian. – Abstr. in Referativ. Zhurn. 51. Astron., 3.51.112 (1975).

042.042 **Limit transformation of the complete integral of the two fixed centres problem.** Yu. N. Isaev.
Pis'ma v Astron. Zhurn., Vol. 1, No. 2, p. 44 - 47 (1975). In Russian.

The possibility is shown to pass from the complete integral of the two fixed centers problem to the complete integral of the limit version of this problem without solving the Hamilton–Jacobi equation.

042.043 **Periodic orbits in Hill's problem.** Eh. A. Vagner.
Pis'ma v Astron. Zhurn., Vol. 1, No. 3, p. 43 - 44 (1975). In Russian.

The existence of periodic satellite-type orbits in Hill's problem is shown by Whittaker's method.

042.044 **A propos de l'équation de Kepler.** J. Meeus.
Ciel et Terre, Vol. 91, 134 - 136 (1975).

042.045 **Resonance phenomena at rotations of artificial and natural celestial bodies.**
V. V. Beletskii (*Beletskij*).
Satellite Dynamics, Symp. 1974, (see 012.007), p. 192 - 232 (1975).

This review describes briefly the history of development and investigation of resonance phenomena in the rotations of natural and artificial celestial bodies.

042.046 **Regularization with Hamilton's principal function.**
K. Zare, V. Szebehely.
Bull. American Astron. Soc., Vol. 7, 342 (1975). – Abstr. AAS.

042.047 **A theory of the Trojan asteroids.** B. Garfinkel.
Bull. American Astron. Soc., Vol. 7, 342 (1975). Abstr. AAS.

042.048 **Some uses of the Jacobi integral for the restricted three-body problem.** T. R. McDonough.
Bull. American Astron. Soc., Vol. 7, 385 (1975). – Abstr. AAS.

042.049 **On fields of force on rotating asteroids.**
G. I. Pokrovskij.
Trudy vos'mykh chtenij, posvyashch. razrab. nauch. naslediya i razvitiyu idej K. Eh. Tsiolkovskogo, Kaluga, 1973. Sekts. "Probl. raket. i kosmich. tekhn". Moskva, 1974, p. 3 - 8. In Russian. – Abstr. in Referativ. Zhurn. 51. Astron., 4.51.127 (1975).

042.050 **Singularities of Newtonian gravitational systems.**
D. G. Saari.
Dynamical Systems, Proc. Sympos. Univ. Bahia, Salvador 1971, p. 479 - 487 (1973).

042.051 **Solutions périodiques et quasi-périodiques du problème restreint. II.**
R. Sergysels, A. Sergysels-Lamy.
Acad. Roy. Belgique, Bull. Cl. Sci., 5e Sér. Vol. 60, 436 - 443 (1974).

042.052 **The secular perturbations of the third order relative to oblateness.** E. I. Timoshkova.
Vestn. Leningr. un-ta, 1974, No. 19, p. 145 - 150. In Russian. Abstr. in Referativ. Zhurn. 51. Astron., 5.51.104; 62. Issled. kosmich. prostranstva, 5.62.334 (1975).

042.053 **On circular orbits in the dynamics of two particles taking into account the lag of interactions.**
V. I. Zhdanov, K. A. Piragas.
Probl. teorii gravitatsii i ehlementarn. chastits. Vyp. (No.) 5. Moskva, Atomizdat, 1974, p. 65 - 80. In Russian. – Abstr. in Referativ. Zhurn. 51. Astron., 5.51.105 (1975).

042.054 **On a method to describe the perturbed motion of a satellite.** M. Yu. Belyaev.
Uch. zap. Tsentr. aehro-gidrodinam. in-ta, Vol. 5, No. 6, p. 48 - 54 (1974). In Russian. – Abstr. in Referativ. Zhurn. 51. Astron., 5.51.106; 62. Issled. kosmich. prostranstva, 5.62.333 (1975).

042.055 **Investigation of the semi-regular precession of a solid body having a fixed point in the Newtonian field of force.** P. M. Burlaka.
Mekh. tverd. tela. Resp. mezhved. sb., 1974, vyp (No.) 7, p.

96 - 103. In Russian. – Abstr. in Referativ. Zhurn. 51. Astron., 5.51.113 (1975).

042.056 On the tides in the n-body problem. II. Quasiperiodic motions of inhomogeneous bodies.
M. S. Volkov.
In-t teor. astron. AN SSSR. Leningrad, 1974. 31 pp. In Russian. – Abstr. in Referativ. Zhurn. 51. Astron., 5.51.114 (1975).

042.057 Adiabatic invariants and phase equilibria for first-order orbital resonances. T. A. Heppenheimer.
Astron. Journ., Vol. 80, 465 - 472 (1975).

In the planar circular restricted three-body problem, the evolution of near-commensurable orbits is studied under change in the mass ratio μ. The evolution involves preservation of two adiabatic invariants. Transition from circulation to libration may occur; such transition are of two types. Type I transition occurs when the evolutionary track in phase space passes through near-zero eccentricity; as in the ordinary case (no transition), pre- and post-evolutionary states are linked by solution of a two-point boundary-value problem. Type II transition occurs when the evolutionary track encounters an unstable phase equilibrium, or periodic orbit.

042.058 The swinging spring – families of periodic solutions and their stability. I. D. L. Hitzl.
Astron. Astrophys., Vol. 40, 147 - 159 (1975).

Eleven families of periodic solutions for a planar spring pendulum system have been identified and their evolution with the energy constant C has been determined. Characteristics for the families are given for three values of a free parameter g (1/4, 1/3, 1/2). Numerous bifurcations are noted. The stability index k is evaluated along each family for the particular choice $g = 1/3$ corresponding to the lowest order resonance $\omega_1 = 2\omega_2$ between the spring (ω_1) and pendulum (ω_2) modes of oscillation.

042.059 The close triple approach. J. Waldvogel.
Celestial Mechanics, Vol. 11, 429 - 432 (1975).

A new method is described for representing the motion in the planar problem of three bodies when all three point masses simultaneously come close to each other. The main results are (1) that the motion during the critical phase of closest approach is intimately connected with triple parabolic escape and (2) that a sufficiently close triple approach generally leads to the escape of one body with arbitrarily high asymptotic velocity.

042.060 The restricted problem: an extension of Breakwell-Perko's matching theory. P. Guillaume.
Celestial Mechanics, Vol. 11, 449 - 467 (1975).

This paper extends Breakwell and Perko's 'first order' matching theory (1965, 1966) to a more general matching theory which is applicable to a wider class of second species solutions. In a first stage, the matching theory is elaborated on the basis of new assumptions on the orders of magnitude of the small parameters. In a second stage, the author constructs a matching theory which takes into account general assumptions which include his assumptions and Breakwell-Perko's.

042.061 Time transformations in the extended phase-space. K. Zare, V. Szebehely.
Celestial Mechanics, Vol. 11, 469 - 482 (1975).

Time transformations involving momenta in addition to the coordinates are studied from the points of view of stabilization and regularization of the equations of motion. The possibility of the stabilization of the equations of motion is investigated similarly to Stiefel's and Baumgarte's recent results. The relation between the original and new independent variables is integrated by a modification of Ebert's theorem

and it is shown that the new independent variable is Hamilton's principal function. Numerical examples illustrate the method and seem to indicate that the computation of close approach trajectories benefit especially by the transformations discussed. The Appendix offers an analytic treatment regarding the stabilization of the constant of energy.

042.062 Investigation of the stability of periodic motions in the neighbourhood of the collinear libration points.
G. I. Shirmin.
Celestial Mechanics, Vol. 11, 483 - 515 (1975). In Russian.

The problem of the stability of three-dimensional periodic orbits in the neighbourhood of a collinear libration point is solved within the limits of the three-dimensional circular restricted three-body problem. Major attention is given to the investigation of stability in the orbital sense, since in the proper sense all orbits are unstable according to Lyapunov's theory. It is shown that in order to resolve the question of stability, it is sufficient to consider the equations in their variational form. Analysis of the roots of the corresponding characteristic equations determines the orbital stability of planar and three-dimensional solutions. Finally, the possibility of conditional stability in the linear approximation is proved.

042.063 Construction of the first approximation of a quasi-periodic solution of the three-body problem by a computer. T. A. Tolmacheva.
Pis'ma v Astron. Zhurn., Vol. 1, No. 5, p. 34 - 38 (1975). In Russian.

The possibility to find a quasiperiodic solution of the three-body problem is considered. The first approximation of a quasiperiodic solution is constructed in numerical-analytical form by a computer.

042.064 On the problem of the three fixed centers.
G. T. Arazov.
Pis'ma v Astron. Zhurn., Vol. 1, No. 6, p. 42 - 45 (1975). In Russian.

The plane problem of motion of a passively gravitating material point in the gravitational field of three fixed points is reduced to quadratures. One of the fixed masses is real and is located in the mass center of the system, while two others may be either real or conjugate-complex, all being located on the same line. The application of the results to the theory of motion of planetary satellites is discussed.

042.065 On the calculation of intermediate motions in triple stellar systems. A. A. Orlov, N. A. Solovaya.
Trudy Gos. Astron. Inst. Shternberga, Vol. 45, 119 - 136 (1974). In Russian.

A method of calculation of intermediate orbits of triple stellar systems is developed. Two systems of calculating formulas accommodate for diverse values of the parameters describing the stellar system are given.

042.066 Some cases of intermediate motions in the stellar three-body problem. N. A. Solovaya.
Trudy Gos. Astron. Inst. Shternberga, Vol. 45, 137 - 153 (1974). In Russian.

An investigation of extreme cases of intermediate orbits of the stellar three-body problem is accomplished. The properties of the extreme orbits, particularly their stability, are analysed. The conditions for the application of general formulas are established.

042.067 On the expansions of Jacobi's function of the three-dimensional circular restricted problem of three bodies into power series in the vicinity of the collinear libration points. G. I. Shirmin.
Trudy Gos. Astron. Inst. Shternberga, Vol. 45, 154 - 178 (1974). In Russian.

Two types of expansions of Jacobi's function in power series in coordinates of the infinitesimal body in the vicinity of the collinear libration points are considered and regions of convergence of these expansions are examined.

042.068 **On calculation of the perturbations of a non-Keplerian orbit in the theory of satellite motion.**
A. A. Orlov.
Soobshch. Gos. Astron. Inst. Shternberga, No. 191, p. 3 - 34 (1974). In Russian.
A group of questions connected with the calculation of the perturbations of non-Keplerian intermediate orbits by the method of Orlov (1970) is considered. The conditions of the applicability of the method are defined; the form of the development of the perturbing function is established; a method of calculations of the perturbations of first order is presented.

042.069 **Beitrag zum Problem der Stabilität.** C.-L. Siegel.
Nachr. Akad. Wiss. Göttingen, II. Math.-Phys. Kl., Jahrgang 1974, No. 3, 36 pp. (1974).

042.070 **'Boomerang' orbits and their numerical determination.** G. I. Tjivanidis, V. V. Markellos.
Astrophys. Space Sci., Vol. 35, 151 - 158 (1975).
A new class of special orbits is introduced which are recurrent in the configuration space only. An algorithm is suggested for their numerical determination in the case of a non-integrable dynamical system.

042.071 **Derivation of a condition on the motion of two deformable bodies.** J. N. Tokis.
Astrophys. Space Sci., Vol. 35, L27 - L32 (1975).
The author attempts to derive the conditions for which the motion of a system of two deformable (fluid or not) bodies can be reduced to the two-body problem. The new condition is discussed for some pairs of such bodies existing in the natural world.

042.072 **The system of equations of the theory of a figure of the fifth approximation.**
V. N. Zharkov, V. P. Trubitsyn.
Astron. Zhurn. Akad. Nauk SSSR, Vol. 52, 599 - 614 (1975). In Russian. English translation in Soviet Astron., Vol. 19, No. 3.
The system of equations of the theory of a figure of the fifth approximation is found. The system of the equations is given in generalized and average radius.

042.073 **Interpretation and application of the problem of four fixed centres.** I. S. Kozlov.
Astron. Zhurn. Akad. Nauk SSSR, Vol. 52, 649 - 656 (1975). In Russian. English translation in Soviet Astron., Vol. 19, No. 3.
Discussion of a new model problem of celestial mechanics – the planar problem of four fixed centres, two of which are located on the $O_3\xi$-axis, the other two on the $O_3\eta$-axis, at a definite distance from one another.

042.074 **Explanation of the symbols used in orbital elements.** S. W. Milbourn.
Journ. British Astron. Ass., Vol. 85, 329 - 331 (1975).

042.075 **Étude des perturbations planétaires directes de la lune.** F. Annabi.
Thèse présentée pour l'obtention du Diplôme de Docteur de 3e Cycle à l'Université de Paris VI. 5 + 85 pp. (1975).

042.076 **Das Dreikörperproblem am Computer-Bildschirm.** K. Steiner.
Sternenbote, 18. Jahrgang, p. 63 - 69 (1975).

042.077 **Positively invariant sets in the general three-body problem.** J. Yoshida.

Publ. Astron. Soc. Japan, Vol. 27, 347 - 356 (1975).
Merman's (1958) and Marchal's (1974) differential inequalities, which characterize the motion of a third distant mass-point, are modified for convenience of treatment, and three positively invariant sets in a 10-dimensional phase space with given values of energy and angular momentum (absolute value) are constructed on the basis of these modified inequalities. The first two of them correspond to escape motions and the last to a motion such that $\rho(t) \geqq kd > kr(t)$ for $t \geqq 0$.

042.078 **A new type of solution of the two-body problem and the non-existence of a solution of analogous structure for the plane restricted circular three-body problem.**
M. N. Kiosa.
In-t teoret. i ehksperim. fiz. ITEhF-37. Moskva, 1974, 14 pp. Price 6 Kop. In Russian. – Abstr. in Referativ. Zhurn. 51. Astron., 6.51.142 (1975).

042.079 **On the development of the perturbing function in the three-body problem.** Eh. A. Borisov.
Izv. vyssh. ucheb. zavedenij. Geod. i aehrofotosemka, 1974, No. 4, p. 59 - 66. In Russian. – Abstr. in Referativ. Zhurn. 51. Astron., 6.51.143 (1975).

042.080 **Geometrical proof of some theorems on the conservation of the integrals of motion.**
A. M. Kovalev.
Mekh. tverd. tela. Resp. mezhved. sb., 1974, vyp. (No.) 7, p. 36 - 40. In Russian. – Abstr. in Referativ. Zhurn. 51. Astron., 6.51.160 (1975).

042.081 **Asymptotic solution of problems of optimization of mass point motion with a generalized law of mass consumption.**
Yu. A. Klikh, O. F. Makarov, N. A. Avramchuk.
Mekh. tverd. tela. Resp. mezhved. sb., 1974, vyp. (No.) 7, p. 114 - 127. In Russian. – Abstr. in Referativ. Zhurn. 51. Astron., 6.51.161; 62. Issled. kosmich. prostranstva, 6.62.312 (1975).

042.082 **Algorithms of celestial mechanics. (Materials for mathematical providing of computers).**
In-t teor. astron. AN SSSR. Preprint 02. Leningrad, 1974. 48 pp. Price 14 Kop. In Russian. – Abstr. in Referativ. Zhurn. 51. Astron., 6.61.170 (1975).

042.083 **The swinging spring – invariant curves formed by quasi-periodic solutions. III.** D. L. Hitzl.
Astron. Astrophys., Vol. 41, 187 - 198 (1975).

042.084 **Periodic orbits in the regularized restricted three-body problem. I. Regularization of the equation of motion.** B. Szczodrowska-Kozar.
Acta Univ. Wratislav. No. 231, Mat., Fiz., Astron. XII, p. 61 - 71 = Wrocław Astron. Obs., Repr. No. 93 (1974).
This article considers the two-dimensional restricted three-body problem and its regularization specially adapted to the study of the collision Trojan orbits (with the sun). With the aid of this regularization the author is going to deal with the short-period Goodrich Trojan orbits. In the article the time regularization, regularization of the coordinates, of the equation of motion, and of the Jacobian integral is described in detail. The problem brought up in such a way be as well applied to another mass ratio.

042.085 **Numerical orbits near the triangular lunar libration points.** J. I. Katz.
Icarus, Vol. 25, 356 - 359 (1975).
The behavior of a test particle placed at a triangular libration point of the earth–moon system is calculated using Newton's equations for the four-body problem, with arbitrarily

chosen initial conditions. If the orbits of the massive bodies have their real eccentricities, then the test particle leaves the vicinity of the libration point in three years, much faster than if the orbits were circular. Very small particles are affected by solar radiation pressure, and may leave even faster.

042.086 On the significance of perihelion shift calculations. M. J. Duff.
General Relativity and Gravitation, (*GB*), Vol. 5, 441 - 452 (1974).
Attempts to resolve a confusion which exists in the literature over whether the perihelion precession of planetary orbits is a linear or non-linear effect. The problem is both one of physics and semantics.

042.087 Symmetries of the positive-energy Kustaanheimo-regularized Kepler problem revisited.
B. R. Sutton.
Nuovo Cimento B, Ser. 11, Vol. 25B, 662 - 668 (1975).
Following previous work of Pirani (1974) for negative-energy Kepler orbits, an $SL_{2,c}$ formalism is developed for the positive-energy case. The group of linear transformations permuting the orbits is found to be locally isomorphic to $SO_{3,3}$.

042.088 New integrals in an elliptic restricted problem of three-bodies in three dimensional co-ordinate system. S. K. Sinha.
Proc. National Acad. Sci. India, Ser. A, Vol. 43, 357 - 371 (1973).
Following the method of Contopoulos (1967), the author determines two nonalgebraic integrals in elliptic restricted problems of three bodies, which are necessary for the complete integration of this problem.

042.089 Gravitational radiation-reaction effects in a star-planet system. J. A. Pelterson.
Phys. Rev. D, Particles and Fields, Vol. 11, 253 - 256 (1975).
The development of the orbit of a planet in the gravitational field of a massive star with a non-zero mass quadrupole moment is investigated. Study of the reaction on the orbit due to the emission of gravitational radiation shows that, in addition to the known decrease in radius, there is also a gradual tilting of the orbit.

042.090 Equations of motion for the perturbed restricted three-body problem. T. A. Heppenheimer.
AIAA Journ., Vol. 12, 1442 (1974).
Various methods for the derivation of the equations of motion for the perturbed restricted three-body problems are discussed. For an exact treatment it has been found necessary to solve analytically a perturbed two-body problem.

Stable and random motions in dynamical systems. See Abstr. 003.104.

On the application of more accurate numerical integration methods to celestial mechanics. See Abstr. 021.004.

Programming system OPAL for analytical theories of celestial mechanics. See Abstr. 031.403.

Non-numeric computation for high eccentricity orbits. See Abstr. 052.007.

On the periodic solutions and resonance of spinning satellites in near-circular orbits. See Abstr. 052.008.

Analytical theories of the motion of artificial satellites. See Abstr. 052.020.

An estimation procedure for orbit determination, using the K.S. transformation. See Abstr. 052.022.

Numerical methods of orbital dynamics. See Abstr. 052.024.

On the comparison of numerical theories of orbital motion. See Abstr. 052.027.

On the dynamic equivalence of the post-Newtonian approximation of the theory of planetary motions in the general theory of relativity and in the inertia-free mechanics with Riemann's gravitation potential. See Abstr. 066.080.

Comparison of sunspot periods with planetary synodic period resonances. See Abstr. 072.050.

Equations of figures of planets. See Abstr. 091.033.

Monte Carlo simulation of asteroid collisions. See Abstr. 098.025.

Conditions suffisantes d'évasion ou de retour dans le cas plan du problème des comètes. See Abstr. 102.011.

Properties of motion in the gravitational field of a rotating bar. See Abstr. 151.018.

043 Astronomical Constants

043.001 **Solar test of Dirac's large numbers hypothesis.**
C. Chin, R. Stothers.
Nature, Vol. 254, 206 - 207 (1975).

Dirac has proposed that the large dimensionless numbers that are constructable from the fundamental constants of physics and astronomy are related to each other, and are simple functions of our present epoch in the universe. Thus he suggests that, as measured by atomic standards, the gravitational constant, G, varies with the time as t^{-1} and the number of nucleons in the universe increases as t^2. In more recent work he has pursued the consequences of assuming different modes of creation of new matter. The authors have found that Dirac's theory of multiplicative creation, but not his theory of additive creation, is not in contradiction with known facts about the sun.

043.002 **Variation of G.** P. A. M. Dirac.
Nature, Vol. 254, 273 (1975).

043.003 **A determination of the rate of change of G.**
T. C. Van Flandern.
Monthly Notices Roy. Astron. Soc., Vol. 170, 333 - 342 (1975).

A new analysis of lunar occultation observations from 1955 to 1974 utilizing Atomic Time gives a value for the empirical part of the secular acceleration of the moon's mean longitude of $(-65'' \pm 18'')/cy^2$. This differs from other determinations which utilized the Ephemeris Time scale. The remaining acceleration has as its most probable cause a decrease in the universal gravitational constant at the rate of $\dot{G}/G = (-8 \pm 5) \times 10^{-11}/yr$. The observed rate is also consistent with the Dirac and the Hoyle–Narlikar cosmological theories, and to a lesser degree, with the Brans–Dicke theory. Other possible interpretations of the observed excess (negative) lunar acceleration are also discussed; but only those with cosmological significance seem at all plausible.

043.004 **New determination of the precession by the principle of the maximum set of stars without proper motions.** B. Thüring.
Astron. Nachr., Vol. 296, 83 - 94 (1975). In German.

The principle of the maximum set of stars without proper motions requires that the corrections of the precession be specified so that the number of the stars "without proper motion" (practically $\mu < 0''10$) becomes a maximum. These stars are named the "Träger" (carriers) of the non-rotating coordinate system. The method avoids any hypothesis about the distribution of the proper motions and hence does not use the method of least squares. The main part of the procedure and the algorithm are described. The method is applied to the proper motions of the catalogue FK4, then to the proper motions of the catalogue N30 and finally to a third catalogue, which consists of the proper motions of N30 reduced to the system of FK4 (N30 → FK4). With regard to the accuracy as well as to the number of the "Träger" (carriers) of the coordinate system free of rotation the results from (N30 → FK4) have the following implications: the correction of lunisolar precession $\Delta p_1 = +0''97 \pm 0''04$ per century; the correction of the motion of equinox $\Delta p_2 = \Delta\lambda + \Delta e = +1''10 \pm 0''04$ per century. The new method has the advantage also, that fewer than half of the stars of a catalogue can influence the result.

043.005 **Recent researches relating to the IAU system of astronomical constants.** J. W. Siry.
Bull. American Astron. Soc., Vol. 7, 342 (1975). – Abstr. AAS.

043.006 **A dynamical determination of the general precession in longitude.** R. E. Laubscher.
Bull. American Astron. Soc., Vol. 7, 342 (1975). – Abstr. AAS.

043.007 **Messung der Lichtgeschwindigkeit nach Ole Roemer.** H.-U. Fuchs.
Orion, 33. Jahrgang, p. 75 - 80 (1975).

043.008 **Determination of precession and galactic rotation from the proper motions of the AGK3.**
G. Asteriadis.
Diss. Nat. Gesamtfakultät, Ruprecht-Karl-Univ., Heidelberg. 3 + 52 pp. (1975).

From the proper motions of 166 179 AGK3 stars of magnitude $m_{pg} \geqslant 8.0$ distributed from declination $-2°5$ to the north pole corrections to Newcomb's precession and Oort's constants of galactic rotation have been determined simultaneously with the solar motion. The result for the correction to Newcomb's lunisolar precession is: $\Delta p_1 = +1''10 \pm 0''05$ (m.e.) per century. The value $\Delta\lambda + \Delta e$ describing the combined correction due to incorrect planetary precession and a non-precessional motion of Newcomb's equinox is $\Delta\lambda + \Delta e = +1''37 \pm 0''07$ (m.e.) per century. The result for Oort's constants of galactic rotation is A = +16.1 ± 1.9 (m.e.) km/s/kpc, B = −9.0 ± 1.9 (m.e.) km/s/kpc.

043.009 **A method for deriving unknown parameters in the problem of determination of the constant of gravitation.** M. U. Sagitov.
Soobshch. Gos. Astron. Inst. Shternberga, No. 187, p. 13 - 19 (1973). In Russian.

043.010 **On the determination of the constant of gravitation with the help of torsion balance by the resonance method.** L. G. Gugel'.
Soobshch. Gos. Astron. Inst. Shternberga, No. 187, p. 34 - 45 (1973). In Russian.

043.011 **Formulas for precession.** H. Kinoshita.
Smithsonian Astrophys. Obs., *Cambridge, Mass.*, Special Report 364, 6 + 25 pp. (1975).

Literal expressions for the precessional motion of the mean equator referred to an arbitrary epoch are constructed. Their numerical representations are obtained. In constructing the equations of motion, the second-order secular perturbation and the secular perturbation due to the long-periodic terms in the motions of the moon and the sun are taken into account. These perturbations contribute more to the motion of the mean equator than does the term due to the secular perturbation of the orbital eccentricity of the sun. In this paper, the author uses the correction $\Delta f = 1''10$ to the speed of Newcomb's lunisolar precession, the correction $\Delta g = -0''03$ to the speed of Newcomb's planetary precession, and the geodesic precession $f_g = 1''915$.

On the torque of mutual gravitation of test masses.
See Abstr. 022.075.

Special cases of expansion of the torque of mutual gravitation of test masses. See Abstr. 022.076.

Precession, universal time, ephemeris time and rotation of the earth. See Abstr. 044.013.

Note on the determination of the 18.6 year nutation terms by the Z-term of the ILS. See Abstr. 045.014.

044 Time, Rotation of the Earth

044.001 Tidal nonuniformity in the earth's rotation.
G. P. Pil'nik.
Astron. Zhurn. Akad. Nauk SSSR, Vol. 52, 178 - 188 (1975).
In Russian. English translation in Soviet Astron., Vol. 19, No. 1
A short review of the investigations of the tidal non-uniformity in the earth's rotation is given. The results of a spectral analysis of the special coordinate system of the astronomical observations compiled from the data of 25 observatories for 20 years are presented.

044.002 Die Sekunde. G. Becker.
PTB Mitt., 85. Jahrgang, p. 14 - 28 (1975).
The development of the time unit and the time scale unit, the second, from the astronomical to the atomic definition is described considering the contributions of the Physikalisch-Technische Bundesanstalt to the resolutions of international scientific and intergovernmental organizations.

044.003 On the earth's axes of rotation and figure.
R. d'E. Atkinson.
Monthly Notices Roy. Astron. Soc., Vol. 171, 381 - 386 (1975).
A rigorous geometrical proof is given for the statement that meridian observations of declination involve the earth's axis of figure, so that nutation should be computed for this axis and not for the instantaneous axis of rotation as has always been done hitherto. A pair of observations of a star at upper and lower transit is analysed to show that the classical expressions for the fundamental declination, and the latitude of date, which had originally been acceptable, needed amendment after the work of Oppolzer, but become valid again if the nutation is changed as suggested.

044.004 Long-period time interval markers from Mercury transits. R. Gerharz.
Journ. British. Astron. Ass., Vol. 85, 252 - 254 (1975).

044.005 Terrestrial timekeeping and general relativity – a discovery. W. H. Cannon, O. G. Jensen.
Science, Vol. 188, 317 - 328 (1975).

044.006 On the annual motion of the poles of rotation and inertia of the earth. I.
A. A. Korsun', S. P. Major, Ya. S. Yatskiv.
Astrometriya i Astrofizika, *Kiev,* vyp. (No.) 24, (see 003.003), p. 26 - 45 (1974). In Russian.
Seasonal components in the motion of the poles of rotation and inertia of the earth derived by means of the polar coordinates 1846 to 1972 are given. When deriving the annual polar motion the mean pole of the epoch of observation was adopted as origin of the coordinate system.

044.007 On the analysis of variations in the rate of the earth's rotation in 1956–1973.
A. A. Korsun', N. S. Sidorenkov.
Astrometriya i Astrofizika, *Kiev,* vyp. (No.) 24, (see 003.003), p. 46 - 52 (1974). In Russian.
Changes in the rate of the earth's rotation are divided by numerical filtering into two components: long-term variations (linear secular motion and long-period waves) and short-term variations. The periods and amplitudes of some waves and components are estimated.

044.008 Measurement of the velocity of the earth's rotation.
N. S. Sidorenkov.
Zemlya i Vselennaya, 1975, No. 2, p. 61. In Russian.

044.009 Reformulation of the relativistic conversion between coordinate time and atomic time. J. B. Thomas.
Astron. Journ., Vol. 80, 405 - 411 (1975).
The relativistic conversion between coordinate time and atomic time is reformulated to allow simpler time calculations relating analysis in solar system barycentric coordinates (using coordinate time) with earth-fixed observations (measuring "earth-bound" proper time or atomic time). After an interpretation in terms of relatively well-known concepts, this simplified formulation, which has a rate accuracy of about 10^{-15}, is used to explain the conventions required in the synchronization of a worldwide clock network and to analyze two synchronization techniques –portable clocks and radio interferometry. Finally, pertinent experimental tests of relativity are briefly discussed in terms of the reformulated time conversion.

044.010 Personal and instrumental error jumps in the photoelectric time service device of the Astronomical Observatory of the Latvian State University in the years 1970 and 1973. A. V. Ivanov.
Uch. Zap. Latv. Gos. Univ., Vol. 220, Astron., vyp. (No.) 11, (see 003.014), p. 17 - 28 (1975). In Russian.
Personal and instrumental error jumps in the work of the photoelectric time service device of the Latvian State University Astronomic Observatory are explained. An analysis of results obtained by the Latvian State University time service shows that photoelectrical transit instruments can give continuous clock-time corrections with an accuracy up to ±0.003 sec.

044.011 Polar motion and irregularities in the earth's rotation (review of papers published in 1970–1973).
Ya. S. Yatskiv, A. A. Korsun', N. T. Mironov.
Astrometriya i Astrofizika, *Kiev,* vyp. (No.) 25, (see 003.015), p. 88 - 97 (1975). In Russian.
The current state of knowledge on the problem of the earth's rotation is summarized. Particular attention is paid to the peculiarities in the earth's rotation which provide the means for studying a number of geophysical problems.

044.012 Time Service of the Mizusawa Observatory. Bulletins, Vol. 17, No. 1 - 12 , 1974.
S. Takagi, M. Aihara, K. Yokoyama, T. Hara, G. Murakami, T. Okuda.
Edited by the International Latitude Observatory of Mizusawa, Mizusawa-Shi, Iwate-Ken, Japan. 2 + 46 pp. (1974).
This Bulletin contains the results of time service and astronomical observations made at the Mizusawa Observatory from 1 January to 31 December, 1972.

044.013 Precession, universal time, ephemeris time and rotation of the earth. W. K. Hristov.
Izv. Tsentr. labor. geod. Blg. AN, 1974, kn. 14, p. 5 - 20.
Abstr. in Referativ. Zhurn. 51. Astron., 4.51.164 (1975).

044.014 Determination of the exact time with long-baseline radio interferometers.
N. S. Blinov, E. N. Fedoseev.
Soobshch. Gos. Astron. Inst. Shternberga, No. 190, p. 20 - 25 (1974). In Russian.
Estimates of the accuracy of determination of point radio source right ascensions and the accuracy of calculation of sidereal time from long-baseline interferometers are given.

044.015 Estimate of the accuracy of quartz clocks of the united Time Service of the Sternberg Astronomical Institute and the General Research Institute of Geodesy and

Cartography. N. S. Blinov, K. G. Belelin, I. N. Makarov. Soobshch. Gos. Astron. Inst. Shternberga, No. 190, p. 33 - 36 (1974). In Russian.

044.016 Il problema della rotazione terrestre.
F. Zagar.
Rend. Seminario Fac. Sci. Univ. Cagliari, Suppl. Vol. 44, (see 012.016), 1 - 18 (1974).

044.017 Methods of observing the earth's rotation from satellite orbit perturbations. K. Lambeck.
Rend. Seminario Fac. Sci. Univ. Cagliari, Suppl. Vol. 44, (see 012.016), 19 - 24 (1974).

044.018 Time and latitude service.
Polish Acad. Sci., Astron. Latitude Station, Borowiec, Circ. Nos. 132, 133, 9 + 11 pp. (1975). — 1974 October — 1975 March.

044.019 Astronomische Zeit- und Breitenbestimmungen. Empfangszeiten von Zeitsignalen.
Edited by Deutsches Hydrographisches Institut, Hamburg. 5 pp. (1974). — 1974 July — September.

044.020 On the amplitude changes of seasonal components in the rate of rotation of the earth. S. Okazaki.
Publ. Astron. Soc. Japan, Vol. 27, 367 - 378 (1975).
In this paper an analysis of seasonal variations in the rate of the earth's rotation is carried out with regard to the amplitude changes particularly. It is found that the annual and semiannual components have peculiar changes in the amplitude, i.e., (1) the annual term has been a tendency of the amplitude enhancement of about 0.10 ms day^{-1} and following decay which occurred rhythmically at 1957.5, 1963.5, and 1969.5 with a 6-yr period and (2) the semi-annual term had a step change of the amplitude by about $+0.13$ ms day^{-1} at the beginning of 1962. The amplitude change of the semi-annual term is proved to be attributed to the difference in $\Delta\alpha_a$ between the fundamental catalogs FK3 and FK4.

044.021 Détermination astronomique de l'heure et de la latitude.
Obs. Neuchâtel, Bull. (B), 1974 September — 1975 March (1975).

044.022 L'heure astronomique définitive de l'Observatoire de Neuchâtel.
Obs. Neuchâtel, Bull. (D), 1974 September — December (1975).

044.023 Astronomische Zeit- und Breitenbestimmungen, Empfangszeiten von Zeitsignalen, Präzisionszeitvergleiche.
Akad. Wiss. DDR, Zentralinst. Phys. Erde, Abt. Geod. Astron., Jahrgang 1974, Nos. 1 - 4, 24 + 17 + 16 + 16 pp. (1975). 1974 January - August.

044.024 Universal time and coordinates of the pole; Emission time of time signals; Universal time (coordinated); Independent local atomic time scales AT (i).

Bureau International de l'Heure, (B.I.H.), Paris, Circ. D98 - D104 (1975). — 1975 January - July.

044.025 Détermination astronomique de l'heure et heures demi-définitives de réception des signaux horaires.
L. Webrová, V. Ptáček.
Acad. Tchécoslov. Sci., Inst. Astron., Station de l'Heure, Prague, Sér. 7, No. 1, 14 pp. (1974). — 1974 January - February.

044.026 International Time and Latitude Service at the Tokyo Astronomical Observatory during 1974.
K. Osawa.
Tokyo Astron. Obs., Time and Latitude Bull., Vol. 48, Nos. 7 - 12, p. 41 - 79, i - vii (1974). — Results of the time determinations 1974 July - December.

044.027 Daily phase values and time differences.
U. S. Naval Obs., Washington, D.C., Time Service Publ., Ser. 4, Nos. 413 - 438 (1975). — 1975 January 1 - June 26.

044.028 Preliminary times and coordinates of the pole.
U.S. Naval Obs., Washington, D.C., Time Service Publ., Ser. 7, Nos. 366 - 391 (1975). — 1975 January 2 - June 26.

044.029 Time Service Announcement. G. M. R. Winkler.
U.S. Naval Obs., Washington D.C., Time Service Publ., Ser. 14, No. 18 (Frequency adjustment).

044.030 Time service. C. Moranzino (Editor).
Oss. Astron. Torino (Pino Torinese), Bull. No. 7, 15 pp. (1974). — Results of the time determinations 1974 January - April.

044.031 On the investigation of the earth's rotation according to results of observations with photographic zenith tubes. M. Meinig.
19th Astrometrical Conference 1972, (see 012.019), p. 259 - 261 (1975). In Russian.

044.032 Investigation of the irregularity of the earth's rotation at the Astronomical Observatory of the Kharkov State University and the Kharkov State Scientific Institute of Standards. V. I. Turenko.
Vestn. Khar'kov. Univ., No. 117, (Ser. Astron., vyp. (No.) 9), p. 51 - 57 (1974). In Russian.

Bureau International de l'Heure. Annual report for 1974. See Abstr. 008.094.

Résultats des observations faites à Alger avec l'astrolabe impersonnel A. Danjon OPL 8. Temps et latitude 1973. See Abstr. 041.023.

Free and forced motions of the earth. See Abstr. 046.029.

045 Latitude Determination, Polar Motion

045.001 The components of the secular motion of the earth pole. E. Fichera, A. Pugliano.
Astron. Nachr., Vol. 296, 41 - 43 (1975). In French.
After homogenizing the declination systems of the ILS Catalogue, the secular motion of the mean pole is studied. Our analysis leads to two components of this motion: a progressive component (already known) and an elliptical one.

045.002 Earth wobble, day length and continental drift. D. W. Hughes.
Nature, Vol. 253, 591 - 592 (1975).

045.003 Secular variations of the earth's gravity field and polar motion. M. I. Yurkina, A. Sh. Fajtel'son.
Astron. Zhurn. Akad. Nauk SSSR, Vol. 52, 171 - 177 (1975). In Russian. English translation in Soviet Astron., Vol. 19, No. 1.
Observational equations are deduced for determining polar motion from repeated astronomical observations of latitudes and longitudes and repeated gravity measurements. Zenith displacements in Kiev, Poltava, Kharkov and Gorky are evaluated using maps of secular variations of gravity. The values obtained are of the same order as the zenith displacements determined by astronomical methods.

045.004 On the occasion of the seventieth anniversary of latitude observations with the ZTF-135 at Pulkovo.
L. D. Kostina, N. R. Persiyaninova, V. I. Sakharov.
Vrashchenie i prilivn. deformatsii Zemli. Vyp. (No.) 6. Kiev, "Nauk. dumka", 1974, p. 105 - 111. In Russian. – Abstr. in Referativ. Zhurn. 51. Astron., 3.51.130 (1975).

045.005 Analysis of variations of the mean latitude of Poltava based on the observations with the Zeiss telescope. R. I. Popova.
Vrashchenie i prilivn. deformatsii Zemli. Vyp. (No.) 6. Kiev, "Nauk. dumka", 1974, p. 111 - 117. In Russian. – Abstr. in Referativ. Zhurn. 51. Astron., 3.51.131 (1975).

045.006 Study on some nutation terms by the ILS Z-term. K. Yokoyama.
Publ. International Latitude Obs. Mizusawa, Vol. 9, 1 - 45 (1973).
The periodic components of the Z-term were statistically investigated with relation to the nutation terms from the liquid core model by making use of the monthly mean latitudes of the International Latitude Service covering the period of three group observations from 1955 to 1971.

045.007 What causes a change of the Chandler motion? S. Takagi.
Publ. International Latitude Obs. Mizusawa, Vol. 9, 47 - 88 (1973).
The author intended to estimate the effect of geophysical phenomena on a change of the Chandler motion. He derived the equations of perturbation for the Chandler motion after Tisserand's method (1891). He derived the formulae of change for both amplitude and phase angle in several cases. He reviewed the geophysical phenomena and divided them into several genres.

045.008 On the nearly diurnal free nutation. M. Ooe.
Publ. International Latitude Obs. Mizusawa, Vol. 9, 133 - 159 (1973).
From the daily mean values of the residual latitudes of the ILS stations from 1962 to 1971 and of other three stations from about 1966 to 1971, a periodic term of 210 mean solar days was derived as to be common to all these stations, and its amplitude was found to be $0''.008 \pm 0''.002$. The period of the nearly diurnal free nutation was derived from the above results as $23^h57^m6^s$ in sidereal time.

045.009 Note on the retrograde sway. M. Ooe.
Publ. International Latitude Obs. Mizusawa, Vol. 9, 161 - 166 (1975).

045.010 On the effect of geoid deformation in latitude observations. I. Okamoto, T. Sasao.
Proc. International Latitude Obs. Mizusawa, No. 14, p. 1 - 8 (1974). In Japanese.
A possibility of explaining non-polar variation of astronomical latitude by means of geoid deformation is examined. A general formula describing latitude variations due to geoid deformation as well as station displacements and polar motion is presented. It is shown that necessary amount of the geoid deformation which could really provide observable latitude variations is rather large, and it should cause remarkable changes in sea level, absolute values of gravitational acceleration and other related quantities.

045.011 The materials for the study of the polar motion before 1891, collected in the library of the Tokyo Astronomical Observatory. N. Sekiguchi.
Proc. International Latitude Obs. Mizusawa, No. 14, p. 9 - 23 (1974). In English and Japanese.

045.012 On the observation errors of latitude observations with the Floating Zenith Telescope.
H. Ōkawa, Y. Gotō.
Proc. International Latitude Obs. Mizusawa, No. 14, p. 39 - 51 (1974). In English and Japanese.
Observing program was changed since 1967.0. The results of the observation made with FZT during 1967–1972 were analysed to see the several kinds of observation errors.

045.013 Spectral analyses of the Z term derived from the several combinations of the ILS stations. H. Ishii.
Proc. International Latitude Ob. Mizusawa, No. 14, p. 96 - 137 (1974). In Japanese.

045.014 Note on the determination of the 18.6 year nutation terms by the Z-term of the ILS.
K. Yokoyama.
Proc. International Latitude Obs. Mizusawa, No. 14, p. 138 - 144 (1974). In Japanese.
Both longitude and obliquity components of the 18.6 year nutation terms were derived for trial by the Z-terms of the ILS three group observations from 1955 to 1966. The results were $9''.203$ for the nutation constant and $-6''.836$ for the longitude component. The former value is in good agreement with the observational ones by other authors, while the latter shows a slightly different value.

045.015 Annual report of the International Polar Motion Service for the year 1972. S. Yumi.
Published for the International Council of Scientific Unions by Central Bureau of the International Polar Motion Service, Mizusawa, Japan. 4 + 193 pp. (1974).
The work of the International Polar Motion Service has progressed with a collaboration of the 56 stations and observatories all over the globe. This volume contains only the results of latitude observations and not those of time related with the polar motion as they were in the previous volumes. They will be published in the later volumes.

045.016 Choix d'un processus aléatoire pour décrire le mouvement du pôle, d'un point de vue cinématique.
F. Nahon.
Comptes Rendus Acad. Sci. Paris, Sér. A, Vol. 280, 1401 - 1403 (1975).

Soit z (t) le processus aléatoire stationnaire à valeurs complexes qui représente les fluctuations du mouvement du pôle. L'auteur a proposé de restreindre le choix des processus admissibles à ceux qui vérifient la condition E z (t) z (t') = 0. Il montre que le processus à accroissements orthogonaux associés vérifie la même condition; il indique les conséquences concernant la représentation empirique de la variation des latitudes.

045.017 Motion of the earth's poles. A. A. Mikhajlov.
Vasiona, Vol. 23, 9 - 13 (1975). In Serbo-Croatian.

045.018 The seasonal components of the motion of the rotation and inertia poles of the earth for 1850 - 1960. L. V. Rykhlova.
Soobshch. Gos. Astron. Inst. Shternberga, No. 190, p. 26 - 32 (1974). In Russian.

The annual components of the rotation pole and the annual motion of the inertia pole are calculated for 1850—1960. The positions of the rotation pole used are referred to the mean pole of the epoch of observations. The results of calculations are tabulated.

045.019 Betrachtungen zum Nachweis von Gezeiteneffekten in geodätisch-astronomischen Breitenbestimmungen. C. Elstner, J. Höpfner.
Akad. Wiss. DDR, Zentralinst. Phys. Erde, Veröff. No. 29, (see 013.012), p. 95 - 107 (1974).

045.020 The secular variation of longitudes and plate tectonic motion. E. Proverbio, V. Quesada.
Rend. Seminario Fac. Sci. Univ. Cagliari, Suppl. Vol. 44, (see 012.016), 37 (1974). — Abstract of the paper cited in 11.045.014.

045.021 Progressi nello studio di un modello matematico del moto di Chandler. P. Zampirollo.
Rend. Seminario Fac. Sci. Univ. Cagliari, Suppl. Vol. 44, (see 012.016), 39 - 45 (1974).

045.022 Mise en évidence du mouvement du pôle à partir d'observations Doppler d'un satellite par une station et des données de trajectoire fournies par le N.W.L. (expérience T.R.A.P.O.L.). N. Capitaine.
Rend. Seminario Fac. Sci. Univ. Cagliari, Suppl. Vol. 44, (see 012.016), 53 - 63 (1974).

045.023 L'introduction des coordonnées du pôle déduites de l'observation Doppler de satellite dans les calculs du Bureau International de l'Heure. N. Capitaine, M. Feissel.
Rend. Seminario Fac. Sci. Univ. Cagliari, Suppl. Vol. 44, (see 012.016), 65 - 73 (1974).

045.024 On some problem in the comparison of the polar motion derived from the classical time and latitude observations and the Doppler satellite method.
C. Sugawa, C. Kakuta.
Rend. Seminario Fac. Sci. Univ. Cagliari, Suppl. Vol. 44, (see 012.016), 117 - 120 (1974).

045.025 Monthly Notes of the International Polar Motion Service.
IPMS Monthly Notes, International Latitude Obs. Mizusawa (Japan). 1974 Nos. 11, 12, p. 85 - 101; 1975 Nos. 1 - 5, p. 1 - 47 (1975). — Announces the values of latitudes observed at the collaborating stations during 1974 November - 1975 May.

045.026 Is the nutation of the solid inner core responsible for the 24-year libration of the pole?
C. Kakuta, I. Okamoto, T. Sasao.
Publ. Astron. Soc. Japan, Vol. 27, 357 - 365 (1975).

045.027 The new observational programme of 12 groups of Horrebow - Talcott pairs for latitude variation determinations in Józefosław for the years 1972 - 1984.
B. Kołaczek, M. Dukwicz-Latka.
Warsaw Techn. Univ., Astron. – Geod. Obs. Józefosław. Latitude Circ. No. 48, 10 pp. (1974).

045.028 Results of the determination of latitude in Józefosław by observations of the Horrebow-Talcott pairs.
L. Pieczyński.
Warsaw Tech. Univ., Astron. – Geod. Obs. Józefosław. Latitude Circ. No. 51, 52 (1974). — 1974 July—December.

Bureau International de l'Heure. Annual report for 1974. See Abstr. 008.094.

Der Einfluß der Saalrefraktion im Meridianhaus des Lohrmann-Observatoriums auf die Breitenbeobachtungen nach Horrebow-Talcott. See Abstr. 031.261.

Résultats des observations faites à Alger avec l'astrolabe impersonnel A. Danjon OPL 8. Temps et latitude 1973. See Abstr. 041.023.

Time and latitude service.
See Abstr. 044.018.

Astronomische Zeit- und Breitenbestimmungen. Empfangszeiten von Zeitsignalen.
See Abstr. 044.019.

Détermination astronomique de l'heure et de la latitude. See Abstr. 044.021.

Astronomische Zeit- und Breitenbestimmungen, Empfangszeiten von Zeitsignalen, Präzisionszeitvergleiche.
See Abstr. 044.023.

International Time and Latitude Service at the Tokyo Astronomical Observatory during 1974.
See Abstr. 044.026.

Eine Lotabweichungskarte Westdeutschlands nach einem geodätisch konsistenten Kolmogorov-Wiener-Modell.
See Abstr. 046.019.

Simultaneous determination of latitude, clock correction and azimuth from observations of two stars in a common vertical. See Abstr. 046.027.

Chandler wobble and viscosity in the earth's core.
See Abstr. 081.017.

Some effects of the local character in the atmospheric structure on the astrometry (II).
See Abstr. 082.078.

Errata

045.901 Errata: 'Dynamics of the pole tide and the damping of the Chandler wobble' [Geophys. Journ. Roy. Astron. Soc., Vol. 39, 539 - 550 (1974)]. C. Wunsch.
Geophys. Journ. Roy. Astron. Soc., Vol. 40, 311 (1975).

046 Astronomical Geodesy, Satellite Geodesy, Navigation

046.001 **An approximate method for latitude determination at high latitudes.**
V. V. Kirichuk, A. S. Lavnikevich.
Geod.,kartogr. i aehrofotosemka. Resp. mezhved. nauch.-tekhn. sb., 1974, vyp. (No.) 20, p. 34 - 37. In Russian.
Abstr. in Referativ. Zhurn. 52. Geodeziya i Aehrosemka, 2.52.82 (1975).

046.002 **Analysis of refractional distortions in the determination of the astronomical azimuth.**
A. A. Akunej.
Geod., kartogr. i aehrofotosemka. Resp. mezhved. nauch.-tekhn. sb., 1974, vyp. (No.) 20, p. 3 - 5. In Russian. – Abstr. in Referativ. Zhurn. 52. Geodeziya i Aehrosemka, 2.52.83 (1975).

046.003 **Rattachement satellitaire des réseaux géodésiques continentaux par la méthode de la détermination de l'ellipsoïde mondiale de référence.** K. A. Czarnecki.
Geodezja Kartografia, Vol. 24, 3 - 21 (1975). In Polish.
L'auteur présente la méthode de la détermination de l'ellipsoïde mondiale de référence à la base des réseaux de triangulation satellitaire et du rattachement des réseaux géodésiques par l'intermédiaire de cette ellipsoïde. Cette méthode peut servir aussi à la détermination de la grande semi-axe de la terre au moyen du procédé géométrique à la base de la triangulation satellitaire.

046.004 **Three-dimensional transformation of coordinates to evaluate the accuracy of a large geodetic triangulation network.** J. Rysz.
Geodezja Kartografia, Vol. 24, 23 - 56 (1975). In Polish.
The astronomical-geodetic triangulation in Poland and the subsequent secondary network executed after the last war now cover the whole country. But both the secondary network and the primary triangulation still require final adjustment: discrepancies in some areas are of the order up to 1 meter, particularly in the centre and in the eastern parts. The author in this work analyses the primary and independently the secondary surveys and in conclusion suggests undertaking the additional linear traverses in $N-S$ and in $E-W$ directions.

046.005 **Sequential filtering applied to the determination of tracking station locations.**
B. E. Schutz, S. P. Condon, B. D. Tapley.
Journ. Geophys. Res., Vol. 80, 823 - 831 (1975).

046.006 **Geodetic uses of Lunar Laser Ranging.** A. Stolz.
Bull. Géod., Nouvelle Sér., No. 115, p. 5 - 16 (1975).
A recent improvement in measuring accuracy together with the development of new techniques suggests that geodetic point coordination to 1 part in 10^8 will become a reality in the near future. As a first step, however, the reference frame in which the points are to be coordinated, must be refined. Lunar Laser Ranging offers a means of solving both of these problems. The necessary equations are developed and the shortcomings of the method are analysed.

046.007 **The effect of systematic differences between reference positions on the second harmonic of the earth's gravitational potential.** P. Brosche.
Monthly Notices Roy. Astron. Soc., Vol. 171, 131 - 133 (1975).
If J_2 is derived from optical satellite tracking data, systematic errors of the order of $\Delta J_2 / J_2 \approx 10^{-7}$ may be expected because of zonal errors in the astrometric coordinate systems used.

046.008 **On the three-dimensional computation of geodetic networks and anomalies.** A. Marussi.
Atti Accad. Nazionale Lincei, Ser. 8, Rend. Cl. Sci. fis., mat., nat., Vol. 55, 460 - 466 (1973/74).

046.009 **Première détermination d'une longue base terrestre par télémétrie laser-lune et localisation du réflecteur de Lunakhod I.** O. Calame.
Comptes Rendus Acad. Sci. Paris, Sér. B, Vol. 280, 551 - 554 (1975).
La communication récente de mesures de distances lunaires issues de la station soviétique de Crimée a permis de déterminer la première très longue base terrestre entre les deux observatoires (McDonald, Texas, U. S. A. et Simeis, Crimée, U. R. S. S.) (9453, 373 km±6 m) ainsi que la localisation du réflecteur de Lunakhod I.

046.010 **Determination of coordinates for Helsinki and Riga stations based on observations of GEOS-B.**
N. A. Sorokin, V. I. Krylov.
Pis'ma v Astron. Zhurn., Vol. 1, No. 3, p. 45 - 47 (1975). In Russian.
Coordinates for Helsinki and Riga stations from observations of artificial earth satellites are determined by the method of chord intersections in space.

046.011 **On the accuracy of astrogeodetic geoid heights.**
E. Groten, H. Schaab.
Boll. Geod. Sci. Affini, Anno 34, p. 21 - 44 (1975).
Accuracy of astrogeodetic method in geoid height computation is studied. Plumb-line deflection, torsion balance observations and astrogravimetric procedures as applied in a test area (Noerdlinger Ries) are discussed.

046.012 **Influence of the atmosphere on the accuracy of coordinate determination from phase measurements of radio signals of artificial earth satellites.**
V. A. Danilin, A. V. Plotnikov.
Kosmich. Issled., Vol. 13, 266 - 271 (1975). In Russian.

046.013 **Geodetic results from ISAGEX data.**
J. G. Marsh, B. C. Douglas, D. M. Walls.
Bull. Géod., Nouvelle Sér., Année 1975, No. 116, p. 117 - 130 (1975).
Laser and camera data taken during the International Satellite Geodesy Experiment (ISAGEX) have been used in dynamical solutions to obtain center-of-mass coordinates for several Astro–Soviet camera sites . The results are accurate to about 20 m in each coordinate. The orbit of PEOLE (i = 15°) has also been determined from ISAGEX data and mean Kepler elements suitable for geodynamic investigations are presented.

046.014 **Satellite Doppler solutions in terms of a single parameter.** M. K. Paul.
Bull. Géod., Nouvelle Sér., Année 1975, No. 116, p. 131 - 142 (1975).
By an appropriate combination of the integrated Doppler counts for a motionless ground station over two consecutive arcs of a satellite path, it is possible to obtain a linear mathematical model relating the coordinates of the ground station to the observations. The method is described and applications are given.

046.015 **Note au sujet de l'article "Satellite Doppler solutions in terms of a single parameter" de M. K. Paul.**

H. M. Dufour.
Bull. Géod., Nouvelle Sér., Année 1975, No. 116, p. 143 - 148 (1975).

046.016   Ortsbestimmung mit Satelliten.
G. Zimmermann.
SuW, Vol. 14, 186 - 188 (1975).

046.017   Determination of the position of a ship from inter-
dependent altitudes of heavenly bodies.
G. V. Makarov.
Sudovozhdenie. Vyp. (No.) 14. Moskva, 1974, p. 128 - 140.
In Russian. – Abstr. in Referativ. Zhurn. 51. Astron., 4.51.185 (1975).

046.018   Verfahren zur Lösung des Oberflächenintegrales für
das Modell der einfachen Schicht in der Satelliten-
geodäsie.   H. Fröhlich.
Deutsche Geod. Kommission Bayer. Akad. Wiss.,Reihe C, No.
207, 47 pp. (1975). – Diss. Hohe Landwirtschaftliche
Fakultät Rheinisch. Friedrich-Wilhelms-Univ. Bonn.

046.019   Eine Lotabweichungskarte Westdeutschlands nach
einem geodätisch konsistenten Kolmogorov-Wiener-
Modell.   E. Grafarend, G. Offermanns.
Deutsche Geod. Kommission Bayer. Akad. Wiss.,Reihe A, No.
82, 60 pp. (1975).
A vertical deflection map of Western Germany is pre-
dicted by the multivariate Kolmogorov–Wiener concept out|
of about 200 astrogeodetic deflections of the Western German
first order net stations. It is assumed that the vertical deflection vector is homogeneously and isotropically distributed, its
covariance matrix has Taylor–Karman structure. The spectrum
of auto- and crosscorrelations for a flat earth model is given.
The model is tested relative to the unvariate Kolmogorov-
Wiener concept of Heitz.

046.020   Ortsbestimmung mit künstlichen Erdsatelliten nach
der Standlinienmethode.   G. Gerstbach.
Österreich. Zeitschr. Vermessungswesen, 62. Jahrgang,
p. 181 - 182 (1975).

046.021   On the determination of elliptic orbits from the
time micrometer observations.   O. Ojanen.
Rep. Finnish Geod. Inst., *Helsinki*, No. 75:3, 15 pp. (1975).
A method used at the Finnish Geodetic Institute to
determine a preliminary orbit from the time micrometer
observations is described. This method is useful in reducing
the satellite's position outside the observed short arcs.

046.022   Station coordinates for GEOS-C altimeter calibra-
tion and experimentation.
J. G. Marsh, B. C. Douglas, D. M. Walls.
Goddard Space Flight Center, Greenbelt, Maryland, GSFC
Document X-921-74-242, Preprint, 6 + 21 pp. (1974).
Station coordinates are given for the C-Band radar
GEOS-C altimeter calibration sites at Bermuda, Merritt, Grand
Turk, and Wallops Islands. Comparisons with other solutions
suggest a relative uncertainty of a few meters in each coordi-
nate.

046.023   Geodetic results from ISAGEX data.
J. G. Marsh, B. C. Douglas, D. M. Walls.
Goddard Space Flight Center, Greenbelt, Maryland, GSFC
Document X-921-74-250, Preprint, 6 + 17 pp. (1974).
Laser and camera data taken during the International
Satellite Geodesy Experiment (ISAGEX) have been used in

dynamical solutions to obtain center-of-mass coordinates for
the four Astro-Soviet camera sites. The results are accurate to
about 20 m in each coordinate. The orbit of PEOLE (i = 15°)
has also been determined from ISAGEX data and mean Kepler
elements suitable for geodynamic investigations are presented.

046.024   Konsequenzen der geodätischen Nutzung von
Laserbeobachtungen künstlicher Erdsatelliten.
L. Stange.
Akad. Wiss. DDR, Zentralinst. Phys. Erde, Veröff. No. 29,
(see 013.012), p. 85 - 93 (1974).
The possibilities taking advantage of the high accuracy
of recent laser instruments for geodetical purposes are in-
vestigated.

046.025   Erweiterung der Satellitenkamera SBG zur Laser-
entfernungsmessung.
H. Fischer, R. Neubert, C. Selke, R. Stecher.
Akad. Wiss. DDR, Zentralinst. Phys. Erde, Veröff. No. 29,
(see 013.012), p. 125 - 135 (1974).

046.026   Geodetic position of Sirahama Hydrographic
Observatory.
Data Rep. Hydrographic Observations, Ser. Astron. Geod.,
Tokyo, No. 9, p. 74 (1975).

046.027   Simultaneous determination of latitude, clock correction and azimuth from observations of two stars
in a common vertical.   V. V. Kirichuk, A. S. Lavnikevich.
Dokl. i nauch. soobshch. L'vov. politekhn. in-t, 1974, No. 3,
p. 163 - 164. In Russian. – Abstr. in Referativ. Zhurn. 51.
Astron., 6.51.210; 52. Geodeziya i Aehrosemka, 6.52.101
(1975).

046.028   On some connections of geophysics and geodesy
with astrometry.   E. Buschmann.
19th Astrometrical Conference 1972, (see 012.019), p. 79 -
81 (1975). In Russian.

046.029   Free and forced motions of the earth.
R. O. Vicente.
Boll. Geod. Sci. Affini, Anno 34, p. 173 - 183 (1975). – Lec-
ture at the International School of Advanced Geodesy, Erice
(Sicily), October 1 - 26, 1974.

Tables K_{11}: Two-star fix without use of altitude difference method. Vol. III N (Lat. 20°–29°30′ N).
See Abstr. 003.022.

Foundations of satellite geodesy.
See Abstr. 003.084.

Astronomy and geodesy. No. 4.
See Abstr. 003.152.

Ergebnisverbesserungen beim Reduktionsverfahren
nach Turner-Vick zum Anschluß von Einzelobjekten an Nachbarsterne.   See Abstr. 031.262.

On a graphical method of determination of the
azimuth of Polaris.   See Abstr. 031.266.

A radio interferometer as geodetical instrument.
See Abstr. 033.033.

On prospects of using lunar ranging measurements.
See Abstr. 094.003.

047 Ephemerides, Almanacs, Calendars

047.001 Mathematics of the calendar (The calendar formula and the permanent calendar).
A. V. Mikhajlovskij.
Nekotor. probl. issled. Vselennoj. No. 1. Leningrad, 1973, p. 184 - 208. In Russian. – Abstr. in Referativ. Zhurn. 51. Astron., 2.51.9 (1975).

047.002 Astronomical calendar of the Observatory in Sofia for the year 1975. D. Rajkova, Z. Krajcheva, N. Bakoev, R. Radkov, edited by A. Bonov.
Izdatelstvo na Blgarskata Akademiya na Naukite. Sofiya. Godina 22, 114 pp. Price 1 Lv. (1975). In Bulgarian.

047.003 Events of 1975 in the graphic time table.
Sky Telescope, Vol. 49, 33 - 35 (1975).

047.004 Anuário Astronômico 1975.
Published by Instituto Astronômico e Geofísico, Universidade de São Paulo, São Paulo, Brasil. 12 + 111 + 180*+ 2 pp. (1974).

047.005 Almanaque Nautico y Aeronautico para el año 1975.
Observatorio Naval, Republica Argentina, Armada Argentina, Servicio de Hidrografia Naval, Buenos Aires. Publ. H. 225, 384 pp. Price Arg $ 36.00 (1974).

047.006 Suplemento al Almanaque Nautico y Aeronautico para el año 1975. Sol, Planetas y Estrellas.
Observatorio Naval, Republica Argentina. Armada Argentina, Servicio de Hidrografia Naval, Buenos Aires. 8 + 133 pp. Price Arg $ 14.00 (1974).

047.007 Almanac for Geodetic Engineers 1975.
Prepared under the supervision of R. L. Kintanar. Republic of the Philippines – Department of National Defense, Philippine Atmospheric, Geophysical and Astronomical Services Administration (Weather Bureau), Quezon City. 11 + 33 pp. Price p 3.00 (1974).

047.008 Tables of Sunrise, Sunset, Twilight, Moonrise, and Moonset 1975.
Prepared under the supervision of R. L. Kintanar. Republic of the Philippines – Department of National Defense, Philippine Atmospheric, Geophysical and Astronomical Services Administration (Weather Bureau), Quezon City. 10 + 57 pp. (1974).

047.009 The Astronomical Ephemeris for the year 1976.
Issued by Her Majesty's Nautical Almanac Office, London; Nautical Almanac Office, United States, Naval Observatory, Washington. Her Majesty's Stationery Office, London. 8 + 574 pp. Price £ 3.80 (1974).

047.010 The Nautical Almanac for the year 1976.
Issued by Her Majesty's Nautical Almanac Office, London; and Nautical Almanac Office, United States, Naval Observatory, Washington. Printed and published by Her Majesty's Stationery Office London. A4 + 276 + 35pp. Price £ 1.60 (1974).

047.011 Connaissance des Temps ou des mouvements célestes pour l'an 1976 à l'usage des astronomes et des navigateurs.
Publiée par le Bureau des Longitudes under the supervision of B. Morando.
Gauthier-Villars Éditeur, Paris. 42 + 495 + A145 pp. Price F 220.00 (1975).

047.012 The American Ephemeris and Nautical Almanac for the year 1976.
Issued by Nautical Almanac Office, United States Naval Observatory, Washington; Her Majesty's Nautical Almanac Office, Royal Greenwich Observatory, London. U. S. Government Printing Office, Washington. 8 + 574 pp. Price $ 10.35 (1974).

047.013 Annuario per il 1975.
Boll. Geod. Sci. Affini, Anno 34, p. 110 - 123(1975).

047.014 Philippine Astronomical Handbook 1975.
Prepared under the supervision of R. L. Kintanar. Republic of the Philippines–Department of National Defense. Philippine Atmospheric, Geophysical and Astronomical Services Administration (Weather Bureau), Quezon City. 11 + 58 pp. (1974).

047.015 The Air Almanac 1975, September - December.
Her Majesty's Stationery Office, London; United States Naval Observatory, Washington. 246 + A84 + F4 pp. Price £ 2.15 (1975).

047.016 The Star Almanac for Land Surveyors for the Year 1976.
Prepared by H. M. Nautical Almanac Office, published by Order of The Science Research Council. Her Majesty's Stationery Office, London. 16 + 76 pp. Price 48p. (1975).

047.017 Japanese Ephemeris 1976.
Compiled under the supervision of A. M. Sinzi, by T. Uniwa, T. Mori, A. Senda, Y. Harada, K. Inoue, K. Nagamori, Y. Suzuki, T. Jojo.
Astronomical Division, Hydrographic Department, Tokyo, Japan. Pub. No. 684, 6 + 464 pp. (1974).

047.018 Apparent Places of Fundamental Stars 1977, containing the 1535 stars in the Fourth Fundamental Catalogue (FK 4).
Edited by Astronomisches Rechen-Institut, Heidelberg, under the supervision of T. Lederle. Published and produced by G. Braun GmbH, Karlsruhe. To be purchased from Verlag G. Braun, Karlsruhe, Germany. 44 + 510 pp. Price DM 42.00 (1975).

047.019 Almanaque Nautico 1976.
Published by Instituto y Observatorio de Marina, San Fernando (Cádiz). Printed in Spain by Observatorio de Marina, San Fernando (Cádiz). 420 + 30 pp. (1975).

047.020 Annuario 1975.
Edited by Osservatorio Astronomico di Torino. Pubbl. Varie Fuori Ser., Oss. Astron. Torino, No. 59, 59 pp. (1974).
Cronologia, Còmputo ecclesiastico gregoriano, feste mobili; Coordinate dell'Osservatorio astronomico di Torino (Pino Torinese); Calendario delle effemeridi del sole e della luna; I pianeti nel 1975; Ecclissi e occultazioni; Attività dell'Osservatorio (*M. G. Fracastoro*); La legge di gravitazione e l'avanzo secolare del perielio di Mercurio (*V. Banfi*); Recenti esplorazioni planetarie (*M. G. Fracastoro*); Insolazione a Pino Torinese (*A. Di Battista*).

047.021 Astronomical ephemeris for the year 1975.
P. M. Đurković.
Vasiona, Vol. 22, 104 - 120 (1974). In Serbo-Croatian.

047.022 Das Himmelsjahr 1975. M. Gerstenberger.

Franckh'sche Verlagshandlung Kosmos-Verlag, Stuttgart. 112 pp. Price DM 7.80 (1975).

047.023 **Astronomical calendar for 1975.** G. Ottewell. Furman University Physics Dept., Greenville, S.C. 48 pp. Price $ 4.95 (1974).

047.024 **Éphémérides Nautiques pour l' an 1976.** Ouvrage publié par le Bureau des Longitudes spécialement à l' usage des marins. Gauthier-Villars Editeur, Paris. 479 pp. (1975).

Compression of ephemerides. See Abstr. 041.006.

Space Research

051 Extraterrestrial Research, Spaceflight Related to Astronomy and Astrophysics

051.001 **Heavy cosmic-ray exposure of Apollo astronauts.**
E. V. Benton, R. P. Henke, J. V. Bailey.
Science, Vol. 187, 263 - 265 (1975).
A comprehensive study of the heavy-particle cosmic-ray exposure received by the individual astronauts during the nine lunar Apollo missions reveals a significant variation in the exposure as a function of shielding and the phase of the solar cycle. The data are useful in planning for future long-range missions and in estimating the expected biological damage.

051.002 **A blueprint for space research.** C. Norman.
Nature, Vol. 253, 151 (1975).

051.003 **Attaining to the asteroids (concerning the problem of opening up the cosmos according to K. Eh. Tsiolkovskij).** M. K. Tikhonravov.
Trudy sed'mykh chtenij, posvyashch. razrab. nauch. naslediya i razvitiyu idej K. Eh. Tsiolkovskogo, Kaluga, 1972. Sekts. Probl. raket. i kosmich. tekhn. Moskva, 1974, p. 58 - 72. In Russian. — Abstr. in Referativ. Zhurn. 62. Issled. kosmich. prostranstva, 2.62.76 (1975).

051.004 **Ten years cosmic era.** V. A. Shatalov.
Chelovek v kosmose. Moskva, Nauka, 1974, p. 23 - 29 (in Russian), p. 30 - 33 (in English).

051.005 **Project Daedalus: the origins and aims of the study.**
A. Bond, A. R. Martin.
Journ. British Interplanet. Soc., Vol. 28, 146 - 149 (1975).
A brief account is given of the origins of Project Daedalus, the BIS Starship Study, and of its aims. The paper is intended to serve as an introduction to the series of technical papers representing the results of the study.

051.006 **An evolution of X-ray astronomy instrumentation.**
N. Jagoda, W. D. Antrim, Jr.
Journ. British Interplanet. Soc., Vol. 28, 299 - 318 (1975).
This paper describes the growth of X-ray astronomy instrumentation from simple, single function, instruments performing survey measurements with crude spectral, spatial and temporal resolution to complex, multiple detector, instruments capable of observing selected phenomena and sources with extremely fine resolution.

051.007 **Investigation of Uranus, its satellites, and distant interplanetary phenomena by spacecraft techniques.**
J. A. van Allen.
Icarus, Vol. 24, 277 - 279 (1975).
A brief digest is given of the principal scientific objectives that can be addressed effectively, and perhaps uniquely, by spacecraft missions to Uranus. Practical considerations favor the launching of such missions in the late 1979, with subsequent swing-bys of Jupiter and arrival at Uranus in late 1986.

051.008 **Automated payload requirements for space shuttle.**
P. E. Culbertson, T. Hagler.
Acta Astronaut., Vol. 1, 1301 - 1314 (1974).

051.009 **The Apollo Telescope Mount on Skylab.**
R. Ise, E. H. Cagle.
Acta Astronaut., Vol. 1, 1315 - 1329 (1974).

051.010 **"Gamma-telescope" experiment. (Preliminary results).**
A. I. Belyaevskij, V. L. Bokov, V. K. Bocharkin, I. F. Bugakov, Yu. G. Derevitskij, B. A. Dmitriev, G. M. Gorodinskij, E. M. Kruglov, G. A. Pyatigorskij, A. M. Romanov, E. I. Chujkin.
Izv. AN SSSR. Ser. fiz., Vol. 38, 1838 - 1841 (1974). In Russian. — Abstr. in Referativ. Zhurn. 51. Astron., 2.51.233; 62. Issled. kosmich. prostranstva, 2.62.100 (1975).

051.011 **Europe in space: the emergence of ESA.**
Sky Telescope, Vol. 49, 284 - 286, 302 (1975).

051.012 **Survivability of microorganisms in space and its impact on planetary exploration.**
M. Frankenberg-Schwager, H. Bücker, H. Wollenhaupt.
Raumfahrtforschung, [DGLR - Deutsche Ges. für Luft- und Raumfahrt, Berlin], Vol. 18, 209 - 212 (1974).
A survey of some investigations on the survivability of microorganisms in space environment and studies on the effect of simulated space factors (extreme temperature, vacuum and UV-radiation) on microorganisms are presented. The results of these investigations indicate that terrestrial microorganisms can tolerate extreme environmental conditions and dispose on repair mechanisms to eliminate damages induced by this extreme environment. The implication of these findings on the exploration of Mars is discussed.

051.013 **A Mars rover concept for future landing missions.**
D. E. Koelle, W. Kokott, W. Schultze.
Raumfahrtforschung, [DGLR - Deutsche Ges. für Luft- und Raumfahrt, Berlin], Vol. 18, 224 - 235 (1974).

051.014 **Die Erforschung der Kometen mit Raumsonden. Vortrag anläßlich der XXIII. HOG-Jahrestagung in Salzburg.** H. Löb.
Astronautik, Jahrgang 12, p. 1 - 5 (1975).

051.015 **Experimental astrophysics.** C. Sagan.
Mercury, (Journ. Astron. Soc. Pacific), Vol. 4, No. 2, p. 18 - 23 (1975).

051.016 **Development of space vehicles in the USSR during 15 years (1957 - 1972).**
A. G. Mrykin, Yu. V. Biryukov.
Trudy sed'mykh chtenij posvyashch. razrab. nauch. naslediya i razvitiyu idej K. Eh. Tsiolkovskogo, Kaluga, 1972. Sekts. "Probl. raket. i kosmich. tekhn". Moskva, 1974, p. 3 - 19. In Russian. — Abstr. in Referativ. Zhurn. 62. Issled. kosmich. prostranstva, 3.62.3 (1975).

051.017 **First rocket experiments for investigation of the upper atmosphere.** B. A. Mirtov, L. A. Vedeshin.
Vestn. AN SSSR, 1974, No. 9, p. 117 - 123. In Russian.
Abstr. in Referativ. Zhurn. 62. Issled. kosmich. prostranstva,

3.62.9 (1975).

051.018 **The work of the cosmonauts A. A. Gubarev and G. M. Grechko aboard the station Salyut 4.**
Zemlya i Vselennaya, 1975, No. 2, p. 2 - 4. In Russian.

051.019 **Aircraft observing program for 100 minutes of totality in the 1977 eclipse.**   R. D. Mercer.
Bull. American Astron. Soc., Vol. 7, 351 (1975). — Abstr. AAS.

051.020 **Extra-atmospheric research in the submillimeter region with satellite-borne telescopes.**
A. E. Salomonovich, A. S. Khajkin.
Trudy fiz. in-ta AN SSSR, Vol. 77, 33 - 35 (1974). In Russian.
Abstr. in Referativ. Zhurn. 51. Astron., 5.51.85 (1975).

051.021 **Baikonur — Canaveral: common program.**
M. F. Rebrov.
Zemlya i Vselennaya, 1975, No. 3, p. 21 - 27. In Russian.

051.022 **Ten years later.**   A. I. Lazarev.
Zemlya i Vselennaya, 1975, No. 3, p. 28 - 31. In Russian.

051.023 **Scientific instrumentation of the Radio-Astronomy-Explorer-2 satellite.**   J. K. Alexander, M. L. Kaiser, J. C. Novaco, F. R. Grena, R. R. Weber.
Astron. Astrophys., Vol. 40, 365 - 371 (1975).
      The RAE-2 spacecraft has been collecting radio astro-nomical measurements in the 25 kHz to 13 MHz frequency range from lunar orbit since June, 1973. This paper presents a summary of the technical aspects of the program and illus-trates the performance of the experiments over the first 18 months of the flight. Among the unique features of the RAE-2 is the capability to observe repeated lunar occultations of strong radio sources at very low frequencies.

051.024 **Test results on the Viking gas chromatograph — mass spectrometer experiment.**   K. Biemann.
Origins of Life, Vol. 5, 417 - 430 (1974).

051.025 **The Copernicus satellite in the new era of space astronomy.**   D. W. Goldsmith.
The heritage of Copernicus, (see 003.020), p. 487 - 507 (1974).

051.026 **Helios/the sungrazer.**   H. O. Ruppe.
Astronaut. Aeronaut., (*USA*), Vol. 12, No. 6, p. 62 - 71 (1974).
      The author discusses the joint German—US space project, Helios, designed for brushing by the sun at 0.3 AU and carry-ing experiments for solar research only.

051.027 **Astronautics in the year 1974.**
M. Grün, P. Koubský.

Ŕíše hvězd, Vol. 56, 87 - 92 (1975). In Czech.

051.028 **Orientation measurement for balloon-borne tele-scopes.**   R. D. Joseph.
Journ. Phys. E, (Sci. Instruments), Vol. 8, 92 - 94 (1975).
      A simple and reliable system has been developed for post-flight measurement of the orientation of balloon-borne tele-scopes. It is easy to calibrate, requires no detailed knowledge of the geomagnetic field, and is adaptable to a variety of in-strumental configurations across the spectrum of balloon astronomy.

051.029 **Strategy for solar system exploration.**
Spaceworld, (*USA*), Vol. K-11-131, p. 4 - 21 (1974).
      The implementation of a solar system exploration pro-gram is reviewed, beginning with missions already approved and scheduled by NASA and suggesting further programs that might be valuable in studying the planets and their satellites, the interplanetary medium, and the comets and asteroids of the solar system.

051.030 **Space report.**
Journ. British Interplanet. Soc., Vol. 28, 284 - 293, 356 - 363, 437 - 440 (1975).

051.031 **Space report.**   Spaceflight, Vol. 17, 16 - 25, 60 - 65, 108 - 114, 139 - 145, 180 - 183, 188, 226 - 230, 237 (1975).

Ariel-V experiments.
Journ. British Interplanet. Soc., Vol. 28, 142 - 144 (1975).

Erfolgreiche HELIOS-Mission.
Umschau, 75. Jahrgang, p. 382 (1975).

A focussing X-ray collector and its response in flight.
See Abstr. 034.082.

On the possibility of using observations of artificial earth satellites in fundamental astrometry.
See Abstr. 041.075.

A Dutch satellite for astronomical research.
See Abstr. 054.018.

Automated life-detection experiments for the Viking mission to Mars.   See Abstr. 097.079.

Organic contamination problems in the Viking molecular analysis experiment.   See Abstr. 097.080.

An automatically-returned Martian sample by 1985?
See Abstr. 097.082.

052 Astrodynamics and Navigation of Space Vehicles

052.001 The thermal regime of "ether towns" at different distances from the sun. A. T. Ulubekov.
Trudy sed'mykh chtenij, posvyashch. razrab. nauch. naslediya i razvitiyu idej K. Eh. Tsiolkovskogo, Kaluga, 1972. Sekts. Probl. raket. i kosmich. tekhn. Moskva, 1974, p. 32 - 41. In Russian. – Abstr. in Referativ. Zhurn. 51. Astron., 2.51.2; 62. Issled. kosmich. prostranstva, 2.62.525 (1975).

052.002 Moments of solar radiation acting upon a satellite.
T. V. Grudnistyj, V. F. Kameko, Yu. T. Reznichenko, Eh. P. Yaskevich.
Kosmich. strela. Moskva, Nauka, 1974, p. 44 - 47. In Russian. Abstr. in Referativ. Zhurn. 62. Issled. kosmich. prostranstva, 2.62.386 (1975).

052.003 Determination of a model of disturbances acting upon a satellite during flight.
V. I. Dranovskij, V. N. Zigunov, L. V. Sokolov.
Kosmich. strela. Moskva, Nauka, 1974, p. 55 - 63. In Russian. Abstr. in Referativ. Zhurn. 62. Issled. kosmich. prostranstva, 2.62.387 (1975).

052.004 A form of differential equations of the motion of artificial satellites of the moon. G. G. Koman.
Astron. Zhurn. Akad. Nauk SSSR, Vol. 52, 207 - 209 (1975). In Russian. English translation in Soviet Astron., Vol. 19, No. 1.
Differential equations for elements of an intermediate orbit of an artificial satellite of the moon are obtained. The intermediate orbit, in contrast to the undisturbed orbit, takes into account perturbations from the second and third zonal harmonics of expansion of the moon's attraction potential.

052.005 Some remarks on the string problem treated by Singh and Demin. P. Hagedorn.
Celestial Mechanics, Vol. 11, 59 - 73, with a correction, p. 530 (1975).
The problem of the motion of a string attached to a satellite on a circular orbit, as treated by Singh and Demin, is reconsidered. In their paper they discuss problems of uniqueness and stability. In particular the radial equilibrium positions were found to be unstable in a certain sense. In the present paper it is shown that: (i) with the stability definition used by Singh and Demin the equilibrium of a string hanging in a uniform gravity field would also be unstable; (ii) a definition of stability more appropriate for continuous systems would establish the stability of the string both in orbit and in a uniform gravity field.

052.006 Project Daedalus: the navigation problem.
G. R. Richards.
Journ. British Interplanet. Soc., Vol. 28, 150 - 160 (1975).
The navigational requirements for an interstellar probe are assessed with reference to the British Interplanetary Society's Project Daedalus study.

052.007 Non-numeric computation for high eccentricity orbits. R. Sridharan, M. L. Renard.
Celestial Mechanics, Vol. 11, 179 - 194 (1975).
Geocentric orbits of large eccentricity ($e = 0.9$ to 0.95) are significantly perturbed in cislunar space by the sun and moon. Between the extremes of high accuracy digital integration of the equations of motion and of using an approximate, but very fast, stability criteria method, this paper is concerned with the development of a method of intermediate complexity using non-numeric computation. The computer is used as the theory generator to generalize Lidov's theory using six osculating elements. Symbolic integration is completely automa-

tized and the output is a set of condensed formulae well suited for repeated applications in launch window analysis. Examples of applications are given.

052.008 On the periodic solutions and resonance of spinning satellites in near-circular orbits.
V. J. Modi, K. C. Pande.
Celestial Mechanics, Vol. 11, 195 - 212 (1975).
Attitude motion of spinning axisymmetric satellites in presence of gravity-gradient and solar radiation pressure torques is studied analytically. The approximate closed-form solution developed for the nonlinear, nonautonomous, coupled fourth-order system proves to be an excellent tool in locating periodic solutions of the system in both circular and noncircular orbits. The variational stability of the periodic motion is examined using the Floquet theory. The resonance analysis suggests the existence of critical combinations of system parameters leading to large amplitude oscillations.

052.009 Théorie analytique programmée du mouvement des satellites artificiels sous l'action gravitationnelle de la terre. X. Berger.
Celestial Mechanics, Vol. 11, 281 - 300 (1975).
In the analytical theory presented here, the developments of the first, second and third orders, different from those given by the direct solution of Lagrange equations, have been got once and for all to avoid literal algebra in the computed program for orbit calculation. So, the literal expressions of all the terms have been established and checked by comparisons of the analytical theory with numerical integration. Furthermore it permits study of the mathematical properties of these developments.

052.010 The problem of the critical inclination revisited.
A. H. Jupp.
Celestial Mechanics, Vol. 11, 361 - 378 (1975).
The behaviour of the argument of the pericentre is investigated for the orbit of an artificial satellite which is moving under a potential V when the inclination of the orbit is close to the critical value $\tan^{-1} 2$. The theory is developed to first order and it is applicable to all values of the eccentricity, with the exception of those in the neighbourhood of zero and unity. Four principal types of behaviour are noted and these are illustrated in appropriate phase-plane diagrams.

052.011 A solution of the motion of an artificial satellite in the vicinity of the critical inclination. L. Liu.
Acta Astron. Sinica, Vol. 15, 230 - 240 (1974). In Chinese.
This article gives the computing methods of all first order perturbations and second order secular perturbations from the oblate earth in the vicinity of the critical inclination.

052.012 Sur l'existence d'une solution périodique dans le mouvement autour du centre de gravité d'un satellite aimanté dans le champ magnétique terrestre.
I. Stellmacher.
Comptes Rendus Acad. Sci. Paris, Sér. A, Vol. 280, 977 - 980 (1975).
En s'appuyant sur la théorie de Malkin relative aux systèmes quasi-linéaires périodiques, on montre comment construire la solution périodique du mouvement autour du centre de gravité.

052.013 Investigation of the equations of satellite motion in a noninertial coordinate system. V. P. Semenko.
Uch. zap. Tsentr. aehro-gidrodinam. in-ta, 1974, tom (Vol.) 5, No. 4, p. 117 - 122. In Russian. – Abstr. in Referativ.

Zhurn. 51. Astron., 3.51.105; 62. Issled kosmich. prostranstva, 3.62.421 (1975).

052.014 Long-periodic perturbations of the orbital elements of artificial earth satellites due to the attraction of the moon.   V. P. Dolgachev.
Vestn. Mosk. un-ta. Fiz., astron., Vol. 15, 591 - 596 (1974). In Russian. – Abstr. in Referativ. Zhurn. 51. Astron., 3.51. 106 (1975).

052.015 Semi-analytical method for computing the motion of artificial resonance satellites with large eccentricity.
A. A. Solov'ev.
In-t prikl. mat. AN SSSR. Preprint No. 87. Moskva, 1974. 39 pp. In Russian. – Abstr. in Referativ. Zhurn. 62. Issled. kosmich. prostranstva, 3.62.420 (1975).

052.016 On the integration of the differential equations of relativistic rocket dynamics.
B. K. Fedyushin, S. I. Makarikhin.
Trudy vos'mykh chtenij, posvyashch. razrab. nauch. naslediya i razvitiyu idej K. Eh. Tsiolkovskogo, Kaluga, 1973. Sekts. "Probl. raket.i kosmich. tekhn." Moskva, 1974, p. 19 - 22. In Russian. – Abstr. in Referativ. Zhurn. 62. Issled. kosmich. prostranstva, 3.62.424 (1975).

052.017 On the motion of an apparatus with two solar sails.
V. V. Malanin, A. V. Rep'yakh.
Probl. mekh. upravlyaemogo dvizheniya. Vyp. (No.) 5. Perm', 1974, p. 99 - 108. In Russian. – Abstr. in Referativ. Zhurn. 62. Issled. kosmich. prostranstva, 3.62.426 (1975).

052.018 The shadow function in the problem of influence of light pressure on the motion of artificial earth satellites.   S. N. Vashkov'yan.
Vestn. Mosk. un-ta. Fiz., astron., Vol. 15, 584 - 590 (1974). In Russian. – Abstr. in Referativ. Zhurn. 62. Issled. kosmich. prostranstva, 3.62.427 (1975).

052.019 Use of computer literal programs to solve the main problem in satellite theory.   A. L. Kutuzov.
Pis'ma v Astron. Zhurn., Vol. 1, No. 2, p. 39 - 43 (1975). In Russian.
Programs have been developed from OPAL package of subroutines to solve analytically the main problem in satellite theory, all gravitational harmonics being zero except J_2. Keplerian elements perturbations of the second order in J_2 have been obtained by successive approximations. Perturbation techniques based on Lie transforms were used to develop the main problem solution up to the third order in J_2. Elimination of short period terms from the Hamiltonian of the main problem is described.

052.020 Analytical theories of the motion of artificial satellites.   G.-I. Hori, Y. Kozai.
Satellite Dynamics, Symp. 1974, (see 012.007), p. 1 - 15 (1975).
The paper gives a general review of various analytical theories of the motion of artificial satellites. The authors review also theories that treat second order sources of perturbations such as tesseral harmonics of the earth's potential, luni-solar attraction, drag of the earth's atmosphere, solar radiation, and so on. The orders of perturbations needed for analytical theories to attain an accuracy of 10^{-6} and 10^{-8} are shown in concluding the review.

052.021 Time elements.   P. E. Nacozy.
Satellite Dynamics, Symp. 1974, (see 012.007), p. 16 - 26 (1975).
Time elements are introduced for use with Sundman time transformations of the type $dt = r^{\alpha} ds$ for satellite equations of motion. Two time elements are presented, one providing maximum accuracy for $\alpha = 1$, the other for $\alpha = 2$. Numerical results show accuracy improvements of more than one order of magnitude when time elements are employed with time transformations in the numerical integration of the satellites equations, compared with using time transformations alone.

052.022 An estimation procedure for orbit determination, using the K.S. transformation.   A. Rios-Neto.
Satellite Dynamics, Symp. 1974, (see 012.007), p. 27 - 34 (1975).
The orbit estimation problem is formulated. Assuming a perturbed two body physical model, the equations of motion in the mathematical model are treated by using the Kustaanheimo-Stiefel transformation in order to get analytical solution in parametric space.

052.023 Importance of the moon—earth coupling effect in the motion of an artificial satellite: semi-analytical solution of this and similar problems.
X. Berger.
Satellite Dynamics, Symp. 1974, (see 012.007), p. 35 - 49 (1975).

052.024 Numerical methods of orbital dynamics.
G. Balmino.
Satellite Dynamics, Symp. 1974, (see 012.007), p. 50 - 97 (1975).   In French.
Different numerical integration algorithms are reviewed in their application to the many body problem, the determination of satellite orbits around the earth, moon and other planets.

052.025 J-adaptive estimation with estimated noise statistics.   A. H. Jazwinski, C. Hipkins.
Satellite Dynamics, Symp. 1974, (see 012.007), p. 98 - 110 (1975).

052.026 Importance of the coupling effects between earth potential harmonics in the motion of an artificial satellite. Computed and checked solution of this coupling problem. The case of J_7.   X. Berger.
Satellite Dynamics, Symp. 1974, (see 012.007), p. 111 - 126 (1975).
The "coupling" effect arising from the simultaneous presence of different harmonics can be great compared with the direct effect of the harmonics. The author shows which coupling effects must be considered for a good representation of the motion of any satellite, and for the determination of zonal harmonics.

052.027 On the comparison of numerical theories of orbital motion.   J. D. Mulholland.
Satellite Dynamics, Symp. 1974, (see 012.007), p. 127 - 135 (1975).
The numerical study of the motion of the closest natural satellite, the moon, provides a means to examine the general concept of a numerical theory, as opposed to a simple ephemeris. Attempts to reconcile a discrepancy between the author's calculations and those of Oesterwinter and Cohen permit some conclusions on the documentation required for such a theory to be complete.

052.028 Stabilization and real world satellite problem.
C. E. Velez.
Satellite Dynamics, Symp. 1974, (see 012.007), p. 136 - 153 (1975).

052.029 Orbit determination in the presence of atmospheric drag errors.   B. D. Tapley, A. R. Neto,

B. E. Schutz.
Satellite Dynamics, Symp. 1974, (see 012.007), p. 154 - 169 (1975).

052.030 Literal algebra for satellite dynamics.
E. M. Gaposchkin.
Satellite Dynamics, Symp. 1974, (see 012.007), p. 170 - 179 (1975).

Analytical developments of a satellite perturbation theory can be accomplished quickly, accurately, and with generality by use of computer algebra. The basic principles of computer algebra, the required and desirable features of computer-algebra programs, and the applications of computer algebra to specific problems in perturbation analysis will be presented.

052.031 A comparative study of the calculation of partial derivatives in space dynamics. H. G. Walter.
Satellite Dynamics, Symp. 1974, (see 012.007), p. 180 - 191 (1975).

052.032 Non-gravitational forces in satellite dynamics.
L. Sehnal.
Satellite Dynamics, Symp. 1974, (see 012.007), p. 304 - 330 (1975).

The effects of different non-gravitational forces on the motion of artificial satellites are described. Drag, direct solar radiation pressure and earth albedo radiation pressure are discussed in detail.

052.033 Solar radiation pressure and balloon type artificial satellites. R. V. de Moraes.
Satellite Dynamics, Symp. 1974, (see 012.007), p. 331 - 341 (1975).

052.034 Optical impulsive interplanetary transfers.
J. P. Gravier, C. Marchal, R. D. Culp.
Satellite Dynamics, Symp. 1974, (see 012.007), p. 342 - 369 (1975).

This study presents the derivation of the equations necessary to establish a practical method for computing optimal transfers in the real case. It makes it possible to compute and use actual optimal transfers for advanced planning of preliminary mission analysis in place of the standard Hohmann transfers.

052.035 On the effect of earth radiation pressure in the motion of Pageos. Comparisons between different models. D. Gambis.
Satellite Dynamics, Symp. 1974, (see 012.007), p. 370 - 376 (1975).

052.036 Short-period and long-period perturbations of a spherical satellite due to direct solar radiation.
K. Aksnes.
Bull. American Astron. Soc., Vol. 7, 341 (1975). — Abstr. AAS.

052.037 Perturbations of a close earth satellite due to sunlight reflected from the earth. D. A. Lautman.
Bull. American Astron. Soc., Vol. 7, 341 (1975). — Abstr. AAS.

052.038 Satellite perturbations with automated Poisson series manipulations. R. Broucke.
Bull. American Astron. Soc., Vol. 7, 341 (1975). — Abstr. AAS.

052.039 On the problem of stability of relative equilibrium of a satellite in a circular orbit.
A. P. Markeev, A. G. Sokol'skij.
Kosmich. Issled., Vol. 13, 139 - 146 (1975). In Russian.

052.040 Periodic solutions of the translatory-rotational motion of a satellite relative to the rotating earth.

V. G. Dëmin, E. B. Bibik.
Kosmich. Issled., Vol. 13, 158 - 162 (1975). In Russian.

052.041 On the stabilization of the height of the pericentre.
K. V. Kholshevnikov, V. E. Markov.
Kosmich. Issled., Vol. 13, 181 - 184 (1975). In Russian.

052.042 On the secular evolution of the orbit of a balloon satellite. Yu. N. Isaev, A. I. Prokof'ev.
Kosmich. Issled., Vol. 13, 185 - 189 (1975). In Russian.

052.043 The observability of navigation of space vehicles in non-linear problems. G. N. Razorenov.
Kosmich. Issled., Vol. 13, 190 - 200 (1975). In Russian.

052.044 Taking into account the influence of constant transversal acceleration in the semi-analytical method for computing the motion of stationary artificial earth satellites. M. A. Vashkov'yak.
Kosmich. Issled., Vol. 13, 272 - 274 (1975). In Russian.
Brief information.

052.045 On the estimate of the momenta of distribution of the height of the pericentre of a nearly circular orbit. Yu. V. Sirotenko.
Kosmich. Issled., Vol. 13, 275 - 276 (1975). In Russian.
Brief information.

052.046 The problem of optimum construction of the elliptic transition with fixed angular distance between circular orbits. V. S. Novoselov.
Probl. mekh. upravlyaemogo dvizheniya. Vyp. (No.) 5. Perm', 1974, p. 54 - 60. In Russian. — Abstr. in Referativ. Zhurn. 51. Astron., 4.51.136 (1975).

052.047 Algorithms of the problems of relativistic rocket dynamics. B. K. Fedyushin.
Trudy vos'mykh chtenij, posvyashch. razrab. nauch. naslediya i razvitiyu idej K. Eh. Tsiolkovskogo, Kaluga, 1973. Sekts. "Probl. raket. i kosmich. tekhn". Moskva, 1974, p. 24 - 31. In Russian. — Abstr. in Referativ. Zhurn. 62. Issled. kosmich. prostranstva, 4.62.322 (1975).

052.048 Semi-analytical method for calculation of the motion of artificial resonance satellites with large eccentricity. 1. Description of the algorithm and of experimental results. Gravitational model. A. A. Solov'ev.
In-t prikl. mat. AN SSSR. Preprint No. 86, Moskva, 1974. 49 pp. In Russian.

052.049 Equations of motion of rockets in the atmosphere.
I. I. Merkulov.
Trudy vos'mykh chtenij, posvyashch. razrab. nauch. naslediya i razvitiyu idej K. Eh. Tsiolkovskogo, Kaluga, 1973. Sekts. "Probl. raket. i kosmich. tekhn". Moskva, 1974, p. 95 - 109. In Russian. — Abstr. in Referativ. Zhurn. 62. Issled. kosmich. prostranstva, 4.62.339 (1975).

052.050 On the plane motion of an artificial satellite.
L. I. Kuznetsov, V. E. Pasynkov, P. E. Tovstik.
Prikl. mekhanika. Vyp. (No.) 1. Leningrad, Leningr. un-t, 1974, p. 32 - 44. In Russian. — Abstr. in Referativ. Zhurn. 62. Issled. kosmich. prostranstva, 4.62.345 (1975).

052.051 Effect of the technique of representation of satellite motion on the accuracy of numerical ephemerides.
M. Yu. Belyaev, V. P. Semenko.
Uch. zap. Tsentr. aehro-gidrodinam. in-ta, Vol. 5, No. 5, p. 126 - 132 (1974). In Russian. — Abstr. in Referativ. Zhurn. 51. Astron., 5.51.107; 62. Issled. kosmich. prostranstva, 5.62.342 (1975).

052.052 **The shadow function in the problem of influence of radiation pressure on the motion of artificial earth satellites.** S. N. Vashkov'yak.
Vestn. Mosk. un-ta. Fiz., astron., Vol. 15, 584 - 590 (1974). In Russian. — Abstr. in Referativ. Zhurn. 51. Astron., 5.51. 109 (1975).

052.053 **Semi-analytical method of calculation of the motion of artificial resonance satellites with large eccentricities.** A. A. Solov'ev.
In-t prikl. mat. AN SSSR. Preprint No. 87. Moskva, 1974, 39 pp. In Russian. — Abstr. in Referativ. Zhurn. 51. Astron., 5.51.111 (1975).

052.054 **Stationary motions of a satellite-gyroscope system.** A. Kh. Akhmetshin.
Kazan. aviats. in-t. Kazan', 1974. 17 pp. In Russian. — Abstr. in Referativ. Zhurn. 62. Issled. kosmich. prostranstva, 5.62. 351; 51. Astron., 6.51.173 (1975).

052.055 **Effect of resonance-oblateness coupling on a satellite orbit.** C. A. Wagner.
Goddard Space Flight Center, Greenbelt, Maryland, GSFC Document X-920-74-334, Preprint, 3 + 25 pp. (1974).
Second order effects of the coupling between geopotential resonance and oblateness on a satellite orbit are calculated. These effects arise from the interaction of resonance with the secular changes of the orbit's node, perigee and mean anomaly. They have the same period and phase as first order resonance perturbations. But their amplitudes are proportional to the square of the period and dominate the first order effects as the orbit becomes commensurate.

052.056 **General relativity and satellite orbits.** D. P. Rubincam.
Goddard Space Flight Center, Greenbelt, Maryland, GSFC Document X-921-75-56, 6 + 35 pp. (1975).
The general relativistic correction to the position of a satellite is found by retaining Newtonian physics for an observer on the satellite and introducing a r^{-3} potential. The potential is expanded in terms of the Keplerian elements of the orbit and substituted in Lagrange's equations. Integration of the equations shows that a typical earth satellite with small orbital eccentricity is displaced by about 17 cm from its unperturbed position after a single orbit, while the periodic displacement over the orbit reaches a maximum of about 3 cm. The moon is displaced by about the same amounts. Application of the equations to Mercury gives a total displacement of about 58 km after one orbit and a maximum periodic displacement of about 12 km.

052.057 **Velocity-space maps and transforms of tracking observations for orbital trajectory state analysis.**
S. P. Altman.
Celestial Mechanics, Vol. 11, 405 - 428 (1975). — Presented at the 23rd congress of the International Astronautical Federation, Vienna, Austria, October 8—15, 1972.
External observations of orbital vehicles, such as provided by optical and radar sensors of tracking systems, are transformable into corresponding velocity state maps, as presented in this paper. These transformations and the consequent state maps are essential for development of the orbit observation matrix used with the unified state matrix, in recursive estimators such as the Kalman filters. Line-of-sight rays and range spheres (or hemispheres) of observations map conformally into orthogonal spherical surfaces in velocity space, as the result of the point-contact transformations. In bi-spherical coordinates, the field of observation maps for a ground-based tracking system site is shown to be a reduced (or degenerate) form of the general field of observation maps for a satellite-based tracking site.

052.058 **Modified multirevolution integration methods for satellite orbit computation.**
O. F. Graf, D. G. Bettis.
Celestial Mechanics, Vol. 11, 433 - 448 (1975).
Multirevolution methods allow for the computation of satellite orbits in steps spanning many revolutions. Modified methods are derived that will integrate exactly products of linear and periodic functions. Numerical examples are given that show that these new methods provide better accuracy for certain satellite problems. It is also shown that information obtained from an approximate analytical solution of the satellite equations of motion may be used to increase the accuracy and/or efficiency of the multirevolution integration.

052.059 **Experimental flight Apollo — Soyuz.** A. A. Bol'shoj.
Priroda, No. 6.75, p. 2 - 7 (1975). In Russian.

052.060 **Semi-analytical method for computing the motion of an artificial lunar satellite.**
M. L. Lidov, V. A. Lyakhova, A. A. Solov'ev.
Kosmich. Issled., Vol. 13, 283 - 310 (1975). In Russian.

052.061 **Optimum parameters of gravitationally-sensitive gyro-elements of orientation of satellites.**
V. A. Sarychev, K. V. Lukanin, S. A. Mirer.
Kosmich. Issled., Vol. 13, 311 - 321 (1975). In Russian.

052.062 **The stability of plane oscillations and rotations of a satellite in a circular orbit.** A. P. Markeev.
Kosmich. Issled., Vol. 13, 322 - 336 (1975). In Russian.

052.063 **Time averaged properties of central orbits.** J. B. Pearce.
Space Sci. Instrum., Vol. 1, 17 - 21 (1975).
Formulae for the amount of time spent at given altitudes and given angles from periapsis for bodies under the influence of a central force are derived. The altitude formula is applied to the case of remote observations from a spin-stabilized earth orbiting vehicle.

052.064 **On the tesseral-harmonics resonance problem in artificial-satellite theory.** B. A. Romanowicz.
Smithsonian Astrophys. Obs., *Cambridge, Mass.*, Special Report 365, 6 + 47 + A4 pp. (1975).
The longitude-dependent part of the geopotential usually gives rise only to short-period effects in the motion of an artificial satellite. However, when the motion of the satellite is commensurable with that of the earth, the path of the satellite repeats itself relative to the earth and perturbations build up at each passage of the satellite in the same spot, so that there can be important long-period effects. In order to take these effects into account in deriving a theoretical solution to the equations of motion of an artificial satellite, it is necessary to select terms in the longitude-dependent part of the geopotential that will contribute significantly to the perturbations. The author has tried to make a selection that is valid in a general case, regardless of the initial eccentricity of the orbit and of the order of the resonance. The solution to the equations of motion of an artificial satellite, in a geopotential thus determined, is then derived by using Hori's method by Lie series, which, by its properties regarding canonical invariance, has proved advantageous in the classical theory.

052.065 **Associated integrals of the equations of dynamical systems in the problems considered by K. Eh. Tsiolkovskij.** V. V. Dobronravov.
Trudy vos'mykh. chtenij, posvyashch. razrabotke nauch. nasledija i razvitiyu idej K. Eh. Tsiolkovskogo, Kaluga, 1973 g. Sekts. "Mekh. kosmich. poleta. Moskva, 1974, p. 3 - 12. In Russian. — Abstr. in Referativ. Zhurn. 51. Astron., 6.51.168;

62. Issled. kosmich. prostranstva, 6.62.342 (1975).

052.066   On the application of group variation method in problems of space research.   Ya. A. Kachmarchik. Trudy vos'mykh chtenij, posvyashch. razrabotke nauch. naslediya i razvitiyu idej K. Eh. Tsiolkovskogo, Kaluga, 1973 g. Sekts. "Mekh. kosmich. poleta". Moskva, 1974, p. 174 - 181. In Russian. − Abstr. in Referativ. Zhurn. 62. Issled. kosmich. prostranstva, 6.62.313 (1975).

052.067   Some qualitative regularities and estimates of evolution of the orbits of artificial earth satellites of Molniya-1 type.   M. L. Lidov, A. A. Solov'ev. In-t prikl. mat. AN SSSR. Preprint No. 6. Moskva, 1975. 28 pp. Price 10 Kop. In Russian. − Abstr. in Referativ. Zhurn. 62. Issled. kosmich. prostranstva, 6.62.314 (1975).

052.068   On the estimate of deviations of the mass center of a space vehicle from programmed motion.   V. S. Ruchinskij. Trudy vos'mykh chtenij, posvyashch. razrabotke nauch. naslediya i razvitiyu idej K. Eh. Tsiolkovskogo, Kaluga, 1973 g. Sekts. "Mekh. kosmich. poleta". Moskva, 1974, p. 53 - 66. In Russian. − Abstr. in Referativ. Zhurn. 62. Issled. kosmich. prostranstva, 6.62.315 (1975).

052.069   Rotational motion of artificial earth satellites and determination of the atmosphere's density.   V. V. Beletskij, V. M. Grigorevskij, S. Ya. Kolesnik. In-t prikl. mat. AN SSSR. Preprint No. 7. Moskva, 1975. 52 pp. Price 15 Kop. In Russian. − Abstr. in Referativ. Zhurn. 62. Issled. kosmich. prostranstva, 6.62.336 (1975).

052.070   Solid body motion around a mass center under the influence of the gravitational, magnetic and aerodynamical momentum.   V. V. Malanin, O. I. Zlotnikov. Sb. nauch. tr. Perm. politekhn. in-t, 1974, No. 152, p. 79 - 88. In Russian. − Abstr. in Referativ. Zhurn. 62. Issled. kosmich. prostranstva, 6.62.339 (1975).

052.071   On the motion of vehicles with small thrust rotating relative to the mass center.   Yu. A. Zakharov. Trudy vos'mykh chtenij, posvyashch. razrabotke nauch. naslediya i razvitiyu idej K. Eh. Tsiolkovskogo, Kaluga, 1973 g. Sekts. "Mekh. kosmich. poleta". Moskva, 1974, p. 67 - 76. In Russian. − Abstr. in Referativ. Zhurn. 62. Issled. kosmich. prostranstva, 6.62.340 (1975).

052.072   Estimate of the boundary of oscillatory motions of artificial earth satellites.   V. A. Sarychev, V. V. Sazonov. In-t prikl. mat. AN SSSR. Preprint No. 130. Moskva, 1974 g. 38 pp. Price 15 Kop. In Russian. − Abstr. in Referativ. Zhurn. 62. Issled. kosmich. prostranstva, 6.62.341 (1975).

052.073   Comparison of statistical orbit determination methods.   B. E. Schutz, J. D. McMillan, B. D. Tapley. AIAA Journ., Vol. 12, 1465 - 1466 (1974). The results obtained in a comparison of the convergence characteristics and accuracy of the batch estimation algorithm and the extended sequential estimation algorithm as applied to the problem of estimating the state of a near-earth satellite in the presence of geopotential modeling errors, are described.

052.074   Satellite motion about an oblate earth.   J. J. F. Liu. AIAA Journ., Vol. 12, 1511 - 1516 (1974). A general theory of the method of averaging is used to study the effect of the earth's oblateness on the motion of an artificial satellite. The first-order harmonic J_2 and the second order harmonics J_3 and J_4 are included in the analysis.

Equations of motion for interconnected rigid and elastic bodies: a derivation independent of angular momentum. See Abstr. 042.030.

Resonance phenomena at rotations of artificial and natural celestial bodies.   See Abstr. 042.045.

053   Lunar and Planetary Probes and Satellites

053.001   **Helios–Chronologie eines Bilderbuchstarts.**
H. Link.
SuW, Vol. 14, 76 - 78 (1975).

053.002   **The resources of the solar system.**
R. C. Parkinson.
Spaceflight, Vol. 17, 124 - 128 (1975).

053.003   **Mariner-Venus-Mercury 1973 project history.
Part 1.**   D. Baker.
Spaceflight, Vol. 17, 131 - 133 (1975).

053.004   **Pioneer 10 and Pioneer 11.**   C. F. Hall.
Science, Vol. 188, 445 - 446 (1975).

053.005   **The Mercury orbiter and its relativistic aspects.**
G. Israel.
Raumfahrtforschung, [DGLR - Deutsche Ges. für Luft- und
Raumfahrt, Berlin], Vol. 18, 218 - 223 (1974).
   Technical means exist for conducting experimental tests
of the gravitational theories inside the solar system. With the
recent discoveries in astrophysics and cosmology, one needs
to improve our confidence in general relativity by more
sophisticated measurements. In addition to the use of large
earth based antennae as a means of accurately evaluating the
relativistic parameters, it is shown how experiments aboard
heliocentric probes like SOREL can be envisaged. Having a satellite in orbit around the planet Mercury can also be the appropriate way of testing the gravitation theories, with the second
benefit of investigating the physics of the closest planet to
the sun. A spacecraft of the Helios type can be proposed for
this purpose.

053.006   **Sonnensonde Helios. Deutsch-amerikanisches Pro-
jekt zur Erforschung der Sonne und des sonnennahen Raumes.**   G. Weiss.
Astronautik, Jahrgang 12, p. 10 - 11 (1975).

053.007   **Long-range communications with Pioneer 10 at
Jupiter.**   L. W. Dickerson, A. J. Siegmeth.
Journ. British Interplanet. Soc., Vol. 28, 371 - 391 (1975).
   The longest two-way telecommunications link ever
established connected Pioneer 10 with its controllers and
experimenters on earth during the first Jovian encounter. This
radio link, operating in the 2.1 and 2.3 gigahertz frequency
bands, maintained high quality communications over a distance of 5-1/2 astronomical units. The data collected by eleven
on-board advanced scientific instruments were transmitted to
the investigators, providing a new wealth of information
about the largest planet of the solar system.

053.008   **Project Helios.**   M. Howard.
Spaceflight, Vol. 17, 184 - 188 (1975).

053.009   **Mariner-Venus-Mercury 1973 project history. Part 2.**
D. Baker.
Spaceflight, Vol. 17, 191 - 194 (1975).

053.010   **Ninety days on Mars.**   Staff of M. Marietta.
Spaceflight, Vol. 17, 202 - 211, 240 (1975).

053.011   **New starts to the planets.**   H. Oja.
Spaceflight, Vol. 17, 215 - 218 (1975).

053.012   **Missions to Salyut 4.**   G. R. Hooper.
Spaceflight, Vol. 17, 219 - 225 (1975).

053.013   **Availability of experimental data from the planetary
space probes at the National Space Science Data
Center.**   W. S. Cameron.
Bull. American Astron. Soc., Vol. 7, 389 (1975). − Abstr. AAS.

053.014   **Jüngste interplanetare Sonden.**
G. Gerstbach, C. Köberl, H. Mucke.
Sternenbote, 18. Jahrgang, p. 46 - 55 (1975).

053.015   **Orbit determination capability analysis for the
Mariner-Jupiter-Saturn 1977 mission.**
G. A. Ransford, C. E. Hildebrand, V. J. Ondrasik.
Journ. Spacecraft and Rockets, (USA), Vol. 11, 658 - 663
(1974).
   A combined earth-based radio tracking/onboard optical
data navigation system designed to meet these goals, is described. Some results of applying this navigation system to the
preliminary Mariner-Jupiter-Saturn 1977 mission trajectories
are presented.

053.016   **Pioneer Venus mission plan for atmospheric probes
and an orbiter.**
J. W. Dyer, R. R. Nunamaker, J. R. Cowley, Jr., R. W. Jackson.
Journ. Spacecraft and Rockets, (USA), Vol. 11, 710 - 715
(1974).
   The 1978 mission concept is described for a spin-stabilized Pioneer spacecraft designed to function as a Venus
orbiter and as a probe carrier on a second flight. The same
'Bus' spacecraft will be the basic vehicle for both flights.

Soviet lunar landing.
Sky Telescope, Vol. 49, 20 (1975).

The Viking Mars lander camera.
See Abstr. 034.085.

Les sondes Viking: À la recherche de la vie sur Mars.
See Abstr. 097.039.

Mariner 9: an instrument of dynamical science.
See Abstr. 097.062.

Pioneer 11: Through the dragon's mouth.
See Abstr. 099.009.

**Pioneer 11 encounter: Preliminary results from the
Ames Research Center plasma analyzer experiment.**
See Abstr. 099.024.

**Pioneer 10 Jovian encounter: radiation dose and
implications for biological lethality.**
See Abstr. 099.074.

Why image Uranus?   See Abstr. 101.006.

Particle and field environment of Uranus.
See Abstr. 101.007.

Particles and fields in the outer solar system.
See Abstr. 106.012.

054 Artificial Earth Satellites

054.001 Laser corner cubes for the Interkosmos AUOS-Z ellips satellite. P. Navara.
Bull. Astron. Inst. Czechoslovakia (BAC), Vol. 26, 18 - 23 (1975).

The Interkosmos satellite AUOS-Z will be equipped with a corner cube panel designed for the laser ranging. The use of corner cubes for satellite laser measurement, the corner cube panel calculation, the actual calculus conditions, are described and reasons are outlined supporting the technical solution. Listed in the conclusion are the AUOS-Z satellite parameters which concern to laser ranging.

054.002 Attitude equilibria and stability of arbitrary gyrostat satellites under gravitational torques.
R. W. Longman.
Journ. British Interplanet. Soc., Vol. 28, 38 - 46 (1975).

054.003 Intercosmos 12. S. A. Nikitin.
Priroda, No. 2.75, p. 101 (1975). In Russian.

054.004 Three-axis stabilised X-ray spacecraft.
J. W. Heaton, G. W. Cocks.
Journ. British Interplanet. Soc., Vol. 28, 326 - 342 (1975).

As scientific experiments in X-ray astronomy become more ambitious the requirements on the spacecraft become more demanding. The aim of this paper is to discuss some of the more critical problems peculiar to advanced X-ray spacecraft and examine possible solutions.

054.005 UHURU − the first X-ray astronomy satellite.
A. C. Fabian.
Journ. British Interplanet. Soc., Vol. 28, 343 - 346 (1975).

054.006 An analytic method to account for drag in the Vinti satellite theory.
J. S. Watson, G. D. Mistretta, N. L. Bonavito.
Celestial Mechanics, Vol. 11, 145 - 177 (1975).

In order to retain separability in the Vinti theory of earth satellite motion when a non-conservative force such as air drag is considered, a set of variational equations for the orbital elements are introduced, and expressed as functions of the transverse, radial, and normal components of the non-conservative forces acting on the system. Results of this technique for the case of the intense air drag satellites San Marco-2 and Air Force Cannonball are given. These results indicate that the satellite ephemerides produced by this theory in conjunction with the Vinti program are of very high accuracy. In addition, since the program is entirely analytic, several months of ephemerides can be obtained within a few seconds of computer time.

054.007 Skylab: the three month vigil. Part 3.
D. Baker.
Spaceflight, Vol. 17, 11 - 15 (1975).

054.008 An orbiting solar power station.
Sky Telscope, Vol. 49, 226 - 228 (1975).

054.009 Analysis of the orbit of Ariel 1, 1962-15A, near 15th-order resonance. D. M. C. Walker.
Planet. Space Sci., Vol. 23, 565 - 574 (1975).

On 8 May 1973 the orbit passed through 15th-order resonance and has been determined, with the RAE orbit refinement program PROP, at eight epochs between February and August 1973 using 500 observations. The orbital inclinations during the time of 15th-order resonance, as given by these eight orbits and 31 U.S. Navy orbits, were fitted with a theoretical curve using the THROE computer program, the best fit giving $10^9 \bar{C}_{15} = -370 \pm 14$ and $10^9 \bar{S}_{15} = -114 \pm 31$. The values of eccentricity were also successfully fitted using THROE, and the results are discussed.

054.010 Skylab. Un laboratoire au service de la terre.
R. Dejaiffe.
Ciel et Terre, Vol. 91, 5 - 40 (1975).

054.011 De AMSAT-OSCAR-7 gelanceerd.
W. L. B. J. Dekker.
Zenit, Vol. 2, 50 - 51 (1975).

054.012 The last OSO satellite. S. P. Maran, R. J. Thomas.
Sky Telescope, Vol. 49, 355 - 358 (1975).

054.013 The rotational motion of a satellite and atmospheric density determination.
V. V. Beletskii (*Beletskij*), V. M. Grigorevsky (*Grigorevskij*), S. Ya. Kolesnik.
Satellite Dynamics, Symp. 1974, (see 012.007), p. 233 - 279 (1975).

054.014 On the stability of a flexible satellite.
L. I. Kuznetsov.
Prikl. mekhanika. Vyp. (No.) 1. Leningrad, Leningr. un-t, 1974, p. 55 - 71. In Russian. − Abstr. in Referativ. Zhurn. 62. Issled. kosmich. prostranstva, 4.62.344 (1975).

054.015 On the evolution of the spin motion of the rocket 1969−51 B. V. Mioc, V. Ciubotaru.
Bull. Astron. Inst. Czechoslovakia, Vol. 26, 149 - 150 (1975).

The spin period evolution of the rocket 1969−1973 B between 1969−1973 is studied. Equations for the spin period variation are deduced. An acceleration of the spin motion in March 1971 is pointed out.

054.016 Magnetometer experiments in the European Space Research Organisation's HEOS satellites.
P. C. Hedgecock.
Space Sci. Instrum., Vol. 1, 61 - 82 (1975).

Details of the design of the magnetometer experiments in the ESRO HEOS-1 and HEOS-2 earth satellites are discussed. Ground measurements and flight data on their performance together with details of their orbits are also presented.

054.017 Considerazioni sulle velocità radiali ottenute da osservazioni radio-ottiche di satelliti artificiali.
L. Buffoni, A. Manara.
Rend. Seminario Fac. Sci. Univ. Cagliari, Suppl. Vol. 44, (see 012.016), 105 - 110 (1974).

A method for calculation of radial velocities of artificial satellites obtained from optical and Doppler observations made in Milano-Brera Observatory is studied. Finally, results obtained with two methods are compared.

054.018 A Dutch satellite for astronomical research.
W. Lulofs.
Philips Telecommun. Rev., (*Netherlands*), Vol. 32, 217 - 224 (1974).

ANS, the first Dutch satellite has been launched into orbit, the on-board equipment serving to investigate the celestial ultra-violet and X-ray radiation. Technicalities with regard to attitude control, telemetry and telecommand, on-board computer and power supply are briefly outlined.

054.019 Satellite digest.

Compiled by R. D. Christy.
Spaceflight, Vol. 17, 35, 76, 115, 152 - 153, 196, 200, 235 (1975).

Some international satellite programs.
Sky Telescope, Vol. 49, 19 - 20 (1975).

On the determination of elliptic orbits from the time micrometer observations.   See Abstr. 046.021.

Variations in air density, satellite drag coefficient and atmospheric rotation rate from analysis of the orbit of 1966-92D.   See Abstr. 082.084.

The ultraviolet experiment onboard the Astronomical Netherlands Satellite – ANS.
See Abstr. 113.022.

055   Observations of Earth Satellites, Lunar and Planetary Probes

055.001   **Visual observations of artificial earth satellites in Finland 1974.**
Prepared under the supervision of A. Tuominen, with an introduction by P. Järvi.
Observations of Satellites, No. 15, (published by the Finnish Meteorological Institute, Helsinki, Finland), 8 + 77 pp. (1975).

055.002   **On the observations of artificial earth satellites.**
A. Drożyner.
Urania Kraków, Vol. 46, 46 - 50 (1975). In Polish.

Theoretical Astrophysics

061 General Theoretical Problems of Astrophysics, Gravitational Instability, Neutrino Astronomy, Infrared, X-Ray, Gamma-Ray Astronomy, Abundances and Origin of Elements

061.001 Rocket observation of energy spectrum of diffuse hard X-rays. Y. Fukada, S. Hayakawa, M. Ikeda, I. Kasahara, F. Makino. Y. Tanaka.
Astrophys. Space Sci., Vol. 32, L1 - L5 (1975).

The energy spectrum of diffuse hard X-rays measured in the range 10–40 keV shows a rather sharp change of slope. The logarithmic derivative of the spectrum changes around 20–30 keV by the increment significantly greater than 0.5 within an interval smaller than 50 keV.

061.002 Primary γ-rays. C. E. Fichtel.
Phil. Trans. Roy. Soc. London, Ser. A, Vol. 277, 365 - 379 (1975).-Conference paper.

Within our Galaxy, cosmic rays can reveal their presence in interstellar space and probably in source regions by their interactions with interstellar matter which lead to γ-rays with a very characteristic energy spectrum. From the study of the intensity of the high energy γ radiation as a function of galactic longitude, it is already clear that cosmic rays are almost certainly not uniformly distributed in the Galaxy and are not concentrated in the centre of the Galaxy.

061.003 Bounds on neutrino burst intensity imposed by the exclusion principle and causality.
S. A. Bludman, M. A. Ruderman.
Astrophys. Journ., (Letters), Vol. 195, L19 - L21 (1975).

Two different limits on the antineutrino (or neutrino) luminosity of a source are derived respectively from the exclusion principle and from the strength and form of weak interactions. Together with causality, the statistics of burst observations, and the maximum total mass in local astronomical objects, our exclusion principle limit on luminosity rules out nearby or nonrelativistic sources as astrophysical models for recently reported antineutrino bursts. More distant sources with relativistically expanding shells are ruled out by our weak interaction limit.

061.004 Observation of the diffuse component of cosmic soft X-rays. S. Hayakawa, T. Kato, Y. Tanaka, K. Yamashita, J. A. M. Bleeker, A. J. M. Deerenberg.
Astrophys. Journ., Vol. 195, 535 - 543 (1975).

A soft X-ray survey by means of thin-window proportional counters on board a spinning rocket covered a large celestial region with a width of about 60° in galactic latitude. In the galactic anticenter side the distribution of the soft X-ray intensity is rather smooth and shows a gradual increase with galactic latitude, whereas in the galactic center side the distribution is irregular and shows several enhanced regions. The overall distribution and the energy spectrum are accounted for in terms of a superposition of the following three components: an extragalactic hard component with a power-law spectrum as established at energies higher than 2 keV, a galactic soft component which has an exponential spectrum with an apparent temperature of 0.18 ± 0.02 keV, and an extragalactic component with a similar exponential spectrum.

061.005 Antineutrino bursts and cosmic-ray air shower experiments. T. C. Weekes, N. A. Porter.
Astron. Astrophys., Vol. 37, 447 - 449 (1974).

The possibility that the recently reported antineutrino burst could have been detected by large cosmic-ray air shower arrays is considered. Cherenkov radiation from positrons produced in the atmosphere by the antineutrinos offers another detection possibility.

061.006 Kosmische Maser. W. H. Kegel.
SuW, Vol. 14, 14 - 17 (1975).

061.007 Astrophysical restrictions on the neutrino-nucleon cross section at energies $E \geqslant 3 \times 10^{17}$ eV.
V. S. Berezinskij, A. Yu. Smirnov.
Izv. AN SSSR. Ser. fiz., Vol. 38, 1834 - 1837 (1974). In Russian. – Abstr. in Referativ. Zhurn. 51. Astron., 2.51.177 (1975).

061.008 Modern concepts on the possibility of formation of superheavy elements in nature.
Eh. E. Berlovich.
Izv. AN SSSR. Ser. fiz., Vol. 38, 1786 - 1790 (1974). In Russian. – Abstr. in Referativ. Zhurn. 51. Astron., 2.51.849 (1975).

061.009 A statistics of γ-radiation bursts.
V. V. Usov, G. V. Chibisov.
Astron. Zhurn. Akad. Nauk SSSR, Vol. 52, 192 - 194 (1975). In Russian. English translation in Soviet Astron., Vol. 19, No. 1.

The dependence of the number of detected γ-bursts on the sensitivity of the detectors is considered.

061.010 A search for VHF radio pulses in coincidence with celestial gamma-ray bursts. G. A. Baird, T. J. Delaney, B. G. Lawless, D. J. Griffiths, J. R. Shakeshaft, R. W. P. Drever, W. P. S. Meikle, J. V. Jelley, W. N. Charman, R. E. Spencer.
Astrophys. Journ., (Letters), Vol. 196, L11 - L13 (1975).

A detailed search for coincident pulses has been made of VHF radio recordings taken at widely spaced stations around the times of 19 celestial γ-ray bursts between 1970 and 1973. No coincident pulses were found above a sensitivity level of the order of 10^{-12} ergs cm^{-2} (event)$^{-1}$ in a 1-MHz bandwidth.

061.011 Observations of cosmic gamma-ray bursts with IMP 7: evidence for a single spectrum.
T. L. Cline, U. D. Desai.
Astrophys. Journ., (Letters), Vol. 196, L43 - L46 (1975).

Spectral observations of nine recent cosmic γ-ray bursts are reported. The average photon number spectra of all nine events are each consistent with a 150-keV exponential from 100 keV to about 400 keV, and a power law of index –2.5 from 400 keV to 1100 keV. The observations also indicate an event rate of 16 in 1972 and 1973, or 8 ± 2 per year, higher than the 5 ± 1 per year initially reported. This corresponds to an approximately 40 percent lower effective intensity threshold, attained by using more sensitive detectors in multiple-satellite coincidence.

061.012 Observational aspects of X-ray halos.
H. Spiegelhauer, J. Trümper.
Journ. British Interplanet. Soc., Vol. 28, 319 - 325 (1975).

061.013 Cosmic neutrinos of ultra-high energies and detection possibility.
V. S. Berezinsky (*Berezinskij*), A. Yu. Smirnov.
Astrophys. Space Sci., Vol. 32, 461 - 482 (1975).

The fluxes and spectra of galactic and extragalactic neutrinos at energy 10^{11} – 10^{19} eV are calculated. In particular, the neutrino flux from the normal galaxies is calculated taking into account the spectral index distribution. The only assumption that seriously affects the calculated neutrino flux at $E_\nu \gtrsim 10^{17}$ eV is the power-like generation spectrum of protons in the entire considered energy region.

061.014 A new shock locus for similarity solutions in one-dimensional unsteady gas dynamics.   R. E. Grundy.
Solar Physics, Vol. 40, 227 - 230 (1975).

This paper deals with shock conditions for the progressing wave (or similarity) solutions of one-dimensional, unsteady gas dynamics. These solutions have hitherto been used to deal with the flow behind shocks moving into stationary atmospheres. By generalising the shock conditions to the case of moving atmospheres, it is shown that the progressing wave solutions can be used to describe a certain class of flows, and a new shock locus can be constructed in the phase plane of the solutions. It is hoped that such solutions will be of use in describing the unsteady flow behind shocks propagating into the ambient solar wind.

061.015 Improved opacity calculations.
N. H. Magee, Jr., A. L. Merts, W. F. Huebner.
Astrophys. Journ., Vol. 196, 617 - 620 (1975).

Inclusion of new atomic data and improvements in the numerical procedure of opacity calculations as well as the increased iron abundance have led to significant increases from previously published astrophysical tables. This increase is at least 50 percent over sizable ranges of temperature and density. The new opacities for Iben XIV mixture with increased iron are presented. The treatment of narrow lines is given as an example of the type of improvements to numerical procedures. The effect of the improvements to other astrophysical mixtures is discussed.

061.016 Speculations on detection of the "neutrino sea".
L. Stodolsky.
Phys. Rev. Letters, Vol. 34, 110 - 112, with a correction, p. 508 (1975).

If there is a high density of ambient neutrinos – a "neutrino sea" – then on the conventional weak-interaction theory two types of possibly measurable effects exist. In one the spin direction of a transversely polarized moving electron rotates in field-free space. In the other the motion of the earth creates a torque on a ferromagnet.

061.017 Effects of primordial fluctuations on the abundances of light elements.
R. I. Epstein, V. Petrosian.
Astrophys. Journ., Vol. 197, 281 - 284 (1975).

Some of the effects of primordial inhomogeneities on the production of ^{2}H, ^{3}He, and ^{7}Li are investigated. For the most part, temperature fluctuations have only small effects on the abundances. The density fluctuations are modeled by a unimodal distribution so that the universe is characterized by the mean density and the relative amplitude of the density variations. The authors have found that the constraint which

^{2}H production imposes on the mean mass density is eased somewhat by allowing for density inhomogeneities. For reasonable estimates of the pregalactic abundances, sufficient ^{3}He and ^{7}Li cannot be produced in the same model universe.

061.018 Jeans' type gravitational instability of finite isothermal gas spheres–II.   S. Yabushita.
Monthly Notices Roy. Astron. Soc., Vol. 171, 85 - 86 (1975).

It has recently been found that there is an extra factor γ in the boundary condition adopted in an earlier analysis of gravitational instability of isothermal gas spheres. It is shown that when the boundary condition is corrected, the general conclusion already arrived at remains true and that there are minor changes in the non-dimensional parameter which specifies neutrally stable configurations.

061.019 Energy spectrum of diffuse component of cosmic soft γ rays.   Y. Fukada, S. Hayakawa, I. Kasahara, F. Makino, Y. Tanaka, B. V. Sreekantan.
Nature, Vol. 254, 398 - 399, with a correction, Vol. 255, 428 (1975).

In this paper the authors present new measurements on the diffuse component of cosmic soft gamma rays in the energy range 0.1–4 MeV obtained with a balloon-borne telescope. These measurements show that the energy spectrum gradually steepens from $E^{-2.3}$ to $E^{-2.8}$ as the energy increases from 100 to 500 keV and becomes less steep thereafter, consistent with the presence of a hump in the MeV region.

061.020 On some assumptions in solving problems of non-equilibrium gas radiation.   A. A. Korovyakovskaya.
Astrofiz. Issled., Izv. Spets. Astrofiz. Obs., Vol. 7, p. 35 - 40 (1975). In Russian.

A possibility of two assumptions is considered which are introduced in solving the problems of nonequilibrium gas radiation, stationarity of the field of radiation in the $L\alpha$ line, and acceptability of the relations for thermodynamic equilibrium. It is shown that in the process of radiation the character of variations of the electron temperature, degree of ionization, populations of the first and second energy levels of hydrogen is independent on the initial number of $L\alpha$-quanta in the medium.

061.021 X-ray astronomy in the Uhuru epoch and beyond.
E. M. Kellogg.
Astrophys. Journ., Vol. 197, 689 - 704 (1975). – Lecture presented at the meeting of the American Astronomical Society, Rochester, New York, 1974 August 21.

A review of results from the Uhuru satellite is presented. An intensive treatment of two subjects is given, rather than a broad review. First, Cyg X-1, a stellar X-ray source and a candidate for a black hole, is discussed; second, the X-ray source in the Perseus cluster of galaxies, which may be a cloud of hot intergalactic gas, is treated. In both cases, the train of logic used in establishing the nature of these objects is presented and evaluated. For both, while alternative explanations cannot be completely eliminated, they become more difficult to sustain when examined in detail, suggesting that the candidate explanations are more likely correct.

061.022 X-ray and gamma-ray detection by means of atmospheric interactions: fluorescence and Čerenkov radiation.   G. G. Fazio.
Astrophysics. Part A, (see 003.002), p. 315 - 359 (1974).

Detection of cosmic X rays by atmospheric fluorescence: the fluorescence process, X-ray induced fluorescence light in the atmosphere, ground-based detection of fluorescence light, sensitivity for detection of cosmic X-ray sources, a search for X-ray pulses from supernovae, a search for X-ray pulses associated with gravitational radiation pulses, background sources of light, conclusions and future experiments; Detection of cosmic gamma rays by atmospheric Čerenkov radiation: Čerenkov radiation in the atmosphere, properties of Čerenkov radiation generated by cosmic-ray air showers, detection techniques for cosmic gamma rays, results and future experiments.

061.023 Diffuse cosmic gamma-ray background in the 28 keV—4.1 MeV range from Kosmos 461 observations.
E. P. Mazets, S. V. Golenetskii (*Golenetskij*), V. N. Il'inskii (*Il'inskij*), Yu. A. Gur'yan, T. V. Kharitonova.
Astrophys. Space Sci., Vol. 33, 347 - 357 (1975).

Diffuse cosmic background and atmospheric gamma-radiation in the range 28 keV—4.1 MeV were studied with a scintillation spectrometer on board of the Kosmos 461 satellite. Separation of the cosmic and atmospheric components was made possible through a reliable determination of the geomagnetic dependences of albedo gamma-radiation. The spectrum of diffuse background in the energy range covered cannot be fitted with a common law. The shape of the high energy component spectrum of the diffuse background constructed using the data of Kosmos 461 and SAS-2 is in agreement with the hypotheses of the cosmological origin of the radiation.

061.024 The mystery of the cosmic boron abundance.
S. Ramadurai, N. C. Wickramasinghe.
Astrophys. Space Sci., Vol. 33, L41 - L44 (1975).

The observed high abundance of boron in type I carbonaceous chondrites may be due to the presence in the primitive solar nebula of graphite grains which have been irradiated by high energy nucleons at some stage of their history. The boron atoms thus produced by spallation reactions are stably locked within interstellar graphite grains and could make a significant contribution to the boron abundance of Cl chondrites.

061.025 The search for superheavy elements in nature.
G. Herrmann.
Phys. Scripta, Vol. 10A, 71 - 76 (1974). — Paper presented at the Nobel Symposium on superheavy elements, Ronneby, Sweden, June 11 - 14, 1974.

A review is given of the methods used in the search for superheavy elements in nature and of the results obtained. It is concluded that there is no convincing evidence for the presence of such elements in nature.

061.026 Superheavies in nature—where and how to look.
M. Nurmia.
Phys. Scripta, Vol. 10A, 77 - 80 (1974). — Paper presented at the Nobel Symposium on superheavy elements, Ronneby, Sweden, June 11 - 14, 1974.

In searching for superheavy elements (SHE) in nature it is usually assumed that the SHE (1) follows the chemistry of its lighter homolog and (2) decays by spontaneous fission. The sensitivity of SHE detection can be substantially improved by replacing these assumptions by a consideration of the geochemical fractionation processes, and by using a mass spectrometer. A good candidate for a search would be element 112, eka-mercury. It is expected to be more volatile and more noble than Hg so that it may behave like a heavy rare gas and be concentrated in the earth's atmosphere.

061.027 Possible paths for synthesis of superheavy elements in nature. W. M. Howard.
Phys. Scripta, Vol. 10A, 138 - 141 (1974). — Paper presented at the Nobel Symposium on superheavy elements, Ronneby, Sweden, June 11 - 14, 1974.

The author discusses the possibility of producing superheavy elements in the astrophysical r-process. Thus, he considers a detailed calculation of fission-barrier heights and neutron separation energies for heavy neutron-rich nuclei. The dependence of the neutron-induced fission cutoff of the r-process on the uncertainty in the nuclear models is discussed in some detail.

061.028 Stability of elements in the r-process region.
R. Bengtsson, R. Boleu, S. E. Larsson, J. Randrup.
Phys. Scripta, Vol. 10A, 142 - 148 (1974). — Paper presented at the Nobel Symposium on superheavy elements, Ronneby, Sweden, June 11 - 14, 1974.

The fission barriers of the nuclei in the r-process region have been calculated with inclusion of axial asymmetry and reflection asymmetry. Neutron separation energies and spontaneous-fission half-lives have been calculated and the possibilities of producing superheavy elements by means of neutron-capture processes are discussed.

061.029 An alternative thermonuclear n-capture path to the superheavy island.
H. W. Meldner, J. Nuckolls, L. Wood.
Phys. Scripta, Vol. 10A, 149 - 155 (1974). — Paper presented at the Nobel Symposium on superheavy elements, Ronneby, Sweden, June 11 - 14, 1974.

061.030 The importance of delayed fission in the production of very heavy and superheavy elements.
C.-O. Wene, S. A. E. Johansson.
Phys. Scripta, Vol. 10A, 156 - 162 (1974). — Paper presented at the Nobel Symposium on superheavy elements, Ronneby, Sweden, June 11 - 14, 1974.

Recent advances in the calculation of fission barriers and in the experimental determination of beta strength functions make it possible to derive reliable estimates of the effect of beta-delayed fission in the decay of very heavy nuclides far from the line of beta-stability. The model presented here is used to calculate the amount of delayed fission during the decay back of nuclides produced in the intense neutron burst from a nuclear explosion and of nuclides produced in astrophysical r-process.

061.031 On the chemical composition of the moon, Jupiter, meteorites and Am stars. E. M. Drobyshevski (j).
Earth Planet. Sci. Letters, Vol. 25, 368 - 378 (1975).

If surface anomalies in the composition of the metallic-line A stars are due to a precipitation of planet-like bodies (planetoids) on them, then one should expect a correlation to exist between the overabundance of heavier-than-iron elements on these stars and their "standard" abundances in the solar system (since chondrites provide the "standard" level for these elements). However, an anticorrelation was revealed. This fact supports the original suggestion on the origin of the metallicism of A stars, and can easily be explained within the author's hypothesis on the formation of the sun from matter escaping from the proto-Jupiter. The Am phenomenon may be seen to result from a precipitation of large geochemically differentiated planetoids onto a star. Such planetoids (including the moon) condense in the cooled envelope of the primary component of a close binary system.

061.032 Some problems and possibilities of nuclear astrophysics. G. E. Kocharov.
Probl. sovrem. fiz. Leningrad, Nauka, 1974, p. 147 - 182. In Russian. — Abstr. in Referativ. Zhurn. 51. Astron., 3.51.152 (1975).

061.033 Metagalactic protons of superhigh energies.
V. S. Berezinskij, S. I. Grigor'eva, G. T. Zatsepin.
Izv. AN SSSR. Ser. fiz., Vol. 38, 1791 - 1795 (1974). In Russian. — Abstr. in Referativ. Zhurn. 51. Astron., 3.51.852 (1975).

061.034 Starke Magnetfelder in der Astrophysik.
J. Schmid-Burgk, H. Pohl.
SuW, Vol. 14, 160 - 163 (1975).

061.035 Distribution of spontaneous fission fragments in

nuclear astrophysical problems. T. Ohnishi.
Astrophys. Space Sci., Vol. 34, 321 - 345 (1975).

Empirical equations for the estimation of fission fragment yield are proposed. These equations can be applied to all nuclides in the nuclidic region of $208 \lesssim A$ and $Z^2/A < 40.2$, and they can explain experimental results within an accuracy of 20 %. A possibility is proposed that the anomalously high abundance of medium weight elements observed in some peculiar A stars may be the accumulation of the fission fragments from the fissioning nuclei in $250 \lesssim A \lesssim 265$.

061.036 On the nature of γ-ray bursts.
O. F. Prilutski (*Prilutskij*), V. V. Usov.
Astrophys. Space Sci., Vol. 34, 387 - 393, 395 - 401 (1975).
In Russian and English.

The hypothesis on the γ-ray burst generation in the process of the collapse of supermassive bodies in the nuclei of active galaxies is considered. It is shown that γ-ray burst properties observed may be interpreted within the frames of the given model. A statistical test for choosing a hypotheses on γ-ray burst nature is discussed.

061.037 On the absorption of neutrinos in cosmic space.
V. A. Krat, I. L. Gerlovin.
Astrophys. Space Sci., Vol. 34, L11 - L12 (1975). – Letter.

061.038 Nucleochronology and chemical evolution.
B. M. Tinsley.
Astrophys. Journ., Vol. 198, 145 - 150 (1975).

Considerations of chemical evolution have been used to generalize Schramm and Wasserburg's formalism for deriving a mean age of the elements in the Galaxy at the time (T) when the solar system formed. It is found that Schramm and Wasserburg's quantity $\Delta^{max} - \Delta$ (which can in principle be evaluated from abundances and nuclear data on the radioactive elements) does not necessarily lie between $T/2$ and T, as it does in the restricted set of models; instead, it may lie anywhere between 0 and T, depending critically on the nature of the evolutionary model. It is concluded that the nucleochronometers provide a model-independent lower limit to the time T, but that derivation of a more precise age of the Galaxy from radioactive time scales will require detailed understanding of its chemical evolution.

061.039 High-energy gamma-ray results from the second Small Astronomy Satellite.
C. E. Fichtel, R. C. Hartman, D. A. Kniffen, D. J. Thompson, G. F. Bignami, H. Ögelman, M. E. Özel, T. Tümer.
Astrophys. Journ., Vol. 198, 163 - 182 (1975).

A high-energy (> 35 MeV) γ-ray telescope employing a 32-level wire spark-chamber system was flown on the second Small Astronomy Satellite. The high-energy galactic γ-radiation is observed to dominate over the general diffuse radiation along the entire galactic plane and is seen to be most pronounced in a region from $l^{II}=335°$ to $l^{II}= 40°$. When examined in detail, the longitudinal and latitudinal distributions seem generally correlated with galactic structural features, and particularly with arm segments. In addition to the general galactic emission, high-energy γ-radiation was seen from the Crab nebula (a significant fraction of which is pulsed at the radio pulsar frequency), Vela X (a supernova remnant whose high-energy γ-radiation possibly provides the first direct experimental evidence associating cosmic rays with supernovae), the general region ($15° < b^{II} < 30°$, $340° < l^{II} < 20°$), and a region a few degrees north of the galactic plane around 190° to 195° in l^{II}.

061.040 New method for recording neutrinos.
G. S. Bisnovatyj-Kogan.
Priroda, No. 4.75, p. 106 (1975). In Russian.

061.041 The effect of gyro-viscosity on the magneto-atmospheric waves. K. M. Srivastava, H. L. Jordan.
Astron. Astrophys., Vol. 39, 345 - 356 (1975).

The stability of a horizontally stratified infinitely conducting compressible medium in the presence of a vertical gravitational field and a horizontal magnetic field has been investigated including the effect of finite Larmor radius through the anisotropic pressure tensor.

061.042 Physical conditions in an optically thick hydrogen gas heated by suprathermal protons. E. Kimmer.
Bull. American Astron. Soc., Vol. 7, 237 (1975). – Abstr. AAS.

061.043 Low temperature Rosseland opacities.
D. R. Alexander.
Bull. American Astron. Soc., Vol. 7, 240 (1975). – Abstr. AAS.

061.044 A general fluid dynamics algorithm for astrophysical applications. L. D. Cloutman, H. M. Ruppel.
Bull. American Astron. Soc., Vol. 7, 255 - 256 (1975). Abstr. AAS.

061.045 Improved opacity calculations.
N. H. Magee, Jr., A. L. Merts, W. F. Huebner.
Bull. American Astron. Soc., Vol. 7, 338 (1975). – Abstr. AAS.

061.046 Thermal effects in beta-processes.
M. A. Rudzinskij, Z. F. Seidov.
Izv. AN AzSSR. Ser. fiz.-tekhn. i mat. n., 1974, No. 4, p. 98 - 103. In Russian. – Abstr. in Referativ. Zhurn. 51. Astron., 4.51.211 (1975).

061.047 On emission lines in the cosmic gamma-ray background. D. D. Clayton, R. A. Ward.
Astrophys. Journ., Vol. 198, 241 - 244 (1975).

The authors calculate the composite spectrum of γ-rays resulting from the decay of ^{56}Ni to ^{56}Co to ^{56}Fe throughout the history of the universe. The results for several cosmological models are presented and compared with the Apollo 15 measurements at low resolution of the cosmic background. The radioactivity background is a significant fraction of the total, and several of its features may be detectable.

061.048 Objects and observations in infrared astronomy.
H. J. Habing.
Space Sci. Rev., Vol. 17, 629 - 643 (1975). – Presented at the workshop on coherent detection in astronomy, held at Rhenen, The Netherlands, 25 and 26 April 1974 – see 012.009.

061.049 Quarks in astrophysics.
D. D. Ivanenko, D. F. Kurdgelaidze, N. Maksyukov.
Vestn. Mosk. un-ta. Fiz., astron., Vol. 15, 667 - 671 (1974).
In Russian. – Abstr. in Referativ. Zhurn. 51. Astron., 5.51.786 (1975).

061.050 A possible mechanism for cosmic gamma ray bursts.
U. Anzer, G. Börner.
Astron. Astrophys., Vol. 40, 123 - 126 (1975).

The recently discovered cosmic γ-ray bursts can be explained by the bremsstrahlung of a beam of relativistic electrons hitting a region of high proton density. Spectra are computed, and estimates on the particle densities and on the geometrical dimensions are derived from the observations. The question of associated optical emission is also discussed.

061.051 Gamma ray astrophysics. F. W. Stecker.
Origin of cosmic rays, (012.012), p. 267 - 334 (1975). – Review paper.

061.052 Observation of celestial gamma rays. K. Pinkau.
Origin of cosmic rays, (012.012), p. 335 - 370

(1975).

This review paper is sub-divided into three parts: (1) Instrumental problems; (2) Observational results on lines, the diffuse flux, the galactic emission and localized sources; (3) Discussion of the Vela supernova remnant.

061.053　Galactic gamma-ray-astronomy.
V. Schönfelder.
Conference on optical observing programs on galactic structure and dynamics, (see 012.013), p. 231 - 252 (1975). In German.

061.054　The X-ray background.　A. C. Fabian.
Observatory, Vol. 95, 80 (1975).

061.055　The gamma-ray background.　R. R. Hillier.
Observatory, Vol. 95, 82 - 83 (1975).

061.056　Neutral currents in astrophysics.　S. Ramadurai.
Bull. Astron. Soc. India, Vol. 3, 9 - 12 (1975).

061.057　Limitations of ground based near and far infrared astronomy.
G. Dall'Oglio, I. Guidi, B. Melchiorri, F. Melchiorri, V. Natale.
Infrared Physics, Vol. 15, 13 - 17 (1975).

The possibilities of ground based infrared astronomy are discussed. It is shown that telescopes used up to now are not optimized and that the diffraction limited detectivity conditions cannot be reached, in the near infrared, by the larger telescopes. The maximum useful diameters for near i.r. and far i.r. telescopes have been evaluated and a new parameter (i. e. visibility) is introduced in order to describe the performances of i.r. ground based telescopes.

061.058　X-ray astronomy.　R. Giacconi, with notes by R. Bland, G. Palumbo.
High energy astrophysics and its relation to elementary particle physics, (see 012.018), p. 201 - 253 (1974).

061.059　On the theory of synchrotron radiation by a charged current-carrying ring.
A. B. Kukanov, G. A. Lavrova.
Izv. vyssh. ucheb. zavedenij. Fizika, 1974, No. 12, p. 84 - 89. In Russian. – Abstr. in Referativ. Zhurn. 51. Astron., 6.51. 951 (1975).

061.060　Introductory remarks (Materials of the VIth international seminar "Particle acceleration and nuclear reactions in space, Leningrad, 19 - 21 August 1974".
S. N. Vernov.
Izv. AN SSSR. Ser. fiz., Vol. 39, 242 - 243 (1975). In Russian.
Abstr. in Referativ. Zhurn. 62. Issled. kosmich. prostranstva, 6.62.200 (1975).

061.061　Electric and magnetic fields originating in the discontinuity of a neutral current layer.
B. V. Somov, S. I. Syrovatskij.
Izv. AN SSSR. Ser. fiz., Vol. 39, 375 - 378 (1975). In Russian.
Abstr. in Referativ. Zhurn. 62. Issled. kosmich. prostranstva, 6.62.217 (1975).

061.062　Particle acceleration in the neighbourhood of the zero-line of a magnetic field.
S. V. Bulanov, P. V. Sasorov.
Izv. AN SSSR. Ser. fiz., Vol. 39, 379 - 382 (1975). In Russian.
Abstr. in Referativ. Zhurn. 62. Issled. kosmich. prostranstva, 6.62.218 (1975).

061.063　Discrete sources of X-rays and gamma-radiation.
Yu. N. Gnedin.
Izv. AN SSSR. Ser. fiz., Vol. 39, 408 - 416 (1975). In Russian.

Abstr. in Referativ. Zhurn. 62. Issled. kosmich prostranstva, 6.62.222 (1975).

061.064　Concluding remarks (Materials of the VIth international seminar "Particle acceleration and nuclear reactions in space, Leningrad, 19 - 21 August 1974").
G. E. Kocharov.
Izv. AN SSSR. Ser. fiz., Vol. 39, 445 - 447 (1975). In Russian.
Abstr. in Referativ. Zhurn. 62. Issled. kosmich. prostranstva, 6.62.226 (1975).

061.065　De heliumabundantie in het heelal.
W. J. Weber, T. de Graaf.
Utrechtse Sterrekundige Overdrukken No. 286, 7 pp. (1974).

061.066　Fe I fluorescence with astrophysical applications.
L. A. M. Willson.
Thesis, Michigan Univ., Ann Arbor (USA). 46 pp. University Microfilms Order No. 74-3750 (1973).

061.067　Neutrino radiation in spherically-symmetric gravitational fields. 2. The structure of the radiation field.
J. B. Griffiths, R. A. Newing.
General Relativity Gravitation, (GB), Vol. 5, 345 - 352 (1974).

061.068　Neutrino radiation in spherically-symmetric gravitational fields. 3. Comparison with photon radiation fields.　J. B. Griffiths.
General Relativity Gravitation, (GB), Vol. 5, 453 - 464 (1974).

061.069　Nickel or iron: the problem of nucleosynthesis at atomic weight 56.　K. L. Hainebach.
Thesis, Rice Univ., Houston, Texas (USA).120 pp. University Microfilms Order No. 74-21,275 (1974).

061.070　Neutrino pair emission by stellar: (a) plasmon-plasmon, (b) plasmon-phonon, and (c) phonon-phonon processes.　V. L. Madhyastha.
Thesis, New York Univ., New York (USA). 145 pp. University Microfilms Order No. 74-16,844 (1973).

061.071　Application of secular stability analysis to several astrophysical problems.　R. A. Siquig.
Thesis, Colorado Univ., Boulder (USA). 230 pp. University Microfilms Order No. 74-22,393 (1974).

061.072　Galactic neutrino sources and cosmic rays.
W. S. Pallister, A. W. Wolfendale.
AIP (American Inst. Phys.) Conference Proc., No. 22, p. 273 - 283 (1974). – Review paper.

061.073　Spatial and spectral anomalies in the soft X-ray background.　P. A. J. de Korte.
Proefschrift Rijksuniversiteit Leiden (Netherlands). 96 pp. (1975).

061.074　Problems and achievements of nuclear astrophysics. IV. Rapid neutron capture and tertiary processes of nucleosynthesis [supernovae].　B. Kuchowicz.
Postępy Fiz., Vol. 25, 475 - 495 (1974). In Polish.

A brief outline of the rapid neutron capture process is given; this process is responsible for the synthesis of heavy nuclides (up to the superheavy ones), and is usually located in supernovae. Two tertiary processes of nucleosynthesis, the p-process and the l-process, are surveyed.

061.075　Escape of photons from magnetized cylinders.
K. D. Krori, J. Barua.
Journ. Phys. A, (Math., Nuclear, General), Vol. 8, 186 - 189 (1975).

The authors have investigated the escape of photons from

the surface of magnetized cylinders. They have considered the escape of photons from a radiating solenoid carrying a current, and a radiating straight wire carrying a current.

061.076 Search for superheavy elements in nature [via neutron induced fission].
C. Stephan, J. Tys, M. Sowinski, E. Cieslak, M. Meunier.
Journ. Physique, Vol. 36, 105 - 112 (1975).
Superheavy elements have been searched for by neutron induced fission of mass separated samples. Various natural materials have been investigated: minerals, manganese nodules, lunar dust, meteoritic materials. Fissioning masses have been collected in the $A = 300$ mass region.

061.077 Variable energy blast wave through self-gravitating gas spheres. P. Chaturani, I. Ram.
Proc. Indian Acad. Sci., Ser. A, Vol. 80, 140 - 159 (1974).
The propagation of a variable energy blast wave through self-gravitating gas spheres has been studied. Similarly methods have been used to obtain the solution in the form of a power series in the non-dimensional shock radius.

061.078 Mass asymmetric fission and the termination of the astrophysical r-process.
R. Bengtsson, W. M. Howard.
Phys. Letters B, (*Netherlands*), Vol. 55B, 281 - 285 (1975).
A recent calculation of the mass-asymmetric fission barriers for heavy neutron-rich nuclei is used to estimate the heavy to light mass fragment ratio resulting from fission at the termination of the astrophysical r-process. The fragments due to mass-asymmetric fission at the termination of the r-process may account for the solar system abundance distribution of neutron-rich nuclei in the mass region $150 \lesssim A \lesssim 170$.

061.079 The abundances of the elements.
P. M. Williams, M. G. Edmunds.
Sci. Progr., (*GB*), No. 243, Vol. 61, 323 - 347 (1974).

061.080 Außerirdisches Leben aus astrophysikalischer Sicht.
K.-H. Schmidt.
Sterne, Vol. 51, 82 - 90 (1975).

Of fundamental electrodynamics and astrophysics.
See Abstr. 022.009.

The influence of plasma turbulence on the radiation of accreting gas. See Abstr. 062.009.

Nuclear and nonnuclear abundance patterns in the manganese stars. See Abstr. 114.020.

On the chemical composition of the sun, Jupiter, meteorites and Am stars. See Abstr. 114.083.

Magnetohydrodynamics: applications to magnetic stars, cosmical gas dynamics, and pulsars.
See Abstr. 116.011.

Dielectronic recombination and abundances near quasars. See Abstr. 141.011.

Extragalactic X-ray sources and the X-ray background. See Abstr. 142.003.

A study of fast time structure within cosmic gamma-ray bursts. See Abstr. 142.091.

Temporal X-ray astronomy with a pinhole camera. See Abstr. 142.095.

Hard X-ray bursts in the energy range of 40–290 keV. See Abstr. 142.097.

Galactic X-ray astronomy. See Abstr. 142.101.

Nuclear gamma ray production by cosmic rays. See Abstr. 143.058.

Light element production by cosmic rays. See Abstr. 143.059.

Nucleosynthesis and matter–antimatter cosmologies. See Abstr. 162.004.

Microphysics, cosmology, and high energy astrophysics. See Abstr. 162.075.

Errata

061.901 Erratum: "On the chemical composition of moon, Jupiter, meteorites and Am stars" [Earth Planet. Sci. Letters, Vol. 25, 368 - 378 (1975)]. E. M. Drobyshevski(j). Earth Planet. Sci. Letters, Vol. 26, 270 (1975). – See 13.061. 031.

062 Hydrodynamics, Magnetohydrodynamics, Plasma

062.001 Numerical simulation of the plasma double layer.
C. K. Goertz, G. Joyce.
Astrophys. Space Sci., Vol. 32, 165 - 173 (1975).

A one-dimensional particle-in-cell computer simulation is used to model the formation of an electrostatic double layer. The conditions for the onset of the layer formation are explored and a relation between the length of the layer and the electrostatic potential difference across is found.

062.002 Population densities and ionization coefficients of fast transient hydrogen plasmas.
M. Cacciatore, M. Capitelli.
Zeitschr. Naturforschung, Vol. 30a, 48 - 54 (1975).

062.003 Force-free pulsar magnetosphere – II. The steady, axisymmetric theory for a normal plasma.
I. Okamoto.
Monthly Notices Roy. Astron. Soc., Vol. 170, 81 - 93 (1975)

A physically reasonable field structure in the force-free pulsar magnetosphere is discussed in the steady, axisymmetric theory for a normal plasma. The null line where the charge density ρ_e vanishes must cross the critical field line near the light cylinder $\tilde{\omega}_L$ from the equator side and approach to it again from the pole side as the axial distance $\tilde{\omega}$ increases to infinity in a plausible field model. One difficulty of the theory is, however, that there must be a mechanism which draws out charges of both sign from the neighbourhood of the critical field line P_c, if all charges are assumed to emanate only from the pulsar surface.

062.004 Optically thin radiating shock waves and the formation of density inhomogeneities.
R. A. Chevalier, J. C. Theys.
Astrophys. Journ., Vol. 195, 53 - 60 (1975).

It is found that the gas flow behind a perturbed, optically thin, radiating shock front tends to create density clumps. A two-dimensional hydrodynamic code is used to follow the shock created by an explosion. The shock is perturbed by a density fluctuation in the ambient medium. The growth of clumps is greatest near the time when radiative cooling first becomes significant. After a clump has formed, it tends to move ahead of the parent shock front due to its large momentum. The magnitude of the density enhancement is approximately proportional to the magnitude of the perturbing fluctuation. The results are discussed, with particular attention to the shock fronts in supernova remnants.

062.005 Thermal fluctuations of potential differences of two separate points in a magnetoactive plasma.
V. M. Kostin.
Izv. vyssh. ucheb. zavedenij. Radiofizika, Vol. 17, 1261 - 1268 (1974). In Russian. – Abstr. in Referativ. Zhurn. 51. Astron., 2.51.206 (1975).

062.006 The equations of hydrodynamics for a thermally conducting, viscous, compressible fluid in special relativity. P. J. Greenberg.
Astrophys. Journ., Vol. 195, 761 - 772 (1975).

The author derives the expressions (expansion, rotation, and shear) describing the kinematics of the fluid as seen by an observer comoving with an element of the fluid and employing three Fermi-transported, local, Cartesian axes. It is furthermore observed that there is the possibility for waves of viscosity to exist in the fluid, and he determines in the case of one-dimensional flow, the condition for the existence of these waves.

062.007 On a nonlinear closure approximation for cosmic-ray diffusion equations. I. Lerche.
Astrophys. Journ., Vol. 195, 783 - 784 (1975).

Owens has recently proposed, and used, a nonlinear closure approximation for the turbulent Liouville equation in order to obtain a cosmic-ray diffusion equation. The author shows that his proposed scheme is forbidden since it demands that a positive definite integral be identically zero.

062.008 On the nonlinear closure approximation for cosmic-ray diffusion. A. J. Owens.
Astrophys. Journ., Vol. 195, 785 - 786 (1975).

Lerche has recently claimed that the nonlinear closure approximation is invalid. His discussion reemphasizes the approximate nature of the model, but gives no indication of its accuracy. His initial equations involve higher-order terms than those retained in the approximation, and hence do not bear on it. His subsequent comments concerning number and energy density fluctuations are shown to apply to Dupree's nonlinear model as well. His conclusions depend critically on α_0 being strictly independent of velocity, which previous discussions have shown is not necessary.

062.009 The influence of plasma turbulence on the radiation of accreting gas. S. A. Kaplan, F. K. Lamb,
D. Pines, C. J. Pethick, V. N. Tsytovich.
Astron. Zhurn. Akad. Nauk SSSR, Vol. 52, 64 - 70 (1975). In Russian. English translation in Soviet Astron., Vol. 19, No. 1.

It is shown that the excitation of plasma turbulence in the process of accretion of gas to dense objects enhances the electromagnetic radiation and accelerates the electrons to ultrarelativistic energy. This leads to an increase of the efficiency of the accretion. The accretion to nonmagnetic objects may generate radiation with sharp maximum in the infrared and the accretion to an object with strong magnetic field generates radiation with the maximum in the X-ray range.

062.010 On the ionization of hydrogen and helium by hydromagnetic shock waves.
E. Daltabuit, J. Andrade, J. Cantó, M. Peimbert.
Rev. Mexicana Astron. Astrofis., Vol. 1, 203 - 209 (1974).

Hydromagnetic shock waves in the 10 to 200 km s^{-1} range including the effect of a magnetic field in the 0 to 1×10^{-4} gauss range perpendicular to the flow, have been computed. The results for $v_0 \leqslant 16$ km s^{-1} are in very good agreement with those of Field et al. (1968). The relevance of the results to some astronomical problems is briefly discussed. In particular the lack of ionized helium in three H II regions near the nucleus of our Galaxy can be easily explained if the ionization is due to hydromagnetic shock waves.

062.011 Energy transfer and superradiance between two high temperature carbon or nitrogen plasmas.
B. A. Norton.
Zeitschr. Naturforschung, Vol. 30a, 263 - 264 (1975).

062.012 Mirroring in the Fokker-Planck coefficient for cosmic-ray pitch-angle scattering in homogeneous magnetic turbulence.
M. L. Goldstein, A. J. Klimas, G. Sandri.
Astrophys. Journ., Vol. 195, 787 - 799 (1975).

The Fokker-Planck coefficient for pitch-angle scattering, appropriate for cosmic rays in homogeneous, stationary magnetic turbulence, is computed from first principles. The general existence of a Dirac δ-function in the pitch-angle scattering coefficient is demonstrated. It is proved in this paper that this delta function is the prediction of the Fokker-Planck equation for pitch-angle scattering due to mirroring in the magnetic field. The conditions under which this δ-function contributes to pitch-angle scattering are determined, and shown to be identical to the conditions under which first-order mirroring occurs in the random field. These conditions are generally fulfilled in interplanetary and probably interstellar space. The implications of the δ-function for the validity of the Fokker-Planck equation are discussed.

062.013 **Theoretical models of magnetic field line merging, 1.**
V. M. Vasyliunas.
Rev. Geophys. Space Phys., Vol. 13, 303 - 336 (1975).

Models of magnetic field line merging that consider processes in a limited region around the magnetic X line, within which the external magnetic fields are roughly uniform and antiparallel, are reviewed.

062.014 **Magnetohydrodynamic equilibrium conditions in the post-Newtonian approximation of general relativity.** N. P. Bondarenko, O. V. Kravtsov.
Astrophys. Space Sci., Vol. 32, 379 - 384 (1975).

A set of equations, which are magnetohydrodynamic equilibrium conditions in the post-Newtonian approximation of general relativity (PNA of GR), is obtained. The given system generalizes the previously obtained magnetohydrodynamic equilibrium conditions of classical mechanics and the hydrodynamic equilibrium conditions in the PNA of GR.

062.015 **The propagation of wave modes in ultrarelativistic magnetoactive plasma—I.** R. D. Blandford.
Monthly Notices Roy. Astron. Soc., Vol. 170, 619 - 632 (1975).

Some results on the propagation of undamped wave modes in fully relativistic, magnetoactive plasma are presented.

062.016 **A unified approach to mean field electrodynamics.**
P. H. Roberts, A. M. Soward.
Astron. Nachr., Vol. 296, 49 - 64 (1975).

Using the first order smoothing approximation and a novel technique (double Fourier transformation and expansion) a number of results, new and old, in the theory of mean field electrodynamics and magnetohydrodynamics are given a systematic and general derivation. They are cast into forms which bring into new prominence the role of the helicity spectrum in induction processes. The situations in which the results may be expected to be accurate are delineated. New expressions are given for the reduction in the mean electromotive force created by the Lorentz forces acting on the microscale turbulence.

062.017 **A model for generation of bow-shock-associated upstream waves.** R. W. Fredricks.
Journ. Geophys. Res., Vol. 80, 7 - 17 (1975).

062.018 **Larmor radius and collisional effects on gravitational instability of a composite rotating plasma.**
K. Prakash, R. C. Sharma.
Astrophys. Space Sci., Vol. 33, 3 - 9 (1975).

Gravitational instability of an infinitely conducting hydromagnetic composite rotating plasma is considered to include simultaneously the finite Larmor radius effects and the frictional effects with neutrals. It is found that Jeans' criterion of instability holds good in the presence of rotation, finite Larmor radius and collisions with neutrals. The particular cases of the above effects on the waves propagated along and perpendicular to the magnetic field have been discussed. The effect of rotation is to decrease the Larmor radius by an amount depending upon the wave number of perturbation.

062.019 **The oscillations and the stability of rotating masses with magnetic fields. IV. Existence of the point of bifurcation.** S. K. Trehan, M. Singh.
Astrophys. Space Sci., Vol. 33, 43 - 48 (1975).

The existence of the point of bifurcation of rotating gaseous masses with toroidal magnetic fields is established from the first variations of the integral properties provided by the second-order virial relations. It is shown that the point of bifurcation, where the Jacobi ellipsoids branch off from the Maclaurin spheroids, is unaffected by the presence of toroidal magnetic fields.

062.020 **The effects of body geometry on the structure in the near wake zone of bodies in a flowing plasma.**
W. A. Oran, N. H. Stone, U. Samir.
Journ. Geophys. Res., Vol. 80, 207 - 209 (1975).

062.021 **Pulsed high-energy radiation from oblique magnetic rotators.** A. Ferrari, E. Trussoni.
Astrophys. Space Sci., Vol. 33, 111 - 126 (1975).

A detailed analysis is reported of charged particle motion in the e.m. fields generated by oblique magnetic rotators. A full treatment of particle trajectories, radiated power and spectrum is given; astrophysical implications of these results in connection with pulsars, mainly NP 0532, are discussed and the possibility of a high energy pulsed radiation model with synchro-Compton emission at the speed-of-light cylinder is investigated.

062.022 **The interaction of an obliquely incident s-polarized plane electromagnetic wave at a warm moving magnetized plasma half-space.**
P. K. Mukherjee, S. P. Talwar.
Astrophys. Space Sci., Vol. 33, 147 - 163 (1975).

The interaction of an s-polarized plane electromagnetic wave incident from a dielectric (or vacuum) region on a warm moving magnetized plasma half-space is considered. Expressions are obtained for the phase and group velocities and the index of refraction for the ordinary mode, as also for power reflection and transmission coefficients. It is found that in contrast to the case of a cold magnetized plasma, the ordinary electromagnetic mode excited in the warm magnetoplasma medium gets modified due to the presence of an external magnetic field. In addition, the various reflection and transmission characteristics for a warm magnetoplasma depend on the velocity of the moving plasma as well as on the strength of the applied magnetic field, as against the case for a cold moving magnetized plasma. Numerical results on the reflection coefficient are presented for several values of the parameters characterizing the electron-plasma temperature, the velocity of the moving medium and the strength of the applied magnetic field.

062.023 **Criterion for propagation of magnetohydrodynamic waves: an equivalence.** Bibhas R. De.
Astrophys. Space Sci., Vol. 33, 235 - 237 (1975).

It is shown that the requirement that a magnetohydrodynamic wave propagates over a distance equal to a large number of wavelengths without substantial attenuation is equivalent to the requirement that the lateral velocity of displacement of the field lines far exceeds a characteristic velocity. This latter velocity is the velocity of slippage of the field lines with respect to the medium.

062.024 **Low-frequency fluctuations in the solar wind. I. Theory.** C. S. Wu, J. D. Huba.
Astrophys. Journ., Vol. 196, 849 - 857 (1975).

Several simple relationships between the power spectra of density and velocity fluctuations and the power spectrum of magnetic field fluctuations are derived within the context of plasma kinetic theory. The theory is restricted to the low-frequency regime (less than the proton cyclotron frequency) where hydromagnetic turbulence is expected to play the most important role. The effects of Alfvén and magnetosonic waves upon the plasma fluctuations are discussed separately. The results are then applied to proton fluctuations in the solar wind, demonstrating a connection between plasma and field fluctuations.

062.025 Nonisothermal magnetostatic equilibria in a uniform gravity field. I. Mathematical formulation.
B. C. Low.
Astrophys. Journ., Vol. 197, 251 - 255 (1975).
The author considers the magnetostatic equilibrium of a two-dimensional plasma in a uniform gravity field and he assumes the ideal gas law. By transforming the pressure and temperature into functions of the magnetic vector potential and the vertical height, he shows that the vector equation of equilibrium in the cases of Cartesian and cylindrical coordinate systems reduces to a single scalar nonlinear elliptic partial differential equation. A discussion is given on the general properties of the equations derived and their physical implications.

062.026 Destabilization of hydromagnetic drift-Alfvén waves in a finite-β, collisional plasma.
J. T. Tang, N. C. Luhmann, Jr., Y. Nishida, K. Ishii.
Phys. Rev. Letters, Vol. 34, 70 - 73 (1975).
The authors have observed experimentally the destabilization of both the Alfvén and the drift branches of the hydromagnetic, coupled drift-Alfvén waves in a steady-state, high-density ($n_0 \approx 10^{13} - 10^{15}$ cm^{-3}) collisional plasma when an axial electron current is drawn along the magnetic field.

062.027 Interaction between the electron-cyclotron emissions at $(n + {}^1/_2) \Omega_e$ and the ring-current protons in space. M. Nambu.
Phys. Rev. Letters, Vol. 34, 387 - 391 (1975).
The interaction between the electron-cyclotron emissions at $(n + {}^1/_2) \Omega_e$ and the ring-current protons is studied by using the linear-response theory of a turbulent plasma. The effective turbulent collisions come from the decay interactions between the electron-cyclotron turbulence and electrostatic ion-cyclotron waves. The ion-cyclotron waves generated by the generalized parametric resonance are effective for the loss of the ring-current protons. The critical electron-cyclotron-turbulence amplitude is consistent with the observed value.

062.028 The oscillations of an incompressible viscous cylinder. T. Ishizawa.
Mem. Fac. Sci., Kyoto Univ., Ser. Phys., Astrophys., Geophys., Chem., Vol. 34, 275 - 280 (1975).
The characteristic frequencies for non-axisymmetric oscillations of an incompressible viscous cylinder are determined. The viscous damping of the oscillations is discussed.

062.029 Electron runaway in turbulent astrophysical plasmas.
M. J. Houghton.
Planet. Space Sci., Vol. 23, 409 - 418 (1975).
An astrophysical electron acceleration process is described which involves turbulent plasma effects: the acceleration mechanism will operate in 'collision free' magnetoactive astrophysical plasmas when ion-acoustic turbulence is generated by an electric field which acts parallel to the ambient magnetic lines of force. It is shown that, in spite of the turbulence, a small fraction of electron population can accelerate freely, i.e. runaway, in the high parallel electric potential.

062.030 Dynamics of the current sheet of a flare. I. Diffusion thickening of the current sheet and estimate of the flare parameters. L. A. Pustil'nik.
Astron. Zhurn. Akad. Nauk SSSR, Vol. 52, 316 - 325 (1975).
In Russian. English translation in Soviet Astron., Vol. 19, No. 2.

A process of magnetic field dissipation in the current sheet of a flare is analysed. It is shown that, due to plasma turbulence, field dissipation during a flare must occur by "splitting" the current sheet and its subsequent diffusion thickening. Estimates of the main characteristics of the flare process are obtained. The suggested mechanism of energy conversion in a flare allows to avoid certain difficulties encountered in the previous models, which are connected with accumulation of the plasma in the current sheet. This flare model, apparently, may be used to explain flare activity observable in various astrophysical objects (chromospheric flare, UV Ceti-type stars, X-ray sources, galactic nuclei, etc.).

062.031 On the acceleration of cosmic particles trapped by plane electromagnetic waves in an inhomogeneous plasma. V. Ya. Davydovskij, A. S. Ukolov.
Izv. Sev.-Kavkaz. nauch. tsentra vyssh. shkoly. Seriya estestv. n., 1974, No. 2, p. 82 - 85. In Russian. — Abstr. in Referativ. Zhurn. 51. Astron., 3.51.176 (1975).

062.032 Thermal instability in plasma with finite Larmor radius. R. C. Sharma, K. Prakash.
Zeitschr. Naturforschung, Vol. 30a, 461 - 465 (1975).
The effects of the finite Larmor radius of the ions on the thermal instability of a plasma are investigated. When the instability sets in as stationary convection, the finite Larmor radius is found to have a stabilizing effect. The conditions for the nonexistence of overstability are investigated. The case with horizontal magnetic field is discussed.

062.033 Electromagnetic instability in collisionless anisotropic plasma contrastreams. S. S. Aggarwal.
Astrophys. Space Sci., Vol. 33, 259 - 264 (1975).
The electromagnetic instability in collisionless, anisotropic magnetized plasma contrastreams for perturbations propagating normally to the ambient magnetic field is investigated. It is found that the parallel temperature stabilizes the configuration for $T_\parallel \leqslant T_\perp$ while it shows destabilizing effect for $T_\parallel > T_\perp$. A comparative study of the growth rates associated with various coexisting modes reveals that the electromagnetic instability may, under certain conditions, be associated with larger growth rates than the electrostatic instability.

062.034 Finite Larmor radius effects on the Rayleigh-Taylor instability of a rotating plasma of variable density.
P. K. Bhatia, J. M. Steiner.
Astrophys. Space Sci., Vol. 34, 459 - 465 (1975).
The Rayleigh-Taylor instability in a rotating plasma of variable density has been investigated to include simultaneously the effects of viscosity and the finiteness of the ion Larmor radius. It is shown that, for a plasma in which the density is stratified along the vertical, the solution is characterized by a variational principle. Making use of this, proper solutions have been obtained for a semi-infinite plasma in which the density varies exponentially.

062.035 On the stability of magnetic fields of active regions.
M. M. Molodenskij.
Solnechnye Dannye 1975 Byull., No. 1, p. 66 - 70 (1975).
In Russian.
A condition for magnetic field stability obtained previously (1974) has been generalized for the case of arbitrary distribution of currents. Configurations of magnetic fields were found to be stable on the condition that the ratio of the current to the field and the gradient of this ratio are small enough.

062.036 Hydromagnetic waves in structured magnetic fields.
L. E. Cram, P. R. Wilson.
Solar Physics, Vol. 41, 313 - 327 (1975).
The authors reformulate the hydromagnetic wave prob-

lem for magnetic fields which vary in one direction perpendicular to the field. The permitted modes of small amplitude hydromagnetic oscillations are considered, first in the case of a single interface between semi-infinite magnetic and non-magnetic compressible regions, and secondly for a magnetic flux sheath of given thickness imbedded in a nonmagnetic region.

062.037 On the character of a force-free magnetic field in streams with spiral symmetry.   G. A. Rubo.
Problems of cosmic physics. Vyp. (No.) 9, (see 003.013), p. 53 - 56 (1974). In Russian.
   The force-free magnetic field character in spirally symmetric streams of a fluid with infinite conduction has been investigated.

062.038 Plasma echo in the ionosphere.
   V. N. Pavlenko, S. M. Revenchuk.
Problems of cosmic physics. Vyp. (No.) 9, (see 003.013), p. 64 - 72 (1974). In Russian.
   It is shown that two successive disturbances — the longitudinal and transverse — applied in various points of space in a collisionless magnetoactive plasma may cause a transverse echo. Some cases have been considered in which the Langmuir waves superimposed on cyclotron waves cause echo waves at electron-cyclotron frequency as well as low frequency echo waves at ion-cyclotron frequency. The possibility of the existence of such waves in the plasma of the ionospheric F- and E-layers is demonstrated.

062.039 Effect of obstacles on the rate of reconnection of magnetic field lines.   P. J. Baum, A. Bratenahl.
Planet. Space Sci., Vol. 23, 813 - 816 (1975).
   Further results of a laboratory magnetic field line reconnection experiment are presented. In particular, it is found that the reconnection rate can be slowed by placing solid obstacles to impede the outflow of plasma from an x-type magnetic neutral point. Without the obstacles the reconnection rate is faster and more impulsive. The fastest reconnection event has strong similarities to solar flares and geomagnetic substorms. It is suggested that more stationary features of solar activity such as prominences may be the result of reconnection slowed by obstacles such as the photosphere.

062.040 Terrestrische und extraterrestrische Plasmen.
   R. Lüst.
Naturwissenschaften, 62. Jahrgang, p. 255 - 263 (1975).
   The cosmic plasma nearest the earth and its interaction with planets can now be studied in detail by observation from satellites and space probes. Among the problems that can be investigated in this way are the plasmaphysical processes that could be significant if thermonuclear fusion were used for the controlled production of energy. Observations from space probes can also contribute to our knowledge of very distant cosmic plasmas that cannot be studied directly.

062.041 Magnetohydrodynamic accretion and the instability of smooth trans-Alfvénic flow.   D. J. Williams.
Monthly Notices Roy. Astron. Soc., Vol. 171, 537 - 549 (1975).
   Radial accretion towards a body, in the presence of a magnetic field, is considered. The results are generalized to show that any ideal aligned-field flow where gas passes smoothly from a super-Alfvénic region into a sub-Alfvénic region will be unstable, without exception. It is concluded that super-Alfvénic flow towards a body must become sub-Alfvénic via a shock, followed by turbulent flow. Super-Alfvénic accretion has thus some similarity to the solar wind's impinging on the earth's magnetosphere.

062.042 Hydromagnetic waves and cosmic-ray diffusion theory.   M. A. Lee, H. J. Völk.
Astrophys. Journ., Vol. 198, 485 - 492 (1975).

Pitch-angle (and energy) diffusion of cosmic rays in hydromagnetic wave fields is considered. The treatment remains strictly within the quasi-linear approximation. It is shown that the popular assumption of an isotropic power spectrum tensor of magnetic fluctuations requires in this case equal forms and magnitudes of Alfvén and magnetosonic wave spectra—a situation which is generally unlikely. The relative contributions to the pitch-angle diffusion coefficient from the cyclotron resonances and Landau resonance due to the different types of waves are evaluated for a typical situation in the solar wind.

062.043 A two-component description of energetic particle scattering in a turbulent magnetoactive plasma.
   G. Morfill.
Journ. Geophys. Res., Vol. 80, 1783 - 1794 (1975).

062.044 The interaction between homogeneous turbulence and an inhomogeneous magnetic field near neutral sheets.   G. Rüdiger.
Astron. Nachr., Vol. 296, 133 - 141 (1975). In German.
   Continuing an investigation concerning the influence of a uniform mean magnetic field on turbulence (Rüdiger, 1974) the author now considers a weak magnetic field changing spatially weakly and containing a neutral sheet. An originally homogeneous and isotropic turbulent field becomes inhomogeneous and anisotropic if such a magnetic field is present. Because of the finite correlation length the turbulent field is also affected in a neutral sheet. For a special class of spectral functions of two- and three-dimensional turbulence the anisotropic damping of the motions is given in the vicinity of the neutral sheet. Furthermore, the author points out the consequence for the mean magnetic field which is affected by such an inhomogeneous turbulent field. Using Bochner's theorem concerning the spectral tensor of the originally homogeneous turbulence he obtains an additional decay of the mean magnetic field.

062.045 Annotation on the paper "On the Reynolds stresses in mean-field hydrodynamics. II. Two-dimensional turbulence and the problem of negative viscosity".
   F. Krause, G. Rüdiger.
Astron. Nachr., Vol. 296, 143 (1975).

062.046. On the relativistic theory of electromagnetic dispersion relations and Poynting's theorem.   I. Lerche.
Astrophys. Journ., Suppl. Ser., No. 279, Vol. 29, 113 - 122 (1975).
   Constitutive relations, and general dispersion relations, are derived for an arbitrary, anisotropic, dispersive and dissipative medium which is moving relative to an inertial observer. The constitutive relations are expressed in terms of the "local" dielectric tensor, magnetic permeability, etc., where "local" refers to the instantaneous rest frame of the medium. The generalization of Poynting's theorem for power flow is also given including the expression for the rate at which the moving medium does not work on the radiation. In view of the current interest in radiation generated in, and passing through, pulsar magnetospheres, it is believed that the general results presented here are, perhaps, not without some astrophysical importance.

062.047 Discontinuities in magnetohydrodynamics with finite electroconductivity and limiting conditions on the magnetopause and flare magnetoplasma.
   K. G. Ivanov.
Geomagn. Aeronom., Vol. 15, 497 - 501 (1975). In Russian.

062.048 Motion of charged particles normal to an irregular magnetic field.   J. R. Jokipii.
Astrophys. Journ., Vol. 198, 727 - 732 (1975).

The motion of charged particles is considered in a fluctuating magnetic field which varies only in directions normal to its mean direction. Such a field would be produced, for example, by an ensemble of magnetosonic waves propagating normal to an ambient magnetic field. The appropriate generalization of gradient drift motion is derived in terms of the magnetic fluctuation spectrum, and an effective diffusion coefficient is derived.

062.049   Introduction to magnetohydrodynamics.
N. O. Weiss.
Magnetohydrodynamics, (see 003.018), p. 1 - 36 (1974).

062.050   Surface density of accumulated electrons on walls in contact with a plasma.   Bibhas R. De.
Astrophys. Space Sci., Vol. 35, L17 - L19 (1975).
It is shown that the surface density of accumulated electrons on a wall in contact with a plasma can be expressed as a simple function of the Debye shielding distance in the plasma. The result may have applications to problems involving objects immersed in a space plasma.

062.051   Non-stationary convection in a fully ionized magnetized plasma.
L. A. Abramov, L. S. Al'perovich.
Astron. Zhurn. Akad. Nauk SSSR, Vol. 52, 667 - 669 (1975). In Russian. English translation in Soviet Astron., Vol. 19, No. 3.
The problem of heating of a fully ionized plasma horizontal layer is solved. The criterion of the convective oscillations instability and the frequency of neutral oscillations for large Hartman numbers is obtained.

062.052   Stochastic acceleration by a single wave in a magnetic field.   G. R. Smith, A. N. Kaufman.
Phys. Rev. Letters, Vol. 34, 1613 - 1616 (1975).

062.053   Plumpton-Ferraro oscillations of a slowly rotating liquid sphere.   P. C. Kendall, J. A. Lawrie.
Geophys. Journ. Roy. Astron. Soc., Vol. 41, 441 - 446 (1975).

062.054   When is quasi-linear theory exact?
F. C. Jones, T. J. Birmingham.
Plasma Phys., Vol. 17, 15 - 22 = Goddard Space Flight Center, Greenbelt, Maryland, Separate print (1975).
The authors use the cumulant expansion technique of Kubo to derive an integrodifferential equation for $\langle f \rangle$, the average one particle distribution function for particles being accelerated by electric and magnetic fluctuations of a general nature. For a very restricted class of fluctuations, the $\langle f \rangle$ equation degenerates exactly to a differential equation of Fokker–Planck type. Quasi-linear theory, including the adiabatic assumption, is an exact theory only for this limited class of fluctuations.

062.055   The developments of condensations in uniform media.   M. Kondo.
Publ. Astron. Soc. Japan, Vol. 27, 215 - 235 (1975).
The fully nonlinear development of self-gravitating density perturbations has been investigated numerically for slab, cylindrical, and spherical configurations, under the assumptions that the equation of state is either adiabatic or isothermal with $\gamma = 5/3$, $4/3$, and 1, and that the unperturbed state is either static or expanding.

062.056   Three-wave interactions involving one whistler.
D. B. Melrose.
Australian Journ. Phys., Vol. 28, 101 - 113 = Separate print Div. Radiophys. C.S.I.R.O. Sydney (1975).
The purposes of this paper are twofold: firstly, to derive specific expressions describing three-wave interactions in which one of the waves is a whistler and the other two are higher

frequency waves, the derivation being based on the theory developed by Melrose and Sy (1972); secondly, to discuss the possible importance of such interactions to plasma radiation (i.e. radiation at or near the local plasma frequency) from the solar corona.

062.057   Improved quantum mechanical treatment of plasma polarization shift of ion lines.   S. Volonté.
Journ. Phys. B, Atomic Molecular Phys., Vol. 8, 1170 - 1176 = Commun. Dép. Astrophys., Fac. Sci. Mons, Mons Astrophys. Papers No. 46 (1975).
An improved quantum mechanical treatment of the plasma polarization shift of ion lines is presented. Calculations using this treatment are made for plasma conditions corresponding to available measurements of the shift of the He II resonance series. Comparisons with the other theories are also discussed.

062.058   Behandlung eines einfachen hydromagnetischen Dynamos mit Hilfe der Gitterpunktmethode.
G. Rüdiger.
Publ. Astrophys. Obs. Potsdam, No. 108, Vol. 32, Heft 1, p. 25 - 39 (1974).
An outline of a computer program is presented which is to provide the solutions of the nonlinear partial differential equation system in case of a simple dynamo model. The nonlinearity arises considering the relation of the induced magnetic field on turbulence. Consequently the eigenvalue-character of the linear problem vanishes. So, the magnetic fields can be determined uniquely for parameters exceeding the smallest eigenvalue of the linear theory. The nonlinear algebraic system that originates using the lattice-point method is solved by Newton's procedure.

062.059   Photoneutrino reaction in a superstrong magneto-active plasma.   C. K. Chou.
Thesis, New York Univ., New York (USA). 97 pp. University Microfilms Order No. 74-13,313 (1973).

062.060   On the theory of large amplitude Alfvén waves.
M. L. Goldstein, A. J. Klimas, F. D. Barish.
Solar wind three, (see 012.020), p. 385 - 387 (1974).

062.061   Numerical models of hydromagnetic dynamos.
S. A. Jepps,
Journ. Fluid Mechanics, Vol. 67, 625 - 646 (1975).
The magnetic induction equation is solved numerically in a sphere for a variety of prescribed fluid flows. The models considered are the so-called '$\alpha \omega$ dynamos', in which both small-scale turbulence and large-scale shearing play a significant role.

062.062   An astronomer's vademecum for plasma turbulent reactors.   C. A. Norman, D. ter Haar.
Fys. Tidsskr., Vol. 72, 84 - 90 (1974).
The authors show how one can test whether an astronomical system with given luminosity and size, or given size and density, could possibly be a plasma turbulent reactor, and in what range of frequencies it can be expected to radiate.

Plasma on the earth and in space.
See Abstr. 003.007.

Elementare Plasmaphysik.
See Abstr. 003.087.

Time-dependent hydrodynamic phenomena in the envelopes of the Be stars.   See Abstr. 064.007.

Thermodynamical approach to current interruption model of solar flares.   See Abstr. 073.015.

Nonisothermal magnetostatic equilibria in a uniform gravity field. II. Sheet models of quiescent prominences. See Abstr. 073.030.

Restricted three-dimensional stellar wind modeling. I. Polytropic case.   See Abstr. 074.015.

The influence of pressure anisotropy of near-solar plasma on the propagation of Alfvén waves. See Abstr. 074.033.

The generation of magnetic fields in astrophysical bodies. X. Magnetic buoyancy and the solar dynamo. See Abstr. 080.026.

Solar wind access to the plasma sheet along the flanks of the magnetotail.   See Abstr. 084.206.

Plasma and fields in the magnetospheric tail.

See Abstr. 084.237.

Hydromagnetic waves in molecular clouds. See Abstr. 131.040.

Hydromagnetic waves and shock waves as an interstellar heat source.   See Abstr. 131.118.

Rotating magnetospheres: frozen-in-flux violation. See Abstr. 141.324.

Self-consistent rotating magnetosphere. See Abstr. 141.329.

Galactic winds driven by cosmic rays. See Abstr. 143.015.

New applications of the equations of stellar hydrodynamics.   See Abstr. 155.006.

063   Radiative Transfer, Scattering

063.001   Improved complete-linearization method for the solution of the non-LTE line transfer problem.
I. Hubený.
Bull. Astron. Inst. Czechoslovakia (BAC), Vol. 26, 38 - 47 (1975).
    A new numerical method is presented for the self-consistent solution of the radiative transfer equation and the equations of statistical equilibrium.

063.002   The inverse source problem in radiative transfer.
    E. W. Larsen.
Journ. Quant. Spectrosc. Radiat. Transfer, Vol. 15, 1 - 5 (1975).

063.003   Relativistic Compton scattering from moving electrons and angular moments.   B. R. Wienke.
Journ. Quant. Spectrosc. Radiat. Transfer, Vol. 15, 151 - 157 (1975).

063.004   Nongray radiative transfer in a semi-infinite medium: H-function.   A. L. Crosbie.
Journ. Quant. Spectrosc. Radiat. Transfer, Vol. 15, 197 - 200 (1975).
    Numerical results are presented for the dimensionless emissive power at the boundary of a nongray semi-infinite medium in radiative equilibrium. The absorption coefficient consists of an array of equal intensity, nonoverlapping bands or lines. Specifically, the rectangular, triangular, exponential, Doppler and Lorentz profiles are considered.

063.005   Non-linear Compton radiative group locking.
    G. S. S. Sweeney, P. Stewart.
Astron. Astrophys., Vol. 37, 201 - 207 (1974).
    Radiative damping can cause charged particles to undergo rapid evolution in a strong low-frequency wave whose phase velocity is greater than 3×10^{10} cm/s. A particle can adjust itself to the group velocity of the strong wave in a matter of

seconds at a distance of about 10^{13} cm from the Crab neutron star.

063.006   On relativistic electron heating by induced Compton scattering.   G. V. Dedkov.
Astron. Zhurn. Akad. Nauk SSSR, Vol. 52, 195 - 197 (1975).
In Russian. English translation in Soviet Astron., Vol. 19, No. 1.
    A formula for the relativistic electron heating by induced Compton scattering of unpolarized isotropic waves is deduced. It is shown that a definite electron density in quasars and nuclei of Seyfert galaxies is necessary to distort the initial radiation spectra due to this effect. An absence of distortions in observed spectra permits to estimate an upper limit of relativistic electron density and efficiency of heating in corresponding sources.

063.007   A note on the 'peaking effect' in spherical-geometry transfer problems.
G. B. Rybicki, D. G. Hummer.
Monthly Notices Roy. Astron. Soc., Vol. 170, 423 - 427 (1975).
    This note presents evidence that the claims advanced by Wilson, Tung and Sen regarding the adequacy of Wilson and Sen's half-range moment method for treating the outward peaking of the radiation field in a spherical system are unjustified. It is suggested that the good values obtained by Wilson et al. for the mean intensity and the Eddington factor arise from their choice of the arbitrary function $A(r)$ to include the known asymptotic forms of the source function.

063.008   Anisotropic scattering in inhomogeneous media–1.
    A new representation formula for the solution of the auxiliary integral equation for the source function.
A. L. Fymat, R. E. Kalaba, E. Zagustin.
Journ. Quant. Spectrosc. Radiat. Transfer, Vol. 15, 259 - 265 (1975).
    A new representation formula for the solution of the auxiliary integral equation for the source function in inhomo-

geneous, anisotropically scattering media is presented. It involves two new functions Φ and Ψ of two variables instead of the original five variables. The present representation for the solution of Fredholm integral equations of second kind with unsymmetric kernels provides a new approach to radiative transfer in anisotropic inhomogeneous media.

063.009 Solution of the transfer equation in a scattering atmosphere with spherical symmetry.
M. Missana.
Astrophys. Space Sci., Vol. 33, 245 - 251 (1975).

An exact formal solution of the n-approximation radiative transfer equations for the Compton scattering in a spherically symmetric atmosphere is obtained. In view of further applications, the simple case of a density $\varrho(r) = \varrho_0/r$ is fully developed and the 20 approximation equations have been studied with the computer.

063.010 Light scattering by a spheroidal particle.
S. Asano, G. Yamamoto.
Applied Optics, Vol. 14, 29 - 49 (1975).

063.011 Strong scintillations in astrophysics. I. The Markov approximation, its validity and application to angular broadening. L. C. Lee, J. R. Jokipii.
Astrophys. Journ., Vol. 196, 695 - 707 (1975).

The Markov approximation to the propagation of waves in an extended, irregular medium is discussed in an astrophysical context. The Markov equation for the angular spectrum is particularly simple, and solutions are discussed for typical turbulence spectra. It is found that the equation for the angular spectrum is very nearly that used by previous authors, and the present discussion shows that these results are much more general than previously thought. A possible observational test for distinguishing between Gaussian and power-law interstellar density spectra is discussed.

063.012 Resonance line transfer with partial redistribution. III. Mg II resonance lines in solar-type stars.
R. W. Milkey, T. R. Ayres, R. A. Shine.
Astrophys. Journ., Vol. 197, 143 - 145 (1975).

The authors discuss the gravity dependence of the Mg II resonance lines calculated including effects of partial redistribution in frequency. Using chromospheric models scaled from a solar model, the increased decoupling of the radiation temperature of the k_1 feature from the minimum electron temperature in lower-gravity models is demonstrated. The limb darkening of the k-line in the main-sequence model is also discussed.

063.013 Line formation in turbulent media. G. Traving.
Problems in stellar atmospheres and envelopes, (see 003.001), p. 325 - 356 (1975).

063.014 Remarks on the relation between emission and absorption coefficients. J. Pfleiderer.
Astron. Astrophys., Vol. 38, 323 - 324 (1975).

A recent paper of Oster (1974) is discussed in terms of conventional notation and some comments are given. − Concerning Abstr. 11.063.036.

063.015 Shifts and asymmetries of lines formed in a thermally driven turbulent medium.
E. Schatzman, C. Magnan.
Astron. Astrophys., Vol. 38, 373 - 380 (1975).

The calculation of the absorption and emission coefficients within a line formed in a microturbulent situation is reconsidered. Both coefficients are usually shifted to the blue, but because the absorption coefficient is shifted to the red with respect to the emission coefficient, the lines may be red-shifted. The theory seems to account well for the observed

asymmetries and shifts of the solar lines and gives especially a residual red-shift at the limb. The asymmetries of the chromospheric lines are equally well reproduced.

063.016 On a problem of uniqueness regarding H-function calculations. C. E. Siewert.
Journ. Quant. Spectrosc. Radiat. Transfer, Vol. 15, 385 - 387 (1975).

It is shown that the developed L equation, especially useful for Chandrasekhar's H-function calculations when ω is close to unity, has a unique solution.

063.017 Invariance principles and internal radiation fields in semi-infinite atmospheres. V. V. Ivanov.
Astron. Zhurn. Akad. Nauk SSSR, Vol. 52, 217 - 226 (1975).
In Russian. English translation in Soviet Astron., Vol. 19, No. 2.

Invariance relations of a new type are formulated for problems of light scattering in semi-infinite atmospheres. All the previously known invariance relations as well as a number of new results are obtained as particular cases.

063.018 Invisible bodies. M. Kerker.
Journ. Optical Soc. America, Vol. 65, 376 - 379 (1975).

An expression is derived for the scattering of electromagnetic radiation by small, nonabsorbing, compound ellipsoids that contain an inner ellipsoidal region and an outer confocal ellipsoidal shell. For certain combinations of dielectric constant, the scattering is zero, thereby rendering the body invisible.

063.019 An approximate solution of the integral equation in radiative transfer. A. V. Piskarev, L. V. Yasnov.
Radioizluchenie Solntsa, Vyp. (No.) 3. Leningrad, Leningr. un-t, 1974, p. 92 - 98. In Russian. − Abstr. in Referativ. Zhurn. 51. Astron., 4.51.459 (1975).

063.020 On the passage of radiation through inhomogeneous, moving media. XII. Polarized waves in a plane, sheared medium. M. A. Lee, I. Lerche.
Astrophys. Journ., Vol. 198, 477 - 484 (1975).

Using the complete set of Maxwell equations, the authors follow the ray paths and Stokes parameters of a polarized wave which passes through a slab of differentially shearing material of variable refractive index. The calculations show that even for simple situations a noticeable polarization variation occurs. Accordingly the calculations reported here are an educative device: they show that it is, perhaps, unwise to assume that the observed polarization properties of pulsar signals are the same as the emitted signals.

063.021 Strahlungstransport in kosmischen Masern.
E. Bettwieser.
Diss. Nat. Gesamtfakultät, Ruprecht-Karl-Univ., Heidelberg. 11 + 85 pp. (1975).

063.022 The albedo of a homogeneous sphere.
V. M. Loskutov.
Vestn. Leningr. un-ta, 1974, No. 19, p. 132 - 135. In Russian. Abstr. in Referativ. Zhurn. 51. Astron., 5.51.211 (1975).

063.023 Numerical solution of radiative transfer equation in extended spherical atmospheres with Rayleigh phase function. A. Peraiah.
Astron. Astrophys., Vol. 40, 75 - 80 (1975).

A numerical solution of radiative transfer equation has been obtained in a spherically symmetric homogeneous medium with Rayleigh's phase function in the framework of discrete space theory of Grant and Hunt (1968) and Peraiah and Grant (1973). The fast doubling algorithm has been used in spherical cases for large optical thicknesses and highly extended spherical shells.

063.024 The use of variable quadrature weights.
S. Frandsen.
Astron. Astrophys., Vol. 40, 213 - 216 (1975).

A technique called the variable weight technique (VWT) is described. The regions where the VWT can be useful is in more complex problems like model atmosphere computations and NLTE multi-line cases. This has been demonstrated by introducing the VWT in a program for calculating model atmospheres using the complete linearization method (Auer and Mihalas, 1969).

063.025 Radiative transfer through spherically-symmetric atmospheres and shells.   J. Schmid-Burgk.
Astron. Astrophys., Vol. 40, 249 - 255 (1975).

The application of an integral equation method to problems of radiative transfer through extended atmospheres and shells of spherical symmetry is discussed under the assumption that the opacities are known functions of radius. Expressions for the integral operators, obtained through the use of spline representations for the source function, are given for linearized non-grey LTE models, for multi-temperature dust shells, for picket fence-models, and for anisotropic scattering. Some results are described for a grey shell model and for a two-step picket-fence atmosphere.

063.026 Finite eddy-size effects on centre-to-limb variations; an alternative to anisotropic microturbulence.
H. Frisch.
Astron. Astrophys., Vol. 40, 267 - 276 (1975).

An attempt is presented to explain the centre-to-limb increase in width of solar lines on the basis of the theory of line formation in a turbulent medium where finite eddy-size is taken into account. Hydrodynamic velocities along the line of sight are represented by a step-wise constant stochastic process with gaussian isotropic velocity distribution and an arbitrary variation of the eddy-size with the continuum optical depth.

063.027 Discrete frequency scattering in spherical layers surrounding a point source.   L. G. Titarchuk.
Astrophys. Space Sci., Vol. 35, 137 - 149 (1975).

The scattered radiation field in homogeneously absorbing and isotropically scattering spherical layers is studied, when the isotropic point source is at the centre. A complete frequency redistribution is assumed. It is shown that on the inner boundary $r = R_0$ of the cavity, when $R_0 \gg 1$ (all radii are expressed in path lengths), the source function $B \sim R_0^{-1} \ln^{-1/2} R_0$ for the Doppler profile and $B \sim R_0^{-3/2}$ for the Voigt and Lorentz profiles. The asymptotical behaviour of the source function $B(r)$ significantly differs from the analogous behaviour of the solution for an infinite medium.

063.028 Acceleration of isolated atoms by radiation pressure.
R. H. Gordon.
Astrophys. Space Sci., Vol. 35, 197 - 202 (1975).

If isolated atoms are accelerated by a point source of radiation with a power law spectrum and an intensity sufficiently large that radiation pressure dominates other forces, and if both acceleration and photoionization of the ions are considered, the resulting spectrum of escape velocities is independent of the intensity of the radiation, depending only on the power law index. For power-law indices typical of astrophysical objects, relativistic velocities are not possible.

063.029 Theory of Thomson scattering in a strong magnetic field. II   T. Hamada.
Publ. Astron. Soc. Japan, Vol. 27, 275 - 286 (1975).

A relativistic quantum theory is formulated for the Compton scattering by electrons in a strong magnetic field.

063.030 Theory of radiative transfer in a strong magnetic field.   S. Kanno.
Publ. Astron. Soc. Japan, Vol. 27, 287 - 306 (1975).

A theory is presented of the radiative transfer in a magnetized plasma with the opacity determined by the Thomson scattering. The Thomson cross section in the magnetic field is highly anisotropic and polarization-dependent. The equation of transfer is established accordingly and approximate solutions are found in the limits of small and large optical thickness. The latter solution is used to find the intensity and the polarization of thermal X-rays from a magnetic dipole star. The concept of mean free path is discussed and also it is shown that the Faraday rotation naturally comes about as a result of the multiple forward scattering.

063.031 Strong radiative reaction in the non-linear Compton process.   P. Stewart.
Astron. Astrophys., Vol. 41, 169 - 174 (1975).

The transient behaviour of the equations of motion of a charged particle undergoing strong radiative reaction under the influence of an intense electromagnetic field of given phase velocity is examined numerically. Analytical solutions are obtained for the steady oscillatory state for arbitrary phase velocity and field amplitude, and the solutions are used to obtain an estimate of the attenuation length scale for a strong wave propagating through a plasma with strong radiative reaction. The results are used to estimate the charge density required for spatial attenuation of the 30 Hz field in the Crab nebula.

063.032 Scattering operator in the theory of discontinuous Markov processes. (Perturbation theory).
J. M. Cook.
Journ. Math. Anal. Appl. (USA), Vol. 47, 578 - 598 (1974).

063.033 Spectral line formation in spherically symmetric and expanding atmospheres.   P. B. Kunasz.
Thesis, Colorado Univ., Boulder (USA). 233 pp. University Microfilms Order No. 74-22,364 (1974).

063.034 Depth of origin and mean depth of a magnetoactive line.   V. E. Stepanov, V. M. Grigor'ev, I. M. Kats.
Dokl. Akad. Nauk SSSR. Ser. Mat. Fiz., Vol. 222, 1057 - 1060 (1975). In Russian.

063.035 Stability of radiative transfer and the problem of choice of boundary conditions.
V. E. Stepanov, V. M. Grigor'ev, I. M. Kats.
Dokl. Akad. Nauk SSSR. Ser. Mat. Fiz., Vol. 222, 1315 - 1317 (1975). In Russian.

063.036 Simulation of pitch angle diffusion of charged particles in a disordered magnetic field.
T. B. Kaiser.
Solar wind three, (see 012.020), p. 203 - 205 (1974).

Improved quantum mechanical treatment of plasma polarization shift of ion lines.   See Abstr. 062.057.

The non-LTE transport equation for polarized radiation in the presence of magnetic fields. I. Formulation. See Abstr. 064.005.

Radiation pressure on grains as a mechanism for mass loss in red giants.   See Abstr. 064.066.

Kinetic equilibrium and line formation of Na I in the solar atmosphere.   See Abstr. 080.004.

X and Y functions for planetary atmospheres with Lambert law reflecting surfaces.   See Abstr. 091.010.

The structure of the Orion A molecular cloud. See Abstr. 131.039.

Line-leaking models for interstellar molecular emission and absorption—I. The anomalous absorption and emission by formaldehyde.   See Abstr. 131.060.

Starlight excitation of permitted lines in the Orion nebula.   See Abstr. 132.014.

Approximate solutions of radiative transfer in dusty nebulae. I. Pure hydrogen nebulae. See Abstr. 132.015.

Polarization of radio sources. VI. An oscillatory behavior of the intensity in a general solution of the radiation transfer problem in a plasma.   See Abstr. 141.022.

Errata

063.901   Errata: 'On the passage of radiation through inhomogeneous moving media. VI. Dispersion effects on phase and ray paths in a plane, differentially shearing medium' [Astrophys. Journ., Vol. 191, 759 - 762 (1974)]. I. Lerche. Astrophys. Journ., Vol. 197, 805 (1975).

064   Stellar Atmospheres, Stellar Envelopes, Mass Loss

064.001   Model atmosphere analysis of the peculiar star 53 Aur. I. The physical atmospheric parameters determination and model atmospheres.   J. Zverko.
Bull. Astron. Inst. Czechoslovakia (BAC), Vol. 26, 58 - 63 (1975).

The physical atmospheric parameters – the effective temperature, the surface gravity and the electron density – are determined. The Balmer line profiles are measured and the Balmer discontinuity deduced.The corresponding values are as follows: T_{eff} = 10 160°K, log g = 3.90 and log N_e = 14.02.

064.002   The radio and infrared spectrum of early-type stars undergoing mass loss.
A. E. Wright, M. J. Barlow.
Monthly Notices Roy. Astron. Soc., Vol. 170, 41 - 51 (1975).

A unified model is presented for the radio and infrared spectrum of early-type stars surrounded by a gaseous ionized envelope resulting from mass loss. The cases of uniform mass loss (1) at constant velocity, and (2) with accelerative effects taken into account, are treated. It is shown that the radio and infrared free–free spectra are predicted to be of the form $S_\nu \propto \nu^{0.6}$ except (typically) in the near infrared where the spectrum will become flatter. Various effects which may cause a deviation from a 0.6 spectral index are considered. Finally, it is shown how radio and infrared flux densities may be used to derive mass loss rates.

064.003   On the scale of photospheric convection in red giants and supergiants.   M. Schwarzschild.
Astrophys. Journ., Vol. 195, 137 - 144 (1975).

An attempt is made to estimate the sizes of the convective elements which dominate the brightness variations on the photospheres of red giants and supergiants. The data assembled permit the extreme hypothesis that these dominant convective elements are so large that only a modest number of them exists at any one time on the entire surface of such a star–in contrast with two million granules on the sun.

064.004   Radiation-driven winds in Of stars.
J. I. Castor, D. C. Abbott, R. I. Klein.
Astrophys. Journ., Vol. 195, 157 - 174 (1975).

The large number of subordinate lines of a representative ion are found to have a dominant effect on the force of radiation on material in O star atmospheres. The force is increased over that due to resonance lines alone so that rates of mass loss are obtained which are 100 times greater than previously thought possible. The force is related to the solution of the line-transfer problem, and it becomes a function of the local velocity gradient. A new stellar wind theory, with a different interpretation of the singular point, is developed to treat this situation. The rate of mass loss, and other properties of the model, are uniquely specified by the luminosity, mass, and radius of the star.

064.005   The non-LTE transport equation for polarized radiation in the presence of magnetic fields. I. Formulation.   L. L. House, R. Steinitz.
Astrophys. Journ., Vol. 195, 235 - 249 (1975).

A formulation of the radiative transfer of polarized light under conditions departing from local thermodynamic equilibrium and accounting for the presence of a magnetic field is presented. The formulation is self-consistent in that the equations of statistical equilibrium for magnetic sublevels are also included. The quantum-mechanical derivation of the absorption matrix, the key to the presentation, is derived in detail utilizing the density matrix approach. The present discussion is restricted to the case in which the magnetic field is sufficiently strong that the natural widths of magnetic sublevels do not overlap producing interference effects.

064.006   A note on the validity of the use of Euler's equation for the motion of the matter in the envelopes of Be stars.   T. H. Morgan.
Astrophys. Space Sci., Vol. 33, 99 - 101 (1975).

Three assumptions which underlie the theoretical treatment of the structure and kinematics of the circumstellar material surrounding Be stars are examined. It is shown that a single-fluid non-viscous hydrodynamical description is a reasonable one for describing the motion of a fluid under typical envelope conditions.

064.007   Time-dependent hydrodynamic phenomena in the envelopes of the Be stars.   T. H. Morgan.
Astrophys. Journ., Vol. 195, 391 - 395 = Rosemary Hill Obs., Univ. Florida, *Gainesville*, Contr. No. 48 (1975).

Temporal and angular variations in the motion and distribution of circumstellar material about Be stars are studied by means of a simple hydrodynamical approach.

064.008   Sodium line formation in Arcturus.   W. L. Kelch.
Astrophys. Journ., Vol. 195, 679 - 687 = Publ. Goethe Link Obs., Indiana Univ., *Bloomington*, No. 163 (1975).

Results of an investigation of the sodium line formation in the K2 IIIp giant Arcturus are given. A model Na I atom of eight bound levels is used, and the radiative transfer problem is solved by the method of complete linearization. The author compares predictions of the D line profiles and central intensities of the $3p–3d$, $3p–5s$, and $3p–4d$ multiplets with Griffin's Photometric Atlas of Arcturus. The local thermodynamic equilibrium (LTE) approximation predicts the D lines and all the subordinate transitions mentioned above to be weaker than as seen in the Arcturus Atlas. It is found that although the resonance lines computed by non-LTE or kinetic equilibrium (KE) differ little from LTE and do not fit the observations well (except in the case of an increase in the microturbulence with depth).

064.009   Circumstellar dust shell models for Alpha Orionis.   H. M. Dyck, T. Simon.
Astrophys. Journ., Vol. 195, 689 - 693 (1975).

The authors present the results from some simple calculations of optically thin dust shell models for α Ori. They show that (1) the mass in the shell is large, $3 \times 10^{-5} - 1 \times 10^{-2} M_\odot$, but compatible with the current picture of mass-loss; and (2) photometry at wavelengths greater than 20μ can help to distinguish among various density distributions in the shell.

064.010   A possible width-luminosity correlation of the Ca II K_1 and Mg II k_1 features.
T. R. Ayres, J. L. Linsky, R. A. Shine.
Astrophys. Journ., (*Letters*), Vol. 195, L121 - L124 (1975).

Existing high resolution stellar profiles of the Ca II and Mg II resonance lines suggest a possible width-luminosity correlation of the K_1 minimum features. The authors show that such a correlation can be simply understood if the continuum optical depth of the stellar temperature minimum is relatively independent of surface gravity as suggested by three stars studied in detail.

064.011   Polarization in Zeeman split absorption lines.
V. N. Sazonov.
Astron. Zhurn. Akad. Nauk SSSR, Vol. 52, 71 - 80 (1975). In Russian. English translation in Soviet Astron., Vol. 19, No. 1.

The paper considers polarization in the absorption line in a magnetic field under the condition of local thermodynamic equilibrium. The polarization and the profiles of the lines considerably depend upon the value of the normal and abnormal dispersions, in particular upon the value of Faraday rotation of the plane of the linear polarization in the magnetoactive plasma in the star's atmosphere.

064.012 Planck mean cross-sections for four grain materials.
R. C. Gilman.
Astrophys. Journ.,Suppl. Ser., No. 268, Vol. 28, 397 - 403 (1974).

Planck mean absorption and radiation-pressure cross-sections have been computed for spherical and spheroidal grains composed of olivine, iron, silicon carbide, or graphite. The spherical grain radii range from 10^{-7} to 10^{-3} cm, while the effective temperature of the Planck function used in calculating these means ranges from $10°$ to $40,000°$K.

064.013 Silicon-rich stellar envelope?
G. A. Gurzadyan, S. S. Rustambekova.
Nature, Vol. 254, 311 - 312 (1975).

During an analysis of Orion-2 data derived from a space astrophysical experiment, there was observed a hot star with a gaseous envelope, SAO077308, of B1e type and nearly ninth magnitude. It is interesting because of an extremely strong emission line at 2,520 Å occurs in its ultraviolet spectrum. Following a series of attempts, the identification of this line with the resonance sextet of neutral silicon (mean wavelength 2,520 Å) seems probable. The components of this sextet, of almost equal strengths, are: 2,507, 2,514, 2,516, 2,519, 2,524 and 2,528 Å; the second of these lines is resonant, the others quasi-resonant in the sense that the lowest levels of these lines are located quite close to the ground level, up to 0.01–0.03 eV. This identification cannot be taken as final and needs further examination.

064.014 A model-atmosphere analysis of the spectrum of Arcturus (basic spectroscopic data).
R. Mäckle, R. Griffin, R. Griffin, H. Holweger.
Astron. Astrophys.,Suppl. Ser., Vol. 19, 303 - 319 (1975).

Five tables supplement the analysis of the Arcturus spectrum by Mäckle et al. (1975). They contain basic observational data and computational results for individual spectral lines.

064.015 Formation of molecular lines in stellar atmospheres.
K. H. Hinkle, D. L. Lambert.
Monthly Notices Roy. Astron. Soc., Vol. 170, 447 - 474 (1975).

Statistical equilibrium of electronic states of diatomic molecules in stellar atmospheres is examined. Atmospheres discussed are representative of the sun, Arcturus (K-giant) and Betelgeuse (M-supergiant). Examination of the equilibrium for excited electronic states demonstrates that the exchange between these states and the ground electronic state is most probably determined by radiative excitation. This result implies that scattering rather than pure absorption is the appropriate mechanism for the formation of lines belonging to these electronic transitions. The scattering hypothesis is given a preliminary check against solar observations.

064.016 Importance of bound-free opacity of OH and CH in solar and stellar atmospheres.
S. P. Tarafdar, P. K. Das.
Monthly Notices Roy. Astron. Soc., Vol. 170, 559 - 568 (1975).

Absorption cross-sections for bound-free transitions of OH and CH have been estimated and their importance in the atmosphere of the sun and of other stars has been examined.

064.017 The Balmer discontinuities of O9–B2 supergiants.
R. E. Schild, F. H. Chaffee.
Astrophys. Journ., Vol. 196, 503 - 513 (1975).

New energy distributions of supergiants at 50 Å resolution from $\lambda\lambda 3200-8000$ show that the Balmer discontinuities are poorly correlated with spectral types. A particularly good example is ϵ Ori, whose Balmer discontinuity is 0.1 mag larger than other O9.5 Ia and B0.5 Ia supergiants. Analysis of available four-color and Hβ photometry shows that whereas Hβ correlates well with luminosity as indicated by MK luminosity class, the Balmer discontinuity does not.

064.018 Models for X-ray illuminated atmospheres.
M. Milgrom, E. E. Salpeter.
Astrophys. Journ., Vol. 196, 583 - 588 (1975).

A numerical scheme has been developed to study plane-parallel stellar atmospheres, with a given surface temperature and gravity, illuminated by a beam of X-rays. For given total flux, spectrum, and angle of incidence of the X-ray beam, the structure of the atmosphere and the characteristics of the outgoing radiation are calculated. The scheme is applicable to binary systems in which one member is a compact X-ray source and the other is a normal star. The results are presented for a few models with different values of the input parameters.

064.019 Line blanketing and opacity probability distribution function. J. van Paradijs, M. S. Vardya.
Astrophys. Space Sci., Vol. 33, L9 - L12 (1975).

Advantages and limitations of using opacity probability distribution function (OPDF) in computations of model atmospheres for cool stars have been examined. It is concluded that a direct approach, in which a large number of frequency points are taken, may be more appropriate relative to OPDF.

064.020 Radiative transfer in gray circumstellar dust envelopes: VY Canis Majoris revisited.
R. D. Schwartz.
Astrophys. Journ., Vol. 196, 745 - 751 = Lick Obs. Bull., No. 683 (1975).

The circumstellar dust model for VY CMa proposed by Herbig is reinvestigated using a generalized form of Huang's theory of radiative transfer. The resultant envelope parameters and the emergent energy distribution are found to be insensitive to the choice of Eddington factor for a given envelope inner boundary temperature. Observed fluxes from 0.43 to 74 μ are incorporated into the model, and problems relating to grain emissivity for $\lambda > 30 \mu$ and grain survival at the indicated inner boundary temperature of $1855°$K are discussed.

064.021 Radiative transfer in spherical circumstellar dust envelopes. II. Is the infrared continuum of Eta Carinae produced by thermal dust emission?
J. P. Apruzese.
Astrophys. Journ., Vol. 196, 753 - 760 (1975).

The radiative transfer techniques of Huang have been further extended in an attempt to construct a specific thermal dust-emission model of η Carinae which fits the observations of both the spectral and the spatial distribution of the infrared radiation. The model which best fits the observations requires infrared emissivities considerably higher than those of normal-size nonsilicate grains. Evidence is presented that the high infrared emissivities are caused by large particles rather than a heavy concentration of infrared-active silicates. The possibility that synchrotron radiation is responsible for some of the infrared continuum is discussed.

064.022 Radiative transfer in spherical circumstellar dust envelopes. III. Dust envelope models of some well known infrared stars. J. P. Apruzese.
Astrophys. Journ., Vol. 196, 761 - 768 (1975).

The radiative transfer techniques described elsewhere by

the author have been employed to construct dust envelope models of several well known infrared stars. The resulting calculations indicate that the infrared emissivity of circumstellar grains generally must be higher than that which many calculations of small nonsilicate grains yield. This conclusion is dependent to some degree on the (unknown) size of the stellar envelopes considered, but is quite firm in the case of the spatially resolved envelope of IRC + 10216. Further observations of the spatial distribution of the infrared radiation from stellar envelopes will be invaluable in deciphering the properties of the circumstellar grains.

064.023 **An alternative mechanism for production of emission features in some infrared objects.**
J. P. Apruzese.
Astrophys. Journ., Vol. 196, 769 - 771 (1975).

Two dust-envelope models of the M supergiant VX Sgr, which exhibits a prominent emission feature at 10 μ, are presented. The models indicate that, for certain envelope sizes, the presence of the observed emission feature does not necessarily indicate that the emitting grains possess a similar feature in their emissivity profile. The mechanism which may in some cases be producing the observed emission feature is discussed.

064.024 **Ionization equilibria in the atmospheres of late-type giants.**   J. R. Auman, J. E. J. Woodrow.
Astrophys. Journ., Vol. 197, 163 - 173 (1975).

Model atmospheres were constructed for late-type giants in which the LTE assumption was relaxed in the calculation of the ionization equilibria of the metals that contribute significantly to the density of free electrons. According to the authors' findings, ionization equilibria are shifted considerably in the atmospheres with effective temperatures less than 3000 K. The electron pressure is increased by a factor of 60 in the model with T_e = 2000 K, and by a factor of 14 in the model with T_e = 2500 K. The approximations and sources of error in the models are discussed, as well as the possible deviations from LTE in the dissociation equilibria of the molecules.

064.025 **Model stellar atmospheres and heavy element abundances.**   E. Böhm-Vitense.
Problems in stellar atmospheres and envelopes, (see 003.001), p. 21 - 56 (1975).

064.026 **Response of a bounded atmosphere to a non-resonant excitation. I: Isothermal case.**
J. Provost.
Solar Physics, Vol. 40, 257 - 273 (1975).

The response of a bounded atmosphere to a non-resonant excitation applied at its basis is studied. It is shown that the essential feature related to this kind of excitation is that the distribution of the energy of the velocity field relatively to the frequency and horizontal wavelength is a function of height and merely depends on the structure of the atmosphere above the level at which it is considered. The preliminary results concerning an isothermal atmosphere are presented and their relevance to the solar case is discussed.

064.027 **Shock propagation through an atmospheric model of an RR Lyrae type star.**   T. Okuda.
Mem. Fac. Sci.,Kyoto Univ., Ser. Phys., Astrophys., Geophys., Chem., Vol. 34, 261 - 273 (1974).

The propagation of shock waves through a model atmosphere of an RR Lyrae type star is examined on the basis of Brinkley-Kirkwood method. The effects of gravity, radiation pressure and variation of specific heat ratio with depth are taken into account. It is supposed that a shock wave is generated below the convective region with the shock strength and shock energy taken as parameters. The numerical results show that the shock strength remains almost constant in the convective region while it abruptly increases

from the top of this region toward the outer surface. A brief discussion on the validity of the calculations of shock propagation is also presented.

064.028 **A-type horizontal-branch stars.**   K. Kodaira.
Problems in stellar atmospheres and envelopes, (see 003.001), p. 149 - 171 (1975).

064.029 **Circumstellar envelopes and mass loss of red giant stars.**   D. Reimers.
Problems in stellar atmospheres and envelopes, (see 003.001), p. 229 - 256 (1975).

064.030 **Radio emission from stellar and circumstellar atmospheres.**   L. Oster.
Problems in stellar atmospheres and envelopes, (see 003.001), p. 301 - 323 (1975).

064.031 **A study of M dwarfs. I. Preliminary model atmospheres.**   J. R. Mould.
Astron. Astrophys., Vol. 38, 283 - 288 (1975).

A sequence of model atmospheres for early M dwarfs is presented. The models contain opacity distribution functions for the most prominent molecular bands in M dwarfs, H_2O and TiO. A mixing-length theory of convection is included, and a test on one model shows no dependence of the emergent flux on the mixing-length to scale height ratio. The emergent fluxes are tabulated for comparison with observations. Fair agreement is evident, but the inclusion of atomic line blanketing seems necessary before a discrepancy in the optical continuum gradient can be remedied.

064.032 **Line blanketing and model stellar atmospheres. II. Interpretation of broad-band photometric observations.**   E. Peytremann.
Astron. Astrophys., Vol. 38, 417 - 434 (1975).

Using the blanketed model atmospheres described in a previous paper (Peytremann, 1974), the author calculates a grid of colour indices including line blocking. These indices are those of the Geneva photometric system. He studies some general properties of this photometric system, in terms of three atmospheric parameters: effective temperature T_{eff}, surface gravity g and scaled solar abundances χ. He also presents and discusses the results of the analysis of about 700 stars of the Geneva catalog (Rufener, 1971), whose colours are in the range of the grid of theoretical colours. Among the results obtained in the course of this investigation, one may mention a correspondence table between MK luminosity classes and gravities log g, abundance determinations for a number of metal-deficient stars, as well as the study of three galactic clusters.

064.033 **Fe I fluorescence in T Tauri stars. II. Clues to the velocity field in the circumstellar envelope.**
L. A. Willson.
Astrophys. Journ., Vol. 197, 365 - 370 (1975).

Radial velocity measurements for the fluorescent iron lines $\lambda\lambda 4063$, 4132 for RW Aur compared with the primary emission lines of hydrogen and calcium show that Fe I $\lambda 3969.26$ has been shifted onto He, as was predicted. Profiles of the fluorescent iron lines indicate that (1) the lines are formed at 3–10 stellar radii from the star, and (2) rotational velocities are equal to or exceed expansion velocities at this distance. The conclusion is drawn that mass-loss rates calculated on the basis of a purely expanding envelope may be in error by factors of 2 to 10 for T Tauri stars. Implications of these results for the envelope structure are also discussed.

064.034 **The continuum flux distribution for Arcturus.**
D. E. Blackwell, R. S. Ellis, P. A. Ibbetson, A. D. Petford, R. B. Willis.

Monthly Notices Roy. Astron. Soc., Vol. 171, 425 - 439 (1975).

A composite of all continuum flux data for Arcturus has been plotted including additional results obtained recently in. Israel. The data are compared with flux distributions calculated from Carbon & Gingerich, and Peytremann grids of models. There is not good agreement between the calculated and observed flux distributions at shorter wavelengths, but the Peytremann model seems considerably better than the Carbon & Gingerich model.

064.035 The spectrum of the free-free radiation from extended envelopes.   N. Panagia, M. Felli.
Astron. Astrophys., Vol. 39, 1 - 5 (1975).

The continuous spectrum from a spherical envelope with $n_e \propto r^{-2}$ is derived by solving the equation of radiative transfer. The solution is also obtained in the general case of $n_e \propto r^{-a}$. It is shown that the optically thin portion of the envelope produces the major part of the emission. The case of $n_e \propto r^{-2}$ is discussed in more detail since it may well represent the radio emission from the envelopes of stars undergoing mass loss. The radio spectrum is found to vary as $v^{0.6}$ and the size of the source as $v^{-0.7}$. It is shown that a measurement of the flux density can provide a good estimate of the mass loss rate. The case of P Cyg is examined, and a value of the mass loss rate of $1.2 \times 10^{-5} M_\odot$/year is derived.

064.036 Colors and ionization equilibria in K dwarfs.   M. N. Perrin, G. Cayrel de Strobel, R. Cayrel.
Astron. Astrophys., Vol. 39, 97 - 106 (1975).

The strong discrepancy found by Oinas for K-dwarfs between the actual ionization ratio and its expected value derived from normal gravities and temperatures obtained by deblanketed scans has been reinvestigated for six stars of his program. It has not been found significant. Therefore it does not seem necessary any longer to look for a new continuous absorber to be introduced in the opacities of K stars in the 5000 - 10800 Å spectral range.

064.037 Blanketed model atmospheres for late-type stars. II.   F. Querci, M. Querci.
Astron. Astrophys., Vol. 39, 113 - 125 (1975).

The authors present a series of models for the atmospheres of cool stars; the temperature range is 2600 °K to 3800 °K, and the gravity is taken as 0.1, 1 or 10. Line blanketing by CO, CN and C_2 is handled by the OPDF (opacity distribution functions) method; the authors' application to cool stars is discussed and compared to other published work.

064.038 The radiation field in photospheric models for extreme supergiants.   C. De Jager, L. Neven.
Astrophys. Space Sci., Vol. 33, 295 - 323 (1975).

On the basis of assumed photospheric temperature models for 36 extreme supergiants (log g_e-values of 1, 0.5 and 0; T_e ranging from approx. 3700–33 000 K) photospheric fluxes were computed for 36 wavelengths ranging from 100 Å to 60 000 Å. The hot models are in perfect radiative equilibrium; the cooler show deviations up to 10%, sometimes even larger. In tables and graphs the authors give for these models radiation fluxes $\pi F(\lambda)$, integrated fluxes, effective temperatures, colours U, B and V, and the Balmer discontinuity D.

064.039 On the best linear approximation of limb-darkening laws for dwarf stars with thin photosphere.
A. A. Rubashevskij.
Astrometriya i Astrofizika, Kiev, vyp. (No.) 24, (see 003.003), p. 88 - 90 (1974). In Russian.

Various linear approximations to non-linear limb darkening laws for stars with thin photospheres are discussed. Discrete analogues of integral relations are given for the determination of linear limb-darkening laws. The linear approximation for which the full flux of stellar radiation is constant

appears to be the best one.

064.040 The dynamics of the envelopes of Be stars in a toroidal magnetic field and stability of the envelope motion.   V. N. Morozov.
Vestn. Leningr. un-ta, 1974, No. 13, p. 137 - 141. In Russian. Abstr. in Referativ. Zhurn. 51. Astron., 3.51.180 (1975).

064.041 Possible existence of the laser effect in stellar atmospheres.   N. N. Lavrinovich, V. S. Letokhov.
Zhurn. ehksperim. i teor. fiz., Vol. 67, 1609 - 1620 (1974). In Russian. – Abstr. in Referativ. Zhurn. 51. Astron., 3.51. 539 (1975).

064.042 Polarization properties of silicate-like grains in circumstellar envelopes of late-type stars due to temperature variations.
J. Svatoš, M. Šolc, V. Vanýsek.
Astrophys. Space Sci., Vol. 34, 149 - 154 (1975). – Paper presented at the Symposium on Solid State Astrophysics, Cardiff, Wales, 9 – 12 July, 1974.

The influence of temperature changes in circumstellar silicate-like envelopes upon the polarization effects is investigated. It is shown that under the assumption that $\Delta T_g > 50°$ and conductivity of silicate grains is indirectly proportional to T_g this mechanism can be responsible for the observed dependence of intensity vs polarization in some late-type stars, e.g. V CVn. The same effects can be produced by dirty ices and graphite grains. It is suggested that irradiation by electrons and/or protons can affect the circumstellar envelopes in a similar way, especially those of early-type stars, and irradiation by neutrons can exert an influence on the envelopes of supernovae.

064.043 On the structure of envelopes of Be stars.   V. G. Gorbatskij.
Pis'ma v Astron. Zhurn., Vol. 1, No. 3, p. 36 - 38 (1975). In Russian.

To obtain a better agreement between theory and observations, one has to assume that envelopes of Be stars consist of separate condensations. Estimates of parameters of the condensations are given.

064.044 The inversion of the Hγ absorption line of a B 1 Ib supergiant, ρ Leonis.   A. G. Hearn, J. N. Holt.
Astron. Astrophys., Vol. 39, 251 - 255 (1975).

The numerical technique for inverting a single line absorption profile is applied to the Hγ profile of the B 1 Ib supergiant ρ Leonis. An estimate of electron density for the atmosphere is obtained which is probably better than one order of magnitude and the variation of the Planck function through the atmosphere is well determined. The inversion shows that microturbulence alone cannot be the main broadening agent for the line profile. Calculations show that the observed line profile is consistent with a rotational velocity of 60 km s^{-1} and thermal Doppler broadening.

064.045 Ultraviolet spectra with line opacities.   E. Peytremann.
Astron. Astrophys., Vol. 39, 393 - 403 (1975).

The author presents a few theoretical ultraviolet spectra that have been calculated with a large number of lines. He discusses the amount of line blocking in several spectral ranges, various effective temperatures, surface gravities and scaled solar abundances. It is shown that for low spectral resolutions (either instrumental or due to stellar rotation) the true continuum can very often not be observed, even for early B stars. As a consequence, traditional spectral analysis will fail if it relies on the definition of a continuum.

064.046 On mass loss by stellar wind in population II red

giants.   F. Fusi-Pecci, A. Renzini.
Astron. Astrophys., Vol. 39, 413 - 419 (1975).

It is shown that the mass loss suggested by many authors to fit the morphology of the horizontal branch of globular clusters may be accounted for by a stellar wind mechanism. It is also shown that a small dispersion within a cluster of the stellar luminosity at the helium flash would produce a significant dispersion in the mass of horizontal branch stars. Furthermore, this mass loss mechanism would greatly emphasize the age dependence of the horizontal branch morphology.

064.047   Theoretical models for Arcturus.
J. G. Collins, H. R. Johnson.
Bull. American Astron. Soc., Vol. 7, 240 (1975). — Abstr. AAS.

064.048   Theoretical colors for cool metal deficient giant stars.   R. A. Bell, B. Gustafsson.
Bull. American Astron. Soc., Vol. 7, 247 (1975). — Abstr. AAS.

064.049   The effect of extended atmospheres on interferometer response functions.
J. P. Cassinelli, N. M. Hoffman.
Bull. American Astron. Soc., Vol. 7, 247 - 248 (1975). Abstr. AAS.

064.050   On the abundance of ^{7}Li and the intermediate mass elements in the envelope of a 7 $M_\odot$ star.
M. Arnould, W. M. Howard, S. A. Lamb.
Bull. American Astron. Soc., Vol. 7, 248 (1975). — Abstr. AAS.

064.051   Stellar winds induced by dynamo-produced magnetic fields.   W. K. Rose, E. H. Scott.
Bull. American Astron. Soc., Vol. 7, 252 (1975). — Abstr. AAS.

064.052   Solution of the co-moving frame equation of transfer in spherically symmetric flows.
D. Mihalas, P. B. Kunasz, D. G. Hummer.
Bull. American Astron. Soc., Vol. 7, 256 - 257 (1975). Abstr. AAS.

064.053   Characteristic relaxation times for non-LTE atmospheres.   A. Skumanich.
Bull. American Astron. Soc., Vol. 7, 257 (1975). — Abstr. AAS.

064.054   On the grey, circumstellar dust model for VY CMa.
R. D. Schwartz.
Bull. American Astron. Soc., Vol. 7, 257 (1975). — Abstr. AAS.

064.055   Atmospheric structure from spectral line intensities.
L. W. Ramsey, H. R. Johnson.
Bull. American Astron. Soc., Vol. 7, 257 - 258 (1975). Abstr. AAS.

064.056   A monochromatic treatment of atomic line blanketing.   B. M. Krupp, H. R. Johnson.
Bull. American Astron. Soc., Vol. 7, 258 (1975). — Abstr. AAS.

064.057   Scattering by dust and the appearance of Eta Carinae.   K. Davidson, M. T. Ruiz.
Bull. American Astron. Soc., Vol. 7, 259 (1975). — Abstr. AAS.

064.058   Stellar upper photosphere models based on the Ca II K-wing. II. The coherent scattering approximation.
T. R. Ayres, J. L. Linsky, R. A. Shine.
Bull. American Astron. Soc., Vol. 7, 359 - 360 (1975). Abstr. AAS.

064.059   Model atmospheres of late-type stars. II. Sources of opacity.   V. E. Panchuk.
Astrometriya i Astrofizika, *Kiev*, vyp. (No.) 25, (see 003. 015), p. 20 - 26 (1975). In Russian.

The results of calculations of opacity in the atmospheres of cool stars are discussed.

064.060   Analysis of theoretical values of linear limb-darkening coefficients of dwarf stars obtained from models of their atmospheres.   A. A. Rubashevskij.
Astrometriya i Astrofizika, *Kiev*, vyp. (No.) 25, (see 003. 015), p. 27 - 38 (1975). In Russian.

On the basis of calculations and tabular comparisons the differences in theoretical coefficients of limb darkening connected with both the methods of their determination and properties of the models themselves are discussed. Dependence of the darkening coefficients on wavelength, temperature, log g and parameters of chemical composition is considered for a set of model atmospheres. For K0—O8 stars in the region of 3646—8205 Å the darkening coefficients are calculated under the condition of conserving the constant radiation flux of the star.

064.061   Stellar atmospheres — the middle man.
L. H. Auer, B. Newell.
Dudley Obs. Rep. No. 9, (see 012.008), p. 1 - 16 (1975).

The basic link between observations and the theoretical HR diagram is the theory of stellar atmospheres. In this paper the authors discuss the parameters that predominantly determine the emergent flux distribution of a star; these parameters are the "observables" of a stellar atmosphere. They conclude by indicating briefly how the theory of stellar atmospheres can be used to design photometric systems and to interpret photometric results.

064.062   Effects of departures from LTE and atmospheric extension on colors and theoretical continuum parameters.   D. Mihalas.
Dudley Obs. Rep. No. 9, (see 012.008), p. 241 - 269 (1975).

The effects of departures from LTE upon continuum parameters and uvbyβ indices for early-type stars are discussed.

064.063   A progress report on theoretical four-dimensional photometry of F, A, and B stars.   R. L. Kurucz.
Dudley Obs. Rep. No. 9, (see 012.008), p. 271 - 296 (1975).

The differential effects of changes in effective temperature, gravity, abundance, and microturbulence on the energy distributions and colors of F, A, and B stars are predicted from a new grid of model atmospheres that includes line opacity.

064.064   General remarks on the variability of spotted stars.   C. Friedemann, J. Gürtler.
Astron. Nachr., Vol. 296, 125 - 132 = Mitt. Univ.-Sternw. Jena No. 122 (1975).

The light variations appearing as a consequence of a dark spot on the photosphere of a star are studied. The shape of the light curve depends on the position of the spot on the stellar surface, the orientation of the rotational axis in relation to the observer, the size of the spot, and the temperature difference between the spot and the photosphere. The spots lead to a shift of the stars in the colour-magnitude diagram.

064.065   Molecular column densities in selected model atmospheres.
H. R. Johnson, R. F. Beebe, C. Sneden.
Astrophys. Journ., Suppl. Ser., No. 280, Vol. 29, 123 - 136 = Publ. Goethe Link Obs., Indiana Univ., *Bloomington*, No. 166 (1975).

Molecular column densities are presented for 35 molecules in a variety of cool stellar model atmospheres. From an examination of the predicted column densities, the following conclusions are drawn: (1) OH might be visible in carbon stars which have been generated from triplet-α burning, but will be absent from carbon stars generated from the CNO bi-cycle. (2)

The TiO/ZrO ratio shows small but interesting variations as C/O is changed and as the effective temperature is changed. (3) The column density of silicon dicarbide (SiC_2) is sensitive to abundance, temperature, and gravity. (4) Unexpectedly, SiC_2 is anticorrelated with C_2. (5) The presence of SiC_2 in a carbon star allows to eliminate the possibility that these stars are both "hot" ($T_{eff} \geqslant 3000$ K) and have been produced through the CNO bi-cycle.

064.066 Radiation pressure on grains as a mechanism for mass loss in red giants.   S. Kwok.
Astrophys. Journ., Vol. 198, 583 - 591 (1975).

A quantitative model is constructed for the process of radiation pressure on dust to explain the mass loss observed in cool giants. Results are given for six sample stars, demonstrating the effects of stellar luminosity, mass, and effective temperature on the mass loss rate. The envelope expansion velocity and average grain size related to the fraction of grain material condensed and the gas density is found to be the key factor in determining the feasibility of mass ejection under this mechanism.

064.067 Magnetic fields and dense chromospheres in dMe stars.   D. J. Mullan.
Astron. Astrophys., Vol. 40, 41 - 54 (1975).

The author examines in a semi-quantitative fashion the hypothesis that dense chromospheres of dMe stars are heated by dissipation of hydromagnetic waves. The waves are thought to be generated in active regions where the (non-spot) field strength can be as large as 5–10 kG, according to the author's starspot model. He proposes that dMe stars are a set of magnetic stars on the lower main sequence with strong fields presumably generated by dynamo action in deep convective envelopes.

064.068 The abundance determination in a stellar atmosphere. 1. LTE experimentation using an artificial non-LTE spectrum.
S. Dumont, N. Heidmann, J. T. Jefferies, J.-C. Pecker.
Astron. Astrophys., Vol. 40, 127 - 132 (1975).

It is shown that the classical LTE analysis of an atomic spectrum of a solar-like star leads to values of the abundance A which may be different from the real values. Various uncertainties affect the results derived from neutral lines. However, proper selection of the observational data, for instance use of ion lines, can lead, in similar cases at least, to good values of A. These conclusions concern the solar-like case; however, they emphasize that, in all cases, the effect of departures from LTE needs very careful discussion.

064.069 The mass loss from the O9.5 Ib supergiant, ζ Orionis, derived from the Hα profile.   A. G. Hearn.
Astron. Astrophys., Vol. 40, 277 - 283 (1975).

The P Cygni type profile of Hα of the O9.5 Ib supergiant, ζ Orionis, measured by Conti and Leep has been interpreted in terms of a uniform spherically symmetrical shell where the excitation and ionization balance is determined by the radiation from the star. The analysis gives a shell radius of 2.1 stellar radii and an electron density in the shell of 1.3×10^{10} cm^{-3}. With the observed expansion velocity of 250 km s^{-1} a mass loss of $1.8 \times 10^{-6} M_\odot$ yr^{-1} is obtained. The observation and the interpretation is consistent with the hypothesis that the mass loss mechanism is a hot coronal wind driven by a corona with a temperature not less than 2.6×10^6 °K.

064.070 The energy balance and mass loss of stellar coronae.   A. G. Hearn.
Astron. Astrophys., Vol. 40, 355 - 364 (1975).

Calculations of the energy losses of a stellar corona by conduction, radiation and mass loss show that for a given coronal base pressure, there is an average coronal temperature for which the energy loss is a minimum. For OB supergiants the main energy loss is by mass loss. The main energy loss from a corona of a Wolf-Rayet star is also by mass loss. The calculations show that the corona which would give the observed mass loss from Wolf-Rayet stars has an opacity from electron scattering of about unity. This would explain the absence of photospheric lines and Balmer discontinuity. It also raises the possibility that the radius of the true photosphere of Wolf-Rayet stars could be much less than the observed radius.

064.071 Extended static stellar atmospheres. I. Principles of computation.
E. Hundt, K. Kodaira, J. Schmid-Burgk, M. Scholz.
Astron. Astrophys., Vol. 41, 37 - 40 (1975).

The principles of computation of extended static stellar atmospheres are discussed, and a computer program is described. Locations in the HR-diagram of stars with extended photospheres are predicted.

064.072 Extended static stellar atmospheres. II. Location in the HR-diagram and some properties of extended atmosphere stars.   J. Schmid-Burgk, M. Scholz.
Astron. Astrophys., Vol. 41, 41 - 45 (1975).

The authors search for combinations of stellar masses, luminosities and radii which may lead to extended static photospheres. Two main groups of such objects are found: (i) Stars with about one to a few solar masses in their final red giant stages (atmospheric gravities log $g_{grav} \lesssim 0$); the temperature drop at the surface caused by the extension may have consequences for the formation of interstellar molecules and grains. (ii) Low-mass stars of high luminosity (log $g_{grav} \simeq 3...1$), including the high-luminosity blue halo stars like Barnard 29 which is briefly discussed.

064.073 Propagation of shock waves in stellar atmosphere.   H. S. Gurm, V. P. Singh.
Bull. Astron. Soc. India, Vol. 2, 40 (1974). – Abstract.

064.074 Coupling of non-LTE radiative transfer with nonlinear hydrodynamics: a study in non-steady stellar atmospheres.   R. I. Klein.
Thesis, Brandeis Univ., Waltham, Mass. (USA). 305 pp. University Microfilms Order No. 74-16,822 (1974).

064.075 Theoretical investigation of stellar atmospheres with specific application to the atmosphere of a carbon star R Coronae Borealis.   K. Kawabata.
Thesis, Pennsylvania State Univ., University Park (USA). 402 pp. University Microfilms Order No. 74-20,931 (1973).

064.076 Spectroscopy of circumstellar shells.   R. R. Treffers.
Thesis, California Univ., Berkeley (USA). 79 pp. University Microfilms Order No. 74-23,976 (1973).

064.077 Mass loss from the A-type supergiant α Cyg.   H. J. G. L. M. Lamers.
Studies on the structure and stability of extended stellar atmospheres. Rijksuniversiteit Leiden (Netherlands). Proefschrift p. 153 - 168 (1974).

064.078 Characteristics of non-thermally heated stellar atmospheres.   H. J. G. L. M. Lamers, M. Kuperus.
Studies on the structure and stability of extended stellar atmospheres. Rijksuniversiteit Leiden (Netherlands). Proefschrift p. 169 - 200 (1974).

064.079 Stellar winds.   P. H. Roberts.
Solar wind three, (see 012.020), p. 231 - 242 (1974).
The first part of this paper is intended to provide a moti-

vation for the study of stellar winds. The second reviews the theory of steady, spherically-symmetric, coronal expansion according to both the one-fluid and the two-fluid continuum pictures.

064.080 Non-LTE-Effekte und Einfluß der Heliumhäufigkeit in Atmosphären von A0 Ia-Ueberriesen.
R. P. Kudritzki.
Diss. Techn. Univ. Berlin, 76 pp. (1973).

Ultraviolet absorption lines arising on metastable states.   See Abstr. 022.073.

A table of semiempirical gf values. Part 1: Wavelengths: 5.2682 nm to 272.3380 nm. Part 2: Wavelengths: 272.3395 nm to 599.3892 nm. Part 3: Wavelengths: 599.4004 nm to 9997.2746 nm.   See Abstr. 022.086.

Improved complete-linearization method for the solution of the non-LTE line transfer problem.
See Abstr. 063.001.

Resonance line transfer with partial redistribution. III. Mg II resonance lines in solar-type stars.
See Abstr. 063.012.

Line formation in turbulent media.
See Abstr. 063.013.

The use of variable quadrature weights.
See Abstr. 063.024.

The effect on the $^{12}C/^{13}C$ ratio of repeated deep mixing to the hydrogen burning shell in a red giant.
See Abstr. 065.003.

On the adiabatic pulsations of the rotating stars with a constant temperature gradient.   See Abstr. 065.015.

Accretion and effluxion of mass and angular momentum.   See Abstr. 065.032.

On the solar curve of growth of titanium.
See Abstr. 071.025.

Restricted three-dimensional stellar wind modeling. I. Polytropic case.   See Abstr. 074.015.

Importance of OH bound-free opacity in the solar and stellar atmosphere.   See Abstr. 080.047.

Supersonic neutral winds in an outer atmosphere. III. A two-dimensional model.   See Abstr. 091.015.

The calibration of uvby photometric indices for population I and II stars in the range $0.5 < \Theta_e < 0.7, 2.0 < \log g < 4.4$.   See Abstr. 113.030.

Colors of metal deficient giant stars.
See Abstr. 113.039.

Atmospheric parameters from four-color photometry.   See Abstr. 113.042.

Wolf-Rayet-Sterne – eine Gruppe außergewöhnlicher Sterne (II. Teil).   See Abstr. 114.015.

Carbon and iron abundances for eleven southern G stars of unusual interest.   See Abstr. 114.029.

Stars of spectral class B with emission lines.
See Abstr. 114.055.

A six-color Q-parameter for yellow supergiants.
See Abstr. 114.056.

Preliminary results of UV spectrophotometry.
See Abstr. 114.057.

Ultraviolet spectra with line opacities.
See Abstr. 114.060.

The Na D lines as surface gravity indicators.
See Abstr. 114.061.

Effective temperatures, surface gravities, and chemical compositions of early B-type stars.
See Abstr. 114.091.

Shell spectrum of the Be star HD 217050, 1966–1972.   See Abstr. 114.310.

The Mg II doublet emissions near 2800 Å observed in Alpha Tauri, Alpha Orionis, and Epsilon Pegasi.
See Abstr. 114.314.

Spectralphotometry and quantitative analysis of the hydrogen-deficient star HD 60344.
See Abstr. 114.317.

Mass loss observed in the ultraviolet spectrum of the A2 supergiant, Alpha Cygni.   See Abstr. 114.328.

Ultraviolet observations of the chromosphere of two M-supergiants.   See Abstr. 114.343.

Recent balloon observations of the chromospheric Mg II lines near 2800 Å.   See Abstr. 114.344.

Evidence for a chromosphere in Vega.
See Abstr. 114.345.

Ultraviolet observations of Capella from Copernicus.
See Abstr. 114.346.

Abundances in the halo cool star HD 128279.
See Abstr. 114.352.

Empirical effective temperature, bolometric corrections, and fundamental stellar properties. See Abstr. 115.008.

Rapidly rotating stars with optically thin stellar winds.   See Abstr. 116.001.

Period study for U Cephei and implications for the mass-transfer theory.   See Abstr. 121.016.

Spectrum of the Delta Scuti variable 20 Canum Venaticorum: a model-atmosphere analysis.
See Abstr. 122.017.

Shock-wave thermalization.   See Abstr. 125.002.

The atmospheres of cool white dwarfs of spectral type DA.   See Abstr. 126.013.

Herbig-Haro objects and T Tauri nebulae.
See Abstr. 132.021.

The IRC+10216 molecular envelope.
See Abstr. 141.612.

065   Star Formation, Stellar Structure and Evolution, Neutron Stars

065.001   The collapse of self-gravitating clouds of pure hydrogen.   W. G. L. Pöppel.
Astrophys. Space Sci., Vol. 32, 175 - 196 (1975).

Exploratory models of the collapse of spherical self-gravitating clouds are studied in relation to the problem of the formation of first generation star systems. The masses which were considered are in the range of 83 to $5.2 \times 10^{10} M_\odot$. The models show the importance of a correct evaluation of the chemical reactions and dissipative mechanisms, which cannot be ignored in a realistic treatment of the collapse of self-gravitating clouds. The influence of the initial conditions on the dynamical and thermal properties during evolution are also analysed.

065.002   The equation of state for neutron stars.   F. Ferrini.
Astrophys. Space Sci., Vol. 32, 231 - 247 (1975).

The author presents the results of an improved analysis of the equation of state of matter from metallic to nuclear density which takes into account: (a) in the region of low densities ($\rho < 10^4$ g cm^{-3}), the electron correlation energy in the lattice; (b) in the region of medium density ($10^4 < \rho < 4.3 \times 10^{11}$ g cm^{-3}) a more refined discussion of the electron energy; (c) in the region of large densities (4.3×10^{11} g cm^{-3} $< \rho$) a new method for determining the surface nuclear energy based on the Thomas-Fermi method. The equation of state which he obtains shows no discontinuity at the boundaries between the different regions.

065.003   The effect on the $^{12}C/^{13}C$ ratio of repeated deep mixing to the hydrogen burning shell in a red giant.
D. S. Dearborn, A. J. C. Bolton, P. P. Eggleton.
Monthly Notices Roy. Astron. Soc., Vol. 170, 7P - 10P (1975).

The effect of thermal instabilities in the hydrogen burning shell on the $^{12}C/^{13}C$ ratio in the envelope is studied. Repeated deep mixing caused by the instability can reduce the $^{12}C/^{13}C$ ratio to values of about 10. This process does not however produce carbon stars.

065.004   Pulsational stability of stars in thermal imbalance. VI. Physical mechanisms and extension to nonradial oscillations.   M. L. Aizenman, J. P. Cox.
Astrophys. Journ., Vol. 195, 175 - 185 (1975).

A clarification of the physical mechanisms operative in the pulsational stability against radial oscillations of spherical stars in thermal imbalance is presented. It is shown that the stability coefficient for such stars can be conceptually separated into distinct "dynamical" and "nonadiabatic" parts. The dynamical part involves no nonadiabatic effects, and is analogous to a slowly time-varying spring constant in a simple mechanical system. The nonadiabatic part arises from phase shifts between the pressure and density variations brought about solely by the small nonadiabatic effects which cannot be neglected in a star in thermal imbalance. It is shown that the formal expression for the stability coefficient for purely radial oscillations also applies identically to the nonradial oscillations of a spherical star in thermal imbalance. We suggest that these thermal imbalance effects as applied to nonradial oscillations may play a role in the instability of the β Cephei stars.

065.005   Carbon-burning nucleosynthesis with convection.   A. S. Endal.
Astrophys. Journ., Vol. 195, 187 - 192 = Rosemary Hill Observatory Univ. Florida, *Gainesville*, Contr. No. 47 (1975).

The effect of convection on carbon-burning nucleosynthesis is explored with a limited network of reactions. Con-

vection is simulated by a series of networks at fixed mass points in the core of an evolving 15 $M_\odot$ star. Complete mixing is always assumed. Comparison to single network calculations show that the "half-energy" approximation of Arnett yields reasonable results, although the abundances of nuclei which are created by β-decays of unstable nuclei tend to be underestimated, by this approximation.

065.006   Modal stability of RR Lyrae stars.   R. F. Stellingwerf.
Astrophys. Journ., Vol. 195, 441 - 466 (1975).

The modal stability of a survey of RR Lyrae models has been investigated using linear nonadiabatic analysis and relaxation to the exact periodic nonlinear limit cycles. With these techniques both the growth rates of the small-amplitude solutions and the mode switching rates of the large-amplitude solutions are obtained, resulting in a complete description of the long-term modal behavior of the model. The systematics of the nonlinear models are described, and an accurate fitting formula for the periods is derived.

065.007   Viscous effects in rapidly rotating stars with application to white-dwarf models. III. Further numerical results.   R. H. Durisen.
Astrophys. Journ., Vol. 195, 483 - 492 = Lick Obs. Bull., No. 670 (1975).

Improved viscous evolutionary sequences of differentially rotating, axisymmetric, nonmagnetic, zero-temperature white-dwarf models are constructed using the relativistically corrected degenerate electron viscosity. The results support the earlier conclusion that angular momentum transport due to viscosity does not lead to overall uniform rotation in many interesting cases.

065.008   Structure and properties of detonation waves. I. Detonation waves in dense stellar material.
S. W. Bruenn, A. Marroquin.
Astrophys. Journ., Vol. 195, 567 - 584 (1975).

Techniques are developed for following the time history of the nuclear abundances and the thermodynamic variables of a fluid element passing through the burning zone of a detonation wave. Application is made here to stellar material at high densities. Assumptions made in prior investigations are examined in detail. It is shown, that a detonation wave fueled by ^{16}O is self-consistent even when crystallization effects are taken into account.

065.009   An analysis of the pulsational stability of very low mass stars on the hydrogen and deuterium main sequence.   J. W. Opoien, A. S. Grossman.
Astron. Astrophys., Vol. 37, 335 - 338 (1974).

The authors have investigated the pulsational stability of the most recent theoretical models of stars in the mass range 0.085 $M_\odot$ to 0.5 $M_\odot$ on the hydrogen main sequence, and in the mass range 0.012 $M_\odot$ to 0.5 $M_\odot$ on the deuterium main sequence. They find all models to be unstable on the hydrogen main sequence. Pulsation periods range from 10 to 41 min on the hydrogen main sequence and from 1 to 20 hrs on the deuterium main sequence.

065.010   Wie sterben die Sterne?   K. Pinkau.
SuW, Vol. 14, 4 - 8 (1975).

065.011   Neutron stars with an anisotropic equation of state: mass, redshift and stability.
H. Heintzmann, W. Hillebrandt.
Astron. Astrophys., Vol. 38, 51 - 55 (1975).

Fully relativistic, anisotropic neutron star models are studied by means of several simple or plausible assumptions about the equation of state at high densities. The authors find that – neglecting stability requirements –: a) there is in principle no limiting mass nor limiting redshift for arbitrarily large anisotropy, b) small deviations from isotropy entail small changes in the mass and redshift of a star, c) estimates with semi-realistic equations of state for neutron star matter show that the maximum mass of a stable neutron star still lies beyond $3-4\,M_\odot$.

065.012 **Steady-state nuclear fusion in accreting neutron-star envelopes.** C. J. Hansen, H. M. Van Horn.
Astrophys. Journ., Vol. 195, 735 - 741 (1975).
Conditions in the envelopes of accreting neutron stars are shown to require thermonuclear burning of hydrogen and helium in thin shells. Numerical models of such envelopes are constructed for neutron stars with M = 0.0925, 0.476, and $1.41\,M_\odot$ and for spherically symmetric accretion rates of $M = 10^{-11}$, 10^{-9}, and $10^{-7}\,M_\odot\mathrm{yr}^{-1}$. Almost all of the shells are found to be thermally unstable, with the growth rate of the instability ranging from milliseconds to months. A crude model of nonspherical accretion is also considered and found to resemble roughly the spherical accretion problem with the same accretion rate per unit area.

065.013 **Slowly braked, rotating neutron stars.** H. Sato.
Astrophys. Journ., Vol. 195, 743 - 749 (1975).
A slowly braked, rotating neutron star is believed to be a star which rapidly rotates, has no nebula, is non-pulsing, and has a long initial braking time of $\sim 10^5$ to 10^6 years because of a low magnetic field. Such an object might be observable as an extended weak source of infrared or radio wave radiation due to the scattering of low-frequency strong-wave photons by accelerated electrons. If these objects exist abundantly in the Galaxy, they would act as sources of relatively low-energy cosmic rays. Pulsars (rapidly braked neutron stars) are shown to have difficulties in providing an adequate amount of cosmic-ray matter, making these new sources seem necessary. It is shown that white dwarfs may be slowly braked stars with braking times longer than $10^{6.5}$ years.

065.014 **Star formation in clouds of molecular hydrogen.** V. C. Reddish.
Monthly Notices Roy. Astron. Soc., Vol. 170, 261 - 280 (1975).
It is shown that the formation of hydrogen molecules on the surfaces of grains in interstellar clouds results in pressure instabilities leading to fragmentation of the clouds and the formation of protostars.

065.015 **On the adiabatic pulsations of the rotating stars with a constant temperature gradient.**
V. Ureche, N. Lungu.
Bull. Astron. Inst. Czechoslovakia, Vol. 26, 116 - 119 (1975).
The equations of the adiabatic pulsations for a star with slow rotation (neglecting the rotational distortion) is considered. For the case when only a certain superficial envelope of the star pulsates, and its temperature gradient is constant, the solution is given in terms of a degenerate hypergeometrical series. The angular velocity ω_{crit}, for which the pulsations become unstable, depends on the thickness of the envelope and the last quantity depends on the temperature gradient. The overtones, above a given limit, are unstable.

065.016 **Formation of population II stars.**
S. Miki, T. Nakano.
Publ. Astron. Soc. Japan, Vol. 27, 147 - 163 (1975).
Thermal and dynamical properties of gas clouds of mass between 1 $M_\odot$ and $10^6\,M_\odot$ with a fractional mass of heavy elements $Z=10^{-3}$ to $Z=10^{-5}$ are investigated by comparing the time scales of cooling, heating, expansion, and contraction

for two cases with and without grains. The conditions under which the gas cloud can contract by its own gravity and the process of contraction are investigated, and the formation of globular cluster stars is discussed.

065.017 **Thermal pulses; p-capture, α-capture, s-process nucleosynthesis; and convective mixing in a star of intermediate mass.** I. Iben, Jr.
Astrophys. Journ., Vol. 196, 525 - 547 (1975).
After extrapolating from the characteristics of a model star of mass 7 $M_\odot$ during the course of 10 thermal relaxation oscillations, the following inferences are drawn: (1) The reaction ^{22}Ne $(\alpha, n)^{25}$Mg serves as a major source of neutrons for s-process nucleosynthesis. (2) This nucleosynthesis takes place at the high temperature base of a convective shell of mass $0.002\,M_\odot$ at and near the peak of a thermal pulse which recurs at intervals of roughly 2500 yr. (3) At maximum size, the convective shell extends from the position of highest helium-burning rate to just below the hydrogen-helium discontinuity, and no further. (4) The ^{22}Ne that is the immediate precursor of neutrons owes its formation to the ^{14}N that has been left behind by the hydrogen-burning shell during the interpulse phase; after two α-captures and a β decay, ^{14}N is converted completely into ^{22}Ne within the convective shell. (5) Following the disappearance of the convective shell, the base of the convective envelope extends into the outer portion of the region previously contained in the convective shell; fresh products of p-capture, α-capture, and s-process nucleosynthesis are then convected to the surface; in particular, surface abundances of ^{4}He, ^{22}Ne, ^{25}Mg, and the s-process progeny of ^{25}Mg and ^{56}Fe are enhanced, and the abundance of ^{12}C may also be enhanced.

065.018 **Neon-22 as a neutron source, light elements as modulators, and s-process nucleosynthesis in a thermally pulsing star.** I. Iben, Jr.
Astrophys. Journ., Vol. 196, 549 - 558 (1975).
If the reaction ^{22}Ne$(\alpha, n)^{25}$Mg is the major source of neutrons in the convective shell of a thermally pulsing star, the fraction of emitted neutrons that can be captured by ^{56}Fe and its neutron-capture progeny is controlled by ^{22}Ne, ^{25}Mg, and the neutron-capture progeny of ^{22}Ne and ^{25}Mg. Adopting currently quoted neutron-capture cross sections and choosing the initial ratio of ^{22}Ne to ^{56}Fe to be within a factor of 2 of the solar system value, it is shown that from three to six neutrons are captured by ^{56}Fe and its progeny for every fresh ^{56}Fe engulfed by the convective shell during a thermal pulse. The resultant distribution of heavy s-process elements resembles the solar system distribution.

065.019 **A realistic lower bound for the maximum mass of neutron stars.**
R. L. Bowers, R. D. Pedigo, A. M. Gleeson, R. L. Zimmerman.
Astrophys. Journ., Vol. 196, 639 - 645 (1975).
An equation of state for neutron matter which includes relativistic pion exchange is used to calculate neutron star masses and moments of inertia. The equation of state sets a reasonable lower bound on the softness of cold superdense matter, and leads to a maximum equilibrium neutron star mass of $0.52\,M_\odot$. Processes which could affect our conclusions such as hyperonization, nonnormal ground states, solidification and local anisotropy, are discussed.

065.020 **The stability of differential rotation with non-axisymmetric perturbations. II: The necessary conditions.** C.-H. Sung.
Astrophys. Space Sci., Vol. 33, 127 - 140 (1975).
A plane-wave analysis on a simplified scheme based on the Boussinesq approximation and shallow convection is used to establish the necessary conditions for stability of a differentially rotating, compressible flow between two coaxial cylinders subject to non-axisymmetric perturbations. To test

the adequateness of this simplification, the sufficient conditions for stability are again established which agree with those obtained by a normal-mode analysis on an exact scheme in an earlier paper by the author. This model is applicable to stellar models with rotation $\Omega = \Omega(\varpi)$, where ϖ is the radial distance from the axis of rotation (the z-axis). A necessary and sufficient condition for stability, in the non-dissipative case, is found.

065.021 **Magnetically distorted polytropes. III: Radial, non-radial, toroidal and transverse shear modes of oscillation.** N. K. Sood, S. K. Trehan.
Astrophys. Space Sci., Vol. 33, 165 - 172 (1975).

The oscillations of a gaseous polytrope with a magnetic field having both a toroidal and a poloidal component are examined using the second-order tensor virial equations on the assumption that the magnetic energy is small compared with the gravitational energy. The frequencies of oscillation of the transverse shear, the toroidal and the coupled pulsation modes are tabulated for polytropic indices $n = 1$, 1.5, 2, 3 and 3.5. It is found that the magnetic field decreases the frequency of oscillation of (i) the transverse shear mode and (ii) the mode which starts as a radial pulsation in the absence of a magnetic field while it increases the frequency of oscillation of (i) the toroidal mode and (ii) the Kelvin mode. In all cases the shift in frequency decreases with increasing n.

065.022 **Vibrational stability of stars in thermal imbalance: a solution in terms of asymptotic expansions. II: The general 'non-isentropic' oscillations.** J. Demaret.
Astrophys. Space Sci., Vol. 33, 189 - 213 (1975).

The solution of the partial differential equation describing the 'non-isentropic' oscillations of a star in thermal imbalance has been obtained in terms of asymptotic expansions up to the first order in the parameter Π/t_s, where Π is the adiabatic pulsation period for the fundamental mode and t_s, a secular time scale of the order of the Kelvin-Helmholtz time. The solution obtained allows one to derive unambiguously a general integral expression for the coefficient of vibrational stability for arbitrary stellar models in thermal imbalance. The physical interpretation of this stability coefficient is discussed and its generality and its simplicity are stressed. Finally, the author emphasizes that the inclusion of the effects of thermal imbalance in the stability calculations of realistic evolutionary sequences of stellar models, not considered up to now by other authors, is quite easy and straightforward with the simple formula derived here.

065.023 **A physically realistic sphere of perfect fluid to serve as a model of neutron stars.**
B. Kuchowicz.
Astrophys. Space Sci., Vol. 33, L13 - L14 (1975).

A new exact solution of Einstein's equations is derived, which constitutes a generalization of the well-known internal Schwarzschild solution, and may be applied to relativistic spheres with a finite density increase toward the centre.

065.024 **Carbon ignition and burning in degenerate stellar cores.** R. G. Couch, W. D. Arnett.
Astrophys. Journ., Vol. 196, 791 - 803 (1975).

Earlier calculations (Couch and Arnett, 1973) have been extended and refined. The conditions at ignition of $^{12}C + ^{12}C$ in degenerate carbon-oxygen cores are explored. Nuclear energy release by carbon burning is found to be balanced initially by neutrino loss from Urca shells. Densities at which thermal runaway occurs are far below those currently thought to be necessary for core collapse and hence neutron star formation. Mathematical and astrophysical uncertainties, and the possible relationship between these models and stars, are discussed.

065.025 **Studies of evolved stars. V. Nucleosynthesis in hot-bottom convective envelopes.**
J. M. Scalo, K. H. Despain, R. K. Ulrich.
Astrophys. Journ., Vol. 196, 805 - 817 (1975).

Convective envelopes fitted to the cores of luminous double shell models have been found to develop high base temperatures, $(40-80) \times 10^6\,°K$, during the period between shell flashes. The authors show that these "hot-bottom" envelopes are the result of the increasing influence of radiation pressure on the radiative and adiabatic temperature gradients. Lower limits on the luminosity ($M_{bol} \approx -5.4$) and total mass ($\sim 1.5\,M_\odot$) required for the phenomenon are derived. Comparison with observations indicates that nucleosynthesis in hot-bottom envelopes is not the primary cause of the abundance anomalies observed in peculiar red giants, except possibly for the large $^{13}C/^{12}C$ ratios and lithium enhancements found in some stars. In particular, mixing of triple-α carbon and s-process elements initiated by the helium shell flashes seems to be required in order to satisfy constraints imposed by the observed line and band strengths and luminosities.

065.026 **Interpretation of the stellar metallicity distribution.**
B. M. Tinsley.
Astrophys. Journ., Vol. 197, 159 - 162 (1975).

The dispersion in metal abundances of stars of a given age is similar to that in the abundance distribution of dwarfs with lifetimes greater than the age of the Galaxy. Thus the latter distribution cannot be correctly interpreted unless the former dispersion is taken into account. With this consideration, dynamical models for the formation of the Galaxy by collapse, and models for chemical evolution that consider accretion of gas shed by halo stars, no longer predict too few moderately metal-poor G–K dwarfs in the solar neighborhood.

065.027 **A variational calculation of the fundamental frequencies of quadrupole pulsation of fluid spheres in general relativity.** S. L. Detweiler.
Astrophys. Journ., Vol. 197, 203 - 217 (1975).

A variational principle is used to calculate, in a relatively simple manner, the frequencies of quadrupole oscillation of relativistic neutron star models for a variety of equations of state and central densities. Typical periods and damping times are on the order of milliseconds and seconds, respectively, in agreement with earlier estimates obtained by Thorne in a different manner. For spheres of incompressible fluids, the analyses indicate that the fundamental quadrupole mode becomes unstable as $2M/R$ attains its limiting value of 8/9.

065.028 **Fossil dust shells around luminous supergiants.**
R. Stothers.
Astrophys. Journ., (*Letters*), Vol. 197, L25 - L27 (1975).

The observed frequency with which infrared excesses appear in F, G, and K supergiants of luminosity class Ia supports the idea that these excesses arise in a "fossil" circumstellar dust shell that was formed during a prior M-supergiant phase of evolution. The required leftward evolution of the star on the H-R diagram would then imply that the Ledoux, rather than the Schwarzschild, criterion for convective mixing is the correct criterion to use in stellar evolution calculations.

065.029 **Axisymmetric α^2-dynamos in the Hayashi-phase.**
M. Schüßler.
Astron. Astrophys., Vol. 38, 263 - 270 (1975).

The evolution of axisymmetric magnetic fields during the Hayashi-phase is investigated on the basis of the α-effect dynamo mechanism. In particular, the influence of the transition from the fully convective to the fully radiative main sequence state (for A stars) is considered for rigid rotation. The fields are calculated by solving the induction equation as an initial value problem. The numerical results show that a dynamo-built magnetic field can "survive" the transition while conserving the main part of its total flux. The topology of the field can change significantly during the evolution. For a certain phase of the transition to the main sequence there are oscillating fields, possibly giving an explanation for magnetic stars with long periods.

065.030 **Instability against nonradial oscillations of models for Beta Cephei stars.** M. L. Aizenman, J. P. Cox, J. R. Lesh.
Astrophys. Journ., Vol. 197, 399 - 404 (1975).

The authors have found nuclear-driven vibrational instability in some of the lower g^+ modes in a 10 $M_\odot$ model evolved off the ordinary hydrogen-burning main sequence, during the initial stages of shell hydrogen burning. It is conceivable that this instability may be the basic excitation mechanism for the β Cephei stars.

065.031 **Theoretical studies of massive stars. I. Evolution of a 15 $M_\odot$ star from the zero-age main sequence to neon ignition.** A. S. Endal.
Astrophys. Journ., Vol. 197, 405 - 413 = Rosemary Hill Obs., Univ. Florida, *Gainesville*, Contr. No. 56 (1975).

The evolution of a 15 $M_\odot$ star from the zero-age main sequence to neon ignition has been computed by the Henyey method. The hydrogen-rich envelope and all shell sources were explicitly included in the models. Energy transfer from the carbon-burning shell to the core by degenerate electron conduction becomes important after the core carbon-burning stage. Neon ignition will occur in a semidegenerate core and will lead to a mild "flash." Detailed numerical results are given.

065.032 **Accretion and effluxion of mass and angular momentum.** R. N. Henriksen, K. C. Heaton.
Monthly Notices Roy. Astron. Soc., Vol. 171, 27 - 34 (1975).

The hydrodynamical accretion and effluxion of isentropic or isothermal material with uniform specific angular momentum, $L \sin \theta$, is solved exactly on a stream line, θ = constant. The critical value for L above which accretion is severely curtailed and effluxion is rotationally driven, is found as the principal new result.

065.033 **Large scale magnetic fields in late-type stars.** D. L. Moss.
Monthly Notices Roy. Astron. Soc., Vol. 171, 303 - 309 (1975).

Models of a late-type zero age main sequence star with a deep surface convection zone are constructed containing a large scale dipolar magnetic field which is assumed to be expelled from the envelope by the action of the turbulence. It is shown that there must be a discontinuity in the transverse component of field at the lower boundary of the envelope, and that for the model considered it is unlikely that the central field can exceed about 10^7 Gauss.

065.034 **Evolution of massive stars.** V. I. Varshavskij, A. V. Tutukov.
Astron. Zhurn. Akad. Nauk SSSR, Vol. 52, 227 - 233 (1975). In Russian. English translation in Soviet Astron., Vol. 19, No. 2.

The structure and evolution of 16 $M_\odot$, 32 $M_\odot$ and 64 $M_\odot$ stars with initial chemical composition $X = 0.602$, $Y = 0.354$ and $Z = 0.044$ ($X_{12} = 0.00619$ and $X_{16} = 0.01847$) were considered from zero-age main sequence up to the core oxygen-depletion in red supergiants. The changes of the helium, carbon and oxygen contents in the interstellar medium were estimated since the Galaxy exists due to the massive stars evolution. Some consequences of stellar rotation and meridional circulation were also considered. The structure and evolution of massive pure hydrogen-helium stars were investigated through the core hydrogen and helium burning stages up to the core carbon ignition.

065.035 **Pulsational stability of neutrino thermal conduc-** tivity stars. A. S. Zentsova, D. K. Nadezhin.
Astron. Zhurn. Akad. Nauk SSSR, Vol. 52, 234 - 242 (1975). In Russian. English translation in Soviet Astron., Vol. 19, No. 2.

The problem of pulsational stability of hot neutron stars nontransparent with respect to electron neutrinos has been considered. Due to inverse dependence of neutrino opacity on temperature the conditions for the excitation of stellar pulsations are created. An expression for the integral of pulsational stability involving the diffusion of lepton charge has been found. The pulsational characteristics of the neutrino stars and the stability integral have been calculated. The neutrino heat conductivity is favourable to the excitation of the pulsations.

065.036 **A self-consistent field method in the theory of rotating stars.** S. I. Blinnikov.
Astron. Zhurn. Akad. Nauk SSSR, Vol. 52, 243 - 254 (1975). In Russian. English translation in Soviet Astron., Vol. 19, No. 2.

A straightforward variant of the self-consistent method is given for determining the equilibrium structure of arbitrarily rotating stars which is more simple for computer programming and requires less storage capacity than that proposed by Ostriker and Mark (1968). The author presents tables of the properties of rigidly rotating polytropes n = 1.5 and n = 3 and compares them with the results obtained by other authors. The accuracy of the energy method of investigation of stability is checked on rotating isentropic polytropes.

065.037 **A process of star formation and discreteness of the dispersions of star velocities.** S. A. Kaplan, R. B. Shatsova.
Astron. Zhurn. Akad. Nauk SSSR, Vol. 52, 260 - 263 (1975). In Russian. English translation in Soviet Astron., Vol. 19, No. 2.

The star formation in spiral arms is due to preliminary compressions of the interstellar medium by spiral shocks. It may lead to the discreteness and multipleness of the dispersions of velocities of stars which are formed in the same gas — dust complex. The discrete character of the dispersions of space velocities is confirmed by observational data for different samples of dwarf and giant stars in the vicinity of the sun.

065.038 **A consideration of the neutron capture time scale in the s-process.** J. B. Blake, D. N. Schramm.
Astrophys. Journ., Vol. 197, 615 - 620 (1975).

The question of the mean neutron capture interval in s-process nucleosynthesis is considered based upon s-process branching at ^{204}Tl and ^{151}Sm. It is concluded that the mean neutron flux during s-process nucleosynthesis was of the order of $10^{15}-10^{16} n$ s^{-1} cm^{-2} with a typical time scale between s-process neutron captures of the order of a few tens of years for nuclei with cross sections ~100 mb.

065.039 **Free-free opacity of dense stellar matter.** G. De Zotti.
Astrophys. Space Sci., Vol. 33, 359 - 367 (1975).

A non-relativistic calculation of the free-free opacity is made in detail for the conditions of intermediate and strong electron degeneracy. Corrections to the Born approximation are included and the ion correlations as well as the electron screening of the nuclear charge are taken into account. The free-free opacity is compared with the conductive one. It is shown that it must be taken into account when the degeneracy is not too strong, that is in the temperature-density regime where the plasma frequency ω_p satisfies the condition $\omega_p \lesssim 2.5 \ kT$.

065.040 **Neutral currents and the cooling of neutron stars.** S. M. Chitre, J. V. Narlikar, S. Ramadurai.
Astrophys. Space Sci., Vol. 33, L45 - L47 (1975).

Although the neutrino cooling rate by purely leptonic processes is altered by the neutral currents, the overall cooling of model neutron stars is shown to remain practically unaffected.

065.041 **How to make metal-poor stars, redden OB associations and grow mantles on grains.**
M. G. Edmunds, N. C. Wickramasinghe.
Astrophys. Space Sci., Vol. 34, 131 - 136 (1975). – Paper presented at the Symposium on Solid State Astrophysics, Cardiff, Wales, 9–12 July, 1974.

Three consequences of the existence of grains with metal-rich ice mantles are considered: (1) The production of metal-poor stars by expulsion of protostellar grains by radiation pressure during star formation. (2) The effects of these expelled grains in reddening massive stars in an OB association. (3) The production of the icy mantles on grains in OB associations.

065.042 **Solid state physics and cooling of neutron stars.**
S. Tsuruta.
Astrophys. Space Sci., Vol. 34, 199 - 208 (1975). – Paper presented at the Symposium on Solid State Astrophysics, Cardiff, Wales, 9–12 July, 1974.

The author shows the possible effect of the 'magnetic' condensation on cooling of neutron stars. Its observational significance (especially for younger pulsars such as the Crab pulsar) is emphasized. Other effects of solid state physics on cooling are also discussed.

065.043 **Neutron star cores.** R. G. Palmer.
Astrophys. Space Sci., Vol. 34, 209 - 222 (1975). Paper presented at the Symposium on Solid State Astrophysics, Cardiff, Wales, 9–12 July, 1974.

Current theories, and the astrophysical implications, of the nature of high density neutron star matter are reviewed. Suggestions are made for a compromise between the alternatives of neutron crystallization and pion condensation.

065.044 **Structure of neutron star cores.**
V. Canuto, B. Datta, J. Lodenquai.
Astrophys. Space Sci., Vol. 34, 223 - 229 (1975). – Paper presented at the Symposium on Solid State Astrophysics, Cardiff, Wales, 9–12 July, 1974.

After reviewing the outer and central regions of a neutron star, the authors discuss the central region and the possibility that the core has a solid structure. They present the work of different groups on the solidification problem, suggesting that the neutron star-cores are indeed solid.

065.045 **A note on critically rotating polytropes.**
P. S. Williams.
Astrophys. Space Sci., Vol. 34, 425 - 430 (1975).

The structure of critically rotating polytropes is calculated using two (related) Roche-type approximation schemes, one of which has been developed for stellar models. Comparisons with other results are made as a validation of these methods in stellar structure calculations.

065.046 **Theoretical isochrones and main sequences for old disk population stars.**
P. Demarque, G. R. Gisler.
Astron. Astrophys., Suppl. Ser., Vol. 20, 237 - 253 (1975).

Theoretical isochrones have been constructed for fourteen mixtures in the ranges of chemical compositions $(0.05 < Y < 0.45)$ and $(0.01 < Z < 0.06)$ and for ages relevant to the old disk population stars. The effects of autoionization absorption as computed by Watson were added to the Cox-Stewart radiative opacities. Comparison with the colour-magnitude diagram of M67 and NGC 188 does not favour supermetallicity, but rather normal abundances. Helium abundance by mass Y in the range of 0.25–0.35 is found. The distance of the Hyades cluster is discussed in terms of main sequence models extended to higher masses.

065.047 **Development of the theory of stellar evolution.**
J. Einasto, M. Jõeveer.
Zemlya i Vselennaya, 1975, No. 2, p. 77 - 81. In Russian.

065.048 **On the pulsational stability of supermassive magnetic stars.** V. V. Usov.
Pis'ma v Astron. Zhurn., Vol. 1, No. 2, p. 33 - 35 (1975). In Russian.

In MHD approximation it is shown that the generation of electromagnetic waves by a pulsating supermassive magnetic star can make it pulsationally stable. For usual parameters of the interstellar medium, the surface magnetic field on a supermassive star necessary for its stabilization is about 10^3 Oe.

065.049 **On the collapse of rotating protostars.**
W. Tscharnuter.
Astron. Astrophys., Vol. 39, 207 - 212 (1975).

Numerical calculations have been made for the early stages of the collapse of axisymmetric protostellar clouds including rotation. The numerical results indicate that a hypothetical solar nebula would evolve with a strong central condensation from the very beginning.

065.050 **Mixed stars from population II giants.**
V. Caloi, V. Castellani.
Astron. Astrophys., Vol. 39, 335 - 339 (1975).

A grid of models has been computed in order to obtain information about the location of (homogeneous and) main sequence models on the H–R diagram for $Z = 10^{-3}$ and $0.10 \leqq X \leqq 0.90$. The authors investigate in this way the expected location of stars which might have originated from chemical mixing of population II red giants. It is shown that fully mixed models cannot account for intermediate temperature subgiants, like the so called "blue stragglers" in galactic globular clusters. Observational evidence for high temperature mixed stars in the halo population is discussed.

065.051 **The $\Delta(3, 3)$ resonance in dense matter: a liquid-solid equation of state and its implications for neutron star structure.** V. R. Pandharipande, R. A. Smith.
Bull. American Astron. Soc., Vol. 7, 240 - 241 (1975). Abstr. AAS.

065.052 **Magnetohydrodynamics of neutron star interiors.**
I. Easson.
Bull. American Astron. Soc., Vol. 7, 241 (1975). – Abstr. AAS.

065.053 **Physics at the magnetospheric boundary of an accreting neutron star and its consequences for models of binary X-ray sources.** R. F. Elsner, F. K. Lamb.
Bull. American Astron. Soc., Vol. 7, 241 (1975). – Abstr. AAS.

065.054 **Hydromagnetic solutions for accretion flows in the magnetosphere of a rotating neutron star.**
P. Ghosh, F. K. Lamb, C. J. Pethick.
Bull. American Astron. Soc., Vol. 7, 241 - 242 (1975). Abstr. AAS.

065.055 **Some effects of atmospheric opacity on the evolution of a 5 $M_\odot$ star.**
R. W. Whitaker, H. R. Johnson.
Bull. American Astron. Soc., Vol. 7, 248 (1975). – Abstr. AAS.

065.056 **The stability of stars containing magnetic fields.**
R. J. Tayler.
Bull. American Astron. Soc., Vol. 7, 252 (1975). – Abstr. AAS.

065.057 **Detailed compositional evolution of first generation stars.** J. G. Eoll.

Bull. American Astron. Soc., Vol. 7, 256 (1975). — Abstr. AAS.

065.058 **Evolution of 15 $M_{\odot}$ and 25 $M_{\odot}$ stars through shell carbon burning.**
W. M. Howard, I. Iben, S. A. Lamb.
Bull. American Astron. Soc., Vol. 7, 256 (1975). — Abstr. AAS.

065.059 **A binary hypothesis for the subdwarf B stars.**
J. G. Mengel, J. Norris. P. G. Gross.
Bull. American Astron. Soc., Vol. 7, 256 (1975). — Abstr. AAS.

065.060 **Das Vogt-Russell-Theorem des inneren Aufbaus der Sterne.** H. Zimmermann.
Astron. in der Schule, 12. Jahrgang, p. 30 - 33 (1975).

065.061 **Quakes of solid cores of neutron stars as the origin of γ-ray bursts.** A. I. Tsygan.
AN SSSR. Fiz.-tekhn. in-t. Preprint 467. Leningrad, 1974. 14 pp. In Russian. — Abstr. in Referativ. Zhurn. 51. Astron., 4.51.659 (1975).

065.062 **Gravitational contraction of protostars. III. Role of heavy elements in the evolution of protostars.**
I. G. Kolesnik.
Astrometriya i Astrofizika, *Kiev*, vyp. (No.) 25, (see 003. 015), p. 10 - 19 (1975). In Russian.
The regularities of gravitational contraction of a protostar with one solar mass and volume energy losses due to neutral atoms of carbon and oxygen were considered. It is shown that after the phase of quasi-free contraction a core formation takes place, where the pressure forces are essential. The effect of finite optical depth of heavy element lines on volume energy losses and the possibility of development of the calculated properties in the evolution of interstellar clouds are discussed.

065.063 **Photometry and the evolutionary tracks in the HR diagram.** A. Maeder.
Dudley Obs. Rep. No. 9, (see 012.008), p. 427 - 436 (1975).
A report is given on the problems encountered in the comparison of cluster sequences in the color-magnitude diagram with theoretical isochrones. It is suggested that further developments in the theory of stellar models are necessary for explaining this basic test of stellar evolution.

065.064 **Stellar evolution at high mass with semiconvective mixing according to the Ledoux criterion.**
R. Stothers, C.-W. Chin.
Astrophys. Journ., Vol. 198, 407 - 417 (1975).
Semiconvective mixing has been included in new evolutionary sequences of models for stars of 10, 15, and 30 $M_{\odot}$, with four different initial chemical compositions. The models have been constructed with the help of the Ledoux criterion both for the definition of convective instability and for the state of convective neutrality that is assumed to be attained in regions with a gradient of mean molecular weight. Allowance for convective overshooting has also been made.

065.065 **On the treatment of convection as a nonradial stellar pulsation.** R. G. Deupree.
Astrophys. Journ., Vol. 198, 419 - 423 (1975).
The problem of nonlinear, nonadiabatic, nonradial stellar pulsation has been examined by a computational approach. A test calculation with a particular RR Lyrae model pulsating in the fundamental mode reveals that convection is present in the second helium ionization zone only when the luminosity amplitude becomes large and even then is present during only part of the pulsation cycle. The convective motion, measured by the nonradial velocity component, grows during rising luminosity, reaches a maximum after maximum luminosity, and decays during the other parts of the cycle. The possible

consequences of this phenomenon are briefly discussed.

065.066 **Double-shell-source evolution at 2 $M_{\odot}$.**
R. A. Gingold.
Astrophys. Journ., Vol. 198, 425 - 429 (1975).
A 2 $M_{\odot}$ population I star was evolved from the core helium-burning phase to the ninth strong helium shell flash. The intershell convective zone did not reach the hydrogen-containing layers, and thus no neutron production by the $^{13}C(\alpha, n)^{16}O$ reaction was possible. Mixing is unlikely to occur during later relaxation cycles. The temperatures are too low for neutron production by the $^{22}Ne(\alpha, n)^{25}Mg$ reaction in the intershell convective zone. The convective envelope fails by a wide margin to reach regions containing processed material. The results indicate that there are difficulties with the suggestion that helium shell flashes explain the existence of low-mass stars having unusual surface abundances.

065.067 **Slowly rotating relativistic stars. V. Static stability analysis of $n = 3/2$ polytropes.**
J. B. Hartle, M. W. Munn.
Astrophys. Journ., Vol. 198, 467 - 476 (1975).
The static stability analysis of slowly and uniformly rotating relativistic stars is applied to the $n = 3/2$ Tooper polytropes corresponding to an ideal nonrelativistic neutron gas. The method tests the stability of a sequence of stellar models against the modes of oscillation which correspond to radial modes in a nonrotating star. It involves a comparison of the uniformly rotating models to be tested with a sequence of models having a specified differential rotation. The equations of structure of these differentially rotating stars are derived and integrated, and the critical central density dividing the stable models from the unstable models is determined.

065.068 **Die Sternentwicklung bis zum Auftreten von Pulsationen.** A. Weigert.
BAV Rundbrief, 24. Jahrgang, p. 1 - 17 (1975). — Abbreviated version of a lecture held during BAV-meeting, Hamburg, 1974 October 27.

065.069 **Intrinsic magnetic polarizability contribution to the susceptibility of dense neutron matter.**
J. Bernabeu, T. E. O. Ericson, C. Ferro Fontan.
Astrophys. Letters, Vol. 16, 57 - 59 (1975).
It is shown that, apart from the neutron interaction contribution to the susceptibility of neutron stars, there is an additional contribution coming from the magnetic polarizability β_n of individual neutrons. For $\beta_n \approx 10^{-4}$ fm^3, this part overwhelms the interaction contribution for $k_F \gtrsim 3.5$ fm^{-1}. No transition to ferromagnetic state is predicted in the interesting density region.

065.070 **Accretion onto neutron stars under adiabatic shock conditions.** S. L. Shapiro, E. E. Salpeter.
Astrophys. Journ., Vol. 198, 671 - 682 (1975).
The accretion of gas onto a neutron star is examined for the case in which a strong, adiabatic shock front forms above the stellar surface to decelerate the incident plasma stream. Steady-state, spherically symmetric flow is considered, all magnetic fields are ignored, and rapid thermalization by plasma instabilities in the shock front is assumed. The dynamical and thermal structure of the emission zone between the surface and the shock front is determined, and the emergent radiation spectrum is calculated.

065.071 **Composition of the neutron star surface in pulsar models.** F. C. Michel.
Astrophys. Journ., Vol. 198, 683 - 685 (1975).
The author examines the question of whether ions are available from the neutron star surface to provide the rela-

tivistic stellar winds from pulsars. Although Ruderman and Sutherland suggest that ions are not available. he is unable to find firm support for that conclusion. The rate at which the ions would be lost is compatible with the total number of ions (helium, he assumes) estimated by Rosen and Cameron to be available and is compatible with the observed distribution of pulsar periods and ages. Consequently it would appear that ion availability from the surfaces of pulsars continues to be a plausible assumption, albeit certainly not an obligatory one.

065.072 Influence of convection on the vibrational stability of stars towards non-radial oscillations.
M. Gabriel, R. Scuflaire, A. Noels, A. Boury.
Astron. Astrophys., Vol. 40, 33 - 39 (1975).

The coupling between convection and non-radial oscillations is discussed. Limitations of the theory come from the lack of a good theory for turbulent convection.

065.073 Dynamo maintenance of magnetic fields in stars.
N. O. Weiss.
Magnetohydrodynamics, (see 003.018), p. 183 - 248 (1974).

065.074 Stellar evolution III: the overshooting from convective cores. A. Maeder.
Astron. Astrophys., Vol. 40, 303 - 310 (1975).

The problem of the boundary of convective cores is examined. An attempt is made to remove in computations of stellar models the local character of the equation of energy transport and also to avoid the use of Schwarzschild's criterion, which is a local one. Convection in stellar cores is calculated explicitly with non-local expressions of the mixing-length formalism and an iterative process is proposed to integrate simultaneously the equations of stellar structure and the equations describing convective motions. The method is applied to a chemically homogeneous star of $2 M_\odot$; the results show that the zone of overshooting typically extends over 15% of a mixing-length.

065.075 Can exploding supermassive stars produce ^{7}Li?
H. Nørgaard, M. Arnould.
Astron. Astrophys., Vol. 40, 331 - 334 (1975).

The nucleosynthesis calculations of Audouze and Fricke (1973) for the implosion-explosion phase of a spherically symmetric $5.2 \times 10^5 M_\odot$ supermassive star are reinvestigated with new reaction rate data and are extended to include also $A < 12$ nuclei, especially in order to examine the possibility of ^{7}Li production if some ^{3}He is supposed to survive at the beginning of the implosion-explosion process.

065.076 Pulsed gamma-ray emission from neutron and collapsing stars and supernovae.
G. S. Bisnovatyi-Kogan (Bisnovatyj-Kogan), V. S. Imshennik, D. K. Nadyozhin (Nadezhin), V. M. Chechetkin.
Astrophys. Space Sci., Vol. 35, 3 - 21, 23 - 41 (1975). In Russian and English.

Three mechanisms generating pulsed γ-ray emission at late stages of stellar evolution are investigated: (1) γ-ray bursts produced by the absorption of neutrino emission of a collapsing star in its envelope; (2) γ-ray bursts of thermal emission when the outer layers of a compact star ($R = 0.01 - 0.1 R_\odot$) are heated up by a powerful shock wave; and (3) γ-ray emission due to ejection of matter from neutron stars at an active stage of their existence.

065.077 Zero-population stars.
V. Castellani, P. Paolicchi.
Astrophys. Space Sci., Vol. 35, 185 - 196 (1975).

The expected characteristics of stars with zero metal content are investigated for a large range of masses and in varying the original helium content. It is shown that in such

an extreme case ($Z = 0$), the main sequence locus is much less sensitive than expected to the He-content; $Y = 0$ models are found to ignite 3α reactions in the zero-age main sequence at masses larger by $\sim 5 M_\odot$ than in the $Y = 0.2$ case. Evolutionary computations confirm that possible survivors of the zero population can develop a giant branch, akin to that of the well-known population II stars.

065.078 On surface perturbations in stellar stability problems.
J. Denis, P. Smeyers.
Meded. Koninkl. Acad. Wet., Letteren Schone Kunsten België, Kl. Wet., Jaargang 37, No. 1. 18 pp. (1975).

The perturbation procedures for the solution of problems involving a surface perturbation are reviewed with reference to their applicability in stellar stability problems. The perturbation procedure of Brillouin for a slight change of the boundary is extended to a class of eigenvalue problems in which one has to deal with a vector wave equation. The method can be applied to obtain the corrected eigenvalues and the corrected eigenfunctions as well.

065.079 On the luminosity of spherical protostars.
I. Appenzeller, W. Tscharnuter.
Astron. Astrophys., Vol. 40, 397 - 399 (1975).

Hydrodynamic model computations have been carried out for a spherically symmetric $1 M_\odot$ protostar. Compared to similar computations by Larson (1969) the authors used a different treatment of the accretion shock front. Their computations basically confirm Larsons results and show that Larson's disputed shock jump conditions have little influence on the protostellar models.

065.080 Loss of stability of low-mass stars under neutronization. M. M. Basko, V. S. Imshennik.
Astron. Zhurn. Akad. Nauk SSSR, Vol. 52, 469 - 480 (1975). In Russian. English translation in Soviet Astron., Vol. 19, No. 3.

The results of numerical calculations of the final stage of evolution and transition to the collapse of the iron cores with masses of $1.21 M_\odot$, $1.4 M_\odot$ and $1.7 M_\odot$ are presented. In the models of $1.7 M_\odot$ and $1.4 M_\odot$ there exists a distinct moment of loss of mechanical stability and transition to the phase of hydrodynamic collapse. In the model of $1.21 M_\odot$ one cannot point out such a moment — the evolution of the core with a time-scale of β-processes gradually accelerates up to the hydrodynamic collapse. In all the models, and especially in the model of $1.21 M_\odot$, the thermodynamically non-equilibrium character of β-processes is of a great significance. The major contribution to the energy loss rate in all the models is due to URCA-neutrinos. The possibility of the formation of low-mass ($M < 1.0 M_\odot$) neutron stars is discussed.

065.081 A note on star formation by gravitational fragmentation. A. A. Suchkov, Yu. A. Shchekinov.
Astron. Zhurn. Akad. Nauk SSSR, Vol. 52, 662 - 664 (1975). In Russian. English translation in Soviet Astron., Vol. 19, No. 3.

The possibility of star formation by gravitational fragmentation is considered. There is a lower limit for the mass of a fragment, which is defined by the cooling mechanism and opacity. The minimum mass obtained coincides with the stellar mass for contemporary conditions as well as for those for the formation of stars of the first generation.

065.082 Hydrodynamic calculations of cocoon stars.
H. W. Yorke.
Conference on optical observing programs on galactic structure and dynamics, (see 012.013), p. 139 - 145 (1975). In German.

065.083 The history of star formation and the presently ob-

served distribution of the heavy elements among stars.   P. Biermann.
Conference on optical observing programs on galactic structure and dynamics, (see 012.013), p. 147 - 149 (1975).

065.084   **A journey to the centre of a neutron star.**
S. M. Chitre.
Bull. Astron. Soc. India, Vol. 1, 19 - 22 (1973).

065.085   **Equation of state at high densities.**   V. K. Garde.
Bull. Astron. Soc. India, Vol. 2, 41 (1974). – Abstract.

065.086   **Evolution of ^{20}Ne core star.**
K. Duorah, H. L. Duorah.
Bull. Astron. Soc. India, Vol. 2, 41 (1974). – Abstract.

065.087   **On the evolution and pulsations of stellar models of higher solar mass.**
H. S. Gurm, H. M. Sukhija.
Bull. Astron. Soc. India, Vol. 2, 41, 43 (1974). – Abstract.

065.088   **Sternentstehung.**   H. Zimmermann.
Astron. in der Schule, 12. Jahrgang, p. 50 - 53 (1975).

065.089   **Thermal instability of helium-burning shell in stars evolving toward carbon-detonation supernovae.**
D. Sugimoto, K. Nomoto.
Publ. Astron. Soc. Japan, Vol. 27, 197 - 213 (1975).
Artificially suppressing the occurrence of thermal pulses, evolution in the phase of a growing carbon-oxygen core was computed through the ignition of carbon burning. More than 4000 thermal pulses take place through the evolutionary phase. The peak energy generation rate is $10^7 L_\odot$ at most, a rate too small to induce any major dynamical effect. After each pulse the convective envelope penetrates into the helium zone, and the products of helium burning, which contain carbon and s-process elements, are mixed into the convective envelope, which thereby develops composition characteristics of carbon stars.

065.090   **Nonradial oscillations of a 10 solar mass star in the main-sequence stage.**   Y. Osaki.
Publ. Astron. Soc. Japan, Vol. 27, 237 - 258 (1975).
The equations of linear adiabatic nonradial oscillations have been solved for realistic models of a 10 solar mass star in the core hydrogen burning stage. A propagation diagram and a phase diagram are introduced to interpret the behavior of eigenfunctions in the interior of stellar models. It is found that the zone with a varying chemical composition left behind by the receding convective core plays an essential role in nonradial oscillations. Pulsational stability is tested using the quasi-adiabatic approximation, and all of the models are found to be stable both for radial and nonradial modes ($l = 2$).

065.091   **X-ray emission from a neutron star with a strong magnetic dipole field.**   H. Inoue.
Publ. Astron. Soc. Japan, Vol. 27, 311 - 323 (1975).
The accretion of matter by a neutron star with a strong magnetic dipole field is considered as a model of X-ray sources. The structure of two narrow magnetic polar cones, along which matter falls down to the stellar surface, has been studied in detail. It is shown that the cone is optically thick and that the photon diffusion time is essential in determining the structure, which depends mainly on the matter accretion rate. Consequently, the black-body spectrum with about 10^7 K is emitted from the surface of the cone for the accretion rate of about $10^{-9} M_\odot$ yr^{-1}. This model accounts for the observed nature of Cen X-3 fairly well.

065.092   **Exotic phases of neutron-star matter.**   N. C. Chao.
Thesis, Washington Univ., Seattle (USA). 132 pp.
University Microfilms Order No. 74-7036 (1973).

065.093   **Nature of the carbon stars.**   S. D. Kilston.
Thesis, California Univ., Los Angeles (USA). 126 pp.
University Microfilms Order No. 74-3978 (1973).

065.094   **Numerical study of protostellar formation.**
C. K. Westbrook.
Thesis, California Univ., Livermore (USA). 93 pp. (1974).

065.095   **Non-linear pulsations and effects of thermal imbalance.**   V. K. Sastri.
Thesis, Nebraska Univ., Lincoln (USA). 111 pp. University Microfilms Order No. 74-13,018 (1973).

065.096   **Many-body plasma corrections to stellar thermonuclear reaction rate.**   C. M. Burgoyne.
Thesis, Pennsylvania State Univ., University Park (USA). 105 pp. University Microfilms Order No. 74-20,989 (1974).

065.097   **Comparison of microscopic calculations of solid neutron star matter.**   V. Canuto, J. Lodenquai, S. M. Chitre.
Nuclear Phys. A, (*Netherlands*), Vol. 233 A, 521 - 528 (1974).

065.098   **Numerical study of protostellar formation.**
C. K. Westbrook.
Thesis, California Univ., Davis (USA). 96 pp. University Microfilms Order No. 74-21,640 (1974).

065.099   **Constraints on the mixing and evolution of the peculiar red giants.**   J. M. Scalo.
Thesis, California Univ., Los Angeles (USA). 292 pp. University Microfilms Order No. 74-12,471 (1974).

065.100   **Advanced evolution of globular cluster stars.**
R. J. Zinn.
Thesis, Yale Univ., New Haven, Conn. (USA). 160 pp. University Microfilms Order No. 74-25,792 (1974).

065.101   **Neutron stars and incompressible fluid spheres in the Jordan-Brans-Dicke theory of gravitation.**
W. Hillebrandt, H. Heintzmann.
General Relativity Gravitation, (*GB*), Vol. 5, 663 - 672 (1974).

065.102   **Cores of neutron stars: are they solid or liquid.**
S. Takacs.
Cesk. Cas. Fys., Vol. 24, 500 - 501 (1974). In Slovak. – Letter.

065.103   **Calculation of the equilibrium structure and oscillations of polytropic stars pervaded by toroidal magnetic fields.**   M. J. Miketinac.
Comput. Phys. Commun., (*Netherlands*), Vol. 7, 410 - 418 (1974).
The program solves the structure equations determining the equilibrium configurations of a self-gravitating, axisymmetric, perfectly conducting polytrope in the presence of a toroidal magnetic field. Parameters such as mass, volume, radius etc., characterising an equilibrium configuration are calculated.

065.104   **The evolution of a $15 M_\odot$ star during the helium burning phase.**   C. de Loore, H. Hensberge.
Meded. Koninkl. Acad. Wet., Letteren. Schone Kunsten België, Kl. Wet., Vol. 36, No. 6, p. 5 - 13 (1974).
The evolution is followed with time steps of a few years for evolutionary phases where the convective regions are increasing or decreasing considerably. It is found that, with the pressure gradient criterion for convective stability and allowing normal mixing, helium ignition occurs when the star crosses

the blue giant region for the first time. The first part of helium burning occurs during the red giant stage, the second part during the blue giant stage.

065.105   Aspects of stellar evolution reappraised.
P. F. Browne.
Contemporary Phys., *(GB)*, Vol. 16, 51 - 68 (1975).
The theory of stellar evolution is reviewed in a simple fashion, with emphasis on topics for which some rethinking may be required. It is suggested that evolution of stars off the main sequence in the H-R diagram into the area occupied by the giants is caused, not by fuel exhaustion, but simply by instability following the ignition of thermonuclear energy release in the interior. The physics of a degenerate gas is discussed.

065.106   Models for stellar interior; pre-main sequence evolution and deuterium burning. E. Toma.
Stud. Cerc. Fiz., *(Rumania)*, Vol. 26, 839 - 868 (1974). In Rumanian.
Evolution tracks are calculated and the effect of chemical composition and mixing length ratio on tracks is discussed. Deuterium burning and the consequent vibrational instability are considered. Comments on the observation of deuterium burning in stars are given.

065.107   Stability of nonradial vibrational modes of relativistic neutron stars. II.
P. Cazzola, L. Lucaroni.
Phys. Rev. D, Particles and Fields, Vol. 10, 2038 - 2040 (1974).
Earlier work is extended to establish the stability of nonradial vibrational modes for multipoles of order $1 = 2, 3, 4$ and 5, and for central densities up to $10^{18} g\ cm^{-3}$.

065.108   Abnormal neutron stars. C.-G. Kallman.
Phys. Letters B, *(Netherlands)*, Vol. 55B, 178 - 182 (1975).
Quasi-classical field theory is used to predict abnormal behaviour of neutron matter in a model with neutrons, scalar mesons and vector mesons. Two stable families of neutron stars are predicted, a conventional one and an unconventional one with surface density in excess of $10^{15} g\ cm^{-3}$, roughly.

065.109   Stellar hydromagnetics. D. L. Moss.
Sci. Progr., *(GB)*, No. 243, Vol. 61, 421 - 442 (1974).

065.110   Positively charged isospin wave softening and proton lattice in neutron stars.
P. W. Anderson, N. Itoh, M. A. Alpar, E. Tosatti, R. G. Palmer.
Nuovo Cimento Lettere, Ser. 2, Vol. 12, 165 - 170 (1975).

Stars, their birth, life and death.
See Abstr. 003.010.

The physics of stellar interiors.
See Abstr. 003.118.

Of atoms, mountains, and stars: a study in qualitative physics. See Abstr. 022.018.

Radiation-driven winds in Of stars.
See Abstr. 064.004.

Fe I fluorescence in T Tauri stars. II. Clues to the velocity field in the circumstellar envelope.
See Abstr. 064.033.

General remarks on the variability of spotted stars.

See Abstr. 064.064.

Slowly rotating relativistic stars. IIIA. The static stability criterion recovered. See Abstr. 066.002.

On the collapse of iron stellar cores.
See Abstr. 066.029.

The black hole that ate the solar neutrinos.
See Abstr. 066.060.

Analytic stellar models in general relativity.
See Abstr. 066.072.

The weak interaction and gravitational collapse.
See Abstr. 066.084.

General relativity, collapse and singularities.
See Abstr. 066.100.

On the stability of the solar core.
See Abstr. 080.012.

The evolution of the solar inner rotation.
See Abstr. 080.022.

Observations of lithium dilution and rotational velocity decay in F and G giant stars. See Abstr. 114.012.

Stellar compositions from narrow-band photometry — V. Barium abundances for 200 evolved stars.
See Abstr. 114.018.

N-type carbon stars and the 3-α process.
See Abstr. 114.034.

The Ap star 108 Aqr. II. — The oblique rotator model. See Abstr. 114.332.

Magnetohydrodynamics: applications to magnetic stars, cosmical gas dynamics, and pulsars.
See Abstr. 116.011.

Photometric elements of the eclipsing binary V444 Cyg and the nature of the Wolf-Rayet star.
See Abstr. 121.007.

A linear, non-adiabatic pulsation analysis of models of dwarf cepheid variable stars. See Abstr. 122.001.

Applications of linear pulsation theory to the cepheid mass problem and the double-mode cepheids.
See Abstr. 122.004.

Classification of intrinsic variables. VI. Ultrashort-period, very small amplitude B-type variables.
See Abstr. 122.045.

Supernova remnants and presupernova models.
See Abstr. 125.007.

The upper mass limit for white dwarf formation as derived from the stellar content of the Hyades cluster.
See Abstr. 126.007.

Analytic surface boundary conditions for white dwarf evolutionary calculations. See Abstr. 126.012.

Evolution of helium white dwarfs in close binaries.
See Abstr. 126.015.

Outer layers of white dwarf stars.
See Abstr. 126.020.

Internal structure and stability of an interstellar cloud heated by an external flux of soft X-rays.
See Abstr. 131.035.

Interstellar molecules and the formation of stars.
See Abstr. 131.094.

Dark clouds, star formation and spiral structure.
See Abstr. 131.138.

Further observations of the Orion nebula cluster.
See Abstr. 132.022.

Envelope ejection to form planetary nebulae.
See Abstr. 133.010.

Discovery of a pulsar in a binary system.
See Abstr. 141.304.

Internal magnetic fields of pulsars, white dwarfs, and other stars.   See Abstr. 141.307.

Theory of pulsars: polar gaps, sparks, and coherent microwave radiation.   See Abstr. 141.316.

Some remarks on the hypothesis of relativistic beaming of pulsar emission.   See Abstr. 141.326.

Acceleration of pulsars to high velocities by asymmetric radiation.   See Abstr. 141.335.

Particle acceleration at pulsar magnetic poles.
See Abstr. 141.350.

Pulsars and high density physics.
See Abstr. 141.358.

PSR 1913 + 16: endpoints of speculation. A critical discussion of possible companions and progenitors.
See Abstr. 141.361.

Spin-down of pulsars.   See Abstr. 141.362.

H_2 and HD infrared lines expected from dense interstellar objects.   See Abstr. 141.613.

Infrared observations of Sharpless 2−106, a possible location for star formation.   See Abstr. 141.619.

Oblique rotators in binary systems.
See Abstr. 142.008.

Model for 1.24 s X-ray pulses in Her X-1.
See Abstr. 142.010.

Anisotropic Thomson scattering for pulse formation in X-ray pulsars.   See Abstr. 142.031.

Neutron star wobble in binary X-ray sources.
See Abstr. 142.068.

Why the number of galactic X-ray stars is so small?
See Abstr. 142.071.

Expected polarization properties of binary X-ray sources.   See Abstr. 142.080.

Collapsed stars, pulsars and the origin of cosmic rays.
See Abstr. 143.073.

On the theoretical determination of gap parameters for cluster color-magnitude diagrams and their comparison with observational data.   See Abstr. 153.015.

Properties of two blue compact galaxies.
See Abstr. 158.041.

Black dwarf stars as missing mass in clusters of galaxies.   See Abstr. 160.001.

066· Relativistic Astrophysics (without Cosmology), Background Radiation, Gravitation Theory

066.001 The theory of transparent gravitational lenses.
R. R. Bourassa, R. Kantowski.
Astrophys. Journ., Vol. 195, 13 - 21 (1975).

The theory of transparent gravitational lenses is presented. All calculations are done within the framework of the linearized Einstein gravitation theory, the geometrical optics approximation, and the small-angle scattering approximation. The authors present a general expression for the bending angle of a light ray that goes around or through a bounded mass. This expression is used to calculate the locations, intensifications, distortions, orientations, and velocities of image seen around or through such masses.

066.002 Slowly rotating relativistic stars. IIIA. The static stability criterion recovered.   J. B. Hartle.
Astrophys. Journ., Vol. 195, 203 - 212 (1975).

The static stability criterion for slowly rotating relativistic stars of Hartle and Thorne is corrected to properly compute the shape of the neutral mode of oscillation of a marginally stable rotating star to zeroth order in the rotation. The correction consists in taking into account the rotation-induced mixing between the neutral radial mode and neutral convective modes. The resulting criterion gives a simple method of analyzing the effect of a slow and rigid rotation on the stability of an isentropic stellar model.

066.003 Time delays for multiply imaged quasars.
J. H. Cooke, R. Kantowski.
Astrophys. Journ., (Letters), Vol. 195, L11 - L14 (1975).

The authors present a general expression for the difference in arrival times of light signals which originate simultaneously at a cosmologically distant source but which travel along different paths due to an intervening gravitational lens. The time delay consists of two terms: the first is due to the difference in geometrical path lengths, and the second is due to the difference in the gravitational potential through which the two light rays travel. The authors conclude that the potential term is significant and cannot be neglected.

066.004 The inertia and the heavy mass of a cosmic body in inertia-free dynamics and the self-absorption of gravity.   H.-J. Treder.
Astron. Nachr., Vol. 296, 9 - 13 (1975). In German.

In the author's 'Machian dynamics' (Treder 1972) without inertia the effective inertial masses of particles are homogeneous functions of the Newtonian potential. He shows that by an 'induction of inertia' the sum of the effective inertial masses of the particles of the star becomes greater than the effective inertial mass of the star itself. The author compares these properties with the effect of self-absorption of gravitation given by his tetrad field theory.

066.005 Response of rigid rotators to gravitational radiation.
G. Dautcourt.
Astron. Nachr., Vol. 296, 25 - 29 (1975).

The response of an axially symmetric rigid rotator to incident gravitational radiation is discussed for particular states of free rotator motion using generalized Eulerian equations, and assuming wavelengths large compared with the rotator dimensions. First, if coincident initially, the rotation and the symmetry axes slightly differ after exposed to a radiation flux which has suitable polarization and propagates perpendicular to the rotation axis. Secondly, the angular velocity of a rotation perpendicular to the symmetry axis is changed in a wave field propagating in the direction of the rotation axis

(Braginski-rotator).

066.006 Cosmological effects of primordial black holes.
G. F. Chapline.
Nature, Vol. 253, 251 - 252 (1975).

Although only black holes with masses $\gtrsim 1.5\,M_\odot$ are expected to result from stellar evolution black holes with much smaller masses may be present throughout the Universe. These small black holes are the result of density fluctuations in the very early Universe. If many small black holes ($M < M_\odot$) exist at the present time then their presence may be revealed because they radiate electromagnetic radiation. During collapse the metric will be changing rapidly on a time scale $\tau \simeq 10^{-5}$ $(M/M_\odot)$ s, so that production of massless particles with energy of order h/τ is expected. Thus masses smaller than about 10^{20} g will radiate X rays and gamma rays when they undergo gravitational collapse.

066.007 Radiation as a source of gravitation.   H. Bondi.
Nature, Vol. 253, 414 - 415 (1975). − Letter.

066.008 The relativistic Roche problem. II. Stability theory.
L. G. Fishbone.
Astrophys. Journ., Vol. 195, 499 - 505 (1975).

The author presents an investigation of the dynamical stability of infinitesimal, incompressible, homogeneous, self-gravitating fluid bodies moving in circular equatorial orbits around a Kerr black hole. For near-photon orbits, instability (tidal disruption) sets in at the Roche-limit point. For the last stable orbit, instability sets in for a certain configuration somewhat more distorted from sphericity than the Roche-limit configuration.

066.009 Cylinders as gravitational radiation telescopes.
R. G. Hier, S. N. Rasband.
Astrophys. Journ., Vol. 195, 507 - 512 (1975).

The multitude of resonance modes of a free elastic cylinder are viewed as a host of resonant detectors for gravitational radiation. The absorption cross sections for all modes using exact cylinder eigensolutions are obtained, and detailed numerical results for the axisymmetric longitudinal modes in the Pochhammer-Chree approximation are presented. The combining of information from all modes excited by a gravitational wave pulse to give spectral and directional information about the incident burst is discussed.

066.010 The Lense-Thirring effect and accretion disks around Kerr black holes.   J. M. Bardeen, J. A. Petterson.
Astrophys. Journ., (Letters), Vol. 195, L65 - L67 (1975).

Astrophysical evidence for the relativistic Lense-Thirring effect could come from its influence on tilted accretion disks around Kerr black holes. The authors show how it causes the gradual transition of the disk into the equatorial plane of the black hole in the region between the radii $10^4 M$ and $10^2 M$. They expect that a considerable part of the radiation emitted in the central part of the disk may be reabsorbed in the transition region, which may lead to observable changes in the X-ray spectrum.

066.011 Gravitational field and metric of collapsing objects.
I. D. Novikov.
New problems of astrophysics, Publ. Astrophys. Winter School, 1972, (see 012.001), p. 75 - 77 (1974). In Russian.

066.012 On gravitational radiation.   R. F. Polishchuk.

New problems of astrophysics. Publ. Astrophys.
Winter School, 1972, (see 012.001), p. 114 - 117 (1974).
In Russian.

066.013 Two kinds of stellar collapse. J. I. Katz.
Nature, Vol. 253, 698 - 699 (1975).
The author shows that the events which produce compact objects may be divided into at least two classes on the basis of the impulse given to a binary system containing the collapsing object. This impulse may be related to the amount and symmetry of mass loss resulting from the collapse event.

066.014 Schwarzschild orbital topography and high Doppler blueshifts. W. R. Stoeger.
Nature, Vol. 254, 276 (1975), with a reply by S. M. Chitre, J. V. Narlikar and R. C. Kapoor.
Chitre, Narlikar and Kapoor discuss the high Doppler blueshift of forward light emission from material particles in circular orbit in a Schwarzschild field at and near $r = 3GM/c^2$, which they claim is the radius of the unstable circular orbit. Massive particles in circular orbits of radii approaching $r = 3GM/c^2$ are not only in an unstable configuration but also highly unbound ($E \gg 1\,mc^2$). In any realistic physical situation they will all move along outward spiral orbits. There seems to be little reason to expect that there could be a thin disk of massive particles executing even approximately circular orbits in the "high blueshift region" near $r = 3GM/c^2$.

066.015 Solar gravitational deflection of radio waves. J. M. Riley.
Nature, Vol. 254, 289(1975).

066.016 Interaction of a photon with a gravitational field. D. F. Crawford.
Nature, Vol. 254, 313 - 314 (1975).
It is usual to analyse the behaviour of photons in gravitational fields as if they were point particles which travel along geodesics. But by taking into account the finite spatial extent of the photon the author shows that it could interact with the transverse gradient of the gravitational field to give a deflection additional to the well known gravitational deflection of General Relativity. He postulates that this deflection, even though it is extremely small, could be the basis for a much more important effect in which the photon can decay into three photons (or possibly several photons with one or more gravitons). If such a mechanism is possible it would appear as a decrease in photon energy and could be of astrophysical importance in explaining many of the anomalous redshifts that have been observed.

066.017 Perturbation of a slowly rotating black hole by a stationary axisymmetric ring of matter. II. Penrose processes, circular orbits, and differential mass formulae. C. M. Will.
Astrophys. Journ., Vol. 196, 41 - 49 (1975).
The author presents a detailed description of the phenomenon of energy extraction ("Penrose process") from a slowly rotating black hole perturbed by a stationary axisymmetric ring of matter, and shows that the gravitational interaction between the ring and the particles used in the Penrose process must be taken into account. A "differential mass formula" relating the total masses of neighboring black-hole-ring configurations is derived.

066.018 Sorte hull. R. Stabell.
Astron. Tidssk., Årg. 8, p. 39 - 43 (1975).

066.019 Alles beim alten? Neues aus der relativistischen Astrophysik. A. Wittmann.
SuW, Vol. 14, 125 - 126 (1975).

066.020 Optically thick accretion onto black holes. S. Tamazawa, K. Toyama, N. Kaneko, Y. Ōno.
Astrophys. Space Sci., Vol. 32, 403 - 421 (1975).
Spherically symmetric, steady-state, optically thick accretion onto a nonrotating black hole with the mass of $M = 1\,M_\odot$ is studied. The gas accreting onto the black hole is assumed to be a fully ionized hydrogen plasma with $n_0 = 10^8$ cm^{-3} and $T_0 = 10^4$ K far from the black hole, and a new approximate expression for the Eddington factor is introduced. The luminosity is estimated to be $L = 1.875 \times 10^{33}$ erg s^{-1}, which primarily arises from the optical surface ($\tau \sim 1$) of $T \sim 10^4$ K. The accretion flow is characterized by $\tau \sim 1$ and $(v/c)\tau \sim 10$. In the optically thin region, the flow remains isothermal, and the increase of temperature occurs at $\tau \sim 1$. The radiative equilibrium is strictly realized at $(v/c)\tau \sim 10$.

066.021 Mach's principle and bundle of geometric quantities. M. Heller.
Astrophys. Space Sci., Vol. 32, L29 - L31 (1975).
It is shown that the bundle of geometric quantities $\sigma(M)$ over a space-time M is a strong anti-Machian element of M.

066.022 Étude de la relation entre le décalage des fréquences vers le rouge et la magnitude apparente lorsqu'une part du décalage observé est dû à un effet de source. G. Le Denmat.
Comptes Rendus Acad. Sci. Paris, Sér. B, Vol. 280, 17 - 19 (1975).
On examine, d'un point de vue phénoménologique, une conséquence de l'hypothèse d'un décalage des raies spectrales vers le rouge qui ne serait pas entièrement dû à un effet Doppler-Einstein mais proviendrait en partie d'un effet intrinsèque à la source émettrice. Pour écrire alors une relation du type relation de Hubble entre le décalage z observé des fréquences et la magnitude apparente de la source il convient donc de distinguer dans le z observé la composante z_c (cosmologique) due à la distance relative et aux états de mouvement de la source et de l'observateur, et la composante intrinsèque z_i due à l'effet de source.

066.023 Étude d'un schéma matière pure isotrope dans le cadre de la méthode de Newman-Penrose. B. Léauté.
Comptes Rendus Acad. Sci. Paris, Sér. A, Vol. 280, 49 - 51 (1975).
L'expression de l'énergie-impulsion totale émise lors d'une transition entre deux états stationnaires successifs confirme la validité de la formule de Bondi-Sachs, donnant l'énergie-impulsion du rayonnement gravitationnel sortant.

066.024 Le mouvement du périhélie et la mécanique invariantive. I. Mihăilă.
Comptes Rendus Acad. Sci. Paris, Sér. A, Vol. 280, 595 - 598 (1975).
En utilisant les équations du mouvement absolu données par la mécanique invariantive, on déduit l'équation du mouvement relatif et finalement la formule du mouvement du périhélie. On établit l'expression de la force qui produit l'avance du périhélie.

066.025 Are very large gravitational redshifts possible? W. B. Bonnor, S. B. P. Wickramasuriya.
Monthly Notices Roy. Astron. Soc., Vol. 170, 643 - 649 (1975).
Solutions of the Einstein-Maxwell equations exist for charged dust in equilibrium under its own gravitational attraction and electrical repulsion. The authors use these to construct solutions for a sphere and a spheroid. In both cases arbitrarily large gravitational redshifts are possible, but as the

spheroid shrinks to a disc, the redshift remains finite. One of the solutions shows clearly how a horizon filters out non-spherically symmetric fields.

066.026 On the vacuum—matter symmetry of gravitation.
W. H. McCrea.
Observatory, Vol. 95, 13 - 15 (1975).

066.027 On the equilibrium figures of an ideal rotating fluid in the post-Newtonian approximation of general relativity. III. Stability of the forms of equilibrium.
K. A. Pyragas, N. P. Bondarenko, A. N. Kryshtal.
Astrophys. Space Sci., Vol. 33, 75 - 97 (1975).

A stability criterion is given for the equilibrium form of an ideal rotating fluid in the post-Newtonian approximation. This generalizes the known Lyapunov criterion in classical dynamics. The sphere stability is also investigated and it is shown that it is stable only when $R > 22.2 R_g$ (R is the relativistic sphere radius, R_g the Schwarzschild radius).

066.028 On recording pulses of gravitational bremsstrahlung from superdense star clusters.
V. I. Panov, V. N. Rudenko.
Dokl. Akad. Nauk SSSR. Ser. Mat. Fiz., Vol. 221, 573 - 576 (1975). In Russian.

066.029 On the collapse of iron stellar cores.
Z. Barkat, G. Rakavy, Y. Reiss, J. R. Wilson.
Astrophys. Journ., Vol. 196, 633 - 638 (1975).

The collapse of iron stellar cores is investigated to see whether the outward shock produced by the bounce at neutron star density is sufficient to burn appreciable amounts of the envelope around the iron core. Several models were tried, and in all cases no appreciable burn took place; hence no explosion results from the collapse of these models.

066.030 Slowly rotating relativistic stars. VIII. Frequencies of the quasi-radial modes of an $n = 3/2$ polytrope.
J. B. Hartle, J. L. Friedman.
Astrophys. Journ., Vol. 196, 653 - 660 (1975).

The expressions derived by Hartle, Thorne, and Chitre and by Chandrasekhar and Friedman for the change in frequency of a radial mode of a spherical star induced by a slow and rigid rotation are applied to the lowest modes of stars constructed from a $n = 3/2$ relativistic polytropic equation of state corresponding to an ideal non-relativistic free neutron gas. The two expressions give identical results.

066.031 Observable effects of a scalar gravitational field in a binary pulsar. D. M. Eardley.
Astrophys. Journ., (Letters), Vol. 196, L59 - L62 (1975).

It is remarked that the Dicke-Brans-Jordan theory of gravitation predicts dipole gravitational radiation from certain binary systems that contain a neutron star or black hole, causing decay of the orbit. Further, it predicts that the true rate of a pulsar clock should vary with distance from a binary companion (other than a black hole). At least one of these effects would be observable in PSR 1913+16.

066.032 Test for the existence of gravitational radiation.
R. V. Wagoner.
Astrophys. Journ., (Letters), Vol. 196, L63 - L65 (1975).

Predictions are presented for the gravitational-radiation-induced decrease of the orbital period and eccentricity of the recently discovered radio pulsar in a binary system, in terms of the mass ratio and angle of inclination of the orbit. It is shown how measurements of the rotation of the periastron and the term of order $(v/c)^2$ in the pulsar period variation would allow these two unknowns to be determined.

066.033 Energy loss of relativistic electrons and positrons traversing cosmic matter. R. J. Gould.
Astrophys. Journ., Vol. 196, 689 - 694 (1975).

Energy loss by Compton scattering, bremsstrahlung, electronic excitations, synchrotron radiation, and pair production is evaluated for relativistic electrons and positrons traversing a neutral or ionized gas of atomic hydrogen and helium with a blackbody radiation field and a magnetic field. Assuming that the observed cosmic X-ray and extragalactic radio background are due to Compton scattering of $2.7°$K blackbody photons and synchrotron radiation, respectively, an rms value $\langle B_o^2 \rangle^{1/2} = 0.9 \times 10^{-7}$ gauss is derived for the intergalactic magnetic field at the present epoch. This value is independent of the cosmological model assumed.

066.034 High efficiency of the Penrose mechanism for particle collisions. T. Piran, J. Shaham, J. Katz.
Astrophys. Journ., (Letters), Vol. 196, L107 - L108 (1975).

It is shown that, by two-particle collisions, rather than particle disintegration, one substantially increases the efficiency of the Penrose mechanism. The process may be relevant when considering both means of extracting energy from black holes and means of producing energetic particles. The process can be operating in close binary systems, such as some compact X-ray sources, of which one of the members is a black hole.

066.035 Electromagnetic radiation from colliding black holes.
F. J. Tipler.
Astrophys. Journ., Vol. 197, 199 - 202 (1975).

It is shown that the collision of two black holes would result in the emission of electromagnetic radiation with a very distinctive wave form. If the gravitational radiation events reported by Weber are produced by black-hole collisions in the galactic center, then the associated electromagnetic pulses would have, in the microwave band, a maximum flux between 4×10^{-6} and 6×10^{-5} Jy. This flux lies at the limit of detectability with present-day radio astronomy technology.

066.036 Uniqueness of the Kerr black hole.
D. C. Robinson.
Phys. Rev. Letters, Vol. 34, 905 - 906 (1975).

The family of Kerr solutions, with $|a| < m$, is shown to be the unique pseudostationary family of black-hole solutions of the Einstein vacuum field equations when the event horizon is assumed to be nondegenerate.

066.037 Die Lichtgeschwindigkeit im Gravitationsfeld der Sonne nach der Relativitätstheorie.
H.-J. Treder.
Sterne, Vol. 51, 17 - 24 (1975).

066.038 Gravitationstheorie – heute. U. Kasper.
Sterne, Vol. 51, 25 - 28 (1975).

066.039 Central gravitational redshifts from static massive objects. P. K. Das, J. V. Narlikar.
Monthly Notices Roy. Astron. Soc., Vol. 171, 87 - 102 (1975).

The possibility of obtaining high gravitational redshifts from static massive objects is investigated in an attempt towards finding a satisfactory explanation of the redshifts of the QSOs. Following Bondi's approach equilibrium configurations of the core and envelope type are discussed within the framework of general relativity. Gravitational redshifts from the centre of the core, from the core-envelope boundary and from the surface of the object are calculated in a variety of cases and their dependence on the equation of state is studied.

066.040 Search for continuous gravitational radiation.
J. Hough, J. R. Pugh, R. Bland, R. W. P. Drever.
Nature, Vol. 254, 498 - 501 (1975).

In an earlier paper the authors have given results of a search for short pulses of gravitational radiation carried out with two gravitational radiation detectors operating in coincidence over a frequency range from 650 to 1,450 Hz. Detectors of that type may also be used within this frequency region to set upper limits to fluxes of gravitational radiation of either continuous or incoherent waveform. The latter may include pulses which occur at a high rate but are too small to be detected individually. The authors describe here a preliminary experiment designed to search for any such signals in a general way.

066.041 **On the ability of current experiments to test π^0-decay gamma-ray background theories.**
T. Montmerle.
Astrophys. Journ., Vol. 197, 285 - 289 (1975).

In some theories for the $1-100$ MeV γ-ray background some definite features are predicted (e.g., matter-antimatter regions annihilating in the case of the Stecker-Puget theory) as a function of redshift. However, the theoretical possibility of observing these features at any redshift up to $z_c \simeq 100$ is, in the case of π^0-decay theories, rather restricted with present experiments, and this is shown by the introduction of a "visibility function" which folds the theoretical γ-ray background intensity as a function of redshift through the efficiency of a given experiment. The study of the spatial structure of the γ-ray background at high redshifts ($\lesssim z_c$) is potentially best accomplished by low-energy γ-ray experiments.

066.042 **The physics and astronomy of black holes.**
D. W. Sciama.
Quarterly Journ. Roy. Astron. Soc., Vol. 16, 1 - 12 (1975).

066.043 **A satellite under the action of gravitational radiation.**
V. N. Rudenko.
Astron. Zhurn. Akad. Nauk SSSR, Vol. 52, 444 - 446 (1975).
In Russian. English translation in Soviet Astron., Vol. 19, No. 2.

The satellite orbit variations under the action of resonant gravitational radiation are calculated. The utilizing of this effect for gravitational wave registering is discussed. The perturbation of an artificial earth satellite by gravitational radiation of known double stars is too small for registering. Estimates for the detectable energy density of the gravitational radiation are obtained.

066.044 **On tidal phenomena in a strong gravitational field.**
B. Mashhoon.
Astrophys. Journ., Vol. 197, 705 - 716 (1975).

A simple framework based on the concept of quadrupole tidal potential is presented for the calculation of tidal deformation of an extended test body in a gravitational field. This method is used to study the behavior of an initially faraway nonrotating spherical body that moves close to a Schwarzschild or an extreme Kerr black hole. In general, an extended body moving in an external gravitational field emits gravitational radiation due to its center of mass motion, internal tidal deformation, and the coupling between the internal and center of mass motions. Estimates are given of the amount of tidal radiation emitted by the body in the gravitational fields considered. The results reported in this paper are expected to be of importance in the dynamical evolution of a dense stellar system with a massive black hole in its center.

066.045 **Post-Newtonian generation of gravitational waves.**
R. Epstein, R. V. Wagoner.
Astrophys. Journ., Vol. 197, 717 - 723 (1975).

An analysis is presented of gravitational radiation from astrophysical bodies, in terms of an expansion in powers of (velocity) $\sim$ (Newtonian potential)$^{1/2}$ within general relativity. Explicit expressions are obtained for the distant radiation field and energy flux from perfect-fluid sources, valid through $^3/_2$ post-Newtonian order (beyond the usual formulae).

066.046 **The N-body problem in general relativity.**
N. Spyrou.
Astrophys. Journ., Vol. 197, 725 - 743 (1975).

The form of the metric tensor in the first and $2^1/_2$ post-Newtonian approximations (PNA) is derived, and the method for its evaluation in the second PNA is presented for a system of N bodies, which are spherical, homogeneous, and rotate uniformly around axes through their centers. The energy, linear-momentum, and angular-momentum integrals up to the first PNA are also derived for the above system. In a particular numerical application the author gives the first PNA correction in the energy integral. This increases with increasing mass, decreasing radius, and decreasing distance of the bodies, and becomes comparable to the Newtonian value of the energy for small distances and for radii approaching the Schwarzschild radius.

066.047 **The stability of nonuniform rotation in relativistic stars.** F. H. Seguin.
Astrophys. Journ., Vol. 197, 745 - 765 (1975).

The stability of axisymmetric, differential rotation in nonmagnetic stars of uniform chemical composition is studied in the context of general relativity theory. Criteria are found for stability against local, linear, axisymmetric perturbations in conducting, viscous stars and in perfect fluid models. Applications of the stability conditions to models of specific astrophysical objects are discussed.

066.048 **On the origin of the microwave background.**
F. Hoyle.
Astrophys. Journ., Vol. 196, 661 - 670 (1975).

The background of microwave radiation is known to be remarkably uniform over the sky, although the regions giving rise to the radiation in widely separated elements of solid angle have, according to the usual cosmological theories, always been out of communication with each other. Using a new approach to the big-bang cosmologies, an explanation of this uniformity is given. The intensity of the background appears to be related to the energy of conversion of hydrogen to helium within galaxies. Yet this circumstance is regarded as coincidental in the usual theories. Here it receives explanation.

066.049 **Roterende zwarte gaten, (6), (7).**
P. Hut, J. Stollman.
Zenit, Vol. 2, 6 - 8, 63 - 66 (1975).

066.050 **On the theory of non-stationary relativistic shock waves in an inhomogeneous cosmic medium.**
A. F. Novak.
Astrometriya i Astrofizika, *Kiev,* vyp. (No.) 24, (see 003.003), p. 53 - 55 (1974). In Russian.

Whitham's method, known in gas dynamics, is generalized for the case of a non-stationary spherical relativistic shock wave. A differential equation of the shock strength is derived. The formulae of the relativistic shock parameters are discussed.

066.051 **Gravitational waves retaining the homogeneity of space.** V. N. Lukash.
Zhurn. ehksperim. i teor. fiz., Vol. 67, 1594 - 1608 (1974). In Russian. – Abstr. in Referativ. Zhurn. 51. Astron., 3.51.865 (1975).

066.052 **On gravitational waves in the field of a collapsing star.** N. R. Sibgatullin, G. A. Alekseev.
Probl. teorii gravitatsii i ehlementarn. chastits. Vyp. (No.) 5. Moskva, Atomizdat, 1974, p. 142 - 153. In Russian. – Abstr. in Referativ. Zhurn. 51. Astron., 3.51.890 (1975).

066.053 **Wave fields outside a collapsing charged star.**
N. R. Sibgatullin, G. A. Alekseev.
Zhurn. ehksperim. i teor. fiz., Vol. 67, 1233 - 1249 (1974).

In Russian. – Abstr. in Referativ. Zhurn. 51. Astron., 3.51. 891 (1975).

066.054 **The motion of spinless bodies in a given space with torsion.**   V. N. Ponomarev.
Vestn. Mosk. un-ta. Fiz. astron., Vol. 15, 541 - 547 (1974). In Russian. – Abstr. in Referativ. Zhurn. 51. Astron., 3.51. 896 (1975).

066.055 **Diffraction of light and lens effect of the stellar gravitation field.**
P. V. Bliokh, A. A. Minakov.
Astrophys. Space Sci., Vol. 34, L7 - L9 (1975).
The purpose of the present communication is to consider the focussing of light by gravitational 'lenses' (stars). Finite estimates of the gain q on the optical axis ($r \rightarrow 0$) can be obtained with an account of diffraction of light.

066.056 **Gravitational lenses and "superlight" velocities.**   L. Kh. Ingel'.
Pis'ma v Astron. Zhurn., Vol. 1, No. 3, p. 18 - 20 (1975). In Russian.
Two images of the same object given by a gravitational lense, when this lense moves, can have a relative velocity above that of light.

066.057 **Charged black holes in the interstellar medium.**   K. G. Suffern.
Astron. Astrophys., Vol. 39, 275 - 279 (1975).
It is shown that an ionized interstellar medium at kinetic temperature T K containing an imbedded magnetic field of strength H Gauss will accrete onto a black hole via the independent particle approximation provided the mass of the hole satisfies the inequality $M/M_\odot \ll 3.4 \times 10^{-8}(T/10^4 \text{ K})^{3/4}(H/10^{-6} \text{ Gauss})^{-1}$.

066.058 **Turbulent heating of the primeval plasma and distortions of the cosmic microwave background radiation spectrum.**   K. L. Chan, B. J. T. Jones.
Bull. American Astron. Soc., Vol. 7, 236 - 237 (1975). Abstr. AAS.

066.059 **Flaring behavior in accretion disks around black holes.**   F. K. Lamb.
Bull. American Astron. Soc., Vol. 7, 241 (1975). – Abstr. AAS.

066.060 **The black hole that ate the solar neutrinos.**   D. D. Clayton, M. J. Newman, R. J. Talbot, Jr.
Bull. American Astron. Soc., Vol. 7, 242 (1975). – Abstr. AAS.

066.061 **290 år gravitation.**   P. E. Kustaanheimo.
Astron. Tidsskr., Årg. 8, p. 56 - 69 (1975).

066.062 **Zwarte gaten (8), (9).**   P. Hut, J. Stollman.
Zenit, Vol. 2, 132 - 135, 181 - 182 (1975).

066.063 **On the problem of determination of gravitationally-inertial radiation.**   A. K. Guts.
Redkollegiya zhurn. "Izv. vyssh. ucheb. zavedenij SSSR. Ser. Fizika". Tomsk, 1974. 11 pp. In Russian. – Abstr. in Referativ. Zhurn. 51. Astron., 4.51.862 (1975).

066.064 **Static scalar and electromagnetic fields in the theory of gravitation.**   K. A. Bronnikov, V. N. Mel'nikov.
Probl. teorii gravitatsii i ehlementarn. chastits. Vyp. (No.) 5. Moskva, Atomizdat, 1974, p. 80 - 91. In Russian. – Abstr. in Referativ. Zhurn. 51. Astron., 4.51.872 (1975).

066.065 **Spectral problem for the Einstein and Klein-Gordon system of equations.**   A. A. Solodov.
Probl. teorii gravitatsii i ehlementarn. chastits. Vyp. (No.) 5.

Moskva, Atomizdat, 1974, p. 92 - 97. In Russian. – Abstr. in Referativ. Zhurn. 51. Astron., 4.51.873 (1975).

066.066 **Some solutions of the equations describing the scalar and electromagnetic field interaction in an exterior gravitational field.**   G. N. Shikin.
Probl. teorii gravitatsii i ehlementarn. chastits. Vyp. (No.) 5. Moskva, Atomizdat, 1974, p. 98 - 101. In Russian. – Abstr. in Referativ. Zhurn. 51. Astron., 4.51.874 (1975).

066.067 **Some non-linear fields in the theory of gravitation.**   G. N. Shikin.
Probl. teorii gravitatsii i ehlementarn. chastits. Vyp. (No.) 5. Moskva, Atomizdat, 1974, p. 101 - 106. In Russian. – Abstr. in Referativ. Zhurn. 51. Astron., 4.51.875 (1975).

066.068 **On quantization of the interval in the Riemann space, gravitational and weak interaction and background radiation.**   K. P. Stanyukovich.
Probl. teorii gravitatsii i ehlementarn. chastits. Vyp. (No.) 5. Moskva, Atomizdat, 1974, p. 106 - 120. In Russian. – Abstr. in Referativ. Zhurn. 51. Astron., 4.51.876 (1975).

066.069 **Geometrical quantization of Schwarzschild's metric.**   V. A. Pilipenko.
Probl. teorii gravitatsii i ehlementarn. chastits. Vyp. (No.) 5. Moskva, Atomizdat, 1974, p. 120 - 122. In Russian. – Abstr. in Referativ. Zhurn. 51. Astron., 4.51.877 (1975).

066.070 **On the "other" physical spaces.**   M. E. Gertsenshtejn, K. P. Stanyukovich.
Probl. teorii gravitatsii i ehlementarn. chastits. Vyp. (No.) 5. Moskva, Atomizdat, 1974, p. 162 - 167. In Russian. – Abstr. in Referativ. Zhurn. 51. Astron., 4.51.878 (1975).

066.071 **Distortions of the microwave background radiation spectrum in the submillimeter wavelength region.**   K. L. Chan, B. J. T. Jones.
Astrophys. Journ., Vol. 198, 245 - 248 (1975).
The authors' numerical calculation shows that the linear approximation of the Compton-distorted radiation spectrum fails seriously in the submillimeter range.

066.072 **Analytic stellar models in general relativity.**   R. C. Adams, J. M. Cohen.
Astrophys. Journ., Vol. 198, 507 - 512 (1975).
In this paper the authors describe a method for deriving analytic stellar models in general relativity. The models are exact solutions to Einstein's equations. By inverting the usual procedure in astrophysics, they reduce Einstein's static field equations to one first-order linear differential equation. Such an equation can always be reduced to quadratures. Two new exact solutions are presented and the rotational properties explored.

066.073 **Some problems of relativistic astrophysics.**   P. R. Amnuehl', O. Kh. Gusejnov.
Izv. vyssh. ucheb. zavedenij. Fizika, 1974, No. 12, p. 69 - 79. In Russian. – Abstr. in Referativ. Zhurn. 51. Astron., 5.51. 772 (1975).

066.074 **The problems of gravitational waves and asymptotic solutions of Einstein's equations.**
N. R. Sibgatullin.
Otchet. o nauch.-issled. rabotakh. vypolnen. v 1973. In-t mekh. Mosk. un-ta. Moskva, Mosk. un-t, 1974, p. 8. In Russian. – Abstr. in Referativ. Zhurn. 51. Astron., 5.51.802 (1975).

066.075 **The increasing role of general relativity in astronomy.**   S. Chandrasekhar.

Ehjnshtejnovsk. sb. 1973. Moskva, Nauka, 1974, p. 207 - 228. In Russian. – Abstr. in Referativ. Zhurn. 51. Astron., 5.51. 803 (1975).

066.076 On the stability of finite motions along a geodetic line in Schwarzschild's metric.
M. F. Shirokov, B. V. Bondarev.
Izv. vyssh. ucheb. zavedenij. Fizika, 1974, No. 12, p. 52 - 55. In Russian. – Abstr. in Referativ. Zhurn. 51. Astron., 5.51.810 (1975).

066.077 Geodetic deviations in Kerr's space.
I. M. Dozmorov, G. V. Lutsenko.
Izv. vyssh. ucheb. zavedenij. Fizika, 1974, No. 12, p. 142 - 145. In Russian. – Abstr. in Referativ. Zhurn. 51. Astron., 5.51.811 (1975).

066.078 On distribution functions which are invariant to transformations from one inertial frame of reference to another. A. E. Kaplan.
Ehjnshtejnovsk. sb. 1973. Moskva, Nauka, 1974, p. 396 - 400. In Russian. – Abstr. in Referativ. Zhurn. 51. Astron., 5.51. 813 (1975).

066.079 On L. Brillouin's book: " Relativity reexamined".
I. I. Gol'denblat.
Ehjnshtejnovsk. sb. 1973. Moskva, Nauka, 1974, p. 401 - 418. In Russian. – Abstr. in Referativ. Zhurn. 51. Astron., 5.51. 820 (1975).

066.080 On the dynamic equivalence of the post-Newtonian approximation of the theory of planetary motions in the general theory of relativity and in the intertia-free mechanics with Riemann's gravitation potential. H.-J. Treder.
Astron. Nachr., Vol. 296, 101 - 107 (1975). In German.

Proof of the dynamical equivalence of the post-Newtonian approximations for the Lagrangian representations of planetary motions according to GRT and according to "Riemannian dynamics with Mach-Einstein doctrine". Finally, the author discusses the equivalence problem for general post-Newtonian Lagrangians of the one-body problem in celestial mechanics.

066.081 The free-fall acceleration for a Schwarzschild metric.
H.-J. Treder, K. Fritze.
Astron. Nachr., Vol. 296, 109 - 110 (1975). In German.

The authors derive the acceleration of free fall for a Schwarzschild metric and consider its dependence on the radial coordinate.

066.082 Mach's principle, gravitation and electromagnetism: the electromagnetic inertia of charged elementary particles. J. F. Woodward, W. Yourgrau.
Astron. Nachr., Vol. 296, 111 - 118 (1975).

The question of the magnitude and behaviour of the electromagnetic contribution to the mass of an elementary charged particle is explored. It is shown that, if the "cut-off" of relativistic quantum electrodynamics is taken as an approximation to the physical process embodied in the postulates: I. That the bare charge of an elementary particle may only emit and absorb virtual photons having energies that are integral multiples of some fundamental energy, and II. that the observed charge of an elementary particle is a constant independent of its "radius", the electromagnetic mass of such a particle is always finite, even when the radius of the particle becomes zero. Alternative postulates that will resolve the cut-off dilemma are briefly discussed. It is noted that the "Machieness" of electromagnetic mass (in terms of Pauli's criterion for "Machieness") depends on the detailed structure of postulate II, and that extant experimental results do not permit one, at present, to decide the issue with certainty.

066.083 Frozen scintillation caused by low-frequency gravitational radiation. G. Dautcourt.
Astron. Nachr., Vol. 296, 119 - 123 (1975).

Intense low-frequency cosmic gravitational radiation in the Megaparsec wave band causes among other effects a small angular displacement of distant light sources. A short treatment of the effect is given.

066.084 The weak interaction and gravitational collapse.
D. N. Schramm, W. D. Arnett.
Astrophys. Journ., Vol. 198, 629 - 639 (1975).

Recent developments in weak interaction physics as well as a more careful examination of previously known phenomena should yield a much clearer picture of the gravitational collapse of an evolved stellar core than has previously been possible. It is shown that the dominant neutrino-emitting process throughout much of the collapse is ordinary electron capture. The effects of neutral currents and the coherent scattering of neutrinos is discussed. A modified neutrino transport supernova model, whereby momentum transfer from neutrino scattering blows off the outer part of a star while leaving a dense remnant, may be quite reasonable.

066.085 Relativity: experiments increase confidence in Einstein. A. L. Robinson.
Science, Vol. 188, 1099 - 1101 (1975).

066.086 A dust model for the cosmic microwave background. N. C. Wickramasinghe, M. G. Edmunds, S. M. Chitre, J. V. Narlikar, S. Ramadurai.
Astrophys. Space Sci., Vol. 35, L9 - L13 (1975).

The cosmic microwave background may be explained on the basis of absorption and reemission of the light from galaxies by graphite whiskers of lengths $l \simeq 0.1-1$ mm. The mass density of such particles required is of the order of 10^{-34} g cm^{-3}.

066.087 On a certain group-kinematical method of generalization of Lorentz transformations in two-dimensional space-time. H. Petryszyn.
Acta Cosmologica, Fasc. 2, p. 97 - 111 (1974/75).

066.088 The scalar-tensor theory of gravitation and the seismic response of the earth on gravitational radiation. L. Eh. Gurevich, S. D. Dynkin.
Astron. Zhurn. Akad. Nauk SSSR, Vol. 52, 521 - 526 (1975). In Russian. English translation in Soviet Astron., Vol. 19, No. 3.

According to the scalar-tensor theory the gravitation radiated by celestial bodies generates "secondary" gravitational waves. The seismic response of the earth on these secondary waves is studied. Several cases are shown in which the secondary radiation and the response on it are comparable or larger than corresponding values for usual (primary) gravitational radiation in the general theory of relativity. It is found that angular dependence and polarization of the seismic response of the earth on the secondary gravitational radiation differ considerably from similar properties in the case of the primary gravitational radiation.

066.089 On the possible form of impulses of gravitational radiation. S. P. Vyatchanin, G. S. Plotnikov.
Astron. Zhurn. Akad. Nauk SSSR, Vol. 52, 527 - 529 (1975). In Russian. English translation in Soviet Astron., Vol. 19, No. 3.

The impulse form of gravitational radiation of a superdense star cluster is considered. The expression for the force acting on the test masses in laboratory is obtained. Numerical estimates are given.

066.090 Gravitational radiation from stellar collapse into a disk. I. D. Novikov.
Astron. Zhurn. Akad. Nauk SSSR, Vol. 52, 657 - 659 (1975).

In Russian. English translation in Soviet Astron., Vol. 19, No. 3.

It is shown that at the collapse of a star into a disk (Thuan-Ostriker model) the main part of energy is radiated during the radiation burst with abrupt stop of the matter in the moment of flattening. The corresponding formulae are given.

066.091 Measurement of the gravitational shift of the frequency in the radio region with the help of an artificial earth satellite. B. M. Chikhachev, N. E. Ivanov, G. M. Fedorenko, A. A. Korchak, I. P. Stakhanov, V. V. Belyj, I. G. Boldovskaya, E. A. Kornitskaya.
Kosmich.Issled., Vol. 13, 381 - 388 (1975). In Russian.

066.092 Black holes. R. Schreiber.
Urania Kraków, Vol. 46, 163 - 167 (1975). In Polish.

066.093 Measurement of the spectrum of the submillimeter cosmic background.
D. P. Woody, J. C. Mather, N. S. Nishioka, P. L. Richards.
Phys. Rev. Letters, Vol. 34, 1036 - 1039 (1975).

The spectrum of the night sky has been measured in the frequency range from 3 to 40 cm^{-1} using a fully calibrated liquid-helium-cooled balloon-borne spectrophotometer at an elevation of 39 km. A model based on the known molecular parameters was used to subtract the atmospheric emission. In the frequency range from 4 to 17 cm^{-1}, the spectrum of the background radiation is that of a blackbody with a temperature of 2.99$^{+0.07}_{-0.14}$ K.

066.094 Unphysical solutions of Yang's gravitational-field equations. R. Pavelle.
Phys. Rev. Letters, Vol. 34, 1114, with a correction, p. 1484 (1975).

A static, spherically symmetric solution of Yang's vacuum gravitational field equations is given which predicts incorrect values for experimental observations. It is argued that Yang's field equations must be supplemented by further restrictions on the class of allowable space-times.

066.095 Superdense matter: neutrons or asymptotically free quarks? J. C. Collins, M. J. Perry.
Phys. Rev. Letters, Vol. 34, 1353 - 1356 (1975).

The authors note the following: The quark model implies that superdense matter (found in neutron-star cores, exploding black holes, and the early big-bang universe) consists of quarks rather than of hadrons. Bjorken scaling implies that the quarks interact weakly. An asymptotically free gauge theory allows realistic calculations taking full account of strong interactions.

066.096 The universal background radiation.
A. S. Webster.
Observatory, Vol. 95, 79 - 80 (1975).

066.097 Observations of the cosmic microwave background.
P. E. Clegg.
Observatory, Vol. 95, 81 - 82 (1975).

066.098 The universal background radiation. Theoretical considerations. M. Rowan-Robinson.
Observatory, Vol. 95, 82 (1975).

066.099 Zwarte gaten als testobjecten voor nieuwe fysische theorieën. (10). P. Hut, J. Stollman.
Zenit, Vol. 2, 213 - 216 (1975).

066.100 General relativity, collapse and singularities.
J. A. Wheeler, with notes by H. G. Hughes, N. Snyderman, J. K. Lawrence, R. K. P. Zia.

High energy astrophysics and its relation to elementary particle physics, (see 012.018), p. 519 - 571 (1974).

066.101 Relativity. R. K. Sachs.
The heritage of Copernicus, (see 003.020), p. 297 - 310 (1974).

066.102 When time slows down. D. Goldsmith.
Mercury, (Journ. Astron. Soc. Pacific), Vol. 4, No. 3, p. 2 - 8 (1975).

066.103 On the cosmological term, compensation and singularities. D. D. Ivanenko.
Izv. vyssh. ucheb. zavedenij. Fizika, 1974, No. 12, p. 35 - 42. In Russian. − Abstr. in Referativ. Zhurn. 51. Astron., 6.51. 900 (1975).

066.104 Post-Newtonian approximation, superdense stars and cosmology in Treder's theory of gravitation.
V. Kasper.
AN USSR. In-t teor. fiz. Preprint ITF-74-140R. Kiev, 1974. 15 pp. Price 4 Kop. In Russian. − Abstr. in Referativ. Zhurn. 51. Astron., 6.51.915 (1975).

066.105 Electromagnetic radiation of a particle in Kerr-Schwarzschild fields.
I. M. Ternov, Y. R. Khalilov, G. A. Chizhov.
Izv. vyssh. ucheb. zavedenij. Fizika, 1974, No. 12, p. 42 - 52. In Russian. − Abstr. in Referativ. Zhurn. 51. Astron., 6.51. 953 (1975).

066.106 Gravitational-inertial waves in a vacuum. Linear approximation.
L. B. Borisova, V. D. Zakharov.
Izv. vyssh. ucheb. zavedenij. Fizika, 1974, No. 12, p. 106 - 113. In Russian. − Abstr. in Referativ. Zhurn. 51. Astron., 6.51.960 (1975).

066.107 On a physical interpretation of the parameters in the generalized Kerr−NUT solution. I.
V. P. Semenov.
Izv. vyssh. ucheb. zavedenij. Fizika, 1974, No. 12, p. 146 - 151. In Russian. − Abstr. in Referativ. Zhurn. 51. Astron., 6.51.979 (1975).

066.108 Relativistic dynamics of a system of gravitating particles. V. S. Brezhnev.
Izv. vyssh. ucheb. zavedenij. Fizika, 1974, No. 12, p. 151 - 152. In Russian. − Abstr. in Referativ. Zhurn. 51. Astron., 6.51.988 (1975).

066.109 Stochastic particle acceleration in gravitational radiation fields of low frequency. G. Dautcourt.
Publ. Astrophys. Obs. Potsdam, No. 108, Vol. 32, Heft 1, p. 5 - 14 (1974).

The motion of particles in gravitational radiation fields with random distribution of amplitudes is studied.

066.110 Structure and appearance of accreting discs around black holes. C. T. Cunningham.
Thesis, Washington Univ., Seattle (USA). 123 pp. University Microfilms Order No. 74-15,510 (1973).

066.111 Perturbations of a rotating black hole.
S. A. Teukolsky.
Thesis, California Inst. Techn., Pasadena (USA). 143 pp. University Microfilms Order No. 74-14,289 (1974).

066.112 Polarization of the blackbody radiation at 3.2 cm.
G. P. Nanos, Jr.
Thesis, Princeton Univ., Princeton, N. J. (USA). 142 pp. Uni-

versity Microfilms Order No. 74-17,481 (1974).

066.113 Note on the motion of black holes.
M. Demianski, L. P. Grishchuk.
General Relativity Gravitation, (*GB*), Vol. 5, 673 - 679 (1974).

066.114 Analysis of the optical data on the deflection of light in the vicinity of the solar limb. P. Merat.
General Relativity Gravitation, (*GB*), Vol. 5, 757 - 764 (1974).

066.115 Spherically symmetric collapse and the naked singularity. A. Banerjee.
Journ. Phys. A, (Math., Nuclear, General), Vol. 8, 281 - 282 (1975). – Brief comment.

066.116 Classification of black holes with electromagnetic fields. D. C. Robinson.
Phys. Rev. D, Particles and Fields, Vol. 10, 458 - 460 (1974).
It is proved that the class of solutions (pseudostationary) of the Einstein–Maxwell equations which determines the exteriors of isolated electromagnetic black holes form discrete families, each depending upon at least one but no more than four parameters.

066.117 Some exact models of inhomogeneous dust collapse.
E. P. T. Liang.
Phys. Rev. D, Particles and Fields, Vol. 10, 447 - 457 (1974).
Some highly symmetric models of irrotational dust collapse are discussed. In the spherical model a black hole is formed. In the plane symmetric case there is either no singularity or the interior solution cannot be joined smoothly onto an exterior static vacuum. In the cylindrical case it is found to be possible to construct a model which, starting with regular initial conditions, will collapse into the naked singularity of an external static field.

066.118 Stability of Reissner-Nordstrom black holes.
V. Moncrief.
Phys. Rev. D, Particles and Fields, Vol. 10, 1057 - 1059 (1974).
Studies the stability of the Reissner-Nordstrom family of black holes against even-parity perturbations, and shows that there are no unstable normal-mode solutions to the wave equations for the gauge-invariant functions describing gravitational and electromagnetic perturbations.

066.119 Electromagnetic scattering from a black hole and the glory effect. B. Mashhoon.
Phys. Rev. D, Particles and Fields, Vol. 10, 1059 - 1063 (1974).
Shows that in the scattering of electromagnetic radiation by a black hole the backward glory effect is absent in the Schwarzschild field and for plane electromagnetic waves incident on a Kerr black hole along its axis of radiation. Estimates the polarisation of the cosmic background radiation due to the cosmological distribution of Kerr black holes.

066.120 Numerical calculation of bound geodesics in the Kerr metric. H. Goldstein.
Zeitschr. Phys., Vol. 271, 275 - 279 (1974).
Timelike geodesics, especially bound orbits in the equatorial plane and spherical orbits, are calculated numerically. The orbits are plotted using the Kerr-Schild coordinate system. The periastron advance and the dragging of nodes have the same values in any coordinate system and can be directly measured by an observer at infinity.

066.121 On the non-linear behavior of nonspherical perturbations in relativistic gravitational collapse.
K. Tomita.
Progr. Theor. Phys. Japan, Vol. 52, 1188 - 1204 (1974).
As a preparation for clarifying the final state of relativistic gravitational collapse, the author examines how nonspheri-

cal perturbations behave in the neighbourhood of the Schwarzschild surface by analyzing the second-order non-linear perturbations as well as the first-order linear ones. It is concluded that, however small the initial nonspherical perturbations are, full non-linearity must be considered, in order to analyze dynamical processes in the neighbourhood of the final state.

066.122 Stationary, axially symmetric perturbations of charged Kerr black holes. S. K. Bose, M. Y. Wang.
Phys. Rev. D, Particles and Fields, Vol. 10, 1675 - 1677 (1974).
Generalizes an earlier result by Carter (1971) to include an electrically charged Kerr black hole, with exterior geometry described by the Newman metric. It is shown that the only permitted axially symmetric, time-independent perturbations around such a black hole are those which correspond to augmenting its mass, charge and angular momentum.

066.123 Black hole in a uniform magnetic field.
R. M. Wald.
Phys. Rev. D, Particles and Fields, Vol. 10, 1680 - 1685 (1974).
Considers a (non-Kerr) axisymmetric black hole, with angular momentum, located in a uniform magnetic field parallel to the hole's symmetry axis.

066.124 Geodesic synchrotron radiation in the Kerr geometry by the method of asymptotically factorized Green's functions. P. L. Chrzanowski, C. W. Misner.
Phys. Rev. D, Particles and Fields, Vol. 10, 1701 - 1721 (1974).
Derives master formulae for all the radiation spectra (gravitational, scalar, and electromagnetic) of a highly relativistic test particle in a circular orbit about a rotating, Kerr black hole. The relevance of the results to the experimental detection of gravity waves is considered.

066.125 The efficiency of the Penrose process.
A. Kovetz, T. Piran.
Nuovo Cimento Lettere, Ser. 2, Vol. 12, 39 - 42 (1975).
The Penrose process (1969) for extracting energy from a Kerr black hole is discussed. The upper limit to the energy that can be gained in a single Penrose process is calculated.

066.126 Zero value of the Schwarzschildian mass of asymptotically Euclidian time-symmetrical gravitational waves. V. N. Folomeshkin.
International Journ. Theor. Phys., (*GB*), Vol. 10, 145 - 151 (1974).
It is shown that a coordinate system with simple coordinate conditions can be chosen such that one can explicitly see that the Schwarzschildian mass of an asymptotically Euclidian time-symmetrical system of gravitational waves is equal to zero.

066.127 Cosmological singularities in a theory of gravitation with second order perturbations.
M. Giesswein, E. Streeruwitz.
Acta Phys. Austriaca, Vol. 41, 41 - 64 (1975). In German.
General relativity is modified by adding quadratic terms in the Riemann tensor to the Lagrangian. These terms are likely to be due to effects of quantization of the matter fields (source fields). The resulting field equations are specialized to homogeneous isotropic cosmologies. Quantum modifications of general relativity of the form considered here are unable to prevent the occurrence of singularities of Friedmann type universes.

066.128 On the gravitational stability of spherical systems.
S. K. Chandra, R. Bondyopadhaya.
Indian Journ. Phys., Vol. 48, 779 - 786 (1974).
Using the collisionless Boltzmann equation in spherical co-ordinates, the perturbation method of Lindstedt has been applied to obtain the dispersion relations. Assuming wave so-

lutions in velocity dependent circular form, a neccessary criterion for gravitational stability has been derived.

066.129 A galactic model. II. The gravitationally radiated angular and linear momenta fluxes. D. J. Booth. International Journ. Theor. Phys., (GB), Vol. 10, 333 - 353 (1974).

Using the linearised theory of general relativity the gravitationally radiated angular and linear momenta from a galactic model of N gravitational radiators are calculated. The results are presented in terms of the lowest order contributing multipole momenta, the orientations of the radiators about a common reference frame, the distances between pairs of radiators and the frequency of each radiator.

066.130 Black and white holes. R. Goldoni. Acta Phys. Austriaca, Vol. 41, 75 - 82 (1975).

066.131 Massive spheres with two constant-density inner layers and a variable density outer layer in general relativity. K. D. Krori, P. Borgohain. Indian Journ. Pure Appl. Phys., Vol. 12, 616 - 621 (1974).

Two classes of spheres with three density distributions are considered in general relativity. Each of them has a constant-density core, a constant density intermediate shell and a variable density outer shell.

066.132 Existence and stability criteria for circular geodesics in the vicinity of a Reissner-Nordstrom black hole. A. Armenti, Jr. Nuovo Cimento B, Ser. 11, Vol. 25B, 442 - 448 (1975).

066.133 Scalar extension of the new theory of gravitation. H. Yilmaz. Nuovo Cimento B, Ser. 11, Vol. 26B, 165 - 170 (1975).

The new theory of gravitation proposed by the author is extendable to accommodate a cosmic field of Dicke-Brans type. It is shown that the predictions are the same as those of the usual Dicke-Brans theory but the cosmic field is not necessarily interpretable as a pure scalar.

066.134 Inertial effects in the gravitational collapse of a rotating shell. L. Lindblom, D. R. Brill. Phys. Rev. D, Particles and Fields, Vol. 10, 3151 - 3155 (1974).

Studies the free-fall expansion or collapse of a slowly rotating spherical shell of matter and solves Einstein's equations to first order in the angular velocity of the shell for shells of arbitrary rest mass and radial velocity, assuming asymptotic flatness and no incoming waves. Discusses the inertial dragging exhibited by this solution in the interior of the shell.

066.135 Diffraction of a plane electromagnetic wave at a Schwarzschild black hole. E. Herlt, H. Stephani. International Journ. Theor. Phys., (GB), Vol. 12, 81 - 93 (1975).

Using the technique of Debye potentials a rigorous solution of the diffraction problem is given as a superposition of an incident wave, strongly connected with the Coulomb scattering wave function, and a scattered wave, which is purely outgoing for large distances. The phase shifts of the partial waves are evaluated in the WBK approximation.

066.136 Magnetic field of a current loop around a Schwarzschild black hole. J. A. Petterson. Phys. Rev. D, Particles and Fields, Vol. 10, 3166 - 3170 (1974).

Uses Maxwell's equations in curved space-time to obtain a multipole expansion for a quasistatic axisymmetric magnetic field in a Schwarzschild background at radii both inside and outside the radius of the source. Calculates the magnetic field of a current loop, symmetrically located in a Schwarzschild background and exhibits the behaviour of the dipole part of this field in various limits.

066.137 Experimental method for the determination of the gravitational interaction velocity in Newtonian gravitational fields. G. Cristea. Inst. Fiz. Atomica, Bucharest,(Romania), FR-117-1974, 12 pp. (1974).

066.138 Gravitationswellen. H.-J. Treder. Sterne, Vol. 51, 69 - 81 (1975).

Radio-wave deflection experiments confirm Einstein. Phys. Today, Vol. 28, No. 4, p. 17, 20 (1975).

Wellenreflexion an rotierenden „Schwarzen Löchern". Phys. Blätter, 31. Jahrg., 91 (1975).

The physics of time asymmetry. See Abstr. 003.049.

Die spezielle Relativitätstheorie. See Abstr. 003.085.

Strong interactions, gravitation and cosmology. See Abstr. 022.085.

Gravitational-wave antenna design to detect random gravitational waves. See Abstr. 031.248.

The relativity effect of deflection of light in astrometry. See Abstr. 041.005.

Terrestrial timekeeping and general relativity – a discovery. See Abstr. 044.005.

Reformulation of the relativistic conversion between coordinate time and atomic time. See Abstr. 044.009.

The Mercury orbiter and its relativistic aspects. See Abstr. 053.005.

The equations of hydrodynamics for a thermally conducting, viscous, compressible fluid in special relativity. See Abstr. 062.006.

Magnetohydrodynamic equilibrium conditions in the post-Newtonian approximation of general relativity. See Abstr. 062.014.

A realistic lower bound for the maximum mass of neutron stars. See Abstr. 065.019.

A variational calculation of the fundamental frequencies of quadrupole pulsation of fluid spheres in general relativity. See Abstr. 065.027.

Slowly rotating relativistic stars. V. Static stability analysis of $n = 3/2$ polytropes. See Abstr. 065.067.

Pulsed gamma-ray emission from neutron and collapsing stars and supernovae. See Abstr. 065.076.

Neutron stars and incompressible fluid spheres in the Jordan-Brans-Dicke theory of gravitation. See Abstr. 065.101.

Stability of nonradial vibrational modes of relativistic neutron stars. II. See Abstr. 065.107.

Abnormal neutron stars.   See Abstr. 065.108.

Photo-Coulomb gravitons and gravitational luminosity of the sun.   See Abstr. 080.006.

The masses of components and the inclination of a binary system with one pulsar determined from relativistic effects.   See Abstr. 117.007.

Ultrashort-period binaries. III. The accretion of hydrogen-rich matter onto a white dwarf of one solar mass. See Abstr. 117.031.

X-ray variability by matter accretion onto a black hole in a detached binary system.   See Abstr. 117.032.

A model of X Persei.   See Abstr. 117.036.

The ultraviolet spectrum of Beta Lyrae. See Abstr. 121.042.

Interstellar H_2CO. I. Absorption studies, dark clouds, and the cosmic background radiation. See Abstr. 131.037.

Massive black holes in extragalactic radio source components?   See Abstr. 141.041.

Massive black holes in extragalactic radio sources. See Abstr. 141.069.

Periastron shifts in the binary system PSR 1913+16: theoretical interpretation.   See Abstr. 141.320.

PSR 1913 + 16: endpoints of speculation. A critical discussion of possible companions and progenitors. See Abstr. 141.361.

The masses of binary X-ray sources. See Abstr. 142.002.

The implausible history of triple star models for Cygnus X-1: evidence for a black hole. See Abstr. 142.011.

Cygnus X-1: an interpretation of the spectrum and its variability.   See Abstr. 142.020.

X-ray-emitting double stars.   See Abstr. 142.024.

Why the number of galactic X-ray stars is so small? See Abstr. 142.071.

Spectrum and polarization of X-rays from accretion disks around black holes.   See Abstr. 142.086.

Collapsed stars, pulsars and the origin of cosmic rays. See Abstr. 143.073.

Possible power source of Seyfert galaxies and QSOs. See Abstr. 158.021.

Does the Local Supercluster rotate? See Abstr. 160.013.

Apparent galaxy clustering caused by low-frequency gravitational radiation.   See Abstr. 160.014.

Scheinbare Virialsatz-Verletzung bei Galaxienhaufen infolge extrem langwelliger kosmischer Gravitationsstrahlung.   See Abstr. 160.034.

Distortions of the 3°K background radiation spectrum: observational constraints on the early thermal history of the universe.   See Abstr. 162.001.

Primeval black holes and galaxy formation. See Abstr. 162.008.

Cosmological absorption of gravitational waves. See Abstr. 162.014.

Large-scale random gravitational waves. See Abstr. 162.024.

Propagation of electromagnetic polarization effects in anisotropic cosmologies.   See Abstr. 162.028.

The porthole effect and rings of fire in finite metagalaxies.   See Abstr. 162.032.

Mach's principle in general relativity. See Abstr. 162.055.

Radar distance in Robertson-Walker space-times. See Abstr. 162.065.

Errata

066.901   Errata: 'Role of f-gravity in cosmological models' [Nature, Vol. 249, 640 - 641 (1974)]. C. Sivaram, K. P. Sinha, E. A. Lord. Nature, Vol. 255, 262 (1975). – See Abstr. 11.066.080.

Sun

071 Solar Photosphere, Spectrum

071.001 **The asymmetry of the Hα absorption coefficient.**
A. Zelenka.
Solar Physics, Vol. 40, 39 - 52 (1975).
Because of fine structure splitting and Lamb shift the seven lines of the Hα transition array do not fall together. The separation between the most intensive components amounts to 0.14 Å, a value which is comparable with the Doppler width representative of the layers where the solar Hα profile originates. Inclusion of this splitting in the line absorption coefficient leads to a better agreement between computed and observed profiles. The transfer problem needs not be solved for each of the seven lines. The new characteristic of these profiles is their asymmetry: the centre of symmetry of the line wings is not equal to that of the line core.

071.002 **Horizontal velocities in the solar photosphere.**
D. Dravins.
Solar Physics, Vol. 40, 53 - 63 (1975).
Horizontal macroscopic velocities V_{hor} in the photosphere are studied. High-resolution spectrograms of quiet regions are analyzed for center-limb variation of rms Doppler shifts. The data are treated to assure that the observed velocities refer to constant size volumes on the sun (800 × 3000 × 250 km), independent of μ. Using known height variation of vertical velocities and calculated line formation heights, the height dependence of $\langle V_{hor} \rangle$ is obtained. From a value around 450 m s⁻¹ it decreases rapidly with increasing height. To study also small-scale velocities, the time evolution of subarcsecond size elements in the photospheric network (solar filigree) is studied on filtergrams. It is concluded that they show proper motions implying $\langle V_{hor} \rangle$ about 1 km s⁻¹.

071.003 **On an empirical determination of the source function for strong Fraunhofer lines.**
R. B. Teplitskaya.
Issled. po geomagnetizmu, aehron. i fiz. Solntsa. Vyp. (No.) 31. Moskva, Nauka, 1974, p. 3 - 12. In Russian. – Abstr. in Referativ. Zhurn. 51. Astron., 2.51.447 (1975).

071.004 **Broadening of some solar Na I lines by atomic hydrogen.** E. Roueff.
Astron. Astrophys., Vol. 38, 41 - 44 (1975).
A critical review of the van der Waals approximation for pressure broadening is given. The theoretical point of view of line broadening is emphasized concerning the interatomic potential and the collision treatment itself. Finally, some solar lines are considered, and the use of more realistic interaction potentials is shown to give large discrepancies from the van der Waals formula; these discrepancies increase with the principal quantum numbers involved in the transition, and also with temperature.

071.005 **On the problem of control of the damping parameter in the Fraunhofer line profiles.**
Eh. A. Gurtovenko.
Solnechnye Dannye 1974 Byull., No. 11, p. 58 - 65 (1975). In Russian.
The methods of theoretical evaluation and empirical determination of the damping constant in the Fraunhofer line profiles are discussed. The profiles of 68 faint lines of different atoms and ions were analysed bv the Voigt functions method. The damping constant deduced empirically is compared with the calculated value resulting from Van der Waals interactions.

071.006 **Investigation of the absorption of the continuum spectrum by Fraunhofer lines. VIII. An influence of the lines on the solar continuum radiation.**
P. P. Kozak, I. V. Baranovskij, L. Yu. Ostroverkhaya.
Solnechnye Dannye 1974 Byull., No. 12, p. 64 - 69 (1975). In Russian.
The integrals of the continuum intensities in the wavelengths under investigation and the effect of Fraunhofer lines on the solar continuum have been calculated. The rate and the character of the lines influence on the continuous absorption with atmospheric depth and also the colour indices in the U, B, V, R and I regions have been determined.

071.007 **Table of solar diatomic molecular lines I. Spectral range: 6100–6600 Å.**
R. Boyer, P. Sotirovski, J. W. Harvey.
Astron. Astrophys., Suppl. Ser., Vol. 19, 359 - 387 (1975).
A table of molecular lines observed in the spectrum of a sunspot in the wavelength range 6100–8300 Å is in preparation. The present publication of results between 6100 and 6600 Å represents the first part of this work. An introductory text gives a detailed description of each column of the table. For each identified line the rotation branch, quantum number and the vibration band are indicated. The authors measured the equivalent widths for all lines with profiles as free from blending as possible, the effect of scattering from the photosphere and the edges of strong Fraunhofer lines being taken into account.

071.008 **Photoelectric speckle interferometry of the solar granulation.** J. W. Harvey, M. Schwarzschild.
Astrophys. Journ., Vol. 196, 221 - 226 (1975).
Using the McMath solar telescope on a morning of average seeing it has been found possible, with the help of photoelectric speckle interferometry, to detect securely the existence of details in the quiet solar granulation up to a wavenumber of 240 × 10⁻⁴ km⁻¹, i.e., a wavelength of one-third of an arc second.

071.009 **Solar granulation and oscillations as spatially random processes.** D. K. Lynch, G. A. Chapman.
Astrophys. Journ., Vol. 197, 241 - 249 (1975).
Using Sheeley and Bhatnagar's technique to separate the slowly varying and oscillatory component of the photospheric velocity field, the authors analyze high spatial resolution λ6102.7 velocitygrams (subtracted spectroheliograms). A new way of interpreting the power spectra is presented. By invoking simple random models of the velocity field, the shape of the power spectra and autocorrelation functions can be explained quantitatively, and the results show that there are no large cells in either field. The effects of seeing are discussed. The interpretation of the results leads to the identification of a restricted region in the k_x-ω diagnostic diagram in which the oscillations fall.

071.010 **On the broadening and shift of spectral lines.**

W. Van Rensbergen, E. De Doncker, G. Deridder.
Solar Physics, Vol. 40, 303 - 315 (1975).

The Van der Waals interaction between a radiator and a light neutral perturber explains the broadening and shift of spectral lines only partly satisfactorily. The authors used a combination of a Smirnov-Roueff (S-R) interaction potential and a Van der Waals (VdW) potential. The agreement between theory and experiment for the shift of the line is – although not completely satisfactory – better than the very poor agreement resulting from a VdW potential. However, the S-R potential may not be used in the case that heavy perturbers are involved. In that case the VdW potential yields the better results. Fairly good agreement is found between theoretically and empirically determined values of γ for the solar centre.

071.011 **High n emission and absorption lines of the sun (II).**
A. Greve.
Solar Physics, Vol. 40, 329 - 331 (1975).

Considering the electronic pressure broadening of high n solar recombination lines, an upper wavelength limit of the observable region of possible lines is derived. The reason that no lines have been observed is discussed.

071.012 **Far-infrared solar brightness measured with a balloon-borne lamellar-grating interferometer.**
P. Stettler, J. Rast, F. K. Kneubühl, E. A. Müller.
Solar Physics, Vol. 40, 337 - 349 (1975).

The solar brightness temperature was measured at wavelengths between $208\,\mu$ and $660\,\mu$ with a lamellar-grating interferometer. In order to avoid terrestrial absorption due to water vapor the measurement of the absolute spectral brightness was performed at altitudes of 30 km and 35 km with a balloon-borne gondola. The methods of observation are explained in detail. The present experimental temperature profile is compared to results obtained from other measurements. The results agree with the empirical HSRA model and with those of other research groups in the spectral range between $208\,\mu$ and $660\,\mu$.

071.013 **One- and multi-component models of the upper photosphere based on molecular spectra. IV: Non-LTE treatment of the CN violet system.**
G. H. Mount, J. L. Linsky.
Solar Physics, Vol. 41, 17 - 33 (1975).

Non-LTE synthetic spectra derived from a detailed analysis of the formation of the CN (0,0) $\lambda3883$ Å spectrum are compared with center-limb photoelectric spectra taken at Kitt Peak National Observatory. Significant non-LTE effects are found and the Kurucz, Altrock-Cannon, Mount-Linsky II, and HSRA models are compared. The authors conclude that a significant decrease is required in the presently accepted value of solar carbon abundance. They specify the regions of formation for the CN (0,0) 3883.35 Å bandhead at disc center and limb.

071.014 **The solar lithium abundance. I: Observations of the solar lithium feature at $\lambda6707.8$ Å.**
J. W. Brault, E. A. Müller.
Solar Physics, Vol. 41, 43 - 52 (1975).

A detailed observational study of the solar photospheric lithium feature has been carried out with emphasis on center-limb observations, continuum location, possible effects of telluric lines, effects of blending by atomic and molecular lines, and decomposition of the solar spectrum around $\lambda6707$ Å.

071.015 **The solar lithium abundance. II: Synthetic analysis of the solar lithium feature at $\lambda6707.8$ Å.**
E. A. Müller, E. Peytremann, R. de la Reza.
Solar Physics, Vol. 41, 53 - 65 (1975).

The solar abundance of lithium and its isotopic abundance ratio were determined by comparing synthetic spectra with the high quality center-to-limb spectra of the solar lithium feature presented in paper I (see 071.014). The following possible effects on the abundance result were considered: deviations from LTE, atmospheric models, blends by molecular and atomic lines, hyperfine structure, large scale velocity fields. It was found that the photospheric lithium abundance is $\log \epsilon_{Li} = 1.0 \pm 0.1$. A mean depth of line formation was determined for the $\mathrm{Li}^7\,\mathrm{I}\,\lambda6707.776$ Å line. The lithium isotopic abundance ratio was found to be $\mathrm{Li}^6/\mathrm{Li}^7 \leqslant 0.01$.

071.016 **The solar niobium abundance.**
Ö. Hauge, N. H. Youssef.
Solar Physics, Vol. 41, 67 - 69 (1975).

The solar Nb abundance is derived from five Nb I and ten Nb II lines in the photospheric spectrum. Equivalent widths are obtained from measurements on spectra recorded at Kitt Peak National Observatory. Synthetic spectrum calculations gave abundances of 2.23 and 2.08 from neutral and ionized lines respectively in the logarithmic $A_H = 12.00$ scale. This gives an average abundance value of $A_{Nb} = 2.13 \pm 0.10$.

071.017 **Acoustic waves and the geometric scale in the solar atmosphere.** F.-L. Deubner.
Solar Physics, Vol. 40, 333 - 335 = Mitt. Fraunhofer Inst., Freiburg, No. 135 (1975). – Research note.

071.018 **Solar line profiles of He I 584 Å and He II 304 Å.**
G. W. Cushman, L. Farwell, G. Godden, W. A. Rense.
Journ. Geophys. Res., Vol. 80, 482 - 486 (1975).

A grazing incidence spectrograph launched in a Black Brant rocket August 30, 1973, at White Sands Missile Range obtained high spectral resolution data of the solar radiation between 200 and 700 Å. Absolute intensities and profiles of the brighter lines in this spectral region were measured. Results for the He I 584-Å and He II 304–Å lines are presented.

071.019 **Measurement of solar disc polarization in a number of Fraunhofer lines and their adjacent continuum.**
E. Wiehr.
Astron. Astrophys., Vol. 38, 303 - 306 (1975).

The linear polarization in the absorption lines Sr^+4077, Ca 4227, Ba^+4554, $\mathrm{Mgb}_{1,2,3}$ and $\mathrm{NaD}_{1,2}$, including their adjacent continua, is measured at a limb distance of 5 arcsec. In addition, the polarization of the violet wings of Ca 4227 and NaD_2 as well as that of the continua close to these two lines is measured as function of the heliocentric angle.

071.020 **Spectroscopic detection of solar ^{3}He.**
D. N. B. Hall.
Astrophys. Journ., Vol. 197, 509 - 512 (1975).

Excess emission in the longward wing of the 10830 Å helium feature in a solar prominence is identified as due to solar ^{3}He. A solar ^{3}He/^{4}He abundance ratio of $4 \pm 2 \times 10^{-4}$ is obtained, and its implications are discussed.

071.021 **A reduced upper limit to the solar boron abundance.**
D. N. B. Hall, O. Engvöld.
Astrophys. Journ., Vol. 197, 513 - 515 (1975).

A new upper limit on the solar boron abundance has been obtained from observations of the photospheric spectrum in the 16240 Å region. The authors find $N_\odot(B)/N_\odot(H) \lesssim 1.2 \pm 0.6 \times 10^{-10}$. The implications of this result for the boron abundance in the primitive solar material are discussed.

071.022 **Ni I transition probabilities and the solar nickel abundance.** W. N. Lennard, W. Whaling,
J. M. Scalo, L. Testerman.
Astrophys. Journ., Vol. 197, 517 - 526 (1975).

Absolute transition probabilities have been measured for 97 lines from 11 levels in Ni I using the beam-foil lifetime and branching-ratio technique. The solar Ni abundance has been computed (1) by matching the equivalent widths of the 10 weakest lines, and (2) by fitting the profiles of 19 lines. Both methods agree within experimental uncertainty and yield an average photospheric Ni abundance of 6.28 ± 0.09.

071.023 The stronger absorption lines in the solar spectrum — an identification list. J. Mitton.
Journ. British. Astron. Ass., Vol. 85, 238 - 241 (1975).

071.024 Stigmatic spectra of the sun between 1200 Å and 2100 Å.
D. Samain, R. M. Bonnet, R. Gayet, C. Lizambert.
Astron. Astrophys., Vol. 39, 71 - 81 (1975).
    Stigmatic spectra of the sun from 1200 Å to 2100 Å with spatial resolution on the disk of 7 arcsec have been obtained. The mean spectral resolution achieved is 0.4 Å. Absolute intensities and center-to-limb variations of the solar UV continuum have been measured on both sides of the temperature minimum. They locate the position of this minimum around 1580 Å with a value of 4430 °K ± 50 °K. It is shown that an extra opacity is required to represent the observations down to, at least, 1580 Å.

071.025 On the solar curve of growth of titanium.
R. Foy.
Astron. Astrophys., Vol. 39, 235 - 239 (1975).
    The purpose in the present work is to investigate whether or not the splitting of the solar curve of growth found in the case of iron applies to other elements. The author shows that, contrary to the solar curve of growth for iron, that for titanium is not split in its damping part; this result emphasizes that reference curves of growth have to be used very carefully in detailed analysis of stellar atmospheres.

071.026 Statistical analysis of solar granulation.
C. Aime, G. Ricort.
Astron. Astrophys., Vol. 39, 319 - 324 (1975). In French.
    The study of statistical properties of solar granulation brightness fluctuation is generally performed by using a scanning spot on the image. The measured signal is one dimensional and to express the physical meaning of the phenomenon, or to correct instrumental and atmospheric effects, it is necessary to compute the two dimensional power spectrum (P. S.). The computation is generally done by applying the Rogerson (1955) and Uberoi (1955) formulae to the one dimensional P. S., or by computing the Hankel transform of the autocorrelation. The authors develop other methods by presenting advantages and disadvantages of each method, limiting their work to those which only impose precision in the obtaining of data along one direction.

071.027 The CO fundamental bands in the solar spectrum.
C. Muller, A. J. Sauval.
Astron. Astrophys., Vol. 39, 445 - 451 (1975).
    The result of a study of CO lines in the $\Delta v = 1$ bands in the solar disk spectrum is presented. Recently, the photospheric spectrum has been recorded using a balloon-borne spectrometer from 40 km altitude. From a comparison between the observed and synthetic solar spectra, an agreement is shown using the best available solar and molecular data. The results are discussed and compared with other investigations and enable the authors to explain an apparent discrepancy between previous results.

071.028 Inhomogeneous model of the photosphere.
P. Turon.
Solar Physics, Vol. 41, 271 - 288 (1975).
    What are the temperature and pressure distributions in the solar photosphere? A new method for the construction of empirical models is developed starting with a first approximation called 'multicolumn'. The old idea of two or three column models is thus kept but completed by the following ideas: (1) A good photograph of granulation is a fair representation of topological features along x and y. (2) Vertical distributions $T(z)$, $P(z)$ cannot be very much different from distributions given by stellar atmosphere models in radiative and hydrostatic equilibrium.

071.029 Equivalent width of molecular lines in stars. I: Lyman and Werner bands of H_2 in the solar atmosphere. K. S. Krishna Swamy.
Solar Physics, Vol. 41, 301 - 312 (1975).
    The expected equivalent widths of the individual rotational lines of the Lyman and Werner bands of the hydrogen molecule from the solar atmosphere have been calculated. These results are used to predict what one expects to observe with a specified wavelength bandpass. These are compared with the observation of Dupree and Reeves.

071.030 Spectral features to be expected from rotational and expansional motions in fine solar structures.
B. Rompolt.
Solar Physics, Vol. 41, 329 - 348 (1975).
    The shapes of spectral features (lines) to be expected from rotating or expanding fine solar filaments, loops, and knots (oriented variously to the spectrograph slit) are given and discussed. Only the appearance of the lines in the spectrum (i.e., their orientation to the direction of dispersion and the shape of their area) is considered. It is shown that the observed inclination to the direction of dispersion of some spectral features can be produced by rotation but not by expansion. The observational evidences are given.

071.031 Is the solar filigree the site of strong photospheric magnetic fields? J. M. Beckers.
Bull. American Astron. Soc., Vol. 7, 346 (1975). – Abstr. AAS.

071.032 A new component of solar magnetism – the inner network fields. W. C. Livingston, J. Harvey.
Bull. American Astron. Soc., Vol. 7, 346 (1975). – Abstr. AAS.

071.033 Evolving force-free magnetic fields in response to photospheric motions. B. C. Low, Y. Nakagawa.
Bull. American Astron. Soc., Vol. 7, 347 (1975). – Abstr. AAS.

071.034 Non-LTE formation in the presence of a magnetic field. L. H. Auer, J. N. Heasley, L. L. House.
Bull. American Astron. Soc., Vol. 7, 349 - 350 (1975). Abstr. AAS.

071.035 Measurements of solar line profiles between 1175 and 3200 Å.
J. L. Kohl, W. H. Parkinson, E. M. Reeves.
Bull. American Astron. Soc., Vol. 7, 360 (1975). – Abstr. AAS.

071.036 Convective flux in the solar photosphere as determined from fluctuations. F. N. Edmonds, Jr.
Bull. American Astron. Soc., Vol. 7, 362 (1975). – Abstr. AAS.

071.037 Observations of velocity and intensity fluctuations in solar granulation. S. Musman.
Bull. American Astron. Soc., Vol. 7, 362 (1975). – Abstr. AAS.

071.038 Physical conditions in granulation.
R. C. Altrock, S. Musman.
Bull. American Astron. Soc., Vol. 7, 362 - 363 (1975). Abstr. AAS.

071.039 Height dependence of horizontal velocities in the

photosphere.   D. Dravins.
Bull. American Astron. Soc., Vol. 7, 363 (1975). — Abstr. AAS.

071.040   Solar sectors in photospheric magnetic fields.
L. Svalgaard.
Bull. American Astron. Soc., Vol. 7, 364 (1975). — Abstr. AAS.

**071.041   High-resolution solar spectra in the 2900 Å and the
1700 Å range.**
H. C. McAllister, P. H. Smith, J. T. Jefferies.
Bull. American Astron. Soc., Vol. 7, 365 (1975). — Abstr. AAS.

**071.042   A phenomenological study of high-resolution granu-
lation photographs.**
B. J. Labonte, G. W. Simon, R. B. Dunn.
Bull. American Astron. Soc., Vol. 7, 366 (1975). — Abstr. AAS.

**071.043   On the investigation of the structure of the solar
surface.**
M. B. Kerimbekov, V. A. Magerramov, M. I. Ovchinnikova.
Izv. AN AzSSR. Ser. fiz.-tekhn. i mat. n., 1974, No. 4, p. 92 -
97. In Russian. — Abstr. in Referativ. Zhurn. 51. Astron.,
4.51.456 (1975).

071.044   The sun in the far infrared and sub-mm region.
C. De Jager.
Space Sci. Rev., Vol. 17, 645 - 654 (1975). — Presented at the
workshop on coherent detection in astronomy, held at Rhenen,
The Netherlands, 25 and 26 April 1974 — see 012.009.

071.045   On line blanketing in solar spectra.   H. Wöhl.
Astron. Astrophys., Vol. 40, 343 - 345 (1975).
The fraction η of absorption lines in several published
solar spectra and tables of solar lines of the photosphere,
umbrae, and penumbrae were computed. The averaged values
of η are given for 100 Å intervals within the wavelength region
from $0.3\,\mu$ to $2.5\,\mu$.

**071.046   A new method of interpretation of lines of the solar
spectrum.   G. F. Sitnik.**
Astron. Zhurn. Akad. Nauk SSSR, Vol. 52, 553 - 560 (1975).
In Russian. English translation in Soviet Astron., Vol. 19, No. 3.
Closely connected triplet lines arising from the general
lower level or from lower sub-levels with small energetic differences are used. It is supposed for such lines that both the
source function in the line and the frequency dependence of
the selective absorption coefficient are equal at any depth of
the region of line formation in the atmosphere and change
equally with this depth. The calculation by formulae allows
to determine the source function in a line, its long-wave dependence inside a line, to construct simultaneously a model
of the solar atmosphere, having determined the source function in the continuous spectrum from observations in the
points of a profile in its middle part and in the wing, to find
the ratio of selective and continuous coefficients of absorption
as a function of the optical depth in the continuous spectrum,
to find long-wave dependences of coefficients of both selective
absorption and selective emission at different optical depths.

**071.047   On the variations of the photospheric lines and con-
tinuum intensities by small changes of the physical
parameters.   O. G. Badalyan, V. N. Obridko.**
Astron. Zhurn. Akad. Nauk SSSR, Vol. 52, 561 - 567 (1975).
In Russian. English translation in Soviet Astron., Vol. 19, No. 3.
Formulae connecting the variations of the solar Fraunhofer lines and continuum intensities by small variations of
the model parameters are derived. Applying these formulae to
observations of faculae spectra confirmed the presence of an
overheated layer in the facula.

071.048   Isotopes of samarium in the sun.

A. Ekeland, Ø. Hauge.
Solar Physics, Vol. 42, 17 - 20 (1975).
Terrestrial samarium consists of seven isotopes. Some
spectral lines from Sm have isotope shifts and hyperfine
structures that will modify the profile of the absorption lines
in the Fraunhofer spectrum. The photospheric spectrum
around the Sm II lines at 4467 and 4519 Å has been studied.
Although it is impossible to derive the solar abundance of each
individual isotope, it is shown that a terrestrial isotopic composition can account for the anomal line width and asymmetry
of the observed solar lines. The solar abundance found from
the two lines is $A(\text{Sm}) = 1.54$ in the logarithmic $A(\text{H}) = 12.00$
scale.

**071.049   Supplementary remarks to 'On the average depth
of formation of weak Fraunhofer lines' by E.
Gurtovenko, V. Ratnikova and C. de Jager.**
E(h). A. Gurtovenko, V. A. Ratnikova.
Solar Physics, Vol. 42, 43 - 45 (1975). — Research note.

**071.050   A model of the supergranulation network and of
active-region plages.   J. O. Stenflo.**
Solar Physics, Vol. 42, 79 - 105 (1975).
Analysis of magnetograph recordings made simultaneously
in different spectral lines have shown that the quiet-region network and active-region plages with average field strengths less
than about 100 G are made up by the same type of elementary
structures, each having the same physical properties. Magnetograph data are used together with continuum, line profile, and
EUV data to derive a model of these subarcsec, spatially unresolved elementary structures.

**Profiles of selected Fraunhofer lines for different
positions center-limb on the solar disk.**
See Abstr. 003.009.

**Absolute transition probabilities for some lines of
neutral calcium.   See Abstr. 022.013.**

On the broadening of iron lines by neutral hydrogen.
See Abstr. 022.014.

**Finite eddy-size effects on centre-to-limb variations;
an alternative to anisotropic microturbulence.**
See Abstr. 063.026.

**On the scale of photospheric convection in red
giants and supergiants.   See Abstr. 064.003.**

Flare model chromospheres and photospheres.
See Abstr. 073.065.

**Study of He I emission lines in the solar atmosphere.
III. The triplet-singlet line intensity ratios in solar prominences.**
See Abstr. 073.077.

The emission of Li-like ions from the solar corona.
See Abstr. 074.043.

**The intensities of helium lines in the solar EUV
spectrum.   See Abstr. 076.007.**

**X-ray bright points, coronal heating and the solar
cycle.   See Abstr. 076.013.**

**Determination of the height of hard X-ray sources
in the solar atmosphere by measurement of photospheric
albedo photons.   See Abstr. 076.053.**

**An anticorrelation between polar and equatorial
rotation of the solar photosphere.   See Abstr. 080.029.**

072 Sunspots, Faculae, Solar Activity Cycles

072.001 Ephemeral active regions in 1970 and 1973.
K. L. Harvey, J. W. Harvey, S. F. Martin.
Solar Physics, Vol. 40, 87 - 102 (1975).

Ephemeral active regions (ER) identified on good quality full-disk magnetograms are studied. A comparison of parameters of ER and regular active regions suggests that ER are the small-scale end of a broad spectrum of active regions. The role of ER in the light of present theories of solar activity is investigated but is not yet clear. Heating of the chromosphere and corona may be significantly affected by ER.

072.002 Recurrent magnetic activity, sunspot number and its rate of decline.
B. N. Bhargava, G. K. Rangarajan.
Solar Physics, Vol. 40, 235 - 239 (1975).

Power spectral densities computed from low-latitude horizontal intensity of the earth's magnetic field over two-year periods of declining phases of solar cycles 16 to 19 show a close relationship with the maximum relative sunspot number of the following solar cycles. The maximum sunspot number shows an exponential rise with the power density near $1/27$ cd^{-1}; maximum R_z, however, increases linearly with power density near $1/14$ cd^{-1}. It is also shown that the rate of decline of sunspot number in a solar cycle is almost exactly related, linearly, to power spectral density for the preceding solar cycle. Power densities near $1/27$ and $1/14$ cd^{-1} in declining phase of solar cycle appear to be satisfactory indices for the maximum relative sunspot number of the following cycle and its rate of decline thereafter.

072.003 CO_2 and HCN in sunspots.
M. C. Pande, V. P. Gaur.
Nature, Vol. 253, 104 (1975). – Letter.

072.004 The beginning of a new cycle of solar activity.
M. Waldmeier.
Nature, Vol. 253, 419 (1975). – Letter.

072.005 Sunspot cycle periodicities.   J. P. Bagby.
Nature, Vol. 253, 482 (1975).

The maximum entropy spectral analysis technique of Burg (unpublished) and Cohen and Lintz has led to the discovery of periodicities in sunspot data and an ability to predict ionospheric reflectivity and climatic features. But the theory that a planetary influence affects sunspot variations still cannot be ruled out, as is shown by a further analysis of the data.

072.006 Is there an ionization anomaly in sunspots?
J. Buurman.
Astron. Astrophys., Vol. 37, 451 - 452 (1974).

Absorption features on umbral spectrograms have been attributed to abnormally strong and magnetically broadened ion lines by Mallia and Petford (1972) and by Blackwell and Mallia (1973). The features are largely of umbral origin indeed. However, the lack of polarization makes the identification as Cr II and Fe II lines questionable.

072.007 The crossover effect in sunspots and the fine structure of the penumbra.   A. A. Golovko.
Issled. po geomagnetizmu, aehron. i fiz. Solntsa. Vyp. (No.) 31. Moskva, Nauka, 1974, p. 24 - 32. In Russian. – Abstr. in Referativ. Zhurn. 51. Astron., 2.51.486 (1975).

072.008 Some problems of the fine structure of 11-year time-latitude diagrams of proper positions of flares in sunspot groups.   V. V. Kasinskij, L. A. Plyusnina.

Issled. po geomagnetizmu, aehron. i fiz. Solntsa. Vyp. (No.) 31. Moskva, Nauka, 1974, p. 44 - 50. In Russian. – Abstr. in Referativ. Zhurn. 51. Astron., 2.51.518 (1975).

072.009 On the solar activity index controlling cosmic ray variations.
L. I. Dorman, I. A. Pimenov, L. F. Churunova.
Izv. AN SSSR. Ser. fiz., Vol. 38, 1951 - 1956 (1974). In Russian. – Abstr. in Referativ. Zhurn. 51. Astron., 2.51.531 (1975).

072.010 On the planetary theory of sunspots.
E. Okal, D. L. Anderson.
Nature, Vol. 253, 511 - 513 (1975).

It has been proposed that sunspot activity is affected by positions of the planets, and calculations have been presented, which purport to show that planetary tides on the sun vary in the same way as the sunspot variations. The authors believe that the apparent agreement of the sunspot cycle with planetary tidal effects is an artefact of the calculation.

072.011 The reality of solar activity forecast in 1972–1973.
P. R. Romanchuk.
Astron. Obs., *Kiev*, Preprint No. 3, 8 pp. (1974). In Russian.

The data characterizing the accuracy of the sunspot numbers forecast given by Astron. Obs. Pulkovo and Kiev have been reduced.

072.012 Systems of latitudinal structures on the sun.
N. I. Kozhevnikov.
Astron. Zhurn. Akad. Nauk SSSR, Vol. 52, 119 - 124 (1975). In Russian. English translation in Soviet Astron., Vol. 19, No. 1.

A Fourier analysis of the distributions of the activity parameters over heliographic latitude was made. It is found that these distributions contain periodic components with periods of 2, 4, 8, 16, 32°.

072.013 An evaluation of the height of formation of emission transformations of the H and K Ca II lines in a sunspot umbra.   R. B. Teplitskaya, S. A. Ehfendieva.
Solnechnye Dannye 1974 Byull., No. 11, p. 66 - 71 (1975). In Russian.

For evaluation of the height of formation of different parts of the H and K lines in a sunspot umbra a simple geometric method is used. The results of the study of seven spectrograms are presented.

072.014 Radio emission in the sunspot groups located over the solar disk near the equator.
L. I. Yurovskaya.
Solnechnye Dannye 1974 Byull., No. 11, p. 94 - 99 (1975). In Russian.

The latitude distributions of sunspot groups over the solar disk accompanied by radio noise storms and sunspot groups without radio emission are considered. The ability of sunspot groups to be accompanied by radio emission is supposed to be independent on the latitude of sunspots on the solar disk.

072.015 On the problem of the Joule dissipation of sunspot magnetic fields.
P. R. Romanchuk, V. N. Krivodubskij.
Solnechnye Dannye 1974 Byull., No. 12, p. 69 - 75 (1975). In Russian.

This paper deals with the problem of the Joule dissipation of sunspot magnetic fields, the solar plasma turbulent conductivity being taken into account. The magnitudes of the solar plasma's turbulent conductivity and the effective velocity of the turbulent motions in the solar atmosphere, which are required for the explanation of the decay of sunspot magnetic fields as a result of Joule dissipation within the time compared with the observed life-time of the spots, are determined.

072.016 **Configuration of the magnetic field and motions in sunspots and sunspot groups.** R. N. Ikhsanov.
Solnechnye Dannye 1974 Byull., No. 12, p. 81 - 89 (1975). In Russian.

The reasons of the observed variations in the configuration of the magnetic field and motions in sunspot groups and separate umbrae in a sunspot are discussed.

072.017 **On the antipodal character of large sunspot groups.** Yu. I. Vitinskij.
Solnechnye Dannye 1974 Byull., No. 12, p. 90 - 94 (1975). In Russian.

The principle characteristics of antipodal large sunspot groups were considered using the Greenwich data for 1874 - 1964. A connection of these groups with the solar cycle and active longitudes on the sun is discussed.

072.018 **Variations of solar activity due to the revolution of Mars about the sun.**
G. Ya. Vasil'eva, A. A. Shpital'naya, N. S. Petrova.
Solnechnye Dannye 1974 Byull., No. 12, p. 94 - 102 (1975). In Russian.

The monthly means of the sunspot relative numbers for twenty 11-year solar cycles have been investigated taking into account the ecliptic longitude of Mars. Variations of the solar activity caused by the revolution of Mars about the sun seem to be present and to depend on the phase of the 22-year solar cycle.

072.019 **Long-term periodicities in the sunspot cycle.**
W. C. Bain.
Nature, Vol. 254, 362, with a reply by T. J. Cohen, P. R. Lintz, p. 362 - 363 (1975). – Concerning Abstr. 12.072.007.

072.020 **Der Einfluss der Szintillation auf den Faktor k der Sonnenfleckenrelativzahl.**
D. Paperlein, H. Pachali.
Applied Optics, Vol. 14, 395 - 399 (1975).

The authors combine their own observations and the results of other authors in order to find the relationship between the factor k of the Wolf numbers and the seeing. To a first approximation this relationship has a linear slope. Furthermore the factor k depends on the Wolf number. At lower activity the factors k are greater and the slope is no longer linear.

072.021 **Recent measurements of the flux excess from solar faculae and the implication for the solar oblateness.**
G. A. Chapman.
Phys. Rev. Letters, Vol. 34, 755 - 758 (1975).

Direct observation of the excess brightness from photospheric faculae are presented. This excess brightness is, at times, large enough to produce an apparent oblateness that exceeds that reported by Dicke and Goldenberg. These results support the Chapman-Ingersoll facular explanation for the excess solar oblateness and support the findings of Hill et al. by offering a possible source for their excess equatorial brightness which, they showed, can produce an apparent, nongeometrical oblateness.

072.022 **The nature of the sunspot phenomenon. III: Energy consumption and energy transport.** E. N. Parker.
Solar Physics, Vol. 40, 275 - 289 (1975).

This paper points out the basic relation between the con-

version of thermal energy into convective fluid motion (Alfvén waves when a strong vertical magnetic field is present) and the convective transport of thermal energy. It is shown that heat transport necessarily accompanies convective driving of fluid motion. Convective motions restricted to a layer whose thickness is a small fraction of the local scale height can divert no more than the same fraction of the energy into Alfvén waves. But if the convecting layer extends over many scale heights, then the convective forces may convert more energy into fluid motion than they transport. Hence the creation of a cool sunspot requires convection extending coherently over several scale heights, at least 500 km. This requirement is basically just the familiar thermodynamic efficiency of an ideal heat engine. The calculations establish that convection need not be much less efficient than the ideal.

072.023 **The nature of the sunspot phenomenon. IV: The intrinsic instability of the magnetic configuration.**
E. N. Parker.
Solar Physics, Vol. 40, 291 - 301 (1975).

It is pointed out that the equilibrium configuration of a sunspot magnetic field, confined to the cool umbra by the pressure of the surrounding photospheric gas, is unstable to the familiar hydromagnetic exchange instability. The characteristic time for dissolution of the sunspot is of the order of one hour. In view of the observed long life of individual sunspots, it is important to understand why the instability does not develop. The author reviews several theoretical possibilities, suggesting, tentatively, that the spot might be stabilized by a suitable redistribution of the cooling in the umbra. In particular, if the cooling extends to greater depth in an elongated portion of the sunspot, the magnetic pressure on the boundary is reduced there, tending to reduce the elongation.

072.024 **Hα oscillations in sunspot umbrae.** G. L. Phillis.
Solar Physics, Vol. 41, 71 - 79 (1975).

Direct measurement of intensity and photographic subtractions of simultaneous Doppler filtergrams at Hα ± 0.3 Å in a time-lapse sequence of McMath 11482 has shown velocity oscillations in sunspot umbrae. These oscillations exhibit the following characteristics: (a) A characteristic oscillation frequency for each umbra, with the periods ranging from 145 ± 10 s to 180 ± 10 s. (b) A duration of at least 8 cycles. (c) A Doppler shift of 10 km s^{-1}. (d) An oscillating element size of $3''$ or less. (e) A visibility to a radius vector of about 0.9.

072.025 **Umbral oscillations and penumbral waves in Hα.**
R. L. Moore, F. Tang.
Solar Physics, Vol. 41, 81 - 88 (1975).

The authors present examples of umbral oscillations observed on Big Bear Hα filtergram movies and investigate the relation between umbral oscillations and running penumbral waves occurring in the same sunspot. They also report and interpret 'dark puffs' which emerge from the edge of the umbra and move outward across the penumbra, and which have the same period as the running penumbral waves. It is suggested that the dark puffs and the running penumbral waves have a common source: photospheric oscillations just inside the umbra.

072.026 **Positionsbestimmung von Sonnenflecken.**
U. Fritz, H. Treutner, O. Vogt.
Orion, 33. Jahrgang, p. 38 - 42 (1975).

072.027 **Polaritätsbestimmungen an Sonnenflecken mit dem Spektrohelioskop.** F. N. Veio.
Orion, 33. Jahrgang, p. 48 - 50 (1975).

072.028 **The role of magnetic forces in sunspot equilibrium.**
M. G. Adam.

Monthly Notices Roy. Astron. Soc., Vol. 171, 287 - 302 (1975).

Detailed measurements have been made to determine the configuration of the magnetic field over a large sunspot, using the Oxford 35-m solar telescope and the 11-m spectroscope. The heliographic direction of the magnetic vector and the field strength have been derived, for more than 200 points specified by solar latitude and longitude, from measurements of Zeeman split lines in the λ 5250 Å region. The horizontal forces in stable regions of sunspots are briefly discussed using the 'magnetic line curvature' approach developed by Jakimiec.

072.029 **Possible mechanism of the solar cycle.**
Yu. V. Vandakurov.
Astron. Zhurn. Akad. Nauk SSSR, Vol. 52, 351 - 358 (1975). In Russian. English translation in Soviet Astron., Vol. 19, No. 2.

A hypothesis is discussed that magnetic fields in stellar convection zones are necessary to compensate some average azimuthal force which is created by convective motions and which is equivalent to a negative viscosity. Some examples confirm the existence of such negative–viscosity effect.

072.030 **Comments on the magnitude of variation of solar activity in different cycles.** J. Mergentaler.
Postępy Astron., Vol. 23, 65 - 68 (1975). In Polish.

Mean differences between the consecutive monthly averages of Wolf numbers were computed and compared with the mean errors computed for cycles VI–XIX by the author and by E. B. Balli. The peculiarity of the XIVth cycle is noted.

072.031 **Sunspots and the solar cycle.** T. G. Cowling.
Nature, Vol. 255, 189 - 190 (1975).

072.032 **Long-term groups in the observational series of solar activity.** P. R. Romanchuk, A. N. Sergeeva.
Astrometriya i Astrofizika, *Kiev*, vyp. (No.) 24, (see 003.003), p. 107 - 114 (1974). In Russian.

Statistical relations are studied between the observed sunspot members W and relative ones formed by long- and short-term groups for 19 and 20 cycles. These relations are used to predict W for the next month.

072.033 **On variations of solar activity caused by the revolution of Jupiter, Saturn and Uranus around the sun.**
G. Ya. Vasil'eva, A. A. Shpital'naya, N. S. Petrova-Pystina.
Solnechnye Dannye 1975 Byull., No. 1, p. 84 - 93 (1975). In Russian.

The monthly and yearly means of the sunspot relative numbers from 1749 to 1973 have been studied, the ecliptic longitudes of Jupiter, Saturn and Uranus being taken into account. The epochs of extrema of solar activity variations caused by the outer planets are associated with the moments of their transits across the projection of the galactic magnetic field ($b^{II} = 0°$, $l^{II} = 70°$) on the ecliptic plane. The existence of the well known double 11-year and the assumed double secular solar cycles is connected with the relative periods of revolution of Jupiter, Saturn, Uranus and Neptune.

072.034 **Observations of weak solar magnetic fields with the Lockheed diode array magnetograph.**
R. C. Smithson.
Bull. American Astron. Soc., Vol. 7, 346 - 347 (1975). Abstr. AAS.

072.035 **Facular magnetic fields: description of the data and preliminary results.** K. P. White III.
Bull. American Astron. Soc., Vol. 7, 347 (1975). – Abstr. AAS.

072.036 **Simultaneous X-ray spectra and X-ray images of an active region.** M. Gerassimenko, J. M. Davis, R. C. Chase, A. S. Krieger, J. K. Silk, G. S. Vaiana.
Bull. American Astron. Soc., Vol. 7, 347 (1975). – Abstr. AAS.

072.037 **The morphology of an active region from the photosphere to the corona.** G. L. Epstein, R. W. Hobbs, R. D. Chapman, W. M. Neupert, R. J. Thomas.
Bull. American Astron. Soc., Vol. 7, 347 (1975). – Abstr. AAS.

072.038 **Correlative studies of a complex active region from X-ray, magnetograph and H-alpha data during the Skylab IV mission.**
N. P. Cumings, M. J. Hagyard, J. B. Smith, S. T. Wu.
Bull. American Astron. Soc., Vol. 7, 347 (1975). – Abstr. AAS.

072.039 **Preliminary results of correlative studies from S-056 X-ray, magnetograph and H-alpha data of the complex active region 137/141 for June 9–18, 1973.**
J. B. Smith, Jr., D. M. Speich, R. M. Wilson, A. C. deLoach, R. B. Hoover, J. P. McGuire.
Bull. American Astron. Soc., Vol. 7, 347 (1975). – Abstr. AAS.

072.040 **High resolution observations of sunspot Hα fibrils.**
V. Gaizauskas.
Bull. American Astron. Soc., Vol. 7, 349 (1975). – Abstr. AAS.

072.041 **The linear polarization of continuum radiation in sunspots.** G. D. Finn, J. T. Jefferies.
Bull. American Astron. Soc., Vol. 7, 349 (1975). – Abstr. AAS.

072.042 **Broad-band circular polarization of sunspots.**
R. M. E. Illing, D. A. Landman, D. L. Mickey.
Bull. American Astron. Soc., Vol. 7, 349 (1975). – Abstr. AAS.

072.043 **Measurement of magnetic fields using Fourier transform techniques.** A. M. Title.
Bull. American Astron. Soc., Vol. 7, 350 (1975). – Abstr. AAS.

072.044 **Course of solar activity during the declining phase of solar cycle 20.** H. W. Dodson, E. R. Hedeman.
Bull. American Astron. Soc., Vol. 7, 364 - 365 (1975). Abstr. AAS.

072.045 **The case of the missing sunspots.** J. A. Eddy.
Bull. American Astron. Soc., Vol. 7, 365 (1975). Abstr. AAS.

072.046 **On the reality of global short-period pulsations of the sun.** V. F. Chistyakov.
Problems of cosmic physics. Vyp. (No.) 9, (see 003.013), p. 19 - 30 (1974). In Russian.

Oscillations with periods of 80–90 years, 22 years, 11 years and monthly fluctuations of solar activity are global long-periodic oscillations at the sun. Besides it is possible to suppose the presence of a second group of global short-periodic oscillations with duration of several dozens of minutes. The results of the observations show that the time variations of the magnetic fields of sunspots and the intensity of the hydrogen flocculi have no definite period. The duration of the oscillations for the magnetic fields of spots is 20–40 minutes and 15.5 ± 1.0 minutes for flocculi. The analyses of the data permit to choose three components of the short-periodical oscillations: a) local oscillations in a definite active region, b) regional oscillations for several active regions, c) global oscillations inherent of the whole sun.

072.047 **On the change of relative intensities of sunspot umbrae during observations with a Halle 0.5 Å Hα-filter.** Eh. P. Surkov, L. D. Surkova.
Problems of cosmic physics. Vyp. (No.) 9, (see 003.013), p. 30 - 42 (1974). In Russian.

Time variations of sunspots contrasts are studied using films taken in the center and in the wings of the Hα line. It was found that the changes of sunspots brightness have periods of 52, 143 and 574 seconds duration. The changes with period

32 minutes are less reliably revealed. Fluctuations of the brightness with 300 sec period are present in the wings of the Hα line. The results are discussed.

072.048 Zur Sonnenaktivität 1973. A. Saul.
SuW, Vol. 14, 206 - 207 (1975).

072.049 Statistical relations between relative numbers and flare indices. E. N. Zemanek.
Astrometriya i Astrofizika, *Kiev*, vyp. (No.) 25, (see 003. 015), p. 39 - 48 (1975). In Russian.
The relation between flare indices and sunspot relative numbers and recurrent relative numbers was studied. The correlation coefficients and regression equations were determined.

072.050 Comparison of sunspot periods with planetary synodic period resonances. R. M. Wood.
Nature, Vol. 255, 312 - 313 (1975).
The author shows not only a correlation between planetary synodic period resonances with solar activity period, but also a rough correlation between the variations of the resonance period resulting from orbit eccentricity and the variations of the sunspot period.

072.051 Rotational motions of sunspots. Š. Knoška.
Bull. Astron. Inst. Czechoslovakia, Vol. 26, 151 - 158 (1975).
On the basis of a study of the rotation of sunspots from the daily drawings of the sun, made at the Mt. Wilson Observatory in the years 1917–1924, it was found that of the 146 sunspots which rotated, 60% rotated in the ENW direction, in the N, as well as in the S hemisphere. The observed rotational motion, concurrent in direction in both hemispheres, contradicts the effect of differential rotation, as well as the action of the Coriolis force. It is probably due to the dynamics of the development of the magnetic field of the group.

072.052 Corrections of the total areas of sunspots in the Greenwich Photoheliographic Results for 1960.
M. Kopecký, F. Kopecká.
Bull. Astron. Inst. Czechoslovakia, Vol. 26, 158 - 159 (1975).
In the "Greenwich Photoheliographic Results" for 1960 the total areas of sunspots are given incorrectly for about one third of the days in the first quarter of 1960. This paper gives the correct values of the sunspot areas for these days, as well as the corresponding correct values of the monthly means, the annual mean and the means by synodic revolutions of the total areas of the sunspots.

072.053 Observations of large-scale moving magnetic features near sunspots.
A. G. Michalitsanos, A. Bhatnagar.
Astrophys. Letters, Vol. 16, 43 - 51 (1975).
High time and spatial resolution magnetograms taken with a longitudinal video magnetograph show the systematic motion of large crescents and ridges of magnetic field at the outer penumbral boundary of a large complex sunspot group. Both the ridges and the crescents are resolved into knots of flux which are typically 2 arc sec to 3 arc sec in extent, and which move in unison with velocities in the range of 0.2 to 0.3 km/sec. Over a four hour period, these ridges of magnetic field, which are predominantly of opposite polarity to the parent sunspot, are observed to move over distances of from 4 arc sec to 6 arc sec, and merge with existing outlying magnetic fields.

072.054 The temperature distribution in sunspot umbrae.
F. G. Rozhavskij.
Astron. Zhurn. Akad. Nauk SSSR, Vol. 52, 568 - 571 (1975). In Russian. English translation in Soviet Astron., Vol. 19, No. 3.

On the basis of monochromatic profile observations the temperature distribution in "small" and "middle" sunspot umbrae has been calculated. The temperature in the sunspot of the same area is lower than it was presented by other authors.

072.055 Vibration-rotation bands of NO in sunspots.
V. P. Gaur.
Bull. Astron. Soc. India, Vol. 2, 39 (1974). — Abstract.

072.056 Verlauf der Sonnenfleckenrelativzahlen ab 1900.
A. Schroll.
Sternenbote, 18. Jahrgang, p. 82 - 86 (1975).

072.057 On a comparison between the predicted and the observed values of the different indices of solar activity for the 20th solar cycle. C. Poulakos.
Praktika Acad. Athens, Vol. 48, 186 - 196 = Res. Center Astron., Applied Math., Acad. Athens, Ser. 1 (Astron.), No. 36 (1973).

072.058 Oscillations in sunspot umbras due to trapped Alfvén waves excited by overstability.
Y. Uchida, T. Sakurai.
Publ. Astron. Soc. Japan, Vol. 27, 259 - 274 (1975).
Oscillations observed in sunspot umbras are interpreted as a vertical motion in the atmosphere induced by a standing Alfvén wave trapped in the region between the overstable layer under the photosphere and the chromosphere-corona transition layer. Some other aspects of the oscillation, such as the relation to the running penumbral waves are also discussed.

072.059 Magnetoactive lines in the medium with the velocity gradient.
V. M. Grigorjev (*Grigor'ev*), J. M. Katz (*I. M. Kats*).
Solar Physics, Vol. 42, 21 - 35 (1975).
The magnetic splitting peculiarities of the absorption lines in the sunspot spectrum are considered. The most common and typical of them is breaking of all Stokes parameter symmetry in regard to the line center. The possible reason of this effect is the macroscopic gas motion with inhomogeneous velocity. Computed contours are given for the line Fe I $\lambda 5250$ Å with various combinations of magnetic and velocity fields. Magneto-optical effects within the line which are connected with the magnetic and velocity field inhomogeneity are discussed. The observation results are discussed for longitudinal magnetic field zero lines. These observations were carried out for the sunspot and photosphere in two spectral lines Fe I $\lambda\lambda 5250$ and 5233 Å.

072.060 On the turbulent decay of strong magnetic fields and the development of sunspot areas.
F. Krause, G. Rüdiger.
Solar Physics, Vol. 42, 107 - 119 (1975).
This paper deals with the conception that two-dimensional turbulence is present in a sunspot where the magnetic field is strong. It is shown that such a two-dimensional turbulence provides for a turbulent decay of the magnetic field. The turbulent decay is studied by investigating a simple model and comparing the results with those deduced by Bumba from the observed decay of sunspot groups areas.

072.061 Comments on the course of solar activity during the declining phase of solar cycle 20 (1970–74).
H. W. Dodson, E. R. Hedeman.
Solar Physics, Vol. 42, 121 - 130 (1975).
In the declining phase of solar cycle 20 (1970–74) three pulses of activity occurred and resulted in two well defined 'stillstands' in the smoothed means of sunspot, 2800 MHz, and calcium plage data. Studies of the latitude distribution of spots and flares show the extent of the dominance of the northern

hemisphere. From mid-1973 to mid-1974 the sun had a relatively inactive hemisphere centered on ~0° longitude. The relationship of certain well defined 'coronal holes' to this inactive hemisphere of the chromosphere is noted.

072.062 **Planetary motion and cyclic activity of the sun: estimate of energies.**
V. A. Vlasov, L. I. Gudzenko, V. E. Chertoprud.
Kratkie soobshch. po fiz., 1974, No. 12, p. 9 - 11. In Russian.
Abstr. in Referativ. Zhurn. 51. Astron., 6.51.465 (1975).

072.063 **Broad-band circular polarization of sunspots: spectral dependence and theory.**
R. M. E. Illing, D. A. Landman, D. L. Mickey.
Astron. Astrophys., Vol. 41, 183 - 185 (1975).

Observations of circular polarization in sunspots in two spectral bands are presented; the two distinct regions of polarization recognized in previous work (uniform and "speckled" spatial dependence) are shown to have different spectral dependences. A simple model is advanced which results in circular polarization of the correct magnitude. The implications of the model for the mass flow and magnetic field structure of sunspots are discussed in the light of our previous observations.

072.064 **Spectroscopic study of dynamic phenomena in sunspots.** R. B. Schultz.
Thesis, Colorado Univ., Boulder (USA). 270 pp. University Microfilms Order No. 74-22,388 (1974).

072.065 **An analysis of the profiles of magnetoactive lines. II. Equivalent widths and degrees of circular polarization of the lines.** A. V. Baranov.
Solnechnye Dannye 1975 Byull., No. 2, p. 52 - 55 (1975). In Russian.

On the basis of an analysis of the equivalent widths and degrees of circular polarization of three triplet lines it is shown that the characteristics of the fields and physical parameters of the matter and their variations with depth are different in different sunspots.

072.066 **Solar activity and the structure of interplanetary matter.**
G. Ya. Vasil'eva, A. A. Shpital'naya, N. S. Pystina.
Solnechnye Dannye 1975 Byull., No. 2, p. 76 - 84 (1975). In Russian.

Variations of the solar activity caused by external planets give evidence to some structure of the interplanetary space. "Active" and "passive" longitudes in space derived from investigations of the monthly and yearly means of Wolf numbers from 1749 to 1973 are organized in four main sectors. The orientation of "active" sectors seems to be determined by the magnetic field of the Galaxy and motion of the sun in space. The form of the solar magnetosphere is supposed to manifest itself through sector structure.

072.067 **On the distribution of the magnetic field and radial velocity in sunspot No. 223 on August 2, 1972.**
L. F. Lazareva, Eh. I. Mogilevskij.
Solnechnye Dannye 1975 Byull., No. 2, p. 97 - 104 (1975). In Russian.

The data were obtained at IZMIRAN on August 2, 1972 by means of the ATB-3 (f = 17 m). The magnetic field and Doppler velocities were determined at the line Fe I λ 6302 Å. H was 3400 gauss. The most probable velocity lies within $0.2 - 0.3$ km/sec.

072.068 **Observation of the sun in Czechoslovakia.**
M. Neubauer.
Říše hvězd, Vol. 56, 7 - 9 (1975). In Czech.

Stray light in solar observations. A method of

numerical integration.   See Abstr. 031.268.

Initial results from a scanning Stokes polarimeter. See Abstr. 034.041.

The theoretical instrumental profile of the combination telescope-pinhole photometer and its effect upon observations of umbra intensities.   See Abstr. 034.075.

Depth of origin and mean depth of a magnetoactive line.   See Abstr. 063.034.

Table of solar diatomic molecular lines I. Spectral range: 6100–6600 Å.   See Abstr. 071.007.

On the variations of the photospheric lines and continuum intensities by small changes of the physical parameters. See Abstr. 071.047.

A model of the supergranulation network and of active-region plages.   See Abstr. 071.050.

A few comments on the two-ribbon flare without sunspots of 29 July 1973.   See Abstr. 073.014.

A morphological study of a solar active region in August 1972. II. A correlation analysis of morphologies of flares and spots.   See Abstr. 073.028.

Étude statistique des associations entre le champ magnétique des groupes de taches solaires et la morphologie des éruptions solaires et des sursauts associés sur ondes décimétriques.   See Abstr. 073.033.

Sunspot motion, flares and type III bursts in McMath 11482.   See Abstr. 073.035.

The relation between flares and the missing energy in spots.   See Abstr. 073.064.

Investigation of the transition region above sunspots based on polarization radio measurements. See Abstr. 073.075.

Relationships – Waldmeier grouping, solar flares. See Abstr. 073.111.

Evolution of solar active regions. See Abstr. 073.113.

The effect of evolution of active regions on high-speed quasi-stationary plasma flows from the sun. See Abstr. 074.007.

Large-scale and solar-cycle variations of the solar wind.   See Abstr. 074.029.

North-south asymmetries in solar wind and solar spottedness in sunspot cycle No. 20. See Abstr. 074.114.

Evolution of the solar Lyman alpha flux during four consecutive years.   See Abstr. 076.001.

X-ray bright points, coronal heating and the solar cycle.   See Abstr. 076.013.

On the connection of local solar radio sources at $\lambda = 6.6$ cm with the characteristics of active regions. I. See Abstr. 077.015.

The metric quiet sun during two cycles of activity and the nature of the coronal holes.   See Abstr. 077.066.

Influence of solar activity on the spatial distribution of cosmic rays in the interplanetary space. See Abstr. 078.002.

Solar magnetic fields and convection. I: Active regions and sunspots.   See Abstr. 080.024.

Variation of the solar constant and solar activity. See Abstr. 080.042.

Information on the third dimension of solar phenomena deduced by statistical procedures. See Abstr. 080.048.

Origin of the sunspot modulation of ozone: its implications for stratospheric NO injection.   See Abstr. 082.010.

Electron content power spectral estimates: periods of 2 days to $^1/_2$ year.   See Abstr. 083.011.

The relationship between sudden ionospheric disturbances and the magnetic structure of sunspot groups. See Abstr. 083.038.

Correlation of the air temperature at Catania and the sunspot cycle.   See Abstr. 085.001.

Solar activity and the weather. See Abstr. 085.002.

Early active sun?: Radiation history of distinct components in fines.   See Abstr. 094.187.

Les comètes, l'activité du soleil et le milieu interplanétaire.   See Abstr. 102.029.

Intensity fluctuations in the head of comet 1973f. See Abstr. 103.100.

073 Solar Chromosphere, Flares, Prominences

073.001 Radiation of chromospheric flares in hydrogen lines.
L. N. Kurochka, E. V. Kurochka, V. A. Ostapenko.
Bull. Astron. Inst. Czechoslovakia (BAC), Vol. 26, 23 - 29 (1975).

The radiation energy of flares of various importances was computed in the Lyman, Balmer, Paschen and Brackett series of hydrogen for different values of the optical thicknesses of the flares. It is shown that the amount of energy, reduced in all emission lines of H, is 4 - 5 times larger than the amount, radiated by flares in H-alpha.

073.002 A comparison of spicules in the Hα and He II (304 Å) lines.
O. K. Moe, O. Engvold, J. M. Beckers.
Solar Physics, Vol. 40, 65 - 68 (1975). – Research note.

073.003 Phenomenology of the subflare; a synthesis of CINOF (Campaign for the Integrated Observation of Solar Flares). C. de Jager.
Solar Physics, Vol. 40, 133 - 140 (1975).

During the CINOF Campaign, June 1972, more than a hundred small chromospheric brightenings were observed. The observational results of about ten of them were investigated in more detail. They show a number of regularities.

073.004 Analysis of the August 7, 1972 white light flare: light curves and correlation with hard X-rays.
D. M. Rust, F. Hegwer.
Solar Physics, Vol. 40, 141 - 157 (1975).

Cinematic, photometric observations of the 3B flare of August 7, 1972 are described in detail. From the close temporal correspondence and from the small distance (3″) separating the layers where the visible emission and the X-rays arose, it is argued that the hard X-ray source must have had the same silhouette as the white light flare and that the emission patches had cross-sections of 3–5″. There was also a correlation between the location of the most intense visible emissions near sunspots and the intensity and polarization of the 9.4 GHz radio emission. The observations are interpreted with the aid of the flare models of Brown to mean that the same beam of non-thermal electrons that was responsible for the hard X-ray bremsstrahlung also caused the heating of the lower chromosphere that produced the white light flare.

073.005 Energy build-up and release mechanisms in solar and auroral flares. T. Obayashi.
Solar Physics, Vol. 40, 217 - 226 = Space Sci. Rev., Vol. 17, 195 - 203 (1975).

Flare phenomena in the solar atmosphere and in the terrestrial magnetosphere exhibit many similarities. The mechanical energy of enhanced photospheric motion is converted and stored in the form of magnetic potential energy in sunspot fields, which is analogous to the case of the growth phase of magnetospheric substorms. The energy release during the explosive phase is initiated by a sudden collapse in the magnetic field topology and the X-type magnetic neutral point is created in the corona. Subsequent electrical discharge takes place in the form of an intense electrojet current flowing in the base of the chromosphere at the altitude where the Cowling conductivity is a maximum. It is suggested that the field-aligned precipitation of hot electrons and the Ohmic heating in the chromosphere result in major features of solar flares.

073.006 Remarks on sequence-of-plasma-instabilities models of solar flares. B. Coppi.
Astrophys. Journ., Vol. 195, 545 - 552 (1975).
The main requirements of solar flare models that explain

their development in terms of an appropriate sequence of plasma instabilities are discussed. This sequence is characterized by a "preheating phase" of the local chromospheric plasma, a "flash phase" resulting from an explosive instability leading to particle acceleration and onset of plasma turbulence, and in the case of large flares by a "hot phase" in which a significant amount of magnetic energy is converted into plasma flow and thermal energy. The relevant evolution of the electron distribution function and the two-dimensional flow and magnetic field configurations that can be consistent with the above-mentioned model are considered.

073.007 On the brightness fluctuation of coarse mottles of the chromospheric network in a line.
V. E. Merkulenko, V. I. Polyakov, V. I. Skomorovskij.
Issled. po geomagnetizmu, aehron. i fiz. Solntsa. Vyp. (No.) 31. Moskva, Nauka, 1974, p. 13 - 23. In Russian. – Abstr. in Referativ. Zhurn. 51. Astron., 2.51.454 (1975).

073.008 A preliminary model of the magnetic field structure of a coarse mottle of the chromospheric network. V. E. Merkulenko, V. I. Polyakov, V. I. Skomorovskij.
Issled. po geomagnetizmu, aehron. i fiz. Solntsa. Vyp. (No.) 31. Moskva, Nauka, 1974, p. 92 - 103. In Russian. – Abstr. in Referativ. Zhurn. 51. Astron., 2.51.455 (1975).

073.009 The mechanism of emergence of chromospheric flares. P. R. Romanchuk, V. N. Krivodubskij.
Astron. Obs., Kiev, Preprint No. 2, 17 pp. (1974). In Russian.

The mechanism of solar chromospheric flares which provides an accumulation of energy in the form of electrical charges in a region of reduced turbulent conductivity of solar plasma, the fast release of this energy and subsequent acceleration of the particles to relativistic velocities have been considered.

073.010 On a mechanism of formation of solar quiescent prominences. P. V. Sasorov.
Astron. Zhurn. Akad. Nauk SSSR, Vol. 52, 106 - 111 (1975). In Russian. English translation in Soviet Astron., Vol. 19, No. 1.

A mechanism of formation of quiescent prominences proposed by Pikel'ner (1971) is further investigated. The prominences are formed by gas flowing from the chromosphere to the region of hollows in magnetic field line arches. The well-known phenomena of the "depletion" of the corona in the neighbourhood of quiescent prominences may be naturally explained with the mechanism discussed.

073.011 Hydrogen emission in a disintegrated prominence with bright Balmer continuum.
N. A. Yakovkin, M. Yu. Zel'dina, A. S. Rakhubovskij.
Astron. Zhurn. Akad. Nauk SSSR, Vol. 52, 112 - 118 (1975). In Russian. English translation in Soviet Astron., Vol. 19, No. 1.

The Balmer lines Hα–H38, many metal and helium lines, the bright Balmer continuum beginning near the H32 are seen in the prominence spectrum. The electron density $n_e = 10^{11}$ cm^{-3} and effective length $l = 3 \times 10^9$ cm were determined by the Stark effect and Saha equations for the highest levels respectively. The electron temperature $T_e = 7200$ K was determined by nine methods. The obtained degree of ionization of hydrogen is less then 10%. It is shown that the Hα emission due to scattering of photospheric radiation is close to 80%. The radiation temperature in the Hα line (4750 K), in the Lα line (6950 K), in Lyman continuum (6100 K) and Balmer continuum (2950 K) were also calculated.

073.012 Laboratory reproduction of a solar X-ray flare

spectrum. Eh. Ya. Gol'ts, I. A. Zhitnik,
Eh. Ya. Kononov, S. L. Mandel'shtam, Yu. V. Sidel'nikov.
Dokl. Akad. Nauk. SSSR. Ser. Mat. Fiz., Vol. 220, 560 - 563
(1975). In Russian.

073.013   Active prominence on July 20, 1972. I. On the rotational motions in a prominence.
T. P. Nikiforova.
Solnechnye Dannye 1974 Byull., No. 11, p. 100 - 103 (1975).
In Russian.

Rotational motions were detected in the active prominence from the screw structure of the H and K lines. Velocities of rotational motions up to 30 km/sec, periods of 36 and
33 min, and radii of 5000 and 9000 km have been found.

073.014   A few comments on the two-ribbon flare without sunspots of 29 July 1973.   L. Hejna.
Bull. Astron. Inst. Czechoslovakia, Vol. 26, 119 - 126 (1975).

The overall development of the active region of the two-
ribbon flare without sunspots is studied in the paper. Common features between this type of flare and proton flares are
sought. The conclusion is drawn that the spotless flares are a
kind of "pure" type of the proton flare with a relatively simple structure, free of the "deformations" due to high intensity
magnetic fields in sunspots.

073.015   Thermodynamical approach to current interruption model of solar flares.   O. Kaburaki.
Publ. Astron. Soc. Japan, Vol. 27, 45 - 69 (1975).

The foundation for the current interruption theory of
solar flares is provided through thermodynamical consideration by taking into account the appearance of anomalous
resistivity in a turbulent plasma. By evaluating the free
energies of a plasma under an electric field with normal and
anomalous conductivities, it is shown that the transition
between these two states is a first order phase transition.
Some thermodynamical relations and properties characterizing the behavior of the plasma-current system under varying electric fields are derived. Possible models of solar flares
are presented and numerical evaluations are given for some
typical cases.

073.016   Forbidden lines of highly ionized iron in solar flare spectra.
G. A. Doschek, U. Feldman, K. P. Dere, G. D. Sandlin, M. E.
VanHoosier, G. E. Brueckner, J. D. Purcell, R. Tousey.
Astrophys. Journ., (Letters), Vol. 196, L83 - L86 (1975).

Forbidden lines of Fe XVIII, Fe XIX, and Fe XXI are
identified at 974, 1118, and 1354 Å, respectively, in NRL
ATM solar flare spectra. These lines are due to magnetic dipole
transitions between levels of the ground configurations. The
widths of the Fe XIX and Fe XXI lines are ~0.5 Å, which is
substantially greater than expected in ionization equilibrium.
The intensity-time behavior and the widths of these lines are
discussed for the 1973 June 15 flare.

073.017   The Compton effect in the chromosphere.
M. Missana.
Astrophys. Space Sci., Vol. 33, 239 - 243 (1975).

The author assumes that the lines of the solar spectrum,
formed in the reversing layer, when travelling for an optical
thickness τ in the chromosphere are subjected to a red shift
$\Delta\lambda(\tau)$ due to multiple Compton scattering. He limits his investigation to the difference of the red shift at the limb–the
red shift at the centre of the disc giving a good agreement between the theoretical results and observations.

073.018   Penetration of protons from solar flares into the polar caps.
I. I. Alekseev, A. P. Kropotkin, V. P. Shabanskij.
Geomagn. Aeronom., Vol. 15, 13 - 19 (1975). In Russian.

073.019   Fe XXIII 263 Å and Fe XXIV 255 Å emission in solar flares.   K. G. Widing.
Astrophys. Journ., (Letters), Vol. 197, L33 - L35 (1975).

A line observed at 263.7 Å in ATM/NRL solar flare
spectra is identified as Fe XXIII $2s^2$ $^1S_0 - 2s2p$ $^3P_1$. The identification is confirmed by observation of the intercombination
line along the isoelectronic sequence, including S XIII, Ar XV,
Ca XVII, and Cr XXI. The predicted intensity of Fe XXIII
263 Å relative to Fe XXIV 255 Å is in qualitative agreement
with the observed. Significant characteristics of the Fe XXIII
and Fe XXIV images observed in the flare of 1973 June 15 are
(1) an archlike spatial structure spanning the magnetic neutral
line and (2) a Doppler-widened emission core, which may pin-
point the site of the energy release.

073.020   The hydrogen Balmer lines and the structure of the quiet solar chromosphere. I: Observations at the limb.   P. Mein, N. Mein.
Solar Physics, Vol. 40, 317 - 328 (1975).

Spectra of Hα, Hβ and Hδ have been taken under good
seeing conditions with the vacuum tower telescope of Sacramento Peak Observatory. Intensity curves are given at various
wavelengths in these lines to permit further comparison with a
theoretical model. Between 2000 and 6000 km above the
limb the average thermal + turbulent velocity of the atoms is
found to increase from 20 km s^{-1} to 30 km s^{-1} and the mean
number of hydrogen atoms per cm^3 in level 2 is given by
$\log n_2 = 4.5 - 0.00056\,(z - 2000)$, z being the altitude above
the limb in km. For line profile computations a new interpolation formula is presented.

073.021   Investigation of five white light flares.
Y. M. Slonim, Z. B. Korobova.
Solar Physics, Vol. 40, 397 - 410 (1975).

On the basis of original observations, five white-light
flares (WLF) are investigated. Evidence is given that their emission is located in two points brightening on either side of the
line $H \parallel = 0$ and lying at the foot points of chromospheric
loops. The area of WLF is $\approx 5 \times 10^{-6}$ hemisphere, i.e. ≈ 0.007
of Hα flare area; the intensity of WLF is sometimes twice that
of the background at the center of the disk. WLF are resolved
into more bright and fine knots of $\approx 2''$ in diameter. The position of WLF coincides with the brightest knots of Hα flares
which are characterized by wide wings with rapid increase and
decrease.

073.022   Short-lived flare-like phenomena in the quiet chro-mosphere.   D. L. Glackin.
Solar Physics, Vol. 41, 115 - 118 (1975).

Objects resembling small and intense flares that last on
the order of tens of minutes and seem to be unrelated to
other activity have been observed. They are strikingly evident
in full-disc Hα cinematograms.

073.023   The relationship between solar flares and solar sector boundaries.   P. H. Dittmer.
Solar Physics, Vol. 41, 227 - 231 (1975).

A superposed epoch analysis of 1964–1970 solar flares
shows a marked increase in flare occurrence within a day (13°
of longitude) of (−+) solar sector boundaries as well as a local
minimum in flare occurrence near (+−) sector boundaries.
This preference for (−+) boundaries is more noticeable for
northern hemisphere flares, where these polarities match the
Hale polarity law, but is not reversed in the south. Plage
regions do not show such a preference.

073.024   The identification of solar flare Fe XVIII to Fe XXIII emission lines from $2s^n 2p^k - 2s^{n-1} 2p^{k+1}$ transitions.   B. C. Fawcett, R. D. Cowan.
Monthly Notices Roy. Astron. Soc., Vol. 171, 1 - 7 (1975).

Solar flare emission lines due to $2s^n 2p^k - 2s^{n-1} 2p^{k+1}$

transitions of Fe XVIII to Fe XXIII are identified in the spectral region between 66 and 140 Å. The published solar flare spectra from the Goddard Space Flight Center experiment on OSO 5 were used for this analysis. Line classifications are based on data acquired from recent laboratory observations and theoretical extrapolations from these data. Previous identifications are revised and observations which may allow the measurement of solar flare densities are discussed.

073.025 **Chromospheric flares or chromospheric aurorae?**
L. D. De Feiter.
Space Sci. Rev., Vol. 17, 181 - 193 (1975). – Review paper – see 012.002.

This paper discusses some of the well-documented flare phenomena and possible analogies with magnetospheric substorm phenomena. Such analogies do exist, but also important differences. The combination of forces from magnetospheric and solar physicists will bring us closer to the understanding of these nearby examples of unstable cosmic plasmas.

073.026 **On emission line profiles in spectra of solar flares.**
S. I. Grachev.
Astron. Zhurn. Akad. Nauk SSSR, Vol. 52, 326 - 331 (1975).
In Russian. English translation in Soviet Astron., Vol. 19, No. 2.

Emission line profiles for the source function $S(\tau) = a(1+B\tau)$ are calculated in case of a one-dimensional medium of thickness τ_0, moving with constant velocity gradient γ. A comparison of theoretical profiles with those in spectra of solar flares is made.

073.027 **Metal emission in a bright disintegrated prominence.**
N. A. Yakovkin, M. Yu. Zel'dina,
A. S. Rakhubovskij.
Astron. Zhurn. Akad. Nauk SSSR, Vol. 52, 332 - 337 (1975).
In Russian. English translation in Soviet Astron., Vol. 19, No. 2.

The abundance of Na, Mg, Ca, Sc, Ti, Fe, Sr and Ba relative to hydrogen in a prominence was determined; it is approximately (except abundance of Na) as in the photosphere and in the chromosphere. The metals are singly ionized; calcium, strontium and barium are some exceptions.

073.028 **A morphological study of a solar active region in August 1972. II. A correlation analysis of morphologies of flares and spots.**
Solar Phys. Division, Yunnan Obs.
Acta Astron. Sinica, Vol. 15, 173 - 189 (1974). In Chinese.

A correlation analysis of the morphology of nine solar flares of a large solar active region in August 1972 and the morphology of the fine structure of sunspot groups in this region has been carried out.

073.029 **A newly observed solar feature: macrospicules in He II 304 Å.**   J. D. Bohlin, S. N. Vogel, J. D. Purcell, N. R. Sheeley, Jr., R. Tousey, M. E. VanHoosier.
Astrophys. Journ., (Letters), Vol. 197, L133 - L135 (1975).

He II 304 Å spectroheliograms, obtained with the NRL extreme-ultraviolet slitless spectrograph during the Skylab mission, show spikelike structures at the sun's polar limb which resemble the familiar Hα spicules. However, the relatively large size and long life of these He II features has led the authors to distinguish them by the name macrospicules. The macrospicules appear as protuberances or jets, ranging from 5″ to over 60″ in length, from 5″ to 30″ in width, and from 5 to over 40 minutes in lifetime. The most radical departure from Hα spicules is that macrospicules occur only within the chromospheric boundaries of coronal holes.

073.030 **Nonisothermal magnetostatic equilibria in a uniform gravity field. II. Sheet models of quiescent prominences.**   B. C. Low.
Astrophys. Journ., Vol. 198, 211 - 217 (1975).

The author uses the mathematical formulation of Low to construct magnetostatic models of the quiescent prominence in the form of a vertical infinite sheet of material which is both denser (at 1.67×10^{-13} g cm^{-3}) and cooler (at 5000 K) than its surrounding. A magnetic field of the order $(8\pi p_0)^{1/2} \sim 2$ gauss serves to support the weight of the prominence and to confine the prominence material in a thin sheet. The gross observed prominence features are reproduced. In particular, he finds that a prominence sheet of thickness 100 km may be approximated by an isobaric model in spite of the presence of the ambient 2 gauss magnetic field. This is a first attempt at considering both questions of magnetostatic support and energy transport of the quiescent prominence.

073.031 **A comparative study of the sudden disappearance of two filaments, with and without flare brightening.**
G. Banos, T. Prokakis.
Astron. Astrophys., Vol. 39, 245 - 250 (1975).

The results of a comparative study of two "disparitions brusques", with and without flare brightening, are presented. It is found that a correlation probably exists between the impulsiveness of the filament's eruption, determined by the maximum apparent (or real) ascent velocity of the material, and the importance of the flare. Probable explanations of the observations are discussed.

073.032 **The active prominence on July 20, 1972. II. Analysis of the spectrum.**
T. P. Nikiforova, T. V. Ustyuzhanina.
Solnechnye Dannye 1975 Byull., No. 1, p. 94 - 102 (1975).
In Russian.

A spectrophotometric study of some bright features of the active prominence (showing screw structure of H and K Ca II lines) was made. Kinetic temperatures of (5600–8200) ±1600°K, turbulent velocities of (8.4–14.8) ±1.9 km/sec, and electron densities $n_e \leqslant 10^{11}$ cm^{-3} were determined. The Balmer lines show an increase in population with the number of the line.

073.033 **Étude statistique des associations entre le champ magnétique des groupes de taches solaires et la morphologie des éruptions solaires et des sursauts associés sur ondes décimétriques.**   A. Koeckelenbergh.
Ciel et Terre, Vol. 91, 119 - 133 (1975).

On the basis of the observations made at Uccle and Humain, one has attempted a statistical study of the associations between descriptive factors of the magnetic field of the sunspots groups and the morphology of the flares and the bursts on decimetric wave range. One points out that three groups of factors correspond each to a phase of the flare processes. One sees the probable part of the microstructures and the general magnetic field. It appears an information on the behavior of the active centers according to their magnitude and the further development of a flare and an associated decimetric burst.

073.034 **On the behaviour of the hydrogen Lyman series in flares.**   L. D. De Feiter, Z. Švestka.
Solar Physics, Vol. 41, 415 - 424 = Mitt. Fraunhofer Inst., Freiburg, No. 122 (1975).

The Lyman spectrum of hydrogen has been computed for a number of flare models, characterized by the column density of hydrogen atoms in the ground state, the electron density and the electron temperature. Broadening by the thermal Doppler effect and by Stark effect has been accounted for. The source functions for the individual lines of the series have been derived from non-LTE calculations of the excitation in hydrogen flares.

073.035 **Sunspot motion, flares and type III bursts in McMath 11482.**   H. Zirin, B. Lazareff.

Solar Physics, Vol. 41, 425 - 438 (1975).

The authors have studied a series of flares in McMath 11482, 1972 August 19−22, with particular reference to the basis for the flares and comparison with dekameter radio data. They find that the flares were produced by rapid (∼1000 km h⁻¹) westward motion of a large new p spot. All flares occurring in front of the spot produce type III bursts. The time of type III emission agrees perfectly with the start of the Hα flare. Thus type III bursts are only produced in favorable configurations.

073.036 Doppler widths of solar chromospheric emission lines. E. C. Bruner, Jr., E. G. Chipman.
Bull. American Astron. Soc., Vol. 7, 255 (1975). − Abstr. AAS.

073.037 The pressure balance and currents in active region loop structures. P. V. Foukal.
Bull. American Astron. Soc., Vol. 7, 346 (1975). − Abstr. AAS.

073.038 Kinematics of sprays, surges, and eruptive prominences.
Y. Nakagawa, S. T. Wu, E. Tandberg-Hanssen.
Bull. American Astron. Soc., Vol. 7, 348 (1975). − Abstr. AAS.

073.039 The high-energy limb event of January 17, 1974.
R. Tousey, J. D. Bohlin, O. K. Moe, J. D. Purcell, N. R. Sheeley.
Bull. American Astron. Soc., Vol. 7, 348 (1975). − Abstr. AAS.

073.040 Eruptive prominences in the EUV: observations with the Harvard spectrometer on ATM.
E. J. Schmahl.
Bull. American Astron. Soc., Vol. 7, 348 - 349 (1975).
Abstr. AAS.

073.041 Spectral and slitjaw observations of an eruptive and untwisting filament. K. Jockers, O. Engvold.
Bull. American Astron. Soc., Vol. 7, 349 (1975). − Abstr. AAS.

073.042 Analysis of eruptive prominence of 30th April 1974.
S. T. Wu, M. Dryer, P. S. McIntosh, E. Reichmann.
Bull. American Astron. Soc., Vol. 7, 349 (1975). − Abstr. AAS.

073.043 Fine structure in quiescent prominences.
J. M. Malville.
Bull. American Astron. Soc., Vol. 7, 350 (1975). − Abstr. AAS.

073.044 Simultaneous hydrogen and helium emission line intensities in prominences.
R. M. E. Illing, D. A. Landman, D. L. Mickey.
Bull. American Astron. Soc., Vol. 7, 350 (1975). − Abstr. AAS.

073.045 The nature of magnetic field reconnection in solar flares: implications of recent observational evidence.
R. L. Moore.
Bull. American Astron. Soc., Vol. 7, 351 - 352 (1975).
Abstr. AAS.

073.046 A non-thermal mechanism for solar flares.
R. P. Lin.
Bull. American Astron. Soc., Vol. 7, 352 (1975). − Abstr. AAS.

073.047 Flare mechanisms based on the current limitation concept. D. S. Spicer.
Bull. American Astron. Soc., Vol. 7, 352 (1975). − Abstr. AAS.

073.048 The super heating instability as a trigger for solar flares. D. A. Tidman, D. S. Spicer, J. Davis.
Bull. American Astron. Soc., Vol. 7, 352 (1975). − Abstr. AAS.

073.049 Time dependent evaporative flare model.

T. Hirayama, F. Endler.
Bull. American Astron. Soc., Vol. 7, 352 - 353 (1975).
Abstr. AAS.

073.050 A general approach to the interpretation of chromospheric mottles. K. B. Gebbie, R. Steinitz.
Bull. American Astron. Soc., Vol. 7, 353 (1975). − Abstr. AAS.

073.051 Calcium II line correlations for the active quiet chromosphere. D. R. Brown.
Bull. American Astron. Soc., Vol. 7, 353 (1975). − Abstr. AAS.

073.052 Height of helium emission in the chromosphere.
S. A. Schoolman, T. Pope.
Bull. American Astron. Soc., Vol. 7, 353 (1975). − Abstr. AAS.

073.053 Observations from Skylab of the density dependent C III multiplet at 1175 Å in active and quiet regions and above the limb. K. Nicolas, G. Brueckner.
Bull. American Astron. Soc., Vol. 7, 353 - 354 (1975).
Abstr. AAS.

073.054 Macro-spicules in He II 304 Å over the sun's polar cap. J. D. Bohlin, S. N. Vogel, J. D. Purcell, N. R. Sheeley, R. Tousey, M. E. Van Hoosier.
Bull. American Astron. Soc., Vol. 7, 354 (1975). − Abstr. AAS.

073.055 A catalogue and classification of flares and flare-like events observed by the ATM XUV spectrograph S082A, and spectroheliograph S082B.
V. E. Scherrer, R. Tousey, G. D. Sandlin.
Bull. American Astron. Soc., Vol. 7, 357 (1975). − Abstr. AAS.

073.056 Transient line emission from ionized iron under flare-like conditions.
J. Davis, P. Kepple, R. Tousey.
Bull. American Astron. Soc., Vol. 7, 357 (1975). − Abstr. AAS.

073.057 Line profiles of the Fe XXIV emission at 192 Å and 255 Å in solar flares.
G. E. Brueckner, O. K. Moe, M. E. Van Hoosier.
Bull. American Astron. Soc., Vol. 7, 357 (1975). − Abstr. AAS.

073.058 On the mechanism of formation of loop prominence systems. B. R. De.
Bull. American Astron. Soc., Vol. 7, 357 (1975). − Abstr. AAS.

073.059 Magnetic structure and flares in active region 331.
J. Sutorik, C. Sawyer.
Bull. American Astron. Soc., Vol. 7, 357 (1975). − Abstr. AAS.

073.060 Characteristics of ³He rich flares. G. J. Hurford.
Bull. American Astron. Soc., Vol. 7, 357 - 358 (1975). − Abstr. AAS.

073.061 The D3 chromosphere, coronal holes, and stellar X-rays. H. Zirin.
Bull. American Astron. Soc., Vol. 7, 359 (1975). − Abstr. AAS.

073.062 Emission lines in the wings of H and K Ca II.
R. E. Stencel.
Bull. American Astron. Soc., Vol. 7, 359 (1975). − Abstr. AAS.

073.063 Solar limb brightening at submillimeter wavelengths.
C. Lindsey, H. S. Hudson.
Bull. American Astron. Soc., Vol. 7, 360 (1975). − Abstr. AAS.

073.064 The relation between flares and the missing energy in spots. D. J. Mullan.
Bull. American Astron. Soc., Vol. 7, 362 (1975). − Abstr. AAS.

073.065 **Flare model chromosphere and photospheres.**
M. E. Machado, J. L. Linsky.
Bull. American Astron. Soc., Vol. 7, 362 (1975). – Abstr. AAS.

073.066 **Time variations in the EUV line emission from the chromosphere and corona.** J. E. Vernazza.
Bull. American Astron. Soc., Vol. 7, 365 (1975). – Abstr. AAS.

073.067 **The work of the diode array.**
D. M. Rust, C. A. Bridges.
Bull. American Astron. Soc., Vol. 7, 365 - 366 (1975).
Abstr. AAS.

073.068 **Diffusion effects on line intensities in the solar transition region.** H. Gerola, R. A. Shine.
Bull. American Astron. Soc., Vol. 7, 366 (1975). – Abstr. AAS.

073.069 **Solar prominences.** Lyu Van Lyong.
Priroda, No. 5.75, p. 30 - 38 (1975). In Russian.

073.070 **Solar flares in August 1972 and connected geophysical phenomena.**
V. I. Bondarenko, A. K. Yukhimuk.
Kosmich. issled. na Ukraine. Resp. mezhved. sb., 1974, vyp.
(No.) 5, p. 14 - 19. In Russian. – Abstr. in Referativ. Zhurn.
51. Astron., 5.51.452 (1975).

073.071 **Flight time of solar fast particles from flares to the earth. Supplement III.** L. Křivský.
Bull. Astron. Inst. Czechoslovakia, Vol. 26, 181 - 185 (1975).

073.072 **Sur une différence d'assombrissement apparent du fond chromosphérique K$_{2,3,2}$ entre le méridien central et l'équateur du soleil.** C. J. Macris.
Comptes Rendus Acad. Sci. Paris, Sér. B, Vol. 280, 637 - 640 (1975).
Les mesures photométriques d'assombrissement apparent faites sur des spectrohéliogrammes K$_{2,3,2}$, pris à l'observatoire d'Arcetri, ont décelé une disymétrie de l'intensité du fond chromosphérique entre le méridien central et l'équateur. La différence de l'intensité augmente en latitude héliographique.

073.073 **A large flare of August 2, 1972.**
J. Kubota, H. Kurokawa, T. Kureizumi.
Rep. Ionosph. Space Res. Japan, Vol. 26, 288 - 290 = Contr.
Kwasan Hida Obs., Univ. Kyoto, No. 208 (1972).

073.074 **Monochromatic observation of a flare on August 4, 1972.** K. Maeda, N. Oda, K. Nakayama.
Rep. Ionosph. Space Res. Japan, Vol. 26, 291 - 294 = Contr.
Kwasan Hida Obs., Univ. Kyoto, No. 209 (1972).

073.075 **Investigation of the transition region above sunspots based on polarization radio measurements.**
G. B. Gel'frejkh, V. P. Nefed'ev.
Pis'ma v Astron. Zhurn., Vol. 1, No. 6, p. 32 - 35 (1975). In Russian.
From the analysis of the 3.2-cm polarized emission variation at the early stage of sunspot development, the thickness of the transition region between chromosphere and corona is estimated.

073.076 **Notes on the chromosphere.** L. B. Nadeau.
Journ. American Ass. Variable Star Observers, Vol. 1, 62 - 63 (1972).

073.077 **Study of He I emission lines in the solar atmosphere. III. The triplet-singlet line intensity ratios in solar prominences.**
J. N. Heasley, E. Tandberg-Hanssen, W. J. Wagner.
Astron. Astrophys., Vol. 40, 391 - 395 (1975).

The authors present observations of the He I spectrum for the quiescent prominence of 4 November 1971. Integrated line-intensity ratios are compared with theoretical calculations to obtain a hydrogen density of $\sim 10^{10}$ cm^{-3}. The widths of He I and H I Balmer lines indicate an electron temperature of ~ 7000 K and a turbulent broading velocity of ~ 8 km/s for this object.

073.078 **The sun: flares, "moustaches", spicules.**
S. B. Pikel'ner.
Priroda, No. 6.75, p. 15 - 19 (1975). In Russian.

073.079 **Determination of physical parameters of prominences with the help of equilibrium equations.**
Ts. Chultemijn (*Chultem*).
Astron. Zhurn. Akad. Nauk SSSR, Vol. 52, 572 - 578 (1975).
In Russian. English translation in Soviet Astron., Vol. 19, No. 3.
A system of statistical equilibrium equations for the continuum and the first 10 levels of the hydrogen atom in prominences is worked out. A method for joint determination of physical parameters of prominences is proposed: optical depth of the prominence in the centre of the H$_\alpha$ line, electron concentration, electron temperature and effective thickness.

073.080 **Optical thickness and electron concentration in the chromospheric flare of August 4, 1972.**
O. N. Mitropol'skaya.
Astron. Zhurn. Akad. Nauk SSSR, Vol. 52, 587 - 592 (1975).
In Russian. English translation in Soviet Astron., Vol. 19, No. 3.
Optical thickness τ ($\Delta\lambda$) was calculated from the deepness of metallic absorption lines situated in wings of H$_\beta$ and H$_\gamma$ lines in the flare and in the photosphere. The dependence of lg τ from the lg $\Delta\lambda$ is calculated by the least-squares method. The electron concentration in the flare n_e and number of hydrogen atoms at the second level n_2 are determined with the theory of the Stark effect for the wings of H$_\beta$ and H$_\gamma$ lines. Profiles of emission lines H$_\beta$ and H$_\gamma$ for the flare are determined taking into account the variation of profiles of hydrogen Fraunhofer lines of the photosphere due to absorption of the radiation in the flare layer.

073.081 **Lifetimes of cells in the solar network.**
R. N. Moses.
Observatory, Vol. 95, 107 - 109 (1975). – Letter.

073.082 **Velocity oscillations of the solar atmosphere at chromospheric heights.** J. C. Bhattacharyya.
Bull. Astron. Soc. India, Vol. 2, 40 (1974). – Abstract.

073.083 **An interpretation of the correlation of the intensity fluctuations in the K-line of Ca II and b$_1$ of Mg I.**
K. R. Sivaraman.
Bull. Astron. Soc. India, Vol. 2, 40 (1974). – Abstract.

073.084 **Photometric research for the K$_{2,3}$ chromospheric flocculi.** C. J. Macris.
Praktika Acad. Athens, Vol. 49, 215 - 237 = Res. Center Astron., Applied Math., Acad. Athens, Ser. 1 (Astron.), No. 37 (1974).
The author estimated the mean value of the intensity of the flocculi relatively to the mean value of the intensity of the chromospheric background at the center of the solar disk.

073.085 **Ultraviolet emission line profiles of flares and active regions.** G. E. Brueckner.
IAU Symposium No. 68, (see 012.017), p. 135 - 151 (1975).

073.086 **Fe XXIV emission in solar flares observed with the NRL/ATM XUV slitless spectrograph.**
K. G. Widing.
IAU Symposium No. 68, (see 012.017), p. 153 - 163 (1975).

073.087 **X-ray and EUV spectra of solar flares and laboratory plasmas.** G. A. Doschek.
IAU Symposium No. 68, (see 012.017), p. 165 - 181 (1975).

073.088 **Studies of the dynamic structure and spectra of solar X-ray flares.** S. W. Kahler, A. S. Krieger, J. K. Silk, R. W. Simon, A. F. Timothy, G. Vaiana.
IAU Symposium No. 68, (see 012.017), p. 185 (1975).
Summary.

073.089 **On the thermal structure of the flare-produced plasma.** I. Craig.
IAU Symposium No. 68, (see 012.017), p. 187 - 189 (1975).
Summary.

073.090 **High time resolution analysis of solar flares observed on the ESRO TD-1A satellite.**
P. Hoyng, J. C. Brown, G. Stevens, H. F. Van Beek.
IAU Symposium No. 68, (see 012.017), p. 233 - 235 (1975).
Summary.

073.091 **Inference of the hard X-ray source dimensions in the 1972, August 7 white light flare.**
D. M. Rust.
IAU Symposium No. 68, (see 012.017), p. 243 (1975).
Summary. — For the full text of the paper see 073.004.

073.092 **Gamma-ray lines from solar flares.**
R. Ramaty, R. E. Lingenfelter.
IAU Symposium No. 68, (see 012.017), p. 363 - 383 (1975).

073.093 **Fast electrons in small solar flares.** R. P. Lin.
IAU Symposium No. 68, (see 012.017), p. 385 - 409 (1975).
This review summarizes both the direct spacecraft observations of non-relativistic solar electrons, and observations of the X-ray and radio emission generated by these particles at the sun and in the interplanetary medium.

073.094 **Nuclei of heavy elements from solar flares.**
C. Y. Fan, G. Gloeckler, D. Hovestadt.
IAU Symposium No. 68, (see 012.017), p. 411 - 421 (1975).

073.095 **Implications of NRL /ATM solar flare observations on flare theories.** C. C. Cheng, D. S. Spicer.
IAU Symposium No. 68, (see 012.017), p. 423 - 424 (1975).
Summary.

073.096 **Nonthermal processes in large solar flares.**
H. S. Hudson, T. W. Jones, R. P. Lin.
IAU Symposium No. 68, (see 012.017), p. 425 - 426 (1975).
Summary.

073.097 **On the acceleration processes in solar flares.**
Z. Švestka.
IAU Symposium No. 68, (see 012.017), p. 427 - 439 (1975).

073.098 **Differential rotation, meridional and random motions of the solar Ca^+ network.**
E. H. Schröter, H. Wöhl.
Solar Physics, Vol. 42, 3 - 16 (1975).
From high precision computer controlled tracings of bright Ca^+ mottles the authors investigated differential rotation, meridional and random motions of these chromospheric fine structures.

073.099 **Height of helium emission in the chromosphere.**
T. Pope, S. A. Schoolman.
Solar Physics, Vol. 42, 47 - 51 (1975).
Filtergrams of the limb show the He I D_3 chromosphere as a shell which is separated from the limb by a gap. The height

of maximum D_3 contribution occurs at about 1350 km above the limb and is independent of the intensity of the D_3 emission. We interpret this effect as the height to which coronal EUV radiation is capable of penetrating the atmosphere.

073.100 **Interpretation of Hα contrast profiles of chromospheric fine structures.** L. E. Cram.
Solar Physics, Vol. 42, 53 - 66 (1975).
Recent observations of the Hα contrast profiles of identifiable chromospheric fine structures are interpreted in terms of an empirical model. Quantitative models for the Hα chromosphere near the limb suggest that the 'dark band' phenomenon is due to low opacity in the neighbourhood of the temperature minimum, while the peculiar appearance of mottle contrast profiles near the limb is explained in terms of foreground absorption.

073.101 **On the spicular density enhancement in the region of formation.** S. K. Sahai, R. J. Bessey.
Solar Physics, Vol. 42, 67 - 70 (1975). — Research note.

073.102 **Emission of helium in prominences and the chromosphere.** N. N. Morozhenko.
Solar Physics, Vol. 42, 71 - 78 (1975).
The excitation of the levels 2^1P, 3^1D, 5^1D, 6^1D and 7^1D of helium in quiescent prominences and the chromosphere is considered using the observations made during the 1952, 1961 and 1970 solar eclipses as well as by means of high altitude coronographs. It is shown that the lower levels of the singlet series of helium, like the lower levels of hydrogen, are excited by resonance ($\lambda = 584$ Å) and photospheric radiation. The higher levels are populated by transitions from the triplet system, so that a relative state LTE is established between the systems for $n \geqslant 6$.

073.103 **Changes of the Hα fibril pattern during solar flares.** A. Bruzek.
Solar Physics, Vol. 42, 215 - 217 (1975).
A pattern of parallel Hα fibrils inside a network cell changed into a pattern of vertical or radial fibrils after a flare had covered the region. This may be evidence for a change of the magnetic configuration during flare occurrence.

073.104 **X-ray heating of a low-temperature region in chromospheric flares.** B. V. Somov.
Solar Physics, Vol. 42, 235 - 246 (1975).
Part of the proper X-ray emission of a flare is absorbed in the chromosphere and heats the region which creates an optical (in particular Hα) flare emission. The heating of chromosphere by X-ray emission may be responsible for the diffuse halo around the flare kernels. The optical emission of flare kernels, whose main sources of heating are energetic particles and/or thermal fluxes, may be also increased. By simple model calculations the present paper discusses the possibility of such effects for the large flare of 1972 August 7.

073.105 **The nuclear composition of solar energetic particles — new source of information about solar flares.** V. N. Lutsenko.
Izv. AN SSSR. Ser. fiz., Vol. 39, 250 - 258 (1975). In Russian.
Abstr. in Referativ. Zhurn. 62. Issled. kosmich. prostranstva, 6.62.202 (1975).

073.106 **On the polarization of X-ray radiation of solar flares.** A. A. Korchak.
Izv. AN SSSR. Ser. fiz., Vol. 39, 281 - 286 (1975). In Russian.
Abstr. in Referativ. Zhurn. 62. Issled. kosmich. prostranstva, 6.62.206 (1975).

073.107 **Generation of hydrogen and helium isotopes and of γ-radiation in solar flares.**

I. A. Ibragimov, G. E. Kocharov.
Izv. AN SSSR. Ser. fiz., Vol. 39, 287 - 303 (1975). In Russian.
Abstr. in Referativ. Zhurn. 62. Issled. kosmich. prostranstva,
6.62.207 (1975).

073.108   Acceleration of charged particles in processes of solar flare-type.   S. I. Syrovatskij.
Izv. AN SSSR. Ser. fiz., Vol. 39, 359 - 374 (1975). In Russian.
Abstr. in Referativ. Zhurn. 62. Issled. kosmich. prostranstva,
6.62.216 (1975).

073.109   A shock wave with emission and the explosive phase of a chromospheric flare.   A. A. Rumyantsev.
Izv. AN SSSR. Ser. fiz., Vol. 39, 383 - 387 (1975). In
Russian. – Abstr. in Referativ. Zhurn. 62. Issled. kosmich.
prostranstva, 6.62.219 (1975).

073.110   Motions in a prominence of April 17, 1972.   O. Engvold, T. Kozar, B. Rompolt.
Inst. Teor. Astrophys., Blindern–Oslo, Rep. No. 40,
2 + 9 pp. (1974).
    The velocity field in a prominence of April 17, 1972, is
studied. An apparently poor spatial correspondence of the
prominence seen in the Ca II K line spectrograms and Hα
filtergrams is discussed.

073.111   Relationships – Waldmeier grouping, solar flares.   M. Taylor.
Journ. American Ass. Variable Star Observers, Vol. 3, 30 - 34
(1974).
    The purpose of this paper is to determine if any relation-
ships exist between solar flares detected with SES equipment,
and the specific sunspot groups classified under the Waldmeier system, when determined visually.

073.112   Asymmetries of the solar Ca II lines.   J. N. Heasley.
Thesis, Yale Univ., New Haven, Conn. (USA). 159 pp. University Microfilms Order No. 74-13,045 (1973).

073.113   Evolution of solar active regions.   R. N. Moses, Jr.
Thesis, Ohio State Univ., Columbus (USA). 171 pp.
University Microfilms Order No. 74-17,795 (1974).

073.114   Photoelectric solar limb scans for determining mean chromospheric structure.   P. E. Barnhart.
Thesis, Ohio State Univ., Columbus (USA). 213 pp. University
Microfilms Order No. 74-24,296 (1974).

073.115   Solar surges: magnetic properties, dynamics and structure.   J.-C. Roy.
Thesis, Univ. Western Ontario, London, (Canada). 221 pp.
[Microfilm copies available from the National Library of
Canada, Ottawa] (1973).

073.116   Dynamic model of solar coronal spicules.   S. K. Sahai.
Thesis, Wyoming Univ., Laramie (USA). 107 pp. University
Microfilms Order No. 74-23,703 (1974).

073.117   An analysis of chromospheric emission lines at the 1962 eclipse.   R. C. Groeneveld.
Proefschrift, Subfaculteit Sterrenkunde, Rijksuniversiteit
Groningen (Netherlands). 106 pp. [This thesis is available from
Rijksuniversiteit, Groningen] (1974).

073.118   Spectral characteristics of flares.   S. Schoolman, A. Title.
Solar wind three, (see 012.020), p. 147 - 149 (1974).

Rotational motions in fine solar structures.

See Abstr. 003.019.

Classification of spectra of K, Ca, Sc, Ti, V and Cr and extrapolation to Fe solar flare lines.
See Abstr. 022.004.

Transitions $2s^2 2p^k - 2s 2p^{k+1}$ of the N I and C I iso-electronic sequences.   See Abstr. 022.029.

Dynamics of the current sheet of a flare. I. Diffusion thickening of the current sheet and estimate of the flare parameters.   See Abstr. 062.030.

Effect of obstacles on the rate of reconnection of magnetic field lines.   See Abstr. 062.039.

Shifts and asymmetries of lines formed in a thermally driven turbulent medium.
See Abstr. 063.015.

Far-infrared solar brightness measured with a balloon-borne lamellar-grating interferometer.
See Abstr. 071.012.

Spectroscopic detection of solar ^{3}He.
See Abstr. 071.020.

Spectral features to be expected from rotational and expansional motions in fine solar structures.
See Abstr. 071.030.

Statistical relations between relative numbers and flare indices.   See Abstr. 072.049.

The microwave structure of coronal condensations and its relation to proton flares.   See Abstr. 074.003.

Direct observations of a flare related coronal and solar wind disturbance.   See Abstr. 074.020.

Limb brightening and dark features observed at 6 cm wavelength.   See Abstr. 074.044.

The effect of flares upon the outer solar corona.
See Abstr. 074.056.

The relationship between coronal bright points and the chromospheric network.   See Abstr. 074.060.

Time variations in coronal active regions.
See Abstr. 074.099.

Interplanetary disturbances: a classification.
See Abstr. 074.126.

X-ray bursts from solar flares behind the limb.
See Abstr. 076.002.

Hard X-ray bursts from flares behind the solar limb.
See Abstr. 076.003.

On the analysis of solar XUV observations and the structure of the chromosphere-corona transition region.
See Abstr. 076.004.

Gamma radiation of solar flares and cosmic ray generation.   See Abstr. 076.005.

Solar flare X-rays in July 1972.
See Abstr. 076.006.

Solar flare X-ray polarisation.
See Abstr. 076.012.

EUV emission, filament activation and magnetic fields in a slow-rise flare.   See Abstr. 076.014.

On the relationships between sfe (crochet) and solar X-ray and microwave bursts.   See Abstr. 076.015.

The location of the site of energy release in an X-ray sub-flare.   See Abstr. 076.018.

The XUV spectrum of He II in quiet regions, a coronal hole, filaments, prominences, and the 7 August 1972 flare.   See Abstr. 076.020.

Estimates of impact polarization of solar flare X-ray lines.   See Abstr. 076.024.

Spatial relationships between X-ray, EUV, and H-alpha sources during a solar flare observed by OSO-7.
See Abstr. 076.025.

General properties of soft X-ray flare images.
See Abstr. 076.026.

Time changes in the structure and spectrum of an X-ray flare.   See Abstr. 076.027.

On the XUV emissions in solar flares observed with the ATM/NRL spectroheliograph.   See Abstr. 076.028.

Solar gamma rays.   See Abstr. 076.033.

Time-dependent 2.2 MeV and 0.5 MeV and 0.5 MeV lines from solar flares.   See Abstr. 076.034.

The origin and implications of gamma rays from solar flares.   See Abstr. 076.035.

Solar gamma-ray lines as probes of accelerated particle directionalities in flares.   See Abstr. 076.036.

X-ray spectra of solar active regions.
See Abstr. 076.041.

X-ray spectroscopy of solar active regions during the third Skylab mission.   See Abstr. 076.042.

Soft X-radiation from single active regions.
See Abstr. 076.045.

Flare-like ultraviolet spectra of active regions.
See Abstr. 076.046.

The structure of solar active regions from EUV and soft X-ray observations.   See Abstr. 076.047.

Association of X-ray flares with solar coronal active regions.   See Abstr. 076.048.

Thermal and nonthermal interpretations of flare X-ray bursts.   See Abstr. 076.051.

Solar flare X-ray measurements and their relation to microwave bursts.   See Abstr. 076.055.

Relation of microwave emission to X-ray emission from solar flares.   See Abstr. 076.056.

X- and γ-ray measurements during the 1972,

August 2 and 7 large solar flares.   See Abstr. 076.057.

High energy gamma-ray radiation above 300 keV associated with solar activity.   See Abstr. 076.058.

Outbursts of γ-radiation observed during the flares of August 4 and 7, 1972 aboard the Prognoz 2 satellite.
See Abstr. 076.064.

3.3 millimeter limb brightening measurements during the 30 June 1973 total solar eclipse.
See Abstr. 077.019.

Some studies on solar microwave bursts in relation to the phases of the associated Hα-flares and their spectral nature.   See Abstr. 077.020.

Solar flare particles: energy-dependent composition and relationship to solar composition.
See Abstr. 078.001.

The variation of solar proton energy spectra and size distribution with heliolongitude.   See Abstr. 078.006.

Solar particle events with anomalously large relative abundance of ^{3}He.   See Abstr. 078.012.

Heavy solar cosmic rays in the January 25, 1971 solar flare.   See Abstr. 078.013.

The relative abundances and energy spectra of solar-flare-accelerated deuterium, tritium, and helium-3.
See Abstr. 078.025.

The acceleration of heavy nuclei in solar flares.
See Abstr. 078.026.

Charge composition of solar cosmic rays in the January 25, 1971 solar flare.   See Abstr. 078.029.

A search for deuterium on the sun.
See Abstr. 080.001.

Kinetic equilibrium and line formation of Na I in the solar atmosphere.   See Abstr. 080.004.

Submillimetre brightness spike at the solar limb.
See Abstr. 080.010.

Dynamics of the solar magnetic field. V. Velocities associated with changing magnetic fields.
See Abstr. 080.018.

Information on the third dimension of solar phenomena deduced by statistical procedures.
See Abstr. 080.048.

Auroral flares and solar flares.   See Abstr. 084.011.

Lunar surface phenomena: solar flare track gradients, microcraters, and accretionary particles.
See Abstr. 094.184.

Lunar microcraters and their solar flare track record.
See Abstr. 094.185.

Charged-particle and micrometeorite impacts on the lunar surface.   See Abstr. 094.186.

Cosmogenic radionuclides in samples from Taurus-Littrow: effects of the solar flare of August 1972.

See Abstr. 094.531.

Solar flare and lunar surface process characterization at the Apollo 17 site. See Abstr. 094.533.

Dynamic modeling of interplanetary disturbances produced by solar flares. See Abstr. 106.033.

Deceleration of flare-generated interplanetary shock waves. See Abstr. 106.038.

A model for the propagation of flare-associated interplanetary shock waves. See Abstr. 106.048.

Analysis of the 0.511 MeV radiation at the OSO-7 satellite. See Abstr. 143.090.

Errata

073.901 Erratum: 'Resonance-line transfer with partial redistribution: a preliminary study of Lyman α in the solar chromosphere' [Astrophys. Journ., Vol. 185, 709 - 726 (1973)]. R. W. Milkey, D. Mihalas. Astrophys. Journ., Vol. 195, 831 (1975).

074 Solar Corona, Solar Wind

074.001 Coronal information from EUV disk spectral line intensities. D. E. Billings, M. Alvarez.
Solar Physics, Vol. 40, 23 - 38 (1975).

Center of disk EUV line intensities from quiet and active regions are used for determining an analytical expression for the variation of temperature with height in the lower corona, including the corona-chromosphere transition region. This approach imposes two coronal temperature regimes in both quiet and active regions. In each case the lower temperature regime is a continuation of the transition region, reaching a maximum of about 1.4 million deg in the quiet and 1.7 million deg in the active region. In the quiet region the high temperature regime, assumed isothermal, has a temperature of about 2.4 million deg, and in the active region, about 4.2 million deg.

074.002 XUV observations of coronal magnetic fields.
N. R. Sheeley, Jr., J. D. Bohlin, G. E. Brueckner,
J. D. Purcell, V. Scherrer, R. Tousey.
Solar Physics, Vol. 40, 103 - 121 (1975).

Spectroheliograms obtained with the Naval Research Laboratory's Extreme Ultraviolet Spectrograph (S082A) on Skylab are compared with Kitt Peak National Observatory magnetograms. A principal result is the characteristic reconnection of flux from an emerging bipolar magnetic region to previously existing flux in its vicinity. Examples of the disappearance of magnetic flux from the solar atmosphere are also shown. The results of a particularly simple, potential field calculation are shown for comparison with the Skylab observations.

074.003 The microwave structure of coronal condensations and its relation to proton flares.
H. Tanaka, S. Énomé.
Solar Physics, Vol. 40, 123 - 131 (1975).

Active regions on the sun in the 20th solar cycle are studied with special reference to their association with proton flares based on microwave interferometric observations at Toyokawa Observatory. It has been reconfirmed that the active regions associated with intense S-component emission with a high 3-cm to 8-cm flux ratio are likely to produce proton flares. About one fourth of 259 active regions during the period investigated are found to have definite features in the spatial distribution of polarization at a wavelength of 3 cm. Active regions with one particular type of polarization pattern have a good correlation with the occurrence of proton flares.

074.004 Magnetic and thermal energies in the solar wind.
L. Diodato, G. Moreno, C. Signorini.
Solar Physics, Vol. 40, 231 - 234 (1975).

Magnetic and thermal energy densities measured in the solar wind by the spacecraft Vela-3/IMP-3, Explorer-34 and HEOS-1 are compared. A linear relation seems to exist, on the average, between these parameters, suggesting a possible equipartition of energy between plasma and magnetic field.

074.005 Oxygen-to-neon abundance ratio in the solar corona.
L. W. Acton, R. C. Catura, E. G. Joki.
Astrophys. Journ., (Letters), Vol. 195, L93 - L95 (1975).

The oxygen-to-neon abundance ratio in the solar corona is determined to be 4.7 ± 1.5 on the basis of a relatively model-independent analysis utilizing 25 separate measurements of the O VIII to Ne IX resonance line ratio. This abundance ratio is smaller than most published results based upon EUV or X-ray observations but is in good agreement with direct measurements of flare cosmic rays.

074.006 Limb-brightening observations from the OSO-7 satellite. I. Electron density and temperature of the non-equatorial corona from EUV lines of Fe XIV and other Fe ions. S. O. Kastner, E. D. Rothe, W. M. Neupert.
Astron. Astrophys., Vol. 37, 339 - 348 (1974).

Spectra in the range 190–300 Å have been observed by the Goddard spectroheliograph on the OSO-7 satellite, at several coronal locations beyond the limb. A comparison is made between the radial variation of Fe XIV emission lines and the expected density dependence of these lines as predicted by Blaha. Excellent agreement is found, and is made use of to study weaker Fe XIV lines and derive preliminary values of electron density. Also the observed intensity ratios of lines of successive iron ions from Fe IX to Fe XV are used, together with the level populations expected in the temperature-dependent ionization equilibrium of iron, to obtain values of electron temperature beyond the limb along solar radii at polar angles of 46° and 8°.

074.007 The effect of evolution of active regions on high-speed quasi-stationary plasma flows from the sun.
V. V. Kasinskij, N. N. Lyakhov.
Issled. po geomagnetizmu, aehron. i fiz. Solntsa, Vyp. (No.) 31. Moskva, Nauka, 1974, p. 51 - 60. In Russian. – Abstr. in Referativ. Zhurn. 51. Astron., 2.51.469; 62. Issled. kosmich. prostranstva, 2.62.276 (1975).

074.008 The solar M-region problem – an old problem now facing its solution? A. Gulbrandsen.
Planet. Space Sci., Vol. 23, 143 - 149 (1975).

The solar M-region problem is briefly reviewed. The Mustel and the Allen-Saemundsson M-region schools are discussed in the light of (a) recent results on coronal structures and solar wind variations, and (b) statistical analyses of coronal-geomagnetic correlations. From this discussion it is suggested that the M-regions should be identified with the central portion of magnetically open solar regions, or coronal holes.

074.009 Detection of coronal holes from λ5303 Fe XIV observations. R. Fisher, S. Musman.
Astrophys. Journ., Vol. 195, 801 - 803 (1975).

The authors use photoelectric measurements of intensity to estimate the distribution of volume emissivity in the coronal green line (λ5303 Fe XIV) over the solar surface. Using this distribution and a calculated dependence of emissivity as a function of T and N_e, it is possible to estimate a coronal density distribution. An example is given which shows a region of drastically reduced emissivity – a coronal hole – which was simultaneously detected by spacecraft.

074.010 Determination of the solar wind velocity from pulsation indices. J. Verö.
Journ. Atmosph. Terr. Phys., Vol. 37, 561 - 564 (1975).
Paper presented at the IAGA-Symposium in Kyoto, September 1973.

An effort is made to deduce solar wind velocities from geomagnetic pulsation indices. It is shown that it is possible to determine the daily solar wind velocity from the pulsation data of a single observatory with smaller mean square error (about ±60 km/sec) than from ΣKp or ΣAp.

074.011 Forbidden-line excitation data for certain coronal lines.
S. J. Czyzak, L. H. Aller, R. N. Euwema.
Astrophys. Journ., Suppl. Ser., No. 272, Vol. 28, 465 - 470 (1974).

Some new data pertinent to forbidden transitions for

coronal lines of S XII, Ca XIII, K XI, Mn XIII, Fe XIV, Co XV, Ni XVI, N XV, Cr IX, Fe X, and Ni XII are presented.

074.012 Waves and instabilities in the solar wind.
J. V. Hollweg.
Rev. Geophys. Space Phys., Vol. 13, 263 - 289 (1975).

The author presents a review of waves and instabilities in the solar wind, concentrating on those aspects that are likely to play important roles in influencing the dynamic and thermodynamic states of the general solar wind expansion. He considers in particular the roles played by various waves and instabilities in influencing the heating and expansion of the solar wind, the angular momentum of the solar wind, the solar wind thermal anisotropy, the heating and flow of alpha particles in the solar wind, the interstellar neutral particles that become ionized in the solar wind, and the 'fluidlike behavior' of the solar wind.

074.013 Solution of three-fluid model equations with anomalous transport coefficients for the quiet solar wind. S. Cuperman, N. Metzler.
Astrophys. Journ., Vol. 196, 205 - 219 (1975).

A three-fluid model consisting of continuity, momentum, and energy equations is used to describe the average behavior of the electrons (e), protons (p), and α-particles (α) in the quiet (undisturbed) expanding solar corona. The numerical solution of the equations is achieved by using a perturbation-method procedure in which a two-fluid solution for electrons and protons is followed by a solution for the α-particles. These model-equations predict the following features for the α-particles at 1 a.u.: (1) equal α-particle and proton streaming velocities, (2) an α-particle to proton temperature ratio of about 3.5, and (3) an α/p density ratio of about 0.035 (which is also less than half of that in the solar corona). These results explain the α-particle characteristics observed at 1 a.u.

074.014 Investigations of the solar plasma near Mars and on the path earth–Mars by means of charged particle traps aboard the Soviet space vehicles 1971 - 1973.
K. I. Gringauz, V. V. Bezrukikh, M. I. Verigin, A. P. Remizov.
Kosmich. Issled., Vol. 13, p. 123 - 128 (1975). In Russian.

074.015 Restricted three-dimensional stellar wind modeling. I. Polytropic case. S. F. Nerney, S. T. Suess.
Astrophys. Journ., Vol. 196, 837 - 847 (1975).

A steady, axisymmetric, nondissipative, three-dimensional solar wind model is investigated through the use of a perturbation expansion in inverse Rossby numbers (the ratio of the Coriolis to the inertial forces). Numerical results are presented for a polytropic gas, although the solutions far from the sun are independent of this approximation. The meridional velocity at large radius is of order $-2 \sin 2\,\theta$ km s^{-1}, where θ is the polar angle. This is consistent with the comet-tail analysis of Brandt, Harrington, and Roosen, who find -2.6 ± 1.2 km s^{-1} at 1 a.u. and $\theta = \pi/4$.

074.016 Interplanetary gas. XX. Does the radial solar wind speed increase with latitude?
J. C. Brandt, R. S. Harrington, R. G. Roosen.
Astrophys. Journ., Vol. 196, 877 - 878 (1975).

The astrometric technique used to derive solar wind speeds from ionic comet-tail orientations has been used to test the suggestion that the radial solar wind speed is higher near the solar poles than near the equator. The authors find no evidence for the suggested latitude variation.

074.017 The reconnection of magnetic field lines in the solar corona. N. R. Sheeley, Jr., J. D. Bohlin, G. E. Brueckner, J. D. Purcell, V. E. Scherrer, R. Tousey.
Astrophys. Journ., (Letters), Vol. 196, L129 - L131 (1975).

Skylab XUV coronal spectroheliograms and photospheric magnetograms are compared. This comparison shows that, as new bipolar magnetic fields emerge through the solar surface into the corona, the new coronal fields interact with the old ones in a manner that suggests the reconnection of the field lines.

074.018 The coronal hole at the 7 March 1970 solar eclipse. M. Waldmeier.
Solar Physics, Vol. 40, 351 - 358 = Astron. Mitt. Sternw. Zürich, No. 337 (1975).

A hole (region of very low emission) appeared in the SW-quadrant of the sun at the total eclipse of March 7, 1970. From the photometry in white light, the density distribution in the hole has been calculated under the assumption of a circular cross section. In the central part, whose diameter is roughly half that of the whole hole, the density is less than 10% of the density of the undisturbed corona at the same distance from the sun. The hole did appear in a region free of solar activity for three consecutive rotations.

074.019 Observed coronal temperatures at $1.37\,R_{\odot}$ in the region of a helmet structure.
D. H. Liebenberg, R. J. Bessey, B. Watson.
Solar Physics, Vol. 40, 387 - 396 (1975).

During the total solar eclipse, 1965 May 30, a 25 cm aperture $f/8.0$ telescope and Fabry-Perot interferometer were operated aboard the USAF-AEC aircraft. High resolution spectra of the Fe XIV emission line, 530.3 nm, were obtained. Deconvolved intensity vs wavelength profiles of the second order fringe overlay a helmet structure on the NM limb at out to $1.37\,R_{\odot}$. The profiles yield coronal temperatures, absolute intensities and Doppler velocities in regions of apparently open magnetic field structure and within the closed field lines of the helmet. Together with white light intensities the observations are interpreted to provide temperatures and turbulent velocities in and around this coronal structure. Comparison is made with a model by Billings and Roberts. A model with radial flow (solar wind) velocities of $\sim$60 km s^{-1} satisfies the observations in the open field line region.

074.020 Direct observations of a flare related coronal and solar wind disturbance. J. T. Gosling, E. Hildner, R. M. MacQueen, R. H. Munro, A. I. Poland, C. L. Ross.
Solar Physics, Vol. 40, 439 - 448 (1975).

Numerous mass ejections from the sun have been detected with orbiting coronagraphs. Here for the first time the authors document and discuss the direct association of a coronagraph observed mass ejection, which followed a 2B flare, with a large interplanetary shock wave disturbance observed at 1 AU. Estimates of the mass (2.4×10^{16} g) and energy content (1.1×10^{32} erg) of the coronal disturbance are in reasonably good agreement with estimates of the mass and energy content of the solar wind disturbance at 1 AU. The energy estimates as well as the transit time of the disturbance are also in good agreement with numerical models of shock wave propagation in the solar wind.

074.021 Solar-interplanetary modeling: 3-d solar wind solutions in prescribed non-radial magnetic field geometries. B. R. Durney, G. W. Pneuman.
Solar Physics, Vol. 40, 461 - 485 (1975).

A model is presented which describes the 3-dimensional non-radial solar wind expansion between the sun and the earth in a specified magnetic field configuration subject to synoptically observed plasma properties at the coronal base. In this paper, the field is taken to be potential in the inner corona based upon the Mt. Wilson magnetograph observations and radial beyond a certain chosen surface. For plasma boundary conditions at the sun, deconvoluted density profiles obtained from synoptic K-coronameter brightness observations are used. The temperature is taken to be 2×10^6 K at the base of closed field lines and 1.6×10^6 K at the base of open field lines. For a sample calculation, the authors employ data taken during the period of the 12 November 1966 eclipse.

074.022 Solar wind: the quasi-radial approximation and its limitations. S. T. Suess, S. F. Nerney.
Solar Physics, Vol. 40, 487 - 499 (1975).

The mathematical basis for approximating the solar wind expansion as nearly radial is examined and defined, removing earlier restrictions thought to occur in the presence of a magnetic field and large variations in latitude. The equations and side conditions governing quasi-radial flow are derived and solved for a simple example to illustrate how this technique can be used for global models of the solar wind.

074.023 On a cold emission in the solar corona. K. V. Alikayeva.
Solar Physics, Vol. 41, 89 - 95 (1975).

The faint emission of hydrogen, helium and metals in the corona which appeared near an active prominence is studied. The calculations showed that the temperature of the emission region is in the limits from 10000 K to 30000 K and the electron density is between $10^9 - 10^{10}$ cm^{-3}, respectively.

074.024 On the identification of Fe IX and Ni XI lines from coronal spectra. F. Magnant-Crifo.
Solar Physics, Vol. 41, 109 - 113 (1975).

Identifications of coronal lines with transitions of Fe IX and Ni XI, as proposed by Wagner and House, and Svensson, Ekberg and Edlén, are checked by means of line intensity ratios, from observations obtained at the 1965 eclipse. All the suggestions of Svensson, Ekberg and Edlén are suitable.

074.025 On coronal streamers with T-type neutral points. B. R. Durney.
Solar Physics, Vol. 41, 233 - 240 (1975).

The gas-magnetic field interaction of an isothermal axisymmetric corona is considered. A method is suggested for solving the MHD equations in the case when a uniform gas pressure and the radial component of the magnetic field (as in a dipole) are specified at the sun's surface. The flux of open field lines (ϕ) can be given arbitrarily, and no reconnection or opening of field lines can take place. If configurations in hydrostatic equilibrium between the regions of open and closed field lines can be found, then the method of solution converges. The equation of hydrostatic equilibrium at the neutral point (assumed to be of the T-type) is written in a simple form, and it is shown that if ϕ is smaller than a certain ϕ_{min}, this equation cannot be satisfied. Configurations in hydrostatic equilibrium between the regions of open and closed field lines are expected to exist for any value of ϕ larger than ϕ_{min}.

074.026 On the effect of latitude dependent base conditions on the structure of the solar wind.
I. W. Roxburgh, C. Singer.
Solar Physics, Vol. 41, 241 - 245 (1975).

The isothermal solar wind equations are solved for the case where the coronal conditions vary with latitude. It is found that the solutions are not uniquely determined by the base density but require knowledge of the injection angle of the fluid. Even for the case of spherically symmetric density at the corona, the solutions are not unique and form a one parameter set, but the latitude variation decreases rapidly with increasing heliocentric distance.

074.027 On the importance of iron in the coronal radiative losses. G. Pineau des Forêts.
Astron. Astrophys., Vol. 38, 457 - 459 (1975).

The author shows iron to be important in the cooling of the corona by comparing the total radiative energy losses of the elements H, He, C, N, O, Ne, Mg, Si and S due to bremsstrahlung, recombination radiation and line emission with that due to line emission of highly ionized iron in the temperature range of maximum abundance of these ions.

074.028 The interpretation of the forbidden emission lines from a coronal condensation. H. E. Mason.
Monthly Notices Roy. Astron. Soc., Vol. 171, 119 - 130 (1975).

Theoretical emission rates have recently been computed for six lines observed in the visible spectrum of the solar corona. These rates are used to analyse the spectra of a coronal condensation observed by Lyot and Aly at the 1952 eclipse. Average electron density and temperature conditions for the condensation are deduced and a specific model is proposed, in which electron density and temperature estimates are defined as a function of position within the condensation.

074.029 Large-scale and solar-cycle variations of the solar wind. M. Neugebauer.
Space Sci. Rev., Vol. 17, 221 - 254 (1975). – Review paper – see 012.002.

This paper summarizes space probe observations relevant to the determination of the large-scale, three-dimensional structure of the solar wind and its solar cycle variations. Observations between 0.6 and 5 AU reveal very little change in the average solar-wind velocity, but a pronounced decrease in the spread of velocities about the average. The velocity changes may be accompanied by a transfer of energy from the electrons to the protons. The variances or noise in the field and plasma have also been measured as a function of radial distance. Dependence on the phase of the solar-activity cycle can be found in the data on the number of high speed streams, the proton density, the percent helium, and the magnetic-field strength and polarity.

074.030 Microstructure of the solar wind. H. J. Völk.
Space Sci. Rev., Vol. 17, 255 - 276 (1975). – Review paper – see 012.002.

In this review microstructure is understood to summarize the dynamical properties of the solar wind that enter the fluid equations, usually employed to describe the overall flow, only in one or the other averaged form. Thus, first a description of the kinetic picture of the solar wind is given; it mainly involves the structure of the particle distributions in velocity space and associated instabilities. Waves and discontinuities are discussed next. Two selected examples are mentioned where the effects of waves and discontinuities play a role. Specific physical properties of the interplanetary magnetic irregularity spectrum and its relevance for cosmic ray propagation are discussed at the end.

074.031 Intensity ratios of forbidden coronal lines of Fe XI at 7892 Å and 3987 Å. J. P. Rozelot, J. C. Noëns.
Journ. Quant. Spectrosc. Radiat. Transfer, Vol. 15, 379 - 384 (1975).

The excitation of the lowest ($3s^2 \ 3p^4$) configuration of Fe XI is discussed and theoretical values for the relative intensities of the Fe XI lines at 7892 Å and 3987 Å are presented.

074.032 L'étude de la couronne blanche à bord de Concorde 001 au cours de l'éclipse totale de soleil du 30 juin 1973. S. Koutchmy.
L'Astronomie, 89e année, p. 149 - 157 (1975).

074.033 The influence of pressure anisotropy of near-solar plasma on the propagation of Alfvén waves.
S. V. Sobolev.
Astron. Zhurn. Akad. Nauk SSSR, Vol. 52, 346 - 350 (1975). In Russian. English translation in Soviet Astron., Vol. 19, No. 2.

The propagation of Alfvén waves in a nonhomogeneous anisotropic plasma is studied. The coefficients of reflection and transmission of Alfvén waves at the contact discontinuity are obtained and the dependence of the laws of reflection and refraction upon the degree of pressure anisotropy of plasma is found. The propagation of Alfvén waves in an isothermal atmosphere with homogeneous gravity field is considered.

074.034 On the nature of irregularities in the solar wind corotating high-speed streams. N. A. Lotova.
Astron. Zhurn. Akad. Nauk SSSR, Vol. 52, 359 - 364 (1975). In Russian. English translation in Soviet Astron., Vol. 19, No. 2.

The peculiarities of the small-scale irregularities which are related with the sector structure in the interplanetary medium were investigated. These are responsible for radio sources scintillation enhancement. Taking into account the characteristics of such streams, it is concluded that probably they are related to plasma magnetosound or anisotropic temperature instabilities.

074.035 Polarization observation of the solar corona at the solar eclipse of 1968 September 22.
Z. Hu, L. Chui.
Acta Astron. Sinica, Vol. 15, 137 - 148 (1974). In Chinese.

At the total solar eclipse on 1968 September 22, observable in Western China, photographic observation of the polarization of the solar corona was made separately on ground and in an airplane. The spectral range covered by photography is 5700–6300 Å. Through photometric measurement, the absolute brightness and polarization of the corona were obtained for the interval $1.1-3.5\,R_\odot$. The corona was of maximum type, but did not have any symmetry. Polarization was nearly radial.

074.036 The effect of asymmetric solar wind on the Lyman α sky background. J. A. Joselyn, T. E. Holzer.
Journ. Geophys. Res., Vol. 80, 903 - 907 (1975).

The Lyman α sky background arises from the scattering of solar Ly α from a spatial distribution of neutral hydrogen in interplanetary space. This distribution is partially determined by the solar wind proton flux, which provides the principal mechanism of loss by charge exchange of the neutral hydrogen By generating isophotal maps of scattered Ly α for several choices of interstellar wind direction and solar wind proton flux distributions, the results show that latitudinal variations of the solar wind proton flux can have a significant effect on the observed location and shape of the Ly α intensity maximum. This fact should aid in the interpretation of Ly α maps and also indicates a possible method for inferring values for the average solar wind proton flux out of the ecliptic plane.

074.037 Alfvén wave refraction in high-speed solar wind streams. J. V. Hollweg.
Journ. Geophys. Res., Vol. 80, 908 - 916 (1975).

The author uses a simple physical theory to calculate the variation of Alfvén wave amplitudes and the wave refraction in a schematic model for a high-speed solar wind stream.

074.038 Multispacecraft study of the solar wind velocity at interplanetary sector boundaries.
E. J. Rhodes, Jr., E. J. Smith.
Journ. Geophys. Res., Vol. 80, 917 - 928 (1975).

The sector structure of the interplanetary magnetic field and the solar wind velocity at the sector boundaries have been compared at two widely separated locations. Magnetic field and solar wind velocity observations made aboard satellites, have been compared at separations of up to 0.3 AU, 50° in heliocentric longitude and 7° in heliographic latitude. The most plausible explanations of the observed velocity differences are (1) a preferential acceleration of sector boundaries with distance, perhaps caused by stream-stream interactions; or (2) a

dependence of the solar wind velocity on heliographic latitude. The data imply a gradient of ≈ 10 km/s per degree of latitude.

074.039 The enhancement of solar wind fluctuations at the proton thermal gyroradius. M. Neugebauer.
Journ. Geophys. Res., Vol. 80, 998 - 1002 (1975).

Average power spectra of solar wind fluctuations at frequencies up to 0.87 Hz are calculated from Ogo 5 measurements of positive ion flux. Although the general spectral trend follows the power law spectrum observed at lower frequencies, a small but statistically significant power enhancement is observed at the frequency $v/2\pi R$, where v is the solar wind velocity and $R = (2mkT)^{1/2}/eB$ is the gyroradius of proton thermal motions in the solar wind. The measured power spectrum is in rough agreement with that deduced from radio scintillation observations.

074.040 Apollo 17 lunar surface cosmic ray experiment – measurement of heavy solar wind particles.
E. Zinner, R. M. Walker, J. Borg, M. Maurette.
Proc. Fifth Lunar Sci. Conference, (see 012.003), Vol. 3, 2975 - 2989 (1974).

074.041 On the propagation of magnetic disturbances in the solar wind. I. Lerche.
Astrophys. Space Sci., Vol. 34, 309 - 319 (1975).

Under the geometrical optics approximation the author discusses the propagation of a polarized magnetic profile, made up of Alfvén waves, in the solar wind. He shows that (1) the profile propagates at an angle to the radial direction (the direction of the solar wind flow), (2) the radial half-width of the profile stays essentially constant, or even diminishes a little, with distance from the sun, (3) the half-width in a direction transverse to the radial direction increases without limit as the magnetic profile moves outward from the sun.

074.042 Cyclotron waves in the solar wind. J.-T. Horng.
Astrophys. Space Sci., Vol. 34, L1 - L5 (1975). Abstract presented at the International Symposium on Solar-Terrestrial Physics, São Paulo, Brazil, 17–22 June, 1974.

Cyclotron waves in the solar wind near 1 AU with frequencies well below the electron cyclotron frequency and wavelengths much larger than the electron cyclotron radius but less than the proton cyclotron radius are considered.

074.043 The emission of Li-like ions from the solar corona.
D. R. Flower, H. Nussbaumer.
Astron. Astrophys., Vol. 39, 295 - 302 (1975).

The authors have analysed the intensities of the solar XUV emission lines of the lithium-like ions O^{+5}, Ne^{+7}, Mg^{+9} and Si^{+11}, observed both by rocket and satellite. The coronal contribution to the observed intensities is shown to be important even for O^{+5}, whose ionisation curve maximises at 3×10^5 K. The coronal electron temperature, as deduced from off-limb OSO 6 observations, is found to be remarkably stable at about 1.6×10^6 K over the period of one month during which the observations were made. From the same spectra, the relative abundances of O, Ne and Mg are deduced.

074.044 Limb brightening and dark features observed at 6 cm wavelength.
F. Chiuderi-Drago, E. Fürst, W. Hirth, P. Lantos.
Astron. Astrophys., Vol. 39, 429 - 434 (1975).

Radio maps of the sun at 6 cm wavelength were obtained using the 100 m-telescope near Bonn (HPBW = 2.6') during 7 days. The most interesting features present on these maps are: a) limb brightening of about 15% without deconvolution, and b) depression zones correlated with filaments. The combination of both observations allows to determine the electron density of the low corona and leads to the conclusion that there is a transition region between filament and the surrounding corona.

074.045 On the correlation of coronal green-line intensity and solar wind velocity.
E. C. Roelof, S. Cuperman, A. Sternlieb.
Solar Physics, Vol. 41, 349 - 366 (1975).

Cross-correlation functions have been computed between green-line intensity (Kislovodsk) and Vela solar wind velocity January—June 1967. The cross-correlation patterns appear to be dominated by two competing effects: a tendency of quasi-stationary green-line emission and solar wind velocity to anti-correlate; and a tendency of transient green-line emission and solar wind velocity enhancements to correlate positively. The authors find evidence for simultaneous (same-day) emission brightenings over 2 to 4 limb quadrants.

074.046 Comment on the paper 'Two-fluid model of the solar corona' by J. W. Knight, C. E. Newman, and P. A. Sturrock.   D. E. Billings.
Solar Physics, Vol. 41, 367 - 369 (1975). — Research note. Concerning 12.074.002.

074.047 Reply to Billings concerning 'Two-fluid model of the solar corona'.
J. W. Knight, C. E. Newman, P. A. Sturrock.
Solar Physics, Vol. 41, 371 (1975). — Research note.

074.048 Population théorique des sous niveaux Zeeman relatifs à la raie 5303 Å de Fe XIV.   J. P. Rozelot.
Solar Physics, Vol. 41, 373 - 379 (1975).

The author gives here under compact tables the relative population of the six first levels arising from the hyperfine structure of Fe XIV. These populations are computed from the best atomic data actually available, so that one can have a well known basis for the interpretation of coronal polarization measurements, both theoretical and observational.

074.049 Dynamic response of an isothermal static corona to finite-amplitude disturbances.
Y. Nakagawa, S. T. Wu, E. Tandberg-Hanssen.
Solar Physics, Vol. 41, 387 - 396 (1975).

The velocity evolution of sprays, surges and fast ejections is characterized by a rapid acceleration followed by a slow-down. In contrast, eruptive prominences show a velocity evolution with a slow increase followed by a rapid acceleration. The authors examine the physical causes which differentiate these two characteristic velocity evolutions, and study the dynamic responses of the solar corona. It is shown that the resultant flow depends strongly on the nature of the disturbing causes. It is shown that for large flare sprays 5×10^{39} particles can be injected into the solar wind.

074.050 Absolute intensities of the outer corona from the June 30, 1973 total solar eclipse.
C. F. Keller, J. E. Tabor, W. Matuska.
Bull. American Astron. Soc., Vol. 7, 234 (1975). — Abstr. AAS.

074.051 Long-term X-ray emitting structures in the solar corona.   J. P. McGuire, R. M. Wilson, A. C.
deLoach, R. B. Hoover, J. B. Smith, Jr., D. M. Speich, S. T. Wu.
Bull. American Astron. Soc., Vol. 7, 235 (1975). — Abstr. AAS.

074.052 A search for solar wind velocity changes between 0.7 and 1 AU.   D. S. Intriligator, M. Neugebauer.
Journ. Geophys. Res., Vol. 80, 1332 - 1334 (1975).

Simultaneous observations of the solar wind velocity as measured at the Pioneer 9 and OGO 5 spacecraft during five solar rotations in 1968 and 1969 are presented. During this time, Pioneer 9 was traveling in toward the sun to approximately 0.7 AU while the earth orbiter OGO 5 was spending long periods in the interplanetary medium. A comparison of the 3-hour averages of solar wind velocity obtained at both spacecraft indicates that the same basic solar wind velocity structure was seen at both spacecraft.

074.053 Differences between the bulk speeds of hydrogen and helium in the solar wind.   K. W. Ogilvie.
Journ. Geophys. Res., Vol. 80, 1335 - 1338 (1975).

This note describes observations made by the Goddard Space Flight Center plasma instruments on Explorer 43 of differences between the bulk speeds of helium and hydrogen. Arguments are given for regarding these differences as a characteristic of the interplanetary medium on the macroscale associated with the whole region of a stream-stream interaction. If the differences were caused by the action of Alfvén waves, then on these occasions they must have been produced closer to the sun than to 1 AU.

074.054 Temperature and density measurements of coronal loops.   R. C. Chase, L. Golub, A. Krieger, J. K.
Silk, G. S. Vaiana, M. Zombeck, A. F. Timothy.
Bull. American Astron. Soc., Vol. 7, 346 (1975). — Abstr. AAS.

074.055 Emerging flux regions observed by S-056.
J. A. Vorpahl.
Bull. American Astron. Soc., Vol. 7, 346 (1975). — Abstr. AAS.

074.056 The effect of flares upon the outer solar corona.
R. H. Munro.
Bull. American Astron. Soc., Vol. 7, 347 - 348 (1975). — Abstr. AAS.

074.057 White light and radio studies of the coronal transient of 14—15 September 1973. I. Observations and emission mechanisms.   R. Robinson, R. M. MacQueen.
Bull. American Astron. Soc., Vol. 7, 348 (1975). — Abstr. AAS.

074.058 White light and radio studies of the coronal transient of 14—15 September 1973. II. The dynamics of the event.   J. T. Gosling, G. A. Dulk.
Bull. American Astron. Soc., Vol. 7, 348 (1975). — Abstr. AAS.

074.059 Temporal and spatial properties of coronal bright points.
L. Golub, A. Krieger, R. Simon, G. Vaiana, A. F. Timothy.
Bull. American Astron. Soc., Vol. 7, 350 (1975). — Abstr. AAS.

074.060 The relationship between coronal bright points and the chromospheric network.   J. G. Timothy.
Bull. American Astron. Soc., Vol. 7, 350 (1975). — Abstr. AAS.

074.061 Spectrophotometric study of a coronal formation observed around a prominence on 23 March 1974.
T. Tsubaki.
Bull. American Astron. Soc., Vol. 7, 350 (1975). — Abstr. AAS.

074.062 Representative temperatures for coronal and transition region ions.   D. E. Billings.
Bull. American Astron. Soc., Vol. 7, 354 (1975). — Abstr. AAS.

074.063 Polar transients observed in the EUV.
G. L. Withbroe, D. Jaffe.
Bull. American Astron. Soc., Vol. 7, 354 (1975). — Abstr. AAS.

074.064 Coronal density distribution from Fe XIV λ 5303 data.   R. Fisher, S. Musman, P. Seagraves.
Bull. American Astron. Soc., Vol. 7, 355 (1975). — Abstr. AAS.

074.065 Brightness distribution in the white light corona $(3-10\,R_\odot)$ from daily OSO-7 observations.
R. A. Howard, M. J. Koomen.
Bull. American Astron. Soc., Vol. 7, 355 - 356 (1975).

Abstr. AAS.

074.066 **Ground observations of the total intensity and polarization of the solar corona at the 1973 eclipse.**
C. Lilliequist, J. Rush, G. Newkirk, L. Lacey, H. Hull.
Bull. American Astron. Soc., Vol. 7, 356 (1975). — Abstr. AAS.

074.067 **Oxygen to neon abundance ratio in the solar corona.**
L. W. Acton, R. C. Catura, E. G. Joki.
Bull. American Astron. Soc., Vol. 7, 356 (1975). — Abstr. AAS.

074.068 **Observations of a long lived coronal streamer.**
A. I. Poland.
Bull. American Astron. Soc., Vol. 7, 356 (1975). — Abstr. AAS.

074.069 **Polar plumes in XUV emission-line corona.**
J. D. Bohlin, J. D. Purcell, N. R. Sheeley, Jr., R. Tousey.
Bull. American Astron. Soc., Vol. 7, 356 (1975). — Abstr. AAS.

074.070 **Coronal loops observed on 9 February 1972.**
R. D. Chapman, W. M. Neupert.
Bull. American Astron. Soc., Vol. 7, 357 (1975). — Abstr. AAS.

074.071 **Solar wind development in the middle corona.**
D. H. Liebenberg, R. Bessey, B. Watson.
Bull. American Astron. Soc., Vol. 7, 358 (1975). — Abstr. AAS.

074.072 **Are there latitudinal variations in solar wind speeds?**
J. C. Brandt, R. S. Harrington, R. G. Roosen.
Bull. American Astron. Soc., Vol. 7, 358 (1975). — Abstr. AAS.

074.073 **The coronal source of recurrent, high speed solar wind streams.**   J. Nolte, A. S. Krieger, D. Webb, G. S. Vaiana, A. J. Lazarus, J. Sullivan, A. F. Timothy.
Bull. American Astron. Soc., Vol. 7, 358 (1975). — Abstr. AAS.

074.074 **A model of coronal holes.**
W. M. Adams, P. A. Sturrock.
Bull. American Astron. Soc., Vol. 7, 358 (1975). — Abstr. AAS.

074.075 **Photoelectric observations of the green coronal line.**
R. M. E. Illing, D. A. Landman, D. L. Mickey.
Bull. American Astron. Soc., Vol. 7, 363 - 364 (1975). Abstr. AAS.

074.076 **On the investigation of accelerated streams of the solar wind.**   N. A. Lotova, I. V. Chashej.
Geomagn. Aeronom., Vol. 15, 193 - 196 (1975). In Russian.

074.077 **Frequency shift of monochromatic cosmic radio emission sources occulted by the corona and the coronal plasma turbulence.**   I. M. Gordon, V. I. Kucherov.
Probl. yader. fiz. i kosmich. luchej. Resp. mezhved. temat. nauch.-tekhn. sb., 1974, vyp. (No.) 1, p. 44 - 53. In Russian.
Abstr. in Referativ. Zhurn. 51. Astron., 4.51.463 (1975).

074.078 **Lunar surface solar wind observations at the Apollo 12 and Apollo 15 sites.**
D. R. Clay, B. E. Goldstein, M. Neugebauer, C. W. Snyder.
Journ. Geophys. Res., Vol. 80, 1751 - 1760 (1975).
Eleven months of simultaneous data were obtained by the ALSEP solar wind spectrometers at the Apollo 12 and 15 sites. There were no observed differences between the properties of the upstream solar wind and the plasma observed at Apollo 15, where the local magnetic field is $3 \pm 3 \gamma$. However, the solar wind flow is often strongly perturbed at the Apollo 12 site, where the field is $\sim 38 \gamma$. The plasma perturbations observed at this site suggest that remanent lunar magnetic fields are the cause of lunar limb compression waves and that these waves should be more noticeable for low solar wind

dynamic pressures. Those physical and chemical properties of the surface layers of the moon which depend on the chemical composition, the energy, or the flux of the bombarding plasma may depend strongly on the strength and scale size of the local magnetic field.

074.079 **An extension of the use of critical conditions in solar wind theory.**
M. S. Gussenhoven, R. L. Carovillano.
Journ. Geophys. Res., Vol. 80, 1761 - 1763 (1975).
This paper is based upon the polytropic Parker model of the solar wind. The critical condition is a regularity requirement for a transonic solution and represents a constraint upon the flow temperature and velocity. This constraint may be imposed at any distance, and this freedom enables us to present the family of all solutions to the flow equations, for each value of the polytrope index γ, in a way that facilitates comparison to experimental solar wind values.

074.080 **A model for the origin of solar wind stream interfaces.**   A. J. Hundhausen, L. F. Burlaga.
Journ. Geophys. Res., Vol. 80, 1845 - 1848 (1975).
The basic variations in solar wind properties that have been observed at 'stream interfaces' near 1 AU are explained by a gas dynamic model in which a radially propagating stream, produced by a temperature variation in the solar envelope, steepens nonlinearly while moving through interplanetary space.

074.081 **On the possibility of experimental testing of the interaction model solar wind — interstellar medium.**
V. B. Baranov, K. V. Krasnobaev, O. G. Onishchenko.
Pis'ma v Astron. Zhurn., Vol. 1, No. 4, p. 29 - 32 (1975). In Russian.
A method based on observations of radio sources scintillations is proposed to test a theoretical model of the solar wind — interstellar medium interaction.

074.082 **Absolute photometry and the structure of the corona during the solar eclipse of June 30, 1973.**
V. Rušin, M. Rybanský.
Bull. Astron. Inst. Czechoslovakia, Vol. 26, 160 - 168 (1975).
The paper gives the geographic co-ordinates of the observation station and the eclipse contacts. Three photographs of the solar eclipse were used to determine the absolute values of the intensity of the white corona for all position angles. The flattening pattern of the isophotes as a function of height is similar to that in the neighbourhood of minimum solar activity. The method of determining the structure of the solar corona from pictures taken without a radial filter, when short exposures are not applied, is also described.

074.083 **On proton acceleration at the boundary of the solar wind with the interstellar medium.**
M. F. Bakhareva.
Geomagn. Aeronom., Vol. 15, 393 - 400 (1975). In Russian.

074.084 **On the helio-latitudinal dependence of solar wind velocity.**   V. I. Vlasov.
Geomagn. Aeronom., Vol. 15, 542 - 543 (1975). In Russian.
Brief information.

074.085 **Confirmation of known numerical solutions for the quiet solar wind equations.**
S. Cuperman, N. Metzler, M. Spiegelglass.
Astrophys. Journ., Vol. 198, 755 - 759 (1975).
A detailed comparison of two integration methods for the quiet solar wind, namely, starting at the "singular critical point" and proceeding outward and then inward (Method I) and starting at "infinity" and proceeding inward (Method II), is given. No consideration is given to interaction with

interstellar gas.

074.086 Solar rotation as marked by extreme-ultraviolet coronal holes. W. J. Wagner.
Astrophys. Journ., (Letters), Vol. 198, L141 - L144 (1975).

Spectroheliograms produced in the light of Fe XV
λ 284 by the Goddard Space Flight Center experiment on
OSO-7 are used to determine solar rotational periods for low-
emission coronal hole features. From 1972 May to 1973
October, daily measures were obtained of the coronal hole
area appearing at the central solar meridian. Autocorrelations
of the coronal-hole-area time series in seven latitude zones
provide synodic rotation periods which indicate a close to
rigid rotation by such features for lag lengths of only one
rotation. For latitudes below about 50°, both the autocorrelation functions for short-lived coronal holes and analyses of
more persistent features over these 18 months suggest a quad-
rupolar distribution of coronal hole longitudes.

074.087 The heating of the solar corona. II. A model based on energy balance.
R. W. P. McWhirter, P. C. Thonemann, R. Wilson.
Astron. Astrophys., Vol. 40, 63 - 73 (1975).

The density and temperature distribution of the solar
corona is calculated assuming an energy balance between
thermal conduction and radiated power loss with the primary
heating of the corona by the dissipation of sound-waves propagated upwards from below the sun's surface. A sharp transition region is found and the calculated results are compared
with observations. A detailed model atmosphere for the transition region and corona is derived using the Harvard Smithsonian Reference Atmosphere (for the chromosphere) as
starting point. Hydrostatic equilibrium is assumed in the calculations but it is also shown that a pressure arises because of
the sound waves which is of comparable magnitude to hydrostatic pressure. The inclusion of this pressure introduces
difficulties that are discussed.

074.088 A compilation of normalized solar wind densities and velocities (1965/1971).
L. Diodato, G. Moreno, C. Signorini.
Astron. Astrophys., Suppl. Ser., Vol. 20, 313 - 362 (1975).

Measurements of the solar wind bulk speed and density
performed at 1 AU by the spacecraft Vela-3, Explorers-33-34-
35 and HEOS-1, in the period from July 1965 to June 1971,
are normalized to similar conditions. The resulting three-
hour averages are shown in tabular form.

074.089 Synoptic tables of the green corona for 1947 - 1970.
J. Sýkora.
Contr. Astron. Obs. Skalnaté Pleso, Vol. 5, 7 - 73 (1973/75).

Synoptic tables of the green corona intensity for the
period 1947 - 1970 are tabulated. Instensities are expressed in
the photometric scale of Pic du Midi and represent the mean
activity in the rectangles with dimensions of 40° (three days)
in longitude and 20° in latitude. Errors which occur during
monochromatic corona observations were taken into account
in the calculations.

074.090 A unified explanation of solar type IV dm con-tinua and zebra patterns. J. Kuijpers.
Astron. Astrophys., Vol. 40, 405 - 410 (1975).

Both type IV dm continuum radiation and zebra patterns
can have their origin in plasma waves at the upper hybrid
frequency, which are excited by a loss-cone distribution of
fast electrons superposed on the thermal background. If the
inverse fractional density of fast electrons surpasses the ratio
of the electron plasma frequency to the electron cyclotron
frequency, instability only exists at those places in the corona
where this ratio is integer, thus giving rise to zebra patterns.
Some observations in favour of the present explanation are

discussed. Upper limits to the magnetic field strength and the
fast particle density in the source are derived from the observations.

074.091 Latitude-dependent nonlinear high-speed solar wind streams.
S. T. Suess, A. J. Hundhausen, V. Pizzo.
Journ. Geophys. Res., Vol. 80, 2023 - 2029 (1975).

The theory of the nonlinear interaction between high-
and low-speed solar wind streams has been limited either to
small amplitudes or to the equatorial plane. It is shown how
both of these restrictions can be removed for flow more than
a few solar radii away from the sun by employing a quasi-radial
approximation to develop a perturbation expansion of the
equations of motion.

074.092 The role of cascade transitions in an exciting initial level of the red coronal line.
M. A. Livshits.
Astron. Zhurn. Akad. Nauk SSSR, Vol. 52, 579 - 586 (1975).
In Russian. English translation in Soviet Astron., Vol. 19, No. 3.

The excitation rate of initial $3p^5$ $^2P_{1/2}$ level of FeX owing
to excitation of upper levels of $3s3p^6$ and $3s^23p^43d$ configurations and followed by downward transition has been
calculated. The interpretation of red coronal line observations
(Waldmeier, 1970) has given T ≤ 1.3 × 10⁶ K and n_e ≈ 10⁸
cm⁻³ near the poles. From the ratio of the green to the red
line intensities for quiet equatorial regions, taking into account
the results of Petrini, 1970, it was found T = 1.6 × 10⁶ °K. The
red line width near the poles is increasing perhaps due to the
radial outward gas flux in the lower coronal lines.

074.093 Some characteristics of the solar wind, position of shock wave, magnetopause and plasmapause during the period with high solar activity in August 1972.
V. V. Bezrukikh, K. I. Gringauz, G. N. Zastenker, M. Z. Khokhlov.
Kosmich.Issled., Vol. 13, 342 - 351 (1975). In Russian.

074.094 Occultation of Crab nebula by solar corona during June 1971. C. V. Sastry, K. R. Subramanian.
Bull. Astron. Soc. India, Vol. 2, 38 (1974). – Abstract.

074.095 Astronomical photography. Part B: Solar corona photography.
R. M. MacQueen, C. L. Ross, R. E. Evans.
Apollo 17 preliminary science report, p. 34-4 - 34-6 = Dudley
Obs., Albany, New York, Repr. No. C81 (1974).

Solar corona photographs obtained on the Apollo 17 mission show coronal streamers that apparently corroborate
earth-based observations; moreover, visual observations by the
crew indicate clearly identifiable streamers which extend to
approximately 100 solar radii.

074.096 Holes in the solar corona. W. M. Glencross.
IAU Symposium No. 68, (see 012.017), p. 19 - 21
(1975). – Summary.

074.097 Classification of new spectral lines of Fe XVII observed in solar active regions.
R. J. Hutcheon.
IAU Symposium No. 68, (see 012.017), p. 69 - 70 (1975).
Summary.

074.098 Interpretation of the X-ray spectra of solar active regions. A. B. C. Walker, Jr.
IAU Symposium No. 68, (see 012.017), p. 73 - 100 (1975).

This paper presents a review of recent analytical studies
of the coronal X-ray spectrum below 25 Å.

074.099 Time variations in coronal active regions.

A. S. Krieger, R. C. Chase, M. Gerassimenko, S. W.
Kahler, A. F. Timothy, G. S. Vaiana.
IAU Symposium No. 68, (see 012.017), p. 103 (1975).
Summary.

**074.100 Observation of a non-uniform component in the
distribution of coronal bright points.**
L. Golub, A. S. Krieger, G. S. Vaiana.
Solar Physics, Vol. 42, 131 - 134 (1975).

The longitude distribution of X-ray bright points shows
very strong variations when plotted in a heliocentric (Carring-
ton) coordinate system. In addition, the latitude distribution
can be interpreted as having two components: a uniformly distributed component and one having a distribution similar to
that of active regions, occurring mostly within ±30° of the
equator.

074.101 The structure and evolution of coronal holes.
A. F. Timothy, A. S. Krieger, G. S. Vaiana.
Solar Physics, Vol. 42, 135 - 156 (1975).

In this preliminary discussion of the structure and evolution of coronal holes the authors investigate typical coronal
configurations which result in the formation of these features.
They also follow the evolution of one particular hole as a function of time and investigate its rotational properties at latitudes
ranging from 20°S to 50°N. Finally, they explore possible
mechanisms which may account for the almost rigid rotational
characteristics displayed by this feature.

**074.102 A coronal hole observed at 10.7 GHz with a large
single dish.** E. Fürst, W. Hirth.
Solar Physics, Vol. 42, 157 - 161 (1975).

On July 24, 1973, a coronal hole was observed at 10.69
GHz (2.8 cm) with the Bonn 100-m telescope. The difference
of the brightness temperature between outside and inside the
hole was about 400 to 500 K. It is shown that this lack of emission can be explained by usually adopted values of the electron
density at the bottom of the corona.

**074.103 The large coronal transient of 10 June 1973. I:
Observational description.**
E. Hildner, J. T. Gosling, R. M. MacQueen, R. H. Munro,
A. I. Poland, C. L. Ross.
Solar Physics, Vol. 42, 163 - 177 (1975).

During the 8.5 month flight of the High Altitude Observatory's white light coronagraph on board Skylab, over 100
coronal transients were observed. In this paper a description of
one well observed loop transient, that of 10 June 1973 is presented. Probable causes, density, mass, energy, shape, typical
and atypical characteristics of the 10 June transient are discussed.

**074.104 On the possibility of deducing interplanetary and
solar parameters from geomagnetic records.**
C. T. Russell.
Solar Physics, Vol. 42, 259 - 269 (1975).

While at present we are able to deduce from ground
records only qualitative properties of the solar wind, in the
future quantitative deductions may be possible, in a statistical
sense, from an examination of polar cap magnetograms
together with records of geomagnetic activity. The qualitative
inferences that are possible now indicate several important
features of the behavior of the solar wind over the last 100
years. There appear to be significant long term changes in
either the solar wind velocity, the magnetic field strength, the
variability of the field or some combination of all three. With
the exception of the most recent solar cycle, there is little
north-south asymmetry in these solar parameters.

**074.105 The "leader center in a cell" method in the three-
dimensional nonstationary problem of interaction**

between the solar wind plasma and a conductive model of the
moon (a numerical experiment). A. S. Lipatov.
AN SSSR. In-t kosmich. issled. Pr-196. Moskva, 1974. 42 pp.
In Russian. – Abstr. in Referativ. Zhurn. 51. Astron., 6.51.45 ḷ,
62. Issled. kosmich. prostranstva, 6.62.192 (1975).

**074.106 Plasma theory of forming of a radar signal
reflected from the sun and the problem of the solar
wind.** I. M. Gordon, N. N. Gerasimova.
Izv. AN SSSR. Ser. fiz., Vol. 39, 334 - 339 (1975). In Russian.
Abstr. in Referativ. Zhurn. 62. Issled. kosmich. prostranstva,
6.62.212 (1975).

074.107 The corona at the solar eclipse of June 20, 1974.
M. Waldmeier.
Astron. Mitt. Eidgenössisch. Sternw. Zürich, No. 331, 24 pp.
(1974). In German.

To observe the corona at the solar eclipse of June 20,
1974 the author has undertaken an expedition to Walpole
(SW-Australia). The shape and structure of the corona are
presented. Pictures of the isophotes are given. The corona's
structure was of intermediate type.

**074.108 The brightness of the corona at the solar eclipse
of June 30, 1973.** J. Dürst.
Astron. Mitt. Eidgenössisch. Sternw. Zürich, No. 335, 17 pp.
(1974). In German.

The solar eclipse of 1973 June 30 has been observed at
Atar, Mauretania. With a camera of 37 cm focal length the
corona and the sky were photographed at an effective wavelength of 6000 Å. Reduction of these photographs yields
brightness distributions of the corona, the coronal aureole
and the sky background up to r = 6.

**074.109 Beobachtungen der Sonnenkorona in den Jahren
1971–1973.** M. Waldmeier.
Astron. Mitt. Eidgenössisch. Sternw. Zürich, No. 341, 19 pp.
(1975).

**074.110 Multicolor polarization measurements of the outer
solar corona.** M. Mattei, J. A. Mattei.
Journ. American Ass. Variable Star Observers, Vol. 2, 80
(1973).

074.111 Search for the solar wind in the lunar soil.
P. H. Payton.
Thesis, California Univ., Los Angeles (USA). 159 pp. Universi-
ty Microfilms Order No. 74-18,799 (1974).

**074.112 Unusual solar wind and solar proton events observed
on the lunar surface.** R. A. Medrano-Balboa.
Thesis, Rice Univ., Houston, Texas (USA). 199 pp. University
Microfilms Order No. 74-21,308 (1974).

074.113 Three-dimensional solar wind modeling.
S. F. Nerney.
Thesis, Colorado Univ., Boulder (USA). 226 pp. University
Microfilms Order No. 74-22,378 (1974).

**074.114 North-south asymmetries in solar wind and solar
spotteddness in sunspot cycle No. 20.**
R. Hartmann.
Journ. Interdiscipl. Cycle Res., Vol. 6, 41 - 46 (1975). – Summary.

**074.115 Latitude distribution of the total emission and of
half-widths of the λ 6374 Å Fe X line.**
N. F. Tyagun, V. E. Stepanov.
Solnechnye Dannye 1975 Byull., No. 2, p. 56 - 64 (1975).
In Russian.

1715 profiles of the λ 6374 Å Fe X line were obtained

by the authors with the 53-cm coronograph–spectrograph at the Sayan Observatory. The behavior of equivalent widths and half-widths has been investigated. The magnitude of the half-widths appears to show the maximum at the latitude of 45° and the minimum within active latitudes. The half-widths in the polar region are, on the average, somewhat greater than those in the equatorial zone. The distribution of the equivalent width values agrees with Waldmeier's results.

074.116 **Influence of a coelostat on polarimetric measurements of the solar corona.** A. K. Kishonkov.
Solnechnye Dannye 1975 Byull., No. 2, p. 64 - 72 (1975). In Russian.

The influence of a coelostat on polarimetric measurements of the solar corona has been considered. Expressions for the degree of polarization and position of the polarization plane of coronal emission after reflection from the mirror have been obtained. For a concrete problem the parameters of the mirror were measured and the corrections to the degree and direction of the polarization plane calculated. An influence of the mirror upon the partially polarized emission obtained by means of a pile has been evaluated. The results are given on graphs.

074.117 **On the stages of evolution of coronal condensations.** V. P. Vasil'ev.
Solnechnye Dannye 1975 Byull., No. 2, p. 72 - 76 (1975). In Russian.

Photometric analyses of the known cases of eclipse observations of coronal condensations in the continuum were investigated. The differences in phenomenologic and model interpretation of these observations are shown to be connected with different stages in the evolution of the center of activity.

074.118 **Electromagnetic waves from solar corona.** S. P. Mishra, A. K. Gwal, K. D. Misra.
Indian Journ. Phys., Vol. 48, 816 - 825 (1974).

The energy radiated by electrons in the Cerenkov mode for plasma waves is numerically obtained with special regard to solar coronal model. An estimate of differential scattering cross section in the solar corona is made. The calculated e.m. wave spectrum is compared with that of synchrotron radiation in the light of available experimental and theoretical results.

074.119 **Coronal and solar wind abundances.** J. Hirshberg.
Solar wind three, (see 012.020), p. 26 (1974). – Summary.

074.120 **Measurement of heavy solar wind particles during the Apollo 17 mission.** E. Zinner, R. M. Walker, J. Borg, M. Maurette.
Solar wind three, (see 012.020), p. 27 - 32 (1974).

074.121 **Relation of solar wind fluctuations to differential flow between protons and alphas.**
M. Neugebauer.
Solar wind three, (see 012.020), p. 33 - 34 (1974).

074.122 **The solar wind as deduced from lunar samples.** P. Eberhardt.
Solar wind three, (see 012.020), p. 58 - 67 (1974).

074.123 **The thermal properties of the ancient solar wind.** J. Borg, A. L. Burlingame, M. Maurette, P. C. Wszolek.
Solar wind three, (see 012.020), p. 68 - 70 (1974).

074.124 **Coronal structure and the solar wind.** E. C. Roelof.
Solar wind three, (see 012.020), p. 98 - 131 (1974).

074.125 **X-ray observations of coronal holes and their relation to high velocity solar wind streams.** A. S. Krieger, A. F. Timothy, G. S. Vaiana, A. J. Lazarus, J. D. Sullivan.
Solar wind three, (see 012.020), p. 132 - 139 (1974).

074.126 **Interplanetary disturbances: a classification.** J. T. Gosling.
Solar wind three, (see 012.020), p. 140 - 143 (1974).

074.127 **Coronal transient phenomena.** S. T. Wu, S. M. Han.
Solar wind three, (see 012.020), p. 144 - 146 (1974).

A hydrodynamical description of outward propagating disturbances from the sun's surface ($\sim 1.043\ R_s$, R_s being the solar radius) in a model solar atmosphere is considered. Use is made of this model to explain the coronal transient phenomenon.

074.128 **Discontinuities in the solar wind.** G. L. Siscoe.
Solar wind three, (see 012.020), p. 151 - 168 (1974).

074.129 **Interpenetrating solar-wind streams.** W. C. Feldman, J. R. Asbridge, S. J. Bame, M. D. Montgomery.
Solar wind three, (see 012.020), p. 179 (1974). – Abstract.

074.130 **Instabilities connected with neutral sheets in the solar wind.** V. Formisano, P. C. Hedgecock, C. T. Russell, J. D. Means.
Solar wind three, (see 012.020), p. 180 - 186 (1974).

A preliminary study of two sets of data is presented taken from HEOS-1 and OGO-5 in the solar wind, that shows the internal structure of two neutral sheets and their two-dimensional structure.

074.131 **Long-term variations of the solar wind proton parameters.**
L. Diodato, G. Moreno, C. Signorini.
Solar wind three, (see 012.020), p. 195 (1974). – Abstract.

074.132 **Effect of magnetic force on solar wind expansion.** T. Yeh.
Solar wind three, (see 012.020), p. 196 - 200 (1974).

A family of critical solutions for isothermal magnetized solar wind with azimuthal motion is calculated, ranging from zero to very large magnetic fluxes. The result indicates that the magnetic force enhances the solar wind expansion and the enhancement is more eminent at lower temperatures.

074.133 **Cosmic ray propagation in the solar wind.** H. J. Volk.
Solar wind three, (see 012.020), p. 202 (1974). – Summary.

074.134 **An effect of cosmic rays on the distant solar wind.** J. R. Jokipii, L. C. Lee.
Solar wind three, (see 012.020), p. 224 - 229 (1974).

074.135 **Radial gradient of solar wind velocity from 1 to 5 AU.** H. R. Collard, J. H. Wolfe.
Solar wind three, (see 012.020), p. 281 - 290 (1974).

074.136 **The radial gradient and the role of turbulence in the solar system.** D. S. Intriligator.
Solar wind three, (see 012.020), p. 294 - 299 (1974).

Pioneer 10 observations from the Ames Research Center Plasma Analyzer experiment between 1 a.u. and 3 a.u. have been used to estimate the power spectra of the solar wind proton streaming speed. The power spectra indicate that significant turbulence on the scale of 10^6 km or more is present throughout this range of heliocentric distances, implying the importance of the role of large scale turbulence between 1 a.u. and 3 a.u.

074.137   The latitude dependencies of the solar wind.
R. L. Rosenberg, C. R. Winge, Jr.
Solar wind three, (see 012.020), p. 300 - 310 (1974)

074.138   Three dimensional modelling.   S. Suess.
Solar wind three, (see 012.020), p. 311 - 317 (1974).
Several recent theoretical advances have made it possible to construct useful global model of the solar wind, where such models are those of more complex geometry than strictly spherical expansion with magnetic field lines lying wholly on cones of half-angle equal to the co-latitude.

074.139   Theoretical predictions of latitude dependencies in the solar wind.
C. R. Winge, Jr., P. J. Coleman, Jr.
Solar wind three, (see 012.020), p. 318 - 320 (1974).

074.140   Variation of the solar wind flux with heliographic latitude, deduced from its interaction with interplanetary hydrogen.   J. Blamont.
Solar wind three, (see 012.020), p. 321 - 328 (1974).

074.141   Evidence of a velocity gradient in the solar wind.
E. J. Smith, E. J. Rhodes, Jr.
Solar wind three, (see 012.020), p. 329 - 331 (1974).

074.142   Waves and instabilities in the solar wind.
J. V. Hollweg.
Solar wind three, (see 012.020), p. 333 (1974). – Abstract.

074.143   Interplanetary heat conduction – IMP 7 results.
W. C. Feldman, M. D. Montgomery, J. R. Asbridge, S. J. Bame, H. R. Lewis.
Solar wind three, (see 012.020), p. 334 - 342 (1974).

074.144   Simulation of colliding solar wind streams with multifluid codes.
K. Papadopoulos, R. W. Clark, C. E. Wagner.
Solar wind three, (see 012.020), p. 343 - 350 (1974).

074.145   The power associated with density fluctuations and velocity fluctuations in the solar wind.
D. S. Intriligator.
Solar wind three, (see 012.020), p. 368 - 372 (1974).

074.146   The enhancement of solar wind fluctuations with scale size near the proton gyroradius.
M. Neugebauer.
Solar wind three, (see 012.020), p. 373 - 374 (1974).

074.147   The fine structure of solar wind velocity.
N. A. Lotova, I. V. Chashey (*Chashej*).
Solar wind three, (see 012.020), p. 375 - 381 (1974).

074.148   Nonlinear hydromagnetic wave evolution in the solar wind.   R. H. Cohen, R. M. Kulsrud.
Solar wind three, (see 012.020), p. 382 - 384 (1974).

074.149   Instability of Alfvén waves in the solar wind.
M. Dobrowolny.
Solar wind three, (see 012.020), p. 388 - 389 (1974).

074.150   The relationship between velocity gradients and magnetic turbulence in the solar wind.
H. B. Garrett.
Solar wind three, (see 012.020), p. 390 - 394 (1974).

074.151   Rigidly rotating component of the solar corona.
E. Antonucci.
Nuovo Cimento B, Ser. 11, Vol. 25B, 513 - 520 (1975).
Analysis of the green-line intensity (Fe XIV) for 1947 - 1970 suggests that the corona is organized on a very large scale. Long-lived structures do exist with the following characteristics: 1) rigid rotation up to ±60° latitude, 2) correlation with the large-scale solar magnetic pattern. They develop essentially as longitudinal persistent features, close to the boundaries of the solar magnetic sectors.

074.152   Solar wind model predictions for the sources of streams observed at 1 AU. Coronal enhancements or coronal holes?   D. E. Jones, R. G. George.
Proc. Utah Acad. Sci. Arts Letters, (*USA*), Vol. 51, 169 - 180 (1974).

The third solar wind conference: a summary.
See Abstr. 011.012.

The excitation of several iron and calcium lines in the visible spectrum of the solar corona.
See Abstr. 022.022.

Wavelengths and transition probabilities for the isoelectronic sequence of oxygen.   See Abstr. 022.024.

A new shock locus for similarity solutions in one-dimensional unsteady gas dynamics.   See Abstr. 061.014.

Low-frequency fluctuations in the solar wind. I. Theory.   See Abstr. 062.024.

Terrestrische und extraterrestrische Plasmen.
See Abstr. 062.040.

Magnetohydrodynamic accretion and the instability of smooth trans-Alfvénic flow.   See Abstr. 062.041.

Hydromagnetic waves and cosmic-ray diffusion theory.   See Abstr. 062.042.

Three-wave interactions involving one whistler.
See Abstr. 062.056.

Simulation of pitch angle diffusion of charged particles in a disordered magnetic field.   See Abstr. 063.036.

Stellar winds.   See Abstr. 064.079.

Observations from Skylab of the density dependent C III multiplet at 1175 Å in active and quiet regions and above the limb.   See Abstr. 073.053.

The D3 chromosphere, coronal holes, and stellar X-rays.   See Abstr. 073.061.

Time variations in the EUV line emission from the chromosphere and corona.   See Abstr. 073.066.

Diffusion effects on line intensities in the solar transition region.   See Abstr. 073.068.

Investigation of the transition region above sunspots based on polarization radio measurements.
See Abstr. 073.075.

On the spicular density enhancement in the region of formation.   See Abstr. 073.101.

Dynamic model of solar coronal spicules.

See Abstr. 073.116.

On the analysis of solar XUV observations and the structure of the chromosphere-corona transition region. See Abstr. 076.004.

X-ray bright points, coronal heating and the solar cycle. See Abstr. 076.013.

Interpreting XUV spectroheliograms in terms of coronal magnetic field structures. See Abstr. 076.017.

The XUV spectrum of He II in quiet regions, a coronal hole, filaments, prominences, and the 7 August 1972 flare. See Abstr. 076.020.

Time variations of solar X-ray bright points. See Abstr. 076.039.

Spatially resolved X-ray spectra of coronal active regions. See Abstr. 076.043.

Association of X-ray flares with solar coronal active regions. See Abstr. 076.048.

On the problem of relationship between the dynamics of the photosphere and coronal structures in X-rays and the solar wind. See Abstr. 076.067.

Polarization reversal during the solar noise storm activity of August 1971. See Abstr. 077.018.

The metric quiet sun during two cycles of activity and the nature of the coronal holes. See Abstr. 077.066.

Étude de l'éclipse solaire partielle du 25 février, 1971 au Pic-du-Midi. See Abstr. 079.102.

Solar wind access to the plasma sheet along the flanks of the magnetotail. See Abstr. 084.206.

Effects of solar wind parameters on the development of magnetospheric substorms. See Abstr. 084.207.

MHD-waves in the solar wind – a possible source of Pc3 geomagnetic pulsations. See Abstr. 084.223.

The solar wind-magnetosphere dynamo and the magnetospheric substorm. See Abstr. 084.251.

Structure of the terrestrial bow shock. See Abstr. 084.270.

Strongly-cooled ionizing plasma flows with application to Venus. See Abstr. 093.011.

Solar wind interaction with Venus: review. See Abstr. 093.038.

Solar wind-Venus interaction observed from magnetic field experiment on Mariner 10. See Abstr. 093.039.

Solar wind sputtering on the lunar surface: equilibrium crater densities related to past and present microparticle influx rates. See Abstr. 094.180.

Solar wind and micrometeorite alteration of the lunar regolith. See Abstr. 094.189.

Interaction of energetic nuclear particles in space with the lunar surface. See Abstr. 094.224.

Observations of moon-plasma interactions by Explorer 35 and Apollo surface and orbital experiments. See Abstr. 094.260.

Electron microscopy of irradiation effects in space. See Abstr. 094.409.

Solar nitrogen: evidence for a secular increase in the ratio of nitrogen-15 to nitrogen-14. See Abstr. 094.431.

Structure and variations of solar wind–Mars interaction region. See Abstr. 097.034.

Jupiter's magnetic field, magnetosphere, and interaction with the solar wind: Pioneer 11. See Abstr. 099.025.

Interaction of a comet with the solar wind. See Abstr. 102.040.

Preliminary results on comet Kohoutek – interactions with the solar wind. See Abstr. 103.100.

Dynamics of partially ionized gas in the gravitational field of the sun. See Abstr. 106.011.

In situ observations of the scale-size of plasma turbulence in the asteroid belt (1.6–3 astronomical units). See Abstr. 106.013.

Direct observations of higher frequency density fluctuations in the interplanetary plasma. See Abstr. 106.016.

Interplanetary shock waves: recent developments. See Abstr. 106.021.

Numerical MHD simulation of interplanetary shock pairs. See Abstr. 106.029.

Interplanetary shock pair disturbances: comparison of theory with space probe data. See Abstr. 106.040.

Interplanetary shock waves and comet brightness fluctuations during June–August 1972. See Abstr. 106.041.

Radial gradients in the interplanetary magnetic field between 1.0 and 4.3 AU: Pioneer 10. See Abstr. 106.051.

Interplanetary scintillations. See Abstr. 106.053.

Evidence of a primordial solar wind. See Abstr. 107.013.

Intensity variations of cosmic radiation and the solar wind. See Abstr. 143.076.

075 Solar Patrol

075.001 Sonnenfleckenbeobachtungen 1973.
P. Völker.
SuW, Vol. 14, 23 - 26 (1975).

075.002 **Prominences.** M. K. V. Bappu.
Quarterly Journ. Roy. Astron. Soc., Vol. 16, 38
(1975).

075.003 **Solar activity during 1973.** V. Barocas.
Journ. British. Astron. Ass., Vol. 85, 255 - 261
(1975). – Report Solar Section.

075.004 **Osservazioni solari (ottiche e radio).** Nos. 23 - 32,
1971 July - 1973 December.
Oss. Astron. Trieste, Pubbl. Nos. 448, 449, 451, 453, 457,
458, 461, 462, 475, 476 (1972/73/74).
Flux density and polarization data at 237 MHz; Interferometric and radiometric data at 408 MHz; Distinctive events observed at 237 and 408 MHz.

075.005 **Observations: Central Section of Solar Observers of
the Polish Amateur Astronomical Society.**
W. Szymański.
Urania Kraków, Vol. 46, 22, 54, 73 - 78, 120, 151, 184
(1975). In Polish.

075.006 **Solar activity in 1974.** J. Mergentaler.
Urania Kraków, Vol. 46, 149 - 151 (1975). In
Polish.

075.007 **Solare Beobachtungsergebnisse (Solar Data).**
E. A. Lauter, C.-U. Wagner, A. Böhme, F. Fürsten-
berg, D. Scholz, S. Böhm.
Zentralinst. für Solar-Terrestrische Physik (Heinrich-Hertz-
Inst.), Akad. Wiss. DDR, HHI Solar Data, Vol. 25, 113 - 130,
November – December (1974); Vol. 26, p. A - F, 2 - 23,
January – March (1975). – Solar radio emission.

075.008 **Solar observations made at Catania Astrophysical
Observatory during 1974.** G. Godoli.
Oss. Astrofis. Catania, Pubbl. Nuova Ser. No. 155, 80 pp.
(1975).
Sunspots; Hα and K faculae; Hα flares; Hα quiescent
prominences; K quiescent prominences; Hα active prominences on disc and at limb; Hα disc and limb patrol hours.

075.009 **Daily Hα chromosphere pictures, daily K_{232} chromosphere pictures, daily white light photosphere pictures.** M. Cimino (Editor).
Photographic Journ. of the Sun, Oss. Astron. Roma, Nos. 87 -
92 (1974). – 1974 May 22 - November 1. – Rotations 1615 -
1620.

075.010 **Solar phenomena.**
M. Cimino, M. Torelli, F. Casamassima, V. Croce.
Oss. Astron. Roma, Monthly Bull. Nos. 200 - 202 (1974/75).
1974 December - 1975 February: Daily total areas of sunspot-
groups; Heliographic position, classification and area of sun-
spot-groups; Longitudinal sunspot magnetic fields; Hours of
K-line cinematographic patrol; Hours of Hα cinematographic
patrol; Sudden cosmic noise absorption S.C.N.A. (18 Mc/s),
sudden enhancement of atmospherics S.E.A. (27 Kc/s);

Explanation.

075.011 **Observations radioélectriques solaires faites sur
600 MHz en 1972 au Laboratoire de Radioastronomie de Humain-Rochefort.** C. Gonze, R. Gonze.
Bull. Astron. Obs. Roy. Belgique, Vol. 8, 142 - 160 (1974).

075.012 **Sunspot relative numbers for 1974.**
M. Waldmeier.
Astron. Mitt. Eidgenössisch. Sternw. Zürich, No. 333, 10 pp.
(1975).

075.013 **Sunspots (sunspot relative-numbers and sunspot-
areas); Synoptic charts of solar magnetic fields**
(Mount Wilson Observatory); **Éruptions chromosphériques
brillantes; Intensité de la couronne solaire; Solar radio emis-
sion.**
M. Waldmeier, R. Howard, G. Olivieri, M. Bernot, H. Tanaka.
Quarterly Bull. Solar Activity (published by Eidgen. Sternw.
Zürich), Nos. 185 - 186, p. 175 - 254 (1975). – 1974 January –
June.

075.014 **Daily maps of the sun and geophysical graphs.**
Solnechnye Dannye 1974 Byull., No. 11, p. 1 - 57;
No. 12, p. 1 - 54; 1975, No. 1, p. 1 - 65; No. 2, p. 1 - 46.
In Russian.

075.015 **Magnetic fields of sunspots.**
Prilozhenie k Byulletenyu "Solnechnye Dannye",
1974, Nos. 11 - 12; 1975, Nos. 1 - 2. In Russian.

075.016 **Indices of geomagnetic activity.**
Journ. Atmosph. Terr. Phys., Vol. 37, 191 - 192,
575 - 576, 851 - 852 (1975). – 1974 August – 1975 January.

075.017 **Solar and solar system activity. Radio Astronomy
Section (BAA).** J. R. Smith, R. J. J. Langton.
Journ. British Astron. Ass., Vol. 85, 161 - 164, 280 - 281,
355 - 357 (1975).

075.018 **Geomagnetic and solar data.**
J. V. Lincoln (Editor).
Journ. Geophys. Res., Vol. 80, 704, 1030, 1379, 1857 - 1860,
2340 (1975). – 1974 October - 1975 February.

075.019 **L'activité solaire.** M.-J. Martres.
L'Astronomie, 89ᵉ année, p. 39 - 43, 86 - 87, 126,
174, 222 - 223, 266 (1975). – Rotations 1614 - 1623.

075.020 **Sunspot numbers.**
Sky Telescope, Vol. 49, 59, 126, 196, 269, 336,
413 (1975).

Synoptic tables of the green corona for 1947 - 1970.
See Abstr. 074.089.

Verification of solar patrol calibration at 15.4 GHz.
See Abstr. 077.055.

Ionospheric spread-F at Huancayo, sunspot
activity and geomagnetic activity. See Abstr. 083.045.

076 Solar UV, X Rays, Gamma Radiation

076.001 **Evolution of the solar Lyman alpha flux during four consecutive years.** A. Vidal-Madjar.
Solar Physics, Vol. 40, 69 - 86 (1975).

Four consecutive years of a quasi-continuous survey of the solar $L\alpha$ line are presented. Absolute calibration and aging correction are evaluated producing higher quality measurements which are: the total $L\alpha$ flux, the central $L\alpha$ flux, the blue wing flux at 0.33 Å from the center, the slope of the blue wing at the same location. Empirical laws are deduced from this large amount of data giving a relation between these different parameters and the flux integrated over the whole line. Furthermore, other empirical laws are obtained between the total $L\alpha$ flux and two solar activity indices.

076.002 **X-ray bursts from solar flares behind the limb.**
J.-R. Roy, D. W. Datlowe.
Solar Physics, Vol. 40, 165 - 182 (1975).

From the UCSD OSO-7 X-ray experiment data, the authors have identified 54 X-ray bursts with 5.1–6.6 keV flux greater than 10^3 photon cm^{-2} keV^{-1} which were not accompanied by visible $H\alpha$ flare on the solar disk. By studying OSO-5 X-ray spectroheliograms, $H\alpha$ activity at the limb and the emergence and disappearance of sunspot groups at the limb, they found 17 active centers as likely seats of the X-ray bursts beyond the limb. The authors present the analysis of 37 X-ray bursts and their physical parameters. They compare their results with those published by Datlowe et al. (1974) for disk events.

076.003 **Hard X-ray bursts from flares behind the solar limb.**
D. L. McKenzie.
Solar Physics, Vol. 40, 183 - 191 (1975).

The determination of the location of the region of origin of hard X-rays is important in evaluating the importance of 10–100 keV electrons in solar flares and in understanding flare particle acceleration. At present only limb-occulted events are available to give some information on the height of X-ray emission. In fifteen months of OSO-7 operation, nine major soft X-ray events had no reported correlated $H\alpha$ flare. The author examines the hard X-ray spectra of eight of these events with good candidate X-ray flare producing active regions making limb transit at the time of the soft X-ray bursts. All eight bursts had significant X-ray emission in the 30–44 keV range, but only one had flux at the 3σ level above 44 keV. The data are consistent with most X-ray emission occurring in the lower chromosphere, but some electron trapping at high altitudes is necessary to explain the small nonthermal fluxes observed.

076.004 **On the analysis of solar XUV observations and the structure of the chromosphere-corona transition region.** D. R. Flower, G. Pineau des Forêts.
Astron. Astrophys., Vol. 37, 297 - 300 (1974).

The method of analysis of solar XUV line intensities is reconsidered in the light of recent observational determinations of the electron and acoustic pressure in the transition region. The assumptions of constant electron pressure and constant thermal conduction flux in the outer transition region are shown to be incompatible with these observations. A model is derived which is consistent with the available observations and which is characterised by (1) a constant electron density and (2) an electron temperature (T_e) gradient varying as $T_e^{-1/2}$ in the outer transition region ($10^5 \lesssim T_e \lesssim 10^6$ K).

076.005 **Gamma radiation of solar flares and cosmic ray generation.** I. M. Gordon.
Izv. AN SSSR. Ser. fiz., Vol. 38, 1855 - 1858 (1974). In Rus-

sian. – Abstr. in Referativ. Zhurn. 51. Astron., 2.51.516 (1975).

076.006 **Solar flare X-rays in July 1972.**
N. N. Evstaf'ev, O. M. Kovrizhnykh, A. S. Melioranskij, N. I. Nazarova, V. M. Pankov, I. R. Rozantsev, I. A. Savenko, L. M. Chupova.
Solnechnye Dannye 1974 Byull., No. 12, p. 55 - 63 (1975). In Russian.

Observations of solar flare X-rays $5<E<30$ keV by the "Prognoz-2" satellite during July 1972 are given. These measurements are discussed in connection with different phenomena of solar activity.

076.007 **The intensities of helium lines in the solar EUV spectrum.** C. Jordan.
Monthly Notices Roy. Astron. Soc., Vol. 170, 429 - 440 (1975).

The present paper points out that the lines of neutral and singly ionized helium in the solar EUV spectrum have anomalously high intensities when compared with lines of other ions formed at similar temperatures. It is suggested that the observed absolute and relative intensities, and in addition line widths, can be accounted for if a mechanism which causes the helium atoms and ions to be excited by electrons with temperatures greater than the ionization equilibrium value is operating.

076.008 **Impulsive solar X-ray bursts. II. Statistical correlation of observations with solar longitude.**
V. Petrosian.
Astrophys. Journ., Vol. 197, 235 - 239 (1975).

It is shown that the model where the impulsive solar X-ray bursts result from interaction of a beam of high-energy electrons with solar plasma, agrees both qualitatively and quantitatively with the existing data on variation with solar longitude of frequency distribution and average value of parameters characterizing these bursts. Concerning center-to-limb variations, the model predicts correctly the softening of the impulsive X-ray burst spectra and the absence of variation in other quantities, such as the burst strength, the soft-to-hard X-ray flux ratio, and the frequency of occurrence of bursts.

076.009 **Ultraviolet solar spectrum (1889–1969 Ångströms).**
H. C. McAllister, P. H. Smith.
Solar Physics, Vol. 41, 3 - 16 (1975).

Six spectrograms of the solar spectrum were obtained in the region from 1800 to 1970 Å at a resolving power of about 2×10^5, using a rocket-borne spectrograph with an echelle as the principal dispersing element. Data reduction has been completed for the region from 1889 to 1969 Å, in which 732 features have been measured and tabulated along with available laboratory wavelengths which correspond reasonably with the solar features. Tracings of the spectra are also presented.

076.010 **Enhancement of the solar X-ray line intensities due to dielectronic recombination as an excitation process.** S. M. R. Ansari, B. Alam.
Solar Physics, Vol. 41, 97 - 107 (1975).

Intensities of resonance lines in X-ray region for Si IX– Si XIV ions are calculated by considering total excitation, i.e., excitation due to electron impact and dielectronic recombination. It is found that the contribution of the latter is quite significant. For consistency, the electron density effect is not only accounted for in the ionization equilibrium but also in the total rate of excitation. It is also found that the contribution of electron density effect is pronounced with the inclusion of dielectronic recombination as an excitation mechanism. The computed average line intensities are compared with the available observations and a table of line flux for various wavelengths of the above-mentioned ions at different temperatures is also given.

076.011 The height distribution of flare hard X-rays in thick and thin target models.
J. C. Brown, A. N. McClymont.
Solar Physics, Vol. 41, 135 - 151 (1975).

Quantitative predictions are made of this height distribution for both models and the results discussed in relation to observations of hard X-ray emission from flares behind the limb. It is concluded that the thick target model is as compatible with such events as the thin target whereas the latter is in general much less satisfactory in terms of energy requirements and of flare observations at other wavelengths. Other source models are also briefly considered.

076.012 Solar flare X-ray polarization.   S. L. Mandel'stam (Mandel'shtam), I. L. Beigman (Bejgman), I. P. Tindo.
Nature, Vol. 254, 462 (1975).

076.013 X-ray bright points, coronal heating and the solar cycle.   M. H. Gokhale.
Solar Physics, Vol. 41, 381 - 386 (1975).

Recent Skylab observations about the bright points in the solar X-ray images seem to confirm an essential prediction of a model proposed by this author for the appearance and the disappearance of the photospheric fields during a solar cycle. The segments of the individually rising strands of the 'fundamental flux-loops' proposed in the model may lead to the X-ray bright points with the observed properties. The emergence of such strands may substantially contribute to the coronal heating at different heights.

076.014 EUV emission, filament activation and magnetic fields in a slow-rise flare.
D. M. Rust, Y. Nakagawa, W. M. Neupert.
Solar Physics, Vol. 41, 397 - 414 (1975).

The evolution of coronal and chromospheric structures is examined together with magnetograms for the 1B flare of January 19, 1972. Soft X-ray and EUV studies are based on the OSO-7 data. The Hα filtergrams and magnetograms came from the Sacramento Peak Observatory. Theoretical force-free magnetic field configurations are compared with structures seen in the soft X-ray, EUV and Hα images.

076.015 On the relationships between sfe (crochet) and solar X-ray and microwave bursts.
J. H. Sastri, B. S. Murthy.
Solar Physics, Vol. 41, 477 - 485 (1975).

Geomagnetic crochets (sfe) observed at Kodaikanal over the period 1966–71 have been studied in relation to solar X-ray bursts observed by NRL satellite (SOLRAD-9) in the 0.5–3 Å, 1–8 Å and 8–20 Å bands and radio bursts observed in the frequency range 1000–17000 MHz. The amplitude of sfe is linearly correlated with the peak intensities of X-ray bursts in the 1–8 Å and 8–20 Å bands. The single frequency correlation of sfe with radio bursts is a flat maximum in the frequency range 2000–3750 MHz.

076.016 The ultraviolet solar spectrum.
L. A. Dreiling, R. A. Bell.
Bull. American Astron. Soc., Vol. 7, 248 (1975). — Abstr. AAS.

076.017 Interpreting XUV spectroheliograms in terms of coronal magnetic field structures.
N. R. Sheeley, Jr., J. D. Bohlin, G. E. Brueckner, J. D. Purcell, V. E. Scherrer, R. Tousey.
Bull. American Astron. Soc., Vol. 7, 346 (1975). — Abstr. AAS.

076.018 The location of the site of energy release in an X-ray sub-flare.   R. D. Petrasso, S. W. Kahler, A. S. Krieger, J. K. Silk, G. S. Vaiana.
Bull. American Astron. Soc., Vol. 7, 352 (1975). — Abstr. AAS.

076.019 X-ray spectral dependence of the relationship between impulsive solar flare X-rays and type III radio bursts.   S. R. Kane.
Bull. American Astron. Soc., Vol. 7, 352 (1975). — Abstr. AAS.

076.020 The XUV spectrum of He II in quiet regions, a coronal hole, filaments, prominences, and the 7 August 1972 flare.   J. L. Linsky, D. L. Glackin, R. D. Chapman, W. M. Neupert, R. J. Thomas.
Bull. American Astron. Soc., Vol. 7, 353 (1975). — Abstr. AAS.

076.021 Extreme ultraviolet solar limb brightening observations of lithium-like ions.
J. T. Mariska, G. L. Withbroe.
Bull. American Astron. Soc., Vol. 7, 354 (1975). — Abstr. AAS.

076.022 Center to limb variations in solar hard X-ray spectra.   D. W. Datlowe, M. J. Elcan, H. S. Hudson, L. E. Peterson.
Bull. American Astron. Soc., Vol. 7, 354 (1975). — Abstr. AAS.

076.023 The shape of the hard solar X-ray spectrum.   M. J. Elcan, D. W. Datlowe, H. S. Hudson, L. E. Peterson.
Bull. American Astron. Soc., Vol. 7, 354 - 355 (1975). Abstr. AAS.

076.024 Estimates of impact polarization of solar flare X-ray lines.   W. A. Brown.
Bull. American Astron. Soc., Vol. 7, 355 (1975). — Abstr. AAS.

076.025 Spatial relationships between X-ray, EUV, and H-alpha sources during a solar flare observed by OSO-7.
R. J. Thomas, W. M. Neupert.
Bull. American Astron. Soc., Vol. 7, 355 (1975). — Abstr. AAS.

076.026 General properties of soft X-ray flare images.   S. W. Kahler, A. S. Krieger, G. S. Vaiana.
Bull. American Astron. Soc., Vol. 7, 355 (1975). — Abstr. AAS.

076.027 Time changes in the structure and spectrum of an X-ray flare.
J. K. Silk, S. W. Kahler, A. S. Krieger, G. S. Vaiana.
Bull. American Astron. Soc., Vol. 7, 355 (1975). — Abstr. AAS.

076.028 On the XUV emissions in solar flares observed with the ATM/NRL spectroheliograph.
C. C. Cheng, K. G. Widing.
Bull. American Astron. Soc., Vol. 7, 356 - 357 (1975). Abstr. AAS.

076.029 Comparison of Skylab X-ray and ground-based helium observations.   J. W. Harvey, A. S. Krieger, J. M. Davis, A. F. Timothy, G. S. Vaiana.
Bull. American Astron. Soc., Vol. 7, 358 (1975). — Abstr. AAS.

076.030 Formation of the solar EUV spectrum.   E. H. Avrett.
Bull. American Astron. Soc., Vol. 7, 360 (1975). — Abstr. AAS.

076.031 Absolute solar UV intensities 1680 Å to 2100 Å.   O. K. Moe, G. E. Brueckner, J.-D. F. Bartoe, M. E. Van Hoosier.
Bull. American Astron. Soc., Vol. 7, 360 (1975). — Abstr. AAS.

076.032 Further measurements of emission line profiles in

the solar ultraviolet spectrum.
B. C. Boland, E. P. Dyer, J. G. Firth, A. H. Gabriel, B. B. Jones, C. Jordan, R. W. P. McWhirter, P. Monk, R. F. Turner. Monthly Notices Roy. Astron. Soc., Vol. 171, 697 - 724 (1975).

A further flight of a high resolution echelle spectrograph has been carried out on a Skylark rocket to measure solar line profiles in the region 1400–2200 Å. Microdensitometer traces for 37 emission lines are presented. Analysis of the profiles leads to a non-thermal mechanical velocity component which increases with temperature over the range $6 \times 10^3 - 10^5$ K, always remaining subsonic. Interpretations are considered in terms of a progressive mechanical energy flux to the corona.

076.033   **Solar gamma rays.**
R. Ramaty, B. Kozlovski, R. E. Lingenfelter. Goddard Space Flight Center, Greenbelt, Maryland, GSFC Document X-660-74-368, 1 + 92 pp. (1974).

The theory of gamma-ray production in solar flares is treated in detail. Both lines and continuum are produced. The strongest line predicted at 2.225 MeV with a width of less than 100 eV and detected at 2.24 ± 0.02 MeV, is due to neutron capture by protons in the photosphere. The strongest prompt lines are at 4.43 MeV from ^{12}C and at ~ 6.2 from ^{16}O and ^{15}N. These lines result from both direct excitation and spallation. The widths of individual prompt lines are determined by nuclear kinematics. Other potentially observable lines are predicted. From the comparison of the observed and calculated intensities of the line at 4.4 MeV to that of the 2.2 MeV line it is possible to obtain information on the spectrum of accelerated nuclei in flares. About 10^{33} protons of energies greater than 30 MeV were produced in the 1972, August 4 flare. The gamma-ray continuum, produced by electron bremsstrahlung, allows the determination of the spectrum and number of accelerated electrons in the MeV region. From the comparison of the line and continuum intensities the authors find a proton-to-electron ratio of about 10 to 10^2 at the same energy for the 1972, August 4 flare.

076.034   **Time-dependent 2.2 MeV and 0.5 MeV lines from solar flares.**   H. T. Wang, R. Ramaty. Goddard Space Flight Center, Greenbelt, Maryland, GSFC Document X-660-75-74, 2 + 36 pp. (1975).

The time dependences of the 2.2 MeV and 0.51 MeV gamma-ray lines from solar flares are calculated and the results are compared with observations of the 1972, August 4 and 7 flares. The time lag between the nuclear reactions and the formation of these two lines are caused, respectively, by capture of the neutrons, and by deceleration of the positrons and decay of the radioactive nuclei.

076.035   **The origin and implications of gamma rays from solar flares.**   R. Ramaty. Goddard Space Flight Center, Greenbelt, Maryland, GSFC Document X-660-75-79, Preprint, 1 + 8 pp. (1975). – Highlights of an invited paper presented at the meeting of the American Physical Society, Washington, D. C., April 1975.

076.036   **Solar gamma-ray lines as probes of accelerated particle directionalities in flares.**
R. Ramaty, C. J. Crannell. Goddard Space Flight Center, Greenbelt, Maryland, GSFC Document X-660-75-96, 1 + 11 pp. (1975).

Anisotropies of charged particles accelerated in solar flares can be studied by observing Doppler shifts of selected gamma-ray lines. The authors have calculated the spectral shape of the 6.1 MeV line of ^{16}O. If the accelerated particles are isotropic, the line remains centered at $E_0 = 6129.4$ keV, and its width is about 100 keV. However, for particle anisotropies that may be produced in solar flares, the line is shifted to lower energies by about 30 to 40 keV.

076.037   **On the role of competing atomic processes and electron density in the line emission of solar X-rays.**
S. M. R. Ansari, Badr-e-Alam, J. Hussain. Bull. Astron. Soc. India, Vol. 2, 40 - 41 (1974). – Abstract.

076.038   **EUV observations of the active sun from the Harvard experiment on ATM.**
R. W. Noyes, P. V. Foukal, M. C. E. Huber, E. M. Reeves, E. J. Schmahl, J. G. Timothy, J. E. Vernazza, G. L. Withbroe. IAU Symposium No. 68, (see 012.017), p. 3 - 17 (1975).

076.039   **Time variations of solar X-ray bright points.**
L. Golub, A. S. Krieger, J. K. Silk, A. F. Timothy, G. S. Vaiana. IAU Symposium No. 68, (see 012.017), p. 23 - 24 (1975). Summary.

076.040   **Solar activity observed in X-rays and the EUV from OSO 7.**   R. J. Thomas. IAU Symposium No. 68, (see 012.017), p. 25 - 42 (1975).

076.041   **X-ray spectra of solar active regions.**
J. H. Parkinson. IAU Symposium No. 68, (see 012.017), p. 45 - 64 (1975).

076.042   **X-ray spectroscopy of solar active regions during the third Skylab mission.**
J. P. Pye, R. J. Hutcheon, J. H. Parkinson, K. A. Pounds. IAU Symposium No. 68, (see 012.017), p. 65 - 66 (1975). Summary.

076.043   **Spatially resolved X-ray spectra of coronal active regions.**   R. C. Catura, L. W. Acton, E. G. Joki, C. G. Rapley, J. L. Culhane. IAU Symposium No. 68, (see 012.017), p. 67 (1975). Summary.

076.044   **Statistical methods in the identification and prediction of the solar X-ray spectral lines.**
S. Yousef. IAU Symposium No. 68, (see 012.017), p. 71 - 72 (1975). Summary.

076.045   **Soft X-radiation from single active regions.**
D. H. Brabban, E. B. Dorling, W. M. Glencross, J. R. H. Herring. IAU Symposium No. 68, (see 012.017), p. 101 - 102 (1975). Summary.

076.046   **Flare-like ultraviolet spectra of active regions.**
G. E. Brueckner. IAU Symposium No. 68, (see 012.017), p. 105 - 107 (1975). Summary.

076.047   **The structure of solar active regions from EUV and soft X-ray observations.**   C. Jordan. IAU Symposium No. 68, (see 012.017), p. 109 - 131 (1975).

076.048   **Association of X-ray flares with solar coronal active regions.**   P. R. Sengupta. IAU Symposium No. 68, (see 012.017), p. 183 (1975). Summary.

076.049   **The relationship between hard and soft X-ray bursts observed by OSO 7.**   D. W. Datlowe. IAU Symposium No. 68, (see 012.017), p. 191 - 208 (1975).

076.050   **Relationship between hard and soft solar X-ray sources observed by OSO-7.**
D. W. Datlowe, H. S. Hudson. IAU Symposium No. 68, (see 012.017), p. 209 (1975).

Summary.

076.051 Thermal and nonthermal interpretations of flare X-ray bursts. S. Kahler.
IAU Symposium No. 68, (see 012.017), p. 211 - 231 (1975).

076.052 Rise time of hard X-ray bursts.
J. Vorpahl, T. Takakura.
IAU Symposium No. 68, (see 012.017), p. 237 - 238 (1975).
Summary.

076.053 Determination of the height of hard X-ray sources in the solar atmosphere by measurement of photospheric albedo photons. J. C. Brown, H. F. Van Beek.
IAU Symposium No. 68, (see 012.017), p. 239 - 241 (1975).
Summary.

076.054 The interpretation of spectra, polarization, and directivity of solar hard X-rays. J. C. Brown.
IAU Symposium No. 68, (see 012.017), p. 245 - 282 (1975).

076.055 Solar flare X-ray measurements and their relation to microwave bursts. L. D. De Feiter.
IAU Symposium No. 68, (see 012.017), p. 283 - 297 (1975).

076.056 Relation of microwave emission to X-ray emission from solar flares. T. Takakura.
IAU Symposium No. 68, (see 012.017), p. 299 - 313 (1975).

076.057 X- and γ-ray measurements during the 1972, August 2 and 7 large solar flares.
R. Talon, G. Vedrenne, A. S. Melioransky (*Melioranskij*),
N. F. Pissarenko (*Pisarenko*), V. M. Shamolin, O. B. Likin.
IAU Symposium No. 68, (see 012.017), p. 315 - 339 (1975).

076.058 High energy gamma-ray radiation above 300 keV associated with solar activity.
E. L. Chupp, D. J. Forrest, A. N. Suri.
IAU Symposium No. 68, (see 012.017), p. 341 - 359 (1975).

076.059 Measurements of a gamma-ray burst above 1 MeV.
R. Koga, G. M. Simnett, R. S. White.
IAU Symposium No. 68, (see 012.017), p. 361 (1975).
Summary.

076.060 Observation of the absolute intensity of the sun in the vacuum ultraviolet region. K. Nishi.
Solar Physics, Vol. 42, 37 - 42 (1975).
The absolute intensity of the solar spectrum between 1550 Å and 1950 Å was measured photoelectrically by a rocketborne spectrometer flown from the Kagoshima Space Centre on 19 February 1973. The spectrometer was a single dispersive type with uniaxial sun-pointer, and the absolute intensity from the whole disk with a 78 Å spectral resolution was measured.

076.061 The analysis of a high resolution X-ray spectrum of a solar active region. J. H. Parkinson.
Solar Physics, Vol. 42, 183 - 207 (1975).
Absolute intensities and wavelengths for 98 X-ray emission lines between 9 and 22.5 Å are reported. The active region spectra were obtained with three rocket-borne crystal spectrometers. Identifications are proposed for 62 of the lines and a model of the emitting plasma is constructed from the intensities of the strongest lines.

076.062 The X-ray line and continuum emission from a solar active region. P. B. Landecker, R. S. Wolff.
Solar Physics, Vol. 42, 209 - 214 (1975).
The X-ray spectrum of the quiet sun in the energy range 2.3–6.9 keV was observed from an Aerobee rocket using an uncollimated graphite crystal spectrometer. These results and spatial measurements made with an onboard modulation collimator are analyzed using solar models. Several methods of estimating coronal temperatures are used in the analysis and all yield results within the range $(4\pm1) \times 10^6$ K.

076.063 Anisotropy and polarization of solar X-ray bursts. J. C. Henoux.
Solar Physics, Vol. 42, 219 - 233 (1975).
The effects of the Compton back-scattered X-ray flux from the photosphere on the directivity and polarization of flare X-rays between 15 keV and 150 keV are computed. The calculations are made with a thin target model for flares of De Jager-Kundu type with electrons spiralling downward around a vertical magnetic field and for an isotropic source. The resulting polarization for an isotropic source is not higher than 4%. The resulting directivity of anisotropic sources is greatly reduced, particularly below 70 keV.

076.064 Outbursts of γ-radiation observed during the flares of August 4 and 7, 1972 aboard the Prognoz 2 satellite. Zh. Verden, O. B. Likin, A. S. Melioranskij,
N. F. Pisarenko, I. A. Savenko, R. Talon, V. M. Shamolin.
Izv. AN SSSR. Ser. fiz., Vol. 39, 272 - 280 (1975). In Russian.
Abstr. in Referativ. Zhurn. 62. Issled. kosmich. prostranstva, 6.62.205 (1975).

076.065 Observation of the solar ultraviolet Mg II doublet, (1). T. Kohno, N. Yajima, Z. Suemoto.
Tokyo Daigaku Uchu Koku Kenkyusho Hokoku., Vol. 10, 166 - 173 (1974). In Japanese.

076.066 Observation of the solar ultraviolet Mg II doublet, (2). N. Yajima, T. Kohno, Z. Suemoto.
Tokyo Daigaku Uchu Koku Kenkyusho Hokoku., Vol. 10, 174 - 182 (1974). In Japanese.

076.067 On the problem of relationship between the dynamics of the photosphere and coronal structures in X-rays and the solar wind. V. M. Tomozov, V. V. Kasinskij.
Solnechnye Dannye 1975 Byull., No. 2, p. 84 - 90 (1975).
In Russian.
The distribution of X-rays over the solar disc (coronal holes and bright regions) is compared with the dynamics of sunspot activity and the solar wind. It is concluded that the regions of minimum X-emission are long living, stable structures under which the sunspot activity is depressed essentially during several rotations of the sun. The probable reason of strengthening and maintaining the high speeds of solar wind lies in rising the new magnetic fields and regeneration of the old ones. The maximum velocities fall within the periphery of coronal holes.

A high resolution decameter multichannel radio spectrograph. See Abstr. 033.032.

Stigmatic spectra of the sun between 1200 Å and 2100 Å. See Abstr. 071.024.

Analysis of the August 7, 1972 white light flare: light curves and correlation with hard X-rays.
See Abstr. 073.004.

Laboratory reproduction of a solar X-ray flare spectrum. See Abstr. 073.012.

Ultraviolet emission line profiles of flares and active regions. See Abstr. 073.085.

Fe XXIV emission in solar flares observed with the NRL/ATM XUV slitless spectrograph.

See Abstr. 073.086.

X-ray and EUV spectra of solar flares and laboratory plasmas.   See Abstr. 073.087.

Studies of the dynamic structure and spectra of solar X-ray flares.   See Abstr. 073.088.

High time resolution analysis of solar flares observed on the ESRO TD-1A satellite.   See Abstr. 073.090.

Inference of the hard X-ray source dimensions in the 1972, August 7 white light flare. See Abstr. 073.091.

Gamma-ray lines from solar flares. See Abstr. 073.092.

Fast electrons in small solar flares. See Abstr. 073.093.

Nonthermal processes in large solar flares. See Abstr. 073.096.

X-ray heating of a low-temperature region in chromospheric flares.   See Abstr. 073.104.

On the polarization of X-ray radiation of solar flares.   See Abstr. 073.106.

Generation of hydrogen and helium isotopes and of γ-radiation in solar flares.   See Abstr. 073.107.

Coronal information from EUV disk spectral line intensities.   See Abstr. 074.001.

Limb-brightening observations from the OSO-7 satellite. I. Electron density and temperature of the non-equatorial corona from EUV lines of Fe XIV and other Fe ions.   See Abstr. 074.006.

Long-term X-ray emitting structures in the solar corona.   See Abstr. 074.051.

Interpretation of the X-ray spectra of solar active regions.   See Abstr. 074.098.

Time variations in coronal active regions. See Abstr. 074.099.

Observation of a non-uniform component in the distribution of coronal bright points. See Abstr. 074.100.

X-ray observations of coronal holes and their relation to high velocity solar wind streams. See Abstr. 074.125.

Atmospheric heating by solar EUV radiation. See Abstr. 082.081.

A cometary hydrogen model: comparison with OGO-5 measurements of comet Bennett (1970 II). See Abstr. 102.009.

077 Solar Radio Radiation

077.001 **Location of radio source at 35 GHz of 2145 UT 2 August 1972 burst.**
H. Ogawa, K.-A. Kawabata.
Solar Physics, Vol. 40, 159 - 163 (1975).

The location of the radio source of a major burst at 2145 UT on 2 August 1972 is determined from observations undertaken by the 35 GHz solar interferometer at Nagoya. The location of the radio source coincides with an Hα brightening.

077.002 **A model exciter for type III solar radiobursts.**
C. C. Harvey.
Solar Physics, Vol. 40, 193 - 216 (1975).

In an earlier paper (Harvey and Aubier, 1973) the large scale radial electron density gradient in the corona and solar wind was shown to cause the phase velocity of plasma waves to decrease as they propagate away from the sun, thus leading to appreciable Landau damping of the plasma waves. It is proposed here that this same phase velocity decrease creates conditions which facilitate the stabilisation of a beam of exciter electrons of finite duration, provided that three conditions are fulfilled. The spatial density of the power converted into plasma waves is calculated as a function of position and time, and is shown to be independent of the nature of the stabilisation mechanism.

077.003 **New microstructure of decametre solar radio bursts.**
H. S. Sawant, S. K. Alurkar, R. V. Bhonsle.
Nature, Vol. 253, 329 - 330 (1975).

Earlier observations of solar radio bursts in the decametre region have revealed some types of fine structure bursts, drift pairs and split pairs being examples of fine temporal and frequency structure bursts respectively. The authors report here a new type of solar decametre bursts which, they believe, has been seen for the first time. This was a result of high resolution (frequency ~5 kHz and time ~10 ms) spectral observations made over a frequency range of only 0.5 MHz near 35 MHz with a scanning rate of 100 Hz.

077.004 **Split-band structure in type II radio bursts from the sun.** S. F. Smerd, K. V. Sheridan, R. T. Stewart.
Astrophys. Letters, Vol. 16, 23 - 28 (1975).

The split-band structure of type II bursts is here attributed to simultaneous plasma-frequency emission from plasma ahead of and behind a type II shock front. The amount of band splitting is then a measure of the shock strength. Since the shock strength is the ratio of the disturbance speed to the Alfvén velocity and since the disturbance speeds can be derived from measured frequency-drift rates and an assumed coronal-density model, it is possible to derive the magnetic field strength along the path of the type II disturbance. Shock strengths in the range $1.2 \lesssim M \lesssim 1.7$, and magnetic fields in the range $0.3 \lesssim H \lesssim 4G$ are thus obtained for nine split-band type II bursts.

077.005 **Characteristic pairs of type III solar radio bursts.**
G. Daigne, B. Møller-Pedersen.
Astron. Astrophys., Vol. 37, 355 - 366 (1974).

The authors report observations of type III solar radio bursts appearing in characteristic pairs and usually considered as fundamental-harmonic emissions from a common source of plasma oscillations. They measured time delay and frequency ratio of the intensity maxima, peak intensities, half power duration and decay time of the intensity time profiles. They have compared peak intensities observed at two harmonic frequencies with a theoretical model based on the

$F-H$ hypothesis (Melrose, 1974).

077.006 **On elliptic polarization in solar radio bursts emission at 3.2 cm wavelength.** V. P. Nefed'ev.
Issled. po geomagnetizmu, aehron. i fiz. Solntsa. Vyp. (No.) 31. Moskva, Nauka, 1974, p. 38 - 43. In Russian. – Abstr. in Referativ. Zhurn. 51. Astron., 2.51.505 (1975).

077.007 **Frequency-time analysis and prediction of average diurnal data for the total flux of solar radio emission.**
S. A. Andrianov, L. V. Yasnov.
Radioizluchenie Solntsa. Vyp. (No.) 3. Leningrad, Leningr. un-t, 1974, p. 55 - 91. In Russian. – Abstr. in Referativ. Zhurn. 51. Astron., 2.51.523 (1975).

077.008 **The effect of electron density fluctuations on the fundamental radiation of type III bursts.**
J. Heyvaerts.
Astron. Astrophys., Vol. 38, 45 - 49 (1975).

The effect of long-wavelength electron density inhomogeneities on the fundamental radiation of type III bursts is discussed. It is shown that the effective radiation transfer coefficients in the source are mainly determined by these fluctuations if the relative electron density of fluctuations exceeds 0.1 %. In this case both the bandwidth in the vicinity of a given altitude and the overall size of the source reflect the properties of the fluctuations of the electronic density.

077.009 **On the "harmonic structure" in type III solar radio bursts.** G. Daigne.
Astron. Astrophys., Vol. 38, 141 - 143 (1975).

The two components of characteristic pairs of type III solar radio bursts have been observed on four different frequencies. Correlations between the peak intensities do not favour the fundamental-harmonic radiation hypothesis generally involved for these burst pairs.

077.010 **Comments on "Polarization fine structure in solar radiobursts of type III on short meter wavelengths" by C. Slottje.** A. C. Riddle.
Astron. Astrophys., Vol. 38, 153 - 155 (1975).

The polarization characteristics of a pair of Type III solar radio bursts observed by Slottje led him to conclude that they were fundamental radiation despite gross differences in other characteristics. It is suggested here that the polarization profile of the second burst can also be explained on the assumption that it is harmonic radiation and then the other differences between the bursts are a natural consequence of the fundamental-harmonic relationship between the bursts.

077.011 **The solar spectrum at 8 mm.**
P. N. Swanson, R. Kuseski.
Nature, Vol. 253, 513 - 514 (1975).

The brightness temperature of the quiet sun has been measured over almost the complete radio spectrum. It ranges from approximately 6,000 K at millimetre wavelengths to over 10^6 K at metre wavelengths. The authors have measured the slope of the brightness-frequency curve near 36 GHz ($\lambda = 8$ mm) to determine whether the slope best matches the van de Hulst model or the observations. The steep slope measured at 36 GHz indicates that the observed disk temperatures in the 20–40 GHz range are probably more reliable than the van de Hulst model calculations.

077.012 **On the directivity of the radiation of local sources of the S-component of the solar radio emission at 3.2 cm.** V. N. Borovik, G. B. Gel'frejkh, B. I. Lubyshev.

Astron. Zhurn. Akad. Nauk SSSR, Vol. 52, 97 - 105 (1975).
In Russian. English translation in Soviet Astron., Vol. 19, No. 1.

18 local sources of the radio emission of the sun at 3.2 cm wavelength were studied to check the directivity of their radiation.

077.013 On the mechanism of fast-drifting radio bursts in absorption. V. V. Zajtsev, A. V. Stepanov.
Solnechnye Dannye 1974 Byull., No. 11, p. 71 - 74 (1975). In Russian.

The type IV continuum with fast-drifting absorption radio bursts is supposed to be due to the high-frequency loss-cone instability in the coronal magnetic trap. The absorption bursts arise as a result of injection into the trap of a beam of superthermal electrons, filling the loss-cone and eliminating instability.

077.014 Connection of the intensity variations of the continuum and bursts of noise storms with active regions. G. P. Chernov.
Solnechnye Dannye 1974 Byull., No. 11, p. 74 - 83 (1975). In Russian.

The results of an investigation of 11 noise storms, recorded in IZMIRAN over the range 151 - 187 MHz during 1971, are given and discussed.

077.015 On the connection of local solar radio sources at $\lambda = 6.6$ cm with the characteristics of active regions.I.
G. B. Gel'frejkh, Z. B. Korobova, N. P. Stasyuk.
Solnechnye Dannye 1974 Byull., No. 11, p. 83 - 88 (1975). In Russian.

The results of reduction of strip-scans of the sun at $\lambda = 6.6$ cm during the 1964 minimum of solar activity are presented. Local radio sources are identified with active regions.

077.016 Fine structure of the continuum solar radio burst on August 22, 1971.
S. T. Akin'yan, Yu. B. Vedeneev, I. M. Chertok.
Solnechnye Dannye 1974 Byull., No. 11, p. 88 - 93 (1975). In Russian.

The radio burst on August 22, 1971, recorded at IZMIRAN and NIRFI with two spectrographs (45 - 90 MHz and 100 - 250 MHz) and several radiometers, is analysed. In this event the pulsations of the radio emission with time scale of 8 - 16 sec have been observed on the background of the metric continuum lasting about four minutes.

077.017 Search for the diffraction structure in meter solar radio bursts. G. P. Chernov.
Solnechnye Dannye 1974 Byull., No. 12, p. 75 - 80 (1975). In Russian.

To discover the diffraction structure in the meter solar bursts, a set of observations of the spectrum within 151 - 187 MHz with high time resolution and of the intensity at the frequency of 169 MHz were made at the IZMIRAN in 1971. The results are presented and discussed.

077.018 Polarization reversal during the solar noise storm activity of August 1971. M. Kurihara.
Publ. Astron. Soc. Japan, Vol. 27, 71 - 79 (1975).

Reversals of the sense of circular polarization of solar radio emission were observed for active type I storms in August 1971. Observations with a 160-MHz interferometer revealed that the reversals were caused by sudden growth and decay of a secondary storm source whose sense of polarization was opposite to that of the long-lasting main source. The time variations of both the associated S-component sources and sunspots are compared with that of the storm sources. The role of the magnetic field, which presumably connects the storm sources, the S-component sources, and the sunspots.

is discussed in relation to the origin of the storm activity.

077.019 3.3 millimeter limb brightening measurements during the 30 June 1973 total solar eclipse.
F. I. Shimabukuro, W. J. Wilson, T. T. Mori, P. L. Smith.
Solar Physics, Vol. 40, 359 - 370 (1975).

Solar limb brightening measurements at a wavelength of 3.3 mm were made during the 30 June 1973 total solar eclipse from a site at Lake Rudolf, Kenya. The results show that at this wavelength there is a limb brightening of about 20%, occurring within one half arc min of the limb.

077.020 Some studies on solar microwave bursts in relation to the phases of the associated Hα-flares and their spectral nature.
S. K. Sarkar, T. Chattopadhyay, M. K. Das Gupta.
Solar Physics, Vol. 40, 411 - 415 (1975).

Occurrences of the flare-associated microwave bursts as well as their peak flux and energy excess spectra have been examined in relation to the pre- and post-maximum phases of the respective flares during the period 1969−72.

077.021 An example of a fundamental type IIIb radio burst.
R. T. Stewart.
Solar Physics, Vol. 40, 417 - 419 (1975). − Research note.

077.022 Type IIIb radio bursts: 80 MHz source position and theoretical model.
T. Takakura, S. Yousef.
Solar Physics, Vol. 40, 421 - 438 = Division Radiophys., CSIRO, Sydney, Radiophys. Publ. RPP 1758 (1975).

The authors present Culgoora spectrograph and radioheliograph observations as well as a model of type IIIb bursts; the latter are defined as chains of striae of slow or no frequency drift, the chain as a whole drifting like a normal type III burst. The 80 MHz source positions are studied for a group of IIIb bursts, a IIIb precursor and harmonic pairs of 1 : 2 frequency ratio. It is found that the IIIb position may vary in a IIIb group. No significant difference was found between the source positions of a IIIb precursor and the following III burst. For one event the authors found that the fundamental IIIb burst showed a high degree of circular polarization ($\sim 46\%$), while its second harmonic, a normal type III burst, was unpolarized.

077.023 High resolution observations of solar bursts at 3.7 and 11.1 cm wavelengths.
C. E. Alissandrakis, M. R. Kundu.
Solar Physics, Vol. 41, 119 - 133 (1975).

Four bursts were observed on August 9, 1973 with the NRAO 3-element interferometer at 3.7 and 11.1 cm. By using a simple source model the authors have calculated the temperature, flux, size and position of the small scale components of the bursts as a function of time. Two of the bursts were found to be right circularly polarized. The existence of burst structures with temperatures of the order of 10^7 K indicates that at least part of the radiation in these bursts is generated by a non-thermal mechanism.

077.024 Time profile of type III bursts.
T. Takakura, Y. Naito, K. Ohki.
Solar Physics, Vol. 41, 153 - 161 (1975).

On the hypothesis that the time profile of a type III burst corresponds directly to the flux of electron beam, the similarity of time profile is shown to be maintained even if the electron velocity decreases with distance provided that the time is normalized to unity at the time of maximum flux. The observed time profiles of type III bursts with simple shape seem to follow the similarity law in almost all frequency range. This evidence may indicate that the time profile, both the rising and decaying phases, of a type III burst should be attributed to a common origin, e.g., the time variation of exciter determined by the initial velocity distribution in the electron beam, instead of attributing the rising time to the beam length and the decay time to the damping of plasma waves after the passage of the electron beam.

077.025 Decameter storm radiation, II.
T. E. Gergely, M. R. Kundu.
Solar Physics, Vol. 41, 163 - 188 (1975).
The physical properties of six decametric storms, observed at Clark Lake Radio Observatory are studied. The occurrence of two distinct classes of type III bursts in storms is discussed: 'off-fringe' and 'onfringe' type III's. A model of the storm region is proposed.

077.026 An intense microwave radio solar burst observed at 1200 MHz on 1974 December 16. A. N. Kelly.
Monthly Notes Astron. Soc. Southern Africa, Vol. 34, 12 - 14, with a correction p. 15 (1975).

077.027 Pulsations of the continuum metre solar radio emission. A. K. Markeev, V. V. Fomichev, I. M. Chertok.
Astron. Zhurn. Akad. Nauk SSSR, Vol. 52, 338 - 345 (1975). In Russian. English translation in Soviet Astron., Vol. 19, No. 2.
The event on May 21, 1973 recorded at Izmiran with spectrographs (45–90 MHz and 140–180 MHz) and polarimeters at 74 MHz and 204 MHz is analysed. In this event pulsations of radio emission with time scale of 10–20 sec have been observed during some hours on the background of the storm continuum.

077.028 On certain features of radiation from local sources on the sun at 2.3–2.7 centimeter wavelengths.
V. M. Bogod, A. N. Korzhavin.
Astrofiz. Issled., Izv. Spets. Astrofiz. Obs., Vol. 7, p. 121 - 133 (1975). In Russian.
Observational data processing results are presented for several local sources on the sun. The data were obtained at 2.3 and 2.7 centimeter wavelengths during July–September 1972. Brightness temperatures of the flocculus component of local sources are estimated.

077.029 Investigation of circular polarization of S-component sources of solar radio emission from high resolution observations. N. G. Peterova.
Astrofiz. Issled., Izv. Spets. Astrofiz. Obs., Vol. 7, p. 134 - 147 (1975). In Russian.
Results are presented of investigations of circularly polarized radio emission from 75 sources of the S-component using the observations of 1967–1970 with the large Pulkovo radio telescope at a wavelength of 4.4 cm, with a resolution of 1.8 min of arc. All the sources practically belonged to unipolar spot groups, ~80 percent of the area of these spot groups contained a magnetic field of one polarity. The results of observations are explained in the framework of the hypothesis on synchrotron origin of the polarized component of radiation from the S-component sources. Some inferences are made on the temperature gradient in the solar atmosphere above an active region.

077.030 Radio observations from Intercosmos-Kopernik 500 satellite. J. Hanasz, R. Schreiber, H. Wełnowski, B. Wikierski, V. I. Aksenov.
Postępy Astron., Vol. 23, 3 - 10 (1975). In Polish.
The space experiment Intercosmos-Kopernik 500 is briefly described. 50 solar type III radio bursts were observed with the aid of a radio-spectrograph in the frequency range of 0.6 to 6.0 MHz. Irregular structure of some type III bursts has been discovered during the increasing phase of bursts. Ionospheric plasma resonances have been continuously recorded.

077.031 Radiobilder der Sonne im cm-Wellenbereich.
O. Hachenberg.
SuW, Vol. 14, 150 - 155 (1975).

077.032 Some features of solar spike burst generation.
V. V. Zheleznyakov, V. V. Zaitsev (*Zajtsev*).
Astron. Astrophys., Vol. 39, 107 - 111 (1975).
The generation process of "spike" bursts is investigated. It is assumed that these bursts are caused by plasma waves excited in the process of the quasi-linear relaxation of the electron stream in the solar corona. The time evolution of bursts and their association with type III radio emission are considered. The possibility of rather short-lived (3×10^{-3} – 3×10^{-2} s) bursts corresponding to the second harmonic of the plasma frequency is pointed out.

077.033 Observations of solar radio emission during August 1972.
I. I. Berulis, A. P. Molchanov, V. P. Olyanyuk, O. Ya. Pudov, R. L. Sorochenko, N. G. Franchuk, L. V. Yasnov.
Radioizluchenie Solntsa. Vyp. (No.) 3. Leningrad, Leningr. un-t, 1974, p. 16 - 44. In Russian. – Abstr. in Referativ. Zhurn. 51. Astron., 3.51.472 (1975).

077.034 The absolute radiation fluxes of the quiet sun in the region 121.6 – 45 nm. A. I. Efremov, S. V. Avakyan, A. L. Podmoshenskij, M. P. Bolgartseva, L. N. Gershun, M. A. Ivanov, V. S. Petrov, I. M. Pribylovskij, G. V. Sazonov.
Issled. po geomagnetizmu, aehron. i fiz. Solntsa. Vyp. (No.) 32. Moskva, Nauka, 1974, p. 162 - 166. In Russian. – Abstr. in Referativ. Zhurn. 51. Astron., 3.51.502; 62. Issled. kosmich. prostranstva, 3.62.180 (1975).

077.035 On a connection of faint local radio sources at λ = 6.6 cm with characteristics of active regions. II.
G. B. Gel'frejkh, Z. B. Korobova, N. P. Stasyuk.
Solnechnye Dannye 1975 Byull., No. 1, p. 70 - 73 (1975). In Russian.
The mean flux of emission from faint local radio sources is shown to be in better correlation with the mean brightness than with the area of flocculi. It was found that about 1/3 – 1/4 flux is connected with a sunspot group and 2/3 – 3/4 with a flocculus.

077.036 On periodic oscillations of the background of noise storms. G. F. Eliseev, P. V. Panov.
Solnechnye Dannye 1975 Byull., No. 1, p. 74 - 80 (1975). In Russian.
The spectral-correlational analysis of the continuum of noise storms, made for 17 sources at 208 MHz frequency, gives evidence that: 1) each local source is characterized by periodic components with amplitude and oscillation periods changing with the development of the source and also due to the presence of non-stationary processes, and 2) the relation between the power of harmonical constants and the values of their periods is of non-monotonous character with maxima within the range of 100–120 sec and 160–180 sec.

077.037 Some comments on the bremsstrahlung mechanism of the S-component of solar radio emission.
E. Ya. Zlotnik, Yu. V. Tikhomirov.
Solnechnye Dannye 1975 Byull., No. 1, p. 80 - 84 (1975). In Russian.
Critical comments are given on the explanation of the S-component as electron bremsstrahlung in the model of a local source with sharp temperature and density gradient over the height. In this model the main mechanism at $\lambda > 4$ cm is shown to be cyclotronic.

077.038 The necessity of fundamental emission in type III bursts. D. F. Smith, W. D. Davis.

Solar Physics, Vol. 41, 439 - 447 (1975).

Observations of some type III radio bursts in the hecto-meter and kilometer wave range are compared with theoretical predictions. It is shown that the burst emission must be near the plasma frequency in the region between $10\,R_\odot$ and $50\,R_\odot$ in order to be consistent with the observed steep rise in brightness temperature for these bursts. The results of Fainberg, Malitson et al., and Haddock and Alvarez are discussed and compared with the interpretation of emission near the plasma frequency.

077.039 **6 cm observations of a solar active region with the Westerbork Synthesis Radio Telescope.**
M. R. Kundu, C. E. Alissandrakis, H. W. van Someren Greve.
Bull. American Astron. Soc., Vol. 7, 235 (1975). – Abstr. AAS.

077.040 **Solar radio bursts at decameter-wave frequencies.**
A. Achong, C. H. Barrow.
Bull. American Astron. Soc., Vol. 7, 338 (1975). – Abstr. AAS.

077.041 **9.1 cm radio maps and sector boundaries.**
W. Graf, R. N. Bracewell.
Bull. American Astron. Soc., Vol. 7, 356 (1975). – Abstr. AAS.

077.042 **The spectrum of the sun at $\lambda = 8$ mm.**
P. N. Swanson, R. Kuseski.
Bull. American Astron. Soc., Vol. 7, 360 - 361 (1975).
Abstr. AAS.

077.043 **Source regions for type II radio bursts.**
J. C. Dodge.
Bull. American Astron. Soc., Vol. 7, 361 (1975). – Abstr. AAS.

077.044 **The amplitude and position distributions of low frequency solar type III bursts.**
R. J. Fitzenreiter, J. Fainberg, H. H. Malitson.
Bull. American Astron. Soc., Vol. 7, 361 (1975). – Abstr. AAS.

077.045 **Center-to-limb distribution of solar radio bursts to 1 AU.** M. L. Kaiser, H. H. Malitson.
Bull. American Astron. Soc., Vol. 7, 361 (1975). – Abstr. AAS.

077.046 **Classification of 2800−2700 MHz solar noise bursts.**
A. E. Covington.
Bull. American Astron. Soc., Vol. 7, 361 - 362 (1975).
Abstr. AAS.

077.047 **Radio observations of the solar 5-min oscillations at 2.0 cm wavelength.** F. L. Wefer.
Bull. American Astron. Soc., Vol. 7, 363 (1975). – Abstr. AAS.

077.048 **Five minute oscillations at 2.8 cm?**
C. J. Greben-Kemper, W. Graf.
Bull. American Astron. Soc., Vol. 7, 363 (1975). – Abstr. AAS.

077.049 **Cinematography of the five minute oscillations; a new aspect of the horizontal propagation.**
D. K. Lynch.
Bull. American Astron. Soc., Vol. 7, 365 (1975). – Abstr. AAS.

077.050 **A nonlinear, time-dependent theory of the type III burst exciter.**
R. A. Smith, M. L. Goldstein, K. Papadopoulos.
Bull. American Astron. Soc., Vol. 7, 365 (1975). – Abstr. AAS.

077.051 **Investigation of the oscillations of continuum of noise storms by a spectral-correlation analysis.**
A. N. Asanova, L. A. Eliseeva, G. F. Eliseev.
Problems of cosmic physics. Vyp. (No.) 9, (see 003.013), p. 42 - 45 (1974). In Russian.

For three noise storms at frequency 208 MHz the results of a spectral-correlation analysis are given. The power spectra of noise storms have indicated quasi-periodic oscillations with average period 5.7, 7.4, 9.4, 12.1, 13.7, 16.8, 25.9 min.

077.052 **On determination of the spectrum slope and spectral index for the radio emission flux of weak sources.**
I. E. Pogodin.
Radioizluchenie Solntsa. Vyp. (No.) 3. Leningrad, Leningr. un-t, 1974, p. 99 - 108. In Russian. – Abstr. in Referativ. Zhurn. 51. Astron., 4.51.507 (1975).

077.053 **Peculiarity of type III solar radio burst generation by stabilized fast particle streams.** V. V. Zajtsev.
Izv. vyssh. ucheb. zavedenij. Radiofizika, Vol. 17, 1438 - 1445 (1974). In Russian. – Abstr. in Referativ. Zhurn. 51. Astron., 4.51.517 (1975).

077.054 **Solar observations at 1.4 and 4.1 mm wavelengths with angular resolutions of $13''$ and $40''$.**
A. G. Kislyakov, Yu. Yu. Kulikov, L. I. Fedoseev, V. I. Chernyshev.
Pis'ma v Astron. Zhurn., Vol. 1, No. 4, p. 24 - 28 (1975). In Russian.

Distributions of radio brightness for the equatorial zone of the sun obtained at 1.4 and 4.1 mm wavelengths with angular resolutions in right ascension of $14''$ and $40''$ show a number of small scale features in solar regions devoid of corresponding optical details. This phenomenon is likely to be caused by chromospheric inhomogeneities analogous to photospheric granulation.

077.055 **Verification of solar patrol calibration at 15.4 GHz.**
J. P. Castelli, D. A. Guidice.
Astrophys. Letters, Vol. 16, 71 - 74 (1975).

The absolute flux of the sun at 15.4 GHz has been determined as 511×10^{-22} sfu for six days during August−September 1974 at the Sagamore Hill Radio Observatory, using an optimum gain horn with hot−cold calibration loads. The maximum uncertainty of this measurement, considering all possible observational and system errors was calculated to be ∼ 4 percent. The normal patrol measurements use the moon as a calibration source and for the corresponding period provided a flux of 518 sfu. The two sets of measurements show a consistency of 1.4 percent, well within the absolute accuracy calculated for the horn-derived values. Critical comments on this and other experiments are reviewed.

077.056 **Linear polarization in meter and decameter solar radiobursts.** A. Boischot, A. Lecacheux.
Astron. Astrophys., Vol. 40, 55 - 61 (1975).

The authors present a new method of detecting linear polarization in solar radiobursts. Observations have been made of many type III bursts in the range 25 to 80 MHz and around 169 MHz with different spectrographs with a time resolution up to 0.02 s and a frequency resolution between 20 kHz and 300 kHz. In none of the spectra did an effect of a linear polarization appear within the sensitivity limits. This result contradicts several other works and the origin of this contradiction is discussed.

077.057 **Radio bursts in the centimeter region as precursors of proton flares.** M. N. Belovskij, M. N. Nazarova, N. K. Pereyaslova, S. G. Frolov.
Dokl. Akad. Nauk SSSR. Ser. Mat., Fiz., Vol. 222, 594 - 595 (1975). In Russian.

077.058 **Polarization of type I bursts.** P. Zlobec.
Proc. 2nd meeting Committee European Solar Radio Astronomers, (C.E.S.R.A.), Trieste 1971, p. 101 - 108 = Oss. Astron. Trieste, Pubbl. No. 450 (1972).

Polarimetric registrations at 237 MHz made in Trieste in

February, March and April of 1970 are compared with the spectrographic results of the observatory of Utrecht.

077.059   Periodicities in solar radio emission.   A. Abrami.
Proc. Summer School on Plasma Physics and Solar Radioastronomy, Ile de Ré 1972, p. 233 - 244 = Oss. Astron. Trieste, Pubbl. No. 466 (1973).

077.060   Drift and polarization behaviour in partially polarized type I bursts.   P. Zlobec.
Proc. 3rd meeting Committee European Solar Radio Astronomers, (C.E.S.R.A.), Bordeaux 1972, p. 151 - 158 = Oss. Astron. Trieste Pubbl. No. 467 (1973).

Partially polarized type I bursts simultaneously recorded on the spectrographic film and on the polarimetric high speed recordings are considered.

077.061   About two type IV events on August 4 and 7, 1972. Polarimetric measurements at 237 MHz.
P. Santin, P. Zlobec.
Report UAG-28—World Data Center A, p. 279 - 282 = Oss. Astron. Trieste, Pubbl. No. 468 (1973).

077.062   Some preliminary results on the distribution of type I bursts at 408 MHz.   A. Abrami.
Proc. 4th meeting Committee European Solar Radio Astronomers, (C.E.S.R.A.), Berne 1974, p. 123 - 130 = Oss. Astron. Trieste, Pubbl. No. 482 (1973).

The positions of type I bursts at 408 MHz are obtained by means of a simple correlation interferometer. Their distribution across the solar disk, for three periods of solar activity, shows the complex structure of the emitting regions and allows the determination of the source heights in the solar corona.

077.063   Determination of Faraday rotation occurring between the burst-source in the solar corona and the earth.   R. V. Bhonsle, S. K. Mattoo.
Bull. Astron. Soc. India, Vol. 2, 37 (1974). – Abstract.

077.064   Radio evidence for an expanding magnetic arch beyond 20 solar radii.   R. G. Stone, J. Fainberg.
Solar Physics, Vol. 42, 179 - 181 (1975). – Research note.

077.065   Type III solar radio bursts and the fundamental-harmonic hypothesis.   H. Rosenberg.
Solar Physics, Vol. 42, 247 - 257 (1975).

The observational evidence is reviewed for the occurrence of type III solar radio bursts in pairs with frequency ratio two to one. The author shows that the observations can be explained under the hypothesis that there is a tendency for a type III burst to be followed by a second burst within approximately one second. The author concludes that in general, type III bursts are emitted at the second harmonic of the plasma frequency and that type III theories should account for this and only under very special circumstances (which are rare) for the emission at the fundamental and the second harmonic.

077.066   The metric quiet sun during two cycles of activity and the nature of the coronal holes.
P. Lantos, Y. Avignon.
Astron. Astrophys., Vol. 41, 137 - 142 (1975).

The radio quiet sun has been studied at 169 MHz (λ = 1.77 m) from 1957 to 1970 with the Nançay interferometer. The comparison with the other UV and radio observations and with the UV models shows that the radio quiet sun corresponds to the coronal holes. The rough constancy of the brightness temperature (T_B = 750000 K) indicates that the physical conditions in the coronal holes do not vary systematically with the cycle of activity. The heating is thus local and independent, at a first approximation, of the neighbouring active regions. The coronal holes appear to be the quiet coronal regions of the sun.

077.067   On the polarization of type IV meter-wave radio radiation.   A. V. Stepanov.
Solnechnye Dannye 1975 Byull., No. 2, p. 47 - 51 (1975). In Russian.

The type IV meter-wave continuum is supposed to be due to the loss-cone instability in the coronal magnetic trap. The observed polarization of the emission corresponding to an ordinary mode is explainable in the dense plasma approach, if a fundamental harmonic is the most essential one in the type IV emission.

077.068   Travelling solar radio bursts.   R. G. Stone.
Solar wind three, (see 012.020), p. 72 - 97 (1974).

077.069   Millimeter wave solar observations.
J. P. Castelli, D. A. Guidice, P. M. Kalaghan.
IEEE Trans. Microwave Theory Techn., Vol. MTT-22, 1292 - 1299 (1974).

Discussions and lists of observatories on both low- and high-resolution millimeter solar observations, AFCRL equipments and some observational results. – *SS*

A new high-speed solar spectrograph for meter and decameter wavelengths.   See Abstr. 033.006.

Radio emission in the sunspot groups located over the solar disk near the equator.   See Abstr. 072.014.

Étude statistique des associations entre le champ magnétique des groupes de taches solaires et la morphologie des éruptions solaires et des sursauts associés sur ondes décimétriques.   See Abstr. 073.033.

Sunspot motion, flares and type III bursts in McMath 11482.   See Abstr. 073.035.

Fast electrons in small solar flares.   See Abstr. 073.093.

On the acceleration processes in solar flares.   See Abstr. 073.097.

Limb brightening and dark features observed at 6 cm wavelength.   See Abstr. 074.044.

A coronal hole observed at 10.7 GHz with a large single dish.   See Abstr. 074.102.

On the relationships between sfe (crochet) and solar X-ray and microwave bursts.   See Abstr. 076.015.

X-ray spectral dependence of the relationship between impulsive solar flare X-rays and type III radio bursts.   See Abstr. 076.019.

Solar flare X-ray measurements and their relation to microwave bursts.   See Abstr. 076.055.

Relation of microwave emission to X-ray emission from solar flares.   See Abstr. 076.056.

Observations of solar eclipses in the radio wave region of 3.2 - 3.4 cm.   See Abstr. 079.001.

078 Solar Cosmic Radiation

078.001 **Solar flare particles: energy-dependent composition and relationship to solar composition.**
H. J. Crawford, P. B. Price, B. G. Cartwright, J. D. Sullivan.
Astrophys. Journ., Vol. 195, 213 - 221 (1975).

Plastic and glass track detectors on rockets and Apollo spacecraft have been used to determine the composition of particles from He to Ni at energies from ~0.1 to ~50 MeV per nucleon in several solar flares of widely varying intensities. The similarity between solar particle abundances and recent abundance data for the photosphere and corona is so close that the few discrepancies (mainly S and Ar) are attributed to errors in the solar abundance tables. The suggested revisions, S/Si = 0.17 and Ar/Si $\leq$ 0.04, would require significant changes in the models of nucleosynthesis of the elements Si to Fe by explosive oxygen and silicon burning.

078.002 **Influence of solar activity on the spatial distribution of cosmic rays in the interplanetary space.**
L. I. Dorman, R. T. Gushchina.
Izv. AN SSSR. Ser. fiz., Vol. 38, 1920 - 1923 (1974). In Russian. – Abstr. in Referativ. Zhurn. 51. Astron., 2.51.470 (1975).

078.003 **Annual cosmic ray variations in 1958 - 1968.**
L. I. Dorman, N. P. Milovidova.
Izv. AN SSSR. Ser. fiz., Vol. 38, 1928 - 1931 (1974). In Russian. – Abstr. in Referativ. Zhurn. 51. Astron., 2.51.484 (1975).

078.004 **Some results of investigations of solar cosmic rays aboard Mars 7.** S. N. Vernov, N. V. Alekseev, P. V. Vakulov, N. I. Vologdin, E. V. Gorchakov, V. A. Iozenas, G. Ya. Kolesov, Yu. I. Logachev, Yu. V. Mineev, N. F. Pisarenko, I. A. Savenko, B. Ya. Shcherbovskij.
Kosmich. Issled., Vol. 13, p. 131 - 135 (1975). In Russian.

078.005 **Monte Carlo model of the highly anisotropic solar proton event of 20 April 1971.**
I. D. Palmer, R. A. R. Palmeira, F. R. Allum.
Solar Physics, Vol. 40, 449 - 460 (1975).

The authors analyse the solar proton event of 20 April 1971 recorded by the 7.6–55 MeV energy channel of the UTD cosmic ray experiment on Explorer 41. The anisotropy for this event remained large ($\gtrsim$ 100%) and field-aligned well into the decay portion of the intensity profile. A Monte Carlo technique is employed to calculate numerically the pitch-angle distributions, anisotropy amplitude, and intensity in two different models of this solar cosmic ray event. With this flexible technique the authors are able to reproduce the anisotropy and intensity profiles in the given event, and infer certain propagation parameters which are consistent with the observations.

078.006 **The variation of solar proton energy spectra and size distribution with heliolongitude**
M. A. I. van Hollebeke, L. S. Ma Sung, F. B. McDonald.
Solar Physics, Vol. 41, 189 - 223 (1975).

A statistical study of the initial phases of 185 solar particle events has been carried out using the data from the Goddard cosmic ray experiments on IMPs IV and V. Special emphasis is placed on the identification of the associated solar flare. The existence of a 'preferred-connection' longitude between 20°W and 80°W is established by examining the heliolongitude of all the flare associated events. It is argued that for heliolongitudes $\lambda_\odot$ = 20–80°W, γ_p, the spectral index determined at the time of maximum particle intensity is representative of the source spectra. For these heliolongitudes

γ_p displays a surprisingly small range. Previous electron measurements provide almost identical average values of the source spectra over similar energy ranges. These results are discussed briefly in terms of Fermi acceleration models. For flare events located further away from the nominal field line connecting the earth and the sun, γ_p becomes progressively steeper.

078.007 **The 1964–1972 quiet-time spectra of protons and helium at 2–20 MeV per nucleon.** R. Zamow.
Astrophys. Journ., Vol. 197, 767 - 780 (1975).

Although normally the fluxes below 20 MeV nucleon^{-1} are highly variable, indicating the dominance of a solar component, there are periods when the fluxes remain at a low and reasonably constant level. During these quiet periods the low-energy fluxes may still be fitted with "turnup" spectral forms proportional to $E^{-\gamma}$, with $\gamma \sim 2.8$. Since the data now span virtually an entire solar cycle, the long-term time-dependence of the spectra over solar cycle 20 has been determined. During the period 1964–1972, the variation of the fluxes is a factor ~7 and is similar to the modulation of the medium-energy cosmic rays. The observed relative abundance of protons and helium is closer to the medium-energy galactic than to the average solar-flare relative abundance. This evidence favors a galactic origin for the low-energy quiet-time turnup.

078.008 **Cosmic ray measurements with the automatic interplanetary station Mars 2.**
S. N. Vernov, B. A. Tverskoj, V. A. Yakovlev, E. V. Gorchakov, P. P. Ignat'ev, G. P. Lyubimov, N. V. Pereslegina, O. A. Marchenko, T. E. Shvidkovskaya, N. N. Kontor, T. I. Morozova, A. G. Nikolaev, Yu. A. Rozental', I. V. Getselev, A. V. Onishchenko, V. I. Tkachenko, E. A. Chuchkov.
Izv. AN SSSR. Ser. fiz., Vol. 38, 1859 - 1862 (1974). In Russian. – Abstr. in Referativ. Zhurn. 51. Astron., 3.51.435; 62. Issled. kosmich. prostranstva, 3.62.283 (1975).

078.009 **Energetic and spatial properties of cosmic ray events in August 1972.** A. M. Altukhov, G. F. Krymskij, A. I. Kuz'min, I. S. Samsonov, Z. N. Samsonova, G. V. Skripin, V. A. Filippov, N. P. Chirkov, G. V. Shafer.
Izv. AN SSSR. Ser. fiz., Vol. 38, 1876 - 1879 (1974). In Russian. – Abstr. in Referativ. Zhurn. 51. Astron., 3.51.436; 62. Issled. kosmich. prostranstva, 3.62.285 (1975).

078.010 **27-day cosmic ray variations associated with non-uniform distribution of solar active regions.**
G. A. Bazilevskaya, V. P. Okhlopkov, T. N. Charakhch'yan.
AN SSSR. Fiz. in-t Preprint No. 119. Moskva, 1974, 34 pp. In Russian. – Abstr. in Referativ. Zhurn. 51. Astron., 3.51. 496 (1975).

078.011 **A possibility of estimating the characteristics of corpuscular electron streams by the parameters of the absorbing layer in the D-region.** T. P. Zhukova.
Issled. po geomagnetizmu, aehron. i fiz. Solntsa. Vyp. (No.) 32. Moskva, Nauka, 1974, p. 153 - 157. In Russian. – Abstr. in Referativ. Zhurn. 51. Astron., 3.51.500 (1975).

078.012 **Solar particle events with anomalously large relative abundance of ^{3}He.**
A. T. Serlemitsos, V. K. Balasubrahmanyan.
Astrophys. Journ., Vol. 198, 195 - 204 (1975).

Experimental results on three solar energetic particle events with extremely high abundance of ^{3}He are reported. The measurements cover an energy range of ~4–80 MeV nucleon^{-1}. In the first of these events $\Gamma(^3$He$/^4$He$) = 1.52 \pm 0.1$ (where Γ is the ratio of the two species in an energy per nucleon representation). In the two subsequent events $\Gamma(^3\mathrm{He}/^4\mathrm{He})$ had values of 0.71 ± 0.06 and 0.35 ± 0.03, respectively. The abundance of protons relative to He nuclei was significantly low in these events. The lower limits $\Gamma(^3\mathrm{He}/^2\mathrm{H})$ obtained for these three events were 300, 250, and 63, respectively, and are much higher than the upper limits expected from theoretical considerations.

078.013 Heavy solar cosmic rays in the January 25, 1971 solar flare. C. J. Pellerin, Jr.
Solar Physics, Vol. 41, 449 - 458 (1975).

A detailed study of the charge composition of heavy solar cosmic rays measured in the January 25, 1971 solar flare including differential fluxes for the even charged nuclei from carbon through argon is presented. The measurements are obtained for varying energy intervals for each nuclear species in the energy range from 10 to 35 MeV nucleon^{-1}. These measurements, when combined with other experimental results, enable the energy dependence of abundance measurements as a function of nuclear charge to be discussed.

078.014 Determination of the upper cutoff of the 1−2 September 1971 proton event from satellite measurements. Dj. Heristchi, G. Trottet.
Solar Physics, Vol. 41, 459 - 460 (1975). − Research note.

078.015 Implications of observed charge states of low-energy solar cosmic rays. J. R. Jokipii, A. J. Owens.
Journ. Geophys. Res., Vol. 80, 1209 - 1212 (1975).

Recent measurements of the charge states of low-energy ($\sim$100 keV/nucleon) solar cosmic rays at 1 AU are discussed. The measurements are consistent with models involving charge equilibrium with neutral matter at the sun only if the particles lose $\sim$90% of their energy owing to adiabatic deceleration in the solar wind. Such an energy loss is shown to be possible only if the diffusion coefficient for 1-MeV/nucleon particles is smaller than 10^{20} cm^2 s^{-1}. The implications of these results for models of solar cosmic ray acceleration are discussed.

078.016 Interplanetary scattering of fast solar electrons deduced from type III burst observed at km-wavelength. H. Alvarez, R. P. Lin.
Bull. American Astron. Soc., Vol. 7, 366 (1975). − Abstr. AAS.

078.017 Anisotropy of solar protons and sectorial structure of the interplanetary field.
Eh. I. Nesmyanovich, A. T. Nesmyanovich.
Problems of cosmic physics. Vyp. (No.) 9, (see 003.013), p. 45 - 53 (1974). In Russian.

The results of a data analysis of 57 proton events (May 1967 − May 1970) and of the sectorial structure of the interplanetary space (1962−1969) are presented in detail.

078.018 Propagation of solar electrons with energies $E_e >$ 30 keV in the interplanetary space.
V. G. Kurt, Yu. I. Logachev, N. F. Pisarenko.
Kosmich. Issled., Vol. 13, 222 - 235 (1975). In Russian.

078.019 Results of measurements of cosmic ray intensity aboard the automatic station Luna 19.
E. A. Chuchkov, G. P. Lyubimov, O. G. Myagchenkova, A. D. Novichkova, N. V. Pereslegina, N. N. Kontor, A. G. Nikolaev.
Kosmich. Issled., Vol. 13, 254 - 265 (1975). In Russian.

078.020 Some unusual features of the cosmic ray storm in August 1972.
J. A. Lockwood, L. Hsieh, J. J. Quenby.
Journ. Geophys. Res., Vol. 80, 1725 - 1734 (1975).

The unusually large cosmic ray disturbance commencing on August 4 had three unusual features. First, a ground level event was seen in neutron monitors at $P_c \lesssim 1.5$ GV beginning at about 1400 UT, almost 8 hours after the large solar flare at 0630 UT. Second, a large precursory increase was associated with the rapid Forbush decrease at 2100−2200 UT. Third, the Forbush decrease had a short time scale, with an overshoot in the initial recovery phase.

078.021 Solar cosmic ray 'square wave' of August 1972.
R. A. Medrano, C. J. Bland, J. W. Freeman, H. K. Hills, R. R. Vondrak.
Journ. Geophys. Res., Vol. 80, 1735 - 1743 (1975).

Three Rice University suprathermal ion detector experiments (Sides) were deployed on the lunar surface during the Apollo 12, 14, and 15 missions. During the exceptional period of solar activity in August 1972, penetrating particles were observed by all Side detectors on the night side of the moon. The penetrating particles are tentatively identified as solar protons with energies ($\sim$ 25 MeV or greater) that were able to penetrate the shielding of all detectors.

078.022 Interplanetary acceleration of low-energy solar protons: a study of the solar particle event of November 18, 1968. P. Venkatarangan, L. J. Lanzerotti.
Journ. Geophys. Res., Vol. 80, 1744 - 1750 (1975).

The spectra of solar proton and alpha particle fluxes from the solar particle event of November 18, 1968, deviate considerably from a power law form for as long as 2 days after the onset of the event. The data are presented and discussed.

078.023 An investigation of cosmic ray variations and solar activity at small frequencies.
E. V. Kolomeets, Ya. E. Shvartsman.
Izv. AN KazSSR. Ser. fiz.-mat., 1974, No. 6, p. 19 - 23. In Russian. − Abstr. in Referativ. Zhurn. 51. Astron., 5.51.438 (1975).

078.024 Increase in the "smoothing" of time-intensity patterns of proton events with distance from the sun. L. Křivský.
Bull. Astron. Inst. Czechoslovakia, Vol. 26, 190 (1975).

The decrease in the occurrence of impulse effects in proton fluxes from the active region of August 1972 with increasing distance in interplanetary space is explained by the effect of "smoothing" as a result of the increasing number of interactions of the new situation of the fields with the preceding situation. The principle of "successive impressions" is used to explain the phase delay of the onset phase of the overall pattern of the proton event with distance.

078.025 The relative abundances and energy spectra of solar-flare-accelerated deuterium, tritium, and helium-3.
J. D. Anglin.
Astrophys. Journ., Vol. 198, 733 - 753 (1975).

The relative abundances and energy spectra of $^2\mathrm{H}$, $^3\mathrm{H}$, and $^3\mathrm{He}$ have been studied in energetic solar particle events with the University of Chicago IMP-5 charged-particle telescope. The outstanding result of these observations is the remarkable scarcity of $^2\mathrm{H}$ and $^3\mathrm{H}$ in several flares which have extremely large abundances of $^3\mathrm{He}$. A detailed comparison of the observations with different nuclear production models is undertaken. It is shown that neither the isotropic production models developed in this paper, nor the model of Ramaty and Kozlovsky, in which $^3\mathrm{He}$ is preferentially produced in the backward direction by a beam of protons directed downward into the photosphere, provides an adequate explanation of all the observational data.

078.026 The acceleration of heavy nuclei in solar flares.
K. Sakurai.
Planet. Space Sci., Vol. 23, 955 - 959 (1975).

The overabundance of heavy nuclei in solar cosmic rays of energy $\lesssim 10$ MeV/nucleon (sometimes up to ~ 30 MeV/nucleon) is explained by taking into account the pre-flare ionization states of these nuclei in the region where they are accelerated. A model is proposed which considers two-step accelerations associated with the initial development of solar flares.

078.027 **Neutrons of solar origin.**
Yu. Dubinskij, Yu. E. Efimov, K. Kudela.
Izv. AN SSSR Ser. fiz., Vol. 39, 259 - 263 (1975). In Russian. Abstr. in Referativ. Zhurn. 62. Issled. kosmich. prostranstva, 6.62.203 (1975).

078.028 **Generation of protons, neutrons and electrons on the sun.** A. B. Bajsakalova, O. A. Bogdanova, T. Z. Iskakov, V. A. Kobzev, E. V. Kolomeets, A. I. Kupchishin.
Izv. AN SSSR. Ser. fiz., Vol. 39, 264 - 271 (1975). In Russian. Abstr. in Referativ. Zhurn. 62. Issled. kosmich. prostranstva, 6.62.204 (1975).

078.029 **Charge composition of solar cosmic rays in the January 25, 1971 solar flare.**
C. J. Pellerin, Jr.
Thesis, Catholic Univ. of America, Washington, D.C. (USA). 141 pp. University Microfilms Order No. 74-19,735 (1974).

078.030 **Relativistic solar cosmic rays.** M. A. Pomerantz.
Journ. Franklin Inst., (*U. S. A.*), Vol. 298, 363 - 383 (1974).
Discusses how ground-based observations of solar cosmic rays are contributing to the understanding of the earth's environs, the interplanetary medium and the sun itself.

078.031 **Nuclear composition of solar cosmic rays.**
D. Hovestadt.
Solar wind three, (see 012.020), p. 2 - 25 (1974).
In this paper three topics will be discussed; (1) Elemental composition of flare particles; (2) Isotopic composition of flare particles and related phenomena; (3) Composition of quiet time cosmic rays at very low energies.

078.032 **Heliocentric cosmic ray gradient 1.0–4.1 A.U.**
M. F. Thomsen.
Solar wind three, (see 012.020), p. 217 - 223 (1974).

On the solar activity index controlling cosmic ray variations. See Abstr. 072.009.

Fast electrons in small solar flares.
See Abstr. 073.093.

Nuclei of heavy elements from solar flares.
See Abstr. 073.094.

The nuclear composition of solar energetic particles – new source of information about solar flares.
See Abstr. 073.105.

The microwave structure of coronal condensations and its relation to proton flares. See Abstr. 074.003.

Unusual solar wind and solar proton events observed on the lunar surface. See Abstr. 074.112.

Measurement of heavy solar wind particles during the Apollo 17 mission. See Abstr. 074.120.

Relation of solar wind fluctuations to differential flow between protons and alphas. See Abstr. 074.121.

Interplanetary disturbances: a classifaction.
See Abstr. 074.126.

Gamma radiation of solar flares and cosmic ray generation. See Abstr. 076.005.

Proton energy deposition in molecular and atomic oxygen and applications to the polar cap.
See Abstr. 082.059.

Dissociative recombination contributions to I (5577) and I(6300) [O I] and $N_m F2$ enhancements resulting from moderate proton PCA events.
See Abstr. 083.069.

Transport of energetic solar particles on closed magnetospheric field lines. See Abstr. 084.202.

Access of solar electrons to the polar regions.
See Abstr. 084.265.

Lunar surface phenomena: solar flare track gradients, microcraters, and accretionary particles.
See Abstr. 094.184.

Primary cosmic radiation on the lunar surface.
See Abstr. 094.414.

Determination of natural and cosmic ray induced radionuclides in Apollo 17 lunar samples.
See Abstr. 094.530.

Cosmogenic radionuclides in samples from Taurus-Littrow: effects of the solar flare of August 1972.
See Abstr. 094.531.

Apollo 17 cosmic-ray experiment: interplanetary heavy nuclei of energies 0.05 to 5.0 MeV per atomic mass unit.
See Abstr. 143.036.

079 Solar Eclipses

079.001 Observations of solar eclipses in the radio wave region of 3.2 - 3.4 cm.   A. P. Molchanov.
Radioizluchenie Solntsa. Vyp. (No.) 3. Leningrad, Leningr. un-t, 1974, p. 3 - 15. In Russian. − Abstr. in Referativ. Zhurn. 51. Astron., 2.51.462 (1975).

079.002 A method for deriving radial brightness distributions from eclipse observations.
J. P. Hagen, P. N. Swanson.
Astrophys. Journ., Vol. 198, 219 - 222 (1975).
   A method for claculating the radial brightness distribution of an eclipsed object by using the slope of the eclipse curve is presented. It is shown that the brightness temperature of a circularly symmetric eclipsed object is related to the slope of the eclipse curve and to the rate of change of the eclipsed area; the rate of change of the eclipsed area may be expressed simply in terms of the sine of a polar angle. The method is applicable to any eclipsing objects with circular cross section. It is particularly useful in the case of radio observations of a solar eclipse.

Two uses of ancient astronomy.
See Abstr. 004.031.

079.100 Solar eclipse 1973 June 30

Sky brightness and polarization during the 1973 African eclipse.   G. E. Shaw.
Applied Optics, Vol. 14, 388 - 394 (1975).
   The absolute intensity, color, and polarization of the sky were measured during the eclipse of 30 June 1973 in Northern Kenya. Zenith sky radiance during totality decreased by a factor of 10^4 from the normal day sky value. The distribution of sky intensity with angle on the celestial hemisphere was approximately symmetrical about the local zenith, with this point having the minimum intensity value. The spectral distribution of zenithal diffuse skylight shifted toward the blue during totality, but the horizon reddened. The polarization ratio P decreased from a normal day value of 0.45 to 0.04. There is evidence that the distribution of polarization ratio is strongly affected by variations in surface albedo. The major results are compatible with predictions based on a radiative transfer model that considers double-scattering processes only.

Observations of short term light variations during the June 30, 1973 solar eclipse.   G. T. Klement.
Astron. Astrophys., Vol. 37, 431 - 433 (1974).
   Electronic registration of shadow bands during the June 30, 1973 solar eclipse is described. A power spectrum of the bands is shown as the final result.

Eclipse determination of the radial brightness distribution of the sun at λ 8 mm and λ 3 mm.
J. P. Hagen, P. N. Swanson.
Bull. American Astron. Soc., Vol. 7, 360 (1975). − Abstr. AAS.

Total solar eclipse of 30 June 1973.
I. Miko, E. Pittich, J. Sýkora.
Kozmos, Vol. 6, 13 - 19, 46 - 51, 78 - 84 (1975). In Slovak.

Observation of the solar eclipse on June 30, 1973 at 3.1 and 3.3 cm wavelengths.
A. P. Molchanov, V. A. Stupin.
Radioizluchenie Solntsa. Vyp. (No.) 3. Leningrad, Leningr. un-t, 1974, p. 50 - 54. In Russian. − Abstr. in Referativ.

Zhurn. 51. Astron., 3.51.426 (1975).

Der Temperaturverlauf bei der totalen Sonnenfinsternis vom 30. Juni 1973.
M. G. Firneis, F. J. Firneis.
Sitzungsber. Österreich. Akad. Wiss., Math.-nat. Kl., Abt. II, Vol. 183, 57 - 70 = Astron. Mitt. Wien, No. 15 (1974).
   During the total solar eclipse of June 30th, 1973 the variation of the air temperature has been recorded. A formula for the temperature decrease ΔT as a function of the daily temperature amplitude A and the mean time t' of the eclipse is derived. The sky brightness decreased by 12.3 magnitudes.

Preliminary report on the observations of the total solar eclipse of 30 June 1973 in Africa.
F. Moriyama, E. Hiei, A. Tokuya, H. Miyazaki.
Tokyo Astron. Obs., Report No. 66, Vol. 17, 389 - 416 (1975). In Japanese.

L'étude de la couronne blanche à bord de Concorde 001 au cours de l'éclipse totale de soleil du 30 juin 1973.
See Abstr. 074.032.

Absolute photometry and the structure of the corona during the solar eclipse of June 30, 1973.
See Abstr. 074.082.

The brightness of the corona at the solar eclipse of June 30, 1973.   See Abstr. 074.108.

Submillimetre brightness spike at the solar limb.
See Abstr. 080.010.

Dayglow of the infrared atmospheric band system of O_2 during a total eclipse of the sun.   See Abstr. 082.045.

079.101 Solar eclipse 1974 December 13

About December's partial solar eclipse.
Sky Telescope, Vol. 49, 123 - 126 (1975).

079.102 Solar eclipse 1971 February 25

Étude de l'éclipse solaire partielle du 25 février, 1971 au Pic-du-Midi.   J. P. Rozelot, G. Ratier.
Solar Physics, Vol. 40, 371 - 385 (1975).
   The scattering in the neighbourhood of the sun is investigated on photographs taken at Pic-du-Midi Observatory during the partial solar eclipse of February 25 1971. The diffraction theory is used to determine the image of a crescent with incoherent illumination, and to compare the 'diffracted' isophote curves with those drawn from the plates. The electronic density in a coronal sector is given and the polarization ratio and the electron density of the first Fe XIV excited level are determined. Some interpretations are compared with those given by other authors.

079.103 Solar eclipse 1970 March 7

Observations of the total solar eclipse of 7 March, 1970.   M. Kanno, T. Tsubaki, H. Kurokawa.

Mem. Fac. Sci., Kyoto Univ., Ser. Phys., Astrophys. Geophys., Chem., Vol. 34, 281 - 292 (1974).

Observations of the flash spectrum of the chromosphere, the slit spectrum of the inner corona, and the direct photograph of the corona were made at the 7 March, 1970 eclipse at Puerto Escondido, Mexico. A detailed description is presented on the instruments, the observations, and the photometric calibration procedures. A discussion is also included of the acquisition in the eclipse observations.

079.104   Solar eclipse 1968 September 22

Observation of the 22 September 1968 solar eclipse at wavelength 11.1 cm.
Solar Eclipse Observation Group, Purple Mountain Obs.
Acta Astron. Sinica, Vol. 15, 113 - 122 (1974). In Chinese.

Measurements were made of the 22 September 1968 solar eclipse by Purple Mountain Observatory in Kashih, Sinkiang at wavelength of 11.1 cm. The association of radio sources with the optical active regions were examined. Source flux densities, associated one-dimensional source sizes, heights and brightness temperatures for these regions are given.

Observation of circular polarization at 3.2 cm wavelength of the solar eclipse of September 22,1968.
Solar Eclipse Observation Group, Purple Mountain Obs.
Acta Astron. Sinica, Vol. 15, 123 - 136 (1974). In Chinese.

This paper gives a general description of the observation of circular polarization at 3.2 cm wavelength of the solar eclipse on September 22,1968. The results obtained are presented.

Polarization observation of the solar corona at the solar eclipse of 1968 September 22.
See Abstr. 074.035.

079.105   Solar eclipse 1972 July 10

Observation of the solar eclipse on July 10, 1972 at 3.2 cm wavelength.
B. Ya. Losovskij, A. P. Molchanov, V. A. Stupin.
Radioizluchenie Solntsa. Vyp. (No.) 3. Leningrad, Leningr. un-t, 1974, p. 45 - 49. In Russian. − Abstr. in Referativ. Zhurn. 51. Astron., 3.51.425 (1975).

079.106   Solar eclipse 1977 October 12

Aircraft observing program for 100 minutes of totality in the 1977 eclipse.   See Abstr. 051.019.

079.107   Solar eclipse 1974 June 20

Sky brightness and temperature at the solar eclipse of June 20, 1974.   M. Waldmeier.
Astron. Mitt. Eidgenössisch. Sternw. Zürich, No. 332, 10 pp. (1974). In German.

Measurements of the brightness of the sky around the zenith have been carried out throughout the eclipse day. The influence of the eclipse upon the air temperature became noticeable 9 min after first contact. Compared to the undisturbed daily variation the largest temperature decrease occurred 29 min after mid-totality and amounted to 7.5°.

The corona at the solar eclipse of June 20, 1974.
See Abstr. 074.107.

079.108   Solar eclipse 1975 May 11

Solar eclipse of 11 May 1975.   J. Bouška.
Říše hvězd, Vol. 56, 75 - 76 (1975).   In Czech.

080 Solar Atmosphere, Figure, Internal Constitution, Neutrinos, Magnetic Fields, Rotation, Miscellanea

080.001 A search for deuterium on the sun. J. M. Beckers.
Astrophys. Journ., (*Letters*), Vol. 195, L43 - L45 (1975).

From observations of the Hα line of deuterium in quiescent solar prominences, the author has determined a new upper limit to the solar D/H ratio of 2.5×10^{-7}.

080.002 On convection in the sun. Yu. V. Vandakurov.
Solar Physics, Vol. 40, 3 - 21 (1975).

Convective motions driven by a superadiabatic temperature gradient in a viscous thermally conductive medium are considered. Approximate linearized equations governing the perturbation are derived under the following conditions: (i) The ratio of the excess temperature gradient over the adiabatic gradient is small compared with the gradient itself. (ii) The perturbation is of low-frequency type. (iii) The rotation is slow. Only the convective mode is described by these equations (as in the Boussinesq approximation), and the equations are valid for compressible configurations with any ratio between the scale heights of the equilibrium and perturbed quantities. Results of a numerical calculation of unstable perturbations for configurations with a large density stratification are given.

080.003 Recent advances in solar physics on the basis of stratospheric observations. V. A. Krat.
Uspekhi fiz. nauk, Vol. 113, 707 - 708 (1974). In Russian. Abstr. in Referativ. Zhurn. 51. Astron., 2.51.443 (1975).

080.004 Kinetic equilibrium and line formation of Na I in the solar atmosphere. T. Gehren.
Astron. Astrophys., Vol. 38, 289 - 302 (1975).

The influence of deviations from local thermodynamic equilibrium (LTE) on the formation of Na I lines under solar conditions is investigated. Based on different prescribed solar model atmospheres and model atoms which include up to 9 energy levels and 14 line transitions, the line transfer problem is solved using the integral equation approach, assuming a stationary, plane-parallel and homogeneous atmosphere, and complete redistribution of line photons. The computed D line cores closely fit observation. Since the formation of the Na D lines is nearly independent of the solar chromospheric temperature, such a fit is obtained for three different solar models. There appears to be no need for the high chromospheric electron densities suggested in previous investigations. The equivalent widths of the Na I lines are affected by deviations from LTE only a few per cent.

080.005 Nuclear reactions on the sun and solar neutrinos. S. S. Vasil'ev, G. E. Kocharov, A. A. Levkovskij.
Izv. AN SSSR. Ser. fiz., Vol. 38, 1827 - 1833 (1974). In Russian. – Abstr. in Referativ. Zhurn. 51. Astron., 2.51.563 (1975).

080.006 Photo-Coulomb gravitons and gravitational luminosity of the sun. D. V. Gal'tsov.
Zhurn. ehksperim. i teor. fiz., Vol. 67, 425 - 427 (1974). In Russian. – Abstr. in Referativ. Zhurn. 51. Astron., 2.51.881 (1975).

080.007 Models of the sun.
P. Demarque, J. G. Mengel, A. V. Sweigart.
Phys. Today, Vol. 28, No. 2, p. 71 (1975).

080.008 Energy transfer and distribution of temperatures on the sun. Eh. E. Dubov.
Priroda, No. 2.75, p. 45 - 51 (1975). In Russian.

080.009 On a general drift of magnetic fields in the solar atmosphere. V. A. Krat.
Pis'ma v Astron. Zhurn., Vol. 1, No. 1, p. 35 - 37 (1975). In Russian.

Spectroscopic observations made from the Solar Stratospheric Observatory are discussed. An upward motion of solar magnetic elements is detected.

080.010 Submillimetre brightness spike at the solar limb. J. E. Beckman, J. C. G. Lesurf, J. Ross.
Nature, Vol. 254, 38 - 39 (1975).

The authors report here the first complete phase of the reduction of data obtained during the solar eclipse of June 30, 1973 from the high altitude moving platform provided by Concorde 001. Submillimetre observations were made at both the second and third contacts, using a rapid-scanning Michelson interferometer. The very limited angular extent of the excess emission, which extends from some 4″ outside the optical limb, where it exhibits a very steep cutoff, to 5″ within the optical limb, with a less steep decline, suggests that the spike can be identified with emission from chromospheric spicules.

080.011 Solar neutrinos and solar rotation. I. W. Roxburgh.
Monthly Notices Roy. Astron. Soc., Vol. 170, 35P - 36P (1975).

Recent criticisms by Monaghan of suggestions that rapid rotation lowers the neutrino flux from the sun are replied to and shown to be invalid. Rapid differential rotation is capable of lowering the neutrino flux provided the ratio of centrifugal force to gravity decreases outwards in the sun.

080.012 On the stability of the solar core. W. Unno.
Publ. Astron. Soc. Japan, Vol. 27, 81 - 99 (1975).

Nonradial g-mode oscillations (g_1-mode of harmonics $l= 3$ or 4) that are effectively trapped in the solar core are shown to be the most likely candidates for types of motions that could possibly destabilize the sun. The gradient of the mean molecular weight in the core is effective for providing favorable conditions for trapping, especially if short range mixing has enhanced the gradient in a narrow layer. In the latter case, overstability appears to be possible, though marginal. The overstability, if possible, is expected to take place intermittently when the ^{3}He burning is activated. It may have some bearing on the solar neutrino deficiency and on the geological ice ages.

080.013 Comparaison de deux observations de déplacements anormaux vers le rouge observés au voisinage du disque solaire (complément).
S. Depaquit, J.-P. Vigier, J.-C. Pecker.
Comptes Rendus Acad. Sci. Paris, Sér. B, Vol. 280, 113 - 114 (1975).

L'observation de W 28 S à 18 cm n'infirme pas les résultats antérieurs obtenus à 21 cm (Tau A) et l'existence d'un «redshift» anormal des objets éclipsés par le soleil pendant cette éclipse.

080.014 Hydrogen-helium inhomogeneities and the solar neutrino problem.
D. J. Faulkner, G. S. Da Costa, A. J. R. Prentice.
Monthly Notices Roy. Astron. Soc., Vol. 170, 589 - 597

(1975).

Sixteen evolutionary sequences with various initial hydrogen-helium distributions have been constructed to test the ideas of Prentice that a configuration with a hydrogen-exhausted core of a few per cent of the sun's mass surrounded by a hydrogen-rich burning region can give a marked reduction in the solar neutrino flux. It is found that such a configuration does give values <2 SNU, but that the configuration is not itself attainable in evolutionary sequences beginning with a central hydrogen deficiency. It is argued, in fact, that no initial abundance inhomogeneity can produce such a configuration at the solar age.

080.015 On solar neutrinos and the method of spherically averaged rotation. A. J. R. Prentice.
Monthly Notices Roy. Astron. Soc., Vol. 170, 67P - 70P (1975).

A critical analysis of the method of spherical averages in rotating stars suggests that the neutrino flux is not nearly as badly underestimated as Monaghan has recently concluded and may even be actually overestimated in the case of stars possessing rapidly rotating cores.

080.016 The rotation of the sun. R. Howard.
Sci. American, Vol. 232, No. 4, p. 106 - 114 (1975).

080.017 The effect of primordial hydrogen/helium fractionation on the solar neutrino flux.
J. C. Wheeler, A. G. W. Cameron.
Astrophys. Journ., Vol. 196, 601 - 605 (1975).

If hydrogen and helium are immiscible below some critical temperature, gravitational separation could occur in the proto-sun, resulting in a nearly pure helium core and a nearly pure hydrogen shell. The authors have constructed solar models according to this scenario and find the neutrino flux reduced to 1.5–3 SNU.

080.018 Dynamics of the solar magnetic field. V. Velocities associated with changing magnetic fields.
R. H. Levine, Y. Nakagawa.
Astrophys. Journ., Vol. 196, 859 - 866 (1975).

Methods of determining horizontal velocities from the magnetic induction equation on the basis of a time series of magnetogram observations are discussed. For the flare of 1972 August 7, it is shown that a previously developed method of predicting positions of likely flare activity provides reasonable agreement with observations. Limitations to this type of solution of the magnetic induction equation are pointed out, and unambiguous solutions, corresponding to phenomenological determinations of velocity patterns under various physical circumstances, are presented for simple magnetic configurations. Implications for the analysis of changes in a series of magnetogram observations are discussed.

080.019 The energy flux of the sun. A critical discussion of standard values for the solar irradiance.
D. Labs.
Problems in stellar atmospheres and envelopes, (see 003.001), p. 1 - 19 (1975).

080.020 Acoustic waves in the lower solar atmosphere.
C. Chiuderi, C. Giovanardi.
Solar Physics, Vol. 41, 35 - 42 (1975).

The propagation and dissipation of acoustic waves in the lower solar atmosphere is studied. The level of shock formation is computed for various initial conditions. It is shown that shocks form rather low in the atmosphere and that this result does not depend critically on the assumed initial conditions.

080.021 Tabulation of the harmonic coefficients of the solar magnetic fields.
M. D. Altschuler, D. E. Trotter, G. Newkirk, Jr., R. Howard.
Solar Physics, Vol. 41, 225 - 226 (1975).

Tables of spherical harmonic coefficients for the global photospheric magnetic field between 1959 and 1974 are now available on microfilm. (These are the same coefficients which were used to construct the maps of the coronal magnetic atlas.)

080.022 The evolution of the solar inner rotation.
T. Sakurai.
Monthly Notices Roy. Astron. Soc., Vol. 171, 35 - 52 (1975).

The evolution of the axisymmetric rotation of the radiative interior of the sun is studied based on the Eddington–Sweet theory of perturbation. The effect of the molecular weight gradient is neglected, but the effects of the eddy viscosity and of the solar wind torque are taken into account. A new formulation of the initial boundary value problem of a system of the potential equation and the higher order non-linear diffusion equation is given, and a reasonably stable method of solution is proposed.

080.023 A solar model with low neutrino emission.
F. Hoyle.
Astrophys. Journ., (Letters), Vol. 197, L127 - L131 (1975).

A model for the sun giving a reaction rate on ^{37}Cl as low as 0.5 SNU is discussed. The low neutrino emission comes about because the sun is taken to have a core containing 0.3 – 0.5 $M_\odot$ with an unusually high concentration of iron-group metals and a low initial concentration of helium. The high opacity caused by the metals makes the core convective, a suggestion which has been made previously by other authors on an ad hoc basis. In such a situation the low neutrino emission arises for the following two reasons: (1) a high hydrogen concentration, $X \approx 0.7$, is maintained at the center; (2) convective mixing of ^{7}Be on a time scale less than the ^{7}Be destruction time minimizes the importance of ^{7}Be$(p, \gamma)^8$B.

080.024 Solar magnetic fields and convection. I: Active regions and sunspots. J. H. Piddington.
Astrophys. Space Sci., Vol. 34, 347 - 362 (1975).

A phenomenological model of solar magnetic fields is developed, which differs drastically from all currently popular (diffuse-field) models. Its acceptance would require a review of a major part of theoretical solar physics including convection.

080.025 Why the sun may appear oblate. K. H. Schatten.
Astrophys. Space Sci., Vol. 34, 467 - 480 (1975).

If the sun loses angular momentum from its core, due to core contraction, into the solar wind at the observed rate, then an ~0.7 day rotational period for the core of the sun is required for temporal equilibrium. The rotational power released in the core contraction process can equal the observed magnetic energy released in the solar activity cycle if the sun's core rotates with a period near 1.4 to 4 days. The rotational power released from a rotating object is $\tau\Omega$, where τ is the torque on the object and Ω is its angular velocity. Fitting this to the solar wind torque and core rotation rate provides an 0.5 to 5 day rotation period for the sun's core. The consistency of the above methods suggests that the sun's observed oblateness is due to a rapidly rotating solar core. The oblateness of the photosphere is estimated to be near 3.4×10^{-5}.

080.026 The generation of magnetic fields in astrophysical bodies. X. Magnetic buoyancy and the solar dynamo.
E. N. Parker.
Astrophys. Journ., Vol. 198, 205 - 209 (1975).

The author shows that magnetic fields can be retained for long periods of time in the stable radiative region beneath the convective zone, but unfortunately the solar dynamo cannot function there because turbulent diffusion is an essential part of its operation. The only possible conclusion appears to be that the dynamo operates principally in the very lowest levels of the convective zone at depths of 1.5×10^5 km or more, where the gas density is 0.1 g cm^{-3}, and the fields are limited to 50 gauss, rather than the usually estimated 10^2 gauss.

080.027 Measurements of solar magnetic fields by Fourier transform techniques. I: Unsaturated lines.
A. M. Title, T. D. Tarbell.
Solar Physics, Vol. 41, 255 - 269 (1975).

If the basic profile shapes of the normal Zeeman triplet do not have zeros in their Fourier transform, the magnetic field splitting can be determined independent of the profile shape. When the ratio of the splitting of the components is greater than the intrinsic FWHM of the component profiles the magnetic splitting can be determined with significantly greater accuracy than the measurement accuracy of the original profile. For Gaussian shaped components and a ratio of magnetic splitting to FWHM of 1.5 the noise reduction factor is 25.

080.028 Long term variation of the solar equatorial velocity and its relation to non-axisymmetric convection.
G. Belvedere, L. Paternò.
Solar Physics, Vol. 41, 289 - 295 (1975).

The long term variations of solar equatorial velocity are considered, as determined by spectroscopic observations of several authors since 1900. By eliminating Storey's observations covering the period 1914–1932 which seem to be affected by casual errors, a computer analysis picks out a period of about 34 yr in the velocity variation. An interpretation is given of this period in the framework of the interaction of non-axisymmetric convection with rotation.

080.029 An anticorrelation between polar and equatorial rotation of the solar photosphere. C. L. Wolff.
Solar Physics, Vol. 41, 297 - 300 (1975).

Published spectroscopic measurements of solar rotation are analyzed to show that when the rotation velocity increases at high latitudes it tends to decrease at low latitudes, and conversely. The high latitude velocities typically vary over only 20% of the range of those near the equator and the smallest variations of all occurred near latitude 60° during the rising portion of the previous solar cycle. The anticorrelation is consistent with a recent suggestion that differential rotation on the sun arises from photospheric wind systems whose strength is determined, ultimately, by oscillations within the sun.

080.030 Long term evolution of solar sector structure.
L. Svalgaard, J. M. Wilcox.
Solar Physics, Vol. 41, 461 - 475 (1975).

The large-scale structure of the solar magnetic field during the past five sunspot cycles (representing by implication a much longer interval of time) has been investigated using the polarity (toward or away from the sun) of the interplanetary magnetic field as inferred from polar geomagnetic observations. It appears that a solar structure with four sectors per rotation persisted through the past five sunspot cycles with a synodic rotation period near 27.0 days. Superposed on this four-sector structure there is another structure with inward field polarity, a width in solar longitude of about 100° and a synodic rotation period of about 28 to 29 days.

080.031 Ultraviolet absorption in the solar atmosphere.
J. N. Dragon, J. P. Mutschlecner.
Bull. American Astron. Soc., Vol. 7, 248 (1975). − Abstr. AAS.

080.032 Calculations of profiles for the Ca II H and K lines including partial redistribution effects.

R. A. Shine, R. W. Milkey, D. Mihalas.
Bull. American Astron. Soc., Vol. 7, 360 (1975). − Abstr. AAS.

080.033 Magneto-atmospheric waves and Moreton's wave phenomenon. A. H. Nye, J. H. Thomas.
Bull. American Astron. Soc., Vol. 7, 361 (1975). − Abstr. AAS.

080.034 What velocities are consistent with the interpretation of supergranulation as penetrative convection?
K. B. Gebbie, J. Toomre.
Bull. American Astron. Soc., Vol. 7, 363 (1975). − Abstr. AAS.

080.035 The theoretical temperature minimum.
W. Kalkofen, P. Ulmschneider.
Bull. American Astron. Soc., Vol. 7, 363 (1975). − Abstr. AAS.

080.036 Observations of the large-scale solar circulation.
P. S. McIntosh.
Bull. American Astron. Soc., Vol. 7, 364 (1975). − Abstr. AAS.

080.037 The internal origin of solar sectors. S. T. Suess.
Bull. American Astron. Soc., Vol. 7, 364 (1975).
Abstr. AAS.

080.038 A comparison of the meridional flows in the sun's convection zone, predicted by theories of the solar dynamo and differential rotation. B. R. Durney.
Bull. American Astron. Soc., Vol. 7, 364 (1975). − Abstr. AAS.

080.039 An anticorrelation between polar & equatorial rotation rates. C. L. Wolff.
Bull. American Astron. Soc., Vol. 7, 364 (1975). − Abstr. AAS.

080.040 Towards building a solar general circulation model.
P. A. Gilman.
Bull. American Astron. Soc., Vol. 7, 364 (1975). − Abstr. AAS.

080.041 Hydrogen lines in the solar atmosphere.
G. Elste.
Bull. American Astron. Soc., Vol. 7, 366 (1975). − Abstr. AAS.

080.042 Variation of the solar constant and solar activity.
V. F. Loginov.
Radiatsion. protsessy v atmosf. i na zemn. poverkhnosti.
Leningrad, Gidrometeoizdat, 1974, p. 158 - 161. In Russian.
Abstr. in Referativ. Zhurn. 51. Astron., 4.51.535 (1975).

080.043 Seasonal variation and magnitude of the solar sector structure−atmospheric vorticity effect.
J. M. Wilcox, L. Svalgaard, P. H. Scherrer.
Nature, Vol. 255, 539 - 540 (1975).

A relationship between the solar sector structure, as swept past the earth by the solar wind, and terrestrial atmospheric vorticity has been reported by Wilcox et al. As observed by spacecraft magnetometers near the earth, the extended solar magnetic field typically consists of four sectors within a 27-d synodic solar rotation period. Within each sector the magnetic field is predominantly either towards or away from the sun. The sectors are separated by thin current sheets that reverse the direction of the field and constitute sector boundaries. Here the authors report evidence of a seasonal variation in this effect.

080.044 Fourier analysis of the five minute oscillations.
D. K. Lynch.
Astrophys. Letters, Vol. 16, 77 - 79 (1975).

The use and limitations of the Fourier transform for analyzing aperiodic signals is discussed with regard to the solar five minute oscillations. It is argued that peaks in the observational $k - \omega$ power spectra do not necessarily represent the true spectral properties of the underlying physical process which is responsible for the data. Alternative interpretations of these peaks are proposed.

080.045 The depth of formation of the CN lines in the solar atmosphere. G. A. Porfir'eva.
Astron. Zhurn. Akad. Nauk SSSR, Vol. 52, 593 - 598 (1975). In Russian. English translation in Soviet Astron., Vol. 19, No. 3.

The depth of formation of the weak bands of the violet system of CN was calculated by the method of weighting functions. Two models of the solar atmosphere were used. Lines with different quantum numbers are formed nearly in the same layer ($\tau_0 \cong 0.05 - 0.06$). The difference between the depths of formation of the centre of the line and of its wing is not great ($\Delta\tau_0 \cong 0.005$). Contribution functions for the solar centre slightly differ from the contribution functions for the limb. The calculations are in good agreement with earlier observations.

080.046 Nuclear reaction on sun. J. Mergentaler.
Urania Kraków, Vol. 46, 34 - 37 (1975). In Polish.

080.047 Importance of OH bound-free opacity in the solar and stellar atmosphere.
S. P. Tarafdar, P. K. Das.
Bull. Astron. Soc. India, Vol. 2, 39 (1974). – Abstract.

080.048 Information on the third dimension of solar phenomena deduced by statistical procedures.
G. Godoli.
Oss. Astrofis. Catania, Pubbl. Nuova Ser. No. 153, 19 pp. (1974). – Communication presented at the Symposium on the sun and solar system in three dimensions, Frascati (Italy), 2 - 3 July 1974.

Geometrical relationships between apparent elements (apparent height, apparent position) and true elements are established for radial phenomena. Applications to polar rays and prominences are reviewed: it is shown how by statistical procedures it is possible to deduce information from the apparent elements on the true ones. Information on the third dimension of spots, faculae and flares deduced from foreshortening laws is also briefly reviewed.

080.049 Some questions of physics and astrophysics of the sun. G. E. Kocharov.
Izv. AN SSSR Ser. fiz., Vol. 39, 244 - 249 (1975). In Russian. Abstr. in Referativ. Zhurn. 62. Issled. kosmich. prostranstva, 6.62.201 (1975).

080.050 Difficulties in the interpretation of a solar neutrino experiment. S. S. Vasil'ev.
Izv. AN SSSR. Ser. fiz., Vol. 39, 304 - 309 (1975). In Russian. Abstr. in Referativ. Zhurn. 62. Issled. kosmich. prostranstva, 6.62.208 (1975).

080.051 Distribution of plasma ions according to velocities and solar neutrinos.
S. S. Vasil'ev, G. E. Kocharov, A. A. Levkovskij.
Izv. AN SSSR. Ser. fiz., Vol. 39, 310 - 315 (1975). In Russian. Abstr. in Referativ. Zhurn. 62. Issled. kosmich. prostranstva, 6.62.209 (1975).

080.052 The role of nuclear reactions in the evolution of matter of the solar system. A. K. Lavrukhina.
Izv. AN SSSR. Ser. fiz., Vol. 39, 395 - 402 (1975). In Russian. Abstr. in Referativ. Zhurn. 62. Issled. kosmich. prostranstva, 6.62.220 (1975).

080.053 Magnetische velden en convectie in de zon.
H. C. Spruit, C. Zwaan.
Nederlands Tijdschr. Natuurkunde, Vol. 40, No. 09, 5 pp. = Utrechtse Sterrenkundige Overdrukken No. 277 (1974).

080.054 The evolution of magnetic fields and unstable solar plasmas. M. Kuperus.
Utrechtse Sterrekundige Overdrukken No. 295, 3 pp. (1974).

080.055 Solar neutrinos. R. K. Ulrich.
AIP (American Inst. Phys.) Conference Proc., No. 22, p. 259 - 272 (1974). – Review paper.

080.056 Magnetic fields in the sun. D. J. Mullan.
Journ. Franklin Inst., (U. S. A.), Vol. 298, 341 - 362 (1974).

The observed properties of solar magnetic fields are reviewed, with particular reference to the complexities imposed on the field by motions of the highly conducting gas.

080.057 On some general properties of the structure of solar magnetic fields. R. N. Ikhsanov.
Solnechnye Dannye 1975 Byull., No. 2, p. 91 - 97 (1975). In Russian.

Three systems of scales in the structure and distribution of the magnetic fields in the solar photosphere are observable. The first system is typical for quiet formations, the second for formations with strong magnetic fields and the third for their internal structure. The regularities, found in the structure of the magnetic field, show that the convection (system) is the main reason of the observed structure.

080.058 Solar neutrinos and experiments to search for the hypothetical level in ^{6}Be.
V. N. Fetisov, Yu. S. Kopysov.
Nuclear Phys. A, (Netherlands), Vol. 239A, 511 - 529 (1975).

The authors discuss the problem of solar neutrinos, as well as the restrictions imposed on the parameters of the hypothetical level in ^{6}Be and experiments to search for new levels in nuclei with $A = 6$.

Stray light in solar observations. A method of numerical integration. See Abstr. 031.268.

Fraunhofer selective diffraction produced by structure elements in sodium vapour. See Abstr. 034.021.

Solar test of Dirac's large numbers hypothesis. See Abstr. 043.001.

The Apollo Telescope Mount on Skylab. See Abstr. 051.009.

Nonisothermal magnetostatic equilibria in a uniform gravity field. I. Mathematical formulation. See Abstr. 062.025.

On the stability of magnetic fields of active regions. See Abstr. 062.035.

Hydromagnetic waves in structured magnetic fields. See Abstr. 062.036.

An approximate solution of the integral equation in radiative transfer. See Abstr. 063.019.

Formation of molecular lines in stellar atmospheres. See Abstr. 064.015.

Importance of bound-free opacity of OH and CH in solar and stellar atmospheres. See Abstr. 064.016.

Response of a bounded atmosphere to a nonresonant excitation. I: Isothermal case. See Abstr. 064.026.

The black hole that ate the solar neutrinos.
See Abstr. 066.060.

The sun in the far infrared and sub-mm region.
See Abstr. 071.044.

Recent measurements of the flux excess from solar faculae and the implication for the solar oblateness.
See Abstr. 072.021.

Solar rotation as marked by extreme-ultraviolet coronal holes.   See Abstr. 074.086.

A unified explanation of solar type IV dm continua and zebra patterns.   See Abstr. 074.090.

Measurement of the angular momentum of Jupiter and the sun by use of the Lense-Thirring effect.
See Abstr. 099.001.

The 11-year cycle of galactic cosmic radiation and the total magnetic field of the sun.   See Abstr. 143.075.

Errata

080.901   Errata: 'Solar oblateness, excess brightness, and relativity ' [Phys. Rev. Letters, Vol. 33, 1497 - 1500 (1974)].   H. A. Hill, P. D. Clayton, D. L. Patz, A. W. Healy, R. T. Stebbins, J. R. Oleson, C. A. Zanoni.
Phys. Rev. Letters, Vol. 34, 296 (1975).

080.902   Erratum: 'Yearly variation in the synodic rotation period of the sun' [Solar Physics, Vol. 37, 257 - 260 (1974)].   W. Graf.
Solar Physics, Vol. 40, 514 (1975).

Earth

081 Figure, Composition, and Gravity of the Earth

081.001　The importance of damping in geophysics.
H. Jeffreys.
Geophys. Journ. Roy. Astron. Soc., Vol. 40, 23 - 27 (1975).

081.002　Changes in the earth's inertia tensor due to earth-
quakes by MacCullagh's formula.
M. Israel, A. Ben-Menahem.
Geophys. Journ. Roy. Astron. Soc., Vol. 40, 305 - 307
(1975). – Research note.

081.003　A comment on 'Normal modes of a rotating, self-
gravitating inhomogeneous earth'.
M. N. Jones, with a reply by P. C. Luh.
Geophys. Journ. Roy. Astron. Soc., Vol. 40, 309 (1975).
Concerning 12.081.006.

081.004　Geopotential harmonics of order 15 and even degree,
from changes in orbital eccentricity at resonance.
D. G. King-Hele, D. M. C. Walker, R. H. Gooding.
Planet. Space Sci., Vol. 23, 229 - 246 (1975).
　　In this paper the changes in eccentricity at resonance for
six satellites in near-circular orbits at inclinations between 56°
and 90° have been analysed to derive 11 pairs of equations
linking the harmonic coefficients of order 15 and (even) degree l, $\bar{C}_{l, 15}$ and $\bar{S}_{l, 15}$ in the usual notation.

081.005　Numerische Probleme der astro-gravimetrischen
Geoidbestimmung.　D. Lelgemann.
Nachr. Karten- und Vermessungswesen, [Verlag Inst. Angew.
Geodäsie, Frankfurt], Sonderheft, p. 87 - 95 (1975).
　　The author selected the procedure of "least squares
collocation" as mathematic basis for a planned astro-gravi-
metrical determination of the geoid of the Federal Republic
of Germany. He describes a series of problems which arose in
the course of the now finished test computations and he
indicates solutions related to general problems of the astro-
gravimetrical determination of the geoid as well as to special
problems of the selected procedure.

081.006　Zur Variation der Gravitationsbeschleunigung.
H.-J. Treder.
Gerlands Beiträge Geophys., Vol. 84, 20 - 24 (1975).
　　According to the analytical formulation of the Mach–
Einstein doctrine about the relativity of inertia (Treder 1972/
73) and according to the reference tetrads theory of gravity
fields with self-absorption (Treder 1967/71) the effective gravitational mass of the earth is a function of the gravity potentials of the universe, the galaxy, and the sun. From these dependences annual and secular variations of the Galilean acceleration g are resulting. The amplitude of the annual variation
is of the order $\Delta g/g \approx \pm 5 \times 10^{-10}$ per year and the amplitude
of the secular variation is of the order $\Delta g/g \approx \pm 10^{-7}$ for 10^8
years.

081.007　Gezeiten.　P. Brosche.
Naturwissenschaften, Vol. 62, 1 - 9 (1975).
　　Tidal forces are present in the whole universe. This interaction is of particular interest for the earth which is subject to
the tidal forces of the moon and the sun. These forces produce
periodic phenomena in the atmosphere, in the solid earth, and
in the oceans. Tides also cause a retardation of the earth's
rotation.

081.008　Is the earth tide phase lag unaffected by anelasticity?
B. Bodri.
Nature, Vol. 254, 314 - 315 (1975). – Letter.

081.009　An almost axially-symmetrical model of the
hydromagnetic dynamo of the earth. I.
S. I. Braginskij.
Geomagn. Aeronom., Vol. 15, 149 - 156 (1975). In Russian.

081.010　15th order resonance terms using the decaying orbit
of TETR-3.　C. A. Wagner, S. M. Klosko.
Planet. Space Sci., Vol. 23, 541 - 549 (1975).
　　The orbit of TETR-3 (1971-83B), inclination: 33°, passed
through resonance with 15th order geopotential terms in
February 1972. The resonance caused the orbit inclination to
increase by 0.015°. Analysis of 48 sets of mean Kepler elements for this satellite in 1971–1972 (across the resonance)
has established strong constraint for high degree, 15th order
gravitational terms.

081.011　Alaskan thermokarst terrain and possible Martian
analog.　L. W. Gatto, D. M. Anderson.
Science, Vol. 188, 255 - 257 (1975).
　　A first-order analog to Martian fretted terrain has been
recognized on enhanced, ERTS-1 (Earth Resources Technology Satellite) imagery of Alaskan Arctic thermokarst terrain.
The Alaskan analog displays flat-floored valleys and intervalley
uplands characteristic of fretted terrain. The thermokarst terrain has formed in a manner similar to one of the processes
postulated for the development of the Martian fretted terrain.

081.012　Processing of altimetry data.　K.-R. Koch.
Bull. Géod., Nouvelle Sér., No. 115, p. 35 - 40
(1975). – Shortened version of a paper presented to the 14th
International Congress of Surveyors in Washington, Sept. 1974.

081.013　Effect of certain anomaly correction terms on
potential coefficient determinations of the earth's
gravitational field.　R. H. Rapp.
Bull. Géod., Nouvelle Sér., No. 115, p. 57 - 63 (1975).

081.014　The free-oscillation equations at the centre of the
earth.　D. J. Crossley.
Geophys. Journ. Roy. Astron. Soc., Vol. 41, 153 - 163 (1975).
　　The well-known equations governing the free oscillations
of an elastic medium possess a singular point at the origin of
the co-ordinate system. The obvious procedure to begin an
integration of the equations at the origin is to use a power-
series expansion, and this appears to have been incorporated
in some free oscillation programs. However, details have never
been published and the method is presented here. A short
discussion of the usefulness of the method is given based on
some computations with a real earth model.

081.015　Secularly unstable Maclaurin ellipsoids and the pre-
fission state of the earth.　J. A. O'Keefe.
Bull. American Astron. Soc., Vol. 7, 384 - 385 (1975).

Abstr. AAS.

081.016 **Marées terrestres 1973.** F. De Meyer.
Annuaire, edited by Institut Royal Météorologique de Belgique, Bruxelles, 84 pp. (1975).

081.017 **Chandler wobble and viscosity in the earth's core.**
M. G. Rochester, Ya. S. Yatskiv, T. Sasao, with a reply by J. Verhoogen.
Nature, Vol. 255, 655 - 656 (1975).

081.018 **High precision measurements for studying the secular variation in gravity in Finland.**
A. Kiviniemi.
Publ. Finnish Geod. Inst., *Helsinki*, No. 78, 68 pp. (1974).

081.019 **On the absolute determination of gravity (I).**
T. Okuda, C. Sugawa, K. Hosoyama, T. Suzuki, T. Sato.
Proc. International Latitude Obs. Mizusawa, No. 14, p. 77 - 90 (1974). In Japanese.

The expectation of many scientists in the world for developing an absolute gravimetry apparatus with an accuracy of few micro-Gals was realized by A. Sakuma. The accuracy is presently approaching to one micro-Gal. The International Latitude Observatory of Mizusawa decided to establish the Sakuma type absolute gravimetry apparatus under the support of the B.I.P.M. till 1977. This note describes briefly the apparatus with a schematic diagram, and reports on the characteristics of the observation building and the structure of the foundations.

081.020 **On the oscillation of the laterally heterogeneous earth, I.** P. Musen.
Goddard Space Flight Center, Greenbelt, Maryland, GSFC Document X-920-75-62, 36 pp. (1975).

081.021 **Goddard earth models (5 and 6).** F. J. Lerch, C. A. Wagner, J. A. Richardson, J. E. Brownd.
Goddard Space Flight Center, Greenbelt, Maryland, GSFC Document X-921-74-145, Preprint, 12 + 100 + A 110 + B 29 pp. (1974). — Results presented at the 55th annual meeting of the American Geophysical Union, April 8 - 12, 1974, Washington, D.C.

A comprehensive earth model has been developed at the Goddard Space Flight Center to satisfy requirements of the National Geodetic Satellite Program. The model consists of two complementary gravitational fields (in spherical harmonics) and center-of-mass locations for 134 tracking stations on the earth's surface. One gravitational field (Goddard Earth Model 5) is derived solely from satellite tracking data. A second (combination) solution (GEM 6) uses this data with 13,400 simultaneous events from satellite camera observations, and 1654 5° X 5° surface gravimetric anomalies.

081.022 **Evaluation and comparisons of recent geopotential solutions.** M. A. Khan.
Goddard Space Flight Center, Greenbelt, Maryland, GSFC Document X-921-74-275, Preprint, 7 + 31 pp. (1974).

081.023 **Precise calculations in some integral transformations of physical geodesy.** Yu. M. Nejman.
Izv. vyssh. ucheb. zavedenij. Geod. i aehrofotosemka, 1974, No. 4, p. 49 - 54. In Russian. — Abstr. in Referativ. Zhurn. 52. Geodeziya i Aehrosemka, 5.52.58 (1975).

081.024 **Determination of the parameters of the earth's gravitational field and perspectives of their further improvement.** L. P. Pellinen.
Materialy Plenuma Komis. po izuch. vrashcheniya Zemli Astron. soveta AN SSSR. Poltava, 1972. Kiev, Nauk. dumka,

1974, p. 9. In Russian. — Abstr. in Referativ. Zhurn. 52. Geodeziya i Aehrosemka, 5.52.63 (1975).

081.025 **Probable characteristics of second order perturbations in gravimetrical measurements on sea.**
V. L. Panteleev.
Soobshch. Gos. Astron. Inst. Shternberga, No. 187, p. 20 - 33 (1973). In Russian.

081.026 **Neue Aspekte zum Geodynamics Project aus hockdruck-festkörperphysikalischer und planetarer Sicht.** F. Frölich.
Akad. Wiss. DDR, Zentralinst. Phys. Erde, Veröff. No. 29, (see 013.012), p. 21 - 31 (1974).

081.027 **Marées terrestres.** P. Melchior (Editor).
Bull. d'Informations, (Obs. Roy. Belgique, Bruxelles), Nos. 70, 71, p. 3921 - 4084 (1975).

081.028 **Étude des marées terrestres et océaniques à partir des perturbations d'orbites de satellites.**
A. Cazenave.
Rend. Seminario Fac. Sci. Univ. Cagliari, Suppl. Vol. 44, (see 012.016), 25 - 29 (1974).

081.029 **On the determination of the potential by satellite observations.** V. Szebehely.
Rend. Seminario Fac. Sci. Univ. Cagliari, Suppl. Vol. 44, (see 012.016), 31 - 35 (1974).

081.030 **Sulla precessione euleriana della terra secondo il modello di Kelvin.** A. Melis.
Rend. Seminario Fac. Sci. Univ. Cagliari, Suppl. Vol. 44, (see 012.016), 47 - 51 (1974).

081.031 **The estimation of 550 km X 550 km mean gravity anomalies.** M. R. Williamson, E. M. Gaposchkin.
Smithsonian Astrophys. Obs., *Cambridge, Mass.,* Special Report 363, 5 + 20 pp. (1975).

The calculation of 550 km X 550 km mean gravity anomalies from 1° X 1° mean free-air gravimetry data is discussed. The block estimate procedure developed by Kaula is used. Estimates for 1452 of the 1654 blocks are obtained.

081.032 **Global detailed geoid computation and model analysis.** J. G. Marsh, S. Vincent.
Geophys. Surveys, [D. Reidel Publ. Company, Dordrecht—Holland], Vol. 1, 481 - 511 = Goddard Space Flight Center, Greenbelt, Maryland, Separate print (1974). — See 12.081. 025.

081.033 **Die innere Struktur der Erde, des Mondes und der Planeten.** H. Stiller.
Akad. Wiss. DDR, Jahrbuch 1974, p. 60 - 63 (1975).

081.034 **Introduction to Molodensky's theory.**
H. Moritz.
Boll. Geod. Sci. Affini, Anno 34, p. 161 - 172 (1975). — Lecture at the International School of Advanced Geodesy, Erice (Sicily), October 1 - 26, 1974.

081.035 **Tidal friction.** H. Jeffreys.
Quarterly Journ. Roy. Astron. Soc., Vol. 16, 145 - 151 (1975).

081.036 **Estimate of the value of variation of the plumb-line deflection.** Yu. Ya. Vashchilov, I. A. Maslov.
Metodika izmereniya gravitats. polej. Moskva — Penza, 1974, p. 50 - 54. In Russian. — Abstr. in Referativ. Zhurn. 52. Geodeziya i Aehrosemka, 6.52.58 (1975).

081.037   **Earth tides.**   P. Melchior.
Geophys. Surveys, Vol. 1, 275 - 303 = Obs. Roy. Belgique, Commun., Sér. B, No. 90 = Sér. Géophys. No. 123 (1974).

The main geometrical characteristics and mechanical properties of bodily tides are described, using the convenient elastic parameters of Love. The problem of the earth's deformation is a problem of spherical elasticity of the sixth order. The importance of earth tides in astronomy and geophysics is emphasized by their relation to the precession-nutation and tesseral tidal problems, the secular retardation of the earth's speed of rotation due to the dissipation of energy in sectorial tides, the periodic variations of the speed of rotation due to zonal tides, the satellite orbit perturbations due to the earth's potential variation, and the radial deformations in laser distance measurements.

Planet earth. (Its past, present and future). See Abstr. 003.006.

On the earth's axes of rotation and figure. See Abstr. 044.003.

Secular variations of the earth's gravity field and polar motion.   See Abstr. 045.003.

Is the nutation of the solid inner core responsible for the 24-year libration of the pole? See Abstr. 045.026.

Geodetic uses of Lunar Laser Ranging. See Abstr. 046.006.

The effect of systematic differences between reference positions on the second harmonic of the earth's gravitational potential.   See Abstr. 046.007.

Importance of the coupling effects between earth potential harmonics in the motion of an aritificial satellite. Computed and checked solution of this coupling problem. The case of J_7.   See Abstr. 052.026.

A planetary radio astronomy discussion of the 1.55 cm microwave emission of the earth. See Abstr. 082.021.

A view of earth and air.   See Abstr. 082.074.

A model of the gravitational differentiation of the interior of planets.   See Abstr. 091.005.

Correlated elements and the bulk composition of the moon, the earth and the parent body of the eucrites. See Abstr. 094.126.

Gravity and magnetic investigations of Meteor crater, Arizona.   See Abstr. 105.085.

082 The Earth's Atmosphere Including Refraction, Scintillation, Extinction, Airglow, Site Testing

082.001 Interior radiances in optically deep absorbing media- III. Scattering from Haze L.
G. W. Kattawar, G. N. Plass.
Journ. Quant. Spectrosc. Radiat. Transfer, Vol. 15, 61 - 85 (1975).

The interior radiances are calculated within an optically deep absorbing medium scattering according to the Haze L phase function. The dependence on the solar zenith angle, the single scattering albedo, and the optical depth within the medium is calculated.

082.002 Measurement of line widths of CO of planetary interest at low temperatures. P. Varanasi.
Journ. Quant. Spectrosc. Radiat. Transfer, Vol. 15, 191 - 196 (1975).

Self-broadened, air-broadened and CO_2-broadened half-widths of lines $R(0)$ through $R(16)$ in the CO fundamental have been measured at $100°K$ (self-broadening only), $200°K$, $250°K$ and $300°K$ using the Ladenburg-Reiche curve-of-growth.

082.003 Polarisation of atmospheric bremsstrahlung.
R. R. Rausaria, R. N. Singh.
Nature, Vol. 253, 28 - 29 (1975). − Letter.

082.004 Statistical investigations of star image vibrations.
I. V. Shvalagin, I. I. Motrunich, M. M. Osipenko.
Atmosf. optika. Moskva, Nauka, 1974, p. 97 - 103. In Russian. − Abstr. in Referativ. Zhurn. 51. Astron., 2.51.103 (1975).

082.005 Probability of clear sky in winter over the territory of the U.S.S.R.
Sh. P. Darchiya, V. I. Ivanov, P. G. Kovadlo.
Issled. po geomagnetizmu, aehron. i fiz. Solntsa, Vyp. (No.) 31. Moskva, Nauka, 1974, p. 61 - 64. In Russian. − Abstr. in Referativ. Zhurn. 51. Astron., 2.51.106 (1975).

082.006 On the study of the dependence of the refraction coefficient on the absolute altitude.
V. I. Vashchenko, B. M. Dzhuman, A. L. Ostrovskij.
Geod. i kartografiya, 1974, No. 10, p. 26 - 27. In Russian. Abstr. in Referativ. Zhurn. 51. Astron., 2.51.144 (1975).

082.007 Photoionization of barium clouds via the ^3D metastable levels. J. L. Carlsten.
Planet. Space Sci., Vol. 23, 53 - 60 (1975).

The photoionization of optically thin barium clouds is analyzed and shown to occur primarily by a two-step process involving the ^3D metastable term as the intermediate state. The equilibrium populations of the ^1D and ^3D metastable levels are calculated and found to differ significantly from the values now in the literature.

082.008 Atmospheric emission measurements at 85 to 118 GHz. C. J. Gibbins, A. C. Gordon-Smith, D. L. Croom.
Planet. Space Sci., Vol. 23, 61 - 73 (1975).

Measurements have been made of the atmospheric gaseous emission spectrum in the region from 85 − 118 GHz (wavelength range 3.5 − 2.5 mm), and the results have been used to estimate the attenuation in the zenith direction over this range, which includes the low frequency side of the oxygen absorption line centered at 118.75 GHz. The results are compared with theoretical predictions based on line shape models in current use, and the discrepancies between experiment and theory are discussed.

082.009 A radio picture of the earth.
W. J. Webster, Jr., T. T. Wilheit, T. C. Chang, P. Gloersen, T. J. Schmugge.
Sky Telescope, Vol. 49, 14 - 16 (1975).

082.010 Origin of the sunspot modulation of ozone: its implications for stratospheric NO injection.
M. A. Ruderman, J. W. Chamberlain.
Planet. Space Sci., Vol. 23, 247 - 268 (1975).

The measured modulation of cosmic rays deposited in the stratosphere over a sunspot cycle produces an oscillating source of stratospheric NO with an 11-yr (quasi) period. The resulting modulation of ozone over this period is calculated and is shown to give good agreement with available measurements of the time lag, the latitude dependence and the magnitude of cyclic variations of ozone.

082.011 Reaction of the oxygen emission at 6300 Å during the onset of pre-dawn conjugate photoelectron precipitation. M. M. Gogoshev.
Planet.Space Sci., Vol. 23, 305 - 310 (1975).

082.012 Atmospheric density from the low altitude satellite 1970-48A: comparison of orbital decay measurements, accelerometer measurements and atmospheric models.
H. R. Rugge, B. K. Ching.
Planet. Space Sci., Vol. 23, 323 - 335 (1975).

Atmospheric densities have been deduced from high resolution radar-determined orbital decay data and from data obtained from a uniaxial accelerometer flown onboard the low altitude satellite 1970-48A. Data were obtained during late June and early July, 1970. The orbital decay-deduced densities, having an effective 6 hr temporal resolution, were determined at an altitude of 143 km, essentially one-half scale height above perigee.

082.013 Comment on 'Superrotation of upper atmosphere by global deposition of meteoroids' [Planet. Space Sci., Vol. 22, 559 - 568 (1974)]. R. L. Hawkes.
Planet. Space Sci., Vol. 23, 379 - 381 (1975).

The meteoric influx explanation of superrotation (Mitra, 1974) is re-examined. It is shown that the excess orbital angular momentum of the meteoroids is transferred to the region below about 110 km, and thus can probably not account for the superrotation of the 150–400 km atmospheric layer.

082.014 The photoabsorption coefficients of CO and CO_2 in the region 350 to 650Å.
W. S. Watson, D. T. Stewart, A. B. Gardner, M. J. Lynch.
Planet. Space Sci., Vol. 23, 384 - 386 (1975). − Research note.

082.015 Measurements of the extraterrestrial solar radiant flux from 2981 to 4000 Å and its transmission through the earth's atmosphere as it is affected by dust and ozone. J. J. DeLuisi.
Journ. Geophys. Res., Vol. 80, 345 - 354 (1975).

Measurements of the directly transmitted ultraviolet solar flux in 2-Å intervals in the range 2981−4000 Å had been obtained at Boulder, Colorado, over a period of 2 years. A spectrum of the extraterrestrial solar flux has been derived from these data by use of a modified Langley method. This spectrum is compared with several spectra measured by other investigators, and differences are discussed. The measurements are also used to obtain total ozone and dust extinction as a function of wavelength. The ozone results are compared with total ozone measured concurrently by the Dobson spectrometer.

082.016 Airborne study of equatorial 6300 Å nightglow.
T. P. Markham, J. Buchau, R. E. Anctil, J. F. Noxon.
Journ. Atmosph. Terr. Phys., Vol. 37, 65 - 74 (1975).

082.017 Low latitude meteor wind observations.
A. J. Scholefield, H. Alleyne.
Journ. Atmosph. Terr. Phys., Vol. 37, 273 - 286 (1975).

082.018 Atmospheric transmission of the 1.27 micron band of oxygen. M. Gadsden, P. C. Wraight.
Journ. Atmosph. Terr. Phys., Vol. 37, 287 - 296 (1975).

Evans et al. (1970) have published calculations of the atmospheric transmission of the (0−0) band in the infrared atmospheric system of molecular oxygen at night time. The authors present results of modified and extended calculations suitable for both nightglow and dayglow conditions and indicate the magnitude of the effect of seasonal changes in atmospheric temperature.

082.019 Methods of calculating atmospheric transmittance and radiance in the infrared. A. J. LaRocca.
Proc. IEEE, Vol. 63, 75 - 94 (1975).

Radiation from remotely sensed objects in the earth's environment is attenuated in its passage through the atmosphere. In the infrared region of the spectrum, this is caused, for clear-sky conditions, mainly through absorption by atmospheric molecular species, i.e., the atmospheric gases. Methods are described for calculating atmospheric transmittance and radiance to correspond to spectral resolutions of those instruments used in remote sensing. These methods involve either numerical integration of the exact line structure over the desired spectral interval, or the use of band models derived from artificially created line structure.

082.020 Long term variations in the albedo and surface temperature of the earth.
A. Sellers, A. J. Meadows.
Nature, Vol. 254, 44 (1975). − Letter.

082.021 A planetary radio astronomy discussion of the 1.55 cm microwave emission of the earth.
W. J. Webster, Jr.,T. C. Chang, L. T. Darby, H. M. Finkelstein.
Icarus, Vol. 24, 143 - 147 (1975).

Using 1.55 cm observations of the earth made by the Electrically Scanned Microwave Radiometer experiment on Nimbus 5, the appearance of the earth from Venus is simulated. A single antenna unable to resolve the earth's disk would give a time averaged disk temperature of 183 K. In one rotation, the disk temperature would vary from 194 K to 172 K. During the 1973 inferior conjunction, a radio telescope with 1 arc sec resolution would resolve most of the major surface features of the earth.

082.022 Dynamics of the chemical evolution of earth's primitive atmosphere. A. Bar-Nun, A. Shaviv.
Icarus, Vol. 24, 197 - 210 (1975).

The course of evolution of earth's primitive reducing atmosphere is shown to possibly have been determined to a large extent by the effect of thunder shock waves, which is comparable to the effect of solar uv radiation. The major chemical reactions occurring during a thunderstorm in the troposphere were pyrolysis of hydrocarbons, their oxidation by water vapor and their reaction with molecular nitrogen. These reactions were studied by the single-pulse shock tube technique and their rates as well as their product distributions

were determined.

082.023 Measurements of the atmospheric transfer function at Mauna Kea, Hawaii.
J. C. Dainty, R. J. Scaddan.
Monthly Notices Roy. Astron. Soc., Vol. 170, 519 - 532 (1975).

Over the 10 nights of observation the seeing was variable, ranging from typical widths of the transfer function at the 25 per cent level of approximately 0.23 cycles/arcsec (2.5 cm in the pupil) to approximately 1.00 cycles/arcsec (11.0 cm in the pupil). The latter values correspond to sub-arcsec seeing. The variation with zenith angle and the form of the atmospheric transfer function were briefly investigated.

082.024 Twilight enhancement of λ5577-Å airglow.
R. C. Schaeffer.
Journ. Geophys. Res., Vol. 80, 154 - 160 (1975).

The measurement of [O I] λ5577-Å airglow intensities in the morning and evening twilights with a Fabry-Perot interferometer has revealed an enhancement with an apparent seasonal variation. The enhancement is found to emanate from the F region of the ionosphere and is well accounted for in terms of the dissociative recombination of O_2^+ and F region electrons. It is shown that no other mechanism for the excitation of $O(^1S)$ atoms can be expected to give rise to a significant twilight enhancement. In addition, the results of earlier ground-based studies of the twilight green line airglow by other authors are reviewed critically with reference to the present results.

082.025 Comment on 'Atomic oxygen densities in the lower thermosphere as derived from in situ 5577-Å night airglow and mass spectrometer measurements' by D. Offermann and A. Drescher. T. M. Donahue, B. Guenther.
Journ. Geophys. Res., Vol. 80, 219 - 220, with a reply by D. Offermann, A. Drescher, p. 221 - 222 (1975). − Concerning Journ. Geophys. Res., Vol. 78, 6690 - 6700 (1973) − see 10.082.057.

082.026 The ratio of primary scattering to total scattering of sky radiance. E. de Bary, G. Eschelbach.
Tellus, Vol. 26, 682 - 690 (1974).

082.027 Ground illumination from a turbid cloudless sky.
M. R. Nagel.
Applied Optics, Vol. 14, 67 - 75 (1975).

An attempt is made to develop a set of empirical equations for the relationship between skylight illuminance, turbidity, and solar altitude under the condition of a cloud-free sky. The maximum possible illuminance from a cloud-free sky is estimated using these equations. The illuminances are stated in relation to the illuminance produced by a Rayleigh sky. An improved approximation for Bemporad's airmass function is used in the computations.

082.028 Possibility of atmospheric temperature sounding by the solar occultation technique.
J. K. G. Watson, P. K. L. Yin.
Applied Optics, Vol. 14, 549 - 552 (1975).

082.029 Diffraction theory of optical scintillations due to turbulent layers. L. S. Taylor, C. J. Infosino.
Journ. Optical Soc. America, Vol. 65, 78 - 84 (1975).

An analysis is presented for the modulation indices and spatial spectra of a plane wave that propagates in a uniform medium after having been randomized by passage through a turbulent layer.

082.030 Real-time wavefront correction through Bragg diffraction of light by sound waves.

V. N. Mahajan.
Journ. Optical Soc. America, Vol. 65, 271 - 278 (1975).

082.031 Measurements of anomalous atmospheric absorp-
tion in the wavenumber range 4 cm^{-1} −15 cm^{-1}.
R. J. Emery, P. Moffat, R. A. Bohlander, H. A. Gebbie.
Journ. Atmosph. Terr. Phys., Vol. 37, 587 - 594 (1975).

Field measurements have been made of atmospheric
absorption in the wavenumber range 4 cm^{-1} to 15 cm^{-1}, using
a 200 metre horizontal transmission path and at a resolution
of 0.2 cm^{-1}. The results are compared with theoretical spectra
based on water vapour monomer absorption, and anomalous
absorption spectra are presented.

082.032 A measurement of the extinction of solar hydrogen
Lyman-α radiation in the summer arctic mesosphere.
E. V. Thrane, A. Johannessen.
Journ. Atmosph. Terr. Phys., Vol. 37, 655 - 661 (1975).

Rocket observations of the extinction of solar hydrogen
Lyman-α radiation have been used to derive pressure, density
and temperature in the mesosphere. Comparison of the derived
atmospheric parameters with the standard atmosphere, with
stratospheric data and with collision frequency measurements,
lead to the conclusion that the laboratory measurements of
the absorption cross section of Lyman-α in molecular oxygen
do not apply to conditions in the upper atmosphere.

082.033 Diurnal variations of atomic hydrogen: observations
and calculations.
B. A. Tinsley, R. R. Hodges, Jr., D. F. Strobel.
Journ. Geophys. Res., Vol. 80, 626 - 634 (1975).

Theoretical calculations of the diurnal variation of atomic
hydrogen in the thermosphere are presented that simultaneously evaluate the effects of thermosphere rotation, thermal
escape, charge exchange with O^+ ions, charge exchange with
hot H^+ ions, transport due to winds, and ballistic fluxes in a
consistent manner.

082.034 Doppler profile measurements of the geocoronal
hydrogen Balmer alpha line.
S. K. Atreya, P. B. Hays, A. F. Nagy.
Journ. Geophys. Res., Vol. 80, 635 - 638 (1975).

Results of high-resolution measurements of the geocoronal hydrogen Balmer alpha line profile made during several
nights in December 1971, May 1972, and October 1972 are
presented. The measured intensities are found to be in general
agreement with the theory and earlier measurements under
essentially similar conditions, and the measured geocoronal
hydrogen temperatures are found to be somewhat lower than
the Jacchia (1971) model exospheric values.

082.035 Observed variations of the exospheric hydrogen
density with the exospheric temperature.
J. L. Bertaux.
Journ. Geophys. Res., Vol. 80, 639 - 642 (1975).

Measurements of exospheric hydrogen densities at a
distance of 3 R_E (earth radii) are presented as a function of
exospheric temperature. They imply that the density n_c at the
exobase level decreases with exospheric temperature T_c but
not enough to keep the Jeans escape flux F_J constant. The
increase of F_J with exospheric temperature may be compensated for by the decrease of F_e, which measures the loss of
hydrogen by charge exchange in the plasmasphere. The two
fluxes would be equal to 6.6 × 10^7 atoms cm^{-2} s^{-1} at a temperature T_c= 1070°K, giving a total escape flux in agreement with
measurements of hydrogen compounds in the region from 30
to 50 km and theoretical mesospheric calculations.

082.036 The earth's upper atmosphere–I, II.
L. G. Jacchia.
Sky Telescope, Vol. 49, 155 - 159, 229 - 232 (1975).

082.037 Polarization measurements of the Hβ line in blue
sky light.   D. Clarke, I. S. McLean.
Planet. Space Sci., Vol. 23, 557 - 559 (1975).

New measurements have been made of the reduced polarization in a Fraunhofer line (Hβ) of blue sky light. The value
of the depolarization agrees well with a previously reported
intensity filling-in effect. There is no evidence of a change of
the azimuth of vibration of the polarization across the line
profile.

082.038 On the production of nitric oxide by cosmic rays
in the mesosphere and stratosphere.
M. Nicolet.
Planet. Space Sci., Vol. 23, 637 - 649 (1975).

Nitric oxide is formed in the atmosphere through the
ionization and dissociation of molecular nitrogen by galactic
cosmic rays. One NO molecule is formed for each ion pair
produced by cosmic ray ionization.

082.039 Probability of a clear night sky on the Highveld.
J. Hers.
Monthly Notes Astron. Soc. Southern Africa, Vol. 34, 28 - 29
(1975). – Research note.

082.040 The earth's upper atmosphere–III.   L. G. Jacchia.
Sky Telescope, Vol. 49, 294 - 299 (1975).

082.041 Determination of the spectral distribution of the
solar radiation in the near infra-red region outside
the earth's atmosphere.   Y. Lin, Y. Hu, L. Shen.
Acta Astron. Sinica, Vol. 15, 149 - 158 (1974). In Chinese.

Spectral measurements of the solar radiation with a
monochromator were made during the spring of 1968 from a
5000 metre altitude site of the Mount Jolmo Lungma region
in Southern Tibet, China. From observations lasting more
than a month, 28 sets of solar spectral irradiance data with
different air mass, corresponding to the four days of best
observational conditions, were carefully processed. Through
extrapolation they were used to determine the spectral distribution of the solar radiation in the near infra-red region outside the earth's atmosphere and the mean atmospheric transmission in the Mount Jolmo Lungma region. The results are
briefly discussed.

082.042 Investigation of the optical atmospheric refraction
for artificial satellites.   K. Huang, Y. Ding.
Acta Astron. Sinica, Vol. 15, 215 - 229 (1974). In Chinese.

A quasi-atmospheric model for computing optical atmospheric refraction is constructed, and formulas for computing
optical refraction for objects inside or outside the atmosphere,
particularly for artificial satellites, are derived. In precision
these formulas are comparable with the numerical integration
method, but more convenient for use.

082.043 Détection au sol de la turbulence stratosphérique
par intercorrélation spatioangulaire de la scintillation stellaire.   J. Vernin, F. Roddier.
Comptes Rendus Acad. Sci. Paris, Sér. B, Vol. 280, 463 - 465
(1975).

On montre qu'il est possible de détecter optiquement, à
partir du sol, la présence de turbulence dans la stratosphère,
par analyse statistique de la scintillation d'une étoile double.

082.044 A rediscussion of the atmospheric extinction and
the absolute spectral-energy distribution of Vega.
D. S. Hayes, D. W. Latham.
Astrophys. Journ., Vol. 197, 593 - 601 (1975).

For both the Lick and the Palomar calibrations of the
spectral-energy distribution of Vega, the atmospheric extinction was treated incorrectly. The authors present a model for
extinction in the earth's atmosphere and use this model to

calculate corrections to the Lick and Palomar calibrations. They also describe a method that can be used to fabricate mean extinction coefficients for any mountain observatory. They combine selected portions of the corrected Lick and corrected Palomar calibrations with the new Mount Hopkins calibration to generate an absolute spectral-energy distribution of Vega over the wavelength range 3300–10,800 Å.

082.045 Dayglow of the infrared atmospheric band system of O_2 during a total eclipse of the sun.
P. C. Wraight, M. Gadsden.
Journ. Atmosph. Terr. Phys., Vol. 37, 717 - 730 (1975).

Observations are reported of the radiation from $^1\Delta_g$ oxygen at 1.27μ and 1.58μ. The observations were made using a tilting filter spectrophotometer in Concorde 001, flying at heights up to 17 km. Results are presented concerning the normal dayglow, and also the variation of radiance during the eclipse; and the observations are compared with the theory of $^1\Delta_g$ oxygen production and quenching.

082.046 Is there a continuum near infra-red dayglow?
P. C. Wraight.
Journ. Atmosph. Terr. Phys., Vol. 37, 731 - 737 (1975).

Observations made from Concorde 001 at wavelengths near 1.27μ and 1.58μ suggest the existence of a continuum background dayglow emission extending from near 1.2μ to longer wavelengths, with a total radiance of 100 MR or more, depending on the spectrum assumed. The possibility that the recombination reaction between atomic oxygen and molecular oxygen may be chemiluminescent is considered.

082.047 Photographing of sunrise or sunset from space.
K. Ya. Kondrat'ev, A. A. Buznikov, A. I. Lazarev, E. V. Khrunov.
Dokl. Akad. Nauk SSSR, Ser. Mat. Fiz., Vol. 221, 832 - 834 (1975). In Russian.

082.048 Spectrophotometering of the earth from the Soyuz 13 manned spaceship. A. A. Buznikov, P. I. Klimuk, K. Ya. Kondrat'ev, V. V. Lebedev, V. M. Orlov.
Dokl. Akad. Nauk SSSR, Ser. Mat. Fiz., Vol. 221, 1310 - 1313 (1975). In Russian.

082.049 Astroclimatic characteristics of Terskol peak. The monochromatic extinction coefficient.
N. S. Komarov, E. A. Depenchuk, R. I. Chuprina.
Astrometriya i Astrofizika, *Kiev*, vyp. (No.) 24, (see 003.003), p. 115 - 119 (1974). In Russian.

082.050 Atmospheric turbulence parameters from visual resolution. M. L. Wesely, Z. I. Derzko.
Applied Optics, Vol. 14, 847 - 853 (1975).

Direct visual observations of image blurring by atmospheric turbulence are used to obtain line averages of the refractive index structure function coefficient along lines of sight in the lower atmosphere.

082.051 Glow in space. A. I. Lazarev, A. G. Nikolaev, V. I. Sevast'yanov.
Zemlya i Vselennaya, 1975, No. 2, p. 9 - 13. In Russian.

082.052 Scattering in the earth's atmosphere: calculations for Milky Way and zodiacal light as extended sources.
H. J. Staude.
Astron. Astrophys., Vol. 39, 325 - 333 (1975).

The author presents the results of detailed calculations on first order Rayleigh- and Mie-scattering in the earth's atmosphere illuminated by the Milky Way and the zodiacal light. The results are compared with the work already done in this area and with the procedures commonly applied in reducing photometric observations of extended sources of the night

sky. It is shown that a substantial part of the discrepancies between the brightness distributions of zodiacal light resp. Milky Way given by different authors is due to an inaccurate treatment of the atmospheric scattered light.

082.053 Chemiluminescence of sodium released at night and its relation to the sodium nightglow.
D. Golomb, H. S. Hoffman, G. T. Best.
Journ. Geophys. Res., Vol. 80, 1363 - 1366 (1975).

Sodium vapor was released in darkness on the downleg of a rocket trajectory, from 150 to 125 km. An intense headglow was observed surrounding the rocket, followed by a short-lived afterglow disappearing within $0.2-0.5$ s. The glow is interpreted in terms of a chemiluminous reaction involving sodium dimers Na_2.

082.054 Twilight enhancement in $O_2(b^1\Sigma_g)$ airglow emission.
J. F. Noxon.
Journ. Geophys. Res., Vol. 80, 1370 - 1373 (1975).

The (0.1) band of the atmospheric system of $O_2(b^1\Sigma_g - X^3\Sigma_g)$ at 8640 Å exhibits a large enhancement at twilight; the morning enhancement greatly exceeds that in the evening. Some contribution to the enhancement arises from fluorescent scattering of sunlight by O_2 in the 80- to 100-km region. Of comparable strength in the evening, but wholly dominant in the morning, is production of $O_2(^1\Sigma)$ by energy transfer from $O(^1D)$ produced in ozone photolysis within the same altitude region.

082.055 A search for lunar tides in the thermosphere.
L. G. Jacchia.
Journ. Geophys. Res., Vol. 80, 1374 - 1375 (1975).

A search for lunar tides in densities derived from satellite drag has yielded spurious oscillations with the period of one lunar day, caused by the '27-day' variation connected with the solar rotation. The real tides remain undetected, with an amplitude smaller than 1 or 2%.

082.056 The abundance of metal chlorides in the surface layers of the earth and Venus. M. H. Hart.
Bull. American Astron. Soc., Vol. 7, 375 (1975). − Abstr. AAS.

082.057 On the possibility of determination of the transparency of the atmosphere by satellite photometry.
M. V. Bratijchuk, Ya. M. Motrunich.
Problems of cosmic physics. Vyp. (No.) 9, (see 003.013), p. 72 - 79 (1974). In Russian.

The possibility of study of the transparency of the atmosphere by means of electrophotometric observations of spherical satellites with specular reflecting coating of the surface was analysed. The coefficients of extinction and atmospherical transparency for south and north parts of the sky have been derived from the results of electrophotometric observations of Explorer 19. They are in agreement with the results of investigations obtained by star photometry.

082.058 Superrotation und zweidimensionale Turbulenz.
G. Rüdiger.
Gerlands Beiträge Geophys., Vol. 84, 92 - 98 (1975).

The earth's atmosphere is supposed to be represented by a rotating spherical fluid layer in which two-dimensional weak turbulent motions perpendicular to the radial direction occur. By using only Bochner's theorem the phenomenon of superrotation is obtained. The horizontal velocities necessary for explaining the measured superrotation rates are of the order of the observed values.

082.059 Proton energy deposition in molecular and atomic oxygen and applications to the polar cap.
B. C. Edgar, H. S. Porter, A. E. S. Green.
Planet. Space Sci., Vol. 23, 787 - 804 (1975).

The authors extend the microscopic approach developed for proton energy deposition in N_2 to the general case of proton bombardment of a polar atmosphere composed of N_2, O_2 and O. They calculate the volume emission rates of various N_2^+, O_2, O_2^+ and OI emissions that would be encountered in a typical PCA event.

082.060   Observations of OI 7774 emission excited by conjugate photoelectrons.   A. B. Christensen.
Planet. Space Sci., Vol. 23, 831 - 842 (1975).

082.061   Systematic winds at heights between 350 and 675 km from analysis of the orbits of four balloon satellites.   J. W. Slowey.
Planet. Space Sci., Vol. 23, 879 - 886 (1975).

082.062   Comment on "Superrotation of the upper atmosphere by global deposition of meteoroids".
B. A. McIntosh.
Planet. Space Sci., Vol. 23, 891 - 892 = Astrophys. Branch, National Research Council of Canada, Ottawa, NRCC No. 14451 (1975). — See 11.082.009.

082.063   Superrotation of the thermosphere by global deposition of meteoroids?
P. W. Blum, D. W. Hughes, K. G. H. Schuchardt.
Planet. Space Sci., Vol. 23, 892 - 896 (1975).

082.064   Results of some airglow observations of internal gravitational waves.
V. I. Krassovsky, K. I. Kuzmin, N. A. Piterskaya, A. I. Semenov, M. V. Shagaev, N. N. Shefov, T. I. Toroshelidze.
Planet. Space Sci., Vol. 23, 896 - 898 (1975).
Data on internal gravitational waves at heights of 90–100 km obtained from the intensities and rotational temperature of atmospheric emissions are compared with data on such waves at heights of 200–400 km obtained by radio methods. The difference in the characteristics of waves at different heights is discussed.

082.065   Dynamics and structure of the quiet thermosphere.
K. S. W. Champion.
Journ. Atmosph. Terr. Phys., Vol. 37, 915 - 926 (1975).
Review paper presented at International Solar-Terrestrial Physics Symposium, São Paulo, 1974 June 17 - 24.

082.066   Optical observations of the thermosphere.
J. Blamont.
Journ. Atmosph. Terr. Phys., Vol. 37, 927 - 938 (1975).
Review paper presented at International Solar-Terrestrial Physics Symposium, São Paulo, 1974 June 17 - 24.
This paper reviews the work performed in the last three years to the optical study of the thermosphere that is, of the region of the atmosphere above 120 km, excluding all phenomena originating at the mesopause and auroral phenomena.

082.067   Energy deposition in the thermosphere caused by the solar wind.   K. D. Cole.
Journ. Atmosph. Terr. Phys., Vol. 37, 939 - 949 (1975).
Review paper presented at International Solar-Terrestrial Physics Symposium, São Paulo, 1974 June 17 - 24.
The subject of joule dissipation of ionospheric currents and movement of the thermosphere by electric fields is briefly reviewed and the depositions of energy into the thermosphere by charged particles and electric fields are compared. A simple explanation of the correlation of magnetic disturbances and weather in the lower atmosphere, propounded earlier by the author, is included here.

082.068   On seasonal and diurnal changes of anomalies of astronomical refraction.   N. A. Vasilenko.

Astrometriya i Astrofizika, Kiev, vyp. (No.) 25, (see 003.015), p. 98 - 110 (1975). In Russian.
On the basis of measuring the astronomical refraction in 1968–1972 at fixed zenith distances of 80–90°, the deviations of the measured refraction from the table values are noted to be of seasonal and diurnal character. These deviations are connected with non-consideration of the changes in density gradients in the lower atmospheric layer, the effect of inclination of the air layers of the same density, and with the discrepancy between the "refraction temperature coefficient" in the formula of reductions for temperature and its true value.

082.069   Extinction parameters on Terskol Peak and investigation of the instrumental system of the photometer.   A. F. Pugach, R. R. Kondratyuk, A. Eh. Rozenbush.
Astrometriya i Astrofizika, Kiev, vyp. (No.) 25, (see 003.015), p. 111 - 114 (1975). In Russian.
Extinction coefficients in the UBV bands for the Terskol Peak (Elbrus, h = 3100 m) were obtained using photoelectric observations of standard stars during 1972–1973. Some remarks concerning application of mean rather than individual values are made.

082.070   Influence of exposure time on spectral properties of turbulence-degraded astronomical images.
C. Roddier, F. Roddier.
Journ. Optical Soc. America, Vol. 65, 664 - 667 (1975).
Mathematical expressions are obtained for the power spectrum of a turbulence-degraded image of a point source as a function of exposure time. Calculations are based on the near-field Rytov-Kolmogorov approximations and the Taylor hypothesis. Comparisons with available data are satisfactory.

082.071   New solutions of the refraction integral.
R. White.
Journ. Optical Soc. America, Vol. 65, 676 - 678 (1975).
In a spherically symmetric inhomogeneous system, certain relations between refractive index and distance from the center of symmetry give closed solutions in elementary functions of the refraction integral. A careful formulation of the fundamental equations reveals that all such solutions reduce to special cases of one of eight standard forms, and that the theory of plane-parallel systems is mathematically identical. One special case gives a better representation of the earth's atmosphere than the traditionally used hypotheses of Cassini and Simpson.

082.072   Twilight airglow. 3. [O I] 6300-Ångström radiation.
D. W. Rusch, W. E. Sharp, P. B. Hays.
Journ. Geophys. Res., Vol. 80, 1832 - 1836 (1975).
A coordinated rocket experiment measuring the red line of atomic oxygen at 6300 Å, the photoelectron flux, and the ionospheric densities of O_2^+ and n_e was flown into the twilight mid-latitude airglow over White Sands, New Mexico. Analysis of the 6300-Å emission rate profile shows that photoelectron impact excitation of atomic oxygen is the main source of $O(^1D)$ atoms above 240 km.

082.073   Spectroscopic measurement of sodium dayglow: absence of a large diurnal variation.
C. R. Burnett, R. W. Lasher, A. S. Miskin, V. L. Sides.
Journ. Geophys. Res., Vol. 80, 1837 - 1844 (1975).

082.074   A view of earth and air.   D. G. King-Hele.
Phil. Trans. Roy. Soc. London, Ser. A, Vol. 278, (No. 1277), p. 67 - 109 (1975). — The Bakerian lecture, 1974.
The theme of this lecture is the earth and its atmosphere as viewed by an interstellar sightseer who arrives in the solar system, able to see to the limits of present human knowledge but no further, and able to turn the clock back but nor forward.

082.075 Two rocket experiments for measuring ionizing
radiation in the upper atmosphere.
T. V. Kazachevskaya, V. V. Selant'ev.
Geomagn. Aeronom., Vol. 15, 565 - 567 (1975). In Russian.
Brief information.

082.076 Spectral analysis of variations of green emission of
the night sky.   K. I. Kuz'min.
Geomagn. Aeronom., Vol. 15, 567 - 568 (1975). In Russian.
Brief information.

082.077 Statistical characteristics of vertical turbulent velocity observed in atmospheric lower boundary layer.
I. Naito.
Publ. International Latitude Obs. Mizusawa, Vol. 9, 89 - 131
(1973).

082.078 Some effects of the local character in the atmospheric structure on the astrometry (II).
T. Gotõ, N. Kikuchi.
Proc. International Latitude Obs. Mizusawa, No. 14, p. 30 -
38 (1974). In Japanese.
Standard deviation of latitude observations with the VZT
at Mizusawa is studied in connection with the coefficient of
stability in the atmosphere. The coefficient of stability in the
atmosphere is obtained by aerological data of Akita.

082.079 On the characteristics of the astronomical refraction
in the northern hemisphere.
C. Sugawa, N. Kikuchi.
Proc. International Latitude Obs. Mizusawa, No. 14, p. 145 -
162 (1974). In English and Japanese.
It has been reconfirmed that the amount of astronomical
refraction closely depends on surface temperature.

082.080 On the nature and mechanisms of 17^+, 18^+, 19^+
and 44^+ ion formation in the upper atmosphere.
G. M. Martynkevich.
Dokl. Akad. Nauk SSSR. Ser. Mat., Fiz., Vol. 222, 72 - 75
(1975). In Russian.

082.081 Atmospheric heating by solar EUV radiation.
R. S. Stolarski, P. B. Hays, R. G. Roble.
Journ. Geophys. Res., Vol. 80, 2266 - 2276 (1975).
The diurnal variation of the neutral gas heating efficiency
due to photo-ionization processes $\lambda \leqslant 1025$ Å is calculated by
using a diurnal model of the mid-latitude ionospheric F-region.
The neutral gas heating efficiency due to EUV solar heating
below 400 km lies between 30 and 35%. The calculated electron gas heating efficiency, below 250 km, where photoelectron heating predominates, is roughly 5%, and the ion heating
efficiency is about 2% with respect to the absorbed EUV solar
energy.

082.082 The N I (5200 Å) dayglow.   D. W. Rusch, A. I.
Stewart, P. B. Hays, J. H. Hoffman.
Journ. Geophys. Res., Vol. 80, 2300 - 2304 (1975).

082.083 Rocket investigations of corpuscular radiation at
different geomagnetic latitudes. I. Complex of the
apparatus for investigations of corpuscular radiation aboard
MR-12 meteorological rockets.
V. F. Tulinov, V. M. Fejgin, Yu. M. Zhuchenko, V. A.
Lipovetskij, L. S. Novikov, V. A. D'yachenko, A. P. Babaev,
T. A. Zhuchenko.
Kosmich.Issled., Vol. 13, 361 - 366 (1975). In Russian.

082.084 Variations in air density, satellite drag coefficient
and atmospheric rotation rate from analysis of the
orbit of 1966-92D.   B. R. Bowman.
Planet. Space Sci., Vol. 23, 1003 - 1010 (1975).

082.085 On the relations between anomalous refraction and
aerological data (I). R. Fukaya.
Tokyo Astron. Obs., Report No. 66, Vol. 17, 357 - 364 (1975).
In Japanese.

082.086 Halo's.   T. Joosten.
Zenit, Vol. 2, 223 - 226 (1975).

082.087 Unusual chemical compositions of noctilucent-
cloud particle nuclei.   C. L. Hemenway.
Evolutionary and physical properties of meteoroids. IAU colloquium No. 13, p. 287 - 290 = Dudley Obs., Albany, New
York, Repr. No. B46 (1973).
On August 8, 1970, two Pandora sounding rocket pay-
loads were launched from the ESRO range in Kiruna, Sweden
during a noctilucent cloud display. Large numbers of sub-mi-
cron particles were collected, most of which appear to be made
up of a high-density material coated with a low-density material. Typical electron micrographs are shown. Particle chemical
compositions have been measured.

082.088 Destruction de l'ozone de la haute atmosphère par
les oxydes d'azote au moment du lever du soleil.
P. Rigaud, O. Steiger, D. Huguenin.
Comptes Rendus Séances, SPHN Genève, Vol. 9, Fasc. 1 - 3,
p. 84 - 90 (1974) = Publ. Obs. Genève, Sér. A, Fasc. 81/II
(1975).

082.089 Atmospheric transmittance in the far infrared at
Testa Grigia.
P. P. Lombardini, F. Melchiorri, G. Salio, L. Dall'Agnola.
Infrared Physics, Vol. 15, 73 - 78 (1975).
Atmospheric transmittance in the band 300 – 2000 μm
was measured in the far i.r. solar observatory of Testa Grigia,
3480 m above standard. These measurements are compared
with meteorological data observed at Plateau Rosa.

082.090 Some results of studying the regularities of change
in optical thicknesses of the atmosphere.
Yu. A. Gongadze, R. G. Indzhgiya, A. E. Mikirov.
Soobshch. AN GruzSSR, Vol. 75, No. 2, p. 321 - 324 (1974).
In Russian. – Abstr. in Referativ. Zhurn. 51. Astron., 6.51.129
(1975).

082.091 Artificial satellites and the upper atmosphere.
I. V. Almar.
Nauka i chelovechestvo. 1975. Moskva, Znanie, 1974, p. 249 -
255. In Russian. – Abstr. in Referativ. Zhurn. 62. Issled.
kosmich. prostranstva, 6.62.250 (1975).

082.092 Investigation of the mechanism of variation of the
density of the atmosphere.
L. S. Khizhak, P. G. Chernyaga.
L'vov. politekhn. in-t. L'vov, 1974. 12 pp. In Russian. – Abstr.
in Referativ. Zhurn. 62. Issled. kosmich. prostranstva, 6.62.
251 (1975).

082.093 La qualité des images téléscopiques à Locarno-
Monti. S. Cortesi.
Astron. Mitt. Eidgenössisch. Sternw. Zürich, No. 334, 16 pp.
(1974).

082.094 Refraction anomalies.   A. I. Nefed'eva.
19th Astrometrical Conference 1972, (see 012.019),
p. 110 - 116 (1975). In Russian.

082.095 The atmospheric extinction at Cerro Tololo,
1969 - 1971.   H. Moreno, J. Maza.
Dep. Astron., Univ. Chile, Obs. Astron. Nacional, Cerro Calán,
Santiago de Chile, Publ. Vol. 2, (No. 4), 157 - 158 (1974).
The values of the monochromatic extinction for the

wavelength interval 3100 Å to 5800 Å are presented. These values have been obtained at the Cerro Tololo Interamerican Observatory for six periods of observations.

Radio meteor investigations of the upper atmosphere circulation.   See Abstr. 003.028.

The physics of mesospheric (noctilucent) clouds. See Abstr. 012.024.

Structure and dynamics of the upper atmosphere. See Abstr. 012.026.

Absolute integrated intensity and individual line parameters for the 6.2 μ band of NO_2. See Abstr. 022.006.

Statistical-band-model analysis and integrated intensity for the 21.8 μm bands of HNO_3 vapor. See Abstr. 022.033.

Photon noise and atmospheric noise in active optical systems.   See Abstr. 031.027.

Die Sichtbarkeit von Sternen und Planeten bei Tageslicht, in der Dämmerung und bei Beobachtungen aus tiefen Schächten.   See Abstr. 031.028.

Automatic solar image motion measurements. See Abstr. 031.229.

A balloon borne liquid nitrogen cooled infrared radiometer.   See Abstr. 034.018.

First rocket experiments for investigation of the upper atmosphere.   See Abstr. 051.017.

Rotational motion of artificial earth satellites and determination of the atmosphere's density. See Abstr. 052.069.

The rotational motion of a satellite and atmospheric density determination.   See Abstr. 054.013.

Der Einfluss der Szintillation auf den Faktor k der Sonnenfleckenrelativzahl.   See Abstr. 072.020.

Zodiacal light photopolarimetry. I. Observations, reductions, disturbing phenomena, accuracy. See Abstr. 106.017.

On the connection of the sectorial structure of the interplanetary magnetic field with the indices of zonal circulation.   See Abstr. 106.036.

Influence of the sectorial structure of the interplanetary magnetic field on the circulation of the earth's atmosphere.   See Abstr. 106.039.

Errata

082.901 Errata: "Fluorescence efficiencies of electrons in second positive bands of N_2 and first negative bands of N_2^+" [Planet. Space Sci., Vol. 21, 1381 - 1387 (1973)]. S. P. Khare, A. Kumar. Planet. Space Sci., Vol. 23, 904 (1975). – See 10.082.019.

083 Ionosphere

083.001 **A large-scale hole in the ionosphere caused by the launch of Skylab.**
M. Mendillo, G. S. Hawkins, J. A. Klobuchar.
Science, Vol. 187, 343 - 346 (1975).

A dramatic ionospheric phenomenon, unique in magnitude and in spatial and temporal extent, occurred along the Atlantic Coast of North America after the launch of the NASA Skylab Workshop on 14 May 1973. The effect was a large and rapid decrease in the total number of ionospheric electrons within a distance of 1000 kilometers of the burning engines of the Saturn V launch vehicle. The observations are interpreted in terms of exceptionally enhanced chemical loss rates due to the molecular hydrogen and water vapor contained in the Saturn second-stage exhaust plume.

083.002 **ATS-6 radio beacon experiment.** H. Soicher.
Nature, Vol. 253, 252 - 254 (1975). – Letter.

083.003 **Storm-time increases in the ionospheric total electron content.**
R. J. Moffett, J. A. Murphy, G. J. Bailey.
Nature, Vol. 253, 330 - 331 (1975). – Letter.

083.004 **Ion composition of the F2-layer and outer ionosphere.** M. N. Fatkullin.
Issled. obl. F i vnesh. ionosfery. Moskva, 1974, p. 8 - 49. In Russian. – Abstr. in Referativ. Zhurn. 51. Astron., 2.51.536 (1975).

083.005 **On the height extent of the seasonal anomaly in the mid-latitude outer ionosphere at high solar activity.**
M. N. Fatkullin, N. M. Boenkova, A. D. Legen'ka.
Issled. obl. F i vnesh. ionosfery. Moskva, 1974, p. 50 - 57. In Russian. – Abstr. in Referativ. Zhurn. 51. Astron., 2.51.537 (1975).

083.006 **Daily and latitudinal development of the seasonal anomaly in the lower part of the F-region at high solar activity.** N. M. Boenkova.
Issled. obl. F i vnesh. ionosfery. Moskva, 1974, p. 58 - 68. In Russian. – Abstr. in Referativ. Zhurn. 51. Astron., 2.51.538 (1975).

083.007 **Theoretical models of the mid-latitude F2-region at summer season during low solar activity.**
M. N. Fatkullin, A. D. Legen'ka, A. Muradov.
Issled. obl. F i vnesh. ionosfery. Moskva, 1974, p. 177 - 206. In Russian. – Abstr. in Referativ. Zhurn. 51. Astron., 2.51.539 (1975).

083.008 **F2-region and outer ionosphere under disturbed conditions.** M. N. Fatkullin.
Issled. obl. F i vnesh. ionosfery. Moskva, 1974, p. 216 - 251. In Russian. – Abstr. in Referativ. Zhurn. 51. Astron., 2.51.540 (1975).

083.009 **Magnetospheric ring current as possible source of disturbance of the ionospheric F2-layer during geomagnetic storms.** N. A. Najdenova.
Issled. po geomagnetizmu, aehron. i fiz. Solntsa. Vyp. (No) 32. Moskva, Nauka, 1974, p. 148 - 152. In Russian. – Abstr. in Referativ. Zhurn. 62. Issled. kosmich. prostranstva, 2.62. 291 (1975).

083.010 **Changes in the ionospheric profile and the Faraday factor M with Kp.**
M. D. Papagiannis, H. Hajeb-Hosseinieh, M. Mendillo.

Planet. Space Sci., Vol. 23, 107 - 113 (1975).

083.011 **Electron content power spectral estimates: periods of 2 days to $^1/_2$ year.** N. C. Low, T. H. Roelofs, P. C. Yuen.
Planet. Space Sci., Vol. 23, 133 - 141 (1975).

083.012 **Theory of photoelectron thermalization and transport in the ionosphere.** G. P. Mantas.
Planet. Space Sci., Vol. 23, 337 - 354 (1975).

083.013 **Atmospheric gravity waves: their effects on ionospheric temperature and composition.**
S. S. Prasad, D. R. Furman.
Journ. Atmosph. Terr. Phys., Vol. 37, 17 - 30 (1975).

083.014 **A new form of representation of the diurnal and solar-cycle variations of ionospheric absorption.**
J. C. Samuel, P. A. Bradley.
Journ. Atmosph. Terr. Phys., Vol. 37, 131 - 141 (1975).

083.015 **On the latitude of the focus of the L current system.**
A. Palumbo.
Journ. Atmosph. Terr. Phys., Vol. 37, 153 - 159 (1975).

The ionospheric current system associated with L, the lunar daily geomagnetic variation, is characterized in the sunlit hemisphere by two northern and two southern foci. An attempt is made to deduce the latitudes of these foci directly from observatory data. This is done by noting the sign and amplitude of the horizontal component of L when the focus is on the meridian.

083.016 **Incoherent scatter measurements of E- and upper D-region ionization changes during three solar flares.**
G. N. Taylor.
Journ. Atmosph. Terr. Phys., Vol. 37, 349 - 357 (1975).

The use of incoherent scatter radar for observing sudden ionospheric disturbances is discussed. Measurements of electron densities at heights between 80 and 130 km during three solar flares are presented, and compared with observations of the simultaneous optical, radio and soft X-ray events. It is not possible to describe the relationship between the magnitudes of the X-ray and ionospheric events uniquely.

083.017 **The equatorial helium ion trough and the geomagnetic anomaly.** S. Chandra.
Journ. Atmosph. Terr. Phys., Vol. 37, 359 - 367 (1975).

083.018 **The twilight D-region and metallic ionization.**
M. P. Gough.
Journ. Atmosph. Terr. Phys., Vol. 37, 565 - 568 (1975).
Short paper.

083.019 **Power spectra of large scintillation signals.**
C. L. Rufenach.
Journ. Atmosph. Terr. Phys., Vol. 37, 569 - 572 (1975).

Scintillation spectra using radio stellar sources at the time of large scintillation levels show a monotonically decreasing spectral density with increasing frequency in contrast to a spectral flatness for $\nu < \nu_f$ previously reported for small levels.

083.020 **Ionospheric solar flare effect observations.**
W. C. Bain, E. Hammond.
Journ. Atmosph. Terr. Phys., Vol. 37, 573 - 574 (1975).

The VLF sudden phase anomaly measured near vertical incidence is shown to be the most efficient measure of flare effects. Its relation to solar X-ray bursts is also described.

083.021 Ionospheric mid-latitude trough and the abrupt scintillation boundary.
L. Kersley, A. P. van Eyken, K. J. Edwards.
Nature, Vol. 254, 312 - 313 (1975). − Letter.

083.022 Wave-particle interactions in the magnetosphere and ionosphere. R. M. Thorne.
Rev. Geophys. Space Phys., Vol. 13, 291 - 302 (1975).

083.023 Relations entre les anomalies brusques de phase en ondes très longues et les sursauts du rayonnement X solaire. A. Kimpara.
Comptes Rendus Acad. Sci. Paris, Sér. B, Vol. 280, 323 - 324 (1975).
L'auteur étudie les relations entre les anomalies brusques de phase (SPA) observées sur les signaux de l'émetteur NWC (22,3 kHz) et les flux dans diverses parties du spectre X solaire pendant l'intervalle géophysique mondial 26 juillet−14 août 1972.

083.024 Equations for magnetospheric convection and a solution for polar cap flows. G. Atkinson.
Journ. Geophys. Res., Vol. 80, 32 - 36 (1975).

083.025 Investigations in the ionosphere with the help of satellite Cosmos 378. 3. Study of electron streams within the energy range $0.5-12$ keV.
A. P. Remizov, M. Z. Khokhlov.
Geomagn. Aeronom., Vol. 15, 3 - 9 (1975). In Russian.

083.026 On a statistical model of geophysical processes with excess and asymmetry of the density probability function. I. S. Vsekhsvyatskaya.
Geomagn. Aeronom., Vol. 15, 46 - 51 (1975). In Russian.

083.027 On the influence of variations of neutral composition of the upper atmosphere on the structure of the lower part of the mid-latitude F-region.
O. A. Mal'tseva, M. N. Fatkullin, T. I. Zelenova.
Geomagn. Aeronom., Vol. 15, 56 - 60 (1975). In Russian.

083.028 Theoretical models of altitude distribution of electron concentrations and various ion components in the lower part of the mid-latitude F-region during negative disturbances. II.
M. N. Fatkullin, O. A. Mal'tseva, T. I. Zelenova.
Geomagn. Aeronom., Vol. 15, 61 - 66 (1975). In Russian.

083.029 Generation of small-scale irregularities in a turbulent plasma and radio aurora.
A. V. Volosevich, V. A. Liperovskij.
Geomagn. Aeronom., Vol. 15, 74 - 77 (1975). In Russian.

083.030 Planetary distribution of the 27-day variation of the electron concentration at the maximum of the F-layer of the ionosphere. A. M. Chkhetiya.
Geomagn. Aeronom., Vol. 15, 160 - 161 (1975). In Russian. Brief information.

083.031 Global evolution of the ionospheric electron content during some geomagnetic storms.
R. P. Kane.
Journ. Atmosph. Terr. Phys., Vol. 37, 601 - 611 (1975).

083.032 Ionospheric electron content and equivalent slab thickness in the equatorial region.
A. Das Gupta, S. Basu, J. N. Bhar, J. C. Bhattacharyya.
Journ. Geophys. Res., Vol. 80, 699 - 701 (1975).
The ionospheric total electron content and equivalent

slab thickness in the equatorial region around 75°E have been studied. It is found that the slab thickness exhibits a strong latitudinal dependence with a maximum at the magnetic equator and a minimum around the crest of the F_2 anomaly, indicating that the F region is thick near the equator and most skewed near the crest. The results are consistent with the electrodynamic drift and diffusion in the equatorial ionosphere.

083.033 Comparative studies of E-region ionospheric drifts and meteor winds.
D. G. Felgate, A. N. Hunter, S. P. Kingsley, H. G. Muller.
Planet. Space Sci., Vol. 23, 389 - 400 (1975).
Ionospheric drifts using total reflections from the E-region have been compared with neutral winds measured by meteor radar. Close agreement was found when both measurements were made in a common volume of atmosphere. It is concluded that the drift technique does measure the movement of the neutral atmosphere in the altitude range $95-120$ km. The agreement between measurements from widely separated regions indicates the horizontal scale of the wind structure is at least 700 km.

083.034 Daytime valley in the F_1-region observed by incoherent scatter.
C. Taieb, G. Scialom, G. Kockarts.
Planet. Space Sci., Vol. 23, 523 - 531 (1975).
Incoherent scatter data obtained at Saint-Santin have been analyzed between 1969 and 1972 in the height interval $120-200$ km. The summer daytime electron density profiles show a maximum around 150 km followed by a minimum around 165 km, when the solar activity is low. The seasonal and solar cycle variations of this phenomenon are discussed. When the day-time valley exists, the maximum density is $ca.$ $20-30$ per cent greater than the minimum value.

083.035 Interpretation of single-site scintillation measurements to estimate F-region drift velocities.
C. L. Rufenach, J. H. Pope.
Planet. Space Sci., Vol. 23, 560 - 562 (1975).
A method of estimating ionospheric drift velocities using single-site scintillation measurements is applied to determine a correlation coefficient of 0.55 between magnetic activity and F-region drift velocity near the auroral ionosphere. This method is based on the relationship between the drift velocity and the scintillation spectral breakpoint.

083.036 Effects of photoelectron heating and interhemisphere transport on day-time plasma temperatures at low latitudes.
G. J. Bailey, R. J. Moffett, W. E. Swartz.
Planet. Space Sci., Vol. 23, 599 - 610 (1975).
The thermal balance of the plasma in the day-time equatorial F region is examined. Steady-state solutions of electron and ion temperatures are obtained, assuming the ions are O^+ and H^+. The theoretical concentrations of O^+ and H^+ and the field-aligned velocity were obtained following Moffett and Hanson (1973), while theoretical photoelectron heating rates of the electron gas were taken from Swartz et al. (1975).

083.037 Ionosphere-magnetosphere coupling. R. A. Wolf.
Space Sci. Rev., Vol. 17, 537 - 562 (1975). − Review paper − see 012.002.
The large-scale electrical coupling between the ionosphere and magnetosphere is reviewed, particularly with respect to behavior on time scales of hours or more. The following circuit elements are included: (1) the magnetopause boundary layer, which serves as the generator for the magnetospheric-convection circuit; (2) magnetic field lines, usually good conductors but sometimes subject to anomalous resistivity; (3) the ionosphere, which can conduct current across magnetic field lines; (4) the magnetospheric particle distributions, in-

cluding tail current and partial-ring currents.

083.038 The relationship between sudden ionospheric disturbances and the magnetic structure of sunspot groups. M. Song.
Acta Astron. Sinica, Vol. 15, 159 - 172 (1974). In Chinese.
Data of the sunspot groups and sudden ionospheric disturbances (SID) during 1959—60 have been statistically analysed. It is found that there is a good statistical relationship between the scattering of magnetic polarities of spot groups and SIDs. This relation can be used to estimate SID activity of spot groups several days ahead. The connection of SIDs with the morphological development of spot groups is also discussed.

083.039 Ionospheric effects of X-ray source Scorpius XR-1.
I. G. Poppoff, R. C. Whitten, D. S. Willoughby.
Journ. Atmosph. Terr. Phys., Vol. 37, 835 - 840 (1975).
A comment is made regarding the problem of matching ionospheric theory with reported X-ray star effects on terrestrial radio propagation. It is concluded that the observations are not readily explained by the X-ray hypothesis.

083.040 Electron production in the lower ionosphere by a diffuse galactic X-ray background.
D. A. Gagliardini, H. Karszenbaum.
Journ. Atmosph. Terr. Phys., Vol. 37, 845 - 849 (1975).
The electron production rates in the lower ionosphere, due to galactic X-rays, are computed. It is shown that the contribution of photons of energies greater than 10 keV is relevant below 80 km.

083.041 On the distribution of electron-ion gas in the F-region of the ionosphere. A. G. Khantadze.
Dokl. Akad. Nauk SSSR, Ser. Mat. Fiz., Vol. 221, 839 - 841 (1975). In Russian.

083.042 Latitude dependence of ionosphere total electron content: observations during sudden commencement storms. L. J. Lanzerotti, L. L. Cogger, M. Mendillo.
Journ. Geophys. Res., Vol. 80, 1287 - 1306 (1975).

083.043 On the importance of doubly charged ions in the auroral ionosphere. S. S. Prasad, D. R. Furman.
Journ. Geophys. Res., Vol. 80, 1360 - 1362 (1975).
Brief report.

083.044 Small ion components in the disturbed ionosphere within heights between 100 and 200 km.
V. N. Dyadichev, S. I. Kozlov.
Kosmich. Issled., Vol. 13, 242 - 248 (1975). In Russian.

083.045 Ionospheric spread-F at Huancayo, sunspot activity and geomagnetic activity. G. G. Bowman.
Planet. Space Sci., Vol. 23, 899 - 903 (1975). — Correction of 12.083.057.

083.046 Investigations in the ionosphere with the help of satellite Cosmos 378. 4. Structure of the regions of recording of electrons with energies of 0.5—12 keV and their convection. M. Z. Khokhlov.
Geomagn. Aeronom., Vol. 15, 207 - 213 (1975). In Russian.

083.047 Investigations in the ionosphere with the help of satellite Cosmos 378. 5. Anisotropy of electron streams of 0.5—12 keV at high latitudes. M. Z. Khokhlov.
Geomagn. Aeronom., Vol. 15, 214 - 220 (1975). In Russian.

083.048 Absorption of radio waves in the ionosphere of temperate latitudes during the cycle of solar activity.
V. V. Belikovich, E. A. Benediktov, A. V. Tolmacheva.

Geomagn. Aeronom., Vol. 15, 251 - 254 (1975). In Russian.

083.049 Variations of concentration of O^+ ions at an altitude of 200 km.
G. S. Ivanov-Kholodnyj, L. A. Antonova.
Geomagn. Aeronom., Vol. 15, 255 - 259 (1975). In Russian.

083.050 Relative ion concentration in the upper ionosphere from data of the artificial earth satellite Intercosmos 5. Ya. I. Likhter, Ya. P. Sobolev.
Geomagn. Aeronom., Vol. 15, 281 - 285 (1975). In Russian.

083.051 Influence of moving irregularities in the ionosphere on the irregular refraction of Cygnus-A.
M. A. Ovsyankin, S. P. Chernysheva, V. M. Sheftel', Eh. G. Shcharenskaya.
Geomagn. Aeronom., Vol. 15, 353 - 354 (1975). In Russian.
Brief information.

083.052 Calculation of N(h) profiles of the ionosphere by the method of quadratic programming.
N. P. Danilkin, P. F. Denisenko, V. V. Sotskij.
Geomagn. Aeronom., Vol. 15, 355 - 357 (1975). In Russian.
Brief information.

083.053 Investigation of the anisotropy of ionospheric irregularities in inclined sounding.
V. D. Gusev, T. A. Gajlit, V. M. Ostrovskij.
Geomagn. Aeronom., Vol. 15, 364 - 366 (1975). In Russian.
Brief information.

083.054 Ionization-recombination cycle of the D-region.
A. D. Danilov.
Journ. Atmosph. Terr. Phys., Vol. 37, 885 - 894 (1975).
Review paper presented at International Solar-Terrestrial Physics Symposium, São Paulo, 1974 June 17 - 24.

083.055 D-region in disturbed conditions, including flares and energetic particles. A. P. Mitra.
Journ. Atmosph. Terr. Phys., Vol. 37, 895 - 913 (1975).
Review paper presented at International Solar-Terrestrial Physics Symposium, São Paulo, 1974 June 17 - 24.

083.056 Effective electron loss coefficient of the disturbed daytime D region. W. Swider, W. A. Dean.
Journ. Geophys. Res., Vol. 80, 1815 - 1819 (1975).
The effective electron loss coefficient of the D region is determined as a function of height for four daytime instances over a 49-hour time span during the November 2 - 5, 1969, solar proton event.

083.057 Techniques, methods and results of complex investigations of the lower ionosphere.
V. A. Misyura, S. S. Shlyuger, Yu. K. Chasovitin, L. A. Piven', L. F. Chernogor, V. G. Somov.
Kosmich. issled. na Ukraine. Resp. mezhved. sb., 1974, vyp. (No.) 5, p. 63 - 67. In Russian. — Abstr. in Referativ. Zhurn. 62. Issled. kosmich. prostranstva, 5.62.140 (1975).

083.058 Investigations in the ionosphere with the help of satellite Cosmos 378. 6. Characteristics of the ionosphere and of precipitating particles and their connection with geophysical phenomena during magnetospheric substorms.
K. I. Gringauz, G. L. Gdalevich, O. P. Kolomijtsev, N. G. Klejmenova, O. M. Raspopov.
Geomagn. Aeronom., Vol. 15, 425 - 433 (1975). In Russian.

083.059 Dynamical processes in the outer ionosphere.
R. V. Gostrem, M. A. Nikitin.
Geomagn. Aeronom., Vol. 15, 434 - 441 (1975). In Russian.

083.060 On day-time variations of VLW-fields, defined by the dependence of electron concentration of the lower ionosphere on the zenith angle of the sun.
G. V. Azarnin, A. B. Orlov, N. N. Sazeeva.
Geomagn. Aeronom., Vol. 15, 462 - 466 (1975). In Russian.

083.061 Variations of $[NO^+]/[O_2^+]$ in the day-time at altitudes of 130 - 200 km.
G. S. Ivanov-Kholodnyj, L. A. Antonova.
Geomagn. Aeronom., Vol. 15, 477 - 482 (1975). In Russian.

083.062 Oxygen emission in the lower ionosphere in the field of a strong electromagnetic wave.
G. M. Milikh.
Geomagn. Aeronom., Vol. 15, 491 - 496 (1975). In Russian.

083.063 Echo 2: a study of electron beams injected into the high-latitude ionosphere from a large sounding rocket.   J. R. Winckler, R. L. Arnoldy, R. A. Hendrickson.
Journ. Geophys. Res., Vol. 80, 2083 - 2088 (1975).

083.064 Variations in ion composition at middle and low latitudes from Isis 2 satellite.
E. L. Breig, J. H. Hoffman.
Journ. Geophys. Res., Vol. 80, 2207 - 2216 (1975).

083.065 A sudden vanishing of the ionospheric F region due to the launch of Skylab.
M. Mendillo, G. S. Hawkins, J. A. Klobuchar.
Journ. Geophys. Res., Vol. 80, 2217 - 2228 (1975).

083.066 Remote sensing of the ionospheric F layer by use of O I 6300-Å and O I 1356-Å observations.
S. Chandra, E. I. Reed, R. R. Meier, C. B. Opal, G. T. Hicks.
Journ. Geophys. Res., Vol. 80, 2327 - 2332 (1975).

083.067 Determination of F region height and peak electron density at night using airglow emissions from atomic oxygen.   B. A. Tinsley, J. A. Bittencourt.
Journ. Geophys. Res., Vol. 80, 2333 - 2337 (1975).

083.068 Streams of high-energy electrons at heights between 200 and 300 km from data of the artificial earth satellite Cosmos 264.   A. M. Gal'per, A. F. Iyudin, B. I. Luchkov, Yu. V. Ozerov, V. T. Samojlenko.
Kosmich. Issled., Vol. 13, 437 - 439 (1975). In Russian.
Brief information.

083.069 Dissociative recombination contributions to I (5577) and I(6300) [O I] and $N_m F2$ enhancements resulting from moderate proton PCA events.
H. S. Porter, B. C. Edgar, A. E. S. Green.
Planet. Space Sci., Vol. 23, 935 - 944 (1975).

083.070 Correlation between the intensity of emission at 6300 Å, temperature and parameters of the ionosphere.   A. I. Semenov, N. N. Shefov.
Planet. Space Sci., Vol. 23, 1013 - 1014 (1975).
The measured and calculated intensities of the 6300 Å emission are compared on the basis of the measured Doppler temperature and intensity of the red emission during night and twilight and the F_2 layer parameters. There is satisfactory correlation between them for the quiet geomagnetic periods ($Kp \leqslant 2$).

083.071 Ferraro's contribution to ionospheric research.
J. E. C. Gliddon.
Geophys. Journ. Roy. Astron. Soc., Vol. 41, 309 - 310 (1975).

083.072 On the theory of diffusion in the ionosphere.
H. Rishbeth.

Geophys. Journ. Roy. Astron. Soc., Vol. 41, 311 - 317 (1975).

083.073 Calculated variations in the H^+ content of the plasmasphere.
J. A. Murphy, G. J. Bailey, R. J. Moffett.
Geophys. Journ. Roy. Astron. Soc., Vol. 41, 319 - 325 (1975).

083.074 On a hydrostatical model of the upper ionosphere and a method of determining the mean profiles of electron concentration and temperature with the help of whistlers.   Ya. L. Al'pert, D. S. Fligel'.
Nizkochastot. volny i signaly vo vnesh. ionosfere. Apatity, 1974, p. 85 - 90. In Russian. – Abstr. in Referativ. Zhurn. 62. Issled. kosmich. prostranstva, 6.62.275 (1975).

083.075 Results of measurements of the electron content of the ionosphere and its horizontal gradients according to the propagation of radio waves from satellites.
V. A. Misyura, I. I. Kapanin, V. A. Podnos, O. F. Tyrnov, A. M. Tsymbal, G. N. Zinchenko, A. K. Surkov, V. M. Trubitsyn, V. N. Podnos.
Kosmich. issled. na Ukraine. Resp. mezhved. sb., 1974, vyp. (No.) 5, p. 8 - 13. In Russian. – Abstr. in Referativ. Zhurn. 62. Issled. kosmich. prostranstva, 6.62.276 (1975).

083.076 Statistical characteristics of fluctuations of effects and inhomogeneities of the ionosphere from observations of signals of artificial earth satellites.
V. A. Misyura, N. P. Svetlichnyj, L. V. Bezrodnaya, V. I. Aboltin, N. D. Gerasimova, L. N. Smirnova, V. M. Mokryj, B. V. Zagvozdkin.
Kosmich. issled. na Ukraine. Resp. mezhved. sb., 1974, vyp. (No.) 5, p. 57 - 63. In Russian. – Abstr. in Referativ. Zhurn. 62. Issled. kosmich. prostranstva, 6.62.277 (1975).

Analysis shows Skylab tore hole in ionosphere. Phys. Today, Vol. 28, No. 5, p. 17 - 18 (1975).

Ionospheric direct measurement techniques. See Abstr. 031.203.

Influence of probe surface properties on plasma measurements in the ionosphere and magnetosphere. See Abstr. 031.252.

The effects of body geometry on the structure in the near wake zone of bodies in a flowing plasma. See Abstr. 062.020.

Plasma echo in the ionosphere. See Abstr. 062.038.

Observations of Birkeland currents at auroral latitudes.   See Abstr. 084.005.

A comparison of satellite observations of Birkeland currents with ground observations of visible aurora and ionospheric currents.   See Abstr. 084.009.

Observations of Birkeland currents. See Abstr. 084.012.

Sq variability and aeronomic structure. See Abstr. 084.210.

The response of the day side magnetosphere-ionosphere system to time-varying field line reconnection at the magnetopause. 1. Theoretical model. See Abstr. 084.259.

The response of the day side magnetosphere-ionosphere system to time-varying field line reconnection at the magnetopause. 2. Erosion event of March 27, 1968. See Abstr. 084.260.

The cause of storm after effects in the middle latitude D-region.   See Abstr. 084.407.

084   Aurorae, Geomagnetic Field, Radiation Belts

Aurorae

084.001   **The continuity of the auroral oval in the afternoon sector.**   A. L. Snyder, S.-I. Akasofu, D. S. Kimball.
Planet. Space Sci., Vol. 23, 225 - 227 (1975).
United States Air Force DMSP satellite auroral images and South Pole auroral all-sky camera data are presented to provide new direct evidence that the afternoon sector auroras lie within a continuous oval belt. An example is also presented to illustrate that afternoon substorm auroras occur within the auroral oval configuration.

084.002   **Dynamics of day and night aurora during substorms.**   V. G. Vorobjev (*Vorob'ev*), G. Gustafsson, G. V. Starkov, Y. I. Feldstein (*Ya. I. Fel'dshtejn*), N. F. Shevnina.
Planet. Space Sci., Vol. 23, 269 - 278 (1975).
The motion of auroral forms on the day- and nightside of the earth has been studied during different substorm phases by means of all-sky camera films.

084.003   **Additional remarks on earlier methods of determining the altitude of auroras.**
W. Schröder.
Journ. Atmosph. Terr. Phys., Vol. 37, 375 - 376 (1975).
Short paper.

084.004   **Observations and theory of the formation of stable auroral red arcs.**   M. H. Rees, R. G. Roble.
Rev. Geophys. Space Phys., Vol. 13, 201 - 242 (1975).

084.005   **Observations of Birkeland currents at auroral latitudes.**   H. R. Anderson, R. R. Vondrak.
Rev. Geophys. Space Phys., Vol. 13, 243 - 262 (1975).

084.006   **On the angular distributions of electrons in 'inverted V' substructures.**   P. Venkatarangan, J. R. Burrows, I. B. McDiarmid.
Journ. Geophys. Res., Vol. 80, 66 - 72 (1975).

084.007   **Intensity correlations of the 4278-Å N_2^+ (0-1) first negative band and 5875-Å emissions in auroras.**
R. D. Sears.
Journ. Geophys. Res., Vol. 80, 215 - 218 (1975).

084.008   **Dynamic relationship between proton and electron auroral substorms.**   H. Fukunishi.
Journ. Geophys. Res., Vol. 80, 553 - 574 (1975).

084.009   **A comparison of satellite observations of Birkeland currents with ground observations of visible aurora and ionospheric currents.**
J. C. Armstrong, S.-I. Akasofu, G. Rostoker.
Journ. Geophys. Res., Vol. 80, 575 - 586 (1975).
Observations of Birkeland (field aligned) currents inferred from magnetic field perturbations observed at ~800-km altitude by the Triad satellite are correlated with simultaneous ground-based auroral and magnetic observations. A detailed study of one satellite pass is presented in which the ground-based data are used to locate the position of the substorm-disturbed region with respect to the region of the upper atmosphere through which Triad passed.

084.010   **Auroral emission of the N_2^+ Meinel bands.**
D. C. Cartwright, W. R. Pendleton, Jr., L. D. Weaver.
Journ. Geophys. Res., Vol. 80, 651 - 654 (1975).
New cross sections for electron impact excitation and quenching of the N_2^+ Meinel bands have been combined with recently determined transition probabilities to yield altitude profiles of the Meinel bands relative to those for the N_2^+ first-negative system.

084.011   **Auroral flares and solar flares.**   T. Nagata.
Space Sci. Rev., Vol. 17, 205 - 220 (1975). – Review paper – see 012.002.
The morphology of development of auroral flares (magnetospheric substorms) for both electron and proton auroras is summarized, based on ground-based as well as rocket-borne and satellite-borne data with specific reference to the morphology of solar flares.

084.012   **Observations of Birkeland currents.**
P. A. Cloutier, H. R. Anderson.
Space Sci. Rev., Vol. 17, 563 - 587 (1975). – Review paper – see 012.002.
Recent measurements of precipitating energetic particles and vector magnetic fields from satellites and sounding rockets have verified the existence of geomagnetically-aligned electric currents at high latitudes in the ionosphere and magnetosphere. The spatial and temporal configuration of such currents, now commonly called Birkeland currents, has delineated their role in providing ionospheric closure of magnetospheric current systems, and gross features of these current systems may be understood in terms of theoretical models of magnetospheric convection. The association of Birkeland currents with auroral features on a very small scale suggests that auroral acceleration may result from the current flow.

084.013   **The $L = 6.6$ Oosik barium plasma injection experiment and magnetic storm of March 7, 1972.**
E. M. Wescott, H. C. Stenbaek-Nielsen, T. N. Davis, W. B. Murcray, H. M. Peek, P. J. Bottoms.
Journ. Geophys. Res., Vol. 80, 951 - 967 (1975).

084.014   **Hiss emitting auroral activity.**   T. Oguti.

Journ. Atmosph. Terr. Phys., Vol. 37, 761 - 768
(1975).

084.015 **The effective recombination coefficient measured in the auroral E-region during a sudden commencement electron precipitation event.** A. Brekke.
Journ. Atmosph. Terr. Phys., Vol. 37, 825 - 833 (1975).

084.016 **Electron density decrease in SAR arcs resulting from vibrationally excited nitrogen.**
G. P. Newton, J. C. G. Walker.
Journ. Geophys. Res., Vol. 80, 1325 - 1327 (1975).

Enhanced vibration of molecular nitrogen is produced in SAR arcs by collisions with hot thermal electrons. The rate coefficient for the reaction between oxygen ions and molecular nitrogen is increased as a result of the vibrational excitation, and this leads to an increase in the rate of loss of F region ionization.

084.017 **The earth as a Jovian-like radio source.**
M. L. Kaiser, R. G. Stone.
Bull. American Astron. Soc., Vol. 7, 390 (1975). – Abstr. AAS.

084.018 **A calculation of auroral hiss with improved models for geoplasma and magnetic field.** K. Maeda.
Planet. Space Sci., Vol. 23, 843 - 865 (1975).

Intensities of auroral hiss generated by the Čerenkov radiation process by electrons in the lower magnetosphere are calculated with respect to a realistic model of the earth's magnetosphere.

084.019 **The topology of the auroral oval as seen by the Isis 2 scanning auroral photometer.**
A. T. Y. Lui, C. D. Anger, D. Venkatesan, W. Sawchuk, S.-I. Akasofu.
Journ. Geophys, Res., Vol. 80, 1795 - 1804 (1975).

Auroral distributions in the polar region viewed by the Isis 2 scanning auroral photometer are studied to examine the topology of the auroral oval. It is shown that a single continuous oval-shaped belt can be defined in which both discrete and/or diffuse auroras lie.

084.020 **Crossed beam measurements of the diffuse radar aurora.**
W. L. Ecklund, B. B. Balsley, R. A. Greenwald.
Journ. Geophys. Res., Vol. 80, 1805 - 1809 (1975).

084.021 **On acceleration of auroral electrons by a stationary double layer.**
E. E. Antonova, B. A. Tverskoj.
Geomagn. Aeronom., Vol. 15, 563 - 565 (1975). In Russian. Brief information.

084.022 **Current-driven plasma instabilities at high latitudes.** F. L. Scarf, R. W. Fredricks, C. T. Russell, M. Neugebauer, M. Kivelson, C. R. Chappell.
Journ. Geophys. Res., Vol. 80, 2030 - 2040 (1975).

084.023 **Energization of auroral electrons by electrostatic shock waves.** J. R. Kan.
Journ. Geophys. Res., Vol. 80, 2089 - 2095 (1975).

Electrostatic shock waves are proposed as a possible mechanism for energizing electrons which are responsible for discrete auroras. It is shown that electrostatic shock solutions can exist in a $T_i \gg T_e$ plasma carrying a field-aligned electron current if the plasma in the high-latitude plasma sheet has an earthward flow component. The model is formulated within the framework of the Vlasov-Poisson equations. It is found that under the high-latitude plasma sheet conditions, a field-aligned potential jump of several kilovolts can be produced by the shock.

084.024 **On the formation of auroral arcs and acceleration of auroral electrons.** D. W. Swift.
Journ. Geophys. Res., Vol. 80, 2096 - 2108 (1975).

It is suggested that the highly structured auroral arc is caused by a current-driven laminar electrostatic shock oblique to the geomagnetic field. Electrons are accelerated by the potential jump associated with the shock. The shock is assumed to be confined to a plane. Self-consistent solutions to the Poisson-Vlasov systems are calculated for the electrostatic potential.

084.025 **Electron fluxes and correlations with quiet time auroral arcs.**
P. F. Mizera, D. R. Croley, Jr., F. A. Morse, A. L. Vampola.
Journ. Geophys. Res., Vol. 80, 2129 - 2136 (1975).

084.026 **Simultaneous measurements of auroral particles and electric currents by a rocket-borne instrument system: introductory remarks.** H. R. Anderson, P. A. Cloutier.
Journ. Geophys. Res., Vol. 80, 2146 - 2151 (1975).

084.027 **Rocket measurement of auroral electron fluxes associated with field-aligned currents.**
P. M. Pazich, H. R. Anderson.
Journ. Geophys. Res., Vol. 80, 2152 - 2160 (1975).

084.028 **Electron currents associated with an auroral band.**
R. J. Spiger, H. R. Anderson.
Journ. Geophys. Res., Vol. 80, 2161 - 2164 (1975).

084.029 **Rocket-based magnetic observations of auroral Birkeland currents in association with a structured auroral arc.** R. T. Casserly, Jr., P. A. Cloutier.
Journ. Geophys. Res., Vol. 80, 2165 - 2168 (1975).

084.030 **Neutral hydrogen flux measured at 100- to 200-km altitude in an electron aurora.**
G. E. Iglesias, H. R. Anderson.
Journ. Geophys. Res., Vol. 80, 2169 - 2171 (1975).

084.031 **On a crossed field two-stream plasma instability in the auroral plasma.** T. N. C. Wang, R. T. Tsunoda.
Journ. Geophys. Res., Vol. 80, 2172 - 2182 (1975).

084.032 **A relationship between synchronous altitude electron fluxes and the auroral electrojet.**
R. D. Sharp, E. G. Shelley, G. Rostoker.
Journ. Geophys. Res., Vol. 80, 2319 - 2324 (1975).

084.033 **The development of the substorm in auroral radio absorption.**
J. K. Hargreaves, H. J. A. Chivers, W. I. Axford.
Planet. Space Sci., Vol. 23, 905 - 911 (1975).

084.034 **Some remaining mysteries in the aurora.**
J. W. Dungey.
Quarterly Journ. Roy. Astron. Soc., Vol. 16, 117 - 127 (1975).

Energy build-up and release mechanisms in solar and auroral flares. See Abstr. 073.005.

Generation of small-scale irregularities in a turbulent plasma and radio aurora. See Abstr. 083.029.

Geomagnetic Field

084.201 **Precambrian geomagnetic field reversal.**
D. K. Bingham, M. E. Evans.
Nature, Vol. 253, 332 - 333 (1975). – Letter.

084.202 **Transport of energetic solar particles on closed magnetospheric field lines.** M. Scholer.
Space Sci. Rev., Vol. 17, 3 - 44 (1975). – Invited lecture, second meeting of the European Geophysical Society, September 1974, Trieste, Italy.

The present review tries to summarize the status of knowledge regarding solar proton behavior on closed magnetospheric field lines. Together with a presentation of recent measurements in the closed field line region relevant theoretical problems are discussed. They fall either under the study of single particle motion in different static magnetospheric configurations (due to different field models or due to real, e.g. ring current induced changes), or under the study of resonant interaction processes as pitch angle scattering and radial diffusion.

084.203 **Electron fluxes during the magnetic storm of December 14 - 15, 1970 according to Cosmos 381 data.** S. V. Avakyan, M. P. Bolgartseva, A. I. Efremov, I. A. Krinberg, A. P. Kulakov, V. S. Petrov, A. L. Podmoshenskij, I. M. Pribylovskij, G. V. Sazonov, Yu. N. Shaulin.
Issled. po geomagnetizmu, aehron. i fiz. Solntsa. Vyp. (No.) 32. Moskva, Nauka, 1974, p. 158 - 161. In Russian. – Abstr. in Referativ. Zhurn. 62. Issled. kosmich. prostranstva, 2.62. 296 (1975).

084.204 **Some properties of the current sheet in the geomagnetic tail.** J. W. Eastwood.
Planet. Space Sci., Vol. 23, 1 - 14 (1975).

The topic of this report is that of the influence of noise, and of the finite length and width of the tail on the behaviour of the current sheet. The presence of a weak magnetic field linking through the current sheet leads to plasma containment and counterstreaming, with the consequence that both the plasma temperature and density are increased in the vicinity of the current sheet. The effect of these changes on the relationship between steady bulk parameters is discussed.

084.205 **Particle precipitation in the South Atlantic geomagnetic anomaly.** D. G. Torr, M. R. Torr, J. C. G. Walker, R. A. Hoffman.
Planet. Space Sci., Vol. 23, 15 - 26 (1975).

084.206 **Solar wind access to the plasma sheet along the flanks of the magnetotail.** M. K. Bird.
Planet. Space Sci., Vol. 23, 27 - 40 (1975).

Low-energy particle trajectories in an idealized magnetotail magnetic field are investigated to determine the accessibility of magnetosheath protons and electrons to the plasma sheet along the flanks of the tail magnetopause. It is suggested that a large fraction of the magnetotail plasma is composed of former solar wind particles which have penetrated the magnetospheric boundary at the tail flanks.

084.207 **Effects of solar wind parameters on the development of magnetospheric substorms.**
T. Murayama, K. Hakamada.
Planet. Space Sci., Vol. 23, 75 - 91 (1975).

Effects of solar wind parameters on the development of substorms during the events of southward interplanetary magnetic field lasting more than one hour were studied.

084.208 **Excitation of magnetosonic waves with discrete spectrum in the equatorial vicinity of the plasmapause.** A. V. Gul'elmi (*Gul'el'mi*), B. I. Klaine (*Klajn*), A. S. Potapov.
Planet. Space Sci., Vol. 23, 279 - 286 (1975).

The purpose of this paper is to present the qualitative theory of excitation of the fast magnetosonic waves in the equatorial region of the plasmapause.

084.209 **A necessary condition for the geodynamo.**
F. H. Busse.
Journ. Geophys. Res., Vol. 80, 278 - 280 (1975).

084.210 *Sq* **variability and aeronomic structure.**
G. M. Brown.
Journ. Atmosph. Terr. Phys., Vol. 37, 107 - 117 (1975).

Evidence is presented for correlations between the day-to-day variability in the phase of the solar diurnal variation of horizontal magnetic intensity on quiet days, $Sq(H)$, and parameters in the stratosphere and in the E-region of the ionosphere.

084.211 **Characteristics of sudden worldwide changes in the geomagnetic field.** B. J. Rigby, J. S. Mainstone.
Journ. Atmosph. Terr. Phys., Vol. 37, 447 - 454 (1975).

084.212 **Comments on some plasma instabilities and their magnetospheric relevance.**
H.-R. Lehmann, R. Treumann.
Gerlands Beiträge Geophys., Vol. 84, 1 - 14 (1975).

Some comments on the gyroresonant and high-frequency electrostatic plasma instabilities in the magnetosphere and their importance for magnetospheric dynamics are presented. The somewhat ambiguous situation of unknown priority of the instabilities is shown. The paper ends with an outlook on the importance of instability study in the magnetosphere for understanding various cosmical problems, and gives a compilation of a number of plasma instabilities which may be of importance in various regions of the magnetosphere.

084.213 **Alternative to the geomagnetic self-reversing dynamo.** K. L. Verosub.
Nature, Vol. 253, 707 - 708 (1975).

The author points out that there is an alternative method of analysing the geomagnetic field which can account for magnetic reversals but which does not require a self-reversing dynamo. The analysis is based on the assumption that the geomagnetic field arises from two separate sources. Each source has a mathematical representation in terms of dipoles, quadrupoles, and so on. The magnetic field which is observed at or above the surface of the earth is the sum of the fields arising from the two sources.

084.214 **The sunspot cycle influence on lunar and solar daily geomagnetic variations.**
S. R. C. Malin, A. Cecere, A. Palumbo.
Geophys. Journ. Roy. Astron. Soc., Vol. 41, 115 - 126 (1975).

The long-standing question of whether L and S respond similarly to the sunspot cycle is re-examined using a very much more extensive data set than hitherto. The global distribution of the phenomenon is discussed in terms of the coefficients of simple spherical harmonic models, including the first such representation of the changes in S and L from sunspot minimum to maximum.

084.215 **Influence de la polarité du champ magnétique interplanétaire sur la variation annuelle et sur la variation diurne de l'activité magnétique.** A. Berthelier.
Comptes Rendus Acad. Sci. Paris, Sĉr. B, Vol. 280, 195 - 198 (1975).

On montre que les maximums des variations annuelles et diurnes de l'activité magnétique s'interprètent de façon cohérente si on fait l'hypothèse que l'activité est favorisée lorsqu'augmente l'angle entre la composante azimuthale du champ magnétique interplanétaire et l'axe du dipôle.

084.216 **Interplanetary magnetic field direction and the configuration of the day side magnetosphere.**
T. W. Hill, M. E. Rassbach.
Journ. Geophys. Res., Vol. 80, 1-6 (1975).
The direction of the interplanetary magnetic field appears to have a significant effect on the equilibrium size of the day side magnetosphere. The qualitative features of this effect are illustrated here by means of a simple vacuum model that ignores solar wind and ionospheric plasma densities as a first approximation. This idealized model produces earthward displacements of the subsolar magnetopause and equatorward displacements of the polar cusp associated with southward-turning interplanetary fields that are about a factor of 2 greater than observed displacements.

084.217 **Electrostatic and electromagnetic turbulence associated with the earth's bow shock.**
P. Rodriguez, D. A. Gurnett.
Journ. Geophys. Res., Vol. 80, 19 - 31 (1975).

084.218 **Simultaneous particle and field observations of field-aligned currents.** F. W. Berko, R. A. Hoffman, R. K. Burton, R. E. Holzer.
Journ. Geophys. Res., Vol. 80, 37 - 46 (1975).

084.219 **Substorm-associated reconfiguration of the dusk side equatorial magnetosphere: a possible source mechanism for isolated plasma regions.**
J. N. Barfield, J. L. Burch, D. J. Williams.
Journ. Geophys. Res., Vol. 80, 47 - 55 (1975).

084.220 **Local time variations of particle flux produced by an electrostatic field in the magnetosphere.**
M. G. Kivelson, D. J. Southwood.
Journ. Geophys. Res., Vol. 80, 56 - 65 (1975).

084.221 **Substorm and interplanetary magnetic field effects on the geomagnetic tail lobes.**
M. N. Caan, R. L. McPherron, C. T. Russell.
Journ. Geophys. Res., Vol. 80, 191 - 194 (1975).

084.222 **Second-order statistical structure of geomagnetic field reversals.** P. S. Naidu.
Journ. Geophys. Res., Vol. 80, 803 - 806 (1975).
From the autocorrelation function of geomagnetic polarity intervals, it is shown that the field reversal intervals are not independent but form a process akin to the Markov process, where the random input to the model is itself a moving average process. The input to the moving average model is, however, an independent Gaussian random sequence. All the parameters in this model of the geomagnetic field reversal have been estimated. In physical terms this model implies that the mechanism of reversal possesses a memory.

084.223 **MHD-waves in the solar wind — a possible source of Pc3 geomagnetic pulsations.**
P. A. Vinogradov, V. A. Parkhomov.
Geomagn. Aeronom., Vol. 15, 134 - 137 (1975). In Russian.

084.224 **Energy spectra and charge states of H, He, and heavy ions observed in the earth's magnetosheath and magnetotail.** C. Y. Fan, G. Gloeckler, D. Hovestadt.
Phys. Rev. Letters, Vol. 34, 495 - 498 (1975).
H, He, and heavy ions of energies $\geqslant 0.12$ MeV/charge were detected in the magnetotail and in the magnetosheath. It was

found that their relative abundances were ~9 : 1 : 4 × 10⁻², their differential energy spectra were ~$1/E^4$, and their atomic electrons were almost completely stripped. These results led the authors to suggest that they were the low-energy "quiet-time" cosmic rays accelerated within the magnetotail and the magnetosheath to the observed energies.

084.225 **Magnetic storm associated changes in the electron content at low latitudes.**
G. W. Prölss, K. Najita.
Journ. Atmosph. Terr. Phys., Vol. 37, 635 - 643 (1975).

084.226 **Structure of the quasi-perpendicular laminar bow shock.** E. W. Greenstadt, C. T. Russell, F. L. Scarf, V. Formisano, M. Neugebauer.
Journ. Geophys. Res., Vol. 80, 502 - 514 (1975).

084.227 **Standing waves at low Mach number laminar bow shocks.** D. H. Fairfield, W. C. Feldman.
Journ. Geophys. Res., Vol. 80, 515 - 522 (1975).

084.228 **A quantitative magnetospheric model derived from spacecraft magnetometer data.**
G. D. Mead, D. H. Fairfield.
Journ. Geophys. Res., Vol. 80, 523 - 534 (1975).
A quantitative model of the external magnetospheric field has been derived by making least squares fits to magnetic field measurements from four Imp satellites. The data set consists of 12,616 vector field averages over half-earth radii intervals between 4 and 17 R_E, taken from 451 satellite orbits between 1966 and 1972.

084.229 **Magnetospheric mapping with a quantitative geomagnetic field model.**
D. H. Fairfield, G. D. Mead.
Journ. Geophys. Res., Vol. 80, 535 - 542 (1975).

084.230 **Substorm-injected protons and electrons and the injection boundary model.**
A. Konradi, C. L. Semar, T. A. Fritz.
Journ. Geophys. Res., Vol. 80, 543 - 552 (1975).

084.231 **A modeling of the magnetospheric substorm.**
F. Yasuhara, Y. Kamide, S.-I. Akasofu.
Planet. Space Sci., Vol. 23, 575 - 578 (1975).
This paper reports some results of an attempt to simulate the large-scale changes of the internal structure of the magnetosphere during the magnetospheric substorm by assuming the growth of two current systems, one in the night-side and the other in the day-side.

084.232 **Weak and intense substorms.** Y. Kamide, S.-I. Akasofu, S. E. Deforest, J. L. Kisabeth.
Planet. Space Sci., Vol. 23, 579 - 587 (1975).
It is shown that there is no qualitative difference between weak and intense substorms. In this paper, the growth of the electrojet across the Canada meridian chain and plasma behaviors at the synchronous distance are compared in detail for a weak and intense substorm which occurred successively on 14 July 1970. The present study, together with earlier ones, shows clearly that it is incorrect to distinguish weak substorms from more intense substorms by calling the former "local" substorms or a growth phase of more intense substorms.

084.233 **Geoactive zones in the solar wind.**
M. S. Bobrov.
Planet. Space Sci., Vol. 23, 627 - 636 (1975).
Three parameters of the solar wind, proton number density n, Z-component of frozen-in magnetic field in solar ecliptic coordinates and magnetic field variability ΔB, may be called geoactive parameters since each of them is responsible

for a certain phase or stage of a geomagnetic storm. The present paper is an attempt to consider qualitatively the problems of geoactive parameters and geoactive zones and to formulate new, more adequate theses on the properties of solar corpuscular streams. These theses will be applied to the analysis of a typical magnetic storm.

084.234 Neutral sheets. J. W. Dungey.

Space Sci. Rev., Vol. 17, 173 - 180 (1975). – Review paper –see 012.002.

With the growth of recognition of the importance of neutral sheets, the complexity of the subject has become apparent. Although the crucial problem concerns the vicinity of a neutral line, where the normal component B_z may be neglected, only a small fraction of satellite crossings can be expected in such a small region. The breakdown of the problem will be discussed and possible trajectories described for particles entering from either the lobes or the magnetosheath.

084.235 Interplanetary field effect on the magnetosphere. A. Nishida.

Space Sci. Rev., Vol. 17, 353 - 389 (1975). – Review paper – see 012.002.

This paper reviews recent developments in the understanding of the solar-wind magnetosphere interaction process in which the interplanetary magnetic field has been found to play a key role.

084.236 Study of magnetospheric dynamics using energetic solar particles. G. Morfill.

Space Sci. Rev., Vol. 17, 391 - 433 (1975). – Review paper – see 012.002.

The author shows the importance of the single particle approximation (trajectories in a reference field) in forming the basis of the understanding of the quiet-time penetration of cosmic rays into the magnetosphere, he considers the 'steady dynamics' such as wave-particle interaction and field line reconnection, which is believed to exist nearly all the time, and finally the author reviews the work which has been done in the much more complex and less well-understood field of 'impulsive dynamics' such as geomagnetic storms and substorms.

084.237 Plasma and fields in the magnetospheric tail. K. Schindler.

Space Sci. Rev., Vol. 17, 589 - 614 (1975). – Review paper – see 012.002.

Recent developments in the physics of the magnetospheric tail are reviewed. Particular emphasis is placed on progress in selfconsistent theory and on the comparison of its results with the observations. It is demonstrated that several of the macroscopic properties of the tail can be understood in terms of a rather simple two-dimensional picture in which the plasma is static and isotropic.

084.238 Rotation of the earth's solid core as a possible cause of declination, drift and reversals of the earth's magnetic field. M. Steenbeck, G. Helmis.

Geophys. Journ. Roy. Astron. Soc., Vol. 41, 237 - 244 (1975).

Calculations carried out on a simple model show that each disturbance of the gyroscope causes a gradually decreasing rotation of the gyroscope in relation to the sphere, the rotation axis falling exactly into the equatorial plane and rotating within the latter. Transferring this concept to the earth's solid core permits one to set up a hypothesis for the explanation of the declination, drift, and reversals of the earth's magnetic field.

084.239 High β plasma instabilities and storm time geomagnetic pulsations. L. J. Lanzerotti, A. Hasegawa.

Journ. Geophys. Res., Vol. 80, 1019 - 1022 (1975).

084.240 Excitation of geomagnetic pulsations of the "serpentine-emission" type in the interplanetary plasma. A. V. Gul'el'mi, B. V. Dovbnya, B. I. Klajn.

Dokl. Akad. Nauk SSSR, Ser. Mat. Fiz., Vol. 221, 1314 - 1317 (1975). In Russian.

084.241 The magnetosheath electron population at lunar distance: general features. P. H. Reiff, D. L. Reasoner.

Journ. Geophys. Res., Vol. 80, 1232 - 1237 (1975).

084.242 Pioneer 7 observations of plasma flow and field reversal regions in the distant geomagnetic tail. R. C. Walker, U. Villante, A. J. Lazarus.

Journ. Geophys. Res., Vol. 80, 1238 - 1244 (1975).

084.243 Double streams of protons in the distant geomagnetic tail. U. Villante, A. J. Lazarus.

Journ. Geophys. Res., Vol. 80, 1245 - 1247 (1975).

084.244 Polar cap currents for different directions of the interplanetary magnetic field in the $Y-Z$ plane. E. Friis-Christensen, J. Wilhjelm.

Journ. Geophys. Res., Vol. 80, 1248 - 1260 (1975).

084.245 Relation of variations in total magnetic field at high latitude with the parameters of the interplanetary magnetic field and with DP 2 fluctuations. R. A. Langel.

Journ. Geophys. Res., Vol. 80, 1261 - 1270 (1975).

084.246 Explorer 45 observations of 1- to 30-Hz magnetic fields near the plasmapause during magnetic storms. W. W. L. Taylor, B. K. Parady, L. J. Cahill, Jr.

Journ. Geophys. Res., Vol. 80, 1271 - 1286 (1975).

084.247 Anisotropic plasma instabilities in the magnetosphere and interplanetary medium. A. K. Yukhimuk, V. I. Bondarenko, I. A. Izmajlov.

Problems of cosmic physics. Vyp. (No.) 9, (see 003.013), p. 56 - 63 (1974). In Russian.

The electromagnetic instabilities in the magnetosphere and in the interplanetary medium are discussed. The instabilities are caused by anisotropic plasma particle velocity distribution. The instability conditions are found. Formulas for resonance particle energies and characteristic sizes of unstable disturbances are obtained. Estimates of the sizes of unstable disturbances in the magnetosphere and in the interplanetary medium and of the flux energies of ejected particles are made.

084.248 Model of the magnetospheric field for slightly disturbed conditions. N. A. Tsyganenko.

Kosmich. Issled., Vol. 13, 215 - 221 (1975). In Russian.

084.249 A physical discussion of secondary effects in the solar daily geomagnetic variation. G. Fanselau, F.-W. Gerstengarbe.

Gerlands Beiträge Geophys., Vol. 84, 107 - 112 (1975).

The solar quiet daily geomagnetic variations are consisting of two components. The principal effect is produced by an ionospheric dynamo and is not discussed here. The phenomenological character and a first physical interpretation of the secondary effect superposed to the primary effect are dealt with.

084.250 On the association of lunar daily variations in H at Alibag with the degree of magnetic activity. D. R. K. Rao, B. R. Arora.

Gerlands Beiträge Geophys., Vol. 84, 113 - 116 (1975).

Presented during the special session of the "Committee on Lunar Effects" – Second General Scientific Assembly of

IAGA, Kyoto, September 1973.

The authors report the association of lunar diurnal variations with the degree of magnetic activity derived from an analysis of a long series of horizontal intensity (H) observations at Alibag.

084.251 **The solar wind-magnetosphere dynamo and the magnetospheric substorm.** S.-I. Akasofu.
Planet. Space Sci., Vol. 23, 817 - 823 (1975).

In a quiet condition, the solar wind kinetic energy is converted into electrical energy. A small part of this energy is dissipated as heat energy in the polar ionosphere. The author identifies at least three types of magnetospheric disturbances which are not associated with an increase of the heat production and calls them reversible disturbances, while the magnetospheric substorm is an irreversible disturbance which is associated with a large increase of the heat production.

084.252 **Corrections to "The compressed geomagnetic field as a function of dipole tilt"** [Planet. Space Sci., Vol. 22, 595 - 608 (1974)].
D. W. Halderson, D. B. Beard, J. Y. Choe.
Planet. Space Sci., Vol. 23, 887 - 890 (1975). – See 11.084.206.

084.253 **Particles and the magnetic field in the outer midday magnetosphere of the earth.**
A. E. Antonova, V. P. Shabanskij.
Geomagn. Aeronom., Vol. 15, 297 - 302 (1975). In Russian.

084.254 **The statistics of inversions of the geomagnetic field.** V. P. Golovkov.
Geomagn. Aeronom., Vol. 15, 383 - 385 (1975). In Russian. Brief information.

084.255 **MHD wave transmission and production near the magnetopause.** A. Wolfe, R. L. Kaufmann.
Journ. Geophys. Res., Vol. 80, 1764 - 1775 (1975).

084.256 **Error enhancement in geomagnetic models derived from scalar data.**
D. P. Stern, J. H. Bredekamp.
Journ. Geophys. Res., Vol. 80, 1776 - 1782 (1975).

084.257 **A net of geomagnetic coordinates in the outer magnetosphere.**
I. I. Alekseev, L. N. Osipova, V. P. Shabanskij.
Geomagn. Aeronom., Vol. 15, 502 - 507 (1975). In Russian.

084.258 **Alfvén waves of compression in the magnetosphere.** Yu. P. Mal'tsev.
Geomagn. Aeronom., Vol. 15, 572 - 573 (1975). In Russian. Brief information.

084.259 **The response of the day side magnetosphere-ionosphere system to time-varying field line reconnection at the magnetopause. 1. Theoretical model.**
T. E. Holzer, G. C. Reid.
Journ. Geophys. Res., Vol. 80, 2041 - 2049 (1975).

084.260 **The response of the day side magnetosphere-ionosphere system to time-varying field line reconnection at the magnetopause. 2. Erosion event of March 27, 1968.**
G. C. Reid, T. E. Holzer.
Journ. Geophys. Res., Vol. 80, 2050 - 2056 (1975).

084.261 **Identifications of the polar cap boundary and the auroral belt in the high-altitude magnetosphere: a model for field-aligned currents.** M. Sugiura.
Journ. Geophys. Res., Vol. 80, 2057 - 2068 (1975).

084.262 **Energization of electrons at synchronous orbit by substorm-associated cross-magnetosphere electric fields.** R. J. Walker, M. G. Kivelson.
Journ. Geophys. Res., Vol. 80, 2074 - 2082 (1975).

084.263 **On the interpretation of low-energy particle access to the polar caps.** F. C. Michel, A. J. Dessler.
Journ. Geophys. Res., Vol. 80, 2309 - 2310 (1975).

084.264 **Differential rotation of the magnetospheric plasma as cause of the Svalgaard-Mansurov effect.**
H. Volland.
Journ. Geophys. Res., Vol. 80, 2311 - 2315 (1975).

Svalgaard, and independently Mansurov, discovered a correspondence between the geomagnetic variations at the geomagnetic poles and the sector polarity of the interplanetary magnetic field (IMF). Heppner noted an asymmetry in the magnetospheric electric convection field observed within the polar caps at ionospheric altitudes which is also related to the sector polarity of the IMF. It is shown that both effects can be consistently explained by differential rotation of the magnetospheric plasma with respect to the earth.

084.265 **Access of solar electrons to the polar regions.**
E. Nielsen, M. A. Pomerantz.
Planet. Space Sci., Vol. 23, 945 - 954 (1975).

The interaction between the geomagnetic and interplanetary magnetic fields is studied through its effects upon the intensities of solar electrons reaching the polar caps during times of strongly anisotropic electron fluxes in the magnetosheath. During the particle event of 18 November 1968, electrons of solar origin were observed outside the magnetopause with detectors aboard OGO-5. Correlative studies of these satellite observations and concurrent measurements by riometers and ionospheric forward scatter systems in both polar regions have revealed that the initial stage of the associated polar cap absorption event is attributable to the prompt arrival of solar electrons. The analysis indicates that an anisotropic electron flux may be isotropized at the magnetopause before propagating into the polar regions.

084.266 **The microstructure of the magnetopause.**
D. M. Willis.
Geophys. Journ. Roy. Astron. Soc., Vol. 41, 355 - 389 (1975).

084.267 **HEOS observations of the configuration of the magnetosphere.**
P. C. Hedgecock, B. T. Thomas.
Geophys. Journ. Roy. Astron. Soc., Vol. 41, 391 - 403 (1975).

The preliminary results of an analysis of magnetic field measurements made over a five-year period in the earth's magnetosphere are presented.

084.268 **Theories of magnetic field annihilation.**
E. R. Priest, B. U. Ö. Sonnerup.
Geophys. Journ. Roy. Astron. Soc., Vol. 41, 405 - 413 (1975).

A critical summary is given of the various theories for magnetic field annihilation. In addition a new exact three-dimensional solution of the magnetohydrodynamic equations is described.

084.269 **Record of observations at Ottawa Magnetic Observatory, July 1968 to December 1969.**
J. Hruska.
Publ. Earth Phys. Branch, Dep. of Energy, Mines and Resources, Ottawa, Canada, Vol. 46, (No. 1), 1 - 83 (1974).

084.270 **Structure of the terrestrial bow shock.**
E. W. Greenstadt.
Solar wind three, (see 012.020), p. 440 - 454 (1974).

084.271 **Variation of the ratio of specific heats across a detached bow shock.**
J. K. Chao, M. J. Wiskerchen.
Solar wind three, (see 012.020), p. 455 - 459 (1974).

084.272 **The terrestrial magnetosphere and comparison with Jupiter's.** F. C. Michel.
Solar wind three, (see 012.020), p. 460 - 471 (1974).

084.273 **Intensity and energy spectrum of electrons accelerated in the earth's bow shock.** K. A. Anderson.
Journ. Geophys., Vol. 40, 701 - 712 (1974).

A model for generation of bow-shock-associated upstream waves. See Abstr. 062.017.

Effect of obstacles on the rate of reconnection of magnetic field lines. See Abstr. 062.039.

Recurrent magnetic activity, sunspot number and its rate of decline. See Abstr. 072.002.

Chromospheric flares or chromospheric aurorae? See Abstr. 073.025.

The solar M-region problem — an old problem now facing its solution? See Abstr. 074.008.

Determination of the solar wind velocity from pulsation indices. See Abstr. 074.010.

Some characteristics of the solar wind, position of shock wave, magnetopause and plasmapause during the period with high solar activity in August 1972. See Abstr. 074.093.

On the possibility of deducing interplanetary and solar parameters from geomagnetic records. See Abstr. 074.104.

A view of earth and air. See Abstr. 082.074.

Magnetospheric ring current as possible source of disturbance of the ionospheric F2-layer during geomagnetic storms. See Abstr. 083.009.

Changes in the ionospheric profile and the Faraday factor $\overline{M}$ with Kp. See Abstr. 083.010.

Wave-particle interactions in the magnetosphere and ionosphere. See Abstr. 083.022.

Ionosphere-magnetosphere coupling. See Abstr. 083.037.

Investigations in the ionosphere with the help of satellite Cosmos 378. 6. Characteristics of the ionosphere and of precipitating particles and their connection with geophysical phenomena during magnetospheric substorms. See Abstr. 083.058.

Correlation between the intensity of emission at 6300 Å, temperature and parameters of the ionosphere. See Abstr. 083.070.

Radiation belt and convection in an artificial magnetosphere. See Abstr. 084.413.

Interplanetary streams and their interaction with the earth. See Abstr. 106.022.

Response of the plasma sheet at $\sim 18 R_E$ to sudden southward turnings of the interplanetary magnetic field. See Abstr. 106.025.

Nonlinear oblique interaction of interplanetary tangential discontinuities with magnetogasdynamic shocks. See Abstr. 106.028.

Some properties of the Svalgaard A/C index. See Abstr. 106.031.

Comment on 'Interplanetary magnetic sector structure, 1926—1971' by L. Svalgaard and 'Correspondence of solar field sector direction and polar cap geomagnetic field changes for 1965' by W. H. Campbell and S. Matsuchita. See Abstr. 106.032.

Interplanetary magnetic field polarity and low-latitude geomagnetic field. See Abstr. 106.044.

Influence of the bounded magnetosphere and the ring current on the geomagnetic cutoff rigidity of cosmic rays. See Abstr. 143.018.

Radiation Belts

084.401 **Energetic He⁺ ions from the radiation belt at low altitudes near the geomagnetic equator.**
M. Scholer, D. Hovestadt, G. Morfill.
Journ. Geophys. Res., Vol. 80, 80 - 85 (1975).
Energetic helium ions trapped in the radiation belt can get lost through charge exchange by collisions with neutral exospheric hydrogen. Since the cross section for conversion of singly ionized He to neutral He is much greater than the cross section for conversion of doubly ionized He to neutral He, the energetic He flux can be considered to originate from the He⁺ population of the radiation belt. At low altitudes the energetic neutral helium atoms are again converted to ionized helium by collisions with the neutral oxygen atoms of the atmosphere.

084.402 **Electromagnetic hiss and relativistic electron losses in the inner zone.** B. T. Tsurutani, E. J. Smith, R. M. Thorne.
Journ. Geophys. Res., Vol. 80, 600 - 607 (1975).

084.403 **Wave-particle interactions and their relevance to substorms.** R. W. Fredricks.
Space Sci. Rev., Vol. 17, 449 - 480 (1975). – Review paper – see 012.002.
Wave-particle effects are implicit in most models of radial diffusion and energization of Van Allen belt particles; they were explicitly used in the wave turbulence model for trapped particle precipitation and trapped flux limitations. In the paper the author reviews the observations and some of the pertinent theoretical interpretations of wave-particle effects as they relate to substorm and storm-time phenomena.

084.404 **Geomagnetically trapped radiation.** M. Schulz.
Space Sci. Rev., Vol. 17, 481 - 536 (1975). – Review paper – see 012.002.
This review covers the major developments in radiation-belt phenomenology of the past four years (1970–1973).

084.405 **Interaction between heavier ions and ring current protons.** N. Brice, C. Lucas.
Journ. Geophys. Res., Vol. 80, 936 - 942 (1975).

084.406 **The quiet time structure of energetic (35–560 keV) radiation belt electrons.**
L. R. Lyons, D. J. Williams.
Journ. Geophys. Res., Vol. 80, 943 - 950 (1975).

084.407 **The cause of storm after effects in the middle latitude D-region.**
W. N. Spjeldvik, R. M. Thorne.
Journ. Atmosph. Terr. Phys., Vol. 37, 777 - 795 (1975).
Fluxes of outer zone radiation belt electrons are considerably enhanced during the main phase of geomagnetic storms. Throughout the storm recovery the electrons radially diffuse to lower L and decay slowly to reform the characteristic two zone structure.

084.408 **Dynamics of the belt of energetic electrons injected into the inner regions of the magnetosphere during the storm of December 16–18, 1971.**
L. M. Kovrygina, Eh. N. Sosnovets, L. V. Tverskaya, O. V. Khorosheva.
Geomagn. Aeronom., Vol. 15, 308 - 312 (1975). In Russian.

084.409 **Light flashes observed by astronauts on Skylab 4.**
L. S. Pinsky, W. Z. Osborne, R. A. Hoffman, J. V. Bailey.
Science, Vol. 188, 928 - 930 (1975).
Two dedicated light flash observing sessions were conducted by one of the crewmen during the Skylab 4 mission. Analyses of his observations reveal a strong correlation between flash frequency and primary cosmic-ray flux, and an even stronger correlation between flash frequency and the South Atlantic Anomaly (SAA) region of the inner belt trapped radiation. Calculations indicate that an all-proton inner belt probably cannot produce the observed SAA flash rate, and they suggest that there may exist a previously unobserved inner belt flux of multiply charged nuclei.

084.410 **An approximate analytic description of plasma bulk parameters and pitch angle anisotropy under adiabatic flow in a dipolar magnetospheric field.**
D. J. Southwood, M. G. Kivelson.
Journ. Geophys. Res., Vol. 80, 2069 - 2073 (1975).

084.411 **Preliminary results of observations of protons of the ring current during magnetic disturbances aboard the satellite Molniya 1.** A. N. Grechin, L. M. Kovrygina, A. S. Kovtyukh, M. I. Panasyuk, I. A. Rubinshtejn, Eh. N. Sosnovets.
Kosmich. Issled., Vol. 13, 352 - 360 (1975). In Russian.

084.412 **A new trapped proton radiation belt model.**
D. M. Sawyer, Jr.
EOS Trans. American Geophys. Union, Vol. 54, 1182 (1973). Abstract.

084.413 **Radiation belt and convection in an artificial magnetosphere.**
I. M. Podgornyj, Eh. M. Dubinin, Yu. N. Potanin.
Izv. AN SSSR. Ser. fiz., Vol. 39, 350 - 353 (1975). In Russian.
Abstr. in Referativ. Zhurn. 62. Issled. kosmich. prostranstva, 6.62.214 (1975).

084.414 **Support for Crand theory from measurements of earth albedo neutrons between 70 and 250 MeV.**
G. Kanbach, C. Reppin, V. Schönfelder.
Journ. Geophys. Res., Vol. 79, 5159 - 5165 = Max-Planck-Inst. Extraterr. Phys., München, Separate print (1974).

New types of motion in Störmer's problem.
See Abstr. 042.002.

Interaction between the electron-cyclotron emissions at $(n + {}^1/_2)\,\Omega_e$ and the ring-current protons in space.
See Abstr. 062.027.

Wave-particle interactions in the magnetosphere and ionosphere. See Abstr. 083.022.

Errata

084.901 **Correction: 'The $O_2(b^1\Sigma_g^+) - X(^3\Sigma_g^-)$ system in aurora'** [Journ. Geophys. Res., Vol. 79, 4821 - 4822 (1974)]. A. V. Jones, R. L. Gattinger.
Journ. Geophys. Res., Vol. 80, 1856 (1975).

085 Solar-Terrestrial Relations

085.001 Correlation of the air temperature at Catania and the sunspot cycle. C. Blanco, S. Catalano.
Journ. Atmosph. Terr. Phys., Vol. 37, 185 - 187 (1975).

Results of the correlation between the air temperature recorded at Catania in the period 1817–1970 and the sunspot number R, are reported. The temperature and the sunspot number, both averaged over the 11 year solar cycle, give a coefficient $r = 0.5$ for a linear correlation. The higher mean temperatures are associated with the more active solar cycles and the lower mean temperatures with cycles of lower mean activity.

085.002 Solar activity and the weather. J. M. Wilcox.
Journ. Atmosph. Terr. Phys., Vol. 37, 237 - 256 (1975).

The attempts during the past century to establish a connection between solar activity and the weather are discussed. Some critical remarks about the quality of much of the literature in this field are given. Several recent investigations are summarized. Use of the solar-interplanetary magnetic sector structure in future investigations may add an element of cohesiveness and interaction to these investigations.

085.003 22-jarige zonnevlekkenperiode in het optreden van winterweer? F. Ijnsen.
Zenit, Vol. 2, 183 - 184 (1975).

085.004 Observations of low-energy electrons upstream of the earth's bow shock. D. L. Reasoner.
Journ. Geophys. Res., Vol. 80, 187 - 190 (1975).

Observations of electron fluxes with a lunar-based electron spectrometer when the moon was upstream of the earth have shown that a subset of observed fluxes are strongly controlled by the interplanetary magnetic field (IMF) direction. The fluxes occur only when the IMF lines connect back to the earth's bow shock. Observed densities and temperatures were in the ranges $2-4 \times 10^{-3}$ cm^{-3} and $1.7-2.8 \times 10^6$ °K. It is shown that these electrons can account for increases in effective solar wind electron temperatures on bow shock connected field lines, which have been observed previously by other investigators. It is further shown that if a model of the bow shock with an electrostatic potential barrier is assumed, the potential can be estimated to be 500 V.

085.005 Magnetic field depression at the earth's surface calculated from the relationship between the size of the magnetosphere and the Dst values.
S.-Y. Su, A. Konradi.
Journ. Geophys. Res., Vol. 80, 195 - 199 (1975).

085.006 Red oxygen emission λ 6300 Å and density of the upper atmosphere.
Yu. L. Truttse, V. D. Belyavskaya.
Geomagn. Aeronom., Vol. 15, 101 - 104 (1975). In Russian.

085.007 On the connection of parameters of solar wind, location of the magnetosphere on the day-side of the earth and the periods of geomagnetic pulsations.
M. S. Kovner, N. M. Rudneva, Ya. I. Fel'dshtejn.
Geomagn. Aeronom., Vol. 15, 124 - 127 (1975). In Russian.

085.008 On interaction between solar wind and the geomagnetic field.
N. V. Erkaev, V. G. Pivovarov.
Geomagn. Aeronom., Vol. 15, 157 - 158 (1975). In Russian.
Brief information.

085.009 Solar conditionality of variations of the length of day and of seismic activity. Yu. D. Kalinin.
Geomagn. Aeronom., Vol. 15, 170 - 171 (1975). In Russian.
Brief information.

085.010 Neutron spectra and connection coefficients within the region of $10^{-2}-10^5$ GeV at solar activity minimum in different atmospheric depths.
L. V. Granitskij, G. A. Novikova, L. E. Rishe.
Issled. po geomagnetizmu, aehron. i fiz. Solntsa. Vyp. (No.) 31. Moskva, Nauka, 1974, p. 130 - 135. In Russian. – Abstr. in Referativ. Zhurn. 51. Astron., 2.51.532 (1975).

085.011 Les bases énergétiques de la paléoclimatique théorique et l'évolution des climats.
E. A. Bernard.
Ciel et Terre, Vol. 91, 41 - 74, 89 - 118, 161 - 219 (1975).

085.012 Sur l'aspect météorologique des relations soleil-terre.
P. Bernard.
L'Astronomie, 89e année, p. 181 - 182 (1975).

085.013 Ice ages and the Galaxy. W. H. McCrea.
Nature, Vol. 255, 607 - 609 (1975).

The passage of the solar system through a dust lane bordering a spiral arm of the Galaxy may cause a temporary variation of the sun's radiation and so lead to an ice epoch on earth.

085.014 Secular cycle of solar activity and pressure field of the earth's northern hemisphere.
I. V. Maksimov, B. A. Sleptsov-Shevlevich.
Probl. Arktiki i Antarktiki. Vyp. (No.) 45. Leningrad, Gidrometeoizdat, 1974, p. 27 - 37. In Russian. – Abstr. in Referativ. Zhurn. 51. Astron., 5.51.453 (1975).

085.015 Solar activity and the concentration of C^{14} in tree rings for 1780 - 1838 measured with a single-channel scintillation device stabilized by light pulse.
G. E. Kocharov, Kh. A. Arslanov, V. A. Dergachev, S. A. Rumyantsev, S. B. Chernov, V. F. Goncharov.
Trudy 5-go Vses. soveshch. po probl. "Astrofiz. yavleniya i radiouglerod", 1973. Tbilisi, 1974, p. 19 - 37. In Russian. Abstr. in Referativ. Zhurn. 51. Astron., 5.51.454 (1975).

085.016 Solar cosmic rays and variations of radiocarbon in the earth's atmosphere.
V. A. Alekseev, A. K. Lavrukhina, Z. K. Mil'nikova, I. V. Smirnov, L. D. Sulerzhitskij.
Trudy 5-go Vses. soveshch. po probl. "Astrofiz. yavleniya i radiouglerod", 1973. Tbilisi, 1974, p. 39 - 46. In Russian. Abstr. in Referativ. Zhurn. 51. Astron., 5.51.455 (1975).

085.017 Relationship between solar (corpuscular) activity and fast global variations of atmospheric pressure.
V. E. Chertoprud.
Pis'ma v Astron. Zhurn., Vol. 1, No. 4, p. 33 - 38 (1975). In Russian.

Statistic analysis of long-term data on atmospheric pressure for the network of meteorological stations at magnetic latitudes from 30° to 70° N reveals with great confidence global influences of solar corpuscular activity on variations of atmospheric pressure for winter and spring seasons.

085.018 Seasonal variations of the electron concentration in years of high solar activity.
M. A. Kutimskaya, O. M. Radzhabova.

Geomagn. Aeronom., Vol. 15, 550 (1975). In Russian. – Brief information.

085.019 Solar activity and precipitation within the zones of latitude 0° − 40° N.
J. Xanthakis, C. Poulakos, B. Tritakis.
Praktika Acad. Athens, Vol. 49, 187 - 214 = Res. Center Astron., Applied Math., Acad. Athens, Ser. 1 (Astron.), No. 38 (1974).

085.020 Influence of the solar radiation on the intensity of the equatorial electrojet. L. N. Yaremenko.
Kosmich. issled. na Ukraine. Resp. mezhved. sb., 1974, vyp. (No.) 5, p. 45 - 48. In Russian. – Abstr. in Referativ. Zhurn. 51. Astron., 6.51.477 (1975).

085.021 Correlation between solar activity and quasi-periodic variations of the gravity of the earth.
G. T. Sobakar.
Geofiz. sb. AN USSR, 1974, vyp. (No.) 62, p. 45 - 47.
In Russian. – Abstr. in Referativ. Zhurn. 52. Geodeziya i Aehrosemka, 6.52.64 (1975).

085.022 Solar activity variations and the radiocarbon content in the earth's atmosphere.
V. A. Dergachev.
Izv. AN SSSR. Ser. fiz., Vol. 39, 325 - 333 (1975). In Russian. – Abstr. in Referativ. Zhurn. 62. Issled. kosmich.

prostranstva, 6.62.211 (1975).

085.023 Solar-terrestrial disturbances of August 1972, 17.
On the relation between the solar activity and the tidal force on the sun induced by the planets. K. Takahashi.
Denpa Kenkyusho Kiho., Vol. 19, 379-383(1973). In Japanese.

085.024 The ancestry of solar-terrestrial research.
M. A. Pomerantz.
EOS Trans. American Geophys. Union, Vol. 55, 955 - 957 (1974).

Solar-terrestrial physics. Part 1.
See Abstr. 003.024.

Solar flares in August 1972 and connected geophysical phenomena. See Abstr. 073.070.

Seasonal variation and magnitude of the solar sector structure−atmospheric vorticity effect.
See Abstr. 080.043.

Absorption of radio waves in the ionosphere of temperate latitudes during the cycle of solar activity.
See Abstr. 083.048.

Interplanetary streams and their interaction with the earth. See Abstr. 106.022.

Planetary System

091 Physics of the Planetary System (Planetary Atmospheres, Figure, Interior, Magnetic Fields, Rotation, etc.)

091.001 Absorption line studies of reflection from horizontally inhomogeneous layers.
J. F. Appleby, D. J. Van Blerkom.
Icarus, Vol. 24, 51 - 69 = Contr. Five College Obs., Univ. Mass., *Amherst*, No. 191 (1975).

A discussion of literature relevant to horizontal inhomogeneities in planetary atmospheres shows this to be an increasingly important yet largely unexplored topic. The authors examine a range of cloud and gas configurations, of line and continuum opacities, and they compare phase variations of bright versus dark limbs. The results in general show trends quite dissimilar to (usually opposite) those predicted by a simple reflecting layer model. Percent equivalent width variations for the tower model are usually somewhat greater for weak than for relatively strong absorption lines, with differences of a factor of about two or three.

091.002 Remarks on the paper "The tidal loss of satellite-orbiting objects and its implications for the lunar surface" by M. J. Reid.   T. Gold.
Icarus, Vol. 24, 134 - 135 (1975).

The paper by Reid, (Icarus, Vol. 20, 240 - 248 (1973) – see Abstr. 10.091.021), suggests that masses may be stored in circumlunar orbits for long periods of time, limited only by tidal dissipation. The real loss may, however, be much faster, due to large changes in the orbit caused by the disturbing field of the earth. It is shown that the example quoted of Jupiter's satellites is inadequate to make the case for stability of such orbits.

091.003 On the gravitational stability of satellite-orbiting objects: a reply to T. Gold.   M. J. Reid.
Icarus, Vol. 24, 136 - 138 = Contr. Div. Geol. Planet. Sci., California Inst. Technology, *Pasadena*, No. 2500 (1975).

The stability of a satellite-orbiting object under the disturbing influence of the parent planet can be assessed by comparison with analogous three-body systems. The changes in the eccentricity and semimajor axis of a satellite-orbiting object (disturbed by the planet) and a planetary satellite (disturbed by the sun) scale equally if the dimensions of the systems are scaled by the sphere of influence of the orbited body. Thus, the apparent gravitational stability of planetary satellites supports theories of the gravitational stability of satellite-orbiting objects.

091.004 Exploring the solar system (V): Atmospheres and climates.   A. L. Hammond.
Science, Vol. 187, 244 - 246 (1975). – Research news.

091.005 A model of the gravitational differentiation of the interior of planets.
V. P. Keondzhyan, A. S. Monin.
Dokl. Akad. Nauk SSSR. Ser. Mat. Fiz., Vol. 220, 825 - 828 (1975). In Russian.

091.006 Über die Bahnen der regulären Satelliten im Planetensystem.   A. Unsöld, T. Gehren.
Naturwissenschaften, Vol. 62, 95 - 96 (1975).

091.007 Magnetochemical properties of bodies in the solar system derived from telescope spectral reflectivity curves.   P. Wasilewski.
Meteoritics, Vol. 9, 416 - 418 (1974). – Abstract.

091.008 Planetary cratering in the early solar system.
G. W. Wetherill.
Meteoritics, Vol. 9, 422 (1974). – Abstract.

091.009 A technique to deduce atmospheric temperature and constituent profiles from a planet's limb radiance profile.   G. Ohring.
Icarus, Vol. 24, 388 - 394 (1975).

The concept is described of deducing the temperature and constituent profile of a planetary atmosphere from orbiter measurements of the planet's ir limb radiance profile. Expressions are derived for the weighting functions associated with the limb radiance profile for a Goody random band model. Analysis of the weighting functions for the Martian atmosphere indicates that a limb radiance profile in the 15 μm CO_2 band can be used to determine the Martian atmospheric temperature profile from 20 to 60 km. Simulation of the Martian limb radiance profile in the rotational water vapor band indicates that Martian water vapor mixing ratios can be inferred from limb radiance observations in a water vapor band.

091.010 X and Y functions for planetary atmospheres with Lambert law reflecting surfaces.
G. S. Sidhu, J. L. Casti.
Astrophys. Journ., Vol. 196, 607 - 612 (1975).

A Riccati factorization technique is employed to derive X and Y functions for a fairly general class of homogeneous, isotropic, plane-parallel atmospheres including those bounded by a Lambert reflector – a situation hitherto not dealt with in any completely satisfactory way. The preliminary results reported illustrate the general scope of the method employed.

091.011 Methane absorption in the visible spectra of the outer planets and Titan.   T. Owen, R. D. Cess.
Astrophys. Journ., (*Letters*), Vol. 197, L37 - L40 (1975).

New spectra of Jupiter, Saturn, and Titan show weak methane bands in the region below 6000 Å which have been known for many years in the spectra of Uranus and Neptune. Adopting the known abundance of methane on Jupiter, the authors have used a band model to determine CH_4 abundances and broadening pressures for the other objects. The results indicate high values of $[CH_4]/[H_2]$ for Uranus and Neptune; for Titan, a surface pressure in excess of 1 atm is implied.

091.012 Origin and evolution of the atmospheres of the terrestrial planets.   A. J. Meadows.
Journ. British. Astron. Ass., Vol. 85, 208 - 212 (1975). Christmas lecture 1974.

091.013 Contributions to planetary exploration by magnetic field measurements.   F. M. Neubauer.
Raumfahrtforschung, [DGLR - Deutsche Ges. für Luft- und

Raumfahrt, Berlin], Vol. 18, 213 - 218 (1974).

It is shown that magnetic observations in the vicinity or on the surface of the planets and satellites of the solar system constitute an important source of information on the interiors, ionospheres and magnetospheres of these bodies. The great scientific potential of magnetic exploration of the planetary system is illustrated by three examples: the general magnetic fields of Jupiter and Saturn, the Jovian satellite Io and the magnetosphere, ionosphere and interior of Mars.

091.014 Erforschung der Planeten mit Raumsonden.
S. Marx.
Astron. in der Schule, 12. Jahrgang, p. 17 - 19 (1975).

091.015 Supersonic neutral winds in an outer atmosphere.
III. A two-dimensional model. N. E. Gilbert.
Australian Journ. Phys., Vol. 28, 85 - 99 (1975).

A steady two-dimensional model that represents the part of a convection cell where the neutral gas constituent is rising and becoming supersonic is presented for both isothermal and variable temperature conditions in an intensely heated region of the outer atmosphere of a planet or star. The stream-lines and orthogonal curves are represented by a system of confocal hyperbolae and ellipses respectively.

091.016 Equation of state of cosmochemical substances and the structure of the major planets.
V. N. Zharkov, V. P. Trubitsyn, I. A. Tsarevskij, A. B. Makalkin.
Izv. AN SSSR. Fiz. Zemli, 1974, No. 10, p. 7 - 18. In Russian.
Abstr. in Referativ. Zhurn. 51. Astron. 3.51.236 (1975).

091.017 The solution to the Titius-Bode rule.
W. E. Greig.
Bull. American Astron. Soc., Vol. 7, 337 (1975). – Abstr. AAS.

091.018 Satellite eclipses: observation and prediction.
C. F. Peters.
Bull. American Astron. Soc., Vol. 7, 343 (1975). – Abstr. AAS.

091.019 Tidal evolution of an accreting satellite.
A. W. Harris.
Bull. American Astron. Soc., Vol. 7, 343 (1975). – Abstr. AAS.

091.020 Quantification of spectral reflectance characteristics for mineralogical and petrological information.
M. J. Gaffey.
Bull. American Astron. Soc., Vol. 7, 370 (1975). – Abstr. AAS.

091.021 Reflections on lunar discoveries.
W. K. Hartmann.
Bull. American Astron. Soc., Vol. 7, 373 (1975). – Abstr. AAS.

091.022 A major perturbing force on small $(1 < R < 10^4$ cm) solar system bodies; the Yarkovsky effect.
C. Peterson.
Bull. American Astron. Soc., Vol. 7, 376 (1975). – Abstr. AAS.

091.023 Photoelectric absorption spectrum of methane in the visible and near infrared. K. A. Dick, U. Fink.
Bull. American Astron. Soc., Vol. 7, 382 (1975). – Abstr. AAS.

091.024 Analysis of earth-based planetary images thru digitization. L. H. Wasserman, W. A. Baum.
Bull. American Astron. Soc., Vol. 7, 389 (1975). – Abstr. AAS.

091.025 Uses and misuses of Minnaert's law.
J. Veverka, M. Noland.
Bull. American Astron. Soc., Vol. 7, 389 (1975). – Abstr. AAS.

091.026 On the use of a finite difference method for solving anisotropic scattering problems. B. R. Barkstrom.
Bull. American Astron. Soc., Vol. 7, 389 - 390 (1975).
Abstr. AAS.

091.027 An initial-value solution of internal radiation field for generalized Chandrasekhar's planetary problem.
S. Ueno, A. P. Wang.
Bull. American Astron. Soc., Vol. 7, 390 (1975). – Abstr. AAS.

091.028 The structure of a thermosphere and its stability.
S. H. Gross.
Bull. American Astron. Soc., Vol. 7, 390 (1975). – Abstr. AAS.

091.029 Brightness temperature measurements at 3 mm wavelength. B. L. Ulich, E. K. Conklin.
Bull. American Astron. Soc., Vol. 7, 391 (1975). – Abstr. AAS.

091.030 Production of organic molecules in the outer solar system by proton irradiation: laboratory simulations.
T. Scattergood, P. Lesser, T. Owen.
Icarus, Vol. 24, 465 - 471 (1975). – Presented at IAU Colloquium No. 28, Cornell Univ., Ithaca, New York, 1974 August 18 - 21.

Preliminary experiments to investigate the formation of colored polymers and other interesting molecules by the irradiation of gas mixtures by protons are discussed. Two to four Mev protons were used, with corresponding beam fluxes (as measured at $6R_J$ from the planet) equivalent to approximately 80 earth years at Jupiter per hour of exposure. An important feature of this work is the presence or absence of absorption at 5 μm in the different materials produced; Titan is quite dark at this wavelength and Io is fairly bright. Such features may provide criteria for accepting or rejecting various materials produced in these experiments as reasonable coloring agents for the outer solar system.

091.031 Elementhäufigkeiten im Sonnensystem und Element-entstehung in der Galaxis. H. Holweger.
SuW, Vol. 14, 189 - 193 (1975).

091.032 Polarization of light reflected from rough planetary surface. M. Wolff.
Applied Optics, Vol. 14, 1395 - 1405 (1975).

A calculation is made of the luminance and polarization of light due to single and double reflections from the faces of particles in a surface composed of random, irregular particles using equations of electromagnetic waves and materials with a complex index of refraction. Some geometric properties of shadows are derived and used. Good agreement is obtained between these results and measurements of polarized light from Mars, Mercury, and the moon. The model can be used to calculate polarization and luminance of rough astronomical bodies and surfaces as a function of the viewing angle.

091.033 Equations of figures of planets.
V. P. Trubitsyn, V. N. Zharkov.
Pis'ma v Astron. Zhurn., Vol. 1, No. 4, p. 39 - 42 (1975).
In Russian.

New equations of the theory of figure are obtained. They permit to calculate the shape and the gravity field for gas-fluid rotating planets with known density distribution.

091.034 Comments on the rotation of the planets.
G. Horedt.
Astrophys. Space Sci., Vol. 35, L15 - L16 (1975). – Letter.

091.035 On the influence of the mesorelief on the brightness distribution over a planetary disk.
L. A. Akimov.
Astron. Zhurn. Akad. Nauk SSSR, Vol. 52, 635 - 641 (1975).
In Russian. English translation in Soviet Astron., Vol. 19, No. 3.

The effect of large, gentle slopes on the brightness distribution over a planetary disk is discussed. The expressions for micro-, meso- and macrorelief are given. The brightness distribution law of an extremely rough planet is obtained. The possible application of this law to the moon is discussed.

091.036 On the mechanism of the magnetic dynamo of planets. Sh. Sh. Dolginov.
Kosmich.Issled., Vol. 13, 367 - 374 (1975). In Russian.

091.037 On the satellites of planets. K. Ziołkowski.
Urania Kraków, Vol. 46, 132 - 135 (1975). In Polish.

091.038 Method for determination of the parameters of axial rotation of a planet. A. L. Abramenko.
Vestn. Leningr. un-ta, 1974, No. 19, p. 151 - 153. In Russian. Abstr. in Referativ. Zhurn. 51. Astron., 6.51.174; 62. Issled. kosmich. prostranstva, 6.62.165 (1975).

091.039 Hypothesis on the resonance structure of the solar system. A. M. Molchanov.
AN SSSR. Nauch. tsentr. biol. issled. n.-i. vychisl. tsentr. Preprint. Pushchino, 1974. 20 pp. In Russian. — Abstr. in Referativ. Zhurn. 51. Astron., 6.51.273 (1975).

091.040 Multiple scattering in planetary atmospheres. W. M. Irvine.
Icarus, Vol. 25, 175 - 204 = Contr. Five Coll. Obs., Univ. Mass., *Amherst*, No. 196 (1975).
Methods for solving radiative transfer problems within the extended visible spectrum in planetary atmospheres are reviewed. Emphasis is placed on rapid, approximate procedures for the determination of such quantities as the plane and spherical (Bond) albedo, surface illumination, absorbed energy, limb darkening, phase curve, and spectra. Precise numerical methods and analytical results are also discussed. Recent approaches to such complications as atmospheric inhomogeneity and reflection from a porous regolith are described briefly.

091.041 Stability of frosts in the solar system. L. A. Lebofsky.
Icarus, Vol. 25, 205 - 217 (1975).
Calculations on the stability of various frosts (against evaporation) for solar system objects in circular and elliptical orbits are made. It is found that the stability of these frosts is dependent on the rate of rotation of the object, the latitude of the area on the object being considered, and the eccentricity of the orbit as well as its mean distance from the sun. These factors greatly influence the amount of solar radiation incident and reradiated from a given area on the object. The likelihood of finding these frosts on the surfaces of objects and the lifetimes of objects composed of these frosts is discussed.

091.042 The role of hydrocarbons in the ionospheres of the outer planets. S. K. Atreya, T. M. Donahue.
Icarus, Vol. 25, 335 - 338 (1975).
The role of hydrocarbons as a possible sink for H^+ and H_3^+ ions in the lower ionosphere of the outer planets is examined. Calculations indicate that H^+ and H_3^+ are efficiently converted to hydrocarbon ions on reaction with methane. The terminal ions, CH_5^+ and $C_2H_5^+$ are rapidly neutralized in dissociative recombination with electrons. Extreme ultraviolet photolysis of hydrocarbons as a potential additional source of lower elevation ions is investigated.

091.043 Realistic models of the giant planets. M. Podolak.
Thesis, Yeshiva Univ., New York (USA). 137 pp. University Microfilms Order No. 74-16,443 (1974).

091.044 Craters on bodies of the solar system. P. Příhoda.
Kozmos, Vol. 6, 70 - 74 (1975). In Czech.

Atlas of the planets. See Abstr. 003.050.

Collision-broadened linewidths of tetrahedral molecules—II. Computations for CH_4 lines broadened by N_2, O_2, He, Ne and Ar. See Abstr. 022.025.

Tidal friction and generalized Cassini's laws in the solar system. See Abstr. 042.005.

Un formulaire pour le calcul des perturbations d'ordres élevés dans les problèmes planétaires. See Abstr. 042.031.

Quasi-periodic intermediate orbits for the major planets and resonances of zero order. See Abstr. 042.032.

Restricted problem stability and the solar system. See Abstr. 042.037.

The present state of the problem of the motion of the major planets. See Abstr. 042.040.

The system of equations of the theory of a figure of the fifth approximation. See Abstr. 042.072.

On variations of solar activity caused by the revolution of Jupiter, Saturn and Uranus around the sun. See Abstr. 072.033.

Interior radiances in optically deep absorbing media—III. Scattering from Haze L. See Abstr. 082.001.

Measurement of line widths of CO of planetary interest at low temperatures. See Abstr. 082.002.

Surface history of Mercury: implications for terrestrial planets. See Abstr. 092.023.

Further comment on Titan's atmospheric scaling. See Abstr. 100.202.

A semi-analytical, long-term solution of Pluto, including the Neptune and Uranus resonance. See Abstr. 101.011.

Flux density measurements of radio sources at 2.14 millimeter wavelength. See Abstr. 141.030.

Errata

091.901 Erratum: 'Particle acceleration in planetary magnetospheres' [Nature, Vol. 251, 205 - 206 (1974)]. M. J. Houghton.
Nature, Vol. 253, 288 (1975).

092 Mercury

092.001   On the thermal regime of the upper layer of Mercury.
V. D. Krotikov, O. B. Shchuko.
Astron. Zhurn. Akad. Nauk SSSR, Vol. 52, 146 - 151 (1975).
In Russian. English translation in Soviet Astron., Vol. 19, No. 1.
    Calculations of the thermal regime of the upper layer of
Mercury are carried out in the framework of the uniform
model. The data of diurnal variations of insolation and the
temperature at various hermographical longitudes and
latitudes are given. The peculiarities of these changes due to
the presence of the inverse motion of the sun on the Mercurian
sky near the perihelion are discussed.

092.002   Le passage de Mercure devant le soleil du 10 novembre 1973.   P. de la Cotardière.
L'Astronomie, 89e année, p. 90 - 92 (1975).

092.003   Mercury revisited by Mariner 10.
Sky Telescope, Vol. 49, 292 - 293 (1975).

092.004   Magnetic field of Mercury confirmed.
N. F. Ness, K. W. Behannon, R. P. Lepping,
Y. C. Whang.
Nature, Vol. 255, 204 - 205 (1975).
    The authors present preliminary results from the NASA
Goddard Space Flight Center magnetometer instrumentation
on Mariner 10. In summary, the observations at the third en-
counter with Mercury by the NASA GSFC magnetic field ex-
periment have confirmed the earlier tentative conclusion that
Mercury has a modest, intrinsic magnetic field, the origin of
which is at present uncertain.

092.005   Preliminary interpretation of plasma electron ob-
              servations at the third encounter of Mariner 10
with Mercury.
R. E. Hartle, K. W. Ogilvie, J. D. Scudder, H. S. Bridge, G. L.
Siscoe, A. J. Lazarus, V. M. Vasyliunas, C. M. Yeates.
Nature, Vol. 255, 206 - 208 (1975).
    The third, and last, encounter took place on March 16,
1975 at which time the spacecraft passed within 330 km of
the surface of the planet. The primary purpose of this close
polar approach was to determine whether the magnetic moment of the planet was intrinsic and to enable further observations to be made inside the magnetosheath and magnetosphere. The results presented are from an electron spectrometer mounted on a scan platform. It accepts electrons from
the anti-solar direction in the energy range 13.4−690 eV in
15 logarithmically spaced channels of width $\Delta E/E = 6.6\%$.
The plasma results strongly support the idea that Mercury's
magnetic field is intrinsic rather than induced, and show that
the magnetosphere of Mercury is remarkably similar to that
of the earth.

092.006   Preliminary geologic-terrain map of Mercury.
N. J. Trask, J. E. Guest.
Bull. American Astron. Soc., Vol. 7, 372 (1975). − Abstr. AAS.

092.007   Magnetic field and interior of Mercury.   N. F. Ness.
              Bull. American Astron. Soc., Vol. 7, 372 (1975).
Abstr. AAS.

092.008   The 3.5-micron polarization of Mercury.
              R. Landau.
Bull. American Astron. Soc., Vol. 7, 372 (1975). − Abstr. AAS.

092.009   New reflectance spectra of Mercury and further
              compositional interpretation.
T. B. McCord, F. Vilas, J. B. Adams.

Bull. American Astron. Soc., Vol. 7, 372 (1975). − Abstr. AAS.

092.010   Mercury's helium exosphere.
              R. E. Hartle, S. A. Curtis, G. E. Thomas.
Bull. American Astron. Soc., Vol. 7, 372 - 373 (1975).
Abstr. AAS.

092.011   Morphological study of Mercury crust.
              S. Miyamoto.
Contr. Kwasan Hida Obs., Univ. Kyoto, No. 220, 9 pp. (1974).
    The photographic observation of Mercury surface by
Mariner 10 showed that the Mercury crust is covered with
craters, and its morphological aspect is very similar to that of
the moon. However, in spite of general similarity, there are
formations characteristic to Mercury. The crater named after
Kuiper, is an example, which is the formation intermediate
between terrestrial type strato-volcano and lunar type crater.
These and other formations are interesting, relating to the
problem of crater formation and crustal evolution of terrestrial
type planets.

092.012   Transit of Mercury, 1973 November 10.
              Compiled by S. Lyttle, D. E. Beesley.
Irish Astron. Journ., Vol. 11, 193 (1974).

092.013   The Mariner 10 pictures of Mercury: an overview.
              B. C. Murray.
Journ. Geophys. Res., Vol. 80, 2342 - 2344 (1975).
    The papers comprising the Mariner 10 imaging team final
report, are brought together, and the salient results of the
Mariner 10 imaging experiment at Mercury are summarized.
Those aspects of the data set acquired which were worked by
the team are identified, and other areas where further work is
needed are designated.

092.014   Preliminary imaging results from the second
              Mercury encounter.
R. G. Strom, B. C. Murray, M. J. S. Belton, G. E. Danielson,
M. E. Davies, D. E. Gault, B. Hapke, B. O'Leary, N. Trask,
J. E. Guest, J. Anderson, K. Klaasen.
Journ. Geophys. Res., Vol. 80, 2345 - 2356 (1975).
    The second Mercury encounter has resulted in the acqui-
sition of about 360 pictures of the south polar regions which
provide a reliable cartographic and geologic tie between the
two sides of the planet photographed on the first encounter.
Stereoscopic coverage of large areas of the southern hemisphere was obtained by combining Mercury 1 and 2 pictures
taken at different viewing angles.

092.015   Acquisition and description of Mariner 10 television
              science data at Mercury.
G. E. Danielson, Jr., K. P. Klaasen, J. L. Anderson.
Journ. Geophys. Res., Vol. 80, 2357 - 2393 (1975).
    The Mariner 10 television science subsystem was an im-
proved version of the Mariner 9 system, using 1500-mm-focal-
length optics. An elaborate picture-taking sequence resulted
in transmission of over 4000 frames back to earth during two
flyby encounters with Mercury. These sequences utilized a
real-time data rate of 117.6 kbit/s, resulting in coverage of
about 75% of the lighted portion of Mercury's surface at a resolution of better than 2 km. The complete set of useful
images, which amounted to about 3000 frames, was processed
with three different types of digital image-processing enhancements.

092.016   IPL processing of the Mariner 10 images of Mercury.
              J. M. Soha, D. J. Lynn, J. J. Lorre, J. A. Mosher,

N. N. Thayer, D. A. Elliott, W. D. Benton, R. E. Dewar.
Journ. Geophys. Res., Vol. 80, 2394 - 2414 (1975).

This paper describes the digital processing performed by the Image Processing Laboratory of the Jet Propulsion Laboratory (JPL) on the images of Mercury returned to earth from Mariner 10. Each image contains considerably more information than can be displayed in a single picture. Several specialized processing techniques and procedures were utilized to display the particular information desired for specific scientific analyses. A principal task was the construction of full disk mosaics as an aid to the understanding of surface structure on a global scale; the detailed steps involved in the production of pictures for these mosaics are described.

092.017 **Mercury rotation period determined from Mariner 10 photography.** K. P. Klaasen.
Journ. Geophys. Res., Vol. 80, 2415 - 2416 (1975).

The rotation period of Mercury has been determined to be 58.661 ± 0.017 (1 sigma) days by using high-resolution photography from the Mariner 10 mission. This value matches the period required for 3/2 synchronism with the orbital period (58.6462 days) within the 1-sigma errors assigned and is consistent with the latest values derived from radar and earth-based telescopic observations.

092.018 **Surface coordinates and cartography of Mercury.**
M. E. Davies, R. M. Batson.
Journ. Geophys. Res., Vol. 80, 2417 - 2430 (1975).

A control net of Mercury has been established photogrammetrically by using the Mariner 10 pictures; coordinates of 1328 points are given. The Mariner 10 coordinate system uses a system of longitudes in which the twentieth meridian passes through the center of the small crater Hun Kal and the spin axis is assumed normal to the orbital plane. A reference mosaic of Mercury has been published, and a series of 1:5,000,000 maps is now being produced.

092.019 **Photometric observations of Mercury from Mariner 10.**
B. Hapke, G. E. Danielson, Jr., K. Klaasen, L. Wilson.
Journ. Geophys. Res., Vol. 80, 2431 - 2443 (1975).

The elimination of the residual image problem which plagued previous Mariner imaging systems allowed photometry of moderately high quality to be carried out on Mercury by Mariner 10. Relative radiance measurements are accurate to about ± 1% within a frame; absolute measurements have an uncertainty of about ± 5%. The conclusions from the photometric analysis are discussed.

092.020 **Some comparisons of impact craters on Mercury and the Moon.** D. E. Gault, J. E. Guest,
J. B. Murray, D. Dzurisin, M. C. Malin.
Journ. Geophys. Res., Vol. 80, 2444 - 2460 (1975).

Although the general morphologies of fresh mercurian and lunar craters are remarkably similar, comparisons of ejecta deposits, interior structures, and changes in morphology with size reveal important differences between the two populations of craters. The differences are attributable to the different gravity fields in which the craters were formed and have significant implications for the interpretation of cratering processes and their effects on all planetary bodies.

092.021 **Preliminary geologic terrain map of Mercury.**
N. J. Trask, J. E. Guest.
Journ. Geophys. Res., Vol. 80, 2461 - 2477 (1975).

A geologic terrain map of Mercury has been constructed by use of the photogeologic methods employed for the moon and Mars. The oldest and most widespread unit, intercrater plains, forms nearly level to rolling surfaces on which are superposed numerous secondary impact craters. This unit may represent a very old surface that predates the last heavy bombardment of the inner planets. The effects of this bombardment are recorded in a second widespread unit, heavily cratered terrain, consisting of closely spaced craters and basins from 30 km to several hundred kilometers in diameter. Younger craters, including some with rays, followed emplacement of the plains; they are much less abundant than the preplains craters. The geologic history of Mercury is remarkably similar to that of the moon.

092.022 **Tectonism and volcanism on Mercury.**
R. G. Strom, N. J. Trask, J. E. Guest.
Journ. Geophys. Res., Vol. 80, 2478 - 2507 (1975).

Mercury appears to have a tectonic framework and diastrophic history not found on other terrestrial planets explored to date (earth, Mars, and the moon). On the part of the planet viewed by Mariner 10, only two localized areas show evidence of tensional stresses, both of which are apparently associated with the Caloris basin. Lobate scarps occur in the remainder of the explored region and appear to be primarily reverse or thrust faults which have resulted from compressive stresses acting on a global scale. Stratigraphic, volumetric, and albedo considerations together with distribution indicate that the majority of smooth plains on Mercury were produced by volcanism which occurred at the close of the period of late heavy bombardment similar to that on the moon and Mars. Several generations of plains are evident; the oldest may have resulted in part from an early differentiation of the planet.

092.023 **Surface history of Mercury: implications for terrestrial planets.**
B. C. Murray, R. G. Strom, N. J. Trask, D. E. Gault.
Journ. Geophys. Res., Vol. 80, 2508 - 2514 (1975).

A working hypothesis of Mercury's history is presented. The authors infer the surface of Mercury to record a sequence of events broadly similar to those recorded on the moon, implying similar histories of impact bombardment. The large lunarlike impact craters on Mercury can be interpreted as part of a distinct episode of bombardment which may have affected all the terrestrial planets about 4 b.y. ago.

092.024 **Mercury in the light of contemporary investigations.**
S. R. Brzostkiewicz.
Urania Kraków, Vol. 46, 100 - 108 (1975). In Polish.

092.025 **Osservazioni del transito di Mercurio del 10 novembre 1973.** M. G. Fracastoro.
Atti Fondaz. G. Ronchi, Vol. 29, 751 - 761 = Contr. Oss. Astron. Torino (Pino Torinese), No. 81 (1974).

092.026 **Merkurov prolaz 10. XI 1973 (Transit of Mercury on 10. XI. 1973).** A. Tomić.
Vasiona, Vol. 22, 86 - 90 (1974).

A refined computation of the perigee angle in Ptolemy's Mercury model. See Abstr. 004.010.

Mariner-Venus-Mercury 1973 project history. Part 2. See Abstr. 053.009.

Erste Ergebnisse von Mariner 10 über Venus und Merkur. See Abstr. 093.040.

Seismic effects from major basin formations on the moon and Mercury. See Abstr. 094.114.

Crater degradation on the moon, Mars, and Mercury. See Abstr. 094.226.

Seismic effects from major basin formations on the moon and Mercury. See Abstr. 094.236.

093 Venus

093.001 Is the four-day "rotation" of Venus illusory?
A. T. Young.
Icarus, Vol. 24, 1 - 10 (1975).

Spectroscopic, photometric, and radiometric evidence against a 4-day atmospheric rotation is reviewed. The bulk of the somewhat contradictory evidence seems to favor slow motions, on the order of 5 m/sec, in the atmosphere of Venus; the 4-day "rotation" may be due to a traveling wavelike disturbance, not bulk motions, driven by the uv albedo differences.

093.002 Theoretical interpretation of the 0.7820 μm CO_2 band and 0.8226 μm H_2O line on Venus.
J. L. Regas, L. P. Giver, R. W. Boese, J. H. Miller.
Icarus, Vol. 24, 11 - 18 (1975).

The authors have analyzed the $P6$, $P8$, and $P10$ lines in the 0.7820 μm CO_2 band of Venus using a scattering model. Their new results compare favorably with previous results from the 1.05 μm CO_2 band. They considered nonabsorbing and absorbing clouds and found that the anisotropic scattering mean free path for both models at the 0.2 atm level is between 0.55 and 0.73 km, a range close to the value of 1 km for terrestrial hazes. The authors used their scattering models to synthesize the 0.8226 μm H_2O line.

093.003 Venus: cloud optical depth and surface albedo from Venera 8. C. Devaux, M. Herman.
Icarus, Vol. 24, 19 - 27 (1975).

The authors have used the measurements of the solar flux obtained by the Venera 8 spacecraft inside the atmosphere of Venus and the values of the Venus spherical albedo to deduce the characteristics of the clouds and of the ground.

093.004 Venus' spectroscopic phase variation: implications of the Mariner 10 photographs.
J. W. Chamberlain.
Astrophys. Journ., Vol. 195, 815 - 817 (1975).

The recent ultraviolet photographs of Venus by Mariner 10 have shown an irregular cloud structure. To illustrate the consequences of subsolar weakening of CO_2 absorption, a mathematically simple (but realistically naïve) model is used. Two conclusions are reached: (1) Efforts to distinguish between single- and double-layer models for the clouds from spectroscopic data alone are not merely ambiguous (as argued earlier by Chamberlain and Smith); with present data they are hopeless. (2) The decrease in the CO_2 absorption close to full phase, as reported by Young et al., could result entirely from an equatorial darkening that is relatively inconsequential at the crescent phase and increasingly predominant for fuller phases.

093.005 Dissipation of the Venus ionosphere.
G. M. Nedyalkova, I. E. Turchinovich.
Astron. Zhurn. Akad. Nauk SSSR, Vol. 52, 139 - 145 (1975).
In Russian. English translation in Soviet Astron., Vol. 19, No. 1.

The paper deals with ion dissipation from the Venus atmosphere. Taking into account electric field interaction and ion deceleration in the plasma the equations of ion motion are considered and solved. The values of critical velocities for H^+, He^+ ions and e are obtained. Dissipation fluxes for H^+, He^+ ions and e from the Venus ionosphere are calculated. The electric field of Venus and the charge of the planet are evaluated provided that the flux of positive particles is equal to that of negative ions.

093.006 Wie ähnlich ist der Planet Venus der Erde?
H. W. Köhler.
Umschau, 75. Jahrgang, p. 86 - 87 (1975).

Considering the latest results of Russian and American space probes to Venus, there seem to be severe differences in the physical data of Venus and earth. This article shows how all of these discrepancies can be explained by the special orbit of Venus.

093.007 The aeronomy of the upper atmosphere of Venus.
S. C. Liu, T. M. Donahue.
Icarus, Vol. 24, 148 - 156 (1975).

Density profiles for CO, O, and O_2 in the Cytherean atmosphere above 90 km are plotted with eddy diffusion coefficient (K) as a parameter, subject to the constraint that the mixing ratios of CO and O_2 approach their observed value or values under the observed upper limit at the lower boundary. It is then shown that the value of K puts upper limits on the amount of hydrogen (in the form of H_2O, HCl, and H_2) the atmosphere near 90 km can contain. The authors suggest either very effective escape mechanisms–despite low exospheric hydrogen densities–or novel excitation mechanisms for $O(3^3S)$ and $O(3^5S)$ in the upper atmosphere.

093.008 Venus wind and temperature structure: the Venera 8 data. J. E. Ainsworth, J. R. Herman.
Journ. Geophys. Res., Vol. 80, 173 - 179 (1975).

An analysis of the Venera 8 measurements yields equatorial morning terminator horizontal and vertical winds that are similar in a number of respects of the winds the authors obtained from their analysis of the Venera 7 measurements. The lower boundary of the horizontal retrograde '4-day' wind is defined by a 50-60% decrease in wind speed in the vicinity of 44 km, and there exists a retrograde wind 'plateau' of 15- to 40-m/s winds extending from 40 km down to the vicinity of 18 km, where the winds decrease rapidly to the order of 0.1 m/s near the surface. Updrafts of 2-5 m/s exist in the vicinity of 20-30 km and are apparently associated with a slightly superadiabatic lapse rate.

093.009 Optical constants of sulfuric acid; application to the clouds of Venus? K. F. Palmer, D. Williams.
Applied Optics, Vol. 14, 208 - 219 (1975).

093.010 Motion of Venusian clouds by the deposition of meteoroids. V. Mitra.
Planet. Space Sci., Vol. 23, 551 - 555 (1975).

The theory of superrotation of the earth's atmosphere by global deposition of meteoroids recently developed by the author (1974) is extended after a slight refinement to explain the rotation period of Venusian clouds. A satisfactory agreement with observations is obtained.

093.011 Strongly-cooled ionizing plasma flows with application to Venus. M. K. Wallis, R. S. B. Ong.
Planet. Space Sci., Vol. 23, 713 - 721 (1975).

On the solar wind's penetration into an atmosphere of hydrogen or helium, symmetric charge exchange interactions give energy and momentum losses as the dominant source terms in the flow equations. One-dimensional, supersonic to subsonic solutions are available if the cooling is strong enough. In a model with transverse field and adiabatic (non-thermal) ions, a range of weakly-shocked solutions with upstream Mach number less than 2.5 are discovered. As in the case of detonation waves, the shock strength is independent of down-stream boundary conditions. The solutions may apply in the solar wind flow into the Venusian atmosphere.

093.012 Venus–Jupiter encounter.

Sky Telescope, Vol. 49, 276 - 279 (1975).

093.013 Structure and dynamics of the thermosphere of Venus. M. N. Izakov, S. K. Morozov.
AN SSSR. In-t kosmich. issled. Pr-176. Moskva,1974, 32 pp. In Russian. − Abstr. in Referativ. Zhurn. 51. Astron., 3.51. 221; 62. Issled. kosmich. prostranstva, 3.62.252 (1975).

093.014 The lower atmosphere of Venus.
D. O. Muhleman, G. Orton, G. L. Berge.
Bull. American Astron. Soc., Vol. 7, 373 (1975). − Abstr. AAS.

093.015 An analysis of the strong zonal circulation within the stratosphere of Venus.
V. Ramanathan, R. D. Cess.
Bull. American Astron. Soc., Vol. 7, 373 (1975). − Abstr. AAS.

093.016 Calculations of the radiative and dynamical state of the Venus atmosphere.
J. B. Pollack, R. E. Young.
Bull. American Astron. Soc., Vol. 7, 373 - 374 (1975). Abstr. AAS.

093.017 Radiation models of Venus in light of the Venera 8 photometer measurements. J. O. Roads.
Bull. American Astron. Soc., Vol. 7, 374 (1975). − Abstr. AAS.

093.018 Hot and cold running weather on Venus.
A. T. Young, L. G. Young.
Bull. American Astron. Soc., Vol. 7, 374 (1975). − Abstr. AAS.

093.019 Dual frequency observations of turbulence in the atmosphere of Venus by Mariner 10.
R. Woo, F. Yang.
Bull. American Astron. Soc., Vol. 7, 374 (1975). − Abstr. AAS.

093.020 The vertical cloud structure on Venus inferred from Mariner 10 infrared measurements. F. W. Taylor.
Bull. American Astron. Soc., Vol. 7, 374 (1975). − Abstr. AAS.

093.021 Vertical structure studies of the atmosphere of Venus using CO_2 line profiles.
E. S. Barker, W. Macy, L. M. Trafton.
Bull. American Astron. Soc., Vol. 7, 374 (1975). − Abstr. AAS.

093.022 Venus: vertical structure of stratospheric hazes from Mariner 10 pictures. B. O'Leary.
Bull. American Astron. Soc., Vol. 7, 374 - 375 (1975). Abstr. AAS.

093.023 He 584 Å airglow emission from Venus: Mariner 10 observations. S. Kumar, A. L. Broadfoot.
Bull. American Astron. Soc., Vol. 7, 375 (1975). − Abstr. AAS.

093.024 Comparison of simultaneous CO_2 measurements of Venus with different telescopes and techniques.
R. A. J. Schorn, L. D. G. Young, E. S. Barker.
Bull. American Astron. Soc., Vol. 7, 375 (1975). − Abstr. AAS.

093.025 Some problems in Venus' aeronomy.
N. D. Sze, M. B. McElroy.
Planet. Space Sci., Vol. 23, 763 - 786 (1975).
Models are presented for the height distribution of various photochemically active gases in Venus' upper atmosphere. Attention is directed to the chemistry and vertical transport of odd hydrogen, odd oxygen, free chlorine, CO, O_2, H_2 and H_2O.

093.026 Venus cloud properties: infrared opacity and mass mixing ratio. R. E. Samuelson, R. A. Hanel,
L. W. Herath, V. G. Kunde, W. C. Maguire.

Icarus, Vol. 25, 49 - 63 (1975).
By using the Mariner 5 temperature profile and a homogeneous cloud model,and assuming that CO_2 and cloud particles are the only opacity sources, the wavelength dependence of the Venus cloud opacity is inferred from the infrared spectrum of the planet between 450 and 1250 cm^{-1}. Volume extinction coefficients varying from 0.5×10^{-5} to 1.5×10^{-5} cm^{-1}, depending on the wavelength, are determined at the tropopause level of 6110 km. By using all available data, a cloud mass mixing ratio of approximately 5×10^{-6} and a particle concentration of about 900 particles cm^{-3} at this level are also inferred. The derived cloud opacity compares favorably with that expected for a haze of droplets of a 75% aqueous solution of sulfuric acid.

093.027 High-dispersion spectroscopic observations of Venus during 1968 and 1969. II. The carbon-dioxide band at 8689 Å.
R. A. J. Schorn, A. Woszczyk, L. D. G. Young.
Icarus, Vol. 25, 64 - 88 (1975).
Thirty well-exposed photographic plates showing the spectrum of the carbon-dioxide band at 8689 Å in the atmosphere of Venus were obtained during 1968 and 1969. The authors find rotational temperatures ranging from 236 to 274 K. The average value of the rotational temperature is 246 ± 1 K (one standard deviation); for their 1967 observations, the rotational temperatures ranged from 222 to 248 K, with an average value of 238 ± 4 K.

093.028 An analysis of the strong zonal circulation within the stratosphere of Venus.
V. Ramanathan, R. D. Cess.
Icarus, Vol. 25, 89 - 103 (1975).
A dynamical model is presented for the observed strong zonal circulation within the stratosphere of Venus. The analysis suggests that propagating internal gravity waves generated by diurnal solar heating of the upper stratosphere induce mean zonal velocities within the upper and lower stratosphere. These velocities increase from zero at the tropopause to about 200 msec^{-1} at the 85 km level. The velocity near the uv-cloud level compares favorably with the observed value of 100 msec^{-1}.

093.029 Model of the Venus atmosphere.
M. Ya. Marov, O. L. Ryabov.
In-t prikl. mat. AN SSSR. Preprint No. 112. Moskva,̇1974. 53 pp. In Russian. − Abstr. in Referativ. Zhurn. 51. Astron., 5.51.253; 62. Issled. kosmich. prostranstva, 5.62.198 (1975).

093.030 Venus: eastern elongation, 1964. D. E. Beesley.
Irish Astron. Journ., Vol. 11, 194 - 196 (1974).

093.031 Structure and dynamics of the Venus thermosphere.
M. N. Izakov, S. K. Morozov.
Kosmich.Issled., Vol. 13, 404 - 414 (1975). In Russian.

093.032 The possibility of organic molecule formation in the Venus atmosphere.
V. A. Otroshchenko, Yu. A. Surkov.
Origins of Life, Vol. 5, 487 - 490 (1974).

093.033 The 1971−72 eastern (evening) apparition of Venus.
J. L. Benton, Jr.
Strolling Astronomer, Vol. 25, 151 - 160 (1975).

093.034 On the weakening of cm-region radio waves by two H_2O phases in the Venus atmosphere.
O. F. Tyrnov.
Kosmich. issled. na Ukraine. Resp. mezhved. sb., 1974, vyp. (No.) 5, p. 13 - 14. In Russian. − Abstr. in Referativ. Zhurn. 51. Astron., 6.51.289; 62. Issled. kosmich. prostranstva, 6.62.173 (1975).

093.035 **High dispersion observations of Venus during 1972:**
the CO_2 band at 7820 Å.
L. D. G. Young, A. T. Young, A. Woszczyk.
Icarus, Vol. 25, 239 - 267 (1975).

Forty-seven well-exposed photographic plates of Venus
which show the spectrum of the carbon dioxide band at
7820 Å were obtained. These spectra showed a semiregular
four-day variation in the CO_2 abundance over the disk of the
planet. The authors also find evidence for temporal variations
in the rotational temperature of this band and temperature
variations over the disk. The average temperature, found from
a curve-of-growth analysis assuming a constant CO_2 line width,
is 249 ± 1.4 K.

093.036 **Observations of Venus water vapor over the disk of**
Venus: the 1972–74 data using the H_2O lines at
8197 Å and 8176 Å.   E. S. Barker.
Icarus, Vol. 25, 268 - 281 (1975).

The Venus water vapor line at 8197.71 Å has been monitored at several positions on the disk of Venus and at phase
angles between 21° and 162°. Variations in the abundance have
been found with spatial location, phase angle and time. Comparisons made between the water vapor abundances and the
CO_2 abundances determined from near-simultaneous observations of CO_2 bands at the same positions on the disk of Venus
show no correlation for the majority of the samples.

093.037 **Semiperiodic variations in CO_2 abundance on Venus.**
E. S. Barker, M. A. Perry.
Icarus, Vol. 25, 282 - 295 (1975).

Photoelectric spectral scans of the P branch of the
8689 Å CO_2 band on Venus were made. The relative CO_2 line
strength was determined and then normalized to remove the
spatial variations leaving only temporal variations. When the
slit positions are referred to the equator of Venus, particularly
near inferior conjunction, the large asymmetries between the
slit positions can be explained by a greater CO_2 line strength
over the polar regions and weaker over the equatorial latitudes.
Simultaneous H_2O measurements during several of the observing runs indicate a lack of correlation in the relative CO_2 line
strengths and the H_2O abundance.

093.038 **Solar wind interaction with Venus: review.**
M. K. Wallis.
Solar wind three, (see 012.020), p. 421 - 427 (1974).

093.039 **Solar wind-Venus interaction observed from**
magnetic field experiment on Mariner 10.
Y. C. Whang, N. F. Ness, K. W. Behannon, R. P. Lepping.
Solar wind three, (see 012.020), p. 428 - 432 (1974).

093.040 **Erste Ergebnisse von Mariner 10 über Venus und**
Merkur.   H. W. Köhler.
Naturwiss. Rundschau [Wiss. Verlagsgesellschaft, Stuttgart],
Vol. 27, 415 - 416 (1974).

093.041 **Interpretation of the polarization of Venus.**
J. E. Hansen, J. W. Hovenier.
Journ. Atmosph. Sci., Vol. 31, 1137 - 1160 (1974).

Analyse de la polarisation linéaire de la lumière réfléchie
par Vénus, par comparaison des observations avec les calculs
de diffusion multiple. Une solution concentrée d'acide sul-
furique ($H_2SO_4 \cdot H_2O$) donne des résultats compatibles avec les
données de polarisation.

093.042 **Venus: vertical transport rates in the visible atmo-**
sphere.   R. G. Prinn.
Journ. Atmosph. Sci., Vol. 31, 1691 - 1697 (1974).

Les particules des nuages sur Vénus sont suffisamment
petites pour que leur distribution verticale soit fortement af-
fectée par la turbulence atmosphérique. La distribution verticale des particules peut être estimée à partir des interprétations
courantes des données de réfraction, de polarisation et de la
bande d'absorption.

093.043 **Phasenbeobachtungen der Venus vor und nach der**
unteren Konjunktion 1974 Januar 23.
P. Ahnert.
Sterne, Vol. 51, 114 - 116 (1975).

Dynamics of the Venus atmosphere.
See Abstr. 003.041.

Physics of planet Venus.
See Abstr. 003.091.

Photographic observations of Venus at the Main
Astronomical Observatory of the Academy of Sciences of the
Ukrainian SSR in 1967 and 1969.   See Abstr. 041.065.

Photographic observations of Venus in Pulkovo with
the 26″ refractor.   See Abstr. 041.066.

Mariner-Venus-Mercury 1973 project history. Part 2.
See Abstr. 053.009.

The abundance of metal chlorides in the surface
layers of the earth and Venus.   See Abstr. 082.056.

Photodissociation of CO_2 and thermal emission
connected with it in the upper atmospheres of Mars and Venus.
See Abstr. 097.061.

094 Moon: Dynamics, Global Properties, Local Properties

Moon, Dynamics

094.001 **Laser observations of the moon: Normal points for 1972.**
P. J. Shelus, J. D. Mulholland, E. C. Silverberg.
Astron. Journ., Vol. 80, 154 - 161 (1975).
   The lunar laser observations taken at the McDonald Observatory during 1972 are presented in the form of compressed normal points, using the technique of an earlier paper. Refinements in the knowledge of the lunar motion have permitted corresponding increases in the ability to discriminate observations contaminated by equipment malfunctions; a list of amendments is given for the 1969–1971 data. The geometry of the telescope must be taken into account in the application of these data.

094.002 **Riddles of the origin and history of the moon.**
B. Yu. Levin, S. V. Maeva.
Zemlya i Vselennaya, 1975, No. 1, p. 22 - 28. In Russian.

094.003 **On prospects of using lunar ranging measurements.**
V. K. Abalakin, V. N. Bojko, Yu. L. Kokurin, V. F. Lobanov, M. A. Fursenko.
Astron. Zhurn. Akad. Nauk SSSR, Vol. 52, 387 - 397 (1975). In Russian. English translation in Soviet Astron., Vol. 19, No. 2.
   The present paper deals with the estimation of expected accuracy of determination of some parameters of astronomical and geodetic interest using lunar laser ranging measurements.

094.004 **Tidal friction and the early history of the moon's orbit.** D. P. Rubincam.
Journ. Geophys. Res., Vol. 80, 1537 - 1548 (1975).
   The effect of strong tidal friction on the moon's orbit in its early history is examined. The results show that the moon could have formed in an equatorial orbit about the earth, whether by fission or accretion, provided that the rheology of the early earth was that of a highly viscous Newtonian liquid and the orbit suffered a perturbation out of the earth's equatorial plane. The present $5°$ inclination of the lunar orbit to the ecliptic can be explained if the moon's orbit was perturbed about $3°$ out of the earth's equatorial plane and the earth's viscosity was not less than 10^{18} P.

094.005 **Bulk compositions of the moon and earth, estimated from meteorites.** R. Ganapathy, E. Anders.
Proc. Fifth Lunar Sci. Conference, (see 012.003), Vol. 2, 1181 - 1206 (1974).

094.006 **The geochemical evolution of the moon.**
S. R. Taylor, P. Jakeš.
Proc. Fifth Lunar Sci. Conference, (see 012.003), Vol. 2, 1287 - 1305 (1974).

094.007 **Lunar chronology.** T. Kirsten.
Raumfahrtforschung, [DGLR - Deutsche Ges. für Luft- und Raumfahrt, Berlin], Vol. 18, 235 - 237 (1974).

094.008 **Dynamics of lunar origin and orbital evolution.**
W. M. Kaula, A. W. Harris.
Rev. Geophys. Space Phys., Vol. 13, 363 - 371 (1975).
   The considerable differences in bulk composition of the moon and the earth have led most investigators to favor the capture hypothesis of lunar origin. However, upon closer examination all forms of the hypothesis still seem much less plausible dynamically than formation by accretion, i.e., acquisition of the moon in many small pieces rather than as predominantly one body. Models of accretion do suggest that the proto-lunar matter had a significantly different history from the proto-earth matter.

094.009 **Analytic expressions for lunar osculating elements.**
T. C. Van Flandern, K. F. Pulkkinen.
Bull. American Astron. Soc., Vol. 7, 343 (1975). – Abstr. AAS.

094.010 **On the cooling of the moon by solid convection.**
P. M. Cassen, R. E. Young.
The Moon, Vol. 12, 361 - 368 (1975).
   If a molten, or partially molten, lunar core exists at present, constraints would be placed on the viscosity of the solid mantle and the distribution of radioactive heat sources. Models in which the heat sources have been concentrated near the surface would rapidly solidify if the effective viscosity was equal to, or less than, 10^{22} $cm^2\,s^{-1}$. Retention of most of the heat sources throughout the mantle would permit present day solid convection to occur without cooling the core.

094.011 **A model of lunar evolution.**
D. W. Strangway, H. N. Sharpe.
The Moon, Vol. 12, 369 - 397 (1975).
   The authors consider a hypothesis in which there is a present day asthenosphere, a heat flow between 24 and 32 ergs $cm^{-2}\,s^{-1}$ and a crust which developed early in the moon's history by melting of the outer 100 to 200 km. They have also introduced a constraint which keeps the deep interior below the Curie point of iron for the first 1 to 1.5 b.y. so that it is able to carry the memory of an early field which magnetized the cold interior. The magnetized mare basalts and breccias cooled in this field from above the Curie point of iron ($\sim 800°$C.) and acquired a thermoremanent magnetization. The consequences that follow from this model are discussed.

094.012 **Parameters of the selenopotential model and the lunar deflections of the vertical.** M. Bursa.
Bull. Astron. Inst. Czechoslovakia, Vol. 26, 140 - 148 (1975).
   Stokes's lunar constants up to degree $n = 13$ (Michael and Blackshear, 1972), the selenocentric gravitational constant, and the angular velocity of the moon's rotation were used to compute the harmonic coefficients of degree n and order k in a development of the selenocentric radius-vector of the equipotential surfaces of the selenopotential, to derive the parameters of the best-fitting tri-axial ellipsoid, the harmonic coefficients in the development of the gravity on an equipotential surface, as well as its heights relative to the best-fitting tri-axial ellipsoid and the components of the deflections of the vertical.

094.013 **The influence of the internal constitution of the moon on its rotation.** K. S. Shakirov.
19th Astrometrical Conference 1972, (see 012.019), p. 261 - 264 (1975). In Russian.

094.014 **Constant rotations of the moon.** A. A. Gorynya.
19th Astrometrical Conference 1972, (see 012.019), p. 277 - 278 (1975). In Russian.

   Stress constraint on the thermal evolution of the moon. See Abstr. 094.219.

   Lunar impact melts and terrestrial analogs: their characteristics, formation and implications for lunar crustal evolution. See Abstr. 094.443.

Apollo 17 petrology and experimental determination of differentiation sequences in model moon compositions. See Abstr. 094.455.

Collisional breakup of particles in a planetary ring. See Abstr. 107.002.

Satellite-sized planetesimals and lunar origin. See Abstr. 107.007.

A co-accretional model of satellite formation. See Abstr. 107.012.

Moon, Global Properties

094.101 **Microwave emission spectrum of the moon: mean global heat flow and average depth of the regolith.**
S. J. Keihm, M. G. Langseth.
Science, Vol. 187, 64 - 66 (1975).

Earth-based observations of the lunar microwave brightness temperature spectrum at wavelengths between 5 and 500 centimeters, when reexamined in the light of physical property data derived from the Apollo program, tentatively support the high heat flows measured in situ and indicate that a regolith thickness between 10 and 30 meters may characterize a large portion of the lunar near side.

094.102 **Bounds on the P velocity for the whole moon.**
E. Nyland, E. J. Roebroek.
Nature, Vol. 253, 179 - 180 (1975).

The hedgehog algorithm suggests that either some Apollo seismic travel time data are in error (some layers dip) or a high velocity lid exists at a depth of 60 km in the moon. The velocity depth curves for the moon derived from the Apollo seismic data are probably not representative of the moon as a whole and lateral inhomogeneity in the near surface layers can affect the generality of measurements of arrival time and amplitude.

094.103 **Lunar microcraters: implications for the micro-meteoroid complex.** F. Hörz, D. E. Brownlee,
H. Fechtig, J. B. Hartung, D. A. Morrison, G. Neukum, E. Schneider, J. F. Vedder, D. E. Gault.
Planet. Space Sci., Vol. 23, 151 - 172 (1975). – Paper presented at the meeting of the Cosmic Dust Panel of COSPAR at Konstanz, Germany 1973.

The contributions of lunar microcrater studies to understand the overall micrometeoroid environment are summarized and compared to satellite data.

094.104 **Lunar magnetic anomalies and the Cayley formation.** D. E. Stuart-Alexander.
Nature, Vol. 253, 658 (1975).

094.105 **On connection of optical and geological-morphological division of the moon into regions.**
G. I. Ammalaimiev, N. N. Evsyukov, V. M. Litvinov.
Astron. Zhurn. Akad. Nauk SSSR, Vol. 52, 205 - 206 (1975).
In Russian. English translation in Soviet Astron., Vol. 19, No. 1.

Practical absence of connection of geological charts with the two-parametric chart constructed on the base of albedo and colour of the lunar surface is shown.

094.106 **A series of maps of optical characteristics of the lunar surface. Scale 1 : 5 000 000. 3 maps, 79 × 94 cm, polychromatic.** N. P. Barabashov.
Compiled by N. N. Evsyukov.
Astronomical Observatory of the Charkov State University, Institute of Space Research of the USSR Academy of Sciences.
Naukova dumka, Kiev. Price 32 Kop. (1973). In Russian.
Review in Astron. Zhurn. Akad. Nauk SSSR, Vol. 52, 210 - 213, 1975 (G. A. Burba).

094.107 **An ancient lunar magnetic dipole field.**
S. K. Runcorn.
Nature, Vol. 253, 701 - 703 (1975). – Letter.

094.108 **On the origin of the lunar smooth-plains.**
V. R. Oberbeck, F. Hörz, R. H. Morrison, W. L. Quaide, D. E. Gault.
The Moon, Vol. 12, 19 - 54 (1975).

If smooth-plains are a result of formation of basins or other distant large craters, then the plains materials are mainly ejecta of secondary craters of these basins or craters with only minor contributions of primary-crater or basin ejecta. This hypothesis is based on synthesis of knowledge of the mechanics of ejection of material from impact craters, photogeologic evidence, remote measurements of surface chemistry, and petrology of lunar samples. Examples of numerous secondary craters observed in and around the Cayley formation and other smooth-plains are presented. Evidence is given for significant lateral transport of highland debris by ejection from secondary craters and by landslides triggered by secondary impact. It is emphasized that the importance of secondary-impact cratering in the highlands has in general been underestimated and that this process must have been important in the evolution of the lunar surface.

094.109 **The great-circle pattern of large circular maria: product of an earth-moon encounter.**
R. J. Malcuit, G. R. Byerly, T. A. Vogel, T. R. Stoeckley.
The Moon, Vol. 12, 55 - 62 (1975).

The circular maria – Orientale, Imbrium, Serenitatis, Crisium, Smythii, and Tsiolkovsky – lie nearly on a lunar great circle. This pattern can be considered the result of a very close, non-capture encounter between moon and earth early in solar-system history. In the case of an encounter with a non-spinning moon, backfalling materials would be distributed along a lunar great circle. However, if the moon is rotating during the encounter, the backfall pattern will deviate from the great circle, the amount depending on the rate and direction of spin. Radiometric dates from mare rocks are consistent with this model of mare formation if the older mare rock dates are considered to date the encounter and younger dates are considered to date subsequent volcanic eruptions on a structurally weakened moon.

094.110 **Crater frequencies on lava-covered areas related to the moon's thermal history.**
C. S. Beals, R. W. Tanner.
The Moon, Vol. 12, 63 - 90 = Earth Phys. Branch, Dept. Energy, Mines and Resources, Ottawa, Contr. No. 523 (1975).

Frequency counts of craters down to 2 km diam as indicators of the relative ages of lunar features, have been made on 264 areas, including 15 terrae, 27 recognized maria, 174 flat-floored craters and 48 lava-covered areas with indefinite boundaries designated as 'marets'. It is considered that the time sequence of separate lava flows represented by the marets may be a reflection of physical processes within the moon responsible for the successive lava flows associated with the larger maria.

094.111 **Comparison of lunar ultraviolet reflectivity with that of terrestrial rock samples.**
J. H. Carver, B. H. Horton, D. G. McCoy, R. S. O'Brien, E. R. Sandercock.
The Moon, Vol. 12, 91 - 100 (1975).

The ultraviolet and visible albedos of a number of terrestrial basalts, gabbros and anorthosites have been investigated over the wavelength range 800 Å to 8000 Å and compared with previously reported measurements of the lunar albedo. For most of the terrestrial samples the albedo changed only slightly between visible and middle ultraviolet wavelengths in striking contrast to the moon where the ultraviolet albedo is about a factor of five or ten less than it is in the visible. The general shape of the lunar ultraviolet albedo may be caused by a layer of anorthositic fragments on the moon such as have been found to be a very abundant component of the Apollo 'coarse-fines'.

094.112 **Lunar crustal density profile from an analysis of Doppler gravity data.** A. W. G. Kunze.
The Moon, Vol. 12, 101 - 112 (1975).

Low altitude line-of-sight gravity data obtained by CSM

and LM radio tracking during several Apollo missions are used to construct an equispaced normalized vertical gravity net 30 km above selected lunar highland regions. Correlation of local vertical gravity anomalies with craters of different depth reveals a density increase with depth in the upper lunar highland crust. Crustal densities determined in this fashion are in good agreement with other, previously published crustal density values. The nature of the density increase implies a lunar crust consisting of fractured rather than competent rock.

094.113   **Some aspects of the minor element chemistry of lunar mare basalts.**   A. E. Ringwood.
The Moon, Vol. 12, 127 - 157 (1975).

The principal minor element (including Ti) characteristics of mare basalts which must be explained by an acceptable theory of petrogenesis are reviewed.These include:(i) The absolute abundances of incompatible elements vary over a twentyfold range yet the relative abundances within this group rarely deviate by more than a factor of two from the chondritic relative abundances. (ii) The sizes of the europium and strontium anomalies show a general trend to decrease as the absolute abundances of incompatible elements decrease. This trend is also one of increasing degree of partial melting and implies that the source region did not possess intrinsic Eu or Sr anomalies. (iii) Titanium seems to behave largely as an incompatible element. (iv) Many mare basalts have Rb/Sr model ages of about 4.5 b.y. whereas their crystallization ages are 3.2 - 3.8 b.y.

094.114   **Seismic effects from major basin formations on the moon and Mercury.**   P. H. Schultz, D. E. Gault.
The Moon, Vol. 12, 159 - 177 (1975).

Grooved and hilly terrains occur at the antipode of major basins on the moon (Imbrium, Orientale) and Mercury (Caloris). Such terrains may represent extensive landslides and surface disruption produced by impact-generated P-waves and antipodal convergence of surface waves.

094.115   **Lunar viscosity as obtained from the selenotherms.**   R. Meissner.
The Moon, Vol. 12, 179 - 191 = Inst. Geophys. Kiel, Contr. No. 82 (1975).

Based on the selenotherms $T(z)$ (=temperature-depth functions) and melting point-depth functions T_m (z) viscosity values $\eta(z)$ are calculated. According to two different creep laws used, two sets of viscosity values are obtained. Viscosities in the outer part of the moon are found to be larger than those anywhere on earth. These high values of η explain the large elasticity Q found in lunar seismograms. Viscosities below about 500 km in depth are so small that, at present, some kind of convection or a flow of matter is possible. Tide-generated moonquakes at depths of around 1000 km seem to be connected with some viscous process. From considerations of viscosities at the time period of mare filling, some selection of ancient selenotherms may be performed.

094.116   **A note on the asymmetric distribution of the impacts which created the lunar mare basins.**
N. A. Barricelli, R. Metcalfe.
The Moon, Vol. 12, 193 - 199 (1975).

In this note the authors are calling attention to the results they have obtained in an earlier computer investigation leading to an interpretation of the asymmetric distribution of the impacts which created many of the lunar mare basins (Barricelli and Metcalfe, 1969; Metcalfe and Barricelli, 1970). These results seem to have been overlooked in the Lunar Science Conference held at Houston last spring, where an investigation of the asymmetric distribution of lunar impacts was proposed. A brief description of the results obtained and the methods used in the investigation is given and a possible extension of the method for the investigation of impacts by

objects which are not earth satellites is discussed.

094.117   **A study of lunar impact crater size-distributions.**
G. Neukum, B. König, J. Arkani-Hamed.
The Moon, Vol. 12, 201 - 229 (1975).

Discrepancies in published crater frequency data prompted this study of lunar crater distributions. Effects modifying production size distributions of impact craters such as surface lava flows, blanketing by ejecta, superposition, infilling, and abrasion of craters, mass wasting, and the contribution of secondary and volcanic craters are discussed. The measured cumulative crater frequencies are used to obtain a general calibration size distribution curve by a normalization procedure. It is found that the lunar impact crater size distribution is largely constant in the size range 0.3 km $\leqslant D \leqslant$ 20 km for regions with formation ages between $\approx 3 \times 10^9$ yr and $\gtrsim 4 \times 10^9$ yr. Deviations of measured size distributions from the calibration distribution are strongly suggestive of the existence of processes having modified the primary impact crater population.

094.118   **Measurement of the lunar neutron density profile.**
D. S. Woolum, D. S. Burnett, M. Furst, J. R. Weiss.
The Moon, Vol. 12, 231 - 250 = Div. Geol. Planet. Sci., California Inst. Techn., *Pasadena,* Contr. No. 2539 (1975).

An in situ measurement of the lunar neutron density from 20 to 400 g cm^{-2} depth below the lunar surface was made by the Apollo 17 Lunar Neutron Probe Experiment using particle tracks produced by the ^{10}B (n, α) ^{7}Li reaction. Both the absolute magnitude and the depth profile of the neutron density are in good agreement with theoretical calculations by Lingenfelter, Canfield, and Hampel. However, relatively small deviations between experiment and theory in the effect of Cd absorption on the neutron density and in the relative ^{149}Sm to ^{157}Gd capture rates reported previously (Russ et al., 1972) imply that the true lunar ^{157}Gd capture rate is about one half of that calculated theoretically.

094.119   **Thicknesses of some lunar mare basalt flows and ejecta blankets based on chemical kinetic data.**
R. Brett.
Meteoritics, Vol. 9, 319 - 320 (1974). − Abstract.

094.120   **Can random impacts cause the observed Ar 39/40 age distribution for lunar highland rocks?**
J. B. Hartung.
Meteoritics, Vol. 9, 349 (1974). − Abstract.

094.121   **Bromine in the lunar atmosphere.**
S. Jovanovic, G. W. Reed, Jr.
Meteoritics, Vol. 9, 357 - 359 (1974). − Abstract.

094.122   **Cosmic ray tracks in lunar soils and lunar soils evolution paths.**   G. Poupeau.
Meteoritics, Vol. 9, 392 - 393 (1974). − Abstract.

094.123   **On the origin of the lunar anorthositic gabbro.**
E. Schonfeld.
Meteoritics, Vol. 9, 401 - 402 (1974). − Abstract.

094.124   **Experimentally reproduced lunar spinels in mare basaltic systems.**   T. M. Usselman.
Meteoritics, Vol. 9, 411 - 412 (1974). − Abstract.

094.125   **Some thoughts on the origin of lunar ANT-KREEP and mare basalts.**
H. Wakita, J. C. Laul, R. A. Schmitt.
Meteoritics, Vol. 9, 412 - 413 (1974). − Abstract.

094.126   **Correlated elements and the bulk composition of the moon, the earth and the parent body of the**

eucrites.   H. Wänke, H. Palme.
Meteoritics, Vol. 9, 414 - 415 (1974). – Abstract.

094.127 The lunar highlands: a mixture of lunar anorthosites and material from the late accumulation stage of the moon.   H. Wänke, H. Palme, H. Kruse, F. Teschke.
Meteoritics, Vol. 9, 415 - 416 (1974). – Abstract.

094.128 Extralunar component, age and origin of the regoliths at the Apollo-14 and Apollo-16 sites.
J. T. Wasson, C.-L. Chou, K. L. Robinson.
Meteoritics, Vol. 9, 421 (1974). – Abstract.

094.129 Lunar 'cataclysm': a misconception?   W. K. Hartmann.
Icarus, Vol. 24, 181 - 187 (1975).
    Models for cataclysms have not been clearly defined; some involve major impacts creating large basins (an idea incorporated in present and pre-Apollo evolutionary theories), while others involve catastrophist theories (abandonment of smooth, monotonic changes of cratering rate during time throughout the size spectrum). The scarcity of old (> 4AE) lunar rocks is here derived as a natural consequence of known paleocratering chronology. Explosive mega-regolith formation prior to 4AE brecciated and heated most earlier material, but no unaccountable cratering episode is evident.

094.130 Lunar microwave brightness temperature observations reevaluated in the light of Apollo program findings.   S. J. Keihm, M. G. Langseth.
Icarus, Vol. 24, 211 - 230 = Lamont–Doherty Geol. Obs. Columbia Univ., Contr. No. 2179 (1975).
    Remote observations of the lunar radiowave emission are reexamined in the light of physical property data accumulated through the Apollo program. It is found that a high heat flow, comparable to the heat flows measured at the Apollo 15 and 17 sites, is required to fit the available 5–20 cm wavelength remote data, and that a lunar surface layer relatively free of large boulders within the upper 10–30 m best fits the observations of a decreasing brightness temperature with wavelength for wavelengths greater than ~ 50 cm.

094.131 Formation of the lunar crust: an electrical source of heating.   C. P. Sonett, D. S. Colburn, K. Schwartz.
Icarus, Vol. 24, 231 - 255 (1975).
    Electrical heating of the outer layers of the moon just after formation is shown to lead to melting in a time of order $10^5 - 10^7$ yr. The heating mechanism is based upon eddy current induction from disordered magnetic fields swept outwards by an intense (T Tauri-like) plasma flow from the sun. Threshold temperature for the development of the intense electrical heating lies between 300–400°C depending upon the bulk electrical conductivity function used. The electrical mechanism is an alternate to intense short period accretion as a source of heat for the evolution of lunar maria and highlands, provided that long-lived radioactives are not swept to the surface from too large a melt volume during the initial thermal episode.

094.132 Formation of an iron-poor moon by partial capture, or: yet another exotic theory of lunar origin.
H. E. Mitler.
Icarus, Vol. 24, 256 - 268 (1975).
    It is shown how it is possible to explain the low abundance of iron and siderophiles in the moon in a natural way. This is done by an extension of Öpik's mechanism, whereby one or more planetoids pass through earth's Roche zone, are broken up, and have part of their material captured. Assuming the planetoids are differentiated, the iron core can easily escape capture. This process does not involve any dissipation mecha-

nisms and goes a long way toward explaining the peculiar bulk composition of the moon.

094.133 Lunar duststorms.   D. W. Hughes.
Nature, Vol. 254, 481 - 482 (1975).

094.134 Dark-haloed craters: their nature and origin.   A. Porter.
Strolling Astronomer, Vol. 25, 120 - 126 (1975).

094.135 On the possibility of cartography of the complex of optical characteristics of the moon.
N. N. Evsyukov.
Astron. Zhurn. Akad. Nauk SSSR, Vol. 52, 398 - 403 (1975).
In Russian. English translation in Soviet Astron., Vol. 19, No. 2.
    The possibility of cartography of the complete complex of optical characteristics of the moon by the method of photographic photometry is shown.

094.136 On the existence of the conducting layer inside the moon.   L. L. Vanyan, I. V. Yegorov (*Egorov*), E. B. Feinberg (*Eh. B. Fajnberg*).
Journ. Geophys. Res., Vol. 80, 1549 - 1550 (1975).
    An attempt is made to reconcile the conductivity profile for the moon derived from the inversion of the day side transfer function with the model fit derived profile gotten from dark side transient response data, corrected for the magnetic field compression in the lunar cavity.

094.137 On the composition of the lunar interior.   D. L. Anderson.
Journ. Geophys. Res., Vol. 80, 1555 - 1557 (1975).
    There is now abundant geophysical and geochemical evidence suggesting that the moon has a thick plagioclase rich outer shell. This is most easily explained by early and extensive melting of a CaO and Al_2O_3 rich moon followed by fractional crystallization involving plagioclase flotation. Melilite is probably an important constituent of the interior. This model explains the seismic velocities, the mean density, and the moment of inertia of the moon. The moon is 73–88 % high-temperature condensate.

094.138 On the origin of the mare basins.   R. B. Baldwin.
Proc. Fifth Lunar Sci. Conference, (see 012.003), Vol. 1, 1 - 10 (1974).

094.139 Ages of the lunar nearside light plains and maria.   J. M. Boyce, A. L. Dial, L. A. Soderblom.
Proc. Fifth Lunar Sci. Conference, (see 012.003), Vol. 1, 11 - 23 (1974).

094.140 Impact cratering models and their application to lunar studies – a geologist's view.   E. C. T. Chao.
Proc. Fifth Lunar Sci. Conference, (see 012.003), Vol. 1, 35 - 52 (1974).

094.141 Fresh lunar impact craters: review of variations with size.   K. A. Howard.
Proc. Fifth Lunar Sci. Conference, (see 012.003), Vol. 1, 61 - 69 (1974).

094.142 Smooth plains and continuous deposits of craters and basins.   V. R. Oberbeck, R. H. Morrison, F. Hörz, W. L. Quaide, D. E. Gault.
Proc. Fifth Lunar Sci. Conference, (see 012.003), Vol. 1, 111 - 136 (1974).

094.143 Problems in the interpretation of lunar mare stratigraphy and relative ages indicated by ejecta from small impact craters.
R. A. Young, W. J. Brennan, D. J. Nichols.

Proc. Fifth Lunar Sci. Conference, (see 012.003), Vol. 1, 159 - 170 (1974).

094.144 Lunar dark-mantle deposits: possible clues to the distribution of early mare deposits.   J. W. Head.
Proc. Fifth Lunar Sci. Conference, (see 012.003), Vol. 1, 207 - 222 (1974).

094.145 Impact-induced fractionation in the lunar highlands.
J. L. Warner, C. H. Simonds, W. C. Phinney.
Proc. Fifth Lunar Sci. Conference, (see 012.003), Vol. 1, 379 - 397 (1974).

094.146 A general model for the textural evolution of lunar soil.   J. F. Lindsay.
Proc. Fifth Lunar Sci. Conference, (see 012.003), Vol. 1, 861 - 878 (1974).

094.147 Grain size and the evolution of lunar soils.
D. S. McKay, R. M. Fruland, G. H. Heiken.
Proc. Fifth Lunar Sci. Conference, (see 012.003), Vol. 1, 887 - 906 (1974).

094.148 The role of horizontal transport as evaluated from the Apollo 15 and 16 orbital experiments.
I. Adler, M. Podwysocki, C. Andre, J. Trombka, E. Eller, R. Schmadebeck, L. Yin.
Proc. Fifth Lunar Sci. Conference, (see 012.003), Vol. 2, 975 - 979 (1974).

094.149 Abundances of the group IVB elements, Ti, Zr, and Hf and implications of their ratios in lunar materials.
W. D. Ehmann, L. L. Chyi.
Proc. Fifth Lunar Sci. Conference, (see 012.003), Vol. 2, 1015 - 1024 (1974).

094.150 Element concentrations from lunar orbital gamma-ray measurements.
A. E. Metzger, J. I. Trombka, R. C. Reedy, J. R. Arnold.
Proc. Fifth Lunar Sci. Conference, (see 012.003), Vol. 2, 1067 - 1078 (1974).

094.151 K and U systematics and average concentrations on the moon.   E. Schonfeld.
Proc. Fifth Lunar Sci. Conference, (see 012.003), Vol. 2, 1135 - 1145 (1974).

094.152 Oxygen isotopic constraints on the composition of the moon.
L. Grossman, R. N. Clayton, T. K. Mayeda.
Proc. Fifth Lunar Sci. Conference, (see 012.003), Vol. 2, 1207 - 1212 (1974).

094.153 Modeling the evolution of Sm and Eu abundances during lunar igneous differentiation.
D. F. Weill, G. A. McKay, S. J. Kridelbaugh, M. Grutzeck.
Proc. Fifth Lunar Sci. Conference, (see 012.003), Vol. 2, 1337 - 1352 (1974).

094.154 Taurus-Littrow chronology: some constraints on early lunar crustal development.
L. E. Nyquist, B. M. Bansal, H. Wiesmann, B.-M. Jahn.
Proc. Fifth Lunar Sci. Conference, (see 012.003), Vol. 2, 1515 - 1539 (1974).

094.155 Chronology of lunar basin formation.
O. A. Schaeffer, L. Husain.
Proc. Fifth Lunar Sci. Conference, (see 012.003), Vol. 2, 1541 - 1555 (1974).

094.156 U−Th−Pb systematics on lunar rocks and inferences about lunar evolution and the age of the moon.
F. Tera, G. J. Wasserburg.
Proc. Fifth Lunar Sci. Conference, (see 012.003), Vol. 2, 1571 - 1599 = Contr. Div. Geol. Planet. Sci., California Inst. Technology, *Pasadena*, No. 2492 (1974).

094.157 Possible effects of ^{39}Ar recoil in ^{40}Ar−^{39}Ar dating.
G. Turner, P. H. Cadogan.
Proc. Fifth Lunar Sci. Conference, (see 012.003), Vol. 2, 1601 - 1615 (1974).

094.158 A study of ^{204}Pb partition in lunar samples using terrestrial and meteoritic analogues.
R. O. Allen, Jr., S. Jovanovic, G. W. Reed, Jr.
Proc. Fifth Lunar Sci. Conference, (see 012.003), Vol. 2, 1617 - 1623 (1974).

094.159 Lunar basins: tentative characterization of projectiles, from meteoritic elements in Apollo 17 boulders.   J. W. Morgan, R. Ganapathy, H. Higuchi, U. Krähenbühl, E. Anders.
Proc. Fifth Lunar Sci. Conference, (see 012.003), Vol. 2, 1703 - 1736 (1974).

094.160 Simulation of lunar carbon chemistry: I. Solar wind contribution.   J. P. Bibring, A. L. Burlingame, J. Chaumont, Y. Langevin, M. Maurette, P. C. Wszolek.
Proc. Fifth Lunar Sci. Conference, (see 012.003), Vol. 2, 1747 - 1762 (1974).

094.161 Simulation of lunar carbon chemistry: II. Lunar winds contribution.   J. P. Bibring, A. L. Burlingame, Y. Langevin, M. Maurette, P. C. Wszolek.
Proc. Fifth Lunar Sci. Conference, (see 012.003), Vol. 2, 1763 - 1784 (1974).

094.162 Loss of oxygen, silicon, sulfur, and potassium from the lunar regolith.
R. N. Clayton, T. K. Mayeda, J. M. Hurd.
Proc. Fifth Lunar Sci. Conference, (see 012.003), Vol. 2, 1801 - 1809 (1974).

094.163 Accumulation and isotopic evolution of carbon on the lunar surface.
J. F. Kerridge, I. R. Kaplan, F. D. Lesley.
Proc. Fifth Lunar Sci. Conference, (see 012.003), Vol. 2, 1855 - 1868 (1974).

094.164 Hydrogen and fluorine in the surfaces of lunar samples.
D. A. Leich, R. H. Goldberg, D. S. Burnett, T. A. Tombrello.
Proc. Fifth Lunar Sci. Conference, (see 012.003), Vol. 2, 1869 - 1884 (1974).

094.165 Deuterium, hydrogen, and water content of lunar material.
L. Merlivat, M. Lelu, G. Nief, E. Roth.
Proc. Fifth Lunar Sci. Conference, (see 012.003), Vol. 2, 1885 - 1895 (1974).

094.166 Concentration-versus-depth profiles of hydrogen, carbon, and fluorine in lunar rock surfaces.
G. M. Padawer, E. A. Kamykowski, M. C. Stauber, M. D. D'Agostino, W. Brandt.
Proc. Fifth Lunar Sci. Conference, (see 012.003), Vol. 2, 1919 - 1937 (1974).

094.167 The association between carbide and finely divided metallic iron in lunar fines.
C. T. Pillinger, P. R. Davis, G. Eglinton, A. P. Gowar, A. J. T. Jull, J. R. Maxwell, R. M. Housley, E. H. Cirlin.

Proc. Fifth Lunar Sci. Conference, (see 012.003), Vol. 2, 1949 - 1961 (1974).

094.168 **Lunar neutron capture as a tracer for regolith dynamics.** D. S. Burnett, D. S. Woolum.
Proc. Fifth Lunar Sci. Conference, (see 012.003), Vol. 2, 2061 - 2074 = Contr. Div. Geol. Planet. Sci., California Inst. Technology, *Pasadena*, No. 2477 (1974).

094.169 **Regolith history from cosmic-ray-produced nuclides.** E. L. Fireman.
Proc. Fifth Lunar Sci. Conference, (see 012.003), Vol. 2, 2075 - 2092 (1974).

094.170 **Titanium spallation cross sections between 30 and 584 MeV and Ar³⁹ activities on the moon.**
F. Steinbrunn, E. L. Fireman.
Proc. Fifth Lunar Sci. Conference, (see 012.003), Vol. 2, 2205 - 2209 (1974).

094.171 **²²Na-²⁶Al chronology of lunar surface processes.** Y. Yokoyama, J. L. Reyss, F. Guichard.
Proc. Fifth Lunar Sci. Conference, (see 012.003), Vol. 2, 2231 - 2247 (1974).

094.172 **Thermal histories and crystal distributions in partly devitrified lunar glasses cooled by radiation.**
R. W. Hopper, P. Onorato, D. R. Uhlmann.
Proc. Fifth Lunar Sci. Conference, (see 012.003), Vol. 3, 2257 - 2273 (1974).

094.173 **The formation of lunar glasses.** D. R. Uhlmann, L. Klein, G. Kritchevsky, R. W. Hopper.
Proc. Fifth Lunar Sci. Conference, (see 012.003), Vol. 3, 2317 - 2331 (1974).

094.174 **Grain orientation in lunar soil.** A. Mahmood, J. K. Mitchell, W. D. Carrier III.
Proc. Fifth Lunar Sci. Conference, (see 012.003), Vol. 3, 2347 - 2354 (1974).

094.175 **Lunar soil density and porosity.** W. N. Houston, J. K. Mitchell, W. D. Carrier III.
Proc. Fifth Lunar Sci. Conference, (see 012.003), Vol. 3, 2361 - 2364 (1974).

094.176 **Mixing of the lunar regolith.** D. E. Gault, F. Hörz, D. E. Brownlee, J. B. Hartung.
Proc. Fifth Lunar Sci. Conference, (see 012.003), Vol. 3, 2365 - 2386 (1974).

094.177 **On the exposure history of the lunar regolith.** T. Gold, G. J. Williams.
Proc. Fifth Lunar Sci. Conference, (see 012.003), Vol. 3, 2387 - 2395 (1974).

094.178 **Micrometeoroid abrasion of lunar rocks: a Monte Carlo simulation.**
F. Hörz, E. Schneider, R. E. Hill.
Proc. Fifth Lunar Sci. Conference, (see 012.003), Vol. 3, 2397 - 2412 (1974).

094.179 **Observation of iron-rich coating on lunar grains and a relation to low albedo.**
T. Gold, E. Bilson, R. L. Baron.
Proc. Fifth Lunar Sci. Conference, (see 012.003), Vol. 3, 2413 - 2422 (1974).

094.180 **Solar wind sputtering on the lunar surface: equilibrium crater densities related to past and present micro**

particle influx rates. J. A. M. McDonnell, R. P. Flavill.
Proc. Fifth Lunar Sci. Conference, (see 012.003), Vol. 3, 2441 - 2449 (1974).

094.181 **The current micrometeoroid flux at the moon for masses ≤10⁻⁷ g from the Apollo window and Surveyor 3 TV camera results.** B. G. Cour-Palais.
Proc. Fifth Lunar Sci. Conference, (see 012.003), Vol. 3, 2451 - 2462 (1974).

094.182 **Lunar microcrater studies, derived meteoroid fluxes, and comparison with satellite-borne experiments.**
H. Fechtig, J. B. Hartung, K. Nagel, G. Neukum, D. Storzer.
Proc. Fifth Lunar Sci. Conference, (see 012.003), Vol. 3, 2463 - 2474 (1974).

094.183 **Lunar surface dynamics: some general conclusions and new results from Apollo 16 and 17.**
G. Crozaz, R. Drozd, C. Hohenberg, C. Morgan, C. Ralston, R. Walker, D. Yuhas.
Proc. Fifth Lunar Sci. Conference, (see 012.003), Vol. 3, 2475 - 2499 (1974).

094.184 **Lunar surface phenomena: solar flare track gradients, microcraters, and accretionary particles.**
G. E. Blanford, R. M. Fruland, D. S. McKay, D. A. Morrison.
Proc. Fifth Lunar Sci. Conference, (see 012.003), Vol. 3, 2501 - 2526 (1974).

094.185 **Lunar microcraters and their solar flare track record.** J. B. Hartung, D. Storzer.
Proc. Fifth Lunar Sci. Conference, (see 012.003), Vol. 3, 2527 - 2541 (1974).

094.186 **Charged-particle and micrometeorite impacts on the lunar surface.** S. A. Durrani, H. A. Khan, R. K. Bull, G. W. Dorling, J. H. Fremlin.
Proc. Fifth Lunar Sci. Conference, (see 012.003), Vol. 3, 2543 - 2560 (1974).

094.187 **Early active sun?: Radiation history of distinct components in fines.**
G. Crozaz, G. J. Taylor, R. M. Walker, M. G. Seitz.
Proc. Fifth Lunar Sci. Conference, (see 012.003), Vol. 3, 2591 - 2596 (1974).

094.188 **Simulated cosmic-ray induced U-fission tracks in artificial lunar soil and implications for the ²³⁸U-fission track dating of lunar surface samples.**
K. Thiel, G. Damm, W. Herr.
Proc. Fifth Lunar Sci. Conference, (see 012.003), Vol. 3, 2609 - 2621 (1974).

094.189 **Solar wind and micrometeorite alteration of the lunar regolith.**
R. M. Housley, E. H. Cirlin, N. E. Paton, I. B. Goldberg.
Proc. Fifth Lunar Sci. Conference, (see 012.003), Vol. 3, 2623 - 2642 (1974).

094.190 **Cosmic ray irradiation pattern at the Apollo 17 site: implications to lunar regolith dynamics.**
J. N. Goswami, D. Lal.
Proc. Fifth Lunar Sci. Conference, (see 012.003), Vol. 3, 2643 - 2662 (1974).

094.191 **Thermal conductivity of Apollo 16 lunar fines.** C. J. Cremers, H. S. Hsia.
Proc. Fifth Lunar Sci. Conference, (see 012.003), Vol. 3, 2703 - 2708 (1974).

094.192 **Ferromagnetic resonance properties of lunar fines**

and comparison with the properties of lunar analogues.   R. A. Weeks, D. Prestel.
Proc. Fifth Lunar Sci. Conference, (see 012.003), Vol. 3, 2709 - 2728 (1974).

094.193   Temperature dependence of the ferromagnetic resonance linewidth of lunar soils, iron and magnetite precipitates in simulated lunar glasses, and nonspherical metallic iron particles.   E. J. Friebele, D. L. Griscom, C. L. Marquardt, R. A. Weeks, D. Prestel.
Proc. Fifth Lunar Sci. Conference, (see 012.003), Vol. 3, 2729 - 2736 (1974).

094.194   Ferromagnetic resonance studies of thermal effects on lunar metallic Fe phases.
F.-D. Tsay, D. H. Live.
Proc. Fifth Lunar Sci. Conference, (see 012.003), Vol. 3, 2737 - 2746 (1974).

094.195   The permanent and induced magnetic dipole moment of the moon.
C. T. Russell, P. J. Coleman, Jr., B. R. Lichtenstein, G. Schubert.
Proc. Fifth Lunar Sci. Conference, (see 012.003), Vol. 3, 2747 - 2760 (1974).

094.196   Iron abundance and magnetic permeability of the moon.   C. W. Parkin, W. D. Daily, P. Dyal.
Proc. Fifth Lunar Sci. Conference, (see 012.003), Vol. 3, 2761 - 2778 (1974).

094.197   Iron distributions and metallic—ferrous ratios for Apollo lunar samples: Mössbauer and magnetic analyses.
G. P. Huffman, F. C. Schwerer, R. M. Fisher, T. Nagata.
Proc. Fifth Lunar Sci. Conference, (see 012.003), Vol. 3, 2779 - 2794 (1974).

094.198   Comparative magnetic studies of some Apollo 17 rocks and soils and their implications.
A. Brecher, W. H. Menke, K. R. Morash.
Proc. Fifth Lunar Sci. Conference, (see 012.003), Vol. 3, 2795 - 2814 (1974).

094.199   Magnetic properties of Apollo samples and implications for regolith formation.
G. W. Pearce, D. W. Strangway, W. A. Gose.
Proc. Fifth Lunar Sci. Conference, (see 012.003), Vol. 3, 2815 - 2826 (1974).

094.200   Magnetic properties of Apollo 11—17 lunar materials with special reference to effects of meteorite impact.   T. Nagata, N. Sugiura, R. M. Fisher, F. C. Schwerer, M. D. Fuller, J. R. Dunn.
Proc. Fifth Lunar Sci. Conference, (see 012.003), Vol. 3, 2827 - 2839 (1974).

094.201   Impact processes and lunar magnetism.
C. S. Cisowski, J. R. Dunn, M. Fuller, M. F. Rose, P. J. Wasilewski.
Proc. Fifth Lunar Sci. Conference, (see 012.003), Vol. 3, 2841 - 2858 (1974).

094.202   Lunar magnetic field palaeointensity determinations on Apollo 11, 16, and 17 rocks.
A. Stephenson, D. W. Collinson, S. K. Runcorn.
Proc. Fifth Lunar Sci. Conference, (see 012.003), Vol. 3, 2859 - 2871 (1974).

094.203   Remanent magnetization directions in a layered boulder from the South Massif.
S. K. Banerjee, K. Hoffman, G. Swits.

Proc. Fifth Lunar Sci. Conference, (see 012.003), Vol. 3, 2873 - 2881 (1974).

094.204   High-frequency lunar teleseismic events.
Y. Nakamura, J. Dorman, F. Duennebier, M. Ewing, D. Lammlein, G. Latham.
Proc. Fifth Lunar Sci. Conference, (see 012.003), Vol. 3, 2883 - 2890 (1974).

094.205   Internal friction in rocks and its relationship to volatiles on the moon.
B. R. Tittmann, R. M. Housley, G. A. Alers, E. H. Cirlin.
Proc. Fifth Lunar Sci. Conference, (see 012.003), Vol. 3, 2913 - 2918 (1974).

094.206   Measurements of lunar atmospheric loss rate.
R. R. Vondrak, J. W. Freeman, R. A. Lindeman.
Proc. Fifth Lunar Sci. Conference, (see 012.003), Vol. 3, 2945 - 2954 (1974).

094.207   Episodic release of ^{40}Ar from the interior of the moon.   R. R. Hodges, Jr., J. H. Hoffman.
Proc. Fifth Lunar Sci. Conference, (see 012.003), Vol. 3, 2955 - 2961 (1974).

094.208   Molecular flow of gases through lunar and terrestrial soils.
A. L. Frisillo, J. Winkler, D. W. Strangway.
Proc. Fifth Lunar Sci. Conference, (see 012.003), Vol. 3, 2963 - 2973 (1974).

094.209   Evidence for a high altitude distribution of lunar dust.   J. E. McCoy, D. R. Criswell.
Proc. Fifth Lunar Sci. Conference, (see 012.003), Vol. 3, 2991 - 3005 (1974).

094.210   Charge transfer in lunar materials: interpretation of ultraviolet—visible spectral properties of the moon.
B. M. Loeffler, R. G. Burns, J. A. Tossell, D. J. Vaughan, K. H. Johnson.
Proc. Fifth Lunar Sci. Conference, (see 012.003), Vol. 3, 3007 - 3016 (1974).

094.211   The application of trend surface analysis to a portion of the Apollo 15 X-ray fluorescence data.
M. H. Podwysocki, J. R. Weidner, C. G. Andre, A. L. Bickel, R. S. Lum, I. Adler, J. I. Trombka.
Proc. Fifth Lunar Sci. Conference, (see 012.003), Vol. 3, 3017 - 3024 (1974).

094.212   The geologic significance of some lunar gravity anomalies.   D. H. Scott.
Proc. Fifth Lunar Sci. Conference, (see 012.003), Vol. 3, 3025 - 3036 (1974).

094.213   Elevation profiles of the moon.
W. E. Brown, Jr., G. F. Adams, R. E. Eggleton, P. Jackson, R. Jordan, M. Kobrick, W. J. Peeples, R. J. Phillips, L. J. Porcello, G. Schaber, W. R. Sill, T. W. Thompson, S. H. Ward, J. S. Zelenka.
Proc. Fifth Lunar Sci. Conference, (see 012.003), Vol. 3, 3037 - 3048 (1974).

094.214   Apollo laser altimetry and inferences as to lunar structure.   W. M. Kaula, G. Schubert, R. E. Lingenfelter, W. L. Sjogren, W. R. Wollenhaupt.
Proc. Fifth Lunar Sci. Conference, (see 012.003), Vol. 3, 3049 - 3058 (1974).

094.215   Temperature and electrical conductivity of the lunar interior from magnetic transient measurements in

the geomagnetic tail.
P. Dyal, C. W. Parkin, W. D. Daily.
Proc. Fifth Lunar Sci. Conference, (see 012.003), Vol. 3,
3059 - 3071 (1974).

094.216 Polarized electromagnetic response of the moon.
C. P. Sonett, B. F. Smith, G. Schubert, D. S. Colburn, K. Schwartz.
Proc. Fifth Lunar Sci. Conference, (see 012.003), Vol. 3,
3073 - 3089 (1974).

094.217 Constraints on lunar structure.
A. M. Dainty, M. N. Toksöz, S. C. Solomon, K. R.
Anderson, N. R. Goins.
Proc. Fifth Lunar Sci. Conference, (see 012.003), Vol. 3,
3091 - 3114 (1974).

094.218 On the origin of mascons and moonquakes.
S. K. Runcorn.
Proc. Fifth Lunar Sci. Conference, (see 012.003), Vol. 3,
3115 - 3126 (1974).

**094.219 Stress constraint on the thermal evolution of the
moon.** J. Arkani-Hamed.
Proc. Fifth Lunar Sci. Conference, (see 012.003), Vol. 3,
3127 - 3134 (1974).

094.220 Zu einigen neuen Erkenntnissen über den Erdmond.
S. Marx.
Astron. in der Schule, 12. Jahrgang, p. 13 - 16 (1975).

094.221 Radar-range equation. S. W. Henriksen.
Proc. IEEE, Vol. 63, 813 - 814 (1975).
The radar-range equation, if not carefully applied, will
indicate that power returned from corner-cube reflectors varies
as the inverse fourth power of the range. Careful attention to
the meaning of each term in the equation leads to an inverse
second-power relationship. This result is confirmed by analysis
according to geometric optics.

**094.222 Dielectric properties of the first 100 meters of the
moon.** G. R. Olhoeft, D. W. Strangway.
Earth Planet. Sci. Letters, Vol. 24, 394 - 404 (1975).
Reviewing 92 measurements of lunar sample dielectric
constant versus density at frequencies above 100 kHz, gives
the relation $K' = (1.93 \pm 0.17)^p$ by regression analysis, where
K' is the dielectric constant of a soil or a solid at a density of
p g/cm^3. This formula is the geometric mean between the
dielectric constant of vacuum and the zero porosity dielectric
constant of lunar material. The dielectric constant increases
smoothly with depth, as a function of the soil compaction
only. The constant does not vary significantly with the temperature observed in a lunar day.

094.223 Mascons and the moon's orientation.
H. J. Melosh.
Earth Planet. Sci. Letters, Vol. 25, 322 - 326 = Div. Geol.
Planet. Sci., California Inst. Techn., Pasadena, Contr. No.
2548 (1975).
This paper reports the discovery of a relation between
the moments of inertia of the mascons (taken about the
moon's center) and the moon's moments of inertia. It is found
that the principal axes of the mascons alone are nearly parallel
to those of the moon. Possible explanations of this parallelism
are discussed. It appears that the mascons have been emplaced
in special sites whose position was controlled by the processes
which produced the farside highlands.

**094.224 Interaction of energetic nuclear particles in space
with the lunar surface.** R. M. Walker.
Annual Rev. Earth Planet. Sci., Vol. 3, (see 003.011), 99 - 128

(1975).

**094.225 The role of ballistic erosion and sedimentation in
lunar stratigraphy.** V. R. Oberbeck.
Rev. Geophys. Space Phys., Vol. 13, 337 - 362 (1975).
Many of the lunar surface formations have been emplaced by impact craters. Two mechanisms have been proposed for transport of crater ejecta; both the base surge and ballistic
transport mechanisms are reviewed in this paper.

094.226 Crater degradation on the moon, Mars, and Mercury.
D. Dzurisin, M. C. Malin.
Bull. American Astron. Soc., Vol. 7, 373 (1975). − Abstr. AAS.

**094.227 Mapping of optical parameters across the lunar sur-
face using silicon vidicon imaging: major results during the past year.** C. Pieters, T. B. McCord.
Bull. American Astron. Soc., Vol. 7, 373 (1975). − Abstr. AAS.

094.228 Electromagnetic induction in the moon.
L. L. Vanyan, I. V. Egorov.
The Moon, Vol. 12, 253 - 275, 277 - 298 (1975). In Russian
and English.
Electromagnetic induction in a stratified moon with a
trailing cavity is discussed. The influence of the moon wake is
studied by using a two-layer lunar model with a perfectly conductive core. The magnetic field is shown to be independent
of the wake length when that quantity is greater than 3 lunar
radii. Regions on the sunlit and dark sides where the magnetic
field may be described in terms of its first spatial harmonic
have been distinguished, together with the corresponding
errors admitted. It is in these regions that the electrical conductivity of the moon can be found with very high accuracy,
by simultaneous observations on the lunar surface and in the
undisturbed solar wind. Results of these observations can be
conveniently related to values of the apparent resistivity.

**094.229 Processes of lunar crater degradation: changes in
style with geologic time.** J. W. Head.
The Moon, Vol. 12, 299 - 329 (1975).
Lunar crater degradation can be divided into two time
periods based on differing styles and rates of crater degradation processes. Comparison of lunar radiometric age scales
and the relative degradation of crater morphologic features
for craters larger than about 5 km diam shows that Period I,
prior to about 3.85–3.95 b.y. ago, is characterized by a high
influx rate and by formation of large, multi-ringed basins.
Period II, from about 3.85–3.95 b.y. to present, is characterized by a much lower influx rate and lack of large multi-
ringed basins.

094.230 Tektites − volcanic ejecta from the moon?
W. S. Cameron, B. E. Lowrey.
The Moon, Vol. 12, 331 - 360 (1975).
The problem of the origin of the enigmatic tektites is
still unsolved. The two leading hypotheses − viz., ejecta from
terrestrial impacts, and ejecta from lunar volcanoes or lunar
impacts, each encounters serious difficulties. The rare
terrestrial strewn tektite fields require restrictive ballistic
trajectories from the moon. Calculations reveal that ellipses
of varying, decreasing sizes which depend on velocity of
vertical ejection from which ejecta will intersect the earth at
low-entrance angles occur on the nearside of the moon.
Reasonable velocities were chosen (2.55 to 3.0 km s^{-1}) and
these ellipses circumscribe areas with longitudes between 30
and 50°east and latitudes between 7°north and south of the
moon's equator. These areas were searched for evidence of
volcanism. These sites are identified and discussed after a review of the manifestations found from the various kinds of
terrestrial volcanism for which lunar counterparts were sought.

094.231 **Seismicity of the earth and moon.**
A. V. Nikolaev, I. N. Galkin.
Priroda, No. 5.75, p. 76 - 86 (1975). In Russian.

094.232 **On the temperature and structure of the moon inferred from electrical conductivity data of its interior.** B. A. Okulesskij.
AN SSSR. In-t kosmich. issled. Pr-167. Moskva, 1974, 16 pp. In Russian. — Abstr. in Referativ. Zhurn. 51. Astron., 4.51.326 (1975).

094.233 **Approximation of the surface of the moon's visible side by spherical functions.**
G. T. Yanovitskaya.
Astrometriya i Astrofizika, *Kiev*, vyp. (No.) 25, (see 003. 015), p. 58 - 65 (1975). In Russian.

The results are presented of a harmonic analysis of the data of the "Compiled catalogue of selenocentric positions of 2580 basic points on the moon" and the lunar marginal zone. The equation approximating the surface of the moon's visible side is obtained. The equation contains 24 terms of expansion of heights by spherical functions. The results are used to compose a hypsometric lunar map.

094.234 **Lunar and terrestrial fields.** P. J. Smith.
Nature, Vol. 255, 525 (1975).

094.235 **Some aspects of the minor element chemistry of lunar mare basalts.** A. E. Ringwood.
The Moon, Vol. 12, 127 - 157 (1975).

The principal minor element (including Ti) characteristics of mare basalts which must be explained by an acceptable theory of petrogenesis are reviewed. Recent hypotheses have proposed that mare basalts formed by equilibrium partial melting of pyroxene-rich cumulates which underlay and were complementary to the anorthositic crust. According to a variant of this category, the residual liquid resulting from fractional crystallization of the highlands and their complementary cumulates segregated to form an intermediate layer between the highlands and the underlying primary cumulates. These hypotheses are examined in detail and are rejected on several grounds. A new hypothesis based upon partial melting under conditions of surface or local equilibrium is proposed. It is assumed that the moon accreted from material which had ultimately formed by fractional condensation from a gas phase of appropriate composition. It is considered that this model is capable of explaining the principal minor element characteristics of mare basalts and is consistent with interpretations of the major element chemistry of their source region based upon experimental petrology.

094.236 **Seismic effects from major basin formations on the moon and Mercury.** P. H. Schultz, D. E. Gault.
The Moon, Vol. 12, 159 - 177 (1975).

Grooved and hilly terrains occur at the antipode of major basins on the moon (Imbrium, Orientale) and Mercury (Caloris). Such terrains may represent extensive landslides and surface disruption produced by impact-generated P-waves and antipodal convergence of surface waves. Order-of-magnitude calculations for an Imbrium-size impact (10^{34} erg) on the moon indicate P-wave-induced surface displacements of 10 m at the basin antipode that would arrive prior to secondary ejecta. Comparable surface waves would arrive subsequent to secondary ejecta impacts beyond 10^3 km and would increase in magnitude as they converge at the antipode.

094.237 **Lunar viscosity as obtained from the selenotherms.**
R. Meissner.
The Moon, Vol. 12, 179 - 191 (1975).

Based on the selenotherms $T(z)$ (= temperature-depth functions) and melting point-depth functions $T_m(z)$ viscosity values $\eta(z)$ are calculated. According to two different creep laws used, two sets of viscosity values are obtained. Viscosities in the outer part of the moon are found to be larger than those anywhere on earth. These high values of η explain the large elasticity found in lunar seismograms.

094.238 **A note on the asymmetric distribution of the impacts which created the lunar mare basins.**
N. A. Barricelli, R. Metcalfe.
The Moon, Vol. 12, 193 - 199 (1975).

In this note the authors are calling attention to the results they have obtained in an earlier computer investigation leading to an interpretation of the asymmetric distribution of the impacts which created many of the lunar mare basins (Barricelli and Metcalfe, 1969; Metcalfe and Barricelli, 1970). These results seem to have been overlooked in the Lunar Science Conference held at Houston last spring, where an investigation of the asymmetric distribution of lunar impacts was proposed. A brief description of the results obtained and the methods used in the investigation is given and a possible extension of the method for the investigation of impacts by objects which are not earth satellites is discussed.

094.239 **A study of lunar impact crater size-distributions.**
G. Neukum, B. König, J. Arkani-Hamed.
The Moon, Vol. 12, 201 - 229 (1975).

Discrepancies in published crater frequency data prompted this study of lunar crater distributions. Effects modifying production size distributions of impact craters such as surface lava flows, blanketing by ejecta, superposition, infilling, and abrasion of craters, mass wasting, and the contribution of secondary and volcanic craters are discussed. The resulting criteria have been applied in the determination of size distributions of unmodified impact crater populations in selected lunar regions of different ages. The measured cumulative crater frequencies are used to obtain a general calibration size distribution curve by a normalization procedure.

094.240 **Measurement of the lunar neutron density profile.**
D. S. Woolum, D. S. Burnett, M. Furst, J. R. Weiss.
The Moon, Vol. 12, 231 - 250 (1975).

An in situ measurement of the lunar neutron density from 20 to 400 g cm^{-2} depth below the lunar surface was made by the Apollo 17 Lunar Neutron Probe Experiment using particle tracks produced by the ^{10}B (n, α) ^{7}Li reaction. Both the absolute magnitude and the depth profile of the neutron density are in good agreement with theoretical calculations by Lingenfelter, Canfield, and Hampel. However, relatively small deviations between experiment and theory in the effect of Cd absorption on the neutron density and in the relative ^{149}Sm to ^{157}Gd capture rates reported previously (Russ et al., 1972) imply that the true lunar ^{157}Gd capture rate is about one half of that calculated theoretically.

094.241 **The moon: 15 years space research.**
V. V. Shevchenko.
Zemlya i Vselennaya, 1975, No. 3, p. 13 - 20. In Russian.

094.242 **Lunar gravity: the first farside map.**
A. J. Ferrari.
Science, Vol. 188, 1297 - 1300 (1975).

A global lunar gravity field has been determined from data on the long-term motion of the Apollo 15 and Apollo 16 subsatellites and Lunar Orbiter 5. The nearside gravity map resolves major mascon basins and, in general, is in excellent agreement with the results of Muller and Sjogren. The farside gravity map is characterized by broad positive gravity in the highland regions with interspersed, localized, negative anomalies corresponding to major ringed basins. A comparison between global gravity and topography indicates that a thicker farside crust could be responsible for these gravitational differences between the two lunar hemispheres.

094.243 On the improvement of the coordinates of craters of the visible hemisphere of the moon from photographic data of the space vehicle Zond 8.   V. S. Kislyuk.
Kosmich.Issled., Vol. 13, 415 - 422 (1975). In Russian.

094.244 On the conductive layer in the interior of the moon.   L. L. Van'yan, I. V. Egorov, Eh. B. Fajnberg.
Kosmich. Issled., Vol. 13, 436 - 437 (1975). In Russian.
Brief information.

094.245 Relative Mondhöhen und Lunar Orbiter Atlas.   L. Fritsch.
Sternenbote, 18. Jahrgang, p. 22 - 28 (1975).

094.246 General problems of volcanism on the moon.   Eh. N. Ehrlikh, I. V. Melekestsev, G. S. Shtejnberg.
Geotektonika, 1975, No. 1, p. 104 - 116. In Russian. — Abstr. in Referativ. Zhurn. 51. Astron., 6.51.367 (1975).

094.247 Applications of a nuclear technique for depth-sensitive hydrogen analysis: trapped H in lunar samples and the hydration of terrestrial obsidian.   D. A. Leich.
Thesis, California Inst. Techn., Pasadena (USA). 198 pp. University Microfilms Order No. 74-21,598 (1974).

094.248 Neutron stratigraphy in the lunar regolith.   G. P. Russ.
Thesis, California Inst. Techn., Pasadena (USA). 343 pp. University Microfilms Order No. 74-17,955 (1974).

094.249 Production of helium, neon, and argon in lunar material by solar cosmic ray protons.   J. R. Walton.
Thesis, Rice Univ., Houston, Texas (USA). 316 pp. University Microfilms Order No. 74-21,347 (1974).

094.250 Classification and possible approaches for solving problems of lunar astrometry.   I. V. Gavrilov.
19th Astrometrical Conference 1972, (see 012.019), p. 59 - 65 (1975). In Russian.

094.251 The gravitational field of the moon and its use for solving some astrometrical problems.
M. U. Sagitov.
19th Astrometrical Conference 1972, (see 012.019), p. 66 - 79 (1975). In Russian.

094.252 On the mixed boundary value problem for determination of the lunar potential.
V. V. Brovar, N. G. Ganifaeva.
19th Astrometrical Conference 1972, (see 012.019), p. 264 - 266 (1975). In Russian.

094.253 On the basic mechanical parameters of the moon.   G. A. Meshcheryakov, P. M. Zazulyak.
19th Astrometrical Conference 1972, (see 012.019), p. 266 - 272 (1975). In Russian.

094.254 Determination of the position of the mass center of the moon. Accuracy of the standard selenographic coordinate system developed at the Sternberg Astronomical Institute.   Yu. N. Lipskij, V. A. Nikonov.
19th Astrometrical Conference 1972, (see 012.019), p. 278 - 282 (1975). In Russian.

094.255 The accuracy of realization of the selenographic system in lunar-astrometrical works.
V. S. Kislyuk.
19th Astrometrical Conference 1972, (see 012.019), p. 282 - 286 (1975). In Russian.

094.256 On the compilation of a catalogue of positions of lunar surface points by referencing to stars.
N. G. Rizvanov.
19th Astrometrical Conference 1972, (see 012.019), p. 286 - 287 (1975). In Russian.

094.257 Investigation of the maps of the lunar limb zone.   L. N. Kizyun.
19th Astrometrical Conference 1972, (see 012.019), p. 287 - 290 (1975). In Russian.

094.258 Perfect and 'aborted' lunar craters.   Z. Carriere.
Ann. Soc. Sci. Bruxelles, (*Belgium*), Ser. 1, Vol. 88, 347 - 355 (1974). In French.
The magnetic theory of the origin of lunar craters postulates a mechanism analogous to the magnetohydrodynamic vorticity of sunspots, i.e. the ejection of a 'plasma' to form the characteristic crater formations. Two types of craters are distinguished: those completely formed and those in which the magnetic forces were insufficient to produce a complete crater.

094.259 Albedo-colour diagrams of the lunar surface.   N. N. Evsyukov.
Vestn. Khar'kov. Univ., No. 117, (Ser. Astron., vyp. (No.) 9), p. 27 - 32 (1974). In Russian.

094.260 Observations of moon-plasma interactions by Explorer 35 and Apollo surface and orbital experiments.   G. Schubert, B. R. Lichtenstein.
Solar wind three, (see 012.020), p. 433 - 439 (1974). — Summary.

094.261 Lunar exploration.   V. C. Juan.
Proc. Geol. Soc. China, No. 17, p. 5 - 11 (1974).

094.262 Experimental petrology of lunar highland basalt composition and applications to models for the lunar interior.   A. Raheim, D. H. Green.
Journ. Geol., (*USA*), Vol. 82, 607 - 622 (1974).

094.263 Lunar mineralogy: a heavenly detective story.   J. V. Smith.
American Mineralogist, Vol. 59, 231 - 243 (1974).

094.264 General peculiarities of lunar volcanism.   E. N. Erlich (*Eh. N. Ehrlikh*), I. V. Melekestsev, G. S. Steinberg (*Shtejnberg*).
Modern Geol., (*GB*), Vol. 5, 31 - 43 (1974).

094.265 Intergrowths in lunar and terrestrial anorthosites with implications for lunar differentiation.
J. V. Smith, I. M. Steele.
American Mineralogist, Vol. 59, 673 - 680 (1974).
Pyroxènes, ilménite apparaissent comme des inclusions dans les plagioclases d'anorthosites lunaires; ils représentent des exsolutions à partir de plagioclases haute température contenant Fe, Mg, Ti et Si en excès dans des conditions anhydres; les inclusions d'amphiboles dans des plagioclases d'anorthosites archéennes sont aussi attribuées à des exsolutions mais en présence d'eau.

094.266 Ilmenite and armalcolite in Apollo 17 breccias.   I. M. Steele.
American Mineralogist, Vol. 59, 681 - 689 (1974).
Étude pétrographique et chimique d'une ilménite riche en Mg(MgO = 6.5 %) et d'armalcolites à Zr et Zr-Cr; les rapports Mg/Fe des armalcolites sont en corrélation positive avec les rapports Mg/Fe des silicates coexistants.

094.267 A gravity map of the moon's far side.
Science News, Vol. 106, 391 (1974).

The map shows gravity to be above normal over the lava-filled basins, but below normal over the unfilled ones. The logical conclusion, which is suggested, is that the mascons are caused by the lava flows, rather than, as has been suggested, by the density of great meteorites lying beneath the basins they created.

094.268 Some results of a reduction of photographs of the moon and Mars.
Yu. N. Lipskij, V. I. Chikmachev.
19th Astrometrical Conference 1972, (see 012.019), p. 290 - 295 (1975). In Russian.

094.269 The gravitational field and figure of the moon from data of artificial lunar stations taking into account the harmonic coefficients to the 7th order. V. V. Buzuk.
19th Astrometrical Conference 1972, (see 012.019), p. 295 - 300 (1975). In Russian.

094.270 Determination of the parameters of lunar mascons from an analysis of the vertical distribution of the gravitational potential and its derivatives. N. A. Chujkova.
19th Astrometrical Conference 1972, (see 012.019), p. 300 - 303 (1975). In Russian.

094.271 Parameters of lunar mascons.
V. V. Brovar, A. P. Yuzefovich.
19th Astrometrical Conference 1972, (see 012.019), p. 303 - 312 (1975). In Russian.

094.272 Representation of the anomalous gravitational field of the moon by a field of point masses.
L. V. Ogorodova, A. P. Yuzefovich, G. N. Bobkov.
19th Astrometrical Conference 1972, (see 012.019), p. 312 - 317 (1975). In Russian.

094.273 Stokes constants of mascons. V. V. Brovar.
19th Astrometrical Conference 1972, (see 012.019), p. 318 - 323 (1975). In Russian.

094.274 Thermal structure within the convecting moon.
J. Iriyama.
Journ. Phys. Earth, Vol. 21, 77 - 96 (1973).
Various types of energy released or consumed during the moon's evolution are analyzed and the energy balance of the moon is discussed. The total heat accumulated during the moon's evolution is about 1.8×10^{36} ergs (assuming no heat escaped by thermal convection). A linear stability problem for the moon's interior is considered. For appropriate values of physical parameters it is found that the moon's interior should be in a convective state. The internal temperature distribution in the presence of the convection current is given.

094.275 Energetics of the development of lunar calderas.
P. Hédervári.
Zeitschr. Geophys., Vol. 39, 1007 - 1021 (1973).
On the base of the morphological likeness between the terrestrial calderas of collapse-origin and many of the lunar ring-mount structures, the author supposes that many of the ring-like structures on the moon— especially among the larger features — really can be regarded as calderas of collapse-origin of gigantic dimensions. However — naturally — there may be crater-like features of impact origin on the moon as well. The author made some calculations regarding the energetics of the development of "lunar calderas." Thermal, collapse and potential energies were calculated for the case of models of different dimensions and on the basis of certain plausible suppositions.

Volcanoes of the earth, moon and Mars.
See Abstr. 003.062.

The voyages of Apollo.
See Abstr. 003.093.

Contribution à la détermination des positions lunaires par l'observation d'occultations au moyen d'un micromètre à double image. See Abstr. 034.131.

Première détermination d'une longue base terrestre par télémétrie laser-lune et localisation du réflecteur de Lunakhod I. See Abstr. 046.009.

On the chemical composition of the moon, Jupiter, meteorites and Am stars. See Abstr. 061.031.

Apollo 17 lunar surface cosmic ray experiment — measurement of heavy solar wind particles.
See Abstr. 074.040.

Lunar surface solar wind observations at the Apollo 12 and Apollo 15 sites. See Abstr. 074.078.

The "leader center in a cell" method in the three-dimensional nonstationary problem of interaction between the solar wind plasma and a conductive model of the moon (a numerical experiment). See Abstr. 074.105.

Remarks on the paper "The tidal loss of satellite-orbiting objects and its implications for the lunar surface" by M. J. Reid. See Abstr. 091.002.

Polarization of light reflected from rough planetary surface. See Abstr. 091.032.

On the influence of the mesorelief on the brightness distribution over a planetary disk. See Abstr. 091.035.

Some comparisons of impact craters on Mercury and the Moon. See Abstr. 092.020.

Multiringed basins — illustrated by Orientale and associated features. See Abstr. 094.434.

Chemistry of Apollo 16 and 17 samples: bulk composition, late stage accumulation and early differentiation of the moon. See Abstr. 094.500.

Provenance of KREEP and the exotic component: elemental and isotopic studies of grain size fractions in lunar soils. See Abstr. 094.504.

U—Th—Pb systematics of some Apollo 17 lunar samples and implications for a lunar basin excavation chronology. See Abstr. 094.508.

Abundances of C, N, H, He, and S in Apollo 17 soils from Stations 3 and 4: implications for solar wind exposure ages and regolith evolution. See Abstr. 094.515.

D/H and $^{18}O/^{16}O$ ratios of H_2O in the "rusty" breccia 66095 and the origin of "lunar water".
See Abstr. 094.518.

Solar flare and lunar surface process characterization at the Apollo 17 site. See Abstr. 094.533.

The interaction of hydrogen with Taurus-Littrow orange soil. See Abstr. 094.538.

Electrical conductivity of lunar surface rocks: laboratory measurements and implications for lunar interior temperatures. See Abstr. 094.546.

Glass production by shock waves in terrestrial and lunar rocks and implications regarding the origin of chondrules. See Abstr. 105.047.

Chemische Aspekte von meteoritischer und lunarer Materie.   See Abstr. 105.147.

Korrelierte Elemente. Ihre Bedeutung für die Erforschung der Fraktionierungsprozesse vor und nach der Bildung planetarer Körper.   See Abstr. 107.009.

Spectrophotometric observations of Mu Cephei and the moon from 4 to 8 microns.   See Abstr. 114.329.

On the variation of the cosmic ray composition in the past.   See Abstr. 143.027.

Moon, Local Properties

094.401 **The oxidation state of europium as an indicator of oxygen fugacity.** M. J. Drake.
Geochim. Cosmochim. Acta, Vol. 39, 55 - 64 (1975).

The distribution of Eu between plagioclase feldspar and magmatic liquid has been determined experimentally for basaltic and andesitic systems as a function of temperature and oxygen fugacity at one atmosphere total pressure. Using the approach of Philpotts the ratios Eu^{2+}/Eu^{3+} in plagioclase and coexisting magmatic liquid have been calculated. Oxygen fugacities for lunar ferrobasalts cluster tightly around $10^{-12.7}$. Data on achondritic meteorites are limited, but calculations indicate oxygen fugacities of two-to-five orders of magnitude lower than lunar ferrobasalts.

094.402 **Light element geochemistry of the Apollo 16 site.**
J. F. Kerridge, I. R. Kaplan, C. Petrowski, S. Chang.
Geochim. Cosmochim. Acta, Vol. 39, 137 - 162 = Inst. Geophys. Planet. Phys., Univ. Calif., *Los Angeles*, Publ. No. 1259 (1975).

Pronounced variations in abundances and isotopic compositions of some light elements in soils from the Apollo 16 site are interpreted in terms of differing degrees of solar wind exposure for an originally, and approximately, homogeneous regolith. Carbon abundances in soils are compatible with a model in which equilibrium is established, after 10^4–10^5 yr, between solar wind input and loss by H stripping. However, this model does not explain the observed C isotopic distribution, suggesting that other sources of C or other processes, or both, are also important.

094.403 **The Big Bertha consortium (lunar breccia 14321).**
Editorial note. S. R. Taylor.
Geochim. Cosmochim. Acta, Vol. 39, 227 (1975).

094.404 **Lunar polymict breccia 14321: a petrographic study.** R. A. Grieve, G. A. McKay, H. D. Smith, D. F. Weill.
Geochim. Cosmochim. Acta, Vol. 39, 229 - 245 (1975).

094.405 **Lunar polymict breccia 14321 : a compositional study of its principal components.** A. R. Duncan, S. M. McKay, J. W. Stoeser, M. M. Lindstrom, D. J. Lindstrom, J. S. Fruchter, G. G. Goles.
Geochim. Cosmochim. Acta, Vol. 39, 247 - 260 (1975).

094.406 **Meteoritic trace elements in lunar rock 14321, 184.**
J. W. Morgan, R. Ganapathy, U. Krähenbühl.
Geochim. Cosmochim. Acta, Vol. 39, 261 - 264 (1975).

094.407 **The life and times of Big Bertha: lunar breccia 14321.** A. R. Duncan, R. A. F. Grieve, D. F. Weill.
Geochim. Cosmochim. Acta, Vol. 39, 265 - 273 (1975).

094.408 **Investigation and simulation of metallic spherules from lunar soils.** P. J. Blau, J. I. Goldstein.
Geochim. Cosmochim. Acta, Vol. 39, 305 - 324 (1975).

Metallic spherules selected from the Apollo 11, 12, 14, 15 and 16 sites were studied by optical techniques as well as the electron probe and scanning electron microscope. The metallic spherules are probably produced from both lunar and meteoritic sources. Impact processes cause localized shock melting of metallic (and non-metallic) constituents at metal-sulfide phase interfaces in surface rocks and in the meteoritic projectile. The major source of metallic spherules is the metal phase present in the lunar rocks and soil. The large variation in spherule bulk compositions is attributed to the different meteoritic projectiles bombarding the moon,

metal phases of differing compositions in the lunar soils and rocks and to the experimental results which indicate that high S, high P alloys form two immiscible liquids when melted.

094.409 **Electron microscopy of irradiation effects in space.**
M. Maurette, P. B. Price.
Science, Vol. 187, 121 - 129 (1975).

Radiation-damaged lunar and meteoritic grains tell us about solar system history and synthesis of molecules in space.

094.410 **Analysis of thermodesorption products from lunar soil.**
Yu. A. Surkov, V. V. Shandor, Yu. P. Toporov, G. P. Vdovykin.
Geokhimiya, 1974, No. 10, p. 1516 - 1522. In Russian.
Abstr. in Referativ. Zhurn. 51. Astron., 2.51.279; 62. Issled. kosmich. prostranstva, 2.62.202 (1975).

094.411 **Oxygen determination in the Luna 16 regolith sample by the activation method using neutrons with 14 MeV energy.**
V. Kliment, T. Vandlik, V. Shchasnar.
Geokhimiya, 1974, No. 10, p. 1571 - 1572. In Russian.
Abstr. in Referativ. Zhurn. 51. Astron., 2.51.280 (1975).

094.412 **Determination of natural radionuclides in a Luna 16 regolith sample.** R. Tykva.
Geokhimiya, 1974, No. 10, p. 1573 - 1574. In Russian.
Abstr. in Referativ. Zhurn. 51. Astron., 2.51.281; 62. Issled. kosmich. prostranstva, 2.62.200 (1975).

094.413 **Lunar anorthosites.** O. A. Bogatikov.
Anortozity SSSR. Moskva, Nauka, 1974, p. 113 - 118. In Russian. – Abstr. in Referativ. Zhurn. 51. Astron., 2.51.283; 62. Issled. kosmich. prostranstva, 2.62.155 (1975).

094.414 **Primary cosmic radiation on the lunar surface.**
A. K. Lavrukhina, V. D. Gorin, L. L. Kashkarov, G. K. Ustinova.
Izv. AN SSSR. Ser. fiz., Vol. 38, 1806 - 1810 (1974). In Russian. – Abstr. in Referativ. Zhurn. 51. Astron., 2.51.481 (1975).

094.415 **Regular laser-radar measurements of the distance of light reflectors brought to the moon by Soviet and American space ships.** Yu. L. Kokurin, V. V. Kurbasov, V. F. Lobanov, A. N. Sukhanovskij.
Phys. in-t. AN SSSR, Preprint No. 121, 22 pp. Moskva (1974). In Russian. – Abstr. in Referativ. Zhurn. 62. Issled. kosmich. prostranstva, 2.62.242 (1975).

094.416 **Results of investigations of the mechanical properties of the lunar ground with the TOR-1 apparatus.**
A. I. Vedenin, V. V. Markachev.
Metrologiya i metody optiko-fiz. izmerenij. Moskva, 1974, p. 28 - 29. In Russian. – Abstr. in Referativ. Zhurn. 62. Issled. kosmich. prostranstva, 2.62.255 (1975).

094.417 **Results of an investigation of drilling instruments working aboard the automatic lunar stations Luna 16 and Luna 20.** V. P. Bulekov, A. S. Buyalo, Yu. V. Makarov, V. V. Markachev.
Metrologiya i metody optiko-fiz. izmerenij. Moskva, 1974, p. 29 - 30. In Russian. – Abstr. in Referativ. Zhurn. 62. Issled. kosmich prostranstva, 2.62.256 (1975).

094.418 **Plutonium-244 fission tracks: an alternative explanation for excess tracks in lunar whitlockites.**
P. Pellas, D. Storzer, I. D. Hutcheon, P. B. Price.
Science, Vol. 187, 862 - 864 (1975).

094.419 'Seasoning' of latent damage trails in lunar samples. H. A. Khan, I. Ahmad.
Nature, Vol. 254, 126 - 127 (1975). – Letter.

094.420 Metal-olivine geothermometry for slow-cooled lunar rocks. R. H. Hewins, J. I. Goldstein.
Meteoritics, Vol. 9, 350 (1974). – Abstract.

094.421 Rare-gas dating of Apollo 16 and 17 rocks. S. B. Kahl, D. Phinney, J. H. Reynolds.
Meteoritics, Vol. 9, 359 (1974). – Abstract.

094.422 Dunite 72417 – a chemical study. J. C. Laul, R. A. Schmitt, H. Wakita.
Meteoritics, Vol. 9, 364 - 365 (1974). – Abstract.

094.423 Trapped noble gases in Apollo 16 rocks: a possible indigenous lunar component.
B. D. Lightner, K. Marti.
Meteoritics, Vol. 9, 366 - 367 (1974). – Abstract.

094.424 Marble-cake clast: a miniature rock complex from the lunar highlands.
U. B. Marvin, D. B. Stoeser, J. Bower.
Meteoritics, Vol. 9, 377 - 379 (1974). – Abstract.

094.425 Ion microprobe analyses of lead and other volatiles in lunar sample 74220.
C. Meyer, Jr., D. S. McKay.
Meteoritics, Vol. 9, 382 - 383 (1974). – Abstract.

094.426 Chemical compositions and origin of Apollo 17 station 7 boulder samples. D. F. Nava, J. A. Philpotts, S. R. Winzer, R. K. L. Lum, S. Schuhmann, C. W. Kouns.
Meteoritics, Vol. 9, 383 - 384 (1974). – Abstract.

094.427 Apollo 14: light element geochemistry. C. Petrowski, I. R. Kaplan.
Meteoritics, Vol. 9, 390 - 392 (1974). – Abstract.

094.428 Excess volatiles in Apollo 17 lunar soils 74240 and 74260. E. Schonfeld.
Meteoritics, Vol. 9, 400 - 401 (1974). – Abstract.

094.429 Solar flare track ages for microcrater pits on lunar basalt 12002. D. Storzer, J. B. Hartung.
Meteoritics, Vol. 9, 409 - 410 (1974). – Abstract.

094.430 Magnetochemistry of returned samples from the Apollo landing sites. P. Wasilewski.
Meteoritics, Vol. 9, 418 - 420 (1974). – Abstract.

094.431 Solar nitrogen: evidence for a secular increase in the ratio of nitrogen-15 to nitrogen-14.
J. F. Kerridge.
Science, Vol. 188, 162 - 164 (1975).
   Solar wind nitrogen, implanted in lunar soil samples, exhibits isotopic variations that are related to the time, although not to the duration, of implantation, with earlier samples characterized by lower ratios of nitrogen-15 to nitrogen-14. An increase in the solar nitrogen-15 content during the lifetime of the lunar regolith is probably caused by spallation of oxygen-16 in the surface regions of the sun.

094.432 An interpretation of volcanic and structural features of crater Aitken.
W. B. Bryan, M.-L. Adams.
Proc. Fifth Lunar Sci. Conference, (see 012.003), Vol. 1, 25 - 34 (1974).

094.433 Thickness of mare material in the Tranquillitatis and Nectaris basins. R. A. De Hon.
Proc. Fifth Lunar Sci. Conference, (see 012.003), Vol. 1, 53 - 59 (1974).

094.434 Multiringed basins – illustrated by Orientale and associated features.
H. J. Moore, C. A. Hodges, D. H. Scott.
Proc. Fifth Lunar Sci. Conference, (see 012.003), Vol. 1, 71 - 100 (1974).

094.435 Structural history of southeastern Mare Serenitatis and adjacent highlands. W. R. Muehlberger.
Proc. Fifth Lunar Sci. Conference, (see 012.003), Vol. 1, 101 - 110 (1974).

094.436 Interpretation of ejecta formations at the Apollo 14 and 16 sites by a comparative analysis of experimental, terrestrial, and lunar craters.
D. Stöffler, M. R. Dence, G. Graup, M. Abadian.
Proc. Fifth Lunar Sci. Conference, (see 012.003), Vol. 1, 137 - 150 (1974).

094.437 A method for estimating the absolute ages of small Copernican craters and its application to the determination of Copernican meteorite flux.
G. A. Swann, V. S. Reed.
Proc. Fifth Lunar Sci. Conference, (see 012.003), Vol. 1, 151 - 158 (1974).

094.438 Orange glass: evidence for regional deposits of pyroclastic origin on the moon.
J. B. Adams, C. Pieters, T. B. McCord.
Proc. Fifth Lunar Sci. Conference, (see 012.003), Vol. 1, 171 - 186 = Planet. Astron. Lab., Mass. Inst. Technology, Cambridge, Mass., MITPAL Publ. No. 107 (1974).

094.439 A study of iron-rich particles on the surfaces of orange glass spheres from 74220.
P. M. Bell, A. El Goresy, H. K. Mao.
Proc. Fifth Lunar Sci. Conference, (see 012.003), Vol. 1, 187 - 191 (1974).

094.440 Apollo 17 orange glass: textural and morphological characteristics of devitrification. S. E. Haggerty.
Proc. Fifth Lunar Sci. Conference, (see 012.003), Vol. 1, 193 - 205 (1974).

094.441 Orange material in the Sulpicius Gallus Formation at the southwestern edge of Mare Serenitatis.
B. K. Lucchitta, H. H. Schmitt.
Proc. Fifth Lunar Sci. Conference, (see 012.003), Vol. 1, 223 - 234 (1974).

094.442 Glass-coated soil breccia 15205: selenologic history and petrologic constraints on the nature of its source region. R. F. Dymek, A. L. Albee, A. A. Chodos.
Proc. Fifth Lunar Sci. Conference, (see 012.003), Vol. 1, 235 - 260 = Contr. Div. Geol. Planet. Sci., California Inst. Technology, Pasadena, No. 2478 (1974).

094.443 Lunar impact melts and terrestrial analogs: their characteristics, formation and implications for lunar crustal evolution.
R. A. F. Grieve, A. G. Plant, M. R. Dence.
Proc. Fifth Lunar Sci. Conference, (see 012.003), Vol. 1, 261 - 273 (1974).

094.444 Electron petrographic study of some Apollo 17 breccias.
A. H. Heuer, J. M. Christie, J. S. Lally, G. L. Nord, Jr.

Proc. Fifth Lunar Sci. Conference, (see 012.003), Vol. 1, 275 - 286 (1974).

094.445 **Petrology and crystal chemistry of poikilitic anorthositic gabbro 77017.**
I. S. McCallum, E. A. Mathez, F. P. Okamura, S. Ghose.
Proc. Fifth Lunar Sci. Conference, (see 012.003), Vol. 1, 287 - 302 (1974).

094.446 **Petrology of clasts in lunar breccia 67915.**
E. Roedder, P. W. Weiblen.
Proc. Fifth Lunar Sci. Conference, (see 012.003), Vol. 1, 303 - 318 (1974).

094.447 **Shock-induced melting in anorthositic rock 60015 and a fragment of anorthositic breccia from the "picking pot" (70052).** C. B. Sclar, J. F. Bauer.
Proc. Fifth Lunar Sci. Conference, (see 012.003), Vol. 1, 319 - 336 (1974).

094.448 **Petrography and classification of Apollo 17 non-mare rocks with emphasis on samples from the Station 6 boulder.**
C. H. Simonds, W. C. Phinney, J. L. Warner.
Proc. Fifth Lunar Sci. Conference, (see 012.003), Vol. 1, 337 - 353 (1974).

094.449 **Petrology of a stratified boulder from South Massif, Taurus-Littrow.** D. B. Stoeser, U. B. Marvin, J. A. Wood, R. W. Wolfe, J. F. Bower.
Proc. Fifth Lunar Sci. Conference, (see 012.003), Vol. 1, 355 - 377 (1974).

094.450 **KREEP basalt: a possible partial melt from the lunar interior.** M. L. Crawford, L. S. Hollister.
Proc. Fifth Lunar Sci. Conference, (see 012.003), Vol. 1, 399 - 419 (1974).

094.451 **Spinel–silicate co-crystallization relations in sample 15555.** J. Dalton, L. S. Hollister.
Proc. Fifth Lunar Sci. Conference, (see 012.003), Vol. 1, 421 - 429 (1974).

094.452 **Igneous rocks from Apollo 16 rake samples.**
E. Dowty, K. Keil, M. Prinz.
Proc. Fifth Lunar Sci. Conference, (see 012.003), Vol. 1, 431 - 445 (1974).

094.453 **The petrology of the Apollo 17 mare basalts.**
J. Longhi, D. Walker, T. L. Grove, E. M. Stolper, J. F. Hays.
Proc. Fifth Lunar Sci. Conference, (see 012.003), Vol. 1, 447 - 469 (1974).

094.454 **Mare basalts from the Taurus-Littrow region of the moon.**
J. J. Papike, A. E. Bence, D. H. Lindsley.
Proc. Fifth Lunar Sci. Conference, (see 012.003), Vol. 1, 471 - 504 (1974).

094.455 **Apollo 17 petrology and experimental determination of differentiation sequences in model moon compositions.** F. N. Hodges, I. Kushiro.
Proc. Fifth Lunar Sci. Conference, (see 012.003), Vol. 1, 505 - 520 (1974).

094.456 **The stability of armalcolite: experimental studies in the system MgO–Fe–Ti–O.**
D. H. Lindsley, S. E. Kesson, M. J. Hartzman, M. K. Cushman.
Proc. Fifth Lunar Sci. Conference, (see 012.003), Vol. 1, 521 - 534 (1974).

094.457 **Equilibria bearing on the behavior of titanate phases during crystallization of iron silicate melts under strongly reducing conditions.** B. R. Lipin, A. Muan.
Proc. Fifth Lunar Sci. Conference, (see 012.003), Vol. 1, 535 - 548 (1974).

094.458 **Experimentally reproduced textures and mineral chemistry of Apollo 15 quartz normative basalts.**
G. Lofgren, C. H. Donaldson, R. J. Williams, O. Mullins, Jr., T. M. Usselman.
Proc. Fifth Lunar Sci. Conference, (see 012.003), Vol. 1, 549 - 567 (1974).

094.459 **Experimental liquid line of descent and liquid immiscibility for basalt 70017.**
M. J. Rutherford, P. C. Hess, G. H. Daniel.
Proc. Fifth Lunar Sci. Conference, (see 012.003), Vol. 1, 569 - 583 (1974).

094.460 **Formational history of lunar rocks: applications of experimental geochemistry of the opaque minerals.**
L. A. Taylor, K. L. Williams.
Proc. Fifth Lunar Sci. Conference, (see 012.003), Vol. 1, 585 - 596 (1974).

094.461 **The grain growth of iron: implications for the thermal conditions in a lunar ejecta blanket.**
T. M. Usselman, G. W. Pearce.
Proc. Fifth Lunar Sci. Conference, (see 012.003), Vol. 1, 597 - 603 (1974).

094.462 **Lunar rocks and glasses: a univariate and multifactorial analysis of chemical data.**
A. Carusi, G. Cavarretta, A. Coradini, O. Fanucci, R. Funiciello, M. Salomone, M. Fulchignoni, A. Taddeucci.
Proc. Fifth Lunar Sci. Conference, (see 012.003), Vol. 1, 605 - 620 (1974).

094.463 **Vapor-phase crystallization of iron in lunar breccias.**
U. S. Clanton, D. S. McKay, R. B. Laughon, G. H. Ladle.
Proc. Fifth Lunar Sci. Conference, (see 012.003), Vol. 1, 621 - 626 (1974).

094.464 **Taurus-Littrow TiO_2-rich basalts: opaque mineralogy and geochemistry.**
A. El Goresy, P. Ramdohr, O. Medenbach, H.-J. Bernhardt.
Proc. Fifth Lunar Sci. Conference, (see 012.003), Vol. 1, 627 - 652 (1974).

094.465 **Metal silicate relationships in Apollo 17 soils.**
J. I. Goldstein, R. H. Hewins, H. J. Axon.
Proc. Fifth Lunar Sci. Conference, (see 012.003), Vol. 1, 653 - 671 (1974).

094.466 **Evidence of extensive chemical reduction in lunar regolith samples from the Apollo 17 site.**
H. K. Mao, A. El Goresy, P. M. Bell.
Proc. Fifth Lunar Sci. Conference, (see 012.003), Vol. 1, 673 - 683 (1974).

094.467 **Ion microprobe mass analysis of plagioclase from "non-mare" lunar samples.**
C. Meyer, Jr., D. H. Anderson, J. G. Bradley.
Proc. Fifth Lunar Sci. Conference, (see 012.003), Vol. 1, 685 - 706 (1974).

094.468 **Opaque mineralogy: Apollo 17, rock 75035.**
H. O. A. Meyer, N. Z. Boctor.
Proc. Fifth Lunar Sci. Conference, (see 012.003), Vol. 1, 707 - 716 (1974).

094.469 Ti^{3+}/Ti^{4+} ratios in lunar pyroxenes: implications to depth of origin of mare basalt magma.
C.-M. Sung, R. M. Abu-Eid, R. G. Burns.
Proc. Fifth Lunar Sci. Conference, (see 012.003), Vol. 1, 717 - 726 (1974).

094.470 Crystal chemical control of element partitioning for coexisting chromite—ulvöspinel and pigeonite—augite in lunar rocks.
H. Takeda, M. Miyamoto, A. M. Reid.
Proc. Fifth Lunar Sci. Conference, (see 012.003), Vol. 1, 727 - 741 (1974).

094.471 β-FeOOH, akaganéite, in lunar rocks.
L. A. Taylor, H. K. Mao, P. M. Bell.
Proc. Fifth Lunar Sci. Conference, (see 012.003), Vol. 1, 743 - 748 (1974).

094.472 Spinel-bearing feldspathic-lithic fragments in Apollo 16 and 17 soil samples: clues to processes of early lunar crustal evolution.
P. W. Weiblen, B. N. Powell, F. K. Aitken.
Proc. Fifth Lunar Sci. Conference, (see 012.003), Vol. 1, 749 - 767 (1974).

094.473 Cation distribution and equilibrium temperature of pigeonite from basalt 15065.
T. Yajima, S. S. Hafner.
Proc. Fifth Lunar Sci. Conference, (see 012.003), Vol. 1, 769 - 784 (1974).

094.474 Petrology of the highlands massifs at Taurus-Littrow: an analysis of the 2—4 mm soil fraction.
A. E. Bence, J. W. Delano, J. J. Papike, K. L. Cameron.
Proc. Fifth Lunar Sci. Conference, (see 012.003), Vol. 1, 785 - 827 (1974).

094.475 Analysis of the grain size-frequency distributions of lunar fines. J. C. Butler, E. A. King, Jr.
Proc. Fifth Lunar Sci. Conference, (see 012.003), Vol. 1, 829 - 841 (1974).

094.476 Petrography of Apollo 17 soils.
G. Heiken, D. S. McKay.
Proc. Fifth Lunar Sci. Conference, (see 012.003), Vol. 1, 843 - 860 (1974).

094.477 Regolith compositions from the Apollo 17 mission.
B. Mason, S. Jacobson, J. A. Nelen, W. G. Melson, T. Simkin, G. Thompson.
Proc. Fifth Lunar Sci. Conference, (see 012.003), Vol. 1, 879 - 885 (1974).

094.478 Apollo 16: core 60004 — preliminary study of <1 mm fines. H. O. A. Meyer, R. H. McCallister.
Proc. Fifth Lunar Sci. Conference, (see 012.003), Vol. 1, 907 - 916 (1974).

094.479 Apollo 17 1—2 mm fines: mineralogy and petrology
I. M. Steele, A. J. Irving, J. V. Smith.
Proc. Fifth Lunar Sci. Conference, (see 012.003), Vol. 1, 917 - 924 (1974).

094.480 Apollo 17: comparative chemistry of olivines, pyroxenes, and plagioclases from regolith samples 74220,87, 74241,40, and 75081,52.
H. C. J. Taylor, J. L. Carter.
Proc. Fifth Lunar Sci. Conference, (see 012.003), Vol. 1, 925 - 933 (1974).

094.481 Preservation of lunar core samples: preparation and interpretation of three-dimensional stratigraphic sections. R. Fryxell, G. Heiken.
Proc. Fifth Lunar Sci. Conference, (see 012.003), Vol. 1, 935 - 966 (1974).

094.482 Stabilization of lunar core samples.
J. S. Nagle, M. B. Duke.
Proc. Fifth Lunar Sci. Conference, (see 012.003), Vol. 1, 967 - 973 (1974).

094.483 Elemental composition of Apollo 17 fines and rocks. A. O. Brunfelt, K. S. Heier, B. Nilssen, E. Steinnes, B. Sundvoll.
Proc. Fifth Lunar Sci. Conference, (see 012.003), Vol. 2, 981 - 990 (1974).

094.484 Descartes Mountains and Cayley Plains: composition and provenance.
M. J. Drake, G. J. Taylor, G. G. Goles.
Proc. Fifth Lunar Sci. Conference, (see 012.003), Vol. 2, 991 - 1008 (1974).

094.485 Chemical and mineralogical composition of Surveyor 3 scoop sample 12029,9.
E. J. Dwornik, C. S. Annell, R. P. Christian, F. Cuttitta, R. B. Finkelman, D. T. Ligon, Jr., H. J. Rose, Jr.
Proc. Fifth Lunar Sci. Conference, (see 012.003), Vol. 2, 1009 - 1014 (1974).

094.486 Primordial radioelement concentrations in rocks and soils from Taurus-Littrow.
J. S. Eldridge, G. D. O'Kelley, K. J. Northcutt.
Proc. Fifth Lunar Sci. Conference, (see 012.003), Vol. 2, 1025 - 1033 (1974).

094.487 Breccia 66055 and related clastic materials from the Descartes region, Apollo 16.
J. S. Fruchter, S. J. Kridelbaugh, M. A. Robyn, G. G. Goles.
Proc. Fifth Lunar Sci. Conference, (see 012.003), Vol. 2, 1035 - 1046 (1974).

094.488 Chemical studies of Apollo 16 and 17 samples.
J. C. Laul, D. W. Hill, R. A. Schmitt.
Proc. Fifth Lunar Sci. Conference, (see 012.003), Vol. 2, 1047 - 1066 (1974).

094.489 Compositional studies of the lunar regolith at the Apollo 17 site. M. D. Miller, R. A. Pacer, M.-S. Ma, B. R. Hawke, G. L. Lookhart, W. D. Ehmann.
Proc. Fifth Lunar Sci. Conference, (see 012.003), Vol. 2, 1079 - 1086 (1974).

094.490 Chemical compositions of some soils and rock types from the Apollo 15, 16, and 17 lunar sites.
D. F. Nava.
Proc. Fifth Lunar Sci. Conference, (see 012.003), Vol. 2, 1087 - 1096 (1974).

094.491 The relationships between geology and soil chemistry at the Apollo 17 landing site.
J. M. Rhodes, K. V. Rodgers, C. Shih, B. M. Bansal, L. E. Nyquist, H. Wiesmann, N. J. Hubbard.
Proc. Fifth Lunar Sci. Conference, (see 012.003), Vol. 2, 1097 - 1117 (1974).

094.492 Chemical composition of rocks and soils at Taurus-Littrow.
H. J. Rose, Jr., F. Cuttitta, S. Berman, F. W. Brown, M. K. Carron, R. P. Christian, E. J. Dwornik, L. P. Greenland.
Proc. Fifth Lunar Sci. Conference, (see 012.003), Vol. 2, 1119 - 1133 (1974).

094.493 Trace element evidence for a two-stage origin of
        some titaniferous mare basalts.   A. R. Duncan,
A. J. Erlank, J. P. Willis, M. K. Sher, L. H. Ahrens.
Proc. Fifth Lunar Sci. Conference, (see 012.003), Vol. 2,
1147 - 1157 (1974).

094.494 The history of lunar breccia 14267.
        G. Eglinton, B. J. Mays, C. T. Pillinger, S. O. Agrell,
J. H. Scoon, J. C. Dran, M. Maurette, E. Bowell, A. Dollfus,
J. E. Geake, L. Schultz, P. Signer.
Proc. Fifth Lunar Sci. Conference, (see 012.003), Vol. 2,
1159 - 1180 (1974).

094.495 Chemical evidence for the origin of 76535 as a
        cumulate.   L. A. Haskin, C.-Y. Shih, B. M. Ban-
sal, J. M. Rhodes, H. Wiesmann, L. E. Nyquist.
Proc. Fifth Lunar Sci. Conference, (see 012.003), Vol. 2,
1213 - 1225 (1974).

094.496 The chemical definition and interpretation of rock
        types returned from the non-mare regions of the
moon.   N. J. Hubbard, J. M. Rhodes, H. Wiesmann,
C.-Y. Shih, B. M. Bansal.
Proc. Fifth Lunar Sci. Conference, (see 012.003), Vol. 2,
1227 - 1246 (1974).

094.497 Possible REE anomalies of Apollo 17 REE patterns.
        A. Masuda, T. Tanaka, N. Nakamura, H. Kurasawa.
Proc. Fifth Lunar Sci. Conference, (see 012.003), Vol. 2,
1247 - 1253 (1974).

094.498 Origin of Apollo 17 rocks and soils.
        J. A. Philpotts, S. Schuhmann, C. W. Kouns,
R. K. L. Lum, S. Winzer.
Proc. Fifth Lunar Sci. Conference, (see 012.003), Vol. 2,
1255 - 1267 (1974).

094.499 The contamination of lunar highland rocks by
        KREEP: interpretation by mixing models.
E. Schonfeld.
Proc. Fifth Lunar Sci. Conference, (see 012.003), Vol. 2,
1269 - 1286 (1974).

094.500 Chemistry of Apollo 16 and 17 samples: bulk com-
        position, late stage accumulation and early differen-
tiation of the moon.
H. Wänke, H. Palme, H. Baddenhausen, G. Dreibus, E. Jagoutz, H. Kruse, B. Spettel, F. Teschke, R. Thacker.
Proc. Fifth Lunar Sci. Conference, (see 012.003), Vol. 2,
1307 - 1335 (1974).

094.501 $^{40}Ar-^{39}Ar$ studies of lunar breccias.
        E. C. Alexander, Jr., S. B. Kahl.
Proc. Fifth Lunar Sci. Conference, (see 012.003), Vol. 2,
1353 - 1373 (1974).

094.502 K−Ar analysis of Apollo 11 fines 10084.
        J. R. Basford.
Proc. Fifth Lunar Sci. Conference, (see 012.003), Vol. 2,
1375 - 1388 (1974).

094.503 Lead isotope systematics of some Apollo 17 soils
        and some separated components from 76501.
S. E. Church, G. R. Tilton.
Proc. Fifth Lunar Sci. Conference, (see 012.003), Vol. 2,
1389 - 1400 (1974).

094.504 Provenance of KREEP and the exotic component:
        elemental and isotopic studies of grain size fractions
in lunar soils.
N. M. Evensen, V. R. Murthy, M. R. Coscio, Jr.

Proc. Fifth Lunar Sci. Conference, (see 012.003), Vol. 2,
1401 - 1417 (1974).

094.505 High resolution argon analysis of neutron-irradiated
        Apollo 16 rocks and separated minerals.
E. K. Jessberger, J. C. Huneke, F. A. Podosek, G. J. Wasserburg.
Proc. Fifth Lunar Sci. Conference, (see 012.003), Vol. 2,
1419 - 1449 = Contr. Div. Geol. Planet. Sci., California Inst.
Technology, *Pasadena*, No. 2483 (1974).

094.506 Chronology of the Taurus-Littrow region III: ages
        of mare basalts and highland breccias and some re-
marks about the interpretation of lunar highland rock ages.
T. Kirsten, P. Horn.
Proc. Fifth Lunar Sci. Conference, (see 012.003), Vol. 2,
1451 - 1475 (1974).

094.507 Equilibration and ages: Rb−Sr studies of breccias
        14321 and 15265.
R. K. Mark, C.-N. Lee-Hu, G. W. Wetherill.
Proc. Fifth Lunar Sci. Conference, (see 012.003), Vol. 2,
1477 - 1485 (1974).

094.508 U−Th−Pb systematics of some Apollo 17 lunar
        samples and implications for a lunar basin excavation
chronology.   P. D. Nunes, M. Tatsumoto, D. M. Unruh.
Proc. Fifth Lunar Sci. Conference, (see 012.003), Vol. 2,
1487 - 1514 (1974).

094.509 On the duration of lava flow activity in Mare Tran-
        quillitatis.
A. Stettler, P. Eberhardt, J. Geiss, N. Grögler, P. Maurer.
Proc. Fifth Lunar Sci. Conference, (see 012.003), Vol. 2,
1557 - 1570 (1974).

094.510 Volatile and siderophilic trace elements in the soils
        and rocks of Taurus-Littrow.
P. A. Baedecker, C.-L. Chou, L. L. Sundberg, J. T. Wasson.
Proc. Fifth Lunar Sci. Conference, (see 012.003), Vol. 2,
1625 - 1643 (1974).

094.511 Volatile-element systematics and green glass in Apol-
        lo 15 lunar soils.
C.-L. Chou, P. A. Baedecker, R. W. Bild, J. T. Wasson.
Proc. Fifth Lunar Sci. Conference, (see 012.003), Vol. 2,
1645 - 1657 (1974).

094.512 Meteoritic and volatile elements in Apollo 16 rocks
        and in separated phases from 14306.
R. Ganapathy, J. W. Morgan, H. Higuchi, E. Anders, A. T.
Anderson.
Proc. Fifth Lunar Sci. Conference, (see 012.003), Vol. 2,
1659 - 1683 (1974).

094.513 Labile and nonlabile element relationships among
        Apollo 17 samples.
S. Jovanovic, G. W. Reed, Jr.
Proc. Fifth Lunar Sci. Conference, (see 012.003), Vol. 2,
1685 - 1701 (1974).

094.514 Composition of the gases associated with the mag-
        mas that produced rocks 15016 and 15065.
C. Barker.
Proc. Fifth Lunar Sci. Conference, (see 012.003), Vol. 2,
1737 - 1746 (1974).

094.515 Abundances of C, N, H, He, and S in Apollo 17 soils
        from Stations 3 and 4: implications for solar wind
exposure ages and regolith evolution.
S. Chang, K. Lennon, E. K. Gibson, Jr.
Proc. Fifth Lunar Sci. Conference, (see 012.003), Vol. 2,

1785 - 1800 (1974).

094.516 The distribution in lunar soil of hydrogen released by pyrolysis.
D. J. DesMarais, J. M. Hayes, W. G. Meinschein.
Proc. Fifth Lunar Sci. Conference, (see 012.003), Vol. 2, 1811 - 1822 (1974).

094.517 Sulfur abundances and distributions in the valley of Taurus-Littrow.
E. K. Gibson, Jr., G. W. Moore.
Proc. Fifth Lunar Sci. Conference, (see 012.003), Vol. 2, 1823 - 1837 (1974).

094.518 D/H and $^{18}O/^{16}O$ ratios of H_2O in the "rusty" breccia 66095 and the origin of "lunar water".
S. Epstein, H. P. Taylor, Jr.
Proc. Fifth Lunar Sci. Conference, (see 012.003), Vol. 2, 1839 - 1854 = Contr. Div. Geol. Planet. Sci., California Inst. Technology, *Pasadena*, No. 2481 (1974).

094.519 Total carbon and sulfur contents of Apollo 17 lunar samples.
C. B. Moore, C. F. Lewis, J. D. Cripe.
Proc. Fifth Lunar Sci. Conference, (see 012.003), Vol. 2, 1897 - 1906 (1974).

094.520 Solar wind nitrogen and indigenous nitrogen in Apollo 17 lunar samples.   O. Müller.
Proc. Fifth Lunar Sci. Conference, (see 012.003), Vol. 2, 1907 - 1918 (1974).

094.521 Light element geochemistry of the Apollo 17 site.
C. Petrowski, J. F. Kerridge, I. R. Kaplan.
Proc. Fifth Lunar Sci. Conference, (see 012.003), Vol. 2, 1939 - 1948 = Publ. Inst. Geophys. Planet. Phys., Univ. California, Los Angeles, No. 1336 (1974).

094.522 Sulfur concentrations and isotope ratios in Apollo 16 and 17 samples.   C. E. Rees, H. G. Thode.
Proc. Fifth Lunar Sci. Conference, (see 012.003), Vol. 2, 1963 - 1973 (1974).

094.523 Noble gases in Apollo 17 fines: mass fractionation effects in trapped Xe and Kr.
D. D. Bogard, W. C. Hirsch, L. E. Nyquist.
Proc. Fifth Lunar Sci. Conference, (see 012.003), Vol. 2, 1975 - 2003 (1974).

094.524 Solar, spallogenic, and radiogenic rare gases in Apollo 17 soils and breccias.
H. Hintenberger, H. W. Weber, L. Schultz.
Proc. Fifth Lunar Sci. Conference, (see 012.003), Vol. 2, 2005 - 2022 (1974).

094.525 Lunar trapped xenon.
B. D. Lightner, K. Marti.
Proc. Fifth Lunar Sci. Conference, (see 012.003), Vol. 2, 2023 - 2031 (1974).

094.526 Lunar breccia 14066: $^{81}Kr-^{83}Kr$ exposure age, evidence for fissiogenic xenon from ^{244}Pu, and rate of production of spallogenic ^{126}Xe.   B. Srinivasan.
Proc. Fifth Lunar Sci. Conference, (see 012.003), Vol. 2, 2033 - 2044 (1974).

094.527 Evidence for solar cosmic ray proton-produced neon in fines 67701 from the rim of North Ray Crater.
J. R. Walton, D. Heymann, J. L. Jordan, A. Yaniv.
Proc. Fifth Lunar Sci. Conference, (see 012.003), Vol. 2, 2045 - 2060 (1974).

094.528 Depth profiles of ^{53}Mn in lunar rocks and soils.
M. Imamura, K. Nishiizumi, M. Honda, R. C. Finkel, J. R. Arnold, C. P. Kohl.
Proc. Fifth Lunar Sci. Conference, (see 012.003), Vol. 2, 2093 - 2103 (1974).

094.529 The saturated activity of ^{26}Al in lunar samples as a function of chemical composition and the exposure ages of some lunar samples.   J. E. Keith, R. S. Clark.
Proc. Fifth Lunar Sci. Conference, (see 012.003), Vol. 2, 2105 - 2119 (1974).

094.530 Determination of natural and cosmic ray induced radionuclides in Apollo 17 lunar samples.
J. E. Keith, R. S. Clark, L. J. Bennett.
Proc. Fifth Lunar Sci. Conference, (see 012.003), Vol. 2, 2121 - 2138 (1974).

094.531 Cosmogenic radionuclides in samples from Taurus-Littrow: effects of the solar flare of August 1972.
G. D. O'Kelley, J. S. Eldridge, K. J. Northcutt.
Proc. Fifth Lunar Sci. Conference, (see 012.003), Vol. 2, 2139 - 2147 (1974).

094.532 Rare gases and trace elements in Apollo 15 drill core fines: depositional chronologies and K−Ar ages, and production rates of spallation-produced ^3He, ^{21}Ne, and ^{38}Ar versus depth.   R. O. Pepin, J. R. Basford, J. C. Dragon, M. R. Coscio, Jr., V. R. Murthy.
Proc. Fifth Lunar Sci. Conference, (see 012.003), Vol. 2, 2149 - 2184 (1974).

094.533 Solar flare and lunar surface process characterization at the Apollo 17 site.
L. A. Rancitelli, R. W. Perkins, W. D. Felix, N. A. Wogman.
Proc. Fifth Lunar Sci. Conference, (see 012.003), Vol. 2, 2185 - 2203 (1974).

094.534 Radioactive rare gases, tritium, hydrogen, and helium in the sample return container, and in the Apollo 16 and 17 drill stems.
R. W. Stoenner, R. Davis, Jr., E. Norton, M. Bauer.
Proc. Fifth Lunar Sci. Conference, (see 012.003), Vol. 2, 2211 - 2229 (1974).

094.535 Examination of Apollo 17 surface fines for porphyrins and aromatic hydrocarbons.   J. H. Rho.
Proc. Fifth Lunar Sci. Conference, (see 012.003), Vol. 2, 2249 - 2250 (1974).

094.536 Uniformity of the uranium content of lunar green and orange glasses.
R. L. Fleischer, H. R. Hart, Jr.
Proc. Fifth Lunar Sci. Conference, (see 012.003), Vol. 3, 2251 - 2255 (1974).

094.537 Some surface properties of Apollo 17 soils.
H. F. Holmes, E. L. Fuller, Jr., R. B. Gammage.
Proc. Fifth Lunar Sci. Conference, (see 012.003), Vol. 3, 2275 - 2285 (1974).

094.538 The interaction of hydrogen with Taurus-Littrow orange soil.   D. A. Cadenhead, W. G. Buergel.
Proc. Fifth Lunar Sci. Conference, (see 012.003), Vol. 3, 2287 - 2300 (1974).

094.539 The interaction of water vapor with a lunar soil, a compacted soil, and a cinder-like rock fragment.
D. A. Cadenhead, J. R. Stetter.
Proc. Fifth Lunar Sci. Conference, (see 012.003), Vol. 3, 2301 - 2316 (1974).

094.540 **Shock compression and adiabatic release of lunar fines from Apollo 17.**
T. J. Ahrens, D. M. Cole.
Proc. Fifth Lunar Sci. Conference, (see 012.003), Vol. 3, 2333 - 2345 = Contr. Div. Geol. Planet. Sci., California Inst. Technology, *Pasadena*, No. 2475 (1974).

094.541 **Optical properties of the Apollo 15 deep core samples.** T. Gold, E. Bilson, R. L. Baron.
Proc. Fifth Lunar Sci. Conference, (see 012.003), Vol. 3, 2355 - 2359 (1974).

094.542 **Auger electron spectroscopy of lunar samples.**
R. W. Grant, R. M. Housley, F. J. Szalkowski, H. L. Marcus.
Proc. Fifth Lunar Sci. Conference, (see 012.003), Vol. 3, 2423 - 2439 (1974).

094.543 **Improved determination of the long-term average Fe spectrum from ~1 to ~460 MeV/amu.**
I. D. Hutcheon, D. Macdougall, P. B. Price.
Proc. Fifth Lunar Sci. Conference, (see 012.003), Vol. 3, 2561 - 2576 (1974).

094.544 **Apollo 17 particle track studies: surface residence times and fission track ages for orange glass and large boulders.**
I. D. Hutcheon, D. Macdougall, J. Stevenson.
Proc. Fifth Lunar Sci. Conference, (see 012.003), Vol. 3, 2597 - 2608 (1974).

094.545 **Electrical properties of sample 70215: low-frequency corrections.** R. Alvarez.
Proc. Fifth Lunar Sci. Conference, (see 012.003), Vol. 3, 2663 - 2671 (1974).

094.546 **Electrical conductivity of lunar surface rocks: laboratory measurements and implications for lunar interior temperatures.**
F. C. Schwerer, G. P. Huffman, R. M. Fisher, T. Nagata.
Proc. Fifth Lunar Sci. Conference, (see 012.003), Vol. 3, 2673 - 2687 (1974).

094.547 **Thermoluminescence and thermal environment of some Apollo 17 fines.**
S. A. Durrani, F. S. W. Hwang.
Proc. Fifth Lunar Sci. Conference, (see 012.003), Vol. 3, 2689 - 2702 (1974).

094.548 **Elastic-wave velocities and thermal diffusivities of Apollo 17 rocks and their geophysical implications.**
H. Mizutani, M. Osako.
Proc. Fifth Lunar Sci. Conference, (see 012.003), Vol. 3, 2891 - 2901 (1974).

094.549 **Rock elastic properties and near-surface structure at Taurus-Littrow.**
R. Trice, N. Warren, O. L. Anderson.
Proc. Fifth Lunar Sci. Conference, (see 012.003), Vol. 3, 2903 - 2911 = Publ. Inst. Geophys. Planet. Phys., Univ. California, Los Angeles, No. 1338 (1974).

094.550 **Implications of elastic wave velocities for Apollo 17 rock powders.**
P. Talwani, A. Nur, R. L. Kovach.
Proc. Fifth Lunar Sci. Conference, (see 012.003), Vol. 3, 2919 - 2926 (1974).

094.551 **Ultrasonic attenuation: Q measurements on 70215,29.**
N. Warren, R. Trice, J. Stephens.

Proc. Fifth Lunar Sci. Conference, (see 012.003), Vol. 3, 2927 - 2938 = Publ. Inst. Geophys. Planet. Phys., Univ. California, Los Angeles, No. 1337 (1974).

094.552 **Shock induced ultra-sound absorption in lunar anorthosite.** C. Herminghaus, H. Berckhemer.
Proc. Fifth Lunar Sci. Conference, (see 012.003), Vol. 3, 2939 - 2943 (1974).

094.553 **Effects of vertical stress, temperature and density on the dielectric properties of lunar samples 72441,12, 15301,38 and a terrestrial basalt.**
A. L. Frisillo, G. R. Olhoeft, D. W. Strangway.
Earth Planet. Sci. Letters, Vol. 24, 345 - 356 (1975).

094.554 **Olivine-matrix reactions in thermally metamorphosed Apollo 14 breccias.**
K. L. Cameron, G. W. Fisher.
Earth Planet. Sci. Letters, Vol. 25, 197 - 207 (1975).
The paper gives a petrologic and chemical study of lunar highland samples 14311, 14304 and 14319 respectively.

094.555 **Magnetic "zig-zag" behavior in lunar rocks.**
K. A. Hoffman, S. K. Banerjee.
Earth Planet. Sci. Letters, Vol. 25, 331 - 337 (1975).
The authors present an extended investigation of lunar olivine basalt sample 15535,28 which illustrates unorthodox alternating field demagnetization behavior as also observed to varying degrees in many lunar rocks. It is suggested that the above behavior is due to the presence of a few planar, multidomain grains representing a local mineral fabric. These "super-grains" do not demagnetize.

094.556 **Field-strength dependence of the viscous remanent magnetization in lunar samples.**
J. G. Carnes, D. W. Strangway, W. A. Gose.
Earth Planet. Sci. Letters, Vol. 26, 1 - 7 (1975).

094.557 **^{87}Rb − ^{86}Sr age of rocks from the Apollo 15 landing site and significance of internal isochrons.**
J. L. Birck, S. Fourcade, C. J. Allegre.
Earth Planet. Sci. Letters, Vol. 26, 29 - 35 (1975).
Internal isochrons for two Apollo 15 rocks give an age of $(3.34 \pm 0.09) \times 10^9$ years. The significance of these results is discussed.

094.558 **Mineralogy and petrology of sample 67075 and the origin of lunar anorthosites.**
I. S. McCallum, F. P. Okamura, S. Ghose.
Earth Planet. Sci. Letters, Vol. 26, 36 - 53 (1975).

094.559 **76535: an old lunar rock.** D. D. Bogard, L. E. Nyquist, B. M. Bansal, H. Wiesmann, C. Y. Shih.
Earth Planet. Sci. Letters, Vol. 26, 69 - 80 (1975).
Total ^{40}Ar − K ages of whole rock and plagioclase range from $4.40 - 4.54 \times 10^9$ years.

094.560 **Some questions of the dynamics of formation of glass particles of regolith.**
Yu. B. Chernyak, M. D. Nusinov.
AN SSSR. In-t kosmich. issled. Pr-189. Moskva, 1974. 27 pp. In Russian. − Abstr. in Referativ. Zhurn. 62. Issled. kosmich. prostranstva, 3.62.224 (1975).

094.561 **Investigation of the mechanical, thermal and electrical properties of lunar rocks and some analogues.**
V. V. Rzhevskij, A. A. Silin, V. V. Shvarev, E. A. Dukhovskoj, N. T. Kruglov, A. R. Golovkin, R. G. Petrochenkov.
Fiz. svojstva gorn. porod i mineralov pri vysokikh davleniyakh i temperaturakh. Tbilisi, 1974, p. 51 - 53. In Russian. − Abstr. in Referativ. Zhurn. 62. Issled. kosmich. prostranstva, 3.62.245;

51. Astron., 4.51.350 (1975).

094.562 **Radar investigation of the moon with cosmic instruments.** N. N. Krupenio.
Zemlya i Vselennaya, 1975, No. 2, p. 21 - 28. In Russian.

094.563 **Rare-gas dating of Apollo 16 and 17 rocks.**
S. B. Kahl, D. Phinney, J. H. Reynolds.
Meteoritics, Vol. 10, 88 (1975). — Abstract.

094.564 **Glass production by shock waves in terrestrial and lunar rocks and implications regarding the origin of chondrules.** S. W. Kieffer.
Meteoritics, Vol. 10, 89 - 90 (1975). — Abstract.

094.565 **Shorty crater, noble gases and chronology.**
P. Eberhardt, O. Eugster, J. Geiss, N. Grögler,
N. Jungck, P. Maurer, M. Morgeli, A. Stettler.
Meteoritics, Vol. 10, 93 - 94 (1975). — Abstract.

094.566 **Spallogenic and "cosmogenic" xenon-yields from barium in Thin- and Thick-Target-experiments.**
W. A. Kaiser, K. P. Rosner, W. Herr.
Meteoritics, Vol. 10, 94 - 95 (1975). — Abstract.

094.567 **Shock reduction and shock melting in norite 78235 from Station 8, Taurus—Littrow valley.**
C. B. Sclar.
Meteoritics, Vol. 10, 97 - 98 (1975). — Abstract.

094.568 **Systematic differences of heights of the lunar marginal zone from charts of Hayn, Nefed'ev, Watts and the catalogue compiled at the Main Astronomical Observatory of the Ukrainian Academy of Sciences (Goloseevo).** A. S. Duma.
Astrometriya i Astrofizika, *Kiev*, vyp. (No.) 25, (see 003. 015), p. 49 - 57 (1975). In Russian.
  The heights in the lunar marginal zone from the charts of Hayn, Nefed'ev and the catalogue of Goloseevo were compared with those from the charts of Watts.

094.569 **Some problems of the dynamics of formation of glass regolith particles.**
Yu. B. Chernyak, M. D. Nusinov.
AN SSSR. In-t kosmich. issled. Pr-189. Moskva, 1974. 27 pp.
In Russian. — Abstr. in Referativ. Zhurn. 51. Astron., 5.51.319 (1975).

094.570 **The nature of V-shaped mountain chains on the moon.** A. V. Bugaevskij.
Zemlya i Vselennaya, 1975, No. 3, p. 37. In Russian.

094.571 **Med månen som mål. VII — Apollo 13 og 14.**
T. S. Ringnes.
Naturen, Årg. 98, No. 3, p. 127 - 144 = Inst. Teor. Astrofys., Blindern—Oslo, Småtrykk No. 82 (1974).

094.572 **Silicate mineral chemistry of selected Apollo 15, 16, and 17 soils.** H. C. Taylor.
Thesis, Texas Univ., Dallas (USA). 285 pp. University Micro-films Order No. 74-18,749 (1973).

094.573 **Petrology and chemistry of some Apollo 15 crystalline rocks and regoliths.**
V. C. Juan, J. C. Chen, C. K. Huang, P. Y. Chen, C. M. Wang Lee.
Proc. Geol. Soc. China, No. 17, p. 13 - 29 (1974).

094.574 **Glass-coated lunar rock fragments.**
H. G. Wilshire, H. J. Moore.
Journ. Geol., (*USA*), Vol. 82, 403 - 417 (1974).

Lunar petrology conference.
See Abstr. 011.043.

Single-domain grain size limits for metallic iron.
See Abstr. 022.017.

Search for the solar wind in the lunar soil.
See Abstr. 074.111.

Grain size and the evolution of lunar soils.
See Abstr. 094.147.

Abundances of the group IVB elements, Ti, Zr, and Hf and implications of their ratios in lunar materials.
See Abstr. 094.149.

Lunar basins: tentative characterization of projectiles, from meteoritic elements in Apollo 17 boulders.
See Abstr. 094.159.

The formation of lunar glasses.
See Abstr. 094.173.

Lunar surface dynamics: some general conclusions and new results from Apollo 16 and 17.
See Abstr. 094.183.

Lunar microcraters and their solar flare track record.
See Abstr. 094.185.

Comparative magnetic studies of some Apollo 17 rocks and soils and their implications. See Abstr. 094.198.

Remanent magnetization directions in a layered boulder from the South Massif. See Abstr. 094.203.

Howardites: surface breccias of parent bodies.
See Abstr. 105.118.

The Nakhlites. See Abstr. 105.119.

095 Lunar Eclipses

095.001 **The November eclipse of the moon.**
Sky Telescope, Vol. 49, 128 - 130 (1975).

095.002 **L'éclipse de lune du 4 juin 1974 observée par nos sociétaires.**   P. de la Cotardière.
L'Astronomie, 89ᵉ année, p. 115 - 117 (1975).

095.003 **Far-east photographs of November's lunar eclipse.**
Sky Telescope, Vol. 49, 190 - 192 (1975).

095.004 **Observing lunar eclipses.**   J. E. Westfall.
Strolling Astronomer, Vol. 25, 85 - 88 (1975).

095.005 **The lunar eclipse of May 25, 1975.**
J. E. Westfall.
Strolling Astronomer, Vol. 25, 88 - 93 (1975).

095.006 **Les éclipses de lune de mai et novembre 1975.**
G. Florsch.
L'Astronomie, 89ᵉ année, p. 131 - 147 (1975).

095.007 **An observer's kit for this month's lunar eclipse.**
Sky Telescope, Vol. 49, 280 - 283 (1975).

095.008 **De maansverduistering op 29 november 1974.**
J. C. Lameer.
Zenit, Vol. 2, 70 - 71 (1975).

095.009 **Delimično pomračenje Meseca 4. VI 1974. godine (Partial lunar eclipse on 4. VI. 1974).**   A. Tomić.
Vasiona, Vol. 22, 94 - 96 (1974).

095.010 **Delimično pomračenje Meseca, 4. VI 1974. godine (Partial lunar eclipse on 4. VI. 1974).**   B. Đorđević.
Vasiona, Vol. 22, 96 - 97 (1974).

095.011 **Pomrčina Mjeseca (Lunar eclipse).**   V. Bermanec.
Vasiona, Vol. 22, 97 - 98 (1974).

095.012 **Observation of the lunar eclipse of 29 November 1974.**   J. Bouška, J. Mojchrovič, M. Dujnič, I. Dupal, P. Rapavý, J. Stuchlík, Z. Machovský, M. Kment.
Říše hvězd, Vol. 56, 33 - 35 (1975). In Czech.

096 Lunar Occultations

096.001 **Photoelectric observations of lunar occultations.**
J. M. Harwood, R. E. Nather, A. R. Walker,
B. Warner, P. A. T. Wild.
Monthly Notices Roy. Astron. Soc., Vol. 170, 229 - 236 (1975).
A list of 58 photoelectric timings of lunar occultations is given. Five diameters have been measured and three double stars discovered. Thirteen recently determined diameters of late-type stars by interferometry and the lunar occultation method are used to show that only a small modification needs to be made to Wesselink's V_0-system surface brightness to $(B-V)_0$ relationship.

096.002 **Lunar occultation summary. I.**
J. J. Eitter, W. I. Beavers.
Astrophys. Journ.,Suppl. Ser., No. 269, Vol. 28, 405 - 412 (1974).
During the period 1970 August 12 to 1972 April 26 111 stars have been observed in 121 separate lunar occultation events. A summary of the results of these routine two-color photoelectric lunar-occultation observations made at Erwin W. Fick Observatory is reported.

096.003 **Grazing occultations, 1975.**
Southern Stars, Vol. 25, 224 - 226 (1974).

096.004 **Beobachtung der streifenden Bedeckung von Zeta Tauri am 24. Januar 1975.**
A. Brömme, W. Kunz.
SuW, Vol. 14, 133 - 134 (1975).

096.005 **Observations of star occultations by the moon at Poltava.** B. F. Sincheskul.
Vrashchenie i prilivn. deformatsii Zemli. Vyp. (No.) 6. Kiev, "Nauk. dumka", 1974, p. 96 - 100. In Russian. – Abstr. in Referativ. Zhurn. 51. Astron., 3.51.147 (1975).

096.006 **The 1975–76 lunar occultations of Beta Scorpii.**
T. C. Van Flandern, P. Espenschied.
Bull. American Astron. Soc., Vol. 7, 338 (1975). – Abstr. AAS.

096.007 **Beobachtung einer streifenden Sternbedeckung.**
I. Reimann.
SuW, Vol. 14, 207 (1975).

096.008 **Photoelectric observations of occultations of the Pleiades and the incidence of duplicity in the cluster.** P. Bartholdi.
Astron. Journ., Vol. 80, 445 - 448 (1975).
Photoelectric observations of occultations of Pleiades stars made with the Geneva telescope at the observatory of Haute Provence in 1972 are reported. In confirmation of other observers Atlas is found to be double, as are several other cluster members. A collation of all data shows that some 40%–50% of the brighter Pleiades are double. A comparison with the diagnosis of duplicity and rotation by the Geneva photometry shows an almost perfect prediction rate for duplicity and a good degree of success for detection of rotation.

096.009 **Occultations of the Pleiades: reappearances observed photoelectrically at McDonald Observatory.**
P. Bartholdi, D. W. Dunham, D. S. Evans, E. C. Silverberg, J. R. Wiant.
Astron. Journ., Vol. 80, 449 - 450 (1975).
Reappearance timings and notes of duplicity for Pleiades stars observed at 6943 Å at McDonald Observatory in 1971 are reported.

096.010 **Radio observations of the lunar occultation of the Crab nebula on September 10, 1974.**
V. S. Artyukh, Yu. V. Volodin, G. I. Dobysh, B. K. Izvekov, V. I. Kostenko, L. I. Matveenko, S. A. Sukhodol'skij, A. P. Timofeev.
Pis'ma v Astron. Zhurn., Vol. 1, No. 6, p. 23 - 27 (1975). In Russian.
The results of observations at 3.5 and 6.4 m are given. The effective radio size of the nebula has been determined. The radio brightness distribution has a gap in the vicinity of the pulsar.

096.011 **Occultation observations in 1973.**
T. Mori, Y. Ganeko, Y. Harada, M. Sasaki, M. Yamaguti.
Data Rep. Hydrographic Observations, Ser. Astron. Geod., *Tokyo*, No. 9, p. 1 - 41 (1975).

096.012 **Occultation of Saturn, 1973 March 2.**
W. Crozier.
Irish Astron. Journ., Vol. 11, 194 (1974).

096.013 **L'observation des occultations.** J. Meeus.
Ciel et Terre, Vol. 91, 220 - 228 (1975).

096.014 **Streifende Sternbedeckung am 22. Feb. 1975.**
J. Meeus.
Sternenbote, 18. Jahrgang, p. 42 - 44 (1975).

096.015 **Four more grazing occultations.** J. Hers.
Monthly Notes Astron. Soc. Southern Africa, Vol. 34, 83 - 92 (1975).

096.016 **Occultations d'étoiles par la lune, observées à l'équatorial de 45 cm en 1972 et 1973.**
J. Dommanget, E. Van Dessel.
Bull. Astron. Obs. Roy. Belgique, Vol. 8, 141 (1974).

096.017 **On organizing observations of grazing occultations of stars by the moon in the USSR and photoelectric observations of occultations at the Engelhardt Astronomical Observatory.**
A. A. Nefed'ev, M. I. Shpekin, A. V. Sergeev.
19th Astrometrical Conference 1972, (see 012.019), p. 272 - 276 (1975). In Russian.

A simultaneous two-channel system for lunar occultation observations. See Abstr. 031.409.

Angular diameters and effective temperatures of red giant stars from lunar occultations with special reference to μ Geminorum. See Abstr. 115.001.

The angular diameter of Mu Geminorum. See Abstr. 115.006.

097 Mars, Mars Satellites

Mars

097.001 On the structure of Mars' atmosphere at 120–220 km.
V. A. Krasnopolsky (*Krasnopol'skij*).
Icarus, Vol. 24, 28 - 35 (1975).

Altitude dependences of $[CO_2]$ and $[CO_2^+]$ are deduced from Mariner 6 and 7 CO_2^+ airglow measurements. CO_2 densities are also obtained from n_e radio occultation measurements. Both $[CO_2]$ profiles are similar and correspond to the model atmosphere of Barth et al. (1972) at 120 km, but at higher altitudes they diverge and at 200–220 km the obtained $[CO_2]$ values are three times less than those of the model.

097.002 Thermal structure of the Martian atmosphere during the dissipation of the dust storm of 1971.
B. J. Conrath.
Icarus, Vol. 24, 36 - 46 (1975).

The secular variation of the thermal structure of the Martian atmosphere during the dissipation phase of the 1971 dust storm is examined, using temperatures obtained by the infrared spectroscopy investigation on Mariner 9. Within the framework of a model which assumes an effective vertical diffusivity K independent of height, a mean dust particle diameter of $\sim 2 \mu m$ is inferred. To provide the necessary vertical mixing, $K \gtrsim 10^7$ cm^{2}sec^{-1} is required in the lower atmosphere.

097.003 An unusual landslide feature on Mars.
J. Veverka, T. Liang.
Icarus, Vol. 24, 47 - 50 (1975).

A flow feature on a crater wall, characteristic of a landslide, has been identified in a Mariner 9 high resolution photograph. Although other evidence of mass wasting is common in Mariner 9 photography, the case presented appears unique. A tentative conclusion is that, at least in some cases, Martian soil exhibits significant internal friction in mass movements.

097.004 Origin of Martian channels:clathrates and water.
S. J. Peale, G. Schubert, R. E. Lingenfelter.
Science, Vol. 187, 273 - 274 (1975).

097.005 Shape of Mars. H. F. Petersons.
Nature, Vol. 253, 103 - 104 (1975).

The geometrical ellipticity of Mars differs considerably from the dynamical ellipticity obtained from the precession of the orbits of the satellites. Study of the Martian centre of mass lead Schubert and Lingenfelter to comment that the location of the Hellas basin near the direction of the thinnest crust, as implied by the offset, might suggest that an impact produced the asymmetric crustal distribution. If such an impact did occur then it is likely to have also affected the shape of Mars. The mass of the impacting body is about 10^{-3} of that of the planet's mass.

097.006 Oberflächenstrukturen und Klimatologie des Planeten Mars. H. Zimmer.
SuW, Vol. 14, 9 - 13 (1975).

097.007 Investigation of the Martian neutral atmosphere by the method of bifrequency radio occultation using the Mars Soviet interplanetary stations.
Yu. N. Aleksandrov, M. B. Vasil'ev, A. S. Vyshlov, V. K. Golovkov, A. I. Danilenko, V. M. Dubrovin, A. L. Zajtsev, A. P. Mestehrton, G. M. Petrov, O. N. Rzhiga, V. A. Samovol, L. N. Samoznaev, A. I. Sidorenko, D. Ya. Shtern.
Dokl. Akad. Nauk. SSSR. Ser. Mat. Fiz., Vol. 220, 548 - 551 (1975). In Russian.

097.008 La planète Mars en 1973. J. Dragesco.
L'Astronomie, 89e année, p. 23 - 34 (1975).

097.009 Martian channels – a classification.
H. Masursky.
Meteoritics, Vol. 9, 379 - 381 (1974). – Abstract.

097.010 On the abundance of NO_2 in the Martian atmosphere. T. Owen, T. Scattergood, J. H. Woodman.
Icarus, Vol. 24, 193 - 196 (1975).

Spectrophotometric scans of Mars and the moon in the region 4000 – 5000 Å were obtained and ratioed. No evidence of any absorption greater than 3% is visible in the Martian spectrum. Using their own laboratory spectra of NO_2 as well as the published work of Hall and Blacet (1952) the authors confirm Marshall's (1964) upper limit of 8 μm atmospheres (0.0008 cm amagat) for the abundance of NO_2 in the atmosphere of Mars.

097.011 Television observations of Martian cloud formations in 1973: preliminary results.
A. N. Abramenko, V. V. Prokof'eva.
Icarus, Vol. 24, 379 - 382 (1975).

During the period July 14–September 11, 1973, about 25000 television pictures of Mars were obtained using a 0.7 m telescope in ten spectral regions. A dust storm in moderate latitudes was recorded. It started before Mars passed perigee. During the observation period the atmosphere was gradually clearing. About 3000 pictures of Mars were taken using a 0.5 m meniscus telescope during the period of November 19, 1973–February 19, 1974. On November 19 the atmosphere was very hazy and heavy dust clouds were recorded. The second dust storm cleared during December through January.

097.012 Water soluble cations and the fluvial history of Mars.
M. P. Silverman, E. F. Munoz.
Icarus, Vol. 24, 383 - 387 (1975).

The electrical conductivity and water soluble Na, K, Ca, and Mg of aqueous solutions of terrestrial soils and finely divided igneous and metamorphic rocks were determined. Similar measurements on multiple samples from the surface of Mars, collected by an automated long-range roving vehicle along a highlands to basin transect at sites with morphological features resembling dry riverlike channels, are suggested to determine the fluvial history of the planet.

097.013 Preliminary results of investigations made aboard the Soviet automatic stations Mars 4, Mars 5, Mars 6 and Mars 7. V. I. Moroz.
Kosmich. Issled., Vol. 13, p. 3 - 8 (1975). In Russian.

097.014 Operating of the reentry vehicle of the automatic interplanetary station Mars 6 in the Martian atmosphere. S. S. Sokolov, V. G. Fokin, V. P. Burtsev, R. S. Romanov, M. K. Rozhdestvenskij, V. P. Karyagin, N. F. Borodin, V. F. Ivannikov, V. I. Shkirina, L. V. Nikolaenko, V. V. Kerzhanovich, B. B. Kotov.
Kosmich. Issled., Vol. 13, p. 9 - 15 (1975). In Russian.

097.015 Experiment on measuring the composition of the Martian atmosphere aboard the reentry vehicle of the space station Mars 6. V. G. Istomin, K. V. Grechnev, L. N. Ozerov, M. E. Slutskij, V. A. Pavlenko, V. N. Tsvetkov.
Kosmich. Issled., Vol. 13, p. 16 - 20 (1975). In Russian.

097.016 The Martian atmosphere in the landing region of

the reentry vehicle of Mars 6 (Preliminary results).
V. S. Avduevskij, Eh. L. Akim, V. I. Aleshin, N. F. Borodin,
V. V. Kerzhanovich, Ya. V. Malkov, M. Ya. Marov, S. F. Moro-
zov, M. K. Rozhdestvenskij, O. L. Ryabov, M. I. Subbotin, V.
M. Suslov, Z. P. Cheremukhina, V. I. Shkirina.
Kosmich. Issled., Vol. 13, p. 21 - 32 (1975). In Russian.

097.017 Preliminary results of measurements of the content
         of water vapour in the atmosphere of the planet
from measurements aboard the automatic interplanetary station Mars 5.   V. I. Moroz, A. Eh. Nadzhip.
Kosmich. Issled., Vol. 13, p. 33 - 36 (1975). In Russian.

097.018 Ozone in the planet's atmosphere from measure-
         ments aboard the automatic interplanetary station
Mars 5.   V. A. Krasnopol'skij, A. A. Krys'ko, V. N. Roga-
chev.
Kosmich. Issled., Vol. 13, p. 37 - 41 (1975). In Russian.

097.019 Measurement of the intensity and spectral charac-
         teristics of radiation in the Lyman α-line in the
Martian upper atmosphere.   J.-L. Bertaux, J. Blamont, S.
I. Babichenko, N. N. Dement'eva, A. V. D'yachkov, V. G.
Kurt, V. A. Sklyankin, A. S. Smirnov, S. D. Chuvakhin.
Kosmich. Issled., Vol. 13, p. 42 - 47 (1975). In Russian.

097.020 Preliminary results of the two-frequency radio-
         graphic inspection of the Martian ionosphere with
the Mars stations in 1974.   M. B. Vasil'ev, A. S. Vyshlov, M.
A. Kolosov, N. A. Savich, V. A. Samovol, L. N. Samoznaev, A.
I. Sidorenko, Yu. N. Aleksandrov, A. I. Danilenko, V. M. Du-
brovin, A. L. Sajtsev, G. M. Petrov, O. N. Rzhiga, D. Ya.
Shtern, L. I. Romanova.
Kosmich. Issled., Vol. 13, p. 48 - 53 (1975). In Russian.

097.021 Results of investigations of the Martian atmosphere
         with the method of radiographic inspection by
means of Mars 2, Mars 4 and Mars 6.   M. A. Kolosov, O. I.
Yakovlev, G. D. Yakovleva, A. I. Efimov, B. P. Trusov, T. S.
Timofeeva, Yu. M. Kruglov, V. A. Vinogradov, V. P. Oreshkin.
Kosmich. Issled., Vol. 13, p. 54 - 59 (1975). In Russian.

097.022 Apparatus and some results of photographic investi-
         gations aboard the automatic stations Mars 4 and
Mars 5.   A. S. Selivanov, M. K. Naraeva, B. A. Suvorov, I. F.
Sinel'nikova, M. I. Bokhonov.
Kosmich. Issled., Vol. 13, p. 60 - 66 (1975). In Russian.

097.023 Results of a geological-morphological analysis of
         some photographs of the Martian surface obtained
from the automatic stations Mars 4 and Mars 5.
K. P. Florenskij, A. T. Bazilevskij, R. O. Kuz'min, I. M. Cher-
naya.
Kosmich. Issled., Vol. 13, p. 67 - 76 (1975). In Russian.

097.024 Infrared radiometry from aboard Mars 5.
         L. V. Ksanfomaliti, V. I. Moroz.
Kosmich. Issled., Vol. 13, p. 77 - 79 (1975). In Russian.

097.025 Spectrophotometry of the planet in the 2–5 mm
         region from aboard the automatic interplanetary
station Mars 5.   V. I. Moroz, N. A. Parfent'ev.
Kosmich. Issled., Vol. 13, p. 80 - 83 (1975). In Russian.

097.026 Pressure and height according to the intensity of
         CO_2 bands from measurements from aboard the
automatic interplanetary station Mars 5 (Preliminary results).
L. V. Ksanfomaliti, B. S. Kunashev, V. I. Moroz.
Kosmich. Issled., Vol. 13, p. 84 - 86 (1975). In Russian.

097.027 Photometry in the 3200–9000 Å region from

aboard the automatic interplanetary station Mars 5.
L. V. Ksanfomaliti, G. N. Krasovskij.
Kosmich. Issled., Vol. 13, p. 87 - 91 (1975). In Russian.

097.028 Polarimetric experiment aboard Mars 5.
         L. V. Ksanfomaliti, V. I. Moroz, A. Dollfus.
Kosmich. Issled., Vol. 13, p. 92 - 98 (1975).

097.029 Analysis of the relief conditions in the landing re-
         gion of the reentry vehicle of the automatic inter-
planetary station Mars 6.   R. B. Zezin, V. P. Karyagin,
I. P. Mamoshina, N. A. Morozov, V. M. Pavlova, M. K. Rozhdestvenskij, V. G. Fokin.
Kosmich. Issled., Vol. 13, p. 99 - 107 (1975). In Russian.

097.030 The magnetic field of the planet Mars from data
         of the Mars 3 and Mars 5 satellites.   Sh. Sh. Dol-
ginov, E. G. Eroshenko, L. N. Zhuzgov.
Kosmich. Issled., Vol. 13, p. 108 - 122 (1975). In Russian.

097.031 First results of ion stream measurements with the
         RIEhP–2801 M instrument aboard the automatic
interplanetary stations Mars 4 and Mars 5.
O. L. Vajsberg, A. V. Bogdanov, V. N. Smirnov, S. A. Romanov.
Kosmich. Issled., Vol. 13, p. 129 - 130 (1975). In Russian.

097.032 On the astronomical theory of Martian climate
         variations.   Sh. G. Sharaf, N. A. Budnikova.
Dokl. Akad. Nauk SSSR. Ser. Mat. Fiz., Vol. 221, 64 - 66
(1975). In Russian.

097.033 Construction of a model of the upper atmosphere
         of Mars based on ionospheric data.
A. V. Mikhajlov, G. S. Ivanov-Kholodnyj.
Geomagn. Aeronom., Vol. 15, 29 - 33 (1975). In Russian.

097.034 Structure and variations of solar wind–Mars inter-
         action region.   A. V. Bogdanov, O. L. Vaisberg.
Journ. Geophys. Res., Vol. 80, 487 - 494 (1975).
    Measurements of ion fluxes in close vicinity to Mars have
been made by the space probes Mars 2 and Mars 3 during
1971–1972. Analysis of measurements confirms the earlier
preliminary conclusion about the existence of bow shock wave
near Mars. Seventeen crossings of the bow shock are used to
obtain the mean location of the bow shock and to estimate
the dimension of the obstacle.

097.035 A new variation on Mars.   G. de Vaucouleurs.
Sky Telescope, Vol. 49, 222 - 223 (1975).

097.036 Recombination rates of O and CO on solid CO_2.
         Implications for the composition of the Martian
atmosphere.   D. S. Sethi, A. L. Smith.
Planet. Space Sci., Vol. 23, 661 - 669 (1975).
    An atomic oxygen flow system and a C^{14} radiochemical
technique have been used to show that the reactions $O + CO \rightarrow$
CO_2 and $O + O \rightarrow O_2$ are heterogeneously catalysed by solid
CO_2 at 77 K. Assuming simple first order kinetics, recombination coefficients are determined. A recombination mechanism involving an intermediate adsorbed CO_3 is proposed. If
the kinetic results are assumed to apply under Martain surface
conditions, then conversion of CO to CO_2 by reaction on the
solid CO_2 at the polar caps occurs at ~10 times the total
column recombination rates for homogeneous reactions previously proposed; night-side CO_2 ice clouds would also constitute an important recombination surface.

097.037 On the figure of Mars.   G. A. Meshcheryakov.
         Astron. Zhurn. Akad. Nauk SSSR, Vol. 52, 374 -
379 (1975). In Russian. English translation in Soviet Astron.,

Vol. 19, No. 2.

On the basis of Stokes constants of Mars, obtained from Mariner 9 tracking data, a triaxial ellipsoid approximating the Martian figure and a Clairaut spheroid are constructed. The flattenings of Mars are discussed.

097.038　Sur la vie à la surface de Mars.
G. de Mottoni y Palacios.
L'Astronomie, 89e année, p. 183 - 191 (1975).

097.039　Les sondes Viking: À la recherche de la vie sur Mars.
P. de La Cotardière.
L'Astronomie, 89e année, p. 192 - 195 (1975).

097.040　On the existence of lateral relative motions on Mars.
V. E. Courtillot, C. J. Allegre, M. Mattauer.
Earth Planet. Sci. Letters, Vol. 25, 279 - 285 (1975).

A preliminary study of photomosaic maps of Mars suggests that large lateral relative motions of a relatively thick lithosphere may have occurred. Transform faults (at two different stages of their evolution) are described, together with a possible triple junction. Diffuse diverging plate boundaries can be found, whereas converging ones remain a problem.

097.041　The Coprates trough assemblage: more evidence for Martian polar wander.
D. C. McAdoo, J. A. Burns.
Earth Planet. Sci. Letters, Vol. 25, 347 - 354 (1975).

Martian surface features such as quasi-circular structures in polar regions have been previously cited as evidence of polar wander (i.e. large-scale relative motion between a body-fixed axis and the rotation pole). Another feature, the Coprates (Valles Marineris) trough assemblage, is proposed as further, plausible evidence of such wander. Simple wander scenarios are constructed to illustrate the hypothesis.

097.042　Soviet automatic interplanetary stations explore Mars.　S. S. Sokolov.
Vestn. AN SSSR, 1974, No. 10, p. 21 - 38. In Russian. Abstr. in Referativ. Zhurn. 51. Astron., 3.51.224; 62. Issled. kosmich. prostranstva, 3.62.255 (1975).

097.043　Hellas and dust storms on Mars.
G. A. Lejkin, E. V. Zabalueva.
Pis'ma v Astron. Zhurn., Vol. 1, No. 2, p. 36 - 38 (1975). In Russian.

The development of slope winds in the circular basin Hellas can generate an anticyclone nearly at the summer solstice of the southern hemisphere, and at that time only. The velocity of wind at the periphery of the anticyclone is sufficient for initiation of a planetwide dust storm. This mechanism may perhaps explain the occurrence of global dust storms at perihelion and a preference of their beginning in the vicinity of Hellas.

097.044　Geology of Mars.　T. A. Mutch.
Bull. American Astron. Soc., Vol. 7, 369 (1975). Abstr. AAS.

097.045　Ancient massive blankets and the surface of Mars.
M. C. Malin.
Bull. American Astron. Soc., Vol. 7, 369 (1975). – Abstr. AAS.

097.046　Mars: two possible impact structures near the south pole.　J. A. Cutts.
Bull. American Astron. Soc., Vol. 7, 369 (1975). – Abstr. AAS.

097.047　Apparent changes in the north polar hood of Mars during dust storms.　L. J. Martin.
Bull. American Astron. Soc., Vol. 7, 369 (1975). – Abstr. AAS.

097.048　Ultraviolet refractive index of Martian dust.
K. Pang, J. Ajello, C. Hord, W. Egan.
Bull. American Astron. Soc., Vol. 7, 369 - 370 (1975). Abstr. AAS.

097.049　Composition of Martian surface regions using reflectance spectroscopy.
J. B. Adams, R. L. Huguenin, T. B. McCord.
Bull. American Astron. Soc., Vol. 7, 370 (1975). – Abstr. AAS.

097.050　Crystal field and charge transfer band assignments in iron (III) oxides and oxyhydroxides: Application to Mars.　R. L. Huguenin.
Bull. American Astron. Soc., Vol. 7, 370 (1975). – Abstr. AAS.

097.051　Martian Bouguer gravity anomalies: interpretation of regional variations.
R. S. Saunders, R. J. Phillips.
Bull. American Astron. Soc., Vol. 7, 370 (1975). – Abstr. AAS.

097.052　Current research on Mars' history and climate.
W. K. Hartmann.
Bull. American Astron. Soc., Vol. 7, 371 (1975). – Abstr. AAS.

097.053　Martian atmospheric water vapor observations. 1972–74 apparition.　E. S. Barker.
Bull. American Astron. Soc., Vol. 7, 371 (1975). – Abstr. AAS.

097.054　Diurnal variation of atmospheric water vapor on Mars.　F. M. Flasar.
Bull. American Astron. Soc., Vol. 7, 371 (1975). – Abstr. AAS.

097.055　Mariner 9 ultraviolet spectrometer experiment: bright limb observations of the lower atmosphere of Mars.　J. Ajello.
Bull. American Astron. Soc., Vol. 7, 371 (1975). – Abstr. AAS.

097.056　Martian atmospheric lee waves.　J. A. Pirraglia.
Bull. American Astron. Soc., Vol. 7, 371 - 372 (1975). – Abstr. AAS.

097.057　Theoretical analysis of the polar cap winds on the planet Mars.
J. B. Pollack, R. M. Haberle, C. B. Leovy.
Bull. American Astron. Soc., Vol. 7, 372 (1975). – Abstr. AAS.

097.058　Mass loss from the region of Mars and the asteroid belt.　S. J. Weidenschilling.
Bull. American Astron. Soc., Vol. 7, 375 - 376 (1975). Abstr. AAS.

097.059　Interpretation of results of measurements of the Martian upper ionosphere by the method of a dispersion interferometer with the help of the Mars 2 station.
A. S. Vyshlov, G. S. Ivanov-Kholodnyj, A. V. Mikhajlov, N. A. Savich.
Kosmich. Issled., Vol. 13, 249 - 253 (1975). In Russian.

097.060　Oppositions of Mars.　V. A. Bronshtehn.
Kvant, 1974, No. 11, p. 12 - 16. In Russian.

097.061　Photodissociation of CO_2 and thermal emission connected with it in the upper atmospheres of Mars and Venus.　A. V. Dembovskij, M. N. Izakov, O. G. Lisin.
AN SSSR. In-t kosmich. issled. Pr-165. Moskva, 1974, 38 pp. In Russian. – Abstr. in Referativ. Zhurn. 51. Astron., 4.51.276 (1975).

097.062　Mariner 9: an instrument of dynamical science.
J. F. Jordan, J. Lorell.
Icarus, Vol. 25, 146 - 165 (1975).

The authors review and evaluate the contributions of Mariner 9 in improving our knowledge of the dynamical characteristics of Mars and its two satellites, Phobos and Deimos.

097.063 Infrared heterodyne spectroscopy of CO_2 on Mars.
A. Betz.
Space Sci. Rev., Vol. 17, 659 (1975). – Presented at the workshop on coherent detection in astronomy, held at Rhenen, The Netherlands, 25 and 26 April 1974 – see 012.009.

097.064 On the choice of a model for the upper Martian atmosphere.
N. I. Burangulov, M. K. Rozhdestvenskij.
Trudy Leningr. gidrometeorol. in-ta, 1974, vyp. (No.) 50, p. 151 - 156. In Russian. – Abstr. in Referativ. Zhurn. 51. Astron., 5.51.261 (1975).

097.065 Analytical study of the secular and seasonal modifications of Martian soil. S. Ebisawa.
Contr. Kwasan Hida Obs., Univ. Kyoto, No. 210, 85 pp. (1973).
Very characteristic behavior of Martian surface changes still remains as a mysterious riddle even at present. Its analytical study has been described here based on the reliable terrestrial observations carried out mainly since 1905. Modifications of the soil are divided into three large categories as follows: (1) secular changes, (2) seasonal changes, (3) ephemeral changes during short period.

097.066 Water vapour and Martian meteorology.
S. Miyamoto.
Contr. Kwasan Hida Obs., Univ. Kyoto, No. 211, 12 pp. (1973).
Recent spectroscopic observations proved the existence, seasonal change, and local concentration of water vapour in Mars. Estimated amount of vapour is enough to explain meteorological phenomena. The Mariner 9 observation of winter north polar region suggests the nature of the cap as ordinary snow.

097.067 Meteorological observations of Mars during the 1973 opposition. S. Miyamoto.
Contr. Kwasan Hida Obs., Univ. Kyoto, No. 217, 64 pp. (1974).
This is the 9-th report of a series about the meteorological observation of Mars during the 1973 apparition. After early July 1973, the author witnessed a sudden darkening of Daedalia-Claritas region, and at the same time, cloud activities around Solis Lacus basin was strengthened. He observed no Noachis cloud in this Martian year, but instead, a mighty yellow cloud appeared over Solis Lacus in middel October 1973.

097.068 Seasonal change of the Martian polar caps.
S. Miyamoto.
Contr. Kwasan Hida Obs., Univ. Kyoto, No. 218, 8 pp. (1974).
Martian polar cap is formed not in mid-winter, but just before the vernal equinox. After reaching the maximum dimension, it starts shrinking from spring to summer. On its way shrinking, the northern polar cap stops or reduces shrinking rate for a while. Observed behaviours of polar caps were interpreted in terms of the ordinary H_2O frost hypothesis of the cap. Momental stop of the cap diminution may correspond to the epoch of transition of the general circulation in the atmosphere from the terrestrial type to the cross-equatorial type. The effects of sand storm and water vapour concentration on the polar cap dimension were discussed.

097.069 Secular changes of Martian albedo feature.
S. Miyamoto.

Contr. Kwasan Hida Obs., Univ. Kyoto, No. 219, 10 pp. (1974).
By the secular change of albedo feature in recent years, Martian surface markings now returned very similar to that of last century. The secular change is accompanied with the change of general circulation pattern in the Martian atmosphere. Martian crust is composed of two layers, terra and maria, and is same in principle with those of moon and earth, but apparently modified by wind erosion.

097.070 Photographic observations of Mars during the 1973 opposition. A. Hattori, T. Akabane.
Contr. Kwasan Hida Obs., Univ. Kyoto, No. 221, 25 pp. (1974).
The authors were able to observe some features of southern summer of Mars. (1) In midsummer of Mars, meridional cloud belts developed, and in late summer, great cloud masses grew up. (2) Daedalia was always dark during the observation. (3) Darkness of a mare belt from Mare Cimmerium to Syrtis Major was always faint. (4) After the outbreak of Solis cloud in October, darkness of Syrtis Major faded, and it combined with Libya to form a triangular feature. Albedo changes of (2), (3), and (4) may be rare phenomena in this century.

097.071 Mars observations and secular change of albedo markings during the 1973 opposition.
S. Ebisawa.
Contr. Kwasan Hida Obs., Univ. Kyoto, No. 223, 24 pp. (1975).
The result of detailed Martian patrol observations mainly at Hida Observatory has been described here and the characteristic feature of secular change and dust storm during the 1973 opposition has been dealt with also.

097.072 Planisphere of Mars showing the general feature of surface markings during the 1973 opposition before the appearance of large dust storm. S. Ebisawa.
Contr. Kwasan Hida Obs., Univ. Kyoto, No. 224, 2 pp. (1975).

097.073 Argon in the Martian atmosphere: Are the results of Mars 6 in agreement with optical and radio refraction measurements? V. I. Moroz.
Pis'ma v Astron. Zhurn., Vol. 1, No. 6, p. 36 - 41 (1975). In Russian.
The space probe Mars 6 has discovered some inert gas, probably argon, in the Martian atmosphere. A combined analysis of known data on infrared spectrometry, radio occultations is given. It shows that all data are in agreement with 25–35% argon abundance.

097.074 "Blue clearing" on Mars in connection with dust clouds in August and September, 1971.
V. V. Prokof'eva, T. A. Chuprakova, S. S. Dzyamko, T. V. Bryzgalova.
Astron. Zhurn. Akad. Nauk SSSR, Vol. 52, 623 - 634 (1975). In Russian. English translation in Soviet Astron., Vol. 19, No. 3.
Photometric measurements of the television pictures obtained with ultraviolet, blue, green and red filters showed that "blue clearing" is connected with the appearance of dust clouds. This confirms the hypothesis on the dust nature of the "blue clearing" published 1972: illusion of the "blue clearing" is caused by increasing of brightness of the light areas when dust clouds appear.

097.075 Infrared temperatures and thermal properties of the Martian surface from measurements aboard the automatic interplanetary station Mars 3.
V. I. Moroz, L. V. Ksanfomaliti, G. N. Krasovskij, V. D. Davydov, N. A. Parfent'ev, V. S. Zhegulev, G. F. Filippov.
Kosmich. Íssled., Vol. 13, 389 - 403 (1975). In Russian.

097.076 **New face of old Mars.**   S. R. Brzostkiewicz.
Urania Kraków, Vol. 46, 37 - 46 (1975). In Polish.

097.077 **The history of Martian cartography.**
S. R. Brzostkiewicz.
Urania Kraków, Vol. 46, 66 - 73 (1975). In Polish.

097.078 **Names of craters on Mars.**   S. R. Brzostkiewicz.
Urania Kraków, Vol. 46, 143 - 146 (1975). In
Polish.

097.079 **Automated life-detection experiments for the
Viking mission to Mars.**   H. P. Klein.
Origins of Life, Vol. 5, 431 - 441 (1974).

097.080 **Organic contamination problems in the Viking
molecular analysis experiment.**
D. A. Flory, J. Oró, P. V. Fennessey.
Origins of Life, Vol. 5, 443 - 455 (1974).

097.081 **Model systems for life processes on Mars.**
M. A. Mitz.
Origins of Life, Vol. 5, 457 - 462 (1974).

097.082 **An automatically-returned Martian sample by 1985?**
G. Eglinton, S. Tonkin.
Origins of Life, Vol. 5, 463 - 482 (1974).
The success of the lunar sample analysis program under-
scores the desirability of a returned Martian sample. A mission
which would bring back 1 kg of soil is outlined. A single soil
sample could be informative as to the general surface composition of Mars. Knowledge of the detailed physics, chemistry
and mineralogy of the Martian sample would be of inestimable
value to planetary studies.

097.083 **Report of the apparition of 1973.**
E. H. Collinson.
Journ. British Astron. Ass., Vol. 85, 336 - 341 (1975).

097.084 **Does Mars have a magnetosphere?**   M. K. Wallis.
Geophys. Journ. Roy. Astron. Soc., Vol. 41, 349 -
354 (1975).
The space-probe plasma and magnetic evidence for Mars
having a significant dipole field is uncertain and contradictory.
If ionization processes in the planet's exosphere are taken into
account, several discrepancies between observation and the
fluid model can be explained. The possibility that some solar
plasma flows directly into the atmosphere and is there absorbed is still open.

097.085 **Observing Olympus Mons in 1975.**
J. E. Westfall.
Strolling Astronomer, Vol. 25, 129 - 130 (1975).

097.086 **Mars 1973−74 apparition −A.L.P.O. Report I.**
R. B. Rhoads, V. W. Capen.
Strolling Astronomer, Vol. 25, 130 - 140 (1975).

097.087 **New Martian charts and their nomenclature.**
V. W. Capen.
Strolling Astronomer, Vol. 25, 167 - 171 (1975).

097.088 **Possibility of improving the gravitational potential
and the structure of the Martian upper atmosphere
from the evolution of the orbit of an artificial Martian satellite.**
N. I. Burangulov, M. K. Rozhdestvenskij.
Trudy Leningr. gidrometeorol. in-ta, 1974, vyp. (No.) 50, p.
137 - 150. In Russian. − Abstr. in Referativ. Zhurn. 51.
Astron., 6.51.179; 62. Issled. kosmich. prostranstva, 6.62.334
(1975).

097.089 **The treatment of results of radar measurements of
Mars from the earth in 1971 in connection with
problems of radio altimetry.**
N. N. Krupenio, V. A. Ladygin.
AN SSSR. In-t kosmich. issled. Pr-182. Moskva, 1974. 52 pp.
In Russian. − Abstr. in Referativ. Zhurn. 51. Astron.,
6.51.308; 62. Issled. kosmich. prostranstva, 6.62.179 (1975).

097.090 **Occultation of ε Geminorum by Mars on 1976
April 8.**   G. E. Taylor.
IAU Circ., No. 2782 (1975).

097.091 **The appearance of Mars from 1907 to 1971: graphic
synthesis of photographs from the I.A.U. Center at
Meudon.**   G. De Mottoni y Palacios.
Icarus, Vol. 25, 296 - 332 (1975).
Using selected high quality plates collected at the I.A.U.
Planetary Data Center in Meudon (Paris), the author has
drawn 32 Mercator charts of the planet Mars for every opposition from 1907 to 1971. This graphic synthesis of albedo
distribution may be used in the study of major surface changes
over the years.

097.092 **Ultraviolet complex refractive index of Martian dust:
laboratory measurements of terrestrial analogs.**
W. G. Egan, T. Hilgeman, K. Pang.
Icarus, Vol. 25, 344 - 355 (1975).
The optical complex index of refraction of four candidate Martian surface materials has been determined between
0.185 and 0.4 μm using a modified Kubelka–Munk scattering
theory. The candidate materials were limonite, andesite,
montmorillonite, and basalt. The effect of scattering has been
removed from the results. Also presented are diffuse reflection
and transmission data on these samples.

097.093 **Standardization of photographic observations of
Mars in 1971.**
Yu. V. Aleksandrov, D. F. Lupishko, T. A. Lupishko.
Vestn. Khar'kov. Univ., No. 117, (Ser. Astron., vyp. (No.) 9),
p. 3 - 11 (1974). In Russian.

097.094 **Results of an absolute surface photometry of Mars
in 1971.**
Yu. V. Aleksandrov, D. F. Lupishko, V. P. Tishkovets.
Vestn. Khar'kov. Univ., No. 117, (Ser. Astron., vyp. (No.) 9),
p. 11 - 19 (1974). In Russian.

097.095 **Mars. Un rendez-vous astronomique: 4 juillet 1976.**
R. Dejaiffe.
Rev. Quest. Sci., (*Belgium*), Vol. 145, 365 - 380 (1974).

Volcanoes of the earth, moon and Mars.
See Abstr. 003.062.

The new Mars: the discoveries of Mariner 9.
See Abstr. 003.077.

New data on Mars. Collection of articles.
See Abstr. 003.103.

Mars as viewed by Mariner 9.
See Abstr. 003.156.

**Nonthermal rotational distribution of $CO(A^1\Pi)$
fragments produced by dissociative excitation of CO_2 by
electron impact.**   See Abstr. 022.031.

**Reduced absorption of the nonthermal $CO(A^1\Pi-
X^1\Sigma^+)$ fourth-positive group by thermal CO and implications
for the Mars upper atmosphere.**   See Abstr. 022.032.

Removal of instrument signature from Mariner 9 television images of Mars.   See Abstr. 031.208.

Test results on the Viking gas chromatograph – mass spectrometer experiment.   See Abstr. 051.024.

Investigations of the solar plasma near Mars and on the path earth–Mars by means of charged particle traps aboard the Soviet space vehicles 1971 - 1973.   See Abstr. 074.014.

Alaskan thermokarst terrain and possible Martian analog.   See Abstr. 081.011.

A technique to deduce atmospheric temperature and constituent profiles from a planet's limb radiance profile. See Abstr. 091.009.

Crater degradation on the moon, Mars, and Mercury. See Abstr. 094.226.

Some results of a reduction of photographs of the moon and Mars.   See Abstr. 094.268.

Mars, Satellites

097.201   The photometric function of Phobos and Deimos.
M. Noland, J. Veverka.
Bull. American Astron. Soc., Vol. 7, 388 (1975). – Abstr. AAS.

097.202   Eerste kaart van marsmaan Phobos.
G. W. E. Beekman.
Zenit, Vol. 2, 198 (1975).

097.203   Viewing Phobos and Deimos for navigating Mariner
9.   T. C. Duxbury, G. H. Born, N. Jerath.
Journ. Spacecraft and Rockets, (U. S. A.), Vol. 11, 215 - 222 (1974).
A new on-board optical navigation data technique has

been successfully demonstrated on Mariner 9. Science TV pictures of Phobos and Deimos against star fields were used in the real time navigation process for inserting Mariner 9 into orbit about Mars.

Mariner 9: an instrument of dynamical science. See Abstr. 097.062.

Errata

097.901   Errata: "Variable features on Mars. IV. Pavonis
Mons" [Icarus, Vol. 22, 24 - 47 (1974)]. C. Sagan,
J. Veverka, R. Steinbacher, L. Quam, R. Tucker, B. Eross.
Icarus, Vol. 24, 271 - 276 (1975).

098 Minor Planets

098.001 The nature of asteroids. C. R. Chapman.
Sci. American, Vol. 232, No. 1, p. 24 - 33 (1975).
The spectra of sunlight reflected by the minor planets
have yielded new clues to their mineralogical composition,
their origin, their evolution and their relationship to meteorites.

098.002 Asteroids: spectral reflectance and color characteristics. T. B. McCord, C. R. Chapman.
Astrophys. Journ., Vol. 195, 553 - 562 (1975).
The authors present visible and near-infrared spectral reflectance curves for 35 asteroids. There is a general, but imperfect, correlation between spectral type and semimajor axis, which is consistent with the hypothesis that meteorites are derived from the asteroid belt. The color-frequency histogram for the 67 objects is bimodal, probably associated with "stony" objects and "carbonaceous" objects. Hirayama family pairs usually show disparate spectral types.

098.003 A photometric study of the minor planet 15 Eunomia. F. Scaltriti, V. Zappalà.
Astron. Astrophys., Suppl. Ser., Vol. 19, 249 - 255 (1975).
A series of photoelectric observations of the asteroid 15 Eunomia were made at the Astronomical Observatory of Torino, during the 1974 opposition. The standard magnitude $V(1, 0)$, the phase coefficient, the albedo, the mean radius, the mass were derived. Furthermore the retrograde sense of rotation was confirmed and the sidereal period of rotation was deduced.

098.004 The asteroids as meteorite parent-bodies. C. R. Chapman.
Meteoritics, Vol. 9, 322 - 324 (1974). – Abstract.

098.005 Great opposition of Eros. V. P. Tsesevich.
Zemlya i Vselennaya, 1975, No. 1, p. 87 - 88.
In Russian.

098.006 The Eros flyby.
Sky Telescope, Vol. 49, 162 - 163 (1975).

098.007 Asteroids: infrared photometry at 1.25, 1.65, and 2.2 microns.
T. V. Johnson, D. L. Matson, G. J. Veeder, S. J. Loer.
Astrophys. Journ., Vol. 197, 527 - 531 (1975).
The authors report the first results of a new program of asteroid photometry at wavelengths of 1.25, 1.65, and 2.2 μ. In this paper the observations of three asteroids are reduced to spectral reflectance. (1) Ceres and (2) Pallas have surfaces which are probably composed of carbonaceous chondritic-type material. A basaltic achondritic composition remains a good candidate for (4) Vesta.

098.008 Amateurs observe the rotation of Eros.
J. Ashbrook.
Sky Telescope, Vol. 49, 331 - 332 (1975).

098.009 Der Lichtwechsel des Kleinplaneten 433 Eros während der Opposition 1975.
W. Herzner, H. Jasicek, G. Klement, E. Nussbaum.
SuW, Vol. 14, 167 - 169 (1975).

098.010 The mass of Pallas. J. Schubart.
Astron. Astrophys., Vol. 39, 147 - 148 (1975).
A new value for the mass of Pallas equals $(1.14 \pm 0.22) \times 10^{-10}$ solar mass. This value results from a differential correction of the orbit of Ceres.

098.011 Asteroids: spectral reflectance and color characteristics. II. T. B. McCord, C. R. Chapman.
Astrophys. Journ., Vol. 197, 781 - 790 (1975).
The authors present new spectrophotometry for 31 asteroids, and improved data for nine previously observed, raising the total sample to 98. Several important new spectral types have been found. Asteroid 349 Dembowska is the first large main-belt asteroid found to resemble ordinary chondritic meteorites in spectral properties. The first two measured Trojan asteroids show unusual spectra not compatible with carbonaceous chondrites or other known meteorites. The spectrum of Mars-crosser 887 Alinda is compatible with unequilibrated chondrites. Most fainter asteroids (especially those in the outer half of the belt) have flat spectra indicating probable carbonaceous composition. Compositional heterogeneity of Hirayama families is common among the 16 families studied to date.

098.012 L'anello degli asteroidi. R. Burchi.
Coelum, Vol. 43, 72 - 77 (1975).

098.013 Il pianetino Eros nel suo avvicinamento alla terra nel gennaio 1975. M. Monaco.
Coelum, Vol. 43, 81 - 82 (1975).

098.014 Planetoïden waarnemen Vesta als voorbeeld.
R. P. A. Huisman, N. A. van der Mey.
Zenit, Vol. 2, 32 - 33 (1975).

098.015 Vergelijkende positiebepalingen tussen Sterrenwacht Leiden en Amateursterrenwacht SON. A. T. Son.
Zenit, Vol. 2, 68 - 69 (1975).

098.016 The program of observations of minor planets at the Goethe Link Observatory.
F. K. Edmondson.
IAU Colloquium No. 22, (see 012.005), p. 23 - 24 (1974).

098.017 The program of minor planets observations at the Crimea Observatory.
N. S. Chernykh, L. I. Chernykh.
IAU Colloquium No. 22, (see 012.005), p. 25 - 27 (1974).

098.018 The program of observation of minor planets and comets at the Bucharest Observatory.
C. Cristescu.
IAU Colloquium No. 22, (see 012.005), p. 29 - 31 (1974).

098.019 Le programme des observations des astéroïdes et des comètes à l'Observatoire de Nice. B. Milet.
IAU Colloquium No. 22, (see. 012.005), p. 33 - 34 (1974).

098.020 Observational selection and statistics of asteroids.
T. Kiang.
IAU Colloquium No. 22, (see 012.005), p. 35 - 38 (1974).

098.021 Photometric astrometry of minor planets.
J. D. Mulholland, R. E. Nather.
IAU Colloqium No. 22, (see 012.005), p. 71 (1974). – Abstract.

098.022 On following up newly-discovered earth-approaching minor planets. B. G. Marsden.
IAU Colloquium No. 22, (see 012.005), p. 73 - 75 (1974).

098.023 De la possibilité pour certains astéroides d'être des satellites lointains de Jupiter. D. Benest.

IAU Colloquium No. 22, (see 012.005), p. 87 - 92 (1974).

098.024   On the fragmentation of asteroids.   B. J. (*Yu.*) Levin.
IAU Colloquium No. 22, (see 012.005), p. 101 - 106 (1974).

098.025   Monte Carlo simulation of asteroid collisions.
W. McD. Napier, R. J. Dodd.
IAU Colloquium No. 22, (see 012.005), p. 107 - 110 (1974).

098.026   A note on a peculiarity in the distribution of the argument of perihelion.   T. Kiang.
IAU Colloquium No. 22, (see 012.005), p. 133 - 134 (1974).

098.027   Structure and evolution of the asteroid belt.
G. A. Chebotarev, M. Ya. Shmakova.
IAU Colloquium No. 22, (see 012.005), p. 135 - 139 (1974).

098.028   Total asteroid mass from comet perturbations.
T. C. Van Flandern.
IAU Colloquium No. 22, (see 012.005), p. 141 (1974). — Abstract.

098.029   On fields of force on rotating asteroids.
G. I. Pokrovskij.
Trudy vos'mykh chtenij, posvyashch. razrab. nauch. naslediya i razvitiyu idej K. Eh. Tsiolkovskogo, Kaluga, 1973. Sekts. "Probl. raket. i kosmich. tekhn". Moskva, 1974, p. 3 - 8. In Russian. — Abstr. in Referativ. Zhurn. 62. Issled. kosmich. prostranstva, 3.62.262 (1975).

098.030   44 Nysa: an iron-depleted asteroid.   B. Zellner.
Astrophys. Journ., (*Letters*), Vol. 198, L45 - L47 (1975).

The minor planet 44 Nysa has a unique combination of photopolarimetric parameters, with the most nearly neutral *UBV* colors, the shallowest negative polarization branch, and by far the highest polarimetric albedo yet obtained for any asteroid. Its surface apparently consists of a low-opacity, iron-free silicate strongly suggestive of enstatite achondrites.

098.031   Asteroid families.   J. G. Williams.
Bull. American Astron. Soc., Vol. 7, 343 (1975). Abstr. AAS.

098.032   Are the present-day asteroids a remnant of a much larger early population?
C. R. Chapman, D. R. Davis.
Bull. American Astron. Soc., Vol. 7, 376 (1975). — Abstr. AAS.

098.033   The collisional evolution of carbonaceous and chemically-differentiated asteroids.   C. R. Chapman.
Bull. American Astron. Soc., Vol. 7, 376 (1975). — Abstr. AAS.

098.034   An angular momentum diagram for asteroid rotations.   J. A. Burns.
Bull. American Astron. Soc., Vol. 7, 376 (1975). — Abstr. AAS.

098.035   Surface properties of asteroids.
B. Zellner, D. Morrison, C. R. Chapman.
Bull. American Astron. Soc., Vol. 7, 376 - 377 (1975). Abstr. AAS.

098.036   Asteroid surface composition from reflection spectroscopy.   T. B. McCord, M. J. Gaffey.
Bull. American Astron. Soc., Vol. 7, 377 (1975). — Abstr. AAS.

098.037   Infrared spectral observations of asteroid (4) Vesta.
H. P. Larson, U. Fink.
Bull. American Astron. Soc., Vol. 7, 377 (1975). — Abstr. AAS.

098.038   Narrowband spectrophotometry of Vesta.
G. L. Veeder, T. V. Johnson, D. L. Matson.
Bull. American Astron. Soc., Vol. 7, 377 (1975). — Abstr. AAS.

098.039   Asteroids: complications for the interpretation of radiometry.   D. L. Matson.
Bull. American Astron. Soc., Vol. 7, 377 (1975). — Abstr. AAS.

098.040   Radiometric studies of Trojan asteroids and Jovian satellites 6 and 7.   D. P. Cruikshank.
Bull. American Astron. Soc., Vol. 7, 377 - 378 (1975). Abstr. AAS.

098.041   Planetoid Eros (433) im Januar 1975.   J. Alean.
Orion, 33. Jahrgang, p. 82 - 83 (1975).

098.042   On observations of Eros.
Kometn. Tsirk., *Kiev*, No. 173 (1975). In Russian.

098.043   (433) Eros.
Kometn. Tsirk., *Kiev*, No. 174 (1975). In Russian.

098.044   Observations of Eros in Tartu.   H. Raudsaar.
Kometn. Tsirk., *Kiev*, No. 175 (1975). In Russian.

098.045   Observations of (433) Eros.
Kometn. Tsirk., *Kiev*, No. 176 (1975). In Russian.
Observations of Uzhgorod (*A. G. Kirichenko, K. A. Kudak*).

098.046   Asteroid (433) Eros.
Kometn. Tsirk., *Kiev*, No. 177 (1975). In Russian.
Observations at the Kleť Observatory (*A. Mrkos*); Photometric observations of Eros at the Solar-Cometary Station in Ussurijsk (*V. A. Golubev, A. S. Konyukhov*).

098.047   Observations of Eros at the Ussurijsk Solar Patrol Station.   V. A. Golubev, A. S. Konyukhov.
Kometn. Tsirk., *Kiev*, No. 178 (1975). In Russian.

098.048   Observations of (433) Eros.
Kometn. Tsirk., *Kiev*, No. 180 (1975). In Russian.
Observations at the Tartu Observatory (*H. Raudsaar*); Observations at the Kleť Observatory (*A. Mrkos*).

098.049   Nutation angles of asteroids.
V. S. Safronov, J. A. Burns.
Izv. AN SSSR. Fiz. Zemli, 1974, No. 11, p. 3 - 10. In Russian.

098.050   Surface properties of asteroids: a synthesis of polarimetry, radiometry, and spectrophotometry.
C. R. Chapman, D. Morrison, B. Zellner.
Icarus, Vol. 25, 104 - 130 (1975).

The surface compositions of 110 asteroids are analyzed from statistically representative data sets of polarimetry as a function of phase angle, broad-band radiometry near 10 and $20\,\mu m$, and visible and near-infrared spectrophotometry. A comparison of albedos and diameters determined by polarimetry and radiometry shows that a modest upward revision of the radiometric albedo scale is needed and that a single law relating the slope of the polarization-phase curve to geometric albedo may not hold for very dark asteroids. The authors sketch some implications for meteoritics and for the early history of the solar system and point to the need for further systematic sampling of smaller and fainter objects by these three observational techniques.

098.051   La conjonction serrée Pallas-Vesta du 6 novembre 1975.   J. Meeus.
L'Astronomie, 89ᵉ année, p. 245 - 249 (1975).

098.052 **Les petites planètes troyennes.** J. Meeus.
L'Astronomie, 89e année, p. 250 (1975).

098.053 **Positions of selected minor planets.**
G. De Sanctis, M. A. Vogliotti, V. Zappalà.
Astron. Astrophys., Suppl. Ser., Vol. 20, 363 - 377 (1975).
The instrumental facilities and the reduction procedure
for the new programme of observations of minor planets at
the Observatory of Torino are briefly described. The precise
positions are given of 25 selected minor planets observed
during the period January 1973 – May 1974.

098.054 **Opozicija male planete 433 Eros (Opposition of the
minor planet (433) Eros).** A. Tomić.
Vasiona, Vol. 23, 14 - 17 (1975).

098.055 **Observations of minor planets at the Skalnaté Pleso
Observatory in the years 1965 - 1971.** M. Antal.
Contr. Astron. Obs. Skalnaté Pleso, Vol. 5, 101 - 141
(1973/75).
The paper contains results of photographic observations
of minor planets made at Skalnaté Pleso during the years
1965 - 1971. A total of 717 accurate positions of 234 asteroids
have been determined. The results are collected in tables.

098.056 **Great opposition of Eros.** M. Pańków.
Urania Kraków, Vol. 46, 7 - 15 (1975). In Polish.

098.057 **Planetoïde met uitzonderlijke baan herontdekt.**
G. W. E. Beekman.
Zenit, Vol. 2, 205 (1975).

098.058 **Mass of Vesta corrected.**
Minor Planet Bull., Vol. 2, 24 (1975).

098.059 **Determination of the absolute magnitude and phase
coefficient of minor planet (887) Alinda.**
E. F. Tedesco.
Minor Planet Bull., Vol. 2, 25 - 27 (1975).
Using available photometric data, primarily from the
1973-74 apparition, the absolute magnitude [B(1,0)] of (887)
Alinda is determined to be 14.98, i.e. approximately 1.3 magnitudes brighter than previously thought. In addition the phase
coefficient [0.047] is approximately twice that of a typical
asteroid.

098.060 **The line of variation in minor planet ephemerides.**
F. Pilcher.
Minor Planet Bull., Vol. 2, 27 - 29 (1975).

098.061 **General report of observations by the A.L.P.O.
Minor Planets Section for the years 1973 and 1974.**
R. G. Hodgson.
Minor Planet Bull., Vol. 2, 34 - 40 (1975).

098.062 **An update on earth-crossing planets.**
R. G. Hodgson.
Minor Planet Bull., Vol. 2, 43 - 44 (1975).

098.063 **Observations de l'astéroïde Eros.**
Pélissier, Ralincourt, Saugère, É. Schweitzer,
Thouet, Védrenne, Verdenet, D. Proust.
A.F.O.E.V. Bull., Vol. 9, 20 - 23 (1975).

098.064 **Minor planets positions.**
H. Wroblewski, L. Panaiotov, S. Vásquez.
Dep. Astron., Univ. Chile, Obs. Astron. Nacional, Cerro Calán,
Santiago de Chile, Publ. Vol. 2, (No. 4), 131 - 142 (1974).
Precise positions of ten minor planets are given. The
dependences and data of the comparison stars are also in-
cluded.

098.065 **Observations photographiques de petites planètes,
effectuées à l'astrographe double de 40 cm au cours
de l'année 1972 (1er semestre).** H. Debehogne.
Bull. Astron. Obs. Roy. Belgique, Vol. 8, 130 - 140 (1974).

098.066 **Photographic photometry of small asteroids.
(Preliminary report).** C.-I. Lagerkvist.
Uppsala Astron. Obs., Rep. No. 6, 25 pp. (1975).
Description of the observational and reductional methods;
The plate material; Some preliminary results: the periods, the
light curves of 352 Gisela and 1789 Dubrovolsky, amplitudes
and magnitudes; Positions of seven asteroids.

098.067 **Minor Planet Circulars, (MPC), Nos. 3779 - 3823**
(1975).
Edited by Cincinnati Observatory under the supervision of
P. Herget.
A repository of nearly all new data for numbered and un-
numbered minor planets: Observations, elements and ephemerides, identifications, newly assigned numbers and names, occultations.

098.068 **Occultation of κ Geminorum A by 433 Eros on
1975 January 24.**
R. S. Harrington, H. L. Giclas, P. Herget.
IAU Circ., No. 2737 (1975).

098.069 **1973 EC.** B. G. Marsden.
IAU Circ., No. 2738 (1975).

098.070 **Occultation of κ Geminorum A by 433 Eros.**
P. Wild, H. L. Giclas, R. S. Harrington, M. Miranian,
H. Crull, M. Mattei, D. Di Cicco, A. R. Upgren, C. A. Whitney,
B. T. O'Leary.
IAU Circ., No. 2741 (1975).

098.071 **1949 OA.** P. Wild, B. G. Marsden.
IAU Circ., No. 2747 (1975).

098.072 **1972 RA.** B. G. Marsden, C. M. Bardwell.
IAU Circ., No. 2753 (1975).

098.073 **A photometric investigation of the asteroid (89)
Julia.** H. J. Schober, G. Lustig.
Icarus, Vol. 25, 339 - 343 (1975).
New photometric data of the light curve of the minor
planet (89) Julia were obtained on nine nights during the 1972
opposition using the 60 cm telescope at Obs. Haute Provence.
A synodic period of $11^h 23^m 14^s \pm 7^s$ and an amplitude of
0.25 mag were derived from the measurements. The light curve
is rather asymmetric and no plausible explanation for this
has been offered so far. The measurements have been carried
out in instrumental V'; the data obtained in B' and U' supplement all conclusions from V' data concerning the rotation of
Julia.

098.074 **Posizioni di pianetini nel 1972 – 1973.**
M. A. Vogliotti, V. Zappalà.
Contr. Oss. Astron. Torino (Pino Torinese), No. 82, 8 pp.
(1974).
162 positions of 45 asteroids are given, as deduced from
plates taken at the Observatory of Torino.

Eros 433.
Journ. Astron. Soc. Western Australia, Vol. 26, Januar 1975,
p. 3 = Perth Observatory Circ. No. 50 (1975).

Bulletin for Eros observers.
Sky Telescope, Vol. 49, 9 (1975).

The asteroids.   See Abstr. 003.109.

On the improvement of star positions from minor planet observations.   See Abstr. 041.014.

Meridian observation of major planets and some minor planets, 1968 – 1973.   See Abstr. 041.041.

On the conversion of photographic observations of selected minor planets into a homogeneous coordinate system. See Abstr. 041.067.

Group method of calculating the perturbations of minor planets.   See Abstr. 042.025.

Sur l'application de la méthode de Laplace-Danjon à la détermination des orbites.   See Abstr. 042.034.

A theory of the Trojan asteroids. See Abstr. 042.047.

Periodic orbits in the regularized restricted three-body problem. I. Regularization of the equation of motion. See Abstr. 042.084.

A major perturbing force on small ($1 < R < 10^4$ cm) solar system bodies; the Yarkovsky effect. See Abstr. 091.022.

Mass loss from the region of Mars and the asteroid belt.   See Abstr. 097.058.

Determination of the masses of Jupiter and Saturn from the motion of Trojans.   See Abstr. 099.035

Recent estimates for the mass of Jupiter based on minor planets.   See Abstr. 099.036.

The comet and minor planet program at Lowell Observatory.   See Abstr. 102.012.

Perihelion longitudes and Jacobi constants of Jupiter group comets as indicators of their possible origin from the two equilateral Trojan clouds.   See Abstr. 102.013.

Observations of comets and minor planets. See Abstr. 103.005.

Observations at the Kleť Observatory. See Abstr. 103.007.

Zodiacal light photopolarimetry. II. Gradients along the ecliptic and the phase functions of interplanetary matter. See Abstr. 106.018.

Pioneer 11 meteoroid detection experiment: preliminary results.   See Abstr. 106.026.

099 Jupiter, Jupiter Satellites

Jupiter

099.001 **Measurement of the angular momentum of Jupiter and the sun by use of the Lense-Thirring effect.**
M. R. Haas, D. K. Ross.
Astrophys. Space Sci., Vol. 32, 3 - 11 (1975).
The authors investigate the feasibility of using the Lense-Thirring effect to measure the rotational angular momentum of Jupiter and the sun. This experiment uses gyroscopes in close Jovian and solar orbits. It is important because it provides direct, unique information. The angular momentum is not derivable from the gravitational moments when non-uniform rotation is present. Analysis shows that this experiment could be done around Jupiter with current technology, but could not be done around the sun for some years.

099.002 **Jovian proton aurora.**
M. G. Heaps, B. C. Edgar, A. E. S. Green.
Icarus, Vol. 24, 78 - 85 (1975).
The planet Jupiter possesses a magnetic field and is surrounded by a magnetosphere. The occurrence of auroral and polar cap phenomena similar to those found on earth is very likely. In this work auroral and polar cap emissions in a model Jovian atmosphere are determined for proton precipitation. Results show that most molecular hydrogen and helium emissions for polar cap precipitation are below the ambient dayglow values. Charge capture by precipitating protons is an important source of Lyman α and Balmer α emissions and offers a key to the detection of large fluxes of low energy protons.

099.003 **Thunderstorms on Jupiter.** A. Bar-Nun.
Icarus, Vol. 24, 86 - 94 (1975).
The presence of a considerable acetylene concentration on Jupiter, despite the fast rate of its photolytic hydrogenation, provides strong evidence for the operation of frequent and powerful thunderstorms in the Jovian atmosphere. The rate of acetylene production by thunder shock waves and the products obtained from its photolytic hydrogenation can account for the large ethane concentration and the absence of ethylene. The yellow-brown acetylene polymer and the ruby-red polymers, obtained from thunder-produced hydrogen cyanide or cyanogen with ammonia, are likely contributors to the Jovian coloration.

099.004 **Note on the beaming of Jupiter's decameter-wave radiation and its effect on radio rotation period determinations.** J. K. Alexander.
Astrophys. Journ., Vol. 195, 227 - 233 (1975).
Comparisons of dynamic spectra of Jupiter's decameter wave radiation for events recorded up to ~12 years apart display a high degree of repeatability in that certain spectral features occur at the same central meridian longitude to within ±10°. The variation in the longitude of the spectral landmarks, which is attributable to a ±10° beaming envelope into which the radiation escapes, places a constraint on the precision to which radio rotation-period determinations can be made from measurement of the apparent longitude drift of dynamic spectra.

099.005 **Recent increase in Jupiter's decimetric radio emission.** M. J. Klein.
Nature, Vol. 253, 102 - 103 (1975).
During an observational programme in support of a study of the linear polarisation properties of Jupiter's decimetric radio emission, it was noted that the planet's total intensity had increased between late summer 1973 and July 1974. The increase is small (~5%) but it seems to be a significant change in the trend of the past decade, during which time Jupiter's decimetric emission has decreased by 20–30% with respect to the flux densities observed before 1967.

099.006 **Jupiter's main magnetic field measured by Pioneer 11.** M. H. Acuna, N. F. Ness.
Nature, Vol. 253, 327 - 328 (1975). – Letter.

099.007 **Spectra of the Jupiter radio bursts.**
G. R. A. Ellis.
Nature, Vol. 253, 415 - 417 (1975).
Between March and September of 1973 and 1974, observations were made of the Jupiter radio bursts with the Llanherne filled-aperture low frequency radio telescope (650 m × 650 m). The signals over the frequency range 4–29 MHz were recorded directly on to six video tape recorders for 5 min each day near transit. Over the observing period, tape records of the Jupiter radio bursts were obtained during about two-thirds of all transits, and so far about 80 h of video tape containing the bursts has been accumulated. Here the author presents a preliminary description of the chief properties of their spectra.

099.008 **Periodic variations of the position of Jovian decameter sources in longitude (system III) and phase of Io.** A. Lecacheux.
Astron. Astrophys., Vol. 37, 301 - 304 (1974).
The author studies the period of rotation of Jupiter and the apparent shifts of the mean positions of the sources of the decameter storms, from data obtained between 1960 and 1971. Previous results for the variations of the jovicentric longitude L of sources A and B and their correlation with the declination D_E of the earth are confirmed.

099.009 **Pioneer 11: Through the dragon's mouth.**
Sky Telescope, Vol. 49, 72 - 78 (1975).

099.010 **Reaction of protons with methane in the Jovian ionosphere.** W. T. Huntress, Jr.
Planet. Space Sci., Vol. 23, 377 - 378 (1975).
Laboratory data shows that the reaction of protons with methane proceeds at thermal ion energies to give both CH_3^+ and CH_4^+ ions in the ratio $CH_3^+/CH_4^+ = 1.5 \pm 0.3$. The thermal rate constant for the reaction is $3.8 \pm 0.3 \times 10^{-9}$ cm^3/sec. This reaction may lead to the formation of hydrocarbon ions in the lower ionosphere of Jupiter.

099.011 **Jupiter en 1973.** C. Botton.
L'Astronomie, 89e année, p. 94 - 108 (1975).

099.012 **Occultation of β Scorpii by Jupiter. V. The emersion of β Scorpii C.**
J. L. Elliot, L. H. Wasserman, J. Veverka, C. Sagan, W. Liller.
Astron. Journ., Vol. 80, 323 - 332 (1975).
The emersion of the C component ($V = +4.92$) of the β Scorpii system, occulted by Jupiter on 13 May 1971, was observed at three wavelengths with a time resolution of 10 msec. A new method for using such multiwavelength observations to remove signal contamination by the Jovian limb is presented. Sharp spikes appearing in the light curve are compared with those observed during the occultation of β Sco AB, and a simple model, containing parameters affecting the spike widths, is developed. The widths of the narrowest spikes are shown to correspond closely to the angular diameters of the occulted stars, suggesting that occultation spikes can be used to determine stellar diameters.

099.013 Report from Jupiter – 2. D. Baker.
Spaceflight, Vol. 17, 102 - 107 (1975).

099.014 Photoelectron escape from the ionosphere of Jupiter.
W. E. Swartz, R. W. Reed, T. R. McDonough.
Journ. Geophys. Res., Vol. 80, 495 - 501 (1975).

Photoelectron escape fluxes and ambient electron heating from the Jovian ionosphere are computed as a function of local time and latitude. Several differences for the fluxes expected from a hydrogen atmosphere, rather than a terrestrial type of atmopshere, are described, including an increase in structure in the energy spectra due to the paucity of ionic states entering the photo-ionization processes and lower escape fluxes above 10 eV than were expected from a simple scaling of earth fluxes.

099.015 The 1972 apparition of Jupiter. P. K. Mackal.
Strolling Astronomer, Vol. 25, 104 - 116, 119 - 120 (1975).

099.016 The occultation of β_1 Sco by Jupiter on May 14, 1971. O. G. Taranova.
Astron. Zhurn. Akad. Nauk SSSR, Vol. 52, 380 - 386 (1975). In Russian. English translation in Soviet Astron., Vol. 19, No. 2.

Results of observation and discussion of the occultation of β_1 Sco by Jupiter on May 14, 1971 are given.

099.017 Observations of Jovian S-bursts with an electro-optical radio spectrograph. J. J. Riihimaa.
Astron. Astrophys., Vol. 39, 69 - 70 (1975).

Medium-resolution spectra of Jovian S-bursts have been recorded with a simple electro-optical radio spectrograph. Over the frequency range 20–22 MHz, the distribution of drift rates of certain S-bursts has a peak at −19 MHz/s in one storm, and −17 MHz/s in the other.

099.018 The Pioneer 10 radio occultation measurements of the ionosphere of Jupiter. G. Fjeldbo,
A. Kliore, B. Seidel, D. Sweetnam, D. Cain.
Astron. Astrophys., Vol. 39, 91 - 96 (1975).

Data from the Pioneer 10 radio occultation measurements are utilized to study the vertical electron number density distribution in the Jovian ionosphere. The detectable portion of the Jovian ionosphere consists of a number of layers distributed over an altitude range of more than 3000 km. The maximum density appears to be on the order of 3×10^5 el/cm³. Assuming that H⁺ is the principal ion in the upper portion of the ionosphere yields a topside plasma temperature of 900 ± 400 °K.

099.019 Study of the atmospheric activity on Jupiter during the years 1955–1973. E. Sarris.
Astron. Astrophys., Vol. 39, 135 - 137 (1975).

A photometric study of the atmospheric activity on Jupiter for the period 1967–1973 is made, and the results are related to previous results on the same subject covering the period 1955–1973, with data of the National Observatory of Athens. A period of 4–5 years in the activity on Jupiter is found; a higher activity in the northern hemisphere is detected again.

099.020 Io-related Jovian decametric bursts and Io sodium. N. L. Cohen.
Journ. Roy. Astron. Soc. Canada, Vol. 69, 89 - 93 (1975).

A correlation between the presence of sodium surrounding Io and Io-related Jovian decametric bursts has been found. This suggests that sodium acts as a conductive mechanism for a potential difference across Io's diameter as it crosses Jovian magnetic field lines. It also removes the need for a low surface resistance to produce the decametric noise bursts.

099.021 Detection of water vapor on Jupiter.
H. P. Larson, U. Fink, R. Treffers, T. N. Gautier III.
Astrophys. Journ., (Letters), Vol. 197, L137 - L140 (1975).

High-altitude (12.4 km) spectroscopic observations of Jupiter at $5\,\mu$ from the NASA 91.5 cm airborne infrared telescope have revealed 14 absorptions assigned to the rotation-vibration spectrum of water vapor. Preliminary analysis indicates a mixing ratio $\sim 10^{-6}$ for the vapor phase of water. Estimates of temperature (>300 K) and pressure (<20 atm) suggest observation of water deep in Jupiter's hot spots responsible for its 5-μ flux. Model atmosphere calculations based on radiative transfer theory may change these initial estimates and provide a better physical picture of Jupiter's atmosphere below the visible cloud tops.

099.022 Polarisation structure of Jupiter's decametre radio bursts. J. J. Riihimaa.
Nature, Vol. 255, 210 - 211 (1975).

The high-resolution dynamic spectra of Jupiter's decameter radio emission exhibit structures consisting of repeated, tilted ridges. The frequency-time slope of these 'diagonal patterns', also referred to as modulation lanes, varies from 50 to 300 kHz s⁻¹. The slope is a function of the System III central meridian longitude of Jupiter, and it may be either positive or negative. Here the author reports a study of polarisation diversity in the spectra of modulation lanes.

099.023 Scientific results from the Pioneer 11 mission to Jupiter. A. G. Opp.
Science, Vol. 188, 447 - 448 (1975).

Magnetosphere, magnetic field, and radiation belts; The planet and atmosphere.

099.024 Pioneer 11 encounter: Preliminary results from the Ames Research Center plasma analyzer experiment.
J. D. Mihalov, H. R. Collard, D. D. McKibbin, J. H. Wolfe, D. S. Intriligator.
Science, Vol. 188, 448 - 451 (1975).

The nature of the plasma transitions across Jupiter's bow shock and magnetopause as observed on Pioneer 10 have also been confirmed on Pioneer 11. However, the northward direction of the Pioneer 11 outbound trajectory and the distance of the final magnetopause crossing (80 Jupiter radii) now suggest that Jupiter's magnetosphere is extremely broad with a half-thickness (normal to the ecliptic plane in the noon meridian) which is comparable to or greater than the sunward distance to the nose.

099.025 Jupiter's magnetic field, magnetosphere, and interaction with the solar wind: Pioneer 11.
E. J. Smith, L. Davis, Jr., D. E. Jones, P. J. Coleman, Jr., D. S. Colburn, P. Dyal, C. P. Sonett.
Science, Vol. 188, 451 - 455 (1975).

The magnetosphere clearly appears to be blunt, not disk-shaped, with a well-defined outer boundary. In the outer magnetosphere, the magnetic field is irregular but exhibits a persistent southward component indicative of a closed magnetosphere. The data contain the first clear evidence in the dayside magnetosphere of the current sheet, apparently associated with centrifugal forces. A revised offset dipole (6-parameter fit) is presented as well as the results of a spherical harmonic analysis (23 parameters). The dipole moment and the composite field appear moderately larger than inferred from Pioneer 10.

099.026 Jupiter revisited: first results from the University of Chicago charged particle experiment on Pioneer 11.
J. A. Simpson, D. C. Hamilton, G. A. Lentz, R. B. McKibben, M. Perkins, K. R. Pyle, A. J. Tuzzolino, J. J. O'Gallagher.
Science, Vol. 188, 455 - 459 (1975).

During the December 1974 Pioneer 11 Jupiter encounter

the experiment provided measurements of Jovian energetic protons and electrons both in the magnetic equatorial zone and at previously unexplored high magnetic latitudes. Many of the observations and conclusions from the Pioneer 10 encounter in 1973 were confirmed, with several important exceptions and new findings. Pioneer 11 data support a model in which the intensity varies with a 10-hour period in phase throughout the sunward side of the magnetosphere and is relatively independent of position within the magnetosphere.

099.027 Pioneer 11 observations of energetic particles in the Jovian magnetosphere.
J. A. Van Allen, B. A. Randall, D. N. Baker, C. K. Goertz, D. D. Sentman, M. F. Thomsen, H. R. Flindt.
Science, Vol. 188, 459 - 462 (1975).

Knowledge of the positional distributions, absolute intensities, energy spectra, and angular distributions of energetic electrons and protons in the Jovian magnetosphere has been considerably advanced by the planetary flyby of Pioneer 11 in November–December 1974 along a quite different trajectory from that of Pioneer 10 a year earlier.

099.028 Jovian protons and electrons: Pioneer 11.
J. H. Trainor, F. B. McDonald, D. E. Stilwell, B. J. Teegarden, W. R. Webber.
Science, Vol. 188, 462 - 465 (1975).

The authors present here a preliminary account of the Pioneer 11 passage through the Jovian magnetosphere as viewed by the particle detector systems of the Goddard Space Flight Center and the University of New Hampshire. They restrict their comments almost entirely to the region well within the Jovian magnetosphere, using data from the low energy telescope (LET-II).

099.029 Radiation belts of Jupiter: a second look.
R. W. Fillius, C. E. McIlwain, A. Mogro-Campero.
Science, Vol. 188, 465 - 467 (1975).

Pioneer 11 data showed: the inner moons of Jupiter are sinks of energetic particles and sometimes sources; a large spike of particles was found near Io; multiple peaks occurred in the particle fluxes near closest approach to the planet; this structure may be accounted for by a complex magnetic field configuration.

099.030 The imaging photopolarimeter experiment on Pioneer 11.
A. L. Baker, L. R. Baker, E. Beshore, C. Blenman, N. D. Castillo, Y.-P. Chen, L. R. Doose, J. P. Elston, J. W. Fountain, T. Gehrels, J. H. Kendall, C. E. KenKnight, R. A. Norden, W. Swindell, M. G. Tomasko, D. L. Coffeen.
Science, Vol. 188, 468 - 472 (1975).

For 2 weeks continuous imaging, photometry, and polarimetry observations were made of Jupiter and the Galilean satellites in red and blue light from Pioneer 11. Measurements of Jupiter's north and south polar regions were possible because the spacecraft trajectory was highly inclined to the planet's equatorial plane. The photometric and polarimetric data are compared with a simple model. The data seem consistent with increased molecular scattering at high latitudes.

099.031 Pioneer 11 infrared radiometer experiment: the global heat balance of Jupiter.
A. P. Ingersoll, G. Münch, G. Neugebauer, D. J. Diner, G. S. Orton, B. Schupler, M. Schroeder, S. C. Chase, R. D. Ruiz, L. M. Trafton.
Science, Vol. 188, 472 - 473 (1975).

Data obtained by the infrared radiometers on the Pioneer 10 and Pioneer 11 spacecraft, over a large range of emission angles, have indicated an effective temperature for Jupiter of $125° \pm 3°K$. The implied ratio of planetary thermal emission to solar energy absorbed is 1.9 ± 0.2, a value not

significantly different from the earth-based estimate of 2.5 ± 0.5.

099.032 Atmosphere of Jupiter from the Pioneer 11 S-band occultation experiment: preliminary results.
A. Kliore, G. Fjeldbo, B. L. Seidel, T. T. Sesplaukis, D. W. Sweetnam, P. M. Woiceshyn.
Science, Vol. 188, 474 - 476 (1975).

Two additional radio occultation measurements of the atmosphere of Jupiter were obtained with Pioneer 11. The combination of two Pioneer 10 measurements and one Pioneer 11 measurement yields an oblateness of 0.06496 at 1 millibar and 0.06547 at 160 millibars. Measurements in the Jovian ionosphere indicate a number of layers distributed over about 3000 kilometers, with a topside temperature of about 750 K.

099.033 Gravity field of Jupiter from Pioneer 11 tracking data.
G. W. Null, J. D. Anderson, S. K. Wong.
Science, Vol.188, 476 - 477 (1975).

Significantly improved values of the zonal gravity harmonic coefficients J_3, J_4 and J_6 of Jupiter have been obtained from a preliminary analysis of Pioneer 11 spacecraft Doppler data taken while the spacecraft was near Jupiter. The new results, which will have an important application as boundary conditions for theoretical models of Jupiter's interior, are consistent with a planet in hydrostatic equilibrium.

099.034 Jupiter-waarnemingsaktie J.W.G. afd. Utrecht.
C. Dullemond.
Zenit, Vol. 2, 77 - 79 (1975).

099.035 Determination of the masses of Jupiter and Saturn from the motion of Trojans.
H. Scholl.
IAU Colloquium No. 22, (see 012.005), p. 93 - 96 (1974).

099.036 Recent estimates for the mass of Jupiter based on minor planets.
W. J. Klepczynski.
IAU Colloquium No. 22, (see 012.005), p. 97 - 100 (1974).

099.037 Spatial and spectral mapping of Jupiter and Saturn with a TV system.
B. Duncan, T. D. Faÿ.
Bull. American Astron. Soc., Vol. 7, 237 (1975). – Abstr. AAS.

099.038 Spectra of Jupiter and Saturn in the ten micron region.
J. D. Bregman, D. M. Rank, J. Lacy.
Bull. American Astron. Soc., Vol. 7, 238 (1975). – Abstr. AAS.

099.039 Aeronomy of the major planets: photochemistry of ammonia and hydrocarbons.
D. F. Strobel.
Rev. Geophys. Space Phys., Vol. 13, 372 - 382 (1975).

The photochemistry of hydrocarbons and ammonia is reviewed for the cold, H_2-dominated atmospheres of the major planets and their satellites. It is concluded that the formation rate of complex hydrocarbons by gas phase kinetics is at most a few percent of the CH_4 dissociation rate. On Jupiter, large concentrations of C_2H_6 and C_2H_2 are predicted, which are in reasonable agreement with observational estimates. A consideration of the Lyman α albedo, the UV albedo between 2000 and 2300 Å, and the NH_3 photochemistry suggests that Jupiter's eddy diffusion coefficient is $> 10^8$ cm^2 s^{-1} near the turbopause but only $\sim 10^4$ cm^2 s^{-1} in the lower stratosphere. Large amounts of C_2H_6 and C_2H_2 may be present on Titan.

099.040 Differential spectra and phase space densities of trapped electrons at Jupiter.
C. E. McIlwain, R. W. Fillius.
Journ. Geophys. Res., Vol. 80, 1341 - 1345 (1975).

Using Pioneer 10 data, the authors have constructed differential spectra and phase space densities of trapped electrons at Jupiter. These quantities should assist in calculating synchrotron radiation from these particles and in evaluating the diffusion mechanisms that accelerate the particles. Absorption by the moons Io and Europa is evident, and injection by Io is demonstrated by a density peak in phase space, which demands a local source. There is also a rapid decrease in density between the moons, which could call for either a local loss mechanism or nonlocal losses fed by diffusion.

099.041　**Early scattering by Jupiter and its collision effects in the terrestrial zone.**
W. M. Kaula, P. E. Bigeleisen.
Bull. American Astron. Soc., Vol. 7, 343 (1975). – Abstr. AAS.

099.042　**The Pioneer 11 images of Jupiter.**
W. Swindell, L. R. Doose, M. Tomasko, J. Fountain.
Bull. American Astron. Soc., Vol. 7, 378 (1975). – Abstr. AAS.

099.043　**Preliminary results on the atmosphere of Jupiter from the Pioneer 11 S-band occultation experiment.**
A. Kliore, G. Fjeldbo, B. L. Seidel, T. T. Sesplaukis, D. N. Sweetnam, P. M. Woiceshyn.
Bull. American Astron. Soc., Vol. 7, 378 (1975). – Abstr. AAS.

099.044　**Oblateness of Jupiter: preliminary results from the analysis of Pioneer 10 and 11 S-band radio occultation measurements.**
P. M. Woiceshyn, A. J. Kliore, T. C. Duxbury, G. Fjeldbo, B. Seidel, T. T. Sesplaukis, D. Sweetnam.
Bull. American Astron. Soc., Vol. 7, 378 (1975). – Abstr. AAS.

099.045　**Photometry of Jupiter at large phase angles.**
M. G. Tomasko, N. D. Castillo.
Bull. American Astron. Soc., Vol. 7, 378 - 379 (1975). Abstr. AAS.

099.046　**The planetary magnetic field and magnetosphere of Jupiter: Pioneer 11 vector helium magnetometer.**
E. J. Smith, L. Davis, Jr., D. E. Jones, P. J. Coleman, Jr., D. S. Colburn, C. P. Sonett.
Bull. American Astron. Soc., Vol. 7, 379 (1975). – Abstr. AAS.

099.047　**The main magnetic field of Jupiter: Pioneer 11.**
M. H. Acuna, N. F. Ness.
Bull. American Astron. Soc., Vol. 7, 379 (1975). – Abstr. AAS.

099.048　**The dynamics of Jupiter's atmosphere.**
P. H. Stone.
Bull. American Astron. Soc., Vol. 7, 379 (1975). – Abstr. AAS.

099.049　**Interactions of the long-enduring white ovals in Jupiter's south temperate current.**　R. F. Beebe.
Bull. American Astron. Soc., Vol. 7, 380 (1975). – Abstr. AAS.

099.050　**The structure of the Jovian atmosphere from ground-based and spacecraft observations in the thermal infrared.**　G. S. Orton.
Bull. American Astron. Soc., Vol. 7, 380 (1975). – Abstr. AAS.

099.051　**The vertical cloud structure of Jupiter from 5 micron measurements.**　R. J. Terrile, J. A. Westphal.
Bull. American Astron. Soc., Vol. 7, 380 (1975). – Abstr. AAS.

099.052　**Limb-darkening scans of Jupiter.**
C. B. Pilcher, T. D. Kunkle.
Bull. American Astron. Soc., Vol. 7, 380 (1975). – Abstr. AAS.

099.053　**Detection of water vapor on Jupiter.**
R. Treffers, H. P. Larson, U. Fink, T. N. Gautier.
Bull. American Astron. Soc., Vol. 7, 380 (1975). – Abstr. AAS.

099.054　**Spatial variations of molecular absorptions on Jupiter.**　J. H. Woodman.

099.055　**On the 23-micron feature in the spectrum of Jupiter.**　K. Fox.
Bull. American Astron. Soc., Vol. 7, 381 (1975). – Abstr. AAS.

099.056　**Photochemistry and kinetics of acetylene in the Jovian atmosphere. Laboratory study of C_2H_2 photodissociation and absolute rate parameters for the reaction $H + C_2H_2$.**　L. J. Stief, W. A. Payne.
Bull. American Astron. Soc., Vol. 7, 381 (1975). – Abstr. AAS.

099.057　**Photochemistry of phosphine in the atmospheres of Jupiter and Saturn.**　R. G. Prinn, J. S. Lewis.
Bull. American Astron. Soc., Vol. 7, 381 (1975). – Abstr. AAS.

099.058　**A comparison of Jovian satellite eclipse results with a preliminary Pioneer 10 Jovian atmosphere.**
D. W. Smith, T. F. Greene.
Bull. American Astron. Soc., Vol. 7, 381 (1975). – Abstr. AAS.

099.059　**Methane absorption in the visible spectra of the outer planets and Titan.**　T. Owen, R. D. Cess.
Bull. American Astron. Soc., Vol. 7, 381 - 382 (1975).
Abstr. AAS.

099.060　**A tentative detection of ammonia emission in the Jovian microwave spectrum.**
M. J. Klein, S. Gulkis, E. T. Olsen.
Bull. American Astron. Soc., Vol. 7, 382 (1975). – Abstr. AAS.

099.061　**Observations of temporal variations in Jupiter's decimetric radio emission.**　M. J. Klein.
Bull. American Astron. Soc., Vol. 7, 382 (1975). – Abstr. AAS.

099.062　**A study of the Jovian atmospheric emission at 3.7 cm wavelength.**　E. T. Olsen, S. Gulkis.
Bull. American Astron. Soc., Vol. 7, 382 (1975). – Abstr. AAS.

099.063　**Calculations of Jupiter's gravitational contraction history.**　J. B. Pollack, A. Grossman, H. Graboske.
Bull. American Astron. Soc., Vol. 7, 382 (1975). – Abstr. AAS.

099.064　**Spectral images of Jupiter and Saturn.**
R. B. Wattson, S. Rappaport, E. Fredericks.
Bull. American Astron. Soc., Vol. 7, 382 (1975). – Abstr. AAS.

099.065　**Interpretation of ground-based polarization observations of Jupiter.**　K. Kawabata, J. E. Hansen.
Bull. American Astron. Soc., Vol. 7, 382 - 383 (1975).
Abstr. AAS.

099.066　**Interior models of Jupiter and Uranus.**
M. Podolak.
Bull. American Astron. Soc., Vol. 7, 383 (1975). – Abstr. AAS.

099.067　**Sodium in the Jovian magnetosphere.**
Yu. Mekler, A. Eviatar, F. V. Coroniti.
Bull. American Astron. Soc., Vol. 7, 387 (1975). – Abstr. AAS.

099.068　**Pioneer 11 infrared radiometer experiment: the global heat balance of Jupiter.**
A. P. Ingersoll, G. Münch, G. Neugebauer, D. J. Diner, G. S. Orton, B. Schupler, S. C. Chase, R. D. Ruiz, L. M. Trafton.
Bull. American Astron. Soc., Vol. 7, 388 - 389 (1975).
Abstr. AAS.

099.069　**Pioneer 11 infrared radiometer experiment: the global heat balance of Jupiter.**
A. P. Ingersoll, G. Münch, G. Neugebauer, D. J. Diner, G. S. Orton, B. Schupler, S. C. Chase, R. D. Ruiz, L. M. Trafton.

Bull. American Astron. Soc., Vol. 7, 391 (1975). – Abstr. AAS.

099.070 Some particularities affecting the movements of Jupiter's Red Spot. F. Link.
Planet. Space Sci., Vol. 23, 805 - 812 (1975).

The fluctuations in longitude of Jupiter's Red Spot are discussed. The long term fluctuations show behaviour similar to the fluctuations of zonal circulations on the earth from 1830–1950. The three-monthly fluctuations have a temporal connection with the inferior conjunction of Mercury 1963–1971. Solar activity may be the key to both phenomena.

099.071 Pioneer 10 ultraviolet photometer observations of the Jovian hydrogen torus: the angular distribution. R. W. Carlson, D. L. Judge.
Icarus, Vol. 24, 395 - 399 (1975). – Presented at IAU Colloquium No. 28, Cornell Univ., Ithaca, New York, 1974 August 18 - 21.

The Pioneer 10 ultraviolet photometer observations of the Jovian hydrogen torus are analyzed to obtain the angular distribution. The cloud is asymmetric about Io, where the atoms presumably originate, with the greater density occurring in the trailing portion. A simple model which assumes Jeans escape from the atmosphere of Io is developed and compared to the observations. The results suggest that the exospheric temperature is high ($\sim$ 3000 K) and that the ionization lifetime of the cloud atoms is $\sim 1 \times 10^5$ sec.

099.072 A theory of the Jovian hydrogen torus. T. R. McDonough.
Icarus, Vol. 24, 400 - 406 (1975). – Presented at IAU Colloquium No. 28, Cornell Univ., Ithaca, New York, 1974 August 18 - 21.

It is shown that the energetic particles observed by Pioneer 10 do not ionize atomic hydrogen sufficiently fast to erode the torus as observed. It is proposed that the reason an incomplete torus exists is the presence of a corotating cold magnetospheric plasma. If this explanation is correct, the angular extent of the fractional torus is a measure of the density of the magnetospheric plasma near Io's orbit, which is found to be $\sim 10^2$ cm^{-3}.

099.073 Spectra of hydrate frosts: their application to the outer solar system. W. D. Smythe.
Icarus, Vol. 24, 421 - 427 (1975). – Presented at IAU Colloquium No. 28, Cornell Univ., Ithaca, New York, 1974 August 18 - 21.

Reflectance spectra from 1 to 6 microns were taken of CH_4 and CO_2 gas hydrates and were found to be very similar to H_2O frost spectra over the entire wavelength region. H_2O clathrates have a gas to H_2O ratio of about 1/6, hence a surface may contain 17% (by number) gas and appear spectroscopically similar to an H_2O frost covered surface. The author concludes that reflectance spectroscopy (especially earthbased) is useful for positive identification of some components of the surface, but does not set stringent limits for spectroscopically active hydrate forming substances in the presence of water frost.

099.074 Pioneer 10 Jovian encounter: radiation dose and implications for biological lethality. M. W. Miller, G. E. Kaufman, H. D. Maillie.
Science, Vol. 187, 738 - 739 (1975).

In its recent Jupiter flyby Pioneer 10 passed through a belt of intense particulate radiation. The radiation dose on the outer surface of the spacecraft was at least 4.9×10^5 rads from electrons plus 2.9×10^6 from protons, sufficient to cause significant microbial decontamination. The radiation dose inside Pioneer 10, approximately 2.8×10^5 to 4.9×10^5 rads, was less likely to cause microbial decontamination but would be lethal to man and to most multicellular biological organisms.

099.075 The interior structure of Jupiter: consequences of Pioneer 10 data. R. Smoluchowski.
Icarus, Vol. 25, 1 - 11 (1975). – Paper presented at the USA–USSR conference on 'Cosmochemistry of the moon and planets', Moscow, June 3–8, 1974.

Models of the Jovian interiors are based on theoretical equations of state of hydrogen and helium supported by a few experimental points and on observed parameters such as oblateness, gravitational coefficients, heat emission, magnetic fields, etc. It appears now that within the limits of error the planet is in a hydrostatic equilibrium. The large heat emission and the need for an efficient source of internal heat is confirmed but the results do not indicate which one of the various possible mechanisms is favored although new evolutionary models suggest that the primordial heat may be insufficient. The presence of a highly eccentric and inclined magnetic field poses new problems, which are related to the pattern of internal convection and to the possibility of a north–south asymmetry of the interior.

099.076 Observations of Jupiter at 26.3 MHz using a large array. M. D. Desch, T. D. Carr, J. Levy.
Icarus, Vol. 25, 12 - 17 (1975).

A 640 element phase-steerable dipole array has been used to make highly sensitive observations of the planet Jupiter during the 1973 apparition. The satellite Io is found to have very little influence at the low flux levels, whereas the definition of sources A and B appears to be relatively flux independent. A two-dimensional analysis of the data in the Jupiter–Io plane has revealed considerable source B activity at low intensities which is not influenced by Io.

099.077 A laboratory atlas of the $5\nu_1$ NH$_3$ absorption band at 6475 Å with applications to Jupiter and Saturn. L. P. Giver, J. H. Miller, R. W. Boese.
Icarus, Vol. 25, 34 - 48 (1975).

The $5\nu_1$ absorption band of NH$_3$ is displayed from 6418 to 6550 Å. The total band intensity has been measured: $S_B = 0.66$ cm^{-1} m^{-1} amagat^{-1}. Line intensities and self-broadening coefficients have been measured for some of the prominent lines. Since the total band absorption has previously been measured by others on moderate resolution photoelectric scans of the spectra of Jupiter and Saturn, one can use the band intensity to derive the NH$_3$ abundance in the atmospheres of these two planets. The NH$_3$ abundances in a single vertical path obtained by this method are about 10m amagat for Jupiter and 2m amagat for Saturn. These results are in agreement with previous results obtained from higher resolution photographic spectra.

099.078 Spectra of high-frequency radio emission from Jupiter. M. J. Groth, R. L. Dowden.
Nature, Vol. 255, 382 - 384 (1975). – Letter.

099.079 Taylor columns in a shear flow and Jupiter's Great Red Spot. C. W. Titman, P. A. Davies, P. M. Hilton.
Nature, Vol. 255, 538 - 539 (1975).

The Great Red Spot (GRS) on Jupiter is not located in a uniform flow but occupies a position at which the zonal wind varies strongly with latitude. The conditions in which Taylor columns form in these circumstances are critically dependent on the sign of the relative vorticity associated with the shear. On Jupiter the GRS is situated at a latitude where conditions are particularly favourable for the formation of a Taylor column, according to this model. Complementary to this theoretical approach the authors report some preliminary results of the first experimental investigation into the effect of shear on the formation of a Taylor column.

099.080 **Observations and analysis of the Jovian spectrum in the 10-micron ν_2 band of NH_3.**
J. H. Lacy, A. I. Larrabee, E. R. Wollman, T. R. Geballe, C. H. Townes, J. D. Bregman, D. M. Rank.
Astrophys. Journ., (Letters), Vol. 198, L145 - L148 (1975).

Observations of the ν_2 band of NH_3 in the Jovian atmosphere have been made at resolutions varying from 4 cm^{-1} to 0.15 cm^{-1}. The observations have been interpreted by computation of synthetic atmospheric spectra. Derived atmospheric parameters include a pressure of 0.5 atm at 145 K and a minimum temperature of 118 K.

099.081 **Jovian protons and electrons: Pioneer 11.**
J. H. Trainor, F. B. McDonald, D. E. Stilwell, B. J. Teegarden, W. R. Webber.
Goddard Space Flight Center, Greenbelt, Maryland, GSFC Document X-663-75-44, 16 pp. (1975).

The authors present a preliminary account of the Pioneer 11 passage through the Jovian magnetosphere as viewed by the particle detector systems of the Goddard Space Flight Center and the University of New Hampshire.

099.082 **The occultation of β Scorpii by Jupiter. III. Discussion of the photometric results.**
J. Berezne, M. Combes, R. Laporte, J. Lecacheux, L. Vapillon.
Astron. Astrophys., Vol. 40, 85 - 90 (1975).

The observation the authors have made of the occultation of β Sco by Jupiter is analysed with respect to its photometric accuracy.

099.083 **Jupiteru u pohode (Visiting Jupiter).**
J. Milogradov-Turin.
Vasiona, Vol. 23, 1 - 4 (1975).

099.084 **On the wavelength dependence of Jovian limb darkening'.** V. G. Tejfel'.
Astron. Zhurn. Akad. Nauk SSSR, Vol. 52, 615 - 622 (1975). In Russian. English translation in Soviet Astron., Vol. 19, No. 3.

The applicability of the empirical formula of Minnaert for convenient analytical and graphic presentations of theoretical calculations of the absolute brightness distribution on the disk of a planet surrounded by a semi-infinite homogeneous atmosphere is shown. The dependence of the limb darkening parameter (k) on the reflectivity of the planetary disk center (r_0) for the isotropic, three-members and Henyey-Greenstein scattering functions is presented. The observed variations of the limb darkening with the wavelength for the equatorial belt of Jupiter in 1962 and 1964 are considered. There was detected a sufficient discrepancy between the observed and calculated relation $k(r_0)$.

099.085 **Life on Jupiter?** W. F. Libby.
Origins of Life, Vol. 5, 483 - 486 (1974).

099.086 **Observations of the magnetic field structure of Jupiter.** D. Stannard.
Geophys. Journ. Roy. Astron. Soc., Vol. 41, 327 - 330 (1975).

Observations are reviewed of the magnetic field structure of the planet Jupiter. The magnetic field geometry deduced from the spacecraft data of Pioneer 10 is in good agreement with that derived from earth-based radio astronomy measurements.

099.087 **Features of Jupiter's trapped particle environment associated with Jovian satellites—Pioneer 10 results and X-ray observations.** J. F. Vesecky.
Geophys. Journ. Roy. Astron. Soc., Vol. 41, 331 - 346 (1975).

Following a brief summary of the relevant details of the Pioneer 10 mission to Jupiter, theoretical considerations regarding the absorption of trapped particles by Jupiter's inner satellites are reviewed.

099.088 **Origin of Jupiter's magnetic field.** R. Hide.
Geophys. Journ. Roy. Astron. Soc., Vol. 41, 347 - 348 (1975).

Plausible theories of non-thermal radio-noise from Jupiter on decimetre and decametre wavelengths invoke a strong and nearly dipolar Jovian magnetic field and an associated system of van Allen-type radiation belts of electrically-charged particles extending beyond and interacting with the first Galilean satellite Io. Data from the recent Pioneer 10 fly-by provide direct evidence for the radiation belts and magnetic field and confirm the values for the strength, position and orientation of the equivalent Jovian magnetic dipole inferred previously by radio astronomers.

099.089 **Jupiter in 1972: rotation periods.** P. W. Budine.
Strolling Astronomer, Vol. 25, 141 - 145 (1975).

099.090 **Large-scale disturbances on Jupiter.** W. K. Wacker.
Strolling Astronomer, Vol. 25, 145 - 150 (1975).

099.091 **Some studies of the Jovian upper atmosphere.**
L. A. Capone.
Thesis, Florida Univ., Gainesville (USA). 156 pp. University Microfilms Order No. 74-10,032 (1973).

099.092 **Io-modulated Jovian decametric radiation.**
R. A. Smith.
Thesis, Maryland Univ., College Park (USA). 228 pp. University Microfilms Order No. 74-16,577 (1973).

099.093 **A second look at Jupiter.** E. Burgess.
New Scient., (GB), Vol. 64, 804 - 806 (1974).

099.094 **Jupiter's magnetosphere as observed with Pioneer 10.**
J. A. Van Allen, R. F. Randall.
Astronaut. Aeronaut., (U. S. A.), Vol. 12, No. 7 - 8, p. 14 - 21 (1974).

The authors describe their exploratory survey of the absolute intensities, energy spectra, and angular distributions of energetic electrons and protons as a function of position along the trajectory of the spacecraft through the magnetosphere of Jupiter.

099.095 **On Jupiter's rate of rotation.** P. H. Stone.
Journ. Atmosph. Sci., (USA), Vol. 31, 1471 - 1472 (1974).

It is suggested that the apparent lag of Jupiter's mean rotation rate in extratropical latitudes behind the rotation rate of Jupiter's radio emissions is caused by the difference between phase speeds and true speeds in extratropical latitudes.

099.096 **On the influence of Io on the Jovian atmosphere.**
M. F. Khodyachikh.
Vestn. Khar'kov. Univ., No. 117, (Ser. Astron., vyp. (No.) 9), p. 32 - 37 (1974). In Russian.

099.097 **The detection of Jovian high energy electrons in interplanetary space $\gtrsim$ 1 A.U. from the planet.**
J. A. Simpson, D. L. Chenette, T. F. Conlon.
Solar wind three, (see 012.020), p. 472 - 474 (1974).

099.098 **Jupiter's bow shock: comparison with theory.**
M. Dryer.
Solar wind three, (see 012.020), p. 475 - 477 (1974).

099.099 **Waves in the Jovian upper atmosphere.**
R. G. French, P. J. Gierasch.
Journ. Atmosph. Sci., Vol. 31, 1707 - 1712 (1974).

On montre que les oscillations du profil des températures observées dans la haute atmosphère de Jupiter par Veverka et. al., peuvent être interprétées par la propagation d'ondes atmosphériques.

099.100 Pioneer 10 measurements of Jupiter's magnetic field. D. E. Jones, E. J. Smith, L. Davis, Jr., D. S. Colburn, P. J. Coleman, Jr., P. Dyal, C. P. Sonett.
Proc. Utah Acad. Sci. Arts Letters, (USA), Vol. 51, 153 - 160 (1974).
Résultats préliminaires des mesures effectuées par Pioneer 10, du 30 novembre au 12 décembre 1973.

099.101 A study of single and dual dipole magnetic field models for Jupiter: Pioneer 10.
D. E. Jones, J. G. Melville.
Proc. Utah Acad. Sci. Arts Letters, (USA), Vol. 51, 161 - 164 (1974).

099.102 Jupiter: the planetless planet? J. Eberhart.
Science News, Vol. 106, 186 - 187 (1974).

Intensity and half-width measurements in the 1.525 μm band of acetylene. See Abstr. 022.026.

Intensity measurements in the 4.5μ fundamental of CH_3D at low temperatures. See Abstr. 022.040.

A high resolution decameter multichannel radio spectrograph. See Abstr. 033.032.

Pioneer 10 and Pioneer 11.
See Abstr. 053.004.

On the chemical composition of the moon, Jupiter, meteorites and Am stars. See Abstr. 061.031.

The terrestrial magnetosphere and comparison with Jupiter's. See Abstr. 084.272.

Methane absorption in the visible spectra of the outer planets and Titan. See Abstr. 091.011.

Production of organic molecules in the outer solar system by proton irradiation: laboratory simulations. See Abstr. 091.030.

Venus—Jupiter encounter.
See Abstr. 093.012.

Measurement of pressure-induced absorption in the atmospheres of Saturn and Jupiter. See Abstr. 100.011.

Speculations on the hydrogen phase diagram. See Abstr. 100.015.

The effects of resonance with Jupiter in the system of short-period comets. See Abstr. 102.016.

Évolution des orbites et des radiants d'un essaim météorique de la famille de Jupiter. See Abstr. 104.018.

Possible formation of meteoritic chondrules and inclusions in the precollapse Jovian protoplanetary atmosphere. See Abstr. 105.124.

Pioneer 11 meteoroid detection experiment: preliminary results. See Abstr. 106.026.

Early scattering by Jupiter and its collision effects in the terrestrial zone. See Abstr. 107.010.

On the chemical composition of the sun, Jupiter, meteorites and Am stars. See Abstr. 114.083.

The occultation of Beta Scorpii by Jupiter. VI. The masses of Beta Scorpii A_1 and A_2. See Abstr. 117.020.

Jupiter, Satellites

099.201 Callisto: disk temperature at 3.71-centimeter wavelength. G. L. Berge, D. O. Muhleman.
Science, Vol. 187, 441 - 443 (1975).
The authors observed the radio emission of Callisto with a three-element interferometer at the time of the 1973 opposition of Jupiter. Special care was taken to remove the residual, unresolved contribution from Jupiter itself in the antenna side lobes. The resulting disk temperature at a wavelength of 3.71 centimeters, assuming a radius of 2500±75 kilometers for Callisto, was 101°±25°K. This temperature is much more consistent with emission from a simple dielectric sphere than the considerably higher temperatures that have been reported for wavelengths of 3.5 and 8.2 millimeters.

099.202 Mutual phenomena of the Galilean satellites in 1973. I. Total and near-total occultations of Europa by Io.
K. Aksnes, F. A. Franklin.
Astron. Journ., Vol. 80, 56 - 63 (1975).
The authors report an analysis of 13 observations of nine occultations of Europa (J2) by Io (J1). Light curves of such occultations can provide the following information: (1) correc-

tions τ_1 and τ_2 to the epochs (ET) of J1 and J2 in Sampson's theory; (2) a correction to the radius R_2 of J2; and (3) surface brightness map(s) of J2. The authors have obtained the mean values $\tau_1 = 0.67 \pm 0.05$ min, $\tau_2 = 0.24 \pm 0.02$ min, at the epoch 1973. 62, and $R_2 = 1521 \pm 27$ km.

099.203 Pioneer 11 photographierte Ganymed.
H. W. Köhler.
SuW, Vol. 14, 57 (1975).

099.204 Sodium D-line emission from Io: synoptic observations from Table Mountain Observatory.
J. T. Bergstralh, D. L. Matson, T. V. Johnson.
Astrophys. Journ., (Letters), Vol. 195, L131 - L135 (1975).
The authors report the first results of a program initiated at Table Mountain Observatory to study the time variation of the sodium D-line emission around Io. The data demonstrate that (a) the sodium emission is highly correlated with Io's orbital position, (b) the gross temporal and amplitude characteristics of the emission are explained if resonance scattering of sunlight is the dominant excitation mechanism, (c) the emission is not directly related to dekametric noise storms from Jupiter, (d) an upper limit of 130 kR is placed on any

constant near-surface (auroral) source of emission, (e) a steady-state source of sodium operated during our 7-week observing interval, (f) the sodium cloud is probably distributed in a partial toroid, possibly with more sodium leading than trailing Io, (g) the D-line emission does not depend strongly on the solar phase angle, α.

099.205 Ganymede from Pioneer 10.
Sky Telescope, Vol. 49, 8 (1975).

099.206 The atmosphere and ionosphere of Io.
M. B. McElroy, Y. L. Yung.
Astrophys. Journ., Vol. 196, 227 - 250 (1975).
A variety of models for Io's atmosphere, ionosphere, surface, and environment are developed and discussed in the context of recent observational data.

099.207 Un treizième satellite de Jupiter.   G. Oudenot.
L'Astronomie, 89e année, p. 108 (1975).

099.208 On the presumed capture origin of Jupiter's outer satellites.   T. A. Heppenheimer.
Icarus, Vol. 24, 172 - 180 (1975).
The problem of the origin of Jupiter's outer satellites is treated within the framework of the theory of capture through collinear libration points. Lower bounds for the satellites' semimajor axes are found from a corrected rederivation of Bailey's capture theory. Upper bounds are found from a new derivation of the stability limit for satellites, based on Floquet stability theory. It is shown that if the bodies had near-zero relative velocity when passing the libration point, direct orbits would lie outside retrograde orbits, which is not the case for Jupiter. It is found that the dimensions and distribution of the direct group are well explained by libration-point capture with Jupiter's mass = 1/1730 solar mass, which is interpreted as indicating capture soon after Jupiter's formation. But ad hoc assumptions are required for this capture model to explain the retrograde group. It is concluded that the direct and retrograde groups may have had different mechanisms of origin.

099.209 A photometric study of mutual phenomena of Galilean satellites.
D. Chen, Z. Wu, X. Yang, C. Wang.
Acta Astron. Sinica, Vol. 15, 190 - 214 (1974). In Chinese.
By use of Aksnes' (1973) revised predictions, six events including two of mutual occultation and four of mutual eclipse have been observed photoelectrically. The equipment used consists of a photoelectric photometer mounted on a 60-cm Cassegrain reflector and a GG11 filter.

099.210 Reuzentelescopen observeren de 'manen van Galilei'.
L. Aerts.
Zenit, Vol. 2, 38 - 39 (1975).

099.211 Recent UBV photometry of the Galilean satellites.
R. L. Millis, D. T. Thompson.
Bull. American Astron. Soc., Vol. 7, 237 - 238 (1975).
Abstr. AAS.

099.212 On the presumed capture origin of Jupiter's outer satellites.   T. A. Heppenheimer.
Bull. American Astron. Soc., Vol. 7, 341 (1975). — Abstr. AAS.

099.213 The orbit of Jupiter XIII.
K. Aksnes, B. G. Marsden.
Bull. American Astron. Soc., Vol. 7, 342 - 343 (1975).
Abstr. AAS.

099.214 Pioneer imaging of the Galilean satellites.
T. C. Duxbury,

Bull. American Astron. Soc., Vol. 7, 379 (1975). — Abstr. AAS.

099.215 The surface of Io: a progress report.
F. P. Fanale, D. B. Nash, T. V. Johnson, D. L. Matson.
Bull. American Astron. Soc., Vol. 7, 386 (1975). — Abstr. AAS.

099.216 Infra-red spectra of Io and Titan and narrow band photometer measurements of the Galilean satellites.
U. Fink, H. P. Larson, G. Howell, G. H. Rieke.
Bull. American Astron. Soc., Vol. 7, 386 (1975). — Abstr. AAS.

099.217 Sodium D-line emission from Io: more synoptic observations from Table Mountain Observatory.
J. T. Bergstralh, D. L. Matson, T. V. Johnson.
Bull. American Astron. Soc., Vol. 7, 386 (1975). — Abstr. AAS.

099.218 Sodium D-line emission from Io: spatial brightness distribution from multislit spectra.
G. Münch, J. T. Bergstralh.
Bull. American Astron. Soc., Vol. 7, 386 (1975). — Abstr. AAS.

099.219 The geometry of Io's sodium cloud.
L. Trafton, W. Macy, Jr.
Bull. American Astron. Soc., Vol. 7, 386 - 387 (1975). Abstr. AAS.

099.220 Io: reported helium emission.
D. P. Cruikshank, C. B. Pilcher, W. M. Sinton.
Bull. American Astron. Soc., Vol. 7, 387 (1975). — Abstr. AAS.

099.221 Ganymede: possibility of an oxygen atmosphere.
Y. L. Yung, M. B. McElroy.
Bull. American Astron. Soc., Vol. 7, 387 (1975). — Abstr. AAS.

099.222 Radar observations of Ganymede.
R. M. Goldstein, G. A. Morris.
Bull. American Astron. Soc., Vol. 7, 387 (1975). — Abstr. AAS.

099.223 Infrared albedos and rotation curves of the Galilean satellites.   O. L. Hansen.
Bull. American Astron. Soc., Vol. 7, 387 (1975). — Abstr. AAS.

099.224 Radiometry of the Galilean satellites.   D. Morrison.
Bull. American Astron. Soc., Vol. 7, 387 (1975). Abstr. AAS.

099.225 On the invertibility of mutual occultation light curves.   R. T. Brinkmann.
Bull. American Astron. Soc., Vol. 7, 387 - 388 (1975). Abstr. AAS.

099.226 Na-D line emission from rock specimens by proton bombardment: implications for emission from Jupiter's satellite Io.
D. B. Nash, D. L. Matson, T. V. Johnson, F. P. Fanale.
Journ. Geophys. Res., Vol. 80, 1875 - 1879 (1975).
The Na-D line emission directly excited during surface sputtering by Jovian magnetospheric protons may possibly be an observable component in the Io emission. In any case, sputtering of sodium-rich surface material is a mechanism which can inject sodium (and other material) into the 'atmosphere' and space about Io.

099.227 The atmosphere of Io from Pioneer 10 radio occultation measurements.
A. J. Kliore, G. Fjeldbo, B. L. Seidel, D. N. Sweetnam, T. T. Sesplaukis, P. M. Woiceshyn.
Icarus, Vol. 24, 407 - 410 (1975). — Presented at IAU Colloquium No. 28, Cornell, Univ., Ithaca, New York, 1974 August 18 - 21.

The occultation of the Pioneer 10 spacecraft by Io (JI) provided an opportunity to obtain two S-band radio occultation measurements of its atmosphere. The dayside entry measurements revealed an ionosphere having a peak density of about 6×10^4 el cm^{-3} at an altitude of about 100 km. The topside scale height indicates a plasma temperature of about 406 K if it is composed of Na$^+$ and 495 K if N$_2^+$ is principal ion. A thinner and less dense ionosphere was observed on the exit (night side), having a peak density of 9×10^3 el cm^{-3} at an altitude of 50 km. The topside plasma temperature is 160 K for N$_2^-$ and 131 K for Na$^+$. Two measurements of its radius were also obtained yielding a value of 1830 km for the entry and 1921 km for the exit.

099.228 Temperature dependence of the water-ice spectrum between 1 and 4 microns: application to Europa, Ganymede and Saturn's rings. U. Fink, H. P. Larson.
Icarus, Vol. 24, 411 - 420 (1975).

Reflection spectra of water ice from 1 to 4 μm are presented as a function of temperature. It is found that a feature at 6056 cm^{-1} changes its intensity sufficiently that it can be used as a spectroscopic measure of the ice temperature. A temperature calibration curve of this feature down to 55 K is developed and is used to determine ice temperatures for the Galilean satellites Europa (95 ± 10 K), Ganymede (103 ± 10 K), and the rings of Saturn (80 ± 5 K).

099.229 Jupiter XIII — neuer Satellit des Riesenplaneten. D. Wattenberg.
Archenhold-Sternw., Berlin-Treptow, 23. Jahrgang, p. 14 (1975).

099.230 Spectres des satellites de Jupiter. I. Données des observations.
D. A. Andrienko, L. A. Ourassine (*L. A. Urasin*).
Ann. Obs. Astron. Alger, Vol. 4, Fasc. 1, p. 19 - 21 (1974).

099.231 Thirteenth satellite of Jupiter.
C. T. Kowal, K. Aksnes, B. G. Marsden, E. Roemer.
Astron. Journ., Vol. 80, 460 - 464 (1975).

The discovery, observations, and attempts to determine the orbit of Jupiter XIII are described. It is found that the orbit is very similar to the orbits of Jupiter VI, VII, and X. An ephemeris is provided for the 1975 opposition.

099.232 Ganymede: observations by radar.
R. M. Goldstein, G. A. Morris.
Science, Vol. 188, 1211 - 1212 (1975).

Radar cross-section measurements indicate that Ganymede scatters to earth 12 percent of the power expected from a conducting sphere of the same size and distance. This compares with 8 percent for Mars, 12 percent for Venus, 6 percent for Mercury, and about 8 percent for the asteroid Toro. Furthermore, Ganymede is considerably rougher than Mars, Venus, or Mercury. Roughness is made evident in this experiment by the presence of echoes away from the center of the disk.

099.233 Satelliti di Giove. M. C. Battaglia.
Coelum, Vol. 43, 109 - 112 (1975).

099.234 Rijpvorming op de satelliet Io. C. de Jager.
Zenit, Vol. 2, 203 - 204 (1975).

099.235 Scans of Io, Europa, and Ganymede in the NaD region.
L. J. Lanzerotti, M. F. Robbins, N. H. Tolk, S. H. Neff.
Publ. Astron. Soc. Pacific, Vol. 87, 449 - 453 (1975).

Scans in the region of the sodium D lines were made of the first three Galilean satellites of Jupiter in late December 1973. Sodium emissions from Io (J I) were observed on three

out of four nights. No evidence of emissions was observed from either Europa (J II) or Ganymede (J III).

099.236 Jupiter XIII. C. T. Kowal, K. Aksnes.
IAU Circ., No. 2742 (1975).

099.237 Jupiter XIII.
IAU Circ., No. 2781 (1975).

099.238 The temperature of Amalthea. G. H. Rieke.
Icarus, Vol. 25, 333 - 334 (1975).

Infrared photometry of Amalthea (JV) indicates that it is at a temperature of 155 ± 15 K and has a radius of 120 ± 30 km. There is no evidence for substantial heating by the Jovian radiation belts.

Jupiter's innermost moon may have crystalline layer.
IEEE Spectrum, Vol. 12, No. 1, p. 22 - 23 (1975).

Io, the anomaly of the solar system.
Nature, Vol. 253, 587 - 588 (1975).

Jupiter: nu 13 manen.
Zenit, Vol. 2, 8 (1975).

Photographic positional observations of the Galilean satellites of Jupiter and Saturn satellites in Pulkovo. See Abstr. 041.062.

The orbital resonance amongst the Galilean satellites of Jupiter. See Abstr. 042.026.

De la possibilité pour certains astéroides d'être des satellites lointains de Jupiter. See Abstr. 098.023.

Radiometric studies of Trojan asteroids and Jovian satellites 6 and 7. See Abstr. 098.040.

Io-related Jovian decametric bursts and Io sodium. See Abstr. 099.020.

Radiation belts of Jupiter: a second look. See Abstr. 099.029.

The imaging photopolarimeter experiment on Pioneer 11. See Abstr. 099.030.

Pioneer 10 ultraviolet photometer observations of the Jovian hydrogen torus: the angular distribution. See Abstr. 099.071.

A theory of the Jovian hydrogen torus. See Abstr. 099.072.

Observations of Jupiter at 26.3 MHz using a large array. See Abstr. 099.076.

Features of Jupiter's trapped particle environment associated with Jovian satellites—Pioneer 10 results and X-ray observations. See Abstr. 099.087.

On the influence of Io on the Jovian atmosphere. See Abstr. 099.096.

Evidence for frost on Rhea's surface. See Abstr. 100.210.

The atmospheres of Titan and the Galilean satellites. See Abstr. 100.216.

100   Saturn, Saturn Satellites

Saturn

100.001   The influence of radiation pressure and drag on Saturn's rings.   P. Vaaraniemi.
Astrophys. Space Sci., Vol. 32, 13 - 24 (1975).
The influence of radiation pressure and drag on the optical thickness of ring C is calculated as a function of the saturnocentric distance. The results are compared with Franklin and Cook's 1958 data. It seems probable that drag exerts more important effects than radiation pressure, at least in inner parts of ring C. The drag effect might also explain the existence of the Cassini division.

100.002   The temperature profile in the upper atmosphere of Saturn from inversion of thermal emission observations.   G. Ohring.
Astrophys. Journ., Vol. 195, 223 - 225 (1975).
The temperature profile in the upper atmosphere of Saturn is inferred from recent observations of the planet's thermal emission spectrum. The temperatures are derived from inversion of the observed brightness temperatures in the 7.7-μ CH_4 band. The inferred profile is characterized by a temperature minimum of about $78°K$ in the tropopause region and an increase of temperature with altitude to a maximum of about $134°K$ in the upper atmosphere.

100.003   Center-to-limb observations of Saturn in the thermal infrared.   F. C. Gillett, G. S. Orton.
Astrophys. Journ., (Letters), Vol. 195, L47 - L49 (1975).
Center-to-limb scans on the disk of Saturn at 11.7 μ show limb brightening and an unexpected enhancement near the south pole. These limb brightening observations confirm the presence of a thermal inversion in the upper atmosphere of Saturn and strongly support the suggestion that the 12-μ feature in the spectrum of Saturn is due to emission from ethane (C_2H_6). The enhancement near the south pole is probably related to the fact that the south pole is presently tipped toward the sun, which increases the insolation at southerly latitudes.

100.004   Ionospheric models of Saturn, Uranus, and Neptune.   S. K. Atreya, T. M. Donahue.
Icarus, Vol. 24, 358 - 362 (1975).
Model ionospheres are calculated for Saturn, Uranus, and Neptune. Protons are the major ions above 150 km altitude measured from a reference level where the hydrogen density is 1×10^{16} molecules cm^{-3}, while below 150 km quick conversion of protons to H_3^+ ions by a three-body association mechanism leads to a rapid removal of ionization in dissociative recombination of H_3^+. Electron density maxima are found at about 260 km for Saturn and Uranus and 200 km for Neptune. Present knowledge of the physical and chemical processes in the atmospheres of these planets suggests that their ionospheres probably will not be Jupiter-like.

100.005   Saturn's $3\nu_3$ methane band: an analysis in terms of a scattering atmosphere.
L. Trafton, W. Macy, Jr.
Astrophys. Journ., Vol. 196, 867 - 876 (1975).
Using parameters recently measured in the laboratory, the authors analyzed the shapes of manifolds belonging to the R-branch of Saturn's $3\nu_3$ CH_4 band in terms of a homogeneous scattering model of Saturn's atmosphere, and compared the results with those of an earlier analysis in terms of a reflecting layer model. The resulting effective pressure is an order of magnitude smaller than for the reflecting layer analysis, and the C/H ratio is an order of magnitude higher than the solar

value. The overall fit of the theoretical profiles to the observed manifold shapes is distinctly worse, especially for $R(7)$. Scattering evidently plays only a minor role in the formation of this band along Saturn's south-central meridian.

100.006   The detection of ethane on Saturn.
A. Tokunaga, R. F. Knacke, T. Owen.
Astrophys. Journ., (Letters), Vol. 197, L77 - L78 (1975).
The ν_9 band of ethane (C_2H_6) has been identified in the spectrum of Saturn.

100.007   The 1972 - 73 apparition of Saturn.
J. L. Benton, Jr.
Strolling Astronomer, Vol. 25, 102 - 103 (1975).

100.008   On the structure of Saturn's rings and the 'real' rotational period for the planet.   I. Ferrín.
Astrophys. Space Sci., Vol. 33, 453 - 457 (1975).
An analysis of the Lowell Observatory photographic plates of Saturn gave the following results: (1) ring A and B show peculiar brightness distributions around the planet, from which the author concludes that both are composed of particles in synchronous rotation. (2) The leading side of the particles in ring A is brighter than the trailing side by about 4%, which may indicate an interaction between such particles and the interplanetary medium. (3) Scans of the rings across the major axis show a small ($\sim 0.3''$) region of enhanced brightness, from which a value of $T_s = 10^h 13^m 8 \pm 5^m 4$ for the actual planetary rotational period of Saturn is derived. (4) In order to explain the synchronous rotation, the particles in ring A have to be at least 42 m in diameter.

100.009   Saturn's rings: the determination of their brightness temperature and opacity at centimeter wavelengths.
J. N. Cuzzi, W. A. Dent.
Astrophys. Journ., Vol. 198, 223 - 227 = Five College Obs., Univ. Mass., Amherst, Contr. No. 189 (1975).
Interferometric observations of Saturn at 3.7 cm wavelength are used to determine the brightness temperature of the B ring, $T_R = 15 \pm 3$ K, and the mean normal optical depth to extinction of the B ring, $\tau_0 = 0.8 \pm 0.1$. When these results are combined with model calculations of the brightness of the rings due to both thermal emission and diffuse multiple scattering of the radiation from the planet, the single scattering albedo of a typical ring particle at $\lambda 3.7$ cm is found to be $\varpi_0 = 0.999$ (+0.001, −0.01).

100.010   The rings of Saturn: a frost coated semiconductor?
W. G. Egan, T. Hilgeman.
Bull. American Astron. Soc., Vol. 7, 238 (1975). − Abstr. AAS.

100.011   Measurement of pressure-induced absorption in the atmospheres of Saturn and Jupiter.
T. Z. Martin, D. P. Cruikshank, C. B. Pilcher, W. M. Sinton.
Bull. American Astron. Soc., Vol. 7, 380 - 381 (1975).

100.012   The detection of ethane on Saturn.
A. Tokunaga, R. F. Knacke, T. Owen.
Bull. American Astron. Soc., Vol. 7, 383 (1975). − Abstr. AAS.

100.013   An interpretation of thermal emission from Saturn.
J. Caldwell.
Bull. American Astron. Soc., Vol. 7, 383 (1975). − Abstr. AAS.

100.014   Tentative observation of Saturn emission near 1 MHz.   L. W. Brown.
Bull. American Astron. Soc., Vol. 7, 383 (1975). − Abstr. AAS.

100.015 Speculations on the hydrogen phase diagram.
R. Smoluchowski.
Bull. American Astron. Soc., Vol. 7, 384 (1975). − Abstr. AAS.

100.016 Microwave scattering by the rings of Saturn.
J. N. Cuzzi, J. B. Pollack.
Bull. American Astron. Soc., Vol. 7, 388 (1975). − Abstr. AAS.

100.017 Infrared thermal models for Saturn's ring.
M. J. Price.
Bull. American Astron. Soc., Vol. 7, 388 (1975). − Abstr. AAS.

100.018 Bistatic-radar study of the rings of Saturn.
E. A. Marouf, G. L. Tyler, V. R. Eshleman.
Bull. American Astron. Soc., Vol. 7, 388 (1975). − Abstr. AAS.

100.019 Saturn's rings are a gravitational "mass-spectrograph". J. Boynton.
Bull. American Astron. Soc., Vol. 7, 389 (1975). − Abstr. AAS.

100.020 Thermal emission from a multiple scattering model of Saturn's rings. Y. Kawata, W. M. Irvine.
Icarus, Vol. 24, 472 - 482 = Contr. Five College Obs., Univ. Mass., *Amherst*, No. 194 (1975). − Presented at IAU Colloquium No. 28, Cornell Univ., Ithaca, New York, 1974 August 18 - 21.

Models of Saturn's B ring have been investigated which include the shadowing mechanism, realistic phase functions for the ring particles, and the effects of multiple scattering and a particle size dispersion. These models are based on the assumption that the rings form a layer many particles thick. A power law relation $dn \propto \rho^{-s}$ is used for the size dispersion law of the ring particles, where dn is the number of particles with radii between ρ and $\rho + d\rho$. In the calculation of the infrared brightness temperature of the rings, the effect of mutual heating among the ring particles is considered quantitatively for the first time.

100.021 Shape of the Cassini division in different colors.
K. Lumme.
Icarus, Vol. 24, 483 - 491 (1975). − Presented at IAU Colloquium No. 28, Cornell Univ., Ithaca, New York, 1974 August 18 - 21.

The true shape of the Cassini division of Saturn's rings in red, yellow, green, and blue light has been derived from photographs of high quality, taking into account the correction due to the spreading of light. It is evident that the division is not completely dark. Its minimum brightness is about 20% of that of ring B and the width of the division $0\rlap{.}''44 \pm 0\rlap{.}''05$.

100.022 Saturn's rings and Pioneer 11. M. J. Price.
Icarus, Vol. 24, 492 - 498 (1975).

Quantitative predictions of the diffuse reflection and transmission properties of Saturn's rings, relevant to the September 1979 Pioneer 11 flyby, are presented. Predictions are based on an elementary anisotropic scattering model. Interparticle separations are considered to be sufficiently large that mutual shadowing is negligible. Likely ranges in both the single scattering albedo and perpendicular optical thickness of the ring are considered. Situations of pronounced backscattering and of isotropic scattering are treated individually.

100.023 Saturn's rings: a photometric study of ring C.
S. Koutchmy.
Icarus, Vol. 25, 131 - 135 (1975).

A simple computational procedure is proposed for determining the true photometric profile of ring C using the spread function obtained from the satellite Dione and also slightly overexposed photographs of Saturn. No trace of a faint additional ring between ring C and the disk was found. The decreasing part, toward the planet, of the recorded photometric profile of ring C exhibits a slight depression tentatively attributed to a new division.

100.024 Illumination of Saturn's ring by the ball. I. Preliminary results. M. J. Price, A. Baker.
Icarus, Vol. 25, 136 - 145 (1975).

Indirect solar illumination of Saturn's ring via scattering from the ball of the planet provides a new ground-based observational technique for studying the single scattering albedo and phase function of the individual particles. Initial application of the technique is reported. The indirect contribution to the radiation scattered from the ring has been marginally detected by electronographic areal photometry. The results have been interpreted using simple scattering models.

100.025 Saturn radio emission near 1 MHz. L. W. Brown.
Astrophys. Journ., (*Letters*), Vol. 198, L89 - L92 (1975).

Emission from the direction of the planet Saturn has been observed by the IMP-6 spacecraft at 15 frequencies between 375 and 2200 kHz during the period 1971 April−1972 October. Initial data reduction has isolated approximately 12 storms whose occurrence corresponded to times in which the spacecraft had an unobstructed view in the direction of Saturn. These events persisted over periods between 1 and 10 minutes. Over the span of 500 days of data another 10−20 Saturnian events may exist. The spectral character of the radiation has been found analogous to that of Jupiter. The detection of this nonthermal radio emission is the first direct evidence for the existence of a Saturnian magnetic field containing energetic particles.

100.026 On the interpretation of Saturn's radio emission.
A. P. Naumov.
Izv. vyssh. ucheb. zavedenij. Radiofizika, Vol. 17, 1755 - 1764 (1974). In Russian. − Abstr. in Referativ. Zhurn. 51. Astron., 5.51.300 (1975).

100.027 Identification of $^{13}CH_4$ in the atmosphere of Saturn.
M. Combes, C. de Bergh, J. Lecacheux, J. P. Maillard.
Astron. Astrophys., Vol. 40, 81 - 84 (1975).

Several lines of the $3\nu_3$ band of $^{13}CH_4$ have been identified from high resolution spectra of Saturn, obtained by Fourier spectroscopy, and their equivalent widths measured. A $^{12}CH_4/^{13}CH_4$ ratio of $60 \pm^{40}_{15}$ has been estimated.

100.028 Saturn. T. Encrenaz, M. Combes, L. Vapillon, J. Berezne, Y. Zeau.
IAU Circ., No. 2743 (1975).

The rings of Saturn. See Abstr. 003.113.

Methane absorption in the visible spectra of the outer planets and Titan. See Abstr. 091.011.

Production of organic molecules in the outer solar system by proton irradiation: laboratory simulations. See Abstr. 091.030.

Determination of the masses of Jupiter and Saturn from the motion of Trojans. See Abstr. 099.035.

Spatial and spectral mapping of Jupiter and Saturn with a TV system. See Abstr. 099.037.

Spectra of Jupiter and Saturn in the ten micron region. See Abstr. 099.038.

Photochemistry of phosphine in the atmospheres of Jupiter and Saturn. See Abstr. 099.057.

Methane absorption in the visible spectra of the outer planets and Titan. See Abstr. 099.059.

Spectral images of Jupiter and Saturn. See Abstr. 099.064.

Spectra of hydrate frosts: their application to the outer solar system. See Abstr. 099. 073.

A laboratory atlas of the $5\nu_1$ NH_3 absorption band at 6475 Å with applications to Jupiter and Saturn. See Abstr. 099.077.

Temperature dependence of the water-ice spectrum between 1 and 4 microns: application to Europa, Ganymede and Saturn's rings. See Abstr. 099.228.

Collisional breakup of particles in a planetary ring. See Abstr. 107.002.

Saturn, Satellites

100.201 **Another look at atmospheric dynamics on Titan and some of its general consequences.** G. S. Golitsyn.
Icarus, Vol. 24, 70 - 75 (1975).

Mean wind velocities, U, and horizontal temperature differences, δT, are estimated for the Titan atmosphere using the similarity theory of the author. It is found that U is of order 1 m/sec and $\delta T \sim 0.1$ K. In the upper part of the Titan atmosphere something like the phenomenon of the 4-day Venus circulation may be developed. It is noted that the analogy between the Titan and Venus atmospheric circulations might be a very close one.

100.202 **Further comment on Titan's atmospheric scaling.** C. B. Leovy, J. B. Pollack.
Icarus, Vol. 24, 76 - 77 = Contr. Dep. Atmosph. Sci., Univ. Washington, *Seattle*, No. 323 (1975).

In discussions of the circulation of planetary atmospheres, the atmospheric efficiency factor η is overestimated when the atmosphere is very thin, or when latent heat transfer is neglected.

100.203 **The morphology of Titan's methane bands. I. Comparison with a reflecting layer model.** L. Trafton.
Astrophys. Journ., Vol. 195, 805 - 814 (1975).

Titan's spectrum at moderate resolution is compared with and contrasted to Saturn's spectrum from 5800 to 12,000 Å. The author finds, that the reflecting layer model cannot be reconciled with the observations of Titan. This is true whether or not results from an analysis of the $R(5)$ manifold of Titan's $3\nu_3$ methane band are considered. This implies that aerosol scattering plays a significant role in Titan's band formation. The most likely candidate for such an aerosol is frozen methane particles.

100.204 **The secular and orbital brightness variations of Titan, 1972–74.** G. W. Lockwood.
Astrophys. Journ., (*Letters*), Vol. 195, L137 - L139 (1975).

A brightening trend evident in Titan since 1972 is continuing in the current season at an annual rate of 0.017 mag per year in y and 0.026 mag per year in b. The mean magnitude at eastern elongation is brighter than that at western elongation by 0.009 mag, suggesting synchronous rotation and a stable system of visible surface features or non-uniform atmospheric clouds.

100.205 **The two faces of Iapetus.** D. Morrison, T. J. Jones, D. P. Cruikshank, R. E. Murphy.
Icarus, Vol. 24, 157 - 171 (1975).

The authors present new 20-μm radiometric observations made in 1971–73 and discuss these together with the photometric studies by Widorn (in 1949), Millis (in 1971), Noland et al. (in 1972–73), and Franklin and Cook (in 1972–74). The linear phase coefficient varies as the satellite rotates from 0.028 to 0.068 mag deg^{-1}. When corrected for this effect, the photometric variations suggest an albedo distribution characterized by a dark area covering most of the leading hemisphere and a bright trailing hemisphere and bright south polar cap. A combined analysis of the photometry and radiometry yields a radius of 800 to 850 km and mean geometric albedos for the light and dark faces of about 0.35 and 0.07, respectively. The average phase integral of the bright hemisphere is between 1.0 and 1.5.

100.206 **Secular and orbital brightness changes of Titan, 1972–1974.** G. W. Lockwood.
Bull. American Astron. Soc., Vol. 7, 384 (1975). – Abstr. AAS.

100.207 **The secular brightening of Titan.** T. J. Jones, D. Morrison.
Bull. American Astron. Soc., Vol. 7, 384 (1975). – Abstr. AAS.

100.208 **The albedo distribution across Titan's disk.** J. L. Elliot, J. Veverka, J. Goguen, T. Dunham.
Bull. American Astron. Soc., Vol. 7, 384 (1975). – Abstr. AAS.

100.209 **A photoelectric color and magnitude of Mimas.** O. G. Franz.
Bull. American Astron. Soc., Vol. 7, 388 (1975). – Abstr. AAS.

100.210 **Evidence for frost on Rhea's surface.** T. V. Johnson, G. J. Veeder, D. L. Matson.
Icarus, Vol. 24, 428 - 432 (1975). – Presented at IAU Colloquium No. 28, Cornell Univ., Ithaca, New York, 1974 August 18 - 21.

The authors have observed Rhea at 1.6 μm and 2.2 μm. The infrared spectral reflectances relative to 0.55 μm are 0.8 ($\pm$ 0.1 p.e.) at 1.65 μm and 0.6 ($\pm$ 0.1 p.e.) at 2.2 μm. The broadband infrared reflectances of Rhea are similar to those of the Galilean satellites Europa and Ganymede and also the rings of Saturn (all of which are known from high resolution scans to have water frosts on their surfaces). Rhea's low density, high albedo and relatively flat reflectance from 0.3 μm to 1.1 μm as well as the low infrared reflectances reported here are consistent with the presence of water ice on Rhea's surface.

100.211 **Photometry of Dione, Tethys, and Enceladus on the UBV system.** O. G. Franz, R. L. Millis.
Icarus, Vol. 24, 433 - 442 (1975). – Presented at IAU Colloquium No. 28, Cornell Univ., Ithaca, New York, 1974 August

18 - 21.

UBV measurements of Dione, Tethys, and Enceladus were made with an area-scanning photometer on several nights during the 1972/73 and 1973/74 apparitions of Saturn. The observed brightness variations have been separated into two components – one a function of orbital position, the other a function of solar phase angle. Dione and Tethys are brightest near greatest eastern elongation and faintest near greatest western elongation. The reverse is true of Enceladus. Opposition surges are observed for Dione and Tethys.

100.212 Near-infrared spectrophotometry of Titan.
L. M. Trafton.

Icarus, Vol. 24, 443 - 453 (1975). – Presented at IAU Colloquium No. 28, Cornell Univ., Ithaca, New York, 1974 August 18 - 21.

Detailed analysis of the $R(5)$ manifold of Titan's $3\nu_3$ CH_4 band confirms that the column abundance of Titan's spectroscopically visible atmosphere is greater than 1.6 km amagats. Recently discovered strong, unidentified absorptions in Titan's spectrum at $1.05-1.06$ μm have been compared with laboratory spectra of a number of gases including CH_4, C_2H_4, C_2H_6, and C_3H_8 with negative results. These comparisons, however, have not excluded the possibility that these features arise from a very large quantity of CH_4 or from an isotope of CH_4. Comparison with Uranus' spectrum suggests that the visible abundance of this molecule in Titan's atmosphere may be much greater than in Uranus' relatively clear, deep atmosphere. Spectra of features at $\lambda 8150.7$ and $\lambda 8272.7$ attributed possibly to H_2 have been obtained.

100.213 Estimates of quasipolar absorption by methane in the atmosphere of Titan. K. Fox.
Icarus, Vol. 24, 454 - 459 (1975). – Presented at IAU Colloquium No. 28, Cornell Univ., Ithaca, New York, 1974 August 18 - 21.

The basis for "quasipolar " absorption (QPA) by CH_4 is the existence of a small electric dipole moment in its ground state. The integrated intensity α_{QPA} at a temperature of 90 K is calculated to be between 4.8×10^{-5} and 1.9×10^{-2} cm^{-2} atm^{-1}. With an assumed mean pressure of 0.1 atm and a relative abundance of $[CH_4]/[H_2] = 1$, it is estimated that the ratio of quasipolar to pressure-induced absorption (PIA) is $0.05 \lesssim \alpha_{QPA}/\alpha_{PIA} \lesssim 18$ for the spectral range from 0 to 300 cm^{-1}. This result suggests that quasipolar absorption may contribute to a weak, CH_4-induced greenhouse in the atmosphere of Titan.

100.214 Infrared observations of the surface and atmosphere of Titan. R. F. Knacke, T. Owen, R. R. Joyce.
Icarus, Vol. 24, 460 - 464 (1975). – Presented at IAU Colloquium No. 28, Cornell Univ., Ithaca, New York, 1974 August 18 - 21.

Infrared photometry of Titan, Saturn, and Saturn's rings at 3.5, 4.9, 17.8, and 18.4 μm is reported. Comparison of the albedo of Titan in the 4.9 μm "window" with the albedo of the rings and with laboratory spectra suggests that frost,

possibly water ice, could be a major constituent. If thick clouds are present they must be very dark at 4.9 μm. The 17.8 and 18.4 μm data are not consistent with a clear, dense molecular hydrogen atmosphere.

100.215 An explanation for Iapetus' asymmetric reflectance.
C. Peterson.

Icarus, Vol. 24, 499 - 503 = Contr. Planet. Astron. Lab., Dep. Earth Planet. Sci., Mass. Inst. Technology, *Cambridge, Mass.,* No. 113 (1975). – Presented at IAU Colloquium No. 28, Cornell Univ., Ithaca, New York, 1974 August 18 - 21.

Cook and Franklin (1970) consider Iapetus originally to have been coated with about a meter of ice. This paper considers an ice deposition mechanism operating more actively on the trailing side. The two main assumptions used are (1) that there are more icy than rocky meteoroids in Saturn's environment, and (2) that some portion of each icy meteoroid will stick to a surface at collision velocities less than 2.4 km sec^{-1}, but will completely vaporize itself at greater velocities. A meteoroid can have the minimum collision velocity of about 1.7 km sec^{-1} with Iapetus only if their velocity vectors are nearly parallel, and under these conditions such collisions would tend to be with the trailing hemisphere. Collisions with the leading hemisphere will tend to be at a much higher velocity.

100.216 The atmospheres of Titan and the Galilean satellites.
S. H. Gross.

Journ. Atmosph. Sci., (*USA*), Vol. 31, 1413 - 1420 (1974).

The stability of various atmospheres on Titan and the Galilean satellites are examined relative to escape. One method consists of the comparison of exospheric temperature and a blow-off temperature defined for the various constituents. Another method consists of the evaluation of outflow based on a polytropic model.

Photographic positional observations of the Galilean satellites of Jupiter and Saturn satellites in Pulkovo. See Abstr. 041.062.

Methane absorption in the visible spectra of the outer planets and Titan. See Abstr. 091.011.

Tidal evolution of an accreting satellite. See Abstr. 091.019.

Aeronomy of the major planets: photochemistry of ammonia and hydrocarbons. See Abstr. 099.039.

Methane absorption in the visible spectra of the outer planets and Titan. See Abstr. 099.059.

Infra-red spectra of Io and Titan and narrow band photometer measurements of the Galilean satellites. See Abstr. 099.216.

Commensurabilities of satellites' apsidal precession periods. See Abstr. 101.001.

101 Uranus, Neptune, Pluto, Transplutonian Planets

101.001 Commensurabilities of satellites' apsidal precession periods.   R. Greenberg.
Monthly Notices Roy. Astron. Soc., Vol. 170, 295 - 303 (1975).

The orientation of the major axis of Saturn's satellite Rhea librates about alignment with Titan's major axis. This behaviour is a result of Titan's gravitational influence on Rhea. Similar effects may be important in the Uranian satellite system if they are enhanced by commensurabilities of apsidal precession periods. The effects of possible stable alignments have not been included in past analyses of the motions of Uranus' satellites, so the results of those studies should be accepted only tentatively.

101.002 On a possible atmosphere on Pluto.
G. S. Golitsyn.
Pis'ma v Astron. Zhurn., Vol. 1, No. 1, p. 38 - 42 (1975).
In Russian.

It is shown that even a very rarefied atmosphere influences substantially the temperature distribution on Pluto. An analysis of observational data shows that the atmospheric pressure is probably of a few tenths of an atmosphere.

101.003 Cosmogonical considerations regarding Uranus.
A. G. W. Cameron.
Icarus, Vol. 24, 280 - 284 (1975).

The cosmogony of Uranus is discussed within the context of a picture in which solid condensed materials accumulate to form a large body, which then acquires significant amounts of gas from the primitive solar nebula. Of prime cosmogonical importance is the tilt of the equatorial plane of the planet and of the plane of the satellite orbits by 98° with respect to the plane of the planetary orbit. The tilt of the planet can easily occur as a result of a major collision during the formation process; it seems most likely that the tilt of the satellite orbits requires that they were formed from a gaseous disc rotating about the planet after the tilt of the planetary rotational axis had occurred. Possible methods for tilting this gaseous disc are discussed.

101.004 Interior structure of Uranus.   W. B. Hubbard.
Icarus, Vol. 24, 285 - 291 (1975).

A mission to Uranus will permit definitive measurements of fundamental parameters of Uranus' interior structure, such as radius, rotation, magnetic moment, atmospheric composition, and gravitational harmonics. The author briefly discusses the utility of such data for constraining interior models.

101.005 The atmosphere of Uranus.   P. H. Stone.
Icarus, Vol. 24, 292 - 298 (1975).

Current knowledge of the atmosphere of Uranus is reviewed and specific objectives are suggested for satellite missions to Uranus. The anomalous composition of Uranus makes determinations of its atmospheric composition particularly valuable for testing theories of solar system evolution. The weakness of its atmospheric heating makes the determination of its atmospheric structure and dynamics particularly valuable for testing theories of atmospheric behavior. The large axial inclination of Uranus implies an anomalous latitudinal variation of temperature and dynamics different from that of the other planets.

101.006 Why image Uranus?
M. J. S. Belton, F. E. Vescelus.
Icarus, Vol. 24, 299 - 310 (1975).

A review of visual and photographic data on the apppearance of Uranus indicates that markings frequently occur on the planet. The featureless images obtained by the Stratoscope II balloon telescope are possibly the result of the broad spectral band that was used. It is proposed that if the choice of an MJU imaging system rests on Uranus objectives alone (i.e., excluding the satellites) then the system should emphasize photo-polarimetric observations between 5500 and 10000 Å. If, however, the total mission objectives are the basis of choice then a high resolution imaging system, based on the Mariner Jupiter—Saturn system, but including a solid state silicon array would be a more suitable choice. The performance of such a system at Uranus is analyzed.

101.007 Particle and field environment of Uranus.
G. L. Siscoe.
Icarus, Vol. 24, 311 - 324 (1975).

In 1985 the spin axis of Uranus points within 10° of the sun and the planet's position is very near the solar apex direction. A Uranus mission with an encounter near 1985 might expect to measure the unusual particle and field configuration of a "pole-on" magnetosphere and also properties of the interstellar medium. The author gives here estimates of the particle and field environment of Uranus based on extrapolation of solar wind data from IAU and on scaling relations for an earth-type magnetosphere. Since the magnetic moment of Uranus is unknown, all magnetospheric parameters are derived as a function of the dipole strength. The onset of special magnetospheric properties are identified as the dipole moment increases from small to large values. A fairly complete set of magnetospheric parameters is given for a specific dipole moment to illustrate the case of a large moment.

101.008 The dynamics of Uranus' satellites.
R. Greenberg.
Icarus, Vol. 24, 325 - 332 (1975).

Current knowledge of the dynamics of Uranus' satellites is reviewed in support of preliminary planning for a mission to that planet. The determination of past and present orbital and rotational behavior is discussed. Improved understanding in this area is important not only for its own sake, but also for the implications with regard to the structure of the planet and to the general dynamical history of the solar system.

101.009 An estimate of the temperature and abundance of CH_4 and other molecules in the atmosphere of Uranus.   M. J. S. Belton, S. H. Hayes.
Icarus, Vol. 24, 348 - 357 (1975).

The authors present a preliminary analysis of CH_4 absorptions near 6800 Å in new high resolution spectra of Uranus. A curve of growth analysis of the data yields a rotational temperature near 100 K and a CH_4/H_2 ratio that is 1 to 3 times that expected for a solar type composition. The analysis of the CH_4 also yields a minimum value for the effective pressure of line formation (~ 2 atm). It is speculated that large amounts of some otherwise optically inert gas is present in the Uranus atmosphere. N_2 is suggested as a possible candidate since there are cosmogonic reasons why Uranus should contain large amounts of N relative to C, He, and H, and also because the pressure-induced pure rotation spectrum of N_2 could possibly account for the low brightness temperatures that have recently been observed at 33 and 350 μm. If N_2 is present the planet probably possesses a surface at the 10—100 atmosphere level.

101.010 On the 1968 occultation of BD — 17°4388 by Neptune.   L. Wallace.
Astrophys. Journ., Vol. 197, 257 - 261 (1975).

The difficulties involved in extracting temperafure information from occultation light curves are emphasized. The least uncertain parameter, the scale height defining the gross variation of refractivity, is determined to be about 49 km with an uncertainty of at least 2 km. Detailed temperature profiles derived from the light curves exhibit considerably greater uncertainty.

101.011   A semi-analytical, long-term solution of Pluto, including the Neptune and Uranus resonance.
P. E. Nacozy, R. E. Diehl.
Bull. American Astron. Soc., Vol. 7, 341 - 342 (1975). Abstr. AAS.

101.012   Neptune's atmosphere: concerning the thermal structure.   W. Macy, Jr., L. Trafton.
Bull. American Astron. Soc., Vol. 7, 383 (1975). − Abstr. AAS.

101.013   Models of the atmospheres of Uranus and Neptune.
R. E. Danielson.
Bull. American Astron. Soc., Vol. 7, 384 (1975). − Abstr. AAS.

101.014   Synchrotron radio emission from Uranus and Neptune.   L. D. Kavanagh, Jr.
Icarus, Vol. 25, 166 - 170 (1975).
    A number of charged-particle radiation belt models for Uranus and Neptune are postulated, and the synchrotron emission spectrum for each is calculated over the frequency range 18.75 to 2400 MHz. Available observations are used in conjunction with the synchrotron calculations to establish a rough upper limit to the size and strength of the planet's magnetic field strength and radiation belt intensity.

101.015   Neptune II (Nereid).   K. Aksnes.

IAU Circ., No. 2771 (1975).

    Astrometric positions of the planet Pluto in the years 1971−1974.   See Abstr. 041.019.

    A discussion of the observations of Neptune 1846−1970.   See Abstr. 041.043.

    The orbital resonance amongst the Galilean satellites of Jupiter.   See Abstr. 042.026.

    Investigation of Uranus, its satellites, and distant interplanetary phenomena by spacecraft techniques.
See Abstr. 051.007.

    Methane absorption in the visible spectra of the outer planets and Titan.   See Abstr. 091.011.

    Interior models of Jupiter and Uranus.
See Abstr. 099.066.

    Spectra of hydrate frosts: their application to the outer solar system.   See Abstr. 099.073.

    Ionospheric models of Saturn, Uranus, and Neptune.
See Abstr. 100.004.

    Near-infrared spectrophotometry of Titan.
See Abstr. 100.212.

    Capture d'une comète par Neptune et son passage à l'orbite d'un satellite de la planète.   See Abstr. 102.017.

102 Comets (Origin, Structure, Atmospheres, Dynamics)

102.001 The volatile fraction of the cometary nucleus.
A. H. Delsemme.
Icarus, Vol. 24, 95 - 110 (1975).

In order to prepare a flyby mission to comet Encke, six different sources of information on the possible chemical composition of the cometary nucleus are compared. These are: the neutral and charged radicals and molecules observed in cometary spectra; the chemical composition of type I carbonaceous chondrites; the meteor spectra; the metallic ions collected in the upper atmosphere and correlated with the meteor shower associated with comet Encke; and finally the volatile molecules observed in a volatile-rich sample of lunar soil, that were interpreted as a possible cometary impact. Possible molecular abundances for the volatile fraction of comet Encke are tentatively proposed.

102.002 Modelling of cometary phenomena.
E. A. Kajmakov.
Probl. sovrem. fiz. Leningrad, Nauka, 1974, p. 277 - 291. In Russian. − Abstr. in Referativ. Zhurn. 51. Astron., 2.51.393 (1975).

102.003 Problems of physics of cometary phenomena.
A. Z. Dolginov.
Probl. sovrem. fiz. Leningrad, Nauka, 1974, p. 292 - 306. In Russian. − Abstr. in Referativ. Zhurn. 51. Astron., 2.51.394 (1975).

102.004 The bias of the distribution of cometary orbits by observational selection. L'. Kresák.
Bull. Astron. Inst. Czechoslovakia, Vol. 26, 92 - 111 (1975).

It is shown that all the large-scale irregularities in the spatial distribution of long-period cometary orbits can be explained by observational bias of a random distribution. Model computation substantiated by statistical tests on all comets observed during the last one hundred years, suggest that the operation of all these effects is extremely sensitive to the perihelion distance. Moreover, the existence of two largely different discovery regimes, switching over near $r = 2$ a.u., is indicated. Contrary to a frequent opinion, there is no observational evidence of a uniform distribution of the perihelia beyond this range. The non-uniformities revealed by earlier ellipsoidal analyses are insignificant in comparison with the effects of observing geometry, and inadequate for upholding any association of cometary orbits with the galactic structure.

102.005 Gas phase chemistry in comets.
M. Oppenheimer.
Astrophys. Journ., Vol. 196, 251 - 259 (1975).

A gas phase model for the formation of molecular species in comets is proposed. It is demonstrated that gas phase reactions rapidly reshuffle parent molecular species and that the interpretation of cometary spectra must take these processes into account. The observed species may be produced by reactions which proceed in steady state following photodissociation and ionization of any one hydrogen-bearing parent molecule. However, C−H bond formation may have a relatively long time scale. Little CH_4 or NH_3 will form in the gas phase. An observational test is proposed employing the density ratio $n(H_2O^+)/n(H_2O)$ to determine whether gas phase formation or sublimation from the nucleus is responsible for the occurrence of H_2O in comets. Densities of molecular species derived for a model of the coma at 10^4 km from the nucleus are in harmony with the observations.

102.006 An argument against a type I cometary tail model.
K. Wurm.
Astrophys. Space Sci., Vol. 32, L19 - L23 (1975).

This letter represents an abstract of a forthcoming publication in which a type I cometary tail model is explained in terms of observational facts without additional assumptions. It is made clear that essential claims and assertions of the currently discussed magneto-hydrodynamical theory of the tails cannot be brought into agreement with this 'experimental' model.

102.007 The physical properties of comets.
A. J. Meadows.
Spaceflight, Vol. 17, 134 - 136 (1975).

102.008 Comets and interstellar masers. M. Oppenheimer.
Nature, Vol. 254, 677 - 678 (1975).

Recent advances in the understanding of the origin of the solar system have led to renewed speculation on the origin of comets. The author suggests that the long-period comets are among the condensed remnants of cooled interstellar masers formed in a dense region at the nebular boundary and he discusses the dynamics of the maser−comet transition and suggests ways in which this relationship may be exploited to expand the understanding of star and planet formation.

102.009 A cometary hydrogen model: comparison with OGO-5 measurements of comet Bennett (1970 II).
H. U. Keller, G. E. Thomas.
Astron. Astrophys., Vol. 39, 7 - 19 (1975).

Highly sensitive observations of the Lyman alpha emission of comet Bennett (1970 II) were made during its perihelion passage in March 1970 by the University of Colorado photometer on board OGO-5. The cometary hydrogen coma extended to more than 30×10^6 km in the antisolar direction. A new density model was developed taking into account the orbital motion of the comet and the gradients of the forces of gravitation and radiation pressure. Exact trajectories of atoms in the orbital plane representing the column densities perpendicular to the plane were calculated. The variation of the hydrogen atom lifetime along the trajectory as well as the solar $L\alpha$ profile were considered. The cometary production rate of

102.010 Catalogue of cometary orbits. B. G. Marsden.
Second Edition. Central Bureau for Astronomical Telegrams, International Astronomical Union.
Smithsonian Astrophysical Observatory, Cambridge, Massachusetts, 83 pp. Price $ 3.00 (1975).

102.011 Conditions suffisantes d'évasion ou de retour dans le cas plan du problème des comètes. F. Nahon.
Comptes Rendus Acad. Sci. Paris, Sér. A, Vol. 280, 1037 - 1040 (1975).

On étudie la variation ΔC du moment cinétique de la comète dans la phase d'éjection au cours d'un mouvement dans le plan des primaires. On établit une majoration simple de ΔC qui ne dépend que de la distance initiale; on en déduit une condition suffisante d'évasion, et une condition suffisante de retour à une distance prescrite a priori.

102.012 The comet and minor planet program at Lowell Observatory. H. L. Giclas.
IAU Colloquium No. 22, (see 012.005), p. 19 - 22 (1974).

102.013 Perihelion longitudes and Jacobi constants of Jupiter group comets as indicators of their possible origin from the two equilateral Trojan clouds. E. Rabe.
IAU Colloquium No. 22, (see 012.005), p. 165 - 169 (1974).

102.014 Nongravitational effects on comets: a progress report. B. G. Marsden.
IAU Colloquium No. 22, (see 012.005), p. 181 - 185 (1974).

102.015 The capture of comets. H. Rickman.
IAU Colloquium No. 22, (see 012.005), p. 187 - 191 (1974).

102.016 The effects of resonance with Jupiter in the system of short-period comets. L. Kresák.
IAU Colloquium No. 22, (see 012.005), p. 193 - 203 (1974).

102.017 Capture d'une comète par Neptune et son passage à l'orbite d'un satellite de la planète.
E. I. Kazimirchak-Polonskaya.
IAU Colloquium No. 22, (see 012.005), p. 205 - 221 (1974).

102.018 Origin and evolution of comets. E. Everhart.
IAU Colloqium No. 22, (see 012.005), p. 223 - 225 (1974).

102.019 On possible mechanisms for dust ejection from comets. O. V. Dobrovolsky (Dobrovol'skij).
IAU Colloquium No. 22, (see 012.005), p. 303 - 305 (1974).

102.020 The vaporization of the volatile fraction in comets. A. H. Delsemme.
IAU Colloquium No. 22, (see 012.005), p. 313 - 322 (1974).

102.021 Formaldehyde polymers in comets.
V. Vanýsek, N. C. Wickramasinghe.
Astrophys. Space Sci., Vol. 33, L19 - L28 (1975).
The possibility that crystalline formaldehyde polymers are present in cometary dust is discussed. The available data concerning the physical properties of comets indicate that polymerized formaldehyde cannot be ruled out as a major constituent of cometary material.

102.022 On the theory of brightness outbursts of comets. L. M. Shul'man.
Astrometriya i Astrofizika, Kiev, vyp. (No.) 24, (see 003.003), p. 91 - 101 (1974). In Russian.
The hypothesis developed is that outbursts of cometary brightness are caused by an explosion in the thin surface layer of the cometary nucleus. Explosive material is supposed to be generated in this layer by chemical reactions induced by solar cosmic rays. The absorbed power of cosmic rays has been calculated. The above-mentioned hypothesis is shown not to contradict the observational data.

102.023 Development of a comet as it pursues its orbit. R. A. Lyttleton.
Astrophys. Space Sci., Vol. 34, 491 - 510 (1975).
The properties of cometary dust-swarms in almost parabolic long-period orbits are examined. These properties exhibit such erratic diversity as to make clear that only a theory involving considerable range of essential parameters can be capable of accounting for them adequately.

102.024 Do comets play a role in galactic chemistry and γ-ray bursts? F. L. Whipple.
Bull. American Astron. Soc., Vol. 7, 343 - 344 (1975). Abstr. AAS.

102.025 The meaning of atomic and molecular abundances in comets. A. H. Delsemme.
Bull. American Astron. Soc., Vol. 7, 385 (1975). – Abstr. AAS.

102.026 Comets, solar activity and interplanetary medium. D. A. Andrienko.
Problems of cosmic physics. Vyp. (No.) 9, (see 003.013), p. 104 - 109 (1974). In Russian.
From the data of Bobrovnikoff's catalogue of comet brightness estimates it is found that the maximum values of Δm took place for comets passing high heliographic latitudes and solar active zones. This fact is interpreted as indication on the intensification of the solar wind in the observed regions of the corona. A suggestion is made for a method of velocity determination of solar corpuscular streams from the time interval between comet flares and geomagnetic disturbances.

102.027 Probable parent molecules in cometary nuclei. E. A. Kajmakov.
Problems of cosmic physics. Vyp. (No.) 9, (see 003.013), p. 141 - 155 (1974). In Russian.
Experimental data pertaining to the problem of parent molecules in cometary nuclei are presented. It is shown that cometary ice may contain some anorganic and organic compounds, in particular vast amounts of aminoacids which can provide the spectral components observed as well as the optically active dust component.

102.028 On some new sources of formation of molecules observed in cometary atmospheres.
V. I. Cherednichenko.
Problems of cosmic physics. Vyp. (No.) 9, (see 003.013), p. 155 - 158 (1974). In Russian.
This paper shows that the molecules H_2CO, CH_2N_2, HC_3N, HCN, CH_3OH, HCOOH, CH_3CN, HN_3 as parent molecules for the molecules CO^+, N_2^+, CH^+, CH, CN, C_2, C_3, OH may be observed in the comets. The dissociative recombination may cause the formation of practically all observed molecules.

102.029 Les comètes, l'activité du soleil et le milieu interplanétaire. D. A. Andrienko.
Ann. Obs. Astron. Alger, Vol. 4, Fasc. 1, p. 22 - 29 (1974).

102.030 Détermination de la variation d'éclat de cinq comètes en fonction de la distance à la terre et au soleil. H. Feijth.
L'Astronomie, 89e année, p. 263 - 265 (1975).

102.031 Resonance scattering from optically thin expanding cometary atmospheres. R. R. Meier.
Astron. Astrophys., Vol. 40, 373 - 380 (1975).
Emission line profiles are computed for optically thin, expanding cometary atmospheres, assuming resonance scattering of sunlight as the excitation mechanism. Application is made to atomic hydrogen, both ignoring and allowing for the effect of radiation pressure. Lyman alpha line profiles are computed for comet Kohoutek both to illustrate the theory and to assess the degree of geocoronal absorption of images taken by Skylab-4. The effect of a varying solar flux on the brightness (the Greenstein Effect) is investigated for atomic oxygen.

102.032 Variations and outbursts of comet brightness. D. O. Mokhnach.
Astron. Zhurn. Akad. Nauk SSSR, Vol. 52, 642 - 648 (1975). In Russian. English translation in Soviet Astron., Vol. 19, No. 3.
A theoretical investigation of the surface brightness distribution in the head of a comet at increase, decrease and outbursts of brightness has been made. The surface brightness distribution was evaluated for initial velocity $v_0 = 3$ km/sec and acceleration due to light pressure $g = 0.1$ cm/sec^2 and $g = 0.5$ cm/sec^2 and also for different conditions of brightness variations.

102.033 The anticipated shape of comets dust tail. A. Pilski.
Urania Kraków, Vol. 46, 15 - 21 (1975). In Polish.

102.034 **On the chemical constitution of cometary nuclei.**
L. Biermann, G. Diercksen.
Origins of Life, Vol. 5, 297 - 301 (1974).

102.035 **Comets.** K. R. Sivaraman.
Bull. Astron. Soc. India, Vol. 1, 35 - 39 (1973).

102.036 **Physical and chemical properties of comets.**
D. A. Mendis.
Bull. Astron. Soc. India, Vol. 2, 63 - 66, 75 (1974). – Paper read at the symposium on 'Some aspects of astrophysics' held at PRL, Ahmedabad, August 1974.

102.037 **Comets as evidence of explosive processes in the solar system and in the Galaxy.**
S. K. Vsekhsvyatskij.
Kosmich. issled. na Ukraine. Resp. mezhved. sb., 1974, vyp. (No.) 5, p. 3 - 8. In Russian. – Abstr. in Referativ. Zhurn. 51. Astron., 6.51.275 (1975).

102.038 **Kometene – deres fysikk og opprinnelse, I. II.**
E. Jensen.
Naturen, Årg. 98, No. 5, p. 215 - 222; No. 6, p. 251 - 254 = Inst. Teor. Astrofys., Blindern–Oslo, Småtrykk No. 83 (1974).

102.039 **A study of the icy tails of the distant comets.**
Z. Sekanina.
Icarus, Vol. 25, 218 - 238 (1975).

The properties of the icy-grain model, formulated recently for the nearly straight, structureless tails of a number of comets with large perihelion distances, are studied. It is established that the transition region between 2 and 3 AU, where water snow starts evaporating rapidly, has a profound effect on the dynamics of the icy tails. It is suggested that the icy (or solid-hydrate) grains, constituting the tails of the distant comets, may be carriers of fine meteoric-dust particles, of micron and submicron sizes, which are set free once the grains start disintegrating by evaporation.

102.040 **Interaction of a comet with the solar wind.**
L. Biermann.
Solar wind three, (see 012.020), p. 396 - 414 (1974).

Die Erforschung der Kometen mit Raumsonden. Vortrag anläßlich der XXIII. HOG-Jahrestagung in Salzburg.
See Abstr. 051.014.

The program of observation of minor planets and comets at the Bucharest Observatory.
See Abstr. 098.018.

Le programme des observations des astéroides et des comètes à l'Observatoire de Nice.
See Abstr. 098.019.

Meteoric storms and formation of meteor streams.
See Abstr. 104.016.

Interplanetary shock waves and comet brightness fluctuations during June–August 1972.
See Abstr. 106.041.

Questions concerning interstellar matter in the Magellanic Clouds. See Abstr. 159.016.

103 Comets: Listed Objects

103.001 Kommende Erscheinungen berühmter Kometen.
V. Kasten.
SuW, Vol. 14, 63 - 64 (1975). − Concerning P/Encke,
P/Halley, Swift-Tuttle.

103.002 Endgültige Bezeichnung der Kometen 1973.
R. Lukas.
SuW, Vol. 14, 68 (1975).

103.003 Die im Jahre 1974 entdeckten Kometen.
R. Lukas.
SuW, Vol. 14, 102 (1975).

103.004 Les comètes brillantes apparues depuis 1956.
C. Bertaud.
L'Astronomie, 89ᵉ année, p. 75 - 83, with a correction p. 158
(1975). − Concerning the comets Arend-Roland (1957 III),
Mrkos (1957 V), Ikeya-Seki (1965 VIII), Tago-Sato-Kosaka
(1969 IX), and Bennett (1970 II).

103.005 Observations of comets and minor planets.
G. Van Biesbroeck, C. D. Vesely, B. G. Marsden.
Astron. Journ., Vol. 80, 246 - 251 (1975).
A series of 138 astrometric observations of comets and
minor planets is presented. The observations were made by the
late G. Van Biesbroeck during 1964−1971.

103.006 Designations of comets of 1973.
Kometn. Tsirk., *Kiev*, No. 174 (1975). In Russian.

103.007 Observations at the Kleť Observatory.
A. Mrkos.
Kometn. Tsirk., *Kiev*, No. 175 (1975). In Russian. − Con-
cerning Honda-Mrkos-Pajdušáková; Schwassmann-Wachmann 1;
Eros.

**103.008 Observations of comets at the Skalnaté Pleso Observ-
atory in the years 1964 - 1971.** M. Antal.
Contr. Astron. Obs. Skalnaté Pleso, Vol. 5, 75 - 99 (1973/75).
The results of 264 photographic positional observations
of 32 comets obtained at the Skalnaté Pleso Observatory are
given. A description of the instruments and the reduction are
given in detail. The results are contained in tables.

103.009 Vier nieuwe periodieke kometen.
G. W. E. Beekman.
Zenit, Vol. 2, 230 (1975).

103.010 Observations of comets.
C. Y. Shao, R. E. McCrosky, G. Schwartz, J. H. Bulger.
IAU Circ., No. 2738 (1975). − Concerning 1925 II, 1973 g,
1973 m, 1974 a, 1974 d.

103.011 Observations of comets. T. Seki, N. Kojima.
IAU Circ., No. 2743 (1975). − Concerning 1969 II,
1973 m, 1974 g.

103.012 Observations of comets. G. R. Kastel', T. M.
Smirnova, N. S. Chernykh, L. V. Zhuravleva,
L. I. Chernykh.
IAU Circ., No. 2749 (1975). − Concerning 1925 II, 1974 a.

103.013 Observations of comets. R. E. McCrosky,
J. H. Bulger, C. Y. Shao, G. Schwartz.
IAU Circ., No. 2753 (1975). − Concerning 1972 IX, 1973 g,
1974 d, 1974 g.

103.014 Observations of comets. R. E. McCrosky.
IAU Circ., No. 2785 (1975). − Concerning 1973 m,
1974 g.

103.015 Observations of comets. C. Torres.
IAU Circ., No. 2790 (1975). − Concerning
1972 XII, 1974 e, 1974 a.

103.016 Observations of comets. E. Roemer, L. M.
Vaughn, C. C. McCarthy, R. A. McCallister.
IAU Circ., No. 2761 (1975). − Concerning 1974 f, 1974 g,
1975 a, 1975 b, 1975 c.

103.100 Comet 1973 XII Kohoutek

Observations visuelles de la comète Kohoutek 1973f.
D. A. Andrienko, L. A. Ourassine (*L. A. Urasin*), J.-C. Pham-
Van, M. A. Svetchnikov (*M. A. Svechnikov*).
Ann. Obs. Astron. Alger, Vol. 4, Fasc. 1, p. 18 (1974).

**Search for H_2CO and excited OH in comet Kohou-
tek (1973f).** R. Schröder, H. J. Wendker, P. Stumpff.
Astron. Astrophys., Vol. 37, 417 - 418 (1974).
Line emission from H_2CO and excited OH was searched
at 6 cm in comet Kohoutek (1973f) with the 100 m-telescope
in Effelsberg. Lines were not detected. Implications of this
result are discussed.

**High resolution Lyman alpha observations of comet
Kohoutek (1973f) near perihelion.**
H. U. Keller, J. D. Bohlin, R. Tousey.
Astron. Astrophys., Vol. 38, 413 - 416 (1975).
The line width of $L\alpha$ was determined from three different
exposures of the comet nuclear region. A simplified analysis of
the optical thickness effects showed that this line width is consistent with the established hydrogen outflow velocity of 8.to
10 km s^{-1}.

**Photometric, colorimetric and polarimetric study
of comet Kohoutek 1973 f, on January 17 and 18, 1974.**
A. Bücher, R. Robley, S. Koutchmy.
Astron. Astrophys., Vol. 39, 289 - 293 (1975).
The photoelectric and photographic observations performed on 17 and 18 January 1974 of the external parts of
the comet Kohoutek show that the color of the tail is much
the same as the solar color, and that the degree of polarization
in the tail and anti-tail is very strong. Observations of the anti-
tail suggest similarity in the properties with the zodiacal cloud
dust.

Intensity fluctuations in the head of comet 1973f.
J. Isserstedt, W. Schlosser.
Astron. Astrophys., Vol. 41, 9 - 13 (1975). In German.
Photoelectric photometry of short time luminosity variations in the coma of comet Kohoutek 1973f is presented. The
authors conclude, that the short period variations are changes
in surface-brightness and thus originate in the coma. They
draw attention to the close agreement with the 300 s period
of the solar photosphere and corona, as well as with similar
variations in the EUV ($P = 4.37$ min).

**Radio search for HC_3N, HCN, OH, and detection
of U8.19 in comet Kohoutek (1973f).**
P. T. Giguere, F. O. Clark.
Astrophys. Journ., Vol. 198, 761 - 764 (1975).

Comet Kohoutek (1973f) was observed with the NRAO 140-foot (43 m) radio telescope in the period 1974 January 4–7 in an attempt to detect the transitions of the molecules: HC_3N, HCN, and OH. All results for these lines were negative. An unidentified line was possibly detected at 8189 MHz.

On the interpretation of the observed cometary scintillations. W.-H. Ip, D. A. Mendis.
Astrophys. Space Sci., Vol. 35, L1 - L4 (1975). – Letter.

Photoelectric observations of comet Kohoutek 1973f. F. Scaltriti, M. A. Vogliotti, V. Zappalà.
Atti Fondaz. G. Ronchi, Vol. 29, 762 - 765 = Contr. Oss. Astron. Torino (Pino Torinese), No. 80 (1974).

A kinematographic study of the tail of comet Kohoutek (1973f).
K. Jockers, R. G. Roosen, D. P. Cruikshank.
Bull. American Astron. Soc., Vol. 7, 385 - 386 (1975). Abstr. AAS.

Spectrophotometry of comet Kohoutek (1973f) in the 2500–7000 cm^{-1} region. L. L. Smith, T. Hilgeman.
Bull. American Astron. Soc., Vol. 7, 391 (1975). – Abstr. AAS.

Observations of comet Kohoutek (1973f) at the satellite station 1147 Ondřejov 2.
G. Karský, J. Kostelecký, I. Synek, J. Vondrák.
Bull. Astron. Inst. Czechoslovakia, Vol. 26, 111 - 114 (1975).

Between October 25, 1973 and February 15, 1974 28 photographic positions of the comet were obtained. A brief description of the tracking camera, the method of observation, evaluation, precision, elements of the orbit and individual determined topocentric positions for the equinox and equator 1950.0 are given. It is shown that even if the focal length of the camera is relatively short (78 cm), it is possible to obtain satisfactory position data with error of less than $2'' - 3''$ which are fit for good orbit determination.

Photometric parameters of the comet Kohoutek 1973f. J. Svoreň, J. Tremko.
Bull. Astron. Inst. Czechoslovakia, Vol. 26, 132 - 138 (1975).

Spectrophotometry of comet Kohoutek (1973f).
G. S. D. Babu.
Bull. Astron. Soc. India, Vol. 2, 35 (1974). – Abstract.

Sodium emission in comet Kohoutek (1973f).
M. K. V. Bappu, M. Parthasarathy, K. R. Sivaraman.
Bull. Astron. Soc. India, Vol. 2, 35 (1974). – Abstract.

Radio observations of comet Kohoutek.
S. Ananthakrishnan, S. M. Bhandari.
Bull. Astron. Soc. India, Vol. 2, 35 - 36 (1974). – Abstract.

The light curve of comet Kohoutek.
R. J. Angione, B. Gates, K. G. Henize, R. G. Roosen.
Icarus, Vol. 24, 111 - 115 (1975).

Visual estimates of total coma brightness define the light curve of comet Kohoutek between November 24, 1973 and February 6, 1974. These data are well fitted by straight lines on the M-log r diagram. The preperihelion value of n is 2.2, and the postperihelion value is 3.8 up to January 16. A standstill in the decline is suspected between January 16 and 19. If the standstill is ignored the postperihelion data can be fitted less precisely by a single line with n = 3.3. From photoelectric measures on four nights between January 5 and 13 after perihelion, n is found to be 4.1.

Narrow band filter photometry of comet Kohoutek.
R. J. Angione, R. G. Roosen, H. Lanning.

Icarus, Vol. 24, 116 - 119 (1975).

Narrow band photoelectric measurements of CN(3870), $CO^+(4250)$, $C_2(4700)$, and $C_2(5120)$ were made on twelve nights in December and January of comet Kohoutek. CN and C_2 appear to be stronger after perihelion, and CO^+ showed a strong post perihelion increase coincident with the first appearance of a strong gas tail and then decreased to a fairly constant level.

An upper limit for methane production from comet Kohoutek by high resolution tilting-filter photometry at 3.3 μm. A. E. Roche, C. B. Cosmovici, S. Drapatz, K. W. Michel, W. C. Wells.
Icarus, Vol. 24, 120 - 127 (1975).

The authors have established an upper limit for the methane production from comet Kohoutek (1973f) by searching for fluorescence radiation from the $P2$, $P3$, and $P9$ lines at 3.3 μm. On January 8, 1974, at 2 UT, their measurements indicated a production rate upper limit of $Q < 10^{29}$ molecules sec^{-1} sr^{-1}.

Upper limit on the radar cross section of the comet Kohoutek. E. J. Chaisson, R. P. Ingalls, A. E. E. Rogers, I. I. Shapiro.
Icarus, Vol. 24, 188 - 189 (1975).

An attempt to observe radar echoes from the comet Kohoutek was made at a radio frequency of 7840 MHz ($\lambda \simeq$ 3.8 cm). The upper limit on the radar cross section is approximately $10^4 B^{1/2}$ km^2, where B is the (unknown) bandwidth of the echo in Hertz. For $B \simeq 100$ Hz, it follows that (i) the nucleus, if a perfect spherical reflector, must be less than 250 km in diameter, and (ii) the density of any millimeter sized particles must be less than 1 m^{-3} for a coma of diameter 10^4 km.

Comet Kohoutek, 1973f.
Kometn. Tsirk., *Kiev*, No. 175 (1975). In Russian. – Observations of the Alpine expedition of the Sternberg Astronomical Institute (*V. M. Kovalenko, O. N. Kovalenko*).

Comet Kohoutek, 1973 XII.
Kometn. Tsirk., *Kiev*, No. 178 (1975). In Russian. – Observations of the Alpine expedition of the Sternberg Astronomical Institute (*V. M. Kovalenko, L. T. Markova, O. N. Kovalenko*).

Observations photographiques et spectrographiques de la comète Kohoutek à l'Observatoire de Haute-Provence.
Y. Andrillat.
L'Astronomie, 89e année, p. 65 - 73 (1975).

Shadow of the head of comet Kohoutek.
J. C. Brandt, R. G. Roosen.
Nature, Vol. 253, 659 (1975).

What came out of Kohoutek's comet?
J. F. James.
Nature, Vol. 254, 282 - 283 (1975).

A possible continuum detection of comet 1973 XII Kohoutek at the wavelength of 4.1 mm.
K. Akabane, Y. Chikada.
Publ. Astron. Soc. Japan, Vol. 27, 101 - 106 (1975).

4.1 mm continuum radiation of comet 1973 XII Kohoutek has been possibly detected. Observations were made for 8 days around the perihelion passage of the comet in December 1973. The averaged flux density for the 8 days of observations was estimated as $(2.0 \pm 1.0) \times 10^{-26}$ w m^{-2} Hz^{-1}. A model of the emitting source is briefly discussed.

The isotope ratio $^{12}C/^{13}C$ in comet 1973 XII Kohoutek. S. Kikuchi, A. Okazaki.

Publ. Astron. Soc. Japan, Vol. 27, 107 - 110 (1975).

Comparing the intensity of the $^{12}C^{13}C$ (1-0) band head with those of the $^{12}C^{12}C$ (1-0) and (2-0) band heads, the authors derive the isotope ratio $^{12}C/^{13}C$ as 95 ± 40. This value is essentially the same as those of comet 1963 I Ikeya and comet 1969 IX Tago-Sato-Kosaka, and is in good agreement with those observed in other objects of the solar system.

Preliminary results on comet Kohoutek — interactions with the solar wind.   J. C. Brandt, S. P. Maran.
Solar wind three, (see 012.020), p. 415 - 420 (1974).

103.101   Comet 1974b Bradfield

Comet Bradfield (1974 b).   C. Y. Shao,
G. Schwartz, B. G. Marsden.
IAU Circ., No. 2762 (1975).

Brightness of comet Bradfield in the B- and V-system.
V. K. Rozenbush.
Kometn. Tsirk., *Kiev,* No. 173 (1975). In Russian.

Visual observations and brightness estimates of comet Bradfield.   J. E. Bortle.
Kometn. Tsirk., *Kiev,* No. 173 (1975). In Russian.

Observations of comet Bradfield, 1974b at the Alpine expedition of the Sternberg Astronomical Institute.
V. M. Kovalenko, O. N. Kovalenko.
Kometn. Tsirk., *Kiev,* No. 177 (1975). In Russian.

La comète Bradfield (1974 b) observée par nos sociétaires.   P. de La Cotardière.
L'Astronomie, 89e année, p. 198 - 200 (1975).

Beobachtungen des Kometen Bradfield (1974 b).
K. Werner.
SuW, Vol. 14, 27 (1975).

103.102   Comet 1969 IX Tago-Sato-Kosaka

Coma diameter of comets — I.
C. S. Morris.
Strolling Astronomer, Vol. 25, 94 - 101 (1975).

103.103   Comet 1974g van den Bergh

Comet van den Bergh 1974g.
S. van den Bergh, B. G. Marsden.
British Astron. Ass. Circ. No. 561 (1975).

Comet van den Bergh (1974 g).   B. G. Marsden.
IAU Circ., No. 2773 (1975).

The discovery of comet 1974g.
S. van den Bergh.
Journ. Roy. Astron. Soc. Canada, Vol. 69, 29 - 31 (1975).

Comet van den Bergh, 1974g.
Kometn. Tsirk., *Kiev,* No. 174 (1975). In Russian.

103.104   Comet 1910 II Halley

The past orbit of Halley's comet.   T. Kiang.
IAU Colloquium No. 22, (see 012.005), p. 171 - 173 (1974).

Observing prospects for Halley's comet.
R. G. Roosen, B. G. Marsden.
Sky Telescope, Vol. 49, 363 - 364 (1975).

The structure of meteor streams associated with comet Halley: Eta Aquarids and Orionids.
See Abstr. 104.019.

103.105   Comet 1970d d'Arrest

On the motion of comet P/d'Arrest.
W. J. Klepczynski.
IAU Colloquium No. 22, (see 012.005), p. 145 - 148 (1974).

103.106   Comet 1969 IV Churyumov-Gerasimenko

Periodic comet Churyumov-Gerasimenko (1969 IV).
B. G. Marsden.
IAU Circ., No. 2783 (1975).

Investigation of the orbital motion of comet Churyumov-Gerasimenko 1969 IV (preliminary investigation).
N. A. Belyaev.
IAU Colloquium No. 22, (see 012.005), p. 149 - 153 (1974).

Elements and ephemeris of comet Churyumov-Gerasimenko, 1969 IV.   N. A. Belyaev.
Kometn. Tsirk., *Kiev,* No. 180 (1975). In Russian.

103.107   Comet 1975f Wolf

Periodic comet Wolf.
E. I. Kazimirchak-Polonskaya, D. K. Yeomans.
IAU Circ., No. 2740 (1975).

Periodic comet Wolf (1975 f).   E. Roemer,
L. M. Vaughn, C. C. McCarthy, C. Y. Shao.
IAU Circ., No. 2784 (1975).

Ephemeris of comet Wolf-Kamieński.
Kometn. Tsirk., *Kiev,* No. 175 (1975). In Russian.

Élimination de la discontinuité de la théorie de la comète Wolf 1 construite par M. Kamieński pour l'intervalle 1884–1967.   E. I. Kazimirchak-Polonskaya.
IAU Colloquium No. 22, (see 012.005), p. 155 - 164 (1974).

103.108   Comet 1973m Borrelly

Periodic comet Borrelly (1973 m).
C. Y. Shao, R. E. McCrosky, G. Schwartz.
IAU Circ., No. 2782 (1975).

Le mouvement de la comète Borrelly de 1904 à 1968 et l'influence des forces non gravitationnelles.
L. M. Beloous (*Belous*).
IAU Colloquium No. 22, (see 012.005), p. 175 - 179 (1974).

103.109   Comet 1925 II Schwassmann-Wachmann 1

Periodic comet Schwassmann-Wachmann 1.
A. Mrkos, H. L. Giclas, M. L. Kantz.
IAU Circ., No. 2745 (1975).

Periodic comet Schwassmann-Wachmann 1.
C. Torres, S. Barros, M. Wischnjewsky.
IAU Circ., No. 2758 (1975).

Periodic comet Schwassmann-Wachmann 1.
IAU Circ., No. 2773 (1975).

Periodic comet Schwassmann-Wachmann 1.
C. Torres, R. E. McCrosky, G. Schwartz, J. H. Bulger.
IAU Circ., No. 2786 (1975).

Outbursts of comet Schwassmann-Wachmann 1, and the cloud of interplanetary boulders.   Z. Sekanina.
IAU Colloquium No. 22, (see 012.005), p. 307 (1974). – Abstract.

Comet Schwassmann-Wachmann 1.
Kometn. Tsirk., *Kiev,* No. 176 (1975). In Russian.

Ephemeris of comet Schwassmann-Wachmann 1, 1925 II.
Kometn. Tsirk., *Kiev,* No. 179 (1975). In Russian.

Comets Schwassmann-Wachmann 1 and Oterma during the years 1957 - 1972.   S. K. Vsekhsvyatskij.
Problems of cosmic physics. Vyp. (No.) 9, (see 003.013), p. 110 - 116 (1974). In Russian.

103.110   Comet 1957 V Mrkos

Comparison of processes in comet Mrkos 1957 V with the solar activity.   V. P. Tarashchuk.
Astrometriya i Astrofizika, *Kiev,* vyp. (No.) 24, (see 003.003), p. 102 - 106 (1974). In Russian.
    The light curve and variations in structure of the comet 1957 V are compared with solar and geophysical activity and with the solar wind velocity during the observational period of the comet in August–September 1957.

103.111   Comet 1965 VIII Ikeya-Seki

Polarization reversal in the tail of comet Ikeya-Seki (1965 VIII).   J. L. Weinberg.
Bull. American Astron. Soc., Vol. 7, 337 (1975). – Abstr. AAS.

103.112   Comet 1958 IV Oterma

Comets Schwassmann-Wachmann 1 and Oterma during the years 1957 - 1972.   See Abstr. 103.109.

102.113   Comet 1970 II Bennett

Photometry of comet Bennett in polarized light.
R. S. Osherov, F. A. Tupieva.
Dokl. AN TadzhSSR, Vol. 17, No. 7, p. 17 - 19 (1974). In Russian. – Abstr. in Referativ. Zhurn. 51. Astron., 4.51.407 (1975).

Investigation of the plasma cloud in the tail of comet Bennett 1970 II.
O. V. Dobrovol'skij, Kh. Ibadinov, L. Zatsepina.
Dokl. AN TadzhSSR, Vol. 17, No. 9, p. 21 - 24 (1974). In Russian. – Abstr. in Referativ. Zhurn. 51. Astron., 6.51.387 (1975).

Photometric study of comet Bennett 1970 II.
Sh. R. Daminov.
Tezisy dokl. Resp. konf. moldykh uchenykh i spetsialistov TadzhSSR. Sekts. fiz. i astrofiz. Dushanbe, Donish, 1974, p. 31 - 32. In Russian. – Abstr. in Referativ. Zhurn. 51. Astron., 6.51.388 (1975).

Detailed photometry of comet Bennett 1969i in polarized light.   S. I. Gerasimenko.
Problems of cosmic physics. Vyp. (No.) 9, (see 003.013), p. 116 - 120 (1974). In Russian.

A cometary hydrogen model: comparison with OGO-5 measurements of comet Bennett (1970 II).
See Abstr. 102.009.

103.114   Comet 1970g Abe

The physical characteristics of the neutral coma of comet Abe 1970g.   V. L. Afanas'ev.
Problems of cosmic physics. Vyp. (No.) 9, (see 003.013), p. 120 - 129 (1974). In Russian.
    Slit spectrograms of the comet Abe with the Alma-Ata observatory 70 cm reflector were obtained. The energy distribution in the spectrum and the distributions of surface brightness for CN 3883 Å, CN 4216 Å, C_2 4737 Å, C_2 5165 Å, C_3 4051 Å emissions in relative units were investigated. The average lifetimes and outflow velocities for the model of a cold collisionless neutral coma were estimated.

Absolute photometry and polarimetry of the head of comet Abe 1970g.   K. I. Churyumov, F. I. Kravtsov.
Problems of cosmic physics. Vyp. (No.) 9, (see 003.013), p. 129 - 134 (1974). In Russian.

103.115   Comet 1911 V Brooks

Equidensitograms of comet Brooks 1911 V.
Yu. E. Migach.
Problems of cosmic physics. Vyp. (No.) 9, (see 003.013), p. 134 - 141 (1974). In Russian.

103.116   Comet 1965 I Tsuchinshan 1

Improved orbits of three periodic comets: Tsuchinshan 1, Tsuchinshan 2 and Kearns-Kwee.   G. Sitarski.
Acta Astron., Vol. 25, 161 - 168 (1975).

103.117   Comet 1965 II Tsuchinshan 2

Improved orbits of three periodic comets: Tsuchinshan 1, Tsuchinshan 2 and Kearns-Kwee.
See Abstr. 103.116.

103.118 Comet 1963 VIII Kearns-Kwee

Association of apparitions of comet Kearns-Kwee 1963 VIII and 1972 IX.   S. Shaporev.
Kometn. Tsirk., *Kiev*, No. 177 (1975). In Russian.

Improved orbits of three periodic comets: Tsuchinshan 1, Tsuchinshan 2 and Kearns-Kwee.
See Abstr. 103.116.

103.119 Comet 1974h Bennett

Comet Bennett (1974 h).   C. Torres, J. Parra.
IAU Circ., No. 2746 (1975).

Comet Bennett (1974 h).   C. Torres.
IAU Circ., No. 2756 (1975).

Comet Bennett, 1974h.
Kometn. Tsirk., *Kiev*, No. 173 (1975). In Russian.

Comet Bennett, 1974h.
Kometn. Tsirk., *Kiev*, No. 174 (1975). In Russian.

103.120 Comet 1974f Honda-Mrkos-Pajdušáková

Periodic comet Honda-Mrkos-Pajdušáková (1974 f).
T. Urata, T. Seki, J. Bortle, B. Comsa, D. Green, C. Sherrod.
IAU Circ., No. 2739 (1975).

Periodic comet Honda-Mrkos-Pajdušáková (1974 f).
R. E. McCrosky, C. Y. Shao, J. H. Bulger, A. Mrkos,
R. Petrovičová, T. Seki, D. Herald, A. C. Gilmore, P. M.
Kilmartin.
IAU Circ., No. 2743 (1975).

Periodic comet Honda-Mrkos-Pajdušáková (1974 f).
D. Herald, W. Rea, P. M. Kilmartin.
IAU Circ., No. 2767 (1975).

Periodic comet Honda-Mrkos-Pajdušáková (1974 f).
E. Roemer, C. C. McCarthy.
IAU Circ., No. 2779 (1975).

Comet Honda-Mrkos-Pajdušáková, 1974f.
Kometn. Tsirk., *Kiev*, No. 173 (1975). In Russian.

Comet Honda-Mrkos-Pajdušáková, 1974f.
Kometn. Tsirk., *Kiev*, No. 174 (1975). In Russian. – Observations of the comet in Stormville (*J. E. Bortle*).

Comet Honda-Mrkos-Pajdušáková, 1974f.
Kometn. Tsirk., *Kiev*, No. 176 (1975). In Russian.

103.121 Comet 1974i Wirtanen

Rediscovery of comet Wirtanen, 1974i.
Kometn. Tsirk., *Kiev*, No. 174 (1975). In Russian.

103.122 Comet 1974a Forbes

Periodic comet Forbes (1974 a).   C. Torres,
S. Barros, M. Wischnjewsky, J. A. Bruwer.

IAU Circ., No. 2746 (1975).

Comet Forbes, 1974a.
Kometn. Tsirk., *Kiev*, No. 174 (1975). In Russian.

103.123 Comet 1960 III Schaumasse

Orbit of comet Schaumasse.   K. P. Matsukov.
Kometn. Tsirk., *Kiev*, No. 174 (1975). In Russian.

103.124 Comet 1975a Boethin

New comet Boethin (1975a).
L. Boethin, B. G. Marsden.
British Astron. Ass. Circ. No. 560 (1975).

Comet Boethin.
L. Boethin, C. Scovil, J. E. Bortle, R. E. McCrosky.
IAU Circ., No. 2745 (1975).

Comet Boethin (1975 a).   T. Urata, R. E.
McCrosky, C. Y. Shao, Saito, H. Kosai, M. Antal, B. G. Marsden
IAU Circ., No. 2748 (1975).

Comet Boethin (1975 a).   L. Boethin, T. Urata,
C. Cristescu, J. E. Bortle, B. G. Marsden.
IAU Circ., No. 2749 (1975).

Comet Boethin (1975 a).   N. Kojima, T. Seki,
K. Tomita, H. Shibasaki.
IAU Circ., No. 2752 (1975).

Comet Boethin (1975 a).   T. Seki.
IAU Circ., No. 2755 (1975).

Periodic comet Boethin (1975 a).   T. Urata,
A. Mrkos, W. Ferreri, H. L. Giclas, M. L. Kantz, R. E. McCrosky,
C. Y. Shao, J. E. Bortle, B. G. Marsden.
IAU Circ., No. 2758 (1975).

Periodic comet Boethin (1975 a).
A. Mrkos, T. Seki, K. Hestir, E. F. Tedesco, B. G. Marsden.
IAU Circ., No. 2765 (1975).

Periodic comet Boethin (1975 a).   K. Suzuki,
T. Urata, T. Seki, E. Roemer, C. C. McCarthy.
IAU Circ., No. 2775 (1975).

Periodic comet Boethin (1975 a).
C. C. McCarthy, E. Roemer, L. Vaughn.
IAU Circ., No. 2787 (1975).

New comet Boethin, 1975a.
Kometn. Tsirk., *Kiev*, No. 175 (1975). In Russian.

Comet Boethin, 1975a.
Kometn. Tsirk., *Kiev*, No. 176 (1975). In Russian.

Periodic comet Boethin, 1975a.
Kometn. Tsirk., *Kiev*, No. 177 (1975). In Russian.

Periodic comet Boethin, 1975a.
Kometn. Tsirk., *Kiev*, No. 178 (1975). In Russian.

103.125   **Comet 1975b West-Kohoutek-Ikemura**

New periodic comet West-Kohoutek-Ikemura 1975b.
L. Kohoutek, T. Ikemura, B. G. Marsden, R. M. West.
British Astron. Ass. Circ. No. 561 (1975).

Comet West.   R. M. West.
IAU Circ., Nos. 2741, 2742 (1975).

Comet West.   R. M. West.
IAU Circ., No. 2751 (1975).

Comet Kohoutek (1975 b).   L. Kohoutek.
IAU Circ., Nos. 2752, 2754 (1975).

Comet Kohoutek-Ikemura (1975 b).
T. Ikemura, N. Kojima, R. E. McCrosky, G. Schwartz,
K. Hurukawa, T. Hirayama.
IAU Circ., No. 2754 (1975).

Comet Kohoutek-Ikemura (1975 b).
L. Kohoutek, R. E. McCrosky.
IAU Circ., No. 2755 (1975).

Periodic comet West-Kohoutek-Ikemura (1975 b).
N. Kojima, H. Kosai, T. Seki, G. Schwartz, C. Y. Shao,
B. G. Marsden.
IAU Circ., No. 2756 (1975).

Periodic comet West-Kohoutek-Ikemura (1975 b).
T. Ikemura, H. Kosai, G. Schwartz, R. E. McCrosky, C. Y.
Shao, H. L. Giclas, M. L. Kantz, B. Milet, J. E. Bortle.
IAU Circ., No. 2759 (1975).

Periodic comet West-Kohoutek-Ikemura (1975 b).
N. Kojima, T. Seki, K. Hestir, E. F. Tedesco, D. Fellers,
J. Bortle.
IAU Circ., No. 2765 (1975).

Periodic comet West-Kohoutek-Ikemura (1975 b).
K. Suzuki, N. Kojima, T. Seki.
IAU Circ., No. 2775 (1975).

Periodic comet West-Kohoutek-Ikemura (1975 b).
IAU Circ., No. 2776 (1975).

Periodic comet West-Kohoutek-Ikemura (1975 b).
K. Hestir, R. S. Harrington, E. Roemer, L. M. Vaughn,
C. C. McCarthy.
IAU Circ., No. 2785 (1975).

New periodic comet West-Kohoutek-Ikemura, 1975b.
Kometn. Tsirk., *Kiev,* No. 176 (1975). In Russian.

Periodic comet West-Kohoutek-Ikemura, 1975b.
Kometn. Tsirk., *Kiev,* No. 178 (1975). In Russian.

103.126   **Comet 1975c Kohoutek**

New periodic comet Kohoutek 1975c.
L. Kohoutek, B. G. Marsden.
British Astron. Ass. Circ. No. 561 (1975).

Comet Kohoutek (1975 c).
L. Kohoutek, R. E. McCrosky, J. H. Bulger.
IAU Circ., No. 2755 (1975).

Periodic comet Kohoutek (1975 c).   C. Y. Shao,
J. H. Bulger, R. E. McCrosky, B. G. Marsden.

IAU Circ., No. 2757 (1975).

Periodic comet Kohoutek (1975 c).   T. Seki,
K. Hestir, E. F. Tedesco, L. Kohoutek, M. Dieckvoss,
B. G. Marsden.
IAU Circ., No. 2766 (1975).

Periodic comet Kohoutek (1975c).
G. Schwartz, C. Y. Shao, T. Seki.
IAU Circ., No. 2770 (1975).

Periodic comet Kohoutek (1975 c).   E. Roemer,
C. C. McCarthy.
IAU Circ., No. 2776 (1975).

Periodic comet Kohoutek (1975 c).   D. E. Sykes,
R. L. Waterfield, E. Roemer, M. A. Daniel, C. C. McCarthy.
IAU Circ., No. 2785 (1975).

New short-periodic comet Kohoutek, 1975c.
Kometn. Tsirk., *Kiev,* No. 176 (1975). In Russian.

Periodic comet Kohoutek, 1975c.
Kometn. Tsirk., *Kiev,* No. 178 (1975). In Russian.

103.127   **Comet 1975d Bradfield**

New comet Bradfield 1975d.
W. A. Bradfield, S. W. Milbourn.
British Astron. Ass. Circ. No. 561 (1975).

Comet Bradfield 1975d.   B. G. Marsden.
British Astron. Ass. Circ., No. 562 (1975).

Comet Bradfield (1975 d).   W. A. Bradfield,
D. Herald, B. Tregaskis.
IAU Circ., No. 2759 (1975).

Comet Bradfield (1975 d).
A. C. Gilmore, P. M. Kilmartin.
IAU Circ., No. 2760 (1975).

Comet Bradfield (1975 d).   J.-P. Swings.
IAU Circ., No. 2761 (1975).

Comet Bradfield (1975 d).   F. Dossin, D. Herald,
S. Stephenson, B. G. Marsden.
IAU Circ., No. 2763 (1975).

Comet Bradfield (1975 d).
D. Herald, T. B. Tregaskis, B. G. Marsden.
IAU Circ., No. 2768 (1975).

Comet Bradfield (1975 d).
D. Herald, P. M. Kilmartin.
IAU Circ., No. 2771 (1975).

Comet Bradfield (1975 d).   P. M. Kilmartin,
D. Herald, J. Hers.
IAU Circ., No. 2776 (1975).

Comet Bradfield (1975 d).   A. C. Gilmore,
P. M. Kilmartin, D. Herald, J. Hers.
IAU Circ., No. 2782 (1975).

Comet Bradfield (1975 d).
D. Herald, J. Hers.
IAU Circ., No. 2786 (1975).

Comet Bradfield (1975 d).
A. C. Gilmore, P. M. Kilmartin, D. Herald.
IAU Circ., No. 2792 (1975).

New comet Herald (= Bradfield), 1975d.
Kometn. Tsirk., *Kiev,* No. 176 (1975). In Russian.

Comet Bradfield, 1975d.
Kometn. Tsirk., *Kiev,* No. 177 (1975). In Russian.

Comet Bradfield, 1975d.
Kometn. Tsirk., *Kiev,* No. 179 (1975). In Russian.

103.128   **Comet 1975e Smirnova-Chernykh**

New periodic comet Smirnova–Chernykh 1975e.
T. M. Smirnova, N. S. Chernykh, G. R. Kastel, R. L. Waterfield, G. H. Rutter.
British Astron. Ass. Circ., No. 562 (1975).

Comet Smirnova-Chernykh (1975 e).
T. M. Smirnova, N. S. Chernykh.
IAU Circ., No. 2764 (1975).

Comet Smirnova-Chernykh (1975 e).
T. Seki, E. Roemer, R. E. McCrosky, B. G. Marsden.
IAU Circ., No. 2769 (1975).

Comet Smirnova-Chernykh (1975 e).   E. Roemer,
C. C. McCarthy, R. E. McCrosky, G. R. Kastel.
IAU Circ., No. 2770 (1975).

Periodic comet Smirnova-Chernykh (1975 e).
T. M. Smirnova, N. S. Chernykh, T. Seki, E. Roemer, R. A.
McCallister, D. Waaland, C. C. McCarthy, G. R. Kastel'.
IAU Circ., No. 2772 (1975).

Periodic comet Smirnova-Chernykh (1975 e).
R. L. Waterfield, G. H. Rutter, N. S. Chernykh, T. M.
Smirnova, T. Seki, G. R. Kastel'.
IAU Circ., No. 2777 (1975).

Periodic comet Smirnova-Chernykh (1975 e).
R. E. McCrosky, C. Y. Shao, G. Schwartz, J. H. Bulger,
R. H. S. South, R. L. Waterfield, N. S. Chernykh, E. Roemer,
L. M. Vaughn, C. C. McCarthy.
IAU Circ., No. 2785 (1975).

New comet Smirnova-Chernykh.
Kometn. Tsirk., *Kiev,* No. 176 (1975). In Russian.

New comet Smirnova-Chernykh, 1975e.
Kometn. Tsirk., *Kiev,* No. 177 (1975). In Russian.

Periodic comet Smirnova-Chernykh, 1975e.
N. S. Chernykh, G. R. Kastel'.
Kometn. Tsirk., *Kiev,* No. 178 (1975). In Russian. – Elements
and ephemeris of comet Smirnova-Chernykh, 1975e (*G. R.
Kastel'*).

Periodic comet Smirnova-Chernykh, 1975e.
G. R. Kastel'.
Kometn. Tsirk., *Kiev,* No. 179 (1975). In Russian.

Periodic comet Smirnova-Chernykh, 1975e.
Kometn. Tsirk., *Kiev,* No. 180 (1975). In Russian. – Observations at the Crimean Astrophysical Observatory (*N. S.
Chernykh, N. A. Bokhan*).

103.129   **Comet 1955 VII Perrine-Mrkos**

Periodic comet Perrine-Mrkos.   B. G. Marsden.
IAU Circ., No. 2768 (1975).

**Ephemeris of comet Perrine-Mrkos, 1896 VII,
1955 VII.**
Kometn. Tsirk., *Kiev,* No. 179 (1975). In Russian.

103.130   **Comet 1969 II Gunn**

Periodic comet Gunn (1969 II).
T. Seki, A. Mrkos, R. Petrovičová.
IAU Circ., No. 2785 (1975).

Periodic comet Gunn, 1969 II.   A. Mrkos.
Kometn. Tsirk., *Kiev,* No. 180 (1975). In Russian.
Contains besides observations at the Kleť Observatory elements
and ephemeris of the comet from the British Astron. Ass.
Handbook.

103.131   **Comet 1974c Lovas**

Comet Lovas 1974c.   B. G. Marsden.
British Astron. Ass. Circ., No. 562 (1975).

Comet Lovas (1974 c).   C. Torres, S. Barros,
M. Wischnjewsky, A. C. Gilmore, P. M. Kilmartin, T. Seki.
IAU Circ., No. 2744 (1975).

Comet Lovas (1974 c).   W. Rea, P. M. Kilmartin,
A. C. Gilmore.
IAU Circ., No. 2767 (1975).

Comet Lovas (1974 c).   B. G. Marsden.
IAU Circ., No. 2776 (1975).

Comet Lovas (1974 c).   A. C. Gilmore.
IAU Circ., No. 2792 (1975).

103.132   **Comet 1972 XII Araya**

Comet Araya (1972 XII).   C. Torres, S. Barros,
M. Wischnjewsky, A. C. Gilmore, P. M. Kilmartin.
IAU Circ., No. 2746 (1975).

103.133   **Comet 1973*l* Schwassmann-Wachmann 2**

Periodic comet Schwassmann-Wachmann 2 (1973 *l*).
A. C. Gilmore, P. M. Kilmartin, T. Seki.
IAU Circ., No. 2750 (1975).

Periodic comet Schwassmann-Wachmann 2 (1973 *l*).
A. C. Gilmore, P. M. Kilmartin, T. Seki.
IAU Circ., No. 2779 (1975).

Periodic comet Schwassmann-Wachmann 2 (1973 *l*).
A. C. Gilmore, L. Krumenaker.
IAU Circ., No. 2792 (1975).

103.134   **Comet 1971 II Encke**

Periodic comet Encke.
P. M. Kilmartin, R. R. D. Austin.
IAU Circ., No. 2751 (1975).

Periodic comet Encke.
IAU Circ., No. 2779 (1975).

103.135   **Comet 1973 X Sandage**

Comet Sandage (1973 X).   T. Seki.
IAU Circ., No. 2753 (1975).

103.136   **Comet 1973 XI Gehrels 2**

Periodic comet Gehrels 2 (1973 XI).   P. Wild.
IAU Circ., No. 2753 (1975).

103.137   **Comet 1974e Cesco**

Comet Cesco (1974 e).   C. Torres, S. Barros,
M. Wischnjewsky.
IAU Circ., No. 2753 (1975).

103.138   **Comet 1944 III du Toit 1**

Periodic comet du Toit 1 (1944 III).   C. Torres.
IAU Circ., No. 2756 (1975).

103.139   **Comet 1972 III Bradfield**

Comet Bradfield (1972 III).   D. Herald.
IAU Circ., No. 2758 (1975).

103.140   **Comet 1973 IX Gibson**

Comet Gibson (1973 IX).
J. Gibson, A. C. Gilmore, P. M. Kilmartin.
IAU Circ., No. 2762 (1975).

103.141   **Comet 1852 IV Westphal**

Periodic comet Westphal.   L. M. Belous.
IAU Circ., No. 2770 (1975).

103.142   **Comet 1896 V Giacobini**

**Periodic comets Giacobini (1896 V) and Metcalf
(1906 VI).**   N. A. Belyaev, N. Yu. Goryajnova,
V. V. Emel'yanenko.
IAU Circ., No. 2780 (1975).

103.143   **Comet 1906 VI Metcalf**

**Periodic comets Giacobini (1896 V) and Metcalf
(1906 VI).**   See Abstr. 103.142.

103.144   **Comet 1975g Longmore**

Comet Longmore (1975 g).   A. J. Longmore,
P. R. Standen, P. Wallace.
IAU Circ., No. 2789 (1975).

104 Meteors, Meteor Streams

104.001 Radio observations of the Geminids 1959 – 1969. Structure. M. Šimek.
Bull. Astron. Inst. Czechoslovakia (BAC), Vol. 26, 1 - 10 (1975).

The fine structure of the shower could be studied by combining the observations from at least three favourably located stations. The influence of the Poynting-Robertson effect should cause the change of the concentration of different particles according to their mass across the stream. The change of the location in $L_\odot$, where the density of particles is maximum, is roughly $0.2°$/magn. in the range interval of $m_r = + 1.2$ to $+ 7.5$.

104.002 A possible new meteor shower. M. Šimek, P. Pecina.
Bull. Astron. Inst. Czechoslovakia (BAC), Vol. 26, 11 - 15 (1975).

From the radar material of Geminid 1959 – 1966 observations a systematic slight drop of the mass distribution index s around December 17 was found. There is an excess of echo rates with durations over 2 sec only and, therefore, the radiant was not in a good position with regard to the antenna beam. Because of this, the right ascension of the supposed radiant can be specified only in the half plane $23° < \alpha < 203°$. The declination is more precisely determined as $15° < \delta < 21°$, and the solar longitude at maximum is $L_\odot = 265.7°$ (1975).

104.003 Photographic data on two EN-fireballs in 1974. Z. Ceplecha, M. Ježková, J. Boček.
Bull. Astron. Inst. Czechoslovakia (BAC), Vol. 26, 16 - 17 (1975).

Geometrical, dynamical, photometrical and orbital data are given for two fireballs photographed by all-sky-cameras of the European Network in March 1974.

104.004 Studies of sounds from meteors. D. O. ReVelle.
Sky Telescope, Vol. 49, 87 - 91 (1975).

104.005 Television observations of faint meteors – I. Mass distribution and diurnal rate variation.
R. L. Hawkes, J. Jones.
Monthly Notices Roy. Astron. Soc., Vol. 170, 363 - 377 (1975).

A low light level television meteor observing system sensitive to stars of magnitude $+ 8.5$ is briefly described. Observations of sporadic meteors with the system yielded a mass distribution index of 2.02 ± 0.04 valid over the absolute magnitude range approximately $+ 3$ to $+ 6$. Diurnal rate of occurrence statistics are also presented.

104.006 Meteorbeobachtungen in der Schweiz. R. Germann.
Orion, 33. Jahrgang, p. 13 - 14 (1975).

104.007 Der Meteorfall vom 30. August 1974. J. Alean.
Orion, 33. Jahrgang, p. 15 (1975).

104.008 Versuch einer Orts- und Höhenbestimmung an einem Geminiden-Meteor. J. Alean.
Orion, 33. Jahrgang, p. 15 - 18 (1975).

104.009 Washougal: a stony meteorite with a retrograde orbit? E. A. Carver, E. Anders.
Journ. Geophys. Res., Vol. 80, 789 - 793 (1975).

The Washougal howardite is alleged to have struck the earth from a retrograde orbit, with a preatmospheric velocity of 55 ± 3 km/s. To check this report, the authors determined the ablation loss of this 0.2-kg meteorite from the angular distribution of cosmic ray tracks, using a new method based on the model of Maurette et al. (1969). The preatmospheric mass was 660 ± 220 kg, which according to meteor theory implies a geocentric velocity of $\leqslant 35$ km/s. This upper limit is well below the minimum value for a retrograde orbit, 48 km/s.

104.010 Television observations of faint meteors–II. Light curves. J. Jones, R. L. Hawkes.
Monthly Notices Roy. Astron. Soc., Vol. 171, 159 - 169 (1975).

Observations of very faint meteors provide a test of the theories of meteroid structure. With a low-light-level television system which the authors have developed they have made such observations and found that the solid compact meteoroid model must be abandoned. Their observations, which are in good agreement with the radio results of Verniani, give a vertical train length of about 6.7 ± 0.7 km.

104.011 Noisy meteors. D. W. Hughes.
Nature, Vol. 254, 384 - 386 (1975).

104.012 Investigation of meteor stream ages. H. Korpikiewicz.
Postępy Astron., Vol. 23, 69 - 71 (1975). In Polish.

The average distance between the orbit of a comet and of a meteorite may serve as a stream age indicator. Computations suggest a genetic relation between the α-Capricornids and the Lexell 1770 I comet, which seems to settle the controversy as to the origin of that stream.

104.013 Zomer 1974: de oogst aan meteoor-waarnemingen. P. A. Koning.
Zenit, Vol. 2, 9 - 11 (1975).

104.014 Plan de campagne 1975 van de Werkgroep Meteoren. B. Apeldoorn.
Zenit, Vol. 2, 16 - 18 (1975).

104.015 Estimations du flux météorique apporté par les essaims. F. Link.
IAU Colloquium No. 22, (see 012.005), p. 229 - 238 (1974).

104.016 Meteoric storms and formation of meteor streams. Z. Sekanina.
IAU Colloquium No. 22, (see 012.005), p. 239 - 267 (1974).

104.017 The system of short period meteor streams. B. A. Lindblad.
IAU Colloquium No. 22, (see 012.005), p. 269 - 281 (1974).

104.018 Évolution des orbites et des radiants d'un essaim météorique de la famille de Jupiter.
E. I. Kazimirchak-Polonskaya, A. K. Terent'eva.
IAU Colloquium No. 22, (see 012.005), p. 283 - 301 (1974).

104.019 The structure of meteor streams associated with comet Halley: Eta Aquarids and Orionids.
A. Hajduk.
IAU Colloquium No. 22, (see 012.005), p. 309 - 311 (1974).

104.020 Résultats d'une étude des perturbations séculaires des orbites de particules météoriques. J. Delcourt.
IAU Colloquium No. 22, (see 012.005), p. 323 - 333 (1974).

104.021 Radar observations of meteor rates at three levels at Frunze in 1973.
K. A. Karimov, V. K. Semenov, S. S. Timofeeva.

Trudy Kirg. un-ta. Ser. fiz. n., 1974, vyp. (No.) 5, p. 3 - 16. In Russian. – Abstr. in Referativ. Zhurn. 51. Astron., 3.51. 348 (1975).

104.022 Verification of some correlations of the physical theory using observations of meteors up to + 12^m. N. V. Novoselova. Problems of cosmic physics. Vyp. (No.) 9, (see 003.013), p. 84 - 92 (1974). In Russian.

Results of mean height and mean deceleration measurements for different velocity intervals are obtained in Kharkov and their linear approximations are given. The theoretical dependences of the mean observed height and deceleration of meteors on the velocity are obtained.

104.023 A possibility of observing faint radio meteors. A. A. Tkachuk. Problems of cosmic physics. Vyp. (No.) 9, (see 003.013), p. 92 - 98 (1974). In Russian.

104.024 On spectrophotometric researches of meteors. I. Inelastic collisions in an elementary meteor coma. I. N. Kovshun. Problems of cosmic physics. Vyp. (No.) 9, (see 003.013), p. 99 - 104 (1974). In Russian.

104.025 Meteors. J. Morgan. Southern Stars, Vol. 26, 1 - 7 (1975).

104.026 Radio wave propagation and ionisation of meteoric constituents. A. K. Sen, B. Saha. Nature, Vol. 255, 313 - 316 (1975). – Letter.

104.027 Parameters of the trajectory and morphological peculiarities of meteors with flares. E. N. Kramer, A. K. Markina. Odessk. un-t. Odessa. 1974. 36 pp. In Russian. – Abstr. in Referativ. Zhurn. 51. Astron., 5.51.351 (1975).

104.028 Parameters of atmospheric trajectories of meteors according to observations in Odessa. E. N. Kramer, V. A. Vorob'eva, N. N. Izraetskaya, A. K. Markina, V. I. Musij, O. A. Rudenko, R. B. Teplitskaya, I. S. Shestaka. Odessk. un-t. Odessa, 1974. 48 pp. In Russian. – Abstr. in Referativ. Zhurn. 51. Astron., 5.51.352 (1975).

104.029 On the possibility of attachment of electrons to NO$_2$-molecules in faint meteors. J. Rajchl. Bull. Astron. Inst. Czechoslovakia, Vol. 26, 138 - 139 (1975).

In this paper it is shown that in meteors of absolute maximum visual magnitude less than −2 a sudden increase in the attachment coefficient can be observed, the basis of which cannot be explained by the process of attachment to the meteoric material itself. Calculations indicate that the increase of this coefficient in faint meteors could be explained by NO$_2$, the attachment coefficient of which is two orders higher, taking part in the attachment process with the meteoric material.

104.030 A meteorite that got away. E. J. Öpik. Irish Astron. Journ., Vol. 11, 165 - 172 (1974).

104.031 Radio observations of the Orionid meteor shower. A. Hajduk. Contr. Astron. Obs. Skalnaté Pleso, Vol. 5, 143 - 157 (1973/75).

Radar echo data obtained by the equipment of the Ondřejov Astronomical Observatory in 1961 - 1969 during the periods of the Orionid shower activity are presented. The hourly rates of echoes are given in four duration classes. The echo amplitude distribution in the range scale and some other characteristics of the shower and the corresponding background are tabulated.

104.032 Temporal fluctuations and anisotropy of the micrometeoroid flux in the earth−moon system measured by HEOS 2. H.-J. Hoffmann, H. Fechtig, E. Grün, J. Kissel. Planet. Space Sci., Vol. 23, 985 - 991 (1975). – Paper presented at the XVIIth COSPAR Meeting at Sao Paulo, Brazil, 1974.

The HEOS detector measures the mass and speed of micrometeoroids in the earth−moon system. They are detected by the plasma produced by particle impacts on the sensor. In this paper the origin of the groups and swarms is discussed. The results imply a lunar origin of the groups, whereas the swarms are correlated with the vicinity of the earth. In addition, the dependence of the cumulative flux upon the detector's viewing direction indicates clearly an anisotropic particle flux.

104.033 On the presence of cosmic dust in the upper atmosphere. F. Link. Planet. Space Sci., Vol. 23, 1011 - 1012 (1975).

The activity of Librids on 4 April 1973 was detected by the rise of the twilight luminosity. The ratio of circumzenithal luminosity in the antisolar to solar azimuth also increases in the presence of cosmic dust in the upper atmosphere.

104.034 De grote vuurbol van 25 april 1975. B. Apeldoorn. Zenit, Vol. 2, 206 - 209 (1975).

104.035 Image-orthicon spectra of Geminids in 1969. P. M. Millman, A. F. Cook, C. L. Hemenway. Evolutionary and physical properties of meteoroids. IAU colloquium No. 13, p. 147 - 151 = Dudley Obs., *Albany, New York,* Repr. No. C84 (1973).

The spectra of 25 meteors, recorded with an image-orthicon technique in December 1969, are studied in relation to similar records made in August of the same year. Of 19 Geminid meteors in the absolute visual magnitude range 0 to +2, only one showed any evidence of the forbidden line of oxygen at 5577 Å, while all Perseid meteors recorded in August exhibited the oxygen line, a result of the large difference in geocentric velocity between the two showers. Atoms identified in faint Geminid meteors include neutral iron, magnesium, calcium and sodium. The molecular bands of nitrogen are also observed.

104.036 An unusual meteor spectrum. A. F. Cook, C. L. Hemenway, P. M. Millman, A. Swider. Evolutionary and physical properties of meteoroids. IAU colloquium No. 13, p. 153 - 159 = Dudley Obs., *Albany, New York,* Repr. No. C85 (1973).

An extraordinary spectrum of a meteor at a velocity of about 18.5 ± 1.0 km s^{-1} (approximate uncertainty) was observed from the Springhill Meteor Observatory with an image-orthicon camera at 1970 August 10.

104.037 Peculiarities of fragmentation of the meteor bodies of the Taurids. P. B. Babadzhanov, V. S. Getman. Dokl. AN TadzhSSR, Vol. 17, No. 9, p. 18 - 20 (1974). In Russian. – Abstr. in Referativ. Zhurn. 51. Astron., 6.51.406 (1975).

104.038 Fireballs of 1975 April 29 and 1975 May 4. British Astron. Ass. Circ., No. 562 (1975).

104.039 Lyridi 1974. godine. Rezultati V meteorske ekspedicije (Lyrids in 1974. Results of the Vth

meteor expedition). K. Pavlovski, S. Kulišić. Vasiona, Vol. 22, 90 - 94 (1974).

104.040 **Bolid Šumava of 4 December 1974.** Říše hvězd, Vol. 56, 56 - 57 (1975). In Czech.

Low latitude meteor wind observations. See Abstr. 082.017.

Comparative studies of *E*-region ionospheric drifts and meteor winds. See Abstr. 083.033.

Asteroids: infrared photometry at 1.25, 1.65, and 2.2 microns. See Abstr. 098.007.

Possible evidence of ablation on cosmic dust particles. See Abstr. 105.137.

Pioneer 11 meteoroid detection experiment: preliminary results. See Abstr. 106.026.

105 Meteorites, Meteorite Craters

105.001 Stable isotope mass fragmentography: identification and hydrogen–deuterium exchange studies of eight Murchison meteorite amino acids.
W. E. Pereira, R. E. Summons, T. C. Rindfleisch, A. M. Duffield, B. Zeitman, J. G. Lawless.
Geochim. Cosmochim. Acta, Vol. 39, 163 - 172 (1975).

105.002 He, Ne and Ar isotopes in inclusions of some iron meteorites. L. K. Levsky (*Levskij*), A. N. Komarov.
Geochim. Cosmochim. Acta, Vol. 39, 275 - 284 (1975).
He^3, He^4, Ne^{21} and Ar^{38} contents were determined in 18 metal, troilite, schreibersite and graphite inclusions of 9 iron meteorites, by total outgassing and stepwise heating. The He^4/He^3 ratio in metal phase ranges from 3.85 to 4.65, but in non-metallic samples, from 6.70 to 30.5. The results for cosmogenic isotopes of helium, neon and argon disagree appreciably with data on accelerator-irradiated targets. It should be noted, however, that some inclusions have lost considerable amounts of gas by diffusion.

105.003 Thermal metamorphism of primitive meteorites–I. Variation of six trace elements in Allende carbonaceous chondrite heated at 400–1000°C.
M. Ikramuddin, M. E. Lipschutz.
Geochim. Cosmochim. Acta, Vol. 39, 363 - 375 (1975).

105.004 Rubidium-87/strontium-87 age of Juvinas basaltic achondrite and early igneous activity in the solar system.
C. J. Allègre, J. L. Birck, S. Fourcade, M. P. Semet.
Science, Vol. 187, 436 - 438 (1975).
A $(4.60\pm0.07) \times 10^9$ year internal isochron has been drawn for the achondrite Juvinas by the rubidium-87/strontium-87 method. Earlier petrographic investigations of achondrites supplemented by a new ion microprobe study of Juvinas strongly suggest an igneous origin for this class of meteorites. The results thus indicate that igneous activity may have rapidly followed the formation of the achondrites' parent body 4.6×10^9 years ago.

105.005 Alternative hypothesis for the origin of CCF xenon.
D. C. Black.
Nature, Vol. 253, 417 - 419 (1975).
The author suggests that the anomalous xenon isotopic composition known as carbonaceous chondrite fission (CCF) xenon is not caused by fission, but is the direct result of a modified r-process nucleosynthesis which produces an abundance peak at $Z = 54$ and the magic neutron number $N = 82$. He further proposes that the xenon so produced ('R xenon') was trapped in dust grains, which were subsequently incorporated in the solar system with minimal degassing. This hypothesis also provides a natural explanation for the intimate association between R xenon and a newly discovered xenon component in primitive meteorites.

105.006 The most primitive objects in the solar system.
L. Grossman.
Sci. American, Vol. 232, No. 2, p. 30 - 38 (1975).
Minerals found in the meteorites known as carbonaceous chondrites represent samples of the solid grains that condensed directly out of the gaseous nebula that gave birth to the sun and the planets.

105.007 Origin of magnetite and pyrrhotite in carbonaceous chondrites.
J. M. Herndon, M. W. Rowe, E. E. Larson, D. E. Watson.
Nature, Vol. 253, 516 - 518 (1975). – Letter.

105.008 On the chemical nature of a superheavy element in meteorites. G. N. Goncharov.
Dokl. Akad. Nauk. SSSR. Ser. Mat. Fiz., Vol. 220, 552 - 555 (1975). In Russian.

105.009 Addendum to :'Possible meteorite crater in México' [Rev. Mexicana Astron. Astrofis., Vol. 1, 81 - 86 (1974)]. L. Maupomé.
Rev. Mexicana Astron. Astrofis., Vol. 1, 269 (1974).

105.010 Effect of thermal metamorphic conditions on mineralogy and trace element retention in the Allende meteorite.
M. Ikramuddin, M. E. Lipschutz, W. R. Van Schmus.
Nature, Vol. 253, 703 - 705 (1975). – Letter.

105.011 The Ellicott meteorite.
E. J. Olsen, G. I. Huss, R. M. Pearl.
Meteoritics, Vol. 9, 263 - 269 (1974).
Two individual specimens (total weight 15.7 kg) of a new medium octahedrite were found near Ellicott, El Paso County, Colorado. The find is only 1.2 km from the find (in 1890) of the Franceville medium octahedrite. Ellicott is a group IA iron while Franceville is in group IIIA.

105.012 The Canyonlands meteorite. D. E. Lange, K. Frost, P. P. Sipiera, C. B. Moore.
Meteoritics, Vol. 9, 271 - 280 (1974).
On the basis of homogeneity and composition of the silicate minerals and the lack of chondrules, the Canyonlands meteorite is classified as an H6 chondrite. It has been heavily shocked as indicated by the presence of blackened silicates and maskelynite.

105.013 Magnetite spherules in miocene versus recent sands of New Jersey. J. H. Puffer.
Meteoritics, Vol. 9, 281 - 288 (1974).

105.014 Magnetism in meteorites.
J. M. Herndon, M. W. Rowe.
Meteoritics, Vol. 9, 289 - 305 (1974).
The authors present a somewhat simplified overview of the subject of magnetism in meteorites. A glossary of magnetism terminology is included, followed by discussion of the various techniques used for magnetism studies in meteorites.

105.015 Paneth's iron, a new group IIIE iron.
V. F. Buchwald, R. Hutchison, J. M. Hall.
Meteoritics, Vol. 9, 307 - 311 (1974).

105.016 Do stony meteorites come from comets?
E. Anders.
Meteoritics, Vol. 9, 313 (1974). – Abstract.

105.017 Record of primordial growth environment in olivine crystals from carbonaceous chondrites.
G. Arrhenius, S. K. Asunmaa, R. W. Fitzgerald, B. K. Kothari, D. Macdougall.
Meteoritics, Vol. 9, 313 - 314 (1974). – Abstract.

105.018 A reinvestigation of the Lodran meteorite.
R. W. Bild, J. T. Wasson.
Meteoritics, Vol. 9, 315 - 316 (1974). – Abstract.

105.019 Comparison of a fusion crust produced by artificial ablation of an olivine with fusion crusts on the Allende and Murchison meteorites.

M. B. Blanchard, G. G. Cunningham, D. E. Brownlee.
Meteoritics, Vol. 9, 316 (1974). – Abstract.

105.020 Kinetic constraints in the formation of chondrites.
M. Blander, L. H. Fuchs.
Meteoritics, Vol. 9, 316 - 317 (1974). – Abstract.

105.021 Cosmic ray-produced nuclides in the Canon City
meteorite.
D. Bogard, R. Clark, R. Davis, J. Keith, R. Stoenner.
Meteoritics, Vol. 9, 317 - 318 (1974). – Abstract.

105.022 Fractionated rare earth abundances in an Allende
inclusion: a sample of a fractionated nebular gas.
W. V. Boynton.
Meteoritics, Vol. 9, 318 (1974). – Abstract.

105.023 On the magnetic systematics of ordinary chondrites.
A. Brecher, R. P. Ranganayaki.
Meteoritics, Vol. 9, 319 (1974). – Abstract.

105.024 Chondritic abundances for interplanetary dust:
evidence from micrometeorite craters.
D. E. Brownlee, F. Hörz, P. W. Hodge.
Meteoritics, Vol. 9, 320 - 321 (1974). – Abstract.

105.025 U-Th-Pb radiometric investigations of the Allende
carbonaceous chondrite.
J. H. Chen, G. R. Tilton.
Meteoritics, Vol. 9, 325 - 326 (1974). – Abstract.

105.026 Fractionation of siderophilic elements among H and
L chondrites.
C.-L. Chou, P. A. Baedecker, J. T. Wasson.
Meteoritics, Vol. 9, 326 - 327 (1974). – Abstract.

105.027 The effect of heating on extractable amino acids in
the Murchison carbonaceous chondrite.
J. R. Cronin, C. B. Moore.
Meteoritics, Vol. 9, 327 - 328 (1974). – Abstract.

105.028 Impact structures from ERTS imagery.
R. S. Dietz, J. McHone.
Meteoritics, Vol. 9, 329 - 333 (1974). – Abstract.

105.029 Kaaba stone: presumably not a meteorite.
R. S. Dietz, J. McHone.
Meteoritics, Vol. 9, 334 (1974). – Abstract.

105.030 St. Mesmin: a petrologic overview.   R. T. Dodd.
Meteoritics, Vol. 9, 334 - 335 (1974). – Abstract.

105.031 Olivine microporphyry in the St. Mesmin chondrite.
R. T. Dodd, E. Jarosewich.
Meteoritics, Vol. 9, 335 - 336 (1974). – Abstract.

105.032 The cosmic ray track record of the Saint-Mesmin
LL chondrite.   A. Ducatel, G. Poupeau.
Meteoritics, Vol. 9, 336 - 337 (1974). – Abstract.

105.033 A chondrite with a regolith fragment containing
non-chondritic material.
R. V. Fodor, M. Prinz, H. G. Brown, K. Keil.
Meteoritics, Vol. 9, 337 - 338 (1974). – Abstract.

105.034 The Bhola stone – a true polymict breccia?
K. Fredriksson, A. Noonan, J. Nelen.
Meteoritics, Vol. 9, 338 - 339 (1974). – Abstract.

105.035 Fusion crust phenomena on some carbonaceous
chondrites.   R. M. Fruland.

Meteoritics, Vol. 9, 339 - 342 (1974). – Abstract.

105.036 Glass inclusions of granitic compositions in ortho-
pyroxenes from three enstatite achondrites.
L. H. Fuchs.
Meteoritics, Vol. 9, 342 (1974). – Abstract.

105.037 Inorganic gas release studies and thermal analysis
investigations on carbonaceous chondrites.
E. K. Gibson, Jr.
Meteoritics, Vol. 9, 343 - 344 (1974). – Abstract.

105.038 High-magnesium glasses associated with North
American microtektites in a Caribbean deep-sea
sediment core.   B. P. Glass.
Meteoritics, Vol. 9, 345 - 347 (1974). – Abstract.

105.039 Fluorine concentrations in carbonaceous chondrites.
R. H. Goldberg, D. S. Burnett, M. J. Furst, T. A.
Tombrello.
Meteoritics, Vol. 9, 347 - 348 (1974). – Abstract.

105.040 Amoeba-shaped olivine aggregates: a new type of
inclusion in the Allende meteorite.
L. Grossman.
Meteoritics, Vol. 9, 348 - 349 (1974). – Abstract.

105.041 Ne in selected materials from the Allende meteorite.
J. C. Huneke, S. P. Smith, R. S. Rajan, G. J.
Wasserburg.
Meteoritics, Vol. 9, 350 - 351 (1974). – Abstract.

105.042 Kakangari – unique member of a distinct chemical
class of chondrite?
R. Hutchison, A. L. Graham.
Meteoritics, Vol. 9, 352 (1974). – Abstract.

105.043 Nitrogen isotope distribution in meteorites.
W. G. Injerd, I. R. Kaplan.
Meteoritics, Vol. 9, 352 - 353 (1974). – Abstract.

105.044 Electron microprobe study of glasses in the
Bununu and Malvern achondritic meteorites.
D. Y. Jerome, C. Desnoyers.
Meteoritics, Vol. 9, 354 - 355 (1974). – Abstract.

105.045 The concentrations of 22 trace elements with
atomic numbers between z = 37 (Rb) and z = 72
(Hf) in the 4 Antarctic meteorites Yamato (a), (b), (c) and
(d) and in 8 other meteorites.
K. P. Jochum, M. Seufert, H. Hintenberger.
Meteoritics, Vol. 9, 355 - 357 (1974). – Abstract.

105.046 Trace element composition of metal during conden-
sation-implications for iron meteorites.
W. R. Kelly, J. W. Larimer.
Meteoritics, Vol. 9, 360 (1974). – Abstract.

105.047 Glass production by shock waves in terrestrial and
lunar rocks and implications regarding the origin
of chondrules.   S. W. Kieffer.
Meteoritics, Vol. 9, 360 - 362 (1974). – Abstract.

105.048 Simultaneous measurement of total N and Li in
silicates and irons by RNAA.
B. K. Kothari, P. N. Shukla, P. A. Goel.
Meteoritics, Vol. 9, 362 - 363 (1974). – Abstract.

105.049 Fragmentation of the Sikhote-Alin meteoric body.
E. L. Krinov.
Meteoritics, Vol. 9, 363 (1974). – Abstract.

105.050 The significance of carbon and oxygen abundances on the condensation of planetary material.
J. W. Larimer.
Meteoritics, Vol. 9, 363 - 364 (1974). – Abstract.

105.051 El Sampal, a new meteorite from Argentina.
C. F. Lewis, C. B. Moore, N. A. Hillar.
Meteoritics, Vol. 9, 365 - 366 (1974). – Abstract.

105.052 Information upon mean orbital elements of stone meteorites by rare gas data.　J. C. Lorin.
Meteoritics, Vol. 9, 367 - 369 (1974). – Abstract.

105.053 Sm-Nd ages: a new dating method.
G. W. Lugmair.
Meteoritics, Vol. 9, 369 (1974). – Abstract.

105.054 Low-energy particle irradiation and possible age indicator for components of carbonaceous chondrites.　D. Macdougall, P. B. Price.
Meteoritics, Vol. 9, 370 - 371 (1974). – Abstract.

105.055 The low temperature fracture behavior of iron-nickel meteorites.　H. L. Marcus, L. H. Hackett, Jr.
Meteoritics, Vol. 9, 371 - 376 (1974). – Abstract.

105.056 Cosmic-ray track record in the San Juan Capistrano meteorite.　B. J. Martinek, D. Lal.
Meteoritics, Vol. 9, 377 (1974). – Abstract.

105.057 The Isna meteorite – a C3 find from Egypt.
R. L. Methot, A. F. Noonan, E. Jarosewich,
A. A. DeGasparis.
Meteoritics, Vol. 9, 381 - 382 (1974). – Abstract.

105.058 "Al-Ca rich chondrules in the Coolidge chondrite".
J. Nelen, A. Noonan, K. Fredriksson.
Meteoritics, Vol. 9, 384 - 385 (1974). – Abstract.

105.059 Microprobe analyses of glassy particles from Howardites.
A. F. Noonan, R. S. Rajan, A. A. Chodos.
Meteoritics, Vol. 9, 385 - 386 (1974). – Abstract.

105.060 Rare-earth abundances in two mineral separates with distinct oxygen isotopic composition from an Allende inclusion.　N. Onuma, T. Tanaka, A. Masuda.
Meteoritics, Vol. 9, 387 - 388 (1974). – Abstract.

105.061 The plutonium-244 fission track record in ordinary chondrites: implications for cooling rates.
P. Pellas, D. Storzer.
Meteoritics, Vol. 9, 388 - 390 (1974). – Abstract.

105.062 The Chassigny meteorite: a relatively iron-rich cumulate dunite.　M. Prinz, P. H. Hlava, K. Keil.
Meteoritics, Vol. 9, 393 - 394 (1974). – Abstract.

105.063 Glassy agglutinate-like objects in the Bununu Howardite.
R. S. Rajan, D. E. Brownlee, G. H. Heiken, D. S. McKay.
Meteoritics, Vol. 9, 394 - 396 (1974). – Abstract.

105.064 Trace element content of the Allende minerals.
E. Rambaldi.
Meteoritics, Vol. 9, 396 - 397 (1974). – Abstract.

105.065 Characteristic features observed in the meteorites Beverbruch and Benthullen, Oldenburg, Germany.
P. Ramdohr, A. El Goresy.

Meteoritics, Vol. 9, 397 - 398 (1974). – Abstract.

105.066 The Macibini meteorite and some thoughts on the origin of basaltic achondrites.　A. M. Reid.
Meteoritics, Vol. 9, 398 - 399 (1974). – Abstract.

105.067 Constraints on magnetic field which magnetized the Farmington meteorite parent body.
M. W. Rowe.
Meteoritics, Vol. 9, 399 (1974). – Abstract.

105.068 Search for extinct ^{146}Sm in meteorites.
N. B. Scheinin, G. W. Lugmair, K. Marti.
Meteoritics, Vol. 9, 399 - 400 (1974). – Abstract.

105.069 Rare gases in the St. Mesmin chondrite.
L. Schultz, P. Signer.
Meteoritics, Vol. 9, 402 - 403 (1974). – Abstract.

105.070 Pressures of formation of iron meteorites from sphalerite compositions.
H. P. Schwarcz, S. D. Scott, S. A. Kissin.
Meteoritics, Vol. 9, 404 (1974). – Abstract.

105.071 Anomalous members of iron meteorite groups.
E. Scott, J. T. Wasson.
Meteoritics, Vol. 9, 404 - 405 (1974). – Abstract.

105.072 The distribution of lithophile elements in separated phases of enstatite chondrites.
M. Shima, M. Honda.
Meteoritics, Vol. 9, 405 - 408 (1974). – Abstract.

105.073 Fission track studies in the Allende carbonaceous chondrite.　J. Shirck.
Meteoritics, Vol. 9, 408 (1974). – Abstract.

105.074 Cosmic ray track production rate for a near surface sample of the Allende meteorite.
M. I. Stapanian, D. S. Burnett.
Meteoritics, Vol. 9, 408 - 409 (1974). – Abstract.

105.075 Host clinohypersthene with exsolved augite and the thermal history of the Juvinas eucrite.
H. Takeda, M. Miyamoto, A. M. Reid.
Meteoritics, Vol. 9, 410 - 411 (1974). – Abstract.

105.076 Some effects of gravitational acceleration on the formation of hypervelocity impact craters.
J. A. Wedekind, D. E. Gault, G. Nakata.
Meteoritics, Vol. 9, 421 - 422 (1974). – Abstract.

105.077 Some studies of the unusual eucrite, Ibitira.
L. L. Wilkening, E. Anders.
Meteoritics, Vol. 9, 422 - 423 (1974). – Abstract.

105.078 A geologic interpretation of the Tunguska event.
M. S. Woyski.
Meteoritics, Vol. 9, 423 (1974). – Abstract.

105.079 Die Narben im Antlitz der Himmelskörper.
W. Gentner.
SuW, Vol. 14, 114 - 119 (1975). – Shortened version of a lecture on the occasion of the general assembly of the Max-Planck-Ges., Münster 1974.

105.080 Do stony meteorites come from comets?
E. Anders.
Icarus, Vol. 24, 363 - 371 (1975).
　　　The place of origin of stony meteorites can be determined from their trapped solar-wind gases. "Gas-rich" meteor-

ites have only 10^{-3}—10^{-4} the solar noble gas content and $\leqslant 10^{-2}$—10^{-4} the surface exposure age of lunar soils. These differences suggest that the gas implantation took place between 1 and 8 AU from the sun, in a region where the cratering rate was 10^2—10^3 times higher than at 1 AU. Both characteristics point to the asteroid belt. It appears that a cometary origin can be ruled out for all stony meteorite classes that have gas-rich members. This includes carbonaceous chondrites.

105.081 The problem of the Tunguska meteorite.
N. V. Vasil'ev.
Zemlya i Vselennaya, 1975, No. 1, p. 29 - 35. In Russian.

105.082 Meteorite Gorlovka. I. T. Zotkin.
Zemlya i Vselennaya, 1975, No. 1, p. 35 - 37
In Russian.

105.083 Tsiolkovskij's bolide. G. T. Chernenko.
Zemlya i Vselennaya, 1975, No. 1, p. 91 - 93.
In Russian.

105.084 A seismic refraction technique used for subsurface investigations at Meteor crater, Arizona.
H. D. Ackermann, R. H. Godson, J. S. Watkins.
Journ. Geophys. Res., Vol. 80, 765 - 775 (1975).

105.085 Gravity and magnetic investigations of Meteor crater, Arizona. R. D. Regan, W. J. Hinze.
Journ. Geophys. Res., Vol. 80, 776 - 788 (1975).

105.086 Meteoritenkrater-Ketten. W. Sandner.
Sterne, Vol. 51, 49 - 50 (1975).

105.087 Rare gases in separated whitlockite from the St. Severin chondrite: xenon and krypton from fission of extinct ^{244}Pu. R. S. Lewis.
Geochim. Cosmochim. Acta, Vol. 39, 417 - 432 (1975).
Helium, neon, argon, krypton and xenon data are presented from stepwise heating of samples of the mineral whitlockite from the chondritic meteorite St. Severin. The xenon is shown to be a uniform mixture derived from ^{244}Pu fission and rare-earth element spallation. The krypton similarly contains spallation products and ^{86}Kr from ^{244}Pu fission.

105.088 Petrography and mineral chemistry of Ca-rich inclusions in the Allende meteorite. L. Grossman.
Geochim. Cosmochim. Acta, Vol. 39, 433 - 454 (1975).

105.089 Purines and triazines in the Murchison meteorite.
R. Hayatsu, M. H. Studier, L. P. Moore, E. Anders.
Geochim. Cosmochim. Acta, Vol. 39, 471 - 488 (1975).

105.090 Interplanetary dust on the earth's surface.
D. W. Sears.
Journ. British. Astron. Ass., Vol. 85, 115 - 119 (1975).

105.091 Invalid 4.01-Gyr model U-Pb 'age' of the Nakhla meteorite.
R. Hutchison, N. H. Gale, J. W. Arden.
Nature, Vol. 254, 678 - 680 (1975). — Letter.

105.092 Australia's Henbury craters.
Sky Telescope, Vol. 49, 287 - 290 (1975).

105.093 ^{26}Al in cores of the Keyes chondrite. P. J. Cressy.
Journ. Geophys. Res., Vol. 80, 1551 - 1554 (1975).
Cosmogenic ^{26}Al was measured in six samples from two cores of the Keyes chondrite, previously analyzed for spallogenic ^{3}He, ^{21}Ne, and ^{38}Ar. Specific activities varied from 68.9 ± 5.4 dpm/kg near the center (depth, ~30 cm) to 50.9 ± 3.7 near the postatmospheric surface (depth, ~3 cm).

105.094 Xenon isotope anomalies in the carbonaceous chondrite Murchison.
P. K. Kuroda, R. D. Sherrill, D. W. Efurd, J. N. Beck.
Journ. Geophys. Res., Vol. 80, 1558 - 1570 (1975).
The isotopic compositions have been measured mass spectrometrically for xenon fractions released from the carbonaceous chondrite Murchison in stepwise heating experiments. Variation of the isotopic ratios was found to be relatively small. It appears that these variations can best be explained as being due to the fact that reservoirs of two isotopically distinct gases (solar and planetary) exist in the meteorite and that mixtures of these gases in various proportions are being released at different temperatures.

105.095 The production of ^{21}Na by low-energy protons in meteorites. G. F. Herzog.
Journ. Geophys. Res., Vol. 80, 1109 - 1112 (1975).
The cross sections for ^{21}Na production induced by 10-, 12-, 14-, and 16-MeV protons on natural Mg targets have been measured as 0.21 ± 0.08, 17.0 ± 2.5, 45.2 ± 6.8, and 44.0 ± 6.6 mb, respectively. These values imply that just below the surface the ^{22}Ne/^{21}Ne ratios found in extraterrestrial samples would be lowered by an excess of low-energy protons.

105.096 Impact structures in Canada: their recognition and characteristics.
P. B. Robertson, R. A. F. Grieve.
Journ. Roy. Astron. Soc. Canada, Vol. 69, 1 - 21 = Contr.
Earth Phys. Branch, Ottawa, No. 430 (1975).
Nine impact sites, some of which do not have an obvious craterlike form due to erosion or heavy sedimentary cover, have been confirmed since 1967 and are described here. An additional nine structures where geological and geophysical data are consistent with an impact origin, but where shock features have not as yet been identified, are considered as probable impact sites. A number of other sites are considered as possible impact structures because of their anomalous circular form.

105.097 A large crater field recognized in Central Europe.
J. Classen.
Sky Telescope, Vol. 49, 365 - 367 (1975).

105.098 Paleomagnetic systematics of ordinary chondrites.
A. Brecher, R. P. Ranganayaki.
Earth Planet. Sci. Letters, Vol. 25, 57 - 67 (1975).

105.099 ^{14}C - ^{39}Ar$_{Me}$ correlations in chondrites and their pre-atmospheric size.
W. Born, F. Begemann.
Earth Planet. Sci. Letters, Vol. 25, 159 - 169 (1975).
Cosmogenic ^{14}C has been measured in 12 chondrites and the stone phase of the mesosiderite Bondoc. In eight cases ^{39}Ar in the metal phase from the same meteorite specimens had been measured previously. The results are combined to derive the pre-atmospheric radii R_0 of the meteorites and depth of burial of the samples investigated. Values of R_0 between 35 and 82 cm are obtained. A compilation of all published C concentrations in chondrites show that the variations between different specimens from the same meteorites are almost as large as those for samples from different meteorites. Thus, there is no need to invoke different orbits of the meteoroids and a strong spatial gradient in the primary cosmic-ray intensity to explain variations of low-energy-produced cosmogenic nuclides in different meteorites.

105.100 Cosmic-ray-produced ^{40}K and ^{50}V in the metal phase of chondrites.
K. Imamura, M. Shima, M. Honda.
Earth Planet. Sci. Letters, Vol. 26, 54 - 60 (1975).

105.101 Thermoluminescence studies and the pre-atmospheric shape and mass of the Estacado meteorite.
D. W. Sears.
Earth Planet. Sci. Letters, Vol. 26, 97 - 104 (1975).

Before it entered the atmosphere Estacado was an elongated shape, approximating to an ellipse of eccentricity 0.8, with a pre-atmospheric mass of at least 8 tons. The results are consistent with laboratory experiments which indicate that secondary radiation plays an extremely important part in the production of thermoluminescence.

105.102 And again ... " beyond the haze and fragrance of the taiga". A. I. Eremeeva.
Zemlya i Vselennaya, 1975, No. 2, p. 63 - 68. In Russian.

105.103 At what heliocentric distances did the different meteorite groups form? J. T. Wasson.
Bull. American Astron. Soc., Vol. 7, 375 (1975). – Abstr. AAS.

105.104 New data about supposed cosmic matter in the region of the Tunguska catastrophe.
N. V. Vasil'ev, Yu. A. L'vov, G. M. Ivanova, Yu. A. Grishin, N. P. Shul'ga, A. S. Salina, S. N. Gryaznova, T. A. Menyavtseva
Problems of cosmic physics. Vyp. (No.) 9, (see 003.013), p. 79 - 83 (1974). In Russian.

Some results of cosmochemical investigations in the region of the Tunguska River catastrophe are discussed.

105.105 Indications for a meteoritic impact of early Cambrian age at Conception Bay, Newfoundland.
W. v. Engelhardt.
Naturwissenschaften, 62. Jahrgang, p. 234 - 235 (1975).

105.106 The Ella Island, Greenland, chondrite.
J. H. Carman, G. R. McCormick.
Meteoritics, Vol. 10, 1 - 8 (1975).

The Ella Island, Greenland, meteorite was found in August of 1971. The meteorite is classified as an L-6 chondrite.

105.107 The magnetic investigation of the Sikhote-Alin iron meteorite shower at the site of fall.
E. S. Gorshkov, E. G. Gus'kova, V. I. Pochtarev.
Meteoritics, Vol. 10, 9 - 19 (1975).

Individual and splinter specimens of the iron meteorite shower of Sikhote-Alin and rock samples from impact craters have been studied magnetically.

105.108 Notes on the Barwise chondrite. L. H. Ahrens.
Meteoritics, Vol. 10, 21 - 22 (1975).

105.109 Constraints on magnetic field which magnetized the Farmington meteorite parent body.
M. W. Rowe.
Meteoritics, Vol. 10, 23 - 30 (1975).

105.110 Gobabeb, a new chondrite: the coexistence of equilibrated silicates and unequilibrated spinels.
R. F. Fudali, A. F. Noonan.
Meteoritics, Vol. 10, 31 - 39 (1975).

Gobabeb, an ordinary chondrite, was found near Gobabeb, South West Africa in 1969. Chemically and petrographically it belongs in the H4 group. But, in addition to almost homogeneous silicates and chromites, it contains rare, non-opaque spinels that vary greatly in composition from grain to grain.

105.111 Mineralogy and chemistry of the Ashmore chondrite. W. B. Bryan, G. Kullerud.
Meteoritics, Vol. 10, 41 - 50 (1975).

The Ashmore olivine-bronzite chondrite is a group H,

type 5 stone which differs from other H5 chondrites mainly in its higher proportion of chromite (0.9 wt %) and in the relatively lower iron and higher magnesium content of the chromite.

105.112 The Clovis (no. 1), New Mexico, meteorite and Ca, Al and Ti-rich inclusions in ordinary chondrites.
A. F. Noonan.
Meteoritics, Vol. 10, 51 - 59 (1975).

105.113 The San Juan Capistrano meteorite.
R. C. Finkel, K. Marti, R. E. Jones.
Meteoritics, Vol. 10, 61 - 66, with a correction, p. 189 (1975).

The San Juan Capistrano chondrite fell on 15 March 1973; the total recovered mass was 56 g. Electron microprobe, chemical and petrographic studies show it to be a member of the H group and of petrologic type 6. Rare gas studies show that only minor radiogenic gas loss has occurred and yield a K-Ar age of 4.6×10^9 years and a Kr-Kr exposure age of 29×10^6 years.

105.114 Compositional study of the Lone Tree, Iowa, chondrite. G. R. McCormick, J. H. Carman.
Meteoritics, Vol. 10, 67 - 74 (1975).

The meteorite was found by Loren Westfall in May 1971. Electron microprobe and petrographic studies reveal its mineral composition to be olivine, low-calcium clinopyroxene, high-calcium clinopyroxene, troilite, kamacite, taenite and iron oxides.

105.115 A new investigation of the Uegit meteorite.
F. Burragato, V. Faccenda.
Meteoritics, Vol. 10, 75 - 84 (1975).

105.116 The Bhola stone – a true polymict breccia?
K. Fredriksson, A. Noonan, J. Nelen.
Meteoritics, Vol. 10, 87 - 88 (1975). – Abstract.

105.117 Al-Ca rich chondrules in the Coolidge chondrite.
J. Nelen, A. Noonan, K. Fredriksson.
Meteoritics, Vol. 10, 90 - 91 (1975). – Abstract.

105.118 Howardites: surface breccias of parent bodies.
T. E. Bunch.
Meteoritics, Vol. 10, 91 - 92 (1975). – Abstract.

105.119 The Nakhlites. T. E. Bunch, A. M. Reid.
Meteoritics, Vol. 10, 92 - 93 (1975). – Abstract.

105.120 Uranium and thorium microdistributions in meteorites. G. Crozaz, D. Burnett.
Meteoritics, Vol. 10, 93 (1975). – Abstract.

105.121 Recent results from studies of organic acids in the Murchison carbonaceous chondrite.
J. G. Lawless, B. Zeitman.
Meteoritics, Vol. 10, 95 (1975). – Abstract.

105.122 Concentrations of eight elements in the light-dark chondrite St. Mesmin.
W. Nichiporuk, C. B. Moore.
Meteoritics, Vol. 10, 95 - 96 (1975). – Abstract.

105.123 White and black inclusions and chondritic host material of the Allende meteorite – a comparative study on Li, Na, K, Al, Ca, Ti, Fe, and Ni contents.
W. Nichiporuk, C. B. Moore.
Meteoritics, Vol. 10, 96 (1975). – Abstract.

105.124 Possible formation of meteoritic chondrules and inclusions in the precollapse Jovian protoplanetary

atmosphere.   M. Podolak, A. G. W. Cameron.
Meteoritics, Vol. 10, 97 (1975). — Abstract.

105.125   Snow clot instead of black hole.
G. V. Korotkevich.
Priroda, No. 4.75, p. 105 (1975) In Russian.

105.126   Search for traces of heavy and superheavy elements in meteorites.
G. N. Flerov, O. Otgonsurehn, V. P. Perelygin.
Obedin. in-t yader. issled. Lab. yader. reakts. Preprint P7-8135. Dubna, 1974. 16 pp. In Russian. — Abstr. in Referativ. Zhurn. 51. Astron. 4.51.437 (1975).

105.127   A contribution to the stable carbon isotope geochemistry of iron meteorites.
P. Deines, F. E. Wickman.
Geochim. Cosmochim. Acta, Vol. 39, 547 - 557 (1975).

105.128   Elemental fractionations among enstatite chondrites.
P. A. Baedecker, J. T. Wasson.
Geochim. Cosmochim. Acta, Vol. 39, 735 - 765 (1975).

105.129   Distribution and significance of chromium in meteorites.   T. E. Bunch, E. Olsen.
Geochim. Cosmochim. Acta, Vol. 39, 911 - 927 (1975). Paper presented at the Geophysical Laboratory, Carnegie Institution of Washington; January 7, 1974.

Chromium is present as a minor element in all meteorite types. Depending on the meteorite type it is lithophile (most frequent), chalcophile (less frequent), or siderophile (rare). Chromium is an indicator of physical and chemical conditions of meteorite formation, especially of the state of oxidation. The Cr contents of meteoritic chromites are related to classification and can be used to distinguish among meteorite types that contain this mineral.

105.130   Offene Fragen zur Tektitenforschung.
W. Gentner, O. Müller.
Naturwissenschaften, 62. Jahrgang, p. 245 - 254 (1975).

Tektites are natural glasses which originated from impacts of large celestial bodies with the earth. After a historical introduction, the authors discuss the age grouping of tektite fields and their correlation with terrestrial craters, the surface structure of tektites, chemical and isotopic analyses and the significance of gas inclusions in bubbles. Unanswered questions in tektite research are given particular attention. The gigantic southeast Asian-Australian tektite field probably originated from a multiple-impact event 0.7 my ago.

105.131   Significance of calcium-rich differentiates in chondritic meteorites.
R. Hutchison, A. L. Graham.
Nature, Vol. 255, 471 (1975).

Knowledge of conditions in the early solar system is based largely on results from the study of meteorites. The authors have found in two chondritic meteorites 'chondrules' or inclusions containing material with the composition of a calcic plagioclase. They believe that this may be evidence of differentiation on a planetary body before compaction of the chondritic parent body(ies) occurred.

105.132   Search for superheavy elements in meteorites.
A. G. Popeko, N. K. Skobelev, G. M. Ter-Akopyan, G. N. Goncharov.
Obedin. in-t yader. issled. Lab. yader. reakts. Preprint R7-8042. Dubna, 1974. 18 pp. In Russian. — Abstr. in Referativ. Zhurn. 51. Astron., 5.51.361 (1975).

105.133   Meteoritic matter in the impactites of the Popigajan crater.   V. L. Masajtis, A. G. Sysoev.

Pis'ma v Astron. Zhurn., Vol. 1, No. 4, p. 43 - 47 (1975). In Russian.

Probably the source of nickel in impactites is the remnant of a cosmic body that formed the Popigaj crater.

105.134   History of a meteorite.
B. Yu. Levin, A. N. Simonenko.
Zemlya i Vselennaya, 1975, No. 3, p. 65 - 70. In Russian.

105.135   Meteorite "not connected with that" bolide.
V. I. Tsvetkov.
Zemlya i Vselennaya, 1975, No. 3, p. 81 - 84. In Russian.

105.136   Amino acids in carbonaceous chondrites.
J. G. Lawless, E. Peterson.
Origins of Life, Vol. 6, 3 - 8 (1975).

105.137   Possible evidence of ablation on cosmic dust particles.   C. L. Hemenway.
Evolutionary and physical properties of meteoroids. IAU colloquium No. 13, p. 255 - 257 = Dudley Obs., *Albany, New York,* Repr. No. B45 (1973).

105.138   Near-earth cosmic dust results from S-149.
C. L. Hemenway, D. S. Hallgren, C. D. Tackett.
AIAA paper No. 74-1226, presented at AIAA/AGU conference on scientific experiments of Skylab, Huntsville, Alabama, 1974, 7 pp. = Dudley Obs., *Albany, New York,* Repr. No. B49 (1975).

Craters and penetration holes have been found ranging from 135 micron diameter to less than 0.5 micron. A cosmic dust flux curve in the mass range from $10^{-16} - 10^{-7}$ grams is presented. Evidence is given concerning the directional characteristics of the particles and their breakup in near-earth space is presented.

105.139   Zoned Ca – Al – rich chondrule in Bali: new evidence against the primordial condensation model.
G. Kurat, G. Hoinkes, K. Fredriksson.
Earth Planet. Sci. Letters, Vol. 26, 140 - 144 (1975).

105.140   The chronology of the Nakhla achondritic meteorite.   N. H. Gale, J. W. Arden, R. Hutchison.
Earth Planet. Sci. Letters, Vol. 26, 195 - 206 (1975).

105.141   Measurement of the specific activity of radiocarbon in meteorites.
A. I. Ivliev, V. A. Alekseev, I. V. Smirnov.
Trudy 5-go Vses. soveshch. po probl. "Astrofiz. yavleniya i radiouglerod", 1973. Tbilisi, 1974, p. 139 - 146. In Russian. Abstr. in Referativ. Zhurn. 51. Astron., 6.51.416 (1975).

105.142   Helium, neon and argon isotopes in troilite and schreibersite inclusions of the Sikhote-Alin meteorite.   L. K. Levskij, A. N. Komarov.
Geokhimiya radiogen. i radioaktivn. izotopov. Leningrad, Nauka, 1974, p. 67 - 79. In Russian. — Abstr. in Referativ. Zhurn. 51. Astron., 6.51.417 (1975).

105.143   Regularities of element distribution in iron meteorites.   A. A. Yavnel'.
Geokhimiya, 1975, No. 2, p. 178 - 189. In Russian. — Abstr. in Referativ. Zhurn. 51. Astron., 6.51.421 (1975).

105.144   On the estimate of the pre-atmospheric size of the Mar'yalakhti meteorite.
V. P. Perelygin, Sh. B. Viik, O. Otgonsurehn.
Obedin. in-t yader. issled. Lab. yader. reakts. Preprint R13-8359. Dubna, 1974. 13 pp. In Russian. — Abstr. in Referativ. Zhurn. 51. Astron., 6.51.429 (1975).

105.145 Search for superheavy elements in meteorites (by spontaneous fission detection).
A. G. Popeko, N. K. Skobelev, G. M. Ter-Akopyan, G. N. Goncharov.
Phys. Letters B, (*Netherlands*), Vol. 52B, 417 - 420 (1974).

By using neutron multiplicity detectors, neutron emission from the samples of the Saratov, Efremovka and Allende meteorites has been measured. The possible relation of the activity observed to spontaneous fission of superheavy elements and anomalies in the Xe content of the meteorites is discussed.

105.146 Observations of the micrometeoroid flux from Prospero. D. K. Bedford.
Proc. Roy. Soc. London, Ser. A, Vol. 343, 277 - 287 (1975).

The micrometeoroid flux for near-earth space is determined by analysis of data gathered during a period of one year by the detector aboard Prospero.

105.147 Chemische Aspekte von meteoritischer und lunarer Materie. O. Müller.
Chemiker-Zeitung, (*Germany*), Vol. 98, 281 - 287 (1974).

Aperçu général des connaissances actuelles sur la nature chimique de la matière des météorites et de la lune. Classement des météorites en 4 classes principales et sous-groupes selon la nature des minéraux trouvés. Aspects chimiques des roches lunaires, leurs différences avec la matière terrestre.

105.148 Diffusion de ^{3}He dans les chondrites par le chauffage solaire. J. C. Lorin de la Grand-Maison.
Comptes Rendus Acad. Sci. Paris, Sér. D, Vol. 278, 1797 - 1800 (1974).

Les variations, observées dans les chondrites, du rapport des isotopes de spallation ^{3}He et ^{21}Ne peuvent être, pour l'essentiel, expliquées par l'existence d'un rapport de production de 10, et de pertes par diffusion de ^{3}He dues au chauffage solaire.

105.149 Composition chimique des verres de la météorite de Bununu. C. Desnoyers, D. Y. Jérome.
Comptes Rendus Acad. Sci. Paris, Sér. D, Vol. 278, 3275 - 3277 (1974).

105.150 Preuve du caractère autochtone des microsphères organiques fluorescentes des météorites carbonées.
Y. Benkheiri, B. Alpern.
Comptes Rendus Acad. Sci. Paris, Sér. D, Vol. 278, 3279 - 3281 (1974).

La disparition de la fluorescence des microsphères organiques dans la croûte des météorites carbonées est comparée aux effets thermiques obtenus en laboratoire. On obtient ainsi la preuve du caractère originel de ces microsphères.

105.151 Bitburg, a group-IB iron meteorite with silicate inclusions. E. Rambaldi, E. Jagoutz, J. T. Wasson.
Mineralog. Mag., (*GB*), No. 305, Vol. 39, 595 - 600 (1974).

Les analyses chimiques et les données minéralogiques indiquent que la météorite de Bitburg n'est pas une pallasite mais une ferrométéorite du groupe IB.

105.152 Détermination des silicates phylliteux des météorites carbonées par microscopie et microdiffraction électroniques. S. Caillere, M. Rautureau.
Comptes Rendus Acad. Sci. Paris, Sér. D, Vol. 279, 539 - 542 (1974).

105.153 Australites of mass greater than 100 grams from Western Australia. W. H. Cleverly.
Journ. Roy. Soc. Western Australia, Vol. 57, No. 3, p. 68 - 80 (1974).

105.154 Condensation time of the solar nebula from extinct

^{129}I in primitive meteorites.
R. S. Lewis, E. Anders.
Proc. National Acad. Sci. U.S.A., Vol. 72, 268 - 273 (1975).

Carbonaceous meteorites. See Abstr. 003.107.

Meteorites of the Caucasus and meteorite showers. See Abstr. 003.157.

The mystery of the cosmic boron abundance. See Abstr. 061.024.

Lunar microcraters: implications for the micrometeoroid complex. See Abstr. 094.103.

Impact cratering models and their application to lunar studies – a geologist's view. See Abstr. 094.140.

Lunar basins: tentative characterization of projectiles, from meteoritic elements in Apollo 17 boulders. See Abstr. 094.159.

Regolith history from cosmic-ray-produced nuclides. See Abstr. 094.169.

Mixing of the lunar regolith. See Abstr. 094.176.

Solar wind sputtering on the lunar surface: equilibrium crater densities related to past and present microparticle influx rates. See Abstr. 094.180.

The current micrometeoroid flux at the moon for masses $\leqslant 10^{-7}$ g from the Apollo window and Surveyor 3 TV camera results. See Abstr. 094.181.

Lunar microcrater studies, derived meteoroid fluxes, and comparison with satellite-borne experiments. See Abstr. 094.182.

Lunar microcraters and their solar flare track record. See Abstr. 094.185.

Charged-particle and micrometeorite impacts on the lunar surface. See Abstr. 094.186.

Solar wind and micrometeorite alteration of the lunar regolith. See Abstr. 094.189.

Tektites – volcanic ejecta from the moon? See Abstr. 094.230.

The oxidation state of europium as an indicator of oxygen fugacity. See Abstr. 094.401.

Electron microscopy of irradiation effects in space. See Abstr. 094.409.

Asteroids: spectral reflectance and color characteristics. See Abstr. 098.002.

The asteroids as meteorite parent-bodies. See Abstr. 098.004.

Asteroids: spectral reflectance and color characteristics. II. See Abstr. 098.011.

44 Nysa: an iron-depleted asteroid. See Abstr. 098.030.

Surface properties of asteroids: a synthesis of polarimetry, radiometry, and spectrophotometry.

See Abstr. 098.050.

Washougal: a stony meteorite with a retrograde orbit? See Abstr. 104.009.

A meteorite that got away. See Abstr. 104.030.

Temporal fluctuations and anisotropy of the micrometeoroid flux in the earth—moon system measured by HEOS 2. See Abstr. 104.032.

On the chemical composition of the sun, Jupiter, meteorites and Am stars. See Abstr. 114.083.

Catalytic reactions in the solar nebula: implications for interstellar molecules and organic compounds in meteorites. See Abstr. 131.137.

On the variation of the cosmic ray composition in the past. See Abstr. 143.027.

106 Interplanetary Matter, Interplanetary Magnetic Field, Zodiacal Light

106.001 Study of γ-radiation in circumterrestrial space.
E. P. Mazets.
Probl. sovrem. fiz. Leningrad, Nauka, 1974, p. 198 - 209. In Russian. − Abstr. in Referativ. Zhurn. 51. Astron., 2.51.419 (1975).

106.002 Motion of interplanetary dust particles in the circumterrestrial space and upper atmosphere.
V. N. Lebedinets, A. V. Manokhina, V. B. Shushkova.
Atmosf. optika. Moskva, Nauka, 1974, p. 214 - 219. In Russian. − Abstr. in Referativ. Zhurn. 51. Astron. 2.51.420 (1975).

106.003 Neutral gases of interstellar origin in interplanetary space.
P. W. Blum, J. Pfleiderer, C. Wulf-Mathies.
Planet. Space Sci., Vol. 23, 93 - 105 (1975).
The distribution of interplanetary neutral gases of interstellar origin is recalculated. Several improvements over earlier solutions of the problem are presented. This is especially true for the upwind and the downwind axes, which are of importance in the analysis of the observations of the back-scattered solar resonance radiation.

106.004 Counterglow from the earth-moon libration points.
J. R. Roach.
Planet. Space Sci., Vol. 23, 173 - 181 (1975). − Paper presented at the meeting of the Cosmic Dust Panel of COSPAR at Konstanz, Germany 1973.
Results from the OSO-6 Rutgers Zodiacal Light Analyzer experiment show photometric perturbations above the background in the anti-sun line of sight. Sixteen successive lunations were examined, and the accumulated perturbations show a maximum value in the direction of the L_4 and L_5 earth-moon libration points. This is interpreted as a counterglow from a cloud of particles at the libration points.

106.005 A source for hyperbolic cosmic dust particles.
H. A. Zook, O. E. Berg.
Planet. Space Sci., Vol. 23, 183 - 203 (1975). − Paper presented at the meeting of the Cosmic Dust Panel of COSPAR at Konstanz, Germany 1973.
Earlier analyses of the Pioneer 8 and 9 experimental meteoroid data have shown that the detectors on these two spacecraft are intercepting meteoroids with hyperbolic orbital parameters. It is shown in this paper that these results are entirely consistent with and, indeed, to be expected from other observations of the interplanetary meteoroid complex. Collisional breakup of meteoroids and post-collision radiation pressure modification of their orbits is found to be a sufficient cause for the observed results. Details of the calculations as well as of the results are presented.

106.006 Spatial and time variations of the interplanetary microparticle flux analysed from deep space probes Pioneers 8 and 9. J. A. M. McDonnell, O. E. Berg, F. F. Richardson.
Planet. Space Sci., Vol. 23, 205 - 214 (1975). − Paper presented at the meeting of the Cosmic Dust Panel of COSPAR at Konstanz, Germany 1973.

106.007 First results of the micrometeoroid experiment S 215 on the HEOS 2 satellite. H.-J. Hoffmann, H. Fechtig, E. Grün, J. Kissel.
Planet. Space Sci., Vol. 23, 215 - 224 (1975). − Paper presented at the meeting of the Cosmic Dust Panel of COSPAR at Konstanz, Germany 1973.
The comparison of both the 'particle rates' (number of detected particles per day) and the cumulative particle flux curves for the earth's apex, antiapex, ecliptic north and south directions shows that the rates vary only within a factor of 2, whereas the particle flux is extremely anisotropic. The flux for the apex direction at 10^{-12} g is 7×10^{-5} m^{-2} s^{-1} which is 1 to 2 orders of magnitude higher than the flux values for the other directions.

106.008 On measurements of interplanetary plasma density variations by radio interference methods.
I. P. Stakhanov.
Dokl. Akad. Nauk SSSR. Ser. Mat. Fiz., Vol. 220, 1310 - 1311 (1975). In Russian.

106.009 Neues vom interplanetaren Staub. H. W. Köhler.
Phys. Blätter, 31. Jahrg., 64 - 68 (1975).

106.010 Local interstellar medium. M. K. Wallis.
Nature, Vol. 254, 202 - 203 (1975). − Letter.

106.011 Dynamics of partially ionized gas in the gravitational field of the sun. E. Ya. Gidalevich.
Astron. Zhurn. Akad. Nauk SSSR, Vol. 52, 92 - 96 (1975). In Russian. English translation in Soviet Astron., Vol. 19, No. 1.
This paper is concerned with the dynamics of interstellar gas partially ionized by the ultraviolet radiation in the gravitational field of the sun. It is shown that the role of the gas ionization reduces to the fact that along with pressure of the neutral component additional partial pressure appears; this leads to acceleration and cooling of the gas stream at distances of about 10 a.u. from the sun. A possible role of this process in deceleration of the solar wind is assumed.

106.012 Particles and fields in the outer solar system.
R. E. Vogt, G. L. Siscoe.
Icarus, Vol. 24, 333 - 347 (1975).
The authors review here the relevant solar system physics, the astrophysics, and the plasma-physics questions associated with the media in the remote solar system, and they indicate what *in situ* measurements are needed to resolve remote sensing ambiguities, to verify or reject theoretical predictions, and to provide data not obtainable by existing methods.

106.013 In situ observations of the scale-size of plasma turbulence in the asteroid belt (1.6−3 astronomical units). D. S. Intriligator.
Astrophys. Journ., (*Letters*), Vol. 196, L87 - L90 (1975).
Pioneer 10 observations from the Ames Research Center Plasma Analyzer experiment between 1 and 3 a.u. in 1972 have been used to estimate the power spectra of the streaming speed of solar wind protons. A power-law spectrum is obtained in the $\sim 10^{-4}$ to $\sim 10^{-3}$ Hz frequency range which is similar to that obtained for the solar wind proton number density and streaming speed at 1 a.u. in 1965 December and 1966 January. The power spectra indicate that significant turbulence on the scale of $\sim 10^6$ km or more is present throughout this range of heliocentric distances, implying the importance of the role of large-scale turbulence between 1 and 3 a.u. The power spectra also present qualitatively information concerning the cosmic-ray diffusion tensor at these extended distances.

106.014 Alfvén waves and directional discontinuities in the

interplanetary medium.
J. W. Belcher, C. V. Solodyna.
Journ. Geophys. Res., Vol. 80, 181 - 186 (1975).
The properties of directional discontinuities occurring in selected periods of interplanetary magnetic field data are investigated. Microscale fluctuations in these selected periods have previously been identified as being predominantly Alfvénic on time scales ranging from 5 min to a few hours. The authors argue that the majority of directional discontinuities in such periods are rotational and outwardly propagating.

106.015 **Studies of cosmic dust.** E. P. Mazets.
Probl. sovrem. fiz. Leningrad, Nauka, 1974, p. 183 - 187. In Russian. − Abstr. in Referativ. Zhurn. 51. Astron., 2.51.418; 62. Issled. kosmich. prostranstva, 2.62.267 (1975).

106.016 **Direct observations of higher frequency density fluctuations in the interplanetary plasma.**
D. S. Intriligator.
Astrophys. Journ., Vol. 196, 879 - 882 (1975).
Direct observations from the Ames Research Center plasma spectrometers on Pioneer 6 at 1 a.u. in 1965 December have been used to obtain the power associated with fluctuations in the number density of solar-wind protons in the 10^{-3} to 10^{-2} Hz frequency range. A power-law spectrum is obtained in this frequency range. The extension of the power-law density spectrum based on direct observations to these higher frequencies is consistent with previous extrapolations of both spacecraft and interplanetary scintillation observations and with the dominance of large-scale turbulence in the solar wind. This result is also consistent with direct observations of the solar-wind proton speed and the interplanetary magnetic field.

106.017 **Zodiacal light photopolarimetry. I. Observations, reductions, disturbing phenomena, accuracy.**
R. Dumont, F. Sánchez.
Astron. Astrophys., Vol. 38, 397 - 403 (1975).
The authors review their observation and reduction procedures 1. to separate airglow continuum from extraterrestrial light, and 2. to minimize and evaluate the starlight component. The various contaminating sources and the r.m.s. errors that they may leave in the results, after corrections, are discussed. Special attention is drawn to disturbances in polarimetry by scattering in the low atmosphere.

106.018 **Zodiacal light photopolarimetry. II. Gradients along the ecliptic and the phase functions of interplanetary matter.** R. Dumont, F. Sánchez.
Astron. Astrophys., Vol. 38, 405 - 412 (1975).
The authors present and discuss the information of their photopolarimetric programme. Brightness and polarization as functions of elongation are given and interpreted in order to derive: a) the phase functions of an elementary volume of interplanetary space, depending upon the law r^{-n} assumed for the decrease of the spatial density; b) the polarization curve, weakly dependent upon the choice of the parameter n. A tentative explanation of the Gegenschein is founded on the similitude between the backscatter range of the authors' phase function, and the phase function of asteroids according to Gehrels and his coworkers.

106.019 **The size dependence of sublimation rates for interplanetary ice particles.**
H. Patashnick, G. Rupprecht.
Astrophys. Journ., (*Letters*), Vol. 197, L79 - L82 (1975).
The sublimation rates for water ice have been computed as a function of particle size for various solar distances. Because of the size dependence of the absorption and emission properties of the particles, a sublimation rate minimum evolves whose depth and position is sensitive to the spectral

absorption properties of the particle in combination with the spectral distribution of solar radiation. As a consequence, a quasi-stable size of interplanetary ice particles is predicted which is independent of solar distance.

106.020 **Collisional heating of interplanetary gas: Fokker− Planck treatment.** M. K. Wallis.
Planet. Space Sci., Vol. 23, 419 - 430 (1975).
Interstellar gas streaming through the solar system undergoes both elastic collisions with solar wind ions and destructive, ionizing processes. The Boltzmann equation is set up, with linear Fokker−Planck terms describing the glancing elastic collisions. Solutions combining the dynamical effects of the central force field and the diffusion in velocity space are derived, appropriate to cool gas.

106.021 **Interplanetary shock waves.** M. Dryer.
Space Sci. Rev., Vol. 17, 277 - 325 (1975). − Review paper − see 012.002.
Direct and indirect observations of interplanetary shock waves have been extended to the study of (1) the shock structure itself; (2) the disturbed solar wind in its wake; (3) additional discontinuities such as reverse shocks and pistons; and (4) the shock's kinematic behavior. Emphasis is presently being placed on numerical modeling of shock-induced disturbances in the solar wind as generated by both flares and stream-stream interactions. The former mechanism is emphasized in this review with several recommendations for further research: (a) further numerical modeling for shocks, starting when they are 'born' within relatively low-Alfvén speed coronal regions; (b) expanded synoptic studies by spacecraft at various heliocentric longitudes, radii, and (eventually) latitudes with coordinated diagnostics; and (c) extended patrol of natural probes, such as comets, augmented with theoretical studies of possible shock-induced mechanical and chemical effects.

106.022 **Interplanetary streams and their interaction with the earth.** L. F. Burlaga.
Space Sci. Rev., Vol. 17, 327 - 352 (1975). − Review paper − see 012.002.
Plasma and magnetic field observations of interplanetary streams near 1 AU are summarized. Two types of streams have been identified − corotating streams and flare-associated, and other flow patterns are present due to interaction among streams. The theory of corotating streams, which attributes them to a high temperature region near the sun, satisfactorily explains many of the effects observed at 1 AU. A correspondingly complete theory of flare-associated streams does not exist. Streams are a key link in the chain that connects solar and geomagnetic activity. The factors that most influence geomagnetic activity are probably related to streams and determined by the dynamics of streams. The evolution of streams on scales of 27 days and 11 years probably determines the corresponding variations of geomagnetic activity.

106.023 **The influence of the dispersion of the velocities of inhomogeneities on the time spectra of interplanetary scintillations.** I. V. Chashej.
Astron. Zhurn. Akad. Nauk SSSR, Vol. 52, 365 - 368 (1975). In Russian. English translation in Soviet Astron., Vol. 19, No. 2.
The problem of the influence of the velocity dispersion of inhomogeneities on the time spectrum of interplanetary scintillations is investigated.

106.024 **On the possibility of measurement of the parameters of large-scale inhomogeneities of the interplanetary plasma by the radio astronomical method.** V. I. Shishov.
Astron. Zhurn. Akad. Nauk SSSR, Vol. 52, 369 - 373 (1975). In Russian. English translation in Soviet Astron., Vol. 19, No. 2.
The possibility of the measurement of the refraction

effect on the large-scale inhomogeneities of the interplanetary plasma by terrestrial radioastronomical observations are considered. The method is based on the measurement of the scintillation picture displacement on two frequencies that is caused by the refraction effect. Estimates of the values of the refraction displacements for the mean electron density gradient and for the inhomogeneities with scale 10^9 cm are given.

106.025 Response of the plasma sheet at $\sim 18R_E$ to sudden southward turnings of the interplanetary magnetic field. A. T. Y. Lui, E. W. Hones, Jr., D. Venkatesan, S.-I. Akasofu, S. J. Bame.
Journ. Geophys. Res., Vol. 80, 929 - 935 (1975).
Simultaneous measurements of the interplanetary magnetic field (IMF) and of the magnetotail plasma are examined. Nineteen cases are found in which the IMF turns southward more than $40°$ ($|\Delta\theta| > 40°$) in less than 10 min and a Vela satellite, operating in a high time resolution mode, is in the plasma sheet. Twelve of these cases show a large decrease of plasma pressure, or the so-called plasma sheet thinning, during the subsequent period of southward IMF. Polar magnetic substorms occur at the earth in conjunction with these twelve plasma sheet thinnings. It is likely that the southward IMF alone does not reduce significantly plasma pressure in the plasma sheet and that a large decrease of plasma pressure is associated with a substorm.

106.026 Pioneer 11 meteoroid detection experiment: preliminary results.
D. H. Humes, J. M. Alvarez, W. H. Kinard, R. L. O'Neal.
Science, Vol. 188, 473 - 474 (1975).
The concentration of meteoroids of mass $\sim 10^{-8}$ gram in interplanetary space, in the asteroid belt, and near Jupiter has been measured.

106.027 Interplanetary scintillation observations at 34.3 MHz with the Cocoa-Cross radio telescope.
F. T. Erskine, W. M. Cronyn, S. D. Shawhan.
Bull. American Astron. Soc., Vol. 7, 238 (1975). – Abstr. AAS.

106.028 Nonlinear oblique interaction of interplanetary tangential discontinuities with magnetogasdynamic shocks. F. M. Neubauer.
Journ. Geophys. Res., Vol. 80, 1213 - 1222 (1975).

106.029 Numerical MHD simulation of interplanetary shock pairs. R. S. Steinolfson, M. Dryer, Y. Nakagawa.
Journ. Geophys. Res., Vol. 80, 1223 - 1231 (1975).
Solar wind disturbances produced by relatively long-lasting solar flares are simulated numerically by using the single-fluid magnetohydrodynamic (MHD) equations with negligible dissipation. The computations are confined to the ecliptic plane of a spherically symmetric flow and are begun when an initial disturbance is introduced near the sun in the ambient solar wind. A comparison with several MHD similarity solutions shows the advantages of the present analysis over the latter theory to be the following: (1) it yields a quantitative prediction of the azimuthally induced plasma flow, (2) it provides for removal of the nonphysical zero pressure at the contact surface, and (3) it allows more flexibility in the specification of initial conditions, which gives the present analysis greater utility in predicting plasma and magnetic data.

106.030 Heliographic latitude dependence of the IMF dominant polarity in 1972–1973 using Pioneer 10 data. R. L. Rosenberg.
Journ. Geophys. Res., Vol. 80, 1339 - 1340 (1975).
The heliographic latitude dependence of the interplanetary magnetic field (IMF) was studied by using Pioneer 10 data taken from March 1972 through June 1973 over Bartels

solar rotation (SR) periods 1896–1913. The results are reported and discussed.

106.031 Some properties of the Svalgaard A/C index.
C. T. Russell, R. K. Burton, R. L. McPherron.
Journ. Geophys. Res., Vol. 80, 1349 - 1351 (1975).
Brief report.

106.032 Comment on 'Interplanetary magnetic sector structure, 1926–1971' by L. Svalgaard and 'Correspondence of solar field sector direction and polar cap geomagnetic field changes for 1965' by W. H. Campbell and S. Matsushita.
P. F. Fougere, C. T. Russell.
Journ. Geophys. Res., Vol. 80, 1376 - 1377, with a reply of W. H. Campbell and S. Matsushita, p. 1378 (1975).

106.033 Dynamic modeling of interplanetary disturbances produced by solar flares.
R. S. Steinolfson, M. Dryer, Z. Smith, Y. Nakagawa.
Bull. American Astron. Soc., Vol. 7, 358 (1975). – Abstr. AAS.

106.034 Zodiakallichtbeobachtungen mit dem Ballonteleskop THISBE. W. Hofmann.
Umschau, 75. Jahrgang, p. 340 - 342 (1975).
The zodiacal light was measured by the balloon-borne telescope THISBE in the wavelength region between 0.35 μm and 2.4 μm. There, the intensity of the zodiacal light shows no significant deviation from the relative solar spectrum. This indicates a rather large particle size of more than one micron. Polarization measurements show that the particles are composed of dielectric material, maybe of quartz. The particle temperature determined from infrared measurements is less than $340°$K near the earth orbit. Probably the particles are produced in the asteroid belt.

106.035 Orbits of neutral atoms entering the heliosphere.
S. Grzędzielski, G. Sitarski.
Acta Astron., Vol. 25, 169 - 175 (1975).
Trajectories of neutral hydrogen atoms of interstellar origin entering the heliosphere were computed for a velocity-dependent radiation pressure force resulting from a double-peaked (self-reversed) solar Lyman-alpha emission line.

106.036 On the connection of the sectorial structure of the interplanetary magnetic field with the indices of zonal circulation. R. N. Kulieva.
Geomagn. Aeronom., Vol. 15, 341 - 343 (1975). In Russian.
Brief information.

106.037 Mariner IV: In-depth analysis of the cosmic dust experiment measurement over a heliocentric range of 1.0–1.6 AU. W. M. Alexander.
Diss. Nat. Gesamtfakultät, Ruprecht-Karl-Univ., Heidelberg. 4 + 117 pp. (1975).
Many techniques have been used to obtain knowledge about the properties of this zodiacal dust cloud. The major ones involve ground-based visual, photographic and radar meteor observations; photometric observations of the solar corona and the zodiacal light; and various types of collection techniques and resulting laboratory analyses. For particle masses greater than a microgram, determinations of vector velocity, mass, density and composition have been obtained from meteor measurements involving collisions of meteoroids with the atmosphere of the earth. In addition, composition, age and gross accretion rates have been studied in laboratory analyses of the remnants of meteoroidal and dust particle material which have survived passage through the atmosphere. For particles with masses less than a microgram, the zodiacal light observations and studies represent the only ground-based technique which has been used to study properties of particles of this size regime in interplanetary space. The main purpose

of this paper is to present the results of an extended in-depth study of the data from the Mariner IV experiment. The sections of the paper present the historical material and data necessary for the study, followed by the results and conclusions derived from the analysis.

106.038 Deceleration of flare-generated interplanetary shock waves.   Š. Pintér.
Bull. Astron. Inst. Czechoslovakia, Vol. 26, 169 - 180 (1975).

This paper deals with the deceleration of flare-generated interplanetary shock waves, drawing on actual observations of these waves in interplanetary space made by probes. For the purposes of the study the author mostly chose such phenomena which were associated with type II radio bursts, in order to be able to compare coronal velocities, calculated from the frequency drift of these bursts, with the initial velocities, derived from the propagation of interplanetary shock waves to 0.75−1.5 AU.

106.039 Influence of the sectorial structure of the interplanetary magnetic field on the circulation of the earth's atmosphere.   R. N. Kulieva.
Geomagn. Aeronom., Vol. 15, 546 - 547 (1975). In Russian. Brief information.

106.040 Interplanetary shock pair disturbances: comparison of theory with space probe data.
R. S. Steinolfson, M. Dryer, Y. Nakagawa.
Journ. Geophys. Res., Vol. 80, 1989 - 2000 (1975).

Three sets of data for shock pair disturbances, which are assumed to have been formed from stream-stream interactions, are modeled by using two time-dependent numerical codes. One code solves the fluid dynamic equations, and the other solves the magnetohydrodynamic equations. The large-scale structure in the disturbance is reproduced if (1) the magnetic field and azimuthal velocity are included (that is, if the code which solves the magnetohydrodynamic equations is used), (2) the time dependence of the initial interaction is appropriately selected, and (3) the observed nonradial shock normals are accounted for in the modeling procedure.

106.041 Interplanetary shock waves and comet brightness fluctuations during June−August 1972.
M. Dryer, A. Eviatar, A. Frohlich, A. Jacobs, J. H. Joseph, E. J. Weber.
Journ. Geophys. Res., Vol. 80, 2001 - 2012 (1975).

A series of anomalous events associated with solar activity occurred in interplanetary space in June and August 1972. These events include non-Io-associated decametric emission from Jupiter; brightness variations of the periodic comets Schwassmann-Wachmann i and Giacobini-Zinner; interplanetary shock waves detected by Pioneer 9 and 10, Heos 2, and Prognoz and Prognoz 2; interplanetary scintillations; and sudden commencements of geomagnetic storms. Trajectories and deceleration characteristics for the shock waves are estimated primarily from observations using solar radio, in situ, and geomagnetic data.

106.042 Collisionless shock waves in space: a very high β structure.   V. Formisano, C. T. Russell, J. D.
Means, E. W. Greenstadt, F. L. Scarf, M. Neugebauer.
Journ. Geophys. Res., Vol. 80, 2013 - 2022 (1975).

106.043 Causes of predecreases and Forbush-decreases in the light of first-hand data on the structure of interplanetary plasma streams.   N. V. Mikerina, K. G. Ivanov.
Kosmich. Issled., Vol. 13, 433 - 436 (1975). In Russian. Brief information.

106.044 Interplanetary magnetic field polarity and low-latitude geomagnetic field.

B. N. Bhargava, G. K. Rangarajan.
Planet. Space Sci., Vol. 23, 929 - 933 (1975).

The relationship between the low latitude magnetic field and the polarity of the interplanetary magnetic field, inferred by Svalgaard for a period of about 4 solar cycles, is derived and discussed for low and high solar activity conditions.

106.045 · Collections of cosmic dust.   C. L. Hemenway.
Dudley Obs., *Albany, New York,* Repr. No. B47, 31 pp. (1974).

Some recent results of collection experiments in the upper atmosphere by rocket techniques, by balloon collections in the intermediate atmosphere, and by recoverable satellites in near-earth space are described. Evidence is presented for the existence of relatively high and variable fluxes of submicron particles entering the atmosphere.

106.046 Astronomical photography. Part A: Zodiacal light photography.
R. D. Mercer, L. Dunkelman, R. E. Evans.
Apollo 17 preliminary science report, p. 34-1 - 34-4 = Dudley Obs., *Albany, New York,* Repr. No. C81 (1974).

Polarized and red- and blue-filter photographs of the zodiacal light were obtained for the first time on the Apollo 17 mission, allowing interpretations of zodiacal dust compositions.

106.047 On the diagnostics of the azimuthal component of the interplanetary magnetic field based on ground observations.   Ya. I. Fel'dshtejn, P. V. Sumaruk, N. F.
Shevnina.
Dokl. Akad. Nauk SSSR. Ser. Mat. Fiz., Vol. 222, 833 - 836 (1975). In Russian.

106.048 A model for the propagation of flare-associated interplanetary shock waves.   J. K. Chao.
Solar wind three, (see 012.020), p. 169 - 174 (1974).

106.049 Numerical simulation of interplanetary shock ensembles.   R. S. Steinolfson, M. Dryer,
Y. Nakagawa.
Solar wind three, (see 012.020), p. 175 - 178 (1974).

106.050 Suprathermal particles.   R. P. Lin.
Solar wind three, (see 012.020), p. 187 - 194 (1974).

Recent observations of suprathermal particles in the interplanetary medium in the energy range from solar wind (≥1 keV) to low energy solar cosmic rays (∼1 MeV) are reviewed.

106.051 Radial gradients in the interplanetary magnetic field between 1.0 and 4.3 AU: Pioneer 10.
E. J. Smith.
Solar wind three, (see 012.020), p. 257 - 280 (1974).

106.052 Mode decay and gradients in the interplanetary magnetic field fluctuation spectrum.
R. H. Cohen.
Solar wind three, (see 012.020), p. 291 - 293 (1974).

106.053 Interplanetary scintillations.
W. A. Coles, B. J. Rickett, V. H. Rumsey.
Solar wind three, (see 012.020), p. 351 - 367 (1974).

Terrestrische und extraterrestrische Plasmen.
See Abstr. 062.040.

On the theory of large amplitude Alfvén waves.
See Abstr. 062.060.

Solar activity and the structure of interplanetary matter. See Abstr. 072.066.

Waves and instabilities in the solar wind. See Abstr. 074.012.

Solution of three-fluid model equations with anomalous transport coefficients for the quiet solar wind. See Abstr. 074.013.

Interplanetary gas. XX. Does the radial solar wind speed increase with latitude? See Abstr. 074.016.

Large-scale and solar-cycle variations of the solar wind. See Abstr. 074.029.

Microstructure of the solar wind. See Abstr. 074.030.

On the nature of irregularities in the solar wind corotating high-speed streams. See Abstr. 074.034.

The effect of asymmetric solar wind on the Lyman α sky background. See Abstr. 074.036.

Multispacecraft study of the solar wind velocity at interplanetary sector boundaries. See Abstr. 074.038.

Instabilities connected with neutral sheets in the solar wind. See Abstr. 074.130.

Cosmic ray propagation in the solar wind. See Abstr. 074.133.

The 1964–1972 quiet-time spectra of protons and helium at 2–20 MeV per nucleon. See Abstr. 078.007.

Anisotropy of solar protons and sectorial structure of the interplanetary field. See Abstr. 078.017.

Interplanetary acceleration of low-energy solar protons: a study of the solar particle event of November 18, 1968. See Abstr. 078.022.

Scattering in the earth's atmosphere: calculations for Milky Way and zodiacal light as extended sources. See Abstr. 082.052.

Interplanetary magnetic field direction and the configuration of the day side magnetosphere. See Abstr. 084.216.

Substorm and interplanetary magnetic field effects on the geomagnetic tail lobes. See Abstr. 084.221.

Interplanetary field effect on the magnetosphere. See Abstr. 084.235.

Polar cap currents for different directions of the interplanetary magnetic field in the $Y-Z$ plane. See Abstr. 084.244.

Relation of variations in total magnetic field at high latitude with the parameters of the interplanetary magnetic field and with $DP\,2$ fluctuations. See Abstr. 084.245.

Anisotropic plasma instabilities in the magnetosphere and interplanetary medium. See Abstr. 084.247.

Differential rotation of the magnetospheric plasma as cause of the Svalgaard-Mansurov effect. See Abstr. 084.264.

Access of solar electrons to the polar regions. See Abstr. 084.265.

Observations of low-energy electrons upstream of the earth's bow shock. See Abstr. 085.004.

The detection of Jovian high energy electrons in interplanetary space $\gtrsim$ 1 A.U. from the planet. See Abstr. 099.097.

Outbursts of comet Schwassmann-Wachmann 1, and the cloud of interplanetary boulders. See Abstr. 103.109.

Chondritic abundances for interplanetary dust: evidence from micrometeorite craters. See Abstr. 105.024.

Catalytic reactions in the solar nebula: implications for interstellar molecules and organic compounds in meteorites. See Abstr. 131.137.

Possible interaction of interstellar particles with the solar and terrestrial environment. See Abstr. 131.140.

Cosmic-ray streaming and anisotropies. See Abstr. 143.002.

The influence of the sectorial structure of the interplanetary magnetic field on the modulation of cosmic rays. See Abstr. 143.008.

Causes of Forbush decreases and other cosmic ray variations. See Abstr. 143.020.

Semidiurnal component of cosmic ray intensity. See Abstr. 143.022.

Intensity of cosmic rays in the interplanetary space. Observational data. Jan. 5 - May 15, 1969. See Abstr. 143.029.

Intensity of cosmic rays in the interplanetary space. Observational data. July 19, 1965 - Jan. 25, 1966. See Abstr. 143.030.

Solar modulation of galactic cosmic ray electrons, protons, and alphas. See Abstr. 143.042.

Cosmic ray intensity variations during 0200–0700 UT, August 5, 1972. See Abstr. 143.043.

Cosmic ray anisotropy in the interplanetary space. See Abstr. 143.078.

Two views of cosmic ray propagation in the solar system. See Abstr. 143.095.

107 Cosmogony of the Planetary System

107.001 **The effect of C/O ratio on the condensation of planetary material.** J. W. Larimer.
Geochim. Cosmochim. Acta, Vol. 39, 389 - 392 (1975).

The condensation temperatures of refractory silicates and oxides in a gas of cosmic composition are strongly dependent on the C/O ratio. As the ratio increases from 0.4 to 0.9 (solar ~ cosmic ~ 0.6), condensation temperatures of compounds such as Al_2O_3, $Ca_2Al_2SiO_7$, $MgAl_2O_4$, Mg_2SiO_4 and $MgSiO_3$ decrease by $50-100°$. As C/O increases from 0.9 to 1.0, these temperatures drop an additional $300-400°$. Other chemical differences result when C/O $\gtrsim 0.9$ include: a new suite of high temperature minerals appears (graphite, CaS, Fe_3C, SiC and TiN); the reaction $CO + 3H_2 \rightarrow CH_4 + H_2O$ proceeds to the right at higher temperatures; and iron, whose condensation temperature is unaffected, condenses at higher temperatures than any silicate or oxide.

107.002 **Collisional breakup of particles in a planetary ring.** A. W. Harris.
Icarus, Vol. 24, 190 - 192 (1975).

Jeffreys (1947) estimated the size of fragments resulting from breakup of a satellite inside the Roche limit, obtaining a result of ~ 100 km. This result does not allow for the further breakup of the fragments due to collisions among themselves, which should reduce the maximum size to $\lesssim 3$ km for rock, or $\lesssim 1$ km for ice. This result affects not only Jeffreys' speculations as to the origin of Saturn's rings, but also recent speculations on the origin of the moon by capture and the possible tidal destruction of satellites of Mercury or Venus.

107.003 **Grains accretion processes in a protoplanetary nebula: conducting and insulating materials.**
A. Carusi, A. Coradini, C. Federico, M. Fulchignoni, G. Magni.
Astrophys. Space Sci., Vol. 33, 369 - 384 (1975).

In a previous work the authors estimated cross-sections for constructive and destructive collisions ('constructive' and 'destructive' cross-sections) related to silica grains embedded in a protoplanetary nebula. In this paper the interaction processes among conducting grains (iron, graphite) and among the insulating (silica) ones have been considered. The following results have been obtained: (1) insulating grains have smaller constructive cross-sections than conducting ones; and (2) conducting grains show very large cross-sections for masses up to about 10^{-9} g.

107.004 **Early scattering by Jupiter and its collision effects in the inner solar system.**
W. M. Kaula, P. E. Bigeleisen.
Bull. American Astron. Soc., Vol. 7, 385 (1975). – Abstr. AAS.

107.005 **Collisional breakup of particles in a planetary ring.** A. W. Harris.
Bull. American Astron. Soc., Vol. 7, 385 (1975). – Abstr. AAS.

107.006 **Cosmogonical problems in W. Kaula's compendium.** S. K. Vsekhsvyatskij.
Problems of cosmic physics. Vyp. (No.) 9, (see 003.013), p. 165 - 173 (1974). In Russian.

Review and analysis of the Russian translation of W. Kaula "An introduction to planetary physics; the terrestrial planets". The various cosmogonical hypotheses which are the basis of this book are seriously critizised and their erroneousness is discussed.

107.007 **Satellite-sized planetesimals and lunar origin.** W. K. Hartmann, D. R. Davis.
Icarus, Vol. 24, 504 - 515 (1975). – Presented at IAU Collo-quium No. 28, Cornell Univ., Ithaca, New York, 1974 August 18 - 21.

Exploratory calculations using accretionary theory are made to demonstrate plausible sizes of second-largest, third-largest, etc., bodies at the close of planet formation in helio-centric orbits near the planets, assuming asteroid-like size distributions at the start of the calculation. Many satellite-sized bodies are found to be available for capture, cratering, or collisional fragmentation. Collision of a large body with earth could eject iron-deficient crust and upper mantle material, forming a cloud of refractory, volatile-poor dust that could form the moon.

107.008 **Fractionation in the solar nebula: condensation of yttrium and the rare earth elements.**
W. V. Boynton.
Geochim. Cosmochim. Acta, Vol. 39, 569 - 584 (1975).

The condensation of Y and the rare earth elements (REE) from the solar nebula may be controlled by thermodynamic equilibrium between gas and condensed solids. Highly fractionated REE patterns may result if condensates are removed from the gas before condensation is complete. It is found that the fractionation is not a smooth function of REE ionic radius but varies in an extremely irregular pattern.

107.009 **Korrelierte Elemente. Ihre Bedeutung für die Erforschung der Fraktionierungsprozesse vor und nach der Bildung planetarer Körper.** H. Wänke.
Naturwissenschaften, 62. Jahrgang, p. 264 - 271 (1975).

The analysis of lunar samples revealed correlations for a number of pairs of elements. It can be shown that in some cases the observed constant element ratios must apply to the whole moon. The ratios of the correlated elements in terrestrial samples deviate considerably from the relevant lunar values whenever two elements of highly different volatility are involved. All refractory elements are more abundant on the moon. In general the correlated elements may become the key to unravel the bulk chemistry of the inner planets.

107.010 **Early scattering by Jupiter and its collision effects in the terrestrial zone.**
W. M. Kaula, P. E. Bigeleisen.
Icarus, Vol. 25, 18 - 33 (1975).

When Jupiter was on the order of three to ten earth masses in size, there undoubtedly was a considerably larger mass of condensed matter in its zone, since Jupiter would have perturbed most of it to other parts of the solar system. Monte Carlo studies indicate a significant portion would have crossed the earth's orbit. If the earth and moon had not yet fully formed, the probability of earth-zone planetesimals being hit by this Jupiter-scattered material was high.

107.011 **The Titius-Bode law and the possibility of recent large-scale evolution in the solar system.**
M. M. Nieto.
Icarus, Vol. 25, 171 - 174 (1975).

Although it is by no means clear that the Titius-Bode law of planetary distances is indeed a "law" (even though there are enticing indications), it is proposed that if one assumes that the law is a "law" and that the planets obey it, then this argues against recent large-scale evolution in the solar system.

107.012 **A co-accretional model of satellite formation.** A. W. Harris, W. M. Kaula.
Icarus, Vol. 24, 516 - 524 (1975). – Presented at IAU Colloquium No. 28, Cornell Univ., Ithaca, New York, 1974 August 18 - 21.

Numerical calculations of a simple accretion model including the effects of tidal friction indicate that coformation is tenable only if the planet's Q is less than about 10^3. The parameter which most strongly affects the final mass ratio of the pair is the time at which the secondary embryo is introduced. The model yields the proper moon—earth mass ratio if the moon embryo is introduced when the earth is only about 1/10 of its final mass. The lunar orbit remains at about 10 earth radii throughout most of the growth.

107.013   Evidence of a primordial solar wind.
            C. P. Sonett.
Solar wind three, (see 012.020), p. 36 - 57 (1974).

The several heat sources for parent body progenitors of meteorites are reviewed.

107.014   Dynamical contraction of infinite plane-symmetric
            gas clouds.   T. Kusaka.
Progr. Theor. Phys. Japan, Vol. 52, 147 - 160 (1974).

Die Kosmogonie Immanuel Kants. II.
See Abstr. 004.014.

Close encounters of small bodies and planets.
See Abstr. 042.019.

Hypothesis on the resonance structure of the solar system.   See Abstr. 091.039.

Surface properties of asteroids: a synthesis of polarimetry, radiometry, and spectrophotometry.
See Abstr. 098.050.

Cosmogonical considerations regarding Uranus.
See Abstr. 101.003.

Comets and interstellar masers.
See Abstr. 102.008.

The most primitive objects in the solar system.
See Abstr. 105.006.

Possible formation of meteoritic chondrules and inclusions in the precollapse Jovian protoplanetary atmosphere.   See Abstr. 105.124.

Significance of calcium-rich differentiates in chondritic meteorites.   See Abstr. 105.131.

Condensation time of the solar nebula from extinct ^{129}**I in primitive meteorites.**   See Abstr. 105.154.

Clumping of interstellar grains during formation of the primitive solar nebula.   See Abstr. 131.002.

Stars

111 Stellar Parallaxes

111.001 **Parallaxes of luminous M, S and carbon stars.**
A. R. Upgren.
Bull. American Astron. Soc., Vol. 7, 240 (1975). – Abstr. AAS.

111.002 **The application of parallaxes and photometry to the lower main sequence.** A. R. Upgren.
Dudley Obs. Rep. No. 9, (see 012.008), p. 437 - 451 (1975).

The results and applications of parallaxes of more than 100 nearby red dwarf stars are described. Combined with BVRI photometry of the same stars and of similar members of the Hyades cluster, the parallaxes indicate a distance modulus of 3.22 ± 0.04 mag. for the cluster. No evidence is found to support a recent contention that overblanketed stars describe a broader sequence than normal stars and thus might be responsible for the greater dispersion of the field stars. The advantage of future parallaxes for subdwarfs is briefly discussed.

111.003 **On the method of statistical parallaxes.**
A. Heck, J. Jung.
Astron. Astrophys., Vol. 40, 323 - 326 (1975).

The way in which the principle of maximum likelihood has been applied by Heck (1972, 1973) in order to improve the luminosity estimates of RR Lyrae stars has recently been criticized by Clube and Jones (1974). The authors show that for the sample of RR Lyrae variables studied by Clube and Jones (1971), and by Heck (1972), arguments against their technique put forward by Clube and Jones and concerning the treatment of the random errors affecting the proper motions cannot account for the discrepancy between the published luminosity estimates.

A new nearby subdwarf M star.
See Abstr. 126.010.

112 Proper Motions, Radial Velocities, Space Motions

112.001 **Wie weit nähern sich die Nachbarn der Sonne?**
M. Hoffmann.
SuW, Vol. 14, 27 (1975).

112.002 **Laboratory exercises in astronomy – Proper motion**
O. Gingerich.
Sky Telescope, Vol. 49, 96 - 98 (1975).

112.003 **Photoelectric radial velocities, paper VI. Heard's IAU standard stars.** R. F. Griffin.
Monthly Notices Roy. Astron. Soc., Vol. 171, 407 - 414 (1975).

Fourteen of the 24 'ninth-magnitude' radial-velocity standards proposed by Heard have been observed. One of them is shown to vary in velocity, in addition to two already rejected by Heard for the same reason. The photoelectric observations give accidental errors substantially smaller than those of the published standard velocities. Criteria appropriate to the selection and adoption of future standard stars are suggested.

112.004 **Absolute proper motions of B stars.**
D. K. Karimova, E. D. Pavlovskaya, M. S. Toropova.
Trudy Gos. Astron. Inst. Shternberga, Vol. 45, 87 - 94 (1974). In Russian.

New proper motions of B stars are given. The proper motions are in the FK4 system and given for the equinox 1950.0.

112.005 **Absolute proper motions of 68 stars.**
D. K. Karimova, E. D. Pavlovskaya, M. S. Toropova.

Soobshch. Gos. Astron. Inst. Shternberga, No. 188, p. 42 - 46 (1974). In Russian.

112.006 **Some thoughts on proper motion.** R. H. Stoy.
Monthly Notes Astron. Soc. Southern Africa, Vol. 34, 38 - 44 (1975).

112.007 **Galactic objects with the largest known radial velocities.** A. D. Thackeray.
Observatory, Vol. 95, 100 - 104 (1975).

A list of galactic objects with radial velocities greater than 250 km/s relative to the sun has been compiled.

112.008 **Lowell Proper Motions XVII. Proper Motion Survey in the southern hemisphere with the 13-inch photographic telescope of the Lowell Observatory.**
H. L. Giclas, R. Burnham, Jr., N. G. Thomas.
Lowell Obs. Bull., *Flagstaff, Arizona,* No. 162, Vol. 8, (No. 2), 9 - 49 (1975).

This survey contains data on 915 stars; 360 in the regular catalog which have a proper motion > 0".19/year, 386 very blue stars with smaller or no motion and 169 very red stars with similar small motions. These two regions contain very few stars with large proper motion; only five with motion > 1".0/year, all of which were previously identified. There are six stars of −1 color in the regular catalog which, from the work of Eggen and Greenstein, have a very high probability of being white dwarfs. In the lists of objects with small motion there are over 100 stars with −1 color that may be added to the lists of new white dwarfs. In the catalog list of motions > 0".20/year there are 48 very red, extreme dwarf stars, and a supplementary list with small or no motion containing 169 such objects. There are four newly recognized moving pairs in the catalog list, and seven interesting close pairs of contrasting color with very small motion added as an appendix.

112.009 Proper motions of variable stars in the Sydney Astrographic Zone. K. P. Sims.
Journ. Proc. Roy. Soc. New South Wales, Vol. 107, 49 - 66 = Sydney Obs. Papers, No. 71 (1974/75).

The relative proper motions of 30 stars determined photographically are given together with absolute measures found by applying corrections for the parallactic motion of the reference stars and for the effects of differential galactic rotation.

112.010 Proper Motion Survey with the forty-eight inch Schmidt telescope. XXXVIII. Binaries with white-dwarf components. W. J. Luyten.
Separate print Univ. Minnesota, Minneapolis, Minnesota, p. 1 - 11 (1974).

In No. XVIII of this series (1969) the author published a list of 125 binaries with white-dwarf components. In this paper the data for 320 pairs are shown in tables.

112.011 On the systematic corrections to the Lowell Proper Motion Survey. W. J. Luyten.
Separate print Univ. Minnesota, Minneapolis, Minnesota, p. 16 - 18 (1974).

112.012 Proper Motion Survey with the 48-inch Schmidt telescope. XXXIX. On the Lick proper motions in the Hyades region. W. J. Luyten.
Separate print Univ. Minnesota, Minneapolis, Minnesota, 4 pp. (1975).

112.013 Radial velocities. M. Barbier.
Centre de Données Stellaires, Inform. Bull. No. 8, (see 002.007), p. 12 (1975).

112.014 Proper motions of stars of the Grachev catalogue in declination. Eh. N. Vorob'eva.
19th Astrometrical Conference 1972, (see 012.019), p. 165 - 166 (1975). In Russian.

112.015 On absolutization of photographic proper motions of stars. A. A. Kiselev, L. I. Yagudin.
19th Astrometrical Conference 1972, (see 012.019), p. 201 - 214 (1975). In Russian.

112.016 Some results of photographic determinations of proper motions of stars relative to galaxies in Pulkovo. A. N. Deutsch, V. V. Lavdovskij, O. N. Orlova.
19th Astrometrical Conference 1972, (see 012.019), p. 214 - 217 (1975). In Russian.

Precise measurements of radial velocity using a Lirepho microphotometer. See Abstr. 031.205.

Programmes for the reduction of radial velocity measurements. See Abstr. 031.412.

Catalogue of differences of proper motions in declination of 34 Washington zenith stars. See Abstr. 041.018.

Herleitung und Erprobung eines erweiterten Verfahrens zur Bestimmung eines Instrumentalsystems von Örtern und Eigenbewegungen von Sternen. See Abstr. 041.027.

Comparison of different methods for investigation of the differences of positions and proper motions of stars. See Abstr. 041.046.

Results of photographing galaxies and first results of determination of absolute proper motions of stars. See Abstr. 041.068.

A catalogue of galactic O stars and the ionization of the low density interstellar medium by runaway stars. See Abstr. 113.008.

Spectral types for proper motion stars. See Abstr. 114.026.

Spectroscopic observations of stars in H II regions. See Abstr. 114.086.

Proper motions of RR Lyrae stars. See Abstr. 122.022.

Absolute magnitudes and motions of RR Lyrae stars. See Abstr. 122.023.

A new nearby subdwarf M star. See Abstr. 126.010.

On the ionization of the intercloud medium by runaway O−B stars. See Abstr. 131.116.

Note on the blue stragglers in NGC 7789. See Abstr. 153.031.

Kinematics of stars selected from box orbit parameters. See Abstr. 155.013.

Structure and age of the local association (Pleiades group). See Abstr. 155.025.

Observations for improving stellar kinematics. See Abstr. 155.046.

Radial velocities of supergiants in the Small Magellanic Cloud. See Abstr. 159.015.

113 Stellar Magnitudes, Colors, Photometry

113.001 Six UBV photoelectric sequences in Vela (l = 257° to 281°). J. Denoyelle.
Astron. Astrophys., Suppl. Ser., Vol. 19, 45 - 55 (1975).

For six sequences along the southern galactic equator, UBV values and an identification chart are presented. Special attention was given to the fields around the galactic longitude 275°, where a possible link between the Carina and the Vela spiral features might be detected. The sequences will be used for photographic photometry on ESO-Schmidt plates, which cover the whole Vela region from 257° to 284°.

113.002 An extreme-ultraviolet search of the north galactic polar region. P. Henry, R. Cruddace, F. Paresce, S. Bowyer, M. Lampton.
Astrophys. Journ., Vol. 195, 107 - 110 (1975).

An area of approximately 1350 square degrees around the north galactic pole has been searched for sources radiating at extreme-ultraviolet wavelengths. Discrete sources within this region were not detected at fluxes above the level set by the instrument sensitivity, 4.3×10^{-8} ergs cm^{-2} s^{-1}, in the 135–475 Å band.

113.003 Photoelectric observations of light and colour variations of the emission star BD + 38° 4062.
L. N. Boldenkova, T. I. Kuznetsova, R. M. Raznik.
Uch. zap. Ul'yanovsk. gos. ped. in-t, Vol. 27, No. 8, p. 151 - 158 (1974). In Russian. – Abstr. in Referativ. Zhurn. 51. Astron., 2.51.606 (1975).

113.004 Stellar flux measurements of early-type stars in the 912–1075 Å band.
B. E. Troy, Jr., C. Y. Johnson, J. M. Young, J. C. Holmes.
Astrophys. Journ., Vol. 195, 643 - 648 (1975).

Stellar flux measurements in the wavelength range 912–1075 Å ($\lambda_0 = 950$ Å) are presented for eight OB type stars. The ultraviolet color index $(m_{950} - V)_c$ derived from the flux values varies from approximately −6 to −4 between spectral classes O8 and B2. Radiation from B3 and later stars was below the detectable limit. The variation of the 950 Å color index with spectral type is similar to, but steeper than, that of the 1115 Å and 1376 Å indices, and more negative in value. The derived color indices are consistently more negative than predictions from model atmosphere calculations.

113.005 Hβ photometry of southern early-type stars and galactic structure away from the plane.
D. Kilkenny, P. W. Hill, T. Schmidt-Kaler.
Monthly Notices Roy. Astron. Soc., Vol. 171, 353 - 374 (1975).

Hβ photoelectric photometry is reported for 165 early-type stars at intermediate and high galactic latitudes. The data are combined with earlier *UBV* and spectroscopic results to determine the stellar space distribution. Stars of type B2 and earlier, at distances of up to 1 kpc from the galactic plane, appear to follow spiral structure in the plane. The available material, particularly the derived colour excesses, is used to select a number of blue stars which may be subluminous.

113.006 Multicolor photometry of metallic-line stars. II. Additional observations of ν Draconis.
S. F. González, T. Gómez, E. E. Mendoza V.
Rev. Mexicana Astron. Astrofis., Vol. 1, 119 - 120 (1974).

The authors have obtained additional and improved UBVRI photometric data for BS 6554 and BS 6555. The program has consisted of 171 observations of standard stars, 370 of comparison stars and 218 of each component of ν Draconis.

113.007 Multicolor photometry of metallic-line stars. III. A photometric catalogue. E. E. Mendoza V.
Rev. Mexicana Astron. Astrofis., Vol. 1, 175 - 201 (1974).

The author compiled over 800 metallic-line stars (Am) and suspected Am. He also presents the UBVRIJHKL photometry of a number of them. Preliminary results indicate that the B−V, V−R and R−I color indices are satisfactorily correlated with the spectral type derived solely from the hydrogen lines.

113.008 A catalogue of galactic O stars and the ionization of the low density interstellar medium by runaway stars. C. Cruz-González, E. Recillas-Cruz, R. Costero, M. Peimbert, S. Torres-Peimbert.
Rev. Mexicana Astron. Astrofis., Vol. 1, 211 - 259 (1974).

A catalogue of 664 galactic O stars is presented. For each object the following characteristics are presented: m_v, B−V, spectral type, distance, radial velocity, radial component of the peculiar velocity, possible multiplicity of the object, whether the O star is inside or outside the faintest H II regions detectable on the Palomar Sky Survey prints and identification of the H II region where the star is projected. From this catalogue the luminosity function for O stars is computed. Out of 386 O stars with known radial component of the peculiar velocities, v_{pr}, the authors have found 72 stars with $|v_{pr}| \geqslant 30$ km s^{-1}; only 19 of them had been proposed previously as runaway candidates. On the other hand, 22 runaway candidates proposed by other authors have $|v_{pr}| < 30$ km s^{-1}. It is estimated that 20 to 30% of the total number of O stars are runaways.

113.009 Study of four stellar rings in the Cygnus constellation. T. A. Uranova, G. S. Tsarevskij.
Soobshch. Spets. Astrofiz. Obs., *Zelenchukskaya*, vyp. (No.) 10, p. 37 - 43 (1973). In Russian.

On the basis of three - colour photographic UBV photometry the properties of four stellar rings: SR 125, SR 127, An 69, and An 70 are studied. It is concluded that there are no arguments to consider the investigated rings as spatial groupings of stars.

113.010 UBV photometry of OB⁺ stars north of 1950.0 declination −15°. J. S. Drilling.
Astron. Journ., Vol. 80, 128 - 130 = Louisiana State Univ. Obs., *Baton Rouge*, Contr. No. 102 (1975).

Photoelectric UBV photometry has been obtained for 164 stars which are brighter than photographic magnitude 12.0, north of 1950.0 declination −15°, and classified as OB⁺ in the Luminous Stars catalogs of the Hamburg and Warner and Swasey Observatories. The U−B, B−V diagram for these stars indicates that they consist primarily of O-type stars and early B-type supergiants.

113.011 Catalogue of early-type stars measured in a narrow-band photometric system.
N. Morguleff, M. Gerbaldi.
Astron. Astrophys., Suppl. Ser., Vol. 19, 189 - 209 (1975).

A compilation of the photoelectric measurements in the Barbier-Morguleff system is presented. The catalogue includes data for 773 stars of spectral type O8 to F6. 706 stars have been measured at least twice.

113.012 Emission-line stars with infrared dust emission: implications of the galactic distribution.
D. A. Allen, I. S. Glass.
Monthly Notices Roy. Astron. Soc., Vol. 170, 579 - 587

(1975).

Near-infrared photometry of a final selection of emission-line stars is presented. A sample of about 700 such stars is analysed and it is shown that those with circumstellar dust have a spiral arm distribution in the Galaxy. The peculiar Be stars with dust are also considered in isolation; these too are found to have a population I distribution. It is therefore unlikely that they are symbiotic stars or the progenitors of normal planetary nebulae; they probably form a distinct type whose nearest relatives are the VV Cephei stars.

113.013   On the optical variability of the helium stars HD 160641 and BD +13°3224.   A. U. Landolt.
Astrophys. Journ., Vol. 196, 789 - 790 = Louisiana State Univ. Obs., *Baton Rouge*, Contr. No. 100 (1975).

UBV photoelectric observations of the helium stars HD 160641 and BD +13°3224 are discussed. A period of 0.107995 days was found for BD +13°3224.

113.014   Photometry of SK160 = SMC-X 1.
J. E. Penfold, P. R. Warren, A. J. Penny.
Monthly Notices Roy. Astron. Soc., Vol. 171, 445 - 455 (1975).

Photometry on the *UBV* system is presented for the SMC B0 I star SK160 which is now identified with the X-ray source SMC-X1. A simple rotating ellipsoid model is shown to be insufficient to represent the data.

113.015   Some theorems of stellar statistics and "paradoxes" of colour excess.   V. V. Radzievskij.
Astron. Zhurn. Akad. Nauk SSSR, Vol. 52, 294 - 298 (1975). In Russian. English translation in Soviet Astron., Vol. 19, No. 2.

Several theorems on arithmetic mean values of magnitude and colour excess of stars for different models of distribution of stellar space density and density of interstellar dust are obtained.

113.016   DDO intermediate-band photometry of moving-group stars.   R. J. Boyle, R. D. McClure.
Publ. Astron. Soc. Pacific, Vol. 87, 17 - 36 (1975).

G and K giant stars in several of Eggen's moving groups have been observed on the DDO system. The cyanogen strength and absolute magnitude calibrations of the system are used to discuss the assignments of the observed stars to moving groups. The authors are able to confirm the assignment of some 50 % of the members of the Hyades moving group and thus to segregate out a more homogeneous sample of Hyades group stars. These stars can be used to recalibrate intrinsic Hyades giant sequences. Less firm conclusions are reached concerning the other, older, groups discussed.

113.017   Photometric standards on the *UBV* and *RI* systems.   O. J. Eggen.
Publ. Astron. Soc. Pacific, Vol. 87, 107 - 109 (1975).

Several thousand *UBV* and *RI* observations have been obtained since 1966 with the 40-inch reflector at Siding Spring Mountain. The standards used in these observations are discussed here.

113.018   Photometry of possible barium stars.   O. J. Eggen.
Publ. Astron. Soc. Pacific, Vol. 87, 111 - 113 (1975).

UBVRI photometry of 42 newly announced Ba stars is discussed. Most of these objects may not be Ba stars in the usual sense of that classification.

113.019   Emission-line effects in Hα, Hβ, Hγ, and *uvby* photometry for B-type stars.
D. L. Crawford, J. V. Barnes, C. L. Perry.
Publ. Astron. Soc. Pacific, Vol. 87, 115 - 121 = Contr. Louisiana State Univ. Obs., *Baton Rouge*, No. 101 (1975).

We have compiled Hα, Hβ, Hγ, and four-color data for

780 B-type stars, and analyzed the relation between indices with respect to possible effects due to emission in the hydrogen lines. It is possible to separate supergiants from Be-type stars by means of an Hα vs Hβ diagram.

113.020   *UBVr* sequences for two η Carinae-like objects.
E. R. Craine, S. Tapia.
Publ. Astron. Soc. Pacific, Vol. 87, 131 - 135 (1975).

Comparison sequences are presented for MWC 645 and IRC 10420 in order to stimulate investigations of the light curves of these objects. Both are reported to be similar to η Car and hence may exhibit significant changes in brightness. Previously unpublished finding charts appear for each object.

113.021   Photometric variations of the B emission star HD 174237.   P. Merlin.
Astron. Astrophys., Vol. 39, 139 - 141 (1975). In French.

The bright B emission star HD 174237 was observed photoelectrically in the *UBV* system over a three weeks interval. It was found that this star displays variations in U, B, V larger than $0^m.10$, and variations larger than $0^m.01$ during a few nights. They show no true periodicity, only a 7–12 days cycle.

113.022   The ultraviolet experiment onboard the Astronomical Netherlands Satellite – ANS.
R. J. van Duinen, J. W. G. Aalders, P. R. Wesselius, K. J. Wildeman, C. C. Wu, W. Luinge, D. Snel.
Astron. Astrophys., Vol. 39, 159 - 163 (1975).

The ultraviolet experiment package onboard the ANS consists of a 22 cm diameter Cassegrain telescope, followed by a five channel intermediate band spectrophotometer. The instrument response function for each bandpass is almost rectangular. The central wavelengths are 155, 180, 220, 250 and 330 nm respectively. The in orbit performance of the instrument is according to expectations. Observational results of some typical objects are summarized to illustrate the capabilities of the instrument.

113.023   Catalogue of stars observed photoelectrically.
C. Jaschek, E. Hernandez, A. Sierra, A. Gerhardt.
Obs. Astron. Univ. Nacional La Plata, Republica Argentina, Ser. Astron., Vol. 38, 2 + 507 pp. (1972).

113.024   Photometric studies of faint stars in the vicinity of the Orion nebula.   V. N. Sincheskul.
Astrometriya i Astrofizika, *Kiev*, vyp. (No.) 24, (see 003.003), p. 75 - 79 (1974). In Russian.

On the basis of an analysis of the $U-B$ and $B-V$ colour index for faint stars in the vicinity of the Orion nebula it is established that as the stellar magnitude weakens the negative $U-B$ colour indices and the number of stars with these indices increase. Such a phenomenon is also characteristic of other open clusters.

113.025   UBV photometry of V1357 Cyg (Cyg X-1).
Kh. F. Khaliullin.
Pis'ma v Astron. Zhurn., Vol. 1, No. 3, p. 30 - 35 (1975). In Russian.

In 1974 during 35 nights 300 photoelectric measurements of the star V1357 Cyg in each *UBV* filter have been obtained. The light curve is symmetrical in respect to phases 0.00, 0.25 and 0.50 and has a double wave. The amplitude of regular light variations is $0^m.045 ± 0^m.005$, the same in all *UBV* filters. Besides the star shows fluctuations of brightness with amplitudes up to $0^m.03$. The analysis of these fluctuations does not show any indications of the triplicity of the system.

113.026   Magnesium b line photoelectric photometry of B0–M6 main-sequence, giant, and supergiant stars.
E. F. Guinan, K. A. Harrison.
Bull. American Astron. Soc., Vol. 7, 234 (1975). – Abstr. AAS.

113.027 **Rotational velocity effects on photometric indices for B-type stars.** W. H. Warren, Jr.
Bull. American Astron. Soc., Vol. 7, 272 (1975). – Abstr. AAS.

113.028 **Calibration of the uvbyβ systems.** D. L. Crawford.
Dudley Obs. Rep. No. 9, (see 012.008), p. 17 - 29 (1975).
Details of the calibrations of the photometry in terms of intrinsic color and absolute magnitude are given, as well as checks on possibly systematic effects.

113.029 **Calibrations and applications of the uvby photometric system. II. Beta index and further extensions.**
M. Breger.
Dudley Obs. Rep. No. 9, (see 012.008), p. 31 - 40 (1975).
The β and c_1 indices are calibrated in terms of T_e and log g for population I A and early F stars. ATLAS model atmospheres and the new $(b-y)_0$, β relations by Crawford have been used. For most A stars within two magnitudes of the main sequence, β is shown to be a function of temperature independent of gravity. The zero points of the uvby system are determined and discussed. A "best" temperature calibration is given for the stars in the spectral range B0 to G0.

113.030 **The calibration of uvby photometric indices for population I and II stars in the range $0.5 < \Theta_e < 0.7$, $2.0 < \log g < 4.4$.** A. G. D. Philip, L. T. Matlock.
Dudley Obs. Rep. No. 9, (see 012.008), p. 45 - 63 (1975).
The atmospheric models of Mihalas (1966) were convolved with the transmission curves of the uvby filters to yield theoretical four-color indices as a function of log g and Θ_e. Blanketing corrections have been determined. The parameters Θ_e and log g can be calculated with rms errors of ± 0.015 and ±0.2 respectively.

113.031 **The calibration of the reddening-free parameters of the Vilnius photometric system in temperatures, surface gravities and metallicities.** V. Straižys.
Dudley Obs. Rep. No. 9, (see 012.008), p. 65 - 71 (1975).
The positions of the filter bands of the Vilnius photometric system were selected in 1962–1965 on the basis of energy distribution curves in order to obtain purely photometric three-dimensional classification of stars reddened by interstellar dust without any additional information from their spectra. For classification of stars by spectral types, luminosities and metallicities of all temperatures from O to M, seven or, still better, eight intermediate band magnitudes are necessary.

113.032 **Photometry of distant K giant stars.** H. L. Helfer, P. A. Jennens.
Dudley Obs. Rep. No. 9, (see 012.008), p. 87 - 97 (1975).
This paper is divided into three parts: the first deals with the importance of avoiding a certain class of systematic photometric errors in doing photometric abundance analyses; the second part deals with the new UBViyz photometric system which is well suited for determining abundances in distant K giants; and the third part deals with some results the authors have obtained which touch upon several interesting problems in galactic structure.

113.033 **The DDO photometric system for late-type stars.** K. A. Janes, R. D. McClure.
Dudley Obs. Rep. No. 9, (see 012.008), p. 99 - 110 (1975).
The use of the DDO (David Dunlap Observatory) system of intermediate-band filter photometry is described, with particular reference to determination of atmospheric parameters T_e, log g, [Fe/H], and CN strength. Some recent results are also discussed.

113.034 **Five-color photometry at the Leiden Southern Station.** J. Lub, J. W. Pel.
Dudley Obs. Rep. No. 9, (see 012.008), p. 133 - 134 (1975).
Short summary.

113.035 **The Geneva photometry for B-type stars.** A. Maeder, N. Cramer.
Dudley Obs. Rep. No. 9, (see 012.008), p. 135 - 141 (1975).
The authors give a brief report on the use of the Geneva system for establishing a luminosity calibration of B-type stars. The proposed method allows one to obtain with an intermediate-band photometry, luminosity criteria which are as sensitive as those obtained by Hβ or Hγ photometry. The principle of the method, which is independent of interstellar reddening, the basis for calibration, the range of application and the accuracy are discussed. It is also shown that a very clear separation of Bp and Ap stars is possible.

113.036 **The d vs. B2−V1 diagram for A−F, Ap and Am stars.** B. Hauck.
Dudley Obs. Rep. No. 9, (see 012.008), p. 143 - 150 (1975).
In the Geneva photometric system the d vs. B2−V1 diagram has the same significance as the HR diagram. One can obtain the absolute magnitude of A stars of luminosity classes V to III and distinguish supergiants. The location in this diagram for the Ap and Am stars is also discussed.

113.037 **Remarks about the use of some color indices as temperature parameters.**
B. Hauck, P. Magnenat.
Dudley Obs. Rep. No. 9, (see 012.008), p. 171 - 175 (1975).
The temperature scale for some color indices in various photometric systems is given. The effects of luminosity and blanketing are discussed.

113.038 **UBV synthetic colors.** D. S. Hayes.
Dudley Obs. Rep. No. 9, (see 012.008), p. 309 - 318 (1975).
The synthesis of UBV colors, using published response functions of the U, B and V magnitudes, depends fundamentally upon the calibration of the spectral energy distribution of Vega. The new Mt. Hopkins calibration has made necessary a re-evaluation of the success of the published response functions in reproducing the UBV photometric system, and has also made necessary the derivation of new values of the constants in the transformation equations relating the "natural" and UBV colors.

113.039 **Colors of metal deficient giant stars.** R. A. Bell, B. Gustafsson.
Dudley Obs. Rep. No. 9, (see 012.008), p. 319 - 340 (1975).
The authors have used the computer programs described by Gustafsson (1971), Gustafsson and Nissen (1972) and Nordlund (1974) to compute a grid of flux constant, line blanketed atmospheres for metal deficient giant stars. These models have been used as the basis for computing synthetic spectra. Finally they have convolved the synthetic spectra with the sensitivity functions of various filter systems.

113.040 **Multivariate analysis of photometric data.** K. A. Janes.
Dudley Obs. Rep. No. 9, (see 012.008), p. 341 - 348 (1975).
Most photometric indices are strongly correlated with one another, a fact which can obscure the more subtle relationships among a set of indices. Using the branch of statistics known as multivariate analysis, such a set of correlated indices can be transformed into a set of linearly independent parameters. This method of analysis has been applied first, to develop transformations of the Copenhagen gnkmf indices into DDO indices and second, to compare the UBVRI, DDO, and gnkmf systems. Among these systems, DDO photometry provides the clearest separation into a three parameter system,

but the analysis of the gnkmf photometry suggests there may be a fourth parameter necessary to describe G and K giants.

113.041 Systematic evaluation of existing and improved multicolor systems. U. W. Steinlin, R. Buser.
Dudley Obs. Rep. No. 9, (see 012.008), p. 349 - 358 (1975).

This paper deals with broad-band photometry (mainly, but by no means exclusively, with three color photometry) and its application to faint stars of all different types in large numbers rather than to a smaller number of (previously coarsely classified) brighter stars within a specific range of spectral types, luminosities or other properties. The essential features for such a system are discussed.

113.042 Atmospheric parameters from four-color photometry.
L. Relyea, L. T. Matlock, A. G. D. Philip.
Dudley Obs. Rep. No. 9, (see 012.008), p. 375 - 382 (1975).

Approximate values of log g, Θ_e and [m/H] have been obtained for groups of field stars (including population II) using Strömgren four-color photometry with blanketing corrections as calibrated by Philip and Matlock. From the results, some inferences are made concerning galactic structure at high galactic latitudes.

113.043 Helium-rich stars in the Strömgren four-color system. D. M. Peterson, C. C. Porco.
Dudley Obs. Rep. No. 9, (see 012.008), p. 383 - 387 (1975).

The authors summarize uvbyHβ measurements of the helium-rich stars and show that they occupy a region of the $[m_1] - [u-b]$ diagram below that of most field B stars. They use this to investigate the lack of such objects in the northern hemisphere and the reality of the cool temperature cutoff found by Osmer and Peterson.

113.044 UBVRI photometry of metallic-line stars.
E. E. Mendoza V.
Dudley Obs. Rep. No. 9, (see 012.008), p. 407 - 412 (1975).

A preliminary analysis of the photometry of 99 metallic line stars indicates: 1) The B−V, V−R, and R−I color indices of unreddened Am stars represent satisfactorily the spectral types obtained solely from the hydrogen lines. 2) Mean colors of δ Del stars are redder than those of classic Am stars. Mean colors of mild and suspected Am stars are bluer than those of classic Am stars. 3) The (U−V, B−V) relationship for Am stars is satisfactorily represented by a straight line.

113.045 Some observational properties of G and K stars.
M. Grenon.
Dudley Obs. Rep. No. 9, (see 012.008), p. 413 - 425 (1975).

The classification of late-type stars by means of the Geneva seven-color photometric system is briefly described. A calibration in terms of absolute magnitude and [Fe/H] is applied to analyze samples of nearby stars. In the resulting HR diagram, the author discusses some observational properties of G and K stars, in connection with their age and evolutionary stage.

113.046 Search for faint red stars. M. F. McCarthy.
Dudley Obs. Rep. No. 9, (see 012.008), p. 453 - 456 (1975).

Searches for faint red stars are important for population studies of the end of the main sequence. To reach beyond limits of objective prism surveys, photographic photometrists usually employ modifications of the UBV, RGU systems. A set of filters is described which permits an extension of observations to the infrared region.

113.047 Stars common to various photometric systems.
A. G. D. Philip, B. Hauck, P. Magnenat.
Dudley Obs. Rep. No. 9, (see 012.008), p. 499 - 513 (1975).

113.048 Intermediate infrared colours of M-dwarf stars.
I. S. Glass.
Monthly Notices Roy. Astron. Soc., Vol. 171, 19P-23P (1975).

M dwarfs are shown to occupy a region of the *JHK* two-colour diagram which is quite distinct from that occupied by M giants. The *JHKL* colours of early M dwarfs are found to be predicted fairly well by a recent model which is dominated by H_2O opacity in the infrared.

113.049 Statistical analysis of infrared color-indices of variable late type stars. J. Krempeć.
Inform. Bull. Variable Stars, (I.A.U. Commission 27), Konkoly Obs., Budapest, No. 977, 6 pp. (1975).

113.050 Photometry of six peculiar A-type stars.
S. C. Wolff, N. D. Morrison.
Publ. Astron. Soc. Pacific, Vol. 87, 231 - 236 (1975).

Four-color (*uvby*) photoelectric observations are presented for six Ap stars. No variations were found for the Hg-Mn star κ Cnc. The periods of $1\overset{d}{.}4450$ for 45 Leo and $2\overset{d}{.}8881$ for HR 5597 found by Winzer (1974) are confirmed. The period of HR 5153 is shown to be $2\overset{d}{.}451$. Possible explanations for the photometric variations of 56 Ari and HR 4369 are discussed.

113.051 On the ratio of total-to-selective absorption.
B. I. Olson.
Publ. Astron. Soc. Pacific, Vol. 87, 349 - 351 (1975).

The variation of the ratio of total-to-selective absorption as a function of a star's intrinsic color and color excess has been examined. A relation is found that allows this variation to be expressed numerically.

113.052 Colors, magnitudes, spectral types, and distances for stars in the field of the X-ray source Cyg X-1 (HDE 226868). J. Bregman, D. Butler, E. Kemper, A. Koski, R. P. Kraft, R. P. S. Stone.
Lick Obs. Bull. No. 647, 4 pp. (1973/74).

113.053 Catalogue of photometric and astrometric data for 4000 stars in the Orion nebula aggregate.
A. D. Andrews.
Bol. Inst. Tonantzintla, Vol. 1, 101 - 187 (1974).

Automatic techniques of measurement are applied to multi-colour Baker-Schmidt plates to provide UBVR magnitudes and colours, equatorial and rectangular coordinates for 4117 stars in the Orion nebula aggregate. The zone covered from 5^h26^m to 36^m, $-3\overset{\circ}{.}6$ to $-6\overset{\circ}{.}2$ (Equinox 1900) excludes the central region of bright nebulosity. Identification lists for 1200 Parenago stars in the range $8 < V < 16$ and key charts to fields surveyed at Tonantzintla and Asiago are given.

113.054 JHKL photometry of late type stars.
R. M. Catchpole.
Monthly Notes Astron. Soc. Southern Africa, Vol. 34, 68 (1975). – Abstract.

113.055 VRI photometry at the S.A.A.O.
A. W. J. Cousins.
Monthly Notes Astron. Soc. Southern Africa, Vol. 34, 68 - 71 (1975).

113.056 Photometry of V1057 Cygni and neighboring stars.
A. U. Landolt.
Publ. Astron. Soc. Pacific, Vol. 87, 379 - 383 = Contr. Louisiana State Univ. Obs., *Baton Rouge*, No. 104 (1975).

UBV photoelectric photometry of stars in the vicinity of the pre-main-sequence object V1057 Cyg has been used to establish a photometric sequence in the magnitude range $5.5 < V < 15.5$. Photographic and photoelectric photometry of V1057 Cyg in the time interval 1971–74 also is reported. The brightness slowly declined from 1971 to 1974.

113.057 **Infrared observations of late-type stars in nebulae.**
M. Cohen.
Publ. Astron. Soc. Pacific, Vol. 87, 421 - 423 (1975).

Two- to 18-μ observations of five late-type stars involved with nebulae reveal no infrared excesses near 10 μ. This fact, coupled with the presence of nebular structures resembling wakes or bow shocks, suggests that the stars have randomly encountered the nebulae.

113.058 **The ratio of color excesses in UBV photometry.**
A. Gutiérrez-Moreno, H. Moreno.
Publ. Astron. Soc. Pacific, Vol. 87, 425 - 432 (1975).

Numerical integrations have been performed in order to obtain the ratio of the color excesses in UBV photometry. The data were analyzed to study the dependence of this ratio and of the value of R on the color of the stars and on the reddening itself. The results show that the ratio of the color excesses is practically independent of the reddening, while R appears to change more with the spectral type than with the total amount of reddening.

113.059 **Infrared observations and the effective temperature of the peculiar star HD 101065.**
A. R. Hyland, J. R. Mould, G. Robinson, J. A. Thomas.
Publ. Astron. Soc. Pacific, Vol. 87, 439 - 441 (1975).

Infrared photometric observations of the peculiar star HD 101065 are presented. The spectral type of HD 101065 is found to be between F5 and F6, and its effective temperature to be $6300° \pm 150°$.

113.060 **Untersuchung der interstellaren Verfärbung an O- und B-Sternen.**
W. Wiemer.
Veröff. Astron. Inst. Bonn, No. 88, 40 pp. (1974).

The paper is based on BVRIJKL-magnitudes of about 350 O- and B-type stars, obtained from literature and from own measurements. With the aid of two-colour-diagrams it is shown, that within the Galaxy there is indeed one mean law of interstellar extinction. Considering the mean error of the measured colours and the scatter of the intrinsic colours, "zones of scatter" in the two-colour-diagrams are given for stars of different spectral types and luminosity classes. In an additional chapter it is shown, that for O- and Of-type stars a larger variation in the intrinsic colours has to be assumed. Mean intrinsic colours have been derived for B-type stars of luminosity class I and V.

113.061 **Méthode photométrique de sélection des étoiles Ap.**
P. Steiger.
Bull. Soc. vaudoise Sci. nat., Vol. 72, Fasc. 2, p. 53 - 59 (1974) = Publ. Obs. Genève, Sér. A, Fasc. 81/I (1975).

In the photometric system of Geneva Observatory, a photometric method is developed to select the Ap stars with spectral type earlier than A5, except the Hg stars. This method is independent of luminosity and interstellar reddening. This method is applied to the selection of Ap stars in about fifteen clusters.

113.062 **Spectrophotometry of Orion stars.**
H. Moreno.
Dep. Astron., Univ. Chile, Obs. Astron. Nacional, Cerro Calán, Santiago de Chile, Publ. Vol. 2, (No. 4), 143 - 155 (1974).

Relative spectral intensity distributions for 60 stars in the region of the Orion aggregate are given. Equivalent widths of Hβ, Hγ and Hδ have been measured. The size and position of the Balmer discontinuity, the gradients in the ultraviolet and blue-green-yellow parts of the spectrum, and some other spectral features were also measured.

113.063 **Bibliographic survey of published photoelectric**
indices and spectral classification.
M. Bischoff.
Centre de Données Stellaires, Inform. Bull. No. 8, (see 002.007), p. 11 (1975).

113.064 **Analyse des correspondances appliquée aux systèmes photométriques.**
D. Egret.
Centre de Données Stellaires, Inform. Bull. No. 8, (see 002.007), p. 13 - 14 (1975).

113.065 **Photométrie uvbyβ et classification spectrale MK.**
E. Oblak, S. Considere, M. Chareton.
Centre de Données Stellaires, Inform. Bull. No. 8, (see 002.007), p. 15 - 16 (1975).

113.066 **Temperatures and luminosities of M type dwarfs from infrared photometry.**
G. J. Veeder, Jr.
Thesis, California Inst. Techn., Pasadena (USA). 120 pp. University Microfilms Order No. 74-14,291 (1974).

113.067 **Properties of two common photometric systems and photometric observations of selected eclipsing binary systems.**
G. G. Spear.
Thesis, Pennsylvania Univ., Philadelphia (USA). 298 pp. University Microfilms Order No. 74-14,143 (1973).

113.068 **Beiträge zur Methodik der photographischen UBV-Photometrie mit Anwendung auf den offenen Sternhaufen NGC 2632 (Praesepe).**
W. Paffhausen.
Diss. Univ. Münster, 73 pp. (1974).

Curvature in Hβ transformations.
See Abstr. 031.206.

Observational technique and data reduction.
See Abstr. 031.216.

A data acquisition programme for photometric measurements.
See Abstr. 031.413.

The instrumentation and techniques of infrared photometry.
See Abstr. 034.030.

Advances in instrumentation for stellar photometry.
See Abstr. 034.050.

The photoelectric photometry with a 60 cm reflector at the Skalnaté Pleso Observatory.
See Abstr. 034.072.

The limiting magnitude of the ESO (B) Survey.
See Abstr. 041.011.

Line blanketing and model stellar atmospheres. II. Interpretation of broad-band photometric observations.
See Abstr. 064.032.

Stellar atmospheres – the middle man.
See Abstr. 064.061.

Extinction parameters on Terskol Peak and investigation of the instrumental system of the photometer.
See Abstr. 082.069.

The application of parallaxes and photometry to the lower main sequence.
See Abstr. 111.002.

Spectroscopic and photometric observations of luminous stars in the Centaurus-Norma ($l = 305°-340°$) section of the Milky Way.
See Abstr. 114.013.

A six-color Q-parameter for yellow supergiants.
See Abstr. 114.056.

Wolf-Rayet stars. VI. The nature of the optical and infrared continua.   See Abstr. 114.070.

How can Ap and Am stars be investigated at large distances?   See Abstr. 114.073.

List of spectroscopic and photometric catalogues lately published or to be published — list V.
See Abstr. 114.080.

Simultaneous observations of variable stars. I. The Be star π Aqr.   See Abstr. 114.315.

On the physical association of the peculiar emission-line stars HD 122669 and HD 122691.   See Abstr. 114.359.

Metal abundances of RR Lyrae stars established from low resolution scanner spectrophotometry.
See Abstr. 122.061.

Theoretical mean colors of pulsating cepheids.
See Abstr. 122.062.

UBVRI photometry of V553 Centauri.
See Abstr. 122.110.

The magnetic field of W Sgr and evidences for a period-magnetic field relation in pulsating variables.
See Abstr. 122.115.

A new nearby subdwarf M star.

See Abstr. 126.010.

Further observations of the Orion nebula cluster.
See Abstr. 132. 022.

UBVr sequences and observations of optically identified radio sources.   See Abstr. 141.048.

Preliminary results concerning DDO photometry of the southern hemispheric globular star cluster NGC 3201, and other remarks.   See Abstr. 154.012.

Four-color photometry of blue horizontal-branch stars in globular clusters.   See Abstr. 154.013.

UBVβ photometry and the galactic distribution of OB stars.   See Abstr. 155.038.

Photoelectric photometry of supergiants in the Large Magellanic Cloud.   See Abstr. 159.002.

Small Magellanic Cloud. First list of probable members.   See Abstr. 159.003.

The intrinsic colours $(B-V)_0$, $(U-B)_0$ and distance moduli of supergiants in the Large Magellanic Cloud.
See Abstr. 159.009.

Some observations of bright Magellanic Cloud stars with a 12-channel scanning photometer.
See Abstr. 159.017.

114 Stellar Spectra, Temperatures, Spectroscopy, Spectra of Individual Stars

Stellar Spectra, Temperatures, Spectroscopy

114.001 Equivalent widths and rotational velocities of southern early-type stars. L. A. Balona.
Mem. Roy. Astron. Soc., Vol. 78, 51 - 72 (1975).

Equivalent widths of the Hγ, some He singlet and triplet lines and the (interstellar) K line of Ca II are presented for 585 early-type stars in the Radcliffe radial velocity programme. The strength of the helium lines and the triplet/singlet intensity ratio is investigated as a function of spectral type and luminosity class. Projected rotational velocities on the Slettebak system are derived from the widths at half intensity of the helium lines.

114.002 Polarization characteristics of Herbig Ae and Be stars. F. J. Vrba.
Astrophys. Journ., Vol. 195, 101 - 106 (1975).

The polarization wavelength dependences of 10 Herbig Ae and Be stars have been determined over the wavelength range 3700–7500 Å. Upon subtraction of an assumed interstellar polarization component, seven of nine objects display an unexplained polarization peak (centered upon the 6500 Å filter) the height of which correlates well with the strength of Hα emission. Four of the objects were studied for polarization time variability, with the result that Z CMa showed significant time variation.

114.003 Ultraviolet energy distributions of luminous early-type stars from TD1 satellite observations.
C. M. Humphries, K. Nandy, E. Kontizas.
Astrophys. Journ., Vol. 195, 111 - 119 (1975).

Ultraviolet spectra are presented for supergiants, giants and main-sequence stars in the spectral type range from B0 to A2. The effect of luminosity has been studied by comparing the ultraviolet energy distributions of luminous and main-sequence stars of the same spectral type. The flux deficiencies which are observed in the energy distributions of luminous stars are interpreted as the result of lower effective temperatures for luminous stars than for corresponding main-sequence stars, and an estimate is given of the effective temperature scale for B-type supergiants.

114.004 Uranium lines in the spectra of peculiar A stars: a search for recent r-process events.
C. R. Cowley, S. J. Adelman.
Astrophys. Letters, Vol. 16, 5 - 7 = Contr. Dominion Astrophys. Obs., Victoria, No. 241 (1975).

Uranium wavelengths in the spectra of Ap stars are studied to see if they give any indication of a recent r-process event. It is concluded that there is no credible evidence for an admixture of uranium-235 in these stars, which would imply such an event. The evidence, though negative, is badly confused by blending of the lines, and a final judgement must wait for an observational clarification of the situation.

114.005 On the continuous energy distributions of peculiar A stars. S. J. Adelman.
Astrophys. Journ., Vol. 195, 397 - 403 (1975).

Spectrophotometric scans which cover the wavelength region λλ3300–7100 of 11 magnetic Ap and 11 normal stars are used in connection with published energy distributions to examine the similarity of the continuous energy distributions of peculiar A and normal main-sequence stars. As ob-

served from the ground, the flux distributions of the Hg-Mn stars empirically match those of the normal stars while those of the magnetic Ap and normal stars match in a gross sense. Many magnetic Ap stars are found to possess broad, continuous absorption features which are most likely produced by bound-free discontinuities of common metals, in particular Si I, neutral iron-peak elements, and the singly ionized rare earths.

114.006 Diffusion and isotope anomalies of Hg in Ap stars.
G. Michaud, H. Reeves, Y. Charland.
Astron. Astrophys., Vol. 37, 313 - 324 (1974).

A model is proposed in which both the mercury overabundances and isotopic anomalies in certain Ap stars are explained in terms of radiation pressure effects in the outer atmosphere of these stars.

114.007 Spectral type and kinematic properties of ApSi 4200 stars. C. Megessier.
Astron. Astrophys., Vol. 37, 439 - 441 (1974).

According to their Balmer lines' equivalent widths ApSi 4200 stars have spectral types comprised between B 5 V and B 8 V. A study of their kinematical properties shows that ApSi 4200 behave similarly as normal B 5 V to B 8 V stars, while ApSr−Eu−Cr behave similarly as normal B 9 V to A 3 V stars. So ApSi 4200 stars are in fact B stars and are as young as normal B 5 V to B 8 V stars.

114.008 Sterne mit anomalen Spektren. H.-H. Voigt.
SuW, Vol. 14, 43 - 48 (1975).

114.009 Wolf-Rayet-Sterne−eine Gruppe außergewöhnlicher Sterne (I. Teil). W. Seggewiss.
SuW, Vol. 14, 84 - 88 (1975).

114.010 The relation of emission properties of several Be stars to the luminosity class and the spectral type.
A. M. Delplace, H. Hubert.
Astron. Astrophys., Vol. 38, 75 - 79 (1975). In French.

In the λλ3900−6600 spectral range, about twenty B 2 type emission line stars are studied. A relation between emission features and their variations, and the luminosity class is investigated. Several Be stars with a different spectral type but the same luminosity class are also studied. The time scale of the emission variation increases toward later spectral types.

114.011 Abundance spots on peculiar A stars and diffusion along the surface. O. Havnes.
Astron. Astrophys., Vol. 38, 105 - 108 (1975).

The author examines the effect of horizontal diffusion along the surface of a magnetic peculiar A star on its surface element distribution. The diffusion is driven by composition gradients in an abundance spot. Such a process will, in the course of the stellar lifetime, cause a considerable redistribution of elements from the abundance spot to surrounding parts of the stellar surface. This will damp the spectral variations in the older of the Ap stars. A completely homogeneous horizontal surface composition cannot be obtained by horizontal diffusion on an abundance spot, if this is repuired, other mechanisms must be found.

114.012 Observations of lithium dilution and rotational velocity decay in F and G giant stars.
W. R. Alschuler.
Astrophys. Journ., Vol. 195, 649 - 660 = Lick Obs. Bull., No. 659 (1975).

Lithium abundances and rotational velocities have been determined for 64 stars between F4 III and G5 III. These results can be compared with theoretical predictions of Li convective dilution and rotational velocity decay for masses 2.4 to 3.0 $M_\odot$. If all these stars are crossing the Hertzsprung gap for the first time since leaving the main sequence, then theory and observation do not entirely agree. A satisfactory agreement with observation could be achieved by adjustments to the models, possibly through a modification of the mixing-length theory of convection. Without such modifications it would be necessary to assume that there is a systematic increase in the sampled mean mass between F4 III and G5 III, which implies that these samples are drawn from different regions of the zero-age main sequence having different initial Li abundances.

114.013 **Spectroscopic and photometric observations of luminous stars in the Centaurus-Norma (l = 305°– 340°) section of the Milky Way.** R. M. Humphreys.
Astron. Astrophys., Suppl. Ser., Vol. 19, 243 - 247 (1975).
New observational data are presented for 111 luminous stars in the Centaurus-Norma region of the Milky Way. The new data include MK spectral types and UBV photometry.

114.014 **Fe I and CH equivalent width measurements for sixty-one F, G and K type stars.** J. B. Hearnshaw.
Astron. Astrophys., Suppl. Ser., Vol. 19, 321 - 336 (1975).
Equivalent width measurements are presented for selected Fe I lines and CH blends in the blue-green region of sixty-one F, G and K stars, observed at high dispersion at the Observatoire de Haute-Provence and at Mt. Stromlo Observatory.

114.015 **Wolf-Rayet-Sterne – eine Gruppe außergewöhnlicher Sterne (II. Teil).** W. Seggewiss.
SuW, Vol. 14, 120 - 123 (1975).

114.016 **Destruction and production of molecules in circumstellar regions of T Tauri stars.** G. F. Gahm.
Astrophys. Space Sci., Vol. 32, 297 - 304 (1975).
It is shown that the lifetimes of molecules against photodissociation in circumstellar envelopes of T Tauri stars are very short. The production rates of CO through gas-phase reactions are not sufficient to keep the equilibrium column densities at an observable level. The absence of molecular features reported for this class of stars is qualitatively understood.

114.017 **Liste et classification d'étoiles M, C et à émission nouvelles.** M. Barbier.
Astrophys. Space Sci., Vol. 32, 423 - 430 (1975).
The author presents a new list of 156 M, C or emission line stars discovered on objective-prism plates taken at Haute Provence Observatory. The good spectra are classified by means of criteria used in previous publications, principally from the ratio of TiO bands.

114.018 **Stellar compositions from narrow-band photometry–V. Barium abundances for 200 evolved stars.** P. M. Williams.
Monthly Notices Roy. Astron. Soc., Vol. 170, 343 - 362 (1975).
Using narrow-band spectrophotometric indices interpreted with synthetic spectra computed from model atmospheres, barium abundances have been measured for 200 stars with uncertainties of about 0.3 in [Ba/H]: Several possible new Ba II stars have been found and the frequency of Ba II stars amongst G and K giants appears to lie between 1 and 4 per cent. It is confirmed that the peculiar colours of many Ba II stars are caused by excess blanketing due to CN, CH and C_2.

114.019 **Photoelectric spectrophotometry of Wolf-Rayet stars.** J. D. R. Bahng.
Monthly Notices Roy. Astron. Soc., Vol. 170, 611 - 618 (1975).
Photoelectric spectrum scans of five southern Wolf-Rayet stars in the spectral range $\lambda\lambda 4600$–4720 were analysed to study the variability of brightness and of emission line strengths. No variations of any kind in short time scale were found. However, in WC stars night-to-night variations of 3 to 4 per cent were detected in the emission line strengths.

114.020 **Nuclear and nonnuclear abundance patterns in the manganese stars.**
C. R. Cowley, G. C. L. Aikman.
Astrophys. Journ., Vol. 196, 521 - 524 = Dominion Astrophys. Obs., Victoria, Contr. No. 243 (1975).
The manganese stars show abundance anomalies that are in conflict with the predictions of nuclear astrophysics. The authors stress the deviations from the odd-even effect that occur for phosphorus, gallium, and yttrium. The abundance anomalies (e.g., the Mn/Fe ratio) in the iron peak are less serious; within the uncertainties of the current determinations, they may be explained in terms of the same processes that predict the solar-system abundances.

114.021 **A comparison of galactic and Large Magellanic Cloud G-type supergiants by a method of spectrum synthesis.** M. A. Fry, L. H. Aller.
Astrophys. Journ., Suppl. Ser, No. 275, Vol. 29, 55 - 75 (1975).
The spectra of two G4 Ia supergiants—HR 8752 and R59, in the Perseus spiral arm and Large Magellanic Cloud, respectively—are analyzed by a method of spectrum synthesis, heretofore applied mostly to the sun. Evidence is presented that both of these stars are reddened by local dust clouds. Consequently, selection of appropriate model atmospheres must depend heavily on line spectrum data. A self-consistent spectrum synthesis analysis suggests that, to within a scatter of ± 0.4 in log A among individual elements, these stars are similar in composition to the sun, although there is evidence that metals are definitely underabundant in the LMC star as compared with the sun.

114.022 **New bright hydrogen-emission stars.** N. J. Irvine.
Astrophys. Journ., Vol. 196, 773 - 775 (1975).
Hα observations of bright, rapidly rotating stars with no history of line emission have led to the discovery of several hydrogen-emission stars and also some possible nonemission shell stars. Suspected variability appears to be a valuable clue in finding new emission stars.

114.023 **Linear polarization of Hα in Be stars.** R. Poeckert.
Astrophys. Journ., Vol. 196, 777 - 787 (1975).
The wavelength dependence of linear polarization across Hα has been measured in 12 Be stars with 2 Å resolution. Shell stars like ζ Tau exhibit a decrease in polarization which is proportional to the strength of the emission component of Hα. Stars considered "pole-on" have no such significant variations in polarization through the emission component. Some stars have substantial interstellar polarization which is indicated by a variation in position angle as well as in percent polarization. The results of this study show that the emission in Hα is unpolarized and therefore must arise in a region of the circumstellar envelope such that the optical depth in electron scattering from that region to the observer is much less than 1.

114.024 **Some interesting bright southern stars of early type.** A. Slettebak.
Astrophys. Journ., Vol. 197, 137 - 138 (1975).

The spectra of 10 bright southern stars of early type, considered to be interesting for a variety of reasons and worthy of further study, are described.

114.025 He I λ4922 profiles in B stars: calculations with an improved line broadening theory.
D. Mihalas, A. J. Barnard, J. Cooper, E. W. Smith.
Astrophys. Journ., Vol. 197, 139 - 142 (1975).

Theoretical profiles for the He I λ4922 line in B star spectra have been computed using the improved broadening theory of Barnard, Cooper, and Smith, and the level populations calculated by Auer and Mihalas from a simultaneous self-consistent solution of the coupled transfer and statistical equilibrium equations. The revised broadening theory yields excellent agreement with laboratory measurements of the width and intensity of the forbidden $(2p\,^1P^0 - 4f\,^1F^0)$ transition. The results of this paper show that stellar profiles computed with the new theory are in excellent agreement with observed profiles.

114.026 Spectral types for proper motion stars.
W. P. Bidelman, S.-G. Lee.
Astron. Journ., Vol. 80, 239 - 244 (1975).

Spectral types compiled from the literature are given for 601 proper motion stars, all of which were noted in the Lowell Observatory survey and are also listed in Luyten's northern hemisphere LTT catalogue.

114.027 Properties and problems of helium stars.
K. Hunger.
Problems in stellar atmospheres and envelopes, (see 003.001), p. 57 - 100 (1975).

114.028 Abundance anomalies in early-type stars.
B. Baschek.
Problems in stellar atmospheres and envelopes, (see 003.001), p. 101 - 148 (1975).

114.029 Carbon and iron abundances for eleven southern G stars of unusual interest. J. B. Hearnshaw.
Astron. Astrophys., Vol. 38, 271 - 282 (1975).

Carbon and iron abundances are presented for eleven stars by differential line blanketed model atmosphere analyses of their CH and Fe I spectra relative to the sun. Nearly all the stars are southern dwarfs or subgiants of spectral type G. Nine are high-velocity stars moving at more than 100 km/s, and three are found to be iron-rich with [Fe/H] = +0.4. It is concluded a) that carbon overabundances relative to iron are a general feature of iron-rich stars, and b) that high-velocity stars have higher (C/Fe) ratios than the average of disk stars with the same iron abundance.

114.030 Ejection of nebulae by BQ radiostars with infrared excess. F. Ciatti, A. Mammano.
Astron. Astrophys., Vol. 38, 435 - 444 (1975).

Peculiar BQ stars characterized by infrared excess, radio emission, and forbidden lines, show absorption bands indicating the presence of late-type components of high luminosity. Mass loss from these stars can cause nebulae, which would be excited by hot companions. The authors discuss the cases of compact or extended nebulosities around V 1016 Cyg, HBV 475, MWC 349, MWC 137, R Aqr, M 2−9 and IRC + 10216. New candidates for radio emission are suggested.

114.031 Ultraviolet spectrophotometry from Gemini 11 of stars in Orion.
T. H. Morgan, G. G. Spear, Y. Kondo, K. G. Henize.
Astrophys. Journ., Vol. 197, 371 - 377 (1975).

Ultraviolet spectrophotometry in the wavelength region 2600–3600 Å is reported for nine bright early-type stars in Orion. The results are in good agreement with other observations, and with the possible exception of the supergiants, are in good agreement with recent line-blanketed model atmospheres. There is evidence that the supergiants possess a small ultraviolet deficiency shortward of 3000 Å relative to main-sequence stars of similar spectral type. The most extreme example of this phenomenon is the star κ Ori.

114.032 Symbiotic stars and dust.
B. L. Webster, D. A. Allen.
Monthly Notices Roy. Astron. Soc., Vol. 171, 171 - 180 (1975).

New members of the class of very high excitation symbiotic stars, with relatively strong lines of [Fe VII], have been found. The strong unidentified emission band at 6830 Å has been detected in two of these and its identification is discussed. There is a subgroup of high excitation objects which also emit near-infrared thermal radiation from dust grains and are distinguished by the properties of their emission-line spectra. It is proposed that the dust exists in neutral gas and that the optical and infrared spectra of this subgroup of symbiotic stars arise because they are radiation bounded either at their perimeter or around condensations.

114.033 Observations of the C III λ8500 $(3s\,^1S - 3p\,^1P)$ line in O and Of stars.
D. Mihalas, S. A. Frost, G. W. Lockwood.
Publ. Astron. Soc. Pacific, Vol. 87, 153 - 161 (1975).

Measurements from coudé infrared image-tube spectra of the equivalent widths of the C III λ8500 $(3s\,^1S - 3p\,^1P^0)$, He II λ10124 $(n = 4 - n = 5)$, and hydrogen $P\delta$ lines have been made for several O and Of stars, and are presented here. The C III λ8500 line strengths provide useful constraints on possible theoretical models, and suggest, on the basis of the calculations of Nussbaumer (1971), that the C III λ5696 $(3p\,^2P^0 - 3d\,^1D)$ and λ8500 lines are formed in relatively extended atmospheres, with a dilution factor ≈ 0.1.

114.034 N-type carbon stars and the 3-α process.
S. Kilston.
Publ. Astron. Soc. Pacific, Vol. 87, 189 - 206 (1975).

High-dispersion yellow-red spectra of eight representative N-type carbon stars have been matched by Minnaert model synthesis, which enables proper allowance for the relative contributions of each atomic and molecular line to observed spectral features. The sources of data are discussed, as well as the calculation of f-values. A brief description of the line-formation model is found. Methods used for determination of physical parameters and abundances are discussed. The synthesis fitting-procedure is described, together with problems encountered for each star. The numerical results, as well as comparisons with previous work are given.

114.035 Effects of line blocking in stars of spectral types O5−G0 and luminosity classes I−V.
A. Ardeberg, B. Virdefors.
Astron. Astrophys., Vol. 39, 21 - 31 (1975).

Line-blocking coefficients have been derived based on selected data from the literature regarding line blocking and line photometry. Stars of spectral types O5−G0 and luminosity classes I−V (pop. I) have been included. The wavelength region λλ 3275 – 10000 has been covered. Pass bands have been selected so as to coincide with those of modern absolute-flux measurements. The line-blocking coefficient depends in all cases clearly on luminosity class as well as on spectral class.

114.036 Linear polarization of Hα in Be stars.
R. Poeckert.
Journ. Roy. Astron. Soc. Canada, Vol. 69, 39 - 40 (1975).
Abstr. Canadian Astron. Soc.

114.037 Ca II H and K reversals in carbon stars.
H. B. Richer.

Astrophys. Journ., Vol. 197, 611 - 614 (1975).
Nine heavily exposed, high-dispersion blue spectrograms of seven different hot carbon stars were obtained. All spectra show a reversal in the Ca II H and K lines. The widths of the reversals were measured and, after establishing a calibration, the absolute visual magnitudes of the seven stars were derived. Four of the stars, all ordinary R stars, have absolute magnitudes and colors which place them close to the normal giant branch in the spectral range G8 to K2. The remaining three stars are all hydrogen-deficient carbon stars. All these stars are supergiants, lying along the Ib supergiant branch in a color-magnitude diagram.

114.038 **Interpretation of the Be stars.** S.-S. Huang.
Sky Telescope, Vol. 49, 359 - 362, 367 (1975).

114.039 **Energy distribution in the spectra of ten stars of B2–G9 type.**
N. S. Komarov, E. A. Depenchuk, R. I. Chuprina.
Astrometriya i Astrofizika, *Kiev*, vyp. (No.) 24, (see 003.003), p. 65 - 70 (1974). In Russian.
The energy distributions E_λ in ten stars of different spectral types ($B2–G9$) in the spectral region from 3250 to 7500 Å are calculated. Values of E are given in absolute units. The results of other authors' data are compared.

114.040 **The effective temperatures of early-type stars derived from TD 1 satellite ultraviolet photometry.**
K. Nandy, E. G. Schmidt.
Astrophys. Journ., Vol. 198, 119 - 125 (1975).
A grid of model atmospheres has been calculated, and the resulting energy distributions have been compared with measured energy distributions of 14 stars of spectral types between B0 and A2. The energy distributions are from a combination of TD 1 satellite data and ground-based photometry, and extend from 1300 Å to 5500 Å. For the star ε Ori, the authors derive the luminosity from their data and the angular diameter, assuming it to be a member of the Orion association. This luminosity implies that the mass of this B0 Ia star is $20–25\,M_\odot$.

114.041 **Isolated Strömgren spheres as a source of galactic Hα emission.** B. G. Elmergreen.
Astrophys. Journ., (*Letters*), Vol. 198, L31 - L35 (1975).
The isolated Hα emission features reported by Reynolds et al. are identified with Strömgren spheres surrounding observed O stars; the root mean square electron densities and Strömgren radii are calculated from the observed emission measure and stellar type. The density has an average value of $2\,cm^{-3}$ but increases to $3.3\,cm^{-3}$ for stars within 200 pc of the galactic plane. An analysis of the Catalogue of Galactic O Stars shows that enhanced Hα emission is observed around all known O stars.

114.042 **Preparation of an atlas of late-type spectra.**
P. C. Keenan.
Bull. American Astron. Soc., Vol. 7, 233 (1975). – Abstr. AAS.

114.043 **A search for technetium in red giant variables.**
I. Little-Marenin, S. J. Little.
Bull. American Astron. Soc., Vol. 7, 234 (1975). – Abstr. AAS.

114.044 **Temperature, gravity and abundance determinations of field horizontal branch stars.** S. C. Danford.
Bull. American Astron. Soc., Vol. 7, 239 (1975). – Abstr. AAS.

114.045 **Carbon and nitrogen abundances in the atmospheres of subgiant and asymptotic giant branch stars in M92.** D. Butler, D. Carbon, R. P. Kraft.
Bull. American Astron. Soc., Vol. 7, 239 - 240 (1975). Abstr. AAS.

114.046 **A search for Ap stars in southern galactic clusters.**
M. R. Hartoog.
Bull. American Astron. Soc., Vol. 7, 270 - 271 (1975). Abstr. AAS.

114.047 **The resonance lines of triply ionized silicon and carbon in the UV spectra of O and B stars.**
R. J. Panek, B. D. Savage.
Bull. American Astron. Soc., Vol. 7, 271 - 272 (1975). Abstr. AAS.

114.048 **Carbon and nitrogen abundances in F and G-type dwarfs.** R. E. S. Clegg, R. A. Bell.
Bull. American Astron. Soc., Vol. 7, 272 (1975). – Abstr. AAS.

114.049 **The spectral classification of dwarf M stars.**
P. C. Boeshaar.
Bull. American Astron. Soc., Vol. 7, 272 (1975). – Abstr. AAS.

114.050 **Spectrophotometry of hot carbon stars: Ba II stars.**
C. E. Gow.
Bull. American Astron. Soc., Vol. 7, 272 (1975). – Abstr. AAS.

114.051 **19 new peculiar A stars.**
A. F. Gulliver, D. A. MacRae.
Astron. Journ., Vol. 80, 402 - 403 (1975).
A list containing 19 new and confirmed peculiar A stars is presented. Most of these stars have not been previously identified as being peculiar or their peculiarity has not been well established. A brief description has been provided for three particularly noteworthy stars, HD 7374, HD 51418, and HD 135679.

114.052 **Observations of lithium dilution and rotational velocity decay in F and G giant stars.**
W. R. Alschuler.
Bull. American Astron. Soc., Vol. 7, 338 - 339 (1975). Abstr. AAS.

114.053 **Spectral flux calibrations (.3 to 1.1 μm) of bright stars for use as standards for planetary astronomy.**
C. Pieters, K. Andersen, M. Gaffey, S. Nygard, F. Vilas.
Bull. American Astron. Soc., Vol. 7, 390 (1975). – Abstr. AAS.

114.054 **Wolf-Rayet stars.**
S. V. Rublev, A. M. Cherepashchuk.
Instationary stars and methods of their investigation. Phenomena of instationarity and stellar evolution, (see 003.012), p. 47 - 124 (1974). In Russian.

114.055 **Stars of spectral class B with emission lines.**
A. A. Boyarchuk.
Instationary stars and methods of their investigation. Phenomena of instationarity and stellar evolution, (see 003.012), p. 125 - 150 (1974). In Russian.

114.056 **A six-color Q-parameter for yellow supergiants.**
S. B. Parsons, R. A. Bell.
Dudley Obs. Rep. No. 9, (see 012.008), p. 73 - 85 (1975).
Using spectrum synthesis with model atmospheres, the authors devise a Q-method for obtaining accurate intrinsic colors and color excesses from the Stebbins and Kron UVBGRI photometric system. Values of E(B–V) obtained earlier by Parsons and Bouw, smaller than those of most other workers, are essentially confirmed.

114.057 **Preliminary results of UV spectrophotometry.**
J. R. W. Heintze, T. M. Kamperman, N. Sakhibullin.
Dudley Obs. Rep. No. 9, (see 012.008), p. 151 - 160 (1975).
An investigation has been started to see whether or not

it will be possible to determine T_e, log g, and microturbulence from a photometric analysis of UV spectra of stars of luminosity III, IV, and V observed by the Utrecht S59 experiment in the TD1A satellite.

114.058 A temperature calibration for B stars.
A. G. D. Philip, B. Newell.
Dudley Obs. Rep. No. 9, (see 012.008), p. 161 - 169 (1975).

A new population I effective temperature calibration of the Strömgren [u–b] parameter is presented in terms of the Hayes (1970) spectrophotometric calibration of Vega. It is shown that [u–b] is sufficiently selective that this population I calibration can be applied to derive reliable effective temperatures for population II blue horizontal-branch stars.

114.059 Metal line blocking in A stars.
S. C. Danford.
Dudley Obs. Rep. No. 9, (see 012.008), p. 177 - 182 (1975).

Coudé spectra of 39 A stars were measured in 25 Å bandpasses in the ultraviolet, blue and red spectral regions to determine light fractions blocked by metal lines. The results are applied to the Oke bandpasses and to the bandpasses of the Strömgren photometric system.

114.060 Ultraviolet spectra with line opacities.
E. Peytremann.
Dudley Obs. Rep. No. 9, (see 012.008), p. 183 - 197 (1975).

The author presents a few theoretical ultraviolet spectra that have been calculated with a large number of lines. He discusses the amount of line blocking in several spectral ranges, various effective temperatures, surface gravities and scaled solar abundances. Finally he shows a few comparisons of his spectra with satellite observations (TD1, S2/68 experiment).

114.061 The Na D lines as surface gravity indicators.
D. F. Gray.
Dudley Obs. Rep. No. 9, (see 012.008), p. 457 - 466 (1975).

Preliminary results are reported for surface gravity measurements in fourteen F, G, and K stars of luminosity classes IV and V. The gravity values are derived from a comparison of photoelectrically measured profiles of the Na D lines with model atmosphere calculations.

114.062 Comparison of spectrophotometric scans.
M. Breger.
Dudley Obs. Rep. No. 9, (see 012.008), p. 481 - 491 (1975).

Over a thousand published and unpublished photoelectric scans of stars have been collected. Typical bandpasses are 50 Å. These observations have been transformed to a uniform absolute calibration of Vega. In some cases additional systematic corrections have been applied to the measured fluxes to bring them to the uniform system. The accuracy of the different observational sources has been evaluated by comparing stars in common between observers as well as comparing synthetic and observed uvby indices.

114.063 Stellar rotation, and violations of the odd-even effect in the manganese stars. C. R. Cowley.
Astrophys. Journ., Vol. 198, 379 - 382 = Dominion Astrophys. Obs., *Victoria*, Contr. No. 256 (1975).

Abundances of phosphorus and yttrium are examined in a number of slowly rotating Ap stars. It is argued that slow rotation is not a sufficient condition to establish the odd-Z abundance anomalies that are frequently observed for these elements. Implications for the diffusion hypothesis are discussed.

114.064 The interstellar lines of the Feige stars.
J. G. Cohen, D. A. Meloy.
Astrophys. Journ., Vol. 198, 545 - 549 (1975).

New measurements of the equivalent widths and radial velocities of the interstellar lines of Ca II and Na I in the spectra of Feige stars are presented. The upper limits to the Na I interstellar line imply that the gas in the halo has a much larger value of the ratio of the column density of Ca II to that of Na I than does the plane, and that this ratio is larger than the intermediate value obtained previously from a group of brighter halo stars. From this the authors deduce that there is gas up to at least 1 kpc above the plane, and that this gas has much more Ca II relative to Na I than does the plane. The velocity measurements imply that this gas is moving slowly toward the plane in a time scale such that replenishment of the halo gas is necessary.

114.065 The Copernicus observations: interstellar or circumstellar material?
G. Steigman, P. A. Strittmatter, R. E. Williams.
Astrophys. Journ., Vol. 198, 575 - 582 (1975).

It is suggested that the sharp absorption lines observed in the ultraviolet spectra of early-type stars by the Copernicus satellite may be entirely accounted for by the circumstellar material in the H II regions and associated transition zones around the observed stars. If this interpretation is correct, the Copernicus results yield little information on the state of any interstellar (as opposed to circumstellar) gas, and in particular shed little light on the degree of element depletion in interstellar space.

114.066 Behavior of λ2800 Mg II in stellar spectra.
G. A. Gurzadyan.
Publ. Astron. Soc. Pacific, Vol. 87, 289 - 299 (1975).

The results of measurements of the equivalent widths of the resonance doublet of ionized magnesium λ2800 Mg II in the spectra of 51 relatively faint stars, up to 10^m, of the spectral classes B1-K5 are presented. The observed material has been obtained by means of the space observatory "Orion-2". Some regularities in the behavior of λ2800 Mg II in stellar spectra are discussed. A well-defined empirical relationship between the equivalent width of λ2800 Mg II and the spectral class of the star has been established.

114.067 Some correlations between abundance anomalies of elements in the atmospheres of Ap stars.
V. L. Khokhlova.
Pis'ma v Astron. Zhurn., Vol. 1, No. 6, p. 28 - 31 (1975). In Russian.

Correlations between abundances of some elements in the atmospheres of magnetic Ap stars are revealed, especially distinct for overabundances of Mn and Zr and also for iron peak elements and rare earths. These correlations may provide a clue to the origin of anomalies in the chemical composition of Ap stars.

114.068 Cyanogen strengths, luminosities, and kinematics of K giant stars. K. A. Janes.
Astrophys. Journ., Suppl. Ser., No. 282, Vol. 29, 161 - 183 (1975).

DDO intermediate-band photometry of 1200 G and K stars is used to derive procedures to estimate the anomalous cyanogen strength and the absolute visual magnitude of a G or K giant star. As expected, the CN-strength index, δCN, is correlated with [Fe/H], but the development of the absolute magnitude calibration leads to several interesting results: (i) A revision is necessary in the Wilson-Bappu calibration of K-line absolute magnitudes, $M_v(K)$. (ii) No correlation is found between δCN and $M_v(K)$, which implies that there is no metallicity-dependence in $M_v(K)$. (iii) The new calibration yields a Hyades distance modulus $(m-M)$ = 3.22 mag. The DDO absolute magnitudes, plus published radial velocities and proper motions, are used to calculate space velocities for 799 giants. The Z velocities (perpendicular to the galactic plane) show the expected correlation with δCN, and the extreme CN-weak stars have high velocities in the plane.

114.069   **Orion 2: ultraviolet spectra of faint stars.**
G. A. Gurzadyan.
Zemlya i Vselennaya, 1975, No. 3, p. 2 - 7. In Russian.

114.070   **Wolf-Rayet stars. VI. The nature of the optical and infrared continua.**
M. Cohen, M. J. Barlow, L. V. Kuhi.
Astron. Astrophys., Vol. 40, 291 - 302 (1975).

Scanner spectrophotometry between 3300 and 11 100 Å of Wolf-Rayet stars is combined with infrared photometry between 1.6 and 11.3 μ to produce composite energy distributions. The dereddened stellar continua are substracted from these distributions yielding "difference spectra" of the infrared excesses. These spectra are matched by either a free-free emission continuum or a blackbody-like continuum. WN stars show only free-free emission whereas only WC stars show dust. Physical parameters of the various emitting regions are derived. New red and near-infrared spectra of WC 9 stars have been obtained and evidence is found for visual extinction by circumstellar dust around the star Ve 2–45. Interpretation of the excesses in WC 9 stars as thermal emission by graphite grains yields estimates of the radii and dust masses of the circumstellar shells. The evolutionary status of the unusual WC 9 star Ve 2–45 is briefly discussed.

114.071   **Spektraluntersuchung von Hα-Sternen im Gebiet des Nordamerikanebels.**   R. Hudec.
MVS, *Sonneberg*, Vol. 6, 171 (1975).

29 Hα-stars from Welin's list were examined for emission lines.

114.072   **The C-classification of spectra of carbon stars, II.**
Y. Yamashita.
Ann. Tokyo Astron. Obs., Second Ser., Vol. 15, (No. 1), 47 - 59 (1975).

The C-spectral types based on the system of Keenan and Morgan are presented for 108 carbon stars observed since 1972 at the Okayama Astrophysical Observatory. In addition, revised spectral types and supplementary data of atomic and molecular line intensities are also given for 27 stars listed in the previous paper.

114.073   **How can Ap and Am stars be investigated at large distances?**   H. M. Maitzen.
Conference on optical observing programs on galactic structure and dynamics, (see 012.013), p. 253 - 256 (1975). In German.

114.074   **Identification of HDE stars.**
F. Ochsenbein, R. Bonnet.
Centre de Données Stellaires, Inform. Bull. No. 8, (see 002. 007), p. 8 (1975).

114.075   **The Eta Carinae complex and the evolution of heavy metal stars.**   M. W. Feast.
Monthly Notes Astron. Soc. Southern Africa, Vol. 34, 76 (1975). – Abstract.

114.076   **Future trends in stellar spectroscopy.**
M. K. V. Bappu.
Bull. Astron. Soc. India, Vol. 2, 23 - 25 (1974). – Presidential address.

114.077   **Emission lines in stellar spectra: observations and interpretation.**   R. E. Gershberg, L. Luud.
ENSV Teaduste Akadeemia W. Struve nimeline Tartu Astrofüüsika Observatoorium, Akademiya nauk Ehstonskoj SSR, Tartuskaya astrofizicheskaya observatoriya imeni W. Struve,

Preprint No. 7, 36 pp. Price 18 Kop. Tartu (1975).

Bright emission lines of hydrogen, calcium, helium, nitrogen, carbon, silicium and other elements have been found in the spectra of more than 6000 stars. The authors do not intend to give a review of the present state of observations and of the theories available for all types of stars with emission lines. They review only the results that have been obtained by the application of Sobolev's (1947) method in analysing the stellar atmospheres of several types: they do not touch on such emission stars as novae, supernovae, U Gemtype, irregular and semiregular stars of high luminosity.

114.078   **CH-like stars.**   Y. Yamashita.
Publ. Astron. Soc. Japan, Vol. 27, 325 - 331 (1975).
CH-like stars are red carbon stars of early types. Their spectra show a close resemblance to those of CH stars, typical high-velocity carbon stars of population II, but their radial velocities and proper motions show no indication of high velocity. The properties of CH-like stars are discussed in relation to other types of carbon stars and other types of peculiar stars of late types. The number of CH-like stars found thus far at Okayama amounts to 16. This number may be compared with 16 CH stars and 33 ordinary carbon stars of C0 – C3.

114.079   **The problem of the faint stars for a data center.**
C. Jaschek.
Centre de Données Stellaires, Inform. Bull. No. 8, (see 002. 007), p. 9 - 10 (1975).

114.080   **List of spectroscopic and photometric catalogues lately published or to be published – list V.**
B. Hauck.
Centre de Données Stellaires, Inform. Bull. No. 8, (see 002.007), p. 17 - 21 (1975). – IAU Commission: Group on spectroscopic and photometric data.

114.081   **Energy distribution in the spectra of 100 stars as a result of independent investigations at the Sternberg Astronomical Institute and the Astrophysical Institute of the Kazakh Academy of Sciences.**
I. N. Glushneva, A. V. Kharitonov, I. B. Voloshina, E. A. Glushkova, V. T. Doroshenko, E. A. Kolotilov, M. F. Novikova, I. G. Petrovskaya, V. T. Rebristyj, V. M. Tereshchenko, T. S. Fetisova, L. D. Frishberg.
Soobshch. Gos. Astron. Inst. Shternberga, No. 197 - 198, 72 pp. (1975). In Russian.

The mean values of the stellar energy distribution for 100 stars of different spectral classes are presented as result of independent investigations at the Sternberg Astronomical Institute and the Astrophysical Institute of Kazakh Academy of Sciences. The observations were carried out by means of photoelectric spectrophotometers in the region λλ 3200–7600 Å. The agreement between the data is good. The differences between these data for the two observatories don't exceed 5% in most cases.

114.082   **Orion 2: first scientific results.**
G. A. Gurzadyan.
Vestn. AN SSSR, 1975, No. 1, p. 13 - 24. In Russian.
Abstr. in Referativ. Zhurn. 51. Astron., 6.51.549; 62. Issled. kosmich. prostranstva, 6.62.145 (1975).

114.083   **On the chemical composition of the sun, Jupiter, meteorites and Am stars.**
Eh. M. Drobyshevskij.
AN SSSR. Fiz.-tekhn. in-t. Preprint 478. Leningrad, 1974, 25 pp. In Russian. – Abstr. in Referativ. Zhurn. 51. Astron., 6.51.579 (1975).

114.084   **Spectral observations of Of and Wolf-Rayet stars in the 0.8–1.1 μ range.**

Y. Andrillat, J. M. Vreux.
Astron. Astrophys., Vol. 41, 133 - 136 (1975). In French.

Spectra of twenty-one O (mainly Of) and ten Wolf-Rayet stars have been obtained between 7000 Å and 12000 Å with a dispersion of 230 Å/mm. The sample of O stars shows that if high temperature and low gravity are highly favorable for emission of He I 10830, they are not however either necessary or sufficient conditions to ensure emission of this line.

114.085 **Classifications of red stars for statistical investigations.** A. G. Velghe.
Reprinted from 'Symposium of the new astronomy', Bloemfontein, South Africa, 1972, 9 pp. = Obs. Roy. Belgique, Commun., Sér. A, No. 25 (1973).

114.086 **Spectroscopic observations of stars in H II regions.** D. Crampton, W. A. Fisher.
Publ. Dominion Astrophys. Obs., Victoria, Vol. 14, (No. 12), 283 - 304 = NRC No. 14395 (1974).

Radial velocities, spectral types and Hγ equivalent widths are given for 57 OB stars in H II regions in the northern hemisphere, and Hγ equivalent widths for 105 stars in southern H II regions.

114.087 **The composition of Beta Coronae Borealis.** R. E. Stencel, C. R. Cowley.
Publ. Dominion Astrophys. Obs., Victoria, Vol. 14, (No. 13), 305 - 317 = NRC No. 14568 (1975).

Elemental identifications for 100 ionic species based on wavelength coincidence statistics for the λλ3600 to 6825 spectrum of β CrB are presented. The possible occurrences of Ne, Zn, P, Os and Li are discussed and upper limits on their atmospheric abundances derived. Magnetic phase related spectral variations of a weak line at 6707.75 Å, which may be Li I, are discussed.

114.088 **Prismatic spectrograms of southern MK stars.** A. R. Condal, H. J. Wood, H. Moreno, L. Campusano B., L. Celis S., P. Córdova V., C. Hollemart V., M. Pedreros A., M. Peña C., C. Sterken, M. Trujillo L..
European Southern Obs., Bull. No. 12, p. 45 - 52 (1975).

The authors present several series of spectrograms of southern stars with MK spectral types which were obtained at Cerro Tololo Inter-American Observatory. The purpose of the paper is to provide, for students and researchers, examples of the MK standards taken with a small telescope and fast prismatic spectrograph. Spectrograms of 38 stars are presented.

114.089 **Variability in Ap stars.** J. M. Huntley.
Journ. American Ass. Variable Star Observers, Vol. 2, 73 - 75 (1973).

114.090 **Comparison of a galactic and a Large Magellanic Cloud G-tye supergiant.** M. A.Fry.
Thesis, California Univ., Los Angeles (USA). 143 pp. University Microfilms Order No. 74-11,528 (1973).

114.091 **Effective temperatures, surface gravities, and chemical compositions of early B-type stars.**
G. J. Peters.
Thesis, California Univ., Los Angeles (USA). 338 pp. University Microfilms Order No. 74-18,801 (1974).

114.092 **Observations of lithium dilution and rotational velocity decay in F and G giant stars.**
W. R. Alschuler.
Thesis, California Univ., Santa Cruz (USA). 46 pp. University Microfilms Order No. 74-20,488 (1974).

114.093 **Carbon and nitrogen abundances in metal-poor stars.** C. A. Sneden.

Thesis, Texas Univ., Austin (USA). 249 pp. University Microfilms Order No. 74-14,768 (1974).

114.094 **Sonneberger Astronomen erforschen jüngste Sterne.** W. Wenzel.
Spektrum, Vol. 5, No. 2, p. 12 - 16 (1974).

Measurement of the HeI 4471 Å profile at an electron density of 10^{15} cm^{-3}. See Abstr. 022.007.

The blackbody as a standard light source. See Abstr. 022.012.

The continuous absorption coefficient of the negative oxygen ion in the infrared. See Abstr. 022.039.

A table of semiempirical gf values. Part 1: Wavelengths: 5.2682 nm to 272.3380 nm. Part 2: Wavelengths: 272.3395 nm to 599.3892 nm. Part 3: Wavelengths: 599.4004 nm to 9997.2746 nm. See Abstr. 022.086.

On digital reduction of stellar spectrograms. See Abstr. 031.264.

The Catalogue of Stellar Identifications, progress report III. See Abstr. 041.036.

On the chemical composition of the moon, Jupiter, meteorites and Am stars. See Abstr. 061.031.

Distribution of spontaneous fission fragments in nuclear astrophysical problems. See Abstr. 061.035.

The radio and infrared spectrum of early-type stars undergoing mass loss. See Abstr. 064.002.

Time-dependent hydrodynamic phenomena in the envelopes of the Be stars. See Abstr. 064.007.

A possible width-luminosity correlation of the Ca II K$_1$ and Mg II k$_1$ features. See Abstr. 064.010.

A study of M dwarfs. I. Preliminary model atmospheres. See Abstr. 064.031.

The radiation field in photospheric models for extreme supergiants. See Abstr. 064.038.

Atmospheric structure from spectral line intensities. See Abstr. 064.055.

The energy balance and mass loss of stellar coronae. See Abstr. 064.070.

Interpretation of the stellar metallicity distribution. See Abstr. 065.026.

Emission lines in the wings of H and K Ca II. See Abstr. 073.062.

Photoelectric radial velocities, paper VI. Heard's IAU standard stars. See Abstr. 112.003.

Emission-line effects in Hα, Hβ, Hγ, and uvby photometry for B-type stars. See Abstr. 113.019.

The calibration of the reddening-free parameters of the Vilnius photometric system in temperatures, surface gravities and metallicities. See Abstr. 113.031.

Colors of metal deficient giant stars.
See Abstr. 113.039.

Some observational properties of G and K stars.
See Abstr. 113.045.

Statistical analysis of infrared color-indices of
variable late type stars.   See Abstr. 113.049.

Colors, magnitudes, spectral types, and distances for
stars in the field of the X-ray source Cyg X-1 (HDE 226868).
See Abstr. 113.052.

Bibliographic survey of published photoelectric
indices and spectral classification.   See Abstr. 113.063.

Photométrie uvbyβ et classification spectrale MK.
See Abstr. 113.065.

Temperatures and luminosities of M type dwarfs
from infrared photometry.   See Abstr. 113.066.

Empirical effective temperature, bolometric corrections, and fundamental stellar properties.
See Abstr. 115.008.

A hypothesis of the binary origin of Be stars.
See Abstr. 117.008.

W Ursae Majoris stars: period-spectral type relation
and period changes.   See Abstr. 121.027.

Spectrum of the Delta Scuti variable 20 Canum
Venaticorum: a model-atmosphere analysis.
See Abstr. 122.017.

Optical interstellar lines in southern supergiants.
See Abstr. 131.046.

The interstellar radiation field between 912 Å and
2740 Å.   See Abstr. 131.051.

The nearby interstellar radiation field between
1750 Å and 504 Å.   See Abstr. 131.052.

Detection of new stellar sources of vibrationally
excited silicon monoxide maser emission at 6.95 millimeters.
See Abstr. 131.053.

Detection of intrinsic linear polarization in emission-
line O stars.   See Abstr. 131.056.

Draft Catalog of Herbig-Haro Objects.
See Abstr. 132.036.

Spectroscopic study of the open cluster NGC 2422.
See Abstr. 153.005.

Spectral types in the open cluster NGC 6475.
See Abstr. 153.030.

Metal abundance of type II systems.
See Abstr. 154.009.

The space distribution of M giants in the Warner and
Swasey Luminosity Function Field LF 14.
See Abstr. 155.004.

Wolf-Rayet stars and galactic structure.
See Abstr. 155.027.

Spectra of Individual Stars

114.301 Equivalent width data for six giants of type G and K. D. Koelbloed, J. van Paradijs.
Astron. Astrophys., Suppl. Ser., Vol. 19, 101 - 113 (1975).

The equivalent widths of spectral lines for six G - and K - type giant stars, derived from high dispersion spectrograms, are presented.

114.302 The composite spectrum and energy distribution of XX Ophiuchi.
G. W. Lockwood, H. M. Dyck, S. T. Ridgway.
Astrophys. Journ., Vol. 195, 385 - 389 (1975).

Broad-band photometry from 0.36 to 10.2 μ, narrow-band photometry and scans from 0.75 to 1.08 μ, and Fourier transform spectrometry from 1.4 to 2.5 μ show that the peculiar Be star XX Oph has a cool companion with a spectral type of M6. The observed energy distribution can be represented by a B0 III star and a M6 III star reddened by 4 mag of visual extinction.

114.303 The spectrum of the supergiant ϵ Ori (B 0 Ia). III. The basic parameters. H. J. G. L. M. Lamers.
Astron. Astrophys., Vol. 37, 237 - 247 (1974).

Spectroscopic and photometric observations and the angular diameter of the star ϵ Ori (HD 37128, B0 Ia) are compared with predictions from a grid of non-LTE model atmospheres. The absolute flux at 4430 Å, the Balmer jump, the equivalent width of Hβ and Hγ and the Si IV/Si III ratio are predicted correctly by a plane parallel non-LTE model atmosphere with T_{eff}= 30500 ± 2000 K and log g = 3.0 ± 0.1. A comparison between these data and published evolutionary tracks shows that the best agreement is reached if ϵ Ori is a star of 40 $M_\odot$, which is in the hydrogen shell burning phase and has an age of 3 to 4 million years.

114.304 Interstellar carbon I lines in ζ Ophiuchi.
K. S. de Boer, D. C. Morton.
Astron. Astrophys., Vol. 37, 305 - 311 (1974).

With the help of two new f-values of C I lines obtained from the spectra of ζ Pup and γ^2 Vel, the authors show that the interstellar C I spectrum in ζ Oph obeys an optical-depth relation similar to the one observed for Na I. From this, the relative strengths for 19 multiplets are determined. From the literature, all processes which excite neutral carbon have been collected and the dominant ones are used to calculate populations of the fine-structure levels as a function of n_H and T.

114.305 Determination of abundances and the surface magnetic field strength of 3 Hya by a curve of growth method. H. Hensberge, C. De Loore.
Astron. Astrophys., Vol. 37, 367 - 373 (1974).

Lick and ESO coudé spectrograms of the peculiar A star 3 Hya (HD 72968, HR 3398) were analysed in the wavelength region $\lambda\lambda 3921-4623$. Absolute abundances were derived with a curve of growth analysis. The strength of the surface magnetic field was estimated from arguments involving the Zeeman intensification of lines on the saturated part of the curve of growth. A surface magnetic field strength of 2800 ± 600 Gauss is inferred.

114.306 Ultraviolet emission lines in the spectrum of Procyon.
R. G. Evans, C. Jordan, R. Wilson.
Nature, Vol. 253, 612 - 613 (1975).

The Princeton instrumentation on the satellite Copernicus has been used to observe the F5 IV star, αCMi (Procyon). In addition to previously observed lines of Mg II, the Lyman-α line of H I (1,216 Å), and the resonance lines of Si III (1,206 Å) and O VI (1,032 Å) have been observed for the first time in an F-type star.

114.307 On the character of change of hydrogen lines in the spectra of magnetic and peculiar stars. I. α^2 CVn and ϵ UMa. R. N. Kumajgorodskaya, N. M. Chunakova.
Soobshch. Spets. Astrofiz. Obs., *Zelenchukskaya*, vyp. (No.) 10, p. 21 - 35 (1973). In Russian.

On the basis of about 50 spectrograms with a dispersion of 15 Å/mm of the magnetic and peculiar stars α^2 CVn, HD 184905, and on the basis of data presented in the literature for several stars conclusions are made on the character of the variation of different parameters of hydrogen lines. The short-period variability (of order of minutes) of the Hδ-line in α^2 CVn and ϵ UMa is studied.

114.308 Ultraviolet variations of ϵ UMa. M. R. Molnar.
Astron. Journ., Vol. 80, 137 - 139 (1975).

OAO-2 spectrometer observations of the Ap variable ϵ UMa indicate that the photometric variations are due to variable ultraviolet absorption from apparently overabundant metals. These data also point out the presence of a secondary maximum of the Fe-group elements, notably Cr, that has not been reported.

114.309 The T Tauri star RU Lupi and its circumstellar surrounding.
G. F. Gahm, H. L. Nordh, S. G. Olofsson.
Icarus, Vol. 24, 372 - 378 (1975).

From simultaneous spectroscopic and photometric observations of the T Tauri star RU Lup, which was followed for nine consecutive nights, it was found that most if not all of the light variations observed on this star were caused by variable circumstellar extinction. The character and the time-scale of the variations imply that the variations are due to dust concentrations of stellar dimension crossing the line of sight to the star. The implications of this interpretation and its possible bearing on problems of protoplanetary systems are discussed.

114.310 Shell spectrum of the Be star HD 217050, 1966–1972. T. Kogure.
Publ. Astron. Soc. Japan, Vol. 27, 165 - 179 (1975).

The outer envelope of HD 217050 is studied through an analysis of the shell absorption spectrum in the Balmer series of hydrogen.

114.311 Changement observé dans le spectre de l'étoile symbiotique CI Cygni. R. Gravina.
Comptes Rendus Acad. Sci. Paris, Sér. B, Vol. 280, 115 - 116 (1975).

L'étoile symbiotique CI Cygni a brusquement augmenté d'éclat le 6 juin 1971 et en décembre 1971 nous observons un spectre de type P Cygni.

114.312 Une explosion de l'étoile HD 200 120.
M. Duval, M. Lacoarret, R. Herman, H. Hubert.
Comptes Rendus Acad. Sci. Paris, Sér. B, Vol. 280, 193 - 194 (1975).

Les auteurs aient observé en décembre 1973 une explosion d'hydrogène dans l'étoile HD 200 120 par son émission dans le spectre d'hydrogène. Cette étoile fait partie d'une série d'étoiles situées dans les bras spiraux de la Galaxie et dont l'âge peut être évalué à 20 X 10^6 ans. Les auteurs publient un tableau d'étoiles de ce type qu'ils observent régulièrement. Le cycle d'activité pour ces étoiles est de l'ordre d'une vingtaine d'années ou plus.

114.313 Neutral helium line strengths. VIII. Line profiles in the weak-helium-line star 3 Scorpii.
J. Norris, P. A. Strittmatter.
Astrophys. Journ., Vol. 196, 515 - 520 (1975).

It is shown that the profiles of the diffuse helium lines

in the weak-helium-line star 3 Sco are anomalously broad and shallow in relation to results obtained for various metal lines. An explanation is proposed in terms of partial gravitational settling which gives rise to a nonuniform distribution of helium either in patches on the surface or in depth. Various constraints and consequences of these alternatives are considered, but neither could be ruled out on the basis of present observational data.

114.314 The Mg II doublet emissions near 2800 Å observed in Alpha Tauri, Alpha Orionis, and Epsilon Pegasi.
Y. Kondo, T. H. Morgan, J. L. Modisette.
Astrophys. Journ., (Letters), Vol. 196, L125 - L128 (1975).

The Mg II doublet emissions at 2795 and 2802 Å were observed in α Tau (K5 III), ε Peg (K2 Ib), and α Ori (M2 Iab). The equivalent-width versus absolute-magnitude relation discussed earlier by Kondo et al. has been updated using these observations and recent observations of α Boo by Moos et al. The asymmetry in the 2795 Å emission together with the 2802 Å component may be an indicator of a cooler shell surrounding supergiants.

114.315 Simultaneous observations of variable stars. I. The Be star π Aqr.
R. Haefner, K. Metz, R. Schoembs.
Astron. Astrophys., Vol. 38, 203 - 207 = Veröff. Univ.-Sternw. München, Vol. 7, No. 18 (1975).

Simultaneous spectroscopic, polarimetric and photometric observations with high time resolution have been carried out for the Be star π Aqr. The results do not show significant periodic or nonperiodic variations in polarization, equivalent width of Hα, Hβ, Hγ, radial velocities or other spectral features. A comparison of the authors' mean values with other published results indicate further slow variations as already known.

114.316 A model-atmosphere analysis of the spectrum of Arcturus.
R. Mäckle, H. Holweger, R. Griffin, R. Griffin.
Astron. Astrophys., Vol. 38, 239 - 257 (1975).

The spectrum of the high-velocity giant α Boo (K 2 III) is analyzed relative to the sun, using observational material based on the Arcturus atlas, and a differential technique employing empirical atmospheric models for both stars. Various photometries together with the comprehensive line data lead to an effective temperature and surface gravity, respectively, of T_{eff} = 4260 ± 50 °K, log g = +0.90 ± 0.35. From this and from the parallax and the diameter as determined by various interferometric observations a very low mass results, lying in the range M = 0.1 to 0.6 $M_\odot$ and implying substantial mass loss. The chemical composition, as determined relative to the sun from atomic, ionic and molecular lines of 32 elements, shows an average underabundance of a factor of ~4, in agreement with what would be expected from the high space velocity of this star. From a discussion of possible evolutionary stages some evidence is found that Arcturus is not in the first giant phase which follows the main sequence, but is a more evolved star which may have experienced the helium flash.

114.317 Spectralphotometry and quantitative analysis of the hydrogen-deficient star HD 60344.
J. P. Kaufmann, K. Hunger.
Astron. Astrophys., Vol. 38, 351 - 357 (1975).

Wavelengths and equivalent widths of 120 absorption lines have been measured in the spectrum of HD 60344, in the spectral range λλ 3720–4720 Å. The spectrum has been fine analysed by use of a grid of flux constant hydrogen-line blanketed models, with T_{eff}, log g and ϵ_H (number fraction of H) as parameters. From the weak color excess and the medium strong interstellar CaK-line, a distance $r \approx 600$ pc is estimated,

from which follows log $L/L_\odot \approx 3.49$, $M \approx 1.3 M_\odot$ and $R \approx 3 R_\odot$. From the locus in the (g, T_{eff})-plane, it is concluded that HD 60344 is either evolving to or evolving away from the helium main sequence towards the red supergiants. No variability is evident from the spectrograms. The rotation velocity is $v \sin i$ = 33 km/s.

114.318 The peculiar near-ultraviolet spectrum of γ Corvi (HD 106625).   R. Faraggiana, K. A. van der Hucht.
Astron. Astrophys., Vol. 38, 455 - 456 (1975).

Spectrograms of γ Crv (B9p(Hg)) obtained with the TD–1A/S59 spectrometer show lines at 2576.5 Å and 2557 Å which are much stronger than in normal stars of the same spectral type. The line at 2576.5 Å is readily explained as a Mn II resonance line. For the blend at 2577 Å the authors offer Mn II (UV 20) as a plausible identification.

114.319 Broad absorption features in a Centauri.
A. B. Underhill, R. P. Fahey, D. A. Klinglesmith.
Astrophys. Journ., Vol. 197, 393 - 398 (1975).

Nightly mean intensity tracings of the spectrum of a Centauri in the neighborhood of lines from multiplets 41 and 43 of Fe I are presented for 10 consecutive nights. On some nights broad, shallow absorption features appear at the positions of the Fe I lines.

114.320 Curve-of-growth analysis of a red giant in M67.
R. Griffin.
Monthly Notices Roy. Astron. Soc., Vol. 171, 181 - 193 (1975).

A coudé spectrogram of a red giant (IV-202) in M67 is analysed, with respect to Arcturus, by the differential curve-of-growth technique. The overall metal abundance is found to be approximately twice that in Arcturus, i.e. half the solar value. There is an indication that the star has subsolar mass. Although these results are quantitatively very sensitive to the temperature assigned to the star, the error limits exclude the interpretation that IV-202 is 'super-metal-rich'.

114.321 Narrow-band photoelectric observations of three Wolf-Rayet binary systems: HD 211853, 190918, 192641.   A. M. Cherepashchuk.
Astron. Zhurn. Akad. Nauk SSSR, Vol. 52, 255 - 259 (1975). In Russian. English translation in Soviet Astron., Vol. 19, No. 2.

Eclipses in the continuum λ 4789 (~0ᵐ.1) and in the He II 4686 emission line (~0ᵐ.4) are discovered in the system HD 211853 (WN6 + BOI). The inclination of the orbital plane is i = 64–71°, the mass of the WN6 star is 9.5–13.5 $M_\odot$. Eclipses are not observed in the systems HD 190918 and 192641. Intrinsic brightness fluctuations are observed in HD 192641.

114.322 On the character of change of hydrogen lines in the spectra of magnetic and peculiar stars. II. Spectrophotometric study of HD 184905.
R. N. Kumajgorodskaya, N. M. Chunakova.
Astrofiz. Issled., Izv. Spets. Astrofiz. Obs., Vol. 7, p. 3 - 12 (1975). In Russian.

The behaviour of phenomenological parameters of the hydrogen lines H_γ–H_{11} and central depths of lines of other elements in the spectrum of the peculiar star HD 184905 during a period is studied.

114.323 Helium emission in the spectrum of κ Canis Majoris.
J. D. R. Bahng, E. Hendry.
Publ. Astron. Soc. Pacific, Vol. 87, 137 - 139 (1975).

Coudé spectra of κ CMa in the red region show Hα and He I λλ5876, 6678 in emission. Each of the lines has two emission components but the helium lines have no detectable absorption feature in between. While the Hα emission peaks are separated by 160 km sec^{-1}, the helium lines are separated by 400 km sec^{-1}. A simple model is proposed to account for the behavior of these emission lines.

114.324   A puzzling new emission-line object in Circinus.
N. Sanduleak.
Publ. Astron. Soc. Pacific, Vol. 87, 147 - 148 (1975).
A new emission-line object, showing unusual spectral characteristics has been discovered in an objective-prism survey of the southern Milky Way.

114.325   New cool and emission-line objects.
L. E. Krumenaker.
Publ. Astron. Soc. Pacific, Vol. 87, 185 - 187 (1975).
Several recent objective-prism discoveries are listed.

114.326   Measurements of the monochromatic flux from Vega in the near-infrared.
D. S. Hayes, D. W. Latham, S. H. Hayes.
Astrophys. Journ., Vol. 197, 587 - 592 (1975).
The authors have measured the monochromatic flux from Vega at several wavelengths in the near-infrared, primarily 10, 400, 8090, and 6800 Å.They have identified a variety of sources of error and estimate that they might accumulate to as much as ±2 percent in the fluxes for Vega. The monochromatic fluxes also provide a good calibration of the size of the Paschen discontinuity in hot stars. The color $m(1/8090) - m(1/10,400) = -0.158$ for Vega should be accurate to ±0.01 mag. For practical applications, the authors have prepared an adopted absolute spectral-energy distribution (published separately) for Vega that combines monochromatic fluxes with previously published spectral-energy distributions.

114.327   The local inhomogeneities of the chemical composition on the surface of the strontium Ap-star HD 140160.
V. L. Khokhlova, T. A. Rjabchikova (*Ryabchikova*).
Astrophys. Space Sci., Vol. 34, 403 - 411 (1975).
The surface inhomogeneities of the chemical composition of HD 140160 (χ Ser) were studied from the spectral line profiles. The method of quantitative analysis of local abundance proposed by Khokhlova (1974) was used. The overabundance of Sr up to 10^3 was found in three 'spots' on the surface of the star and normal abundance outside the spots. Iron was found to be overabundant up to 10 times in four spots and deficient by 10 times outside the spots. Some arguments in favour of a supernova hypothesis are mentioned to explain the deficiency of iron.

114.328   Mass loss observed in the ultraviolet spectrum of the A2 supergiant, Alpha Cygni.
Y. Kondo, T. H. Morgan, J. L. Modisette.
Astrophys. Journ., (*Letters*), Vol. 198, L37 - L39 (1975).
High-resolution observations of the Mg II resonance lines and nearby subsidiary absorption features in the spectrum of α Cyg have been obtained. The resonance lines show a broad, asymmetric absorption extending almost 3 Å shortward of the reference wavelength, indicating a large-scale mass loss and a slight asymmetry in the line bottom which could be due to emission. Comparison with existing theory suggests that the added absorption arises in an extended envelope or shell maintained by a mass loss of the order of 10^{-8} to $10^{-10} M_\odot$ yr^{-1}.

114.329   Spectrophotometric observations of Mu Cephei and the moon from 4 to 8 microns.
R. W. Russell, B. T. Soifer, W. J. Forrest.
Astrophys. Journ., (*Letters*), Vol. 198, L41 - L43 (1975).
The authors have obtained the first $4-8\,\mu$ spectrophotometric observations ($\Delta\lambda/\lambda \sim 0.01$) of μ Cep and the moon, using the NASA Airborne Infrared Observatory. The lunar spectrum shows nongray behavior from 6.5 to $8\,\mu$. The spectrum of μ Cep, an M2 Ia star with circumstellar emission,

shows no evidence for circumstellar excess emission from 4 to $8\,\mu$; the authors conclude that silicates provide the only infrared-active component of the circumstellar material.

114.330   Line identification list of 3 Hya.
H. Hensberge, C. De Loore.
Astron. Astrophys., Suppl. Ser., Vol. 20, 183 - 197 (1975).
Line identifications, equivalent widths and atomic data used in another study of 3 Hya are given.

114.331   Ultraviolet spectrophotometry of the emission star SAO 040183.   G. A. Gurzadyan.
Astron. Astrophys., Vol. 39, 213 - 216 (1975).
The results of spectrophotometric processing of three ultraviolet spectrograms of the emission star SAO 040183 of spectral class B2e, obtained with the help of the space observatory "Orion-2" in the range of wavelengths 1950–5000 Å, are given. The observed energy distribution has been derived in the continuous spectrum of this star as well as the absorption curve in the ultraviolet, up to 2000 Å; the latter differs from that which was known earlier through data of other observations. The absorption and emission lines which are probably present in the ultraviolet part of the spectrum of SAO 040183 are noted.

114.332   The Ap star 108 Aqr. II. – The oblique rotator model.   C. Mégessier.
Astron. Astrophys., Vol. 39, 263 - 273 (1975).
The observations made on the Ap Si 4200 star 108 Aqr can be interpreted according to the oblique-rotator model. The mathematical formalism given by Deutsch (1970) was used to describe the distribution of the elements on the stellar surface. The abundance maxima of Cr II and Fe II are distinct from those of Ti II, but lie on the same great circle on the star. The indications concerning the signs of the magnetic field of 108 Aqr given by Babcock (1958) allowed to localise the magnetic pole and the magnetic equator. It can be concluded that CR II, Fe II, and Ti II spots lie near the magnetic equator.

114.333   The symbiotic binary V 1016 Cygni, early stage of a planetary nebula.   A. Mammano, F. Ciatti.
Astron. Astrophys., Vol. 39, 405 - 412 (1975).
New spectroscopic observations of the radio-star with infrared excess V1016 Cyg in the range 3800–10900 Å have been obtained. The physical characteristics indicate that a very hot star excites a nebula which is ejected by a long-period variable. This latter may be responsible for the strong infrared excess. An evolutionary trend is suggested, linking symbiotic stars and BQ[] stars to V 1016 Cyg, leading next to the formation of a compact planetary nebula.

114.334   High resolution profiles of sodium and potassium lines in α Orionis.
L. Goldberg, L. Ramsey, L. Testerman, D. Carbon.
Bull. American Astron. Soc., Vol. 7, 233 (1975). – Abstr. AAS.

114.335   Low resolution airborne spectra of Alpha Orionis, Alpha Bootis and Alpha Scorpii in the 1.2 to 4.3 micron spectral region.   G. C. Augason, E. F. Erickson,
D. Goorvitch, F. C. Witteborn, L. J. Caroff, W. L. Bailey,
D. W. Strecker.
Bull. American Astron. Soc., Vol. 7, 233 (1975). – Abstr. AAS.

114.336   The ultraviolet spectrum of Gamma Cygni.
H. M. Johnson.
Bull. American Astron. Soc., Vol. 7, 233 - 234 (1975).
Abstr. AAS.

114.337   Variable blue object with peculiar spectrum.
A. Elvius.

Bull. American Astron. Soc., Vol. 7, 251 - 252 (1975).
Abstr. AAS.

114.338 Rapid variations of Hα in ζ Tauri. J. D. R. Bahng.
Bull. American Astron. Soc., Vol. 7, 252 (1975).
Abstr. AAS.

114.339 Photoelectric study of the bright Be stars γ Cas, 48 Per, ψ Per, and φ Per.
S. L. Baliunas, M. A. Ciccone, E. F. Guinan, P. Miskinis.
Bull. American Astron. Soc., Vol. 7, 252 (1975). – Abstr. AAS.

114.340 He II lines in the spectrum of ζ Puppis.
M. A. J. Snijders, A. B. Underhill.
Bull. American Astron. Soc., Vol. 7, 257 (1975). – Abstr. AAS.

114.341 Copernicus observations of the Ap star ε UMa.
A. D. Mallama, M. R. Molnar.
Bull. American Astron. Soc., Vol. 7, 270 (1975). – Abstr. AAS.

114.342 Spectral types for four suspected carbon stars of the Two-Micron Sky Survey. C. B. Stephenson.
Astron. Journ., Vol. 80, 404 (1975).
Four stars noted by Baumert (1974) to be in the Two-Micron Sky Survey and reported by him, on the basis of the literature, as possible carbon stars are, on the basis of new observations plus published literature, M stars. Three of them were never actually published as possible carbon stars.

114.343 Ultraviolet observations of the chromosphere of two M-supergiants. A. P. Bernat, D. L. Lambert.
Bull. American Astron. Soc., Vol. 7, 359 (1975). – Abstr. AAS.

114.344 Recent balloon observations of the chromospheric Mg II lines near 2800 Å.
Y. Kondo, T. H. Morgan, J. L. Modisette, D. R. White.
Bull. American Astron. Soc., Vol. 7, 359 (1975). – Abstr. AAS.

114.345 Evidence for a chromosphere in Vega.
F. Praderie, E. Simonneau, T. P. Snow, Jr.
Bull. American Astron. Soc., Vol. 7, 359 (1975). – Abstr. AAS.

114.346 Ultraviolet observations of Capella from Copernicus.
A. K. Dupree.
Bull. American Astron. Soc., Vol. 7, 359 (1975). – Abstr. AAS.

114.347 The Mt. Hopkins calibration of the spectral energy distribution of Vega.
D. S. Hayes, D. W. Latham.
Dudley Obs. Rep. No. 9, (see 012.008), p. 215 - 219 (1975).
The authors have measured the monochromatic flux from Vega at 6800, 8090, and 10400 Å in the near infrared. These values, when extrapolated back to 5556 Å, give a flux at that wavelength only 3% above the result reported by Oke and Schild (1970).

114.348 97-color photometry of the holmium star HD 51418. J. Hardorp.
Dudley Obs. Rep. No. 9, (see 012.008), p. 467 - 476 (1975).
Photoelectric scans in adjacent 50 Å bands from 3250 to 8050 Å were carried out over the period of light variation. All light curves are in phase, except between 3800 and 4100 Å where they are in antiphase. An oblique rotator model qualitatively explains why the light curves look different from those of HD 125248.

114.349 A search for H⁻ in the shell surrounding χ Ophiuchi.
T. P. Snow, Jr.
Astrophys. Journ., Vol. 198, 361 - 367 (1975).
The author has explored some of the consequences of a recent suggestion that the infrared excesses in a specific class

of Be stars, called extreme Be stars, are produced by H⁻ free-bound emission. The star χ Oph is shown to have an infrared flux distribution typical of extreme Be stars, and it is estimated from the observed infrared excess that the column density of H⁻ in the shell surrounding χ Oph should be on the order of a few times 10^{15} cm⁻². The discrepancy between the observed and expected column densities does not necessarily imply that the adopted interpretation of the infrared excesses in extreme Be stars is incorrect, although this is a possibility.

114.350 The circumstellar shell of Alpha Orionis from a study of the Fe II emission lines.
A. M. Boesgaard, C. Magnan.
Astrophys. Journ., Vol. 198, 369 - 378 (1975).
Nine spectrograms exposed for the 3100–3300 Å region in α Orionis at 3.4 and 6.7 Å mm⁻¹ have been obtained to study the Fe II emission lines. A model of a spherical envelope moving with a uniform velocity, which gives an effective velocity gradient of geometrical origin, has been derived to explain the observed features. This model accounts fairly well for the observed redshifts, asymmetries, widths, intensities, and shapes of most of the lines. The velocity field of the model appears to be too crude to explain all the phenomena, but the random motions which give rise to the broad lines are intimately associated with the mean motion of the infalling material. This can perhaps be explained by motions of large-scale bubbles or convective cells.

114.351 Heavy elements in the peculiar A star HD 25354.
D. M. Pyper, M. R. Hartoog.
Astrophys. Journ., Vol. 198, 555 - 559 (1975).
The spectrum of the Sr-Cr-Eu Ap star HD 25354 shows strong lines of Eu II, Dy II, Ce II, Nd II, and Sm II. Lines of Hg II and Pt II are probably present, and U II is possibly present but very weak. The authors' results disagree with an earlier line identification of HD 25354 by Jaschek and Brandi in that they find no Th, Am, or Cm lines, and in that Ho lines are not unusually strong on their spectrograms. The apparent overabundances of Hg and Pt in HD 25354 as well as in several other Ap stars of the Cr-Eu type may imply that these stars represent a transition between two groups: Ap stars having abundance anomalies in Si, Fe-peak elements, and the rare earth elements; and the Hg-Mn Ap stars.

114.352 Abundances in the halo cool star HD 128279.
F. Spite, M. Spite.
Astron. Astrophys., Vol. 40, 141 - 146 (1975).
The metal content of this star turns out to be 1% of the solar metal content. All metals analysed show the same deficiency, including the heavy elements produced by the "s" process. From the atmosphere analysis as well as from other data an absolute magnitude $M_V = 3$ is estimated. The parameters of the stellar galactic orbit are then computed. The eccentricity is small, but the component W of the space velocity is large. Interstellar lines are measured and discussed. The relative abundances of the elements are compared with abundances of stars with similar deficiencies and briefly discussed.

114.353 On the cool "Am" star in the Pleiades.
D. J. Stickland.
Astron. Astrophys., Vol. 40, 195 - 197 (1975).
A rough atmospheric analysis of the strong-lined late A type Pleiades star HD 23325 indicates that it probably exhibits overabundances of all the elements studied including Mg, Ca and possibly Sc; it is plausibly not a classical Am star.

114.354 Five new B emission-line stars.
J. Boulon, V. Doazan, N. Letourneur.
Astron. Astrophys., Vol. 40, 203 - 205 (1975). In French.
The authors give the description of the spectra of five new B emission-line stars. Previous observations show they

present at least one of the characteristics frequently observed among Be stars: radial velocity variations, diffuse lines, broad lines. One of these stars is a known spectroscopic binary.

114.355 The effect of line blocking on the light curves of the Ap star HR 5355 (HD 125248).
C. A. Pilachowski, W. K. Bonsack.
Publ. Astron. Soc. Pacific, Vol. 87, 221 - 229 (1975).

Line-blocking coefficients were measured from coudé spectrograms covering the blue and green parts of the spectra, throughout the 9^{d}3 cycle of HR 5355. The coefficient for the v band is found to vary through a range of 0^{m}06 of the continuum in a double wave, with maxima corresponding to the maxima of both the Eu II and the Cr spectrum variation, which occur in antiphase. This line blocking cannot account for the strong single-wave light variation in the v band, which has a minimum at rare-earth maximum, but may be responsible, by redistribution of flux, for the double-wave character of the b- and y-band light variations. It does not appear to be possible to interpret the basic light variations of HR 5355 by the mechanisms recently proposed for this and other Ap stars.

114.356 Ultraviolet continuous spectra of Gamma Cassiopeiae.
G. A. Gurzadyan.
Astron. Astrophys., Vol. 40, 447 - 450 (1975).

Ultraviolet spectrograms of γ Cas in the wavelength region of 2000–3800 Å have been obtained by means of space observatory "Orion-2". The distribution of the energy in the ultraviolet for γ Cas is derived. The most important feature of the spectrum of γ Cas is considered the occurrence of a powerful and very wide depression in its continuum at 2800 Å. A possible extraordinary role of magnesium absorption lines, 2800 Mg II and 2852 Mg I, in the spectrum of γ Cas, an emission line star, is indicated.

114.357 Ultraviolet spectrum of the WN 6 star HD 192163.
A. Cucchiaro, D. Macau-Hercot, C. Jamar.
Astron. Astrophys., Vol. 40, 459 - 460 (1975).

The ultraviolet spectra of the WN 6 star HD 192163 supplied by the S2/S68 experiment has been analyzed. A comparison with OAO 2 photometer results for this star and with the S2/S68 spectra of HD 50896 and 191765 has been made.

114.358 The current shell spectrum of Pleione.
T. Prabhu, M. K. V. Bappu.
Bull. Astron. Soc. India, Vol. 2, 36 (1974). – Abstract.

114.359 On the physical association of the peculiar emission-line stars HD 122669 and HD 122691.
R. F. Garrison, W. A. Hiltner, N. Sanduleak.
Publ. Astron. Soc. Pacific, Vol. 87, 369 - 371 (1975).

Spectroscopic and photometric observations indicate a physical association between the peculiar early-type emission-line stars HD 122669 and HD 122691. The latter has undergone a drastic change in the strength of its emission lines during the past 20 years. There is some indication that both stars vary with shorter time scales.

114.360 Two G giants with strong CH, HR 7606 and HR 8626.
S. R. Baird, W. J. Roberts, T. P. Snow, G. Wallerstein.
Publ. Astron. Soc. Pacific, Vol. 87, 385 - 396 (1975).

Two G0 stars, HR 7606 and HR 8626, are found to have greatly enhanced CH features but no enhancement of the lines of CI. C_2 is strong and CN moderately so. The authors conclude that these stars with strong CH are evolving to the right for the first time in the H-R diagram, while the vast majority of F and G Ib stars have already been giants and have altered their surface compositions by mass loss and mixing.

114.361 The origin of the 6379 Å absorption in SC and CS stars.
R. M. Catchpole.
Publ. Astron. Soc. Pacific, Vol. 87, 397 - 400 (1975).

The A and B bands of calcium hydride are identified at high dispersion in the SC star, UY Cen. In the SC stars, the B band is largely responsible for the λ6379 absorption noted by Rybski (1973). CaH is also present in the CS star TT Cen, but may not be responsible for the λ6379 absorption in WZ Cas.

114.362 HD 186058 and other stars with unconfirmed H emission ("Wer Dornen sucht der findet sie"!).
D. Hoffleit.
Monthly Notes Astron. Soc. Southern Africa, Vol. 34, 81 - 82 (1975).

114.363 A spectrophotometric study of the Be star HD 184279.
S. N. Svolopoulos.
Astron. Astrophys., Vol. 41, 199 - 202 (1975).

Spectra of the Be star HD 184279 have been studied. The H$_\alpha$-line appears in emission and its profile is given. The equivalent widths and the intensities of the measured absorption lines are also given. The lines observed in HD 184279 have lead the author to estimate the central star's atmospheric parameters, as well as those of its surrounding shell.

114.364 Shell of DV Aquarii.
W. R. Beardsley, M. W. King.
IAU Circ., No. 2790 (1975).

114.365 HD 87643.
W. Wamsteker.
IAU Circ., No. 2794 (1975).

114.366 New analysis of the $A^2\Delta-X^2\Pi$ system of CH and the $^{12}C/^{13}C$ ratio in Arcturus.
B. M. Krupp.
Thesis, Maryland Univ., College Park (USA). 88 pp. University Microfilms Order No. 74-17,050 (1973).

114.367 The spectrum of the supergiant Epsilon Orionis (B0 Ia). 3. The basic parameters.
H. J. G. L. M. Lamers.
Studies on the structure and stability of extended stellar atmospheres. Rijksuniversiteit Leiden (Netherlands). Proefschrift p. 43 - 72 (1974).

Radiative lifetimes for the $A\ ^1\Pi$ and $B\ ^1\Delta$ states of the CH$^+$ molecule with application to the CH$^+$ abundance in Zeta Ophiuchi. See Abstr. 022.020.

Term analysis of Fe VI. See Abstr. 022.028.

Model atmosphere analysis of the peculiar star 53 Aur. I. The physical atmospheric parameters determination and model atmospheres. See Abstr. 064.001.

Sodium line formation in Arcturus. See Abstr. 064.008.

A model-atmosphere analysis of the spectrum of Arcturus (basic spectroscopic data). See Abstr. 064.014.

The continuum flux distribution for Arcturus. See Abstr. 064.034.

The inversion of the Hγ absorption line of a B 1 Ib supergiant, ρ Leonis. See Abstr. 064.044.

A rediscussion of the atmospheric extinction and the absolute spectral-energy distribution of Vega. See Abstr. 082.044.

Photometry of six peculiar A-type stars.

See Abstr. 113.050.

Infrared observations and the effective temperature of the peculiar star HD 101065. See Abstr. 113.059.

On the effective temperature of Alpha Herculis A. See Abstr. 115.010.

The magnetic field, spectrum and light variations of the Ap star HD 49976. See Abstr. 116.004.

CoD −44°3318 – a peculiar luminous F star. See Abstr. 122.029.

The magnetic field of W Sgr and evidences for a period-magnetic field relation in pulsating variables. See Abstr. 122.115.

Interstellar absorption lines in the spectrum of Zeta Ophiuchi. See Abstr. 131.045.

Far-ultraviolet extinction in σ Scorpii. See Abstr. 131.074.

Spectral types in Trumpler 10. See Abstr. 153.010.

Errata

114.901 **Erratum: 'New catalogue of A stars with peculiar spectra (Ap) and with metallic lines (Am)'** [Astron. Astrophys., Suppl. Ser., Vol. 16, 71 - 153 (1974)].
C. Bertaud, M. Floquet.
Astron. Astrophys., Suppl. Ser., Vol. 20, 83 (1975).

115 Stellar Luminosities, Masses, Diameters, HR-Diagrams and Others

115.001 **Angular diameters and effective temperatures of red giant stars from lunar occultations with special reference to μ Geminorum.** D. W. Dunham, D. S. Evans, S. S. Vogt.
Astron. Journ., Vol. 80, 45 - 47 (1975).
Angular diameter measures of μ Geminorum made at occultation are reported leading to a mean value near 13 arc msec. This result is shown to be consistent with measures for stars of closely similar type. The calibration of effective temperature is in need of revision for normal M giants because of defects in the effective temperature-bolometric correction relationship hitherto adopted.

115.002 **Welche Fragen beantworten uns die Untersuchungen der sonnennahen Sterne?** W. Gliese.
SuW, Vol. 14, 79 - 83 (1975).

115.003 **Michelson and the problem of stellar diameters.** D. H. DeVorkin.
Journ. History Astron., Vol. 6, 1 - 18 (1975).

115.004 **The masses of O stars implied by their luminosities and temperatures.** P. S. Conti, M.-L. Burnichon.
Astron. Astrophys., Vol. 38, 467 - 470 (1975).
The positions in a theoretical H−R diagram of 72 O-stars uniformly classified and belonging to clusters with "well established" distances, have been compared with the behavior during main sequence lifetimes of 15 $M_\odot$, 30 $M_\odot$, 60 $M_\odot$, 120 $M_\odot$ models. If one excepts the O3 stars, which the authors discuss separately, there is a remarkable agreement between the lower boundary of observed stars and the theoretical ZAMS. The O star masses lie between ~18 $M_\odot$ and ~120 $M_\odot$.

115.005 **Pre-main-sequence masses and evolution in NGC 2264.** B. J. McNamara.

Publ. Astron. Soc. Pacific, Vol. 87, 97 - 101 = Lick Obs. Bull., No. 665 (1975).
The masses of nine pre-main-sequence stars in NGC 2264 have been determined by employing flux measurements extending from 0.36 μ−3.4 μ in conjunction with model atmospheres. The resultant masses are in good agreement with theoretical predictions although there appears to be a systematic trend in that those stars located near the main sequence are more massive than expected from evolutionary calculations of nonrotating stars. In addition, a probable age spread of at least 2 × 10⁶ years is shown to exist in NGC 2264.

115.006 **The angular diameter of Mu Geminorum.** M. R. Nelson.
Astrophys. Journ., Vol. 198, 127 - 129 (1975).
The angular diameter of μ Geminorum (Sp M3 III) has been measured by lunar occultation yielding a value of 15.6 milli-arcseconds for a uniformly illuminated disk or 16.5 for a fully darkened disk.

115.007 **Infrared occultation observations.** S. T. Ridgway, D. C. Wells.
Bull. American Astron. Soc., Vol. 7, 248 (1975). – Abstr. AAS.

115.008 **Empirical effective temperature, bolometric corrections, and fundamental stellar properties.** A. D. Code.
Dudley Obs. Rep. No. 9, (see 012.008), p. 221 - 240 (1975).
In this paper observational data on stellar angular diameters and essentially complete flux curves are presented. These observations provide the necessary data to determine empirical effective temperatures for thirty-two stars of spectral types earlier than the sun and bolometric corrections for many more. For those stars with reliable parallaxes fundamental luminosities, radii, and surface gravities have been determined.

115.009 **A comparison of spectrographic and DDO photo-electric luminosities and composition indices.**
K. M. Yoss, L. D. Deming.
Dudley Obs. Rep. No. 9, (see 012.008), p. 359 - 365 (1975).

A comparison is made between absolute magnitudes and CN anomalies derived through the David Dunlap Observatory intermediate-band photometry and those obtained from the literature.

115.010 **On the effective temperature of Alpha Herculis A.**
S. L. Knapp, D. G. Currie, K. M. Liewer.
Astrophys. Journ., Vol. 198, 561 - 562 (1975).

Measurements of α Her A with the amplitude interferometer yield a uniform disk angular diameter of $0\overset{''}{.}058 \pm 0\overset{''}{.}009$. The fundamental effective temperature, 2450 K, obtained from this measurement is in good accord with those of luminous stars of neighboring spectral types, and differs markedly from previous determinations.

115.011 **On the population of the instability strip on the H−R diagram.** N. N. Yakimova.
Soobshch. Gos. Astron. Inst. Shternberga, No. 189, p. 21 - 42 (1974). In Russian.

The author has compared qualitatively curved border lines of the instability strip (IS) − theoretical ones and those observed in the Galaxy and Magellanic Clouds. The monotonous change (along the IS) of mean unsmoothed parameters for cepheids may be explained as the consequence of penetration of F-, G-, H-turning points of the evolution tracks into the IS.

115.012 **On the distances and velocities of M supergiants associated with OH and H_2O emission sources.**
R. M. Humphreys.
Publ. Astron. Soc. Pacific, Vol. 87, 433 - 437 (1975).

New radial velocities for S Per, PZ Cas, and VY CMa are included in a discussion of the distances and velocities of M supergiants which are also OH and H_2O emission sources. It is shown that S Per and PZ Cas are probably associated with Per OB1 and Cas OB5, respectively. The peculiar star VY CMa may be associated with NGC 2362 and τ CMa or the more distant cluster NGC 2439. The available data on the velocities and distances of these M supergiants appear to support theoretical models for the maser emission in which the true velocity of the source is at or near the center of the OH velocity structure.

115.013 **The Weistrop Watergate.** W. J. Luyten.
Separate print Univ. Minnesota, Minneapolis, Minnesota, p. 12 - 15 (1974).

Zooming in on Betelgeuse.
Sci. American, Vol. 232, No. 2, p. 42 - 43 (1975). − Short contribution.

Stellar atmospheres − the middle man.
See Abstr. 064.061.

On the luminosity of spherical protostars.
See Abstr. 065.079.

Photoelectric observations of lunar occultations.
See Abstr. 096.001.

The application of parallaxes and photometry to the lower main sequence. See Abstr. 111.002.

The d vs. B2−V1 diagram for A−F, Ap and Am stars. See Abstr. 113.036.

Some observational properties of G and K stars.
See Abstr. 113.045.

Temperatures and luminosities of M type dwarfs from infrared photometry. See Abstr. 113.066.

Ca II H and K reversals in carbon stars.
See Abstr. 114.037.

The effective temperatures of early-type stars derived from TD 1 satellite ultraviolet photometry.
See Abstr. 114.040.

Temperature, gravity and abundance determinations of field horizontal branch stars. See Abstr. 114.044.

Cyanogen strengths, luminosities, and kinematics of K giant stars. See Abstr. 114.068.

Absolute magnitudes and motions of RR Lyrae stars.
See Abstr. 122.023.

Luminous stars in galactic supernova remnants.
See Abstr. 125.021.

Errata

115.901 Errata: 'The luminosity law for late-type main-sequence stars in the solar neighborhood' [Publ. Astron. Soc. Pacific, Vol. 86, 697 - 741 (1974)].
O. J. Eggen.
Publ. Astron. Soc. Pacific, Vol. 87, 352 (1975).

115.902 Errata: 'High-luminosity red stars in or near galactic clusters. Paper I' [Publ. Astron. Soc. Pacific, Vol. 86, 960 - 977 (1974)]. O. J. Eggen.
Publ. Astron. Soc. Pacific, Vol. 87, 352 (1975).

116 Stellar Magnetic Field, Figure, Rotation

116.001 **Rapidly rotating stars with optically thin stellar winds.** J. M. Marlborough, M. Zamir.
Astrophys. Journ., Vol. 195, 145 - 155 (1975).

The analysis of Cassinelli and Castor, concerning the role of the radiation field in the heating, cooling, and transfer of momentum to the atmospheres of early-type stars, is extended to the case of rapidly rotating stars in general, and of Be stars in particular. The general equations for steady flow are presented and the solution discussed for flow in the equatorial plane and in the region near the pole under the assumptions of an optically thin wind and radially streaming radiation. Limiting values for the ratio of velocity of escape to mean thermal speed at the sonic point are obtained if transonic flow is to occur. The observational evidence presently available is considered and is shown to support the theoretical predictions.

116.002 **Investigations of the magnetic Ap stars 53 Cam and γ Equ.** G. Scholz.
Astron. Nachr., Vol. 296, 31 - 39 (1975). In German.

The radial velocity of 53 Cam shows a dependence on the excitation, resp. ionization potential in the way that evidently lines with higher excitation potential have higher radial velocity. The author cannot explain this result with the rotator model without an essential modification of that. The amount and the time variation of the effective magnetic field agrees only approximately with that determined by Babcock. The investigation of γ Equ yielded the following results: The best way to represent both the radial velocity and the effective magnetic field strength is a period of 1786 days given by Steinitz and Pyper.

116.003 **Rotational velocities of marginal metallic-line stars.** H. A. Abt.
Astrophys. Journ., Vol. 195, 405 - 409 (1975).

Projected rotational velocities are estimated for the 44 marginal metallic-line stars (Am:) listed by Cowley et al. These rotational velocities are slightly larger than for the more pronounced Am stars. Analysis of published data on binaries indicates that the Am: stars have the same high frequency of short-period binaries as do the Am stars and that in both cases the low rotational velocities are caused mostly by tendencies toward synchronous rotation in binaries.

116.004 **The magnetic field, spectrum and light variations of the Ap star HD 49976.**
C. A. Pilachowski, W. K. Bonsack, S. C. Wolff.
Astron. Astrophys., Vol. 37, 275 - 279 (1974).

The Sr II Ap variable HD 49976 is found to have a magnetic field which varies cyclically between the limits of ± 2.0 kilogauss with a period of 2^d976. Both Sr II and Ca II K line equivalent widths and $uvby$ photometric observations vary as double waves with the same period, but the synchronous line and light maxima appear to lag the magnetic maximum by 0.15 in phase. The oblique rigid rotator model is used to interpret these results.

116.005 **Mechanisms of light variability of magnetic stars.** K. Stępień.
Postępy Astron., Vol. 23, 21 - 32 (1975). In Polish.

Recent photometric data (in particular obtained in the far ultraviolet) are discussed. Possible mechanisms of light variability are reviewed. It is shown that variations of effective temperature have no firm theoretical foundations and are not able to explain the observations. Also silicon patches cannot be the only cause of light variations. The most promising mechanism seems to be a combination of variable absorption in far ultraviolet and variable blanketing in the whole spectrum caused by nonuniform distribution of many elements.

116.006 **Magnetic star properties from spectral and photometric studies.**
I. A. Aslanov, V. L. Khokhlova, V. Shënajkh.
Uspekhi fiz. nauk, Vol. 114, 375 - 377 (1974). In Russian.
Abstr. in Referativ. Zhurn. 51. Astron., 3.51.574 (1975).

116.007 **The structure of surface magnetic fields in Ap-stars.** I. A. Aslanov.
Pis'ma v Astron. Zhurn., Vol. 1, No. 3, p. 39 - 42 (1975). In Russian.

From the analysis of radial velocities and magnetic fields of some Ap-stars, the conclusion is made that the region of spot formation and magnetic activity is situated along the equator of rotation.

116.008 **Broad-band polarization expected in magnetic M dwarfs.** D. J. Mullan, R. A. Bell.
Bull. American Astron. Soc., Vol. 7, 271 (1975). – Abstr. AAS.

116.009 **HD 151965 – a new bright Bp : Si variable star.** R. W. Hilditch.
Monthly Notices Roy. Astron. Soc., Vol. 171, 25P - 27P (1975).

New photometric observations are presented which show that HD 151965 is a variable of amplitude ∼ 0^m05 in V and period of either 1^d85 or 4^d2. The light variations, together with the published spectral classification, indicate that the star might be a magnetic variable.

116.010 **Measurement of the magnetic field of β CrB.** I. A. Aslanov, Yu. S. Rustamov.
Pis'ma v Astron. Zhurn., Vol. 1, No. 4, p. 21 - 23 (1975). In Russian.

In the Ap star β CrB a correlation is found between the strength of the magnetic field deduced from different lines and the effective optical depths of layers in which these lines are formed.

116.011 **Magnetohydrodynamics: applications to magnetic stars, cosmical gas dynamics, and pulsars.**
L. Mestel.
Magnetohydrodynamics, (see 003.018), p. 37 - 182 (1974).

116.012 **A system of standard stars for rotational velocity determinations.** A. Slettebak, G. W. Collins II, P. B. Boyce, N. M. White, T. D. Parkinson.
Astrophys. Journ., Suppl. Ser., No. 281, Vol. 29, 137 - 159 (1975).

Profiles for the He I λ 4471, Mg II λ 4481, and Fe I λ 4476 lines were measured in the spectra of 217 bright northern and southern stars of types O9—F9, using photoelectric scans and photographic coudé spectrograms obtained at the Cerro Tololo Inter-American Observatory, Kitt Peak National Observatory, and Lowell Observatory. Half-intensity widths of the observed line profiles were compared with the corresponding quantities in a set of theoretical rotationally-broadened line profiles computed using the model-atmosphere approach. The resulting $v \sin i$'s are used to establish a system of standard rotational velocity stars, for use in estimating $v \sin i$ directly from visual inspection of spectrograms. An atlas of rotationally broadened spectra is presented, showing the effects of rotation on spectra of representative B-, A-, and F-type stars.

116.013 **Magnetic field measurements in ξ Bootis A.**
A. M. Boesgaard, D. Chesley, G. W. Preston.

Publ. Astron. Soc. Pacific, Vol. 87, 353 - 355 (1975).

Four Zeeman spectrograms from Lick Observatory of ξ Boo A and two of ι Peg at 2 Å mm^{-1} have been measured to determine if a weak magnetic field is present in ξ Boo A. The results indicate that the field is too weak to be measured by this technique on these spectrograms, although remeasurements of spectrograms from Mauna Kea at 3.4 Å mm^{-1} still give a positive field of 170 gauss.

116.014 **Magnetic fields in dMe stars: how effective is the battery mechanism?**   D. J. Mullan.
Publ. Astron. Soc. Pacific, Vol. 87, 455 - 459 (1975).

Although the author agrees with Worden (1974) that the surfaces of dMe stars may be the sites of magnetic fields as large as 1−100 kilogauss, he disagrees with his conclusion that Biermann's battery mechanism is responsible for generating such large fields. The author believes that it is more probable that field generation occurs in convective conditions, where the battery cannot operate efficiently. He also disagrees with the argument used by Worden to relate magnetic energies with observed properties of flare stars.

116.015 **Aligned rotating magnetospheres.**
E. T. Scharlemann.
Thesis, Cornell Univ., Ithaca, N.Y. (USA). 107 pp. University Microfilms Order No. 74-10,199 (1973).

Large scale magnetic fields in late-type stars.
See Abstr. 065.033.

A self-consistent field method in the theory of rotating stars.   See Abstr. 065.036.

On the pulsational stability of supermassive magnetic stars.   See Abstr. 065.048.

The stability of stars containing magnetic fields.
See Abstr. 065.056.

Dynamo maintenance of magnetic fields in stars.
See Abstr. 065.073.

Equivalent widths and rotational velocities of southern early-type stars.   See Abstr. 114.001.

Observations of lithium dilution and rotational velocity decay in F and G giant stars.   See Abstr. 114.012.

New bright hydrogen-emission stars.
See Abstr. 114.022.

Observations of lithium dilution and rotational velocity decay in F and G giant stars.   See Abstr. 114.052.

Observations of lithium dilution and rotational velocity decay in F and G giant stars.   See Abstr. 114.092.

Determination of abundances and the surface magnetic field strength of 3 Hya by a curve of growth method.
See Abstr. 114.305.

On the character of change of hydrogen lines in the spectra of magnetic and peculiar stars. I. α^2 **CVn and** ϵ **UMa.**
See Abstr. 114.307.

Broad absorption features in a Centauri.
See Abstr. 114.319.

On the character of change of hydrogen lines in the spectra of magnetic and peculiar stars. II. Spectrophotometric study of HD 184905.   See Abstr. 114.322.

On the possibility of magnetic starspots on the primary components of W Ursae Majoris type binaries.
See Abstr. 117.033.

Rotational velocities in IC 2602.
See Abstr. 153.002.

117 Binary and Multiple Stars, Theory

117.001 Outburst in U Cephei. M. Plavec, R. S. Polidan. Nature, Vol. 253, 173 - 174 (1975).

Observations of U Cephei show that the mass transfer and disk formation may be a very variable process. Batten and coworkers have tried to detect emission in this system, but found either no emission or at best a marginal phenomenon. On August 8, 1974, however, the authors observed very strong Hα emission during a primary eclipse. This agrees with the model of a rotating ring around the primary component, originally proposed by Joy for RW Tauri. For the first time, however, the dispersion and time resolution have been so good as to permit a quantitative analysis of the profiles, which is under way.

117.002 Outburst of U Cephei. A. H. Batten, W. A. Fisher, B. W. Baldwin, C. D. Scarfe. Nature, Vol. 253, 174 - 176 (1975).

During the eclipse of September 7, 1974, very strong emission was observed in the course of a routine programme of observation of this and similar systems. The emission was clearly present in all Balmer lines from Hβ to H18 and also in the H and K lines of Ca II, λ4481 of Mg II, and possibly in some lines of Fe II and He I. The changing intensities of the red and violet components during the course of the eclipse suggest that the emission arises in a disk that rotates around the primary star with an average velocity close to 250 km s^{-1}.

117.003 TT Arietis: an evolved, very short period binary. A. P. Cowley, D. Crampton, J. B. Hutchings, J. M. Marlborough. Astrophys. Journ., Vol. 195, 413 - 421 (1975).

A spectroscopic investigation of the rapid variable TT Ari reveals it to be a binary system with an orbital period of 0^d1375. The spectrum is nearly continuous, with weak hydrogen emission lines superposed on shallow broad absorption. The presence of a nonuniform, rotating ring surrounding the primary is inferred from the H emission lines. The complex light variations show a broad maximum during each orbital period, in addition to rapid flickering. This variation may be due to a hot spot in a ring or disk around the primary excited by infalling matter from the secondary. The inferred model of the system is similar to that derived for old novae.

117.004 On supernova explosion in a binary system. Yu. G. Khabazin. Astron. Zhurn. Akad. Nauk SSSR, Vol. 52, 57 - 63 (1975). In Russian. English translation in Soviet Astron., Vol. 19, No. 1.

The calculations available on the evolution of close binary systems in which one component can explode show the exploding component to be of mass less than the other mass; the other star must be a young massive star on the main sequence. Binary system orbit perturbations caused by the explosion of a component are considered. The eccentricity of Cyg X-1 is notably larger than that of massive close binaries, and it is likely that there was an explosion in the Cyg X-1 binary system.

117.005 The structure of synchronously rotating close binaries built on polytropic model ν = 3. L. C. Green, E. K. Kolchin. Astrophys. Journ., Suppl. Ser., No. 271, Vol. 28, 449 - 463 (1974).

The purpose of the present paper is to develop mathematical procedures which are capable of providing an exact solution for the structure of synchronously rotating close binaries built on polytropic model ν = 3 with a view to treating, in the future, cases of small separation which may be of significance for an understanding of the fission problem. Methods are described for computing the variation with depth of the rotational and tidal deformation of the "level" surfaces. Results are discussed for separations of centers of 10.0, 4.0, 3.0, and 2.8 solar radii and for four truncation points in the series expansions, such that all relevant spherical harmonics through those of degree 2, 4, 6, and 8 are included. Only binaries with identical components are treated in the present paper but no limitation to such systems appears to be inherent in the mathematical procedures or the computational programs.

117.006 The kinematics of trapezium systems. C. Allen, A. Poveda, C. E. Worley. Rev. Mexicana Astron. Astrofis., Vol. 1, 101 - 118 (1974).

An analysis has been made of all the observations available of Ambartsumian's trapezia, with the aim of establishing the kinematic properties of this interesting class of multiple stars. It was found that there is no evidence for a systematic expansion in any of the 46 trapezia that were studied. In particular, the Orion trapezium, the best documented case, shows no expansion. However, a sizeable fraction of the trapezia show one or two stars with relative motions. A statistical test and a proper motion test have been applied to establish membership of these stars in their trapezia. Some of these stars have transverse velocities larger than 30 km sec^{-1}. This behaviour is important in understanding the dynamics of trapezia.

117.007 The masses of components and the inclination of a binary system with one pulsar determined from relativistic effects. V. A. Brumberg, Ya. B. Zel'dovich, I. D. Novikov, N. I. Shakura. Pis'ma v Astron. Zhurn., Vol. 1, No. 1, p. 5 - 9 (1975). In Russian.

The first order Doppler shift of the pulsar period in a binary system can be used to determine the mass function involving two masses, the inclination and eccentricity of the orbit. In order to determine separately the masses of the pulsar and of the companion star, frequency shift effects of the second order − gravitational redshift and quadratic Doppler effect − must be included.

117.008 A hypothesis of the binary origin of Be stars. S. Kříž, P. Harmanec. Bull. Astron. Inst. Czechoslovakia, Vol. 26, 65 - 81 (1975).

The good agreement between the predicted properties of certain categories of mass-exchanging binaries and the observed properties of Be stars is demonstrated. Most of the observed spectral changes of Be stars can be explained satisfactorily as a consequence of different modes of mass transfer between components. In particular, this applies to the long-termed variations of these objects which have remained unexplained so far. It seems reasonable, therefore, to suppose that a large portion of Be stars (if not all!) are in fact interacting binaries.

117.009 The detection of extrasolar planetary systems: Part III; Review of recent developments. A. R. Martin. Journ. British Interplanet. Soc., Vol. 28, 182 - 190 (1975).

This note, the third in a series of review papers, discusses results which have, in the main, been presented in the period following the literature search upon which the first two papers were based. New data on low mass stars, and their evolution, is mentioned.

117.010 **Spectroscopic investigation of X Persei
(= 2U 0352 + 30?).**
J. B. Hutchings, D. Crampton, R. O. Redman.
Monthly Notices Roy. Astron. Soc., Vol. 170, 313 - 324 (1975).

Radial velocity and line intensity measurements are reported from recent high dispersion spectrograms of X Persei. The binary interpretation of the previously reported 580-day periodicity is discussed and revised orbital parameters are presented. Line intensity changes occur but are not found to be correlated with this cycle. Rapid line profile changes have been observed with an Image Isocon camera. The emission line profiles are discussed in terms of a binary model.

117.011 **The optical properties of binary star systems with
accretion discs.** D. N. C. Lin.
Monthly Notices Roy. Astron. Soc., Vol. 170, 379 - 392 (1975).

This paper derives a correlation between the period and the optical visibility of a hot spot on an accretion disc in a close binary system. X-ray binaries and rapid blue variables are considered. For the HZ Her–Her X-1 system, the mass ratio is estimated and an explanation is given for the observed marching of the dips in the X-ray light curve.

117.012 **On angular momentum transfer in binary systems.**
R. E. Wilson, R. Stothers.
Monthly Notices Roy. Astron. Soc., Vol. 170, 497 - 501 (1975).

The limiting efficiency is considered with which orbital angular momentum can be converted into rotational angular momentum, J_{rot}, of the mass-gaining component in binary systems which undergo mass exchange. This limit, $(dJ/dM)_{max}$, then specifies the maximum extent to which the observed rates of period change, dP/dt, can be affected by such reduction of orbital angular momentum. Upon integrating $(dJ/dM)_{max}$ over the entire accretion process, we find that the maximum accumulated rotational angular momentum is larger than the amount implied by the observed underluminosities of stars in certain extreme binary systems, by factors of 3 to 4. Shell stars and emission-line stars in binary systems may be produced when core angular momentum is later transferred into an envelope which already has nearly the limiting J_{rot}.

117.013 **Parallax and orbital motion of the astrometric
binary BD + 6° 398.** G. E. Martin, P. A. Ianna.
Astron. Journ., Vol. 80, 321 - 322 (1975).

Plates taken with the McCormick 26-in. refractor of the astrometric binary BD +6° 398 from 1915 to 1974 have been analyzed for parallax, proper motion, and orbital motion. An orbital period of 60 yr satisfies the observations; the mass of the unseen companion is on the order of 0.12 solar masses. The data suggests a large Δm.

117.014 **The dimensions of circumstellar discs in binary
systems.** S. L. Piotrowski.
Acta Astron., Vol. 25, 21 - 28 (1975).

The examination of the dependence of the velocity of rotation of the disc on the period of revolution of the binary star as well as the statistical analysis of the scatter of observed points about the mean relation confirm the conclusion that the radius of the ring (disc) increases with the increase of the mass of the component surrounded by the gaseous stream. There are indications that generally similar dependence of the dimensions of the ring-like structure on mass ratio q and $K/V \sin i$ should be expected when no direct recourse is made to the point-mechanical treatment.

117.015 **Dynamical instabilities and mass exchange in binary
systems.** G. T. Bath.
Monthly Notices Roy. Astron. Soc., Vol. 171, 311 - 328 (1975).

A hydrodynamic scheme is presented by means of which the dynamical behaviour of stars transferring mass in semi-detached binary systems may be investigated. Cool stellar envelopes are found to be dynamically unstable when in contact with the Roche lobe, but only so long as the recombination energy is sufficient to overcome the gravitational potential change to the surface. Such unstable states are followed by phases in which the photosphere shrinks inside the Roche lobe, readjusting to thermal equilibrium. The quasi-periodic mass transfer that results is suggested as being responsible for dwarf nova eruptions, and may be of importance in X-ray sources, Algol systems, and binary symbiotic variables.

117.016 **Classification et nombre des étoiles doubles.**
P. Baize.
L'Astronomie, 89e année, p. 159 - 164 (1975).

117.017 **Gas motion in the close binary system V 444 Cyg.**
Yu. P. Korovyakovskij, Yu. V. Sukharev.
Astrofiz. Issled., Izv. Spets. Astrofiz. Obs., Vol. 7, p. 19 - 26 (1975). In Russian.

Motion of matter is considered in the close binary system V 444 Cyg, one of the components of which is a star with an extended envelope (WN 5.5 + 06). The calculations are made allowing for light pressure and gas-dynamic effects. Matter motion paths are obtained for different initial velocities of matter outflow and bolometric luminosities of the Wolf-Rayet component of the system.

117.018 **Dynamical theory of tides. II. Isentropic motion of
matter.** Yu. P. Korovyakovskij.
Astrofiz. Issled., Izv. Spets. Astrofiz. Obs., Vol. 7, p. 27 - 34 (1975). In Russian.

This paper continues the investigation of nonlinear tidal phenomena in the envelope of a satellite in a dwarf-star system started earlier (1972) for a more general case: isentropic motions of matter. Conditions for creation of a stream of matter from the satellite to the main star are investigated on the basis of numerical integration of a set of nonlinear equations of the theory of dynamic tides.

117.019 **The undersize subgiants: addendum.** D. S. Hall.
Acta Astron., Vol. 25, 95 (1975).

Concerning 11.117.033. It is shown that KO Aql no longer appears to have an undersize subgiant.

117.020 **The occultation of Beta Scorpii by Jupiter. VI.
The masses of Beta Scorpii A$_1$ and A$_2$.**
J. L. Elliot, K. Rages, J. Veverka.
Astrophys. Journ., (Letters), Vol. 197, L123 - L126 (1975).

From images of the spectroscopic binary β Sco A, formed by the Jovian atmosphere during its 1971 May 13 occultation by Jupiter, the angular separation of the binary components was found to be $(1.496 \pm 0.018) \times 10^{-3}$ arc sec. The combination of this measurement, the elements of the spectroscopic orbit, and the parallax determination of Bertiau yields masses of $21.1 \pm 3.2\ M_\odot$ for the primary (B0.5 V) and $12.7 \pm 1.9\ M_\odot$ for the secondary. New observations are suggested for improving the accuracy of the mass determinations.

117.021 **Photometric detection of extrasolar planets using
L.S.T.-type telescopes.**
A. J. Fennelly, G. L. Matloff, G. Frye.
Journ. British Interplanet. Soc., Vol. 28, 399 - 404 (1975).

Detection of planets similar to the earth and Jupiter that might be circling nearby stars is possible with the Large Space Telescope (LST), or modifications of this instrument. The use of the moon as an occulter to increase the signal to noise ratio S/N, and the expected photon fluxes from hypothetical planetary companions to α Centauri A and B, τ Ceti, and ϵ Eridani, are discussed.

117.022 Trapezium in de grote Orionnevel. M. Drummen.
Zenit, Vol. 2, 101 - 106 (1975).

117.023 On the origin of nova-like binary systems.
V. G. Gorbatsky (*Gorbatskij*).
Astrophys. Space Sci., Vol. 33, 325 - 332 (1975).
Nova-like binary systems are similar to W UMa-systems in their basic physical characteristics. Outwardly such systems are different – nova-like systems contain a white dwarf as a component, while both components of a W UMa-system are near the main sequence. A hypothesis is proposed, seeking the origin of contact W UMa-type systems in a fission of rapidly-rotating helium isothermal core of an evolved giant star. The contraction of the more massive component leads to the formation of a white dwarf and, consequently, to a transformation of a W UMa-type system into a nova-like system.

117.024 Computer simulations of gas flow around close binary systems.
S.-A. Sørensen, T. Matsuda, T. Sakurai.
Astrophys. Space Sci., Vol. 33, 465 - 480 (1975).
A gas-dynamical model of gas streams around close binary systems is given. The velocity field and the density distribution are determined for different parameter ranges. The results succeed in explaining the formation of a ring and a disk around the accreting component. The models furthermore reveal the existence of a tongue of matter extending from the inner Lagrangian point and a jet perpendicular to the system axis.

117.025 The solution parameters of very close binary systems. H. Mauder.
Astrophys. Space Sci., Vol. 34, 297 - 308 = Veröff. Remeis Sternw. Bamberg, Astron. Inst. Univ. Erlangen-Nürnberg, Vol. 10, No. 112 (1975).
The problems encountered in the derivation of elements from the light curves of very close binary systems are discussed. It is shown that the rectification procedure is critically dependent on the underlying model. The assumption of a contact configuration leads to better coincidence of theoretical and observed light curves. The Fourier analysis method is used to derive the solution parameters of the examples YY Eri and AM Leo.

117.026 The binary frequency among solar-type stars.
H. A. Abt, S. G. Levy.
Bull. American Astron. Soc., Vol. 7, 268 (1975). – Abstr. AAS.

117.027 Characteristics and evolution of close binary systems.
M. A. Svechnikov, L. I. Snezhko.
Instationary stars and methods of their investigation. Phenomena of instationarity and stellar evolution, (see 003.012), p. 181 - 230 (1974). In Russian.

117.028 Onzichtbare begeleiders van sterren.
M. Drummen.
Zenit, Vol. 2, 129 - 131 (1975).

117.029 Zèta Cancri, meervoudige ster met 'onzichtbare' witte dwerg. M. Drummen.
Zenit, Vol. 2, 171 - 174 (1975).

117.030 Gas dynamics of semidetached binaries.
S. H. Lubow, F. H. Shu.
Astrophys. Journ., Vol. 198, 383 - 405 (1975).
The authors analyze the gas dynamics of semidetached binary systems within the context of the Roche model. Using this concept, they demonstrate the following by semianalytical methods. (1) The escape of material from the surface of the contact component is accomplished by a highly nonisotropic stellar wind which reaches sonic velocities in a neighborhood of the inner Lagrangian point, L1. (2) This wind throttles into a narrow stream of material which makes a prescribed angle with respect to the line joining the stellar centers ranging from $19°.5$ to $28°.4$ for the full range of possible stellar mass ratios. (3) The stream width remains nearly constant over the part of the stream which is nearly straight, and narrows somewhat as the stream curves toward the detached component. (4) If the detached component is smaller than a certain specified size, the stream results in the formation of a disk of material of prescribed size orbiting the detached component in a direct sense.

117.031 Ultrashort-period binaries. III. The accretion of hydrogen-rich matter onto a white dwarf of one solar mass. R. E. Taam, J. Faulkner.
Astrophys. Journ., Vol. 198, 435 - 438 = Lick Obs. Bull., No. 684 (1975).
The authors have followed the thermonuclear runaways which develop when white dwarfs of $1 \, M_\odot$ accrete hydrogen-rich material at $10^{-10} M_\odot$ per annum, a canonical rate of accretion which would accompany the emission of gravitational radiation in ultrashort-period binaries. The existence of such thermonuclear effects is a necessary but not sufficient criterion for gravitational radiation to play an important role in the evolution of these binaries.

117.032 X-ray variability by matter accretion onto a black hole in a detached binary system.
A. F. Illarionov, R. A. Syunyaev.
Pis'ma v Astron. Zhurn., Vol. 1, No. 4, p. 11 - 15 (1975). In Russian.
The angular momentum of gas captured by a black hole in a detached binary system is quite small. The formation of an accreting disk emitting X-rays around a black hole is impossible. However, velocity and density fluctuations of stellar wind matter may give rise to a short-time disk formation, leading to strong X-ray variability.

117.033 On the possibility of magnetic starspots on the primary components of W Ursae Majoris type binaries. D. J. Mullan.
Astrophys. Journ., Vol. 198, 563 - 573 (1975).
The author examines the hypothesis that magnetic starspots occur in W UMa stars. Crude estimates of toroidal field strength in these rapidly-rotating stars suggest that the fields may be large enough (2−10 kilogauss) to permit spots to form. The probability of spot formation is larger on the primary component than on the secondary, especially in Rucinski's W type systems. It is shown that the existence of starspots on the primary component in W UMa systems can account for the apparent temperature excess of the secondary relative to the primary in W type systems, and for the much smaller variations in the light curve exhibited by Rucinski's A type systems. The author predicts upper limits on the amplitude of the distortions which can occur in the light curve of these systems due to starspot activity. The presence of starspots leads to flare activity.

117.034 The frequency of contact binaries and its consequence on their evolution. F. Van't Veer.
Astron. Astrophys., Vol. 40, 167 - 174 (1975). In French.
The author tries to estimate the number of contact binaries (CB) limited to a certain apparent magnitude. The number of CB's must be deduced from the number of observed W UMa type eclipsing binaries. He found it tempting to postulate that the secondary component is transformed into a protoplanetary cloud having the primary star as central condensation. The next step could be the formation of planets and satellites out of this orbiting material.

117.035 Effect of ellipticity and the parameters of X-ray

binaries Cyg X-1 and Cen X-3.
N. G. Bochkarev, E. A. Karitskaya, N. I. Shakura.
Pis'ma v Astron. Zhurn., Vol. 1, No. 6, p. 12 - 17 (1975). In Russian.

Possible masses of the components and the orbit inclinations of the X-ray binary systems Cyg X-1 and Cen X-3 were found from comparison of calculated and observed light curves. Estimate of the mass of the X-ray source Cyg X-1 gives $7 M_\odot \lesssim M \lesssim 15 M_\odot$. This confirms the existence of a black hole in this system. For Cen X-3 the observed light amplitudes larger than $0.^m07 - 0.^m08$ require an additional source of optical variability.

117.036 **A model of X Persei.** B. Paczyński, J. Ziółkowski.
Astron. Astrophys., Vol. 40, 351 - 354 (1975).

A possible model for the close binary system X Persei is discussed. The authors present evidence that the primary is a post mass loss star of about 5 solar masses, and the secondary is a rapidly rotating star of about 25 solar masses, surrounded by a gaseous disc accreted in the recent phase of mass exchange.

117.037 **On the R- and S-modes in a synchronously rotating component of a binary.** J. Denis, P. Smeyers.
Astron. Astrophys., Vol. 40, 411 - 414 (1975).

The perturbation procedure derived in a previous paper is applied to obtain the frequencies of the R- and S-modes in a synchronously rotating component of a binary. It is found that a tidal action increases the rotational splitting between the frequencies of the R- and S-modes. This effect is more pronounced when the central concentration of the model is low. It does not seem possible to explain the multiple periodicities of some β Canis Majoris stars in terms of R- and S-oscillations in the context of uniform rotation.

117.038 **The masses of components and inclination of a binary system with one pulsar determined from relativistic effects.**
V. A. Brumberg, Ya. B. Zel'dovich, I. D. Novikov, N. I. Shakura.
In-t prikl. mat. AN SSSR. Preprint No. 121, Moskva, 1974. 10 pp. In Russian. – Abstr. in Referativ. Zhurn. 51. Astron., 6.51.736 (1975).

117.039 **Planetary systems and extraterrestrial life.**
S. S. Kumar.
Origins of Life, Vol. 5, 491 - 495 (1974).

The paper reviews the present status of the problem of the existence of other planetary systems in the Galaxy. Observational data and theoretical results are presented to show that the occurrence of planetary orbits is not a universal phenomenon. Comments are made on the existence of extraterrestrial life in the solar system and around other stars in the Galaxy.

117.040 **Generation of a magnetic field and acceleration of charged particles in binary systems.**
A. Z. Dolginov.
Izv. AN SSSR. Ser. fiz., Vol. 39, 354 - 358 (1975). In Russian. Abstr. in Referativ. Zhurn. 62. Issled. kosmich. prostranstva, 6.62.215 (1975).

117.041 **Binary stellar winds.**
G. L. Siscoe, M. A. Heinemann.
Solar wind three, (see 012.020), p. 243 - 255 (1974).

On the calculation of intermediate motions in triple stellar systems. See Abstr. 042.065.

Some cases of intermediate motions in the stellar three-body problem. See Abstr. 042.066.

Two kinds of stellar collapse. See Abstr. 066.013.

Occultations of the Pleiades: reappearances observed photoelectrically at McDonald Observatory. See Abstr. 096.009.

Occultation of β Scorpii by Jupiter. V. The emersion of β Scorpii C. See Abstr. 099.012.

The occultation of β_1 Sco by Jupiter on May 14, 1971. See Abstr. 099.016.

The occultation of β Scorpii by Jupiter. III. Discussion of the photometric results. See Abstr. 099.082.

Proper Motion Survey with the forty-eight inch Schmidt telescope. XXXVIII. Binaries with white-dwarf components. See Abstr. 112.010.

Narrow-band photoelectric observations of three Wolf-Rayet binary systems: HD 211853, 190918, 192641. See Abstr. 114.321.

Rotational velocities of marginal metallic-line stars. See Abstr. 116.003.

Interpretation of BM Orionis. See Abstr. 121.003.

Fourier analysis of the light curves of eclipsing variables, I. See Abstr. 121.028.

Masses and orbital elements of the binary system Phi Persei. See Abstr. 121.034.

The systemic velocity of β Lyrae. See Abstr. 121.060.

On the model of outburst of a U Gem star. See Abstr. 122.012.

The location of the hot spot in cataclysmic variable stars as determined from particle trajectories. See Abstr. 122.015.

Evolution of helium white dwarfs in close binaries. See Abstr. 126.015.

Discovery of a pulsar in a binary system. See Abstr. 141.304.

On the nature of the binary system containing the pulsar PSR 1913+16. See Abstr. 141.308.

Some implications of period changes in the first binary radio pulsar. See Abstr. 141.309.

First radio pulsar in a binary system. See Abstr. 141.310.

Properties of the Hulse-Taylor binary pulsar system. See Abstr. 141.319.

Periastron shifts in the binary system PSR 1913+16: theoretical interpretation. See Abstr. 141.320.

Timing effects in pulsed binary systems. See Abstr. 141.325.

On the origin of the binary pulsar PSR 1913 + 16.
See Abstr. 141.338.

Upper limit of electron concentration in the vicinity of the pulsar PSR 1913 + 16 in the binary system.
See Abstr. 141.341.

On the binary system containing PSR 1913+16: I. Constraints on the nature of the companion imposed by geometry and the classical apsidal motion test.
See Abstr. 141.342.

On the binary system containing PSR 1913+16: II. Classical and relativistic effects and the determination of the orbital elements.   See Abstr. 141.343.

Binary pulsar PSR 1913+16: model for its origin.
See Abstr. 141.357.

Search for an optical counterpart of the binary pulsar PSR 1913+16.   See Abstr. 141.359.

The masses of binary X-ray sources.
See Abstr. 142.002.

The implausible history of triple star models for Cygnus X-1: evidence for a black hole.
See Abstr. 142.011.

Physical conditions on the stream of Cyg X-1 produced by the X-ray flux.   See Abstr. 142.023.

Circular polarisation in HDE226868.
See Abstr. 142.026.

Calculation of the light curves of HZ Herculis.
See Abstr. 142.035.

Radial velocities of Scorpius X-1.
See Abstr. 142.048.

Is Cir X-1 a runaway binary?
See Abstr. 142.053.

Binary systems with an X-ray component.
See Abstr. 142.054.

Massieve röntgendubbelsterren: ontstaan en evolutie.
See Abstr. 142.063.

Neutron star wobble in binary X-ray sources.
See Abstr. 142.068.

Sco X-1 and Cyg X-2 as binary systems.
See Abstr. 142.072.

Mass determination for the degenerate member of binary X-ray sources.   See Abstr. 142.078.

X-ray sources in binary systems.
See Abstr. 142.079.

Expected polarization properties of binary X-ray sources.   See Abstr. 142.080.

Modes of mass transfer and classes of binary X-ray sources.   See Abstr. 142.094.

X-ray binaries and asymmetry of supernova outbursts.   See Abstr. 142.096.

A model of the X-ray source Her X-1 and nature of the 35-day cycle.   See Abstr. 142.099.

Supernova explosions in close binary systems. II. Runaway velocities of X-ray binaries.
See Abstr. 142.103.

Numerical studies of mass transfer and accretion in X-ray binary systems.   See Abstr. 142.133.

118 Visual Double and Multiple Stars

118.001 Rotational velocities and spectral types for a sample of binary systems. H. Levato.
Astron. Astrophys., Suppl. Ser., Vol. 19, 91 - 99 (1975).
Rotational velocities and MK spectral types are determined for the primaries and secondaries of 59 visual binaries and for 88 spectroscopic and eclipsing binaries.

118.002 Parallax and motions of the Capella system. W. D. Heintz.
Astrophys. Journ., Vol. 195, 411 - 412 (1975).
A combined parallax of $0\overset{''}{.}079 \pm 0\overset{''}{.}005$, relative proper motions, and the orbital motion of the faint double companion are derived from Sproul astrometric plates.

118.003 Micrometric measures of double stars. G. Van Biesbroeck.
Astrophys. Journ., Suppl. Ser., No. 270, Vol. 28, 413 - 448 (1974).
Approximately 2200 visual micrometer measures are given for 700 double-star systems, mostly with small angular separations and measured during 1965–1973.

118.004 Radial velocity measurements of visual binaries. I. R. F. Griffin.
Astron. Journ., Vol. 80, 245 (1975).
Doubt is cast upon the validity of some of Bakos's observations and therefore of his conclusion that the primaries of visual binary stars are mostly spectroscopic binaries.

118.005 Parallax, mass, and orbit of the 9 Puppis binary system. L. A. Breakiron, G. Gatewood.
Astron. Journ., Vol. 80, 318 - 320 (1975).
The parallax, orbit, and masses of the 9 Puppis binary system are redetermined. The weighted mean absolute parallax of all determinations is $+0.062 \pm 0.003$ (standard error). This parallax, with the visual orbit determined by Douglass, indicates masses of $0.86 \pm 0.17\,M_\odot$ and $0.99 \pm 0.19\,M_\odot$ for the primary and secondary, respectively. These values fall within the scatter of the empirically determined mass-luminosity relationship.

118.006 Combined-light UBV photometry of 103 bright southern visual doubles. P. R. Hurly.
Monthly Notes Astron. Soc. Southern Africa, Vol. 34, 7 - 11 (1975).
Combined-light UBV photometry of 103 bright southern close visual doubles is presented. Most of the pairs have separations between 1 and 10 seconds of arc. All the observations were made during 1974 using the 50 cm and 100 cm reflectors at Sutherland. Three different photometers were used. E region stars (Cousins 1973) were used as standards.

118.007 The file on an "ordinary" star: GC 21827. J. Ashbrook.
Sky Telescope, Vol. 49, 299 - 300 (1975).

118.008 Étoiles doubles et type spectral. P. Baize.
L'Astronomie, 89e année, p. 209 - 214 (1975).

118.009 The parallax and orbit of Kuiper 37. L. A. Breakiron.
Bull. American Astron. Soc., Vol. 7, 266 (1975). – Abstr. AAS.

118.010 Area scanner measurements of the astrometric-visual binary Ross 614 and the unresolved astrometric binary BD +6°398. B. Atwood, D. R. Curott.
Bull. American Astron. Soc., Vol. 7, 337 - 338 (1975).
Abstr. AAS.

118.011 DDO photometry of visual binaries. D. Deming.
Bull. American Astron. Soc., Vol. 7, 338 (1975).
Abstr. AAS.

118.012 Les binaires visuelles dont les éléments orbitaux ont été calculés. P. Baize.
L'Astronomie, 89e année, p. 253 - 262 (1975).

118.013 Micrometric measures of visual double stars. F. Holden.
Publ. Astron. Soc. Pacific, Vol. 87, 253 - 257 = Lick Obs. Bull., No. 697 (1975).

118.014 New double stars (12th series) discovered at Nice. P. Couteau.
Astron. Astrophys., Suppl. Ser., Vol. 20, 379 - 389 (1975).
In French.
A list of 155 double stars discovered at the 50 and 74 cm refractors is given.

118.015 Measurements of double stars made at Nice. P. Couteau.
Astron. Astrophys., Suppl. Ser., Vol. 20, 391 - 410 (1975).
In French.
683 measurements of 198 binaries made at the 74 and 50 cm refractors are given. These are mostly very close binaries discovered recently. The two components of the spectroscopic binary β CrB have been seen for the first time.

118.016 Orbit of ADS 3303 = HU 1082. P. Couteau.
Astron. Astrophys., Suppl. Ser., Vol. 20, 411 - 412 (1974). In French.
The orbital elements and the residuals of ADS 3303 $(4^h31^m5 + 39°03'\,(1950)\,K5\,MVT = 8{,}7\,\Delta M = 0{,}5)$ are given.

118.017 The orbits of visual binary stars ADS 3182, ADS 4890, I 7, ADS 10585, ADS 15267. G. A. Starikova.
Soobshch. Gos. Astron. Inst. Shternberga, No. 188, p. 47 - 55 (1974). In Russian.

118.018 Orbites nouvelles. Étoiles doubles nouvelles. Costa-Morales, W. D. Heintz.
Circ. d'Inform. (U.A.I. Commission des Étoiles Doubles), Grasse, France, No. 64 (1975).

118.019 Étoiles doubles découvertes à Nice (Lunette de 50 cm). P. Couteau, P. Muller.
Circ. d'Inform. (U.A.I. Commission des Étoiles Doubles), Grasse, France, No. 64 (1975).

118.020 Orbites nouvelles. R. L. Walker, J. Dommanget.
Circ. d'Inform. (U.A.I. Commission des Étoiles Doubles), Grasse, France, No. 65 (1975).

118.021 Étoiles doubles nouvelles, Belgrade (Lunette de 65 cm), Nice (Lunette de 50 cm). G. M. Popovic, P. Couteau.
Circ. d'Inform. (U.A.I. Commission des Étoiles Doubles), Grasse, France, No. 65 (1975).

118.022 The first general catalogue of double-star observations made in Belgrade, 1951 – 1971. G. M. Popović.

Publ. Obs. Astron. Beograd, No. 19, 235 pp. (1974).

At the end of 1951 the systematic double-star observations with the 65/1055 cm Zeiss refractor began at Belgrade Observatory. This is the summary of the 21 series of measures and 3 supplements which are published in the 20 years of observations.

118.023 Trajectoires relatives rectilignes des composantes de sept systèmes stellaires visuels. J. Dommanget.
Bull. Astron. Obs. Roy. Belgique, Vol. 8, 164 - 175 (1974).

Calcul des trajectoires rectilignes relatives des sept systèmes ADS 639; ADS 1440 (AB); ADS 1805; ADS 2690; ADS 2735 (A-BC); ADS 5080 (AB) et ADS 9870. Calcul des résidus et des éphémérides ainsi que comparaison des mouvements propres individuels avec les mouvements propres relatifs calculés. Application de l'un de nos critères d'opticité.

118.024 Orbits of the visual binaries ADS 7044 and ADS 8446. R. Van de Wiele.
Bull. Astron. Obs. Roy. Belgique, Vol. 8, 176 - 182 (1974).

The orbital elements of the visual binaries ADS 7044 and ADS 8446 have been computed by the method of Thiele-Innes. Ephemeris, dynamical parallaxes and physical constants are given for both. There seems to exist a third body in the system ADS 8446, but nothing definite can be said now.

118.025 Über die Genauigkeit visuell gemessener Farbdifferenzen bei engen Doppelsternen.
K. Ferrari d'Occhieppo.

Anzeiger Österreich. Akad. Wiss., Math.-nat. Kl., Vol. 111, 105 - 109 = Astron. Mitt. Wien, No. 16 (1974).

118.026 Photographic measures of double stars.
F. J. Josties, C. C. Dahn, V. V. Kallarakal, M. Miranian, G. G. Douglass, J. W. Christy, A. L. Behall, R. S. Harrington.
Publ. United States Naval Obs., *Washington*, Second Ser., Vol. 22, Part 6, 105 pp. (1974).

The series of photographic measures of double stars reported here covers approximately a 5 1/2 year period and is complete through the winter observing season 1972–73. This is a continuation of two earlier series (Franz, et al., 1963; Kallarakal, et al., 1969). 2942 measures of $\Delta\alpha\cos\delta$ and $\Delta\delta$ on 2693 individual plates for 398 double stars are listed.

Results of an investigation of the automatic coordinate measuring machine in measurements of photographic binaries. See Abstr. 034.108.

La représentation des positions observées d'un mobile animé d'un mouvement linéaire uniforme. Application aux cas des couples stellaires optiques et des passages des planètes inférieures sur le soleil. See Abstr. 041.042.

Lunar occultation summary. I.
See Abstr. 096.002.

On K. D. Rakos' photoelectric measurements of Sirius B. See Abstr. 126.018.

119 Spectroscopic Binaries

119.001 Spectroscopic binary orbits from photoelectric radial velocities. Paper I: HD 45088.
R. F. Griffin, B. Emerson.
Observatory, Vol. 95, 23 - 27 (1975).

119.002 CW Cephei: an important close binary member of the III Cephei association. I.-S. Nha.
Astron. Journ., Vol. 80, 232 - 238 (1975).

Yellow and blue photoelectric observations of CW Cephei are presented and new orbital elements are derived. Variation of the orbital period indicates a period of apsidal motion of about 39 yr. The absolute dimensions of this system contribute to the empirical mass-luminosity and mass-radius relations. CW Cep suggests a value of 3.0 for the ratio of total to selective absorption in the region of III Cep association.

119.003 Ultraviolet observations of HD 77581 (= 2U 0900 – 40). K. Nandy, W. McD. Napier, G. I. Thompson.
Monthly Notices Roy. Astron. Soc., Vol. 171, 259 - 262 (1975).

The absolute flux distribution in the spectral range from 1350 Å to 2740 Å of HD 77581 is presented here. The companion of this star is the X-ray source 2U 0900 – 40. The ultraviolet spectrum has been obtained with the S2/68 Ultraviolet Sky Survey telescope in the TD1 satellite. Its flux distribution, corrected for interstellar reddening determined from the observed ultraviolet colour ($m_{2190} - m_{2500}$), is found to be similar to that of a B0 supergiant star, ϵ Ori (HD 37128).

119.004 The spectroscopic binary system HD 158320

(= 3U 1727–33?).
A. J. Penny, J. E. Penfold, L. A. Balona.
Monthly Notices Roy. Astron. Soc., Vol. 171, 387 - 393 (1975).

HD 158320 has been studied spectroscopically and photometrically. It is shown to be a B0.5 II spectroscopic binary of 38.10 day period, with a companion of minimum mass 1.7 $M_\odot$, and to be a member of a physical visual triple system. No photometric variations are found. It is a possible candidate for the X-ray source 3U 1727–33.

119.005 The orbit of the spectroscopic binary HD 11291.
J. F. Heard, A. Krautter.
Journ. Roy. Astron. Soc. Canada, Vol. 69, 22 - 24 (1975).

Orbital elements are recorded for the single-line spectroscopic binary HD 11291 (2 Per), based on 42 spectrograms taken between 1941 and 1974.

119.006 The orbit of the double-line spectroscopic binary HD 153720. J. F. Heard, R. J. Hurkens.
Journ. Roy. Astron. Soc. Canada, Vol. 69, 25 - 28 (1975).

Orbital elements are presented for the double-line F0-type spectroscopic binary HD 153720, based on forty-six spectrograms.

119.007 Period finding for single-lined and double-lined spectroscopic binary stars. G. W. Jones.
Bull. American Astron. Soc., Vol. 7, 266 - 267 (1975).
Abstr. AAS.

119.008 Variation of Hα and H & K emission in late-type

spectroscopic binaries. E. J. Weiler.
Bull. American Astron. Soc., Vol. 7, 267 (1975). − Abstr. AAS.

119.009 **Mid-UV spectra of four binary systems observed by the S59 spectrometer.**
T. J. Herczeg, Y. Kondo, K. A. Van der Hucht.
Bull. American Astron. Soc., Vol. 7, 338 (1975). − Abstr. AAS.

119.010 **UBV and JHKL photometry of the radio star UX Ari = HD 21242.**
D. S. Hall, R. E. Montle, H. L. Atkins.
Acta Astron., Vol. 25, 125 - 132 (1975).

The authors present 1972 *UBV* and *JHKL* photometry of the spectroscopic binary radio star UX Ari = HD 21242. The light seemed to vary with the 6^d43791 orbital period determined spectroscopically by Carlos and Popper. The minimum occurred around 0^P1 at all wavelengths, but the amplitude (max to min) is greater at longer wavelengths, ranging from 0^m1 at U to 0^m2 at L. The light curve changed somewhat during the year, with minimum deeper in late 1972.

119.011 **A new spectroscopic orbit for Capella.**
A. H. Batten, V. Erceg.
Monthly Notices Roy. Astron. Soc., Vol. 171, 47P - 50P (1975).

Spectrograms of Capella at a dispersion of 2.4 Å mm^{-1} were obtained at Victoria in the fall of 1973 and the spring of 1974. From them the following orbital elements were derived for the primary component: $P = 104^d023$ (assumed), $T_0 =$ JD 2 442 119.352, $\omega = 292°$, $e = 0.014$, $K_1 = 26.06$ km s^{-1}, $V_0 = 29.48$ km s^{-1}. There is no indication of an increase in period that had been suggested by an analysis of earlier observations.

119.012 **Binary O star HR 8281.**
D. Crampton, R. O. Redman.
Astron. Journ., Vol. 80, 454 - 457 (1975).

Recent high-dispersion spectroscopic observations of HR 8281 establish the presence of the spectrum of a secondary companion. The system appears to be composed of an O6 and O9 star of normal mass and hence it is unlikely that it is related to the X-ray source Cep X-4.

119.013 **Helium λ10830 in Alpha Virginis A and B.**
D. D. Meisel, R. A. Berg.
Astrophys. Journ., Vol. 198, 551 - 553 (1975).

Line profiles and velocities of He I λ10830 obtained at maximum velocity separation of the spectroscopic binary Spica (α Vir) are presented. The authors derive a mass ratio, M_A/M_B, of 1.4 and an equatorial velocity of 88 km s^{-1} for α Vir A. A meaningful discussion of observed equivalent widths of the He I λ10830 line awaits a more certain determination of the spectral classification of α Vir B.

119.014 **Discovery of a new spectroscopic binary: 21 Hydrae.** M.-T. Chauville.
Astron. Astrophys., Vol. 40, 207 - 211 (1975). In French.

The star 21 Hydrae, HD 79193, classified as Am by Strömgren, is a spectroscopic binary with a period of 7.75 days and a mass ratio $m_1/m_2 = 1.14$. The primary star seems to belong to giant class III and the secondary may be a dwarf close to A0 type.

119.015 **Radial velocity variation of the K giant HD 107325.**
C. T. Bolton, A. Young, G. Wicks, R. B. Jones.
Publ. Astron. Soc. Pacific, Vol. 87, 259 - 264 (1975).

New spectroscopic observations of the K giant HD 107325 from three observatories are reported. These observations, which extend from 1970 February to 1974 June, rule out the large-amplitude velocity variations of one-half day period that were reported by Fehrenbach, but suggest that a small-amplitude long-period velocity variation is present. The

old observations are rediscussed.

119.016 **On the reality of the spectroscopic orbit for R Aquarii.** T. S. Jacobsen, G. Wallerstein.
Publ. Astron. Soc. Pacific, Vol. 87, 269 - 271 (1975).

New velocities for the emission lines in R Aqr show that the system has not followed the orbit with a period of 26.7 years tentatively suggested by Merrill. The binary nature of the star and its enormous mass function of $16.5 M_\odot$ are probably unreal. Evidence for mass loss with ionization and velocity increasing outward is presented.

119.017 **A revised orbit for HR 5361.**
C. D. Scarfe, S. Alers.
Publ. Astron. Soc. Pacific, Vol. 87, 285 - 288 (1975).

Radial velocities from 26 coudé spectrograms and from five coudé scanner observations have been used to redetermine the elements of this single-line K-giant binary.

119.018 **The double-line spectroscopic binary HD 159176.**
P. S. Conti, A. P. Cowley, G. B. Johnson.
Publ. Astron. Soc. Pacific, Vol. 87, 327 - 332 (1975).

From coudé observations at Kitt Peak and Cerro Tololo we derive the orbital elements of the system HD 159176. Both stars are practically identical O7 stars, each already evolved from the ZAMS. The close separation, probable synchronous rotation, and near filling of their Roche lobes suggest photometric variability will be observable. Eclipses are unlikely.

119.019 **Spectroscopic binary orbits from photoelectric radial velocities. Paper 2: HD 9313.**
R. F. Griffin, B. Emerson.
Observatory, Vol. 95, 98 - 100 (1975).

119.020 **The spectroscopic binary−b Persei.**
R. Rajamohan, M. Parthasarathy.
Bull. Astron. Soc. India, Vol. 2, 36 (1974). − Abstract.

119.021 **On galactic orientation of orbits of spectrum binaries.**
Eh. F. Brazhnikova, M. M. Dagaev, V. V. Radzievskij.
Astron. Zhurn. Akad. Nauk SSSR, Vol. 52, 546 - 552 (1975). In Russian. English translation in Soviet Astron., Vol. 19, No. 3.

The orientation of orbital apsidal lines of spectrum binaries contained in Bečvář's and Batten's catalogues is investigated. It is established that the periastrons of the stars with periods P > 20^d and radial velocity V$_r$ < 0 are mostly concentrated behind the tangent plane. The galactic orientation of the apsidal lines of these stars is discovered.

Photoelectric observations of occultations of the Pleiades and the incidence of duplicity in the cluster.
See Abstr. 096.008.

Five new B emission-line stars.
See Abstr. 114.354.

Rotational velocities and spectral types for a sample of binary systems. See Abstr. 118.001.

Étoiles doubles et type spectral.
See Abstr. 118.008.

Measurements of double stars made at Nice.
See Abstr. 118.015.

The systemic velocity of β Lyrae.
See Abstr. 121.060.

The distance and absolute magnitude of the super-

lithium S star T Sagittarii.   See Abstr. 122.002.

Photometric properties and evidence of duplicity for SZ Lyncis.   See Abstr. 122.003.

Observations of the red-dwarf emission-line objects LP101-15, BY Draconis, GT Pegasi, and FF Andromedae. See Abstr. 122.030.

Duplicity of some T Tauri stars and related variables. See Abstr. 122.041.

HD 206267, a candidate star for the transient X-ray source Cepheus X—4?   See Abstr. 142.057.

Cygnus X-1: discovery of variable circular polariza- tion.   See Abstr. 142.092.

Spectroscopic study of the open cluster NGC 2422. See Abstr. 153.005.

Spectral types in the open cluster NGC 6475. See Abstr. 153.030.

Errata

119.901   Erratum: 'Orbite spectroscopique et éléments absolus de deux étoiles doubles W Gru et UX Men' [Astron. Astrophys., Vol. 32, 429 - 434 (1974)]. M. Imbert. Astron. Astrophys., Vol. 39, 487 (1975).

120   Variable Stars: Catalogues, Ephemerides, Miscellanea

120.001   RV Tauri Sterne – ein internationales Beobachtungs programm.   E. Heiser. BAV Rundbrief, 24. Jahrgang, p. 17 - 18 (1975).

120.002   60th name-list of variable stars. B. V. Kukarkin, P. N. Kholopov, N. P. Kukarkina, N. B. Perova. Inform. Bull. Variable Stars, (I.A.U. Commission 27), Konkoly Obs., Budapest, No. 961, 15 pp. (1975).

120.003   Les cartes des étoiles du programme A.F.O.E.V. É. Schweitzer. A.F.O.E.V. Bull., Vol. 9, 12 - 19 (1975).

Notes on observing methods and programs for new observers.   See Abstr. 031.218.

Methods for search of variable star periods. See Abstr. 031.263.

121 Eclipsing Variables

121.001 **Spectrographic observations of β Lyr during the international campaign of 1971.**
U. Flora, M. Hack.
Astron. Astrophys., Suppl. Ser., Vol. 19, 57-89 (1975).

The results of the spectrographic observations of β Lyr during the international campaign of July-August 1971 are given, i.e. radial velocity, profile and intensity variation of several significant lines. An asymmetrical distribution of matter in the envelope surrounding the whole system and having higher density in front of the preceding hemisphere of the B9 star is indicated by the observations. Mass exchange is suggested.

121.002 **Ultraviolet photometry from the Orbiting Astronomical Observatory. XVIII. The 1972 eclipse of 31 Cygni.** L. R. Doherty, A. F. Jung.
Astrophys. Journ., Vol. 195, 121 - 125 (1975).

Wide-band filter photometry of the 1972 eclipse of 31 Cyg has been reduced for seven filters with effective wavelengths 4250–1430 Å. Comparison of the light curves with earlier OAO eclipse observations of the similar system 32 Cyg suggests that both K supergiant atmospheres can be described with a single density profile $\rho/\rho_0 = f(r/r_0)$. Line absorption appears to be the principal source of opacity at all ultraviolet wavelengths.

121.003 **Interpretation of BM Orionis.** S.-S. Huang.
Astrophys. Journ., Vol. 195, 127 - 135 (1975).

The entire light curve of the BM Ori system both inside and outside primary and secondary eclipses has been examined on the basis of two models for the disk around the secondary component: one with the luminous energy of the disk coming entirely from the secondary, and another with the luminous energy coming at least partly from the primary. It has been found that if the disk is highly opaque, as is suggested by the fitting of the light curve, there exist in the first model discrepancies between what has been derived from the luminosity consideration for the secondary component and what has been derived from the radius consideration. Hence the second model is accepted. Based on this model the nature of both component stars has been examined from a consideration of the luminosity and the dimensions of the disk.

121.004 **A spectrometric study of the Lyman-alpha line of Beta Persei.** K.-Y. Chen, F. B. Wood.
Astrophys. Journ., (*Letters*), Vol. 195, L73 - L76 = Rosemary Hill Obs., Univ. Florida, *Gainesville*, Contr. No. 52 (1975).

A brief description is given of the Lα line based on observations made by the Copernicus satellite. No emission was detected. No significant differences were found between observations outside eclipse and those during primary minimum. Some of the blended lines in this spectral region are identified.

121.005 **A photometric study of U Cephei. Part I.**
D. S. Hall, K. Walter.
Astron. Astrophys., Vol. 37, 263 - 273 = Mitt. Astron. Inst. Univ. Tübingen No. 130 (1974).

Three photoelectric light curves (Tschudovitchev; Khozov and Minaev; Catalano and Rodono) are analyzed and a photometric solution found which allows for an irregular surface brightness distribution on the B star and for gas stream absorption. Then 104 light curves (between 1880 and 1970) are discussed to see how changes in the shape during totality correlate with period variations. One set of geometrical elements ($i=83°14$, $b_B = 0.1665$, $b_G = 0.3340$) could be shown to satisfy all three photoelectric light curves.

121.006 **Apsidal motion in the FT Orionis system.**
B. Grønbech.
Astron. Astrophys., Vol. 37, 435 - 437 (1974).

Apsidal motion with a period of 520 ± 100 years has been detected in the FT Orionis binary system. This leads to a mean density concentration coefficient $\bar{k}_2 = 0.0046$, which is in good agreement with theoretical values. The spectrum is double-lined at the phase of maximum radial velocity separation.

121.007 **Photometric elements of the eclipsing binary V444 Cyg and the nature of the Wolf-Rayet star.**
A. M. Cherepashchuk.
Astron. Zhurn. Akad. Nauk SSSR, Vol. 52, 81 - 91 (1975). In Russian. English translation in Soviet Astron., Vol. 19, No. 1.

It is shown that the Wolf-Rayet star in the system V444 Cyg is a helium star on the final evolution stage after the main sequence.

121.008 **Photometric study of AE Phoenicis.**
R. M. Williamon.
Astron. Journ., Vol. 80, 140 - 144 = Rosemary Hill Obs., Univ. Florida, *Gainesville*, Contr. No. 53 (1975).

The bright W Ursae Majoris-type eclipsing binary system AE Phoenicis was observed with UBV filters during 1970 at the Cerro Tololo Inter–American Observatory. A significant asymmetry in maximum light was observed and evidence for a small orbital eccentricity was detected. The eclipses are complete, and solutions of the light curves based on the Russell method give $k = 0.57$.

121.009 **A photometric study of U Cep. Part II (tables).**
D. S. Hall, K. Walter.
Astron. Astrophys., Suppl. Ser., Vol. 19, 337 - 350 = Mitt. Astron. Inst. Univ. Tübingen, No. 131 (1975).

Part I of the authors' study of U Cep is published in Astronomy and Astrophysics. This Part II contains the tables giving explicit numerical representations of photometric solutions, new *UBV* observations of the variable, and the compilation of data between 1880 and 1970 concerning the slope of intensity during the phase of total eclipse.

121.010 **A new spectroscopic orbit of the eclipsing binary 68u Her.** B. J. Kovachev, W. Seggewiss.
Astron. Astrophys., Suppl. Ser., Vol. 19, 395 - 402 (1975).

New orbital elements of 68u Her have been determined from Crimea spectrograms obtained in 1962. The new values of the velocity amplitudes and the γ velocity differ only slightly from earlier results. The discrepancy between photometric and spectroscopic orbit is discussed.

121.011 **Hydrogen profiles, helium line strengths, and surface gravities of eclipsing binary stars.**
E. C. Olson.
Astrophys. Journ., Suppl. Ser., No. 274, Vol. 29, 43 - 54 (1975).

Coudé spectroscopy and four-color photometry are used to show that surface gravities of eclipsing binary stars determined from hydrogen line profiles and from theoretical profile grids with gravities based on binary light and radial velocity solutions are in agreement. Agreement is obtained with both the Griem and the Vidal-Cooper-Smith line-broadening theories. Agreement is also found for gravities determined from neutral helium line strengths for $N(\text{He})/N(\text{H}) = 0.1$.

121.012 **Three-colour photometry of the eclipsing binary EI Cephei.** T. D. Padalia, R. K. Srivastava.

Astrophys. Space Sci., Vol. 32, 285 - 290 (1975).

The photoelectric elements of the system EI Cephei have been determined in U, B, V colours. The absolute dimensions have been determined and some evolutionary aspects have been discussed. A modified period of $8\overset{d}{.}439334$ has been obtained. The system is a detached one.

121.013 Three-colour photometry of TW Cas.
C. D. Kandpal.
Astrophys. Space Sci., Vol. 32, 291 - 295 (1975).

Photoelectric elements of the system TW Cas have been determined in U, B and V filters. A refined period of $1\overset{d}{.}428\ 324\ 77$ has been given though no change in period is noticed. Spectroscopic elements given by Struve have been used to get the absolute elements. The system is found to be a detached one.

121.014 On the period variations of RT Persei.
S. Mancuso, L. Milano.
Astrophys. Space Sci., Vol. 32, 385 - 401 (1975).

The observed minima of RT Persei up to 1971 are collected and treated by the method of least squares to establish periodic terms with a fundamental period of $40\overset{y}{.}0$ and, furthermore, a quadratic term. It is shown that the periodic terms might be explained by the presence of a third body, but this hypothesis seems unlikely, and could be definitively ruled out by new spectroscopic observations. A dynamical instability of the system may constitute a more likely explanation of the observed period changes.

121.015 Period changes in eruptive binaries.
J. E. Pringle.
Monthly Notices Roy. Astron. Soc., Vol. 170, 633 - 642 (1975).

The evidence for period changes in close eclipsing binary systems is investigated from a statistical viewpoint. It is found that the evidence is not clear cut as has been previously suggested.

121.016 Period study for U Cephei and implications for the mass-transfer theory.   D. S. Hall.
Acta Astron., Vol. 25, 1 - 20 (1975).

The $O-C$ curve between 1880 and 1970 is represented with 10 upward-curving parabolic segments. This scheme is the simplest consistent with the period-change model of Biermann and Hall. The residual scatter of normal points is small: $\sigma = \pm 0\overset{d}{.}002$. The effect of light curve scatter is negligible. The average period decrease is $\Delta P/P = -5.5 \times 10^{-5}$ and the average subsequent increase is $\Delta P/P = +6.5 \times 10^{-5}$. The intervals between are not equal but average 9.1 yr.

121.017 The variability of period of Beta Lyrae. II.
Z. Klimek, J. M. Kreiner.
Acta Astron., Vol. 25, 29 - 37 (1975).

This paper contains photoelectric observations obtained in Cracow in 1973, a supplement to the list of times of minima published in an earlier paper and a short discussion of periodicity in the $O-C$ curve.

121.018 A computer program for the analysis of the 1972 chromospheric spectrum of 31 Cygni.
C. L. Morbey, K. O. Wright, R. G. Carlberg.
Journ. Roy. Astron. Soc. Canada, Vol. 69, 40 (1975). – Abstr. Canadian Astron. Soc.

121.019 A luminous circumstellar ring in the eclipsing binary RX Geminorum.
D. S. Hall, K. Walter.
Astron. Astrophys., Vol. 38, 225 - 237 = Mitt. Astron. Inst. Univ. Tübingen, Nr. 30 (1975).

Analyzing their UBV observations of this eclipsing

binary obtained between 1967 and 1971, the authors searched for a consistent photometric solution. A conventional solution, based on a conventional rectification, yielded several serious inconsistencies. A solution modified to allow for various gas streaming effects, and based on a short-region rectification, yielded $b_A = 0.0779$, $b_K = 0.2316$, and $i = 81°.18$, valid for all three colours. Absolute dimensions are derived by assuming the secondary star fills its Roche lobe and the primary obeys the mass-luminosity relation. The results $(2.72 \pm 0.56\ M_\odot$ and $2.56 \pm 0.25\ R_\odot$ and $0.46 \pm 0.10\ M_\odot$ and $7.60 \pm 0.67\ R_\odot$ respectively) are consistent with the radial velocity curve of Gaposchkin if allowance is made for gas stream distortion.

121.020 Ultraviolet photometry from the Orbiting Astronomical Observatory. XIX. Atmospheric properties of the detached binaries VV Orionis and MR Cygni.
J. A. Eaton.
Astrophys. Journ., Vol. 197, 379 - 391 (1975).

A series of OAO 2 light curves for VV Ori covering the wavelength range 1300–4300 Å is presented. Analysis of these light curves using light-curve synthesis techniques leads to a mass ratio for the system consistent with radial-velocity observations and to limb darkening consistent with theory. Analysis of light curves for MR Cyg, a system similar to VV Ori, indicates that the gravity darkening of MR Cyg could be as low as that found for VV Ori. The mass-luminosity relation applied to this system, however, implies that the primary is probably fully gravity darkened or that the stars may not be adequately represented by Roche potentials.

121.021 The G-type eclipsing binary TY Pyxidis.
J. Andersen, D. M. Popper.
Astron. Astrophys., Vol. 39, 131 - 134 (1975).

The 7th magnitude G-type eclipsing binary TY Pyx, with a period of $3\overset{d}{.}2$ and with the H and K lines in emission in both components, is shown to consist of two approximately equal components with masses $1.2\ M_\odot$, radii $1.65\ R_\odot$, and spectral type G 5. No other binary system with similar properties is known to the authors and they encounter some difficulty in understanding its history. The motion of TY Pyx implies that it belongs to the older disc population.

121.022 An outburst of U Cephei.   A. H. Batten,
W. A. Fisher, B. W. Baldwin, C. D. Scarfe.
Journ. Roy. Astron. Soc. Canada, Vol. 69, 40 (1975). – Abstr. Canadian Astron. Soc.

121.023 The accuracy of visually determined times of primary minima of β Lyrae-type and W Ursae Majoris-type variables.   A. D. Mallama.
Journ. American Ass. Variable Star Observers, Vol. 3, 49 - 51 (1974).

121.024 Minima of eclipsing binary stars–I.   M. E. Baldwin.
Journ. American Ass. Variable Star Observers, Vol. 3, 60 - 69 (1974).

This report continues previous lists of minima published in the IAU Inform. Bull. Variable Stars, Nos. 111, 114, 119, 129, 154, 180, 221, 247, and 795. This report contains 441 observed heliocentric minima of 50 eclipsing binary stars.

121.025 Six-color observations of Algol, 1949–1951.
J. Stebbins, K. C. Gordon.
Astrophys. Space Sci., Vol. 33, 481 - 486 (1975).

Observations of Algol in six colors, covering both eclipses, are presented. Primary minimum occurred at JD 2433 228.8684.

121.026 On the dynamical evolution of binary systems of the W Ursae Majoris type.   D. Poshanova.
Matematika i mekhanika. Vyp. (No.) 8. Alma-Ata, 1973,

p. 206 - 210. In Russian. – Abstr. in Referativ. Zhurn. 51. Astron., 3.51.743 (1975).

121.027 W Ursae Majoris stars: period-spectral type relation and period changes. A. Yamasaki.
Astrophys. Space Sci., Vol. 34, 413 - 424 (1975).

The relation between period and spectral type is examined for 33 W Ursae Majoris stars for which accurate observations have enabled us to clearly classify their eclipse types at the primary minimum (transit (A) or occultation (W)). About a half of the examined stars are of A-type, and the rest correspond to W-type. Periods of W-type systems are found to fall within 0.25–0.5 days, while periods of A-type systems range between 0.25–0.9 days. For A-type systems certain period-spectral type relations seem to hold, but for W-type systems no definite relation could be found.

121.028 Fourier analysis of the light curves of eclipsing variables, I. Z. Kopal.
Astrophys. Space Sci., Vol. 34, 431 - 457 (1975).

The aim of the paper will be to pioneer a new approach to the analysis of the light changes of eclipsing binary systems in the frequency domain, and to point out its merits in comparison with a conventional treatment of the same problem in the time-domain which has been developed so far.

121.029 Periods and period variations of some eclipsing variables.
A. Guarnieri, A. Bonifazi, P. Battistini.
Astron. Astrophys., Suppl. Ser., Vol. 20, 199 - 225 (1975).

For the eclipsing binaries AR Aur, BF Aur, AH Cep, EK Cep, V1143 Cyg, u Her, AR Lac available data have been collected and discussed with regard to period study. Seventeen new times of minimum are given. For AR Aur, BF Aur, AH Cep, AR Lac period changes are put into evidence. 1672 photoelectric measurements of these systems are listed.

121.030 UBV photometry of the Algol system RX Gem. D. S. Hall, K. Walter.
Astron. Astrophys. Suppl. Ser., Vol. 20, 227 - 235 (1975).

We present UBV observations of this eclipsing binary obtained between 1967 and 1971 at three different observatories: 325 in V, 316 in B, 182 in U. Normal points of the individual observations are also presented: 72 in V, 74 in B, 46 in U.

121.031 Spectrophotometry of the eclipsing binary systems U Cep, U Sge, and SX Cas.
C. G. Rhombs, J. D. Fix.
Bull. American Astron. Soc., Vol. 7, 267 (1975). – Abstr. AAS.

121.032 Theory of the ratio of depths of primary and secondary eclipses and its application.
S.-S. Huang, D. A. Brown.
Bull. American Astron. Soc., Vol. 7, 267 (1975). – Abstr. AAS.

121.033 U Cep: photometry during the recent outburst. E. C. Olson.
Bull. American Astron. Soc., Vol. 7, 267 (1975). – Abstr. AAS.

121.034 Masses and orbital elements of the binary system Phi Persei. E. M. Hendry.
Bull. American Astron. Soc., Vol. 7, 268 (1975). – Abstr. AAS.

121.035 Photoelectric observations of U Cephei. N. L. Markworth.
Bull. American Astron. Soc., Vol. 7, 338 (1975). – Abstr. AAS.

121.036 The eclipsing binary CW Eridani. K.-Y. Chen.
Acta Astron., Vol. 25, 89 - 101 = Rosemary Hill Obs., Univ. Florida, *Gainesville*, Contr. No. 50 (1975).

Three-color photoelectric observations of CW Eridani

during 1970–1973 are presented. The period is determined to be $2^d.72837$. Solutions based on the Russell model yield average values $r_g = 0.183$, $r_s = 0.129$ and $i = 86°3'$, for the system.

121.037 Photoelectric light curve of the eclipsing variable GK Cephei. T. Z. Dworak.
Acta Astron., Vol. 25, 103 - 116 (1975).

This paper presents the results of B and V photoelectric observations of the eclipsing binary GK Cephei. The light curve based on the normal points is given. The period has been redetermined on the basis of the photoelectric observations of both minima. The $O-C$ diagram is given. Some physical properties are discussed.

121.038 UBV photometry of the eclipsing binary OX Cassiopeiae, a possible member of NGC 381.
T. H. Frazier, D. S. Hall.
Acta Astron., Vol. 25, 117 - 124 (1975).

New UBV photometry shows that the period of Reim (1957) should be doubled and leads to the new ephemeris JD (hel.) = $2441269.6355 + 2^d.4893427 E$. Secondary minimum falls at $0^P.512$. Use of the mass-luminosity relation indicates a distance of 2100 ± 500 pc, consistent with but not conclusive evidence of membership in NGC 381.

121.039 Spectroscopic evidence for extrastellar matter in u Her. B. J. Kovachev, M. Reinhardt.
Acta Astron., Vol. 25, 133 - 151 (1975).

The authors present the phase variations of the equivalent widths of the Balmer lines, He I lines, Mg II 4481, Ca II 3933, C II 4267 lines in the spectrum of the eclipsing binary u Her. The strong asymmetry of the equivalent widths with respect to the secondary and primary photometric minimum provides strong evidence for the existence of a gaseous stream in the system. A qualitative model of the spatial distribution of extrastellar matter is given which is able to explain the observed phase variations of the equivalent widths.

121.040 Spectrophotometric study of the continuous spectrum in the eclipsing binary system UV Vir.
M. B. Babaev, S. M. Azimov.
Izv. AN AzSSR. Ser. fiz.-tekhn. i mat. n., 1974, No. 4, p. 104 - 110. In Russian. – Abstr. in Referativ. Zhurn. 51. Astron., 4.51.771 (1975).

121.041 Four-color photometry and spectroscopic gravities of early-type eclipsing binary systems.
E. C. Olson.
Dudley Obs. Rep. No.9, (see 012.008), p. 41 - 43 (1975).

Four-color photometry and coudé spectroscopy of early-type eclipsing binaries with known surface gravities are used to demonstrate agreement between theoretical model atmospheres and observations. Continuum color indices, hydrogen line profiles, and helium line strengths are internally consistent for 41 eclipsing systems covering the spectral range B0 to A3. The value N(He)/N(H) = 0.1 is consistent with observations of the B-type binaries.

121.042 The ultraviolet spectrum of Beta Lyrae.
M. Hack, J. B. Hutchings, Y. Kondo, G. E. McCluskey, M. Plavec, R. S. Polidan.
Astrophys. Journ., Vol. 198, 453 - 465 = Dominion Astrophys. Obs., *Victoria*, Contr. No. 249 (1975).

Details are presented of low-resolution (0.2–0.4 Å) scans of the spectrum of β Lyrae in the region 1000–3000 Å, obtained from the Copernicus OAO. The continuous spectrum appears to correspond to type B5 or later while the line spectrum is largely dominated by very strong emission of lines characteristic of earlier spectral types. Small velocity variations, similar to those of the shell lines in the visible spectrum, are seen. A discussion of models for the system is given, and it is considered that present evidence suggests that the secondary object is a mass-accreting black hole.

121.043 Minima of eclipsing variables.
B. A. Krobusek, A. D. Mallama.
Inform. Bull. Variable Stars, (I.A.U. Commission 27), Konkoly Obs., Budapest, No. 954, 3 pp. (1975).

121.044 Photometric observations of the suspected VV Cephei star BD +61°219. P. Tempesti.
Inform. Bull. Variable Stars, (I.A.U. Commission 27), Konkoly Obs., Budapest, No. 955, 2 pp. (1975).

121.045 RZ Eri. B. Grønbech.
Inform. Bull. Variable Stars, (I.A.U. Commission 27), Konkoly Obs., Budapest, No. 956, p. 1 - 2 (1975).

121.046 Times of minima for V523 Sgr and V526 Sgr.
B. Grønbech.
Inform. Bull. Variable Stars, (I.A.U. Commission 27), Konkoly Obs. Budapest, No. 956, p. 3 (1975).

121.047 Photoelectric lightcurve and minima of the eclipsing binary CW Cas.
R. Burchi, R. de Santis.
Inform. Bull. Variable Stars, (I.A.U. Commission 27), Konkoly Obs., Budapest, No. 962, 4 pp. (1975).

121.048 HD 169454: a possible Zeta Aur system.
C. Bartolini, S. Scardovi.
Inform. Bull. Variable Stars, (I.A.U. Commission 27), Konkoly Obs., Budapest, No. 963 (1975).

121.049 Photoelectric minima of ER Ori and XY Leo.
R. Burchi, F. Zavatti.
Inform. Bull. Variable Stars, (I.A.U. Commission 27), Konkoly Obs., Budapest, No. 964 (1975).

121.050 The next minimum of the long period eclipsing binary EE Cep. L. Meinunger.
Inform. Bull. Variable Stars, (I.A.U. Commission 27), Konkoly Obs., Budapest, No. 965 (1975).

121.051 A new eclipsing variable. R. Szafraniec.
Inform. Bull. Variable Stars, (I.A.U. Commission 27), Konkoly, Obs., Budapest, No. 971 (1975).

121.052 Period increase in RW Persei confirmed.
D. S. Hall, T. Stuhlinger.
Inform. Bull. Variable Stars, (I.A.U. Commission 27), Konkoly Obs., Budapest, No. 972 (1975).

121.053 Revised elements of eclipsing stars. P. Ahnert.
Inform. Bull. Variable Stars, (I.A.U. Commission 27), Konkoly Obs., Budapest, No. 978, p. 1 (1975).

121.054 Photographic minima of eclipsing stars.
P. Ahnert.
Inform. Bull. Variable Stars, (I.A.U. Commission 27), Konkoly Obs., Budapest, No. 978, p. 2 (1975).

121.055 Photographic photometry of BD +60°2289, a new bright eclipsing binary. F. Gieseking.
Inform. Bull. Variable Stars, (I.A.U. Commission 27), Konkoly Obs., Budapest, No. 980, 3 pp. (1975).

121.056 New bright eclipsing binary. W. Strohmeier.
Inform. Bull. Variable Stars, (I.A.U. Commission 27), Konkoly Obs., Budapest, No. 984 (1975).

121.057 v^1 Ori A − a new eclipsing binary in the Trapezium. E. Lohsen.
Inform. Bull. Variable Stars, (I.A.U. Commission 27), Konkoly Obs., Budapest, No. 988, 4 pp. (1975).

121.058 UBV and spectral data on RV Pictoris.
R. H. Méndez.
Inform. Bull. Variable Stars, (I.A.U. Commission 27), Konkoly Obs., Budapest, No. 1000, 4 pp. (1975).

121.059 High-resolution observations of the radio emission from Beta Persei.
B. G. Clark, K. I. Kellermann, D. Shaffer.
Astrophys. Journ., (Letters), Vol. 198, L123 - L124 (1975).
The angular size of the radio emission from β Persei (Algol) was measured during a flare and found to be about 4 milli-arcsec equivalent Gaussian diameter, corresponding to linear dimensions of 0.1 AU and mean brightness temperature 4×10^8 K. The observed change in the interferometer fringe visibility in a few hours corresponds to a mean apparent expansion velocity of 500–1000 km s^{-1}, or to a stationary, slightly elliptical source.

121.060 The systemic velocity of β Lyrae.
A. H. Batten, J. M. Fletcher.
Publ. Astron. Soc. Pacific, Vol. 87, 237 - 244 (1975).
Radial velocities of β Lyr determined at Victoria from 1966 to 1974 are all well satisfied by an assumed circular orbit with $V_0 = -17.8$ km sec$^{-1} \pm 0.8$ km sec^{-1}, and $K_1 = 184.0$ km sec$^{-1} \pm 0.7$ km sec^{-1}, provided account is taken of the changing period in the calculation of phases, and the velocity measurements are made only from the lines of Si II. Previous work indicating variations in the value of V_0 is discussed. The use of unreliable lines for velocity measurement, small systematic differences between spectrographs, and possible occasional distortion of the velocity curve can account for the variations reported.

121.061 Spectroscopy of the massive eclipsing binary HD 163181 (=V453 Scorpii). J. B. Hutchings.
Publ. Astron. Soc. Pacific, Vol. 87, 245 - 251 (1975).
New coudé spectrograms of the OB binary HD 163181 are analyzed. An absorption spectrum attributed to the secondary is reported for the first time, and it is deduced that the secondary is more massive and less luminous than the supergiant primary. The primary shows phase-dependent mass-loss characteristics and the present light curve suggests that it is near its Roche limiting radius. Nitrogen and oxygen abundances are unusual in the primary spectrum. The nature of the secondary is discussed.

121.062 AZ Cassiopeiae at the 1956−57 eclipse.
R. H. Méndez, G. Münch, J. Sahade.
Publ. Astron. Soc. Pacific, Vol. 87, 305 - 310 (1975).
The spectrum of AZ Cas at the time of the 1956−57 eclipse can be described as a combination of an F8 Ib star and a main-sequence B0-1 companion. The latter star is ejecting matter with approximately spherical symmetry at some 50–60 km sec^{-1}, while the former has a very extended atmosphere. The whole system seems to be also surrounded by an expanding envelope.

121.063 Element abundances in Algol-type binaries.
S. A. Naftilan.
Publ. Astron. Soc. Pacific, Vol. 87, 321 - 326 (1975).
The results of an abundance analysis of both the primaries and secondaries of six eclipsing binary stars are presented. The analysis was done by adjusting the parameters used to generate synthetic spectra until the best fit with the observed spectra was obtained. The secondaries of three "classical" Algol-type binaries, U Cep, TT Hya, and ZZ Cnc,

were found to be very slightly metal deficient in agreement with photometric results. The subgiant secondaries of RS CVn and RW UMa, which do not fill their Roche lobes, are found to be moderately metal deficient. The secondary of the widely separated supergiant system, RZ Oph, shows no metal deficiency. All of the primaries are either normal or very slightly metal rich. Several theories of binary-star evolution are considered, but none explain these results satisfactorily.

121.064 **Emission lines and radial velocities of the VV Cephei type star BD +63°3.** M. Barbier.
Astron. Astrophys., Suppl. Ser., Vol. 20, 305 - 312 (1975). In French.

Three spectrographic plates at a dispersion of 20 Å mm^{-1} and one spectrographic plate at a dispersion of 12 Å mm^{-1} have been obtained for the VV Cephei type star BD +63°3. 93 emission lines are identified, principally of [Fe II] and also of H, O II, [S II], S II, [Cr II], Fe II, [Ni II], Ni II, Cu II. From the emission [Fe II] lines a mean temperature of 6300 ± 1500°K has been derived. The radial velocity of the emission lines of the various elements differs by 6 km s^{-1} from the velocity determined for the absorption lines of the late type component. An upper limit for the absolute magnitude of BD +63°3 has been determined on the basis of the velocity of the interstellar calcium line: $M_v > -7.9$.

121.065 **A study of W UMa-type eclipsing binaries by the method of light curve synthesis: application to V 566 Oph and AB And.** E. Berthier.
Astron. Astrophys., Vol. 40, 237 - 248 (1975). In French.

Recently several authors proposed methods of light curve synthesis in order to determine the parameters of eclipsing binary systems from observed light curves. This approach is particularly well suited in the case of W UMa systems, for which the proximity effects are important. In the present article the author develops a highly automatic numerical method for the analysis of W UMa systems. The method is applied to V 566 Oph and AB And.

121.066 **MWC 930, AS 299 and AS 341: three probable VV Cephei stars.** D. A. Allen.
Astron. Astrophys., Vol. 40, 335 - 336 (1975).

Spectra and near-infrared photometry of MWC 930, AS 299 and AS 341 are described. All three have previously been classified Be, but are probably VV Cephei stars.

121.067 **Fourier analysis of the light curves of eclipsing variables, II.** Z. Kopal.
Astrophys. Space Sci., Vol. 35, 159 - 170 (1975).

The aim of the present paper is to extend the Fourier methods of analysis of the light curves of eclipsing binaries, outlined in a previous communication (1975) in connection with systems whose components would appear as uniformly bright discs, to systems whose components exhibit discs characterized by an arbitrary radially-symmetrical distribution of brightness – i.e., an arbitrary 'law of darkening' towards the limb – be it linear or nonlinear. Fundamental equations are set up which govern the light changes arising from the mutual eclipses of limb-darkened stars – be such eclipses total, partial or annular. A closed algebraic solution for the elements of the occultation eclipses terminating in total phase is given.

121.068 **Fourier analysis of the light curves of eclipsing variables, III.** Z. Kopal.
Astrophys. Space Sci., Vol. 35, 171 - 183 (1975).

The aim of the present paper is to extend the author's new methods of analysis of the light curves of eclipsing binary systems, consisting of spherical components, by Fourier approach to eclipses of transit type – which arise when the eclipsing component happens to be smaller of the two. His

present principal concern are transit eclipses, terminating in annular phase, of stars characterized by arbitrary radially-symmetrical distribution of brightness over their apparent discs – a phenomenon which will cause the light of the system to vary continuously during annular phase.

121.069 **Neue Elemente von RT Andromedae.** P. Ahnert.
MVS, *Sonneberg*, Vol. 6, 172 - 181 (1975).

New mean elements as well as instantaneous ones have been derived from the short period Algol type variable star RT Andromedae.

121.070 **Über die Periode von RV Piscium.** P. Ahnert.
MVS, *Sonneberg*, Vol. 6, 181 - 183 (1975).

From 40 minima of the time between 1925 and 1973 new elements of this scarcely observed Algol star were derived. The period seems to be remarkably constant.

121.071 **Konstanz der Periode des W-Ursae-Majoris-Sterns U Pegasi.** P. Ahnert.
MVS, *Sonneberg*, Vol. 6, 184 - 189 (1975).

An investigation of the minima observed since 1955 confirms the elements given by Rigterink. Despite of the abnormally large dispersion of the O–C values the period seems to have been constant during the last twenty years.

121.072 **Die Periodenänderung von BX Andromedae.** P. Ahnert.
MVS, *Sonneberg*, Vol. 6, 189 - 193 (1975).

The period of BX Andromedae was constant from 1899 to 1950. About 1951 the period became longer by a quarter of a second. Then it remained constant again from 1952 till now.

121.073 **The period of RZ Cassiopeiae.** A. J. Stokes.
Journ. American Ass. Variable Star Observers, Vol. 1, 17 - 19 (1972).

121.074 **BL Telescopii – a rare eclipsing binary.** J. H. Akyuz.
Journ. American Ass. Variable Star Observers, Vol. 1, 50 - 51 (1972).

121.075 **A recent photoelectric minimum of RZ Cassiopeiae.** A. J. Stokes.
Journ. American Ass. Variable Star Observers, Vol. 1, 54 - 55 (1972).

121.076 ***UBV* photoelectric photometry of V 453 Scorpii.** B. F. Madore.
Astron. Astrophys., Vol. 40, 451 - 453 (1975).

UBV photoelectric photometry is presented for the eclipsing system V 453 Sco = HD 163181. The system is seen to be of the β-Lyrae type.

121.077 **Three colour photometry of TW Cas.** C. D. Kandpal.
Bull. Astron. Soc. India, Vol. 2, 31 (1974). – Abstract.

121.078 **Studio preliminare di un minimo della binaria ad eclisse EE Cephei.** L. Baldinelli, C. Tubertini.
Giorn. Ass. Astrofili Bolognesi, No. 37, p. 3 - 5 (1975).

Photographic observations of the 1969 minimum of the EE Cep eclipsing binary are reported. From the light curve the date of the minimum at J.D. 2440493 ± 1 is derived.

121.079 **Osservazioni fotoelettriche di "β Lyrae".** L. Baldinelli, S. Ghedini.
Giorn. Ass. Astrofili Bolognesi, No. 37, p. 6 - 7 (1975).

121.080   EE Cephei, une algolide à très longue période.
A. Brun.
A.F.O.E.V. Bull., Vol. 8, 34 - 35 (1974).

121.081   Times of minima and light curve for V382 Cygni.
A. U. Landolt.
Publ. Astron. Soc. Pacific, Vol. 87, 409 - 416 = Contr. Louisiana State Univ. Obs., *Baton Rouge,* No. 106 (1975).

Five new photoelectrically determined times of minima are given for the massive, double-lined eclipsing binary V382 Cyg. *UBV* photoelectric observations gathered in the interval 1964–71 are tabulated.

121.082   Period variation in the white-dwarf eclipsing binary BD +16°516.   A. Young, H. H. Lanning.
Publ. Astron. Soc. Pacific, Vol. 87, 461 - 464 (1975).

Photometric observations of eclipse contacts over the past 4.7 years (3300 cycles) are discussed, revealing systematic variations in the period which suggest active mass transfer in progress. The period is found to have increased and decreased over this interval, and current models are not adequate to account for such variations when one component is a white dwarf. Some new and improved data for the system are also discussed.

121.083   The new massive eclipsing binary HR 6773.
A. Young, P. Etzel.
Publ. Astron. Soc. Pacific, Vol. 87, 471 - 477 (1975).

A light curve, velocity curve, and preliminary orbit elements and structure are presented for the early B-star eclipsing binary HR 6773. The evidence suggests that the more massive (primary) component is in early post-main-sequence evolution. Transient behavior is exhibited by the light curve which is suggestive of mass transfer activity.

121.084   A search for polarization in U Cephei.
G. V. Coyne.
Ric. Astron. Specola Vaticana, *Castel Gandolfo*, Vol. 8, (No. 24), 475 - 479 (1974).

Observations of U Cephei near primary eclipse, from just before first contact to just after third contact, show no intrinsic polarization, although polarization should be expected from the predicted densities and configuration of the circumstellar material. This negative result is probably due to the fact that the gas stream from the secondary and the flattened disk about the primary are not permanent features, a fact also evidenced by spectroscopic observations.

121.085   V 2283 Sgr, an eclipsing star with a rotating apse.
H. H. Swope.
Ric. Astron. Specola Vaticana, *Castel Gandolfo,* Vol. 8, (No. 25), 481 - 490 (1974).

121.086   Photometric orbit and apsidal motion of V 2283 Sagittarii.   D. J. K. O'Connell.
Ric. Astron. Specola Vaticana, *Castel Gandolfo*, Vol. 8, (No. 26), 491 - 497 (1974).

A list of minima is given, derived from observations on plates taken at Riverview Observatory between 1932 and 1952. A photometric orbit was determined from the analysis of Miss Swope's photoelectric light curve. The orbital eccentricity is 0.487 and the period of apsidal rotation about 560 years.

121.087   Photoelectric narrow-band photometry on Hβ and Hγ of β Lyrae system.   N. Güdür.
Sci. Rep. Fac. Sci. Ege Univ., *Izmir*, No. 218 (Astron. No. 16), 12 pp. (1975).

121.088   Lists of minima of eclipsing binaries.
Compiled by P. Carnevali, R. Diethelm, A. Figer, R.

Germann, J.-F. Le Borgne, K. Locher, N. Mauron, A. Marot, H. Peter, P. Ralincourt, J. Remis, C. Sanchez, C. Domec, Z. Hevesi, R. Rolland, T. Roudier, M. Behagle, A. Royer, G. Dumarchi.
BBSAG Bull., No. 19, p. 1 - 4; No. 20, p. 1 - 4, 5; No. 21, p. 1 - 4; No. 22, p. 1 - 4, 5 (1975). — 52nd - 55th list of Swiss Astronomical Society's Eclipsing Variable Observers.

121.089   Probable long period eclipsing binary BV 1616 Lep: minimum lasts at least 14 days.   K. Locher.
BBSAG Bull., No. 19, p. 4 (1975).

121.090   BV 1616 Lep: minimum lasts at least 44 days.
J. Lienhard, K. Locher.
BBSAG Bull., No. 20, p. 4 (1975).

121.091   AC Tauri: duration and magnitude at totality.
K. Locher.
BBSAG Bull., No. 20, p. 5 (1975).

121.092   DE Hydrae: duration and magnitude at totality.
K. Locher.
BBSAG Bull., No. 20, p. 5 (1975).

121.093   EQ Tauri: translation of the previous results to the wholly unrelated new elements of the GCVS 1974.
K. Locher.
BBSAG Bull., No. 20, p. 5 (1975).

121.094   Index of star names for BBSAG Bulletin 1 through 20.
BBSAG Bull., No. 20, p. 6 - 9, No. 21, p. 5 (1975).

121.095   AC UMa: the amplitude is 4^m instead of the 1^m catalogued.   K. Locher.
BBSAG Bull., No. 21, p. 5 (1975).

121.096   Minimum brightness of V 640 Ori.
R. Diethelm, K. Locher.
BBSAG Bull., No. 21, p. 5 (1975).

121.097   On the visual brightness of OS Ori.
R. Diethelm, K. Locher.
BBSAG Bull., No. 21, p. 5 (1975).

121.098   NW Aur: an unsuccessful attempt to improve its period.   K. Locher.
BBSAG Bull., No. 21, p. 6 (1975).

121.099   BV 1616 Lep: 70 days' minimum ended 1975 March 2.   J. Lienhard, K. Locher.
BBSAG Bull., No. 21, p. 6 - 7 (1975).

121.100   Improved period of SZ Librae.   K. Locher.
BBSAG Bull., No. 22, p. 4 (1975).

121.101   The totality duration of SX Hya.   K. Locher.
BBSAG Bull., No. 22, p. 5 (1975).

121.102   The totality duration of TY Lib.
H. Peter, K. Locher.
BBSAG Bull., No. 22, p. 5 (1975).

121.103   Note on a recent minimum of UV Lyn.
R. Diethelm.
BBSAG Bull., No. 22, p. 5 (1975).

121.104   UBV photometry of 32 Cygni during the 1974 eclipse.   M. Saitō, H. Sato, E. Watanabe, K. Okida, H. Ogata, C. Hukusaku, H. Sugai.
Tokyo Astron. Bull., Second Ser., No. 237, p. 2773 - 2780

(1975).

32 Cygni is one of Zeta Aurigae stars and its period is about 1147 days. Photoelectric observations of this star were made between November 1974 and January 1975 at four observatories in Japan. All these observations are reported, although only incomplete light curves were obtained.

121.105 β Persei.   D. Gibson.
IAU Circ., No. 2739 (1975).

121.106 RT Lacertae.
D. M. Gibson, R. M. Hjellming, F. N. Owen.
IAU Circ., No. 2789 (1975).

121.107 New elements for some eclipsing binary stars.
M. E. Baldwin.
Journ. American Ass. Variable Star Observers, Vol. 2, 7 - 13 (1973).

121.108 The variable period of W Delphini.   D. S. Hall.
Journ. American Ass. Variable Star Observers,
Vol. 2, 20 - 22 (1973).

121.109 A visual light curve of EG Cephei.
B. Keel, R. E. Montle, D. S. Hall.
Journ. American Ass. Variable Star Observers, Vol. 2, 23 - 25 (1973).

121.110 Differential UBV photometry of β Lyrae, IV.
H. J. Landis, L. P. Lovell, T. H. Frazier, D. S. Hall.
Journ. American Ass. Variable Star Observers, Vol. 2, 67 - 70 (1973), with a correction, Vol. 3, 35 (1974).

121.111 A photoelectric study of the eclipsing binary system
V566 Ophiuchi.   R. H. Kaitchuck, N. G. Sprague.
Journ. American Ass. Variable Star Observers, Vol. 3, 1 - 5 (1974).

121.112 A visual light curve of the eclipsing variable 143025
AD Bootis.   D. Van Buren.
Journ. American Ass. Variable Star Observers, Vol. 3, 6 - 10 (1974).

121.113 The accuracy of visually determined times of
primary minima of Algol-type variables.
A. D. Mallama.
Journ. American Ass. Variable Star Observers, Vol. 3, 11 - 14 (1974).

121.114 New elements for some eclipsing binary stars, II.
M. E. Baldwin.
Journ. American Ass. Variable Star Observers, Vol. 3, 24 - 29 (1974).

121.115 Photoelectric investigations of AA Ceti and UZ Puppis.   R. H. Bloomer, Jr.
Thesis, Florida Univ., Gainesville (USA). 189 pp. University Microfilms Order No. 74-9534 (1973).

A method for deriving radial brightness distributions from eclipse observations.   See Abstr. 079.002.

Properties of two common photometric systems and photometric observations of selected eclipsing binary systems.   See Abstr. 113.067.

Wolf-Rayet stars.   See Abstr. 114.054.

The composite spectrum and energy distribution of XX Ophiuchi.   See Abstr. 114.302.

Outburst in U Cephei.   See Abstr. 117.001.

Outburst of U Cephei.   See Abstr. 117.002.

The solution parameters of very close binary systems.   See Abstr. 117.025.

On the possiblity of magnetic starspots on the primary components of W Ursae Majoris type binaries.   See Abstr. 117.033.

Rotational velocities and spectral types for a sample of binary systems.   See Abstr. 118.001.

Mid-UV spectra of four binary systems observed by the S59 spectrometer.   See Abstr. 119.009.

Three variable stars in Cygnus.   See Abstr. 122.040.

Duplicity of some T Tauri stars and related variables.   See Abstr. 122.041.

RW Arietis, a short period pulsating star, one component of an eclipsing binary.   See Abstr. 122.096.

The linear polarization of the white-dwarf binary BD +16° 516.   See Abstr. 126.017.

Photoelectric observations of Centaurus X-3.   See Abstr. 142.005.

A search for optical pulsations from Centaurus X-3.   See Abstr. 142.006.

Optical studies of UHURU sources. XI. A probable period for Scorpius X-1 = V818 Scorpii.   See Abstr. 142.007.

Nature of Her X-1.   See Abstr. 142.009.

Model for 1.24 s X-ray pulses in Her X-1.   See Abstr. 142.010.

Search for millimeter-wave emission from Uhuru X-ray sources and radio binary stars.   See Abstr. 142.015.

Physical parameters of the Centaurus X-3 system.   See Abstr. 142.018.

Photoelectric observations of Krzeminski's star, the companion of Centaurus X-3.   See Abstr. 142.019.

Hα observations and the distribution of circumstellar material in the HD 77581 system (3U 0900−40).   See Abstr. 142.021.

Origin of the optical emission from Sco X-1.   See Abstr. 142.025.

Observations of six binary X-ray sources with the UCSD OSO-7 X-ray telescope.   See Abstr. 142.041.

Binary systems with an X-ray component.   See Abstr. 142.054.

Further optical observations of HZ Herculis.   See Abstr. 142.056.

Ellipsoidal light variations and masses of X-ray binaries.   See Abstr. 142.059.

Limits on the soft X-ray and extreme ultraviolet flux from RX Andromedae and U Geminorum. See Abstr. 142.060.

X-ray heating and the optical light curve of HZ Herculis.   See Abstr. 142.066.

Optical pulsations in HZ Herculis. III. Discovery of pulsed emission lines.   See Abstr. 142.090.

Determination of the distances of the nearest galaxies by method of parallaxes of eclipsing binaries. See Abstr. 158.117.

Errata

121.901   Erratum: 'Rediscussion of eclipsing binaries. X. The B stars AG Persei and CW Cephei'. [Astrophys. Journ., Vol. 188, 559 - 565 (1974)].   D. M. Popper. Astrophys. Journ., Vol. 198, 515 (1975).

122 Intrinsic Variables, Flare Stars, Pulsation Theory

122.001 **A linear, non-adiabatic pulsation analysis of models of dwarf cepheid variable stars.** J. R. Percy.
Monthly Notices Roy. Astron. Soc., Vol. 170, 155 - 163 (1975).

A linear, non-adiabatic pulsation analysis of models of dwarf cepheid variable stars has been carried out. The models include convection, radiation pressure, surface boundary condition determined from model atmospheres, and opacity determined either analytically or by quadratic interpolation in tables. Best agreement between theory and observations occurs if the hydrogen content by weight is between 0.70 and 0.75. The problem of the evolutionary state of these stars is discussed briefly.

122.002 **The distance and absolute magnitude of the super-lithium S star T Sagittarii.**
R. B. Culver, P. A. Ianna.
Astrophys. Journ., (Letters), Vol. 195, L37 - L38 (1975).

The results of photoelectric and spectroscopic observations of the S star T Sgr at minimum light are presented. A distance to the system of 1.0 kpc, an absolute visual magnitude of -2.4 at maximum light, and a lower mass limit of $1.5\,M_\odot$ are derived for T Sgr.

122.003 **Photometric properties and evidence of duplicity for SZ Lyncis.** T. G. Barnes III, T. J. Moffett.
Astron. Journ., Vol. 80, 48 - 55 (1975).

UBV photometry with a time resolution of 20 sec is reported for the AI Vel variable, SZ Lyn. The reported irregularities in height of maximum light are confirmed and found to have a total amplitude ~ 0.04 mag. No periodic amplitude modulation is observed. It is demonstrated that the colors of SZ Lyn are normal for its spectral type and luminosity class. The periodic variation from a linear ephemeris found by van Genderen is confirmed and its parameters improved. The most plausible interpretation of this periodicity is a light-travel-time effect, leading to the prediction that SZ Lyn is a single-lined spectroscopic binary with period 3.14 yr and total velocity amplitude 19 km/sec.

122.004 **Applications of linear pulsation theory to the cepheid mass problem and the double-mode cepheids.**
D. S. King, C. J. Hansen, R. R. Ross, J. P. Cox.
Astrophys. Journ., Vol. 195, 467 - 474 (1975).

Linear pulsation constants and transition lines for cepheids of composition $X = 0.7, 0.8$ and $Z = 0.02$ are presented. These results are applied to the problem of the cepheid mass discrepancy where it is found that self-consistent masses may be obtained if a color-temperature scale is used which reduces cepheid effective temperatures by $300°-650°\,K$ below those usually assumed.

122.005 **Comparison of UV Ceti flares with solar flares.**
W. Haupt, W. Schlosser.
Astron. Astrophys., Vol. 37, 219 - 223 (1974).

During several periods, totalling 26 hrs, 94 flares of UV Ceti were observed using a photometer with 1 s time resolution. Using solar flare terminology, all flares of UV Ceti observed can be characterized by a typical light curve consisting of a pre-flare, a flash and a slow phase. UV Ceti flare flash phases are compared with solar "white light" emission during the flash phase and it is suggested that during all UV Ceti flares strong particle emission occurs, as for solar white light flares.

122.006 **Pulsational instability of massive stars in the**
regions of Beta Cephei variables. C. Chiosi.
Astron. Astrophys., Vol. 37, 281 - 284 (1974).

In the light of recent results on the vibrational stability of non radial adiabatic oscillations in massive stars during the central H-burning phase, the problem of the Beta Cephei variables is briefly discussed and a speculative suggestion for their instability mechanism is made.

122.007 **Three-colour photometry of the RR Lyrae star YZ Bootis.**
W. Gieren, F. Gieseking, M. Hoffmann.
Astron. Astrophys., Vol. 37, 443 - 445 (1974).

A photoelectric UBV-photometry of the RR_s-variable YZ Bootis was carried out. Three subsequent cycles were observed and distinct changes in shape, amplitude, and mean luminosity could be detected. This explains the scatter found previously in the light curve by other authors who combined observations of different nights. A revised ephemeris has been obtained.

122.008 **Beobachtungsdaten von Mirasternen 1974.**
R. Lukas.
SuW, Vol. 14, 102 (1975).

122.009 **The classification of intrinsic variables. V. The large-amplitude red variables.** O. J. Eggen.
Astrophys. Journ., Vol. 195, 661 - 678 (1975).

$(UBVRI)$ observations covering several cycles of 10 large-amplitude red field variables and three probable group members are discussed.

122.010 **A possible γ-flares creation mechanism on UV Ceti stars.** E. A. Karitskaya.
Astron. Zhurn. Akad. Nauk SSSR, Vol. 52, 189 - 192 (1975). In Russian. English translation in Soviet Astron., Vol. 19, No. 1.

It is suggested that γ-flares recorded on spacecraft "Vela" are connected with flare activity of UV Ceti stars. A possible model of creation of γ-flares is proposed.

122.011 **The light curves of Delta Scuti stars HR 1170 and HR 7563.**
J. Warman, Z. Malacara, M. Breger.
Rev. Mexicana Astron. Astrofis., Vol. 1, 143 - 150 (1974).

Light curves of δ-Scuti type stars HR 1170 and HR 7563 are presented. Preliminary results of a multiperiod analysis are mentioned.

122.012 **On the model of outburst of a U Gem star.**
V. G. Gorbatskij.
Pis'ma v Astron. Zhurn., Vol. 1, No. 1, p. 23 - 28 (1975). In Russian.

It is shown that all phenomena observed during the outburst of a U Gem type star may be considered as consequences of thermal instability of the secondary. The brightening of the secondary is accompanied by a considerable increase of inflow of matter into the disklike envelope of the primary. Therefore radiation from the disk and particularly from its central region becomes predominant during the light maximum.

122.013 **Observations of two δ Scuti stars: HR 1225 and HR 1298.**
H. E. Jørgensen, H. U. Nørgaard-Nielsen.
Astron. Astrophys., Suppl. Ser., Vol. 19, 235 - 241 (1975).

New observations of HR 1225 and HR 1298 are discussed. Variability of the δ Scuti type is confirmed and the dominating periods are derived. For HR 1225 we find $P =$

$0\overset{d}{.}1562$ and for HR 1298, $P = 0\overset{d}{.}081$.

122.014 UV Ceti flare stars: observational data.
T. J. Moffett.

Astrophys. Journ., Suppl. Ser., No. 273, Vol. 29, 1 - 42 (1974).

Photometric observations of UV Ceti flare stars during the period 1971 – 1972 are presented. Most observations were obtained using integration times of 1 s or less. Four-hundred and nine flares were detected during 469.2 hours of flare monitoring. The mean $(B - V)$ of flare light for 77 flares was found to be +0.34 ± 0.44, and the mean $(U - B)$ for 153 flares was –0.88 ± 0.31. A discussion of the usefulness and limitation of various flare parameters is presented.

122.015 The location of the hot spot in cataclysmic variable stars as determined from particle trajectories.
B. P. Flannery.

Monthly Notices Roy. Astron. Soc., Vol. 170, 325 - 331 = Contr. Lick Obs., No. 384 (1975).

Warner and Peters using particle trajectories in the restricted three-body approximation, have previously published a set of models for the hot spot location in cataclysmic variable stars. This paper presents a new set of models, based on their assumptions, which corrects an error in their calculations. With the new models the spot radius is moved substantially closer to the blue star for most mass fractions, and other parameters are changed accordingly. It is pointed out that the well-defined picture for the hot spot geometry produced in this model is probably not directly applicable to real stars.

122.016 On the colour excesses of the long-period cepheids.
E. G. Schmidt.

Monthly Notices Roy. Astron. Soc., Vol. 170, 39P - 44P (1975).

Photometric measurements have been made on the $uvby\beta$ system of stars located in the fields near five long-period cepheids. The interstellar reddening of these field stars is compared with published data for the cepheids. It appears that for the longest period cepheids (periods greater than about 15 days) systematic errors exist in the methods of finding colour excesses which seriously affect the pulsational masses.

122.017 Spectrum of the Delta Scuti variable 20 Canum Venaticorum: a model-atmosphere analysis.
M. Ishikawa.

Publ. Astron. Soc. Japan, Vol. 27, 1 - 33 (1975).

The author has analyzed the high dispersion spectra of 20 CVn by using model atmospheres. By fitting computed continua, hydrogen lines, ionization equilibria etc. to the observed quantities, he derives a reciprocal effective temperature $\theta_{eff} = 0.64 \pm 0.02$, a surface gravity $\log g = 3.8 \pm 0.3$, and a microturbulent velocity $\xi = (3.5 \pm 0.5)$ km s^{-1} (depth-independent). The abundances of 26 elements relative to the sun are determined. The abundances of the chemical elements relative to Fe in δ Scuti stars are compared with those in Am stars. The general abundance pattern of δ Scuti stars suggests that in δ Scuti stars Am anomalies are less pronounced (often becoming normal) than in Am stars. The results can be explained by the diffusion hypothesis.

122.018 SY Fornacis and the Mira Ceti B phenomenon.
M. W. Feast.

Observatory, Vol. 95, 19 - 23 (1975).

The spectrum of SY For indicates that the object consists of an M type star (SR variable) and a peculiar emission object strongly resembling the hitherto unique hot subluminous companion to Mira. The high radial velocity (−158 km/s) at $b = -62°$ suggests membership of the halo population. Velocity measurements also suggest that the SY For and Mira AB systems are being viewed from different aspects. The high mass loss rate for Mira variables deduced by Gehrz and Woolf

is good support for Deutsch's suggestion that in the wide Mira system, Mira B is heated by accretion of matter flowing outwards from Mira A.

122.019 Observations of large-amplitude red variables.
O. J. Eggen.

Astrophys. Journ., Suppl. Ser., No. 276, Vol. 29, 77 - 86 (1975).

The $(UBVRI)$ observations of large-amplitude red variables presented in this supplement are discussed in Astrophys. Journ., Vol. 195, 661 - 678 (1975). – See Abstr. 13.122.009.

122.020 On the intrinsic properties of cepheids in the Galaxy, in Andromeda, and in the Magellanic Clouds.
I. Iben, Jr., R. S. Tuggle.

Astrophys. Journ., Vol. 197, 39 - 54 (1975).

With opacities and opacity derivatives fitted by spline interpolation to fine-grid Cox-Stewart opacity tables, pulsation calculations have been performed in the linear nonadiabatic approximation to provide pulsation parameters for cepheid models covering a range of masses, luminosities, and compositions required for a discussion of population I variables. The results of pulsation calculations are used in conjunction with observational data to estimate (1) absolute magnitudes, (2) color-temperature conversions, (3) masses, and (4) composition parameters characterizing cepheids in three extragalactic complexes: SMC, LMC, and M31.

122.021 UBV photometry of the cepheid V367 Scuti in the open cluster NGC 6649.
B. F. Madore, S. van den Bergh.

Astrophys. Journ., Vol. 197, 55 - 65 (1975).

UBV photometry has been obtained for the cepheid V367 Sct and for the cluster NGC 6649 in which it is located. The cluster is found to have $(m-M)_V = 15.4 \pm 0.2$ and suffers a reddening $E_{B-V}(B0) = 1.37$. For the cepheid V367 Sct, $\langle V \rangle = 11.58$ and $\langle B \rangle - \langle V \rangle = 1.76$, so that $M_{\langle V \rangle} = -3.8 \pm 0.2$ and $\langle B \rangle_0 - \langle V \rangle_0 = 0.49$. A period of 5.255 ± 0.002 days is derived for V367 Sct. NGC 6649 also contains a nonvariable star in the cepheid instability strip. Furthermore, the cluster appears to contain two red giants, one of which is a variable.

122.022 Proper motions of RR Lyrae stars.
M. K. Hemenway.

Astron. Journ., Vol. 80, 194 - 198 (1975).

New proper motions of 65 RR Lyrae field stars are derived using the McCormick 26-in. refractor to obtain plate pairs with an average time base of 39 yr. An estimate of accuracy is made by comparing these results to previously published values. Comments are made on the determination of the absolute motion of the reference frame.

122.023 Absolute magnitudes and motions of RR Lyrae stars.
M. K. Hemenway.

Astron. Journ., Vol. 80, 199 - 207 (1975).

Absolute magnitudes are found by the use of statistical parallaxes for RR Lyrae stars which have been grouped by period and by metal type. The Bailey c-type stars are treated independently. Solar motion solutions are made using velocity data. Parameters of galactic orbits obtained from space velocities show a relation to period. For the total a-star group the resultant mean light visual absolute magnitude is 0.49 mag, which corresponds to M_B of 0.63 mag. Using this figure, the distance to the galactic center is 9.3 kpc.

122.024 A search for variable stars in globular clusters.
A. Terzan, B. Rutily.

Astron. Astrophys., Vol. 38, 307 - 312 (1975). In French.

Photographic observations of the globular clusters of our Galaxy have been undertaken with an image-tube at E.S.O. (Chile) since 1972. They have resulted in the discovery of 57 variable stars in 5 globular clusters: 4 variables in NGC 4590, 22 in NGC 6401, 26 in NGC 6638, 1 in NGC 7099 and 4 in Terzan 1. The results also show that all significant variables in NGC 1904 and NGC 6304 have now been discovered.

122.025 The unusual period distribution of RR Lyrae variables in the globular cluster IC 4499.
C. Coutts, R. J. Dickens, E. Epps, M. Read.
Astrophys. Journ., (Letters), Vol. 197, L45 - L47 (1975).

New periods for 35 RR Lyrae variables in the distant globular cluster IC 4499 have been determined. The mean period of 28 ab-type variables is 0.582 ± 0.012, showing that the cluster more closely resembles globular clusters of Oosterhoff group I ($\langle P_{ab} \rangle$ ~0^{d}55) than those of group II ($\langle P_{ab} \rangle$ ~0^{d}65). However, attention is drawn to the possibility and implications of a closer similarity to nearby dwarf spheroidal systems such as Leo II and Draco whose variables have mean periods intermediate between the two Oosterhoff groups.

122.026 Frequency analysis of the three Delta Scuti stars BS 3265, 1653 and 242.
M. J. Smyth, R. S. Stobie, R. R. Shobbrook.
Monthly Notices Roy. Astron. Soc., Vol. 171, 143 - 157 (1975).

Photometric observations, and some radial velocity data, for the three delta scuti stars BS 3265, 1653 and 242 were Fourier analysed for their component frequencies.

122.027 The mode of pulsation of Mira variables.
P. R. Wood.
Monthly Notices Roy. Astron. Soc., Vol. 171, 15P - 16P (1975).

The variation of the period of Mira variables along the old disk giant branch observed by Eggen is used to produce a (period, Q) diagram. A comparison with theoretical calculations shows that the Mira variables of the old disk population are first overtone pulsators.

122.028 Investigation of the light curve of SS Cyg.
N. F. Vojkhanskaya, G. N. Alekseev.
Astrofiz. Issled., Izv. Spets. Astrofiz. Obs., Vol. 7, p. 13 - 18 (1975). In Russian.

The light curve of SS Cyg of the period from 1896 to 1971 is investigated. Relative energies of the flares are determined. It is shown that according to the amount of energy the flares may be divided into large and small ones. The properties of each kind of flares are considered. A notion of the energetic cycle with the help of which an attempt is made to explain certain peculiarities of the light curve is introduced.

122.029 CoD −44°3318 − a peculiar luminous F star.
N. J. Irvine.
Publ. Astron. Soc. Pacific, Vol. 87, 87 - 88 (1975).

The unnamed variable CoD −44°3318 has been classified F0 Ipe. It is probably one of a rare group of luminous objects related to the T Tauri stars.

122.030 Observations of the red-dwarf emission-line objects LP101-15, BY Draconis, GT Pegasi, and FF Andromedae.   D. H. Martins.
Publ. Astron. Soc. Pacific, Vol. 87, 163 - 171 = Rosemary Hill Obs., Univ. Florida, Gainesville, Contr. No. 51 (1975).

Photoelectric UBVRI photometry of LP101-15, BY Dra, GT Peg, and FF And is presented, and implications with respect to the "center of activity" model are discussed. Apparently the most physically acceptable model to date, facularlike areas and flare activity are included in an attempt to explain all aspects of the behavior of this class of stars.

122.031 Short time-scale photometric variability of the shell star EW Lacertae.   D. F. Lester.

Publ. Astron. Soc. Pacific, Vol. 87, 177 - 184 (1975).

The quasi-periodic variability of the shell star EW Lac is considered. Four-color photometry is presented which defines the time scale (~0^{d}7) and the wavelength dependence of the variations. The observed variations are compared with those expected from several possible models for variability.

122.032 Pulsation properties of four RR Lyrae stars.
T. S. van Albada, K. S. de Boer.
Astron. Astrophys., Vol. 39, 83 - 90 (1975).

Four RR Lyrae stars have been observed over the complete cycle in the uvby Hβ filter system with the ESO 1-meter telescope. After correction for reddening and blanketing the phase dependent parameters T_{eff} and g have been derived, and from these the equilibrium properties. The analysis uses theoretical colors for static model atmospheres.

122.033 On the absolute magnitudes of RV Tauri variables.
D. L. DuPuy, T. G. Barnes III.
Journ. Roy. Astron. Soc. Canada, Vol. 69, 40 - 41 (1975). Abstr. Canadian Astron. Soc.

122.034 Period changes of RR Lyrae variables in M14.
A. Wehleu, J. Conville, H. Hogg.
Journ. Roy. Astron. Soc. Canada, Vol. 69, 41 (1975). − Abstr. Canadian Astron. Soc.

122.035 Two-colour observations of RR Lyrae variables.
J. E. Penfold.
Observatory, Vol. 95, 44 - 50 (1975).

One of the programmes being carried out at the South African Astronomical Observatory is the observation of RR Lyrae stars whose periods are listed as unknown in the General Catalogue of Variable Stars. The programme is being carried out in an attempt to find stars in the intermediate period ranges, viz (1) ~ 0.3 day and (2) between 1 and 3 days. Light curves, periods and ephemerides for four stars are presented here, along with finding charts for identification.

122.036 SS Cygni.   J. A. Mattei.
Journ. American Ass. Variable Star Observers, Vol. 3, 43 - 48 (1974).

122.037 New period-determinations for eight variable stars.
J. Manella.
Journ. American Ass. Variable Star Observers, Vol. 3, 52 - 54 (1974).

122.038 Six long-period variables in Sagittarius.
P. D. Owensby.
Journ. American Ass. Variable Star Observers, Vol. 3, 55 - 56 (1974).

122.039 Updating the period of V Comae Berenices.
J. Johnston.
Journ. American Ass. Variable Star Observers, Vol. 3, 59 (1974). − Abstract.

122.040 Three variable stars in Cygnus.   L. Dexter.
Journ. American Ass. Variable Star Observers, Vol. 3, 59 (1974). − Abstract.

122.041 Duplicity of some T Tauri stars and related variables.
G. Romano.
Astrophys. Space Sci., Vol. 33, 487 - 490 (1975).

A number of double stars of which at least one component is a T Tauri star or related variable were measured in three fields: φ and χ Tauri (Taurus dark cloud), NGC 7023 and NGC 2264.

122.042 Investigation of pulsating variables. III. Monochrom-

atic light gradients of cepheids.
I. G. Kolesnik, V. G. Krivdik.
Astrometriya i Astrofizika, *Kiev*, vyp. (No.) 24, (see 003.003), p. 56 - 64 (1974). In Russian.

Using a series of monochromatic magnitudes for two classical cepheids — η Aql, δ Cep — and two variables of RR Lyrae type — RR Lyr, SU Dra —, monochromatic light gradients are calculated. The authors showed that monochromatic light gradients give information on the relative amplitudes of radial variations at different levels of stellar atmospheres. On the basis of the obtained data the properties of wide-band gradient diagrams for cepheids are discussed.

122.043 Autocorrelation analysis of the brightness of Z And and AG Dra. E. I. Lenderman.
Astrometriya i Astrofizika, *Kiev*, vyp. (No.) 24, (see 003.003), p. 71 - 74 (1974). In Russian.

The autocorrelation function of the brightness of Z And with τ_{max} = 800^d and two autocorrelation functions of the brightness of AG Dra for two parts of Robinson's observations with τ_{max} = 500^d were obtained. The functions indicate a period of 600^d for Z And and a period of 380^d for AG Dra.

122.044 On the variability of three stars in the vicinity of the open cluster NGC 6830.
B. L. Shaganyan, L. M. Subbotina.
Astrometriya i Astrofizika, *Kiev*, vyp. (No.) 24, (see 003.003), p. 80 - 87 (1974). In Russian.

As a part of investigation of all the variable stars in the vicinity of the open cluster NGC 6830 the results are given of photographic observations of OQ, PR and NT Aql. Maps of surroundings, photographic magnitudes of comparison stars and epochs of maximum brightness are given for each variable.

122.045 Classification of intrinsic variables. VI. Ultrashort-period, very small amplitude B-type variables.
O. J. Eggen.
Astrophys. Journ., Vol. 198, 131 - 138 (1975).

These very young disk-population variables are found to be distributed in a very narrow instability strip, extending from M_{bol} = −9 mag to −3 mag and including stars with a range of at least 25 to 5 solar masses. Some new variables, for which periods are not yet available, are suggested as possibly filling the existing gaps in the luminosity distribution and extending the instability region to still lower masses.

122.046 Spectra and colour indices of bright irregular variables in M31 and M33.
A. S. Sharov, V. F. Esipov, V. M. Lyutyj.
Pis'ma v Astron. Zhurn., Vol. 1, No. 2, p. 15 - 19 (1975). In Russian.

Spectra ($\lambda\lambda$ 5500–7500 Å) and photoelectric UBV colours of three bright irregular variables in M31 (AE And, AF And–V19 and Var A–1), and also the spectrum of Var C in M33 were obtained. Strong Hα emission line was found in all stars. On the two-colour $(U-B)-(B-V)$ diagram the variables are in the region of reddened early B-stars. New data on colours show them to belong to S Dor type.

122.047 On the light curves of OH-Mira variables.
P. F. Bowers.
Astron. Astrophys., Vol. 39, 473 - 476 (1975).

The shapes of the mean light curves of a number of OH-Mira stars have been examined. It is shown that, for a given period, the light curves of OH stars generally have steeper ascending branches relative to their descending branches than the light curves of a sample of nearby Miras for which no OH has been detected. The above result is tentative but offers the possibility of a distance-independent criterion for the detection of new OH-Mira variables.

122.048 High dispersion observations of the Hα region in the flare stars BY Draconis and EQ Virginis.
C. M. Anderson, B. W. Bopp.
Bull. American Astron. Soc., Vol. 7, 235 (1975). – Abstr. AAS.

122.049 Radiofrequency polarimetry of flares from UV Ceti stars.
S. R. Spangler, J. M. Rankin, S. D. Shawhan.
Bull. American Astron. Soc., Vol. 7, 235 (1975). – Abstr. AAS.

122.050 Decimetric flares of UV Ceti stars.
S. R. Spangler, S. D. Shawhan, J. M. Rankin.
Bull. American Astron. Soc., Vol. 7, 235 (1975). – Abstr. AAS.

122.051 Spectra of AE And and the Hubble-Sandage variables in M31 and M33. R. M. Humphreys.
Bull. American Astron. Soc., Vol. 7, 238 - 239 (1975). Abstr. AAS.

122.052 Metal abundances of RR Lyrae stars in galactic globular clusters. D. Butler.
Bull. American Astron. Soc., Vol. 7, 239 (1975). – Abstr. AAS.

122.053 A search for variable continuum radio emission from red giant stars. J. D. Fix, S. R. Spangler.
Bull. American Astron. Soc., Vol. 7, 249 (1975). – Abstr. AAS.

122.054 Double mode cepheids and U TrA.
D. S. King, A. N. Cox, D. D. Eilers, J. P. Cox.
Bull. American Astron. Soc., Vol. 7, 251 (1975). – Abstr. AAS.

122.055 π Aqr: a pulsating Be star? J. D. Fernie.
Bull. American Astron. Soc., Vol. 7, 251 (1975). Abstr. AAS.

122.056 Satellite observations of β Cephei stars. J. R. Lesh.
Bull. American Astron. Soc., Vol. 7, 252 (1975). Abstr. AAS.

122.057 Variable stars of young clusters and of the galactic disc. Yu. N. Efremov.
Instationary stars and methods of their investigation. Phenomena of instationarity and stellar evolution, (see 003.012), p. 13 - 46 (1974). In Russian.

122.058 Variable stars of the galactic halo and of stellar aggregates which have formed long ago.
B. V. Kukarkin.
Instationary stars and methods of their investigation. Phenomena of instationarity and stellar evolution, (see 003.012), p. 151 - 180 (1974). In Russian.

122.059 UBV photometry of the β Cephei type variable stars. IV. V986 Ophiuchi (HD 165174).
M. Jerzykiewicz.
Acta Astron., Vol. 25, 81 - 87 (1975).

Six BV and one UBV light-curves of V986 Ophiuchi are presented. The light-variation is very nearly the same in B and V, the range amounting to about 0^{m}015. There is no strict periodicity.

122.060 Statistical analysis of the brightness of R CrB, RY Sgr and SU Tau. G. U. Koval'chuk,
E. I. Lenderman, F. I. Lukatskaya, A. Eh. Rozenbush.
Astrometriya i Astrofizika, *Kiev*, vyp. (No.) 25, (see 003. 015), p. 3 - 9 (1975). In Russian.

The autocorrelation functions and distribution functions of brightness are obtained for R CrB, RY Sgr and SU Tau using sequences of visual observations. The distributions of time intervals between the fixed magnitudes are also calculated. The brightness gradients are obtained using UBV observations during minima.

122.061 Metal abundances of RR Lyrae stars established from low resolution scanner spectrophotometry.
D. Butler, R. P. Kraft.
Dudley Obs. Rep. No. 9, (see 012.008), p. 121 - 132 (1975).
Using the Wampler-Robinson-Miller image-tube image dissector scanner operated at the 120-inch Cassegrain focus, the authors have reestablished the Preston ΔS system for RR Lyrae stars. ΔS is calibrated as a function of [Fe/H] from coarse analysis of coudé spectrograms. The system is applied to obtain [Fe/H] for RR Lyrae stars in globular clusters and various parts of the Galaxy. Plans for future observations are outlined.

122.062 Theoretical mean colors of pulsating cepheids.
A. N. Cox, C. G. Davis.
Dudley Obs. Rep. No. 9, (see 012.008), p. 297 - 307 (1975).
The paper concerns the method of taking a mean of the color variations of cepheids over their pulsational cycle. It is demonstrated that the mean color depends on the type of mean employed. Thus color observations of cepheids can be interpreted by a color-effective temperature relation to give different T_e values for each kind of mean color. Here, theoretical colors from numerical integrations of cepheid pulsations are used to determine the proper method of taking the color mean in order to get, by the color–T_e relation, the correct, non-pulsating T_e.

122.063 Spectral scans of five short period field variable stars. I. Epstein, A. E. Abraham de Epstein.
Dudley Obs. Rep. No. 9, (see 012.008), p. 477 - 480 (1975).

122.064 B–R bei 103 langperiodisch Veränderlichen.
E. Heiser.
BAV Rundbrief, 24. Jahrgang, Sonder-Rundbrief, 39 pp. = BAV Mitt. No. 27 (1975).
The author presents 103 O–C diagrams of long period variable stars, essentially based on AAVSO observational material. The Ludendorff-classification of light curves is compared with quantitative parameters of mean light curves. Obviously, no positive correlation can be introduced between form of the mean light curve and O–C pattern.

122.065 Pulsating stars. J. R. Percy.
Sci. American, Vol. 232, No. 6, p. 66 - 75 (1975).
A star that varies regularly in brightness is vibrating like the air in an organ pipe. The fundamental vibration or its harmonics are strong clues to the star's composition and internal architecture.

122.066 The light curve of 132554 BV Centauri.
F. M. Bateson.
Publ. Variable Star Section, Roy. Astron. Soc. New Zealand, No. 2 (C 74), p. 1 - 12 (1974).
Observations of BV Cen are presented as a light curve for the interval J.D. 2,434,855 to 2,442,189. It is shown that outbursts fall into at least eight types, differing from one another by the shape of the curve at maximum, as well as by the presence, or absence, of standstills and dips on the rise. The average cycle between successive outbursts is 149.4 days.

122.067 A new RR Lyrae variable in Coma Berenices.
D. Hoffleit.
Inform. Bull. Variable Stars, (I.A.U. Commission 27), Konkoly Obs., Budapest, No. 957 (1975).

122.068 Blazhko effect in the RR Lyrae type star WY Draconis. D. Chiş, G. Chiş, I. Mihoc.
Inform. Bull. Variable Stars, (I.A.U. Commission 27), Konkoly Obs., Budapest, No. 960, 3 pp. (1975).

122.069 Flare observations of V1216 Sagittarii.
A. H. Jarrett, G. Grabner.
Inform. Bull. Variable Stars, (I.A.U. Commission 27), Konkoly Obs., Budapest, No. 968, 4 pp. (1975).

122.070 A flare star in Lupus. S. Suryadi.
Inform. Bull. Variable Stars, (I.A.U. Commission 27), Konkoly Obs., Budapest, No. 975 (1975).

122.071 Further observations of UV Ceti.
A. H. Jarrett, J. B. Gibson.
Inform. Bull. Variable Stars, (I.A.U. Commission 27), Konkoly Obs., Budapest, No. 979, 4 pp. (1975).

122.072 Photometric observations of suspected small-amplitude cepheids. J. R. Percy.
Inform. Bull. Variable Stars, (I.A.U. Commission 27), Konkoly Obs., Budapest, No. 983, 4 pp. (1975).

122.073 Blue-infrared behaviour of some variables of Mira type. P. Maffei.
Inform. Bull. Variable Stars, (I.A.U. Commission 27), Konkoly Obs., Budapest, No. 986, 4 pp. (1975).

122.074 Note on the RRab star AT Andromedae.
K. Oláh.
Inform. Bull. Variable Stars, (I.A.U. Commission 27), Konkoly Obs., Budapest, No. 987, p. 1 (1975).

122.075 On the period variation and Blažko effect of XZ Cygni. V. Pop.
Inform. Bull. Variable Stars, (I.A.U. Commission 27), Konkoly Obs., Budapest, No. 990, 4 pp. (1975).

122.076 Proposed ground-base observations of UV Ceti flare stars in coordination with the MIT/SAS-C satellite. T. J. Moffett.
Inform. Bull. Variable Stars, (I.A.U. Commission 27), Konkoly Obs., Budapest, No. 995 (1975).

122.077 Flare-ups in stars of the Pleiades field.
G. Szécsényi-Nagy.
Inform. Bull. Variable Stars, (I.A.U. Commission 27), Konkoly Obs., Budapest, No. 996, 4 pp. (1975).

122.078 Optical observations of UV Ceti flare stars simultaneous with radio coverage. T. J. Moffett.
Inform. Bull. Variable Stars, (I.A.U. Commission 27), Konkoly Obs., Budapest, No. 997, 12 pp. (1975).

122.079 Flare activity of YZ CMi. B. B. Sanwal.
Inform. Bull. Variable Stars, (I.A.U. Commission 27), Konkoly Obs., Budapest, No. 998, 2 pp. (1975).

122.080 On the period of the secondary cycle in XZ Cygni.
P. Kunchev.
Inform. Bull. Variable Stars, (I.A.U. Commission 27), Konkoly Obs., Budapest, No. 1001, 2 pp. (1975).

122.081 Flare stars in the region of NGC 7000. III.
M. K. Tsvetkov, L. K. Erastova, K. P. Tsvetkova.
Inform. Bull. Variable Stars, (I.A.U. Commission 27), Konkoly Obs., Budapest, No. 1002, 4 pp. (1975).

122.082 On the period-luminosity relation of the Delta Scuti stars. T. Z. Dworak, S. Zieba.
Inform. Bull. Variable Stars, (I.A.U. Commission 27), Konkoly Obs., Budapest, No. 1005, 6 pp. (1975).

122.083 Flare activity on UV Ceti, 1975.01.
W. E. Kunkel, N. Zárate.

Inform. Bull. Variable Stars, (I.A.U. Commission 27), Konkoly Obs., Budapest, No. 1006, 3 pp. (1975).

122.084 Note on AX Sgr.   J. D. Fernie.
Astron. Journ., Vol. 80, 458 - 459 (1975).
BVRI photometry of AX Sgr on six nights in September 1974 is reported. No variability in excess of 0.01 mag was found, which is considered surprising if the star is a long-period Cepheid, as has been suggested. The location of the star in the $(B-V)$ vs $(R-I)$ diagram favors the G8 Ia spectral type of Morgan and Roman rather than M2 as suggested by Neckel.

122.085 X-ray emission from U Geminorum and the radio spur at $l \simeq 200°$.   R. Novick, L. Woltjer.
Astrophys. Letters, Vol. 16, 67 - 69 (1975).
Rocket X-ray data on U Geminorum obtained at phase 0.062 during a quiescent period are shown to yield an upper limit of 7×10^{31} ergs sec^{-1} for the X-ray luminosity in the 150 to 280 eV band. This limit is compared with the recent reported X-ray observation of SS Cygni during an active period and the prediction of Warner's model for U Gem. It is suggested, in agreement with Berkhuijsen, that a soft X-ray feature at $l \simeq 200°$, $b \simeq + 20°$ arises from a radio spur at that location.

122.086 Coarse photospheric convection and the ejection of dust by R Coronae Borealis.   T. J. Wdowiak.
Astrophys. Journ., (Letters), Vol. 198, L139 - L140 (1975).
It is suggested that 10 to 100 convection cells exist in a coarse structure across the surface of R CrB and are responsible for the formation of dust patches that are ejected asymmetrically by radiation pressure.

122.087 On the ionization zones around flare stars.   V. M. Tomozov.
Pis'ma v Astron. Zhurn., Vol. 1, No. 5, p. 31 - 33 (1975). In Russian.
The development of an ionization zone (H II) around a flare UV Cet-type star and subsequent recombination Hα emission after powerful flares is considered. Possibilities to observe these zones are discussed.

122.088 Hβ and continuum photometry of BW Vulpeculae.   T. A. Cherewick, A. Young.
Publ. Astron. Soc. Pacific, Vol. 87, 311 - 316 (1975).
The β CMa star BW Vul was observed in the Strömgren violet color, and an Hβ index was observed to search for variations which might correlate with the continuum radiation, and to note any changes in the equivalent width of the Hβ line which might indicate pressure changes. We observed no variation and no correlation of the Hβ index with the phase (contrary to the findings of Kubiak (1971)). A new period study confirms a systematic increase in period.

122.089 Infrared light curves for V1057 Cygni (1971-74).   T. Simon.
Publ. Astron. Soc. Pacific, Vol. 87, 317 - 320 (1975).
Broadband infrared photometry at 5μ, 10μ, and 20μ is presented for V1057 Cyg (=LkHα 190). The observations extend from 1971 July through 1974 July, during which time the 5μ brightness has remained essentially constant and the long-wavelength fluxes have decreased by $\sim 0\overset{m}{.}8$. The origin of the flux excess observed at wavelengths between 1μ and 5μ is uncertain; remnant circumstellar dust, which is radiatively heated by a central point source of declining luminosity, provides a satisfactory explanation of the 10μ and 20μ light curves.

122.090 The classical cepheid program J. D. 2,440,000 - 2,441,000.   T. A. Cragg.
Journ. American Ass. Variable Star Observers, Vol. 1, 9 - 15 (1972). – Progress report on the AAVSO classical cepheid program.

122.091 Metallic-line equivalent widths in the spectra of thirteen RR Lyrae stars.   D. Butler.
Lick Obs. Bull. No. 693, 8 pp. (1974/75).

122.092 Progressi nell'interpretazione e studio delle stelle variabili (XVIII, XIX).   L. Rosino.
Coelum, Vol. 43, 93 - 108 (1975).

122.093 A method of determining the period of short-period variable stars with application to EL Comae and three new variables.   D. J. Henry.
Journ. American Ass. Variable Star Observers, Vol. 1, 29 - 38 (1972).

122.094 The spectra of Mira variable stars.   R. F. Garrison.
Journ. American Ass. Variable Star Observers, Vol. 1, 39 - 42 (1972).

122.095 New sequences for two cepheid variables.   H. J. Landis, C. E. Scovil.
Journ. American Ass. Variable Star Observers, Vol. 1, 52 - 53 (1972).

122.096 RW Arietis, a short period pulsating star, one component of an eclipsing binary.   E. J. Woodward.
Journ. American Ass. Variable Star Observers, Vol. 1, 68 - 69 (1972).

122.097 A note on some flare stars in Ophiuchus and Scorpius.   G. Haro, E. Chavira.
Bol. Inst. Tonantzintla, Vol. 1, 189 - 192 (1974).
Using the Tonantzintla Schmidt camera and centering in $\alpha = 16^h 23^m$ and $\delta = -24°20'$ the authors obtained 210 different exposures with an effective observational time of $43^h 5^m$. In this photographic material they found 4 different stars showing conspicuous flare-ups.

122.098 Some recent observations relating to stellar populations.   T. Lloyd Evans.
Monthly Notes Astron. Soc. Southern Africa, Vol. 34, 51 (1975). – Abstract.

122.099 Flare stars.   D. S. Evans.
Monthly Notes Astron. Soc. Southern Africa, Vol. 34, 51 - 52 (1975).

122.100 S Doradus variables.   A. D. Thackeray.
Monthly Notes Astron. Soc. Southern Africa, Vol. 34, 55 (1975).

122.101 The period of Alpha Lupi.   A. van Hoof.
Monthly Notes Astron. Soc. Southern Africa, Vol. 34, 73 - 75 (1975).

122.102 Period of pulsating variables.   M. Breger.
Monthly Notes Astron. Soc. Southern Africa, Vol. 34, 76 (1975). – Abstract.

122.103 SU Ursae Majoris, 1955 - 69.   J. E. Isles.
Journ. British Astron. Ass., Vol. 85, 346 - 352 (1975).

122.104 Discovery of flare activity on the dM5e star Gliese 268.   B. R. Pettersen.
Astron. Astrophys., Vol. 41, 87 - 90 (1975).
The discovery of flare activity on Gliese 268 is reported and a preliminary flare energy spectrum is derived. Predicted activity level for some flare star candidates, based on the relation between Hα-intensity and flare activity, is given.

122.105 **Discovery of flare activity on BD +66°34 (= Gliese 22 A).** B. R. Pettersen.
Astron. Astrophys., Vol. 41, 113 (1975). – Research note.

122.106 **Nu Centauri – a suspected β Canis Majoris type variable.** R. Rajamohan, V. Natarajan.
Bull. Astron. Soc. India, Vol. 2, 31 (1974). – Abstract.

122.107 **Photoelectric study of X Arietis.** H. S. Mahra.
Bull. Astron. Soc. India, Vol. 2, 31, 34 (1974). Abstract.

122.108 **Energy distribution of five β Scuti stars.** S. C. Joshi, B. S. Rautela.
Bull. Astron. Soc. India, Vol. 2, 34 (1974). – Abstract.

122.109 **Les étoiles variables à longue période.** R. Fillit, D. Proust.
A.F.O.E.V. Bull., Vol. 8, 74 - 82 (1974).

122.110 *UBVRI* **photometry of V553 Centauri.** A. U. Landolt.
Publ. Astron. Soc. Pacific, Vol. 87, 373 - 377 = Contr. Louisiana State Univ. Obs., *Baton Rouge,* No. 105 (1975).

UBVRI photoelectric observations are presented for the population II cepheid V553 Cen. Its period of $2\overset{d}{.}06051$ is confirmed.

122.111 **The Blazhko effect in observations of XZ Cygni.** H. A. Smith.
Publ. Astron. Soc. Pacific, Vol. 87, 465 - 470 (1975).

An analysis was made of the Blazhko effect in the RR Lyrae star XZ Cyg using 410 times and brightnesses of primary maxima obtained between 1905 and 1973. A secondary periodicity was found to exist for XZ Cyg, the period of this variation being about $57\overset{d}{.}38$ until JD2438500 (1964) and thereafter about $58\overset{d}{.}7$. Mean curves are derived for the $57\overset{d}{.}38$ and $58\overset{d}{.}7$ variations. It is shown that the increase in the secondary period can be explained under the assumption that the secondary period is a beat period involving interference between the primary periodicity and another periodicity.

122.112 **Distribution and ages of Magellanic cepheids.** C. H. Payne-Gaposchkin.
Smithsonian Contr. Astrophys., *Cambridge, Mass.,* No. 16, 2 + 32 pp. (1974).

The distribution and ages of the Magellanic cepheids are discussed under four major topics: (1) The Large Cloud is shown to contain components with a variety of ages. A time scale is set up. (2) Changes of period for galactic and Magellanic cepheids are used to test the time scales derived for cepheids. (3) Evidence bearing on possible differences of composition is reviewed. (4) The period-amplitude and period-frequency relations, and their differences, are shown to be compatible with differences in the time of formation of stars that are now cepheids in the two Clouds and in the Galaxy. The discussion under these four topics suggests that the observed differences between the three systems can be interpreted in terms of differences in the timetable of star formation, without the need to invoke initial differences of composition.

122.113 **Period, color, and luminosity for cepheid variables.** C. H. Payne-Gaposchkin.
Smithsonian Contr. Astrophys., *Cambridge, Mass.,* No. 17, 2 + 10 pp. (1974).

The cepheids of the Galaxy and the Magellanic Clouds display a gradation in properties, but when amplitude as well as period is taken into account, their intrinsic colors are sensibly similar. The period-luminosity relations in the three systems are probably parallel, but not necessarily coincident.

122.114 **UBV photometry of R Coronae Borealis during the brief minimum of 1973 - 1974.** M. Nakagiri.
Publ. Astron. Soc. Japan, Vol. 27, 379 - 383 (1975).

R CrB was observed in UBV colors during its minimum beginning at the end of 1973. B – V increased by 0.7 mag during the drop to minimum and decreased slowly in the recovery phase, while the star showed a small variation in B – V color during the brief minima of 1966 and 1972. U – B decreased during the descent, increased very rapidly a little after the minimum, and decreased slowly to the normal value during the recovery phase.

122.115 **The magnetic field of W Sgr and evidences for a period-magnetic field relation in pulsating variables.** W. W. Weiss, H. J. Wood.
Astron. Astrophys., Vol. 41, 165 - 168 (1975).

Four Zeeman plates of the cepheid W Sgr were measured. In three plates, a very weak magnetic field was measured. Simultaneous four-color and Hβ photometry is also presented on four of the nights of spectroscopy. For pulsating variable stars, a possible correlation between the maximum magnetic field and the period is found.

122.116 **Variable star BB Cam.** M. Kazarian, A. Terzan.
Astron. Astrophys., Vol. 41, 217 - 221 (1975).

The photometric study of the variable star BB Cam results of a long series of photographic observations (m_{pg} and m_r). The authors come to the conclusion that: (a) BB Cam is a red variable star with a long period ($\bar{P} = 142j \pm 1$) and $CI = \overline{m_{pg} - m_r} \cong + 2^m 25$. (b) Period and amplitude change from one cycle to the other. (c) Sometimes, the brightness changes very quickly, particularly in blue light. (d) The star is redder in the maximum than in the minimum. (e) BB Cam seems to be a pulsating semi-regular variable star type SRb.

122.117 **Outburst of CI Cygni.** W. Lowder, J. Bauer, E. Hayden.
IAU Circ., No. 2788 (1975).

122.118 **The 1972 anomaly of XZ Cygni.** M. E. Baldwin.
Journ. American Ass. Variable Star Observers, Vol. 2, 14 - 19 (1973).

122.119 **The RV Tauri variables.** J. A. Mattei.
Journ. American Ass. Variable Star Observers, Vol. 2, 26 - 28 (1973).

122.120 **SS Aurigae – an anomaly.** C. J. Hurless.
Journ. American Ass. Variable Star Observers, Vol. 2, 35 - 37 (1973).

122.121 **Observations of RT Aurigae – a short period classical cepheid.** B. F. Small.
Journ. American Ass. Variable Star Observers, Vol. 2, 40 (1973).

122.122 **VX Sagittarii: a variable at many wavelengths.** H. Dinerstein.
Journ. American Ass. Variable Star Observers, Vol. 2, 52 - 59 (1973).

122.123 **A study of some flare stars in a Coma field.** S. Nygard.
Journ. American Ass. Variable Star Observers, Vol. 2, 60 - 62 (1973).

122.124 **Visual observations of short-period cepheids.** B. F. Small.
Journ. American Ass. Variable Star Observers, Vol. 2, 76 - 79 (1973).

122.125 **Visual observations of eight cepheids in Cassiopeia.**
B. F. Small.
Journ. American Ass. Variable Star Observers, Vol. 3, 15 - 18
(1974).

122.126 **S Persei — a semi-regular variable with two periods.**
H. A. Smith.
Journ. American Ass. Variable Star Observers, Vol..3, 20 - 23
(1974).

122.127 **Variable stars as observed in infrared and visual**
radiation. E. B. Weston.
Journ. American Ass. Variable Star Observers, Vol. 3, 35
(1974). — Abstract.

122.128 **3.5 micron light curves of long period variable stars.**
D. W. Strecker.
Thesis, Minnesota Univ., Minneapolis (USA). 260 pp. University Microfilms Order No. 74-10,595 (1973).

122.129 **Spectroscopic study of flare stars.** B. W. Bopp.
Thesis, Texas Univ., Austin (USA). 147 pp. University Microfilms Order No. 74-14,672 (1974).

122.130 **Studies of the Z Cam type variable stars.**
E. L. Robinson, Jr.
Thesis, Texas Univ., Austin (USA). 185 pp. University Microfilms Order No. 74-5316 (1973).

122.131 **Apsidal constant and pulsational characteristics of**
Alpha Virginis. A. P. Odell.
Thesis, Wisconsin Univ., Madison (USA). 54 pp. University Microfilms Order No. 74-10,259 (1974).

122.132 **Modal stability of RR Lyrae stars.**
R. F. Stellingwerf.
Thesis, Colorado Univ., Boulder (USA). 143 pp. University Microfilms Order No. 74-22,395 (1974).

On methods of high-accuracy electrophotometric
observations of variable stars. See Abstr. 031.204.

Calcul de la période d'un phénomène cyclique.
See Abstr. 031.245.

The photoelectric photometry with a 60 cm reflector
at the Skalnaté Pleso Observatory. See Abstr. 034.072.

Dynamics of the current sheet of a flare. I. Diffusion
thickening of the current sheet and estimate of the flare parameters. See Abstr. 062.030.

Shock propagation through an atmospheric model
of an RR Lyrae type star. See Abstr. 064.027.

Fe I fluorescence in T Tauri stars. II. Clues to the
velocity field in the circumstellar envelope.
See Abstr. 064.033.

General remarks on the variability of spotted stars.
See Abstr. 064.064.

Magnetic fields and dense chromospheres in dMe
stars. See Abstr. 064.067.

Theoretical investigation of stellar atmospheres with
specific application to the atmosphere of a carbon star R
Coronae Borealis. See Abstr. 064.075.

Pulsational stability of stars in thermal imbalance.
VI. Physical mechanisms and extension to nonradial oscillations.

See Abstr. 065.004.

Modal stability of RR Lyrae stars.
See Abstr. 065.006.

Instability against nonradial oscillations of models
for Beta Cephei stars. See Abstr. 065.030.

On the treatment of convection as a nonradial stel-
lar pulsation. See Abstr. 065.065.

Die Sternentwicklung bis zum Auftreten von
Pulsationen. See Abstr. 065.068.

Nonradial oscillations of a 10 solar mass star in the
main-sequence stage. See Abstr. 065.090.

UBVr **sequences for two η Carinae-like objects.**
See Abstr. 113.020.

A search for technetium in red giant variables.
See Abstr. 114.043.

The T Tauri star RU Lupi and its circumstellar sur-
rounding. See Abstr. 114.309.

Simultaneous observations of variable stars. I. The
Be star π Aqr. See Abstr. 114.315.

The magnetic field, spectrum and light variations
of the Ap star HD 49976. See Abstr. 116.004.

HD 151965 — a new bright Bp : Si variable star.
See Abstr. 116.009.

Veränderliche in einem Feld um ν Andromedae.
See Abstr. 123.032.

Observations of rapid blue variables — XV. VW
Hydri. See Abstr. 124.001.

The gravity oscillations of white dwarfs.
See Abstr. 126.004.

Circumstellar grains and the intrinsic polarization of
starlight. See Abstr. 131.012.

On the nature of maser sources in infrared stars.
See Abstr. 141.617.

A possible candidate for LMC X-3 and observations
of cepheids in the galactic centre. See Abstr. 142.102.

On the stellar origin of low energy cosmic rays.
See Abstr. 143.053.

Tentative membership of the 11-day cepheid TW
Normae in the open cluster Lyngå 6. See Abstr. 153.009.

UBV **photometry of NGC 6649.**
See Abstr. 153.021.

The distance to the galactic center derived from
RR Lyrae variables, the distribution of these variables in the
Galaxy's inner region and halo, and a rediscussion of the
galactic rotation constants. See Abstr. 155.051.

Errata

122.901 Errata: 'The light variation of I Monocerotis'

[Monthly Notices Roy. Astron., Soc., Vol. 169, 643 - 661 (1974)]. R. R. Shobbrook, R. S. Stobie. Monthly Notices Roy. Astron. Soc., Vol. 171, 235 (1975).

122.902 Errata: 'On the beat phenomenon in the Beta Cephei stars' [Astrophys. Journ., Vol. 190, 631 - 636 (1974)]. R. G. Deupree. Astrophys. Journ., Vol. 198, 777 (1975).

123 Variable Stars: Lists of Observations, Individual Observations

123.001 **CH Ursae Majoris.** I. D. Howarth.
Journ. British. Astron. Ass., Vol. 85, 120 - 123 (1975).

123.002 **Two RV Tauri variables: U Mon, 1964 - 73 and RV Tau, 1950 - 73.** J. E. Isles.
Journ. British. Astron. Ass., Vol. 85, 156 - 160 (1975). — Report of Variable Star Section.

123.003 **RU Pegasi, 1927 — 69.** I. D. Howarth.
Journ. British. Astron. Ass., Vol. 85, 271 - 277 (1975). — Report of Variable Star Section.

123.004 **Verdachte variabele ontmaskerd?** M. Drummen.
Zenit, Vol. 2, 74 - 75 (1975).

123.005 **Amatörobservationer av EM Cygni.** H. Bengtsson.
Astron. Tidssk., Årg. 8, p. 74 (1975).

123.006 **De veranderlijke van de maand: R Draconis.** H. Feijth.
Zenit, Vol. 2, 168 - 170 (1975).

123.007 **Observations of variable stars July - December 1974. Report No. 27.** L. Plaut, H. Feijth.
Nederlandse Vereniging voor Weer- en Sterrenkunde. Kapteyn Astron. Lab. Groningen—Netherlands. 13 pp. (1975).
This report gives 4490 visual observations of 192 Mira-type stars, 5 semi-regular type stars, 7 U Gem and Z Cam type stars, and 5 miscellaneous type stars.

123.008 **Observations of CR Muscae — a Mira variable star.** B. F. Marino, W. S. G. Walker.
Southern Stars, Vol. 26, 11 - 14 (1975).

123.009 **BH Crucis.** F. M. Bateson.
Publ. Variable Star Section, Roy. Astron. Soc. New Zealand, No. 2 (C 74), p. 20 - 25 (1974).
Visual estimates of the extremely red variable, BH Cru, are compared with photoelectric V magnitudes to determine systematic deviations for individual observers. Observed dates for maxima and minima are listed. From a corrected curve a period of 421 days is determined, with the Epoch of Primary Minimum J.D. 2,440,858.

123.010 **Light curves.** F. M. Bateson.
Publ. Variable Star Section, Roy. Astron. Soc. New Zealand, No. 2 (C 74), p. 26 - 41 (1974).
Light curves are reproduced for T & V Aps; R, T, RT, RV and RX Cen; W Hya; R & Y Lup; R & T Nor; RR, RS and RZ Sco. A list of observed maxima and minima is given, together with the elements used and remarks on these variables.

123.011 **Some predictions for 1975.** F. M. Bateson.
Publ. Variable Star Section, Roy. Astron. Soc. New Zealand, No. 2 (C 74), p. 42 - 43 (1974).
Predictions of maxima and minima for a number of Mira variables are given.

123.012 **Seventeen long period variables in Sagittarius.** D. Hoffleit.
Inform. Bull. Variable Stars, (I.A.U. Commission 27), Konkoly Obs., Budapest, No. 958, 3 pp. (1975).

123.013 **CSV 7917.** C. B. Stephenson.
Inform. Bull. Variable Stars, (I.A.U. Commission 27), Konkoly Obs., Budapest, No. 966, 2 pp. (1975).

123.014 **Nine new variable stars in a field around μ Cep.** E. H. Geyer, F. Gieseking.
Inform. Bull. Variable Stars, (I.A.U. Commission 27), Konkoly Obs., Budapest, No. 967, 2 pp. (1975).

123.015 **v^1 Puppis, a pulsating B3 Ve star.** A. van Hoof.
Inform. Bull. Variable Stars, (I.A.U. Commission 27), Konkoly Obs., Budapest, No. 969, p. 1 (1975).

123.016 **Note on v^2 Puppis.** A. van Hoof.
Inform. Bull. Variable Stars, (I.A.U. Commission 27), Konkoly Obs., Budapest, No. 969, p. 2 (1975).

123.017 **27 Vir: a possible Delta Scuti star.** C. Bartolini, A. Piccioni, P. Silveri.
Inform. Bull. Variable Stars, (I.A.U. Commission 27), Konkoly Obs., Budapest, No. 981 (1975).

123.018 **Oklahoma variable number 30.** O. C. S. Cyr, Jr.
Inform. Bull. Variable Stars, (I.A.U. Commission 27), Konkoly Obs., Budapest, No. 982, 2 pp. (1975).

123.019 **New variable stars in the field of M16—M17.** P. Maffei.
Inform. Bull. Variable Stars, (I.A.U. Commission 27), Konkoly Obs., Budapest, No. 985, 6 pp. (1975).

123.020 **A note on V CrA and W Men.** L. A. Milone.
Inform. Bull. Variable Stars, (I.A.U. Commission 27), Konkoly Obs., Budapest, No. 989 (1975).

123.021 **V 450 Lyr.** K. Häussler.
Inform. Bull. Variable Stars, (I.A.U. Commission 27), Konkoly Obs., Budapest, No. 991 (1975).

123.022 **Observations of 26, 27 and 28 CMa.**
A. van Hoof.
Inform. Bull. Variable Stars, (I.A.U. Commission 27), Konkoly Obs., Budapest, No. 992, 4 pp. (1975).

123.023 **HD 163868, a new bright southern variable.**
E. J. Woodward.
Inform. Bull. Variable Stars, (I.A.U. Commission 27), Konkoly Obs., Budapest, No. 993, 3 pp. (1975).

123.024 **New faint southern variable stars.** B. S. Carter.
Inform. Bull. Variable Stars, (I.A.U. Commission 27), Konkoly Obs., Budapest, No. 994, 6 pp. = Veröff. Remeis-Sternw. Bamberg, Astron. Inst. Univ. Erlangen-Nürnberg, Vol. 10, No. 115 (1975).

123.025 **New variable stars in Perseus.** G. Romano.
Inform. Bull. Variable Stars, (I.A.U. Commission 27), Konkoly Obs., Budapest, No. 999 (1975).

123.026 **Improved period of the Ap variable HR 5153.**
Z. Mikulášek.
Inform. Bull. Variable Stars, (I.A.U. Commission 27), Konkoly Obs., Budapest, No. 1003 (1975).

123.027 **Optical variations of Cen X-3.**
A. R. Walker, P. R. Hurly.
Inform. Bull. Variable Stars, (I.A.U. Commission 27), Konkoly Obs., Budapest, No. 1004 (1975).

123.028 **Index to Variable Star Section Circulars Nos. 1 - 199.** C. W. Venimore.
Publ. Variable Star Section, Roy. Astron. Soc. New Zealand, No. 2 (C 74), p. 43 - 48 (1974).

123.029 **Photographische Beobachtungen von Veränderlichen auf Platten der Sonneberger Himmelsüberwachung.**
E. Splittgerber.
MVS, *Sonneberg*, Vol. 6, 193 - 196 (1975).

123.030 **Identität.** I. Meinunger.
MVS, *Sonneberg*, Vol. 6, 201 (1975).
The variable star S 8050 = XY Com is identified with S 8488 = BX Com.

123.031 **Sternverzeichnis – MVS Band 6.**
MVS, *Sonneberg*, Vol. 6, 203 - 205 (1975).

123.032 **Veränderliche in einem Feld um ν Andromedae.**
L. Meinunger.
MVS, *Sonneberg*, Vol. 7, 1 - 21 (1975).
On plates of a field around ν Andromedae taken with the 40/195 cm astrograph of Sonneberg, 21 new variable stars of the following types have been found: 9 RR Lyr; 4 E; 4 irregular variable blue or uncoloured objects; 1 Mira; 1 U Gem; 1 UV Ceti; nova in M 31. For these objects and some other variable stars the type of variability, light curves, coordinates, and charts are given.

123.033 **Revised and updated periods for twelve variables in Sagittarius.** C. Day.
Journ. American Ass. Variable Star Observers, Vol. 1, 43 - 46 (1972).

123.034 **Two stars in Sagittarius with changing periods.**

P. R. Knight.
Journ. American Ass. Variable Star Observers, Vol. 1, 56 - 59 (1972).

123.035 **Three long period variables in Sagittarius.**
B. Hatfield.
Journ. American Ass. Variable Star Observers, Vol. 1, 70 (1972).

123.036 **Association Française des Observateurs d'Étoiles Variables A.F.O.É.V. Année 1974, janvier – décembre.**
A.F.O.E.V. Bull., Vol. 8, 9 - 33, 46 - 73, 95 - 124 (1974); Vol. 9, 29 - 58 (1975). – 17209 observations of 1155 variables are reported.

123.037 **Identification of variable stars in the Magellanic Clouds.** C. J. Butler, P. A. Wayman.
Commun. Dublin Inst. Advanced Studies, Ser. C. Dunsink Obs. Publ., Vol. 1, (No. 7), 193 - 206 (1974).
Two regions of the Large Magellanic Cloud have been searched for variable stars and 307 new identifications of suspected variables are made.

123.038 **Maxima and minima of long period variables.**
J. A. Mattei.
American Ass. Variable Star Observers, AAVSO, Bull. No. 38, 9 pp. (1975). – 1975 annual predictions.

123.039 **SU Tauri.** E. Mayer.
IAU Circ., No. 2739 (1975).

123.040 **R CrB variables.** M. V. Jones, A. F. Jones, U. Surawski, D. P. Elias, W. E. Pennell, E. Mayer, C. Hurless, L. C. Peltier, C. Scovil.
IAU Circ., No. 2751 (1975).

123.041 **AL Comae Berenices.** C. Scovil
IAU Circ., No. 2760 (1975).

123.042 **SU Tauri.** G. Comello, E. Mayer, J. Bortle, U. Hopp.
IAU Circ., No. 2767 (1975).

123.043 **CH Ursae Majoris.** J. E. Bortle, R. Annal.
IAU Circ., No. 2773 (1975).

123.044 **AL Comae Berenices.** C. Scovil, L. Rosino.
IAU Circ., No. 2782 (1975).

123.045 **Al Comae Berenices.** H. E. Bond, A. U. Landolt.
IAU Circ., No. 2786 (1975).

123.046 **A photoelectric light curve of X Cygni.**
H. J. Landis.
Journ. American Ass. Variable Star Observers, Vol. 2, 38 - 39 (1973).

123.047 **Five variable stars in Cygnus.** B. Buratti.
Journ. American Ass. Variable Star Observers, Vol. 2, 80 - 81 (1973).

123.048 **Variable star notes.** J. A. Mattei.
Journ. Roy. Astron. Soc. Canada, Vol. 69, 53 - 56, 101 - 104 (1975).

Proper motions of variable stars in the Sydney Astrographic Zone. See Abstr. 112.009.

124 Novae

124.001 Observations of rapid blue variables — XV. VW Hydri. B. Warner.
Monthly Notices Roy. Astron. Soc., Vol. 170, 219 - 228 (1975).

The Southern Hemisphere dwarf nova VW Hyi is shown, on the basis of repetitive humps in its light curve, to be a binary with a period of 106.95 min. Observations made during two supermaxima show a hump periodicity of 110.03 min, the cause of which has not been explained.

124.002 Eruptive variable stars: initial stages of the outburst of VW Hydri. W. S. G. Walker, B. F. Marino.
Southern Stars, Vol. 25, 214 - 219 (1974).

124.003 Novae, supernovae, and neutron sources. S. Starrfield, J. W. Truran, W. M. Sparks.
Astrophys. Journ., (Letters), Vol. 198, L113 - L117 (1975).

The authors have evolved thermonuclear runaways in 1.1 $M_\odot$ and 1.25 $M_\odot$ white dwarfs where they have assumed extremely enhanced ^{12}C in the envelope. A "super" nova-type outburst results, in which peak temperatures exceeding 2×10^9 K and peak burning rates of 3×10^{24} ergs g^{-1} s^{-1} are achieved. A shock forms and ejects more than 10^{29} g moving with speeds up to 90,000 km s^{-1}. The subsequent β^+-decays provide enough additional energy to eject nearly 2×10^{30} g. The peak luminosity is $10^{10} L_\odot$ at an effective temperature of nearly 3×10^5 K. One model produces a substantial neutron flux for a short time, sufficient to drive an intermediate neutron-capture process.

124.004 The state of ionization in nova shells. II. Integration of the ionization equation for optically thick shells. P. B. Bosma.
Astron. Astrophys., Vol. 40, 175 - 184 (1975).

To investigate new explanations for the light curves of novae which show a marked minimum like DQ Herculis 1934, the ionization equation for expanding, spherically symmetric homogeneous hydrogen shells, surrounding a central star (derived by the author in an earlier paper, 1972), is integrated over time. From the resulting dependence of the degree of ionization on position in the shell and time, the output of visible radiation between 5400 and 5500 Å is calculated. From this visible radiation, "light curves" are constructed for a number of shells in which the mass, the electron temperature, the initial radius and the expansion velocity are varied. Considerations lead to restricting the integration to optically thick shells.

124.005 The state of ionization in nova shells. III. A new explanation for the light curve of DQ Herculis 1934. P. B. Bosma.
Astron. Astrophys., Vol. 40, 185 - 189 (1975).

A new interpretation of the light curve of nova DQ Herculis 1934 is given on the basis of a study of the time integration of the ionization equation for expanding shells surrounding a central star, performed in an earlier paper (1975).

124.006 Two southern dwarf novae. B. Warner.
Monthly Notes Astron. Soc. Southern Africa, Vol. 34, 54 (1975). — Abstract.

124.007 Orbital period and qualitative model for the exnova RR Pictoris. N. Vogt.
Astron. Astrophys., Vol. 41, 15 - 20 (1975).
New photoelectric observations of RR Pic in UBV and integrated light, combined with previously published data, yield a precise value of the orbital period ($0^{d}1450255$). Light curves and color variations as a function of orbital phase are discussed. The spectrum of RR Pic shows a strong continuum and mostly weak emission lines of H, He II and C III. A qualitative model of a close binary with an extended hot area on a gas disc surrounding the blue component is suggested. The gas disc seems to suffer a partial eclipse by the red, cool companion. This interpretation is also supported by UV photometry carried out with the OAO-2.

124.008 Nova or other eruptive variable. P. Wild.
IAU Circ., No. 2788 (1975).

Novae and supernovae. See Abstr. 003.004.

Nova-like stars and novae. See Abstr. 003.008.

Outburst of U Cephei. See Abstr. 117.002.

TT Arietis: an evolved, very short period binary. See Abstr. 117.003.

On the origin of nova-like binary systems. See Abstr. 117.023.

Period changes in eruptive binaries. See Abstr. 121.015.

The location of the hot spot in cataclysmic variable stars as determined from particle trajectories. See Abstr. 122.015.

Progressi nell'interpretazione e studio delle stelle variabili (XVIII, XIX). See Abstr. 122.092.

Veränderliche in einem Feld um ν Andromedae. See Abstr. 123.032.

124.100 Nova Delphini 1967

A spectrophotometric study of nova Delphini 1967 (HR Del). Y. Yamashita.
Ann. Tokyo Astron. Obs., Second Ser., Vol. 15, (No. 1), 1 - 46 (1975).

Twenty-six high-dispersion spectrograms and twenty-three low-dispersion plates of nova Delphini 1967 were obtained at the Okayama Astrophysical Observatory during the period of July 15, 1967 through July 20, 1970. In addition, photoelectric spectral scanning observations were also carried out on two occasions in July and October, 1967. Series of microphotometer traces in several portions are shown to give the development of the spectrum of this nova, and several characteristic features are discussed. Equivalent widths measured for emission and absorption components are listed. The results of curve-of-growth analysis and the characteristic spectral features are interpreted in terms of resonance fluorescence of diluted photospheric radiation.

Spectrophotometric study of nova Delphini 1967. M. B. Babaev.
Astron. Zhurn. Akad. Nauk SSSR, Vol. 52, 48 - 56 (1975). In

Russian. English translation in Soviet Astron., Vol. 19, No. 1.
The results of information processing of the spectral data (obtained from 24 October, 1967 to 24 November, 1969) are presented.

Echelle spectroscopy of nebular emission lines in nova Del 1967.   C. M. Anderson, J. S. Gallagher.
Bull American Astron. Soc., Vol. 7, 242 - 243 (1975). Abstr. AAS.

124.101   Nova Herculis 1934

The period and light curve of the 71-second variation in DQ Herculis.   M. R. Nelson.
Astrophys. Journ., (*Letters*), Vol. 196, L113 - L116 (1975).
The 71^s signal in the old nova DQ Herculis has been observed during the 1974 season. The period at the present epoch allows more accurate estimates of the secular decrease in period. The period decrease appears to be nonlinear. Evidence is also presented to show that the true period of the signal is 142^s.

Isotopic abundance ratios for carbon and nitrogen in nova Herculis 1934.   C. Sneden, D. L. Lambert.
Monthly Notices Roy. Astron. Soc., Vol. 170, 533 - 540 (1975).
Spectra of nova Herculis 1934 taken during the episode of intense CN absorption have been analysed to obtain limits on the isotopic abundance ratios $^{12}C/^{13}C$ and $^{14}N/^{15}N$. The results, $^{12}C/^{13}C \gtrsim 1.5$, $^{14}N/^{15}N \gtrsim 2$, are compared with recent predictions based on models of thermonuclear runaways in hydrogen-rich envelopes of white dwarfs.

Harmonic analysis of the light of DQ Herculis.
A. L. Kiplinger, R. E. Nather.
Nature, Vol. 255, 125 (1975).
The authors report a study of the harmonic components of the 71-s periodicity, and show that the results place severe limits on the available models of the variation.

The state of ionization in nova shells. III. A new explanation for the light curve of DQ Herculis 1934.
See Abstr. 124.005.

124.102   Nova T Coronae Borealis

Periodic fluctuations in the recurrent nova T CrB.
J. Bailey.
Journ. British. Astron. Ass., Vol. 85, 217 - 223 (1975).

124.103   Nova Tucanae 1974

Probable member stars in the western extremity of the Small Magellanic Cloud.   See Abstr. 159.010.

124.104   Nova Persei 1974

Nova Persei 1974.   M. Kiehl, J. Bortle, K. Simmons, P. O. Taylor, M. J. Taylor, E. Mayer.
IAU Circ., No. 2741 (1975).

Nova Persei 1974.   D. Jones, M. Besley, M. Jaques, G. Hurst, S. Anderson.

IAU Circ., No. 2742 (1975).

Nova Per 1974.   U. Hopp, D. Böhme, K. Simmons, M. J. Taylor.
IAU Circ., No. 2750 (1975).

Nova Per 1974.   U. Surawski, P. Taylor.
IAU Circ., No. 2761 (1975).

Identification chart for nova Persei 1974.
N. Sanduleak.
Inform. Bull. Variable Stars, (I.A.U. Commission 27), Konkoly Obs., Budapest, No. 959 (1975).

Beobachtungen der Nova Persei 1974.
W. Wenzel.
MVS, *Sonneberg,* Vol. 6, 201 (1975).

Nova Persei 1974.   A. U. Landolt.
Publ. Astron. Soc. Pacific, Vol. 87, 407 - 408 = Contr. Louisiana State Univ. Obs., *Baton Rouge,* No. 107 (1975).
UBV photometry and a finding chart are presented for nova Persei 1974.

124.105   Nova RS Ophiuchi

RS Oph.   U. Surawski.
IAU Circ., No. 2769 (1975).

Nova RS Ophiuchi as a semiregular variable.
P. Tempesti.
Inform. Bull. Variable Stars, (I.A.U. Commission 27), Konkoly Obs., Budapest, No. 974, 2 pp., with a correction, No. 987 (1975).

124.106   Nova Sagittarii 1969

Pre-maximum brightness of nova Sgr 1969.
I. Radiman, B. Hidajat.
Inform. Bull. Variable Stars, (I.A.U. Commission 27), Konkoly Obs., Budapest, No. 976 (1975).

124.107   Nova V 1017 Sagittarii

VRI photometry of the recurrent nova V 1017 Sagittarii.   A. U. Landolt.
Publ. Astron. Soc. Pacific, Vol. 87, 265 - 267 = Contr. Louisiana State Univ. Obs., *Baton Rouge,* No. 103 (1975).
VRI photometric observations of V1017 Sgr in 1973 and 1974 are discussed.

124.108   Nova GK Persei

GK Persei (=nova Persei 1901).
K. Hirosawa, Y. Hirasawa, J. A. Bailey.
British Astron. Ass. Circ. No. 560 (1975).

GK Persei.   K. Hirosawa, Y. Hirasawa, C. Hurless, E. Mayer, R. Annal.
IAU Circ., No. 2740 (1975).

GK Persei.   P. O. Taylor, M. J. Taylor, C. Scovil, J. E. Bortle, E. Mayer.

IAU Circ., No. 2743 (1975).

GK Per.   C. E. Sullivan, K. Simmons, G. Kelley, D. Böhme, M. J. Taylor, J. Bortle, J. Morgan.
IAU Circ., No. 2750 (1975).

GK Per.   U. Hopp, E. Mayer, G. Comello, J. Bortle.
IAU Circ., No. 2761 (1975).

GK Per.   U. Surawski.
IAU Circ., No. 2769 (1975).

Errata: 'Some spectroscopic properties of nova GK Persei' [Publ. Astron. Soc. Pacific, Vol. 86, 952 - 956 (1974)].
J. S. Gallagher, V. Oinas.
Publ. Astron. Soc. Pacific, Vol. 87, 352 (1975). — See 12.124.112.

124.109   Nova RR Pictoris

Orbital period and qualitative model for the exnova RR Pictoris.   See Abstr. 124.007.

124.110   Nova Scuti 1975

Nova Scuti 1975.
R. Mewis, J. Muirden, W. E. Pennell.
British Astron. Ass. Circ., No. 563 (1975).

Probable nova in Scutum.   P. Wild.
IAU Circ., No. 2791 (1975).

Nova Scuti 1975.   A. J. Weitenbeck, S-G. Lee, L. E. Krumenaker, J. Mattei, J. H. Bulger, P. Garnavich, G. Schwartz, C. A. Whitney, R. E. McCrosky, C. Y. Shao.
IAU Circ., No. 2792 (1975).

124.111   Nova Sagittarii 1962

1962 nova discovered in Sagittarius.
H. Dinerstein.
Journ. American Ass. Variable Star Observers, Vol. 2, 71 - 72 (1973).

125 Supernovae, Supernova Remnants

125.001 The physical conditions in the shells of type I supernovae. E. R. Mustel (*Eh. R. Mustel'*), N. N. Chugay (*Chugaj*).
Astrophys. Space Sci., Vol. 32, 25 - 38, 39 - 53 (1975).
In Russian and English.

Two models of the origin of the spectrum of type I supernovae are analysed: (I) the photosphere of the 'central remnant' and the expanding shell are separated by a density cavity; (II) the 'photosphere' (the layer which produces the continuous spectrum) is the inner part of the expanding shell. The arguments are given in favour of model I. From the analysis of the spectra of supernova 1972e a lower limit for the mass of the shell is obtained $(M > 10^{31}$ g). Then from the fact of absence of a detectable H-absorption line and a simultaneous presence of strong Si II absorption lines (6347, 6371 Å) in the spectrum of supernova 1972e it follows that the ratio Si/H is at least two or three orders greater than that for the 'normal' stars.

125.002 Shock-wave thermalization. S. A. Colgate.
Astrophys. Journ., Vol. 195, 493 - 498 (1975).
The cooling of the high-ion-temperature precursor of the shock wave in the presumed low-density envelope of a type II supernova is examined in greater detail for the galactic production of deuterium and some of the other light elements. When the combined effects of a modified ion energy loss rate, modification and inclusion of electron bremsstrahlung, and the truncation of the bremsstrahlung at the Planck limit are taken into account, the cooling time becomes $\sim 10 T_i^{1/2} + 30$ Thomson scattering periods, where T_i is the kinetic energy (in MeV per nucleon) of fluid flow behind the shock.

125.003 Emission from supernova remnants. I. Thermal bremsstrahlung in the Sedov-Taylor phase.
W. C. Straka, C. J. Lada.
Astrophys. Journ., Vol. 195, 563 - 566 (1975).
Calculations of total emission and isophotal contours due to thermal bremsstrahlung in the Sedov-Taylor phase of a supernova remnant have been carried out for two ages and two initial energies. The maximum surface brightness occurs inside the shock front, with the position of the maximum moving inward at higher frequencies. The thickness of the observed ring also increases with frequency. At radio frequencies, the surface brightness increases with age, but is independent of initial energy. The total brightness, however, is greater for greater initial energy at a given age due to the greater radius. A formula for scaling of the results is given.

125.004 A possible relation between spectral index and z-distribution for supernova remnants.
R. H. Becker, M. R. Kundu.
Astron. Astrophys., Vol. 38, 149 - 152 (1975).
A statistical relation has been found between spectral index and the mean distance from the galactic plane for supernova remnants. Using z-distances calculated from the Σ–D relationship and the observed spectral indices at radio frequencies for 93 galactic supernova remnants, the mean z-distance is found to vary from ~ 175 pc to ~ 60 pc for remnants with flat and steep spectra respectively. A possible correlation between flat-spectra supernova remnants and pulsars is discussed.

125.005 On the cold H I shell around the supernova remnant W 44. R. H. Cornett, P. E. Hardee.
Astron. Astrophys., Vol. 38, 157 - 160 (1975).
Recent observations of the supernova remnants W 44 and

HB 21 suggest the existence of neutral hydrogen shells expanding away from the supernova remnants with a velocity of a few km s^{-1} and a radius several times that of the radio continuum source. It appears that the object associated with W 44 can be explained as hydrogen set in motion by pressure gradients resulting from X-ray or cosmic ray heating by the supernova or its remnant.

125.006 The interaction between the blast wave of a supernova remnant and interstellar clouds.
C. F. McKee, L. L. Cowie.
Astrophys. Journ., Vol. 195, 715 - 725 (1975).
When an interstellar (or circumstellar) cloud is overtaken by a blast wave, a shock is driven into the cloud. The authors derive the velocity of the shock as a function of the cloud density and discuss the effects of thermal conduction between the blast wave and the shocked cloud. It is shown that many of the observed features of the quasi-stationary flocculi in Cas A can be accounted for by this model. Some aspects of the evolution of supernova remnants in a cloudy interstellar medium are discussed. Application of the results to the Cygnus Loop shows that Woltjer's interpretation of the optical and X-ray observations is self-consistent. Vela X and Shajn 147 are also briefly discussed.

125.007 Supernova remnants and presupernova models.
W. D. Arnett.
Astrophys. Journ., Vol. 195, 727 - 733 (1975).
Preliminary models of stellar evolution and neutron-star formation, taken with simple models of the Crab nebula and its pulsar, and of Cas A, can give a coherent picture of the origin of these objects. Energetics, abundances of elements in the nebulae, and characteristics of the neutron star are examined.

125.008 Optical nebulae − supernova remnants.
T. A. Lozinskaya.
Astron. Zhurn. Akad. Nauk SSSR, Vol. 52, 39 - 47 (1975). In Russian. English translation in Soviet Astron., Vol. 19, No. 1.
In 1970 - 1973 interferometric observations of two extremely faint nebulae identified with supernova remnants VRO 42.05.01 and HB9 were carried out. The results are presented in detail. A survey of supernova remnants with known expansion velocity is presented. The nebulae are presented in an evolutionary sequence of uniform objects expanding into interstellar medium of different density. A type of supernovae responsible for observed remnants is discussed.

125.009 On gas distribution in shells of type I and II supernovae near brightness maximum. N. N. Chugaj.
Astron. Zhurn. Akad. Nauk SSSR, Vol. 52, 197 - 199 (1975).
In Russian. English translation in Soviet Astron., Vol. 19, No. 1.
A conclusion on the noticeable difference in the behaviour of Doppler shifts of absorptions in type I and II supernovae is made. This difference is due to the different density distribution in the shells of type I and II supernovae.

125.010 Fluctuations of interstellar medium density − the cause of difference between optical and X-ray velocities of supernova remnants.
K. V. Bychkov, S. B. Pikel'ner.
Pis'ma v Astron. Zhurn., Vol. 1, No. 1, p. 29 - 34 (1975).
In Russian.
The optical emission of SNR is formed in fluctuations of density of cloudlets in which the main shock wave that pro-

duces thermal X-ray emission is slowed down and converges into the fluctuations. The gas here has enough time to cool down to optical temperatures.

125.011 Copernicus: the X-ray spectrum of Puppis-A.
P. A. Charles, J. L. Culhane, A. C. Fabian, R. J. Mitchell, J. C. Zarnecki.
Monthly Notices Roy. Astron. Soc., Vol. 170, 61P - 65P (1975).

The MSSL X-ray telescopes on Copernicus have observed Puppis-A in the energy ranges 0.5−1.4 keV and 1.6−4.7 keV. The measured flux in the higher energy range is used with the spectral data from the lower energy range to determine the temperature of the remnant to be 8±1 million degrees. A spectrum which includes line emission was used to establish this result, which is in agreement with the Pup-A observations of Gorenstein, Harnden & Tucker who use a similar spectrum. A power law gives a very poor fit to the data. The authors find no evidence for changes in spectrum across the source.

125.012 Neutral hydrogen associated with supernova remnants. I. The Cygnus Loop.   L. K. DeNoyer.
Astrophys. Journ., Vol. 196, 479 - 487 (1975).

Observations of H I near the Cygnus Loop show that (1) the remnant is encountering interstellar clouds with densities 5−10 cm^{-3} adjacent to its optical filaments, and (2) the remnant does not contain the cool H I shell as predicted by current evolution theories which equate the present shock velocity with the optical expansion velocity. It is suggested that the remnant is adiabatically expanding into a two-phase medium, and the original energy of the explosion was 10^{51} ergs.

125.013 Thermal instability in supernova shells.
R. McCray, R. F. Stein, M. Kafatos.
Astrophys. Journ., Vol. 196, 565 - 570 (1975).

Thermal instability in the radiative cooling region behind a shock will cause upstream density fluctuations to collapse into thin sheets aligned parallel to the shock front. A linearized calculation demonstrates the development of this instability. Thermal conduction suppresses the development of small-scale perturbations. Estimates of the scale sizes for the fully developed condensations agree roughly with the scale sizes of fine structure observed in supernova shells such as the Cygnus Loop.

125.014 Cosmic-ray production in the Cassiopeia A supernova remnant.   J. S. Scott, R. A. Chevalier.
Astrophys. Journ., (Letters), Vol. 197, L5 - L8 (1975).

A model for the production of high-energy particles in the supernova remnant Cas A is considered. The ordered expansion of the fast moving knots produces turbulent cells in the ambient interstellar medium. The turbulent cells act as magnetic scattering centers, and charged particles are accelerated to large energies by the second-order Fermi mechanism. Model predictions are shown to be consistent with the observed shape and time dependence of the radio spectrum, and with the scale size of magnetic field irregularities. Assuming a galactic supernova rate of one per 50 years, this mechanism is capable of producing the observed galactic cosmic-ray flux and spectrum below 10^{16} eV per nucleon. Several observed features of galactic cosmic rays are shown to be consistent with model predictions.

125.015 Neutral currents and supernovas.
D. N. Schramm, W. D. Arnett.
Phys. Rev. Letters, Vol. 34, 113 - 116 (1975).

It is shown that if more accurate neutrino opacities (including effects of electron degeneracy) are used in a gravitational collapse calculation, then the effects of neutral currents and coherent scattering may be considerably greater than was previously thought. It is also shown that a careful inclusion

of the electron-capture neutrinos should increase the importance of the region near densities of $\sim 2 \times 10^{11}$ g/cm^3.

125.016 The envelopes of type II supernovae.
R. P. Kirshner, J. Kwan.
Astrophys. Journ., Vol. 197, 415 - 424 (1975).

Absolute spectral energy distributions of type II supernovae have been used to derive a consistent physical picture of their ejected envelopes. Continuum observations are consistent with a photosphere that expands to a radius of 10^{15} cm and then begins to shrink at nearly constant temperature. The hydrogen recombination implies an electron density which decreases from 10^{10} to 2×10^7 cm^{-3} in a time of 400 days. A total hydrogen mass in excess of 1 $M_\odot$ is required to prevent rapid recombination in the envelope. A quantitative analysis of line optical depths confirms the identifications of Na I $\lambda 5890$, Mg I $\lambda 5174$, and the Fe II blends near $\lambda\lambda 4600$ and 5100. A possible identification of K I $\lambda 7677$ is suggested.

125.017 Copernicus: the X-ray spectrum of Cassiopeia A.
P. A. Charles, J. L. Culhane, J. C. Zarnecki, A. C. Fabian.
Astrophys. Journ., (Letters), Vol. 197, L61 - L63 (1975).

The MSSL X-ray telescopes on Copernicus have obtained the spectrum of the supernova remnant Cas A in the energy range 0.5−7.5 keV. The observations may be explained by a two-component thermal spectrum with temperatures of 8 and 30 million degrees K. This result can be broadly interpreted in terms of McKee's reverse shock wave model of young remnants where the lower temperature is attributed to the reverse shock and the higher one to the initial blast wave. A 3σ upper limit of 0.02 photons cm^{-2} s^{-1} was also derived for the emission from a line feature at 2 keV.

125.018 Indirect methods for detecting γ-ray bursts from supernovae.
A. A. Belyaev, V. V. Guzhavin, I. P. Ivanenko.
Nature, Vol. 254, 461 - 462 (1975).

125.019 The radio remnant of Kepler's supernova.
S. F. Gull.
Monthly Notices Roy. Astron. Soc., Vol. 171, 237 - 242 (1975).

A high-resolution radio map of Kepler's SNR has been constructed from 5 GHz observations made at Cambridge and at Owens Valley. The remnant is shown to have an irregular shell structure of 170″ arc diameter. The remnant has a close resemblance to Tycho's SNR, but is rather more irregular.

125.020 The X-ray, optical and radio properties of young supernova remnants.   S. F. Gull.
Monthly Notices Roy. Astron. Soc., Vol. 171, 263 - 278 (1975).

The present work demonstrates the close connection between several different phenomena in young supernova remnants. The formation of filaments, a dramatic increase in X-ray luminosity, the energy available from the Rayleigh−Taylor instability and the generation of magnetic field are all in turn consequences of the energy stored in the velocity dispersion of the ejecta. This amounts to 2 or 3 per cent of the expansion energy and is made available at times between 10^9 and 8×10^9 s for the author's standard supernova.

125.021 Luminous stars in galactic supernova remnants.
H. M. Johnson.
Publ. Astron. Soc. Pacific, Vol. 87, 89 - 95 (1975).

A discussion of reasons to expect luminous stars in the vicinity of supernova remnants is followed by a list of them from the Luminous Stars in the Northern − and Southern − Milky Way catalogs inside the areas of 24 SNR in An Optical Atlas of Galactic Supernova Remnants and in distances that are consistent with the SNR distances. This is supplemented by remarks on stars and other data of seven more optical counterparts of possible SNR that are not in the Atlas, including new spectroscopic data of S 104 and S 188.

125.022 A multifrequency study of the radio structure of 3C 10, the remnant of Tycho's supernova.
R. M. Duin, R. G. Strom.
Astron. Astrophys., Vol. 39, 33 - 42 (1975).

Synthesis observations of 3C 10 have been made at 6 cm and 50 cm using the Westerbork telescope. Combining these results with earlier 21 cm Westerbork maps, the authors have made detailed comparisons between the three wavelengths. The total intensity maps are remarkably similar. By comparing their 6 cm map with 2.8 cm and 10.4 cm results they are able to construct a rotation measure map, and confirm that the intrinsic position angles are tangentially aligned, with an average rotation measure of -249 rad m^{-2}. The authors conclude that the ordered component of the magnetic field is radially directed, and suggest that the mechanism for producing such a configuration can be found in the Sedov solution for shock waves in a fluid medium.

125.023 W44: a buoyant supernova remnant.
E. M. Jones.
Astron. Astrophys., Vol. 39, 143 - 146 (1975).

If the H I shell observed around supernova remnant W44 is identified as the fossil blast wave, W44 is 7 million years old. The apparent 13 pc buoyant displacement of the non-thermal source has been produced by a 0.7 pc Myr^{-2} = 2.1×10^{-9} cm s^{-2} local gravitational field. This acceleration is twice the mean Oort value at the altitude of W44. The acceleration vector is inclined about 30 degrees from the normal vector to the galactic plane.

125.024 The supernova remnant DA530 (4C(T)55.38.1).
R. S. Roger, C. H. Costain.
Journ. Roy. Astron. Soc. Canada, Vol. 69, 42 (1975). — Abstr. Canadian Astron. Soc.

125.025 Les supernovae de type I et l'anisotropie de la «constante» de Hubble.
G. Le Denmat, J.-P. Vigier.
Comptes Rendus Acad. Sci. Paris, Sér. B, Vol. 280, 459 - 461 (1975).

L'observation de supernovae de type I dans des galaxies lointaines permet de disposer d'un échantillon de sources très homogènes pour évaluer leur distance réelle. L'examen de leur répartition sur le fond du ciel permet de confirmer de façon significative l'anisotropie de la «constante» de Hubble déjà observée par Rubin, Rubin et Ford.

125.026 Rapid nuclear reactions in supernovae and cosmic rays. E. R. Hilf, W. Hillebrandt, K. Takahashi, M. F. El Eid, T. Kodama.
Phys. Scripta, Vol. 10A, 132 - 137 (1974). — Paper presented at the Nobel Symposium on superheavy elements, Ronneby, Sweden, June 11 - 14, 1974.

A synopsis of the transmutations of matter in a supernova to the final stages of neutron-star matter, r-nuclei, p-nuclei and cosmic rays is given. In particular discussed are (1) the freezing-out of nuclear reactions during cooling, (2) the consequences of an improved β-decay and β-delayed neutron emission systematics for the calculations of the r-process element abundances and of its application to the estimation of the nucleosynthesis-age of the galaxy, (3) the production mechanism of the high-energy light-element cosmic rays, which the authors propose to originate from the neutron star's surface.

125.027 Revision of the expansion velocity of IC 443.
T. A. Lozinskaya.
Pis'ma v Astron. Zhurn., Vol. 1, No. 2, p. 25 - 28 (1975). In Russian.

Additional interferometric measurements were made of the optical supernova remnant IC 443. Mainly the weak central region was observed. Some filaments have velocities of 160–170 km/s. The corresponding age of IC 443 is 25000 years. The agreement with the modern concept of shock wave propagation in supernova remnants is discussed.

125.028 High resolution measurements of the supernova HB 21 at a frequency of 2.7 GHz.
C. G. T. Haslam, N. J. Keen, W. E. Wilson, D. A. Graham, P. Thomasson.
Astron. Astrophys., Vol. 39, 453 - 454 (1975).

Total brightness temperature measurements of HB 21 at 2.7 GHz show a structure very similar to that reported by Hill at 1.4 GHz at similar angular resolution. The spectral index over large parts of the source is approximately 2.8.

125.029 The use of supernovae for determining the Hubble constant and estimating extragalactic distances.
B. W. Rust.
Bull. American Astron. Soc., Vol. 7, 236 (1975). – Abstr. AAS.

125.030 Radio studies of supernova remnants. J. R. Dickel.
Bull. American Astron. Soc., Vol. 7, 245 (1975). Abstr. AAS.

125.031 Observations of X-ray emissions from supernova remnants. P. Gorenstein.
Bull. American Astron. Soc., Vol. 7, 245 - 246 (1975). Abstr. AAS.

125.032 The next galactic supernova. S. van den Bergh.
Bull. American Astron. Soc., Vol. 7, 246 (1975). Abstr. AAS.

125.033 An anomaly in the flux of Cassiopeia A at 38 MHz.
W. C. Erickson, R. A. Perley.
Bull. American Astron. Soc., Vol. 7, 246 (1975). – Abstr. AAS.

125.034 The soft X-ray spectra of Cas A and Tycho's supernova remnant.
R. W. Hill, G. A. Burginyon, F. D. Seward.
Bull. American Astron. Soc., Vol. 7, 246 (1975). – Abstr. AAS.

125.035 Recent Copernicus results on the X-ray emission from supernova remnants. P. A. Charles, J. L. Culhane, C. G. Rapley, J. C. Zarnecki, A. C. Fabian.
Bull. American Astron. Soc., Vol. 7, 246 (1975). – Abstr. AAS.

125.036 Spectrum and structure of X-ray emission from Puppis A. R. C. Catura, L. W. Acton.
Bull. American Astron. Soc., Vol. 7, 246 - 247 (1975). Abstr. AAS.

125.037 A possible relation between z-distribution and spectral index for galactic supernova remnants.
R. H. Becker, M. R. Kundu.
Bull. American Astron. Soc., Vol. 7, 247 (1975). – Abstr. AAS.

125.038 Calculation of SNR interaction. E. M. Jones.
Bull. American Astron. Soc., Vol. 7, 247 (1975). Abstr. AAS.

125.039 Supernovae. Yu. P. Pskovskij.
Instationary stars and methods of their investigation. Phenomena of instationarity and stellar evolution, (see 003. 012), p. 261 - 337 (1974). In Russian.

125.040 The evolution of supernova remnants. III. Thermal waves. R. A. Chevalier.
Astrophys. Journ., Vol. 198, 355 - 359 (1975).

The effect of heat conduction on the evolution of supernova remnants is investigated. A thermal wave, or electron conduction front, can travel more rapidly than a shock wave during the first thousand years of the remnant's evolution. A self-similar solution describing this phase has been found by Barenblatt. Numerical computations verify the solution and give the evolution past the thermal wave phase. While shell formation is not impeded, the interior density and temperature profiles are smoothed by the action of conduction.

125.041 Electromagnetic pulse from supernovae.
S. A. Colgate.
Astrophys. Journ., Vol. 198, 439 - 445 (1975).

The author has calculated the upper and lower limits to the radiated electromagnetic pulse from a supernova assuming that the mass fraction of the matter expanding inside the dipole magnetic field shares energy and maintains pressure balance in the process. This results in a pulse of 10^{46} ergs in a width of $\lambda/2 \approx 150$ cm. Circumstellar matter like a corona would not affect the pulse. Similarly, electron self-radiation effects appear to be small. The principal attenuation of the pulse occurs due to interstellar matter.

125.042 Low-frequency radio maps and spectra of supernova remnants. J. R. Dickel, L. K. DeNoyer.
Astron. Journ., Vol. 80, 437 - 444 (1975).

Low-frequency radio maps of 18 supernova remnants have been constructed from observations made at the Arecibo Observatory. The integrated flux densities have been combined with others in the literature to show that most of the sources have simple power-law spectra. None of the sources show features which would suggest changes in the spectrum spatially across the given source.

125.043 The light of the supernova outburst. II. The case of supernova 1972e.
B. C. Chiu, P. Morrison, L. Sartori.
Astrophys. Journ., Vol. 198, 617 - 628 (1975).

The optical output of the bright type I supernova 1972e is best studied of any, both for quantitative spectrophotometry and for the range of spectral band measured. The authors examine it in detail as a test of the fluorescence theory of supernova light. The observed light curve and spectra allow the partition of the output into two distinct portions: one, the rather red quasi-Planckian continuum with its accompanying emission and absorption lines which comes directly from the expanding gas shell ejected in the explosion; the other, wide fluorescent emission lines and a weak blue two-photon continuum. The fluorescence model implies a much greater energy output in XUV light than in all optical emission, so that the early optical dominance of the direct shell radiation in this case does not affect the main conclusion: the total radiative output of a type I supernova explosion is at least tens of times greater than all one can observe in the infrared and visible bands.

125.044 Type I supernovae. I. The He II, He I, H I spectrum, 30 days after the explosion. C. Gordon.
Astrophys. Journ., Vol. 198, 765 - 773 (1975).

The author has computed the He II, He I, and H I spectra (lines and continuum) for a model of type I supernovae consisting of a spherically expanding shell with a stratification of ionization. Helium is the most abundant element, but hydrogen must be present in the bulk of the ejected material if the author wants to explain the observed stratification of ionization. The relative abundance of H and He varies with the density law adopted. For all the density models the temperature T_e in the inner layers of the shell must be close to 20,000 K. The expansion of the shell cannot be adiabatic, and the model requires the presence of a neutron star inside the shell.

125.045 The radio structure of the supernova remnant IC 443. R. M. Duin, H. van der Laan.
Astron. Astrophys., Vol. 40, 111 - 122 (1975).

High-resolution radio observations of the supernova remnant IC 443 at three wavelengths ($\lambda = 50, 21, 6$ cm) are presented. A very detailed correlation between the optical filaments and the small scale radio features was found. The filaments are interpreted as regions that were formed by unstable cooling of the hot gas behind the shock front. Condensation modes perpendicular to the magnetic field lines can provide the non-thermal volume emissivity enhancement needed to account for the total radio flux of IC 443 if the relativistic electrons and the magnetic fields have an interstellar origin.

125.046 Interferometric observations of the supernova remnant CTB 1.
T. A. Lozinskaya, G. P. Pustovojt.
Pis'ma v Astron. Zhurn., Vol. 1, No. 5, p. 24 - 30 (1975).
In Russian.

Interferometric observations of the faint nebula identified with the supernova remnant CTB 1 were carried out by a contact image converter. The mean radial velocity $- 33 \pm 2$ km/sec determined relative to the L.S.R. permits to calculate an improved supernova remnant distance of 2.6 ± 0.5 kpc. The corresponding nebula radius is 15 pc. The object is located inside a neutral hydrogen cloud and near an H II region. The expansion velocity of the shell is evaluated.

125.047 Chemical potential effects on neutrino diffusion in supernovae. T. J. Mazurek.
Astrophys. Space Sci., Vol. 35, 117 - 135 (1975).

Typically collapsing supernova hydrodynamic computations assume LTE neutrino transport and impose photon-like behavior, i.e., fix the neutrino chemical potentials at zero. The validity of the latter condition is investigated in the diffusion approximation to transport. A coupled system of diffusion equations for energy and lepton number is solved in a collapsing supernova ambience. The results indicate a substantial growth in the neutrino chemical potential for densities above 10^{12} gm cm^{-3}. It is found that the photon-like condition on neutrino transport may misrepresent supernova energetics substantially.

125.048 Supernovae in clusters of galaxies. P. Flin.
Acta Cosmologica, Fasc. 2, p. 21 - 32 (1974/75).

This paper is an attempt to investigate whether parent galaxies in which supernovae were observed belong to clusters of galaxies. 74% of them can be identified as members of clusters.

125.049 Hypothesis for the type I supernova light curve.
M. Leventhal, S. L. McCall.
Nature, Vol. 255, 690 - 692 (1975).

In a type I supernova event the observed luminosity rises very rapidly to a maximum absolute visual magnitude of about -19.0 ± 0.3. After about 20 d the light curve can be characterised by two exponential decays with half lives τ_1 and τ_2 where τ_1 is of the order of several days and τ_2 varies approximately from 40 to 60 d. The integrated observed luminosity is similar for all type I supernova and is about 3.6×10^{49} erg. The authors suggest that a model involving ^{56}Ni white dwarf can explain these changes.

125.050 Optical investigations of supernova remnants: the nebula near γ Cyg. T. A. Lozinskaya.
Astron. Zhurn. Akad. Nauk SSSR, Vol. 52, 515 - 520 (1975).
In Russian. English translation in Soviet Astron., Vol. 19, No. 3.

Detailed optical observations of the nebula near γ Cyg identified with the supernova remnant DR 4 were carried out with a 125-cm reflector. Filter images of the object in H$_a$,

[N II], [S II], [O III] lines were obtained. A series of interferometric observations of the nebula and a nearby H II region were carried out in H_α and 6584 Å [N II] lines. Spectrograms of the nebula in the region 6000–7000 Å were obtained with an image converter diffraction spectrograph. The results are presented in detail.

125.051 **Historical searches for supernovae.**
F. R. Stephenson.
Origin of cosmic rays, (012.012), p. 399 - 424 (1975).
The object of the paper is to discuss historical records of supernovae which have appeared in our Galaxy other than the well known supernovae of 1054, 1572 and 1604.

125.052 **Supernovae and the origin of cosmic rays (I).**
S. A. Colgate.
Origin of cosmic rays, (012.012), p. 425 - 445 (1975).
Review paper.

125.053 **Supernovae and the origin of cosmic rays (II), a model of cosmic ray production in supernovae.**
S. A. Colgate.
Origin of cosmic rays, (012.012), p. 447 - 466 (1975).

125.054 **Polarization of supernova remnants: internal Faraday effects and derived magnetic fields.**
T. Velusamy, M. R. Kundu.
Bull. Astron. Soc. India, Vol. 2, 38 (1974). – Abstract.

125.055 **Supernova remnants.** T. Velusamy.
Bull. Astron. Soc. India, Vol. 2, 67 - 69 (1974).

125.056 **The 1974 Palomar supernova search.**
C. T. Kowal, W. L. W. Sargent, J. Huchra.
Publ. Astron. Soc. Pacific, Vol. 87, 401 - 403 (1975).

125.057 **Supernovae: the origin of the chemical elements, cosmic rays, neutron stars, and maybe even black holes.** D. N. Schramm, W. D. Arnett.
Mercury, (Journ. Astron. Soc. Pacific), Vol. 4, No. 3, p. 16 - 22 (1975).

125.058 **Supernova in anonymous galaxy.** M. Lovas.
IAU Circ., Nos. 2755, 2789 (1975).

125.059 **Suspected supernova.** J. P. Huchra.
IAU Circ., No. 2760 (1975).

125.060 **Supernovae.** J. R. Dunlap, Y. Dunlap.
IAU Circ., No. 2782 (1975).

125.061 **Spectrum of supernova in anonymous galaxy.**
J. Liebert, H. Spinrad.
IAU Circ., No. 2790 (1975).

125.062 **Evolution of supernova remnants.** R. A. Chevalier.
Thesis, Princeton Univ., N.J. (USA). 110 pp. University Microfilms Order No. 74-9668 (1973).

Novae and supernovae. See Abstr. 003.004.

Explosive events in the universe.
See Abstr. 004.081.

High-energy gamma-ray results from the second Small Astronomy Satellite. See Abstr. 061.039.

Observation of celestial gamma rays.
See Abstr. 061.052.

Problems and achievements of nuclear astrophysics. IV. Rapid neutron capture and tertiary processes of nucleosynthesis [supernovae]. See Abstr. 061.074.

Optically thin radiating shock waves and the formation of density inhomogeneities. See Abstr. 062.004.

Structure and properties of detonation waves. I. Detonation waves in dense stellar material.
See Abstr. 065.008.

Pulsed gamma-ray emission from neutron and collapsing stars and supernovae. See Abstr. 065.076.

Thermal instability of helium-burning shell in stars evolving toward carbon-detonation supernovae.
See Abstr. 065.089.

The weak interaction and gravitational collapse.
See Abstr. 066.084.

On supernova explosion in a binary system.
See Abstr. 117.004.

Novae, supernovae, and neutron sources.
See Abstr. 124.003.

Internal motions in H II regions. II. The radial velocity field of IC 443. See Abstr. 131.509.

Observations of radio recombination lines toward supernova remnants at 428 MHz. See Abstr. 131.514.

The radio spectra of the γ-Cygni source and the IC 1318b, c nebular complex. See Abstr. 141.031.

The nonthermal radio sources at $l = 74°9$ and $b = +1°2$. See Abstr. 141.039.

The collision of a strong shock with a gas cloud: a model for Cassiopeia A. See Abstr. 141.051.

Luminosity of quasi-stellar objects.
See Abstr. 141.052.

5 GHz polarization observations of 33 galactic radio sources. See Abstr. 141.060.

G2.4 + 1.4, a supernova remnant or ring nebula around a peculiar star. See Abstr. 141.061.

A multiple pulsar model for quasi-stellar objects and active galactic nuclei. See Abstr. 141.093.

Statistical acceleration of relativistic particles in an assembly of spherical electromagnetic waves.
See Abstr. 141.094.

Connection between pulsars and supernova remnants. See Abstr. 141.312.

Is Cir X-1 a runaway binary?
See Abstr. 142.053.

Gamma-ray lines: a ^{22}Na radioactive diagnostic of young supernovae. See Abstr. 142.065.

X-ray binaries and asymmetry of supernova outbursts. See Abstr. 142.096.

X-ray emission of Cas A and Tycho.
See Abstr. 142.100.

Supernova explosions in close binary systems.
II. Runaway velocities of X-ray binaries.
See Abstr. 142.103.

The region of cosmic rays and the Vela gamma-ray
excess.   See Abstr. 143.037.

Galactic propagation of cosmic rays below 10^{14} eV.
See Abstr. 143.070.

Cloud structure in the galactic plane: a cosmic
bubble bath?   See Abstr. 155.022.

125.100 Supernova in NGC 5253

Search for high-frequency optical variations in the
supernova 1972E.
R. H. Miller, B. M. Lasker, J. E. Hesser, S. B. Bracker.
Astrophys. Journ., Vol. 196, 121 - 123 (1975).
A search for pulsar-like optical variations in the light of
the supernova 1972E near NGC 5253 was conducted with
null results as follows: No periodic variations within the frequency range 1/8 to 500 Hz were observable to a limit of
1 percent of the brightness of the supernova, and no aperiodic
variability was detectable with a limit of about 2– 4 percent
over the same frequency interval.

The physical conditions in the shells of type I
supernovae.   See Abstr. 125.001.

The light of the supernova outburst. II. The case
of supernova 1972e.   See Abstr. 125.043.

125.101 Supernova in NGC 4414

Spectre de la supernova découverte en 1974 dans
NGC 4414.   Y. Andrillat.
Comptes Rendus Acad. Sci. Paris, Sér. B, Vol. 280, 605 - 608
(1975).
L'auteur a obtenu un spectre de la supernova SN 1974–
NGC 4414 dans la région 3 100–5 000 Å avec une dispersion de
67 Å mm^{-1}, à une période très proche du maximum d'éclat. Ce
spectre montre nettement les caractéristiques des supernovae
de type I: larges bandes d'absorption, grande vitesse d'éjection
et diminution très nette du rayonnement en deçà de 3 600 Å.

The 1974 supernova in NGC 4414.   I. D. Howarth.
Journ. British Astron. Ass., Vol. 85, 352 - 354 (1975).

125.102 Supernova in NGC 2207

Supernova in NGC 2207.
J.R. Dunlap, Y. Dunlap.
IAU Circ., No. 2738 (1975).

Supernova in NGC 2207.   R. Green.
IAU Circ., No. 2743 (1975).

Supernova in NGC 2207.   N. V. Vidal.
IAU Circ., No. 2753 (1975).

125.103 Supernova in NGC 4102

Supernova in NGC 4102.
J. R. Dunlap, Y. Dunlap.
IAU Circ., No. 2776 (1975).

Supernovae.   See Abstr. 125.060.

126 Low-luminosity Stars, Subdwarfs, White Dwarfs

126.001 Hot vibrating white dwarf models of pulsating X-ray sources. A. G. W. Cameron.
Astrophys. Space Sci., Vol. 32, 215 - 229 (1975).

A number of white dwarf models have been calculated which correspond to various radial and nonradial modes of vibration with eigenfrequencies in agreement with the observed pulsation frequencies of the X-ray sources Hercules X-1 and Centaurus X-3. It is concluded that if the pulsating X-ray sources are hot white dwarfs, the mass of Cen X-3 probably lies in the range $0.7-1.2\,M_\odot$, and the mass of Her X-1 probably lies in the range $1.1-1.25\,M_\odot$.

126.002 Identification of G227-35 as a strongly polarized magnetic white dwarf. J. R. P. Angel, P. Hintzen, J. D. Landstreet.
Astrophys. Journ., *(Letters)*, Vol. 196, L27 - L29 (1975).

G227-35, a white dwarf suspect with high proper motion from the Lowell survey, is found to be a magnetic white dwarf with no detectable absorption features. The circular polarization spectrum has been measured both with broad-band filters and at 160 to 360 Å resolution. It shows a maximum of 3.3 percent at ~ 4500 Å, and is increasing again strongly in the infrared, reaching ~ 8 percent at 1 μ. No variability was detected over a 5-month period, and no linear component of polarization is found.

126.003 The polarization of optical radiation from the magnetic white dwarfs.
V. N. Sazonov, V. V. Chernomordik.
Astrophys. Space Sci., Vol. 32, 339 - 353, 355 - 369 (1975).
In Russian and English.

It is pointed out that, because of the large Faraday rotation an outlet of linear polarization from the photosphere of a white dwarf is hampered. In accordance with this fact it is proposed to distinguish two types of magnetic white dwarfs. The first type (its representative is Grw 70°8247) has a linear polarization which is comparable in magnitude with the circular one. Polarization of radiation from the white dwarfs of the first type cannot arise in the photosphere. The white dwarfs of the second type (its representative is G 99-37) have their linear polarization much smaller than the circular one. Polarization of these white dwarfs can arise as a result of transfer of radiation in the nonisothermal photosphere.

126.004 The gravity oscillations of white dwarfs.
A. J. Brickhill.
Monthly Notices Roy. Astron. Soc., Vol. 170, 405 - 421 (1975).

The second order differential equations governing the non-radial adiabatic oscillations of stars have been solved for a variety of white dwarf models. The eigenfunctions of hot white dwarf models which include a hydrogen burning shell in equilibrium show that the oscillations have a large amplitude only in the hydrogen envelope. The relevance of these results to the observation of periodic variations in cataclysmic variables is discussed.

126.005 The polarization spectrum and magnetic field strength of the white dwarf Grw +70°8247.
J. D. Landstreet, J. R. P. Angel.
Astrophys. Journ., Vol. 196, 819 - 825 (1975).

The wavelength dependence of the circular and linear polarization of the magnetic white dwarf Grw +70°8247 has been obtained with 80–360 Å resolution in the range 0.3 – 1.1 μ. There is no strong evidence for variability of either circular or linear polarization. The wavelength dependence of circular polarization is interpreted as being due to magnetic circular dichroism in a helium atmosphere. The mean longitu-

dinal field strength estimated from both the circular polarization longward of 4000 Å, which is due to He⁻ opacity, and from the strong ultraviolet polarization feature, caused by the bound-free opacity of He°, is $4-5 \times 10^7$ gauss. The linear polarization is probably due to a transverse field, estimated at 5×10^7 to 10^8 gauss, again on the basis of He⁻ opacity. New observations of the absorption feature at 5855 Å are presented, and this is identified with the quadratically shifted π component of the 5876 line of He originating in a lower field region of 1.5×10^7 gauss.

126.006 A further list of degenerate stars. VIII.
J. L. Greenstein.
Astrophys. Journ., *(Letters)*, Vol. 196, L117 - L120 (1975).

A statistical survey of proper-motion stars with the multichannel spectrophotometer yields 44 further degenerate stars. A considerable number are cool; two composite stars are found, as well as a double white dwarf. Use of the University College, London, digital spectrograph reveals some interesting features visible only at its higher resolution, notably a hot, metallic-line degenerate (Gr 346); and several suspected red degenerates proved to be weak-lined subdwarfs.

126.007 The upper mass limit for white dwarf formation as derived from the stellar content of the Hyades cluster. E. P. J. van den Heuvel.
Astrophys. Journ., *(Letters)*, Vol. 196, L121 - L123 (1975).

The observed number of white dwarfs in the Hyades cluster (between 11 and 14) sets an upper limit m_w smaller than $6\,M_\odot$ (and almost certainly smaller than $5\,M_\odot$) to the masses of stars which finish life as white dwarfs. This result is obtained using the total number of cluster stars brighter than $m_v = 8.4$, together with a turnoff mass of $2.0 \pm 0.3\,M_\odot$, as obtained from the fitting of theoretical timelines to the upper part of the cluster main sequence. The most likely value for m_w thus derived is $3-4\,M_\odot$. This limit is of importance for questions regarding carbon detonation and pulsar formation. If all stars more massive than $4\,M_\odot$ leave remnants, at least some 10 to 20 Hyades stars must have evolved into neutron stars or black holes.

126.008 A search for circular polarization in the white dwarfs Ox+25° 6725 and L1512-34B.
J. D. Landstreet, J. R. P. Angel, R. M. E. Illing.
Acta Astron., Vol. 25, 39 - 41 (1975).

Circular polarization observations have been made of the DA white dwarfs Ox+25° 6725, in which Shulov and Belokon (1972) report periodically variable broad-band circular polarization, and of L1512-34B, in which a possible variable polarization is reported. No significant polarization was found in either star, although observational errors were substantially smaller than the reported effects.

126.009 White dwarfs: composition, mass budget and galactic evolution. V. Weidemann.
Problems in stellar atmospheres and envelopes, (see 003.001), p. 173 - 203 (1975).

126.010 A new nearby subdwarf M star.
A. R. Upgren, E. W. Weis.
Astrophys. Journ., *(Letters)*, Vol. 197, L53 - L54 (1975).

A recent parallax determination shows that the nearby red dwarf AC +54° 1646−56 lies at a distance of only 6.7 pc. Photoelectric photometry, also recently obtained, shows the star to lie about 1 magnitude below the $(M_v, R-I)$ main sequence. Its colors appear similar to those of 20 other M subdwarfs with trigonometric parallaxes. Its tangential velocity is unusually low for a subluminous star, being only 15 km s⁻¹. Its radial velocity has not been determined but would be expected to be high.

126.011   **High dispersion Hβ spectroscopy of six southern white dwarfs.**   M. S. Bessell, D. T. Wickramasinghe.
Monthly Notices Roy. Astron. Soc., Vol. 171, 11P - 14P (1975).

Twenty Å mm⁻¹ coudé spectra of the white dwarfs EG 141, CD−38° 10980, EG 62, EG 21, EG 99 and EG 131 have been obtained at Hβ. All except the latter two have sharp non-LTE line cores similar to those found in Wolf 1346 and 40 Eri B suggesting slow rotation. A number of low dispersion spectra of EG 131, previously classified as DA$_{wk}$, have failed to reveal any evidence of hydrogen line absorption.

126.012   **Analytic surface boundary conditions for white dwarf evolutionary calculations.**
G. Fontaine, H. M. Van Horn.
Astrophys. Journ., Vol. 197, 647 - 650 (1975).

Analytic fits for the pressure, temperature, density, and fractional radius at the base of white dwarf envelopes are given for three chemical compositions: X_H = 0.999, X_{He} = 0.999, and X_C = 0.999. The fractional mass is chosen large enough so that convection has stopped and ionization is complete at the base of an envelope. The fits are accurate to a few percent and are valid in the range $0.22 \leqslant M/M_\odot \leqslant 1.22$ and $10^{-4} \leqslant L/L_\odot \leqslant 10^{-1.5}$.

126.013   **The atmospheres of cool white dwarfs of spectral type DA.**   R. Wehrse.
Astron. Astrophys., Vol. 39, 169 - 175 (1975).

In order to interpret the Strömgren *uvby* colors of the cool hydrogen-rich DA white dwarfs observed by Graham (1972), model atmospheres have been constructed in the range $7000 \lesssim T_{eff} \lesssim 12\,000$ K and $7 \lesssim \log g \lesssim 8.5$. Metal abundances $\epsilon_M = \epsilon_M^\odot$ and $\epsilon_M = \epsilon_M^\odot/100$ were assumed. The models include energy transport by convection and line-blanketing from Balmer and metal lines; all are corrected to constant energy flux. It is shown that the hydrogen-rich white dwarfs under consideration have gravities given by $\log g = 8.3$, and that the metal abundances are reduced by at least a factor of 100 compared to the solar value.

126.014   **Spectrophotometry of stars near the cool end of the white dwarf sequence.**   G. Wegner.
Monthly Notices Roy. Astron. Soc., Vol. 171, 529 - 536 (1975).

The spectra of five stars near the cool end of the white dwarf sequence are described. From the spectra of the five stars, a few hypotheses are made regarding the appearances of the spectra of cool white dwarfs and their evolution.

126.015   **Evolution of helium white dwarfs in close binaries.**
R. F. Webbink.
Monthly Notices Roy. Astron. Soc., Vol. 171, 555 - 568 (1975).

Stars of 0.10, 0.15, 0.20, 0.25, 0.35 and 0.50 $M_\odot$ have been evolved from the zero-age main sequence to extinction of the hydrogen-burning shell as a white dwarf, or to the onset of the helium flash. Shell-burning models are discussed as close binary remnants. Only p−p shell burning configurations exist for 0.10 and 0.15 $M_\odot$ stars. More massive stars display strong thermal pulses at the onset of shell degeneracy and at the reversion from CNO-cycle to p−p chain in the shell. All shells are thermally unstable along the white dwarf track; only close binary systems which reach the second period of mass transfer before remnants reach this stage are therefore likely to avoid nova outbursts.

126.016   **A spectroscopic survey of southern hemisphere white dwarfs − V. Data for fifteen additional stars and the nature of the W 219 group.**   G. Wegner.

Monthly Notices Roy. Astron. Soc., Vol. 171, 637 - 646 (1975).

Spectroscopic data on 13 new southern white dwarfs are described. None of these stars have been previously observed spectroscopically. Spectral types, line profiles, and equivalent widths for the Hβ or K line of Ca II, depending on which is visible in the spectrum are given. Five new stars with peculiar spectra and which appear to belong to the W 219 group are described. The kinematical and spectroscopic properties of these stars are discussed.

126.017   **The linear polarization of the white-dwarf binary BD +16°516.**   J. C. Kemp, R. J. Rudy.
Publ. Astron. Soc. Pacific, Vol. 87, 301 - 303 (1975).

The eclipsing white-dwarf binary BD +16°516 has been found to have a small but finite linear polarization, $p \simeq 0.15\%$, $\theta \simeq 50°$, as averaged over orbital phase and wavelength. While this is probably interstellar, there is also marginal evidence for a component in the ultraviolet which varies with phase. Further work is needed.

126.018   **On K. D. Rakos' photoelectric measurements of Sirius B.**   I. W. Lindenblad.
Astron. Astrophys., Vol. 41, 111 - 112 (1975).

An analysis of Rakos' technique and results (see 11.118. 002), does not support his conclusions.

126.019   **A search for linear polarization in white dwarfs.**
G. V. Coyne.
Ric. Astron. Specola Vaticana, *Castel Gandolfo*, Vol. 8, (No. 27), 499 - 502 (1974).

A search has been made for linear polarization in 15 white dwarf stars. No intrinsic polarization has been detected.

126.020   **Outer layers of white dwarf stars.**   G. Fontaine.
Thesis, Rochester Univ., New York (USA). 236 pp. University Microfilms Order No. 74-22,578 (1974).

126.021   **Hydrogen and helium spectra in magnetic white dwarf stars.**   S. B. Kemic.
Thesis, Colorado Univ., Boulder (USA). 166 pp. University Microfilms Order No. 74-12,384 (1973).

126.022   **Evolution of pure ¹²C white dwarfs.**
D. Q. Lamb, Jr.
Thesis, Rochester Univ., New York (USA). 356 pp. University Microfilms Order No. 74-22,601 (1974).

The broadening of calcium II H and K lines by helium.   See Abstr. 022.019.

Viscous effects in rapidly rotating stars with application to white-dwarf models. III. Further numerical results. See Abstr. 065.007.

Ultrashort-period binaries. III. The accretion of hydrogen-rich matter onto a white dwarf of one solar mass. See Abstr. 117.031.

Period variation in the white-dwarf eclipsing binary BD +16°516.   See Abstr. 121.082.

Novae, supernovae, and neutron sources. See Abstr. 124.003.

Internal magnetic fields of pulsars, white dwarfs, and other stars.   See Abstr. 141.307.

Limits on the soft X-ray and extreme ultraviolet flux from RX Andromedae and U Geminorum. See Abstr. 142.060.

Interstellar Matter, Gaseous Nebulae, Planetary Nebulae

131   Interstellar Matter, Polarization of Starlight, H I, H II Regions

Interstellar Matter, Polarization of Starlight

131.001   The electron density in the direction of ζ Oph.
T. P. Stecher, D. A. Williams.
Astrophys. Space Sci., Vol. 32, 211 - 213 (1975).
It is shown that photoionization of vibrationally excited H_2 and photodissociation of the H_2^+ ions produced thereby constitute a significant electron production route in high UV flux situations. A significant fraction of the electron density in the direction of ζ Oph (−15 km s^{-1} cloud) deduced from observations may be expected to arise in this way.

131.002   Clumping of interstellar grains during formation of the primitive solar nebula.   A. G. W. Cameron.
Icarus, Vol. 24, 128 - 133 (1975).
The author has previously shown that a considerable amount of clumping of interstellar grains is likely to take place during the free-fall collapse phase of an interstellar cloud which is forming the primitive solar nebula, with the assumption of sonic turbulence in the gas. A more realistic calculation was now carried out in which it was assumed that clumps of grains would tend to stick together if their collisions were approximately head-on, but that they would tend to fragment into smaller pieces if the collisions were more tangential.

131.003   Polyoxymethylene polymers as interstellar grains.
N. C. Wickramasinghe.
Monthly Notices Roy. Astron. Soc., Vol. 170, 11P - 16P (1975).
Formaldehyde molecules in interstellar dust clouds condense on interstellar silicate grains as polyoxymethylene whiskers. A significant fraction of interstellar O and C atoms may be frozen on to grains in this form, and this material could be responsible for a major part of the observed extinction and polarization of starlight at optical wavelengths.

131.004   A molecular cloud in IC 1396.
R. B. Loren, W. L. Peters, P. A. Vanden Bout.
Astrophys. Journ., Vol. 195, 75 - 79 (1975).
Observations of the large cometary nebula in IC 1396 show it to be a strong source of molecular line emission. The $^{12}C^{16}O$ and $^{13}C^{16}O$ ($J = 1-0$) line intensities have been mapped and are well correlated with the optical appearance of the cloud. Emission from SO, HCN, and CS has been detected at the position of peak CO line intensity. Analysis of the CO line intensity maps reveals differing distributions of CO excitation temperature and column density. A heating source is required for the cloud, and a search for an infrared source is suggested.

131.005   Formation of molecules on small interstellar grains.
M. Allen, G. W. Robinson.
Astrophys. Journ., Vol. 195, 81 - 90 (1975).
A model is proposed for the formation of molecules on small (radius ≲ 0.04 μ) interstellar grains. It is suggested that the energy liberated when a chemical bond is formed between two atoms on a grain is transferred to the lattice vibrations of the grain, heating up the grain. The "hot" grain, during the

time that it is radiatively recooling, may then liberate its adsorbed volatiles. The time scales for molecular desorption from a grain are calculated for OH and CO for four different grain compositions. The results provide some tentative explanations of astronomical observations.

131.006   Maser radiometer observations of water vapor and OH in weak galactic OH sources.
K. S. Yngvesson, A. G. Cardiasmenos, J. F. Shanley, O. E. H. Rydbeck, J. Elldér.
Astrophys. Journ., Vol. 195, 91 - 99 (1975).
Observations of water vapor and ground-state hydroxyl lines in weak galactic OH sources were performed with maser-equipped radio telescopes. In general, the weak sources exhibit simpler spectra than the strong sources, and detailed correlation studies are facilitated. New water vapor sources have been detected in W37, W43 N, NGC 2438, and ON-3. W12 and W37 are examples of sources in which H_2O mases while OH is seen only in absorption. In other sources, e.g., W31 and W43 N, the velocities of H_2O and OH correlate well. Interferometer measurements of the positions of these fairly weak H_2O sources are required for further correlation between OH and H_2O.

131.007   Interstellar absorption and an apparent anisotropy in the Hubble expansion.   F. D. A. Hartwick.
Astrophys. Journ., (*Letters*), Vol. 195, L7 - L9 (1975).
Residuals from the magnitude-redshift relation for first ranked E and S0 galaxies suggest a nonrandom distribution on the sky similar to the effect (though of smaller amplitude) recently found for ScI galaxies by Rubin et al. It is tentatively concluded that interstellar absorption in a large region of the sky at galactic longitudes 90° ≲ l ≲ 200° is responsible for at least part of this effect.

131.008   Ion-atom charge-transfer reactions and a hot intercloud medium.   G. Steigman.
Astrophys. Journ., (*Letters*), Vol. 195, L39 - L41 (1975).
The ratios of C III/C II and C II/C I observed toward several unreddened stars are inconsistent with radiative recombinations in a hot, partially ionized intercloud medium. The observed underabundance of C III and overabundance of C I may be due to previously neglected charge-transfer reactions. It is shown that, if the reaction $C^{++} + H \rightarrow C^+ + H^+$ proceeds rapidly ($k \approx 10^{-11}$ cm^3 s^{-1}), then the observed C III underabundance can be understood.

131.009   Vibrationally excited SiO: a new type of maser source in the millimeter wavelength region.
N. Kaifu, D. Buhl, L. E. Snyder.
Astrophys. Journ., Vol. 195, 359 - 366 (1975).
The recently discovered strong emission lines at 86.24 GHz (~3.48 mm) from the $J = 2-1$ rotational transition of SiO in its first excited vibrational state have been found to be a maser source associated with infrared stars. The authors have detected 12 maser sources, and almost all are late M-type Mira or semiregular variables. Many of the known OH/H_2O/IR sources are also SiO maser sources, and good velocity correlations between SiO and H_2O emission profiles were found. Characteristics of these new SiO maser sources are discussed.

131.010 **Observations of formaldehyde toward M17.**
C. Lada, E. J. Chaisson.
Astrophys. Journ., Vol. 195, 367 - 377 (1975).

Observations of 6-cm H_2CO toward M17 show evidence for a wide, weak line near a radial velocity of 19 km s^{-1}, and a narrow, intense feature at about 24 km s^{-1}. The former, conspicuously associated with a lane of obscuring dust west of the H II region, exhibits a systematic line-width gradient that indicates a collapse of part of the H_2CO cloud near a region of high density and intense molecular emission. Dynamical arguments suggest $M \gtrsim 4 \times 10^4 M_\odot$ for the collapsing region, while considerations of optical extinction and H_2CO density imply $M \simeq 8700 M_\odot$ for the remaining part of the extended dust lane. The 24 km s^{-1} cloud, of relatively low density and mass, is apparently a foreground object.

131.011 **The diffuse interstellar features studied in HD 21389 by polarimetry and spectrophotometry.**
P. G. Martin, J. R. P. Angel.
Astrophys. Journ., Vol. 195, 379 - 383 (1975).

If the diffuse interstellar features originate in grains which cause interstellar polarization, then a polarization change proportional to the change in optical depth at the features is expected. HD 21389 was chosen as an optimum candidate for a polarization study of the $\lambda4430$ and $\lambda6284$ diffuse features, because of its strong interstellar polarization and diffuse features and because of the reported blue emission wing at $\lambda4430$. While a consistent explanation of these observations in terms of small, poorly aligned grains is possible, the authors suggest that a molecular or atomic origin for the features is more attractive.

131.012 **Circumstellar grains and the intrinsic polarization of starlight.**
W. J. Forrest, F. C. Gillett, W. A. Stein.
Astrophys. Journ., Vol. 195, 423 - 440 (1975).

Twenty-five long-period variable stars exhibiting intrinsic variable polarization have been monitored over the range $3.5 - 11 \mu$ for several cycles. No conclusive evidence for gross changes in amount of circumstellar grains has been found. Thus circumstellar infrared emission is attributed to the total abundance of grains surrounding the star, which does not change by a large amount with time, while intrinsic polarization is attributed to more localized scattering and absorption effects. Spectrophotometry with $\Delta\lambda/\lambda \approx 0.015$ over the $8-14 \mu$ wavelength range of several stars with different chemical compositions indicates excess emission characteristic of 3 types of grains: (1) "blackbody" grains, (2) silicate grains, and (3) silicon carbide grains.

131.013 **Radiative trapping and hyperfine structure: HCN.**
J. Kwan, N. Scoville.
Astrophys. Journ., (*Letters*), Vol. 195, L85 - L88 (1975).

The anomalous weakness of the $F = 1 \rightarrow 1$ hyperfine component in the $J = 1 \rightarrow 0$ emission of interstellar HCN can be caused by radiative trapping in the $J = 2 \rightarrow 1$ lines. The anomaly is readily produced if the $J = 1$ levels are populated largely by collisional excitation from $J = 0$ to $J = 2$ followed by radiative decay to $J = 1$ with the $J = 2 \rightarrow 1$ lines optically thick. Regions where the anomaly is found probably have H_2 densities less than 10^5 cm^{-3} and optical depths in the $J = 1 \rightarrow 0$ lines greater than 50.

131.014 **The visible photodissociation spectrum of ionized methane.** E. S. Ensberg, K. B. Jefferts.
Astrophys. Journ., (*Letters*), Vol. 195, L89 - L91 (1975).

Photodissociation of CH_4^+, observed as a decrease in population in a Paul ion trap, occurs with a cross section of the order of 10^{-19} cm^2, decreasing monotonically with wavelength from 3600 Å to 6000 Å. Any predissociative structure is smaller than about 2×10^{-20} cm^2. These results offer no evidence to support Herzberg's suggestion that predissociative absorption in CH_4^+ may be responsible for diffuse interstellar lines.

131.015 **Far-infrared properties of interstellar grains.**
C. D. Andriesse.
Astron. Astrophys., Vol. 37, 257 - 262 (1974).

Physical arguments are discussed which predict a quadratic law for the frequency dependence of the absorption coefficient of cold interstellar grains in the far infrared. The strength of this absorption is derived using an oscillator model with resonance frequencies in the near infrared, and from an estimate of the probable static dielectric constant of the grains. The quantitative relation found for the far-infrared absorption coefficient is applied to calculate the interstellar grain temperature, and to analyse a measurement of the $100\mu m$ galactic background radiation.

131.016 **On the anisotropy of the absorption and emission coefficient in cosmic masers.**
E. V. Bettwieser, W. H. Kegel.
Astron. Astrophys., Vol. 37, 291 - 295 (1974).

The emission and absorption profiles in cosmic masers are investigated assuming pure Doppler broadening. When the maser medium saturates in an anisotropic radiation field, the absorption coefficient and the source function become anisotropic.

131.017 **Cloud collisions and element depletion.**
H. W. Yorke.
Astron. Astrophys., Vol. 37, 375 - 382 (1974).

A simple time dependent depletion model with two free parameters is presented which takes into account the accretion of the heavier elements ($Z \geqslant 6$) on the surfaces of dust grains and the evaporation of these elements back into the interstellar gas. The equilibrium properties of the model are discussed as a function of the ratio of the sticking factor s and evaporation parameter β. The evolution of two large (3.8 pc) colliding clouds is also discussed as a function of s and β.

131.018 **The 4830 MHz formaldehyde absorption in the direction of galactic radio sources.**
J. B. Whiteoak, F. F. Gardner.
Astron. Astrophys., Vol. 37, 389 - 404 (1974).

Observations have been made at Parkes of the 4830 MHz H_2CO absorption in the direction of about 280 sources or source components. For the average absorbing cloud the line-to-continuum ratio is 0.052, the half-intensity width of the associated absorption feature is 5.2 km s^{-1}, and the projected density of H_2CO molecules is 3×10^{13} cm^{-2}. High values of the line-to-continuum ratio occur for clouds which are near either H II regions or OH emission sources, and these values may be associated with transition temperatures about 1 K below the cosmic background temperature.

131.019 **Forced diffusion: a mechanism to create dust clouds.**
P. W. J. L. Brand.
Astrophys. Space Sci., Vol. 33, 231 - 234 (1975).

Photodesorption from dust grains by an anisotropic radiation field will lead to much greater momentum transfer than that associated with radiation pressure. This can cause the segregation of dust to the centres of moderately dense interstellar clouds.

131.020 **Das Linienspektrum des interstellaren Gases. I. Die kühle Komponente.** S. Marx.
Sterne, Vol. 51, 5 - 16 (1975). – Lecture on the occasion of the 65th birthday of H. Lambrecht, 1973 October 3.

131.021 **On the interpretation of Copernicus observations of interstellar absorption lines in front of ξ Persei.**

J. Gómez-Gonzàlez, J. Lequeux.
Astron. Astrophys., Vol. 38, 29 - 39 (1975).

Observations by the satellites Copernicus and TD 1 of the equivalent widths of ultraviolet interstellar absorption lines in the spectrum of ξ Per are analyzed using models of the velocity distribution of interstellar matter in front of this star, derived from known profiles of optical interstellar lines. Observations can be explained only if there exist in the interstellar medium not only clouds, but a large-velocity dispersion gas containing 4–35 % of the matter. The authors show that Na, K and possibly S and Ar are probably not much depleted in the interstellar gas with respect to their "cosmic" abundances, whilst Mg, Si, P, Ca and Ti have various degrees of depletion. They obtain some quantitative data on the degree of ionization in the medium, and also show that the degree of ionization in the clouds is $n_e/n_H < 3$ to 9×10^{-4}, thus requiring no ionizing agent other than UV stellar light if carbon, the main supplier of electrons, is not depleted.

131.022　**H_2CO and H I observations of dark clouds in NGC 2264.**　Y. K. Minn, J. M. Greenberg.
Astron. Astrophys., Vol. 38, 81 - 85 (1975).

The 6-cm H_2CO and the 21-cm H I lines have been surveyed at positions of dark clouds in NGC 2264. The cloud boundaries are well defined by the H_2CO distributions. The H_2CO and OH line intensities are generally correlated but not strictly proportional. The average value of $[H_2CO]/[OH]$ obtained is rather large, namely $\sim 2 \times 10^{-2}$. At a few positions the H_2CO line widths are observed to be ~ 3.5 km s^{-1} which is much larger than the typical value of 1 km s^{-1} for local dust clouds. At two positions a velocity difference of more than 3 km s^{-1} between the H_2CO and OH lines was measured. It is suggested that at these positions the H_2CO and OH clouds do not coexist. A good correlation exists between the velocities of a sharp depression in the H I line profiles and molecular lines. At a position where several other molecules have been detected there is a good correlation with H_2CO in both velocities and line widths.

131.023　**Positions of OH sources discovered at 1612 MHz.**　A. Winnberg, Nguyen-Q-Rieu, L. E. B. Johansson, W. M. Goss.
Astron. Astrophys., Vol. 38, 145 - 148 (1975).

Right ascensions of 42 newly discovered OH sources have been determined with the Nançay radio telescope. The r.m.s. errors of the measurements are $\lesssim 1$ s of time.

131.024　**The detection of interstellar vinyl cyanide (acrylonitrile).**　F. F. Gardner, G. Winnewisser.
Astrophys. Journ., (*Letters*), Vol. 195, L127 - L130 (1975).

Interstellar vinyl cyanide, $H_2C=CH–CN$, has been discovered in its $2_{11}–2_{12}$ (1372 MHz) transition in Sgr B2. The line is in emission with a peak antenna temperature of 0.036° K. It probably results from maser amplification of the continuum. The total molecular content is uncertain.

131.025　**Spatial distribution of OH absorption in the source NGC 6334.**　V. I. Slysh.
Astron. Zhurn. Akad. Nauk SSSR, Vol. 52, 26 - 33 (1975). In Russian. English translation in Soviet Astron., Vol. 19, No. 1.

Maps of the NGC 6334 region were produced in the 1667 MHz OH line in both circular polarizations and at 1665 MHz in the left-hand circular polarization only. Right ascension scans through the center of the source were made also in 1612 and 1720 MHz satellite lines. Two absorption features in the OH spectrum of NGC 6334 were found to have different spatial distribution.

131.026　**The diffuse interstellar bands. IV. The region 4400–6850 Å.**　G. H. Herbig.
Astrophys. Journ., Vol. 196, 129 - 160 = Lick Obs. Bull., No.

669 (1975).

Between 4400 and 6850 Å a total of 39 diffuse absorption features are described which are certainly or probably of interstellar origin, together with seven more which also may qualify. Equivalent widths of the 17 bands for which the data are most extensive correlate well ($r \approx 0.9$) with $E(B–V)$, but even better with $E(V–I)$ and other long-wavelength indices which avoid the bend near 4400 Å in the interstellar extinction curve. The internal correlation of one line with another is in most cases excellent.

131.027　**Carbon monoxide observations of a dust cloud in the Orion region: L1630.**
A. S. Milman, G. R. Knapp, F. J. Kerr, S. L. Knapp, W. J. Wilson.
Astron. Journ., Vol. 80, 93 - 100, 165 (1975).

The authors have observed the $J = 1 - 0$ rotational transitions of $^{12}C^{16}O$ and $^{13}C^{16}O$ at 115 and 110 GHz in the dust cloud L1630, in the Orion region. L1630 contains several reflection nebulae, the Horsehead nebula, and many young stars. The CO emission is strongest at the positions of the reflection nebulae NGC 2023 and NGC 2068, and may be enhanced near individual young stars. The shapes of the ^{12}CO and ^{13}CO profiles provide evidence that the excitation temperature of the ^{12}CO line is different at different radial velocities. The possibility that the excitation temperature of the ^{13}CO line is smaller than that of the ^{12}CO line is discussed. If this is so, one can only calculate a lower limit to the ^{13}CO optical depth, and an upper limit to the CO column density. Two alternative models that involve turbulent motions of the gas are briefly explored.

131.028　**Carbon monoxide observations of 34 dust clouds.**
A. S. Milman, G. R. Knapp, S. L. Knapp, W. J. Wilson.
Astron. Journ., Vol. 80, 101 - 110 (1975).

The authors have made observations in the $J = 1 - 0$ rotational lines of $^{12}C^{16}O$ and $^{13}C^{16}O$ in the directions of 34 dust clouds. ^{12}CO emission was detected from 26 of these dust clouds; no ^{12}CO emission was seen in the directions away from dust clouds. The ^{12}CO and ^{13}CO profiles are presented for all the positions where they have observations of both lines. The CO excitation temperatures, column densities, and the ^{13}CO optical depths are discussed; the authors conclude that, in most of the clouds they observed, the ^{13}CO line probably has a moderate optical depth, and the neutral-particle density is probably smaller than that necessary to thermalize the $J = 0$ and $J = 1$ levels of CO collisionally. They have found that the CO emission is often enhanced near young stars embedded in dust clouds.

131.029　**Charged dust grains and excitation of rotational levels of interstellar molecular hydrogen.**
S. P. Tarafdar, N. C. Wickramasinghe.
Nature, Vol. 254, 203 - 205 (1975).

A process which could contribute to non-thermal excitation of rotational levels involves encounters of charged dust grains with interstellar hydrogen. The authors have examined this mechanism and find that it could be of comparable, if not greater, importance in relation to other processes which have been proposed.

131.030　**Detection of giant dust complexes in the direction of NGC 7538 in the Perseus arm.**
Y. K. Minn, J. M. Greenberg.
Astrophys. Journ., Vol. 196, 161 - 165 (1975).

The authors have found a giant dust region in the Perseus arm, in the direction of NGC 7538, which contains three large dust complexes. These complexes are not apparent optically as dark clouds but are detected by 6-cm H_2CO absorption.

131.031 Wavelength dependence of interstellar polarization and ratio of total to selective extinction.
K. Serkowski, D. S. Mathewson, V. L. Ford.
Astrophys. Journ., Vol. 196, 261 - 290 (1975).

Wavelength dependence of interstellar linear polarization has been observed for about 180 stars, mostly southern, in the $UBVR$ spectral regions. A multichannel polarimeter-photometer, in which spectral regions are separated by dichroic filters, was used. Wavelength λ_{max} is well correlated with the ratios of color excesses E_{V-K}/E_{B-V}, E_{V-K}/E_{V-R}, and E_{V-I}/E_{V-R}. These correlations indicate that the ratio R of total to selective interstellar extinction can be found for any individual star from the relationship $R = 5.5 \lambda_{max}$. Polarimetry seems to be the most practical method of estimating R. A map of distribution of λ_{max} on the sky, based on values for about 350 stars, indicates several well defined regions with λ_{max}, and hence R, clearly larger (or smaller) than the median value $\bar{\lambda}_{max} = 0.545 \mu$, corresponding to $R = 3.0$.

131.032 Observations of the $J = 2 \to 1$ transition of carbon monoxide in interstellar clouds. P. F. Goldsmith, R. L. Plambeck, R. Y. Chiao.
Astrophys. Journ., (Letters), Vol. 196, L39 - L42 (1975).

The authors have observed the $J = 2 \to 1$ rotational transitions ($\lambda = 1.3$ mm) of ^{12}CO and ^{13}CO in a number of dense molecular clouds associated with H II regions. The intensities of the $J = 2 \to 1$ lines are roughly equal to the intensities previously measured for the $J = 1 \to 0$ transition. The velocities and line widths of the $J = 2 \to 1$ and $J = 1 \to 0$ transitions are in close agreement. They find that the ^{13}CO $J = 2 \to 1$ line is optically thick in these sources.

131.033 Growth of interstellar grains by coagulation.
S. Simons, I. P. Williams.
Astrophys. Space Sci., Vol. 32, 493 - 498 (1975).

A discussion is given of the spontaneous coagulation of interstellar grains due to their natural Brownian motion. It is shown that the predicted present day size of such grains, assuming such a mechanism being operative, is in reasonable agreement with observations for three different environments that have been considered: namely, normal interstellar gas clouds, molecular clouds and the atmospheres of cool stars.

131.034 On the absence of detectable formaldehyde in the direction of Centaurus A.
R. W. O'Connell, W. C. Saslaw, B. E. Turner.
Monthly Notices Roy. Astron. Soc., Vol. 170, 29P - 30P (1975).

The authors report an upper limit to formaldehyde absorption or emission in the direction of the prominent dust lane of NGC 5128. Since it is conceivable that the lane is a foreground object, they have searched at both local and systemic velocities. The upper limit to the mean optical depth in cold H_2CO toward the central continuum source of the galaxy is 0.02.

131.035 Internal structure and stability of an interstellar cloud heated by an external flux of soft X-rays.
Y. Sabano, M. Tosa.
Publ. Astron. Soc. Japan, Vol. 27, 137 - 145 (1975).

The authors study the properties of an interstellar gas cloud which is heated by an external flux of soft X-rays and has a uniform pressure distribution. The heating flux is significantly attenuated inside the cloud even for a rather small cloud and the central region of the cloud is much cooler and denser than that heated uniformly, and hence the cloud can be compressed easier. The stability of such a gas cloud and its implications for the process of star formation are discussed on the basis of the two-phase model of the interstellar medium. The large scale galactic shock seems important as a triggering mechanism for the formation of a dense cloud and for the gravitational collapse leading to star formation.

131.036 Interstellar formaldehyde production.
T. J. Millar, D. A. Williams.
Monthly Notices Roy. Astron. Soc., Vol. 170, 51P - 55P (1975).

Several mechanisms for forming interstellar formaldehyde are reviewed. Three new mechanisms involving neutral species are suggested. Calculations show that mechanisms of this type may explain observations of formaldehyde in dense clouds, but not in the less dense and more typical regions.

131.037 Interstellar H_2CO. I. Absorption studies, dark clouds, and the cosmic background radiation.
N. J. Evans II, B. Zuckerman, G. Morris, T. Sato.
Astrophys. Journ., Vol. 196, 433 - 456 (1975).

The $2_{11} \leftarrow 2_{12}$ transition at 2-cm wavelength of interstellar formaldehyde has been observed in absorption against galactic continuum sources and the 2.7° K cosmic background radiation. In addition, the $1_{10} \leftarrow 1_{11}$ transition at 6-cm wavelength of the rare isotopic species $H_2^{13}C^{16}O$ has been detected in absorption against the cosmic background radiation in at least two clouds. Observations toward Cas A indicate that the temperature of the cosmic background radiation at 2-mm wavelength is $\lesssim 2.9°$ K.

131.038 The nucleation and expulsion of carbon particles formed in stellar atmospheres.
R. G. Tabak, J. P. Hirth, G. Meyrick, T. P. Roark.
Astrophys. Journ., Vol. 196, 457 - 463 (1975).

The conditions for significant nucleation of carbon particles to occur in several late-type stellar atmospheres are considered. The theoretical treatment includes the contributions from the translational, rotational, and vibrational degrees of freedom of the nuclei which have been omitted in previous considerations. It is shown that the range of possible surface free energies of the particles, for which nucleation can be achieved in these atmospheres, is small. The effects of electrostatic drag on escaping particles are discussed. Under the conditions considered here, these effects are shown to be of secondary importance to those of viscous drag.

131.039 The structure of the Orion A molecular cloud.
H. Gerola, S. Sofia.
Astrophys. Journ., Vol. 196, 473 - 478 (1975).

A consistent model of the Orion A molecular cloud is obtained by making use of the observed brightness temperature distributions of the $J = 2 \to 1$ and the $J = 1 \to 0$ transitions of the CO molecule, and the central component ($F = 2 \to 1$) of the $J = 1 \to 0$ transition of HCN, as well as the observed line profiles of the $J = 2 \to 1$ transition of CO, and the $J = 1 \to 0$ transition of HCN. The authors find that Orion A is strongly gravitationally bound and contracting, and that it can maintain the observed temperature distribution only by virtue of internal energy sources other than the contraction. This last conclusion is reached by computing the radiative losses due to the CO and HD cooling, as well as the losses due to inelastic collisions between the gas and the dust. Their results show that while the contraction rate is just about sufficient to balance the rate of radiation by CO, it is less than one-tenth of the rate at which energy is radiated by HD, and less than 0.001 of that at which energy could be lost to cool grains through totally inelastic collisions.

131.040 Hydromagnetic waves in molecular clouds.
J. Arons, C. E. Max.
Astrophys. Journ., (Letters), Vol. 196, L77 - L81 (1975).

The authors suggest that the observed large widths of CO emission lines may be due to the presence of moderate-amplitude hydromagnetic waves in molecular clouds. For parameters typical of CO emitting regions, this hypothesis is viable if such clouds contain systematic magnetic fields of strength $B_0 \gtrsim 40$ microgauss. The damping rate of the waves is discussed, as is the energy required if the observed line widths are due to hydromagnetic waves. To illustrate how wave energy may be supplied, a simple model is outlined for the emission of magnetoacoustic waves by H II regions embedded in magnetized molecular clouds.

131.041 Interstellar carbon monosulfide.
H. S. Liszt, R. A. Linke.
Astrophys. Journ., Vol. 196, 709 - 717 (1975).

The authors have measured intensities of the $J = 3-2$, $2-1$, and $1-0$ transitions of carbon monosulfide in five molecular clouds associated with galactic H II regions. The two higher-lying lines were mapped in each source, and the lowest line, $J = 1-0$, was observed at the map centers. The observed intensity ratios are shown to be dominated by effects of instrumental resolution. The observations are corrected for these effects and discussed in terms of recently published calculations for radiative transfer in collapsing clouds.

131.042 Distribution of 73-GHz para-formaldehyde line emission in the Orion nebula.
N. Kaifu, T. Iguchi, M. Morimoto.
Astrophys. Journ., Vol. 196, 719 - 722 (1975).

The $1_{01}-0_{00}$ transition of para-formaldehyde (72.8381 GHz) has been observed for 10 points in Ori A by the 6-m millimeter-wave telescope at Mitaka. The emission has been found to be extended to the north direction compared with the 2-mm lines of formaldehyde; the north–south extension of the 73-GHz line emission is about 6'. The distribution of the excitation temperature of para-formaldehyde in the Orion molecular cloud is discussed. The column density of para-formaldehyde is estimated to be $2 \sim 5 \times 10^{13}$ cm^{-2}.

131.043 Detection of interstellar trans-ethyl alcohol.
B. Zuckerman, B. E. Turner, D. R. Johnson, F. O. Clark, F. J. Lovas, N. Fourikis, P. Palmer, M. Morris, A. E. Lilley, J. A. Ball, C. A. Gottlieb, M. M. Litvak, H. Penfield.
Astrophys. Journ., (Letters), Vol. 196, L99 - L102 (1975).

Three transitions of trans-ethyl alcohol (CH_3CH_2OH) were detected in emission toward the Sagittarius B2 molecular cloud. The $6_{06}-5_{15}$ transition at 85,265.46 MHz, the $4_{14}-3_{03}$ transition at 90,117.51 MHz, and the $5_{15}-4_{04}$ transition at 104,808.58 MHz were observed at $V_{lsr} \sim 60$ km s^{-1}. The abundance of ethyl alcohol in the Sgr B2 cloud is comparable to that of many molecules previously detected there. Careful comparison of the abundance of ethyl alcohol with chemically related molecules such as its isomer dimethyl ether, acetaldehyde, and methyl alcohol may provide important information about mechanisms for molecular formation in the interstellar medium.

131.044 The interstellar medium near the sun: the line of sight to Lambda Scorpii. D. G. York.
Astrophys. Journ., (Letters), Vol. 196, L103 - L106 (1975).

A preliminary analysis of the column densities and ionization structure in the line of sight to λ Scorpii shows that at least three, and possibly four, regions can be identified by physical properties or by velocity separation in the line profiles. These include a region near $T = 2 \times 10^5$ °K, containing the O VI lines; one or more H II regions; and an H I region, which, however, may contain some S III and C III. Two weak lines of molecular hydrogen are detected, but their velocities are not consistent with their being in the H I region. The abundances of the elements in the H I region show a depletion pattern quite similar to, though somewhat less pronounced than, that found by Morton for the line of sight to ζ Oph.

131.045 Interstellar absorption lines in the spectrum of Zeta Ophiuchi. D. C. Morton.
Astrophys. Journ., Vol. 197, 85 - 115 (1975).

Extensive high-resolution scans with the ultraviolet spectrometer on the Copernicus satellite have been combined with the available ground-based data on interstellar lines to obtain temperatures, densities, and abundances in H I clouds and H II regions in the direction of ζ Oph. Column densities have been obtained for 21 elements in various stages of ionization. A new determination for CO and the results for H_2, HD, CH, CH$^+$, and CN from other authors also have been included. In addition, upper limits have been reported for five elements and 11 molecules. The H_2, HD, and most of the neutral atoms are concentrated in one cloud at a heliocentric velocity of -14.4 km s^{-1}, while N I, O I, Ar I, and most of the first ions are distributed over at least two of these clouds. The velocities of the higher ion states implied there are H II regions in addition to the Strömgren sphere. Calculations of the ionization equilibrium for C, Mg, S, and Ca have shown that the electron density $n_e \sim 0.7$ cm^{-3} in the -14.4 km s^{-1} cloud. Relative to hydrogen, most of the elements in the H I clouds are depleted by factors of 3 to 4000 compared with the solar system abundances. Only sulfur and zinc are present in the gas with near normal abundances. Several elements, particularly Al, Si, and Fe, appear to be depleted in the H II regions as well, but nitrogen is normal.

131.046 Optical interstellar lines in southern supergiants.
J. G. Cohen.
Astrophys. Journ., Vol. 197, 117 - 122 (1975).

Photoelectric photometry of 35 southern supergiants reveals a good correlation of the central residual intensity of the diffuse interstellar line at 4430 Å with color excess. From spectra of the brighter southern supergiants, the author has derived column densities of Na I and Ca II and, in a few cases, of CH and CH$^+$. Models for the gas in the halo are derived, and the author obtained the mass of gas in the halo (from 1×10^7 to $6 \times 10^7 M_\odot$) and a maximum value of the metal depletion of the halo gas relative to the disk gas (from 300 to 90, depending on whether grains exist in the halo). This maximum value implies that the gas in the halo has been significantly contaminated by material which has undergone nuclear processing.

131.047 Observations of interstellar CH and a study of its chemistry and excitation.
B. Zuckerman, B. E. Turner.
Astrophys. Journ., Vol. 197, 123 - 136 (1975).

The 140-foot telescope of the National Radio Astronomy Observatory was used to observe the three Λ-doublet transitions in the $^2\Pi_{1/2}$, $J = {}^1/_2$ state of interstellar CH. Improved rest frequencies for the three CH hyperfine components are obtained. These are discussed in terms of the structure of the molecule. The three hyperfine lines are always observed in emission with relative intensities often far from those expected in local thermodynamic equilibrium. The CH ground state Λ-doublet population is generally inverted in the Galaxy, probably via special collisions with hydrogen molecules. The nonequilibrium hyperfine ratios probably result from radiative processes. The formation rate of CH is inadequate to explain the anomalous excitation in terms of selection of particular states during formation. Comparison is made between observed abundances and those predicted by some ion-molecule reaction schemes.

131.048 Discovery of interstellar methyl formate.
R. D. Brown, J. G. Crofts, F. F. Gardner, P. D. Godfrey, B. J. Robinson, J. B. Whiteoak.
Astrophys. Journ., (Letters), Vol. 197, L29 - L31 (1975).

The $1_{10}-1_{11}$ A-state transition of methyl formate HCOOCH$_3$ has been detected in emission in the spectrum of Sgr B2. With a laboratory determination of 1610.249 MHz for the rest frequency of the transition, the radial velocity of the observed line is 53 ± 6 km s^{-1}. The nearby E-state transition at 1610.906 MHz may also have been detected. It is probable that the 1_{10} and 1_{11} levels are inverted and that the continuum emission of Sgr B2 is being amplified.

131.049 Cosmic masers.   W. H. Kegel.
Problems in stellar atmospheres and envelopes, (see 003.001), p. 257 - 299 (1975).

131.050 Lower limits for the interstellar reddening at the galactic poles.   I. Appenzeller.
Astron. Astrophys., Vol. 38, 313 - 314 (1975).
Interstellar polarization data are used to derive lower limits for the interstellar reddening at the galactic poles. The mean colour excess $E(B - V)$ is found to be > 0.011 at the north galactic pole ($b > 70°$) and > 0.016 at the south galactic pole ($b < -70°$).

131.051 The interstellar radiation field between 912 Å and 2740 Å.   P. M. Gondhalekar, R. Wilson.
Astron. Astrophys., Vol. 38, 329 - 333 (1975).
The average intensity of the interstellar diffuse radiation field between 912 Å and 2740 Å has been calculated on the basis of a galactic model but using, as far as possible, observational inputs for ultraviolet stellar spectra and the properties of the interstellar medium. An additional interstellar effect included in the calculations is the strong absorption in the Lyman series of atomic hydrogen. A comparison of the computed flux with wide field observations of the ultraviolet background gives reasonable agreement.

131.052 The nearby interstellar radiation field between 1750 Å and 504 Å.   M. Grewing.
Astron. Astrophys., Vol. 38, 391 - 396 (1975).
On the assumption that the ionization of the intercloud gas in the vicinity of the sun is indeed due to direct photoionization, the author derives for unreddened stars a self-consistent model which allows to infer directly the interstellar photon flux in the range from 1750 Å to 504 Å. The results suggest that, in the immediate vicinity of the sun, interstellar hydrogen that is not lumped up in concentrations will be roughly 95% ionized. The spectral distribution of the flux indicates that it originates from early type B-stars.

131.053 Detection of new stellar sources of vibrationally excited silicon monoxide maser emission at 6.95 millimeters.   L. E. Snyder, D. Buhl.
Astrophys. Journ., Vol. 197, 329 - 340 (1975).
The authors have searched 32 selected sources for $v = 1$, $J = 1 \to 0$ SiO maser emission at 43,122.0 MHz (~ 6.95 mm), and detected 16 of them. They detected eight of the SiO $v = 1$, $J = 2 \to 1$ SiO maser sources observed by Kaifu, Buhl, and Snyder at 86,243.27 MHz and found that the $v = 1, J = 1 \to 0$ maser is often the most intense transition, and hence promises to be the best frequency to use in searching for new maser sources. Most of the detected maser sources reported in this paper can readily be associated with known late-type stars. The proposed maser mechanisms are discussed in the context of current observations.

131.054 Interstellar molecules: direct formation on graphite grains.   A. Bar-Nun.
Astrophys. Journ., Vol. 197, 341 - 345 (1975).
It is demonstrated experimentally that ground-state atoms of oxygen, hydrogen, and sulfur, thermalized to 78 K, can, upon impinging on graphite grains, produce carbon monoxide and carbon dioxide, a variety of hydrocarbons, and carbon disulfide. Since interstellar extinction curves indicate the existence of bare graphite surfaces in the clouds, it is suggested that direct chemical reactions of atoms with graphite grains supply a considerable number of these molecules to the clouds. In clouds where the temperatures are not much below 78 K, the carbon oxides and most of the hydrocarbons will evaporate from the grains and will not contribute to mantle growth.

131.055 The C$^+$—CO transition in interstellar clouds.   A. E. Glassgold, W. D. Langer.
Astrophys. Journ., Vol. 197, 347 - 350 (1975).
Ionized carbon recombines as the ultraviolet radiation field is attenuated inside interstellar clouds. The authors propose that C$^+$ is transformed into CO, the CO being formed by ion-molecule reactions of C$^+$ with OH and H_2O. The characteristic column density of hydrogen required for the transition is $N \approx 3 \times 10^{21}$ cm^{-2}.

131.056 Detection of intrinsic linear polarization in emission-line O stars.   D. P. Hayes.
Astrophys. Journ., (*Letters*), Vol. 197, L55 - L56 (1975).
Preliminary results of a search for variable linear polarization in ten Of stars are presented. Four of these stars have been found to have variable polarization. It is tentatively concluded that these results may be understood on the basis of spectral classification schemes which attempt to identify the presence of extended envelopes, but further work remains to be done before a correlation between variable polarization and spectral class can be established.

131.057 Microwave observations of the Rho Ophiuchi dark cloud.   E. J. Chaisson.
Astrophys. Journ., (*Letters*), Vol. 197, L65 - L68 (1975).
Observations of the 158α recombination-line spectrum toward the dark cloud ρ Oph show evidence for two features, which, if the cloud is quiescent as seems evident from molecular observations, are most reasonably attributed to carbon and probably sulfur. An observation of Zeeman splitting of OH suggests that if a uniform, longitudinal magnetic field exists, it must be $\lesssim 0.5$ milligauss over much of the central part of the ρ Oph cloud.

131.058 Observations of circular polarization of the $J = 2-1$, $v = 1$ transition of the SiO maser.
D. R. Johnson, F. O. Clark.
Astrophys. Journ., (*Letters*), Vol. 197, L69 - L72 (1975).
Circular polarization has been observed in the $J = 2-1$, $v = 1$ SiO emission from the Orion A molecular source (OMC 1). The net polarization was observed at two different times separated by three days.

131.059 Observations of the $J = 1 \to 0$ transitions of the ^{13}C isotopic species of cyanoacetylene (HCCCN) in the direction of Sagittarius B2.   F. F. Gardner, G. Winnewisser.
Astrophys. Journ., (*Letters*), Vol. 197, L73 - L76 (1975).
The isotopic species HC^{13}CCN and HCC^{13}CN of cyanoacetylene have been detected in Sgr B2 at 9.06 GHz. An upper limit to the abundance ratio ^{12}C/^{13}C of 36 ± 5 is obtained for the gas in front of the continuum source. The ratio of the $F = 1 \to 1$ to the $F = 2 \to 1$ hyperfine transitions is lower than the LTE value of 0.6, a similar result to that found previously for the ^{12}C species.

131.060 Line-leaking models for interstellar molecular emission and absorption—I. The anomalous absorption and emission by formaldehyde.
M. A. Pelling, D. ter Haar.
Monthly Notices Roy. Astron. Soc., Vol. 171, 103 - 118 (1975).
The authors sketch a line-leaking (radiative transfer) theory with special reference to the absorption and emission by interstellar formaldehyde. The model, in which population inversions or anti-inversions are produced because radiation has leaked out of some of the frequencies corresponding to the transitions between relevant levels, is applied to several formaldehyde sources. On the whole reasonable agreement with observational data is obtained.

131.061 Molecules in space. A. H. Cook.
Quarterly Journ. Roy. Astron. Soc., Vol. 16, 21 - 37 (1975). – Larmor lecture of the Cambridge Philosophical Society 1974 October 14.

131.062 Origin of broad interstellar feature at 1.6 μm^{-1}.
P. G. Manning.
Nature, Vol. 255, 40 - 41 (1975).
The author suggests that a discontinuity at ~1.6 μm^{-1}, which he sees as a minor feature on the mean extinction curve for several reddened stars, represents $Fe^{2+} \rightarrow Fe^{3+}$ charge transfer absorption in interstellar grains.

131.063 The interstellar reddening in front of globular clusters. D. L. Crawford, J. V. Barnes.
Publ. Astron. Soc. Pacific, Vol. 87, 65 - 76 (1975).
The authors have observed 187 field stars in galactic fields containing globular clusters. For 15 of the 28 fields, they derive color excesses that should be essentially the same as for the globular clusters.

131.064 Hot hydrogen in prebiological and interstellar chemistry. C. Sagan, R. S. Becker.
Science, Vol. 188, 72 - 73 (1975). – Technical comment.

131.065 Evidence for maser action in the 1.2 cm transitions of methanol in Orion.
R. Hills, V. Pankonin, T. L. Landecker.
Astron. Astrophys., Vol. 39, 149 - 153 (1975).
Observations of the 25 GHz emission from CH_3OH in Orion, with resolution 5.4 kHz and beamwidth 35″, show that the lines consist of at least 4 components each ~ 30 kHz (0.36 km/s) wide arising from positions differing by ~ 10″. The upper limits on the sizes of the emitting regions are between 10″ and 30″. The brightness temperatures are at least 3 times the kinetic temperature of ~ 100 K derived from the widths of the features. The emitting molecules are therefore not in LTE and it is most likely that the emission is produced by weak maser action.

131.066 Column densities of interstellar molecular hydrogen.
L. Spitzer, Jr., W. D. Cochran, A. Hirshfeld.
Astrophys. Journ., Suppl. Ser., No. 266, Vol. 28, 373 - 389 (1974).
Equivalent widths of some 50 lines in the 0–0 to 5–0 Lyman bands of H_2 are reported in the spectra of 28 stars. Curves of growth are given and column densities for levels from $J = 0$ to $J = 5$ are tabulated, with a few values and upper limits for $N(6)$ and $N(7)$, together with values for b, the velocity spread parameter. In three Orion stars and in ρ Leo pairs of components are detected, the difference in radial velocity determined, and column densities measured or estimated; tentative identifications are made with the components observed by Hobbs in the Na D-lines. Column densities for HD are given for 13 stars. Upper limits for column densities in the first and second vibrational levels are listed for several stars. Data are presented showing an apparent increase of velocity dispersion with J for a number of stars, as measured both from the curves of growth and from line widths.

131.067 The CO line receiver at the U.B.C. telescope.
W. L. H. Shuter, W. H. McCutcheon.
Journ. Roy. Astron. Soc. Canada, Vol. 69, 39 (1975). – Abstr. Canadian Astron. Soc.

131.068 Six-centimeter H_2CO emission from the Orion nebula. B. Zuckerman, P. Palmer, L. J. Rickard.
Astrophys. Journ., Vol. 197, 571 - 573 (1975).
With the Max Planck Institute's 100-m antenna at Effelsberg, the 1_{11}–1_{10} transition of H_2CO has been detected in emission in the Orion molecular cloud. Thus, the otherwise ubiquitous cooling of this 6-cm transition is quenched at the H_2 densities ~ 10^5 cm^{-3} deduced to exist in the central regions of this cloud.

131.069 Interstellar clouds containing optically thin H_2.
M. Jura.
Astrophys. Journ., Vol. 197, 575 - 580 (1975).
The author considers clouds which contain H_2 that is optically thin in the Lyman and Werner bands; and he develops models to explain the observed populations in the excited rotational levels. He argues that a considerable fraction of all the H_2 formed on grains is created with some degree of rotational excitation, and that the formation rate on grains, R, is such that $10^{-17} \leqslant R \leqslant 3 \times 10^{-17}$ cm^3 s^{-1}. If R is assumed to be uniform throughout the interstellar gas, it would seem that $R = 3 \times 10^{-17}$ cm^3 s^{-1}, but his results are not sufficiently accurate to determine if R does vary. The author infers the radiation field within clouds and whether the cloud is near an early-type star. For example, the neutral hydrogen observed toward τ Sco is probably at the outer boundary of the H II region surrounding that star.

131.070 Interstellar clouds containing optically thick H_2.
M. Jura.
Astrophys. Journ., Vol. 197, 581 - 586 (1975).
The author considers interstellar clouds in which the H_2 is observed to be optically thick in the Lyman bands. He finds that if H_2 is formed on grains with rate R, and if n is the total density of hydrogen nuclei, then the product of these two quantities can be determined quite simply from the observations. The author estimates the radiation field in each cloud from the populations of the H_2 rotational levels. Of the 10 clouds studied, five are illuminated by an ultraviolet field appropriate to within a factor of 2 to that in the solar neighborhood, four appear to be near the stars against which they are observed, and one is irradiated by a very low intensity field. The results also suggest that there are substantial pressure variations in the interstellar gas, even among those clouds not near H II regions.

131.071 La materia interstellare. P. Tempesti.
Coelum, Vol. 43, 45 - 71 (1975).

131.072 On the ratio of the total to selective absorption.
W. A. Sherwood.
Astrophys. Space Sci., Vol. 34, 3 - 10 (1975). – Paper presented at the Symposium on Solid State Astrophysics, Cardiff, Wales, 9–12 July, 1974.
The ratio of total to selective absorption, R, has been found to remain constant as dust is processed in clouds from low to high density, through H II regions and open clusters, and returned to the interstellar medium. R has the same value in dense dust clouds as it has in H II regions of different ages. Variations in R values obtained from stars in H II regions may be due to errors in special type classification. Globular cluster diameters show no tendency to increase with distance from the sun when $R = 3.2$ is used. Large grains evidently do not exist in the interstellar medium. There is no evidence for neutral extinction in the Galaxy at large.

131.073 Features of the interstellar extinction curve.
D. H. Morgan.
Astrophys. Space Sci., Vol. 34, 11 - 17 (1975). – Paper presented at the Symposium on Solid State Astrophysics, Cardiff, Wales, 9–12 July, 1974.
The extinction curves for spherical particles are subject to the errors of the particle material's refractive index. Their sensitivity to these errors has been investigated and is found to be dependent upon wavelength. For graphite, significant errors are produced in the far ultraviolet part of the extinction curve; for silicates, in the near ultraviolet; while for iron the

error is relatively small. The wavelength dependence of the 10 μm and 20 μm absorption bands of small silicate spheroids upon their shape and alignment has been studied. It is found that the bands can be displaced by ~1 μm towards longer wavelengths from their positions for corresponding spheres: and that a further, though small, displacement can be super-imposed upon this by their subsequent alignment.

131.074 Far-ultraviolet extinction in σ Scorpii.
T. P. Snow, Jr., D. G. York.
Astrophys. Space Sci., Vol. 34, 19 - 22 (1975). – Paper pre-sented at the Symposium on Solid State Astrophysics, Cardiff, Wales, 9–12 July, 1974.

It was found earlier from OAO-2 data (Bless and Savage, 1972) that considerable variability with direction in space is present in both the shape and level (relative to $B - V$ color excess) of the interstellar extinction curve in the far ultra-violet. The authors have obtained UV data on σ Sco using Copernicus (OAO-3), which has an entrance slit on the order of 10^3 times smaller in projected area than that of OAO-2, so that the contribution to the signal from scattered nebular light would be correspondingly smaller. They find very good agree-ment with the extinction curve of Bless and Savage, confirm-ing the low UV extinction in the line of sight to σ Sco.

131.075 Diffuse interstellar band formation in dense clouds.
T. P. Snow, Jr., J. G. Cohen.
Astrophys. Space Sci., Vol. 34, 33 - 38 (1975). – Paper pre-sented at the Symposium on Solid State Astrophysics, Cardiff, Wales, 9–12 July, 1974.

Measurements of the strengths of the diffuse interstellar bands at 4430, 5780 and 5797 Å show that the bands tend to be weak with respect to extinction in dense interstellar clouds. Data on 10 stars in the ρ Ophiuchi cloud complex show further that the diffuse band-producing efficiency of the grains de-creases systematically with increasing grain size. It is concluded that the diffuse bands are not formed in the mantles which accrete on the grains in interstellar clouds, but that they could be produced in the cores of grains or in some molecular species.

131.076 Interstellar extinction and diffuse absorption features.
J. Dorschner.
Astrophys. Space Sci., Vol. 34, 39 - 47 (1975). – Paper pre-sented at the Symposium on Solid State Astrophysics, Cardiff, Wales, 9–12 July, 1974.

The equivalent width of the λ 2175 Å band, W_{2175}, well known as the big bump in the interstellar extinction curves, has been found to be closely correlated with the colour excess E_{B-V} as well as with the extinction differences E_{8-6} and E_{9-7} defined to characterize quantitatively the steep slopes of the extinction curves in the far ultraviolet. The equivalent widths of the $\lambda\lambda$ 5780 and 5797 Å diffuse lines show good correlation' with E_{B-V}. The correlations of W_{5780} and W_{5797} with E_{8-6} resp. E_{9-7} are, however, rather weak. Correlations between W_{2175} and W_{5780} and between W_{2175} and W_{5797} are indicated. The re-sults have been qualitatively interpreted in favour of the dust model consisting of a mixture of small silicate grains and larger silicate grains coated by molecular mantles.

131.077 Physical adsorption of hydrogen on interstellar graphite grain surfaces.
R. F. Willis, B. Fitton.
Astrophys. Space Sci., Vol. 34, 57 - 71 (1975). – Paper pre-sented at the Symposium on Solid State Astrophysics, Cardiff, Wales, 9–12 July, 1974.

The authors review existing single-particle theories con-cerning parameters of importance which determine the kinetics of hydrogen molecule formation and ejection from cold ($T_{gr} \lesssim 20$ K) graphite grain surfaces. The nature of the single-particle quantum states of low mass gas atoms and molecules in a periodic surface lattice potential is considered.

131.078 Effects of charged dust grains.
S. Hayakawa.
Astrophys. Space Sci., Vol. 34, 73 - 79 (1975).
Paper presented at the Symposium on Solid State Astrophysics, Cardiff, Wales, 9–12 July, 1974.

Dust grains expelled by radiation pressure of stars are charged to potentials in the range 30–40 V in H I clouds. These grains may be responsible for the following phenomena which are otherwise hardly explicable. (1) A considerable fraction of electrons knocked-out by charged grains of high speeds have energies around 15 eV and produce singly ionized ions but not doubly ionized ones in accord with an ultraviolet observation of interstellar atoms and ions. (2) Transverse momentum transferred to grains by Coulomb scattering of ambient electrons and protons is greater than that by multiple scattering of cosmic ray protons, thus the former being more effective for the grain alignment than the latter. (3) At a shock front charge separation due to a large inertial mass of grains produces an electric field, thus accelerating charged particles and causing a drift of interstellar matter.

131.079 The influence of grain mantles on the formation of hydrogen molecules on grain surfaces.
T. J. Lee.
Astrophys. Space Sci., Vol. 34, 123 - 130 (1975). – Paper presented at the Symposium on Solid State Astrophysics, Cardiff, Wales, 9–12 July, 1974.

The physical adsorption energy, E, of hydrogen mole-cules on various substrates at temperatures between 5 and 30 K and at the lowest practicable gas densities has been meas-ured. Values of E/k are for condensed CO 340 K, CO_2 800 K, H_2O 850 K and for 'dirty' graphite 980 K and 'dirty' copper 800 K. From these measurements temperature ranges in which H atoms might combine on the surface to form H_2 molecules are estimated.

131.080 UV radiation fields in dark clouds.
A. P. Whitworth.
Astrophys. Space Sci., Vol. 34, 155 - 173 (1975). – Paper pre-sented at the Symposium on Solid State Astrophysics, Cardiff, Wales, 9–12 July, 1974.

If interstellar extinction at UV wavelengths is mainly due to scattering with a strongly forward throwing phase-function, the interior of a dark cloud may be much better illuminated at UV wavelengths than its measured extinction would suggest. The author considers the penetration of radiation into a dark cloud against scattering and absorption by grains and he de-fines a new group property for interstellar grains. Computa-tions are made of the radiation fields in (1200, 4500) Å, at the centres of dark clouds with measured visual extinctions. It is found that even in very dark clouds, the radiation energy density in (1200, 1800) Å may be significant, due to the high grain albedo at these short wavelengths.

131.081 Optical properties of particulates.
D. R. Huffman.
Astrophys. Space Sci., Vol. 34, 175 - 184 (1975). – Paper pre-sented at the Symposium on Solid State Astrophysics, Cardiff, Wales, 9–12 July, 1974.

Optical properties of small particles of olivine (less than 0.1 μ) have been studied in the ultraviolet as an example of an insulating solid. Measured extinction of small olivine particles in the infrared agrees with calculations based on newly meas-ured optical constants, but dominant sharp structure in the 10 μ region still presents a bit of a problem in explaining 'silicate' features in astronomical data.

131.082 The plausibility of silicate-core ice-mantle grains.
M. J. Dempsey, N. C. Wickramasinghe.
Astrophys. Space Sci., Vol. 34, 185 - 189 (1975). – Paper pre-sented at the Symposium on Solid State Astrophysics, Cardiff, Wales, 9–12 July, 1974.

Extinction curves for silicate-core ice-mantle grains are

computed and compared with the infrared spectral data on the Becklin-Neugebauer object in the Orion nebula. A ratio of outer mantle to core radius of 1.3 which best fits this data suggests that silicate-core ice-mantle grains are unlikely to contribute a major part to the total visual extinction coefficient of interstellar material.

131.083 Considerations about the absorption efficiency of dust particles in the infrared.
E. Bussoletti, A. Borghesi, G. Leggieri, A. Blanco.
Astrophys. Space Sci., Vol. 34, 191 - 197 (1975). – Paper presented at the Symposium on Solid State Astrophysics, Cardiff, Wales, 9–12 July, 1974.

Analytical approximations used often in the literature for calculating energy rates emitted by dust grains in infrared are discussed. Comparisons with correct complete formulations are made for three grain models: (1) pure graphite, (2) ice mantle-graphite core, (3) silicates. λ^{-1} and λ^{-2} dependences for the average effective emissivity of such grains are used. The authors find that for silicate and graphite grains the simplified approximations are valid only when accuracies between 10% and 50% are required and only for grain temperatures higher than 80 K. At lower temperatures the validity of the approximations fails for the graphite particle while it is variable for the silicate dust grain. The ice core mantle particles can instead be treated with approximated formulae without introducing appreciable errors.

131.084 Interstellar CN at radio wavelengths.
B. E. Turner, R. H. Gammon.
Astrophys. Journ., Vol. 198, 71 - 89 (1975).

Seven of the nine hyperfine lines in the 2.6-mm $K = 1 \rightarrow 0$ transition of CN have been observed in many interstellar sources. CN is detected in the direction of infrared sources (with and without associated H II regions), toward at least one star which shows interstellar lines of CH and CH^+, toward several objects regarded as protostellar in nature, but not in dark dust clouds or in several strong sources of HCN and HC_3N. Derived total particle densities are typically 10 times less than have been estimated from excitation studies of other molecules (CS, HCN). This is understandable if radiative trapping is important in exciting CS and HCN. CN abundances appear inversely correlated with total densities in the range $\sim 10^4 \leqslant n_{H_2} \leqslant 10^6 \, cm^{-3}$. This, and the relative abundances of CN, CS, CO, and n_{H_2}, appear to fit well the predictions of ion-molecule processes for forming and destroying interstellar molecules.

131.085 OH Zeeman observations of interstellar dust clouds.
R. M. Crutcher, N. J. Evans II, T. Troland, C. Heiles.
Astrophys. Journ., Vol. 198, 91 - 93 (1975).

The authors have attempted to measure Zeeman splitting of OH emission lines in two dust clouds. The upper limits on the Zeeman splitting indicate that the large-scale magnetic fields are less than 50 microgauss, a value significantly less than that expected if the magnetic field is frozen in during collapse.

131.086 Empirical relation between interstellar X-ray absorption and optical extinction.   P. Gorenstein.
Astrophys. Journ., Vol. 198, 95 - 101 (1975).

An empirical relation between interstellar X-ray absorption and optical extinction is derived from the correlation of measurements made on objects of large intrinsic diameter. The result is $A_v = 4.5 \times 10^{-22} N_H$ mag, with the principal error being largely systematic in origin, where N_H represents the column density of interstellar matter in the Brown and Gould model for the X-ray absorption coefficient. The value of A_v/N_H determined from the X-ray measurements is compatible with the $L\alpha$ results from OAO 2. It is in accord with an interstellar dust-to-mass ratio of $\sim 10^{-2}$.

131.087 X-ray absorption, interstellar reddening, and elemental abundances in the interstellar medium.
C. Ryter, C. J. Cesarsky, J. Audouze.
Astrophys. Journ., Vol. 198, 103 - 109 (1975).

X-ray absorption and interstellar reddening along the line of sight to five supernova remnants are shown to yield the relationship $N_x/E_{B-V} = (6.8 \pm 1.6) \times 10^{21}$ equivalent H atoms cm^{-2} mag^{-1}. It is also shown that within 1 to ~ 5 kpc of the sun, the "universal" abundances hold in the interstellar medium. The quoted value of the ratio N_x/E_{B-V} may also apply to the galactic center X-ray source; N_x and E_{B-V} individually are high, probably due to the presence of a large column of molecular hydrogen and/or an abundance gradient of metals (mainly C, N, O) in the Galaxy.

131.088 Study of statistical properties of emission of OH maser sources.   E. E. Lekht, G. M. Rudnitskij, O. Franquelin, J.-P. Drouhin.
Pis'ma v Astron. Zhurn., Vol. 1, No. 2, p. 29 - 32 (1975). In Russian.

Measurements of the probability density function of intervals between zero-crossings of the signal of 18-cm hydroxyl lines from W3 and VY Canis Majoris have been made at the Nançay radio telescope. The results obtained confirm the conclusion made earlier by other authors that the OH maser lines signal is noise-like.

131.089 On the interstellar abundance of sodium and potassium.   J. Lequeux.
Astron. Astrophys., Vol. 39, 257 - 261 (1975).

The author discusses the column densities of interstellar neutral sodium and potassium in front of 15 stars for which detailed profiles of the $K^0 \lambda$ 7699-line and of the Na^0 D-line exist. The total column densities of Na and K thus obtained imply depletions by factors 3–10, and more for two stars in Scorpius. It is not certain that depletions by factors 3–10 are real: some atomic parameters used in the study may be incorrect; alternatively, it may be that most of the strongly-depleted calcium is in the boundaries of the clouds, which may be more ionized than the central parts where most Na and K are found.

131.090 The 4.8 GHz formaldehyde absorption in the direction of Cygnus X-3.   Y. K. Minn.
Astron. Astrophys., Vol. 39, 303 - 305 (1975).

Formaldehyde observations at 4.83 GHz have been made toward the X-ray and radio source Cygnus X-3 ($l = 79.83°$, $b = 0.70°$) during and after its most recent outburst. Three H_2CO absorption lines at 13, 2 and -46 km s^{-1} were observed when the flux of Cygnus X-3 was enhanced. The -46 km s^{-1} line confirms the large kinematic distance of Cygnus X-3 previously estimated from H I absorption line measurements. From the comparison of H_2CO line profiles taken toward Cygnus X-3 during and after the outburst, the author estimates that the optical depths of the H_2CO clouds are small, with values of 0.026 for the 2 and 13 km s^{-1} features and of 0.029 for the -46 km s^{-1} feature, and that the cloud filling factors lie between 0.5 and 1.0. The H_2CO and H I column densities are compared. The linear dimensions and the total mass of the cloud at -46 km s^{-1} are derived.

131.091 Maps of formaldehyde absorption at 4.8 GHz in three dust clouds.
A. Sume, D. Downes, T. L. Wilson.
Astron. Astrophys., Vol. 39, 435 - 444 (1975).

Heiles' Clouds 1 and 2, and Lynds' cloud L 134 have been mapped in the 4.8 GHz line of formaldehyde with the Onsala 25.6-m telescope. Average number densities of H_2CO have been estimated from the maps, and are of the order of 10^{-5} cm^{-3}. From the absence of recombination lines, the r.m.s. electron density is estimated to be < 0.2 cm^{-3}. H_2CO spectra taken with long integration times yield excitation temperatures of 2.3 ± 0.2 K for a position of strong absorption in Cloud 2 and < 1.0 K in L 134.

131.092 A study of the near ultraviolet interstellar extinction curve from selected stars observed with the orbiting stellar spectrophotometer S 59.
R. Viotti, H. J. G. L. M. Lamers.
Astron. Astrophys., Vol. 39, 465 - 471 (1975).
   The S 59 scans were used to select pairs of reddened-unreddened stars with similar optical and ultraviolet spectra. It is confirmed that the interstellar extinction curves are not the same in shape. This suggests that towards different stars the interstellar grains differ both in composition and in size distribution.

131.093 Structure of the hot regions in the interstellar medium. B. W. Smith.
Bull. American Astron. Soc., Vol. 7, 242 (1975). – Abstr. AAS.

131.094 Interstellar molecules and the formation of stars.
P. Palmer, B. Zuckerman.
Bull. American Astron. Soc., Vol. 7, 244 - 245 (1975).
Abstr. AAS.

131.095 Observations of very broadband structure in the interstellar reddening curve. K. H. Rex, D. S. Hayes.
Bull. American Astron. Soc., Vol. 7, 249 (1975). – Abstr. AAS.

131.096 Ultraviolet interstellar extinction based on OAO-2 filter photometry. A. D. Code.
Bull. American Astron. Soc., Vol. 7, 249 (1975). – Abstr. AAS.

131.097 Observations of carbon α and β recombination lines at 22 cm. W. L. Boughton.
Bull. American Astron. Soc., Vol. 7, 250 (1975). – Abstr. AAS.

131.098 Shock-driven implosion of interstellar gas clouds.
P. R. Woodward.
Bull. American Astron. Soc., Vol. 7, 258 (1975). – Abstr. AAS.

131.099 High resolution profiles of the narrow diffuse interstellar features. B. D. Savage.
Bull. American Astron. Soc., Vol. 7, 260 (1975). – Abstr. AAS.

131.100 Detection of the Zeeman effect in 21-cm emission lines. T. H. Troland, C. E. Heiles.
Bull. American Astron. Soc., Vol. 7, 260 - 261 (1975).
Abstr. AAS.

131.101 Evidence for Zeeman splitting in 1720-MHz OH line emission. K. Y. Lo, R. C. Walker, B. F.
Burke, J. M. Moran, K. J. Johnston, M. S. Ewing.
Bull. American Astron. Soc., Vol. 7, 261 (1975). – Abstr. AAS.

131.102 Observations of the $1_2 - 1_1$ 13 GH$_3$ transition of SO.
F. O. Clark, D. R. Johnson.
Bull. American Astron. Soc., Vol. 7, 264 (1975). – Abstr. AAS.

131.103 Detection of the $3_{13} - 3_{12}$ transitions of thioformaldehyde and acetaldehyde in Sgr B2.
J. M. MacLeod, L. H. Doherty.
Bull. American Astron. Soc., Vol. 7, 265 (1975). – Abstr. AAS.

131.104 Cyanoacetylene in dense interstellar clouds.
M. Morris, B. E. Turner, P. Palmer, B. Zuckerman.
Bull. American Astron. Soc., Vol. 7, 265 (1975). – Abstr. AAS.

131.105 The C^+–CO transition in molecular clouds.
A. E. Glassgold, W. D. Langer.
Bull. American Astron. Soc., Vol. 7, 265 (1975). – Abstr. AAS.

131.106 The carbon monoxide abundance in low temperature interstellar clouds. W. D. Langer.
Bull. American Astron. Soc., Vol. 7, 265 (1975). – Abstr. AAS.

131.107 A study of carbon monoxide in dark clouds.
R. L. Dickman.
Bull. American Astron. Soc., Vol. 7, 265 (1975). – Abstr. AAS.

131.108 Infrared photometry of the Mon R2 molecular cloud and detection of the $3\,\mu$ ice band in a highly embedded source. R. B. Loren, P. A. Vanden Bout.
Bull. American Astron. Soc., Vol. 7, 266 (1975). – Abstr. AAS.

131.109 Detection of intrinsic linear polarization in emission-line stars. D. P. Hayes.
Bull. American Astron. Soc., Vol. 7, 271 (1975). – Abstr. AAS.

131.110 X-ray absorption and interstellar reddening.
C. Ryter, C. Cesarsky, J. Audouze.
Bull. American Astron. Soc., Vol. 7, 337 (1975). – Abstr. AAS.

131.111 Ionization waves in the interstellar medium.
I. A. Klimishin.
Problems of cosmic physics. Vyp. (No.) 9, (see 003.013), p. 3 - 18 (1974). In Russian.
   An investigation on the classification of ionization fronts moving in the interstellar medium, on studying their structure, stability and motion is made. General questions of H II-regions dynamics and evolution of globules immersed in the ionized hydrogen are discussed.

131.112 Interstellar molecules in the Galaxy.
G. Winnewisser.
Naturwissenschaften, 62. Jahrgang, p. 200 - 210 (1975).
   Within the last four years more than 40 different molecules in the interstellar gas clouds of our Galaxy have been identified from their radiofrequency spectra. The number of different molecules, the large size of many of them, and the variety of degrees of excitation reveal a totally unexpected complexity in the gas component of interstellar clouds.

131.113 Observations of H$_2$O sources with the very long base-line radio interferometer Pushchino-Simeiz.
V. S. Ablyazov, V. A. Alekseev, M. A. Antonets, V. I. Ariskin, V. P. Vekshin, Eh. D. Gatehlyuk, V. V. Demin, V. A. Efanov, B. G. Kutuza, B. N. Lipatov, L. I. Matveenko, S. M. Mkrtchyan, I. G. Moiseev, V. N. Nikonov, V. A. Oganesyan, V. A. Sanamyan, A. S. Sizov, R. L. Sorochenko, V. P. Sosnin, V. S. Troitskij, B. P. Fateev, A. I. Chikin, B. V. Shchekotov.
Izv. vyssh. ucheb. zavedenij. Radiofizika, Vol. 17, 1431 - 1437 (1974). In Russian. – Abstr. in Referativ. Zhurn. 51. Astron., 4.51.720 (1975).

131.114 On the structure of hydroxyl maser sources.
A. H. Cook.
Monthly Notices Roy. Astron. Soc., Vol. 171, 605 - 618 (1975).
   Recent interferometric observations of hydroxyl maser sources are analysed and it is shown that radiation is observed only along isolated lines of sight characterized by a specific Doppler shift of frequency, a unique sense of circular polarization and a single transition. Some implications of this result for possible properties of pumping processes or for solutions of the equations of transfer are indicated and it is argued that the observations could be accounted for by a filter mechanism depending on a correlation between mass velocity and magnetic field in the source region.

131.115 On the nature of the intercloud medium.
J. K. Hill, J. Silk.
Astrophys. Journ., Vol. 198, 299 - 306 (1975).
   Earlier interpretations of the Copernicus observations of

interstellar absorption lines in the spectra of unreddened stars are reviewed in the context of steady-state ionization models of the low-density (intercloud) component of the interstellar medium. The authors conclude that the observations can be adequately explained if between 10 and 30 percent of a representative line of sight passes through a photoionized medium of density $0.1-0.2$ cm^{-3}; the bulk of the gas along the line of sight is H I, and can either be at a similar density, in which case the mean primary soft X-ray ionization rate cannot exceed $1 \times 10^{-18} s^{-1}$ (H atom)$^{-1}$, or else it must be in shielded clumps or clouds.

131.116 On the ionization of the intercloud medium by runaway O—B stars. T. X. Thuan.
Astrophys. Journ., Vol. 198, 307 - 329 (1975).

The H II regions around high-velocity stars of type O5 to B1 are studied, with "on the spot" reabsorption assumed for the diffuse ultraviolet radiation, but with time-dependent cooling and recombination behind the star considered. Ratios of column densities, integrated through the H II regions are given for stationary and moving stars. Global properties of the intercloud medium are discussed. It is found that the presence of clouds may reduce the volume ionized by runaway stars by as much as a factor of 7 and that as much as 30 percent of the total ionizing radiation from early-type stars may leak out of the Galaxy. The Copernicus observations are consistent with the lines of sight crossing only the H II regions around the observed stars and a neutral ICM with the existence of a "hole" in the distribution of O—B5 stars in the vicinity of the sun.

131.117 Molecule-dust correlations in dark cloud Khavtassi 3. P. C. Myers.
Astrophys. Journ., Vol. 198, 331 - 348 (1975).

Mapping observations of microwave spectral lines of H_2CO, OH, and H toward the dark cloud Kh 3 are reported. For each species, comparison is made between column density N, excitation temperature T_{ex}, and red extinction A, determined from star counts. Stars were counted in resolution elements which coincide with beam size and position for each microwave spectrum taken. It is found that for T_{ex} assumed independent of A, N increases with A strongly for H_2CO, increases moderately for H, and is independent of A for OH. For N assumed independent of A, T_{ex} decreases strongly with A for H_2CO, decreases moderately with A for H, and is indepentent of A for OH.

131.118 Hydromagnetic waves and shock waves as an interstellar heat source. J. Silk.
Astrophys. Journ., (Letters), Vol. 198, L77 - L80 (1975).

The dissipation of hydromagnetic waves in diffuse interstellar matter is investigated as a heat source that does not itself produce any significant nonthermal ionization. In the low-density interstellar medium, shock waves, both hydromagnetic and hydrodynamic, can provide appreciable heating and yield either a neutral intercloud medium ($T \sim 10^4$ K) or a tenuous hot medium ($T \sim 10^6$ K) that is in pressure equilibrium with interstellar clouds. Quantitative estimates are given of the heat input by these various mechanisms, and the predicted parameters and properties of the interstellar medium are discussed.

131.119 Radio detection of interstellar sulfur dioxide.
L. E. Snyder, J. M. Hollis, B. L. Ulich, F. J. Lovas, D. R. Johnson, D. Buhl.
Astrophys. Journ., (Letters), Vol. 198, L81 - L84 (1975).

Interstellar sulfur dioxide (SO_2) has been detected in emission from the direction of the Orion nebula molecular cloud and from Sgr B2. SO_2 is the heaviest interstellar molecule detected to date, and the only nonlinear triatomic molecule which does not contain hydrogen. The remarkable Orion emission profiles suggest that two components are supporting the

SO_2 emission: a dense circumstellar-type envelope, which may be in maser emission, and a warm galactic cloud component.

131.120 Excitation of interstellar methylamine.
N. Kaifu, K. Takagi, T. Kojima.
Astrophys. Journ., (Letters), Vol. 198, L85 - L88 (1975).

The $2_{02}-1_{01}$ transition of CH_3NH_2 (88.67 GHz) has been detected in Sgr B2. The observed intensity of the $2_{02}-1_{01}$ line confirms an anomalously large population of the 2_{02} level as suggested by Fourikis et al. The collisional excitation of CH_3NH_2 is discussed, and it is concluded that an optically thin source with a particle density of $\sim 10^5$ cm^{-3} can explain well the observed anomalies in intensity.

131.121 Radio observations of interstellar CH. Part I.
O. E. H. Rydbeck, E. Kollberg, Å. Hjalmarson, A. Sume, J. Elldér, W. M. Irvine.
Res. Lab. Electronics, Onsala Space Obs., Chalmers Univ. Techn., Gothenburg, Sweden, Res. Rep. No. 120, 6 + 247 + A11 + B7 + C9 + R16 pp. (1975).

This paper reports in detail on observations performed at Onsala Space Observatory, of the 9 cm CH $^2\Pi_{1/2}$, J = 1/2 ground state hyperfine transitions (F = 1—0, upper satellite; F = 1—1, main line; F = 0—1, lower satellite), in the directions of, and in many cases in extended areas around the following regions: Cas A, W 1, W 3, Ori A, W 12, W 28 (SNR), W 31, W 43, W 44 (H II), W 49 A, W 51, M 17, Cyg X (including e.g. DR 5, DR 15, DR 21, DR 22), ON 1, ON 2, ON 3 (K 3—50), NGC 2068, NGC 7000, NGC 7538, L 134, L 134 N, L 1036, L 1082, L 1399, L 1500/3C 123, L 1630, Heiles' Clouds 1 and 2, and the Per OB 2 dust cloud (including NGC 1333 and IC 348). The observations were made with a travelling-wave maser preamplifier, in the Onsala 25.6 m telescope. The CH seems to exhibit weak maser characteristics almost everywhere towards H I and H II regions. Theoretical considerations of excitation temperature and optical depth calculations are presented. No good agreement is found between the observations and existing theories of interstellar CH formation, which seem to underestimate the CH abundance.

131.122 Radio observations of interstellar CH. Part II.
Å. Hjalmarson, A. Sume, J. Elldér, O. E. H. Rydbeck, E. Moore, R. Huguenin, A. Sandqvist, P. O. Lindblad, P. Lindroos.
Res. Lab. Electronics, Onsala Space Obs., Chalmers Univ. Techn., Gothenburg, Sweden, Res. Rep. No. 124, 1 + 36 + 62 pp. (1975).

The Onsala Space Observatory 25.6 m telescope, equipped with a travelling wave maser preamplifier, has been used to study emission in the three hyperfine transitions of the $^2\Pi_{1/2}$, J = 1/2, Λ doublet state of interstellar CH. The main line (F = 1—1) has been detected towards more than one hundred positions in optically dark nebulae. All three lines have been observed in many directions with relatively strong main line emission, including the Sharpless H II region S 159, the open cluster NGC 2264, as well as 3C 353. Observations on and off 3C 123, which lies behind Lynds' cloud L 1500, indicate that the three CH transitions are probably inverted in dark dust clouds. The main line excitation temperature is found to be about —9 K in L 1500. A statistical analysis of the data indicates that the antenna temperature increases linearly with the cloud opacity class, or the cloud density. An increase is also found for the CH column density, but it seems that the ratio between the CH and the total hydrogen column densities decreases for more opaque clouds. Available optical data agree well with the radio results.

131.123 On the absence of interstellar ring molecules.
J. H. Fertel, B. E. Turner.
Astrophys. Letters, Vol. 16, 61 - 64 (1975).

The authors have searched unsuccessfully for several benzene derivatives and one 4-membered ring molecule. It is argued that a relative under-abundance of such ring molecules compared to non-ring molecules is consistent with the formation of interstellar molecules by gas-phase ion-molecule reactions, and may be inconsistent with their formation on the surfaces of interstellar grains.

131.124 **Time variations and spectral structure of the methanol maser in Orion A.**
A. H. Barrett, P. Ho. R. N. Martin.
Astrophys. Journ., (*Letters*), Vol. 198, L119 - L122 (1975).

Temporal variations have been detected in the CH_3OH emission from Orion A over the period from 1973 December to 1974 December, in the $J = 4, 5, 6$, and 7 transitions. Temporal variations are one of the classical symptoms of astrophysical masers. Observations with a resolution of 0.12 km s^{-1} show that these transitions are composed of at least six features.

131.125 **Microwave detection of interstellar HDO.**
B. E. Turner, B. Zuckerman, N. Fourikis, M. Morris, P. Palmer.
Astrophys. Journ., (*Letters*), Vol. 198, L125 - L128 (1975).

The $1_{11}-1_{10}$ transition of singly deuterated water has been detected in emission at the wavelength of 3.72 mm toward the Kleinmann—Low nebula (KL) in Orion A. This is probably the first observation of nonmaser (i.e., low brightness temperature) emission from interstellar water molecules. Analysis indicates that most of the oxygen in KL is probably contained in water molecules if $[HDO]/[H_2O] \approx [D]/[H]$, but not if $[HDO]/[H_2O] \gg [D]/[H]$. The HDO line shape is different from that of other molecules previously observed toward KL.

131.126 **Studies on high-velocity clouds of neutral hydrogen.**
A. N. M. Hulsbosch.
Astron. Astrophys., Vol. 40, 1 - 25 (1975).

A general study has been made of the properties of high-velocity neutral hydrogen. New detailed observations are presented in a region with coordinates $100° \leq l \leq 130°$ and $0° \leq b \leq +25°$, and in a few complexes of high-velocity gas around $l, b = 100°, +50°$.

131.127 **On cosmic rays and final equilibrium states for the Parker instability.**   T. C. Mouschovias.
Astron. Astrophys., Vol. 40, 191 - 194 (1975).

If the Parker instability develops in a system consisting of a conducting interstellar gas and a magnetic field in a galactic gravitational field, the tension of the field lines will eventually stop their inflation. Even if cosmic rays are present, final equilibrium states of the system are still possible. Based on this formalism one may predict the synchrotron emission along spiral arms in which magneto-gravitational instability develops.

131.128 **Molecules in astrophysics.**   G. Winnewisser.
Computational Techniques in Quantum Chemistry and Molecular Physics, Proc. NATO Advanced Study Institute 1974, [D. Reidel Publishing Company, Dordrecht—Holland], p. 529 - 568 = Max-Planck-Inst. Radioastron., Bonn, Sonderdruck, Ser. A, No. 59 (1975).

131.129 **On the presence of phyllosilicate minerals in the interstellar grains.**
A. Zaikowski, R. F. Knacke, C. C. Porco.
Astrophys. Space Sci., Vol. 35, 97 - 115 (1975).

The composition of the interstellar silicate dust is investigated. Condensation or alteration of silicate grains at temperatures of a few hundred degrees, in the presence of H_2O, would result in hydrous or phyllosilicates, the silicate type most abundant in the type I carbonaceous chondrites. Laboratory spectra of several phyllosilicates give better agreement as does the spectrum of a carbonaceous chondrite. The authors propose that the silicates in the interstellar grains are predominantly phyllosilicates and suggest additional spectral tests for this hypothesis.

131.130 **A southern dark globule.**   I. J. Danziger, M. Dennefeld, R. J. Havlen, H. E. Schuster.
Astron. Astrophys., Vol. 40, 455 - 457 (1975).

The identification of a dark globule is reported at galactic position, $l^{II} = 315°$ and $b^{II} = -4°.9$. An associated star causing reflection from the globule provides a distance estimate of 450 parsecs. Comparisons are made with other known dark clouds and globules.

131.131 **Ultraviolet observations of associations in the Large Magellanic Cloud.**
J. Borgman, R. J. van Duinen, J. Koornneef.
Astron. Astrophys., Vol. 40, 461 - 465 (1975).

Ultraviolet observations in a region close to 30 Doradus, made with the Netherlands Astronomical Satellite , are presented. The results indicate that part of the observed reddening is internal to the Large Magellanic Cloud and that the extinction law differs from the average extinction law in the solar neighbourhood.

131.132 **Table of scattering for spherical particles by Mie theory.**   S. Isobe.
Ann. Tokyo Astron. Obs., Second Ser., Vol. 14, (No. 3), 141 - 226 (1975).

Tables of the four coefficients of scattering and of the amplitude function for axially symmetric scattering are given for ice, graphite, silicate, and graphite core-ice mantle particles with radius less than one micron.

131.133 **Application of Mie theory to the problems of interstellar three micron absorption band.**
S. Isobe.
Ann. Tokyo Astron. Obs., Second Ser., Vol. 14, (No. 4), 238 - 246 (1975).

The causes of non-existence or weakness of 3 μ absorption band by ice grains or ice mantle grains are examined for 7 stars which have been observed. The possible causes are: 1) The strength of 3 μ absorption band relative to visual absorption depends on grain radius. 2) Since all of these stars are supergiants, ice mantles of circumstellar grains were evaporated by the strong radiation of the stars.

131.134 **Application of Mie theory to the problems of albedo and phase parameter of interstellar grains.**
S. Isobe.
Ann. Tokyo Astron. Obs., Second Ser., Vol. 14, (No. 4), 247 - 257 (1975).

It is shown that mixtures of graphite core-ice mantle grains with large and small mean sizes and graphite grains provide good fits not only to the extinction curve but also to the phase parameter and the wavelength dependence of albedo.

131.135 **Application of Mie theory to the problems of time dependence of the λ2200 Å bump.**   S. Isobe.
Ann. Tokyo Astron. Obs., Second Ser., Vol. 14, (No. 4), 258 - 269 (1975).

Friedemann (1974) has shown that the color excess ratio $E(\lambda 2180-\lambda 5500)/E(B-V) = E_1$ depends on the age of associations. The author calculates the time dependence of the value $E(\lambda 2200-\lambda 5500)/E(B-V)$ considering the radiation pressure from the central stars and the frictional force due to the gas surrounding the grains.

131.136 Ice mantle growth in interstellar clouds and inter-cloud regions. M. Crézé, S. Isobe.
Ann. Tokyo Astron. Obs., Second Ser., Vol. 14, (No. 4), 270 - 280 (1975).

The ratios R of total to selective extinction in the arm and inter-arm regions are derived to be 4 and 1.6 from the interstellar Ca line absorptions, respectively. Moreover, it is shown that growth of ice mantles is possible even in the normal interstellar radiation field and the existence of such grains in the intercloud and inter-arm regions is proposed. Mixtures of graphite grains and graphite core-ice mantle grains provide good fits to the observed values of R in the arm and inter-arm regions.

131.137 Catalytic reactions in the solar nebula: implications for interstellar molecules and organic compounds in meteorites. E. Anders, R. Hayatsu, M. H. Studier.
Origins of Life, Vol. 5, 57 - 67 (1974).

131.138 Dark clouds, star formation and spiral structure. I. Appenzeller.
Conference on optical observing programs on galactic structure and dynamics, (see 012.013), p. 131 - 138 (1975). In German.

131.139 Population of rotationally excited levels of interstellar H_2. P. Joshi, S. P. Tarafdar.
Bull. Astron. Soc. India, Vol. 2, 39 (1974). − Abstract.

131.140 Possible interaction of interstellar particles with the solar and terrestrial environment.
J. M. Greenberg.
Evolutionary and physical properties of meteoroids. IAU colloquium No. 13, p. 375 - 377 = Dudley Obs., *Albany, New York,* Repr. No. C86 (1973).

The possibility for detection of interstellar particles in the earth's environment is considered on the basis of the passage of the solar system through the interstellar medium. Among the forces which inhibit interstellar particle penetration, the deflection by the solar magnetic field and the repulsive force due to the radiation from the sun are by far the most important.

131.141 Aperture synthesis observations of OH absorption in the galactic center. J. H. Bieging.
Thesis, California Inst. Techn., Pasadena (USA). 187 pp. University Microfilms Order No. 74-17,938 (1974).

131.142 Kinetics of interstellar grain orientation. E. G. Derringh.
Thesis, Rensselaer Polytechnic Inst., Troy, N.Y. (USA). 153 pp. University Microfilms Order No. 74-12,782 (1973).

131.143 Interstellar molecules.
G. Winnewisser, P. G. Mezger, H. D. Breuer.
Fortschr. Chem. Forsch., No. 44, p. 1 - 81 (1974). − Review paper.

131.144 Circumstellar grains and the intrinsic polarization of starlight. W. J. Forrest.
Thesis, California Univ., San Diego (USA). 357 pp. University Microfilms Order No. 74-23,966 (1974).

131.145 Heat balance state of the interstellar gas. Dependences on the amount of cooling- and heating-agencies. T. Sato.
Progr. Theor. Phys. Japan, Vol. 52, 1174 - 1187 (1974).

The properties of the interstellar gas in which the heating balances the cooling are investigated for the temperature range $30\,K < T < 10^8\,K$ taking the abundance of heavy elements and the flux of low energy cosmic rays as parameters. The tempera-ture and the pressure of the gas in a heat balance state are investigated as functions of the density for various values of these parameters.

131.146 Molecular millimeter wave astronomy.
L. E. Snyder.
IEEE Trans. Microwave Theory Techn., Vol. MTT-22, 1299 - 1300 (1974).

Emphasis is placed on the newer areas of research in astronomy and astrophysics which have been facilitated by radio observations of galactic molecules in the millimeter wavelength range of the spectrum.

On OH formation in collision between H and O^-
See Abstr. 022.015.

Microwave absorption spectrum of the CO^+ ion.
See Abstr. 022.036.

The microwave frequencies, line parameters, and spectral constants for $^{14}NH_3$. See Abstr. 022.045.

Ground state centrifugal distortion constants of vinyl isocyanide, CH_2-CH-NC, from the microwave and milli-meter wave rotational spectra. See Abstr. 022.064.

Kosmische Maser. See Abstr. 061.006.

The mystery of the cosmic boron abundance.
See Abstr. 061.024.

Strong scintillations in astrophysics. I. The Markov approximation, its validity and application to angular broadening. See Abstr. 063.011.

Strahlungstransport in kosmischen Masern.
See Abstr. 063.021.

Polarization properties of silicate-like grains in circumstellar envelopes of late-type stars due to temperature variations. See Abstr. 064.042.

Spectroscopy of circumstellar shells.
See Abstr. 064.076.

Star formation in clouds of molecular hydrogen.
See Abstr. 065.014.

Formation of population II stars.
See Abstr. 065.016.

How to make metal-poor stars, redden OB associations and grow mantles on grains. See Abstr. 065.041.

Gravitational contraction of protostars. III. Role of heavy elements in the evolution of protostars.
See Abstr. 065.062.

Charged black holes in the interstellar medium.
See Abstr. 066.057.

On the possibility of experimental testing of the interaction model solar wind − interstellar medium.
See Abstr. 074.081.

Electron microscopy of irradiation effects in space.
See Abstr. 094.409.

Comets and interstellar masers.
See Abstr. 102.008.

Do comets play a role in galactic chemistry and γ-ray bursts?   See Abstr. 102.024.

Local interstellar medium.
See Abstr. 106.010.

Grains accretion processes in a protoplanetary nebula: conducting and insulating materials.
See Abstr. 107.003.

On the ratio of total-to-selective absorption.
See Abstr. 113.051.

The ratio of color excesses in UBV photometry.
See Abstr. 113.058.

Polarization characteristics of Herbig Ae and Be stars.   See Abstr. 114.002.

Linear polarization of $H\alpha$ in Be stars.
See Abstr. 114.023.

Symbiotic stars and dust.
See Abstr. 114.032.

Linear polarization of $H\alpha$ in Be stars.
See Abstr. 114.036.

Isolated Strömgren spheres as a source of galactic $H\alpha$ emission.   See Abstr. 114.041.

The interstellar lines of the Feige stars.
See Abstr. 114.064.

Interstellar carbon I lines in ζ Ophiuchi.
See Abstr. 114.304.

A search for H^- in the shell surrounding χ Ophiuchi.
See Abstr. 114.349.

On the distances and velocities of M supergiants associated with OH and H_2O emission sources.
See Abstr. 115.012.

The interaction between the blast wave of a supernova remnant and interstellar clouds.   See Abstr. 125.006.

Fluctuations of interstellar medium density — the cause of difference between optical and X-ray velocities of supernova remnants.   See Abstr. 125.010.

Search for continuous fluorescence in reflection nebulae.   See Abstr. 132.002.

Emission-line shifts and broadening for Herbig-Haro objects.   See Abstr. 132.023.

Large-scale ionization fronts and the nature and distribution of light scattering particles in the Orion nebula.
See Abstr. 132.024.

Observations of carbon monoxide emission in NGC 5367.   See Abstr. 132.025.

Application of Mie theory to the problems of the scattered continuum light in diffuse nebulae.
See Abstr. 132.037.

Decimeter-wavelength studies of hydrogen and carbon recombination lines toward galactic nebulae.
See Abstr. 132.040.

The underabundance of gaseous iron in the planetary nebula NGC 7027.   See Abstr. 133.001.

Radio observations at 5 GHz of southern planetary nebulae.   See Abstr. 133.012.

Thermal emission spectra of silicates from planetary nebulae.   See Abstr. 133.020.

Interstellar scintillation of extragalactic radio sources.   See Abstr. 141.035.

Pulsar scintillation on inhomogeneities of the interstellar plasma.   See Abstr. 141.315.

The velocity of pulsars and interstellar irregularities in the scintillation pattern.   See Abstr. 141.321.

Measurements of neutral-hydrogen absorption in the spectra of five pulsars and parameters of the Gum nebula.
See Abstr. 141.356.

The velocity of pulsars and inhomogeneities of the interstellar medium in the scintillation pattern.
See Abstr. 141.360.

Spin-down of pulsars.   See Abstr. 141.362.

Silicate absorption at $18\,\mu m$ in two peculiar infrared sources.   See Abstr. 141.601.

M78: an active region of star formation in the dark cloud Lynds 1630.   See Abstr. 141.609.

Infrared emission from OH 284.2−0.8.
See Abstr. 141.610.

H_2 and HD infrared lines expected from dense interstellar objects.   See Abstr. 141.613.

Energy dependence of the Si/Fe ratio in the galactic cosmic rays.   See Abstr. 143.023.

A measurement of cosmic-ray positron and negatron spectra between 50 and 800 MV.   See Abstr. 143.041.

Nuclear gamma ray production by cosmic rays.
See Abstr. 143.058.

Light element production by cosmic rays.
See Abstr. 143.059.

The Galaxy and interstellar medium.
See Abstr. 143.063.

Photometric study of the Chamaeleon T-association.
See Abstr. 152.002.

Further study of the stellar cluster embedded in the Ophiuchus dark cloud complex.   See Abstr. 153.008.

The distribution of interstellar matter in NGC 654.
See Abstr. 153.013.

The distribution of interstellar matter in IC 5146.
See Abstr. 153.014.

Dark matter in open clusters.
See Abstr. 153.032.

The observability of ionized interstellar gas in

globular clusters.   See Abstr. 154.005.

Sources of excitation of the interstellar gas and galactic structure.   See Abstr. 155.005.

Study of galactic gas and dust using observations of elliptical galaxies.   See Abstr. 155.016.

Is the galactic disk well mixed?
See Abstr. 155.018.

Cloud structure in the galactic plane: a cosmic bubble bath?   See Abstr. 155.022.

Investigation of low-latitude hydrogen emission in terms of a two-component interstellar gas model.
See Abstr. 155.039.

Fine-scale structure of carbon monoxide emission observed in the direction of the galactic nucleus.
See Abstr. 155.041.

Neutral hydrogen in the galactic centre region – I. The observations.   See Abstr. 157.008.

Finding list of bright galaxies behind the SMC.
See Abstr. 158.005.

The galactic extinction from the apparent distribution of galaxies.   See Abstr. 158.050.

An atlas of dust and H II regions in galaxies.
See Abstr. 158.056.

The gas-to-dust ratio in M 31.
See Abstr. 158.126.

Obscuration in and around the Small Magellanic Cloud.   See Abstr. 159.001.

Questions concerning interstellar matter in the Magellanic Clouds.   See Abstr. 159.016.

H I, H II Regions

131.501   21 cm observations of hot neutral intercloud gas.
R. D. Davies, E. R. Cummings.
Monthly Notices Roy. Astron. Soc., Vol. 170, 95 - 113 (1975).
Observations of the 21 cm neutral hydrogen absorption spectra of the strong radio sources Vir A, Cyg A and Cas A have been taken. The kinetic temperature range is T_k = 770 to $\lesssim$ 10000 K. There is evidence for an increase of T_k with height above the galactic plane. Some implications of these first direct observations of the temperature of the neutral intercloud gas are discussed.

131.502   The distribution of the H134α recombination line in W33.
F. F. Gardner, T. L. Wilson, P. Thomasson.
Astrophys. Letters, Vol. 16, 29 - 33 (1975).
The H II region W33 has been mapped in the H134α line (2702.8 MHz) with an angular resolution of 4.8 arc min. The line emission has velocity peaks near 32 and 58 km/sec, and these occur over an appreciable part of the nebula. The observations are interpreted in terms of a source expanding at a velocity of about 15 km/sec. The clouds containing OH and H_2CO are believed to surround the nebula and to be pushed ahead of the expanding ionized shell.

131.503   The chemical composition of selected H II regions in the Magellanic Clouds.   R. J. Dufour.
Astrophys. Journ., Vol. 195, 315 - 332 (1975).
The results of an investigation of the optical spectra of three SMC and 11 LMC H II regions are presented. From the emission-line strengths corrected for interstellar reddening, the relative abundances of H, He, N, O, Ne, S, Cl, and Ar are calculated. The effects of temperature and density fluctua-

tions on the abundance results are also discussed.

131.504   H I absorption of nine pulsars.   D. A. Graham, U. Mebold, K. H. Hesse, D. L. Hills, R. Wielebinski.
Astron. Astrophys., Vol. 37, 405 - 410 (1974).
Observations have been made of the 21 cm H I line absorption in the direction of nine pulsars, six of them being new measurements. Investigations of all the parameters which affect the distance determination have been made.

131.505   A decimeter-wavelength recombination line study of W 51 A and W 49 A.
V. Pankonin, A. Parrish, Y. Terzian.
Astron. Astrophys., Vol. 37, 411 - 416 (1974).
The authors investigate the results of new decimeter-wavelength recombination line observations toward W 51 A and W 49 A. They find the H 221α and H 247, 248α lines from both sources are probably emitted from a low electron density medium ($N_e \lesssim$ 50 cm⁻³). These same lines toward W 49 A also may originate in a medium with an electron temperature less than 4000 K.

131.506   Aperture synthesis observations of galactic H II-regions.
J. E. Wink, W. J. Altenhoff, W. J. Webster, Jr.
Astron. Astrophys., Vol. 38, 109 - 128 (1975).
Aperture synthesis observations at 2.695 GHz and 8.085 GHz of the H II regions NRAO 591, NGC 6857, NGC 7538, M 8, W3 and W49A made with the NRAO interferometer are presented. A set of gaussian functions is derived to describe the radiation distribution of each H II region at each frequency. Fine structure is found in all regions. With the exception of the extended source in W49A, all previously known sources with high excitation parameters are resolved into smaller sources. The electron densities of individual sources range from 10^2 cm⁻³ up to 10^5 cm⁻³. In NGC 6857, NGC 7538 and

W49A, continuum sources smaller than 4 arc sec and with electron densities exceeding 10^4 cm^{-3} are found close to the known class II OH emission sources. In M 8, the authors find a ring which seems to be split at the position of the O-star Herschel 36.

131.507 Dust absorption in the Lyman continuum.
E. Krügel.
Astron. Astrophys., Vol. 38, 129 - 132 (1975).

The properties of H II regions containing both dust and He are calculated for a variety of density distributions, exciting objects, dust absorption coefficients for the He- and H-ionizing photons, and gas temperatures. The relation between the emitted Lyman-continuum flux N_c and the ratio R of the volume of the He$^+$-region to that of the H$^+$-region is discussed. For $N_c > 10^{52}$ photon/s, the He-recombination lines should be much weaker than the H-lines. The computations reveal a sharp relation between R and the infrared excess ϵ_{IR}, defined as the factor by which the total IR-luminosity exceeds the energy contributed by the Lα-photons alone. The effects of higher gas temperatures (due to an underabundance of cooling ions) are also briefly discussed.

131.508 The neutral hydrogen in the molecular-dust cloud in the direction of the nebula NGC 6618.
I. V. Gosachinskij, I. A. Rakhimov.
Astron. Zhurn. Akad. Nauk SSSR, Vol. 52, 34 - 38 (1975). In Russian. English translation in Soviet Astron., Vol. 19, No. 1.

It is found that the optical depth of the absorption line in the dust cloud covering the west part of NGC 6618 is reduced by a factor 2. The mean radial velocity of this phenomenon is close to that of the formaldehyde line detected in this cloud. Taking into account the probable decreasing of the kinetic temperature, this corresponds to the reducing of the H I density by a factor about 10 in the dust cloud relative to the undisturbed regions.

131.509 Internal motions in H II regions. II. The radial velocity field of IC 443. P. Pişmiş, M. Rosado.
Rev. Mexicana Astron. Astrofis., Vol. 1, 121 - 127 (1974).
Paper presented at the Second European Regional Meeting of the IAU held in Trieste, Italy, September 1974.

The radial velocities determined from the Hα emission line on two Fabry-Pérot interferograms at nearly 170 points in the supernova remnant IC 443 have been presented. The mean velocity of the SNR is -3 km s^{-1}, while the velocity of expansion is around 60 km s^{-1}. The discussion of some specific features indicates that there are appreciable relative mass motions within the assembly of filaments over the supernova shell. In the northern edge the recession from the observer of two filaments is detected. Along another of the features there is a marked bifurcation, the difference in radial velocity of the two branches reaching up to 76 km s^{-1}. It is suggested that a magnetic field along the supernova shell may account for the observed relative motions.

131.510 The ionization structure of H II regions of different helium content.
L. F. Rodríguez, S. Torres-Peimbert, M. Peimbert.
Rev. Mexicana Astron. Astrofis., Vol. 1, 161 - 173 (1974).

The authors have computed the ionization structure corresponding to H II regions and ionizing stars with $y = N$ (He)/N(H) in the 0.07 to 0.30 range and T_* in the 30000 to 45000°K range. They have adopted a pregalactic value of $y = 0.07$. $Z = 0$ and an evolutionary dependence of $\Delta Y = 3\Delta Z$. From stellar structure considerations it is found that an increase in y, for a star of a given mass, produces a significant increase in the stellar luminosity and effective temperatures; therefore for a higher value of y, and for a constant mass function, on the average the H II regions would have a higher degree of ionization of helium. These effects are very im-

portant in the study of emission line gradients in external galaxies; and in particular, in the study of the lack of ionized helium in the nucleus of our galaxy. Several explanations for the lack of ionized helium in the center of our galaxy are reviewed.

131.511 A very high velocity H I cloud in the direction of M31. R. D. Davies.
Monthly Notices Roy. Astron. Soc., Vol. 170, 45P - 49P (1975).

A neutral hydrogen cloud with a velocity of -447 km s^{-1} relative to the local standard of rest has been discovered near the direction of the Andromeda nebula (M31). Two alternative theories of the origin of this cloud are discussed.

131.512 Study of a neutral hydrogen feature previously observed by Cugnon.
I. F. Mirabel, W. G. L. Pöppel, E. R. Vieira.
Astrophys. Space Sci., Vol. 33, 23 - 41 (1975).

An anomalous velocity cloud near $l=349°$, $b=+3°$, was investigated by Cugnon (1968). The authors made a new set of observations in order to obtain a more complete picture of the feature, including the region originally out of Cugnon's limit of observation. A comparison with optical and radio observations was made and several possibilities of interpretation as to the nature of the object were analyzed.

131.513 An almost complete survey of 21 cm line radiation for $|b| \geqslant 10°$. IV. The H I column density as a function of position of the sky. C. Heiles.
Astron. Astrophys., Suppl. Ser., Vol. 20, 37 - 55 (1975).

A contour map of H I column density for velocities between -93 to $+75$ km s^{-1} is presented.

131.514 Observations of radio recombination lines toward supernova remnants at 428 MHz.
V. Pankonin.
Astron. Astrophys., Vol. 38, 445 - 450 (1975).

Recombination lines at 428 MHz have been observed toward the supernova remnants 3C391 and W49B and the galactic plane at $l^{II} = 44.2$, $b^{II} = 0.0$. These results are related to existing line data in order to resolve the problem of the origin of this emission. The measurements indicate that the recombination lines arise in sources with physical properties similar to those of low brightness H II regions.

131.515 Are the electron temperatures of H II regions a function of galactic radius?
E. Churchwell, C. M. Walmsley.
Astron. Astrophys., Vol. 38, 451 - 454 (1975).

The electron temperatures of H II regions, as derived from radio recombination line surveys of the galactic plane, are analyzed as a function of distance from the galactic center. A weak correlation is found in the sense that H II region electron temperature increases with increasing galactic radius. This might be due to one of a variety of selection effects but a more likely possibility appears to be that an abundance gradient of heavy elements exists in our Galaxy similar to that found by optical observers in some nearby spirals.

131.516 Temperature and emission-line structure at the edges of H II regions. D. C. V. Mallik.
Astrophys. Journ., Vol. 197, 355 - 363 (1975).

Models of ionization fronts located at the edges of expanding H II regions are presented. The emission in [O II] and [N II] lines is greatly enhanced because of the high temperature at the front. The emission in these and other important lines is calculated and compared with Hβ. Effects of different velocities of flow, of different exciting stars, and of different gas densities on the structure of the fronts are also investigated. Observations of bright rims of H II regions where enhanced emission of [N II] lines has been reported are discussed on the basis of these models.

131.517 Temperature conditions in ionized hydrogen regions.
T. B. Pyatunina.
Astrofiz. Issled., Izv. Spets. Astrofiz. Obs., Vol. 7, p. 101 - 120 (1975). In Russian.

The distributions of the electron temperature and density for 44 H II regions are obtained. It is shown that the temperature gradient is, as a rule, negative in the central dense part of the H II zone, and is probably positive in the tenuous envelope. All the regions investigated are supposed to be divided into two groups: high-excitation regions with central electron temperature $T_{ec} > 7000°$K and temperature of the exciting stars $T_{eff} \geqslant 38000°$K, and low-excitation regions with $T_{ec} < 7000°$K and $T_{eff} \leqslant 35000°$K. It is shown that non-LTE effects in the emission of recombination radio lines are negligible.

131.518 Infrared forbidden lines in H II regions and planetary nebulae. J. P. Simpson.
Astron. Astrophys., Vol. 39, 43 - 60 (1975).

Emissivity coefficients as functions of electron density and temperature are calculated for the forbidden lines between 2μ and 300μ. The elements and ionization stages considered are those most abundant in H II regions and planetary nebulae. These emissivity coefficients are used to predict the infrared line spectrum of the Orion nebula through the use of the model described by Simpson (1973). The effect of self-absorption on the infrared line fluxes is discussed, and a method is given for estimating when self-absorption may be important. Further predictions are made for infrared line strengths in planetary nebulae. The abundance of sulfur and the sulfur/oxygen ratio is discussed for H II regions and planetary nebulae.

131.519 A review of recent observations of dust in H II regions. L. F. Smith.
Astrophys. Space Sci., Vol. 34, 49 - 55 (1975). – Paper presented at the Symposium on Solid State Astrophysics, Cardiff, Wales, 9–12 July, 1974.

The upper limit for the absorption cross section σ_H^{ext}, of dust in H II regions in the wavelength range 912–504 Å derived by Mezger et al. (1974), is compatible with that expected for large dust grains, and a gas-to-dust ratio equal to that in the general interstellar medium. The albedo of the small grains must be high for $\lambda > 504$ Å. This restriction is lifted if the visual extinction cross section of the grains in H II regions is less than that for grains in the general interstellar medium. New observations of the Orion nebula indicate that the visual extinction cross section is within a factor 2 of the value in the general interstellar medium.

131.520 Grain charging in H II regions.
A. F. M. Moorwood, B. Feuerbacher.
Astrophys. Space Sci., Vol. 34, 137 - 147 (1975). – Paper presented at the Symposium on Solid State Astrophysics, Cardiff, Wales, 9–12 July, 1974.

Equilibrium grain potentials have been calculated as a function of radial position and for a wide range of electron densities in H II regions ionized by stars of spectral type O5 and B0. Results are presented for both graphite (low yield photoemitter) and aluminium oxide (high yield photoemitter) for which laboratory photoemission data has been obtained. The results for aluminium oxide should approximate the behaviour expected of dielectric grains – e.g., silicates which may be present in H II regions. The importance of charging is discussed in relation to the growth and motion of grains in these regions.

131.521 Thermal bremsstrahlung radiospectra for inhomogeneous objects, with an application to MWC 349.

F. M. Olnon.
Astron. Astrophys., Vol. 39, 217 - 223 (1975).

Radio spectra of very small H II regions are often interpreted in terms of homogeneous density distributions. In this paper the author calculates radio spectra for objects where the density gradually decreases outward. He demonstrates that for these models the optical thick spectral index can be considerably lower than +2. Moreover, he shows that the actual value gives information on the steepness of the density gradient. Finally, the radio data on MWC 349 (spectral index +0.69) are interpreted as being due to a density structure of the form $n_e \propto r^{-2.1}$.

131.522 Young stellar clusters in diffuse nebulae.
Yu. I. Glushkov, E. K. Denisyuk, Z. V. Karyagina.
Astron. Astrophys., Vol. 39, 481 - 485 (1975).

The authors present some results of spectrophotometric studies of compact H II regions. From optical observations they find compact components in 13 diffuse nebulae. The authors believe that many of the investigated nebulae actually consist of several components each containing a stellar cluster, in different evolutionary stages.

131.523 Westerbork synthesis observations of the H 109α recombination line in DR21 and W3.
W. T. Sullivan III, K. J. Wellington, W. M. Goss, H. J. Matthews.
Bull. American Astron. Soc., Vol. 7, 258 - 259 (1975). Abstr. AAS.

131.524 Dust in the H II region NGC 2024. C. L. Sarazin.
Bull. American Astron. Soc., Vol. 7, 259 - 260 (1975). – Abstr. AAS.

131.525 New compact H II regions in dark clouds.
W. Gilmore, R. L. Brown, B. Zuckerman.
Bull. American Astron. Soc., Vol. 7, 260 (1975). – Abstr. AAS.

131.526 The spectrum of portions of H II regions shadowed from the central star. J. S. Mathis.
Bull. American Astron. Soc., Vol. 7, 260 (1975). – Abstr. AAS.

131.527 H I observations in cloud 2.
Y. K. Minn, T. L. Wilson.
Bull. American Astron. Soc., Vol. 7, 339 (1975). – Abstr. AAS.

131.528 G0.55–0.85, an H II region-molecular cloud complex. F. F. Gardner, J. B. Whiteoak.
Monthly Notices Roy. Astron. Soc., Vol. 171, 29P - 31P (1975).

OH, H_2CO and $H110\alpha$ lines have been observed in the direction of G0.55–0.85, an H II region identified with RCW 142. The results are interpreted in terms of a class I OH source embedded in a dense molecular cloud, probably no more distant than the Sagittarius arm.

131.529 Maser radiometer observations of water vapor emission from H II regions and IR stars.
B. T. Cato, B. O. Rönnäng, P. T. Lewin, O. E. H. Rydbeck, K. S. Yngvesson, A. G. Cardiasmenos, J. F. Shanley.
Res. Lab. Electronics, Onsala Space Obs., Chalmers Univ. Techn., Gothenburg, Sweden, Res. Rep. No. 123, 78 pp. (1975).

The spatial structure of water vapor microwave line emission has been investigated with moderate angular resolution in the H II regions W 28, W 43, ON 2 (G 75.8 + 0.3), W 75, and NGC 7538. Several new H_2O sources have been found. Three of them are preliminarily identified with infrared (IR) sources. One of these sources, IRC-20411, has been investigated at optical wavelengths. In NGC 7538 new high velocity features have been discovered. Two new weak water vapor masers, G 30.1–0.7 and G 32.8–0.3, have been detected in a search among eight Class II OH/IR sources. H_2O emission coinciding with the low velocity OH features of VY Canis Majoris has also been detected. A search for LTE water vapor line emission in molecular clouds associated with H II regions is also reported. No line was detected with the utilized sensitivity. The physical implications of this are discussed.

131.530 A study of H I absorption using Karhunen-Loève series. D. Pelat.
Astron. Astrophys., Vol. 40, 285 - 290 (1975).

An attempt is made to use Karhunen-Loève series for the interpretation of an absorption spectrum. The method allows to separate the absorption lines and to identify the lines being blended in the original.

131.531 The two basic components of the interstellar neutral hydrogen and its relation with the galactic structure. R. J. Quiroga.
Astrophys. Space Sci., Vol. 35, 67 - 80 (1975).

The two basic components of the neutral hydrogen, cool dense clouds merged in a hotter tenuous medium, are studied using 21 cm absorption data of the Parkes Survey. The mean parameters obtained for the typical clouds next to the galactic plane are τ_p = 1.7, velocity half-width = 3.3 km s^{-1}. Their temperatures are $T_{sc} \geqslant 40$ K with a mean T_{sc} = 63 ± 12 K and the obtained hot gas density is n_{HH} = (0.15 ± 0.05) atom cm^{-3}. Theoretical analysis following Giovanelli and Brown (1973) reveals that the pressure equilibrium condition ($n_{HH} + 2n_e$) $T_{SH} \approx n_{HC} T_{sc}$ is compatible with the quoted values if it is assumed that the cosmic abundances in the interstellar medium are below the adopted normal solar abundances. The same pressure condition leads to a mean cool cloud density of $n_{HC} \sim 30$ atom cm^{-3} and a hot gas temperature of $T_{SH} \sim$ 10500 K. Comparison with 21 cm emission data shows that the galactic cloud layer is only about a quarter as thick as the hot gas layer.

131.532 A survey of the unusual motions in M17 with a Fabry-Perot monochromator.
K. H. Elliott, J. Meaburn.
Astrophys. Space Sci., Vol. 35, 81 - 96 (1975).

The results of the most extensive survey ever undertaken of the profiles of the [O III] line over M17 are presented. Some very unusual velocity features have been revealed. Several regions were found to be emitting lines split by motions of 20 km s^{-1} whereas others produce lines with three or more separate velocity components which stretch up to -70 km s^{-1} from the mean radial velocity of the nebula. Some speculative explanations for these motions are offered.

131.533 The structure of the H I, H II, molecular and dust complex of M17. J. Meaburn.
Astrophys. Space Sci., Vol. 35, L5 - L8 (1975).

Some modifications to the model of M17 proposed by Elliott and Meaburn (1975) are suggested. – See also 13.131.532.

131.534 An improved 1665 MHz position for W3 (OH).
W. M. Goss, I. A. Lockhart, E. B. Fomalont.
Astron. Astrophys., Vol. 40, 439 - 440 = Publ. Div. Radiophys., C.S.I.R.O., Sydney, RPP 1849 (1975).

The OH position of W3 (OH) at 1665 MHz has been remeasured with the Owens Valley interferometer. To within the accuracy of several arc seconds the centroid of the OH source coincides with the compact H II region. In this respect W3 (OH) belongs to the class of type I OH masers which coincide with compact H II regions (Habing et al., 1974).

131.535 Optical study of the Carina nebula.
L. Deharveng, M. Maucherat.

Astron. Astrophys., Vol. 41, 27 - 36 (1975).

The authors present monochromatic photographs in Hβ, [O III] λ 5007, Hα, [N II] λ 6584, [S II] $\lambda\lambda$ 6717–6731, of two peculiar regions of the Carina nebula: a bright circular ring around the radio peak Car II, and bright rims near Car I. They show small scale structure and steep variations in the ionization conditions. A kinematical study of the nebula is obtained with interference techniques and high angular and spectral resolution in Hα and [N II]. This confirms the existence of a general expansion of the nebula, with non-spherical symmetry. The expansion velocity of the ionized gas is $\geqslant 25$ km s^{-1} although that of the neutral surrounding gas is much smaller.

131.536 Fine structure in M8 at 327 MHz.
Gopal-Krishna, G. Swarup.
Bull. Astron. Soc. India, Vol. 2, 37 (1974). – Abstract.

131.537 Temperature at the edges of H II regions.
D, C. V. Mallik.
Bull. Astron. Soc. India, Vol. 2, 39 (1974). – Abstract.

131.538 Improved calculation of the ionization equilibrium of Ca and Na in H I region.
A. Qaiyum, S. M. R. Ansari.
Bull. Astron. Soc. India, Vol. 2, 41 (1974). – Abstract.

131.539 A study of the nebulae S 206 and S 209.
C. M. Walmsley, E. Churchwell, I. Kazès, A. M. Le Squéren.
Astron. Astrophys., Vol. 41, 121 - 132 (1975).

Radio continuum maps, radio recombination, H_2CO and H I absorption line measurements are presented for the optically visible nebulae S 206 (NGC 1491) and S 209 as well as for the neighbouring radio source NRAO 150. Distances to the nebulae and physical parameters are obtained. Negative results from the formaldehyde observations suggest that there is no molecular cloud associated with these nebulae.

131.540 Molecular observations of the Sharpless H II region S 228. R. Lucas, P. J. Encrenaz.
Astron. Astrophys., Vol. 41, 233 - 235 (1975).

Carbon monoxyde and formaldehyde observations of the Sharpless H II region S 228 are presented. A cold ($T \sim 10$ K) molecular cloud of density $n_{H_2} \sim 350$ cm^{-3} and mass $M \sim 2000\ M_\odot$ has been mapped next to this H II region, which itself coincides with a higher density and hotter region. A CO profile in the direction of S 288 is also presented.

131.541 Chemical composition of selected H II regions in the Magellanic Clouds. R. J. Dufour.
Thesis, Wisconsin Univ., Madison (USA). 150 pp. University Microfilms Order No. 74-16,203 (1974).

131.542 Carbon monoxide studies of H II regions.
H. S. Liszt.
Thesis, Princeton Univ., Princeton, N. J. (USA). 84 pp. University Microfilms Order No. 74-25,961 (1974).

On the ionization of hydrogen and helium by hydromagnetic shock waves. See Abstr. 062.010.

A catalogue of galactic O stars and the ionization of the low density interstellar medium by runaway stars. See Abstr. 113.008.

The Copernicus observations: interstellar or circumstellar material? See Abstr. 114.065.

Spectroscopic observations of stars in H II regions. See Abstr. 114.086.

On the ionization zones around flare stars.
See Abstr. 122.087.

W44: a buoyant supernova remnant.
See Abstr. 125.023.

Interstellar absorption lines in the spectrum of Zeta Ophiuchi.   See Abstr. 131.045.

Effects of charged dust grains.
See Abstr. 131.078.

On the ionization of the intercloud medium by runaway O–B stars.   See Abstr. 131.116.

Radio observations of interstellar CH. Part I.
See Abstr. 131.121.

Radio observations of interstellar CH. Part II.
See Abstr. 131.122.

Recombination lines near 8.9 GHz of strong sources in the southern Milky Way.   See Abstr. 132.032.

Photometry of a bright filament in the Orion nebula.
See Abstr. 132.035.

The Carina nebula at 3.4 and 6 cm.
See Abstr. 132.039.

Chemical abundances of planetary nebulae.
See Abstr. 133.004.

High resolution observations of planetary nebulae at 5 GHz.   See Abstr. 133.008.

Far-infrared observations of W51 with high spatial resolution.   See Abstr. 141.605.

Detection of H_2O emission from the far-infrared source AFCRL (UOA) 19.   See Abstr. 141.607.

Infrared observations of Sharpless 2–106, a possible location for star formation.   See Abstr. 141.619.

A search for ionized hydrogen in globular clusters.
See Abstr. 154.007.

The alignment of interstellar dust clouds and the differential z-field of the Galaxy.   See Abstr. 155.003.

Non-linear density-wave theory and the latitude distribution of HI gas.   See Abstr. 155.033.

The distribution of optical H II regions in our Galaxy.   See Abstr. 155.043.

Observations of H 110α radio recombination-line emission associated with distributed ionized hydrogen. See Abstr. 157.002.

An atlas of dust and H II regions in galaxies.
See Abstr. 158.056.

The relative abundance of nitrogen and sulphur in three spiral galaxies: M 33, M 101, M 51.
See Abstr. 158.073.

A search for water vapour emission from extra-galactic nebulae.   See Abstr. 158.074.

An H I study of Scd galaxies.   See Abstr. 158.104.

The giant spiral galaxy M 101: radio observations of H II regions in external galaxies. II. Radio continuum emission from the H II regions and the nonthermal disc of M 101.   See Abstr. 158.119.

Continuum observations of M 81 at 4.8 GHz.
See Abstr. 158.128.

Distribution and kinematics of neutral hydrogen in the spiral galaxy M81.   See Abstr. 158.138.

A neutral hydrogen study of the integral-sign galaxy MCG 12-7-28.   See Abstr. 158.308.

Monochromatic photographs of giant H II region in the Magellanic Clouds.   See Abstr. 159.012.

Errata

131.901   Erratum: 'Helium abundance in galactic H II regions' [Astron. Astrophys., Vol. 32, 283 - 308 (1974)].
E. Churchwell, P. G. Mezger, W. Huchtmeier.
Astron. Astrophys., Vol. 38, 479 (1975).

132 Emission Nebulae, Reflection Nebulae

132.001 Observations of carbon recombination-line emission in the reflection nebula M78. R. L. Brown, G. R. Knapp, T. B. H. Kuiper, E. N. Rodriguez Kuiper. Astrophys. Journ., (Letters), Vol. 195, L23 - L25 (1975).

The C 166α and 142α recombination lines have been detected in the reflection nebula M78. The line emission arises in a cold ($T \approx 20°$ K) C II Strömgren sphere ionized by HD 38563 N and S, the exciting stars of M78 that are embedded in the dark cloud L1630. The observed lines are extremely narrow, the full width at half-maximum $\Delta V < 1.3$ km s^{-1}, making these the narrowest radio recombination lines yet detected.

132.002 Search for continuous fluorescence in reflection nebulae. W. F. Rush, A. N. Witt. Astron. Journ., Vol. 80, 31 - 36, 73 - 76 (1975).

Photometric and spectrophotometric observations have been made of the reflection nebulae NGC 1435, NGC 2068, NGC 7023, and IC 1287 in an attempt to detect continuous fluorescence by dust grains. Several effects of importance for observations of such faint objects are discussed, including instrumental light scattering, a photographic effect, and a time-delay effect which can occur if the illuminating star is a spectrum variable. It is found that continuous fluorescence by interstellar grains is not likely to exist and that it cannot account for more than 10% of the total surface brightness of these reflection nebulae. No evidence of diffuse interstellar features is found in the spectra of these nebulae.

132.003 Albedo of particles in reflection nebulae. W. F. Rush. Astron. Journ., Vol. 80, 37 - 44 (1975).

The relation between the apparent angular extent of a reflection nebula and the apparent magnitude of its illuminating star derived by Hubble (1922) has been reconsidered under a less restrictive set of assumptions. A computational technique has been developed which permits the use of fits to the observed m-log a values to determine the albedo of particles composing reflection nebulae, providing only that a phase function and average optical thickness are assumed. Multiple scattering, anisotropic phase functions, and illumination by the general star field are considered, and the albedo of reflection nebular particles appears to be the same as that for interstellar particles in general. The possibility of continuous fluorescence contributions to the surface brightness is also considered.

132.004 A large southern reflection nebula at high galactic latitude. I. J. Danziger, M. Dennefeld, D. Kunth, H. E. Schuster. Astron. Astrophys., Vol. 37, 419 - 423 (1974).

A large new reflection nebula of low surface brightness has been identified on ESO Schmidt plates at galactic position $l^{II} = 314°$, $b^{II} = -22°.6$. The study of an associated star illuminating a small part of the nebula suggests a distance of 77 parsec, and a size, projected on the sky, of 13.5 parsec. It is suggested that the nebula is illuminated by the integrated light of the Galaxy.

132.005 T Tauri nebulae and Herbig-Haro nebulae: evidence for excitation by a strong stellar wind. R. D. Schwartz. Astrophys. Journ., Vol. 195, 631 - 642 = Lick Obs. Bull., No. 673 (1975).

Radial velocities of the emission nebulae and NGC 1555 (Hind's reflection nebula) which are associated with T Tau suggest the presence of extended mass outflow which is supersonic with respect to the medium in which the star is embedded. The Herbig-Haro nebulae, if produced by the same mechanism, must involve central objects with a considerably higher rate of mass outflow than is indicated for T Tau.

132.006 Line splitting across the massive neutral intrusions in M16. K. H. Elliott, J. Meaburn. Monthly Notices Roy. Astron. Soc., Vol. 170, 237 - 239 (1975).

Variable splitting of the [O III] line has been observed over the neutral intrusions in the core of M16 with a Fabry—Perot monochromator. A simple model is proposed to explain this.

132.007 The mysterious "Egg nebula" in Cygnus. E. P. Ney. Sky Telescope, Vol. 49, 21 - 23 (1975).

132.008 Mosaics of a southern nebula. E. W. Miller, J. C. Muzzio. Sky Telescope, Vol. 49, 94 - 95, 98 (1975).

132.009 Ionization structure of gaseous nebulae: sulphur, nitrogen and helium. M. Peimbert, L. F. Rodríguez, S. Torres-Peimbert. Rev. Mexicana Astron. Astrofis., Vol. 1, 129 - 141 (1974).

Detailed ionization structure models of gaseous nebulae, for different electron densities and ionizing radiation fields are presented. The observed [S II]/[N II] line intensity ratios in the Orion nebula are greater than those predicted by our models. Possible explanations for this discrepancy are analyzed. It is found that the ionization structure of the heavy elements critically depends on the electron density. From the computations it follows that in order to infer by observation the amount of neutral helium in a given H II region, it is necessary that not only the degree of ionization of heavy elements should be obtained, but also that its electron density should be known.

132.010 Physical conditions in the Cygnus Loop. K. V. Bychkov. Soobshch. Spets. Astrofiz. Obs., Zelenchukskaya, vyp. (No.) 10, p. 3 - 19 (1973). In Russian.

A model for a shock wave with the velocity of gas 210 > $> v_g >$ 170 km/s behind the front can explain optical, X-ray, and 'coronal' radiation. Observational tests to check up the model are presented. The mean gas column density in the direction of the Cygnus Loop is obtained.

132.011 Radio recombination line observations of the C II region NGC 2023. G. R. Knapp, R. L. Brown, T. B. H. Kuiper. Astrophys. Journ., Vol. 196, 167 - 177 (1975).

The reflection nebula NGC 2023 has been observed at four frequencies in a study of its carbon recombination line emission. The observations show the recombination line region to be about 2 pc in diameter, with an average electron density of ~0.2 cm^{-3} and electron temperature of ~20°K. The electron density increases exponentially near the B1.5 star HD 37903, the illuminating star of NGC 2023. The most reasonable interpretation of the observational results is that the carbon emission lines arise in a cold C II region surrounding HD 37903. The coexistence of the C II region with a region of enhanced CO emission suggests that the CO in NGC 2023 may be excited by electron collisions.

132.012 Copernicus: soft X-ray emission from certain features of the Cygnus Loop. P. A. Charles, J. L. Culhane, J. C. Zarnecki. Astrophys. Journ., (Letters), Vol. 196, L19 - L22 (1975).

The MSSL 0.5–1.5 keV X-ray telescope Copernicus has examined several features of previous X-ray maps of the Cygnus Loop. The authors derive fluxes and spectra for the north and northeastern filamentary regions, together with an upper limit for the X-ray flux from the radio source labeled CL4 by Keen et al. Correlation between the X-ray and radio structure is established.

132.013 **UBV photometry of the stars in the fields of emission nebulae. I. M 20.** K. Ogura, K. Ishida.
Publ. Astron. Soc. Japan, Vol. 27, 119 - 135 (1975).

Three-color photometry is presented for 320 stars in the field of the emission nebula M 20 = NGC 6514. An average $(B-V)$ color excess of 0.23 mag is found for the stars. Analysis of the UBV data confirms that most of them belong to the open cluster NGC 6514 which is located at a distance of 1.4 kpc. A contraction age of 7×10^6 yr is obtained from the color-magnitude diagram of the cluster.

132.014 **Starlight excitation of permitted lines in the Orion nebula.** S. A. Grandi.
Astrophys. Journ., Vol. 196, 465 - 472 (1975).

From an idealized model of the Orion nebula and from an analysis of line ratios it is shown that direct starlight excitation of the permitted O I lines dominates over recombination and Lyman line fluorescence. The line strengths predicted by this mechanism agree reasonably well with those observed in the Orion nebula. The application of direct starlight excitation to other ions is also discussed.

132.015 **Approximate solutions of radiative transfer in dusty nebulae. I. Pure hydrogen nebulae.**
V. Petrosian, R. A. Dana.
Astrophys. Journ., Vol. 196, 733 - 744 (1975).

This paper discusses nebulae containing dust and hydrogen. Approximate solutions, accurate to within 50 percent, are obtained for spherical nebulae containing purely absorbing dust. Less accurate solutions are presented for the general case of scattering and absorbing dust. The results are presented in a form convenient for calculating the required number of ionizing photons, the fractional ionization of hydrogen, and the fractions of ionizing photons absorbed by dust and gas in terms of size, hydrogen density, dust albedo, and optical depth. Explicit relations between these quantities are obtained for uniform nebulae. Simple integrals are presented for calculating the above quantities for spherically symmetric nebulae with nonuniform gas and dust distribution.

132.016 **Evidence for a compact source of soft X-rays in the Cygnus Loop.** R. D. Bleach, R. C. Henry, J. F. Meekins, G. Fritz, S. D. Shulman, H. Friedman.
Astrophys. Journ., (*Letters*), Vol. 197, L13 - L17 (1975).

Data obtained in a 3-s observation of the Cygnus Loop (44–60 Å) indicate a 21.4 ms periodicity. When the pulsed fraction of 50–60 percent is removed, the residual intensity agrees well with the mean Loop intensity reported by others. The evidence therefore suggests that a compact object is present within the nebula.

132.017 **Catalogue of southern stars embedded in nebulosity.** S. van den Bergh, W. Herbst.
Astron. Journ., Vol. 80, 208 - 211, 255 - 262 (1975).

The Curtis Schmidt telescope of the Cerro Tololo Observatory was used for a survey of southern reflection nebulae. The present paper presents a catalogue and identification charts for 136 southern stars embedded in nebulosity. A detailed photometric and spectroscopic study of these stars is given in a following paper [Herbst (1975). Astron. Journ., Vol. 80, 212 - 216].

132.018 **R associations I. UBV photometry and MK spectro-** scopy of stars in southern reflection nebulae.
W. Herbst.
Astron. Journ., Vol. 80, 212 - 226 (1975).

Photoelectric UBV photometry and MK spectral types are presented for stars in van den Bergh and Herbst's (1975) catalogue of southern reflection nebulae. These data are used to identify and to derive distances to 20 new R associations. The galactic distribution of these associations is briefly discussed. The effects of possible contamination of the photometry by nebular light is considered in some detail.

132.019 **Dust in the reflection nebula surrounding VY Canis Majoris.** M. Jura.
Astron. Journ., Vol. 80, 227 - 231 (1975).

The author shows that the observations of the polarization and colors in the reflection nebula surrounding VY CMa can be roughly reproduced by a simple spherically symmetric model. To account for both the near greyness of the scattered light and the very high polarization, he suggests both that the dust grains are somewhat larger than those in the normal interstellar gas and that the grains have a large imaginary component of their index of refraction. Using these same grains, he sketches a model to explain the observed infrared linear and circular polarization.

132.020 **Strukturanalysen des Orion-Nebels mit photographischen Detailfilterverfahren.**
E. Lau, N. Richter.
Sterne, Vol. 51, 1 - 4 (1975).

132.021 **Herbig-Haro objects and T Tauri nebulae.**
K.-H. Böhm.
Problems in stellar atmospheres and envelopes, (see 003.001), p. 205 - 228 (1975).

132.022 **Further observations of the Orion nebula cluster.** M. V. Penston, J. K. Hunter, A. O'Neill.
Monthly Notices Roy. Astron. Soc., Vol. 171, 219 - 234 (1975).

New spectral types for 16 stars, UBVRI photometry for 13 stars and UBVRIHKL photometry for 35 stars in the Orion nebula cluster are presented. Analysis of this data together with those of Penston (1973) shows the reddening law in the cluster is normal. Previous claims for high ratios of total to selective absorption are caused by mistaking infrared emission for a hole in the absorption at that wavelength. The distance modulus of M42 is 8.0±0.1 mag. Star formation in the cluster has been in progress over the last 10^7 yr. The infrared colours of cluster stars are correlated with their range of variation in the optical.

132.023 **Emission-line shifts and broadening for Herbig-Haro objects.** M. Friedjung.
Observatory, Vol. 95, 51 - 52 (1975). – Letter.

132.024 **Large-scale ionization fronts and the nature and distribution of light scattering particles in the Orion nebula.** M. A. Dopita, S. Isobe, J. Meaburn.
Astrophys. Space Sci., Vol. 34, 91 - 121 (1975). – Paper presented at the Symposium on Solid State Astrophysics, Cardiff, Wales, 9–12 July 1974.

Image-tube filter photographs calibrated against photoelectric filter photometry have been used to give maps of M42 in absolute flux units over the central 15 arc min of the nebula in Hα, [N II] (λ 6584 Å), Hβ and continuum at λ 4700 Å. Maps of the ratios Hα/[N II] and (for the first time) of continuum/Hβ have been produced with unprecedented spatial resolution. These show that the gas to dust ratio is high near the exciting stars and falls strongly in the vicinity of large scale ionization fronts marked by minima in the Hα/[N II] ratio. These results are interpreted in terms of detailed shell models containing either ice or graphite or silicate scattering particles. A schematic model of the Orion nebula is presented to attempt to explain the large scale phenomena observed here. It demonstrates that simple shell models for this nebula are dubious.

132.025 Observations of carbon monoxide emission in NGC 5367. H. Van Till, R. Loren, J. Davis.
Astrophys. Journ., Vol. 198, 235 - 239 (1975).

Observations of the $J = 1 - 0$ transitions of $^{12}C^{16}O$, $^{13}C^{16}O$, and $^{12}C^{18}O$ in NGC 5367 are reported. The CO line intensity contours lead the authors to conclude that the kinetic temperature peaks at the positions of two stars embedded in this molecular cloud, indicating that the gas is being heated by these stars. From the CO line intensities the mass of the molecular cloud is estimated to be $M_{cloud} \simeq 30\,M_\odot$. Assuming that these two embedded stars recently formed in this cloud, NGC 5367 is then an example of a relatively low mass cloud in which stellar formation may take place.

132.026 The radio recombination line spectrum of Orion A: observation and analysis.
F. J. Lockman, R. L. Brown.
Bull. American Astron. Soc., Vol. 7, 249 (1975). – Abstr. AAS.

132.027 Fifty and 100-micron maps of the Orion nebula with one arc minute resolution.
M. Werner, I. Gatley, E. Becklin, L. Cheung, A. Harper, R. Loewenstein, C. Telesco, H. Thronson.
Bull. American Astron. Soc., Vol. 7, 249 - 250 (1975). Abstr. AAS.

132.028 High velocity emission features in the Orion nebula.
T. R. Gull.
Bull. American Astron. Soc., Vol. 7, 250 (1975). – Abstr. AAS.

132.029 Photography of the Gum nebula.
J. C. Brandt, R. G. Roosen, J. Thompson.
Bull. American Astron. Soc., Vol. 7, 250 - 251 (1975). Abstr. AAS.

132.030 On calculation of spectra in gaseous nebulae.
V. V. Golovatyj, O. N. Zhukov.
Problems of cosmic physics. Vyp. (No.) 9, (see 003.013), p. 162 - 164 (1974). In Russian.

Results are given of the calculation of the function for the transition from the relative abundance of heavy elements to corresponding intensities of forbidden lines. The calculations were carried out with new more precise values of the probabilities of forbidden transitions and of cross-sections of collision excitation.

132.031 On the intensities of decimetric-wavelength radio recombination lines.
A. Parrish, V. Pankonin.
Astrophys. Journ., Vol. 198, 349 - 354 (1975).

The authors summarize the intensity results of some of the 221 and 248α recombination-line observations taken with the Arecibo telescope, and report additional results including 166α observations from the NRAO 300-foot (91 m) telescope. The brightness temperatures of these lines increase sharply with wavelength. It is shown that these results require that the upper levels of the recombining atoms be overpopulated with respect to LTE conditions. The most reasonable interpretation of the results is that the line emission at these decimetric wavelengths is stimulated by a background source of continuum radiation.

132.032 Recombination lines near 8.9 GHz of strong sources in the southern Milky Way.
R. X. McGee, L. M. Newton, R. A. Batchelor.
Australian Journ. Phys., Vol. 28, 185 - 207 (1975).

Seventeen intense nebulae in the southern Milky Way have been surveyed for their radio recombination lines of hydrogen and helium and of elements heavier than helium, X90α. The H 90α line for 30 Doradus in the Large Magellanic Cloud was also observed. Data on source size, flux density, continuum temperature, line half-width and radial velocity are used to derive information about the sources. This information includes electron temperatures and turbulent velocities, the abundance ratio of singly ionized helium to ionized hydrogen, and the intensity ratios of β and γ lines to α lines. The lines from elements heavier than helium are discussed.

132.033 On the luminosity source of the northern part of the nebula M20. N. V. Voshchinnikov.
Pis'ma v Astron. Zhurn., Vol. 1, No. 5, p. 15 - 18 (1975). In Russian.

Polarimetric and photometric observations carried out by the author indicate the star HD 164492 to be considered as the radiation source for the northern part of the nebula M20.

132.034 Optical polarization of a Herbig-Haro object in Corona Austrina.
F. J. Vrba, S. E. Strom, K. M. Strom.
Publ. Astron. Soc. Pacific, Vol. 87, 337 - 339 (1975).

Optical polarization measurements suggest that H-H 100 in the CrA dark cloud represents a reflection nebula illuminated by the infrared source discovered by Strom, Strom, and Grasdalen (1974).

132.035 Photometry of a bright filament in the Orion nebula.
R. J. Dufour, J. S. Mathis.
Publ. Astron. Soc. Pacific, Vol. 87, 345 - 348 (1975).

The spectrum of a bright filament in the outer portion of the Orion nebula, 14' from the Trapezium, was studied. From this study it is concluded that the filament is photoionized, presumably by the Trapezium, rather than a shock front, as had originally been suspected.

132.036 Draft Catalog of Herbig-Haro Objects.
G. H. Herbig.
Lick Obs. Bull. No. 658, 11 + 10 pp. (1974).

This Draft Catalog of Herbig-Haro Objects gives accurate coordinates, rough indication of apparent brightness, a summary of the spectroscopic information available, and identifications for 41 small nebulae that certainly or probably are members of that class. Most of the Lick data given here were collected in the years 1946–1965 but remained largely unpublished.

132.037 Application of Mie theory to the problems of the scattered continuum light in diffuse nebulae.
S. Isobe.
Ann. Tokyo Astron. Obs., Second Ser., Vol. 14, (No. 4), 227 - 237 (1975).

The ratios of surface brightness of continuum light scattered by ice, graphite, or silicate grains to Hβ light are calculated for NGC 6514, 6523, 6611, and the values are compared with the observations by O'Dell, Hubbard, and Peimbert.

132.038 Optical colours and polarization of model reflection nebula. G. A. Shah.
Bull. Astron. Soc. India, Vol. 2, 38 (1974). – Abstract.

132.039 The Carina nebula at 3.4 and 6 cm.
W. K. Huchtmeier, G. A. Day.
Astron. Astrophys., Vol. 41, 153 - 164 (1975).

The Carina nebula has been mapped at 3.4 cm with the 64 m Parkes radio telescope. At this wavelength the half-power beamwidth is 2.6 arcmin. Radio recombination line observations of the nebula have been obtained at 6 and 3.4 cm (H 109 α and H 90 α). In the vicinity of Carina II double line profiles with a separation of $\sim$45 km s^{-1} are seen. These are interpreted in terms of an expanding sphere of ionized hydrogen. Carina I shows single profiles. The mean radial velocity of the observed lines is approximately constant over the whole nebula. Centered on the expanding region are two young galactic clusters: Trümpler 14 and Trümpler 16. It is suggested that they are responsible for the observed expansion.

132.040   Decimeter-wavelength studies of hydrogen and carbon recombination lines toward galactic nebulae.
V. L. Pankonin.
Thesis, Cornell Univ., Ithaca, N.Y. (USA). 237 pp. University Microfilms Order No. 74-10,892 (1973).

Astrophysics of gaseous nebulae.
See Abstr. 003.112.

Discovery of Herbig-Haro objects by [S II] interference photography.   See Abstr. 031.260.

Photometric studies of faint stars in the vicinity of the Orion nebula.   See Abstr. 113.024.

Catalogue of photometric and astrometric data for 4000 stars in the Orion nebula aggregate.
See Abstr. 113.053.

Infrared observations of late-type stars in nebulae.
See Abstr. 113.057.

Spectrophotometry of Orion stars.
See Abstr. 113.062.

Trapezium in de grote Orionnevel.
See Abstr. 117.022.

A molecular cloud in IC 1396.
See Abstr. 131.004.

Observations of the $J = 2 \rightarrow 1$ transition of carbon monoxide in interstellar clouds.   See Abstr. 131.032.

Distribution of 73-GHz para-formaldehyde line emission in the Orion nebula.   See Abstr. 131.042.

Observations of circular polarization of the $J = 2-1$, $\nu = 1$ transition of the SiO maser.   See Abstr. 131.058.

Six-centimeter H$_2$CO emission from the Orion nebula.   See Abstr. 131.068.

Observations of carbon α and β recombination lines at 22 cm.   See Abstr. 131.097.

Radio detection of interstellar sulfur dioxide.
See Abstr. 131.119.

A southern dark globule.
See Abstr. 131.130.

Infrared forbidden lines in H II regions and planetary nebulae.   See Abstr. 131.518.

Young stellar clusters in diffuse nebulae.
See Abstr. 131.522.

A survey of the unusual motions in M17 with a Fabry-Perot monochromator.   See Abstr. 131.532.

The structure of the H I, H II, molecular and dust complex of M17.   See Abstr. 131.533.

Optical study of the Carina nebula.
See Abstr. 131.535.

Fine structure in M8 at 327 MHz.
See Abstr. 131.536.

A study of the nebulae S 206 and S 209.
See Abstr. 131.539.

Radio emission from the peculiar nebulae M2-9.
See Abstr. 133.002.

G2.4 + 1.4, a supernova remnant or ring nebula around a peculiar star.   See Abstr. 141.061.

Measurements of neutral-hydrogen absorption in the spectra of five pulsars and parameters of the Gum nebula.
See Abstr. 141.356.

Studies of the infrared source CRL 2688.
See Abstr. 141.615.

Spectroscopic observations of CRL 2688.
See Abstr. 141.616.

Pulkovo sky survey in the interstellar neutral hydrogen line. II. Neutral hydrogen in the region of Loop II.
See Abstr. 157.006.

Chemical composition of nebulosities in the Magellanic Clouds.   See Abstr. 159.019.

Errata

132.901   Errata: 'The continuous spectrum of Herbig-Haro objects' [Astrophys. Journ., Vol. 193, 353 - 357 (1974)].   K.-H. Böhm, R. D. Schwartz, W. A. Siegmund. Astrophys. Journ., Vol. 197, 805 (1975).

133 Planetary Nebulae

133.001 The underabundance of gaseous iron in the planetary nebula NGC 7027. G. A. Shields.
Astrophys. Journ., Vol. 195, 475 - 478 (1975).

Computer models which take into account the density fluctuations in NGC 7027 are used to derive an iron abundance $\log [N(\text{Fe})/N(\text{H})] = -6.1 \pm 0.4$ in this nebula. This abundance is below the solar value by more than an order of magnitude whereas the abundances of lighter elements appear normal. The iron depletion factor in NGC 7027 is close to that in the interstellar gas and may result from the condensation of iron into grains.

133.002 Radio emission from the peculiar nebula M2-9. C. R. Purton, P. A. Feldman, K. A. Marsh.
Astrophys. Journ., Vol. 195, 479 - 481 (1975).

Observations at a wavelength of 2.8 cm have yielded a flux density of $(45 \pm 7) \times 10^{-29}$ W m^{-2} Hz^{-1} for the nebula M2-9. When combined with the available optical data, this value is found to be surprisingly low and suggests the existence of condensations in the wings of the nebula, together with a dense gaseous core of emission measure $\sim 10^{10}$ cm^{-6} pc.

133.003 The creation of planetary nebulae. A. Finzi, R. Finzi, G. Shaviv.
Astron. Astrophys., Vol. 37, 325 - 334 (1974).

In the picture presented here, the progenitor of a planetary nebula becomes degenerate almost up to the surface before it ejects its hydrogen-rich outer shell. However, just before degeneracy reaches the surface, the hydrogen there begins to burn fiercely, and the hydrogen-rich shell is gradually ejected since the large pressure gradient in the outer envelope cannot be fully balanced by the force of gravitation. The authors have found a 2-parameter set of time-independent models representing mass-ejecting stars.

133.004 Chemical abundances of planetary nebulae. G. O. Boeshaar.
Astrophys. Journ., Vol. 195, 695 - 704 (1975).

A procedure for measuring chemical abundances from emission-line spectra is presented that accounts for small-scale changes in ionization, temperature, and density within the source. Applied to the spectral data for six planetary nebulae, this method appears consistent in the allowance for both filamentary structure and unobserved ionization stages. Applications to the study of H II regions and the gas content of other galaxies are discussed.

133.005 Identification of the 890 cm^{-1} carbonate signature in NGC 7027. J. D. Bregman, D. M. Rank.
Astrophys. Journ., (Letters), Vol. 195, L125 - L126 = Lick Obs. Bull., No. 687 (1975).

High-resolution spectroscopic measurements of the 11 μ continuum emission from NGC 7027 indicate that $MgCO_3$ is a constituent of the dust in the nebula. Other carbonate compounds may also be present in concentrations approximately one order of magnitude below that of $MgCO_3$. Dust-grain temperatures combined with the abundance of $MgCO_3$ require that most of the material composing the grains must be carbon or C, N, O compounds.

133.006 The optical spectrum of the nebula YM 29. E. M. Leibowitz.
Astrophys. Journ., Vol. 196, 191 - 194 (1975).

New spectrograms of the nebula YM 29 lend some support to the identification of this object as a low-excitation planetary nebula. The abundances of oxygen, nitrogen, and probably sulfur in this nebula are close to the mean values for these elements in planetary nebulae.

133.007 Spectroscopic studies of very old hot stars. I. NGC 246 and its exciting star. S. R. Heap.
Astrophys. Journ., Vol. 196, 195 - 204 (1975).

A spectroscopic study has been made of the planetary nebula NGC 246 and its exciting star. The only lines definitely present in the stellar spectrum are lines of C IV and O VI, although lines of He II are probably present. Nitrogen lines, in any ionization state, are absent. The stellar spectrum, therefore, suggests that the triple-α process has operated in this star. The author derives the following probable values for the stellar parameters: $T_{\text{eff}} \sim 200,000°$K, $\log g \sim 6.8$, and $R/R_\odot \sim 0.05$. Comparison of the properties of the star and surrounding nebula with Paczynski's evolutionary tracks for planetary nuclei suggests that the star is very near the point of exhaustion of its nuclear fuels.

133.008 High resolution observations of planetary nebulae at 5 GHz. P. F. Scott.
Monthly Notices Roy. Astron. Soc., Vol. 170, 487 - 495 (1975).

Five planetary nebulae have been observed at a frequency of 5 GHz with 2″ arc resolution. For two of the nebulae, NGC 6543 and NGC 7662, the radio structures are well determined and closely resemble those found optically. The radio spectra imply hydrogen temperatures ~ 6000 K. An alternative classification of the nebula M1−78 as a compact H II region is discussed briefly.

133.009 Some misclassified planetary nebulae. D. A. Allen, R. A. E. Fosbury.
Observatory, Vol. 95, 15 - 17 (1975).

Several objects classified by Kohoutek [Bull. Astron. Inst. Czechoslovakia, Vol. 16, 221 (1965)] as stellar planetary nebulae have been shown to have the 2 μm continua characteristic of M stars. These were thought by Allen to be symbiotic stars. The authors have secured low-dispersion spectra of a sample of five of these objects using an image-tube spectrograph at a dispersion of 210 Å/mm on the Isaac Newton telescope. The five objects are quite definitely neither planetary nebulae nor symbiotic stars: all are M stars and only one of them shows emission lines.

133.010 Envelope ejection to form planetary nebulae. P. E. Stry.
Astrophys. Journ., Vol. 196, 559 - 563 (1975).

Previous calculations at $L/L_{\text{crit}} \approx 0.5$ are extended to higher luminosities to determine whether partial, small-scale ejections of mass continue to be the general result. The $0.2 M_\odot$ envelope of a $1.2 M_\odot$ star was perturbed by raising the interior boundary condition luminosity 10 percent to $L \approx L_{\text{crit}}$. After a brief period of oscillation the entire envelope was ejected. The numerical methods include time dependent convection (not necessarily efficient) and a simplified version of a spherically extended atmosphere.

133.011 Stratification effects in the planetary nebula NGC 7009. L. H. Aller, H. W. Epps.
Astrophys. Journ., Vol. 197, 175 - 184 (1975).

Spectrophotometric scans ($\lambda\lambda 4000-8600$ Å) have been obtained. Profound excitation differences noted by Berman, by Aller, and by Ford and Rubin are substantiated, and an attempt is made to derive ionic concentrations, electron temperatures, and densities. Particular attention is called to the large variations in the Hα/[N II] ratio, which ranges from around 2.0 in the ansae to about 0.03 in the bright ring, i.e., a range of nearly two orders of magnitude. The data illustrate the difficulties in attempting to derive elemental abundances from nebular-line intensities with either radiative models or empirical corrections.

133.012 Radio observations at 5 GHz of southern planetary nebulae.   D. K. Milne, L. H. Aller.
Astron. Astrophys., Vol. 38, 183 - 196 = Div. Radiophys., C.S.I.R.O., *Sydney*, Radiophys. Publ. RPP 1638 (1975).

High-sensitivity radio observations for 165 planetary nebulae are presented and compared with the Hβ intensities to obtain extinction coefficients. A radio distance scale is derived and the mean optical absorption within the galactic disk is estimated to be 1.3 ± 0.8 mag kpc⁻¹. Comparison is made with the stellar absorption in the direction of these nebulae.

133.013 The infrared continuum of the compact planetary nebula NGC 6210.   I. J. Danziger.
Astron. Astrophys., Vol. 38, 475 - 478 (1975).

A study of the continuous radiation from NGC 6210 between 0.37 and 10μ shows contributions from the central star, the nebula, and the dust in the nebula, each characterised by different temperatures. An excess peaking at 1.4μ requires a different explanation, and the possible roles of H⁻ and a faint red star are discussed. The helium abundance $N(He^+)/N(H^+)$ in the nebula is 0.11.

133.014 Electronographic measurements of the density structure of NGC 6543 using the O⁺ and Cl⁺⁺ ions.
M. A. Dopita, A. H. Gibbons.
Monthly Notices Roy. Astron. Soc., Vol. 171, 73 - 83 (1975).

Measurements have been made with the electronographic image tube spectrograph on the Isaac Newton telescope of the [O II] doublet λ3726, 3729 Å and the [Cl III] doublet λ 5537, 5517 Å at 80 and 140 positions respectively in the planetary nebula NGC 6543. A temperature is found from the ratio of the Balmer jump to the sum of H15, H16 and H17. The electron density as indicated by the O⁺ ions is found to vary from 1500 to 5000 cm⁻³ and that indicated by the Cl⁺⁺ ions to vary from 3000 to 6500 cm⁻³. No correlation between the two density structures was found.

133.015 Monochromatic isophotometry of planetary nebulae – I. Hβ observations of NGC 6210, 6543, 6826, 7009, 7027 and 7662.
C. I. Coleman, N. K. Reay, S. P. Worswick.
Monthly Notices Roy. Astron. Soc., Vol. 171, 415 - 423 (1975).

The spectracon electronographic detector, with its advantages over photography of large dynamic range, linearity and high detective quantum efficiency, is used in a program of monochromatic surface photometry of planetary nebulae. The instrumentation and data reduction techniques are described. Data in the form of Hβ (λ 4861 Å) isophote maps are presented for NGC 6210, 6543, 6826, 7009, 7027 and 7662. Each object is discussed with particular reference to faint outer envelopes and maximum overall dimensions.

133.016 Orientation of planetary nebulae within the Galaxy.   G. Melnick, M. Harwit.
Monthly Notices Roy. Astron. Soc., Vol. 171, 441 - 444 (1975).

The authors have analysed planetary nebulae shown in the catalogue of Perek & Kohoutek. They find that, among elliptical planetaries, there is a correlation between the orientation of the major axis of the planetary nebula and the direction of the galactic equator.

133.017 A study of planetary nebulae in the galactic centre direction.   B. A. Vorontsov-Vel'yaminov,
E. B. Kostyakova, O. D. Dokuchaeva, V. P. Arkhipova.
Astron. Zhurn. Akad. Nauk SSSR, Vol. 52, 264 - 273 (1975).
In Russian. English translation in Soviet Astron., Vol. 19, No. 2.

The results of current study of a large planetary nebulae group in the direction of the galactic centre are given. The monochromatic energy flux in series of emission lines was determined for 47 planetaries. For the most interesting objects the results of detailed analysis of the spectra are given. 3 new planetary nebulae were discovered.

133.018 Neon III in planetary nebulae.   F. J. Ahern.
Astrophys. Journ., Vol. 197, 635 - 637 (1975).

The author shows that there is remarkably little variation in the Ne III/O III abundance ratio among the bright planetary nebulae. This makes it reasonable to use the intensity ratio of the nebular transitions Ne III λ3869/O III λ5007 to aid in estimating the electron density and temperature where suitable auroral to nebular intensity ratios are not available.

133.019 Absolute emission-line intensities for V1016 Cygni.   F. J. Ahern.
Astrophys. Journ., Vol. 197, 639 - 645 (1975).

Photometric and spectrophotometric observations have been made of V1016 Cygni to determine the absolute intensities of the brighter emission lines. The Balmer decrement is much less steep than it was when measured in 1966 by O'Dell. Since the total brightness of the nebula has not increased since 1966, this cannot be interpreted as a decrease in reddening. The author postulates an additional source of Hα emission which has since disappeared. A stratification model is presented for the case of a sphere of uniform density and temperature, and the relative intensities of He II λ4686/Hβ imply a blackbody effective temperature of $T_* = 164,000 \pm 3600$ K for the central star. If the Balmer line optical depth arises from hydrogen in the 2s state, he infers a radius of 1.0×10^{16} cm and a mass of $0.027\,M_\odot$. Photometric observations relative to NGC 7027 indicate an Hβ flux of $3.69 \pm 0.20 \times 10^{-11}$ ergs cm⁻² s⁻¹.

133.020 Thermal emission spectra of silicates from planetary nebulae.
E. Bussoletti, J. P. Baluteau, N. Epchtein.
Astrophys. Space Sci., Vol. 34, 81 - 89 (1975). – Paper presented at the Symposium on Solid State Astrophysics, Cardiff, Wales, 9–12 July, 1974.

Calculations of the grain equilibrium temperature and of the expected infrared spectra of IC 418, BD+30° 3639, NGC 6572 and NGC 7027 have been performed using dielectric constants of lunar silicates. The results have been compared with previous work on pure graphite and ice-mantle grains. Lα heating of dust followed by thermal re-emission is consistent with the large infrared excesses detected in planetary nebulae. An extra source of heating is, nevertheless, necessary to fit correctly the experimental results. It appears from the calculations that, for each object, it is possible to define theoretically the most probable nature of the emitting dust.

133.021 On the nebular mass of planetary nebulae.   M. Perinotto.
Astron. Astrophys., Vol. 39, 383 - 391 (1975).

The best available optical and radio informations on forty-three planetary nebulae have been used to investigate the problem of the individual determination of the nebular masses. Their average mass is found to be 0.15 ± 0.30 $M_\odot$ using optical fluxes (31 objects), and 0.23 ± 0.52 $M_\odot$ from radio continuum fluxes (30 objects). A discussion of the errors associated with the various quantities is given.

133.022 New observations of VV8 and M3-27.   T. F. Adams.
Bull. American Astron. Soc., Vol. 7, 243 (1975). – Abstr. AAS.

133.023 Internal dust in planetary nebulae.   J. R. Doughty.
Bull. American Astron. Soc., Vol. 7, 243 (1975). Abstr. AAS.

133.024   **The physical conditions in NGC 7027.**
J. B. Kaler, L. H. Aller, S. J. Czyzak, H. W. Epps.
Bull. American Astron. Soc., Vol. 7, 243 (1975). – Abstr. AAS.

133.025   **The peculiar objects He2-467, M1-2 and IC 2120.**
J. Lutz.
Bull. American Astron. Soc., Vol. 7, 243 - 244 (1975).
Abstr. AAS.

133.026   **The structure of planetary nebulae.**   Y. Terzian.
Bull. American Astron. Soc., Vol. 7, 244 (1975).
Abstr. AAS.

133.027   **The infrared emissivity profile of the dust of**
NGC 7027.   J. P. Apruzese.
Bull. American Astron. Soc., Vol. 7, 244 (1975). – Abstr. AAS.

133.028   **Spectrophotometric studies of gaseous nebulae.**
XXIV. The amorphous low-excitation planetary
IC 4593.   S. J. Czyzak, E. G. Buerger, L. H. Aller.
Astrophys. Journ., Vol. 198, 431 - 433 (1975).

Photoelectrically calibrated line intensity measurements
are presented for IC 4593, whose spectrum has also been
observed photographically. An estimate of the chemical composition from ionic concentrations, based on extrapolation
procedures of the type suggested by Seaton, is compared with
results of a model calculation by Buerger. It is suggested that
great caution must be used with extrapolation of [N II] lines
to estimate nitrogen abundance in gaseous nebulae. It is
concluded that nitrogen abundances estimated from extrapolation of [N II] line data can sometimes be in error.

133.029   **Physical conditions and structure in NGC 7293,**
the "Helix".   J. W. Warner, V. C. Rubin.
Astrophys. Journ., Vol. 198, 593 - 603 (1975).

Image-tube interference filter photographs and spectra
are presented for NGC 7293, the "Helix" planetary nebula.
Extinction values, and electron temperatures and densities,
are derived for 12 points in the central region and in the
bright SW filament. The structure of the nebula probably
consists of two or three rings of low-ionization material surrounding a higher ionization zone. The appearance of the
object in the light of [O I] 6300 Å and the lack of adequate
evidence from velocity data suggest that a true helical interpretation is not realistic.

133.030   **Calculation of Balmer continuum radiation for**
planetary nebulae considering their physical con-
ditions.   E. B. Kostyakova.
Soobshch. Gos. Astron. Inst. Shternberga, No. 188, p. 3 - 14
(1974). In Russian.

The Balmer continuum radiation was calculated for
planetary nebulae with considering their different physical
conditions. The results were compared with the author's observations of 16 planetary nebulae in the near ultraviolet
(1968, 1970, 1971).

133.031   **Kinematics of southern planetary nebulae.**
A. Acker.
Astron. Astrophys., Vol. 40, 415 - 420 (1975). In French.

Radial velocities of forty planetary nebulae have been
determined. The qualitative intensity of measured lines allows
the determination of the excitation class of the PN. In addition, internal velocities have been evaluated for some of the
PN. The agreement between the observed velocities and those
calculated from purely circular motion was analysed. A mean
residual velocity of 36 km/s was found for the PN belonging
to the disc population. Some of the large differences observed
can be explained in terms of overestimated distances (optically thick nebulae), underestimated distances (stellar like
nebulae) and elliptical orbits (objects located in the direction
of the galactic center).

133.032   **Spectra of some planetary nebulae between 8000**
and 11000 Å.
Y. Andrillat, A. Baranne, L. Houziaux.
Astron. Astrophys., Vol. 41, 99 - 102 (1975). In French.

Spectra of some planetary nebulae between 8000 and
11000 Å have been secured. The strongest features belong to
[S III] and He I spectra. [S II], [N I], [C I], [Cl II], He II, O I
ions have been identified. A weak feature at 8830 Å appears
in some planetaries and might be due to [Fe III] or [Cr II].

Ejection of nebulae by BQ radiostars with infrared
excess.   See Abstr. 114.030.

The symbiotic binary V 1016 Cygni, early stage of
a planetary nebula.   See Abstr. 114.333.

Infrared forbidden lines in H II regions and
planetary nebulae.   See Abstr. 131.518.

134   Crab Nebula

134.001   **Measurement of circular polarisation in the Crab nebula at 1,415 MHz.**   K. W. Weiler.
Nature, Vol. 253, 24 - 25 (1975).

The author reports a measurement of the distribution of circular polarisation (Stokes parameter V) in the Crab nebula at 1,415 MHz obtained during a series of high accuracy full polarisation measurements with the Westerbork Synthesis Radio Telescope in November 1971. The final accuracy of the measurement of the distribution of circular polarisation in the Crab nebula was approximately 0.02% of the total intensity.

134.002   **Measured offset between the Crab pulsar and Tau X-1.**
P. J. N. Davison, J. L. Culhane, L. V. Morrison.
Nature, Vol. 253, 610 - 612 (1975).

The authors have used the MSSL X-ray detector on board the Copernicus spacecraft to measure the effective diameter of the Crab nebula and to measure the location of the centroid of the emission with respect to the accurately known pulsar position.

134.003   **Wave zone structure of NP 0532 and infrared radiation excess of Crab nebula.**   V. V. Usov.
Astrophys. Space Sci., Vol. 32, 371 - 373, 375 - 377 (1975). In Russian and English.

At the distance $r \gtrsim 10^{15}$ cm from NP 0532 the plasma concentration decreases so that the intense low-frequency wave ($\nu = 30$ Hz) can propagate. The interaction of this wave with the electrons ejected from the pulsar should result in the IR radiation with $F_\nu \sim 10^2$ fu at $\lambda \sim 10\mu$. This flux is the order of the excess IR radiation from the Crab nebula.

134.004   **The absorption of soft X-rays from the Crab nebula.**
V. S. Iyengar, S. Naranan, B. V. Sreekantan.
Astrophys. Space Sci., Vol. 32, 431 - 446 (1975).

The X-ray spectrum of the Crab nebula has been determined in the energy range 0.5−10 keV using thin window proportional counters carried aboard a Centaur IIA rocket launched from TERLS, India. The spectrum can be well represented by a power law with an exponent −2.1 beyond 2 keV. The absorption of the soft X-ray component below 2 keV is clearly seen in the experiment. It is concluded that it is not necessary to consider anomalous abundance of elements like carbon or neon either in the source or in the interstellar medium as suggested by some authors. The absorption of X-rays in the interstellar dust in the light of current dust models is presented.

134.005   **Change in the high-energy radiation from the Crab.**
K. Greisen, S. E. Ball, Jr., M. Campbell, D. Gilman, M. Strickman, B. McBreen, D. Koch.
Astrophys. Journ., Vol. 197, 471 - 479 (1975).

On 1973 July 23, the Cornell γ-ray telescope had a second exposure to the Crab nebula and pulsar. On the previous flight, 1971 October 6, the signal from the pulsar had been unmistakably clear, expecially at the highest energies (>800 MeV), and an unpulsed component of similar average power had also been discernible. On the second flight, the pulsed signal was barely detectable and the DC component not observable at all. The drop in intensity seems to increase with energy, being only a factor of 2 at 200 MeV but an order of magnitude at 1 GeV. There is some indication (not compelling) that the γ-ray flux may have changed even within the 6-hour exposure. The first flight may have viewed the Crab in a state of enhanced high-energy emission correlated with the glitches that occurred on 1971 August 1 and October 25.

134.006   **High-energy X-ray observations of a lunar occultation of the Crab nebula.**
G. R. Ricker, A. Scheepmaker, S. G. Ryckman, J. E. Ballintine, J. P. Doty, P. M. Downey, W. H. G. Lewin.
Astrophys. Journ., (Letters), Vol. 197, L83 - L87 (1975).

X-ray observations of both immersion and emersion for the lunar occultation of 1974 August 13, were conducted from a balloon-borne X-ray telescope. The authors find that ~70 percent of 20−150 keV (average photon energy ~50 keV) emission from Tau X-1 (3U 0531 + 21) originates in a region with dimensions $24'' \pm 7''$ along position angle (P.A.) = $130°$, and $49'' \pm 7''$ along P.A. = $244°$. The centroid of the X-ray emitting region is offset $10'' \pm 4''$ to the northwest of the pulsar NP 0532.

134.007   **Crab nebula X-ray lunar occultation.**
A. Toor, T. Palmieri, F. Seward.
Bull. American Astron. Soc., Vol. 7, 247 (1975). – Abstr. AAS.

134.008   **Brightness distribution and flux density of the Crab nebula in the 1.7−3.7 mm wavelength band.**
V. F. Zabolotnyj, V. I. Kostenko, V. I. Slysh.
Astron. Zhurn. Akad. Nauk SSSR, Vol. 52, 665 - 667 (1975). In Russian. English translation in Soviet Astron., Vol. 19, No. 3.

The method and results of measurements of brightness distribution of the Crab nebula in the 1.7−3.7 mm wavelength band with $2'$ resolution are presented.

134.009   **A model of the X-ray structure of the Crab nebula.**
B. Aschenbach, W. Brinkmann.
Astron. Astrophys., Vol. 41, 147 - 151 (1975).

The authors propose a model of the Crab nebula explaining the results of recent X-ray measurements which have been obtained by means of the lunar occultation techniques. The proposed spatial orientation of the pulsar leads to an elliptically shaped configuration of the X-ray source elongated in a north-east-to-southwest direction. A motion of the pulsar and its magnetosphere relative to the centroid of the nebula may produce an enhanced emission northwest of the pulsar by magnetic field amplification.

High-energy gamma-ray results from the second Small Astronomy Satellite.   See Abstr. 061.039.

Strong radiative reaction in the non-linear Compton process.   See Abstr. 063.031.

Radio observations of the lunar occultation of the Crab nebula on September 10, 1974.   See Abstr. 096.010.

Supernova remnants and presupernova models. See Abstr. 125.007.

Optical radiation from the Crab pulsar. See Abstr. 141.317.

The alignment of the Crab pulsar magnetic axis. See Abstr. 141.327.

Crab pulsar.   See Abstr. 141.364.

Attempt to monitor pulsed X-rays from the Crab pulsar (NP 0532) utilizing atmospheric fluorescence. See Abstr. 141.367.

Lunar occultation of the hard X-ray source in the Crab nebula.   See Abstr. 142.087.

Radio Sources, Quasars, Pulsars, Infrared, X-Ray, Gamma-Ray Sources, Cosmic Radiation

141 Radio Sources, Quasars, Pulsars, Infrared Sources

Radio Sources, Quasars

141.001 Optical monitoring of radio sources – IV. (Results up to 1973 April).
R. A. Selmes, K. P. Tritton, R. W. Wordsworth.
Monthly Notices Roy. Astron. Soc., Vol. 170, 15 - 40 (1975).
Results for a period ending in 1973 April are presented from the Herstmonceux programme of optical monitoring of radio sources and compact galaxies. Thirty-six quasars, galaxies and BL Lac objects are listed.

141.002 The structures and properties of 4C radio sources in Abell clusters – I. J. M. Riley.
Monthly Notices Roy. Astron. Soc., Vol. 170, 53 - 79 (1975).
To investigate the nature of the radio sources and their associated galaxies in rich clusters of galaxies, observations have been made with the Cambridge One-Mile telescope at 408 and 1407 MHz of the 25 sources in the 4C catalogue in the declination range 20° to 40° which lie in Abell clusters. The results of the observations of 16 of these sources are presented here in tabular form and as contour maps.

141.003 The high frequency radio spectra of secondary standard sources. H. N. Ross, E. R. Seaquist.
Monthly Notices Roy. Astron. Soc., Vol. 170, 115 - 119 (1975).
Eight sources commonly used as standards were compared at 3.24, 6.63 and 10.63 GHz with 3C 274, a source that has been established recently on an absolute scale. As a result, accurate high frequency spectra for these eight sources have been obtained.

141.004 An investigation of the optical fields of 35 3CR radio sources to faint limiting optical magnitudes.
M. S. Longair, J. E. Gunn.
Monthly Notices Roy. Astron. Soc., Vol. 170, 121 - 138 (1975).
The fields of 35 3CR radio sources, for which detailed radio structural information is available, have been investigated using the 200-in. Hale telescope to a uniform limiting apparent magnitude of about 23.5 or 24. All the identifications fainter than apparent magnitude 20 are radio galaxies. Various problems which hamper the identification of radio sources are discussed in the light of these results. It is concluded that, while some of the non-identifications may be due to gross asymmetries of the radio source structures, some of the unidentified sources must still lie beyond the present plate limits.

141.005 High-resolution radio observations of the sources near K3–50. S. Harris.
Monthly Notices Roy. Astron. Soc., Vol. 170, 139 - 153, with a correction, Vol. 171, 235 (1975).
The group of radio sources near the optical object K3–50 has been observed at 5 GHz with the Cambridge 5-km telescope. Two of the sources contain high-density compact

structure, and the one coincident with K3–50 itself has an unusually high brightness temperature. The new radio data are compared with infrared and optical observations of these sources and the anomalous extinction is discussed in terms of a scattering model.

141.006 The flux-density variations of the radio emission from Cygnus X-3. M. McEllin.
Monthly Notices Roy. Astron. Soc., Vol. 170, 1P - 6P (1975).
The Cambridge 5-km telescope has been used to monitor the flux density variations of the radio emission from Cygnus X-3 over the period 1972 September–1974 August. The results have been combined with those of other groups to provide a fairly complete picture of the behaviour of Cygnus X-3 at radio frequencies since the 1972 outburst.

141.007 High frequency radio observations of the Stephan's Quintet region.
M. A. Kaftan-Kassim, J. W. Sulentic, G. Sistla.
Nature, Vol. 253, 176 - 177 (1975). – Letter.

141.008 Radio spectra and red shifts of 179 QSOs.
J. D. Kraus, M. R. Gearhart.
Astron. Journ., Vol. 80, 1 - 8 (1975).
A comparison is made between radio spectral type and red shift for all known QSOs having adequate radio and optical data. A ratio of CE (centimeter excess) to N (normal) spectrum sources is defined and its variation with red shift is given. A radio color-color versus red shift diagram for the 179 QSOs is also presented. The CE/N excess is found to increase rapidly for $z > 2$ and the significance of this trend is discussed.

141.009 The spectrum of OH 471 (0642 + 44).
R. F. Carswell, P. A. Strittmatter, R. E. Williams, E. A. Beaver, R. Harms.
Astrophys. Journ., Vol. 195, 269 - 277 (1975).
Results of image-tube and Digicon spectroscopy of OH 471 (0642 + 44) are reported. Wavelengths of 89 stronger absorption features in the range 4000–6000 Å are given, and a number of absorption redshift systems are suggested. The lack of radiation shortward of 4000 Å is attributed to Lyman continuum absorption in the gas giving rise to the various absorption-line systems.

141.010 Coherent curvature radiation and low-frequency variable radio sources.
W. J. Cocke, A. G. Pacholczyk.
Astrophys. Journ., Vol. 195, 279 - 283 (1975).
A model of low-frequency variable radio sources such as 3C 454.3 based on coherent curvature emission by high-energy electrons in the magnetosphere of a spinar is presented and discussed. This model is applicable to sources at cosmological distances and yields inverse Compton fluxes compatible with present observational data.

141.011 Dielectronic recombination and abundances near quasars. K. Davidson.

Astrophys. Journ., Vol. 195, 285 - 291 (1975).

Major obstacles to determining certain elemental abundances associated with some quasars are (1) a lack of accurately measured helium line intensities, and (2) uncertainty about the extent to which dielectronic recombination $C^{+3} + e^- \to C^{+2}$ is suppressed at the relevant gas and radiation densities. Some calculated results are presented in an attempt to reduce the latter obstacle.

141.012 A third 1415 MHz survey with the Westerbork Synthesis Radio Telescope: the 5C2 region (part I).
P. Katgert.
Astron. Astrophys., Vol. 38, 87 - 104 (1975).

The 5C2 region, observed originally with the Cambridge One-Mile telescope at 408 MHz, has been reobserved at 1415 MHz. The resulting source list contains 238 sources with attenuated flux densities exceeding the catalogue limit of 6.25 m.f.u. Out of a total of 190 5C2 sources (i.e. all 5C2 sources within the 10dB attenuation contour of the present survey) 128 were detected with flux densities above the catalogue limit. Another 22 5C2 sources were detected with flux densities below the catalogue limit.

141.013 A model for simultaneous synchrotron and inverse Compton fluxes.
F. W. Peterson, C. King III.
Astrophys. Journ., Vol. 195, 753 - 759 (1975).

Optical and radio fluxes, possibly correlated, from two extragalactic variable radio sources are analyzed on the assumption that the radio flux is synchrotron radiation whereas the optical flux is inverse Compton radiation resulting from scattering of the synchrotron photons. Detailed analysis of an outburst in 3C 120 shows that a description of the evolution of the outburst, based on this interpretation, is compatible with its description on the basis of the adiabatic expansion model in all respects but the rate of falloff of the optical flux. Compton optical and X-ray fluxes are calculated for the 1972 radio flare in Cygnus X-3.

141.014 On the structure and evolution of a shock wave model of QSO's. E. Daltabuit, J. Cantó.
Rev. Mexicana Astron. Astrofis., Vol. 1, 151 - 159 (1974).

Some rough analytical approximations are proposed to describe the behavior of the shock wave model of QSO's proposed by Daltabuit and Cox (1972) for $10^7 \text{ cm}^{-3} \leqslant n \leqslant 10^8$ cm^{-3} and 700 km sec$^{-1} \leqslant v \leqslant 1000$ km sec^{-1}. The possible effect of radiation pressure on the evolution of this model is studied.

141.015 Flux density measurements of some radio sources in the 1 - 5 mm wavelength range.
V. F. Zabolotnyj, V. I. Kostenko, N. Ya. Nikolaev, V. I. Slysh, V. A. Soglasnova.
Pis'ma v Astron. Zhurn., Vol. 1, No. 1, p. 14 - 17 (1975).
In Russian.

Method and results of flux density measurements in 1 - 5 millimeter wavelength range for eight galactic and extragalactic sources are presented.

141.016 26.3-MHz radio source survey. I. The absolute flux scale. M. R. Viner.
Astron. Journ., Vol. 80, 83 - 92 (1975).

The absolute flux densities of Cyg A, Cas A, Tau A, and Vir A have been measured with an accuracy of about ±5% at 26.3 MHz using a method well suited for use at low radio frequencies where ionospheric scintillation is a limiting factor. No evidence for time variations in any of the four sources has been found.

141.017 Non-variable 13.5 mm flux in the strong millimetre component of Centaurus A.

W. G. Fogarty, N. J. Schuch.
Nature, Vol. 254, 124 - 125 (1975).

It has been suggested that the peculiar galaxy and strong radio source Cen A has a strong and possibly variable component in the millimetre wavelength region. The authors' observations do not support this claim.

141.018 Optical variation of 3C 371. L. Ourassine (Urasin).
Nature, Vol. 254, 125 (1975).

While studying the nucleus of 3C446 Cannon and Penston put forward a hypothesis involving a small continuous source of constant luminosity which may be temporarily obscured by absorbing clouds. This hypothesis is supported by observations of the nucleus of the N-galaxy 3C371, which reveal variations of its luminosity during a characteristic time of $t = 40$ min, or 2.4×10^3 s.

141.019 High resolution radio observations of unidentified radio sources from the B2 catalogue.
G. Grueff, M. Vigotti.
Astron. Astrophys., Suppl. Ser., Vol. 19, 117 - 141 (1975).

The authors present high resolution radio observations of 199 sources selected from the B2 catalogue. Most of them are empty fields, namely, no optical object is visible at their position on the Palomar Sky Survey prints. Positions accurate to a second of arc, and structure information with a resolution of 6 arc sec are obtained. The observations have been made with the Westerbork Synthesis Radiotelescope, operating at a frequency of 5 GHz.

141.020 Westerbork six centimeter observations of a complete sample of quasars from the B2 catalogue.
C. Fanti, R. Fanti, A. Ficarra, L. Formiggini, G. Giovannini, C. Lari, L. Padrielli.
Astron. Astrophys., Suppl. Ser., Vol. 19, 143 - 187 (1975).

A list of 116 QSS's candidates from the B2 catalogue is given in an area of 0.142 steradians, between 8^h and 16^h in right ascension and declination around 29°30'. The identifications are made on the basis of the excess in UV colour, using 48'' Schmidt multicolour plates taken at Palomar. The suggested identifications have been observed at 6 cm at Westerbork and accurate radio positions, radio structure and spectra are given for them.

141.021 Fine structure of 25 extragalactic radio sources.
J. J. Wittels, C. A. Knight, I. I. Shapiro, H. F. Hinteregger, A. E. E. Rogers, A. R. Whitney, T. A. Clark, L. K. Hutton, G. E. Marandino, A. E. Niell, B. O. Rönnäng, O. E. H. Rydbeck, W. K. Klemperer, W. W. Warnock.
Astrophys. Journ., Vol. 196, 13 - 39 (1975).

25 extragalactic radio sources were observed interferometrically at 7.8 GHz ($\lambda \simeq 3.8$ cm) with five pairings of antennas. These sources exhibit a broad variety of fine structures from very simple to complex. Although the structure and the total power of some of these sources have remained unchanged within the sensitivity of the measurements during the year of observations, both the total flux and the correlated flux of others have undergone large changes in a few weeks.

141.022 Polarization of radio sources. VI. An oscillatory behavior of the intensity in a general solution of the radiation transfer problem in a plasma.
A. G. Pacholczyk, T. L. Swihart.
Astrophys. Journ., Vol. 196, 125 - 127 (1975).

The solution for a general form of the transfer equation is given. The total intensity is found to oscillate under general conditions, although the oscillations vanish under most special cases discussed previously in the literature. The oscillations are a manifestation of the Faraday effects and their interaction with polarized absorption.

141.023 Quasars: le mystère s'épaissit. L. Gouguenheim.
L'Astronomie, 89e année, p. 59 - 64 (1975).

141.024 **Some new radio sources associated with the Sersic peculiar galaxies.**
P. Kaufmann, P. M. dos Santos, M. A. Bráz, R. M. Borges.
Astrophys. Space Sci., Vol. 32, L25 - L27 (1975).

141.025 **The spatial distribution and cosmological evolution of scintillating radio sources.**
A. C. S. Readhead, M. S. Longair.
Monthly Notices Roy. Astron. Soc., Vol. 170, 393 - 404 (1975).

An analysis of the scintillation properties of complete samples of extragalactic radio sources indicates: (i) There exists a correlation between compact physical structure and high radio luminosity and redshift once allowance is made for several observational selection effects. (ii) Strongly scintillating radio sources exhibit strong cosmological evolution of the form inferred for quasars and powerful radio sources in general. The strong evolutionary effects are found in samples of both 3CR and 4C radio sources. (iii) Strongly scintillating radio galaxies exhibit strong cosmological evolution as previously inferred by Schmidt.

141.026 **A new variable radio source in Cygnus.**
R. L. Adgie, H. P. Palmer, M. V. Penston.
Monthly Notices Roy. Astron. Soc., Vol. 170, 31P - 34P (1975).

Observations made at NRAO, Greenbank, at a wavelength of 3.7 cm of a source in Cygnus revealed a flux variation from 2.2 to 6.3 Jy during 1973. Although the field was very crowded because of the low galactic latitude, accurate radio and optical position measurements have yielded a provisional identification with a blue stellar object which had a magnitude of 19.5 on the Sky Survey. The identification is confirmed by recent optical observations showing that the object had brightened to 18.5 by 1973 November. A spectroscopic examination has revealed an emission feature near 4300 Å, and the object is almost certainly a quasar.

141.027 **Investigations into reported anisotropies in radio source counts and spectra at 1421 MHz.**
A. R. Gillespie.
Monthly Notices Roy. Astron. Soc., Vol. 170, 541 - 549 (1975).

New surveys at Cambridge show that there is no evidence for any significant anisotropies in the spectral index distributions or counts of radio sources at 1421 MHz, contrary to reports by other observers.

141.028 **Search for optical identifications in the 5C3-radio survey. II. Statistical treatment and results.**
G. A. Richter.
Astron. Nachr., Vol. 296, 65 - 81 (1975).

On plates of the large Schmidt-telescope of Karl-Schwarzschild-Observatory Tautenburg 139 radio sources of the 5C3 area were inspected for possible identifications with optical objects. The results are published in paper I (Astron. Nachr., Vol. 295, 19 - 26 (1974). — See Abstr. 11.141.042) and in the appendix of the present paper. A detailed analysis of these objects showed a relatively large number of about 65 real identifications up to the utmost plate limit $B \approx 21^m7$, which corresponds to an identification rate of about 47%. The individual reliability of each possible optical identification is estimated. Apparently the identified objects are galaxies, "blue" and "neutral" quasars, and one H II-region of the Andromeda nebula. It is of great cosmological interest that no "red" quasars could be found.

141.029 **Close pairs of quasars – reality or projection effects?** B. V. Komberg.
Zemlya i Vselennaya, 1975, No. 1, p. 56 - 57. In Russian.

141.030 **Flux density measurements of radio sources at 2.14 millimeter wavelength.**
J. R. Cogdell, J. H. Davis, B. T. Ulrich, B. J. Wills.
Astrophys. Journ., Vol. 196, 363 - 368 (1975).

Flux densities of galactic and extragalactic sources, and planetary temperatures, have been measured at 2.14 mm wavelength (140 GHz). Results are presented for OJ 287; the galactic sources DR 21, W3, and Orion A; the extragalactic sources PKS 0106+01, 3C 84, 3C 120, BL Lac, 3C 216, 3C 273, 3C 279, and NGC 4151; and the sun, Venus, Mars, and Jupiter. Also presented is the first measurement of the 2.14-mm temperature of Uranus. The spectra of some of these sources are discussed. The flux density scale was calibrated absolutely. The measurements were made with a new continuum receiver on the 4.88-m radio telescope of The University of Texas.

141.031 **The radio spectra of the γ-Cygni source and the IC 1318b, c nebular complex.** C. Goudis.
Astrophys. Space Sci., Vol. 33, 103 - 109 (1975).

The radio spectra of both the γ-Cygni radio source and the IC 1318b, c nebular complex are established from all the available data. The non-thermal nature of the γ-Cygni source is established (sp. index -0.93 ± 0.30). The IC 1318b, c shows an optically thin thermal spectrum. The possibility of the γ-Cygni source being a SNR is discussed.

141.032 **Optical identifications of radio sources selected from the B2 catalogue. IV.** G. Grueff, M. Vigotti.
Astron. Astrophys., Suppl. Ser., Vol. 20, 57 - 82 (1975).

An attempt has been made to push the optical identifications of radio sources beyond the limit of the Palomar Sky Survey. A total of 115 fields, which were empty on the Palomar Sky Survey, have been photographed with the 48″ Schmidt telescope, and 200″ telescope of Hale Observatories. Reliable identifications are proposed for 44 radio sources, or 38% of the starting sample. Some implications of these results are discussed briefly. In particular, it is found that the number of identifications with galaxies increases very sharply when one goes deeper than $m_v \sim 19.5$, while almost no quasars are found fainter than 20^m5.

141.033 **Lyman-β and Fe II emission from QSOs and peculiar galaxies.** T. F. Adams.
Astrophys. Journ., Vol. 196, 675 - 682 (1975).

The scattering of Lβ in an idealized QSO envelope is examined, including the effects of noncoherent scattering. Approximate analytic results are presented for plane-parallel and spherical geometries. Detailed numerical results for the plane-parallel case are also presented to show that the approximate treatment is reasonably accurate. It is shown that appreciable errors can be made if the observations of Lβ in QSOs are interpreted in terms of an analysis based on coherent scattering. Similar formulae are then derived for the scattering of multiplet 3 UV of Fe II. The presence of optical Fe II emission (in this case, multiplet 42) is shown to indicate the presence of large amounts of matter in the region around the central nonthermal source. The effects of inhomogeneities in the scattering medium are considered briefly.

141.034 **The spectra of 3C 273 and PKS 0736+01.**
J. A. Baldwin.
Astrophys. Journ., (*Letters*), Vol. 196, L91 - L93 = Lick Obs. Bull., No. 690 (1975).

The spectra of the quasi-stellar objects 3C 273 and PKS 0736+01 are found to be almost identical, having strong Fe II, hydrogen, and helium permitted lines, but no forbidden lines. These spectra are quite distinct from typical low-redshift QSO spectra, and may be difficult to explain with models where photoionizations are the principal ionization mechanism. This may point to a different ionization and heating mechanism for the gas in 3C 273-type QSOs than in normal QSOs.

141.035   Interstellar scintillation of extragalactic radio
          sources.   J. J. Condon, D. C. Backer.
Astrophys. Journ., Vol. 197, 31 - 38 (1975).

Observations of 12 compact extragalactic sources were
made at 2695 and 8085 MHz in order to detect weak intensity
fluctuations caused by interstellar scintillation. Pulsar data are
used to estimate the parameters of the interstellar medium
needed to interpret the measured upper limits in terms of
source angular diameters which are much larger than the scintillation cutoff diameter. It is shown that the observed source
rms brightness temperatures are less than 10^{15} K and 10^{14} K at
2695 and 8085 MHz, respectively, making self-absorbed pro-
ton-synchrotron radiation and high-brightness coherent mechanisms unlikely. If the sources are composed of "point"
components, each source can contain no fewer than 10^4 such
components.

141.036   Time dependence of the integrated Stokes para-
          meters of compact radio sources at 5 GHz.
G. A. Seielstad, G. L. Berge.
Astron. Journ., Vol. 80, 271 - 281 (1975).

The authors have monitored the integrated 5-GHz radio
emission, including the complete polarization state as well as
the flux density, of 15 compact radio sources for a period of
27 months. In the data presented herein there are several
examples of interesting time variations. However, the diverse
behavior and other properties of the sources defy any satisfactory generalizations regarding mechanisms. Instead the
authors are led to a discussion of the individual sources separately.

141.037   On the errors in measurements of Ohio V radio
          sources in the light of the GB survey.
J. Machalski.
Acta Astron., Vol. 25, 43 - 61 (1975).

Positions and flux densities of 405 OSU V radio sources
surveyed at 1415 MHz down to 0.18 f.u. (Brundage et al.,
1971) have been examined in the light of data from the GB
survey made at 1400 MHz (Maslowski, 1972). An identification
analysis has shown that about 56% of OSU sources reveal themselves as single, 18% − as confused, 20% − as unresolved and
6% − having no counterparts in the GB survey down to 0.09
f.u. − seem to be spurious. The OSU V completeness is 67%
at 0.18 f.u. and 79% at 0.25 f.u.

141.038   A complete sample of radio sources identified with
          elliptical galaxies: radio luminosity function and
other properties.   G. Colla, C. Fanti, R. Fanti, I. Gioia,
C. Lari, J. Lequeux, R. Lucas, M.-H. Ulrich.
Astron. Astrophys., Vol. 38, 209 - 223 (1975).

In this paper the authors describe and discuss a homogeneous and complete sample of 54 bright elliptical galaxies
($m_{pg} \lesssim 15.7$) identified with radiosources of the B 2 catalogue. The radio luminosity function of elliptical galaxies
(using also 3 CR radiogalaxies) is derived and its changes with
optical absolute luminosity are studied.

141.039   The nonthermal radio sources at $l = 74°9$ and
          $b = +1°2$.
R. M. Duin, F. P. Israel, J. R. Dickel, E. R. Seaquist.
Astron. Astrophys., Vol. 38, 461 - 465 (1975).

Radio observations over a large frequency range and high
resolution mapping at 1415 MHz of the source G 74.9 + 1.2
show an extended nonthermal source; some 7 arcmin west a
pointsource is found, which seems to be extragalactic. The
pointsource is weakly polarized and has a flat spectrum which
shows a turnover at 7500 MHz. At high frequencies no variability is detected over a period of 26 months. The extended
component is identified as a supernova remnant. Its spectrum
and polarization are quite normal for a supernova remnant, but
a definite shell structure is not apparent from the map pre-

sented here.

141.040   Source counts at high spatial densities from pencil
          beam observations of background fluctuations.
J. V. Wall, D. J. Cooke.
Monthly Notices Roy. Astron. Soc., Vol. 171, 9 - 25 (1975).

The Parkes 64-m telescope has been used at 2700 and
5000 MHz to determine frequency distributions of the spatial
fluctuations in the background continuum. These distributions reflect the number−flux density relations for the weaker
sources. The very limited observations at 5000 MHz demonstrate the feasibility of the technique for determining the
number−flux density relation at this frequency to ∼10^6
sources per steradian. Brief comparison is made between the
relations derived here and the relations previously found at
lower frequencies.

141.041   Massive black holes in extragalactic radio source
          components?   M. J. Rees, W. C. Saslaw.
Monthly Notices Roy. Astron. Soc., Vol. 171, 53 - 57 (1975).

Mechanisms are discussed whereby massive black holes,
ejected from galactic nuclei, might conceivably travel through
the rarefied intergalactic gas, and release the necessary energy
over relevant time scales to explain extragalactic double radio
sources.

141.042   Quasar absorption spectra: radiative interactions
          between absorbing clouds and the origin of redshift
'doublets'.   M. J. Rees.
Monthly Notices Roy. Astron. Soc., Vol. 171, 1P - 5P (1975).

Some implications of recent high-dispersion spectra of
PKS 0237−23 are discussed. In particular, a mechanism is proposed whereby the observed redshift doublets with characteristic splitting $\Delta z \simeq 0.0012$ could arise.

141.043   Neutrino processes and QSOs.   K. Tennakone.
          Nature, Vol. 254, 399 - 400 (1975).

Peculiar jets of matter emanate from some galactic nuclei
and QSOs, including the galaxy M87 and the QSO 3C273. It is
generally believed that these jets are matter ejected from the
central regions of the object. As a result of non-conservation
of parity in weak interactions, the neutrino processes associated with gravitational collapse of a galactic nucleus in the
presence of a strong magnetic field can lead to ejection of matter from the nucleus in one direction. The author suggests that
the jets observed in M87 and 3C 273 are the result of this
mechanism.

141.044   On the distribution of quasars according to the
          degree of linear polarization and according to the
degree of homogeneity of the magnetic field.
V. M. Kolobov, V. N. Sazonov.
Astron. Zhurn. Akad. Nauk SSSR, Vol. 52, 274 - 277 (1975).
In Russian. English translation in Soviet Astron., Vol. 19, No. 2.

The observed distribution of the number of quasars n
according to the degree of linear polarization p_l at wavelength
λ 49 cm can be fitted by $n(p_l) \sim \exp(-p_l/\bar{p}_l)$, where $\bar{p}_l = 1.5\%$.
It is shown that such an exponential distribution can be ob-
tained in the frame of the following model. The radio emitting
area of a quasar consists of a great number $N \simeq \bar{p}_l^{-1}$ of regions
in each of which the magnetic field is constant and has random
orientation.

141.045   The decrease of intensity of Cassiopeia A at the
          frequency of 13 MHz.   E. A. Benediktov.
Astron. Zhurn. Akad. Nauk SSSR, Vol. 52, 441 - 442 (1975).
In Russian. English translation in Soviet Astron., Vol. 19, No. 2.

Antenna pattern passages of the source are recorded. The
signal level of the source is about 25% of the background level.
The data of the autumn months of 1963−1972 are used to
determine the decrease of the source intensity which is equal

to 0.4 ± 65% per year.

141.046 Significance of the angular diameter-redshift relation. R. C. Roeder.
Nature, Vol. 255, 124 - 125 (1975).

Hewish et al. (1974) have reported that interplanetary scintillation techniques reveal an absence of small-diameter sources at the largest redshifts. They have discussed the cosmological implications of the result using standard, homogeneous Friedmann models of the universe. Here the author points out the important effects on the angular diameter-redshift relation caused by small-scale inhomogeneities.

141.047 Accurate positions and identifications for eleven Ohio survey sources.
J. W. Warner, G. E. Assousa, B. Balick, E. R. Craine.
Publ. Astron. Soc. Pacific, Vol. 87, 103 - 106 (1975).

Accurate radio positions and optical identifications are presented for some sources from the Ohio survey.

141.048 *UBVr* sequences and observations of optically identified radio sources.
E. R. Craine, K. Johnson, S. Tapia.
Publ. Astron. Soc. Pacific, Vol. 87, 123 - 130 (1975).

Photoelectric observations of eight optically identified radio sources and comparison stars in nearby fields are presented. Six of these objects have continuous spectra and may belong to the BL Lac class of variables. The reported observations indicate possible short-term, small-amplitude variations in some of the optical counterparts of the radio sources.

141.049 On the identifications of the radio sources 3C73, 3C282, 3C321 and 3C383. J. E. V. Bystedt.
Astron. Astrophys., Vol. 39, 155 - 156 (1975).

New data on the structure and position of radio sources at 1415 MHz by Högbom and Carlsson (1974) has led to the identification of 3C73 and 3C321 with galaxies. An earlier identification of 3C383 has been confirmed, while that earlier suggested for 3C282 is probably not correct.

141.050 The optical emission-line spectrum of Cygnus A.
D. E. Osterbrock, J. S. Miller.
Astrophys. Journ., Vol. 197, 535 - 544 = Lick Obs. Bull., No. 686 (1975).

Spectrophotometric measurements are reported of the radio galaxy 3C405 = Cyg A, made using the image-tube image-dissector scanner on the Lick 120-inch telescope. The measurements were reduced to energy units by comparison with scans of standard stars made with the same system on the same nights. The emission lines and continuum were measured in the spectral region $\lambda\lambda3346-6731$ (in the rest system of Cyg A). The corrected line strengths are discussed for the information they contain on the physical conditions and the energy-input mechanism to the ionized gas. Published calculations of photoionization by a synchrotron spectrum, extending far into the ultraviolet, approximately match the observed emission-line spectrum.

141.051 The collision of a strong shock with a gas cloud: a model for Cassiopeia A. A. G. Sgro.
Astrophys. Journ., Vol. 197, 621 - 634 (1975).

The result of the collision of the shock with the cloud is a shock traveling around the cloud, a shock transmitted into the cloud, and a shock reflected from the cloud. The structure of such a cloud and its expected appearance to an observer are discussed and compared with the quasi-stationary condensations of Cas A. A model in which the soft X-radiation of Cas A is due to a collection of such clouds is discussed. The faint emission patches to the north of Cas A are interpreted as preshocked clouds which will probably become quasi-stationary condensations after being hit by the shock.

141.052 Luminosity of quasi-stellar objects.
S. A. Colgate, J. D. Colvin, A. G. Petschek.
Astrophys. Journ., (*Letters*), Vol. 197, L105 - L108 (1975).

A model of quasi-stellar objects is described in which infrared and optical photons are produced by the following sequence: supernova ejecta produce both plasma oscillations and a high electron temperature. The oscillations in turn produce transverse oscillations. These photons are then heated, first by interaction with the plasmons and later by inverse Compton collisions. A luminosity of 10^{47} ergs s^{-1} peaking near 100 μ can be achieved in a dimension of 5×10^{16} to 1×10^{17} cm, given a galactic nucleus of that dimension containing 10^8 stars and a gas density of 10^8 electrons cm^{-3}, and either supernovae or energetic stellar collisions with 50 $M_\odot$ ejecta of several MeV per nucleon energy.

141.053 Milli-arcsecond structure of 3C 84, 3C 273, and 3C 279 at 2 centimeter wavelength.
A. E. Niell, K. I. Kellermann, B. G. Clark, D. B. Shaffer.
Astrophys. Journ., (*Letters*), Vol. 197, L109 - L112 (1975).

Long baseline interferometer observations of 3C 84, 3C 273, and 3C 279 show structure on a scale of 1 milli-arcsec which is more complex than can be described by any simple one- or two-component models. Furthermore, for 3C 84 and 3C 273 the structure is not collinear.

141.054 Observations of compact radio nuclei in Cygnus A, Centaurus A, and other extended radio sources.
K. I. Kellermann, B. G. Clark, A. E. Niell, D. B. Shaffer.
Astrophys. Journ., (*Letters*), Vol. 197, L113 - L116 (1975).

Observations of Cygnus A show a compact radio core 2 milli-arcsec in extent oriented in the same direction as the extended components. Other large double- or multiple-component sources, including Centaurus A, have also been found to contain compact radio nuclei with angular sizes in the range 1–10 milli-arcsec.

141.055 An interpretation of the redshift problem of quasar-galaxy pairs. O. Obregón, H. Dehnen.
Astrophys. Space Sci., Vol. 34, 481 - 489 (1975).

By use of a theoretical scheme presented in a previous publication it is possible to calculate the functional relations among the actual cosmological redshift, the peculiar velocity, and the change of the angular separation for quasar-galaxy pairs. Applying the results to the quasi-stellar object Markarian 205 which seems to be connected with the spiral galaxy NGC 4319 the authors find, under the assumption of two different limiting cases, that the distance of the object Markarian 205 is reduced drastically, and that the velocity of one or the two objects remains under 0.1 of the velocity of light if the angle δ between the direction of motion of the quasar and the earth quasar-galaxy line lies in the range $120° \lesssim \delta \lesssim 180°$.

141.056 The identification of absorption redshift systems in quasar spectra.
M. Aaronson, C. F. McKee, J. C. Weisheit.
Astrophys. Journ., Vol. 198, 13 - 30 (1975).

The published spectra of eight quasars have been reexamined: 4C 05.34, PHL 957, PKS 0237−23, 1331+170, Ton 1530, 3C 191, PHL 938, and Mar 132. Fifteen systems appear well established, while nine more are suggested as possible. The existence of many previously identified redshift systems has been confirmed, but the authors' results indicate that the total number of well-established redshifts is substantially smaller than previously suspected. The distribution of absorption redshifts z_a in well-established Class II systems (those with z_a significantly less than the emission redshift z_e) is consistent with the statistical predictions of the cosmological hypothesis for both steady state and Friedmann cosmologies; on the other hand, certain values of $(1 + z_e)/(1 + z_a)$ appear favored and are suggestive of "line locking". Alternative explanations for the large fraction of unidentified lines are considered. The authors conclude that a number of these lines are probably members of additional redshift systems.

141.057 Doublet structure in the absorption redshifts in the spectrum of PKS 0237−23.
A. Boksenberg, W. L. W. Sargent.
Astrophys. Journ., Vol. 198, 31 - 43 (1975).

The spectrum of the quasar PKS 0237−23 has been observed over the wavelength range $\lambda\lambda 3730-4300$ at a resolution of 0.71 Å. Two new redshift systems, a complex of redshifts near $z_{abs} = 1.55$ and a system at $z_{abs} = 2.1758$, were found. The systems $z = 1.51, 1.59, 1.65$, and 1.67 discovered by earlier workers were all found to be double or multiple; in addition, the system $z_{abs} = 1.36$ may be double. A constant splitting, $\Delta z = 0.0012$, between adjacent redshifts, corresponding to 141 ± 9 km s^{-1} in velocity, occurs at least seven times. Most of the absorption lines are unresolved and have widths corresponding to velocity spreads of less than 50 km s^{-1} in the rest frames of the absorbing clouds. There is evidence for absorption line-locking involving superpositions of the lines of the C IV resonance doublet in nearby redshift systems.

141.058 Emission-line strengths and the chemical compositions of quasi-stellar objects.
Y.-W. T. Chan, E. M. Burbidge.
Astrophys. Journ., Vol. 198, 45 - 55 (1975).

Estimates of emission-line strengths in 160 QSOs are listed. They are used, together with those for 60 objects published earlier, to compile a table of rest-frame equivalent widths for a composite QSO containing 28 lines seen in objects with z from 0.06 to 3.53. Models are computed for matter partially filling a spherical volume, ionized by ultraviolet flux from a central object and optically thick with absorption from heavy elements taken into account. The results are very sensitive to the slope of the continuum for $\lambda < 912$ Å. There is no unique set of fitting parameters; ranges are possible. Values of the parameters μ and Y_{He} which give a reasonably good fit are $\mu = -0.8$ to -1.2, and $Y_{He} = 0.03$ to 0.08, where μ is the spectral index of the ionizing radiation and $Y_{He} = $ He/H by number.

141.059 Time-dependent effects in the radially streaming particle model. R. Hubbard.
Astrophys. Journ., Vol. 198, 57 - 62 (1975).

The radially streaming particle model for broad quasar and Seyfert galaxy emission features is modified to include sources of time dependence. The results seem to harmonize with reported observations of multiple components, variability, and transient features in the wings of Seyfert and quasi-stellar emission lines.

141.060 5 GHz polarization observations of 33 galactic radio sources. D. K. Milne, J. R. Dickel.
Australian Journ. Phys., Vol. 28, 209 - 230 (1975).

Polarization observations have been made of 33 galactic radio sources, mostly supernova remnants, at a frequency of 5 GHz using the 64 m telescope at Parkes. Maps of the observed polarization vectors superposed upon total intensity isotherms are presented for each source.

141.061 G2.4+1.4, a supernova remnant or ring nebula around a peculiar star. H. M. Johnson.
Astrophys. Journ., Vol. 198, 111 - 118 (1975).

G2.4+1.4 is a probable nonthermal radio source and an optical nebula which appears to be a supernova remnant (SNR). It also contains an O VI sequence star of great excitation. The author presents new radiofrequency-continuum and (nil) H 92α observations, optical spectroscopy, and Fabry-Perot scanner observations of the nebula. The object distance (5 kpc?), origin of gas kinematics (SNR expansion?), and

mode of excitation of the gas (photoexcitation and/or shock wave?) remain uncertain. The possible roles of the O VI star as "runaway" in a SNR, as a source of photoexcitation, and as an ejector of a "counterfeit" SNR are discussed.

141.062 The nature of quasars. S. van den Bergh.
Astrophys. Journ., (Letters), Vol. 198, L1 - L2 (1975).

Most elliptical galaxies occur in relatively rich clusters whereas the majority of Seyferts are field galaxies. It is pointed out that this difference might be used to find out whether quasars are generically related to giant ellipticals or to Seyfert galaxies. The fragmentary data that are so far available do not favor the view that quasars occur in cD galaxies.

141.063 Two new quasars near galaxies.
H. Arp, J. A. Baldwin, E. J. Wampler.
Astrophys. Journ., (Letters), Vol. 198, L3 - L5 = Lick Obs. Bull., No. 694 (1975).

A quasar of magnitude $V = 19.2$ and redshift $z = 1.94$ is reported at a distance of $1\rlap{.}'6$ from a galaxy whose apparent diameter is about $2\rlap{.}'0$. A second quasar of magnitude $V = 17.6$ and redshift $z = 0.71$ is reported at a distance of $1\rlap{.}'3$ from a galaxy whose apparent diameter is about $1\rlap{.}'5$. Both quasars were discovered by noticing blue stellar objects close to galaxies which were companions of larger nearby galaxies.

141.064 The orientation of double radio sources associated with elliptical galaxies. D. M. Gibson.
Astron. Astrophys., Vol. 39, 377 - 382 (1975).

The comparison of the relative orientations of oblate galaxies and their associated double radio sources with a distribution of position angle differences predicted theoretically provides a sensitive test of models in which the ejection has a preferred direction. The observed distributions are found to be consistent with random ejection; they are also found to be consistent with planar ejection provided the dispersion about the rotation plane is large ($\sim 45°$). Additionally, there is no statistical difference in the distributions of position angle difference when the close components are compared to the more distant components.

141.065 Quasar absorption lines − a review.
W. L. W. Sargent.
Bull. American Astron. Soc., Vol. 7, 261 (1975). − Abstr. AAS.

141.066 Quasar absorption lines: the effects of relativistic winds and radiation pressure. M. J. Rees.
Bull. American Astron. Soc., Vol. 7, 261 (1975). − Abstr. AAS.

141.067 Circular polarization of extragalactic variable radio sources at 8 GHz.
P. E. Hodge, H. D. Aller, T. V. Seling.
Bull. American Astron. Soc., Vol. 7, 262 (1975). − Abstr. AAS.

141.068 Optical polarization models of quasi-stellar objects and BL Lac objects. K. H. Nordsieck.
Bull. American Astron. Soc., Vol. 7, 262 (1975). − Abstr. AAS.

141.069 Massive black holes in extragalactic radio sources.
M. J. Rees, W. C. Saslaw.
Bull. American Astron. Soc., Vol. 7, 262 (1975). − Abstr. AAS.

141.070 Observations of variable radio sources at 18 cm wavelength.
J. C. Webber, L. K. Denoyer, K. S. Yang.
Bull. American Astron. Soc., Vol. 7, 262 (1975). − Abstr. AAS.

141.071 A quasar model revisited.
M. F. Barnothy, J. M. Barnothy.
Bull. American Astron. Soc., Vol. 7, 269 (1975). − Abstr. AAS.

141.072 On the significance of periodicities in the observed quasar redshifts and in the intrinsic redshift components as computed from Bell and Fort's quasar model.
G. J. Corso, J. M. Barnothy.
Bull. American Astron. Soc., Vol. 7, 269 (1975). – Abstr. AAS.

141.073 3C 273 as a galactic object. Y. P. Varshni.
Bull. American Astron. Soc., Vol. 7, 269 (1975).
Abstr. AAS.

141.074 Positions and flux densities of 1075 radio sources at 365 MHz. J. R. Sharp, F. N. Bash.
Astron. Journ., Vol. 80, 335 - 352 (1975).

This paper presents the positions and fringe amplitudes of 1075 discrete radio sources at 365 MHz obtained at the University of Texas Radio Astronomy Observatory using the Bandwidth Synthesis Interferometer. The average standard deviation of the positions is about 2 arcsec in right ascension and 1.5 arcsec in declination. This paper completes the analysis of the data taken with the Texas interferometer in its original form, providing a total of 3467 position measurements of 2779 radio sources.

141.075 Brightness distribution data on 2918 radio sources at 365 MHz.
W. D. Cotton, F. N. Owen, F. D. Ghigo.
Astron. Journ., Vol. 80, 353 - 378 (1975).

This paper is the second in a series describing the results of a program attempting to fit models of the brightness distribution to radio sources observed at 365 MHz with the Bandwidth Synthesis Interferometer operated by the University of Texas Radio Astronomy Observatory. Results for a further 2918 radio sources are given. An unresolved model and three symmetric extended models with angular sizes in the range 10–70 arcsec were attempted for each radio source. In addition, for 348 sources for which other observations of brightness distribution are published, the reference to the observations and a brief description are included.

141.076 Instationary objects of non-stellar nature.
Yu. P. Pskovskij.
Instationary stars and methods of their investigation. Phenomena of instationarity and stellar evolution, (see 003.012), p. 338 - 369 (1974). In Russian.

141.077 The 5C 5 survey of radio sources.
T. J. Pearson.
Monthly Notices Roy. Astron. Soc., Vol. 171, 475 - 505 (1975).

The 5C 5 survey, made with the Cambridge One-Mile telescope, covers an area about 4° in diameter at 408 MHz centred at $\alpha = 09^h 40^m$, $\delta = 47°00'$ to a limiting flux density of 8.7×10^{-29} W m^{-2} Hz^{-1} at the centre, and a concentric area of diameter about 1° at 1407 MHz to a limiting flux density of 1.8×10^{-29} W m^{-2} Hz^{-1}. The positions and flux densities of 230 sources observed at 408 MHz, and of 52 observed at 1407 MHz, are listed with suggested optical identifications for some of the sources.

141.078 Ultraviolet colours of quasars.
D. H. Morgan, K. Nandy.
Monthly Notices Roy. Astron. Soc., Vol. 171, 599 - 604 (1975).

The ultraviolet continuum colours of quasars have been computed attributing the observed redshift variation of the visual quasar colours corrected for the effects of emission lines, to (a) luminosity evolution, (b) a general curvature of spectrum, and (c) reddening due to intergalactic dust. For the dust model, graphite, silicate, and dust which obeys the observed interstellar extinction law have each been considered. The near ultraviolet colours are shown to provide a definite test of whether or not the visual colour-redshift variation is due to dust, and if appropriate, a differentiation between the dust models proposed.

141.079 The absorption-line spectrum of the quasi-stellar object PHL 957. D. W. Wingert.
Astrophys. Journ., Vol. 198, 267 - 279 (1975).

Six new high-dispersion spectra of the quasi-stellar object PHL 957 ($z_{em} = 2.69$) have been obtained with the Princeton integrating television system at the coudé focus of the 200-inch (5 m) telescope. Together with earlier data, 179 absorption lines were found, mostly between 3550 and 4500 Å The absorption redshift system at $z = 2.3099$ is most prominent, with 26 lines, including a very strong Lα line. Another system, close to the emission redshift, was found to be double, with $z = 2.6638$ and 2.6624. The redshift system at $z = 2.3099$ resembles interstellar material in ionization and abundance ratios, and constitutes strong evidence for absorption originating in a galaxy along the line of sight to the quasar. The double system at $z = 2.663$ has higher ion states and most likely is associated with the quasar.

141.080 The nebulosity around 3C 48. E. J. Wampler, L. B. Robinson, E. M. Burbidge, J. A. Baldwin.
Astrophys. Journ., (Letters), Vol. 198, L49 - L52 = Lick Obs. Bull., No. 701 (1975).

Spectrophotometric measurements of the nebulosity around 3C 48 have been made with the Lick image-tube scanner on the 3-m telescope. The forbidden emission lines [O III] $\lambda\lambda 5007$, 4959, [Ne III] $\lambda 3869$, and [O II] $\lambda 3727$ were detected in the nebulosity on the north side, while the Balmer lines were absent. The forbidden lines are narrower but have larger equivalent widths than those present in the spectrum of the central QSO, and have a slightly larger redshift, 0.370, than that of the QSO (0.368), thus ruling out the gravitational redshift hypothesis for this object. There is no evidence concerning the presence or absence of a galaxy of stars associated with 3C 48.

141.081 A comparison of some radio and optical properties of quasi-stellar sources and BL Lacertae objects.
P. D. Usher.
Astrophys. Journ., (Letters), Vol. 198, L57 - L60 (1975).

An empirical relation between radio spectral index and amplitude of optical variability for quasi-stellar sources with continuous, featureless optical spectra (QC sources) is extended to include sources with optical spectral lines. The radio spectral index at 5 GHz is found to correlate with the ratio of radio to optical luminosity in a clear-cut fashion when optical and radio variability is taken into account. QC sources are found to represent a tightly knit group vis-à-vis the general population. The predictive capabilities of the correlations are discussed. When optical variability is taken into account, the Hubble diagram provides further evidence for a cosmological interpretation of redshift.

141.082 Comments on Gott and Gunn's solution of double quasars. J. M. Barnothy.
Astrophys. Journ., (Letters), Vol. 198, L61 - L62 (1975).

The explanation offered by Gott and Gunn for the double quasar 1548+115—namely, that the gravitational lens brightening of faraway quasars by foreground quasars increases the chance of finding more pairs of quasars—is examined and shown to be incorrect.

141.083 Optical condensations and filaments in the northeast radio lobe of NGC 5128.
V. M. Blanco, J. A. Graham, B. M. Lasker, P. S. Osmer.
Astrophys. Journ., (Letters), Vol. 198, L63 - L64 (1975).

Extended diffuse and filamentary optical features are shown to exist in the outer northeast component of the radio source associated with NGC 5128. Some spectroscopic observations are reported, and preliminary remarks are made about the nature of these features.

141.084   The Parkes 2700 MHz Survey (seventh part). Supple-
mentary catalogue for the declination zone −4° to
−30°.   J. G. Bolton, A. J. Shimmins, J. V. Wall.
Australian Journ. Phys., Astrophys. Suppl. No. 34, p. 1 - 32 =
Separate print Div. Radiophys., C.S.I.R.O. Sydney (1975).

A catalogue of 348 radio sources is presented, covering
right ascensions 08^h00^m to 17^h00^m and 19^h30^m to 06^h30^m between declinations −4° and −30°. The regions omitted are
within ~10° of the galactic plane. The catalogue was compiled
from a 'fast' finding survey at 2700 MHz aimed at detecting
sources stronger than 0.5 Jy. Subsequent measurements of
flux density and position were made on all the sources which
were not in the Parkes 408 MHz catalogue, and on some
sources in the 408 MHz catalogue for which only data of low
accuracy were available. Positions of the sources were determined to an accuracy of 7″ arc r.m.s. in both coordinates.
Flux densities for all sources were also measured at 5009 MHz.
Identifications are suggested for 50 galaxies, 100 quasi-stellar
objects and 4 planetary nebulae.

141.085   The Parkes 2700 MHz Survey (eighth part). Cata-
logue for the declination zone −65° to −75°.
J. G. Bolton, P. W. Butler.
Australian Journ. Phys., Astrophys. Suppl. No. 34, p. 33 - 53 =
Separate print Div. Radiophys. C.S.I.R.O. Sydney (1975).

A catalogue of 515 radio sources is presented for the
region between declinations −65° and −75°. The survey is
believed to be complete to 0.22 Jy at 2700 MHz (~900
sources per steradian). Flux densities at 5009 MHz have been
measured for most of the sources stronger than 0.35 Jy at
2700 MHz.

141.086   The Parkes 2700 MHz Survey (ninth part). Supple-
mentary catalogue for the declination zone −45°
to −65°.   J. V. Wall, A. J. Shimmins, J. G. Bolton.
Australian Journ. Phys., Astrophys. Suppl. No. 34, p. 55 - 62 =
Separate print Div. Radiophys. C.S.I.R.O. Sydney (1975).

A catalogue of 166 radio sources is presented, covering an
area of 0.63 sr with right ascensions 00^h to 03^h, 04^h to 08^h30^m,
17^h30^m to 19^h and 20^h to 23^h between declinations −45° and
−65°. The regions omitted are either close to the galactic plane
or are covered in the third part of the Parkes 2700 MHz survey.

141.087   The Parkes 2700 MHz Survey (tenth part). Supple-
mentary catalogue for the declination zone +4° to
+25°.   A. J. Shimmins, J. G. Bolton, J. V. Wall.
Australian Journ. Phys., Astrophys. Suppl. No. 34, p. 63 - 83 =
Separate print Div. Radiophys. C.S.I.R.O. Sydney (1975).

A catalogue of 181 radio sources is presented covering
the declination zone +4° to +25° and all right ascensions,
omitting two regions within approximately 10° of the galactic
plane. The positions of sources were determined to an accuracy
of 10″ arc r.m.s. in both coordinates. Flux densities for most of
the sources were also measured at 5009 MHz. Identifications
are suggested for 17 galaxies and 62 quasi-stellar objects.

141.088   Radio brightness distribution of 3C 273.
R. G. Conway, D. Stannard.
Nature, Vol. 255, 310 - 312 (1975).

The authors provide here detailed information of the
radio brightness distribution of components A and B and
the jet of 3C 273.

141.089   On an attempt at an explanation of quasar stability.
I. Nedyalkov.
Godishn. Vissh. tekhn. uchebni. zaved. Fiz., Vol. 9, No. 2, p.
37 - 44 (1972/1974). In Russian. − Abstr. in Referativ. Zhurn.
51. Astron., 5.51.793 (1975).

141.090   Possible effect of misidentification of QSOs on the
redshift distribution.   D. Basu.
Astrophys. Letters, Vol. 16, 53 - 55 (1975).

The emission lines of QSOs on entering the U, B, V
filters may change the (U−B) and (B−V) color indices of the
objects such that these indices have values similar to those of
main sequence stars. Some QSOs will thus be lost due to misidentification and may account for some of the "holes" in the
redshift (z) distribution. It is shown that this effect is important at $z > 0.7$.

141.091   Close associations of pairs of objects at different
distances.   J. F. Dolan.
Astrophys. Letters, Vol. 16, 65 - 66 (1975).

The equation of Burbidge et al. (1974) used to calculate
chance coincidences between pairs of objects at different distances is observationally investigated and found to be valid.
The result supports their inference that the presently known
cases of small angular separation can neither prove nor dis-
prove the cosmological distances of QSOs with statistical confidence.

141.092   Theoretical emission line profiles in QSOs and
Seyfert galaxies.
G. R. Blumenthal, W. G. Mathews.
Astrophys. Journ., Vol. 198, 517 - 526 = Lick Obs. Bull.,
No. 689 (1975).

The gas responsible for the observed emission line
spectra in QSOs and Seyfert galaxies is assumed to be in the
form of small clouds which can be accelerated to high velocities by ultraviolet radiation from a small central source. The
authors show that for lines produced by recombination, the
outward motion of the cloud system generates a logarithmic
emission line profile which is independent of the mass distribution of the clouds, the change in the gas density and
temperature in the clouds, and the initial location of the
cloud-forming region. The logarithmic profile agrees very well
with three symmetric profiles recently observed by Baldwin,
except (as expected) in the cores of the lines and occasionally
in the far wings. They discuss how the basic logarithmic line
profile is influenced by drag, gravity, and chemical composition, dust, and other effects, some of which may account for
the deviations of the observed profiles from the logarithmic
form.

141.093   A multiple pulsar model for quasi-stellar objects
and active galactic nuclei.
J. Arons, R. M. Kulsrud, J. P. Ostriker.
Astrophys. Journ., Vol. 198, 687 - 707 (1975).

The authors have attempted to develop a unified, self-
consistent, and reasonably complete model for QSOs and
active galactic nuclei that encompasses both thermal and nonthermal emission processes, both steady state and fluctuations about the mean; and they have attempted to follow,
however roughly, matter and energy from their liberation
during pulsar-forming supernova explosions to their transformation into the ultimately observable spectrum.

141.094   Statistical acceleration of relativistic particles in an
assembly of spherical electromagnetic waves.
R. M. Kulsrud, J. Arons.
Astrophys. Journ., Vol. 198, 709 - 715 (1975).

It is shown that if a charged particle moves among a large
number of antennae, each of which radiates an electromagnetic
wave, the energy and momentum of the particle is increased in
a stochastic manner. This acceleration is a resonant process: at
various points on its orbit, the particle encounters regions in
which the phase of the beats between a pair of waves encounters the frame in which the particle is, on average, at rest. The
energy and momentum of a particle change in a secular manner
at these points, and this phenomenon, when summed over all

resonances, leads to a stochastic acceleration if the antennae are randomly located. This process may be significant in models of quasi-stellar objects and active galactic nuclei which rely upon strong electromagnetic waves as the basic energy input.

141.095 B 2.1028 + 31: a possible association of a quasar with a group of galaxies.
P. Battistini, A. Braccesi, L. Formiggini, C. Lari.
Astron. Astrophys., Vol. 40, 217 - 219 (1975).
        The radio source B 2.1028 + 31 is identified with a quasar of redshift $z = 0.177$, surrounded by some galaxies in a group. Radio structure and colours of the quasar are given. Photoelectric colours of one of the galaxies indicate that the quasar is at the same distance as the group.

141.096 The Hubble diagram for the brightest quasars.
R. E. Hills, J. N. Bahcall.
Ann. New York Acad. Sci., Vol. 224, 58 - 64 = Max-Planck-Inst. Radioastron., Bonn, Sonderdruck, Ser. A, No. 18 (1973).

141.097 Double radio sources: energetic evidence that galaxies remember. W. D. Metz.
Science, Vol. 188, 1289 - 1292 (1975).

141.098 A spherical subsystem of galactic radio sources.
A. G. Gorshkov, M. V. Popov.
Astron. Zhurn. Akad. Nauk SSSR, Vol. 52, 538 - 545 (1975). In Russian. English translation in Soviet Astron., Vol. 19, No. 3.
        A considerable concentration of sources with flat spectra towards the galactic center was obtained as the result of a statistical investigation of data of the Ohio Survey (1415 MHz). Quantitative evaluations of the phenomena testified that these sources form a spherical subsystem of the Galaxy. The parameters of this subsystem are close to those of the old population of the Galaxy, such as globular clusters and RR Lyrae stars. The radio luminosity of the object is 10^{33} erg/sec, the total number of sources is about 7000.

141.099 On detection of QSOs by ultra low dispersion spectroscopy.
M. K. V. Bappu, M. Parthasarathy.
Bull. Astron. Soc. India, Vol. 2, 36 - 37 (1974). – Abstract.

141.100 QSO's, observations, and the redshift problem.
H. C. Arp, with notes by M. Shara, G. Cavallo, J. Kormendy, C. Bloch.
High energy astrophysics and its relation to elementary particle physics, (see 012.018), p. 1 - 76 (1974).

141.101 Theoretical problems of high energy astrophysics.
G. R. Burbidge, with notes by R. Epstein, K. Brecher, H. G. Hughes.
High energy astrophysics and its relation to elementary particle physics, (see 012.018), p. 173 - 199 (1974).

141.102 Evidence for discordant redshifts. H. Arp.
        The redshift controversy, (see 003.021), p. 15 - 58 (1973).

141.103 The linear polarization of radio sources. I. Observations at wavelengths of 6, 11, 18 and 21 cm.
F. F. Gardner, J. B. Whiteoak, D. Morris.
Australian Journ. Phys., Astrophys. Suppl., No. 35, 36 pp. (1975).
        The results of surveys made at Parkes of linear polarization at wavelengths of 6, 11, 18 and 21 cm are presented for a total of 1121 small-diameter radio sources or source components. All the objects are south of declination +27°, and the source sample is believed to be virtually complete down to a limiting flux density of 0.8 f.u. at 2700 MHz.

141.104 Culgoora-2 list of radio source measurements at 80 MHz. O. B. Slee, C. S. Higgins.
Australian Journ. Phys., Astrophys. Suppl., No. 36, 60 pp. (1975).
        The 80 MHz survey of radio sources made with the Culgoora radioheliograph has been extended by a further 1748 previously catalogued sources in the declination range −48° to +35°. Positions, flux densities and beam broadening measured with this 3'.7 arc resolution instrument are given for 1291 sources, while 457 undetected sources are given in a separate list. Sucess rates for the detection of various classes of radio sources at 80 MHz are listed and discussed.

141.105 0846 + 51 W1. H. C. Arp, A. G. Willis, H. de Ruiter.
IAU Circ., No. 2750 (1975).

141.106 3C 249.1. L. J. Eachus, W. Liller.
IAU Circ., No. 2784 (1975).

141.107 26.3 MHz radio source survey with an absolute flux scale. M. R. Viner.
Thesis, Maryland Univ., College Park (USA). 247 pp. University Microfilms Order No. 74-17,066 (1973).

141.108 Structure of compact radio sources at 10.7 GHz.
D. B. Shaffer.
Thesis, California Inst. Techn., Pasadena (USA). 106 pp. University Microfilms Order No. 74-14,285 (1974).

141.109 Absorption-line spectrum of the quasi-stellar object PHL 957. D. W. Wingert.
Thesis, Princeton Univ., Princeton, N. J. (USA). 58 pp. University Microfilms Order No. 74-17,501 (1974).

141.110 Possibility of a gravitational effect in the spectra of quasi-stellar objects. 2. M. C. Durgapal.
Journ. Phys. A, (Math., Nuclear, General), Vol. 7, 2236 - 2247 (1974).

141.111 Emission and absorption lines of QSOs and properties of the gas that produces them. Y. W. T. Chan.
Thesis, California Univ., San Diego (USA). 231 pp. University Microfilms Order No. 74-19,605 (1974).

141.112 Extragalactic double radio sources. The current observational and theoretical position.
R. D. Blandford, M. J. Ress.
Contemporary Phys., (GB), Vol. 16, 1 - 16 (1975).

141.113 Six years of pulsars. B. Onderlička.
Cesk. Cas. Fys., Vol. 24, 497 - 500 (1974). In Czech. – Letter.

141.114 Comment on the quark burning model of quasars.
T. Toiya, K. Terasaki.
Progr. Theor. Phys. Japan, Vol. 52, 713 - 715 (1974).
        The authors calculate the burning life of a quasar due to three quark collisions $q_1 + q_1 + q_1 \rightarrow + q_3$ assuming pair type collisions to be suppressed.

141.115 Modified Hoyle-Fowler model for quasars.
M. C. Durgapal.
Indian Journ. Pure Applied Phys., Vol. 12, 457 - 458 (1974).
        The Hoyle-Fowler model has been modified by taking two density distributions. Central redshifts much larger than 3.5 are possible in the proposed modified model; it can also account for the large width of emission lines.

Die Quasare.   See Abstr. 003.143.

The determination of the electron to proton inertial mass ratio via molecular transitions.   See Abstr. 022.010.

Ultraviolet absorption lines arising on metastable states.   See Abstr. 022.073.

A cross-dispersed echelette spectrograph and a study of the spectrum of the QSO 1331 + 170. See Abstr. 034.010.

Scientific instrumentation of the Radio-Astronomy-Explorer-2 satellite.   See Abstr. 051.023.

A search for VHF radio pulses in coincidence with celestial gamma-ray bursts.   See Abstr. 061.010.

The radio and infrared spectrum of early-type stars undergoing mass loss.   See Abstr. 064.002.

The spectrum of the free-free radiation from extended envelopes.   See Abstr. 064.035.

Central gravitational redshifts from static massive objects.   See Abstr. 066.039.

Influence of moving irregularities in the ionosphere on the irregular refraction of Cygnus-A.   See Abstr. 083.051.

UBV and JHKL photometry of the radio star UX Ari = HD 21242.   See Abstr. 119.010.

High-resolution observations of the radio emission from Beta Persei.   See Abstr. 121.059.

A search for variable continuum radio emission from red giant stars.   See Abstr. 122.053.

A multifrequency study of the radio structure of 3C10, the remnant of Tycho's supernova. See Abstr. 125.022.

21 cm observations of hot neutral intercloud gas. See Abstr. 131.501.

Thermal bremsstrahlung radiospectra for inhomogeneous objects, with an application to MWC 349. See Abstr. 131.521.

A study of the nebulae S 206 and S 209. See Abstr. 131.539.

The Carina nebula at 3.4 and 6 cm. See Abstr. 132.039.

Radio observations at 5 GHz of southern planetary nebulae.   See Abstr. 133.012.

Sources of infrared and submillimeter radiation. See Abstr. 141.602.

Search for millimeter-wave emission from Uhuru X-ray sources and radio binary stars.   See Abstr. 142.015.

Soft X-ray spectroscopy of three extragalactic sources.   See Abstr. 142.043.

The X-ray, optical, and radio behavior of Scorpius X-1: the 1971 coordinated observations. See Abstr. 142.046.

The X-ray, optical, and radio behavior of Scorpius

X-1: the 1972 coordinated observations. See Abstr. 142.047.

Highly polarized radio outburst from Cygnus X-3. See Abstr. 142.058.

Nuclei of galaxies and quasars. See Abstr. 158.019.

Possible power source of Seyfert galaxies and QSOs. See Abstr. 158.021.

A radio survey of Markarian galaxies at 6 centimeters.   See Abstr. 158.026.

Radio and optical data on a complete sample of radio faint galaxies.   See Abstr. 158.031.

The structure of the radio galaxy NGC 1265. See Abstr. 158.040.

Broad Balmer emission lines in radio galaxies. See Abstr. 158.045.

Physical conditions in active nuclei − I. The Balmer decrement.   See Abstr. 158.048.

Projection des galaxies Sc et des radiosources quasi-stellaires (QSS) sur les plans de coordonnées supergalactiques. See Abstr. 158.059.

3C 411: a newly discovered N galaxy with a large redshift.   See Abstr. 158.066.

On the emission line variability in nuclei of Seyfert galaxies and quasars.   See Abstr. 158.072.

The fate of gas in elliptical galaxies and the density evolution of radio sources.   See Abstr. 158.083.

Variability and circular polarisation in the nucleus of NGC 5128 at 10.7 GHz.   See Abstr. 158.095.

Observational problems of high energy astrophysics. See Abstr. 158.132.

Extragalactic observational astronomy. See Abstr. 158.134.

Redshifts as distance indicators. See Abstr. 158.135.

Four-color photometry of OJ 287 during its recent three-magnitude decline.   See Abstr. 158.305.

Variability and optical-radio properties of BL Lacertae objects.   See Abstr. 158.312.

Radio properties of BL Lac type objects. See Abstr. 158.313.

Pencil beam observations of Abell clusters of galaxies. I. 2695 MHz.   See Abstr. 160.011.

Sur l'existence de la supergalaxie locale et la distribution de galaxies et des radiosources quasistellaires (QSSs) dans cet objet.   See Abstr. 160.016.

The cosmological redshift.   See Abstr. 162.013.

Kinematical description of quasi-stellar objects. See Abstr. 162.071.

Pulsars

141.301 Pulsar PSR 1919 + 21: notches, drifting subpulses, microstructure, and other emission.
J. M. Cordes.
Astrophys. Journ., Vol. 195, 193 - 202 (1975).

The radiofrequency dependence of intensity fluctuations of PSR 1919 + 21 is discussed. An intensity model that includes drifting subpulses and an independent zero- and low-frequency (steady) intensity is shown to be quantitatively consistent with the salient features of fluctuation spectra. Drifting subpulses are modeled as Gaussian envelopes that modulate a random intensity, a representation that permits estimation of the separate drifting subpulse and steady contributions to the mean intensity profile. Quantitative evaluation of the model shows that the steady intensity dominates the pulse character above 200 MHz. The results also suggest that microstructure and notches in mean pulse profiles are associated with the drifting subpulses. It is suggested that the drifting subpulses and the steady emission originate from different radii in the pulsar magnetosphere.

141.302 Diffraction effects in pulsating radio sources.
F. A. Hopf, W. J. Cocke, N. D. Lubart.
Astrophys. Letters, Vol. 16, 35 - 38 (1975).

A pulsar radio emission model is developed to explore the effects of diffraction on pulsar beaming. At very low frequencies ($\nu \lesssim 2n$ MHz, where n is the number of pulse components), the authors predict an increase of pulse width and the discovery of new pulsars. The pulse width increase might be difficult to untangle from interstellar scintillation effects. Observation of these diffraction effects would place restrictions on the size and radius of curvature of the emission regions.

141.303 Observations of pulsar radio emission. I. Total-intensity measurements of individual pulses.
J. H. Taylor, R. N. Manchester, G. R. Huguenin.
Astrophys. Journ., Vol. 195, 513 - 528 (1975).

The total intensities of individual pulses and subpulses were measured for 16 pulsars by using a multi-channel, de-dispersing receiver and the NRAO 300-foot telescope. Fluctuation statistics are analyzed as a function of pulsar longitude, and cross-correlation analyses show that fluctuations observed in different parts of the pulse window are sometimes, but not always, closely related. These results are consistent with a pulsar model in which the radio emission arises from charged-particle bunches streaming along curved field lines above the pulsar magnetic pole.

141.304 Discovery of a pulsar in a binary system.
R. A. Hulse, J. H. Taylor.
Astrophys. Journ., (*Letters*), Vol. 195, L51 - L53 (1975).

The authors have detected a pulsar with a pulsation period that varies systematically between $0\overset{s}{.}058967$ and $0\overset{s}{.}059045$ over a cycle of $0\overset{d}{.}3230$. No eclipses are observed. The authors infer that the unseen companion is a compact object with a mass comparable to that of the pulsar. In addition to the obvious potential for determining the masses of the pulsar and its companion, this discovery makes feasible a number of studies involving the physics of compact objects, the astrophysics of close binary systems, and special- and general-relativistic effects.

141.305 Pulsar extinction. F. C. Michel.
Astrophys. Journ., (*Letters*), Vol. 195, L69 - L71 (1975).

The author shows that the plasma from a very slowly rotating pulsar can escape without forcing open the field lines. His suggestion is that such opening is essential for the radio

emission, and hence the consequence for slow pulsars is extinction of the pulsar phenomenon. The critical period is estimated to be about 3 seconds for typical pulsar magnetic moments (10^{20} weber-m), in reasonable agreement with the data, which suggest a cutoff in this vicinity. Such slow 'pulsars' nevertheless should continue to emit relativistic particles as long as surface ions (e.g., iron) are available.

141.306 Observations of pulsar spectra at 1420 MHz.
A. Wolszczan, K. H. Hesse, W. Sieber.
Astron. Astrophys., Vol. 37, 285 - 290 (1974).

Using a correlation spectrometer, narrow-band spectra integrated over short time intervals have been measured at 1420 MHz for four pulsars. The observed large scatter of decorrelation widths is explained in terms of a simple theory based on a ray approximation. The authors conclude that the statistical properties of this scatter are not due to experimental errors, but depend on the number of interfering beams reaching the observer as well as on the pulsar dispersion measure.

141.307 Internal magnetic fields of pulsars, white dwarfs, and other stars. R. F. O'Connell.
Astrophys. Journ., Vol. 195, 751 - 752 (1975).

The possibility that the very early magnetic fields in pulsars and other stars have a large number of nodes m in the radial function, has been considered by some authors. However, these analyses have not taken into account the fact that, as the author shows here, such large m values imply large internal magnetic fields. Such large fields could affect the interior dynamics and may also give rise to stability problems.

141.308 On the nature of the binary system containing the pulsar PSR 1913 + 16.
A. R. Masters, D. H. Roberts.
Astrophys. Journ., (*Letters*), Vol. 195, L107 - L111 (1975).

The authors consider the geometry and apsidal motion of the binary system containing the pulsar PSR 1913 + 16 and show that the observations already made by Hulse and Taylor strongly suggest that the companion to the pulsar is a compact star or, possibly, a helium main-sequence star. General-relativistic properties are discussed. They note that further observations of this system may provide the first direct measurement of the mass of a pulsar.

141.309 Some implications of period changes in the first binary radio pulsar. K. Brecher.
Astrophys. Journ., (*Letters*), Vol. 195, L113 - L115 (1975).

The absence of a large advance of the periastron of the first binary pulsar is shown to imply that the unseen companion is also collapsed. The existence of a lower limit to the mass of the observed pulsar of about 0.3 $M_\odot$ requires a minimum general relativistic periastron advance of about $2°yr^{-1}$. Other causes of apparent and true pulse period changes are also briefly considered.

141.310 First radio pulsar in a binary system.
G. S. Bisnovatyj-Kogan.
Priroda, No. 3.75, p. 100 - 101 (1975). In Russian.

141.311 On the possibility of a false detection of pulsar velocities. I. V. Chashej, V. I. Shishov.
Pis'ma v Astron. Zhurn., Vol. 1, No. 1, p. 18 - 22 (1975). In Russian.

It is shown that the temporal shift of the pulsar scintillations cross-correlation function can be due not only to the drift of diffraction pattern but also to noise of the receiving apparatus.

141.312 Connection between pulsars and supernova remnants. G. S. Tsarevskij.

New problems of astrophysics, Publ. Astrophys. Winter School, 1972, (see 012.001), p. 118 - 136 (1974). In Russian.

141.313 On particle emission by a pulsar. V. V. Usov.
New problems of astrophysics, Publ. Astrophys. Winter School, 1972, (see 012.001), p. 137 - 144 (1974). In Russian.

141.314 Cyclotron radiation of relativistic electrons in pulsars. V. M. Charugin.
New problems of astrophysics, Publ. Astrophys. Winter School, 1972, (see 012.001), p. 145 - 151 (1974). In Russian.

141.315 Pulsar scintillation on inhomogeneities of the interstellar plasma. L. A. Pustil'nik.
New problems of astrophysics, Publ. Astrophys. Winter School, 1972, (see 012.001), p. 152 - 165 (1974). In Russian.

141.316 Theory of pulsars: polar gaps, sparks, and coherent microwave radiation.
M. A. Ruderman, P. G. Sutherland.
Astrophys. Journ., Vol. 196, 51 - 72 (1975).
The huge magnetic fields characteristic of pulsars cause the nuclei (largely iron) of the stellar surface to form a tightly bound condensed state. Except for the very young Crab pulsar, theory and observation both support the view that the stellar surface is not hot enough to sustain an outflow of positive ions to balance the outflow of electrons as charge leaves the magnetosphere through the light cylinder along the open magnetic field lines. Adopting the conventional assumption that electrons do not return to the neutron star by coming back through the light cylinder along the open field lines, the authors are led to the following consequences for a neutron star whose magnetic moment tends to be antiparallel (as opposed to parallel) to its spin angular momentum: A polar magnetospheric gap is formed that spans the open field lines from the stellar surface up to an altitude of about 10^4 cm. The gap continually breaks down (sparking) by forming electron-positron pairs on a time scale of a few microseconds.

141.317 Optical radiation from the Crab pulsar.
P. A. Sturrock, V. Petrosian, J. S. Turk.
Astrophys. Journ., Vol. 196, 73 - 82 (1975).
Possible mechanisms for producing the optical radiation from the Crab pulsar are proposed and discussed. There are severe difficulties in interpreting the radiation as being produced by an incoherent process, whether it be synchrotron radiation, inverse-Compton radiation, or curvature radiation. It is proposed, therefore, that radiation in the optical part of the spectrum is coherent. Calculations, which involve a number of simplifying assumptions, indicate that the optical radiation from the Crab pulsar can be understood in this way if the mass of the star is approximately 0.3 $M_\odot$. Various consequences of this model, which may be subjected to observational test, are discussed.

141.318 Observations of pulsar radio emission. II. Polarization of individual pulses.
R. N. Manchester, J. H. Taylor, G. R. Huguenin.
Astrophys. Journ., Vol. 196, 83 - 102 (1975).
Observations of the complete polarization characteristics of individual pulses at frequencies between 110 and 450 MHz are presented for a total of 12 pulsars, together with linear polarization characteristics at 1400 MHz for four pulsars. Subpulses are often highly polarized, with linear polarization generally dominating over circular.

141.319 Properties of the Hulse-Taylor binary pulsar system.
L. W. Esposito, E. R. Harrison.
Astrophys. Journ., (Letters), Vol. 196, L1 - L2 (1975).
Various elementary calculations are summarized concerning the close-binary-system pulsar PSR 1913+16, discovered by Hulse and Taylor. The results are approximate and indicate only the relative importance of various relativistic and nonrelativistic effects.

141.320 Periastron shifts in the binary system PSR 1913+16: theoretical interpretation. C. M. Will.
Astrophys. Journ., (Letters), Vol. 196, L3 - L5 (1975).
Measurements of the shift in the periastron of a recently discovered close binary system containing a pulsar may be interpreted in several ways: as a test of relativistic gravity, as a means of measuring the total mass of the system, or as a means of setting limits on the size and structure of the pulsar's companion. The author presents formulae for the relativistic periastron shift and for the shift produced by the bodies' quadrupole moments, and discusses each interpretation in detail.

141.321 The velocity of pulsars and interstellar irregularities in the scintillation pattern.
N. A. Lotova, I. V. Chashey (Chashej).
Astrophys. Space Sci., Vol. 32, 331 - 338 (1975).
The scintillation theory is developed for application to the interstellar medium taking into account both the movement of the pulsars and the movement of the interstellar irregularities. It is shown that the velocity of the drifting pattern differs essentially from that for the pulsars. In contrast with the interplanetary scintillation, the asymmetry of the form of the cross-correlation function of the interstellar scintillations is caused not only by the motion of the interstellar irregularities, but also by the movement of the source itself.

141.322 Amplification of radiation by relativistic particles in a strong magnetic field. R. D. Blandford.
Monthly Notices Roy. Astron. Soc., Vol. 170, 551 - 557 (1975).
The possibility of maser action by relativistic charged particles confined to one-dimensional motion in a strong magnetic field is considered. It is shown that if the field lines are curved to form circular arcs, genuine maser action is impossible. Under more complex electrodynamical conditions when, for instance, points of inflection and torsion may be present in the field geometry or particle creation, electrostatic acceleration or plasma effects may be occurring, maser action is not impossible. Simple estimates suggest that this type of mechanism cannot be excluded from being responsible for the coherent radio emission from pulsars.

141.323 Pulsar geometries. I. Basic model. L. Oster.
Astrophys. Journ., Vol. 196, 571 - 577 (1975).
A "geometrical" model for pulsar emissions is presented according to which they occur perpendicular to organized magnetic field structures, but in finite sets of narrow beams that rotate with constant speed in the plane perpendicular to the field; such a "wagon wheel" pattern might arise out of interference among wedge-shaped patterns of coherent synchrotron radiation near the surface of the neutron star, or it may be related to standing plasma waves in a circumstellar magnetic field. No further requirements are necessary with regard to this emission mechanism to interpret the existence and preferred location of interpulses, fast and slow pulse-amplitude modulations, and drifting subpulses in terms of the inclination of magnetic field and rotational axes of the neutron star, and the accidental latitude position of the observer.

141.324 Rotating magnetospheres: frozen-in-flux violation.
F. C. Michel.
Astrophys. Journ., Vol. 196, 579 - 582 (1975).
The author shows that previous solutions for the distant magnetic field structure about a rotating axisymmetrically (dipole) magnetized object are internally inconsistent. It ap-

pears necessary to allow for an effective (nonuniform) rotation rate $\omega(f)$ which differs from that of the star itself and which varies with magnetic latitude. As a result, the frozen-in-flux approximation must fail somewhere in the system. With this modification, the resultant equations seem self-consistent.

141.325 Timing effects in pulsed binary systems.
J. C. Wheeler.
Astrophys. Journ., (Letters), Vol. 196, L67 - L70 (1975).

An analysis of a binary stellar system is made in terms of arrival-time data. This is the form which pertains most directly to systems in which one star pulses with an accurately known period, as is the case with the recently discovered binary pulsar and the pulsed binary X-ray sources. Measurement of differences of order 100 μs in time of arrival across the orbit might in principle serve to determine both masses and the inclination angle in the case of the binary pulsar, independent of the rotation of the periastron. Measurement to $\lesssim$ 1 ms could serve to determine the mass of the primary in Cen X-3.

141.326 Some remarks on the hypothesis of relativistic beaming of pulsar emission.
V. V. Zheleznyakov, V. E. Shaposhnikov.
Astrophys. Space Sci., Vol. 33, 141 - 145 (1975).

The hypothesis of the relativistic beaming of pulsar emission is discussed in connection with objections advanced by Manchester et al. (1973). It is shown that the same pulse duration of some pulsars with the non-power-law spectrum over a wide frequency interval may be associated either with large source sizes in longitude, or with location of the emitting source at various distances from the rotation axis of the neutron star (depending on the emission frequency).

141.327 The alignment of the Crab pulsar magnetic axis.
P. B. Jones.
Astrophys. Space Sci., Vol. 33, 215 - 230 (1975).

The magnetic distortion is estimated for neutron stars in which the matter in the interior consists of superfluid neutrons and superconducting protons. It is shown that for the Crab pulsar PSR 0531+21, which is considered to have the greater part of its mass in the form of superfluid neutrons and superconducting protons, the magnetic distortion is almost certainly more important than the elastic energy of the outer shell in determining the departure of the inertia tensor from its spherically symmetric form. With the assumption that internal and external magnetic fields have the same symmetry axis, the external field dipole moment of the Crab pulsar is predicted to be approximately perpendicular to the spin direction, in agreement with a number of published interpretations of observational data.

141.328 Amplitude-modulated noise: an empirical model for the radio radiation received from pulsars.
B. J. Rickett.
Astrophys. Journ., Vol. 197, 185 - 191 (1975).

The radio radiation received from pulsars shows erratic variations over time and frequency. Experimentally these have been characterized by the autocorrelation function of the observed intensity and radio power spectrum. This paper gives theoretical expressions for these functions, under the assumption that the received radiation is described by random Gaussian noise that has been amplitude modulated. The conclusions are a specific test for unresolved pulse components, which are easily confused with system noise, and a test for the significance of pulsar spectrum variations.

141.329 Self-consistent rotating magnetosphere.
F. C. Michel.
Astrophys. Journ., Vol. 197, 193 - 197 (1975).

We show that, to a good approximation, the emission of particles from the poles of an axisymmetrically (dipole) mag-netized rotating star consists of particles of one sign from essentially the entire magnetic polar region ("cap"). Particles of the opposite sign are emitted in a thin sheath ("auroral zone") at the perimeter of the polar cap. This flow pattern determines the torque on the star and the distribution of plasma and fields at large distances from the star. Thus the basic first-order axisymmetric rotator problem seems to be solved in the limit that the particle inertia is small.

141.330 Synchrotron emission from an oblique rotator and its application to pulsars.
B. Miller, B. J. Eastlund.
Phys. Rev. Letters, Vol. 34, 901 - 905 (1975).

A method is presented which permits rapid interpretation of the synchrotron patterns emitted by a charged particle moving in the uniform magnetic field of an oblique rotator. The analysis predicts a variety of intensity and polarization patterns that are compared with data from a number of pulsars. It also predicts the observed frequency dependence of the peak-to-peak separation of two double-peaked pulsars.

141.331 Simultaneous interferometric and spectrometric observations of pulsar 0329+54.
J. A. Galt, N. W. Broten, T. H. Legg.
Astron. Journ., Vol. 80, 311 - 317, 333 (1975).

Observations of the pulsar with an interferometer of lobe spacing 5.5×10^{-2} arcsec were made in the frequency band 408–412 MHz. The rms image motion of the pulsar was found to be $\leqslant 2.5 \times 10^{-3}$ arcsec, a result consistent with thin-screen-scattering theory and with the assumed distance to the pulsar. Dynamic spectra with a frequency resolution of 10 kHz show fine structure in the distribution of pulse energy. Measurements made on features in the dynamic spectra are combined with the image motion limit to derive crude estimates of the pulsar's distance.

141.332 Observations of pulsars at high frequencies.
W. Sieber, R. Reinecke, R. Wielebinski.
Astron. Astrophys., Vol. 38, 169 - 182 (1975).

Observations of mean pulse profiles have been made of a large sample of pulsars at 2.7, 4.9 and 10.7 GHz. The observed pulsars are compared with similar observations at lower frequencies showing that in nearly all cases the mean pulse shape changes with frequency. It is shown that in the majority of cases, pulse widths decrease with increasing frequency, whereas the component distances may either increase or decrease.

141.333 Drifting subpulse behaviour of PSR's 0943 + 10 and 2303 + 30. W. Sieber, L. Oster.
Astron. Astrophys., Vol. 38, 325 - 327 (1975).

The parameters P_2 and P_3 are derived for the drifting-subpulse patterns, the ambiguity in drift direction is discussed on a theoretical basis, and a decision suggested in the case of PSR 0943 + 10. For PSR 2303 + 30 two systems are found and compared with previous results.

141.334 Pulsar fluctuation spectra and the generalized drifting-subpulse phenomenon. II.
D. C. Backer, J. M. Rankin, D. B. Campbell.
Astrophys. Journ., Vol. 197, 481 - 487 (1975).

Pulsar observations at meter wavelengths have been analyzed to investigate pulse-to-pulse variations and to identify regions of the pulse profile which display distinct statistical properties. Fluctuation spectra of eight recently discovered pulsars are presented along with a description of the drifting-subpulse phenomenon in several objects with quasi-periodic responses in their fluctuation spectra. A quantitative analysis of the drifting-subpulse phenomenon in PSR 0031−07 and PSR 0809 + 74 is given which emphasizes the broad-band nature of the phenomenon. A summary of the memory exhibited by pulsars with time scales of ~50 periods is given.

141.335 Acceleration of pulsars to high velocities by asymmetric radiation.   E. Tademaru, E. R. Harrison.
Nature, Vol. 254, 676 - 677 (1975).

The authors propose that pulsars are accelerated by the emission of asymmetric low frequency electromagnetic radiation and rapidly attain characteristic velocities exceeding 100 km s^{-1}.

141.336 PSR 1911 + 03: a new pulsar with a large duty-cycle.   D. K. Mohanty, V. Balasubramanian.
Monthly Notices Roy. Astron. Soc., Vol. 171, 17P - 18P (1975).

A new pulsar with a duty-cycle of about 21 per cent has been detected at 327 MHz using the Ooty radio telescope. In this letter the authors report the observations of the pulsar and discuss the implications of its large duty-cycle for Sturrock's polar-cap model.

141.337 On the polarization of radio radiation of pulsars.   O. G. Onishchenko.
Astron. Zhurn. Akad. Nauk SSSR, Vol. 52, 278 - 281 (1975). In Russian. English translation in Soviet Astron., Vol. 19, No. 2.

The strong linear polarization of the radiation from the pulsar PSR 0833−45 may be accounted for by the influence of an ultrarelativistic magnetoionic plasma in the magnetosphere of the pulsar on the propagation of radio radiation.

141.338 On the origin of the binary pulsar PSR 1913 + 16.   B. P. Flannery, E. P. J. van den Heuvel.
Astron. Astrophys., Vol. 39, 61 - 67 (1975).

The authors suggest that the progenitor of the binary pulsar PSR 1913 + 16 could have been a cataclysmic variable binary or an evolved massive X-ray binary. In both cases a range of satisfactory initial binary periods and masses can be found if one postulates that a small kick (~ 200 km/s) is imparted to the remnant pulsar by a slightly asymmetric supernova explosion. In the second case such an asymmetric explosion is not absolutely required.

141.339 Periodic pulse intensity modulation of pulsars.   K. H. Hesse.
Journ. Roy. Astron. Soc. Canada, Vol. 69, 43 (1975). − Abstr. Canadian Astron. Soc.

141.340 On the measurement of the mass of PSR 1913+16   R. Blandford, S. A. Teukolsky.
Astrophys. Journ., (Letters), Vol. 198, L27 - L29 (1975).

In an analysis of pulse arrival times from the binary pulsar PSR 1913+16, the frequency shift due to the combined transverse Doppler effect and gravitational redshift cannot be distinguished from that due to the first-order Doppler effect as long as there is no secular variation of the orbital elements. However, the effects can be separated in the presence of apsidal motion, allowing a measurement of the masses of PSR 1913+16 and its companion. The authors estimate that at least 5 years of regular observations will be required for a 10 percent measurement of the component masses.

141.341 Upper limit of electron concentration in the vicinity of the pulsar PSR 1913 + 16 in the binary system.   L. M. Ozernoj, V. I. Shishov.
Pis'ma v Astron. Zhurn., Vol. 1, No. 3, p. 21 - 24 (1975). In Russian.

A method is developed to estimate the electron concentration near a pulsar using the pulse width produced by scattering of radio waves in a turbulent plasma. Its application to the pulsar PSR 1913 + 16 gives $N_e < 5 \times 10^5 \text{ cm}^{-3}$.

141.342 On the binary system containing PSR 1913+16: I. Constraints on the nature of the companion im-

posed by geometry and the classical apsidal motion test.   A. R. Masters, W. D. Arnett, D. H. Roberts.
Bull. American Astron. Soc., Vol. 7, 263 (1975). − Abstr. AAS.

141.343 On the binary system containing PSR 1913+16: II. Classical and relativistic effects and the determination of the orbital elements.   D. H. Roberts, A. R. Masters.
Bull. American Astron. Soc., Vol. 7, 263 (1975). − Abstr. AAS.

141.344 A pulsar model revisited.   J. M. Barnothy.
Bull. American Astron. Soc., Vol. 7, 263 - 264 (1975). − Abstr. AAS.

141.345 Phenomenological model for radio pulsars.   L. F. Oster, W. Sieber.
Bull. American Astron. Soc., Vol. 7, 264 (1975). − Abstr. AAS.

141.346 Pulsar microstructure: a model for the emission region.   J. M. Cordes.
Bull. American Astron. Soc., Vol. 7, 264 (1975). − Abstr. AAS.

141.347 Morphology of pulsar microstructure.   T. H. Hankins, B. J. Rickett, J. M. Cordes.
Bull. American Astron. Soc., Vol. 7, 264 (1975). − Abstr. AAS.

141.348 Near-zone and far-zone structure of aligned rotator magnetospheres.   F. C. Michel.
Bull. American Astron. Soc., Vol. 7, 336 (1975). − Abstr. AAS.

141.349 The period derivatives of pulsars.   A. G. Lyne, R. T. Ritchings, F. G. Smith.
Monthly Notices Roy. Astron. Soc., Vol. 171, 579 - 597 (1975).

The period derivatives for 56 pulsars have been obtained from two sets of timing observations 1 yr apart. A total of 84 period derivatives are now known. Statistical relationships between period, period derivative, magnetic field strength, luminosity, integrated pulse width and apparent age are examined. The distribution of values of period and period derivative is explained by postulating a secular decay of magnetic dipole moment. The cut-off of radio emission appears to be related to the magnetic field near the velocity of light cylinder.

141.350 Particle acceleration at pulsar magnetic poles.   N. J. Holloway.
Monthly Notices Roy. Astron. Soc., Vol. 171, 619 - 635 (1975).

Attempts have been made to explain the pulsar phenomenon in terms of a model in which electrons and ions are accelerated by large electric fields near the magnetic poles of rotating neutron stars. The following considerations indicate that these mechanisms are not possible in the axisymmetric situations for which they were invoked, and are unlikely to exist in other situations.

141.351 On the nature of the companion of the radio pulsar PSR 1913 + 16 in the binary system.   L. M. Ozernoj, M. Reinhardt.
Pis'ma v Astron. Zhurn., Vol. 1, No. 4, p. 16 - 20 (1975). In Russian.

Some geometrical and physical constraints on the parameters of the binary system containing the radio pulsar PSR 1913 + 16 are considered. It is pointed out that a neutron star of moderate or small mass may be paired with a red dwarf (or a subdwarf) stripped of its outer layers by a supernova explosion.

141.352 Contribution from pulsars.   J. Wdowczyk.
Phil. Trans. Roy. Soc. London, Ser. A, Vol. 277, 443 - 451 (1975). Conference paper.

A detailed review of the information on the primary cosmic-ray spectrum in the energy interval $10^{12} − 10^{16} \text{ eV}$ is given. Methods based on the analysis of the secondary cosmic rays in the atmosphere are described and the results are discussed. By using quantitative predictions based on the mechanism suggested by Gunn & Ostriker, and experimental information about pulsars, the energy spectrum of primary cosmic rays is obtained.

141.353 Pulsars and the origin of cosmic rays.
T. Gold.
Phil. Trans. Roy. Soc. London, Ser. A, Vol. 277, 453 - 461 (1975).-Conference paper.
The distribution of pulsars and their main properties are discussed. Spin and magnetic fields provide them with the means of radiating radio pulses, and there are indications that these are due to relativistic particles radiating in magnetic fields. A contribution to the cosmic rays may be produced at the same time, expecially by the youngest pulsars.

141.354 The binary pulsar 1913 + 16.   S. van den Bergh.
Astrophys. Letters, Vol. 16, 75 (1975).
The a priori probability that the only known binary radio pulsar should also have the second shortest period known is only ~ 0.02. This suggests that the spin-down of PSR 1913 + 16 may have been slowed down or even reversed by mass transfer.

141.355 Observations of pulsar radio emission. III. Stability of integrated profiles.
D. J. Helfand, R. N. Manchester, J. H. Taylor.
Astrophys. Journ., Vol. 198, 661 - 670 (1975).
An investigation of the manner in which individual pulsar pulses combine to form a stable integrated profile is presented. The stabilization time is found to be closely related to pulsar type, with sources that exhibit drifting subpulses stabilizing most rapidly and the multicomponent sources requiring the largest number of pulses to establish a constant profile shape. The mode-changing phenomenon is investigated in detail for PSR 0329 + 54 over the frequency range 150−1400 MHz. Evidence for mode changes with a short memory time scale is presented for other multicomponent sources. Secular changes in integrated profile shape over periods of days to years were searched for but not found.

141.356 Measurements of neutral-hydrogen absorption in the spectra of five pulsars and parameters of the Gum nebula.   K. J. Gordon, C. P. Gordon.
Astron. Astrophys., Vol. 40, 27 - 31 (1975).
Absorption was found in the spectra of the two pulsars, PSR 0835−41 and PSR 1642−03. Upper limits of $\tau < 0.3$, < 0.5, and < 0.16 were found for the absorption towards PSR 0809+74, PSR 1508+55, and PSR 2045−16, respectively. For PSR 0835−41, the velocity of absorption results in the distance limits $2400 < d < 5000$ pc. The column density of neutral hydrogen in cool clouds is $N_H = 1.0$ and 0.5×10^{21} cm^{-2} toward PSR 0835−41 and 1642−03, respectively, for $\langle 1/T_s \rangle^{-1} = 50$ °K. Values of $\langle n_e \rangle$ in the Gum nebula are derived for the models of the nebula as a fossil Strömgren sphere surrounding the Vela pulsar and as the H II region excited by ζ Pup and γ^2 Vel.

141.357 Binary pulsar PSR 1913+16: model for its origin.
H. M. Van Horn, S. Sofia, M. P. Savedoff, J. G. Duthie, R. A. Berg.
Science, Vol. 188, 930 - 933 (1975).
The existing observational data for the binary pulsar PSR 1913+16 are sufficient to give a rather well-defined model for the system. On the basis of evolutionary considerations, the pulsar must be a neutron star near the upper mass limit of 1.2 solar masses ($M_\odot$). The orbital inclination is probably high, $i \gtrsim 70°$, and the mass of the unseen companion probably lies close to the upper limit of the range $0.25 \, M_\odot$ to $1.0 \, M_\odot$. The secondary cannot be a main sequence star and is probably a

degenerate helium dwarf.

141.358 Pulsars and high density physics.   A. Hewish.
Science, Vol. 188, 1079 - 1083 (1975). − This article is the lecture which the author delivered in Stockholm, Sweden on 12 December 1974 when he received the Nobel Prize in Physics.

141.359 Search for an optical counterpart of the binary pulsar PSR 1913+16.   P. L. Bernacca, F. Ciatti, L. Guzzi, G. Sedmak, I. E. Campisi, A. Treves.
Astron. Astrophys., Vol. 40, 327 - 329 (1975).
An optical counterpart of the binary pulsar PSR 1913+16 was searched in the error box of the radio pulsar. No evidence of a counterpart was found, in agreement with theoretical expectations.

141.360 The velocity of pulsars and inhomogeneities of the interstellar medium in the scintillation pattern.
N. A. Lotova, I. V. Chashej.
Astron. Zhurn. Akad. Nauk SSSR, Vol. 52, 530 - 537 (1975). In Russian. English translation in Soviet Astron., Vol. 19, No. 3.
The scintillation theory applied to the interstellar medium is developed with account taken both of inhomogeneities of the interstellar medium and pulsar motion. It is found that the velocity of the drifting pattern differs essentially from that of the pulsars. The difference is connected with the effect of the extended medium and with the motion of the inhomogeneities. Formulae have been obtained from which by means of the known parameters of the cross-correlation function of scintillations one can get the pulsar velocity and the parameters characterizing the motion of the inhomogeneities.

141.361 PSR 1913 + 16: endpoints of speculation. A critical discussion of possible companions and progenitors.
R. F. Webbink.
Astron. Astrophys., Vol. 41, 1 - 8 (1975).
The author attempts a comprehensive discussion of the nature and origin of the binary pulsar PSR 1913 + 16. It is shown that the system must have been binary before the creation of the pulsar. Main-sequence stars, white dwarfs, helium stars, neutron stars and black holes are considered as possible secondaries. It is shown that the pulsar component must have been a helium star before collapse. The companion is probably a neutron star, although a helium star or black hole cannot be excluded.

141.362 Spin-down of pulsars.
G. Chanmugam, M. Maheswaran.
Publ. Astron. Soc. Japan, Vol. 27, 307 - 310 (1975).
It is suggested that after a pulsar has slowed down to a period of the order of a few seconds, the interstellar material plays the dominant role in further braking.

141.363 Binary pulsar.   H. W. van Someren Greve, H. van der Laan, J. W. M. Baars.
IAU Circ., No. 2740 (1975).

141.364 Crab pulsar.   S. Ryckman, G. Ricker, A. Scheepmaker, J. Ballintine, J. Doty, P. Downey, W. Lewin.
IAU Circ., No. 2760 (1975).

141.365 Correlation of pulsar positions and the arrival directions of air showers of energies $10^{17}−10^{18}$ eV observed at Chacaltaya.   C. Aquirre.
Journ. Phys. A, (Math., Nuclear, General), Vol. 7, 1474 - 1482 (1974).

141.366 Pulsar electrodynamics.   S. Hinata.
Thesis, Illinois Univ., Urbana (USA). 67 pp. Univer-

sity Microfilms Order No. 74-5592 (1973).

**141.367   Attempt to monitor pulsed X-rays from the Crab
pulsar (NP 0532) utilizing atmospheric fluorescence.**
E. E. Frederick.
Thesis, Ohio Univ., Athens (USA). 215 pp. University Microfilms Order No. 74-7637 (1973).

141.368   Small stars raise large problems [pulsars].
A. Hewish.
Phys. Bull., (GB), Vol. 25, 459 - 461 (1974).
     The author discusses the structure and models of neutron
stars. It is shown that it is necessary to have a better understanding of these pulsars in order to be able to tackle larger
problems in radio galaxies and quasars.

141.369   Electrohydrodynamics of the pulsar magnetosphere.
H. Heintzmann, W. Kundt, J. P. Lasota.
Phys. Letters A, (Netherlands), Vol. 51A, 105 - 106 (1975).
     The dispersion relations for weak waves in a cold, charge-
separated plasma (due to a strong rotating magnetic field)
show that radio waves, and even low frequency waves can
propagate through a (one-component) pulsar magnetosphere.

**141.370   Die Abbremsung eines Pulsars durch elektroma-
gnetische Strahlung.   J. Pfarr.**
Diss. Univ. Köln, 66 pp. (1974).

**141.371   The broadening of pulses due to multipath pro-
pagation of radiation [pulsar observations].**
I. P. Williamson.
Proc. Roy. Soc. London, Ser. A, Vol. 342, 131 - 147 (1975).
     Observations of pulsars at low radio frequencies (ca.
100 MHz) show that the pulses received are broadened in
time presumably because of multipath propagation in the
interstellar medium. Uscinski's (1974) recent analysis of this
effect, by means of a diffraction theory, is here reworked
into a form which is readily compared with the results of a
stochastic ray-path theory, and the two approaches are found
to agree exactly. The reasons for this are investigated. The
general problem for a non-uniform medium is also discussed.

Force-free pulsar magnetosphere — II. The steady,
axisymmetric theory for a normal plasma.
See Abstr. 062.003.

Pulsed high-energy radiation from oblique magnetic
rotators.   See Abstr. 062.021.

On the relativistic theory of electromagnetic disper-
sion relations and Poynting's theorem.
See Abstr. 062.046.

Non-linear Compton radiative group locking.
See Abstr. 063.005.

On the passage of radiation through inhomogene-
ous, moving media. XII. Polarized waves in a plane, sheared
medium.   See Abstr. 063.020.

Wie sterben die Sterne?   See Abstr. 065.010.

Solid state physics and cooling of neutron stars.
See Abstr. 065.042.

Composition of the neutron star surface in pulsar
models.   See Abstr. 065.071.

Observable effects of a scalar gravitational field in
a binary pulsar.   See Abstr. 066.031.

Test for the existence of gravitational radiation.
See Abstr. 066.032.

Analytic stellar models in general relativity.
See Abstr. 066.072.

Magnetohydrodynamics: applications to magnetic
stars, cosmical gas dynamics, and pulsars.
See Abstr. 116.011.

The masses of components and the inclination of a
binary system with one pulsar determined from relativistic
effects.   See Abstr. 117.007.

The masses of components and inclination of a
binary system with one pulsar determined from relativistic
effects.   See Abstr. 117.038.

A possible relation between spectral index and z-
distribution for supernova remnants.   See Abstr. 125.004.

Supernova remnants and presupernova models.
See Abstr. 125.007.

H I absorption of nine pulsars.
See Abstr. 131.504.

Measured offset between the Crab pulsar and
Tau X-1.   See Abstr. 134.002.

Wave zone structure of NP 0532 and infrared radia-
tion excess of Crab nebula.   See Abstr. 134.003.

Change in the high-energy radiation from the Crab.
See Abstr. 134.005.

A model of the X-ray structure of the Crab nebula.
See Abstr. 134.009.

A multiple pulsar model for quasi-stellar objects
and active galactic nuclei.   See Abstr. 141.093.

Pulsars and the origin of cosmic rays.
See Abstr. 143.001.

Collapsed stars, pulsars and the origin of cosmic rays.
See Abstr. 143.073.

Remarks on the possibility of pulsar-induced bump
in the cosmic ray spectrum at $10^{13}-10^{16}$ eV/particle.
See Abstr. 143.091.

Infrared Sources

141.601 Silicate absorption at 18 μm in two peculiar infrared sources. T. Simon, H. M. Dyck.
Nature, Vol. 253, 101 - 102 (1975).

Two broad emission features often observed in the infrared spectra of cool stars near 10 and 18 μm are commonly attributed to thermal radiation by silicate dust grains. The authors report here the detection of 18-μm absorption features in the spectra of AFCRL809−2992 and an infrared object associated with the microwave source OH 26.5+0.6. They attribute these features to the Si−O−Si bending mode in silicates.

141.602 Sources of infrared and submillimeter radiation. G. B. Sholomitskij.
New problems of astrophysics, Publ. Astrophys. Winter School, 1972, (see 012.001), p. 77 - 90 (1974). In Russian.

141.603 Optical identifications of AFCRL rocket infrared sources. M. Cohen.
Astron. Journ., Vol. 80, 125 - 127, 167 - 174 (1975).

The positions of almost 700 unidentified infrared sources from the AFCRL rocket sky survey have been searched photographically for objects of exceptional redness. Forty-four optical identifications are suggested and finding charts are given.

141.604 The peculiar object HD 44179 ("The red rectangle") M. Cohen, C. M. Anderson, A. Cowley, G. V. Coyne, W. M. Fawley, T. R. Gull, E. A. Harlan, G. H. Herbig, F. Holden, H. S. Hudson, R. O. Jakoubek, H. M. Johnson, K. M. Merrill, F. H. Schiffer III, B. T. Soifer, B. Zuckerman.
Astrophys. Journ., Vol. 196, 179 - 189 (1975).

A strong infrared source detected in the AFCRL sky survey is confirmed, and is identified with the binary star HD 44179, embedded in a peculiar nebula. The complex nebular structure is discussed on the basis of photographs through narrow-band continuum and emission-line filters. The polarization data support the suggestion of a disk containing some large particles. No radio continuum emission is detected.

141.605 Far-infrared observations of W51 with high spatial resolution. P. M. Harvey, W. F. Hoffmann, M. F. Campbell.
Astrophys. Journ., (*Letters*), Vol. 196, L31 - L34 (1975).

Observations of the core of W51 have been made at 55 μ. These observations suggest the presence of a small ($\lesssim 20''$) source at 55 μ surrounded by a much broader diffuse background. In the center of the source the optical depth at 55 and 100 μ is on the order of 0.1−0.2.

141.606 Very long baseline interferometric observations of OH/IR stars. M. J. Reid, D. O. Muhleman.
Astrophys. Journ., (*Letters*), Vol. 196, L35 - L37 (1975).

Very long baseline interferometric observations of five type II OH/IR objects indicate that the apparent size of the masing components is greater than $\sim 0''.1$. The present observations, combined with current theories, indicate that the 1612-MHz OH maser associated with Mira variables originates in the outer regions of the circumstellar dust.

141.607 Detection of H_2O emission from the far-infrared source AFCRL (UOA) 19.
G. J. White, L. T. Little, E. A. Parker, P. S. Nicholson, G. H. Macdonald, F. Bale.
Monthly Notices Roy. Astron. Soc., Vol. 170, 37P - 38P (1975).

The authors have detected 22.2 GHz radiation from interstellar water vapour in the far-infrared source AFCRL (UOA) 19. The H_2O source does not appear to be dynamically associated with the CO, HCN and CS clouds already detected in AFCRL (UOA) 19.

141.608 Infra-red sources near CoD−42°11721. I. S. Glass, D. A. Allen.
Observatory, Vol. 95, 27 - 30 (1975).

There may be a very dense dust cloud near the star CoD−42°11721.

141.609 M78: an active region of star formation in the dark cloud Lynds 1630.
K. M. Strom, S. E. Strom, L. Carrasco, F. J. Vrba.
Astrophys. Journ., Vol. 196, 489 - 501 (1975).

Several luminous stars in the dark cloud (L1630) containing M78 (NGC 2068) are found to have ages as young as a few times 10^5 years. The properties of the stellar population apparently associated with this dark cloud are investigated; it appears that star formation has taken place over a period of several million years. A significant number of heavily obscured ($A_v \gtrsim 10$ mag) stars has been found in the course of a 2-μ survey of selected regions in the L1630 cloud complex. These objects appear best interpreted as stars of late B and early A spectral type embedded within the dense cloud material. The ratio of total to selective extinction achieves values as high as 5.8 in L1630; furthermore, the ratio of λ4430 band strength to visible extinction is anomalously low in this dense cloud complex.

141.610 Infrared emission from OH 284.2−0.8. J. A. Frogel, S. E. Persson.
Astrophys. Journ., Vol. 197, 351 - 353 (1975).

OH 284.2−0.8, a strong, peculiar OH emitter, has been found to be a bright infrared source less than $4''$ in diameter, and to be associated with a knot of optical nebulosity. The general appearance of the infrared energy distribution resembles that of a compact H II region, although the emission shortward of 3.5 μ is relatively blue, and may contain a contribution from reddened starlight. There is an absorption feature at 3.1 μ, but no apparent emission or absorption features near 10 or 20 μ.

141.611 Recent revelations of infrared astronomy. W. A. Stein.
Publ. Astron. Soc. Pacific, Vol. 87, 5 - 16 (1975). – Invited review presented at 143rd meeting of American Astronomical Society, Rochester, New York, August 1974.

141.612 The IRC+10216 molecular envelope. M. Morris.
Astrophys. Journ., Vol. 197, 603 - 610 (1975).

Microwave emission lines from eight linear molecules have been reported in the direction of the carbon star, IRC+10216. The author proposes that the observed excitation of the rotational energy levels of these molecules is a result of absorption of infrared radiation into excited vibrational states and the subsequent decay back to the rotational levels of the ground vibrational state. A model of the source is constructed. The abundances of CS, SiS, SiO, and HC_3N are derived by fitting their observed lines to the predictions of the model. The molecular abundances yield an estimate for the total mass of the envelope of $(2-35) \times 10^{-3} M_\odot$. Conditions which determine the shapes of emission lines from expanding clouds are discussed, including the geometrical effect of excitation by the anisotropic infrared radiation field, the effect of finite optical depth, and the effect of partial resolution of the cloud.

141.613 H_2 and HD infrared lines expected from dense interstellar objects. E. Bussoletti, G. Stasinska.
Astron. Astrophys., Vol. 39, 177 - 184 (1975).

Expected intensities of rotational H_2 and HD lines are computed for several dense clouds and presumed protostars. It is shown that some lines are detectable with present techniques. Their observation would provide the first direct estimate of masses of dense objects.

141.614 Variability of R CrB and NML Cyg at 3.5 μ.
D. W. Strecker.
Astron. Journ., Vol. 80, 451 - 453 (1975).
3.5-μ photometry of R CrB and NML Cyg from 1968 through 1974 is tabulated and plotted.

141.615 Studies of the infrared source CRL 2688.
E. P. Ney, K. M. Merrill, E. E. Becklin, G. Neugebauer, C. G. Wynn-Williams.
Astrophys. Journ., (Letters), Vol. 198, L129 - L134 (1975).
Infrared, optical, and radio observations are described of a newly discovered galactic infrared source. Most of the radiation comes from a 1".5 diameter infrared source at a temperature of about 150 K, but some visible emission in the form of a symmetrical highly polarized reflection nebulosity is also seen. The object could represent either a very early or a very late stage in stellar evolution.

141.616 Spectroscopic observations of CRL 2688.
D. Crampton, A. P. Cowley, R. M. Humphreys.
Astrophys. Journ., (Letters), Vol. 198, L135 - L137 (1975).
Spectra of the visible objects at the position of the infrared source CRL 2688 indicate that these are nebulosities reflecting the light of an obscured F5 Ia star. Apart from this apparently normal spectrum, strong absorption features identified as molecular C_3 and emission lines identified as molecular C_2 and [S II] are observed. The authors' observations suggest that an F5 Ia star surrounded by a dense shell is at the position of the infrared source.

141.617 On the nature of maser sources in infrared stars.
V. V. Burdjuzha (Burdyuzha), T. V. Ruzmaikina (Ruzmajkina).
Astron. Astrophys., Vol. 40, 233 - 236 (1975).
It is shown that behind the shock front in the shell of an infrared star, thermal instability conditions are fulfilled, which results in fragmentation of the medium. In the clouds thus formed, conditions for ignition of OH and H_2O masers may be fulfilled. The 1612-MHz spatial structure of OH sources in NML Cyg and VY CMa is discussed, as well as the problems of the pumping mechanisms of OH and H_2O masers. The peculiarities of microwave spectra are briefly considered.

141.618 An infrared point-source in Sharpless 149.
J. Bergeat, F. Sibille, M. Lunel.
Astron. Astrophys., Vol. 40, 347 - 349 (1975).
The Sharpless 2-148 and 149 regions were explored at 2.2μ, and a bright point-source (17 f.u.) was detected at the northern rim of Sh 2-149. Within guiding errors this infrared source is shown to coincide with an extremely red star-like object on the POSS prints. The authors have obtained further photometric (0.55 - 3.4 μ), spectroscopic and polarimetric, observations of this object. They interpret these in terms of a highly reddened hot star which could be the exciting star of the Sh 2-149 region.

141.619 Infrared observations of Sharpless 2–106, a possible location for star formation.
F. Sibille, J. Bergeat, M. Lunel, R. Kandel.
Astron. Astrophys., Vol. 40, 441 - 446 (1975).
This paper reports discovery and observations, in the wavelength range 1.25–3.4 μ, of a bright infrared source in Sharpless 2–106. Results are interpreted by a stellar core, surrounded by an emitting dust shell at about 2000 °K, a cooler external dust layer being responsible for the strong local absorption. This object is found near the center of a dense H II knot surrounded by a compact H II region.

Radiative transfer in gray circumstellar dust envelopes: VY Canis Majoris revisited. See Abstr. 064.020.

Radiative transfer in spherical circumstellar dust envelopes. II. Is the infrared continuum of Eta Carinae produced by thermal dust emission? See Abstr. 064.021.

Radiative transfer in spherical circumstellar dust envelopes. III. Dust envelope models of some well known infrared stars. See Abstr. 064.022.

An alternative mechanism for production of emission features in some infrared objects. See Abstr. 064.023.

Ejection of nebulae by BQ radiostars with infrared excess. See Abstr. 114.030.

The symbiotic binary V 1016 Cygni, early stage of a planetary nebula. See Abstr. 114.333.

On the light curves of OH-Mira variables. See Abstr. 122.047.

Coarse photospheric convection and the ejection of dust by R Coronae Borealis. See Abstr. 122.086.

Infrared light curves for V1057 Cygni (1971-74). See Abstr. 122.089.

Vibrationally excited SiO: a new type of maser source in the millimeter wavelength region. See Abstr. 131.009.

The plausibility of silicate-core ice-mantle grains. See Abstr. 131.082.

Considerations about the absorption efficiency of dust particles in the infrared. See Abstr. 131.083.

Time variations and spectral structure of the methanol maser in Orion A. See Abstr. 131.124.

The distribution of the H134α recombination line in W33. See Abstr. 131.502.

Maser radiometer observations of water vapor emission from H II regions and IR stars. See Abstr. 131.529.

An improved 1665 MHz position for W3 (OH). See Abstr. 131.534.

T Tauri nebulae and Herbig-Haro nebulae: evidence for excitation by a strong stellar wind. See Abstr. 132.005.

Optical polarization of a Herbig-Haro object in Corona Austrina. See Abstr. 132.034.

Further study of the stellar cluster embedded in the Ophiuchus dark cloud complex. See Abstr. 153.008.

350-micron mapping of Sagittarius B2. See Abstr. 155.008.

The nucleus of NGC 253. See Abstr. 158.033.

The origin of infrared emission from the nucleus of NGC 1068. See Abstr. 158.042.

A search for water vapour emission from extra-galactic nebulae.   See Abstr. 158.074.

Nonthermal continuum radiation in three elliptical galaxies.   See Abstr. 158.099.

Errata

141.901   Erratum: 'New observations of the angular dia-meter-redshift relation for radio sources'
[Nature, Vol. 252, 657 - 659 (1974)].
A Hewish, A. C. S. Redhead, P. J. Duffett-Smith.
Nature, Vol. 253, 378 (1975).

142 X-Ray, Gamma-Ray Sources

142.001 **Long term variability of Cyg X-1 in hard X-rays.**
F. Frontera, F. Fuligni, C. Cavani.
Astrophys. Space Sci., Vol. 32, 197 - 203 (1975).
Results of the analysis of four balloon observations of Cyg X-1 in the energy range 20−200 keV are reported. Long term time variability of the flux is confirmed and is estimated to amount to a factor 1.5−2. The source appears to remain in a state of lower flux for longer times. No obvious correlation with the low energy behaviour following the transition detected by Uhuru is found.

142.002 **The masses of binary X-ray sources.**
J. A. de Freitas Pacheco.
Astrophys. Space Sci., Vol. 32, 205 - 210 (1975).
From optical and X-ray data available for the binary X-ray sources, the masses of the components were determined univocally in three systems. A discussion about the masses in other three systems is also presented. Besides Cyg X-1, SMC X-1 is another black-hole candidate.

142.003 **Extragalactic X-ray sources and the X-ray background.** M. Rowan-Robinson, A. C. Fabian.
Monthly Notices Roy. Astron. Soc., Vol. 170, 199 - 217 (1975).
Identifications with a relatively complete sample of high latitude X-ray sources, including four new identifications with Abell clusters, are used to establish the luminosity function for extragalactic sources, and to determine the maximum rates of evolution for different populations of source consistent with the observed integrated X-ray background. A significant limit is obtained on any evolution that operates on all sources. Both inverse Compton and thermal Bremsstrahlung models of cluster sources are investigated and severe restrictions can be placed on models of Silk & Tarter, in which a strong correlation of X-ray luminosity with richness is predicted. The authors are sceptical of reported correlations of the X-ray luminosity of clusters with radio luminosity or morphological type.

142.004 **Gamma-ray bursts observed by a hard X-ray experiment aboard OSO-6.**
G. Pizzichini, G. G. C. Palumbo, A. Spizzichino.
Astrophys. Journ., (*Letters*), Vol. 195, L1 - L5 (1975).
The data of the Bologna hard X-ray experiment aboard the OSO-6 satellite have been searched for events in coincidence with the 14 cosmic γ-ray bursts reported by Strong, Klebesadel, and Olson which occurred during the OSO-6 lifetime (1969 August to 1972 January). Three events have been detected.

142.005 **Photoelectric observations of Centaurus X-3.**
H. Mauder.
Astrophys. Journ., (*Letters*), Vol. 195, L27 - L30 (1975).
Photoelectric observations in the V band have yielded a light curve for the optical counterpart of Cen X-3. The amplitude of the light variations is 0.14 mag, which rules out the possibility of a very low mass ratio. It is shown that the X-ray source, which is believed to be a neutron star, must have a lower mass limit close to 3 $M_\odot$. The photometric properties indicate a distance of about 7 kpc. Besides a double wave due to the tidal distortion of the star, the light variations also show some intrinsic activity on a time scale of fractions of a minute. A radial velocity variation with an amplitude of 60 km s^{-1} is predicted.

142.006 **A search for optical pulsations from Centaurus X-3.**
B. A. Peterson, J. Middleditch, J. Nelson.

Astrophys. Journ., (*Letters*), Vol. 195, L31 - L32 (1975).
No optical pulses were detected from Cen X-3 at the 0.2 percent level. The observations were made during the binary phase expected to produce optical pulses of the greatest amplitude.

142.007 **Optical studies of UHURU sources. XI. A probable period for Scorpius X-1= V818 Scorpii.**
E. W. Gottlieb, E. L. Wright, W. Liller.
Astrophys. Journ., (*Letters*), Vol. 195, L33 - L35 (1975).
1068 magnitudes of Sco X-1 = 3U 1617−15 = V818 Sco have been obtained from blue plates in the Harvard collection taken between 1889.5 and 1974.3. We have analyzed these data for periods from 0.25 to 100 days; the best period found, $0\overset{d}{.}787313 \pm 0\overset{d}{.}000001$, gives a sinusoidal light curve with full amplitude of 0.22 ± 0.03 mag. This period is confirmed by published postdiscovery photometry.

142.008 **Oblique rotators in binary systems.**
K. M. V. Apparao.
Nature, Vol. 253, 27 - 28 (1975).
Radio pulsar PSR 1913+16, which has a period of about 59 ms, has been identified as a member of a binary system. The X-ray 'oscillars' Her X-1 and Cen X-3, which are also members of binary systems, are not radio pulsars. These facts, together with the 34-d periodicity of Her X-1 and the lack of this periodicity in Cen X-3, can be explained by using the rotating neutron star hypothesis for the compact object in all of these systems. As is common in pulsar theories, the author assumes that rotating neutron stars are oblique rotators, that they have magnetic fields and that the magnetic axes are not along the axes of rotation.

142.009 **Nature of Her X-1.** Y.-M. Wang.
Nature, Vol. 253, 249 - 250 (1975).
Iyengar et al. have reported observations of hard X rays from Her X-1, which seem to set an upper limit of 10% on the pulsating component in the energy range 20 to 45 keV. The author gives a simple explanation that does not require the additional source of X rays postulated. He refers to the model widely used to explain the X-ray pulsing from members of binary systems (specifically Her X-1 and Cen X-3), in which a rotating neutron star with a non-aligned magnetic field accretes material from its companion, the magnetic field channelling the infalling matter toward a 'hot spot' near the surface of a neutron star.

142.010 **Model for 1.24 s X-ray pulses in Her X-1.**
E. D. Feigelson.
Nature, Vol. 253, 250 - 251 (1975).
The 1.24 s X-ray pulses observed in the 2−20 keV region from Her X-1 are thought to arise from matter being accreted on to a rotating, magnetic neutron star. The pulsed emission is usually thought to be produced by a process analogous to that of radio pulsars: X rays are emitted in a directed fan beam from the magnetic poles of the neutron star, which scans a circle in the sky as the star rotates. The earth is in the line-of-sight of the beam only during a portion of each 1.24 s period. The author proposes that the 1.24 s pulses originate from luminosity variations at the source arising from time-varying accretion rather than from the passage of a beam of constant intensity past the earth.

142.011 **The implausible history of triple star models for Cygnus X-1: evidence for a black hole.**
H. L. Shipman.
Astrophys. Letters, Vol. 16, 9 - 12 (1975).

Models for Cygnus X-1 as a triple star system have been proposed as a way of avoiding the presence of a black hole. Consideration of the likely past history of these models indicates that the supernova explosion which preceded the formation of the neutron star would have caused the system to have become a runaway. Tidal evolution and mass exchange do not seem to offer much hope of avoiding the presence of a black hole.

142.012   Search for 4.8-hour modulation of the Cyg X-3 intensity in hard X-rays.
L. Feretti, F. Frontera, F. Fuligni.
Astrophys. Letters, Vol. 16, 13 - 15 (1975).

Results of the analysis of a balloon observation of the Cygnus region have been analyzed to search for the modulation of the intensity of Cyg X-3. An upper limit for the modulation ratio of 0.4 was found. It is concluded that this ratio is a decreasing function of energy.

142.013   A photographic study of HDE 226868 (Cygnus X-1) based on archive plates.
T. J. Herczeg, J. K. Sutton.
Astrophys. Letters, Vol. 16, 17 - 18 (1975).

The optical object identified with Cygnus X-1 was measured on 60 short-exposure astrographic plates in the collection of the University of Oklahoma Observatory. The interval covered is April–November, 1949. There is indication of a small amplitude, periodic light variation, similar to the variability recently found by photoelectric observers. No evidence for secular changes of the brightness has been found.

142.014   Copernicus: new positions for Cygnus X-1 and 3U 0352+30 (X Per?).
F. J. Hawkins, K. O. Mason, P. W. Sanford.
Astrophys. Letters, Vol. 16, 19 - 22 (1975).

Positions obtained with the Copernicus instrumentation confirm identification of Cyg X-1 with HDE 226868 and 3U 0352+30 with the X Per–ADS 2859B system.

142.015   Search for millimeter-wave emission from Uhuru X-ray sources and radio binary stars.
G. H. MacDonald.
Astrophys. Letters, Vol. 16, 39 - 42 (1975).

A search has been made for radio emission at wavelengths of 9.6 and 3.5 mm from 28 X-ray sources listed in the Uhuru catalogue and the three radio binary stars, β Persei (Algol), β Lyrae and Antares B. No emission was detected from any object to a flux density limit of 10^{-26} W m^{-2} Hz^{-1} at either wavelength.

142.016   An investigation of the optical spectrum of Scorpius X-1.   D. M. Chesley.
Astrophys. Journ., Vol. 195, 529 - 533 (1975).

A table of features found in the optical spectrum of Sco X-1 with suggested identifications is presented. Plots of the $\lambda\lambda 4640{-}4650$ blend as a function of the brightness of Sco X-1 are displayed.

142.017   Detection of γ-radiation from Cyg X-3.
A. M. Gal'per, V. G. Kirillov-Ugryumov, A. V. Kurochkin, B. I. Luchkov, Yu. T. Yurkin.
Izv. AN SSSR. Ser. fiz., Vol. 38, 1801 - 1805 (1974). In Russian. – Abstr. in Referativ. Zhurn. 51. Astron., 2.51.647 (1975).

142.018   Physical parameters of the Centaurus X-3 system.
P. S. Osmer, W. A. Hiltner, J. A. J. Whelan.
Astrophys. Journ., Vol. 195, 705 - 707 (1975).

Photographic spectra of Cen X-3 show that the primary star has a spectral type near O6.5 with weak, variable emission at $\lambda\lambda 4640$ and 4686. No orbital motion of the emission or absorption lines is detected; for the latter the upper limit is $\sim \pm 50$ km s^{-1}. Analysis of the available data indicates that the primary is a factor of 2–3 less massive than expected from normal evolutionary models while the X-ray source has a mass near 1.5 $M_\odot$.

142.019   Photoelectric observations of Krzeminski's star, the companion of Centaurus X-3.   L. D. Petro.
Astrophys. Journ., Vol. 195, 709 - 713 (1975).

V-magnitudes of Krzeminski's star and their interpretation are presented. The author finds $M_{primary} = 20.5 \pm 3.5\ M_\odot$, $M_x = 2.5 \pm 1.1\ M_\odot$, $\sin i = 0.985 \pm 0.02$, and $q = 0.12 \pm 0.05$.

142.020   Cygnus X-1: an interpretation of the spectrum and its variability.   K. S. Thorne, R. H. Price.
Astrophys. Journ., (Letters), Vol. 195, L101 - L105 (1975).

X-ray observations of Cygnus X-1 are compared with the "standard" model (accretion disk surrounding a rotating black hole). Attention focuses on the shape of the X-ray spectrum, the fluctuations in X-ray intensity, and the 1971 transition in the spectrum. There are several encouraging agreements of model and data, but a detailed comparison awaits further theoretical and observational studies. Several predictions of the standard model are listed as suggestive guides for future work.

142.021   Hα observations and the distribution of circumstellar material in the HD 77581 system (3U 0900–40).
M. S. Bessell, N. V. Vidal, D. T. Wickramasinghe.
Astrophys. Journ., (Letters), Vol. 195, L117 - L120 (1975).

Coudé spectra taken at Hα over many cycles show a nearly constant emission component with some periodic changes in the blue-shifted absorption components. It is suggested that the absorption components arise from an asymmetric outflow of gas (with respect to the line centers) from the vicinity of the inner Lagrangian point L_1 and perhaps the accretion disk of the secondary. The persistent outflow from L_1 seen in Hα during the phases immediately following primary minimum may be responsible for distortion of the light and radial-velocity curves.

142.022   X-ray sources and their optical counterparts – III.
C. Jones, W. Forman, W. Liller.
Sky Telescope, Vol. 49, 10 - 13 (1975).

142.023   Physical conditions on the stream of Cyg X-1 produced by the X-ray flux.
G. F. Bisiacchi, D. Dultzin-Hacyan, C. Firmani, S. Hacyan.
Rev. Mexicana Astron. Astrofis., Vol. 1, 261 - 267 (1974).

The authors have studied the interaction of an X-ray flux with the gas in the stream that flows from the primary star to the secondary object in the system Cyg X-1 (HDE 226868). From the analysis of thermal and ionization equilibrium, it turned out that according to the X-ray characteristics, the gas might present instability conditions that induce phase transitions. The possibility of using this mechanism to explain the observations of the He λ 4686 emission line was considered.

142.024   X-ray-emitting double stars.
H. Gursky, E. P. J. van den Heuvel.
Sci. American, Vol. 232, No. 3, p. 24 - 35 (1975).

An analysis of certain very powerful X-ray sources suggests that the radiation emanates from a binary system where a superdense collapsed star is orbiting closely around a massive normal star.

142.025   Origin of the optical emission from Sco X-1.
J. A. de Freitas Pacheco.
Nature, Vol. 253, 699 - 701 (1975). – Letter.

142.026 **Circular polarisation in HDE226868.**
J. J. Michalsky, J. B. Swedlund, R. W. Avery.
Nature, Vol. 254, 39 - 40 (1975). − Letter.

142.027 **Galactic gamma rays from the superposition of X-ray sources.** R. Fusco-Femiano.
Nature, Vol. 254, 125 - 126 (1975).
The author suggests that the gamma-ray flux observed by SAS-II and OSO-III is the result of a superposition of sources rather than to cosmic ray collisions. If X-ray sources are distributed along the spiral arms, their gamma-ray flux would have the same geometry as that resulting from cosmic ray collisions. The model could be tested by searching for time variations in the gamma-ray counts (one can reasonably expect them if gamma rays arise from flares) and by obtaining gamma-ray data with better angular resolution to bring into evidence peaks in correspondence to individual X-ray sources.

142.028 **The helium lines in the spectrum of Scorpius X-1.**
S. E. Chesley.
Astrophys. Journ., Vol. 196, 103 - 106 (1975).
Equivalent widths and relative line intensities for He II λ 4686, He I λ5876, and He I λ5015.7 are determined as a function of B magnitude for Sco X-1. These are used to investigate the temperature variations of the line-forming region.

142.029 **An increase in the X-ray flux from Centaurus A.**
P. J. N. Davison, J. L. Culhane, R. J. Mitchell, A. C. Fabian.
Astrophys. Journ., (*Letters*), Vol. 196, L23 - L25 (1975).
The X-ray spectrum of Cen-A has been studied in the energy range 0.5−7.5 keV. The value of absorbing gas column density and power law index are consistent with those determined by Uhuru, but the 2.5−7.5 keV luminosity is found to have increased by a factor of 4.

142.030 **Copernicus: X-ray observations of 3C 390.3.**
P. A. Charles, M. S. Longair, P. W. Sanford.
Monthly Notices Roy. Astron. Soc., Vol. 170, 17P - 22P (1975).
The MSSL X-ray telescopes on OAO-Copernicus have detected 2.5−7.5 keV X-rays from the direction of the N galaxy 3C 390.3. Upper limits are placed on the flux in the 0.6−3.1 keV range from 3C 390.3 itself. The detected signal is at least a factor of four greater than that expected from the nearby X-ray source 3U 1825+81.

142.031 **Anisotropic Thomson scattering for pulse formation in X-ray pulsars.** T. Daishido.
Publ. Astron. Soc. Japan, Vol. 27, 181 - 189 (1975).
It is shown that a hot neutron star with an intense dipole magnetic field (10^{12} G) appears as an X-ray pulsar. The mean free path of X-ray photons depends on the direction of the propagation vector with respect to the magnetic field. When the neutron star is viewed from varying directions according to its rotation, pulses of X-rays are observed. Profiles of the pulses are computed for different geometries, energies of photons and intensities of the magnetic field. Fitting the computed pulses with the observed profiles of Her X-1 and Cen X-3, values of those parameters such as the intensity of the surface magnetic field, the inclination of the spin axis and the obliquity of the magnetic dipole relative to the spin axis are determined. The validity of this model will be checked, when one observes the polarization profile of the pulse and/or the energy dependence of the pulsed component relative to the steady component.

142.032 **Large outburst of Cyg X-3 in May, 1974.**
N. Kawano, N. Kawajiri.
Publ. Astron. Soc. Japan, Vol. 27, 191 - 193 (1975).
A large outburst of Cyg X-3 was observed at 4.2 GHz in May, 1974. The observations indicate a violent time variation in flux density and a high degree of linear polarization. The maximum flux density reached about 10 f.u. on May 19 and the degree of linear polarization varied between 5 and 9% during the observational period of 10 days. Its mean position angle was about 0° and it seems to have changed from about 25° to 160° during the same period.

142.033 **Southern hard X-ray sources.** R. M. Thomas, P. J. N. Davison, M. C. Clancy, G. Buselli.
Monthly Notices Roy. Astron. Soc., Vol. 170, 569 - 577 (1975).
A re-analysis of University of Adelaide balloon data supports the conclusion that the dominant hard X-ray source in the galactic centre is GX1+4. Data are also presented for GX354−5. The hard source Ara XR-1 is probably the Uhuru object 3U1705−44 and is variable by a factor of at least 1.7 times (20−40 keV). Nor XR-1 is difficult to identify and in common with the rocket object Nor X-1, no convincing Uhuru candidate exists. Ambiguities in pre-Uhuru nomenclature are pointed out with regard to these sources.

142.034 **A possible identification of the X-ray source 3U 0400−59.** M. V. Penston, L. S. Sparke.
Observatory, Vol. 95, 17 - 18 (1975).
On examining the first plates of the ESO−SRC Sky Survey the authors found, overlapping the north end of the error box of the source 3U 0400−59, the NGC 1566 cluster of nearby (∼19 Mpc[4] if H_0 = 55 km/s/Mpc) galaxies which is thus a candidate for identification.

142.035 **Calculation of the light curves of HZ Herculis.**
M. Milgrom, E. E. Salpeter.
Astrophys. Journ., Vol. 196, 589 - 592 (1975).
The consequences are tested of the assumption that HZ Herculis is exposed to X-ray illumination from Her X-1, with intensity extrapolated from the X-rays observed terrestrially during the peak of the 35-day cycle. Further assumptions are that HZ Her fills its Roche lobe and that we are practically in the orbital plane. The authors compute the magnitude of HZ Her, at different wavelengths, as a function of the orbital phase. The computed excess luminosity, caused by the X-ray heating, agrees very well with the observed value, and so does the wavelength dependence of this quantity. The scheme adopted cannot reproduce the sharp minimum in the light curve. They also provide some results for the structure of the X-ray affected atmosphere.

142.036 **Upper limits of hard gamma-ray emission from six X-ray sources.** M. Campbell, J. Alexander, S. E. Ball, Jr., K. Greisen, D. Koch, B. McBreen.
Astrophys. Journ., Vol. 196, 593 - 595 (1975).
The Cornell 100-inch γ-ray telescope has been used to set upper limits for the γ-ray flux above 400 MeV from the X-ray sources Cyg X-1, Cyg X-2, Cyg X-3, Cas A, Cas B, and Per X-1. An upper limit for the diffuse flux of the galactic disk was set which is consistent with OSO-3 and SAS-2 data. A new measurement of the atmospheric γ-ray generation rate is presented.

142.037 **Energy spectrum variability of Cygnus X-1 in hard X-rays.** F. Frontera, F. Fuligni.
Astrophys. Journ., Vol. 196, 597 - 599 (1975).
Energy spectrum analysis of a balloon observation of Cygnus X-1 is reported. During a period of flaring activity of the source, a much softer spectrum than usual, fitted by a power law with number index 2.7, has been found. The "normal status" of flux is attained in a rather smooth way. During this last period the energy spectrum is well represented by a power law with number index 1.84, in agreement with other observations.

142.038   **Evidence for X-ray emission from Capella.**
R. C. Catura, L. W. Acton, H. M. Johnson.
Astrophys. Journ., (*Letters*), Vol. 196, L47 - L49 (1975).

X-ray emission in the range from 0.2 to 1.6 keV has been detected from an area of the sky which contains the binary star system Capella. The X-ray source is at most a few arc minutes in extent and shows no spectral turnover at low energy, consistent with a nearby source. The authors suggest Capella as the source of this emission and that this object belongs to a new class of galactic X-ray sources with a luminosity of $10^{31}-10^{34}$ ergs s^{-1}. Emission from this class of objects is variable, predominantly below 2 keV, and originates from nearby stellar objects.

142.039   **Soft X-ray observations of Centaurus X-3 from Copernicus.**
B. Margon, K. O. Mason, F. J. Hawkins, P. W. Sanford.
Astrophys. Journ., (*Letters*), Vol. 196, L51 - L53 (1975).

The authors have detected soft X-ray emission from Centaurus X-3 in the 0.6–1.9 keV band, using the focusing telescope aboard OAO Copernicus. The flux is compatible with an extrapolation of the harder X-ray spectrum, attenuated by $3-4 \times 10^{22}$ atoms cm^{-2} of interstellar and/or circumstellar matter. The data are consistent with the distance estimate of 5–10 kpc derived from the spectroscopic modulus of the optical component, and obviate the need to postulate the primary to be an anomalously subluminous hot star. There is currently no compelling evidence that such models must be invoked to explain any of the observed compact X-ray sources.

142.040   **Observations of X-ray sources in the Large Magellanic Cloud by the OSO-7 satellite.**
T. H. Markert, G. W. Clark.
Astrophys. Journ., (*Letters*), Vol. 196, L55 - L58 (1975).

Observations of the Large Magellanic Cloud with the 1–40 keV X-ray detectors on the OSO-7 satellite are reported. Results include the discovery of a previously unreported source LMC X-5, measurements of the spectral characteristics of four sources, and observations of their variability on time scales of months.

142.041   **Observations of six binary X-ray sources with the UCSD OSO-7 X-ray telescope.**   M. P. Ulmer.
Astrophys. Journ., Vol. 196, 827 - 835 (1975).

The author reports observations of six binary X-ray sources with the UCSD OSO-7 X-ray telescope: SMC X-1, Vel XR-1 (2U 0900-40), Cen X-3, Her X-1, Cyg X-1, and Cyg X-3.

142.042   **A search for the Zeeman effect in the X-ray star candidates θ^2 Orionis and X Persei, the X-ray source Cygnus X-1, and the B2 Ib star HD 31327.**   E. F. Borra.
Astrophys. Journ., (*Letters*), Vol. 196, L109 - L111 (1975).

Observations of the circular polarization in the wings of the He I line $\lambda6678$ in the X-ray star candidates θ^2 Orionis A and X Persei, the X-ray star HDE 226868 (Cygnus X-1), and the B2 Ib star HD 31327 do not show evidence of the longitudinal Zeeman effect.

142.043   **Soft X-ray spectroscopy of three extragalactic sources.**
B. Margon, M. Lampton, S. Bowyer, R. Cruddace.
Astrophys. Journ., Vol. 197, 25 - 29 (1975).

Spectra for 3C 273, M87, and NGC 4151 are presented in the 0.5–5 keV band. General techniques for error estimation in parameters of X-ray spectra are discussed, and the results are applied to the spectra previously reported for these objects. It is shown that the χ^2+1 technique which has been used in these analyses can drastically underestimate the magnitude of these errors. The authors conclude that the suggested division of extragalactic X-ray sources into two groups characterized by the magnitude of absorption in their spectra is premature.

142.044   **A search for stellar soft X-ray sources.**
M. J. Vanderhill, R. J. Borken, A. N. Bunner, P. H. Burstein, W. L. Kraushaar.
Astrophys. Journ., (*Letters*), Vol. 197, L19 - L22 (1975).

A sensitive soft X-ray survey of about one-tenth of the sky, carried out as part of the Skylab program and flown on 1973 July 28, revealed no evidence for soft X-ray emission from stellar sources. The authors present upper limits to the soft X-ray luminosity of about 50 nearby stars. With nominal assumed values for stellar space densities, they show it unlikely that emission from giants, main sequence stars, white dwarfs, or stars in known binary systems can account for the soft X-ray diffuse background.

142.045   **Results of X-ray and optical monitoring of Scorpius X-1 in 1970.**   D. E. Mook, R. J. Messina,
W. A. Hiltner, R. Belian, J. Conner, W. D. Evans, I. Strong, V. M. Blanco, J. E. Hesser, W. E. Kunkel, B. M. Lasker, J. C. Golson, J. Pel, N. R. Stokes, K. Osawa, K. Ichimura, K. Tomita.
Astrophys. Journ., Vol. 197, 425 - 441 (1975).

Scorpius X-1 was monitored at optical and X-ray wavelengths from 1970 April 26 to 1970 May 20. There was a tendency for the object to show greater variability in X-ray emission when the object was optically bright. The intensity histograms for both the optical and X-ray observations are discussed, as well as periodic variations in the optical intensity.

142.046   **The X-ray, optical, and radio behavior of Scorpius X-1: the 1971 coordinated observations.**
H. V. Bradt, L. L. E. Braes, W. Forman, J. E. Hesser, W. A. Hiltner, R. Hjellming, E. Kellogg, W. E. Kunkel, G. K. Miley, G. Moore, J. W. Pel, J. Thomas, P. Vanden Bout, C. Wade, B. Warner.
Astrophys. Journ., Vol. 197, 443 - 455 (1975).

Scorpius X-1 has been monitored at radio, optical, and X-ray wavelengths for 21 days during the period 1971 February 23 through 1971 April 2. The X-ray intensity was found to fluctuate rapidly (time constants <6 minutes) by a factor of about 2 during periods when the object was in a bright and flaring optical state. The X-ray flux was constant to within ~5 percent during the faint and relatively quiescent optical periods. No correlation between the X-ray/optical activity and radio flaring was noted.

142.047   **The X-ray, optical, and radio behavior of Scorpius X-1: the 1972 coordinated observations.**
C. R. Canizares, G. W. Clark, F. K. Li, G. T. Murthy, D. Bardas, G. F. Sprott, J. H. Spencer, D. E. Mook, W. A. Hiltner, W. L. Williams, T. J. Moffett, G. Grupsmith, P. A. Vanden Bout, J. C. Golson, C. Irving, A. Frohlich, A. M. van Genderen
Astrophys. Journ., Vol. 197, 457 - 466 (1975).

The authors present the results of a 10-day observation of Scorpius X-1 involving seven optical stations, NRAO, and the MIT experiment on the OSO 7 satellite. The source exhibits all of its characteristic variability, including flares and quiescent periods. There is clear evidence for an X-ray–optical correlation in that X-ray activity occurs only where the object is brighter than $m_B \approx 12.7$. However, detailed comparisons of X-ray and optical flares do not show a strict correlation between the intensities, with the exception of one ~20-minute X-ray flare which lags 2 to 3 minutes behind a triple optical flare of similar overall duration. There is no apparent correspondence between two radio flares observed and the shorter wavelength activity.

142.048   **Radial velocities of Scorpius X-1.**
D. Crampton, A. P. Cowley.

Astrophys. Journ., Vol. 197, 467 - 469 (1975).

Radial velocity and intensity measurements are presented for the following emission lines of Scorpius X-1: He II λ4686, the C III blend near λ4650, Hβ, and Hδ. Analysis of these data suggests that a periodicity near $0^d\!.27$ may be present, but the data do not confirm the periods ($3^d\!.931$, $3^d\!.7400$, and $0^d\!.7873$) recently suggested. If the velocity variation represents orbital motion, a model similar to an old nova is inferred. The X-ray flux might be accounted for by a rapid rate of mass transfer.

142.049 Centaurus X-3: new low- and medium-energy X-ray observations.
K. Long, P. C. Agrawal, G. Garmire.
Astrophys. Journ., (Letters), Vol. 197, L57 - L59 (1975).

Low- and medium-energy X-ray observations of Cen X-3 during two pointed sounding rocket experiments are discussed. The lack of low-energy emission from Cen X-3 is consistent with the identification of Krzeminski's star as the optical counterpart of Cen X-3 and normal reddening, in contrast to an earlier measurement by Bleeker et al.

142.050 Distance to Cygnus X-1.
B. Margon, S. Bowyer, R. Kraft.
Nature, Vol. 254, 461 (1975).

142.051 Variable X-ray source near Cen X-3.
C. J. Eyles, G. K. Skinner, A. P. Willmore, F. D. Rosenberg.
Nature, Vol. 254, 577 - 578 (1975). – Letter.

142.052 Observations of a transient X-ray source with regular periodicity of 6.75 min.
J. C. Ives, P. W. Sanford, S. J. Bell Burnell.
Nature, Vol. 254, 578 - 580 (1975).

During planned observations, by the satellite Ariel V, of the X-ray source Cen X-3 (3U1118-60), a previously unreported source nearby was detected by the collimated proportional counter of 100 cm². The source was monitored over a period of 39 d between December 19, 1974, and January 27, 1975. Here the authors report measurements of the light curve, spectrum and of a regular variability with period of 6.755 ± 0.010 min.

142.053 Is Cir X-1 a runaway binary?
D. H. Clark, J. H. Parkinson, J. L. Caswell.
Nature, Vol. 254, 674 - 676 (1975).

Runaway stars are produced when there is a supernova explosion in a close binary system. Usually such an explosion disrupts the binary system, but Gott has suggested that in a significant fraction of such supernovae the collapsed residue of the explosion may be retained in orbit around the runaway second star, the resultant system being a binary with high eccentricity and large space velocity. The authors suggest here that Cir X-1 is such a system.

142.054 Binary systems with an X-ray component.
T. Jarzębowski.
Postępy Astron., Vol. 23, 33 - 53 (1975). In Polish.

This article presents a review of all hitherto known X-ray binaries.

142.055 Extragalactic X-ray sources.   T. Kwast.
Postępy Astron., Vol. 23, 55 - 60 (1975). In Polish.

The Uhuru catalogue of X-ray sources is shortly described. The basic features of some extragalactic X-ray sources are given and three mechanisms of generation of X-rays are discussed. More detailed observations are necessary to decide which process dominates in any given source.

142.056 Further optical observations of HZ Herculis.

J. N. Bahcall, N. A. Bahcall, T. J. Herczeg, P. C. Joss, E. M. Leibowitz, A. Segalovitz, S. Stolero, M. Véron, P. Véron, P. A. Wehinger, D. Weistrop, S. Wyckoff.
Publ. Astron. Soc. Pacific, Vol. 87, 141 - 145 (1975).

Optical observations of HZ Her made in the spring and summer of 1973 are presented. Some of the observations are practically coincident with estimated turn-on times for Her X-1 and others are simultaneous with X-ray observations made using the Copernicus satellite. The long-term variations for the epoch 1953–60 reported by Jones, Forman, and Liller (1973) are confirmed. An interesting theoretical explanation by Fabian, Pringle, and Rees (1973) for the persistence of large optical variations throughout the 35-day X-ray period is ruled out by simultaneous UHURU (X-ray) and Wise Observatory (optical) observations.

142.057 HD 206267, a candidate star for the transient X-ray source Cepheus X-4?
G. Hensberge, R. H. Hammerschlag.
Astron. Astrophys., Vol. 39, 157 - 158 (1975).

The spectrum of the 3.7 day spectroscopic binary HD 206267 shows no evidence for large mass loss. No emission lines could be detected in the wavelength region λλ 3750–6680 Å. The orbital elements of the system suggest a massive secondary which can be interpreted as a normal non-degenerate star.

142.058 Highly polarized radio outburst from Cygnus X-3.
E. R. Seaquist, P. C. Gregory, R. A. Perley, R. H. Becker, J. B. Carlson, M. R. Kundu, R. C. Bignell, J. R. Dickel.
Journ. Roy. Astron. Soc. Canada, Vol. 69, 42 (1975). – Abstr.
Canadian Astron. Soc.

142.059 Ellipsoidal light variations and masses of X-ray binaries.   Y. Avni, J. N. Bahcall.
Astrophys. Journ., Vol. 197, 675 - 688 (1975).

The ellipsoidal light curves of four X-ray binaries are analyzed assuming the light variations are caused by gravity- and limb-darkening effects of a tidally distorted star. Some of the uncertainties regarding the validity of this interpretation are pointed out, including possible departures from repeatability in the light curve as well as various theoretical questions. It is shown by numerical calculations of light curves that (a) the secondary in Cyg X-1 can be either a black hole or a normal early-type star, with $M_2 \geqslant 9 M_\odot$; (b) for SMC X-1, M_X can be $\sim 2.5 M_\odot$; (c) for Cen X-3, $0.6 \leqslant M_X \leqslant 1.1 M_\odot$; and (d) no significant constraints can be derived from the available data for Her X-1.

142.060 Limits on the soft X-ray and extreme ultraviolet flux from RX Andromedae and U Geminorum.
P. Henry, R. Cruddace, M. Lampton, F. Paresce, S. Bowyer.
Astrophys. Journ., (Letters), Vol. 197, L117 - L121 (1975).

Soft X-ray and extreme-ultraviolet observations of RX And during a flare and U Gem during a quiescent state have been made. The 3 σ limits to the flux from RX And are 2.7×10^{-11} ergs cm^{-2} s^{-1} (44–125 Å) and 6.3×10^{-7} ergs cm^{-2} s^{-1} (145–515 Å); for U Gem the 3 σ limit is 7.5×10^{-11} ergs cm^{-2} s^{-1} (44–85 Å). A model of the soft X-ray emission from these systems is developed, and, when combined with these data, constrain the mass accretion rate onto the white dwarf to be less than $\sim 1 \times 10^{-8} M_\odot$ yr^{-1}. For the U Gem system this implies that the accretion efficiency for the white dwarf is less than 10 per cent and that the observed increase in orbital period is due to mass loss from the system rather than mass transfer within the system.

142.061 Possible identification of Ariel 1118–61.
A. C. Fabian, J. E. Pringle, R. F. Webbink.
Nature, Vol. 255, 208 (1975).

The transient X-ray source Ariel 1118–61 has been discovered by Ariel V. The authors point out the presence in the error box of the long period Mira-type variable RS Cen.

142.062 Observations of low energy gamma-ray bursts with SAS-2.
H. Ögelman, C. E. Fichtel, D. A. Kniffen.
Nature, Vol. 255, 208 - 210 (1975).

Low energy celestial gamma-ray bursts were discovered in 1973 and the detection later confirmed by several groups. In this paper the authors report the low energy gamma-ray bursts observed by the plastic scintillator anticoincidence dome of the Small Astronomy Satellite-2 (SAS-2) gamma-ray telescope.

142.063 Massieve röntgendubbelsterren: ontstaan en evolutie.
C. de Loore.
Zenit, Vol. 2, 44 - 46 (1975).

142.064 Possible association of a hard X-ray source with a cosmic gamma-ray burst.
S. Biswas, R. K. Manchanda, B. V. Sreekantan.
Astrophys. Space Sci., Vol. 33, L15 - L18 (1975).

Recently Palumbo et al. (1974) reported the results of the observation of a hard X-ray burst in the OSO-6 satellite wheel experiment that occurred in coincidence with the cosmic γ-ray burst event 69-2 on 7 October 1969 detected by the Vela satellites. In this note, the authors wish to point out the possible association of the X-ray and γ-ray burst event of 7 October 1969 with the hard X-ray source observed by them with X-ray telescopes flown in balloons.

142.065 Gamma-ray lines: a ^{22}Na radioactive diagnostic of young supernovae. D. D. Clayton.
Astrophys. Journ., Vol. 198, 151 - 152 (1975).

The nuclear γ-ray lines emitted following the decay of ^{22}Na should be detectable for roughly a decade following galactic explosions of massive stars in which the helium shell is heated sufficiently to synthesize ^{19}F and ^{21}Ne from seed ^{14}N.

142.066 X-ray heating and the optical light curve of HZ Herculis. S. C. Perrenod, G. A. Shields.
Astrophys. Journ., Vol. 198, 153 - 160 (1975).

The authors discuss theoretically the optical light curve of HZ Her, the binary companion of the pulsed X-ray source Her X-1. Using model stellar atmospheres, they construct light curves that are in agreement with *UBV* photometry of HZ Her except for the sharpness of the minimum. They find that heating of the photosphere of HZ Her by the observed X-ray flux is sufficient to explain the amplitude of the light variations in each color, if the X-ray emission persists at HZ Her throughout the 35-day on-off cycle. The authors propose that the extra light is caused by a stellar wind that electron-scatters optical light emitted by the photosphere of the hot side of the star.

142.067 An upper limit on optical pulsations from Sanduleak 160.
C. R. Pennypacker, B. M. Lasker, S. B. Bracker, J. E. Hesser.
Astrophys. Journ., Vol. 198, 161 - 162 (1975).

SK 160, the optical counterpart of the X-ray star SMC X-1, has been searched for periodic optical pulsations from 0.1 to 500 Hz, setting an upper limit of roughly 0.1 percent on the modulation of the total optical flux.

142.068 Neutron star wobble in binary X-ray sources.
D. Q. Lamb, F. K. Lamb, D. Pines, J. Shaham.
Astrophys. Journ., (*Letters*), Vol. 198, L21 - L25 (1975).

Calculations of the influence of accreting-matter torques on the free precession of neutron stars in close binary systems are summarized. To the extent that the flow of accreting matter is modulated by the wobble motion of the neutron star,

there will be components of the accretion torque locked in phase with this motion. Such components can be very effective in exciting or damping stellar wobble. It is suggested that large-amplitude stellar wobble, which can result from certain matter flow patterns and types of accretion gate, may be responsible for some of the high-low X-ray states observed in binary X-ray sources, including the 35^d cycle in Her X-1.

142.069 Bursts of hard X-ray radiation on June 25 and 26, 1971.   O. P. Babushkina, L. S. Bratolyubova-Tsulukidze, M. I. Kudryavtsev, A. S. Melioranskij, I. A. Savenko, B. Yu. Yushkov.
Pis'ma v Astron. Zhurn., Vol. 1, No. 2, p. 20 - 24 (1975).
In Russian.

From Cosmos—428 on June 25 and 26, 1971 two bursts of hard X-ray radiation in the band 40–290 keV were observed Integrated energy fluxes are of the order of 3.6×10^{-6} erg/cm^2.

142.070 On the variability of Cyg X-3 gamma-radiation.
B. M. Vladimirskij, A. M. Gal'per, V. G. Kirillov-Ugryumov, A. V. Kurochkin, B. I. Luchkov, Yu. I. Neshpor, A. A. Stepanyan, V. P. Fomin, Yu. T. Yurkin.
Pis'ma v Astron. Zhurn., Vol. 1, No. 3, p. 25 - 29 (1975).
In Russian.

The results of Cyg X-3 gamma-radiation measurements for energies above 4×10^7 eV obtained with a spark chamber during a balloon flight and above 2×10^{12} eV with a Cherenkov detector of extensive air showers are presented. Comparison of results shows gamma-radiation in both energy regions to be variable and possibly periodical with a period of 4^h8 equal to the period of X-ray and infrared radiation of Cyg X-3.

142.071 Why the number of galactic X-ray stars is so small?
A. F. Illarionov, R. A. Sunyaev (*Syunyaev*).
Astron. Astrophys., Vol. 39, 185 - 195 (1975).

The number of galactic X-ray sources is by $4 - 7$ orders of magnitude less than the expected number of neutron stars and black holes in binary systems. It is shown that the main condition for the appearance of X-ray source in a binary system including relativistic star is that the size of the normal component must be close to the size of its critical Roche lobe.

142.072 Sco X-1 and Cyg X-2 as binary systems.
J. I. Katz.
Astron. Astrophys., Vol. 39, 241 - 244 (1975).

Sco X-1 and Cyg X-2 resemble old novae spectroscopically and photometrically, which has led to the suggestion that they are binary systems. The energy distribution and emission-line spectrum of Sco X-1 are very different from those of the well-known binary X-ray sources. These differences may be accounted for by a small inclination angle, a small orbit size, the observed softer X-ray spectrum, and a companion star of low intrinsic luminosity in the Sco X-1 system. Cyg X-2 has the properties expected of a cooler version of HZ Her, viewed from a small inclination angle.

142.073 Supercritical accretion disks around compact X-ray sources.   P. Meszaros, M. J. Rees.
Bull. American Astron. Soc., Vol. 7, 242 (1975). – Abstr. AAS.

142.074 Soft X-ray emission from the region containing the north polar spur.
R. G. Cruddace, G. Fritz, S. Shulman.
Bull. American Astron. Soc., Vol. 7, 336 (1975). – Abstr. AAS.

142.075 X-ray spectra for Cyg X-1, Cyg X-3 and Her X-1.
P. Serlemitsos, E. A. Boldt, S. S. Holt, R. Rothschild.
Bull. American Astron. Soc., Vol. 7, 336 (1975). – Abstr. AAS.

142.076 The X-ray rotation modulation collimator onboard

Ariel 5.
F. D. Rosenberg, A. C. Newton, A. P. Willmore.
Bull. American Astron. Soc., Vol. 7, 336 (1975). – Abstr. AAS.

142.077 Observations of X-ray sources with Ariel 5 experiment C. J. C. Ives, P. W. Sanford.
Bull. American Astron. Soc., Vol. 7, 336 (1975). – Abstr. AAS.

142.078 Mass determination for the degenerate member of binary X-ray sources. S. Sofia, R. E. Wilson.
Bull. American Astron. Soc., Vol. 7, 336 - 337 (1975).
Abstr. AAS.

142.079 X-ray sources in binary systems.
R. A. Syunyaev, N. I. Shakura.
Instationary stars and methods of their investigation. Phenomena of instationarity and stellar evolution, (see 003.012), p. 231 - 260 (1974). In Russian.

142.080 Expected polarization properties of binary X-ray sources. M. J. Rees.
Monthly Notices Roy. Astron. Soc., Vol. 171, 457 - 465 (1975).
X-rays emitted by compact sources involving neutron stars or black holes would be highly polarized. X-ray polarimetry could yield important evidence on (a) the beaming and radiation mechanism for accreting neutron stars, and (b) the structure of accretion discs around black holes. Interesting information could also be derived from observations of the highly polarized X-rays resulting from scattering by material in the binary system.

142.081 Galactic γ-rays and cosmic ray origin.
D. Dodds, A. W. Strong, A. W. Wolfendale.
Monthly Notices Roy. Astron. Soc., Vol. 171, 569 - 577 (1975).
Gamma-ray emission from the Galaxy in directions away from the galactic centre supplies a test for models for the origin of cosmic ray nuclei of a few GeV. Extragalactic origin is shown to be improbable, while a plausible distribution of galactic sources gives a good explanation of the observations.

142.082 Observation of an accretion wake and pre-eclipse dips in Centaurus X-3. I. R. Tuohy, A. M. Cruise.
Monthly Notices Roy. Astron. Soc., Vol. 171, 33P - 39P (1975).
The authors report the results of seven Copernicus observations of Centaurus X-3 in the energy range 2.5 - 7.5 keV, between 1973 April and 1974 June. An abrupt absorption event was observed at phase 0.5 which is interpreted as an accretion wake sweeping through the line of sight. The wake appears to be associated with an increase in the gas density prior to an extended low state. The data support a stellar wind model for the mass transfer in the Cen X-3 system.

142.083 Copernicus – X-ray observations of 3U 0750–49.
J. N. Bahcall, P. A. Charles, P. J. N. Davison, P. W. Sanford, E. Kellogg, D. York.
Monthly Notices Roy. Astron. Soc., Vol. 171, 41P - 46P (1975).
It has been suggested that there is an association between X-ray sources and bright stars (Gursky 1972). One such possible association is between 3U 0750–49 and the bright variable star V Pup. The authors have analysed approximately five days worth of X-ray data from Copernicus on V Pup and HD 64740, two bright stars close to the relatively weak X-ray source 3U 0750–49. They have derived an error box perpendicular to the 3U error box and of similar dimensions. Based on both error boxes the variable stars most likely to be the X-ray source are V Pup, CN Pup and CSV 1199.

142.084 Observations of the Circinus X-1 region.
W. A. Baity, M. P. Ulmer, L. E. Peterson.
Astrophys. Journ., Vol. 198, 447 - 452 (1975).

The UCSD X-ray telescope on board the OSO-7 satellite has provided observations of the Cir X-1 region for over 110 days between 1971 December and 1973 March. There is a positive correlation between intensity and spectral shape, with a steeper spectrum at times of peak intensity. Although power-law, exponential, and blackbody spectral shapes are equally good fits to the data alone, examination of all the available data appears to rule out a simple blackbody spectrum, at least at times of maximum intensity.

142.085 Variability of the X-ray sources in the Magellanic Clouds. I. R. Tuohy, C. G. Rapley.
Astrophys. Journ., (Letters), Vol. 198, L69 - L72 (1975).
The authors have conducted further Copernicus observations of the SMC and LMC X-ray sources. The SMC X-1 data provide an improved X-ray period of $3\overset{d}{.}89217 \pm 0\overset{d}{.}00012$ and establish a new phase of JD 2,442,276.15 ± 0.04 for the eclipse center. Several aspects of source variability are discussed. The LMC data confirm the variability of LMC X-1, LMC X-2, and LMC X-3. In addition recent data concerning optical counterparts of these sources are reviewed.

142.086 Spectrum and polarization of X-rays from accretion disks around black holes.
A. P. Lightman, S. L. Shapiro.
Astrophys. Journ., (Letters), Vol. 198, L73 - L75 (1975).
The emergent radiation spectrum and polarization of X-rays from accretion disks around black holes are calculated. A linear polarization of a few percent should be observed above ~ 1 keV from compact binary X-ray sources containing black holes if a disk exists. The observed, hard X-ray component of the Cyg X-1 spectrum cannot be explained by the standard, instability-free accretion disk model.

142.087 Lunar occultation of the hard X-ray source in the Crab nebula.
Y. Fukada, I. Kasahara, S. Hayakawa, F. Makino, Y. Tanaka, H. Akiyama, J. Nishimura, M. Matsuoka, M. Oda, M. Nakagawa, H. Sakurai, V. S. Iyengar, R. K. Manchanda, P. K. Kunte, B. V. Sreekantan.
Nature, Vol. 255, 465 - 466 (1975).
Hard X rays from the Crab nebula were observed at a lunar occultation on 24 January 1975 with two sets of scintillation counters on board two balloons. The result obtained by one of the balloon observations is reported.

142.088 Possible models for some transient X-ray sources.
F. Pacini, S. L. Shapiro.
Nature, Vol. 255, 618 - 619 (1975).
A transient X-ray source flared up in Centaurus during December 1974 and was observed by the satellite Ariel V in the energy interval 3–30 keV. The significance of this source is that it exhibits a definite periodicity, P, of 6.75 ± 0.03 min. This may shed some light on the nature of the object and (possibly) of other transient sources as well. The authors explore here two alternative models for this phenomenon and suggest in both cases that the source may flare again in the near future; although there is no clear evidence for repetitive outbursts, repetitive behaviour cannot be ruled out.

142.089 On the origin of $\lambda\lambda 4640-4650$ emission in X-ray stars. J. E. McClintock, C. R. Canizares, C. B. Tarter.
Astrophys. Journ., Vol. 198, 641 - 652 (1975).
A prominent and rapidly variable emission feature near $\lambda\lambda 4640-4650$ has been observed in six of the eight optical counterparts of compact galactic X-ray sources. It appears very likely that the emission is generated in the vicinity of the compact source by the interaction of X-rays with nearby gaseous matter. The results of a detailed computational model of an X-ray source surrounded by a uniform distribution of

matter are presented. The computed temperature, ionization structure, electron density, and radiation fields are used to discuss three possible origins of the $\lambda\lambda 4640-4650$ emission: (1) nonselective emission by one or more of the ions C III, N III, and O II; (2) selective emission of N III $\lambda\lambda 4634-4641$ in the "Of star process"; (3) selective emission of N III $\lambda\lambda 4634-4641$ in the Bowen fluorescence process.

142.090 Optical pulsations in HZ Herculis. III. Discovery of pulsed emission lines.
A. Davidsen, B. Margon, J. Middleditch.
Astrophys. Journ., Vol. 198, 653 - 660 (1975).

The authors have detected periodic optical pulses from HZ Herculis in a 100 Å band containing the emission features He II $\lambda 4686$ and the N III $\lambda 4640$ multiplet. The data indicate that one or both lines originate at HZ Her, rather than near the compact X-ray source or in the gas stream as suggested by other observers. The authors propose that radiative recombinations of He II and dielectronic recombinations of N III can explain the observed pulsed emission line strengths and that the pulsed lines are formed in a dense region of the HZ Her atmosphere. The occurrence of optical pulses before X-ray turn-on proves that pulsed X-rays are at times present in the HZ Her system even when they are unobservable at earth.

142.091 A study of fast time structure within cosmic gamma-ray bursts.
W. L. Imhof, G. H. Nakano, R. G. Johnson, J. R. Kilner, J. B. Reagan, R. W. Klebesadel, I. B. Strong.
Astrophys. Journ., Vol. 198, 717 - 725 (1975).

Data taken with a fast time resolution germanium spectrometer surrounded by a large area plastic scintillator, on board the low-altitude polar orbiting satellite 1972–076B, and with the γ-ray sensors on the Vela satellite system have been used to study the fine time structure of two cosmic γ-ray bursts. For the 1972 December 18 event, both energy spectra and intensity measurements were obtained, whereas due to depletion of the cryogen for the germanium sensor, only intensity information was obtained for the 1973 July 21 event, all on a fast time scale. Several microbursts with time widths of the order of 60 ms or less occurred at various times during each of these γ-ray bursts.

142.092 Cygnus X-1: discovery of variable circular polarization.
J. J. Michalsky, J. B. Swedlund, R. A. Stokes.
Astrophys. Journ., (Letters), Vol. 198, L101 - L104 (1975).

HDE 226868, the optical counterpart of Cyg X-1, has been observed for circular polarization during 1974. Observations in five colors suggest that circular polarization results from an interstellar effect. Measurements of the blue polarization reveal circular polarization variations synchronous with the $5^d_.6$ orbital period. The circular polarization variation appears to be similar to the blue intensity variation.

142.093 Evidence for long-period sporadic pulsations in the hard X-ray flux of Cygnus X-1.
F. Frontera, F. Fuligni.
Astrophys. Journ., (Letters), Vol. 198, L105 - L108 (1975).

Power spectral density estimates performed on the data of three balloon observations of Cyg X-1 show a peak at a frequency of 5.75×10^{-2} Hz. This is interpreted as evidence for the presence of strong correlations in the hard X-ray flux of the source. That peak, however, appears to be present, in the single observation, only for limited periods of time, outside which it tends to disappear or to be displaced. Pulsations of limited duration, about 1 hour, cannot be excluded as a possible cause of the detected phenomenon.

142.094 Modes of mass transfer and classes of binary X-ray sources.
E. P. J. van den Heuvel.

Astrophys. Journ., (Letters), Vol. 198, L109 - L112 (1975).

It is shown that if one assumes that compact stars (neutron stars or black holes) can be born as companions to main-sequence stars of any mass M, powerful ($L_x = 10^3 - 10^5 L_\odot$) binary X-ray sources with lifetimes $\gtrsim 10^3$ yr are preferably expected to occur for $M < 2.1 M_\odot$ and $M \gtrsim 20 M_\odot$. Systems in the first mass range become X-ray sources when the normal star overflows its Roche lobe due to evolutionary expansion. Systems in the second mass range become X-ray sources when the normal star becomes a blue supergiant or an Of star, and develops a strong stellar wind before it overflows its Roche lobe. In the mass range $2.1 \lesssim M/M_\odot \lesssim 20$ practically no long-lived X-ray binaries are expected since evolutionary stages with strong winds are absent and the accretion rates resulting from Roche-lobe overflow are in general so large (10^{-6} to $10^{-3} M_\odot$ yr^{-1}) that any X-ray emission from a companion will soon be extinguished.

142.095 Temporal X-ray astronomy with a pinhole camera.
S. S. Holt.
Goddard Space Flight Center, Greenbelt, Maryland, GSFC Document X-661-75-114, 23 + 15 pp. (1975). − Invited paper presented at the COSPAR Symposium on Fast Transients in X- and Gamma-Rays, Varna, Bulgaria.

The first preliminary results from the Ariel-5 All-Sky X-Ray Monitor are presented, along with sufficient experiment details to define the experiment sensitivity. Periodic modulation of the X-ray emission is investigated from three sources (Cyg X-3, Cyg X-1, Sco X-1) with which specific periods have been associated. The light curve of a bright nova-like transient source in Triangulum is presented, and compared with previously observed transient sources.

142.096 X-ray binaries and asymmetry of supernova outbursts.
P. R. Amnuehl', O. Kh. Gusejnov, Sh. Yu. Rakhamimov.
Pis'ma v Astron. Zhurn., Vol. 1, No. 5, p. 19 - 23 (1975). In Russian.

Only $\sim 10^{-3}$ of all close binary systems remain after the collapse of one of the components. Therefore there may be ~ 100 strong and ~ 1000 weak X-ray sources in the Galaxy. The cause of binary system disruption is the asymmetry of supernova outbursts, a relativistic star receiving $0.05-0.15$ of the total impulse of the supernova envelope.

142.097 Hard X-ray bursts in the energy range of 40−290 keV.
O. P. Babushkina, L. S. Bratolyubova-Tsulukidze, R. N. Izrailovich, M. I. Kudryavtsev, A. S. Melioranskij, I. A. Savenko, V. M. Shamolin.
Pis'ma v Astron. Zhurn., Vol. 1, No. 6, p. 6 - 11 (1975). In Russian.

The bursts of hard X-rays with energy flux within $3 \times 10^{-7} - 3.5 \times 10^{-6}$ erg/cm^2 were found by means of a scintillation spectrometer on-board Cosmos 428 in the range of 40−290 keV. The integral distribution of the bursts number versus energy flux does not contradict $N (P \geqslant P_0) = kP^{-3/2}$, nearly independent on galactic latitude.

142.098 Sonneberger Beobachtungen der Röntgenquelle HZ Herculis.
W. Wenzel, H. Geßner.
MVS, Sonneberg, Vol. 6, 196 - 200 (1975).

142.099 A model of the X-ray source Her X-1 and nature of the 35-day cycle.
G. S. Bisnovatyj-Kogan, B. V. Komberg.
Astron. Zhurn. Akad. Nauk SSSR, Vol. 52, 457 - 468 (1975). In Russian. English translation in Soviet Astron., Vol. 19, No. 3.

A model of the binary system with the X-ray source Her X-1 is considered. The central star is supposed to have a sufficiently strong magnetic field, so the outflow of matter occurs in jets. The motion of the neutron star relative to the jet occurs due to nonsynchronization between the orbital motion and rotation of the central star. If the duration of the whole circle around the central star is 35 days and the dimension of the jet is about 1/3 of the orbit, then such a model explains a feature connected with the 35-day cycle. The change of the 35-day period is connected with the motion of the magnetic poles of the central star. The authors predict that the long duration of constancy of the 35-day period must be followed by a long time of absence of X-ray emission.

142.100 X-ray emission of Cas A and Tycho.
K. V. Bychkov, T. G. Sitnik.
Astron. Zhurn. Akad. Nauk SSSR, Vol. 52, 505 - 514 (1975). In Russian. English translation in Soviet Astron., Vol. 19, No. 3.
On the basis of a previously suggested model, which explains the emission from knots of Cas A, it is shown that these knots can emit the observed X-radiation from the nebula in the photon energy range $1 < E < 10$ keV. Observational tests are given. Both the absolute magnitude and the spectrum of the observed X-radiation from Tycho can be explained by the emission from swept-up interstellar material ($E < 4$ keV) and by nonrelativistic gas of the SNR ($E > 4$ keV) by assuming that the distance to Tycho is equal to 1.5 kpc.

142.101 Galactic X-ray astronomy. K. P. Beuermann.
Conference on optical observing programs on galactic structure and dynamics, (see 012.013), p. 191 - 229 (1975). In German.

142.102 A possible candidate for LMC X-3 and observations of cepheids in the galactic centre. P. R. Warren.
Monthly Notes Astron. Soc. Southern Africa, Vol. 34, 75 (1975). – Abstract.

142.103 Supernova explosions in close binary systems. II. Runaway velocities of X-ray binaries.
W. Sutantyo.
Astron. Astrophys., Vol. 41, 47 - 52 (1975).
The effects of a spherically symmetric explosion on the runaway velocity of a close binary system with an initial circular orbit is considered. It is shown that the runaway velocity is completely determined by the final orbital parameters regardless of the initial condition. For Cyg X-1, the upper mass limit of the exploded star is found to be $\sim 16\,M_\odot$. For $M_2 = 30\,M_\odot$ these upper limit becomes ~ 9–$10\,M_\odot$ and $19\,M_\odot$ respectively. An argument is given suggesting that the explosion in Her X-1 cannot have been triggered by mass transfer and that the system most likely originated from a population I object.

142.104 X-ray astronomy in 1974. L. E. Peterson.
Bull. Astron. Soc. India, Vol. 2, 51 - 58 (1974).

142.105 On the nature of γ-radiation from the region of the galactic center. A. A. Stepanyan, B. M.
Vladimirskij, Yu. I. Neshpor, V. P. Fomin.
Izv. AN SSSR. Ser. fiz., Vol. 39, 417 - 423 (1975). In Russian. Abstr. in Referativ. Zhurn. 62. Issled. kosmich. prostranstva, 6.62.223 (1975).

142.106 γ-radiation from the central region of the Galaxy and search for flare activity at energies of E $\geqslant 100$ MeV A. I. Belyaevskij, V. L. Bokov, V. K. Bocharkin,
G. M. Gorodinskij, E. M. Kruglov, E. V. Myakinin,
E. I. Chujkin.
Izv. AN SSSR. Ser. fiz., Vol. 39, 424 - 434 (1975). In Russian. Abstr. in Referativ. Zhurn. 62. Issled. kosmich. prostranstva, 6.62.224 (1975).

142.107 Cygnus X-3. M. Ryle.

IAU Circ., No. 2737 (1975).

142.108 Cygnus X-3. N. Kawajiri.
IAU Circ., No. 2742 (1975).

142.109 MX 0513–40. N. V. Vidal, K. C. Freeman.
IAU Circ., No. 2744 (1975).

142.110 New X-ray source. C. J. Eyles, F. D. Rosenberg,
G. K. Skinner, A. P. Willmore.
IAU Circ., No. 2752 (1975).

142.111 Transient X-ray source.
W. A. Wheaton, W. A. Baity, L. E. Peterson.
IAU Circ., No. 2761 (1975).

142.112 X-ray sources. T. H. Markert, H. V. Bradt,
G. W. Clark, W. H. G. Lewin, F. K. Li, H. W.
Schnopper, G. F. Sprott, G. F. Wargo.
IAU Circ., No. 2765 (1975).

142.113 Hercules X-1. B. A. Cooke.
IAU Circ., No. 2770 (1975).

142.114 X-ray sources. C. J. Eyles, G. K. Skinner,
A. P. Willmore, F. D. Rosenberg, R. Berthelsdorf,
E. W. Gottlieb, W. Liller.
IAU Circ., No. 2774 (1975).

142.115 X-ray sources. H. Gursky, J. Grindlay,
H. Schnopper, E. Schreier, D. Parsignault, A. C.
Brinkman, J. Heise, J. Schrijver, R. Mewe, E. Gronenschild,
A. den Boggende, C. Chevalier, S. A. Ilovaisky.
IAU Circ., Nos. 2778, 2779 (1975).

142.116 Cygnus X-1. C. C. Wu, P. R. Wesselius, R. J. van
Duinen, K. S. de Boer, J. W. G. Aalders, J. Heise,
A. C. Brinkman, R. Mewe, J. Schrijver, A. G. F. den
Boggende, E. Gronenschild, R. M. Hjellming, D. M. Gibson,
F. N. Owen, C. T. Bolton.
IAU Circ., No. 2779 (1975).

142.117 Prime candidate for the transient X-ray source A0535 + 26. W. Liller.
IAU Circ., No. 2780 (1975).

142.118 Centaurus A. R. C. Haymes.
IAU Circ., No. 2780 (1975).

142.119 X Persei. B. Margon, S. Bowyer.
IAU Circ., No. 2781 (1975).

142.120 A0535 + 26. P. Murdin.
IAU Circ., No. 2784 (1975).

142.121 A0535 + 26. C. J. Eyles, G. K. Skinner,
A. P. Willmore, F. D. Rosenberg, W. Mayer,
L. A. Higgs.
IAU Circ., Nos, 2787, 2788 (1975).

142.122 X-ray increase of Aql X-1. J. Buff.
IAU Circ., No. 2788 (1975).

142.123 New soft X-ray source. D. R. Hearn, J. A.
Richardson, J. Condon, J. Liebert, H. Spinrad.
IAU Circ., No. 2790 (1975).

142.124 A0535 + 26 and HDE 245770.
S. Rössiger, W. Wenzel.
IAU Circ., No. 2790 (1975).

142.125   CD −33°12119 and X-ray source 3U 1727−33.
         G. Hensberge, E. J. Zuiderwijk.
IAU Circ., No. 2791 (1975).

142.126   **Aquila X-1.**   A. Davidsen, P. Sanford,
         P. Davison, K. Mason.
IAU Circ., No. 2793 (1975).

142.127   **X-ray observations of Cyg X-1 from Salyut 4.**
         E. Shaffer, E. Moskalenko.
IAU Circ., No. 2793 (1975).

142.128   **3U 0900−40.**   S. Rappaport, J. McClintock.
         IAU Circ., No. 2794 (1975).

142.129   **Cygnus X-3.**   E. Kendziorra, W. Pietsch,
         R. Staubert, J. Trümper.
IAU Circ., No. 2794 (1975).

142.130   **Accretion flows onto galactic X-ray sources.**
         J. S. Buff.
Thesis, Colorado Univ., Boulder (USA). 208 pp. University
Microfilms Order No. 74-22,323 (1974).

142.131   **Outbursts of cosmic gamma rays.**   T. Montmerle.
         Recherche, (*France*), No. 47, Vol. 5, 666 - 668
(1974).

142.132   **Cosmic gamma-ray bursts.**
         A. Fabian, J. Pringle.
New Scient., (*GB*), Vol. 65, 313 - 315 (1975).

142.133   **Numerical studies of mass transfer and accretion in
         X-ray binary systems.**   M. L. Alme.
Thesis, California Univ., Livermore (USA). 200 pp. (1974).

142.134   **Extended X-ray sources and missing masses in
         clusters of galaxies.**   R. Hoshi.
Progr. Theor. Phys. Japan, Vol. 52, 1392 - 1393 (1974).
   Based on the observation of soft X-rays in the Coma
cluster, the existence of hot gases is assumed. For the case of
the gases being bound by the missing masses distributed in
them, the missing mass is calculated to be the same as that
required by the dynamics of the galaxies.

142.135   **Spiral structure and galactic gamma radiation.**
         R. Schlickeiser, K. O. Thielheim.
Phys. Letters B, (*Netherlands*), Vol. 53B, 369 - 372 (1974).
   The intensity of high energy galactic gamma rays is calculated using a model for the distribution of interstellar hydrogen and cosmic ray gas coupled through the interstellar magnetic field. Results are compared with recent empirical data
from SAS-2.

142.136   **Models of compact X-ray sources.**   J. I. Katz.
         Thesis, Cornell Univ., Ithaca, N.Y. (USA). 204 pp.
University Microfilms Order No. 74-6382 (1973).

A focussing X-ray collector and its response in flight.
See Abstr. 034.082.

**Observation of the diffuse component of cosmic
soft X-rays.**   See Abstr. 061.004.

**Observations of cosmic gamma-ray bursts with
IMP 7: evidence for a single spectrum.**   See Abstr. 061.011.

X-ray astronomy in the Uhuru epoch and beyond.
See Abstr. 061.021.

**Diffuse cosmic gamma-ray background in the 28
keV−4.1 MeV range from Kosmos 461 observations.**
See Abstr. 061.023.

On the nature of γ-ray bursts.
See Abstr. 061.036.

**On emission lines in the cosmic gamma-ray back-
ground.**   See Abstr. 061.047.

Gamma ray astrophysics.   See Abstr. 061.051.

Observation of celestial gamma rays.
See Abstr. 061.052.

The X-ray background.   See Abstr. 061.054.

X-ray astronomy.   See Abstr. 061.058.

Models for X-ray illuminated atmospheres.
See Abstr. 064.018.

**Accretion and effluxion of mass and angular
momentum.**   See Abstr. 065.032.

**Physics at the magnetospheric boundary of an ac-
creting neutron star and its consequences for models of binary
X-ray sources.**   See Abstr. 065.053.

**Accretion onto neutron stars under adiabatic shock
conditions.**   See Abstr. 065.070.

**X-ray emission from a neutron star with a strong
magnetic dipole field.**   See Abstr. 065.091.

**The Lense-Thirring effect and accretion disks around
Kerr black holes.**   See Abstr. 066.010.

**High efficiency of the Penrose mechanism for
particle collisions.**   See Abstr. 066.034.

**On the ability of current experiments to test π^0 -
decay gamma-ray background theories.**
See Abstr. 066.041.

**The D3 chromosphere, coronal holes, and stellar
X-rays.**   See Abstr. 073.061.

Ionospheric effects of X-ray source Scorpius XR-1.
See Abstr. 083.039.

UBV photometry of V1357 Cyg (Cyg X-1).
See Abstr. 113.025.

**Colors, magnitudes, spectral types, and distances for
stars in the field of the X-ray source Cyg X-1 (HDE 226868).**
See Abstr. 113.052.

**Spectroscopic investigation of X Persei
(= 2U 0352 + 30?).**   See Abstr. 117.010.

**The optical properties of binary star systems with
accretion discs.**   See Abstr. 117.011.

**X-ray variability by matter accretion onto a black
hole in a detached binary system.**   See Abstr. 117.032.

**Effect of ellipticity and the parameters of X-ray
binaries Cyg X-1 and Cen X-3.**   See Abstr. 117.035.

A model of X Persei.   See Abstr. 117.036.

Ultraviolet observations of HD 77581 (= 2U 0900 −
40).   See Abstr. 119.003.

The spectroscopic binary system HD 158320
(= 3U 1727−33?).   See Abstr. 119.004.

X-ray emission from U Geminorum and the radio
spur at $l \simeq 200°$.   See Abstr. 122.085.

Copernicus: the X-ray spectrum of Puppis-A.
See Abstr. 125.011.

Copernicus: the X-ray spectrum of Cassiopeia A.
See Abstr. 125.017.

Observations of X-ray emissions from supernova
remnants.   See Abstr. 125.031.

Recent Copernicus results on the X-ray emission
from supernova remnants.   See Abstr. 125.035.

Spectrum and structure of X-ray emission from
Puppis A.   See Abstr. 125.036.

Hot vibrating white dwarf models of pulsating X-ray
sources.   See Abstr. 126.001.

Empirical relation between interstellar X-ray absorption and optical extinction.   See Abstr. 131.086.

X-ray absorption, interstellar reddening, and
elemental abundances in the interstellar medium.
See Abstr. 131.087.

The 4.8 GHz formaldehyde absorption in the direction of Cygnus X-3.   See Abstr. 131.090.

Copernicus: soft X-ray emission from certain features of the Cygnus Loop.   See Abstr. 132.012.

Evidence for a compact source of soft X-rays in the
Cygnus Loop.   See Abstr. 132.016.

The absorption of soft X-rays from the Crab
nebula.   See Abstr. 134.004.

Change in the high-energy radiation from the Crab.
See Abstr. 134.005.

High-energy X-ray observations of a lunar occultation of the Crab nebula.   See Abstr. 134.006.

A model of the X-ray structure of the Crab nebula.
See Abstr. 134.009.

The flux-density variations of the radio emission
from Cygnus X-3.   See Abstr. 141.006.

A model for simultaneous synchrotron and inverse
Compton fluxes.   See Abstr. 141.013.

Timing effects in pulsed binary systems.
See Abstr. 141.325.

On the origin of the binary pulsar PSR 1913 + 16.
See Abstr. 141.338.

Attempt to monitor pulsed X-rays from the Crab
pulsar (NP 0532) utilizing atmospheric fluorescence.
See Abstr. 141.367.

The origin of cosmic rays and the Vela gamma-ray
excess.   See Abstr. 143.037.

K_a X-rays from cosmic ray oxygen.
See Abstr. 143.057.

Evidence for the detection of gamma rays from
Centaurus A AT $E_\gamma \geq 3 \times 10^{11}$ eV.   See Abstr. 158.035.

Soft X-ray survey of the Large Magellanic Cloud.
See Abstr. 159.004.

The velocity dispersion of the cluster of galaxies
Abell 1060 (3U 1044−30).   See Abstr. 160.010.

The energy spectrum of the Perseus cluster of
galaxies.   See Abstr. 160.019

The soft X-ray spectrum of the Perseus cluster.
See Abstr. 160.020.

Redshift dispersion in the X-ray cluster of galaxies
A1367.   See Abstr. 160.021.

Southern galaxy clusters identified with 3U X-ray
sources.   See Abstr. 160.026.

Effects of the reheating of the intergalactic gas on
the collapse of late density perturbations.
See Abstr. 161.006.

143 Cosmic Radiation

143.001 Pulsars and the origin of cosmic rays.
T. N. Rengarajan.
Astrophys. Space Sci., Vol. 32, 55 - 75 (1975).
The capabilities and limitations of pulsars as sources of cosmic rays are reviewed in the light of experimental observations. Pulsars can supply the cosmic ray power if they have rotational velocities in excess of 700 rad s^{-1} at birth. The existence of a steady flux of cosmic rays of energy greater than 10^{17} eV demands acceleration of particles to last over fifty years, the time interval between supernovae outbursts, whereas the expected period of activity is less than a few years. Finally, the problem of anisotropy with relevance to pulsars as sources and the possibility of observing pulsar accelerated particles from galactic clusters is considered.

143.002 Cosmic-ray streaming and anisotropies.
M. A. Forman, L. J. Gleeson.
Astrophys. Space Sci., Vol. 32, 77 - 94 (1975).
The principal result of this paper is the demonstration that in interplanetary space the electric-field drifts and convective flow parallel to the magnetic field of cosmic-ray particles combine as a simple convective flow with the solar wind. In addition there are diffusive currents and transverse gradient drift currents. With this interpretation direct reference to the interplanetary electric-field drifts is eliminated and the study of steady-state and transient cosmic-ray anisotropies is both more systematic and simpler.

143.003 Propagation of cosmic rays in the Galaxy.
R. R. Daniel, S. A. Stephens.
Space Sci. Rev., Vol. 17, 45 - 158 (1975).
(1) Introduction: Models of cosmic ray confinement; The galactic model for cosmic ray confinement; Scope of the present review. (2) General description of the Galaxy: Dimensions and density; Structure; Interstellar gas; Magnetic fields; Radiation fields. (3) Cosmic ray propagation in the Galaxy: theoretical aspects: Quasi-stationary state and spatial homogenity of cosmic rays; Models of diffusion of cosmic rays; Diffusion process and isotropy of cosmic rays; General formulation of the propagation of cosmic rays in interstellar space; Quasi-equilibrium models. (4) Interpretation of the observed data: Matter traversed by cosmic rays; The chemical composition at the source; Modification of energy spectrum at low energies; Modification of energy spectrum at high energies; Propagation of the electron component and its source characteristics; Leakage lifetime of cosmic rays; Antinuclei in cosmic rays. (5) Cosmogenic electromagnetic radiations in the Galaxy: Cosmic ray interaction with interstellar gas; Cosmic ray interactions with radiation fields; Synchrotron radiation. (6) The role of cosmic rays in galactic dynamics: Hydrostatic equilibrium of the gaseous component of the Galaxy; Stability of self-gravitating gas and the formation of clouds; Thermal equilibrium of interstellar gas.

143.004 Estimate of the ^{3}He/(^{3}He + ^{4}He) ratio in primary cosmic rays at energies ~ 1 GeV/nucleon.
M. G. Iodko, V. A. Romanov.
Izv. AN SSSR. Ser. fiz., Vol. 38, 1813 - 1815 (1974). In Russian. − Abstr. in Referativ. Zhurn. 51. Astron., 2.51.704 (1975).

143.005 Cosmic ray modulation in various solar activity periods.
R. R. Ashirov, A. G. Zusmanovich, E. V. Kolomeets.
Izv. AN SSSR. Ser. fiz., Vol. 38, 1932 - 1936 (1974). In Russian. − Abstr. in Referativ. Zhurn. 51. Astron., 2.51.705 (1975).

143.006 Anisotropic flux of cosmic radiation in different periods of solar activity. S. D. Asylbaeva,
G. A. Gonchar, E. V. Kolomeets, N. V. Slyunyaeva.
Izv. AN SSSR. Ser. fiz., Vol. 38, 1917 - 1919 (1974). In Russian. − Abstr. in Referativ. Zhurn. 62. Issled. kosmich. prostranstva, 2.62.279 (1975).

143.007 Cosmic ray fluxes at minimum and maximum solar activity. G. F. Krymskij, A. I. Kuz'min,
A. M. Altukhov, P. A. Krivoshapkin, V. P. Mamrukova, Z. N. Samsonova, G. V. Skripin, N. P. Chirkov.
Izv. AN SSSR. Ser. fiz., Vol. 38, 1912 - 1916 (1974). In Russian. − Abstr. in Referativ. Zhurn. 62. Issled. kosmich. prostranstva, 2.62.280 (1975).

143.008 The influence of the sectorial structure of the interplanetary magnetic field on the modulation of cosmic rays.
G. A. Bazilevskaya, V. P. Okhlopkov, T. N. Charakhch'yan.
Izv. AN SSSR. Ser. fiz., Vol. 38, 1895 - 1898 (1974). In Russian. − Abstr. in Referativ. Zhurn. 62. Issled. kosmich. prostranstva, 2.62.281 (1975).

143.009 Preliminary results of an investigation of the effect of cosmic ray increase by the flare of December 10, 1966. L. I. Dorman, N. S. Kaminer, A. E. Kuz'-micheva, Yu. I. Okulov, T. S. Khadakhanova.
Izv. AN SSSR. Ser. fiz., Vol. 38, 1880 - 1883 (1974). In Russian. − Abstr. in Referativ. Zhurn. 62. Issled. kosmich. prostranstva, 2.62.282 (1975).

143.010 Fluctuation phenomena in cosmic ray propagation in interplanetary space.
L. I. Dorman, S. E. Kats.
Izv. AN SSSR. Ser. fiz., Vol. 38, 1961 - 1965 (1974). In Russian. − Abstr. in Referativ. Zhurn. 62. Issled. kosmich. prostranstva, 2.62.285 (1975).

143.011 Results and perspectives of cosmic ray investigation.
S. N. Vernov, G. B. Khristiansen.
Probl. yader. fiz. i kosmich. luchej. Resp. mezhved. temat. nauch.-tekhn. sb., 1974, vyp. (No.) 1, p. 3 - 13. In Russian. Abstr. in Referativ. Zhurn. 62. Issled. kosmich. prostranstva, 2.62.286 (1975).

143.012 The dependence of diffusion of galactic cosmic radiation on energy.
S. V. Bulanov, V. A. Dogel', V. S. Ptuskin.
Izv. AN SSSR. Ser. fiz., Vol. 38, 1796 - 1800 (1974). In Russian. − Abstr. in Referativ. Zhurn. 62. Issled. kosmich. prostranstva, 2.62.288 (1975).

143.013 Cosmic rays and the Galaxy.
A. M. Hillas, M. Ouldridge.
Nature, Vol. 253, 609 - 610 (1975). − Letter.

143.014 The confinement of galactic cosmic rays by Alfvén waves. J. A. Holmes.
Monthly Notices Roy. Astron. Soc., Vol. 170, 251 - 260 (1975).
A leaky-box model of cosmic ray confinement is investigated in an axially-symmetric model of the Galaxy. The cosmic rays are confined by the Alfvén waves which they excite in the magnetic field. The effect of non-linear damping of the Alfvén waves is discussed, and the residence time and path length of cosmic rays in the Galaxy are calculated as functions of rigidity. Observations of the electron spectrum

and of pulsar scintillation are discussed in terms of this model.

143.015 Galactic winds driven by cosmic rays.
F. M. Ipavich.
Astrophys. Journ., Vol. 196, 107 - 120 (1975).

Cosmic rays traveling through a magnetized plasma with mean velocity greater than the Alfvén speed can become coupled to the plasma by the emission of magnetohydrodynamic waves. Using a time-independent, spherically symmetric, hydrodynamic treatment, the author shows that cosmic rays escaping from a galaxy can carry along thermal gas and produce a "galactic wind." Solutions indicate that a typical galaxy can lose mass at a rate $\sim 1-10\,M_\odot\,\mathrm{yr}^{-1}$ and energy at a rate $\sim 10^{40}-10^{42}$ ergs s^{-1}. The possible role of galactic winds in radio and X-ray sources is explored. A model is proposed to explain the X-ray emission observed from clusters of galaxies.

143.016 Decaying nuclei and the age of cosmic rays in the Galaxy.
V. L. Prishchep, V. S. Ptuskin.
Astrophys. Space Sci., Vol. 32, 257 - 263, 265 - 271 (1975). In Russian and English.

The propagation of radioactive nuclei of cosmic rays in a flat diffusion galactic model (sources and the main gaseous mass are concentrated in the galactic disc) is considered. The corresponding results are not reducible to the results of a simple homogeneous model. It is shown that the recent data on the Be^{10} nuclei abundance in cosmic rays do not contradict the occurrence of a large cosmic ray halo.

143.017 Scattering of cosmic rays on a statistically uneven surface.
L. I. Dorman, V. Kh. Shogenov.
Geomagn. Aeronom., Vol. 15, 10 - 12 (1975). In Russian.

143.018 Influence of the bounded magnetosphere and the ring current on the geomagnetic cutoff rigidity of cosmic rays.
L. G. Glikman, V. P. Shabanskij.
Geomagn. Aeronom., Vol. 15, 20 - 23 (1975). In Russian.

143.019 The spectrum of cosmic electrons with energies between 6 and 100 GeV.
C. A. Meegan, J. A. Earl.
Astrophys. Journ., Vol. 197, 219 - 233 (1975).

The spectral intensity of primary cosmic-ray electrons in particles $\mathrm{m}^{-2}\,\mathrm{s}^{-1}\,\mathrm{sr}^{-1}\,\mathrm{GeV}^{-1}$ was found to have the following power-law dependence upon the electron energy E in GeV; $dJ/dR = (800 \pm 60)E^{-3.4 \pm 0.1}$. Similarly, the ground level spectrum of secondary cosmic-ray electrons was found to be $dJ/dE = 1.1\,E^{-2.9 \pm 0.1}$. The steepness of the spectrum of cosmic electrons relative to that of nuclei implies one of the following conclusions: either the injection spectrum of electrons is steeper than that of nuclei, or the electron spectrum has been steepened by Compton/synchrotron losses in the energy range covered by the experiment.

143.020 Causes of Forbush decreases and other cosmic ray variations.
E. Barouch, L. F. Burlaga.
Journ. Geophys. Res., Vol. 80, 449 - 456 (1975).

The relationship between neutron monitor variations and the intensity variations of the interplanetary magnetic field is studied by using Deep River data and Imp series satellite data. In over 80% of the cases studied in 1968, identifiable depressions of the cosmic ray intensity are associated with magnetic field enhancements of several hours duration and intensity above 10 γ.

143.021 Modulation of low-energy cosmic rays.
J. W. Sari.
Journ. Geophys. Res., Vol. 80, 457 - 469 (1975).

The objectives of this work are to test directly the effects of the predicted diffusion coefficients on the propagation of low-energy cosmic rays as well as to determine the relative contributions of the discontinuities and the fluctuations between them to the scattering process.

143.022 Semidiurnal component of cosmic ray intensity.
R. P. Kane.
Journ. Geophys. Res., Vol. 80, 470 - 481 (1975).

In this communication the author examines the behavior of the semidiurnal component vis-à-vis specific orientations of the interplanetary magnetic field.

143.023 Energy dependence of the Si/Fe ratio in the galactic cosmic rays.
M. Garcia-Munoz, E. Juliusson, G. M. Mason, P. Meyer, J. A. Simpson.
Astrophys. Journ., Vol. 197, 489 - 493 (1975).

Recent measurements by Israel et al. and Webber have indicated that the silicon-to-iron ratio in the galactic cosmic radiation increases rapidly with decreasing energy over the range ~ 0.4 to 1 GeV per nucleon. Measurements made with the University of Chicago instrumentation on satellites and high altitude balloons over the energy range 35 MeV to 4 GeV per nucleon do not support this result. Energy spectra for Si and Fe are presented, and the corresponding local interstellar spectra are calculated using a standard model of solar modulation. It is found that the local interstellar spectra can be fitted by current models of galactic propagation.

143.024 A time-dependent diffusion-convection model for the long-term modulation of cosmic rays.
J. J. O'Gallagher.
Astrophys. Journ., Vol. 197, 495 - 507 (1975).

Incorporation of the effects of time-dependent diffusive propagation of galactic cosmic rays inside a modulating region whose basic parameters are slowly changing in time leads to a new prediction for the modulated density expected to be observed at a given time. This model provides a natural physical explanation for observed rigidity-dependent phase lags in modulated spectra sometimes referred to as cosmic-ray "hysteresis". If all of the phase lag observed between 500-MeV protons and the Deep River neutron intensity is attributed to the effects described here, the average distance to the modulating boundary during the last solar cycle is estimated to be 45−55 AU.

143.025 Ultraheavy cosmic rays.
M. H. Israel, P. B. Price, C. J. Waddington.
Phys. Today, Vol. 28, No. 5, p. 23 - 27, 29 - 31 (1975).

The distribution of the high-Z nuclei, the youngest elements in the Galaxy, can help us understand the synthesis of matter in exploding stars and its interactions en route to earth.

143.026 Narrow-angle cosmic-ray anisotropies.
S. Barrowes.
Nature, Vol. 255, 171 (1975).

Speller et al. (1972) have presented a measurement of the diurnal cosmic-ray anisotropy, as found with underground muons in London. An additional, but quite different, interpretation of their data is, however, possible.

143.027 On the variation of the cosmic ray composition in the past.
N. Bhandari, J. T. Padia.
Proc. Fifth Lunar Sci. Conference, (see 012.003), Vol. 3, 2577 - 2589 (1974).

143.028 On the requirements for acceleration mechanisms to produce flattened heavy particle cosmic ray energy spectra.
J. R. Wayland.
Astrophys. Space Sci., Vol. 33, 459 - 463 (1975).

The recently observed flattening of the energy spectra of high Z particles is investigated in terms of the limits it places on acceleration mechanisms. The possibility of a new mechanism is suggested.

143.029 Intensity of cosmic rays in the interplanetary space. Observational data. Jan. 5 - May 15, 1969.
Mezhduved. geofiz. kom. pri Prezidiume AN SSSR. Materialy Mirovogo tsentra dannykh B. Moskva. 54 pp. Price 11 Kop. (1974). In Russian. − Abstr. in Referativ. Zhurn. 62. Issled. kosmich. prostranstva, 3.62.279 (1975).

143.030 Intensity of cosmic rays in the interplanetary space. Observational data. July 19, 1965 - Jan. 25, 1966.
Mezhduved. geofiz. kom. pri Prezidiume AN SSSR. Materialy Mirovogo tsentra dannykh B. Moskva. 55 pp. Price 14 Kop. (1974). In Russian. − Abstr. in Referativ. Zhurn. 62. Issled. kosmich. prostranstva, 3.62.280 (1975).

143.031 Cosmic ray anomalies during magnetic storms.
Kh. Z. Aldagarova, A. G. Zusmanovich, E. V. Kolomeets, M. A. Musabaev.
Izv. AN SSSR. Ser. fiz., Vol. 38, 1899 - 1903 (1974). In Russian. − Abstr. in Referativ. Zhurn. 62. Issled. kosmich. prostranstva, 3.62.281 (1975).

143.032 Parameters of cosmic ray variations determined by the spectrographic method.
L. I. Dorman, V. M. Dvornikov, A. A. Luzov, A. V. Sergeev, A. L. Yanchukovskij.
Izv. AN SSSR. Ser. fiz., Vol. 38, 1942 - 1945 (1974). In Russian. − Abstr. in Referativ. Zhurn. 62. Issled. kosmich. prostranstva, 3.62.282 (1975).

143.033 Search for antiprotons in primary cosmic radiation.
V. A. Romanov.
Probl. sovrem. fiz. Leningrad, Nauka, 1974, p. 265 - 276. In Russian. − Abstr. in Referativ. Zhurn. 62. Issled. kosmich. prostranstva, 3.62.291 (1975).

143.034 Some investigations of the multiply charged component of cosmic rays aboard satellites and automatic interplanetary stations. N. S. Ivanova.
Probl. sovrem. fiz. Leningrad, Nauka, 1974, p. 245 - 264. In Russian. − Abstr. in Referativ. Zhurn. 62. Issled. kosmich. prostranstva, 3.62.294 (1975).

143.035 Investigations of the corpuscular radiation aboard satellites and space ships.
A. A. Kolchin, V. V. Lebedev, G. P. Skrebtsov.
Probl. sovrem. fiz. Leningrad, Nauka, 1974, p. 231 - 244. In Russian. − Abstr. in Referativ. Zhurn. 62. Issled. kosmich. prostranstva, 3.62.295 (1975).

143.036 Apollo 17 cosmic-ray experiment: interplanetary heavy nuclei of energies 0.05 to 5.0 MeV per atomic mass unit. R. T. Woods, H. R. Hart, Jr., R. L. Fleischer.
Astrophys. Journ., Vol. 198, 183 - 194 (1975).
Glass detectors were exposed to interplanetary heavy ions for 45.5 hr on the moon during 1972 December 12 and 13. In the energy range 0.05−5.0 MeV amu⁻¹, the differential flux of iron-group nuclei was found to be a power-law spectrum of the form E^{-3}. The observations of solar-related anisotropy and the E^{-3} spectra, and the factor of ∼ 10 enhancement in the Fe/α ratio, imply strongly that these heavy ions are solar in origin.

143.037 The origin of cosmic rays and the Vela gamma-ray excess. J. C. Higdon, R. E. Lingenfelter.
Astrophys. Journ., (Letters), Vol. 198, L17 - L20 (1975).
The recent observation of excess γ-ray emission from the region around the Vela supernova remnant provides the most direct evidence so far that supernovae produce sufficient energy in relativistic particles to be the source of cosmic rays.

The authors show that an estimate of the cosmic ray energy around Vela suggests that either only a fraction of galactic supernovae are the source of most cosmic rays, or only a fraction of the matter traversed by cosmic rays is in the interstellar medium.

143.038 Two classes of cosmic ray decrease.
H. J. Verschell, R. B. Mendell, S. A. Korff, E. C. Roelof.
Journ. Geophys. Res., Vol. 80, 1189 - 1201 (1975).
From an analysis of the time variations during 1968−1971 of the fast neutron flux in the upper atmosphere versus those of ground-based neutron monitors we have identified two classes of transient intensity decrease on the basis of differences in their spectral responses, time histories, and flare associations. Type I events are found to be classic Forbush decreases, sharp declines accompanying a geomagnetic storm sudden commencement, following by 1−3 days a large optical flare with radio noise and energetic particle production, whereas type II events are more symmetric in their time histories and are therefore not associated with a particular flare.

143.039 The evaluation of cutoff rigidities and reentrant albedo calculations for Palestine, Dallas, and Midland, Texas. M. A. Shea, D. F. Smart.
Journ. Geophys. Res., Vol. 80, 1202 - 1208 (1975).
By using the trajectory-tracing technique, cutoff rigidities for Palestine, Dallas, and Midland, Texas, have been calculated as a function of various zenith and azimuth angles. Extensive analysis of the trajectory calculations shows that there is a systematic uncertainty involved in computing the lowest allowed rigidity, and this uncertainty may be a significant fraction of the penumbral width. Continuation of the trajectory-tracing process below the Stormer cutoff allows an evaluation of the reentrant albedo, showing that the average invariant latitude of the guiding center of the trajectory at the albedo origin is the same as the average invariant latitude of the guiding center of the particle trajectory at the detection point.

143.040 Search for antiprotons in primary cosmic radiation.
V. A. Romanov.
Probl. sovrem. fiz. Leningrad, Nauka, 1974, p. 265 - 276. In Russian. − Abstr. in Referativ. Zhurn. 51. Astron., 4.51.741 (1975).

143.041 A measurement of cosmic-ray positron and negatron spectra between 50 and 800 MV.
J. K. Daugherty, R. C. Hartman, P. J. Schmidt.
Astrophys. Journ., Vol. 198, 493 - 505 (1975).
A balloon-borne spark chamber magnetic spectrometer has been used to measure the spectra of cosmic-ray positrons and negatrons at energies between 50 and 800 MV. The present results indicate that the dominance of negatrons from primary sources, found in earlier experiments above 200 MV, extends down to at least 50 MV. Solar modulation of the positron component is found to be consistent with that of the total electron spectrum, assuming that the positron component is entirely attributable to collisions between cosmic-ray nuclei and the interstellar gas.

143.042 Solar modulation of galactic cosmic ray electrons, protons, and alphas. G. J. Fulks.
Journ. Geophys. Res., Vol. 80, 1701 - 1714 (1975).
Over the 5-year period 1968−1972 the author has measured the energy spectrum of primary cosmic ray electrons from 20 MeV to 20 GeV using a balloon-borne absorption spectrometer. During the same period, other investigators have inferred the interstellar electron spectrum mainly from radio observations and have directly measured the electron spectrum at low energy, proton and alpha energy spectra, neutron monitor counting rates, radial particle gradients, the velocity

of the solar wind, and the power spectrum of interplanetary magnetic field irregularities. In this paper the author presents observations of the electron spectrum, examines the simple spherically symmetric model of modulation involving convection, diffusion, and adiabatic deceleration, and compares the model with the various measurements. The model is substantially capable of reproducing the observations above 40 MeV/nucleon.

143.043 Cosmic ray intensity variations during 0200—0700 UT, August 5, 1972.
D. Venkatesan, T. Mathews, L. J. Lanzerotti, D. H. Fairfield, C. O. Bostrom.
Journ. Geophys. Res., Vol. 80, 1715 - 1724 (1975).

A detailed investigation using interplanetary magnetic field measurements and particle data from ground-based neutron monitors, lunar sensors, and satellite-borne detectors has been made of the marked increase in cosmic ray intensity early on August 5, 1972. The similarities in the structure of the event as observed by different detectors, the changes in the helium to proton ratios, the time delay of ~ 9 min between observations at Explorer 41 and those at the moon, and a north-south asymmetry in the enhancement observed by the neutron monitors are all explained in terms of this model.

143.044 Origin of ultra high energy cosmic rays.
P. Kiraly, J. L. Osborne, M. White, A. W. Wolfendale.
Nature, Vol. 255, 619 - 620 (1975). – Letter.

143.045 Cosmic-ray nuclei up to 10^{10} eV/u in the Galaxy.
M. M. Shapiro, R. Silberberg.
Phil. Trans. Roy. Soc. London, Ser. A, Vol. 277, 319 - 348 (1975). – Conference paper.

(1) The energy spectrum of the arriving nuclei after its modulation in the solar system; (2) The composition of the arriving cosmic rays below and above 1 GeV/u; (3) Calculated abundances of cosmic rays at the sources, and comparisons with those in the sun and nearby stars; (4) Some ways in which the source composition might arise; (5) The relative abundances as a function of energy; (6) Transformation and propagation of cosmic-ray nuclei in space; (7) Isotopic composition of cosmic rays – observation and calculations; (8) Abundances and spectra of ultra-heavy nuclei.

143.046 Composition and spectra of primary cosmic-ray electrons and nuclei above 10^{10} eV. P. Meyer.
Phil. Trans. Roy. Soc. London, Ser. A, Vol. 277, 349 - 363 (1975).-Conference paper.

Recent experiments have extended the knowledge of the flux and energy spectra of individual cosmic-ray components to much higher energies than had previously been accessible. Both electron and nuclear components show a behaviour at high energy which is unexpected, and which carries information regarding the sources and the propagation of particles between sources and observer.

143.047 The search for cosmic-ray anisotropies.
H. Elliot.
Phil. Trans. Roy. Soc. London, Ser. A, Vol. 277, 381 - 393 (1975).-Conference paper.

At the present time there is no generally accepted evidence for any statistically significant anisotropy in the energy range $10^{17} - 10^{19}$ eV. The upper limits on the possible anisotropy provide strong evidence that these particles are extragalactic. In that part of the cosmic-ray magnetic rigidity spectrum below ca. 2×10^{11} V the interplanetary magnetic field effectively prevents the detection of anisotropies in interstellar space and the only isotropies measured are associated with the solar wind and its associated magnetic field. In the range of magnetic rigidities extending from 10^{11} to 10^{12} V the cosmic-

ray intensity shows evidence for a small anisotropy of about 2×10^{-4} which can be explained as the result of solar motion relative to the average galactic rotation in our neighbourhood.

143.048 Long-term variations in the cosmic-ray flux.
D. Lal.
Phil. Trans. Roy. Soc. London, Ser. A, Vol. 277, 395 - 411 (1975).-Conference paper.

The present-day information on the temporal and spatial variations in the flux and chemical composition of cosmic-ray protons and multicharged nuclei during certain intervals of time, since the beginning of the solar system, is discussed.

143.049 Survey of data on primary cosmic-ray nuclei above 10^{14} eV. A. M. Hillas.
Phil. Trans. Roy. Soc. London, Ser. A, Vol. 277, 413 - 428 (1975).-Conference paper.

Primary cosmic-ray particles, detected by means of the extensive cascades they generate in the atmosphere, have been observed over a continuous range of energies up to 10^{20} eV, and apparently somewhat higher. At energies such that the radius of curvature of their trajectories, if they are protons, as expected, is comparable to our distance from the galactic centre, the arrival directions of 84 observed particles are distributed randomly over the sky.

143.050 Explanations of the spectral shape in the energy range $10^{14} - 10^{20}$ eV. A. W. Wolfendale.
Phil. Trans. Roy. Soc. London, Ser. A, Vol. 277, 429 - 442 (1975).-Conference paper.

There is evidence suggesting an increase in slope of the energy spectrum of primary cosmic-ray nuclei at about 3×10^{15} eV. Alternative explanations are advanced for this change, related to diffusion of particles of galactic origin or black body cut-off effects for particles of extra-galactic origin.

143.051 On the origin of cosmic rays. V. L. Ginzburg.
Phil. Trans. Roy. Soc. London, Ser. A, Vol. 277, 463 - 479 (1975).-Conference paper.

The origin of the main part of the cosmic rays observed near the earth is discussed. This includes first of all the choice between galactic and metagalactic models and the source problem. Some remarks about other related topics also are made expecially in connexion with the prospects for the future research.

143.052 The extra-galactic contribution to the primary cosmic-ray flux. G. R. Burbidge.
Phil. Trans. Roy. Soc. London, Ser. A, Vol. 277, 481 - 487 (1975).-Conference paper.

In previous studies it has been shown that a good case can be made for supposing that a large fraction of the primary cosmic rays is of extra-galactic origin. These ideas are reviewed here, and the most recent observations and theoretical suggestions bearing on the problem are described.

143.053 On the stellar origin of low energy cosmic rays.
B. Lovell.
Phil. Trans. Roy. Soc. London, Ser. A, Vol. 277, 489 - 501 (1975).-Conference paper.

Large solar flares are associated with the production of low energy cosmic rays, and given the identity of the flare mechanism on the M type stars it seems reasonable to assume that these stars will also produce cosmic rays during the flare phase. The purpose of this paper is to make estimates of the probable contribution of the M type and K type stars to the total cosmic-ray flux in the Galaxy.

143.054 Dependence of the north-south anisotropy of cosmic rays on the orientation of interplanetary shock waves. S. A. Rumyantsev, V. S. Smirnov.

Geomagn. Aeronom., Vol. 15, 401 - 404 (1975). In Russian.

143.055   A spectrographic method of investigation of cosmic
ray variations taking into account the penumbra.
L. I. Dorman, G. Sh. Shkhalakhov.
Geomagn. Aeronom., Vol. 15, 405 - 411 (1975). In Russian.

143.056   Concerning the article of R. B. Salimzibarov "About
the meteor variation of cosmic rays".
S. A. Bel'skij.
Geomagn. Aeronom., Vol. 15, 543 - 545 (1975). In Russian.
Brief information.

143.057 K_α X-rays from cosmic ray oxygen.
S. H. Pravdo, E. A. Boldt.
Goddard Space Flight Center, Greenbelt, Maryland, GSFC
Document X-661-75-39, 9 + A4 + 6 pp. (1975).
Equilibrium charge fractions are calculated for subrela-
tivistic cosmic ray oxygen ions in the interstellar medium.
These are used to determine the expected flux of K_α rays
arising from atomic processes for a number of different postulated interstellar oxygen spectra. Relating these results to the
diffuse X-ray background measured at the appropriate energy
(i.e. ~ 0.6 keV) suggests an observable broadened line feature.

143.058   Nuclear gamma ray production by cosmic rays.
M. Meneguzzi, H. Reeves.
Astron. Astrophys., Vol. 40, 91 - 98 (1975).
The rate of nuclear gamma ray emission from the interaction of cosmic rays with thermal nuclei in the interstellar
medium and the consequent expected flux at the earth are
calculated. The same calculation is made in the case of the
existence of a high flux of low energy cosmic rays, which
could be responsible for the ^{7}Li production in the Galaxy.
The authors also present the flux expected from supernova
remnants, in particular if the galactic ^{7}Li is produced in such
objects by low energy cosmic rays.

143.059   Light element production by cosmic rays.
M. Meneguzzi, H. Reeves.
Astron. Astrophys., Vol. 40, 99 - 110 (1975).
The yield of the isotopes of Li, Be and B (L isotopes)
through cosmic particle bombardment of the interstellar gas
is reexamined in detail in the light of the most recent data on
nuclear cross sections and elemental abundances. The possibility that ^{7}Li is produced by low energy cosmic rays in interstellar space or in supernova remnants is investigated.

143.060   Anisotropy of cosmic radiation in the Galaxy.
T. Gombosi, J. Kota, A. J. Somogyi, A. Varga,
B. Betev, L. Katsarsky, S. Kavlakov, I. Khirov.
Nature, Vol. 255, 687 - 689 (1975).
The authors report a measurement which proves the
existence of an anisotropy in galactic cosmic radiation. It was
detected by examining extensive air showers produced by
cosmic-ray primaries of energies ~6×10^{13} eV (which are
certainly not affected by solar modulation).

143.061   Origin of very high-energy cosmic rays.
S. A. Colgate.
Phys. Rev. Letters, Vol. 34, 1177 - 1180 (1975).
It is suggested that a single source of cosmic rays (super-
novas) occurring in all galaxies can produce the observed
spectrum and the observed anisotropy, and is predictable from
supernova shock theory. Below 10^{13-14} eV the source and observed spectrum are the same, $N(> E) \propto E^{-a}$, $a \simeq 1.75$. Above
10^{14} eV the author predicts a source with $a \simeq 1.2$. Galactic
leakage above 10^{15} eV is linear so that $a \simeq 2.2$ as observed.
Above 10^{19} eV cosmic rays fill the metagalaxy to a flux
several times the anisotropic residual flux from a few events
in our galaxy as observed.

143.062   Introductory cosmic rays.   A. W. Wolfendale.
Origin of cosmic rays, (012.012), p. 1 - 12 (1975).

143.063   The Galaxy and interstellar medium.
J. L. Osborne.
Origin of cosmic rays, (012.012), p. 13 - 24 (1975).
The purpose of this paper is to describe the galactic
setting for the propagation and possible origin of cosmic rays.
A description of the overall structure and constituents of our
Galaxy is followed by an account of the physical properties
and distribution of the interstellar medium. An attempt is
made to solve the problem of covering such a broad subject in
a brief review by concentrating on those features which are
judged to have a direct bearing on the origin and propagation
of cosmic rays.

143.064   Extragalactic cosmic rays.   G. R. Burbidge.
Origin of cosmic rays, (012.012), p. 25 - 36 (1975).
A survey of the components of the extragalactic universe
is given. Much of the discussion centres about the various
types of extragalactic non-thermal sources.

143.065   Sidereal daily variations in cosmic ray intensity and
their relationships to solar modulation and galactic
anisotropies.   T. Thambyahpillai.
Origin of cosmic rays, (012.012), p. 37 - 59 (1975).

143.066   Energy spectrum and mass composition of cosmic
ray nuclei from 10^{12} to 10^{20} eV.   A. A. Watson.
Origin of cosmic rays, (012.012), p. 61 - 95 (1975).

143.067   Nuclear mass composition at 'low' energies (i.e.
$< 10^{12}$ eV/nucleon).   I. L. Rasmussen.
Origin of cosmic rays, (012.012), p. 97 - 133 (1975).
In this paper the author gives a review of present and
proposed methods for the measurement of the masses of
cosmic rays, and discusses some of the results already obtained.

143.068   Nucleosynthesis and galactic cosmic rays.
H. Reeves.
Origin of cosmic rays, (012.012), p. 135 - 164 (1975).

143.069   Galactic structure, magnetic fields and cosmic ray
containment.   K. O. Thielheim.
Origin of cosmic rays, (012.012), p. 165 - 201 (1975).

143.070   Galactic propagation of cosmic rays below 10^{14} eV.
J. L. Osborne.
Origin of cosmic rays, (012.012), p. 203 - 220 (1975).
This paper is concerned with the question of whether the
observed anisotropy and lifetime of cosmic rays can be reconciled with the hypothesis of galactic origin in discrete sources
such as supernovae or supernova remnants. The author considers the nuclear component only. In the energy region above
10^{14} eV the energy density of the cosmic rays and the pressure
exerted by them are so low that they must have a negligible
effect on the magnetic fields through which they propagate.
At the lower energies considered here the mutual interaction
between the particles and the field has to be taken into
account.

143.071   Possible explanations of the spectral shape.
A. W. Wolfendale.
Origin of cosmic rays, (012.012), p. 221 - 231 (1975).
A satisfactory model for the origin of cosmic rays must
explain the many facts which are known about these particles
and quanta. In the present work the nucleonic component is
the main concern, particularly those particles above 10^{12} eV.
The form of the primary spectrum at these energies is briefly
examined and then the relative merits of a number of origin
models are studied. Attention is confined to those models with

which the author has been concerned.

143.072   The cosmic ray electron component.   P. Meyer.
Origin of cosmic rays, (012.012), p. 233 - 266
(1975).
The paper summarises the present knowledge of the
electron spectrum over a wide range of energies from about
100 keV up to almost 1000 GeV.

143.073   Collapsed stars, pulsars and the origin of cosmic rays
F. Pacini.
Origin of cosmic rays, (012.012), p. 371 - 397 (1975).
The paper deals with various equilibrium configurations
which are possible at the end of stellar evolution and the
pulsar phenomenon from an observational point of view. The
author finally introduces the basic electrodynamics of pulsars
and the mechanisms which have been proposed for the
acceleration of particles.

143.074   Surface and underground measurements of long-
term changes in the cosmic ray solar diurnal variation.   T. Thambyahpillai, R. D. Speller.
Planet. Space Sci., Vol. 23, 961 - 971 (1975).

143.075   The 11-year cycle of galactic cosmic radiation and
the total magnetic field of the sun.
S. N. Vernov, A. N. Charakhch'yan, Yu. I. Stozhkov,
T. N. Charakhch'yan.
Izv. AN SSSR. Ser. fiz., Vol. 39, 316 - 324 (1975). In Russian.
Abstr. in Referativ. Zhurn. 62. Issled. kosmich. prostranstva,
6.62.210 (1975).

143.076   Intensity variations of cosmic radiation and the
solar wind.   S. N. Vernov, B. A. Tverskoj,
G. P. Lyubimov, N. V. Pereslegina, N. N. Kontor, E. A.
Chuchkov.
Izv. AN SSSR Ser. fiz., Vol. 39, 340 - 349 (1975). In Russian.
Abstr. in Referativ. Zhurn. 62. Issled. kosmich. prostranstva,
6.62.213 (1975).

143.077   On the chemical composition of cosmic radiation
in the Galaxy.   V. S. Ptuskin.
Izv. AN SSSR Ser. fiz., Vol. 39, 403 - 407 (1975). In Russian.
Abstr. in Referativ. Zhurn. 62. Issled. kosmich. prostranstva,
6.62.221 (1975).

143.078   Cosmic ray anisotropy in the interplanetary space.
S. D. Asylbaeva, G. A. Gonchar, E. V. Kolomeets,
L. A. Mirkin, N. V. Slyunyaeva.
Materialy Itog. nauch. konf. prof.-prepodavat. sostava.
Kazakhsk. un-t. Alma-Ata, 1974, p. 255. In Russian. − Abstr.
in Referativ. Zhurn. 62. Issled. kosmich. prostranstva, 6.62.
233 (1975).

143.079   Cosmic ray intensity variation because of the varia-
tion of geomagnetic cutoff rigidity.
Kh. Z. Aldagarova, E. V. Kolomeets, V. T. Pivneva.
Materialy Itog. nauch. konf. prof.-prepodavat. sostava.
Kazakhsk. un-t. Alma-Ata, 1974, p. 259 - 260. In Russian.
Abstr. in Referativ. Zhurn. 62. Issled. kosmich. prostranstva,
6.62.234 (1975).

143.080   Modulation of galactic cosmic radiation.
A. A. Ajtmukhambetov, R. R. Ashirov, O. A.
Bogdanova, A. G. Zusmanovich, E. V. Kolomeets.
Materialy Itog. nauch. konf. prof.-prepodavat. sostava.
Kazakhsk. un-t. Alma-Ata, 1974, p. 260. In Russian. − Abstr.
in Referativ. Zhurn. 62. Issled. kosmich. prostranstva, 6.62.
235 (1975).

143.081   North-south asymmetry of cosmic radiation.

A. Kh. Bykovskaya, E. V. Kolomeets.
Materialy Itog. nauch. konf. prof.-prepodavat. sostava.
Kazakhsk. un-t. Alma-Ata, 1974, p. 260 - 261. In Russian.
Abstr. in Referativ. Zhurn. 62. Issled. kosmich. prostranstva,
6.62.236 (1975).

143.082   The nature of intensity increases of cosmic radia-
tion before Forbush effects.
E. V. Kolomeets, M. A. Musabaev.
Materialy Itog. nauch. konf. prof.-prepodavat. sostava.
Kazakhsk. un-t. Alma-Ata, 1974, p. 261 - 262. In Russian.
Abstr. in Referativ. Zhurn. 62. Issled. kosmich. prostranstva,
6.62.237 (1975).

143.083   Anisotropy of the one-third day variation of cosmic
rays.   T. Kanno, Y. Ishida, T. Saito.
.Uchusen Kenkyu., Vol. 18, 127 - 136 (1973). In Japanese.

143.084   Spherical zonal components of cosmic rays in inter-
planetary space.
H. Takahashi, N. Yahagi, K. Nagashima.
Uchusen Kenkyu., Vol. 18, 106 - 112 (1973). In Japanese.

143.085   Time correlation of heavy cosmic ray nuclei.
S. E. Walker.
Thesis, California Univ., Riverside (USA). 101 pp. University
Microfilms Order No. 74-13,582 (1973).

143.086   Satellite measurement of cosmic-ray abundances and
spectra in the charge range $2 \leq Z \leq 10$.
J. W. Brown.
Thesis, California Inst. Techn., Pasadena (USA). 116 pp. University Microfilms Order No. 74-14,264 (1974).

143.087   Charge composition of high energy heavy primary
cosmic ray nuclei.   R. D. Price.
Thesis, Catholic Univ. of America, Washington, D.C. (USA).
175 pp. University Microfilms Order No. 74-15,995 (1974).

143.088   Composition of cosmic rays with $Z \geq 12$.
R. C. Maehl.
Thesis, Washington Univ., St. Louis, Mo. (USA). 261 pp. Uni-
versity Microfilms Order No. 74-22,535 (1974).

143.089   Measurement of the primary cosmic-ray electron
spectrum from 6 GeV to 100 GeV.   C. A. Meegan.
Thesis, Maryland Univ., College Park (USA). 221 pp. University Microfilms Order No. 74-17,054 (1973).

143.090   Analysis of the 0.511 MeV radiation at the OSO-7
satellite.   P. P. Dunphy.
Thesis, New Hampshire Univ., Durham (USA). 138 pp. University Microfilms Order No. 74-21,092 (1974).

143.091   Remarks on the possibility of pulsar-induced bump
in the cosmic ray spectrum at 10^{13}–10^{16} eV/parti-
cle.   J. W. Elbert, M. O. Larson, G. H. Lowe, J. L.
Morrison, G. W. Mason, R. L. Spencer.
Journ. Phys. A, (Math., Nuclear, General), Vol. 8, L13 - L17
(1975). − Letter.

143.092   Contribution of an increasing proton−proton cross
section to steepening of the cosmic ray energy spectrum.
T. K. Gaisser, C. J. Noble, G. B. Yodh.
Journ. Phys. G, Vol. 1, L9 - L12 (1975). − Letter.

143.093   Cosmic rays and particle physics at extremely high
energies.   T. K. Gaisser.
Journ. Franklin Inst., (U. S. A.), Vol. 298, 271 - 287 (1974).
Cosmic ray extensive air shower observations can give

clues about the properties of high-energy particle interactions above 10^{15} eV, in addition to giving information about the primary cosmic ray spectrum at these energies. A brief review of recent high-energy accelerator data and its theoretical interpretation is given to explain the necessity for obtaining information from cosmic rays about hadronic interactions beyond machine energies. In particular, it is shown that recent measurements of the energy dependence of total cross-sections and of production cross-sections suggest some sort of threshold behavior in the 1000 GeV region. The use of air shower data for this purpose is illustrated by comparing shower size vs depth measurements with the expectations of various models of strong interactions.

143.094 **The cosmic-ray electron and positron spectra and their modulation.** S. Cecchini, C. Winkler. Nuovo Cimento Lettere, Ser. 2, Vol. 12, 86 - 90 (1975).

A study is presented of the modulation of the electron cosmic-ray component and want to emphasize that similar constraints exist in the choice both of the diffusion coefficient and of the shape of the galactic electron spectrum for low-energy electrons (below 200 MeV).

143.095 **Two views of cosmic ray propagation in the solar system.** E. Barouch. Solar wind three, (see 012.020), p. 206 - 213 (1974).

143.096 **Progress report on the radial gradients of cosmic ray nuclei 0.5 MeV per nucleon to relativistic energies and electrons 6 to 30 MeV.** J. A. Simpson, T. F. Conlon, J. J. O'Gallagher, R. B. McKibben, A. J. Tuzzolino. Solar wind three, (see 012.020), p. 214 - 216 (1974).

Cosmic rays from the Galaxy. Nature, Vol. 253, 588 - 589 (1975).

Cosmic rays: variations and space explorations. See Abstr. 003.054.

Cosmic ray physics. Part 2. Astrophysical aspect. See Abstr. 003.079.

Primary γ-rays. See Abstr. 061.002.

Gamma ray astrophysics. See Abstr. 061.051.

Galactic neutrino sources and cosmic rays. See Abstr. 061.072.

On a nonlinear closure approximation for cosmic-ray diffusion equations. See Abstr. 062.007.

On the nonlinear closure approximation for cosmic-ray diffusion. See Abstr. 062.008.

Mirroring in the Fokker-Planck coefficient for cosmic-ray pitch-angle scattering in homogeneous magnetic turbulence. See Abstr. 062.012.

Hydromagnetic waves and cosmic-ray diffusion theory. See Abstr. 062.042.

Motion of charged particles normal to an irregular magnetic field. See Abstr. 062.048.

Slowly braked, rotating neutron stars. See Abstr. 065.013.

Cosmic ray propagation in the solar wind. See Abstr. 074.133.

An effect of cosmic rays on the distant solar wind. See Abstr. 074.134.

The 1964—1972 quiet-time spectra of protons and helium at 2—20 MeV per nucleon. See Abstr. 078.007.

Results of measurements of cosmic ray intensity aboard the automatic station Luna 19. See Abstr. 078.019.

Heliocentric cosmic ray gradient 1.0 — 4.1 A.U. See Abstr. 078.032.

Interaction of energetic nuclear particles in space with the lunar surface. See Abstr. 094.224.

Cosmic-ray production in the Cassiopeia A supernova remnant. See Abstr. 125.014.

Rapid nuclear reactions in supernovae and cosmic rays. See Abstr. 125.026.

Supernovae and the origin of cosmic rays (I). See Abstr. 125.052.

Supernovae and the origin of cosmic rays (II), a model of cosmic ray production in supernovae. See Abstr. 125.053.

On cosmic rays and final equilibrium states for the Parker instability. See Abstr. 131.127.

Contribution from pulsars. See Abstr. 141.352.

Pulsars and the origin of cosmic rays. See Abstr. 141.353.

Correlation of pulsar positions and the arrival directions of air showers of energies 10^{17}–10^{18} eV observed at Chacaltaya. See Abstr. 141.365.

Galactic γ-rays and cosmic ray origin. See Abstr. 142.081.

Stellar Systems

151   Kinematics and Dynamics of Stellar Systems

151.001   **Shock wave cascades and the formation of proto-galaxies.**   A. A. Rumyantsev, A. D. Chernin.
Astrophys. Space Sci., Vol. 32, L15 - L18 (1975).
   Modern cosmogonical theories attribute to primeval potential or vortex hydrodynamic motions a prominent role in the formation of galaxies. The authors consider here this problem on the basis of the concept of hydrodynamic instability in shock waves.

151.002   **On the nonlinear time development of gas flow in spiral density waves.**   P. R. Woodward.
Astrophys. Journ., Vol. 195, 61 - 73 (1975).
   Time-dependent calculations of the interstellar gas flow in a tightly wound spiral density wave have been performed. The gravitational potential of the gas is ignored, and the flow development is driven by the gravitational potential of the stars. For sufficiently large wave amplitudes, shocks form within one or two transits of the gas through the spiral pattern. The mechanism for steepening of the wave form is discussed by analyzing the roles played by terms in the flow equations as the flow develops.

151.003   **Dynamical method to estimate the relative masses of stars in spherical clusters.**   A. S. Baranov.
Vestn. Leningr. un-ta, 1974, No. 13, p. 122 - 130. In Russian.
Abstr. in Referativ. Zhurn. 51. Astron., 2.51.770 (1975).

151.004   **The ejection of massive objects from galactic nuclei: gravitational scattering of the object by the nucleus.**
W. C. Saslaw.
Astrophys. Journ., Vol. 195, 773 - 781 (1975).
   When a compact massive object moves through a galactic nucleus or stellar system, its gravity produces coherent fluctuations in the stellar density. The massive object is scattered by these self-induced fluctuations and deflected from its initial orbit. This process can be used to probe the gravitational response of stellar systems, and it may be related to the observed misalignment of double extragalactic radio sources.

151.005   **Dynamics and evolution of galaxies.**   J. Einasto.
New problems of astrophysics, Publ. Astrophys.
Winter School, 1972, (see 012.001), p. 4 - 31 (1974).
In Russian.

151.006   **Vibrations of inhomogeneous non-rotating gravitational systems.**   G. Severne, A. Kuszell.
Astrophys. Space Sci., Vol. 32, 447 - 459 (1975).
   Dispersion relations are obtained and analysed for a non-uniform, non-rotating gravitational system. A restriction to short wavelengths makes it possible to consider a linearized form of the collisionless Boltzmann equation, differing from that for homogeneous systems by the appearance of a term expressing the effect of the mean self-gravitational field upon the motion. The mean field affects the radially directed wave perturbations, with a breaking of symmetry. Inwardly and outwardly directed modes have quite different propagation characteristics, inward modes being preferentially propagated. Locally, the stability of the system is found to be enhanced due to the effect of the mean field.

151.007   **Small vertical oscillations in the field of a rotationally symmetrical potential.**   L. P. Osipkov.
Dokl. Akad. Nauk SSSR. Ser. Mat. Fiz., Vol. 221, 309 - 311 (1975). In Russian.

151.008   **Density wave theory and the classification of spiral galaxies.**   W. W. Roberts, Jr., M. S. Roberts, F. H. Shu.
Astrophys. Journ., Vol. 196, 381 - 405 (1975).
   Axisymmetric models of disk galaxies taken together with the density wave theory allow the authors to distinguish and categorize spiral galaxies by means of two fundamental galactic parameters: the total mass of the galaxy, divided by a characteristic dimension; and the degree of concentration of mass toward the galactic center. These two parameters govern the strength of the galactic shocks in the interstellar gas and the geometry of the spiral wave pattern. In turn, the shock strength and the theoretical pitch angle of the spiral arms play a major role in determining the degree of development of spiral structure in a galaxy and its Hubble type. The application of these results to 24 external galaxies demonstrates that the categorization of galaxies according to this theoretical framework correlates well with the accepted classification of these galaxies within the observed sequences of luminosity class and Hubble type.

151.009   **Is there a gravothermal catastrophe?**
L. G. Taff, H. M. Van Horn.
Astrophys. Journ., (Letters), Vol. 197, L23 - L24 (1975).
   A study of the thermodynamics of the finite (confined) isothermal sphere led Lynden-Bell and Wood to predict a gravothermal collapse for such systems when the density contrast exceeds a critical value. Here the authors show that their analysis becomes invalid at precisely this critical point, and that the existence of a gravothermal catastrophe has thus not been proven in their investigation.

151.010   **Dynamical models of elliptical galaxies.**
C. P. Wilson.
Astron. Journ., Vol. 80, 175 - 187 (1975).
   Self-consistent dynamical models are constructed for elliptical galaxies by solving Poisson's equation. They possess rotational symmetry and differential rotation. The density is obtained by integrating over all velocities a near-Gaussian distribution function that depends on energy and angular momentum with two free parameters. Observable properties are predicted by projecting the three-dimensional models onto the plane of the sky. The models give radial intensity profiles and isophote shapes that agree quite well with observations of NGC 3379, and they also predict a rotation curve consistent with the one observed.

151.011   **Numerical investigation of galactic tidal effects on spherical stellar systems.**
D. W. Keenan, K. A. Innanen.
Astron. Journ., Vol. 80, 290 - 302 (1975).
   A numerical investigation is presented which seeks to extend our knowledge of tidal effects on spherical stellar systems. The approach used was essentially the solution of the three-body equations of motion. Three different systems

were computed. In each, the orbits of many test particles were computed in the field of a smooth, spherically symmetric cluster model, which, in turn, moved in its own orbit around a model galaxy. The effect of the tidal fields on the stellar orbits was investigated by observing the evolution of the orbits in energy and in angular momentum. Shock disruption times of a tidal field were also investigated. Observations that support these effects are presented. Finally, the authors investigate the galactic orbits of stars which escape from their parent systems due to tidal forces.

151.012 **On Freeman's collisionless stellar systems.**
M. Nishida, T. Ishizawa.
Mem. Fac. Sci., Kyoto Univ., Ser. Phys., Astrophys. Geophys., Chem., Vol. 34, 353 - 360 (1974).

The structure of Freeman's two dimensional elliptical collisionless stellar systems is restudied in the revised range of the parameters b/a and $\Omega^2/2\pi G\rho$ (the axial ratio and the angular velocity of the cylinder). It is found that Freeman's systems contain rigidly rotating circular (Maclaurin) and elliptical (Jacobi) cylinders and that, if the angular velocity is less than that of the Jacobi elliptical cylinder, the mean circulation is in the direction of rotation and otherwise it is opposite to the direction of rotation.

151.013 **Instability in rotating gravitating systems with radial perturbations.** I. L. Genkin, V. S. Safronov.
Astron. Zhurn. Akad. Nauk SSSR, Vol. 52, 306 - 315 (1975).
In Russian. English translation in Soviet Astron., Vol. 19, No. 2.

The conditions of instability with radial perturbations are considered for four models of rotating gravitating systems: infinite uniform medium, infinitely thin disc and the sheet of finite thickness with rigid and Keplerian rotation.

151.014 **Correlation of gravitational force in a homogeneous stellar system.** V. Yu. Terebizh.
Astron. Zhurn. Akad. Nauk SSSR, Vol. 52, 442 - 444 (1975).
In Russian. English translation in Soviet Astron., Vol. 19, No. 2.

The correlation matrices of components of the gravitational force resulting from randomly distributed field stars are found.

151.015 **The evolution of galaxies. IV. Highly flattened disks.**
R. J. Talbot, Jr., W. D. Arnett.
Astrophys. Journ., Vol. 197, 551 - 570 (1975).

The authors present computations of the structure and evolution of models of the disk component of galaxies. The models display the qualitative characteristics and the statistically dominant trends observed in late-type spirals. The radial distributions of neutral hydrogen, of H II regions, and of colors are well reproduced. Composition gradients across the faces of disk models are found. Using the observed supernova rates in spirals, these models place fairly severe constraints upon the progenitors of supernova. If azimuthally averaged, high-resolution observations of neutral hydrogen will provide interesting quantitative tests of models of this type. Equations governing the structure and evolution of the models of the disk components are discussed in detail, as are the uncertainties in input physics and astronomy.

151.016 **Finite, two-component isothermal spheres. I. Equilibrium models.**
L. G. Taff, H. M. Van Horn, C. J. Hansen, R. R. Ross.
Astrophys. Journ., Vol. 197, 651 - 666 (1975).

The equilibria of finite (confined) isothermal gas spheres composed of particles with two different masses are studied. For given mass ratio μ and dimensionless radius ξ_0 there exists a one-parameter family of solutions. It is found that for $\mu > 3/2$ the total mass of the heavier particles remains finite. Numerical calculations are carried out both for sequences with fixed central density ratio λ and for sequences with fixed

total mass ratio but variable λ. In both cases the sequences display closed loops in the (U, V) phase plane. This suggests interesting, but quite disparate, stability properties for the two types of sequences. A comparison with the two-component dynamical calculations of Spitzer and Hart is also given, and confirms the equilibration of the cores of their models at later evolutionary times.

151.017 **The two-time autocorrelation function for force in bounded gravitational systems.**
L. Cohen, A. Ahmad.
Astrophys. Journ., Vol. 197, 667 - 673 (1975).

The correlation of the force at the same point but at two different times is calculated for bounded gravitational systems. Uniform density and a Gaussian distribution of velocity are assumed. The autocorrelation function decreases as $1/t^5$ for $t \to \infty$. This is in contrast to Chandrasekhar's result for infinite systems, where it decreases only as $1/t$. The authors also show that for an arbitrary distribution of speeds the decrease with time is generally $1/t^4$ and in the case where the distribution is a function of the square of the velocity the decrease is $1/t^5$. The mean square velocity change is calculated via the autocorrelation function.

151.018 **Properties of motion in the gravitational field of a rotating bar.** M. Michalodimitrakis.
Astrophys. Space Sci., Vol. 33, 421 - 440 (1975).

A qualitative study of the properties of motion (equilibrium points, regions of motion, periodic orbits) of a test particle in the gravitational field of a uniformly rotating solid bar is made. Two different models are used for the bar, a homogeneous ellipsoid and a homogeneous rectangular parallelepiped, and the dependence of the properties of motion on the specific choice of the model is investigated. It is found that stability properties, especially those of the equilibrium points on the long axis of the bar, are more pronounced in the case of the parallelepiped.

151.019 **Spiral modes in cold cylindrical systems.**
H. Robe.
Astron. Astrophys., Vol. 39, 455 - 459 (1975).

The linearized hydrodynamical equations governing the non-axisymmetric free modes of oscillation of cold cylindrical stellar systems are separated in cylindrical coordinates and solved numerically for two models. Short-wavelength unstable modes corresponding to tight spirals do not exist; but there exists an unstable growing mode which has the form of trailing spirals which are quite open.

151.020 **Computer models of encounters of galaxies.**
G. G. Byrd.
Bull. American Astron. Soc., Vol. 7, 344 (1975). – Abstr. AAS.

151.021 **The application of anaglyphs to N-body graphics.**
R. S. Harrington, M. Miranian.
Bull. American Astron. Soc., Vol. 7, 344 (1975). – Abstr. AAS.

151.022 **Escape of stars from clusters by the action of tidal fields.** D. W. Keenan, K. A. Innanen.
Bull. American Astron. Soc., Vol. 7, 344 (1975). – Abstr. AAS.

151.023 **The construction of models for elliptical galaxies.**
C. Hunter.
Bull. American Astron. Soc., Vol. 7, 344 (1975). – Abstr. AAS.

151.024 **On the rotation and shape of a star cluster.**
W. H. Jefferys.
Bull. American Astron. Soc., Vol. 7, 344 (1975). – Abstr. AAS.

151.025 **Binary evolution in stellar systems.** R. H. Miller.
Bull. American Astron. Soc., Vol. 7, 344 - 345

(1975). – Abstr. AAS.

151.026 Adiabatic regime for stellar dynamics in thin disk galaxies. I. Drifting epicyclic orbits.
J. W-K. Mark, R. H. Berman.
Bull. American Astron. Soc., Vol. 7, 345 (1975). – Abstr. AAS.

151.027 Adiabatic regime of stellar dynamics in thin disk galaxies. II. Hydrodynamic equations.
R. H. Berman, J. W-K. Mark.
Bull. American Astron. Soc., Vol. 7, 345 (1975). – Abstr. AAS.

151.028 Liouville's theorem and the third integral of motion for steady-state stellar systems. III. Hydrodynamical analogy and some special cases. L. P. Osipkov.
Vestn. Leningr. un-ta, 1974, No. 19, p. 136 - 144. In Russian. Abstr. in Referativ. Zhurn. 51. Astron., 5.51.697 (1975).

151.029 Stability of a spherical stellar system.
M. Ya. Pal'chik, A. Z. Patashinskij, V. K. Pinus.
Zhurn. prikl. mekh. i tekhn. fiz., 1974, No. 6, p. 63 - 73. In Russian. – Abstr. in Referativ. Zhurn. 51. Astron., 5.51. 710 (1975).

151.030 Capture of stars by rotating homogeneous spherical clusters. A. S. Baranov.
Celestial Mechanics, Vol. 11, 517 - 528 (1975).
Changes of the orbit of a star passing through a homogeneous spherical cluster have been estimated. Before entering the cluster the star is supposed to move in a Keplerian parabolic orbit. Due to dynamical friction the energy of the star becomes negative which leads to the elliptic-type motion of the star after leaving the cluster and to capturing the star by the cluster. The formulae for the changes of the star orbit in the cluster are given. Numerical estimates show that open clusters transform star orbits more noticeably than do globular clusters.

151.031 Collisionless generation of galactic rotation.
A. A. Ruzmajkin.
Pis'ma v Astron. Zhurn., Vol. 1, No. 5, p. 10 - 14 (1975). In Russian.
Thomson's theorem is shown to fail in a collisionless system, and generation of vorticity is possible. This process would certainly take place at the early stage of galaxy formation, and also in present-day galaxies with nonequilibrium function of velocity distribution, leading to equilibrium rigid-body rotation.

151.032 On dynamics and cosmogony of galactic coronae.
V. A. Antonov, A. D. Chernin.
Pis'ma v Astron. Zhurn., Vol. 1, No. 6, p. 18 - 22 (1975). In Russian.
A simple stellar dynamical model of galactic coronae with almost radial motions and matter density $\rho \propto r^{-2}$ is considered. The early stage of evolution is discussed for these systems when they were in gaseous state and undergone a quasistationary contraction and became spherical.

151.033 About the structure of spherical galaxies and of clusters of galaxies. I. V. Petrovskaya.
Acta Cosmologica, Fasc. 2, p. 87 - 96 (1974/75).
At the early evolution stages in the spherical galaxies and spherical clusters of galaxies the velocity distribution is supposed to have been significantly radially prolated. In due course at the centre of the systems a region with a spherical velocity distribution is formed which is quasistationary with respect to the irregular force-field. That region gradually expands. This evolution scheme is confirmed by comparing the theoretical models with the observed density distribution in the Coma cluster of galaxies, in the galaxy NGC 3379 and in dwarf galaxies.

151.034 Investigation of homogeneous evolution models of stellar systems. E. R. Astaf'ev.
Astron. Zhurn. Akad. Nauk SSSR, Vol. 52, 498 - 504 (1975). In Russian. English translation in Soviet Astron., Vol. 19, No. 3.
Changes in time of the physical character of models are studied. Models are alike liquid ellipsoids. The initial parameters are: masses $10^9 - 10^{12} M_\odot$, densities $100 - 0.05 M_\odot/\text{pc}^3$, initial ratios of axes 0.9–0.5.

151.035 Spiral tracers, density-wave theory and optical observations. R. Wielen.
Conference on optical observing programs on galactic structure and dynamics, (see 012.013), p. 59 - 74 (1975). In German.

151.036 Present theories of spiral structure: general review and crucial observations. T. Schmidt-Kaler.
Conference on optical observing programs on galactic structure and dynamics, (see 012.013), p. 75 - 96 (1975). In German.

151.037 Momentum transfer and group velocity of a density wave excited by a central bar.
J. V. Feitzinger.
Conference on optical observing programs on galactic structure and dynamics, (see 012.013), p. 105 - 106 (1975). In German.

151.038 Computer experiments in stellar dynamics.
A. Ahmad.
Bull. Astron. Soc. India, Vol. 2, 29 (1974). – Abstract.

151.039 Formation of double galaxies by tidal capture.
K. S. Sastry, A. Potdar, S. M. Alladin.
Bull. Astron. Soc. India, Vol. 2, 29 - 30 (1974). – Abstract.

151.040 On the determination of the masses of disk galaxies using the density-wave model of their spiral structure. B. Basu.
Bull. Astron. Soc. India, Vol. 2, 30 (1974). – Abstract.

151.041 Gravitational attraction of a pair of disk galaxies.
G. M. Ballabh.
Bull. Astron. Soc. India, Vol. 2, 30 (1974). – Abstract.

151.042 Gravitational interactions between galaxies.
S. M. Alladin.
Bull. Astron. Soc. India, Vol. 3, 5 - 8 (1975).

151.043 On the kinetics of excitation of density drift waves by the Doppler effect. M. N. Maksumov.
AN TadzhSSR, In-t astrofiz. Dushanbe, Donish, 1974, 21 pp. Price 12 Kop. In Russian. – Abstr. in Referativ. Zhurn. 51. Astron., 6.51.785 (1975).

151.044 The collisionless generation of galactic rotation.
A. A. Ruzmajkin.
In-t prikl. mat. AN SSSR. Preprint No. 4. Moskva, 1975. 11 pp. In Russian. – Abstr. in Referativ. Zhurn. 51. Astron., 6.51.791 (1975).

151.045 The third integral of motion and the velocity field for the quasi-Newtonian potential. II.
A. A. V'yuga.
Vestn. Leningr. un-ta, 1974, No. 19, p. 126 - 131. In Russian. Abstr. in Referativ. Zhurn. 51. Astron., 6.51.918 (1975).

151.046 Numerical experiments on expanding gravitational systems. G. Janin, M. J. Haggerty.

Journ. Comput. Phys., (*U. S. A.*), Vol. 16, 76 - 92 (1974).

The formation of galaxies, clusters of galaxies, and super-clusters in an expanding cloud of small lumps of material may be due to relatively close gravitational binary interactions between the lumps. An experimental study of expanding gravitational systems of N bodies is performed by means of computer experiments. The results are compared with the evolution of similar systems with more discrete particles placed outside.

151.047   A numerical code for multiple 'water bag' gravitational systems.   S. Cuperman, A. Harten.
Comput. Phys. Commun., (*Netherlands*), Vol. 8, 307 - 319 (1974).

The computer code developed to investigate one-dimensional, collisionless systems of stars, consisting of regions of constant density matter (in phase space) is described.

151.048   Collective instabilities of self-gravitating systems. Infinite homogeneous case.
S. Ikeuchi, T. Nakamura, F. Takahara.
Progr. Theor. Phys. Japan, Vol. 52, 1807 - 1818 (1974).

The instability modes of self-gravitating stellar systems are investigated in comparison with electron oscillations in a plasma for the following four cases; one system of a collisionless or a hydrodynamical description, two systems with a relative velocity of a collisionless and a hydrodynamical one, two collisionless ones and two hydrodynamical ones.

'Boomerang' orbits and their numerical determination.   See Abstr. 042.070.

The collapse of self-gravitating clouds of pure hydrogen.   See Abstr. 065.001.

On tidal phenomena in a strong gravitational field. See Abstr. 066.044.

Dark clouds, star formation and spiral structure. See Abstr. 131.138.

Numerical integration methods for galactic orbit computations.   See Abstr. 155.014.

Numerical calculations on the dynamics of the Galaxy under the influence of spiral arms and the Magellanic Clouds.   See Abstr. 155.040.

On the three-dimensional structure of the spiral arms.   See Abstr. 155.049.

The velocity field of the stars and gas in NGC 2903. See Abstr. 158.006.

Virial tests for fourteen nearby groups of galaxies: a case for the dynamical stability of some groups. See Abstr. 158.009.

Formation of satellites by fragmentation of galaxies.   See Abstr. 158.010.

The role of thermal instability in the formation of galaxies.   See Abstr. 158.016.

Interaction of protoclusters of galaxies with intergalactic matter.   See Abstr. 158.017.

The bar-like objects in the centres of galaxies as a possible generator of spiral density waves. I. See Abstr. 158.018.

A remarkable regularity in galaxy systems; its dynamical and cosmogonical significance. See Abstr. 158.063.

On a possibility of radio astronomical study of the generation of galaxies.   See Abstr. 158.064.

Rejection of the hypothesis on disintegration of galaxy groups due to continuous mass loss. See Abstr. 158.070.

A re-analysis of the dynamics of the nearby groups of galaxies.   See Abstr. 160.006.

152 Stellar Associations

152.001 **A study of the stellar association Canis Major OB 1.** J. J. Clariá.
Astron. Astrophys., Vol. 37, 229 - 236 (1974).

An analysis of the photometric and spectroscopic data of the OB stars in Canis Major confirms the existence of the stellar association Canis Major OB 1. The age of the association is estimated to be 3.0×10^6 years. CMa OB 1 is physically related to the reflection nebula association CMa R 1. The open cluster NGC 2353 appears to be the nucleus of the star association while the open clusters NGC 2343 and NGC 2335, located in the same area, are not related with it.

152.002 **Photometric study of the Chamaeleon T-association.** G. Grasdalen, R. Joyce, R. F. Knacke, S. E. Strom, K. M. Strom.
Astron. Journ., Vol. 80, 117 - 124 (1975).

Optical and infrared photometry was carried out for a large sample of stars in the Chamaeleon dark cloud complex. The distance derived for the complex is 115 pc which makes it the nearest known dark cloud. A study of the stellar population contained within the cloud reveals a wide range of types ranging from a main-sequence A0 star to a large number of Orion population objects. The Chamaeleon complex may represent a stellar cluster intermediate in character between a T-association and an OB-association. The reddening law derived for this dark cloud suggests a value of R, the ratio of total-to-selective extinction, of 5.5; this result is similar to that found for other high gas and dust density regions. An HR diagram was constructed after correcting the observed stellar spectral energy distributions for the effects of reddening and circumstellar emission. The resulting diagram suggests a wide spread in stellar formation time. Finally, the optical and IR emission characteristics of the circumstellar envelopes are outlined briefly.

152.003 **Les magnitudes stellaires en trois couleurs des étoiles des anneaux stellaires en Cygnus et de l'amas NGC 7127.** T. A. Uranova, G. S. Tsarevskij.
Soobshch. Gos. Astron. Inst. Shternberga, No. 188, p. 33 - 41 (1974). In Russian.

Une liste des magnitudes stellaires photographiques U, B et V des étoiles de quatre "anneaux stellaires" et de l'amas NGC 7127 est publiée.

How to make metal-poor stars, redden OB associations and grow mantles on grains. See Abstr. 065.041.

Study of four stellar rings in the Cygnus constellation. See Abstr. 113.009.

CW Cephei: an important close binary member of the III Cephei association. See Abstr. 119.002.

R associations I. UBV photometry and MK spectroscopy of stars in southern reflection nebulae. See Abstr. 132.018.

Pulkovo sky survey in the interstellar neutral hydrogen line. I. Neutral hydrogen in the neighbourhood of stellar associations λ Orionis and Monoceros I. See Abstr. 157.005.

153 Galactic Clusters

153.001 **Uniform survey of clusters in the southern Milky Way.** S. van den Bergh, G. L. Hagen.
Astron. Journ., Vol. 80, 11 - 16 (1975).

The Curtis-Schmidt telescope of the Cerro Tololo Observatory has been used to make a two-color survey of a ~ 12° side strip of the southern Milky Way extending from $l \simeq 250°$ to $l \simeq 360°$. A total of 262 clusters, 63 of which had not previously been catalogued, were found. The apparent distribution on the sky of the clusters in this uniform survey of the southern Milky Way is discussed.

153.002 **Rotational velocities in IC 2602.** H. Levato.
Astrophys. Journ., Vol, 195, 825 - 827 (1975).

Rotational velocities for 20 members of IC 2602 are determined. Their values show that they are, on the average, fast rotators.

153.003 **The kinematic parameters of the star stream Hyades-Praesepe.** E. M. Nezhinskij, L. P. Osipkov.
Astron. Zhurn. Akad. Nauk SSSR, Vol. 52, 203 - 204 (1975). In Russian. English translation in Soviet Astron., Vol. 19, No. 1.

This article deals with kinematic parameters of the open cluster Hyades-Praesepe. The results obtained have been compared with those by Ogorodnikov and Latyshev (1968, 1969).

153.004 **UBV photometry of the southern open cluster NGC 5822.** Ş. Bozkurt.
Rev. Mexicana Astron. Astrofis., Vol. 1, 89 - 100 (1974).

Photoelectric and photographic UBV photometry, to V = 15.50, is carried out for 424 stars in and around NGC 5822. A mean color excess of +0.18, corresponding to a visual absorption of A_v = +0.54. was found. The cluster has a pronounced giant branch. The true distance modulus and the distance of the cluster were found to be 9.33 and 735 pc, respectively. The angular diameter is 54', corresponding to a linear diameter of 11.5 pc. An age of 2.76×10^8 years, using the Lindoff method, was obtained.

153.005 **Spectroscopic study of the open cluster NGC 2422.** M. M. Dworetsky.
Astron. Journ., Vol. 80, 131 - 133 (1975).

Rotational velocities, MK spectral types, and radial velocities are given for the 15 brightest stars in the open cluster NGC 2422. Two of these are probably foreground objects. The brightest member was previously classified B2 IV? e?; no emission lines were seen in 1972–1973. The cluster contains one Si star and another Be star. The mean $V_e \sin i$ of the cluster members is very low; NGC 2422 may contain a large percentage of spectroscopic binaries.

153.006 Membership and photometry of the open cluster IC 4756.

A. D. Herzog, W. L. Sanders, W. Seggewiss.
Astron. Astrophys., Suppl. Ser., Vol. 19, 211 - 234 (1975).

Probabilities of membership, based on relative proper motions, for 464 stars in the field of IC 4756 are given. The cluster proper motion dispersion (m.e.) of $0''09$ yields 173 probable members. Photographic UBV magnitudes of 471 stars in the field are presented. The colour excess of IC 4756 is $E(B-V) = 0^m19$; the apparent distance modulus $V-M_v$ equals 8^m5, corresponding to the true distance of 400 pc. The membership probabilities of 6 blue stragglers and of the supergiant HD 172365 (F9 Ib) are discussed.

153.007 UBV photometry of NGC 2439. S. D. M. White.
Astrophys. Journ., Vol. 197, 67 - 75 (1975).

Photoelectric and photographic measures of 183 stars in the field of NGC 2439 show the cluster to be $\sim 2 \times 10^7$ years old and 4.45 kpc from the sun. Three supergiant stars, including the G0 Ia star R Puppis, are found to be very probable members on the basis of photometric, spectral and radial-velocity data. An extended group of high-luminosity B stars may also be associated with the cluster.

153.008 Further study of the stellar cluster embedded in the Ophiuchus dark cloud complex.

F. J. Vrba, K. M. Strom, S. E. Strom, G. L. Grasdalen.
Astrophys. Journ., Vol. 197, 77 - 84 (1975).

An extension of a 2-μ map of the Ophiuchus dark cloud carried out by Grasdalen, Strom, and Strom has brought to 67 the total number of detected 2-μ point sources. Additional infrared photometry has considerably strengthened the authors' previous conclusion that these sources represent the brighter members of a young cluster obscured by dark-cloud material. The luminosity function derived for the Ophiuchus cluster is virtually identical with that observed for other young clusters. The reddening law in this dense, dark-cloud region, derived from the photometry, is significantly different from the interstellar mean; R, the ratio of total to selective extinction, is larger, suggesting a particle-size distribution peaked toward larger sizes. Furthermore, the ice absorption feature at 3.05 μ has been observed for two sources in this region. The observed strength of this feature seems consistent with the range found for other dense, dark clouds.

153.009 Tentative membership of the 11-day cepheid TW Normae in the open cluster Lyngå 6.

B. F. Madore.
Astron. Astrophys., Vol. 38, 471 - 473 (1975).

UBV photoelectric observations are presented for the brightest stars in the galactic cluster Lyngå 6. The data suggest a reddening to the cluster of $E(B-V) = 1.37 \pm 0.03$ mag (A.D.) and an apparent distance modulus of 16.2 ± 0.5 mag. In the line-of-sight of Lyngå 6 is the 11-day classical cepheid TW Normae, whose magnitude and colours are consistent with it being in the cluster. $M_v = -4.5 \pm 0.5$ mag is provisionally derived for the cepheid, making it the intrinsically brightest cepheid found to be in a cluster.

153.010 Spectral types in Trumpler 10.
H. Levato, S. Malaroda.
Publ. Astron. Soc. Pacific, Vol. 87, 173 - 175 (1975).

MK types for 25 stars in the field of Trumpler 10 are given. The true distance modulus of Trumpler 10 is found to be 8^m2, corresponding to a distance of 440 pc.

153.011 On the question of the uniformity of chemical

composition of stars in clusters. C. R. Cowley.
Observatory, Vol. 95, 55 - 56 (1975). − Letter.

153.012 Spectral and photoelectric observations of stars in the cluster NGC 6913. R. M. Raznik.
Uch. zap. Ul'yanovsk. gos. ped. in-t, Vol. 27, No. 8, p. 142 - 150 (1974). In Russian. − Abstr. in Referativ. Zhurn. 51. Astron., 3.51.590 (1975).

153.013 The distribution of interstellar matter in NGC 654.
W. B. Samson.
Astrophys. Space Sci., Vol. 34, 363 - 376 (1975).

The young cluster NGC 654 is studied using UBV photographic photometry with a view to determining the distribution of interstellar matter in a region where star formation recently occurred. NGC 654 is found to be enclosed in a shell of interstellar matter of mass 1500 $M_\odot$. The mass of all stars in the cluster is 4000 $M_\odot$.

153.014 The distribution of interstellar matter in IC 5146.
W. B. Samson.
Astrophys. Space Sci., Vol. 34, 377 - 386 (1975).

The very young star cluster IC 5146 is studied using star counts, with a view to determining the distribution of interstellar matter in a region where star formation recently occurred. IC 5146 is embedded in a dark nebula which is very dense near its centre. The total mass of interstellar dust in the nebula is found to be about 4.5 $M_\odot$. Comparison of radio and optical observations of the region indicates that gas and dust are not separated to any great degree by radiation from the embedded stars. A gas/dust ratio of about 150/1 by mass is found. This ratio varies with the dust grain model used.

153.015 On the theoretical determination of gap parameters for cluster color-magnitude diagrams and their comparison with observational data. R. Mitalas.
Astrophys. Journ., Vol. 198, 139 - 144 (1975).

A new procedure to determine the theoretical gap parameters of clusters has been developed. The method takes into account the existence of a range of masses in the hydrogen exhaustion phase. It is demonstrated that the mass range of stars in the hydrogen exhaustion phase must be taken into account before theory and observation can be compared.

153.016 Southern open star clusters IV. UBV-Hβ photometry of 26 clusters from Monoceros to Vela.
A. F. J. Moffat, N. Vogt.
Astron. Astrophys., Suppl. Ser., Vol. 20, 85 - 124 (1975).

The photoelectric photometry (summarized in a table) reveals the reality of 21 clusters of which 7 were previously uncatalogued, 11 have spectral types earlier than B3, and 8 contain 9 supergiants. The distances range out to 6 kpc from the sun. Four clusters coincide with interesting stars: 2 cepheids, one Wolf-Rayet, and one U Gem type star.

153.017 Southern open star clusters V. UBV-Hβ photometry of 20 clusters in Carina.
A. F. J. Moffat, N. Vogt.
Astron. Astrophys., Suppl. Ser., Vol. 20, 125 - 153 (1975).

The photoelectric photometry (summarized in a table) reveals the reality of 16 clusters of which 4 were previously uncatalogued, 12 have spectral types earlier than B3 and 6 contain 7 supergiants. The distances range out to 5 kpc from the sun. Two clusters coincide with 2 cepheid variables.

153.018 Southern open star clusters VI. UBV-Hβ photometry of 18 clusters from Centaurus to Sagittarius.
A. F. J. Moffat, N. Vogt.
Astron. Astrophys., Suppl. Ser., Vol. 20, 155 - 182 (1975).

The photoelectric photometry (summarized in a table) reveals the reality of 14 clusters of which 2 were previously uncatalogued, 9 have spectral types earlier than B3 and 8 contain 10 supergiants. The distances range out to 2.3 kpc from the sun. One cluster coincides with a cepheid variable.

153.019   A study of the motion, membership, and distance of the Hyades cluster.   R. B. Hanson.
Astron. Journ., Vol. 80, 379 - 401 = Lick Obs. Bull., No. 685 (1975).

New absolute proper motions referred directly to external galaxies, photographic photometry, and cluster membership probabilities have been determined for over 600 faint stars in the Hyades region. These have been used to investigate the Hyades convergent point and distance. A detailed review of the foundations of the convergent-point method, its applicability to the Hyades, the several mathematical variations of the method, and its possible role in producing the apparent discrepancy between previous Hyades proper motion results and "secondary" Hyades distance indicators, has been carried out. The convergent-point method, as applied to the Hyades, is entirely valid, and cannot be the source of any significant error in the Hyades distance. The resulting Hyades distance modulus is $m - M = 3.42 \pm {}^{0.21}_{0.19}$, significantly greater than all previous values from proper motions, and in agreement with predictions of the Hyades distance from stellar structure theory, and with the results of "secondary" distance indicators. Significant effects of the increased Hyades distance on the cosmic distance scale and other questions of general astronomical importance are briefly discussed.

153.020   A new determination of the distance of the Hyades cluster by the convergent-point method.
T. E. Corbin, D. L. Smith, M. S. Carpenter.
Bull. American Astron. Soc., Vol. 7, 337 (1975). – Abstr. AAS.

153.021 UBV photometry of NGC 6649.   F. D. Talbert.
Publ. Astron. Soc. Pacific, Vol. 87, 341 - 344 (1975).

This paper reports UBV photographic photometry of NGC 6649 and the analysis to derive a minimum distance modulus, $(m - M)_0$, of $11.^m4$ which indicates the cluster is well within the Carina-Sagittarius-Scutum arm of the Milky Way. Distance and age of the cepheid V367 Sct near the cluster center are consistent with cluster membership.

153.022   The stellar groups Ba 12, Ba 13, Ba 14 and Ba 15.   S. M. Hassan.
Astron. Astrophys., Suppl. Ser., Vol. 20, 255 - 267 (1975).

A three colour photometric study of four stellar groups in the UBV system is represented. The colour-magnitude diagrams of the four groups indicate the existence of a main sequence and of a giant branch in contrast to groups of stars produced by random fluctuations in a star field. The distances, the colour excesses, and other parameters are determined for the four groups.

153.023   On the nature of the Puppis cluster NGC 2483.   M. P. FitzGerald, A. F. J. Moffat.
Astron. Astrophys., Suppl. Ser., Vol. 20, 289 - 304 = Contr. Univ. Waterloo Obs. No. 39 (1975).

The photographic UBV photometry ($V \leq 15.7$) and the spectral types presented here for NGC 2483 and a nearby field do not support the suggestion of Lindoff and Johansson (1968) that NGC 2483 is a unified cluster with pre-main-sequence evolution. In an appendix the authors present photoelectric UBV photometry to $V = 15.7$ for 103 stars in a 4° X 4° field centred on NGC 2483. Objective prism classification for 95 of these is also given.

153.024   Stellar density distribution in wide surroundings of the Pleiades and position of the cluster center.
N. M. Artyukhina, P. N. Kholopov.

Soobshch. Gos. Astron. Inst. Shternberga, No. 188, p. 15 - 24 (1974). In Russian.

By counts of stars brighter than $15.^m5$ pg the stellar density distribution in the region with radius of 5°5 around the center of the Pleiades cluster is studied. The center of the cluster determined as the point of concentration of possible cluster members brighter than $16.^m7$ B (selected by their proper motions) lies 13°5 to the west and 4' to the north of Alcyone.

153.025   Stellar density distribution in the cluster NGC 7789.   N. M. Artyukhina, P. N. Kholopov.
Soobshch. Gos. Astron. Inst. Shternberga, No. 188, p. 25 - 32 (1974). In Russian.

The density distribution of stars brighter than $15.^m5$ B in wide surroundings of NGC 7789 has been studied. The radius of the cluster is 1°5 (49 pc). The cluster contains about 2100 stars brighter than $M_v = +2.^m7$, 34% of them being in the nucleus of the system. The radius of the nucleus is 19°5 (10.5 pc).

153.026   The Hyades program at the Figl-Observatory (Vienna).   W. W. Weiß, W. Primik.
Conference on optical observing programs on galactic structure and dynamics, (see 012.013), p. 257 - 261 (1975). In German.

153.027   Dynamics of the Hyades.   R. F. Griffin.
Monthly Notes Astron. Soc. Southern Africa, Vol. 34, 50 (1975). – Abstract.

153.028   Photoelectric photometry of the open cluster NGC 1778.   U. C. Joshi, R. Sagar, P. Pandey.
Bull. Astron. Soc. India, Vol. 2, 34 (1974). – Abstract.

153.029   Micro-spectra of stars in the extremely young open cluster NGC 6530.   M. Parthasarathy.
Bull. Astron. Soc. India, Vol. 2, 36 (1974). – Abstract.

153.030   Spectral types in the open cluster NGC 6475.   H. A. Abt.
Publ. Astron. Soc. Pacific, Vol. 87, 417 - 419 (1975).

Spectral classification of 27 of the brightest cluster members shows (1) and earliest type of B5 IV, (2) a hot Am or Sirius-type star, (3) three Ap stars of the Si or Si-Cr kind, all with weak Ca II K lines, (4) two additional stars with weak K lines, and (5) one "classical" Am star. It is noticed that among the eight spectroscopic binaries in this cluster, the ones with broad lines have normal K-line strengths whereas nearly all of those with sharp lines have weak K lines.

153.031   Note on the blue stragglers in NGC 7789.   E. S. Pendl.
Astron. Astrophys., Vol. 41, 239 - 240 (1975).

A proper motion survey of the star cluster NGC 7789 revealed twelve stars beyond and blueward the turn-off point to be probable members of this cluster. A comparison of the present investigation with a radial velocity study by Strom and Strom (1970) establish the membership of at least four stars with near certainty. The most remarkable result is the long extent of the blue straggler sequence, which reaches to stars about five magnitudes brighter than the luminosity at the turn-off point.

153.032   Dark matter in open clusters.   Å. Wallenquist.
Nova Acta Regiæ Soc. Sci. Upsaliensis, Ser. V:A, Vol. 2, 1 - 98 = Uppsala Astron. Obs. Annaler, Band 5, No. 8 (1975).

The present paper is an attempt to trace dark matter in open clusters and in their surroundings on the basis of the star count method. More than 700 000 stars were counted within about 50 000 square fields which made up the basic material for the study of dark matter within 83 open clusters. The results of this investigation are presented chiefly in a series of distribution maps and diagrams.

Line blanketing and model stellar atmospheres. II. Interpretation of broad-band photometric observations. See Abstr. 064.032.

Theoretical isochrones and main sequences for old disk population stars. See Abstr. 065.046.

Photoelectric observations of occultations of the Pleiades and the incidence of duplicity in the cluster. See Abstr. 096.008.

Occultations of the Pleiades: reappearances observed photoelectrically at McDonald Observatory. See Abstr. 096.009.

The application of parallaxes and photometry to the lower main sequence. See Abstr. 111.002.

DDO intermediate-band photometry of moving-group stars. See Abstr. 113.016.

Beiträge zur Methodik der photographischen UBV-Photometrie mit Anwedung auf den offenen Sternhaufen NGC 2632 (Praesepe). See Abstr. 113.068.

A search for Ap stars in southern galactic clusters. See Abstr. 114.046.

Curve-of-growth analysis of a red giant in M67. See Abstr. 114.320.

On the cool "Am" star in the Pleiades. See Abstr. 114.353.

Pre-main-sequence masses and evolution in NGC 2264. See Abstr. 115.005.

UBV **photometry of the cepheid V367 Scuti in the open cluster NGC 6649.** See Abstr. 122.021.

On the variability of three stars in the vicinity of the open cluster NGC 6830. See Abstr. 122.044.

Variable stars of young clusters and of the galactic disc. See Abstr. 122.057.

The upper mass limit for white dwarf formation as derived from the stellar content of the Hyades cluster. See Abstr. 126.007.

H_2CO and H I observations of dark clouds in NGC 2264. See Abstr. 131.022.

Young stellar clusters in diffuse nebulae. See Abstr. 131.522.

UBV **photometry of the stars in the fields of emission nebulae. I. M 20.** See Abstr. 132.013.

Dynamical method to estimate the relative masses of stars in spherical clusters. See Abstr. 151.003.

Les magnitudes stellaires en trois couleurs des étoiles des anneaux stellaires en Cygnus et de l'amas NGC 7127. See Abstr. 152.003.

Galactic structure based on young southern open star clusters. See Abstr. 155.029.

Photoelectric photometry of star clusters in the Andromeda nebula. See Abstr. 158.098.

Errata

153.901 Errata: 'NGC 6259: southern image of M 11'
[Monthly Notices Roy. Astron. Soc., Vol. 169, 539 - 544 (1974)]. T. G. Hawarden.
Monthly Notices Roy. Astron. Soc., Vol. 171, 235 - 236 (1975).

154 Globular Clusters

154.001 Die mittleren Geschwindigkeiten der Sterne in 58 kugelförmigen Sternhaufen. W. Lohmann.
Astrophys. Space Sci., Vol. 32, 153 - 163 = Astron. Rechen-Inst. Heidelberg, Mitt. Ser. A (1975).

The mean velocities of the stars in 58 globular clusters are derived from the photometric measurements of Kron and Mayall (1960). The mass-visual brightness ratio M/H = 0.935 $M_\odot/H_\odot$ is used. The velocities are falling in the interval $2.9 \leqslant \bar{v} \leqslant 11.6$ km s^{-1}, their mean value is 6.75 km s^{-1}.

154.002 Globular cluster colors and the cosecant law. D. Burstein, L. H. McDonald.
Astron. Journ., Vol. 80, 17 - 30 = Lick Obs. Bull., No. 664 (1975).

All available integrated UBV colors and spectral types of globular clusters are analyzed to determine whether one can obtain the reddening from the integrated parameters. Reddening determinations by various authors for individual clusters, from field stars, comparison of color-magnitude diagrams, etc., are collected and made consistent with one another. Using these data, the intrinsic color-spectral type relationships for globular clusters in the galaxy are found to agree with those for M31, allowing for a reddening of M31 of $E(B-V) = 0.08 \pm 0.04$. The same color-spectral type relationships are found for early-type globular clusters by an analysis of the csc law of galactic absorption. From these determinations of the reddening, the presence of a hole in the dust layer in the vicinity of the sun is confirmed.

154.003 The asymptotic giant branch in NGC 6397. E. A. Mallia.
Monthly Notices Roy. Astron. Soc., Vol. 170, 57P - 60P (1975). – Short communication.

154.004 A photometric study of NGC 2419. R. Racine, W. E. Harris.
Astrophys. Journ., Vol. 196, 413 - 432 (1975).

Photometry to $V = 22.2$ and $B = 23.7$ is reported for the outer-halo globular cluster NGC 2419. The color-magnitude diagram of the cluster is similar to that of the classic metal-poor cluster M92, and indicates a very low metallicity $Z \cong 1.5 \times 10^{-4}$. The reddening $E(B-V)$ is 0.03 ± 0.01 mag, and the apparent distance modulus is $(m - M)_V = 19.87 \pm 0.09$, leading to a galactocentric distance of $R_g = 100 \pm 5$ kpc. The galactic orbit of NGC 2419 is determined. The cluster is gravitationally bound to the Galaxy, traveling on an orbit of eccentricity 0.62 with a period of 3.4×10^9 yr, and is presently near its apogalacticon. It is argued that the cluster was born close to its perigalacticon distance of 24 kpc. A possible gravitational encounter between NGC 2419 and the Magellanic Clouds is mentioned briefly. Finally it is shown that NGC 2419, like many metal-poor halo clusters, possesses an orbit of large angular momentum per unit mass h, and that globular clusters with the largest h are among the metal poorest.

154.005 The observability of ionized interstellar gas in globular clusters. E. H. Scott, W. K. Rose.
Astrophys. Journ., Vol. 197, 147 - 153 (1975).

Recent radio observations have set upper limits to the amounts of neutral and ionized gas in globular clusters. In an attempt to understand these observations, the authors calculate "stellar wind" type solutions for the outflow of ionized gas from globular clusters. The models, which are based on the mass distributions of Peterson and King, predict only about 1 $M_\odot$ or less of ionized gas within a few parsecs of the cluster centers in most cases and thus lead to less free-free emission than the observed upper limits of Hills and Klein. However,

the results are sensitive to the total stellar mass, its spatial distribution, and the metal abundance of the gas. The authors find that the mass of ionized gas for one cluster, NGC 6388, may be sufficient for observability.

154.006 The mass and tidal radius of Omega Centauri. A. Poveda, C. Allen.
Astrophys. Journ., Vol. 197, 155 - 157 (1975).

Using Harding's velocity measurements and rotation curve, the authors obtain two new values for the mass of ω Centauri: (1) $M \gtrsim 3.23 \times 10^6 M_\odot$, from the rotation curve; (2) $M \gtrsim 3.24 \times 10^6 M_\odot$, from the virial theorem. The discrepancy by a factor of 5 between these values and those of Dickens and Woolley is discussed. With the new mass, the inconsistency found by Keenan, Innanen, and House between the observed and the calculated values for the tidal radius vanishes. The resulting mass-luminosity ratio $(M/L)_B \gtrsim 3.2$ is significantly larger than the M/L ratio for other globular clusters.

154.007 A search for ionized hydrogen in globular clusters. J. W. Erkes, A. G. D. Philip.
Astrophys. Journ., Vol. 197, 533 - 534 (1975).

No evidence for radio emission at 3 and 6 cm from ionized hydrogen was found in the globular clusters investigated.

154.008 A color-magnitude diagram for the strong-line globular cluster NGC 6553. F. D. A. Hartwick.
Publ. Astron. Soc. Pacific, Vol. 87, 77 - 81 (1975).

A preliminary color-magnitude (CM) diagram is obtained for NGC 6553, one of the most strong-lined globular clusters known. The diagram shows features characteristic of other metal-rich globular clusters. The position in the CM diagram of an RR Lyrae star spatially superimposed on the cluster suggests that it is a member. A tentative distance modulus for NGC 6553 of $(m-M)_0 = 13\overset{m}{.}4 \pm 0.3$ is derived.

154.009 Metal abundance of type II systems. R. Canterna.
Bull. American Astron. Soc., Vol. 7, 239 (1975).
Abstr. AAS.

154.010 Dark patches in globular clusters. S. P. Kanagy II, S. P. Wyatt.
Bull. American Astron. Soc., Vol. 7, 259 (1975). – Abstr. AAS.

154.011 Infrared observations in the field of the globular cluster M15. V. Castellani, A. Martini, R. Petitti.
Acta Astron., Vol. 25, 153 - 159 (1975).

Usefulness of the infrared photometry for determining giants membership of a globular cluster is confirmed in the case of M15. Application of the infrared criterion to the stars in the neighbourhood of this cluster shows that all but two are certainly not cluster members.

154.012 Preliminary results concerning DDO photometry of the southern hemispheric globular star cluster NGC 3201, and other remarks. R. E. White.
Dudley Obs. Rep. No. 9, (see 012.008), p. 367 - 370 (1975).

154.013 Four-color photometry of blue horizontal-branch stars in globular clusters. A. G. D. Philip.
Dudley Obs. Rep. No. 9, (see 012.008), p. 371 - 374 (1975).

Blue horizontal-branch stars have been observed in the globular clusters M3, M4, M5, M13, and M55. The observations have been transformed to the Θ_e, log g plane and are compared to the evolutionary models of Sweigart and

Gross to obtain coarse limits on model parameters such as helium abundance, total mass, and core mass.

154.014 The gas and horizontal branch star content of globular clusters. R. J. Tayler, P. R. Wood.
Monthly Notices Roy. Astron. Soc., Vol. 171, 467 - 474 (1975).

The observed number of horizontal branch stars in eight globular clusters is combined with an estimate of the minimum time since the clusters last crossed the galactic plane and the assumption that stars lose ~0.2 $M_\odot$ on the first ascent of the giant branch to derive the amount of gas which should be present in the clusters. As the time since a typical cluster crossed the galactic plane is likely to be significantly greater than the minimum time and as all the observations are upper limits rather than definite measurements, it appears that there is a discrepancy between the theoretical estimates and observations and some possible explanations of this result are discussed. If the numbers of horizontal branch stars are used in conjunction with estimates of the total cluster mass and the assumption that the initial mass function was of Salpeter type, the mass of the least massive stars at present in the clusters can be derived. It appears that stars less massive than ~0.25 $M_\odot$ have preferentially escaped from the clusters.

154.015 Near-infrared photometry of globular clusters — IV. The metal poor cluster NGC 6656 (M 22).
T. Lloyd Evans.
Monthly Notices Roy. Astron. Soc., Vol. 171, 647 - 657 (1975).

Photographic photometry in the BVI_K system is presented for 157 stars within 7' of the centre of NGC 6656. The giant branch is steeper than indicated by earlier work, in better agreement with spectroscopic and photometric evidence of low metal abundance. The weakness of the asymptotic giant branch is confirmed.

154.016 The structure of star clusters. VI. Observed radii and structural parameters in globular clusters.
C. J. Peterson, I. R. King.
Astron. Journ., Vol. 80, 427 - 436 (1975).

Observations have been collected from a wide range of sources, to determine structural parameters for globular clusters. Core radii are given for 101 clusters, limiting radii for 43, and all three parameters for 37 clusters. Distance moduli, reddening values, and integrated magnitudes are also collected. Through fitting to previous theoretical models, values are calculated wherever possible for the density, relaxation time, velocity dispersion, and escape velocity at the center of each cluster.

154.017 UBV and uvby photometry of globular clusters in the Large Magellanic Cloud. A. Bernard.
Astron. Astrophys., Vol. 40, 199 - 202 (1975). In French.

Integrated magnitudes and colors on the UBV system are given and discussed for 35 red globular clusters in the Large Magellanic Cloud (LMC) and for 24 of them on the uvby system.

154.018 Membership of the globular cluster NGC 6397.
E. A. Mallia.
Astron. Astrophys., Vol. 41, 103 - 105 (1975).

Radial velocities of stars in the globular cluster NGC 6397 were measured on medium dispersion spectra. From these velocities and comparisons with spectra of standard stars it proved possible to determine cluster membership unambiguously for the late type giants. There is an indication that stars on the asymptotic giant branch have weaker lines than those on the red giant branch with similar colours. Of two early type stars observed one is a foreground object. The other is probably a cluster member with a surface temperature close to 50000 °K. The Mira variable in the direction of the cluster is almost certainly a non-member.

A-type horizontal-branch stars.
See Abstr. 064.028.

On mass loss by stellar wind in population II red giants. See Abstr. 064.046.

Mixed stars from population II giants.
See Abstr. 065.050.

Advanced evolution of globular cluster stars.
See Abstr. 065.100.

A search for variable stars in globular clusters.
See Abstr. 122.024.

The unusual period distribution of RR Lyrae variables in the globular cluster IC 4499. See Abstr. 122.025.

Metal abundances of RR Lyrae stars in galactic globular clusters. See Abstr. 122.052.

The interstellar reddening in front of globular clusters. See Abstr. 131.063.

Dynamical method to estimate the relative masses of stars in spherical clusters. See Abstr. 151.003.

Numerical investigation of galactic tidal effects on spherical stellar systems. See Abstr. 151.011.

Photoelectric photometry of star clusters in the Andromeda nebula. See Abstr. 158.098.

Color-magnitude diagrams for four rich clusters of the Large Magellanic Cloud. See Abstr. 159.007.

Black dwarf stars as missing mass in clusters of galaxies. See Abstr. 160.001.

155 Structure and Evolution of the Galaxy

155.001 A numerical study of local stellar motions.
F. C. House, K. A. Innanen.
Astrophys. Space Sci., Vol. 32, 139 - 151 (1975).

The orbits of over 10000 stars are integrated in a steady-state model of the Galaxy for a time 6.0×10^8 yr. The mean velocities and dispersions of stars within 1 kpc of an 'observer' moving at the circular velocity are calculated as functions of time. The quantities show a strong time-dependence with oscillations of period 10^8 yr. A vertex deviation of the velocity ellipsoid, an asymmetric drift and a K-effect occur as natural consequences of the oscillations. Attempts to apply the Oort method for density determinations in the galactic plane are also influenced by the oscillations. Spiral density waves appear to have a small effect on the motions of the test stars.

155.002 Laboratory plasmas and the galactic spiral arms.
J. C. Cataldo, A. J. Skalafuris.
Astrophys. Space Sci., Vol. 32, L11 - L14 (1975).

The aim of this letter is to draw attention to a plasma experiment which may clarify the physical origin of our galactic spiral arms. The authors have observed a Kelvin-Helmholtz instability in a rotating cylindrical plasma column which takes the form of two spiral arms. The laboratory phenomenon to which they wish to suggest a galactic analogy is shown.

155.003 The alignment of interstellar dust clouds and the differential z-field of the Galaxy.
M. J. Disney, P. B. Hopper.
Monthly Notices Roy. Astron. Soc., Vol. 170, 177 - 184 (1975).

The gradient of the z component of the gravitational field of the galactic disc exerts a differential force across a dust cloud which will be sufficient to squeeze the clouds into disc-like shapes with their long axes parallel to the galactic plane. The authors suggest this as a possible explanation for the observed alignments of dust clouds.

155.004 The space distribution of M giants in the Warner and Swasey Luminosity Function Field LF 14.
G. Vleeming.
Astron. Astrophys., Suppl. Ser., Vol. 19, 21 - 44 (1975).

In the present study, which is an extension of a study of the Warner and Swasey Luminosity Function Field LF 15 (Thé et al. 1974), a survey of M–type stars has been made in square field of $3° \times 3°$ in LF 14. Using Schalén's method space densities have been calculated for the M2–M4, M5–M6.5 and M7–M9 groups. The results were then corrected for interstellar absorption by means of Seeliger's method. The infrared interstellar absorption as function of distance was determined using data of M2, M3 and M4 stars and those published by Neckel (1966). A short discussion of the results is given.

155.005 Sources of excitation of the interstellar gas and galactic structure.
J. J. Cowan, M. Kafatos, W. K. Rose.
Astrophys. Journ., Vol. 195, 47 - 51 (1975).

The excitation of the interstellar gas is discussed in the light of recent evidence from γ-ray, molecular, and 21-cm line observations. Previous studies of the excitation of the interstellar gas have not taken into account the substantial density contrast that exists between spiral arms and interarm regions. The authors examine the role played by the galactic distribution of three sources of excitation (supernovae, OB stars, and ultraviolet stars) in determining the physical state of the interstellar gas in arm and interarm regions.

155.006 New applications of the equations of stellar hydro-
dynamics. P. O. Vandervoort.
Astrophys. Journ., Vol. 195, 333 - 341 (1975).

Formulae are derived which enable the determination of the curvature in the law of galactic rotation and a characteristic scale length of galactic structure parallel to the galactic plane from a knowledge of the velocity of galactic rotation, the slope in the law of rotation, and the moments of the velocity distribution. When supplemented with considerations bearing on the stability of the galactic disk, these formulae provide a practical basis for an approximate determination of the gradients of the density and velocity dispersion of the common stars in the solar neighborhood. Accordingly, the asymmetric drift of the common stars can be determined by inverting the conventional application of the standard hydrodynamic equations.

155.007 The third and fourth moments of the local stellar velocity distribution. R. R. Erickson.
Astrophys. Journ., Vol. 195, 343 - 358 (1975).

The velocity moments through the fourth order were calculated for a sample of 870 nearby stars chosen to be representative of the common stars of the galactic disk. The moment variances and covariances were also calculated, taking into account both the size of the sample of stars and the observational errors in the data. The structure of the Galaxy was investigated by utilizing the moments in a hydrodynamic model to evaluate such parameters as the curvature in the law of rotation, the gradient in the galactic plane of the number density, the gradient in the galactic plane of the velocity dispersion in the galactic radial direction, and the asymmetric drift.

155.008 350-micron mapping of Sagittarius B2.
G. Righini, M. Simon, R. R. Joyce, D. Y. Gezari.
Astrophys. Journ., (Letters), Vol. 195, L77 - L79 (1975).

The authors have mapped at 350μ a $15' \times 15'$ region of Sgr B2 with $3''.5$ resolution, and the central $2' \times 3'$ region with $56''$ resolution. The observations show an extended low-surface-brightness region which is coincident with the regions of diffuse 100-μ, radio continuum, and molecular line emission, and a bright compact submillimeter source whose intensity distribution is similar to that of the core of the molecular source. The peak brightness temperature in the compact source is $23°K$, which together with the gas temperature inferred from the CO observations suggests that the 350-μ optical depth is of order unity in the central $1'$ of the source.

155.009 Surface brightness and continuity of spiral arm –I in the near infrared. W. Schlosser.
Astron. Astrophys., Vol. 38, 133 - 135 (1975). In German.

Near-infrared image tube photographs of the region around the dark cloud Khavtassi No. 494 have been obtained. The differential absorption of the cloud at the effective wavelength (8000 Å) amounts to about $0^m.7$, whereas the visual absorption is about $4^m.5$ (FitzGerald and Schmidt-Kaler, 1975).

155.010 Models for the inner regions of the Galaxy. I. An elliptical streamline model. W. L. Peters III.
Astrophys. Journ., Vol. 195, 617 - 629 (1975).

A new model is proposed for the inner regions ($R < 4$ kpc) of the Galaxy. This model involves gas flowing along concentric elliptical streamlines such as might be caused by a barlike perturbation in the gravitational field of the inner regions of the Galaxy. It is found that such a model can produce not only the so-called 3 kpc arm and the 135 km s^{-1} arm features, but other high-velocity features as well. This model does not require a highly energetic explosion at the galactic nucleus as does the usual expanding spiral-arm model, because there is no net flow of matter out of the inner regions. This model suggests that the outer spiral-arm density wave might be associated with a barlike density wave in the inner regions of the Galaxy.

155.011 **Statistical principles of galactic optical astronomy. Part IV.** H. Eelsalu.
Akademiya Nauk Ehstonskoj SSR. Tartuskaya Astrofiziches-kaya Observatoriya im.W. Struve, 103 pp. = Tartu Astron. Obs., Teated, No. 50 (1975). In Russian.

155.012 **Statistical principles of galactic optical astronomy. Supplements and registers.** H. Eelsalu.
Akademiya Nauk Ehstonskoj SSR. Tartuskaya Astrofiziches-kaya Observatoriya im. W. Struve, 64 pp. = Tartu Astron. Obs., Teated No. 51 (1975). In Russian.

155.013 **Kinematics of stars selected from box orbit para-meters.** D. K. Karimova, E. D. Pavlovskaya.
Astron. Zhurn. Akad. Nauk SSSR, Vol. 52, 201 - 203 (1975). In Russian. English translation in Soviet Astron., Vol. 19, No. 1.

It is shown that kinematics is practically independent on spectral type for stars selected from eccentricity and box angle. The values of $V_\odot$, σ_3, σ_2, σ_1 and σ_1/σ_3 increases with the values i and e. The percentage of circular orbits is larger among emission-line stars.

155.014 **Numerical integration methods for galactic orbit computations.**
F. C. House, K. A. Innanen, D. W. Keenan.
Celestial Mechanics, Vol. 11, 3 - 11 (1975).

Seven algorithms are investigated for integrating stellar orbits in axisymmetric and time-independent galactic models. The authors find that for this purpose, impressive gains over older methods are possible with higher order Runge-Kutta methods and variable order methods.

155.015 **Galactic orbits near the sun.** J. Palouš, L. Perek.
Bull. Astron. Inst. Czechoslovakia, Vol. 26, 82 - 89 (1975).

Results of a systematic exploration of m-periodic orbits in the potential of the Schmidt model are presented. Families of symmetric periodic orbits with ($m \leqq 5$) are represented in the (ϖ, Z) plane by their characteristics. The stability of peri-odic orbits was investigated using a method by Hénon. Stabili-ty curves were constructed and critical points of various types were found. Non-periodic orbits were investigated by the method of invariant curves. The authors tried to define the boundary of the region of dissipated orbits (ergodic region).

155.016 **Study of galactic gas and dust using observations of elliptical galaxies.** G. R. Knapp.
Astron. Journ., Vol. 80, 111 - 116 (1975).

The gas-to-dust ratio has been measured in the directions of 55 elliptical galaxies by comparing 21-cm H I column densities with color excesses and absorptions for the galaxies. The value of the gas-to-dust ratio found in this study is similar to previous measurements, but the data suggest sig-nificant small-scale structure of the interstellar medium.

155.017 **RGU three colour photometry in the large Sagit-tarius cloud.** P. Gschwind.
Astron. Astrophys.,Suppl. Ser., Vol. 19, 281 - 302 (1975).

A field of 0.14 sq. degrees containing 2220 stars, in the direction of the galactic centre has been measured on 17 48″ Palomar Schmidt plates in the RGU system down to a limit-ing magnitude of 16.5 mag. in G. A comparison between the late-type giants and the main-sequence stars of the same absolute magnitude shows the existence of two different groups of late-type giants with normal positions at the left and at the right side of the late main-sequence branch, as has been

found also in other directions. The density functions of the main-sequence stars show a maximum, within the limit of completeness, lying some hundred parsecs nearer than the one of the late-type giants. These density maxima coincide approximately with the next inner spiral arm.

155.018 **Is the galactic disk well mixed?** M. G. Edmunds.
Astrophys. Space Sci., Vol. 32, 483 - 491 (1975).

A consideration of the mixing processes for heavy ele-ments in the interstellar medium suggests that the Galaxy is well-mixed locally. Inhomogeneities would be insufficient to explain the observed abundance spread of old disk stars, or to allow the theory of metal enhanced star formation to account for the paucity of metal poor dwarfs in the solar neighbourhood. Differential abundance effects during the formation of stars or clusters, or the existence of a radial abundance gradient may have to be invoked to explain the observations.

155.019 **The north—south asymmetry in the galactic rota-tion curves.** S. Manabe, M. Miyamoto.
Publ. Astron. Soc. Japan, Vol. 27, 35 - 44 (1975).

It is shown that the large scale north—south asymmetry in the galactic rotation curves obtained from observations of neutral hydrogen can be explained as a consequence of non-circular galactic rotation caused by a non-axisymmetric mass distribution in the central region of the Galaxy, i.e., a bar-like structure, and by a radial motion of the local standard of rest at the sun. In the authors' model the bar is rotating uniformly with an angular velocity 8 km s^{-1} kpc^{-1} and the local standard of rest at the sun is moving with a velocity 3 km s^{-1} toward the galactic center. The attitude and the elongation of their model bar resemble those of an elliptical dispersion ring pro-posed by Shane (1972) and Simonson and Mader (1973). No attempt is made to describe how the bar-like structure is formed.

155.020 **Nouveau schéma général de la structure spirale de notre Galaxie à partir des données stellaires.**
Y. M. Georgelin.
Comptes Rendus Acad. Sci. Paris, Sér. B, Vol. 280, 349 - 352 (1975).

L'auteur a établi un schéma cohérent de la structure spirale de notre Galaxie à partir de l'ensemble des données optiques et radio. Le modèle comporte quatre bras symé-triques deux à deux. Ces bras passent au milieu des régions H II dont les distances sont connues (plus de 90 % de ces ré-gions sont centrées sur ces bras).

155.021 **Local mass density of population II and halo stars.** D. Weistrop.
Astron. Journ., Vol. 80, 303 - 306 (1975).

Kinematic criteria are used to identify halo and popula-tion II members in the Gliese catalogue. A mass-luminosity relation is adopted and the local mass densities of the two types of stars are calculated. Assuming a model for the mass distribution in the halo, the total halo mass is estimated. The result is considerably less than the mass required by Ostriker and Peebles to stabilize the galactic disk.

155.022 **Cloud structure in the galactic plane: a cosmic bubble bath?** P. W. J. L. Brand, W. J. Zealey.
Astron. Astrophys., Vol. 38, 363 - 371 (1975).

An examination of photographs of the Milky Way has given strong indications of large scale loop-shaped dust and gas structures. A list of these, many of them already noted by others, is provided. In addition it is suggested that a large part of the interstellar medium is in the form of filaments, and that these filaments may be related to the loop structures. It is shown that much of the interstellar medium may have been processed through supernova blast waves.

155.023 **Random motions of neutral hydrogen clouds in the Perseus spiral arm.** I. V. Gosachinskij.
Astrofiz. Issled., Izv. Spets. Astrofiz. Obs., Vol. 7, p. 96 - 100 (1975). In Russian.

Results of radial-velocity dispersion measurements, using the data of Maryland Catalogue, are presented for random motions of H I clouds within complexes of 50 X 200 pc in the Perseus spiral arm of the Galaxy. The mean dispersion is found to be 2.9 ± 0.1 km/sec.; any dependence on the angle between the line of sight and the spiral arm is not detected. The character of the motion of the complexes in the arm probably indicates that the arm consists of at least two branches.

155.024 **Absorption of soft X-rays by material within Gould's Belt.** C. G. Rapley.
Nature, Vol. 255, 41 - 42 (1975).

It seems likely that the effects of soft X-ray absorption by material in Gould's Belt will contribute significantly to the understanding of the galactic sources of diffuse soft X-ray emission and may permit the uncertainty concerning the existence of a halo or extragalactic component of the flux to be resolved.

155.025 **Structure and age of the local association (Pleiades group).** O. J. Eggen.
Publ. Astron. Soc. Pacific, Vol. 87, 37 - 63 (1975).

The available photometric and motion parameters for some 500 early-type stars brighter than $M_V = -1^m$ are used to compute space-motion vectors. It is concluded that about one-third of the stars are members of the local association (Pleiades group) which are easily isolated from the other inter-arm stars because $V = -25$ km sec^{-1}. The structure and age of the local association (or Gould Belt) is investigated.

155.026 **Chemical evolution of the Galaxy during its contraction.** M. Kaufman.
Astrophys. Space Sci., Vol. 33, 265 - 293 (1975).

Models for the chemical evolution of the Galaxy are constructed in which the time evolution is imposed by the contraction rate of the Galaxy and present observations of stellar metal abundances as a function of height above the galactic plane.

155.027 **Wolf-Rayet stars and galactic structure.** B. Stenholm.
Astron. Astrophys., Vol. 39, 307 - 318 (1975).

A 15° wide strip along the galactic equator between longitudes 250° and 360° has been searched for Wolf-Rayet stars. A sample of 154 WR stars has been treated statistically. The physical plane formed by these objects is tilted about one degree to the galactic plane and the tilt is upwards in the Cygnus direction. This result is also received by a least squares solution of the objects when given in rectangular coordinates. The WR star sample is regarded as fairly complete out to a distance of 5 kpc.

155.028 **The space density of M dwarfs in the region of the south galactic pole.** J. F. Dolan.
Astron. Astrophys., Vol. 39, 463 - 464 (1975).

The stars used by Thé and Staller (1974) to derive the space density of M2 to M4 dwarfs in the region of the south galactic pole are reanalysed using the matrix method (Dolan, 1974). The associated statistical uncertainties indicate that the space density of M dwarfs derived by Thé and Staller in that region is not necessarily in disagreement with the higher-density results derived by others in the region of the north galactic pole.

155.029 **Galactic structure based on young southern open star clusters.** N. Vogt, A. F. J. Moffat.
Astron. Astrophys., Vol. 39, 477 - 479 (1975).

New observations of 43 southern young open star clusters out to $\sim$ 8 kpc are briefly discussed in relation to galactic structure. The main new results are: (1) extension of the outer arm $+II$ from $l \sim 105°$...$180°$ to $215°$ and possibly to $l \sim 245°$; (2) a probable elongation of the local arm (0) out to at least 4 kpc in the direction $l \sim 245°$; (3) definition of the Carina arm out to $d \sim$ 8 kpc.

155.030 **On the pregalactic helium to hydrogen abundance ratio.** M. Peimbert, S. Torres-Peimbert.
Bull. American Astron. Soc., Vol. 7, 237 (1975). – Abstr. AAS.

155.031 **Preliminary results of a high-sensitivity OH mapping survey in the galactic center.** N. Kaifu.
Bull. American Astron. Soc., Vol. 7, 251 (1975). – Abstr. AAS.

155.032 **The motion of our Galaxy and the local group of galaxies.** V. C. Rubin, W. K. Ford, Jr.
Bull. American Astron. Soc., Vol. 7, 253 (1975). – Abstr. AAS.

155.033 **Non-linear density-wave theory and the latitude distribution of H I gas.** D. L. Ball.
Bull. American Astron. Soc., Vol. 7, 258 (1975). – Abstr. AAS.

155:034 **Isolated Strömgren spheres as a source of galactic Hα emission.** B. G. Elmergreen.
Bull. American Astron. Soc., Vol. 7, 259 (1975). – Abstr. AAS.

155.035 **Large-scale distribution of carbon monoxide in the plane of the Galaxy.**
T. M. Bania, W. B. Burton, M. A. Gordon, F. J. Lockman.
Bull. American Astron. Soc., Vol. 7, 266 (1975). – Abstr. AAS.

155.036 **Galactic rotation derived from radial velocities and fundamental proper motions of stars.** W. Fricke.
Bull. American Astron. Soc., Vol. 7, 342 (1975). – Abstr. AAS.

155.037 **On the local standard of rest and the spiral structure of the Galaxy.** C. C. Lin, C. Yuan.
Bull. American Astron. Soc., Vol. 7, 344 (1975). – Abstr. AAS.

155.038 **UBVβ photometry and the galactic distribution of OB stars.** R. J. Havlen.
Dudley Obs. Rep. No. 9, (see 012.008), p. 389 - 398 (1975).

An extensive survey of OB stars identified in the 'Luminous Stars in the Southern Milky Way' catalogue has been carried out at ESO. Selected regions in Ara, Norma, and Puppis have been chosen to complement other investigations and to direct attention to areas of particular structural interest in the galactic plane. Hβ indices serve as luminosity indicators in the delineation of the young stellar distribution as well as for the isolation and analysis of young clusters and associations.

155.039 **Investigation of low-latitude hydrogen emission in terms of a two-component interstellar gas model.**
P. L. Baker, W. B. Burton.
Astrophys. Journ., Vol. 198, 281 - 297 (1975).

The high-resolution 21-cm hydrogen line observations at low galactic latitude of Burton and Verschuur have been analyzed to determine the large-scale distribution of galactic hydrogen. The distribution parameters are found by model fitting. The analysis permits the determination of values, corrected for optical depth effects, for the thickness of the galactic hydrogen disk between half density points—280 pc for the intercloud medium and 190 pc for the cloud medium for R < 10 kpc – the total observed neutral hydrogen mass of the Galaxy—$1.2 \times 10^9 M_\odot$ of the background gas and $1.4 \times 10^9 M_\odot$ of clouds– and the central number density of the intercloud hydrogen atoms—0.17 cm^{-3}.

155.040 Numerical calculations on the dynamics of the Galaxy under the influence of spiral arms and the Magellanic Clouds. J. Feitzinger.
Diss. Fachbereich Physik Eberhard-Karls-Univ. Tübingen. 4 + 81 pp. (1973). In German.

Hydrodynamical models of our stellar system are calculated to find an explanation for the north-south asymmetry of the galactic rotation curves. It is assumed that the Magellanic Clouds are moving in a circular orbit, inclined 45° to the galactic plane. Their gravitational disturbance and perturbations due to the spiral arms cause deviations from circular orbits of the gas in the galactic plane. After 5×10^9 years there is a difference in the north and south rotation curves of the observed order of $10 - 15$ km/s. This difference disappears after a further 0.5×10^9 years. In the framework of this model the observed velocity difference of the rotation curves can be interpreted as an oval distortion in the rotation velocity.

155.041 Fine-scale structure of carbon monoxide emission observed in the direction of the galactic nucleus.
H. S. Liszt, R. H. Sanders, W. B. Burton.
Astrophys. Journ., Vol. 198, 537 - 544 (1975).

Preliminary results are reported of a fine-scale survey of the Sgr A molecular cloud complex in the 2.6-mm line of carbon monoxide. The observations contain no evidence that galactic rotation is the dominant factor in determining the kinematic characteristics of CO located within a few arc-minutes of the galactic center; similarly no evidence is found here for a rapidly rotating disk of molecular gas in the core of the Galaxy. The distribution of the major CO emission features correlates well with that of the 6-cm H_2CO features observed by Whiteoak et al.; the structure of the Sgr A molecular cloud complex is discussed in terms of this correlation. Several of the CO spectra show a feature at zero velocity which the authors interpret as due to absorption by a cold (~2.7 K) molecular cloud.

155.042 _RGU_ three-colour photometry of an anticentre field near M 35. L. Topaktas.
Astron. Astrophys., Suppl. Ser., Vol. 20, 269 - 288 (1975).

A galactic anticentre field of 2.32 square degrees with 2400 stars near M 35 has been studied photometrically in the _RGU_ system down to a limiting magnitude of $G = 15.5$. The interstellar reddening function, the density functions for different groups of stars and luminosity functions have been determined. The reddening is caused by three interstellar clouds with distances of 360, 800 and 2560 pc. Two types of late-type giants have been identified forming branches at both sides of the main sequence of late-type dwarfs. The density functions reaching larger distances show almost a constant density from the sun to the outer border of the local spiral arm followed by a continuous decrease which is partly caused by incompleteness as revealed by a comparison with results found by Hersperger. For stars with $M(G) > 5$ the density gradients are affected by incompleteness already at distances smaller than 1 kpc. With few exceptions the density gradients of different groups extrapolated to zero distance conform with the space density as accepted for the solar neighbourhood.

155.043 The distribution of optical H II regions in our Galaxy. D. Crampton, Y. M. Georgelin.
Astron. Astrophys., Vol. 40, 317 - 321 = Dominion Astrophys. Obs., _Victoria_, Contr. No. 239 = NRC No. 13913 (1975).

The spiral structure of our Galaxy is mapped by the use of spectrophotometric and kinematic distances of H II regions to distances from the sun up to 8 kpc. An analysis of the kinematics of H II regions gives values for the solar motion components $U_0 = -7.2 \pm 0.7$ km s^{-1}, $V_0 = +14.7 \pm 0.8$ km s^{-1} and Oorts constant $A = 14.4 \pm 0.4$ km s^{-1} kpc^{-1}.

155.044 The Milky Way in comparison to other galaxies.

G. A. Tammann.
Conference on optical observing programs on galactic structure and dynamics, (see 012.013), p. 1 - 58 (1975). In German.

155.045 Radio-astronomical observations and optical structure. K. Rohlfs.
Conference on optical observing programs on galactic structure and dynamics, (see 012.013), p. 107 - 123 (1975). In German.

155.046 Observations for improving stellar kinematics.
W. Fricke, W. Gliese.
Conference on optical observing programs on galactic structure and dynamics, (see 012.013), p. 151 - 156 (1975). In German.

155.047 Present state of the Heidelberg OB-star program.
G. Klare, T. Neckel.
Conference on optical observing programs on galactic structure and dynamics, (see 012.013), p. 157 - 160 (1975). In German.

155.048 Observations on the stellar content of the Milky Way. W. Seggewiß.
Conference on optical observing programs on galactic structure and dynamics, (see 012.013), p. 161 - 164 (1975). In German.

155.049 On the three-dimensional structure of the spiral arms. W. Schlosser.
Conference on optical observing programs on galactic structure and dynamics, (see 012.013), p. 165 - 171 (1975). In German.

155.050 Galactic research at high latitudes. A. Blaauw.
Monthly Notes Astron. Soc. Southern Africa, Vol. 34, 72 - 73 (1975).

155.051 The distance to the galactic center derived from RR Lyrae variables, the distribution of these variables in the Galaxy's inner region and halo, and a rediscussion of the galactic rotation constants. J. H. Oort, L. Plaut.
Astron. Astrophys., Vol. 41, 71 - 86 (1975).

The results of recent surveys of faint RR Lyrae variables in fields near the galactic center are discussed. The density distributions show sharp maxima, and half-widths between 2 and 3 kpc. They indicate that the distance to the center is 8.7 kpc, with a mean error of about ±0.6 kpc. The density of the variables varies as R^{-3}, and is about 260 per kpc^3 at $R = 1.5$. There may be an extra concentration near the center, with a density of about 12000 per kpc^3 at $R = 0.6$. The R^{-3} variation for the density corresponds to a Maxwellian velocity distribution with a dispersion of 125 km/s in one coordinate. In all fields the average absorption in front of the distant variables is about 4 times less than that which has been inferred from galaxy counts.

155.052 Variation of the galactic force law in the region of the sun. K. D. Abhyankar.
Bull. Astron. Soc. India, Vol. 2, 28 (1974). – Abstract.

155.053 Gravitational stability of our galactic disk.
R. Bondyopadhaya.
Bull. Astron. Soc. India, Vol. 2, 28 (1974). – Abstract.

155.054 The effect of magnetic field on the dynamics of the Galaxy. A. K. Ray.
Bull. Astron. Soc. India, Vol. 2, 28 (1974). – Abstract.

155.055 Plasma instability of galactic disk.

S. N. Paul, R. Bondyopadhaya.
Bull. Astron. Soc. India, Vol. 2, 29 (1974). – Abstract.

155.056 A simple model of the Galaxy with a mixed density
law.   K. Bandyopadhyay.
Bull. Astron. Soc. India, Vol. 2, 29 (1974). – Abstract.

155.057 A comprehensive model of the Galaxy using three
heterogeneous spheroids and a mass point.
B. Basu, G. Saha.
Bull. Astron. Soc. India, Vol. 2, 29 (1974). – Abstract.

155.058 Models of the inner regions of the Galaxy.
W. L. Peters.
Thesis, Texas Univ., Austin (USA). 282 pp. University Micro-
films Order No. 74-14,748 (1974).

155.059 The history of the chemical evolution of our
Galaxy.   B. Stroemgren.
Fys. Tidssk., Vol. 72, 135 - 142 (1974). In Danish.

155.060 Radial flow from the nucleus of the Galaxy.
R. Bondyopadhaya.
Indian Journ. Phys., Vol. 48, 585 - 592 (1974).

An expression for the radial flow velocity of gaseous
materials from the nucleus of the Galaxy has been derived. It
is found that mainly gravitational force has the capability of
controlling this flow. It is suggested that an accelerating
mechanism (most probably nuclear explosion) caused the 3-
kpc arm to move.

155.061 Zur Interpretation der Nord-Süd Asymmetrie der
galaktischen Rotationskurven.   J. Feitzinger.
Kleinheubacher Ber., [Fernmeldetechn. Zentralamt, Darm-
stadt], No. 17, p. 511 - 513 (1974).

Harlow Shapley and the discovery of the center of
our Galaxy.   See Abstr. 004.079.

Determination of precession and galactic rotation
from the proper motions of the AGK3.
See Abstr. 043.009.

Nucleochronology and chemical evolution.
See Abstr. 061.038.

High-energy gamma-ray results from the second
Small Astronomy Satellite.   See Abstr. 061.039.

Galactic gamma-ray-astronomy.
See Abstr. 061.053.

On the ionization of hydrogen and helium by
hydromagnetic shock waves.   See Abstr. 062.010.

Interpretation of the stellar metallicity distribution.
See Abstr. 065.026.

A process of star formation and discreteness of the
dispersions of star velocities.   See Abstr. 065.037.

Electromagnetic radiation from colliding black holes.
See Abstr. 066.035.

Elementhäufigkeiten im Sonnensystem und Element-
entstehung in der Galaxis.   See Abstr. 091.031.

Comets as evidence of explosive processes in the
solar system and in the Galaxy.   See Abstr. 102.037.

Hβ photometry of southern early-type stars and

galactic structure away from the plane.
See Abstr. 113.005.

A catalogue of galactic O stars and the ionization
of the low density interstellar medium by runaway stars.
See Abstr. 113.008.

Emission-line stars with infrared dust emission:
implications of the galactic distribution.
See Abstr. 113.012.

Photometry of distant K giant stars.
See Abstr. 113.032.

Atmospheric parameters from four-color photo-
metry.   See Abstr. 113.042.

Spectroscopic and photometric observations of
luminous stars in the Centaurus-Norma ($l = 305°-340°$) sec-
tion of the Milky Way.   See Abstr. 114.013.

A comparison of galactic and Large Magellanic
Cloud G-type supergiants by a method of spectrum synthesis.
See Abstr. 114.021.

The interstellar lines of the Feige stars.
See Abstr. 114.064.

Cyanogen strengths, luminosities, and kinematics of
K giant stars.   See Abstr. 114.068.

Classifications of red stars for statistical investiga-
tions.   See Abstr. 114.085.

On the intrinsic properties of cepheids in the Gal-
axy, in Andromeda, and in the Magellanic Clouds.
See Abstr. 122.020.

Variable stars of young clusters and of the galactic
disc.   See Abstr. 122.057.

Distribution and ages of Magellanic cepheids.
See Abstr. 122.112.

A possible relation between spectral index and z-
distribution for supernova remnants.   See Abstr. 125.004.

Lower limits for the interstellar reddening at the
galactic poles.   See Abstr. 131.050.

The interstellar radiation field between 912 Å and
2740 Å.   See Abstr. 131.051.

Studies on high-velocity clouds of neutral hydrogen.
See Abstr. 131.126.

The ionization structure of H II regions of differ-
ent helium content.   See Abstr. 131.510.

Are the electron temperatures of H II regions a func-
tion of galactic radius?   See Abstr. 131.515.

The two basic components of the interstellar neu-
tral hydrogen and its relation with the galactic structure.
See Abstr. 131.531.

Orientation of planetary nebulae within the Galaxy.
See Abstr. 133.016.

A spherical subsystem of galactic radio sources.
See Abstr. 141.098.

Galactic gamma rays from the superposition of X-ray sources.   See Abstr. 142.027.

On the nature of γ-radiation from the region of the galactic center.   See Abstr. 142.105.

γ-radiation from the central region of the Galaxy and search for flare activity at energies of E⩾100 MeV. See Abstr. 142.106.

Spiral structure and galactic gamma radiation. See Abstr. 142.135.

Propagation of cosmic rays in the Galaxy. See Abstr. 143.003.

The Galaxy and interstellar medium. See Abstr. 143.063.

Galactic structure, magnetic fields and cosmic ray containment.   See Abstr. 143.069.

On the chemical composition of cosmic radiation in the Galaxy.   See Abstr. 143.077.

Density wave theory and the classification of spiral galaxies.   See Abstr. 151.008.

Numerical investigation of galactic tidal effects on spherical stellar systems.   See Abstr. 151.011.

Spiral modes in cold cylindrical systems. See Abstr. 151.019.

On the nature of the Puppis cluster NGC 2483. See Abstr. 153.023.

A photometric study of NGC 2419. See Abstr. 154.004.

Observations of H 110α radio recombination-line emission associated with distributed ionized hydrogen. See Abstr. 157.002.

The radio halo of the Galaxy. See Abstr. 157.004.

An atlas of galactic hydrogen. I. The region $0° \leqslant l \leqslant 12°, +3° \leqslant b \leqslant +17°$.   See Abstr. 157.010.

On the masses and relative velocities of galaxies. See Abstr. 158.023.

The galactic extinction from the apparent distribution of galaxies.   See Abstr. 158.050.

A possible new segment of the Magellanic Stream in the northern sky.   See Abstr. 159.005.

Radial velocities of supergiants in the Small Magellanic Cloud.   See Abstr. 159.015.

Errata

155.901   Errata: 'A new look at the interstellar hydrogen through a very-wide-field photographic Hα survey of the whole Milky Way' [Astron. Astrophys., Suppl. Ser., Vol. 16, 163 - 172 (1974)].   J. P. Sivan. Astron. Astrophys., Suppl. Ser., Vol. 19, 257 (1975).

155.902   Erratum: 'Faint M-type stars in the south galactic pole region' [Astron. Astrophys., Vol. 36, 155 - 161 (1974)].   P. S. Thé, R. F. A. Staller. Astron. Astrophys., Vol. 40, 225 (1975).

156   Galactic Magnetic Field

156.001   **Interpretation of Faraday rotation in the galactic magnetic field.**   D. Nissen, K. O. Thielheim. Astrophys. Space Sci., Vol. 33, 441 - 451 (1975).

The autocorrelation function of Faraday rotation measures is discussed in terms of different types of galactic field configurations. The autocorrelation function evaluated from published data of 139 radio galaxies and quasars is found to resemble a form typical for a quasi-longitudinal field, whereas the autocorrelation function of 38 pulsars turns out to be of the form expected for a longitudinal field. These observations are interpreted with respect to the position of the solar system relative to the neutral sheet in a quasi-longitudinal field configuration.

156.002   **Generation of a galactic magnetic field by spiral shocks.**
Yu. N. Mishurov, V. M. Peftiev, A. A. Suchkov: Pis'ma v Astron. Zhurn., Vol. 1, No. 4, p. 8 - 10 (1975). In Russian.

It is shown that spiral shocks realized in galaxies in the presence of density waves generate a galactic magnetic field with spiral structure.

156.003   **Structure of irregular galactic magnetic fields.**   A. J. Somogyi. Nature, Vol. 255, 689 - 690 (1975).

The author describes a method which, supposing the diffusion approximation of cosmic-ray propagation to be valid, is suitable for determining the structural composition of random galactic fields on the basis of cosmic-ray measurements, down to structures with characteristic lengths $\sim 0.001 - 1$ pc.

The confinement of galactic cosmic rays by Alfvén waves.   See Abstr. 143.014.

Galactic structure, magnetic fields and cosmic ray containment.   See Abstr. 143.069.

A high-resolution neutral-hydrogen study of the galaxy M81.   See Abstr. 158.003.

157 Galactic Radio Radiation

157.001 Radio emission from the central region of the Galaxy ($l^{II} = 351 - 8°$) at 7700 MHz. V. I. Abramenko, A. V. Ipatov, N. M. Lipovka, A. K. Obolenskij. Soobshch. Spets. Astrofiz. Obs., *Zelenchukskaya,* vyp. (No.) 12, p. 43 - 50 (1974). In Russian.

Isophotes are obtained for the galactic background radiation at 7700 MHz for the longitudes $351° < l^{II} < 8°$. Brightness and spectral index variations with longitude are investigated.

157.002 Observations of H 110α radio recombination-line emission associated with distributed ionized hydrogen. P. D. Jackson, F. J. Kerr. Astrophys. Journ., Vol. 196, 723 - 731 (1975).

Nine Milky Way positions have been observed for H 110α radio recombination-line emission associated with ionized hydrogen lying outside those large H II regions detectable as discrete sources on published radio continuum surveys. Line emission was detected for seven of the nine positions. The rapid change of profile shape with position, the lack of narrow-frequency-width profile components, and the correlation of line radiation with continuum radiation at the same positions lead to a conclusion that most of the line radiation originates in small H II regions along the line of sight, over at least the galactic longitude range $l = 24°–33°$. An electron temperature of $4400° \pm 600°$K is associated with the line emission for this range. Other interpretations of the kind of regions emitting the line radiation are briefly discussed.

157.003 Hydrogen 21-cm line temperature scale. R. H. Harten, G. Westerhout, F. J. Kerr. Astron. Journ., Vol. 80, 307 - 310 (1975).

The recent publication of two extensive, high-resolution, galactic 21-cm line surveys, covering both the northern and southern galactic plane regions, draws attention to the long-standing temperature scale problem. The new Maryland-Green Bank survey, the new Parkes survey, and the Berkeley survey are shown to agree with each other to within 5%, which allows to establish a common scale for these surveys.

157.004 The radio halo of the Galaxy. A. Webster. Monthly Notices Roy. Astron. Soc., Vol. 171, 243 - 257 (1975).

A new way of analysing radio observations for evidence of a radio halo around the Galaxy is presented. When applied to appropriate measurements clear evidence for a weak radio halo with a steep spectrum is found. This radio halo is much weaker than expected on the hypothesis that the galactic cosmic rays fill the halo region.

157.005 Pulkovo sky survey in the interstellar neutral hydrogen line. I. Neutral hydrogen in the neighbourhood of stellar associations λ Orionis and Monoceros I. N. V. Bystrova, I. A. Rakhimov. Astrofiz. Issled., Izv. Spets. Astrofiz. Obs., Vol. 7, p. 70 - 86 (1975). In Russian.

Seven drift curves obtained with the large Pulkovo radio telescope at a bandwidth of 20 kHz are given for declinations between $+05°36'$ and $+14°16'$ and right ascensions from 4^h to 8^h. The curves are presented in antenna temperatures for 10 radial velocities fixed relative to the local standard of rest.

157.006 Pulkovo sky survey in the interstellar neutral hydrogen line. II. Neutral hydrogen in the region of Loop II. N. V. Bystrova, I. A. Rakhimov. Astrofiz. Issled., Izv. Spets. Astrofiz. Obs., Vol. 7, p. 87 - 95 (1975). In Russian.

Eight drift curves obtained with the large Pulkovo radio

telescope at a bandwidth of 20 kHz are given for declinations between $0°00'$ and $+17°30'$ and right ascensions from 1^h to 5^h for 10 radial velocities.

157.007 Loops of galactic radio radiation. T. A. Lozinskaya. Zemlya i Vselennaya, 1975, No. 2, p. 33 - 36. In Russian.

157.008 Neutral hydrogen in the galactic centre region – I. The observations. R. J. Cohen. Monthly Notices Roy. Astron. Soc., Vol. 171, 659 - 696 (1975).

A survey of the 21-cm line of neutral hydrogen has been made at Jodrell Bank with the Mark II radio telescope (beam-width $31' \times 35'$) and the Mark IA radio telescope (beamwidth $13' \times 13'$), covering the region $355° \leqslant l \leqslant 10°$, $-5° \leqslant b \leqslant 5°$, and the velocity range -530 km s^{-1} to $+530$ km s^{-1}. The data are presented in the form of contour maps of antenna temperature as a function of galactic latitude and velocity, and also as a function of galactic longitude and velocity. Six new high-velocity emission features have been detected, all lying out of the galactic plane.

157.009 A consistent scheme of definitions of polarisation brightness temperature and brightness temperature. E. M. Berkhuijsen. Astron. Astrophys., Vol. 40, 311 - 316 (1975).

New definitions of polarisation brightness temperature and brightness temperature are proposed and discussed. In the new system T_b^p is one half times T_b^p as defined in Westerhout et al. (1962). Relevant literature and conversion factors for T_b^p to the new system are given in a table. A consistent scheme of definitions is given in the appendix.

157.010 An atlas of galactic hydrogen. I. The region $0° \leqslant l \leqslant 12°, +3° \leqslant b \leqslant +17°$. W. G. L. Pöppel, E. R. Vieira. Publ. Carnegie Institution, Washington, No. 633, 46 pp. = Contr. Inst. Argentino Radioastron., No. 42 (1974). In English and Spanish.

Observational data in the 21 cm neutral hydrogen line are presented for the region $0° \leqslant l \leqslant 12°, +3° \leqslant b \leqslant +17°$. The radial velocity interval extends from -100 to $+100$ km/s.

157.011 Radiocontinuum emission of the Galaxy compared with other galaxies. R. Wielebinski. Conference on optical observing programs on galactic structure and dynamics, (see 012.013), p. 125 - 130 (1975).

Are the electron temperatures of H II regions a function of galactic radius? See Abstr. 131.515.

The nonthermal radio sources at $l = 74°9$ and $b = +1°2$. See Abstr. 141.039.

Investigation of low-latitude hydrogen emission in terms of a two-component interstellar gas model. See Abstr. 155.039.

Radio-astronomical observations and optical structure. See Abstr. 155.045.

157.901 Erratum: 'An almost complete survey of 21 cm line radiation for $|b| \geqslant 10°$. I. Atlas of contour maps' [Astron. Astrophys., Suppl. Ser., Vol. 14, 1 - 555 (1974)]. C. Heiles, H. J. Habing. Astron. Astrophys., Suppl. Ser., Vol. 19, 115 (1975).

158 Single and Multiple Galaxies, Peculiar Objects

Single und Multiple Galaxies

158.001 **On the determination of the masses of disk galaxies using the density-wave model of their spiral structure. M 31 and M 33.** B. Basu.
Bull. Astron. Inst. Czechoslovakia (BAC), Vol. 26, 48 - 57 (1975).

Using the density-wave solutions of the gas-dynamical equations in two-dimensional cylindrical coordinates, the surface density distributions of matter can be expressed in terms of several galactic parameters in spiral galaxies. Using the numerical values of the parameters involved that are appropriate for each separate galaxy, the density values can be evaluated along the radial distances except at regions quite close to the centre. These density values can than be fitted to suitable algebraic laws over different appropriate regions of each galaxy and the masses of the corresponding regions can be evaluated separately by integration. The masses of M 31 and M 33 have been evaluated to be between $3.8 \times 10^{11} M_\odot$ to $4.53 \times 10^{11} M_\odot$ and between $1.45 \times 10^{10} M_\odot$ to $1.66 \times 10^{10} M_\odot$ respectively.

158.002 **Optical positions of bright galaxies. III.**
L. Gallouët, N. Heidmann, F. Dampierre.
Astron. Astrophys., Suppl. Ser., Vol. 19, 1 - 19 (1975).

The authors are publishing here the third and last part of their catalogue of optical positions of the 2118 galaxies at declination higher than $-33°$ in the Reference Catalogue of Bright Galaxies, by G. and A. de Vaucouleurs (1964). The present list, which contains 557 galaxies, covers the range $-33°$ to $-2°$ in declination. The accuracy is, as before, of a few seconds of arc.

158.003 **A high-resolution neutral-hydrogen study of the galaxy M81.** S. T. Gottesman, L. Weliachew.
Astrophys. Journ., Vol. 195, 23 - 45 (1975).

A 2′ synthesis of M81 has been made in the 21-cm neutral-hydrogen emission line. In addition to the interferometric data, a single-dish measurement provided information about the large-scale structure. One-third of the H I observed appears to be divorced from the main body of the galaxy and shows large non-circular motions. A rotation curve was established for M81, and within 20 kpc of the center a mass of $1.1 \times 10^{11} M_\odot$ is found. The neutral-hydrogen mass in M81 is $4.9 \times 10^9 M_\odot$. A density-wave model is discussed which appears to be in better agreement with the structure of the galaxy than the model using older optical data proposed by Shu et al. The kinematics of M81, as derived from the study, do not show the systematic pattern of noncircular velocities expected from the density-wave theory.

158.004 **Stellar content of the nuclei of elliptical galaxies determined from 2.3-micron CO band strengths.**
J. A. Frogel, S. E. Persson, M. Aaronson, E. E. Becklin, K. Matthews, G. Neugebauer.
Astrophys. Journ., (Letters), Vol. 195, L15 - L18 (1975).

Observations of the luminosity-sensitive CO absorption band at 2.3 μ are presented for the nuclei of 16 E and S0 galaxies. Preliminary data on the radial variation of the CO index in six galaxies are also given. The data show that the 2-μ radiation from the nuclei is dominated by high-luminosity stars of integrated spectral type later than K5. There is a small variation in the CO absorption from galaxy to galaxy. The data can be used to synthesize stellar populations for elliptical galaxy nuclei.

158.005 **Finding list of bright galaxies behind the SMC.**
P. W. Hodge, T. P. Snow.
Astron. Journ., Vol. 80, 9 - 10, 71 (1975).

A finding list and chart identify 30 probable galaxies with $V \gtrsim 15$, seen behind the main body of the SMC. These objects could be useful for photoelectric study of the effects of the SMC interstellar dust.

158.006 **The velocity field of the stars and gas in NGC 2903.** S. M. Simkin.
Astrophys. Journ., Vol. 195, 293 - 313 (1975).

Detailed velocities are reported for the ionized gas and the stars contributing to the H10 absorption line in NGC 2903. Large, noncircular velocities and radial streaming motions are found in both the gas and the stars. These seem to be associated with the spiral arms. The position angle of the line of nodes and orientation of this object are established, and its systemic velocity is discussed.

158.007 **Redshift differences of galaxies in nearby groups.** E. R. Harrison.
Astrophys. Journ., (Letters), Vol. 195, L61 - L63 (1975).

It is reported that galaxies in nearby groups exhibit anomalous nonvelocity redshifts. In this discussion, (a) four classes of nearby groups of galaxies are analyzed, and no significant nonvelocity redshift effect is found; and (b) it is pointed out that transverse velocities (i.e., velocities transverse to the line of sight of the main galaxy, or center of mass) contribute components to the redshift measurements of companion galaxies. The redshifts of galaxies in nearby groups of appreciable angular size are considerably affected by these velocity projection effects. The transverse velocity contributions average out in rich isotropic groups, and also in large samples of irregular groups of low membership, as in the four classes referred to in (a), but can introduce apparent discrepancies in small samples as studied by Arp of nearby groups of low membership.

158.008 **On the distribution and kinematics of the ionized gas in the inner region of NGC 4151.**
K. J. Fricke, M. Reinhardt.
Astron. Astrophys., Vol. 37, 349 - 353 (1974).

Image tube spectra have been taken in 8 different position angles across the nucleus of NGC 4151. Emission in the line [O III] λ 5007 can be traced out to approximately 40″ (4 kpc) distance from the nucleus in the NE and SW directions. The observations seem to rule out purely rotational motions. The distribution and the velocity field of the ionized gas can be interpreted by a superposition of rotational motions with a projected axis in P.A. 120° and radial motions in the NE–SW and the E–W directions.

158.009 **Virial tests for fourteen nearby groups of galaxies: a case for the dynamical stability of some groups.**
J. Materne, G. A. Tammann.
Astron. Astrophys., Vol. 37, 383 - 388 (1974).

Fourteen nearby groups of galaxies, as defined by de Vaucouleurs (1965), are tested for dynamical stability. Improved redshifts and allowance for their observational errors (Materne, 1974) reduce considerably the virial masses previously indicated. Monte Carlo calculations suggest that the results are not decisively altered by variations of the adopted masses. It is concluded that the evidence for the existence of bound groups is significant.

158.010 **Formation of satellites by fragmentation of galaxies.** B. A. Vorontsov-Velyaminov.
Astron. Astrophys., Vol. 37, 425 - 429 (1974).

The author describes the main results of a study of 2000 interacting galaxies in the Palomar Sky Atlas. Among them are found systems of the "twofold M 51 type". Many systems and much of the data lead to the concept that the tight "nest of galaxies" disrupt into looser nests and that many large galaxies, by fragmentation, create satellites of various types. The latter develop inside their bodies as do the satellites of the M 51 systems.

158.011   Anomalies in the 18-centimeter OH lines in the galaxy NGC 4945.   J. B. Whiteoak, F. F. Gardner.
Astrophys. Journ., (Letters), Vol. 195, L81 - L84 (1975).

The OH clouds in the central regions of NGC 4945 produce anomalous emission and absorption features in the satellite-line profiles at 1612 and 1720 MHz. The anomalies are consistent with perturbations of the relative populations of the $F = 1$ and $F = 2$ hyperfine levels of the ground state of OH, which could be caused by infrared pumping.

158.012   Kinematic and dynamic study of M31 from observations of emission regions.
J. M. Deharveng, A. Pellet.
Astron. Astrophys., Vol. 38, 15 - 28 (1975). In French.

Radial velocities of Hα were measured at 1636 points in the arms of M31. A Fabry-Perot interference method was used with mean dispersions of 25 Å/mm and 15.5 Å/mm, giving a mean accuracy of ±11 km s^{-1}. The choice of the systemic velocity is discussed in terms of the nuclear and minor axis velocities. The lack of significant differences in radial velocity along this axis indicates that expansion motions are not present. The authors obtain a peak in the rotational velocity of 220 km s^{-1} at 1.'5 from the center, due to a rapidly rotating disk of 350 pc radius and this is followed by a minimum. No velocities could be measured near $R = 2$ kpc due to interference from night sky emission, so the minimum in the rotation curve could not be confirmed. The authors have calculated the distribution of mass, according to the method of Burbidge and Prendergast, using two possible rotation curves, and by varying the axial ratios of the spheroids. The mass inside a 20 kpc radius is $(1.63 + 0.15) \, 10^{11} M_\odot$. The mass to luminosity ratio shows large variations with a minimum at 2.'5 from the center and a maximum at 25', although the mean ratio is close to that of previous determinations ($\simeq 12$). The results found at the center are not in agreement with those of Spinrad.

158.013 $(V-K)$ colors of galaxies: statistical differences between spirals and ellipticals and the color-diameter relation for elliptical galaxies.   G. L. Grasdalen.
Astrophys. Journ., Vol. 195, 605 - 610 (1975).

New K ($2.2 \, \mu$) observations are presented for a survey of galaxies. These are combined with available measures of V-magnitude to form $(V-K)$ colors. The $(V-K)$ colors of bright elliptical galaxies are found to have a very small cosmic dispersion, while for a heterogeneous sample of spiral galaxies the dispersion is quite large. There is no evidence that the $(V-K)$ color of bright ellipticals depends on absolute luminosity. The $(V-K)$ color of elliptical galaxies is found to have a strong radial gradient. This conclusion is verified by detailed observations of M87. The relation of the $(V-K)$ color to other metal-content indicators is briefly discussed.

158.014   Rotation properties of the infrared spiral galaxy NGC 3675.   P. C. van der Kruit.
Astrophys. Journ., Vol. 195, 611 - 615 (1975).

Measurements of the rotation curve of NGC 3675 from slit spectra in three position angles are reported. There is no evidence for strong noncircular motions. The rotation curve rises steeply in the inner part, and the corresponding central concentration of mass might be related to the presence of the relatively strong infrared (and perhaps radio) source in the nucleus. The rotation curve is consistent with the presence of

a rather massive (about 20 percent of the total mass) nuclear bulge, in agreement with the optical appearance. The mass-luminosity ratio is probably small.

158.015   Observations of M87 at 5 GHz with the 5-km telescope.   B. D. Turland.
Monthly Notices Roy. Astron. Soc., Vol. 170, 281 - 294 (1975).

M87 has been observed at 5 GHz with the 5-km telescope at Cambridge. Maps of total intensity and polarized flux were made of the core component. Compact structure, corresponding to knots in the optical jet, was found in the Np component. The knots are significantly polarized (~ 10 per cent) at 5 GHz and models of the jet with particle densities $\geqslant 10^8$ m^{-3} are shown to be no longer tenable. The Sf component is more extended and contains no compact features. The nuclear source of flux density 4.0×10^{-26} W m^{-2} Hz^{-1} was unresolved, but has an extension in the direction of the jet, and possibly in p.a. 170°. The X-ray emission from M87 is most easily explained as thermal bremsstrahlung from an extended hot gas.

158.016   The role of thermal instability in the formation of galaxies.   L. Eh. Gurevich, A. D. Chernin.
Astron. Zhurn. Akad. Nauk SSSR, Vol. 52, 3 - 8 (1975). In Russian. English translation in Soviet Astron., Vol. 19, No. 1.

The authors suppose that at the epoch of formation of galaxies large-scale hydrodynamic motions existed that generate shock waves. Cosmic plasma became heated and ionized at their fronts and subsequently lost its thermal energy because of Compton scattering and free-free radiation. The plasma was compressed behind the fronts under approximately constant pressure. When free-free radiation dominated thermal instability occurred, and plasma fragmentated into dense regions with masses close to the masses of normal galaxies.

158.017   Interaction of protoclusters of galaxies with intergalactic matter.
A. G. Doroshkevich, S. F. Shandarin.
Astron. Zhurn. Akad. Nauk SSSR, Vol. 52, 9 - 14 (1975). In Russian. English translation in Soviet Astron., Vol. 19, No. 1.

The problem of the origin of galaxies and clusters of galaxies is closely associated with the existence of intergalactic matter and its properties. Any theory of the formation of galaxies should explain the following facts: 1) the low density of H I in the intergalactic space at $z \sim 2-4$ resulting from observed spectra of distant QSOs, 2) the low ultra-violet and X-ray background, 3) the high X-ray radiation of clusters of galaxies. In this paper the above questions are discussed in the frame of the theory of formation of galaxies based on the nonlinear theory of gravitational instability.

158.018   The bar-like objects in the centres of galaxies as a possible generator of spiral density waves. I.
V. I. Korchagin, L. S. Marochnik.
Astron. Zhurn. Akad. Nauk SSSR, Vol. 52, 15 - 25 (1975). In Russian. English translation in Soviet Astron., Vol. 19, No. 1.

The possibility of generation of the two-armed spiral pattern by the bar-like structure in a uniform and uniformly rotating disk of gas is shown. The sense of the winding of the spiral pattern is determined by the relation of the angular speeds of the bar and the disk. The trailing two-armed spiral pattern is generated by the bar rotating more rapidly than the disk. The slow bar rotation generates a leading spiral pattern.

158.019   Nuclei of galaxies and quasars.   L. M. Ozernoj.
New problems of astrophysics, Publ. Astrophys. Winter School, 1972, (see 012.001), p. 32 - 74 (1974). In Russian.

158.020   Ring galaxies and intergalactic gas clouds.
C. D. Mackay.

Nature, Vol. 254, 181 (1975).

158.021 Possible power source of Seyfert galaxies and QSOs.
J. G. Hills.
Nature, Vol. 254, 295 - 298 (1975).

The possible presence of massive black holes in the nuclei of galaxies has been suggested many times. In addition, there is considerable observational evidence for high stellar densities in these nuclei. The author shows that the tidal breakup of stars passing within the Roche limit of a black hole initiates a chain of events that may explain many of the observed principal characteristics of QSOs and the nuclei of Seyfert galaxies.

158.022 Velocities with grating and Fabry-Perot spectrographs of the ionized gas of M31.
J. M. Deharveng, A. Pellet.
Astron. Astrophys., Suppl. Ser., Vol. 19, 351 - 358 (1975). In French.

A Fabry-Perot interference method (mean dispersion 25 Å/mm and 15.5 Å/mm) has given 1389 radial velocities of Hα in the arms of M31. Mean rotation velocities are computed by grouping measured points according to the morphology.

158.023 On the masses and relative velocities of galaxies.
G. Burbidge.
Astrophys. Journ., (Letters), Vol. 196, L7 - L10 (1975).

Contrary to the results obtained by Einasto et al. and Ostriker et al., the author shows that there is no unambiguous dynamical evidence which demonstrates that galaxies have very massive halos. The results that $M(R) \propto R$ suggested by Ostriker et al. is entirely dependent on the assumptions (a) that all systems studied are physical systems, and (b) that they are bound. In many cases these assumptions are in doubt. It is shown that the scale of internal velocities in galaxies and relative velocities between galaxies is remarkably constant with a radial component $\sim$ 300 km s^{-1}. Significantly larger values of this quantity are only found in rich clusters of galaxies.

158.024 The integrated neutral hydrogen properties of nearby galaxies. J. F. Dean, R. D. Davies.
Monthly Notices Roy. Astron. Soc., Vol. 170, 503 - 518 (1975).

The H I emission from 17 nearby spiral and irregular galaxies has been studied with the MK II radio telescope (beamwidth 31' × 33') to provide integrated H I flux densities to an accuracy of 10 per cent or better. An indicative total mass, M_T, was estimated directly from the H I observations which agreed well with the value derived by other observers from rotation curves. This provides a simple H I technique for measuring M_T for any spiral or irregular galaxy. These new results gave values of M_H/L, M_H/M_T and M_T/L which showed less scatter within each morphological type as compared with previous investigations.

158.025 Systematic errors in the velocities of galaxies.
B. M. Lewis.
Mem. Roy. Astron. Soc., Vol. 78, 75 - 100 (1975).

The 21-cm estimate V_{21} of the systemic velocities of 202 galaxies are compared with all available optical estimates for the same objects. The average difference between these measurements is shown to depend on the type of the galaxy. An average estimate V_A of the optical velocity is derived by combining estimates from different observers. The differences $(V_{21} - V_A)$ are normally distributed about zero, showing that despite the difference in precision, both the optical and 21-cm techniques produce statistically equivalent results. There is no dependence of the results on the absolute luminosity of a galaxy and no support is found for the existence of an anomalous redshift.

158.026 A radio survey of Markarian galaxies at 6 centimeters. R. A. Sramek, H. M. Tovmassian.
Astrophys. Journ., Vol. 196, 339 - 345 (1975).

All 506 galaxies in Markarian's first five lists were surveyed at 6 cm wavelength with a 3 σ detection limit of about 30 mJy. Twenty-eight galaxies were detected. The Markarian galaxies with Seyfert characteristics have higher radio luminosities. Among these, radio emission is predominantly detected from galaxies of color type II and from those having the more diffuse d nuclei. Among the non-Seyfert Markarian galaxies, those without any spectral lines are more often detected, and these tend to have the more condensed s-type nuclei. There is a tendency for galaxies which are double, or which have ejections, tails, or some form of optical peculiarity, to also have detectable radio emission.

158.027 Millimeter-wave observations of elliptical galaxies.
D. S. Heeschen, E. K. Conklin.
Astrophys. Journ., Vol. 196, 347 - 349 (1975).

Millimeter-wavelength observations of elliptical galaxies are described. At least one of the galaxies is variable, on a time scale of the order of a year.

158.028 The formation of the nuclei of galaxies. I. M31.
S. D. Tremaine, J. P. Ostriker, L. Spitzer, Jr.
Astrophys. Journ., Vol. 196, 407 - 411 (1975).

Globular clusters passing near the center of M31 interact with the background stars through dynamical friction and spiral in to the center of the galaxy, where they are tidally disrupted by interactions with the growing nucleus. By this process a distinct high-density central nucleus of mass $\sim 5 \times 10^7 M_\odot$ will be formed in 10^{10} years. These predictions are consistent with recent Stratoscope balloon observations of the nucleus of M31.

158.029 Statistical analysis of catalogs of extragalactic objects. VI. The galaxy distribution in the Jagellonian field. P. J. E. Peebles.
Astrophys. Journ., Vol. 196, 647 - 652 (1975).

The Jagellonian catalog lists more than 10,000 galaxies in a 6° × 6° area. Estimates of the two-point and three-point correlation functions for the distribution are presented. The results for both functions agree very well with the correlation functions derived from the Zwicky catalog of galaxies and scaled to take account of the greater depth in the Jagellonian survey.

158.030 Studies of galaxies with peculiar nuclei.
M. G. Pastoriza.
Astrophys. Space Sci., Vol. 33, 173 - 188 (1975).

Photometric, morphological and spectral grounds suggest that the peculiar nuclei regions of NGC 1097, 1672, 2997, 5236, and 7552 are composed basically of a normal nucleus surrounded by several large but otherwise normal ($T_e \sim 8000$ K, $N_e \sim 1000/cm^3$) H II regions. The stellar components are also normal although in some cases (NGC 1097, 1672, 1808 and 7552) there is a larger contribution of late type stars.

158.031 Radio and optical data on a complete sample of radio faint galaxies.
G. Colla, C. Fanti, R. Fanti, I. Gioia, C. Lari, J. Lequeux, R. Lucas, M. H. Ulrich.
Astron. Astrophys., Suppl. Ser., Vol. 20, 1 - 36 (1975).

Radio sources from the B2 catalogue have been identified with galaxies in the "Catalogue of galaxies and of clusters of galaxies" (Zwicky and Herzog, 1963, 1966; Zwicky and Kowal 1968). 54 ellipticals, 3 peculiar galaxies and 10 spiral galaxies were found. All these objects have been observed at 1415 MHz at Nançay, 42 have been observed at Westerbork and a few at Nançay at 5000 MHz. These observations allow the determination of the radio spectrum and a study of the radio structure. Spectroscopic observations have also been made at McDonald Observatory and Kitt Peak National Observatory providing redshifts for 41 galaxies and as a result the redshifts are known for all the galaxies except 3 spirals.

158.032 The emission-line spectrum of NGC 1068.
G. A. Shields, J. B. Oke.
Astrophys. Journ., Vol. 197, 5 - 16 (1975).

The authors report new photoelectric observations of the nucleus of NGC 1068 and an analysis of the emission-line intensities. Their results support the hypothesis that the emission-line region of NGC 1068 is ionized by ultraviolet radiation rather than cloud collisions. Assuming a solar oxygen abundance, a nitrogen abundance $N(N)/N(H) = 4 \times 10^{-4}$ is found, which is larger than the solar value, but similar to the high value in the nuclei of normal galaxies.

158.033 The nucleus of NGC 253.
G. H. Rieke, F. J. Low.
Astrophys. Journ., Vol. 197, 17 - 23 (1975).

Detailed mapping and photometry of the nucleus of NGC 253 has been carried out from 1.6 through 34 μ. Most of the 10-μ flux comes from a region ~45 pc in diameter; the 2.2 μ source is more extended. The 10-μ core is visible in near-infrared photographs and is at the position of a compact radio source. The "silicate" absorption feature at 10 μ varies in strength across the source, indicating strong, nonuniform interstellar extinction. The ratio of the optical depth in the feature to the visual extinction is similar to that found toward our galactic center. The M/L ratio in the nucleus is less than ~0.1; therefore, thermonuclear burning of the mass currently in the nucleus could not have sustained the present level of emission for the life of the galaxy.

158.034 An improbable coincidence in NGC 985: a ring galaxy with a Seyfert nucleus.
G. de Vaucouleurs, A. de Vaucouleurs.
Astrophys. Journ., (Letters), Vol. 197, L1 - L4 (1975).

The compact spheroidal component of the ring galaxy NGC 985 = VV 285 was found to possess a class I Seyfert nucleus having narrow [O III] lines and broad Balmer lines (half-width $\simeq$ 5500 km s^{-1}) with narrow cores (half-width < 200 km s^{-1}). The mean redshift of the system is $z = 0.0432$ corresponding to a Hubble distance $\Delta = 130h^{-1}$ Mpc (h = H/100). The absolute magnitude about -22, and the ring diameter $D \approx 22$ kpc (for $h = 1$) make it one of the largest and brightest known ring systems.

158.035 Evidence for the detection of gamma rays from Centaurus A at $E_\gamma \geq 3 \times 10^{11}$ eV.
J. E. Grindlay, H. F. Helmken, R. Hanbury Brown, J. Davis, L. R. Allen.
Astrophys. Journ., (Letters), Vol. 197, L9 - L12 (1975).

Results of extended observations of the active galaxy NGC 5128 (Cen A) at energies > 10^{11} eV are presented. The atmospheric Cerenkov technique was used, together with partial cosmic-ray rejection, to search for γ-ray-initiated extensive air showers from the direction of Cen A. A 4.5 σ (time-averaged) excess over background was detected. Some implications of this probable γ-ray flux are discussed.

158.036 Television surface photometry of the edge-on spiral galaxies NGC 3987 and NGC 5907. M. Davis.
Astron. Journ., Vol. 80, 188 - 193, 253 (1975).

Recent dynamical calculations suggest that the masses of spiral galaxies might have been systematically underestimated and that they could contain a halo of low intrinsic luminosity. With the goal of detecting a spherical halo component the author has obtained numerousexposures of two edge-on spiral galaxies with an SEC vidicon camera in colors roughly B, V, and R. His results constrain the average surface brightness of a halo with twice the surface area of the disc to be less than 3% of the night sky level, and to contribute less than 25% of the total luminosity of each galaxy.

158.037 Luminosity function of nearby galaxies.
C. G. Christensen.
Astron. Journ., Vol. 80, 282 - 289 (1975).

The luminosity function of nearby field galaxies has been determined from a sample of 231 "normal" galaxies with $m_B \lesssim 11.85$. The sample has been confined to galactic latitudes above $\pm 20°$ and excludes the Virgo cluster region. Absolute magnitudes for galaxies with radial velocities exceeding 400 km sec^{-1} have been based on the Hubble law with H = 50 km sec^{-1} Mpc^{-1}. Other distance criteria have been used for systems of lower velocity.

158.038 Size vs absolute magnitude relation for Haro and Im galaxies.   K. Wakamatsu, S. Oka, K. Sakka.
Mem. Fac. Sci.,Kyoto Univ., Ser. Phys., Astrophys., Geophys., Chem., Vol. 34, 243 - 245 (1974).

The size vs absolute magnitude relation for Haro galaxies has been re-examined. It does not appear to differ significantly from the relation for Im galaxies.

158.039 Photographic photometry of bright galaxies— II: NGC 3384.
R. Barbon, M. Capaccioli, M. Tarenghi.
Astron. Astrophys., Vol. 38, 315 - 321 (1975).

Photographic photometry in the B system for the SB0 galaxy NGC 3384 is presented here and compared with previous work. The luminosity distribution and several standard photometric parameters are derived. From the equivalent profile, three different components have been distinguished: the nucleus, the lens and the exponential disk. A possible interaction with the nearby E galaxy NGC 3379 is finally discussed.

158.040 The structure of the radio galaxy NGC 1265.
G. K. Miley, K. J. Wellington, H. van der Laan.
Astron. Astrophys., Vol. 38, 381 - 390 (1975).

New data are presented on the 5 GHz polarization distribution of the head tail radio galaxy NGC 1265 in the Perseus cluster. Polarization reaches 60% in places indicating that magnetic energy is dominant. The results are in agreement with the predictions of the magnetospheric trail model for head tail radio sources. The physical conditions inside and outside the source are discussed within the framework of several assumptions. The derived magnetic field strengths (~3–7μG), electron densities (~10^{-3} to 10^{-4} cm^{-3}) and temperatures (~10^7K) are similar to those previously reported for 3C129.

158.041 Properties of two blue compact galaxies.
H. Arp, R. W. O'Connell.
Astrophys. Journ., Vol. 197, 291 - 296 (1975).

An analysis of spectroscopic and photometric data for two blue, low-luminosity compact galaxies similar to the "isolated extragalactic H II regions" studied by Sargent and Searle is presented. The observed ionization and spectral energy distributions of both systems are consistent with a young stellar population dominated by hot stars $\lesssim 10^8$ years old. Inferred rates of star formation and supernova explosions imply that both galaxies may be considered to be rapidly evolving compared with normal galaxies.

158.042 The origin of infrared emission from the nucleus of NGC 1068. T. W. Jones, W. A. Stein.
Astrophys. Journ., Vol. 197, 297 - 307 (1975).

Recent infrared observational results for the nucleus of the Seyfert galaxy NGC 1068 are reviewed and analyzed in terms consistent with information available at other wavelengths. It is concluded that the infrared and optical data imply that $\gtrsim 85$ percent of the infrared emission at 10μ is radiation from dust grains in the nucleus. The grains are heated radiatively by an underlying source or sources of radiation also responsible for ionizing the emission-line–producing gas. The underlying source could be nonthermal, or it could be a hot plasma. Physical constraints on each of these models are derived.

158.043 The galaxy M82: nuclear luminosity as determined by the polarization pattern.
A. B. Solinger, T. Markert.
Astrophys. Journ., Vol. 197, 309 - 315 (1975).

The polarization pattern of M82 is considered anew and is shown to reflect the ratio of nuclear to disk luminosity. The M82 galactic disk is found to make a significant contribution to the scattered light, independent of type of scatterers assumed. The major portion of the scattered continuum beyond the M82 plane originates in the M82 galactic disk, and not in the nucleus. This, in turn, means that the polarized Hα line observed being scattered in the filaments may originate in the disk, leaving open the possibility that M82 is a Seyfert galaxy.

158.044 Surface photometry of galaxies. I. The SB0 NGC 2950. P. Crane.
Astrophys. Journ., Vol. 197, 317 - 328 (1975).

Surface photometry of NGC 2950, a classic example of the SB0 type galaxy, is presented. The analysis shows that except for the rudimentary bar the galaxy closely resembles an S0 galaxy. The disk component has an exponential light curve with a central surface brightness of $B(0) = 21.65$ per square second of arc and a scale size of 4.3 kpc. The nucleus, bar, and disk components contain 29, 22, and 48 percent, respectively, of the light. Integral and differential colors are presented and discussed in terms of the three components mentioned above.

158.045 Broad Balmer emission lines in radio galaxies.
D. E. Osterbrock, A. T. Koski, M. M. Phillips.
Astrophys. Journ., (*Letters*), Vol. 197, L41 - L44 = Lick Obs. Bull., No. 696 (1975).

The radio galaxies 3C 382, 3C 390.3 and 3C 445 have very broad Balmer emission lines, with irregular profiles that are probably due to massive motions in the ionized gas. Detailed spectrophotometric results for 3C 390.3 show that its forbidden-line spectrum is similar to that of Cyg A, except that the mean density is higher in 3C 390.3. The density is much higher ($N_e \gtrsim 10^{8.5}$ cm^{-3}) in the broad Balmer-line region. Approximately one-fourth of the emission-line radio galaxies observed to date in the program have broad Balmer lines. The phenomenon is probably similar to that in Class 1 Seyfert galaxies, but more intensive in the radio galaxies.

158.046 A study of 3000 faint galaxies. R. J. Dodd,
D. H. Morgan, K. Nandy, V. C. Reddish, H. Seddon.
Monthly Notices Roy. Astron. Soc., Vol. 171, 329 - 351 (1975).

This paper reports measurements of the positions, sizes, shapes and orientations of 3000 images of galaxies in an area of 2 (arcdeg)^2 near the south galactic pole. The measurements extend to large numbers of images as small as 3.3 arcsec diameter. The data are analysed to examine such matters as the possible existence of intergalactic dust clouds, the clustering and pairing of the galaxies, the ratio of spirals to ellipticals and the number-diameter relationship.

158.047 On the possibility of radioastronomical investigation of the birth of galaxies.
R. A. Sunyaev (*Syunyaev*), Ya. B. Zel'dovich.
Monthly Notices Roy. Astron. Soc., Vol. 171, 375 - 379 (1975).

During the prestellar epoch, protogalaxies were composed mainly of neutral hydrogen with mass approximately 100 times the mass of interstellar gas in galaxies today. Protogalaxies and protoclusters of galaxies therefore radiated strongly in the 21-cm line of neutral hydrogen. Due to cosmological redshift this line will be shifted to the metre wavelength band, so detection of redshifted 21-cm lines from protogalaxies may provide the possibility of finding the epoch of galaxy formation and the properties of protogalaxies.

158.048 Physical conditions in active nuclei – I. The Balmer decrement. H. Netzer.
Monthly Notices Roy. Astron. Soc., Vol. 171, 395 - 405 (1975).

The Balmer decrement in Seyfert galaxies and QSO's is calculated under self-absorption conditions, using a 'mean escape probability' technique. It is found that for densities of over 10^8 cm^{-3} and large optical depth in Hα, absorption from level 3 becomes significant. In such a case a large deviation from previous self-absorption calculations by Capriotti is obtained, and a steep decrement is common in most cases. Results for different optical depths in Lyα and different electron densities are given. Comparison with observational results is made, and a possible photoionization model introduced.

158.049 Photoelectric observations of compact galaxies in the region of the Andromeda nebula.
A. S. Sharov, V. M. Lyutyj, V. F. Esipov.
Astron. Zhurn. Akad. Nauk SSSR, Vol. 52, 282 - 286 (1975). In Russian. English translation in Soviet Astron., Vol. 19, No. 2.

Photoelectric observations of some compact galaxies in the region of the Andromeda nebula M31 are made at the G. A. Shajn telescope of the Crimean Astrophysical Observatory. It is found that systematic differences between the photoelectric and photographic magnitudes of galaxies in the V system lead to the overestimation of $B-V$ color in the photographic photometry. According to the photoelectric measurements some compact galaxies are red objects with color $B-V \sim 1^m2-1^m3$. Measured colors $B-V$ and $U-B$ may be connected with peculiar stellar population of those galaxies.

158.050 The galactic extinction from the apparent distribution of galaxies. B. I. Fesenko.
Astron. Zhurn. Akad. Nauk SSSR, Vol. 52, 287 - 293 (1975). In Russian. English translation in Soviet Astron., Vol. 19, No. 2.

The fluctuations of apparent galaxies distribution are considered, the Lick Observatory counts being used. The clustering effect of galaxies is excluded. The effect of variable observing conditions is taken into account by the statistical method. For the galactic extinction fluctuations the cosecant law is confirmed. Parameters of interstellar clouds are derived. An approximated law for the galactic extinction distribution is derived.

158.051 *UBV* photographic surface photometry of bright galaxies in the neighbourhood of NGC 1068.
A. I. Shapovalova.
Astrofiz. Issled., Izv. Spets. Astrofiz. Obs., Vol. 7, p. 41 - 57 (1975). In Russian.

Results of photographic surface photometry of five bright galaxies ($m_B \leqslant 13^m$) in the neighbourhood of NGC 1068 are presented. Integral stellar magnitudes of the galaxies and their colour indices $U-B$, $B-V$ are determined. Distributions of the surface brightness and colour indices $U-B$, $B-V$ along the major axes are obtained. Main features of the morphology, structure and composition of the galaxies are pointed out.

158.052 *UBV* photometry of the galaxy M 82.
B. P. Artamonov, F. Börngen, P. Notni.
Astrofiz. Issled., Izv. Spets. Astrofiz. Obs., Vol. 7, p. 58 - 64 (1975). In Russian.

UBV photometry is performed of the galaxy M 82 from exposures obtained with the 2-meter Tautenburg Schmidt telescope. The central and outer regions of the galaxy are studied. An effect of asymmetry in the distribution of color indices relative to the major axis is detected, which in a first approximation is explained by different stellar composition and different degree of reddening.

158.053 Equidensitometry of galaxies in the group M 81.
L. G. Antropova, B. P. Artamonov, F. Börngen.
Astrofiz. Issled., Izv. Spets. Astrofiz. Obs., Vol. 7, p. 65 - 69 (1975). In Russian.

Through equidensitometry in four colors — U, B, V, R — the dependence is obtained of the ratios for the visible galactic axes of M 81, M 82, NGC 3077, NGC 2976 on the distance from the center.

158.054 Observations of M 31 and M 33 at 1.4 and 2.7 GHz.
B. Dennison, T. J. Balonek, Y. Terzian, B. Balick.
Publ. Astron. Soc. Pacific, Vol. 87, 83 - 86 (1975).

Radio contour maps of M 31 and M 33 at 1410 MHz and 2695 MHz are presented. At these frequencies most of the radio emission appears to be associated with the nuclear and spiral arm regions, and halo emission is not observed. The radio spectra of these galaxies are discussed.

158.055 Pairs consisting of one Markarian and one normal galaxy.
C. Casini, J. Heidmann.
Astron. Astrophys., Vol. 39, 127 - 129 (1975).

The authors show statistically that there are close physical pairs made up of one Markarian and one normal galaxy. The distribution of the separations of two such components is less concentrated than for Markarian-Markarian pairs.

158.056 An atlas of dust and H II regions in galaxies.
B. T. Lynds.
Astrophys. Journ., Suppl. Ser., No. 267, Vol. 28, 391 - 396 (1974).

Image-tube photographs of 41 galaxies are reproduced as an atlas. On the basis of photographs taken in the light of Hα and in the red continuum, H II regions are identified and plotted on a sketch which also shows the distribution of dust in the galaxy. Detailed comments are made for each galaxy.

158.057 The radio emission from the nucleus of M82.
P. P. Kronberg.
Journ. Roy. Astron. Soc. Canada, Vol. 69, 42 (1975). — Abstr. Canadian Astron. Soc.

158.058 The classification of normal galaxies.
S. van den Bergh.
Journ. Roy. Astron. Soc. Canada, Vol. 69, 57 - 76 (1975).

This paper presents a review of galaxy classification systems. Special emphasis is placed on the relations between galaxy classification, galactic evolution and stellar content of galaxies.

158.059 Projection des galaxies Sc et des radiosources quasistellaires (QSS) sur les plans de coordonnées supergalactiques.
H. Karoji.
Comptes Rendus Acad. Sci. Paris, Sér. B, Vol. 280, 455 - 457 (1975).

Les Sc et QSS sont portés sur les plans Y-Z, X-Y des coordonnées supergalactiques de de Vaucouleurs. Comme dans les histogrammes, les distributions dans l'espace des Sc et des QSS «locaux» se ressemblent bien.

158.060 Binary galaxy orbit statistics. I. Fixed mass and major axis.
P. D. Noerdlinger.
Astrophys. Journ., Vol. 197, 545 - 550 (1975).

Averages for the projected separation ρ, squared radial velocity difference $\dot z^2$, and the product of these $\rho \dot z^2$, are presented for a binary galaxy system of fixed mass and major axis,

but any orbital eccentricity ϵ. The average of $\rho \dot z^2$ varies by a factor ~ 3 in the range $0 < \epsilon < 1.0$. The mutual regressions of the velocity and separation on each other are calculated, and are presented in such a way as to exhibit the relative likelihood of occupation of the different parts of the regression curves. "Isopleths" for the probability distribution in $(\rho, |\dot z|)$-space are presented for a few values of the eccentricity. It is concluded that previous analyses were inadequate for failing to take into account that the regression curves represent in many cases a distribution that is almost wholly depopulated through most of the range.

158.061 Sterrenstelsels als M33 lijken (en zijn) roterende formaties.
M. Drummen.
Zenit, Vol. 2, 19 - 22 (1975).

158.062 An early photograph of Fornax A.
P. W. Hodge.
Sky Telescope, Vol. 49, 354 (1975).

158.063 A remarkable regularity in galaxy systems; its dynamical and cosmogonical significance.
L. M. Ozernoj.
AN SSSR. Fiz. in-t. Otd. teor. fiz. Preprint No. 124. Moskva, 1974. 26 pp. In Russian. — Abstr. in Referativ. Zhurn. 51. Astron., 3.51.796 (1975).

158.064 On a possibility of radio astronomical study of the generation of galaxies.
Ya. B. Zel'dovich, R. A. Syunyaev.
In-t prikl. mat. AN SSSR. Preprint No. 94. Moskva, 1974. 9 pp. In Russian. — Abstr. in Referativ. Zhurn. 51. Astron., 3.51.820 (1975).

158.065 The mass-angular momentum relation of normal spiral galaxies.
T. Aikawa, Z. Hitotuyanagi.
Sci. Rep. Tôhoku Univ., First Ser., Vol. 57, 121 - 130 (1974).

The mass-angular momentum relation is rediscussed of normal spiral galaxies by applying the Brandt model. For twenty-six galaxies, masses and angular momenta are determined from optical rotation curves. A power law $P \propto M^{7/4}$ thus obtained is shown to be nearly independent of the parameters of the model. Previous arguments against the Brandt model are briefly discussed.

158.066 3C 411: a newly discovered N galaxy with a large redshift.
H. Spinrad, H. E. Smith, R. Hunstead, M. Ryle.
Astrophys. Journ., Vol. 198, 7 - 11 (1975).

3C 411 (PKS 2019+09), normal extragalactic radio source at low galactic latitude, is identified with a variable, 20th magnitude N galaxy. It has a rich emission-line spectrum yielding a redshift $z = 0.469$. The authors discuss quantitative spectroscopic criteria which tend to place 3C 411 between QSOs and normal radio galaxies, similar to the nearer N systems. A high-resolution synthesis radio map at 5 GHz shows a double-lobed source with a low-brightness connecting bridge, and a nuclear source centered over the optical object.

158.067 Empirical evidence for galaxy evolution.
P. Crane.
Astrophys. Journ., (Letters), Vol. 198, L9 - L12 (1975).

An analysis of the spectral energy distributions of distant galaxies suggests that galaxy evolution has been detected as a redshift-dependent change in the rest-frame spectra. The measured rate of color evolution agrees with previous limits, and is in qualitative agreement with synthesized galaxy models.

158.068 Radio variations in a normal galaxy.
R. A. Sramek.

Astrophys. Journ., (*Letters*), Vol. 198, L13 - L15 (1975).

The radio radiation from M89 (NGC 4552) was found to vary at centimeter wavelengths on a time scale of several years. These variations are consistent with an expanding, self-absorbing synchroton source, and are very similar to the radio variations seen in quasars. The velocity of the expanding cloud must be $v \sim 0.005c$; this low velocity is probably due to confinement of the cloud by interstellar matter.

158.069 Comet-like radio galaxies. V. N. Kuril'chik.
Zemlya i Vselennaya, 1975, No. 2, p. 29 - 32.
In Russian.

158.070 Rejection of the hypothesis on disintegration of galaxy groups due to continuous mass loss.
L. M. Ozernoj.
Pis'ma v Astron. Zhurn., Vol. 1, No. 2, p. 9 - 14 (1975).
In Russian.

Some expected properties of galaxy systems, following from the hypothesis of their instability due to continuous or quasi-continuous mass loss, are analyzed. It is shown that these properties are not confirmed by observational data on groups of galaxies. This is regarded as an argument against the instability of galaxy groups due to mass loss.

158.071 On the nature of coronas around spiral galaxies.
B. V. Komberg, I. D. Novikov.
Pis'ma v Astron. Zhurn., Vol. 1, No. 3, p. 3 - 7 (1975).
In Russian.

The hypothesis that massive coronas around spiral galaxies are composed of hot ionized gas is likely to contradict to observations of the background and M31 in soft X-rays and to the estimate of ultraviolet emission from the corona, based on the fact that neutral hydrogen is present on the periphery of M31.

158.072 On the emission line variability in nuclei of Seyfert galaxies and quasars.
N. G. Bochkarev, S. P. Pudenko.
Pis'ma v Astron. Zhurn., Vol. 1, No. 3, p. 12 - 17 (1975).
In Russian.

As calculations show, the optical emission spectrum and emission line variability of Seyfert galaxies and quasars nuclei can be explained by heating and ionization of gas by variable 100−500 eV X-rays. The required X-ray intensity agrees with observational data.

158.073 The relative abundance of nitrogen and sulphur in three spiral galaxies: M 33, M 101, M 51.
G. Comte.
Astron. Astrophys., Vol. 39, 197 - 205 (1975).

The author presents photographic spectrophotometric surveys of the forbidden lines of [N II] λλ 6548, 6584 and [S II] λλ 6717, 6731 in H II regions of three galaxies: M 33, M 101 and M 51. Since under normal excitation conditions, these lines are emitted in common volume elements of the interstellar ionized gas, the intensity ratio I ([N II])/([S II]), which is practically electron temperature independent and reddening-free, is an indicator of the chemical abundance of nitrogen relative to sulphur.

158.074 A search for water vapour emission from extragalactic nebulae.
B. H. Andrew, M. B. Bell, N. W. Broten, J. M. MacLeod.
Astron. Astrophys., Vol. 39, 421 - 428 (1975).

The authors have searched for water vapour emission at 132 independent positions in seven external galaxies. No emission was detected, with the possible exception of a line seen one night out of four at the position of the nucleus of M 82.

158.075 Photoelectric B and V photometry of ScI galaxies.
J. A. Graham.
Bull. American Astron. Soc., Vol. 7, 253 (1975). − Abstr. AAS.

158.076 Photographic and photoelectric photometry of M51.
M. S. Burkhead, J. K. Kalinowski.
Bull. American Astron. Soc., Vol. 7, 253 (1975). − Abstr. AAS.

158.077 Dust in dwarf ellipticals? W. P. Bidelman.
Bull. American Astron. Soc., Vol. 7, 253 (1975).
Abstr. AAS.

158.078 Detection of extragalactic carbon monoxide at 2.6 mm. L. J. Rickard, P. Palmer, M. Morris, B. Zuckerman, B. E. Turner.
Bull. American Astron. Soc., Vol. 7, 253 - 254 (1975).
Abstr. AAS.

158.079 Rotation of the barred spiral galaxy NGC 7723.
R. A. Chevalier.
Bull. American Astron. Soc., Vol. 7, 254 (1975). − Abstr. AAS.

158.080 Rotation of the nuclear region of M31.
C. J. Peterson.
Bull. American Astron. Soc., Vol. 7, 254 (1975). − Abstr. AAS.

158.081 OAO-2 observations of M31 and M33. J. E. Davis.
Bull. American Astron. Soc., Vol. 7, 254 (1975).
Abstr. AAS.

158.082 cD galaxies in poor clusters.
W. W. Morgan, S. E. Kayser, R. A. White.
Bull. American Astron. Soc., Vol. 7, 254 (1975). − Abstr. AAS.

158.083 The fate of gas in elliptical galaxies and the density evolution of radio sources. G. R. Gisler.
Bull. American Astron. Soc., Vol. 7, 263 (1975). − Abstr. AAS.

158.084 Detection of Fe II emission in the Seyfert galaxy NGC 7469. G. M. MacAlpine, S. A. Shectman.
Bull. American Astron. Soc., Vol. 7, 268 (1975). − Abstr. AAS.

158.085 Physical conditions in Seyfert galaxy nuclei.
R. Ptak, R. Stoner.
Bull. American Astron. Soc., Vol. 7, 268 - 269 (1975).
Abstr. AAS.

158.086 Red halos around spiral galaxies? S. M. Simkin.
Bull. American Astron. Soc., Vol. 7, 339 (1975).
Abstr. AAS.

158.087 Variation d'éclat du noyau de 3C 371.
L. A. Ourassine (*L. A. Urasin*).
Ann. Obs. Astron. Alger, Vol. 4, Fasc. 1, p. 15 - 17 (1974).

158.088 Concealed mass in galaxies.
J. E. Einasto, A. D. Chernin, M. M. Jõeveer.
Priroda, No. 5.75, p. 39 - 43 (1975). In Russian.

158.089 UBVRI surface photometry of four edge-on Sb galaxies. S. M. Simkin.
Dudley Obs. Rep. No. 9, (see 012.008), p. 401 - 406 (1975).
UBVRI surface photometry of NGC 2683, 4192, 4216, and 5746 based on direct plates calibrated with photoelectric sequences in the galaxies show: (1) The "halo" colors for these objects are notably bluer than the nuclear colors. (2) There is a tendency towards redder colors in the outer regions of NGC 4216 but not in the other three objects.

158.090 Absolute spectrophotometry in M31 and M32.

J. B. Oke, M. Schwarzschild.
Astrophys. Journ., Vol. 198, 63 - 70 (1975).

For a number of places in the bulge of M31, and for two places in M32, photometric scans from $\lambda = 3300$ Å to $\lambda = 10,600$ Å have been obtained. The scans show that in both objects the color temperature (particularly shortward of 5000 Å) decreases toward the center, and that the strength of the CN bands increases toward the center in both objects, in agreement with other, earlier observations. The new data can all be interpreted in terms of an increase of heavy-element abundance toward the center in both objects by a factor probably less than 2, and by an excess of heavy elements in M31 compared with M32 by a factor probably greater than 2, in qualitative agreement with earlier conclusions of other observers.

158.091 The broad component of hydrogen emission lines in nuclei of Seyfert galaxies: comments on a charge exchange model. A. Katz.
Astrophys. Journ., Vol. 198, 255 - 260 (1975).

A model to account for the broad hydrogen line emission from the nuclei of Seyfert galaxies based on charge exchange and collisional processes, as proposed by Ptak and Stoner, is investigated. The model consists of a source of fast ($E \sim 10^5$ eV) protons streaming through a medium of quiescent gas. One of the major problems that results from such a model concerns the strong narrow hydrogen line core that would be produced, in direct conflict with the observations. The energy balance and energy requirements of such a model are investigated, and it is found that an energy equal to or greater than the total luminosity of most Seyfert galaxies is required to produce the hydrogen lines alone. The gas must be mostly neutral and dense ($N \sim 10^7$) if a reasonable temperature is to be maintained.

158.092 Oddballs and galaxy formation. J. Binney.
Nature, Vol. 255, 275 - 276 (1975).

158.093 How to study the birth of galaxies using radio-astronomical techniques. J. Gribbin.
Nature, Vol. 255, 276 (1975).

158.094 Normal redshifts in Markarian galaxies.
J. Huchra, with a reply by P. Teerikorpi.
Nature, Vol. 255, 430 (1975).

158.095 Variability and circular polarisation in the nucleus of NGC 5128 at 10.7 GHz.
K. M. Price, M. A. Stull.
Nature, Vol. 255, 467 - 468 (1975).

The galaxy NGC 5128 (Centaurus A), has been found to contain a compact radio source in its nucleus. The results which the authors report place important constraints on its structure and magnetic field strength. The nucleus of NGC 5128 is probably variable at 10.7 GHz; however, the time scale for this variability may be significantly longer than that reported at millimetre wavelengths. The authors failed to detect any circularly polarised flux and from the fluctuations of the measured crossed-horn visibilities they calculate a 5σ upper limit of 2.5% to the fraction which might be present.

158.096 Galaxies with hot spots in central regions. II.
L. S. Nazarova.
Kazan. un-t. Kazan', 1974. 35 pp. In Russian. – Abstr. in Referativ. Zhurn. 51. Astron., 5.51.739 (1975).

158.097 Hot spots in central regions of galaxies. III.
L. S. Nazarova.
Kazan. un-t. Kazan', 1974. 38 pp. In Russian. – Abstr. in Referativ. Zhurn. 51. Astron., 5.51.740 (1975).

158.098 Photoelectric photometry of star clusters in the Andromeda nebula.
A. S. Sharov, V. M. Lyutyj, V. F. Esipov.
Pis'ma v Astron. Zhurn., Vol. 1, No. 4, p. 3 - 7 (1975). In Russian.

The results of photoelectric UBV observations of 44 star clusters in M 31 are presented.

158.099 Nonthermal continuum radiation in three elliptical galaxies. M.-H. Ulrich, T. D. Kinman, C. R. Lynds, G. H. Rieke, R. D. Ekers.
Astrophys. Journ., Vol. 198, 261 - 266 (1975).

Optical, infrared, and radio observations are presented for three elliptical galaxies (NGC 6454, B2 1652 + 39, and B2 1101 + 38) which have flat high-frequency radio spectra. Two of these galaxies emit optical nonthermal continuum radiation, as evidenced by linear polarization of the ultraviolet radiation of the nuclei. No emission lines were detected in the optical spectra of the galaxies, and in that respect they may be related to the BL Lacertae class.

158.100 Observations of M82 and NGC 253 at 8–13 microns.
F. C. Gillett, D. E. Kleinmann, E. L. Wright, R. W. Capps.
Astrophys. Journ., (Letters), Vol. 198, L65 - L68 (1975).

Observations of the 8–13 μ spectrum ($\Delta\lambda/\lambda = 0.02$) of the nuclear regions of M82 and NGC 253 are presented. The emission features seen in both spectra are very similar to those of the planetary nebula BD +30°3639. This similarity supports a thermal reradiation interpretation of the infrared sources in M82 and NGC 253. In addition, there is a strong, broad depression around 10 μ. The bulk of this depression can be identified with "silicate" grain absorption.

158.101 Photométrie photographique de 3C 120.
C. Bertaud, B. Dumortier, C. Pollas.
Inform. Bull. Variable Stars, (I.A.U. Commission 27), Konkoly Obs., Budapest, No. 970, 2 pp. (1975).

158.102 Observations photographiques de 3C 120 (BW Tau).
L. A. Ourassine (Urasin), I. A. Ourassina (Urasina).
Inform. Bull. Variable Stars, (I.A.U. Commission 27), Konkoly Obs., Budapest, No. 973 (1975).

158.103 Intensity profiles and color distributions in the non-nuclear regions of Sb galaxies. S. M. Simkin.
Astron. Journ., Vol. 80, 415 - 426, 473 - 475 (1975).

Intensity profiles, isophotes, and color distributions for a sample of four nearby Sb galaxies indicate: (1) The "exponential disk" in this type of object is uniform in color out to a radius of ~ 15 kpc, where it appears to either change abruptly or merge into the halo. (2) The color indices and outer isophotes of the regions beyond this "merger point" are consistent with a metal-poor "halo" population enriched with BHB stars and possibly some M dwarfs. (3) Observed "red excesses" in the outer regions of one of these spiral galaxies may conceivably be explained by tidal interaction with nearby galaxies.

158.104 An H I study of Scd galaxies.
G. S. Shostak.
Astrophys. Journ., Vol. 198, 527 - 536 (1975).

H I line profiles, together with derived redshifts and integral properties, are presented for 46 Scd galaxies. The profile shapes are quite regular, with a double-peaked structure characteristic of galaxies with flat rotation curves. The ratios M_{HI}/M_T and M_{HI}/L_{pg}, which are known to vary with Hubble type, have observed dispersions of ± 80 percent within this classification, with intrinsic dispersions being probably less than ± 50 percent. This establishes that integral property ratios are substantially homogeneous within the Scd classification. No indication of H I optical-depth effects with inclination is observed. Evidence is presented that distances derived solely from redshifts have an average uncertainty of <0.8 mag in the distance modulus.

158.105 The corona of the giant spiral galaxy NGC 253.
K. C. Freeman, D. W. Carrick, J. L. Craft.
Astrophys. Journ., (Letters), Vol. 198, L93 - L96 (1975).

Giant spirals may have coronae with masses of order 10^{12} to $10^{13} M_\odot$ and radii up to about 1 Mpc. Deep surface photometry of the giant spiral NGC 253 shows that the total photographic luminosity L of its corona is unlikely to exceed 3×10^9 $L_\odot$. If its corona is typical of those postulated by Ostriker et al., then its M/L ratio exceeds 600. If the picture proposed by Einasto et al. is more appropriate, then its M/L exceeds 5800.

158.106 Les galaxies de Markarian et l'anisotropie angulaire de la «constante» de Hubble.
H. Karoji, M. Moles.
Comptes Rendus Acad. Sci. Paris, Sér. B, Vol. 280, 609 - 612 (1975).

L'ensemble observé des galaxies de Markarian permet de disposer d'un échantillon de sources assez nombreuses pour étudier leurs déplacements vers le rouge. L'examen de leur répartition sur le fond du ciel permet de confirmer de façon significative l'anisotropie de la «constante» de Hubble observée pour la première fois par Rubin, Rubin et Ford; anisotropie récemment corroborée par Le Denmat et Vigier à l'aide de supernovae de type I.

158.107 Study of principal catalogues of diameters of galaxies. G. Paturel.
Astron. Astrophys., Vol. 40, 133 - 139 (1975). In French.

The author calculates the limiting surface brightness defining the optical diameters of galaxies in the catalogues by Nilson and Vorontsov - Vel'yaminov based on the limiting surface brightness given by Holmberg and de Vaucouleurs. The internal dispersions in the surface brightness are given. Grouping the galaxies in two classes of morphological types, he derives statistical relations between the diameter systems. He then obtains the internal dispersions of the diameters for Holmberg's, Vorontsov - Vel'yaminov's, Nilson and de Vaucouleurs' catalogues. Finally, he justifies the correction due to the inclination effect on the apparent major axis and calculates the coefficient used in this correction for the different catalogues.

158.108 The structure of groups of galaxies.
J. Einasto, A. Kaasik, P. Kalamees, J. Vennik.
Astron. Astrophys., Vol. 40, 161 - 165 (1975).

Dwarf galaxies and hidden matter are located around giant galaxies, forming gravitationally bound systems. Giant galaxies form a multiplet at the centre, dwarf galaxies are distributed over the whole volume of the system. Total masses of groups of galaxies range from 10^{12} to 3×10^{13} solar masses, the limiting radii being close to $1-2$ Mpc. The densities of both the hidden matter and the system of galaxies decrease with distance from the centre according to the inverse square law. Dwarf galaxies of different morphological types are separated from each other, elliptical galaxies are located in the inner regions, while non-elliptical galaxies populate the peripheral ones.

158.109 Radio tail source associated with NGC 7385.
R. T. Schilizzi, R. D. Ekers.
Astron. Astrophys., Vol. 40, 221 - 224 (1975).

Observations of 2247 + 11, a further example of the "head-tail" class of radio source, are reported. The emission is associated with NGC 7385, a member of cluster Zwicky 2247.3 + 1107. The brightness distributions at 408 and 1415 MHz and the spectral index variations along the tail are discussed. Based on the magnetospheric model for radio tail

sources of Jaffe and Perola, the density of the intra-cluster gas in this region is $\sim 10^{-27}$ gm cm^{-3} and its temperature is $\sim 10^6$ K.

158.110 A group of galaxies containing five Markarian galaxies.
I. D. Karachentsev, V. E. Karachentseva.
Pis'ma v Astron. Zhurn., Vol. 1, No. 5, p. 3 - 6 (1975). In Russian.

A scattered group of galaxies (9^h43^m, $+32°25'$) is noteworthy because 6 of its 14 members with measured radial velocities are blue dwarf galaxies. The mean radial velocity of group members is +1465 km/s with a dispersion of 100 km/s, the mean linear radius is 0.90 Mpc, the virial mass-to-luminosity ratio is $(200-270) M_\odot/L_\odot$. It is suggested that a considerable number of galaxies regarded usually as "background" may belong to such scattered groups.

158.111 "Hidden" mass in galaxies. J. Einasto.
Zemlya i Vselennaya, 1975, No. 3, p. 32 - 36. In Russian.

158.112 Results of observations of Seyfert galaxies.
B. A. Vorontsov-Vel'yaminov, G. Ivanišević.
Soobshch. Gos. Astron. Inst. Shternberga, No. 189, p. 3 - 20 (1974). In Russian.

The list of objects presented aims to facilitate the observations that should be made in order to complete the data (colour, shape and spectra) of Seyfert galaxies, and on a homogeneous scale.

158.113 On possible systematic redshifts across the disks of galaxies.
T. Jaakkola, P. Teerikorpi, K. J. Donner.
Astron. Astrophys., Vol. 40, 257 - 266 (1975).

Velocity observations in 25 galaxies have been examined for possible systematic redshifts across their disks: a possible origin for the redshifts could be the radiation fields. Velocities increase towards the far side in most cases. Deviation of the kinematic major axis from the optical axis is found for 10 galaxies and in 9 of these the largest velocities occur in the far side. In the central regions of four galaxies are found large velocity gradients in the same direction.

158.114 Spectroscopic observations of extremely compact galaxies. M.-H. Ulrich.
Astron. Astrophys., Vol. 40, 337 - 338 (1975).

Low-dispersion spectrograms have been obtained for 22 galaxies noted as "extremely compact" in the Catalogue of Galaxies and of Clusters of Galaxies published by Zwicky and collaborators and radial velocities are given for 14 of them. Ten galaxies have emission lines spectra and one of them (0934+01) is a new Seyfert-type galaxy with broad hydrogen lines and narrow forbidden lines. K and H lines at nearly zero velocity are seen on the spectra of two galaxies and presumably originate from galactic stars projected on the galaxy nuclei.

158.115 The nuclear region of NGC 1097. J. J. Rickard.
Astron. Astrophys., Vol. 40, 339 - 341 (1975).

Image tube photographs of the barred-spiral galaxy NGC 1097 reveal a tightly wound spiral structure in its nucleus. Dust lanes extend from within 350 pc of the nucleus out to the full extend of the bar. Spectral observations are used to redetermine the line of nodes, rotation velocity, and mass of the nuclear region.

158.116 Request for visual observations of Seyfert galaxies.
R. J. Weymann.
Journ. American Ass. Variable Star Observers, Vol. 1, 16 - 17 (1972).

158.117 Determination of the distances of the nearest

galaxies by method of parallaxes of eclipsing
binaries.   T. Z. Dworak.
Acta Cosmologica, Fasc. 2, p. 13 - 19 (1974/75).

In this paper the distances of the three nearest galaxies
(Small Magellanic Cloud, Large Magellanic Cloud and Andromedae nebula) are computed by using Gaposchkin's method
of determination of parallaxes of eclipsing binaries, modified
by the writer. The method may be used as an independent
method of determining distances to nearby galaxies.

**158.118   The color gradient in spiral galaxies: application to
M81.   A. Segalovitz.**
Astron. Astrophys., Vol. 40, 401 - 404 (1975).

The calculated development of the color of a star cluster
(Searle et al., 1973) is used to predict the expected color
evolution, as a function of radius, in a spiral galaxy. It is assumed that the fraction of gas which is converted into stars
during a spiral arm passage (consumption fraction) is a function of radius only. Applying this model to M81, it is shown
that the observed color and mass distributions can be explained by an initial disk-like gas distribution proportional to
the inverse square of the radius and a consumption fraction
which is an increasing function of radius.

**158.119   The giant spiral galaxy M101: radio observations
of H II regions in external galaxies. II. Radio continuum emission from the H II regions and the nonthermal
disc of M101.   F. P. Israel, W. M. Goss, R. J. Allen.**
Astron. Astrophys., Vol. 40, 421 - 438 (1975).

The authors have used the Westerbork Synthesis Radio
Telescope to observe the galaxy M101 in the continuum near
$\lambda\lambda$ 6, 21 and 49 cm. The results show: 1.) The flux density
spectral indices of the radiation from several individual positions in the galaxy have a mean value of -0.4 between λ 21
and λ 49 cm, similar to the integrated spectral index of M101.
2.) Six H II region complexes have been detected with certainty. 3.) For some of the detected H II regions, at least 50%
of the radio flux density comes from the bright dense cores
seen on short-exposure optical photographs. 4.) The cores
have masses of the order of several $\times 10^6 M_\odot$; the envelopes
have masses less than of the order of $10^7 - 10^8 M_\odot$. 5.) Several
hundred early-type stars are needed for the ionization of the
core components. 6.) Comparison with optical observations
suggest the presence of large amounts of internal dust in these
H II region complexes, producing on average more than 1^m
internal extinction. 7.) The radio luminosity distribution
function of the H II regions in M101 is shown to differ substantially in shape from that of M 33.

**158.120   Variability of the emission line spectrum of the
nucleus of the Seyfert galaxy NGC 7469.**
I. I. Pronik.
Astron. Zhurn. Akad. Nauk SSSR, Vol. 52, 481 - 490 (1975).
In Russian. English translation in Soviet Astron., Vol. 19,
No. 3.

It is shown that the intensities and profiles of emission
Balmer lines of the gaseous envelope of the nucleus of NGC
7469 are variable within about 20 days. It is shown also that
the hydrogen envelope of the nucleus of NGC 7469 is inhomogeneous in density. Some of the physical parameters of the
inner dense hydrogen envelope were estimated.

158.121   An interacting pair of galaxies NGC 5194-5.
B. A. Vorontsov-Vel'yaminov.
Astron. Zhurn. Akad. Nauk SSSR, Vol. 52, 491 - 497 (1975).
In Russian. English translation in Soviet Astron., Vol. 19,
No. 3.

The author shows that NGC 5195 is an active young
galaxy of SB 0/a type with a hot nucleus, is an infrared and
radio emitter, compact and very rich with dust. It has a
luminosity $\leqslant -20$ and a mass possibly larger than NGC 5194.

**158.122   An investigation of predominant orientation of
galaxies in pairs and in scattered groups.**
I. D. Karachentsev, B. I. Fesenko.
Astron. Zhurn. Akad. Nauk SSSR, Vol. 52, 659 - 662 (1975).
In Russian. English translation in Soviet Astron., Vol. 19, No. 3.

For isolated pairs of galaxies in the catalogue of Karachentsev (1972) the distribution on the position angles and
in the direction of twisting of spiral structure doesn't show
signs of predominant mutual orientation. There is a slight
positive correlation between apparent ellipticities of galaxies
in pairs and in scattered groups that may indicate a correlation
of true ellipticities of members of systems.

158.123   Spatial distribution of galaxies.   W. Zonn.
Urania Kraków, Vol. 46, 2 - 7 (1975).   In Polish.

**158.124   Optical investigations of galactic nuclei.
K. Fricke.**
Conference on optical observing programs on galactic structure and dynamics, (see 012.013), p. 97 - 103 (1975). In German.

**158.125   A note on the velocity-distance relationship for
nearby galaxies and galaxy groups.**
P. Teerikorpi.
Observatory, Vol. 95, 105 - 107 (1975). – Note.

158.126   The gas-to-dust ratio in M31.   S. van den Bergh.
Astron. Astrophys., Vol. 41, 53 - 54 (1975).

The mean gas-to-dust ratio in M31 is similar to that observed near the sun. No evidence is found for a radial gradient
in the gas-to-dust ratio within the disk of the Andromeda
nebula.

**158.127   Optical and neutral hydrogen study of Markarian
galaxies.**
L. Bottinelli, R. Duflot, L. Gouguenheim, J. Heidmann.
Astron. Astrophys., Vol. 41, 61 - 69 (1975).

Nine non-Seyfert Markarian galaxies have been observed
in the 21-cm line of neutral hydrogen. The authors confirm
that this class of Markarian galaxies is not homogeneous,
neither from the point of view of the integral properties nor
from the optical line intensities. Using their integral properties,
a morphological type can be ascribed to seven of them,
ranging from early spiral to irregular, if they are assumed to
be overluminous; however, for two of them no morphological
type is acceptable. The dynamical properties of two close pairs
of Markarian galaxies are investigated: Ma 7–8 is unstable and
Ma 325–326 is bound.

**158.128   Continuum observations of M 81 at 4.8 GHz.
A. von Kap-herr, B. B. Jones, R. Wielebinski.**
Astron. Astrophys., Vol. 41, 115 - 117 (1975).

The galaxy NGC 3031 (M 81) has been mapped at the
wavelength of 6 cm using the 100-m radio telescope of the
Max-Planck-Institut für Radioastronomie. At this wavelength
the beamwidth is 2.'6 which gives sufficient resolution to see
some of the structure of the galaxy. The authors' map is compared with a similar 21 cm map of M81 to investigate the
spectral index and hence the contribution of H II regions.

**158.129   Elliptical galaxies with compact nuclei.
M. K. V. Bappu, M. Parthasarathy.**
Bull. Astron. Soc. India, Vol. 2, 37 (1974). – Abstract.

**158.130   M81: voor amateurs en Westerbork.
M. Drummen.**
Zenit, Vol. 2, 217 - 222 (1975).

**158.131   Drie dwergsterrenstelsels bij M31 ontdekt.
G. W. E. Beekman.**

Zenit, Vol. 2, 235 (1975).

158.132 Observational problems of high energy astrophysics.
E. M. Burbidge, with notes by R. Chevalier,
M. Rosenberg, T. Thuan.
High energy astrophysics and its relation to elementary particle physics, (see 012.018), p. 109 - 172 (1974).

158.133 Structure and dynamics of galaxies.
K. H. Prendergast, with notes by K. Brecher,
G. Cavallo.
High energy astrophysics and its relation to elementary particle physics, (see 012.018), p. 415 - 439 (1974).

158.134 Extragalactic observational astronomy.
W. L. W. Sargent, with notes by J. Kormendy,
G. Cavallo.
High energy astrophysics and its relation to elementary particle physics, (see 012.018), p. 453 - 518 (1974).

158.135 Redshifts as distance indicators. J. N. Bahcall.
The redshift controversy, (see 003.021), p. 61 - 121,
with a response of H. Arp, p. 123 - 129 (1973).

158.136 UBV surface photometry of M33 and M101.
V. J. Lee.
Thesis, Indiana Univ., Bloomington (USA). 228 pp. University Microfilms Order No. 74-9433 (1973).

158.137 Study of possible models for Balmer emission line regions in Seyfert galaxies. W. M. Adams.
Thesis, Stanford Univ., Stanford, Calif.. (USA). 98 pp. University Microfilms Order No. 74-13,594 (1974).

158.138 Distribution and kinematics of neutral hydrogen in the spiral galaxy M81. A. H. Rots.
Proefschrift, Subfaculteit Sterrenkunde, Rijksuniversiteit Groningen (Netherlands). 69 pp. [This thesis is available from Rijksuniversiteit, Groningen] (1974).

Morphological catalogue of galaxies. V. Catalogue of 1637 galaxies from $-33°$ to $-45°$ declination.
See Abstr. 003.016.

The simultaneous 'discovery' of internal motions in spiral nebulae. See Abstr. 004.072.

On the nature of γ-ray bursts.
See Abstr. 061.036.

On the intrinsic properties of cepheids in the Galaxy, in Andromeda, and in the Magellanic Clouds.
See Abstr. 122.020.

Spectra and colour indices of bright irregular variables in M31 and M33. See Abstr. 122.046.

Spectra of AE And and the Hubble-Sandage variables in M31 and M33. See Abstr. 122.051.

Interstellar absorption and an apparent anisotropy in the Hubble expansion. See Abstr. 131.007.

On cosmic rays and final equilibrium states for the Parker instability. See Abstr. 131.127.

Optical monitoring of radio sources — IV. (Results up to 1973 April). See Abstr. 141.001.

High frequency radio observations of the Stephan's Quintet region. See Abstr. 141.007.

Lyman-β and Fe II emission from QSOs and peculiar galaxies. See Abstr. 141.033.

A complete sample of radio sources identified with elliptical galaxies: radio luminosity function and other properties. See Abstr. 141.038.

Massive black holes in extragalactic radio source components? See Abstr. 141.041.

Neutrino processes and QSOs.
See Abstr. 141.043.

The optical emission-line spectrum of Cygnus A.
See Abstr. 141.050.

Milli-arcsecond structure of 3C 84, 3C 273, and 3C 279 at 2 centimeter wavelength. See Abstr. 141.053.

An interpretation of the redshift problem of quasar-galaxy pairs. See Abstr. 141.055.

Time-dependent effects in the radially streaming particle model. See Abstr. 141.059.

The nature of quasars. See Abstr. 141.062.

Two new quasars near galaxies.
See Abstr. 141.063.

The orientation of double radio sources associated with elliptical galaxies. See Abstr. 141.064.

Instationary objects of non-stellar nature.
See Abstr. 141.076.

Optical condensations and filaments in the northeast radio lobe of NGC 5128. See Abstr. 141.083.

Theoretical emission line profiles in QSOs and Seyfert galaxies. See Abstr. 141.092.

A multiple pulsar model for quasi-stellar objects and active galactic nuclei. See Abstr. 141.093.

Statistical acceleration of relativistic particles in an assembly of spherical electromagnetic waves.
See Abstr. 141.094.

QSO's, observations, and the redshift problem.
See Abstr. 141.100.

Theoretical problems of high energy astrophysics.
See.Abstr. 141.101.

Evidence for discordant redshifts.
See Abstr. 141.102.

Copernicus: X-ray observations of 3C 390.3.
See Abstr. 142.030.

Soft X-ray spectroscopy of three extragalactic sources. See Abstr. 142.043.

The ejection of massive objects from galactic nuclei: gravitational scattering of the object by the nucleus.
See Abstr. 151.004.

Dynamics and evolution of galaxies.
See Abstr. 151.005.

Density wave theory and the classification of spiral galaxies. See Abstr. 151.008.

Dynamical models of elliptical galaxies. See Abstr. 151.010.

About the structure of spherical galaxies and of clusters of galaxies. See Abstr. 151.033.

The Milky Way in comparison to other galaxies. See Abstr. 155.044.

Radio-astronomical observations and optical structure. See Abstr. 155.045.

Radiocontinuum emission of the Galaxy compared with other galaxies. See Abstr. 157.011.

Nuclear magnitudes for galaxies in the Coma and Perseus clusters. See Abstr. 160.002.

Study of three southern groups of galaxies. See Abstr. 160.005.

Binary galaxies and the problem of masses of clusters of galaxies. See Abstr. 160.008.

Possible systematic magnitude error in the Zwicky Catalogue of Galaxies. See Abstr. 160.029.

Spectrophotometry of faint cluster galaxies and the Hubble diagram: an approach to cosmology. See Abstr. 162.005.

Steps toward the Hubble constant. V. The Hubble constant from nearby galaxies and the regularity of the local velocity field. See Abstr. 162.023.

Steps toward the Hubble constant. VI. The Hubble constant determined from redshifts and magnitudes of remote Sc I galaxies: the value of q_0. See Abstr. 162.031.

Peculiar Objects

158.301 The nature of BL Lacertae. J. A. Baldwin, E. M. Burbidge, L. B. Robinson, E. J. Wampler. Astrophys. Journ., (Letters), Vol. 195, L55 - L59 = Lick Obs. Bull., No. 682 (1975).

Spectrophotometric measurements of the nebula around BL Lacertae have been made with the Lick image-tube scanner on the 3-m telescope. No absorption features were seen, and, although the nebular spectrum is curved and steeper than that of BL Lac itself, it is not a good match to the energy distribution of a normal elliptical galaxy. Thus the authors do not confirm the observations by Oke and Gunn in which absorption features were seen corresponding to those in the redshifted spectrum of an elliptical galaxy. They have found no evidence that the nebula around BL Lac is a normal galaxy of stars.

158.302 A study of the variable radio source BL Lacertae in total intensity and linear and circular polarization. R. D. Ekers, K. W. Weiler, J. M. van der Hulst. Astron. Astrophys., Vol. 38, 67 - 73 (1975).

During the period from March 1971 to September 1972, the variable radio source VRO 42.22.01 (BL Lacertae) was observed approximately twice per week at 1.4 GHz ($\lambda = 21$ cm) with the Westerbork Synthesis Radio Telescope. These observations provide the first extensive series of measurements of the source at wavelengths longer than a few centimeters. The total intensity varies from 4.8 to 6.6×10^{-26} Wm^{-2} Hz^{-1} with a form very similar to that seen at higher frequencies two months earlier. The linearly polarized flux is relatively small but is clearly variable between 1.5 and 3 % of the total intensity. The position angle of the linear polarization undergoes one clear jump of $-30°$ and either a monotonic rotation or further series of jumps in the same direction during the observing period. The circular polarization is variable in both sense and magnitude with time but never exceeds 0.4 %. There is a possible correlation between the sense (sign) of the circular polarization and the phase of the total intensity outburst. A short

series of observations at 5 GHz ($\lambda = 6$ cm), carried out on 2 and 8 December 1972, reveal no short time scale total intensity variations exceeding 2 %.

158.303 Redshifts of BL Lac objects. R. F. Carswell. Nature, Vol. 253, 589 - 590 (1975).

158.304 The surface brightness of the nebulosity in BL Lacertae. T. D. Kinman. Astrophys. Journ., (Letters), Vol. 197, L49 - L51 (1975).

UBV observations of BL Lac through apertures with diameters in the range $7''-20''$ show a surface brightness distribution which is consistent with that expected for a giant elliptical galaxy with redshift 0.07 but not 0.02. The evidence from the change of color with aperture is inconclusive, probably because the assumption that the nonthermal component has a power-law distribution is only correct for the average of many observations. New photographs of the object are given.

158.305 Four-color photometry of OJ 287 during its recent three-magnitude decline. A. G. Smith, R. L. Scott, R. J. Leacock, B. Q. McGimsey, P. L. Edwards, R. L. Hackney, K. R. Hackney. Publ. Astron. Soc. Pacific, Vol. 87, 149 - 151 = Rosemary Hill Obs., Univ. Florida, Gainesville, Contr. No. 54 (1975).

The lacertid OJ 287 has undergone a five-year optical outburst from which it is still declining. Photographic UBV photometry throughout the declining phase has shown that the color of OJ 287 remained remarkably constant despite a 3-magnitude drop in brightness. There were, however, significant short-term excursions about the mean color, which is represented by the indices $(U-B) = -0^m65$ and $(B-V) = 0^m39$. A smaller number of infrared measurements show the same general trend.

158.306 Nine new enigmatic objects. B. A. Vorontsov-Vel'yaminov. Pis'ma v Astron. Zhurn., Vol. 1, No. 2, p. 3 - 6 (1975). In Russian.

Photographs of nine interesting enigmatic objects, probably extragalactic ones, are presented. The existence of faint streams of stars and dark matter between NGC 5194 and its companion is revealed. Such streams are not predicted by the tidal model.

158.307 Photographic photometry of five BL Lacertae-type objects. P. Véron, M. P. Véron.
Astron. Astrophys., Vol. 39, 281 - 288 (1975).

Photographic B magnitudes of five BL Lacertae objects are given. B 2 1101 + 38 is shown to be variable. The variations of PKS 0735 + 17 are confirmed. The known ranges of variation of OJ 287 and B 2 1215 + 30 are increased.

158.308 A neutral hydrogen study of the integral-sign galaxy MCG 12-7-28.
L. Bottinelli, L. Gouguenheim.
Astron. Astrophys., Vol. 39, 341 - 343 (1975).

The "integral-sign" galaxy MCG 12-7-28 has been studied in the 21-cm line of neutral hydrogen, with the Nançay radio telescope. Its integral properties are consistent with the following results: the galaxy has an Sb or Sbc type, it is located at a distance of about 18 Mpc and it is seen edge-on.

158.309 Optical variability of BL Lacertae: 1973–1975.
D. W. Wingert, H. R. Miller, G. H. Folsom, R. M. Williamon.
Bull. American Astron. Soc., Vol. 7, 262 (1975). – Abstr. AAS.

158.310 Optical behavior of 3C 66A. G. H. Folsom, H. R. Miller, D. W. Wingert, R. M. Williamon.
Bull. American Astron. Soc., Vol. 7, 262 (1975). – Abstr. AAS.

158.311 The color variations of six newly identified Lacertids. S. Tapia, E. R. Craine, K. Johnson.
Bull. American Astron. Soc., Vol. 7, 339 (1975). – Abstr. AAS.

158.312 Variability and optical-radio properties of BL Lacertae objects. J. T. Pollock.
Astrophys. Journ., (Letters), Vol. 198, L53 - L56 (1975).

The optical variability of new candidates for the BL Lacertae class of objects and of two known members of this class has been examined using the Harvard plate collection. A correlation is indicated between the maximum known range of optical variability and the observed radio spectral index at 5 GHz. Light curves for PKS 0735+17 and 2254+07 (OY 091) are presented.

158.313 Radio properties of BL Lac type objects.
D. R. Altschuler, J. F. C. Wardle.
Nature, Vol. 255, 306 - 310 (1975).

New radio observations of nine BL Lac type objects are discussed. It is shown that in their radio properties, these objects range from highly variable sources with a strong centimetre excess to non-variable sources with comparatively steep spectra. All the evidence strongly suggests that the most variable BL Lac type objects are extremely young.

158.314 Low frequency variability of BL Lac.
D. Stannard, A. M. Treverton, R. W. Porcas, R. J. Davis.
Nature, Vol. 255, 384 - 385 (1975).

The unusual object BL Lacertae has attracted attention because of its large and erratic variability at optical, millimetre and centimetre wavelengths. The general form of the varia-

tions at short radio wavelengths is compatible with the models of Van der Laan and Ozernoy and Sazonov. The authors report observations of the variability of BL Lac at relatively long radio wavelengths which are in serious conflict with these models.

158.315 Optical investigation of the peculiar spiral galaxy NGC 2146.
P. Benvenuti, M. Capaccioli, S. D'Odorico.
Astron. Astrophys., Vol. 41, 91 - 98 (1975).

Photometric and spectroscopic investigation of the peculiar spiral galaxy NGC 2146 is presented. Among the spirals this galaxy exhibits one of the brightest central radio sources with a non-thermal spectrum. Photographs in four color bands (from UV to IR) lead to a detailed discussion of the complex morphology, revealing features that cannot be explained by a simple model of spiral structure. An absorption of 5 ± 1 mag is estimated for the highly obscured central region. From the photometric and kinematic data, the mass within the last observed point of the rotation curve gives 4.8 $\times 10^{10} M_\odot$, from which a lower limit of 6.2 for the mass-to-light ratio is derived.

158.316 Les "lacertides". M. Duruy.
A.F.O.E.V. Bull., Vol. 9, 1 - 10 (1975).

158.317 BL Lacertae-type object. P. Wild.
IAU Circ., No. 2791 (1975).

Explosive events in the universe.
See Abstr. 004.081.

Optical monitoring of radio sources – IV. (Results up to 1973 April). See Abstr. 141.001.

Some new radio sources associated with the Sersic peculiar galaxies. See Abstr. 141.024.

Flux density measurements of radio sources at 2.14 millimeter wavelength. See Abstr. 141.030.

Optical polarization models of quasi-stellar objects and BL Lac objects. See Abstr. 141.068.

Instationary objects of non-stellar nature. See Abstr. 141.076.

A comparison of some radio and optical properties of quasi-stellar sources and BL Lacertae objects. See Abstr. 141.081.

Nonthermal continuum radiation in three elliptical galaxies. See Abstr. 158.099.

Observational problems of high energy astrophysics. See Abstr. 158.132.

Errata

158.901 Errata: "Further observations of the nuclear rotation in M31" [Publ. Astron. Soc. Pacific, Vol. 86, 861 - 866 = Lick Obs. Bull., No. 674 (1974)]. M. F. Walker.
Publ. Astron. Soc. Pacific, Vol. 87, 479 (1975).

159 Magellanic Clouds

159.001 Obscuration in and around the Small Magellanic Cloud. H. T. MacGillivray.
Monthly Notices Roy. Astron. Soc., Vol. 170, 241 - 249 (1975).

The detailed distribution of the obscuration in and around the Small Magellanic Cloud is discussed in the light of recently obtained galaxy counts. Several regions of high absorption are detected. Star masking is taken into account in the determination of the absorption profile across the Cloud. Comparison with the neutral hydrogen profile reveals that the extinction coefficient is less than in the Galaxy, indicating a lower ratio of dust to gas.

159.002 Photoelectric photometry of supergiants in the Large Magellanic Cloud. J. Isserstedt.
Astron. Astrophys., Suppl. Ser., Vol. 19, 259 - 269 (1975). In German.

Photoelectric photometry for 392 supergiants in the Large Magellanic Cloud is presented. The catalogue contains almost all stars of luminosity class I and almost all OB stars with $m_{pg} \leqslant 12^{m}.5$ taken from the spectral survey of Sanduleak (1969), which had not already been measured by Ardeberg et al. (1972).

159.003 Small Magellanic Cloud. First list of probable members. M. Azzopardi, J. Vigneau.
Astron. Astrophys., Suppl. Ser., Vol. 19, 271 - 279 (1975).

An objective-prism survey in the direction of the Small Magellanic Cloud has enabled the authors to detect many new stars which show high luminosity spectral characteristics. Use of interferential filters has allowed to reach fainter stars, by cutting down the background fog, and to limit to a minimum the overlappings. The authors give a first list of 63 stars for which they have photoelectric photometry measurements in the UBV system.

159.004 Soft X-ray survey of the Large Magellanic Cloud. S. Rappaport, A. Levine, R. Doxsey, H. V. Bradt.
Astrophys. Journ., (Letters), Vol. 196, L15 - L18 (1975).

A soft X-ray survey (0.15 – 1.5 keV) of the Large Magellanic Cloud, during a sounding rocket flight, reveals possible emission from the centrally located barlike structure. No evidence was seen for absorption of the soft X-ray 'background' in the direction of the LMC. Both of these observations support the hypothesis of a local production of the soft X-ray background.

159.005 A possible new segment of the Magellanic Stream in the northern sky.
R. J. Cohen, R. D. Davies.
Monthly Notices Roy. Astron. Soc., Vol. 170, 23P - 27P (1975).

A new high-velocity neutral hydrogen feature has been discovered at RA = 02^h30^m, Dec. $+9°.5$ ($l = 160°$, $b = -46°$) at a velocity of -334 km s^{-1} relative to the local standard of rest. Its properties are similar to those of the adjacent parts of the Magellanic Stream and the authors suggest that this new feature is part of the same phenomenon.

159.006 Recherche au prisme objectif de nouveaux membres du Petit Nuage de Magellan.
M. Azzopardi, J. Vigneau.
Comptes Rendus Acad. Sci. Paris, Sér. B, Vol. 280, 87 - 89 (1975).

La recherche au prisme objectif d'étoiles très lumineuses en direction du Petit Nuage de Magellan a conduit les auteurs à mettre en évidence de nombreux nouveaux membres. L'utilisation de filtres interférentiels a permis d'atteindre des étoiles plus faibles en coupant la lumière du ciel nocturne et en limitant le nombre des superpositions. L'appartenance est ensuite confirmée par des mesures photométriques UBV.

159.007 Color-magnitude diagrams for four rich clusters of the Large Magellanic Cloud.
P. J. Flower, P. W. Hodge.
Astrophys. Journ., Vol. 196, 369 - 380 (1975).

Color-magnitude diagrams are presented for four very rich clusters of the Large Magellanic Cloud. Comparisons with theoretical color-magnitude diagrams show good agreement in the location and distribution of stars in the giant regions. All four have nearly identical ages of approximately 5×10^7 years. Each cluster has at least one anomalously luminous giant of intermediate color and $M_v \cong -5.7$, over a magnitude brighter than both the observed and predicted giant branch.

159.008 Le Nubi di Magellano. F. Bònoli.
Coelum, Vol. 43, 1 - 13 (1975).

159.009 The intrinsic colours $(B-V)_0$, $(U-B)_0$ and distance moduli of supergiants in the Large Magellanic Cloud.
J. Isserstedt.
Astron. Astrophys., Vol. 39, 225 - 234 (1975). In German.

The intrinsic colours $(B-V)_0$, $(U-B)_0$ and the distance moduli of the 342 physical members of the Large Magellanic Cloud for which photoelectric UBV photometry and spectral classification on the MK-system are available in the catalogues of Ardeberg et al. (1972), Isserstedt (1975) and Sanduleak (1969), are discussed.

159.010 Probable member stars in the western extremity of the Small Magellanic Cloud. N. Sanduleak.
Astron. Astrophys., Vol. 39, 461 - 462 (1975).

Five new probable member stars in the Small Magellanic Cloud have been identified on objective-prism plates. They appear to be associated with a neutral hydrogen cloud situated near the western boundary of the SMC. Nova Tucanae 1974 also lies in the same region.

159.011 A deep spectral survey of the Magellanic Clouds. A. G. D. Philip.
Bull. American Astron. Soc., Vol. 7, 254 - 255 (1975). Abstr. AAS.

159.012 Monochromatic photographs of giant H II regions in the Magellanic Clouds.
L. H. Aller, S. J. Czyzak.
Bull. American Astron. Soc., Vol. 7, 255 (1975). – Abstr. AAS.

159.013 The abundances of helium in the Large Magellanic Cloud. I. J. Danziger.
Bull. American Astron. Soc., Vol. 7, 255 (1975). – Abstr. AAS.

159.014 De Magellaense stroom – restant van een kosmische ontmoeting? F. P. Israel.
Zenit, Vol. 2, 189 - 193 (1975).

159.015 Radial velocities of supergiants in the Small Magellanic Cloud. P. Dubois.
Astron. Astrophys., Vol. 40, 227 - 231 (1975).

This paper presents the kinematic results of a study comparing Small Magellanic Cloud supergiants to galactic supergiants. The stars were chosen from the lists of Florsch (1972) and Sanduleak (1968). A rough correlation is found between the radial velocities of the supergiants and the radial velocities of neutral hydrogen. In some cases a better relation exists with the radial velocities of the H II regions.

159.016 **Questions concerning interstellar matter in the Magellanic Clouds.** W. H. McCrea.
Monthly Notes Astron. Soc. Southern Africa, Vol. 34, 45 - 49 (1975).

Questions are proposed that may be answered by observations of the Magellanic Clouds from outside better than by observations of the Galaxy from inside. The answers would be of much cosmogonic significance, as illustrated here by some current work on comets.

159.017 **Some observations of bright Magellanic Cloud stars with a 12-channel scanning photometer.**
J. H. Walraven, T. Walraven.
Monthly Notes Astron. Soc. Southern Africa, Vol. 34, 56 - 67 (1975).

159.018 **The structure and space orientation of the Large Magellanic Cloud from the distribution and distance moduli of the supergiants.** J. Isserstedt.
Astron. Astrophys., Vol. 41, 21 - 26 (1975). In German.

The distribution of supergiants in the Large Magellanic Cloud (LMC) is discussed. The following results were found: The most luminous supergiants $m_{bol} < 10\overset{m}{.}0$ and the bluest supergiants $(U - B)_0 \lesssim -0\overset{m}{.}60$ have a strongly structured distribution with concentrations in the form of longish filaments. The dimensions of these filaments, of the order $(2-3)$ kpc $\times$ 0.4 kpc, are similar to those of spiral arms in the solar neighbourhood. No centre of symmetry of the structure can be recognized. There is no relation between the structure of the LMC derived from the distribution of supergiants and the position and orientation of the bar. The distribution of supergiants is different from the less filamentary but more cloudy distribution of the neutral hydrogen. The space orientation of the LMC is discussed using the distance moduli of 170 supergiants, for which MK-classifications from slit spectra and photoelectric UBV photometry are available. It seems that the LMC is rotating in the direction of increasing position angle, i.e. with leading spiral filaments.

159.019 **Chemical composition of nebulosities in the Magellanic Clouds.**
L. H. Aller, S. J. Czyzak, C. D. Keyes, G. Boeshaar.
Proc. National Acad. Sci. USA, Vol. 71, 4496 - 4499 = Contr. Perkins Obs., Ohio State, Ohio Wesleyan Univ., Ser. II, No. 48 (1974).

From photoelectric spectrophotometric data secured at Cerro Tololo Interamerican Observatory the authors have attempted to derive electron densities and temperatures, ionic concentrations, and chemical abundances of He, C, N, O, Ne, S, and Ar in nebulosities in the Magellanic Clouds. A comparison with the Orion nebula suggests He, N, Ne, O, and S may all be less abundant in the Magellanic Clouds.

159.020 **Interstellar reddening in the Large Magellanic Cloud.** J. Isserstedt.
Astron. Astrophys., Vol. 41, 175 - 182 (1975). In German.

Interstellar reddening has been determined individually for 702 supergiants in the Large Magellanic Cloud (LMC) from the two-colour-diagram (for early type stars), and from the intrinsic colours $(B - V)_0$ using MK spectral types (for intermediate and late type stars).

159.021 **The interstellar gas to dust ratio in the Large Magellanic Cloud.** J. Isserstedt, T. Schmidt-Kaler.
Astron. Astrophys., Vol. 41, 241 - 243 (1975). In German.

The gas to dust ratio in the Large Magellanic Cloud has been investigated using newly determined reddening values of supergiants from the catalogues of Ardeberg et al. (1972), Brunet et al. (1973) and Isserstedt (1975). The resulting value $N_{H\,I}/E_{B-V} = 6 \times 10^{21}$ cm^{-2} mag^{-1} is in excellent agreement with those derived in our Galaxy.

Photometry of SK160 = SMC-X 1.
See Abstr. 113.014.

A comparison of galactic and Large Magellanic Cloud G-type supergiants by a method of spectrum synthesis.
See Abstr. 114.021.

Comparison of a galactic and a Large Magellanic Cloud G-type supergiant. See Abstr. 114.090.

On the intrinsic properties of cepheids in the Galaxy, in Andromeda, and in the Magellanic Clouds.
See Abstr. 122.020.

Distribution and ages of Magellanic cepheids.
See Abstr. 122.112.

Period, color, and luminosity for cepheid variables.
See Abstr. 122.113.

Identifications of variable stars in the Magellanic Clouds. See Abstr. 123.037.

Ultraviolet observations of associations in the Large Magellanic Cloud. See Abstr. 131.131.

The chemical composition of selected H II regions in the Magellanic Clouds. See Abstr. 131.503.

Chemical composition of selected H II regions in the Magellanic Clouds. See Abstr. 131.541.

Observations of X-ray sources in the Large Magellanic Cloud by the OSO-7 satellite.
See Abstr. 142.040.

Variability of the X-ray sources in the Magellanic Clouds. See Abstr. 142.085.

UBV and uvby photometry of globular clusters in the Large Magellanic Cloud. See Abstr. 154.017.

Finding list of bright galaxies behind the SMC.
See Abstr. 158.005.

160 Clusters of Galaxies

160.001 Black dwarf stars as missing mass in clusters of galaxies.   W. McD. Napier, B. N. G. Guthrie.
Monthly Notices Roy. Astron. Soc., Vol. 170, 7 - 14 (1975).

The missing mass in clusters of galaxies may be largely in the form of degenerate low-mass stars (black dwarfs) with masses down to $\sim 10^{-2} M_\odot$. Clusters of galaxies retain most of their black dwarfs, so that the virial mass-to-light ratios of clusters are much higher than the mass-to-light ratios of the nuclear regions of individual elliptical galaxies.

160.002 Nuclear magnitudes for galaxies in the Coma and Perseus clusters.   D. W. Weedman.
Astrophys. Journ., Vol. 195, 587 - 592 (1975).

Multiaperture BV photometry is presented for 72 galaxies in the Coma cluster and 44 galaxies in the Perseus cluster. Apparent correlations similar to those reported by Tifft are found between the galactic nuclear magnitudes and galaxy redshifts in the Coma cluster but not in the Perseus cluster. The author does not presently conclude that the data provide confirming evidence for noncosmological redshifts in these clusters. However, it is found that either the absolute nuclear magnitudes of the Perseus cluster galaxies are about 1 mag brighter than those of Coma or that the Coma cluster is more distant relative to Perseus than its redshift implies.

160.003 The 1400-MHz luminosity function for Abell clusters of galaxies.   F. N. Owen.
Astrophys. Journ., Vol. 195, 593 - 603 (1975).

New observations have been used to derive the radio luminosity function of Abell clusters of galaxies at 1400 MHz. A relatively narrow, approximately Gaussian luminosity function is found. Correlations of detection probability with optical properties are examined. No strong correlation with cluster richness is found; however, a correlation of detection above a constant absolute luminosity with distance is detected. A correlation with Rood and Sastry cluster type is also found. The mutual relation of these correlations is discussed. The relation of the cluster luminosity function to field galaxies detections and previously derived radio luminosity functions is examined.

160.004 The missing mass.   B. Margon.
Mercury, Vol. 4, No. 1, p. 2 - 6 (1975).

160.005 Study of three southern groups of galaxies.
G. A. Welch, G. Chincarini, H. J. Rood.
Astron. Journ., Vol. 80, 77 - 82, 163 - 164 (1975).

New radial velocities, accurate to about 100 km/sec, are presented for 39 bright galaxies in a region including de Vaucouleurs groups No. 21, 53 (Fornax cluster), and 31. The ratio of virial mass to luminous mass for the Fornax cluster is $M_{vt}/M_L = 6$. The virial velocity dispersion in kilometers per sec is approximately 200, 400, and 200 for groups No. 21, 53, and 31, respectively.

160.006 A re-analysis of the dynamics of the nearby groups of galaxies.
G. Chincarini, H. J. Rood, G. A. Welch.
Monthly Notices Roy. Astron. Soc., Vol. 170, 441 - 445 (1975).

A new analysis of the dynamics of the groups of galaxies listed by de Vaucouleurs has been made using new radial velocities and corrections for observational errors. The authors confirm the finding of Turner & Sargent that groups tend to cluster around two peaks on an expansion-rate histogram. If these classes are identified as representing bound and unbound groups, then the similarity in slope of their regression lines on

a plot of log (M_{VT}/M_L) *vs* log V_D provides evidence that this correlation lacks cosmological significance.

160.007 On the "Seyfert Sextet," VV 115.
G. Chincarini, D. Martins.
Astrophys. Journ., Vol. 196, 335 - 337 (1975).

Radial velocities in a field east of the Seyfert Sextet indicate that the Seyfert's compact group is a condensation in an extended group of galaxies.

160.008 Binary galaxies and the problem of masses of clusters of galaxies.   T. W. Noonan.
Astrophys. Journ., Vol. 196, 683 - 686 (1975).

The question of whether galaxies' masses M and luminosities L may be related by a power law $M \propto L^{2.5B}$ rather than a direct proportionality is raised in relation to two problems. First, the effect of $B \neq 0.4$ on the problem of galaxy-cluster masses is examined. Second, a preliminary investigation of binary elliptical galaxies suggests $B = 0.3 \pm 0.1$ (s.e.). A value of $B = 0.3$, with a mass-luminosity ratio of 260 for the brightest cluster member as suggested by Jenner's work, indicates a typical cluster mass-luminosity ratio of 580.

160.009 Shahbazian 123: a new distant compact group of compact galaxies.
L. V. Mirzoyan, J. S. Miller, D. E. Osterbrock.
Astrophys. Journ., Vol. 196, 687 - 688 = Lick Obs. Bull., No. 681 (1975).

Image tube scanner spectrograms were taken of seven objects in two Shahbazian compact groups of compact objects. The three brightest objects in Shahbazian 78 are all late-type stars. On the other hand, three of the brightest objects in Shahbazian 123 are luminous compact galaxies with $z = 0.115$, while a fourth is likely also to be a galaxy in the group.

160.010 The velocity dispersion of the cluster of galaxies Abell 1060 (3U 1044−30).
N. V. Vidal, B. A. Peterson.
Astrophys. Journ., (*Letters*), Vol. 196, L95 - L97 (1975).

Radial velocity measurements of 15 galaxies in the cluster A1060 give an average velocity of 3233 km s^{-1} and a dispersion velocity of 799 km s^{-1}. The X-ray luminosity is 1.5×10^{43} ergs s^{-1}. The correlation of velocity dispersion and X-ray luminosity is reviewed.

160.011 Pencil beam observations of Abell clusters of galaxies. I. 2695 MHz.   F. N. Owen.
Astron. Journ., Vol. 80, 263 - 270 (1975).

Observations at 2695 MHz of radic sources in the direction of Abell clusters of galaxies are reported. Positions, flux densities, and angular sizes are given for 226 sources.

160.012 The influence of the Scott effect on the determination of q_0.   A. Kruszewski, I. Semeniuk.
Acta Astron., Vol. 25, 63 - 78 (1975).

The statistical model for taking into account the Scott effect was constructed. The suggestion that clusters with exceptionally bright first-ranked cluster member possess fainter than average second and third-ranked galaxies is not substantiated by raw observational data. The first-ranked galaxies are brighter and less cluster richness dependent than expected from the statistical model. The bias due to the Scott effect may increase the deceleration parameter by up to 0.5 but with proper care it should be possible to take it into account even without employing complicated statistical models.

160.013 Does the Local Supercluster rotate?

G. Dautcourt.
Astron. Astrophys., Vol. 38, 335 - 339 (1975).

Intense ultra low-frequency intergalactic gravitational radiation produces an additional component in the measured redshifts of extragalactic sources. It is noted that the near-field of gravitational waves generates, for sources with distances not exceeding one wavelength, an anisotropic and distance-dependent redshift component. This redshift field, which would have to be added to the normal redshift caused by the Hubble expansion, has some similarity with the observed redshift distribution for nearby galaxies, usually attributed to a differential rotation and expansion of the Local Supercluster.

160.014 Apparent galaxy clustering caused by low-frequency gravitational radiation. G. Dautcourt.
Astron. Astrophys., Vol. 38, 341 - 344 (1975).

Low frequency intergalactic gravitational radiation with wavelengths λ megaparsec produces a frozen scintillation of distant ($D \gg \lambda$) extragalactic sources: The objects are seen in a slightly distorted position, and the deviations may reach minutes of arc. The effect could lead to an apparent clustering of faint and distant galaxies. The author discusses whether this clustering can be detected by using the dispersion-subdivision curve analysis introduced by Zwicky.

160.015 Studies of rich clusters of galaxies—III. Photometry of the cluster A1930 and the $m^* - z$ relation.
T. B. Austin, J. G. Godwin, J. V. Peach.
Monthly Notices Roy. Astron. Soc., Vol. 171, 135 - 142 (1975).

Photometry is presented for objects in the central region of the cluster A1930. There is no evidence of radial luminosity or colour segregation. The discontinuity in the integrated luminosity function occurs at a total absolute magnitude $M_v^* = -21.0$. The extension of Bautz & Abell's data to higher redshift made by this cluster and A1413 confirms that M_v^* is a better distance indicator than the apparent magnitude of the brightest cluster member. The low dispersion of M_v^* is used directly to obtain a value for the Hubble constant at the Virgo cluster of $H_0(VC) = 62 \pm 19$ km s^{-1} Mpc^{-1}.

160.016 Sur l'existence de la supergalaxie locale et la distribution de galaxies et des radiosources quasistellaires (QSSs) dans cet objet. H. Karoji.
Comptes Rendus Acad. Sci. Paris, Sér. B, Vol. 280, 421 - 423 (1975).

L'histogramme du nombre de galaxies en fonction de la distance au centre de la supergalaxie locale est établie.

160.017 High resolution observations of Abell 2256.
C. H. Costain, R. S. Roger.
Journ. Roy. Astron. Soc. Canada, Vol. 69, 42 (1975). − Abstr. Canadian Astron. Soc.

160.018 Absorption-line redshifts of galaxies in remote clusters obtained with a sky-subtraction spectrograph using an SIT television detector.
J. A. Westphal, J. Kristian, A. Sandage.
Astrophys. Journ., (Letters), Vol. 197, L95 - L98 (1975).

A prism spectrograph with an associated SIT television camera, operating as a two-dimensional detector with digital readout, has been used at the 5-m Hale telescope. The system was tested for its ability to subtract the sky spectrum, and was found to produce difference spectra that are essentially photon noise limited. Redshifts of 14 galaxies in clusters with $0.01 \leq z \leq 0.4$ were obtained, each with exposure times of 90 minutes or less. Nine of the redshifts are new. Redshifts for the remaining five agree with previous values to within the measuring errors. The speed and sky-subtraction capabilities of the instrument are sufficient to begin routine measurement

of absorption-line redshifts for remote cluster galaxies in an effort to extend the Hubble diagram.

160.019 The energy spectrum of the Perseus cluster of galaxies. R. S. Wolff, H. Helava, M. C. Weisskopf.
Astrophys. Journ., (Letters), Vol. 197, L99 - L103 (1975).

A one-dimensional focusing X-ray telescope has been used to measure the spatial structure and energy spectrum of the Perseus cluster in the 0.5−4.5 keV X-ray range. The authors observe a significant variation in the X-ray spectrum from different regions of the cluster, indicating different possible emission mechanisms. The spectrum from the region centered on NGC 1275 appears harder and more cut off at low energy than that of the remainder of the cluster.

160.020 The soft X-ray spectrum of the Perseus cluster.
A. Davidsen, S. Bowyer, M. Lampton, R. Cruddace.
Astrophys. Journ., Vol. 198, 1 - 6 (1975).

The X-ray spectrum of the Perseus cluster in the range 0.1−4.0 keV has been observed. No large flux of soft X-rays was found. The X-ray spectrum from 0.1 to 56 keV is consistent with bremsstrahlung from an intracluster gas with $T \approx 10^8$ K, attenuated by the interstellar column density inferred from 21-cm observations. The mass of hot intracluster gas required to produce the observed X-ray flux is a small fraction of the gravitational binding mass.

160.021 Redshift dispersion in the X-ray cluster of galaxies A1367. W. G. Tifft, M. Tarenghi.
Astrophys. Journ., (Letters), Vol. 198, L7 - L8 (1975).

The redshift dispersion in the cluster of galaxies A1367 is shown to be consistent with the X-ray luminosity-redshift dispersion correlation.

160.022 On the determination of distances of the Abell clusters of galaxies.
M. Kalinkov, K. Stavrev, I. Kaneva.
Pis'ma v Astron. Zhurn., Vol. 1, No. 2, p. 7 - 8 (1975).
In Russian.

An improved regression equation to determine distances of the Abell clusters of galaxies from the magnitude of the 10th (according to brightness) galaxy, is derived. It is based on data for 75 clusters with observed redshifts.

160.023 On the infrared radiation of clusters of galaxies.
S. A. Pustil'nik.
Pis'ma v Astron. Zhurn., Vol. 1, No. 3, p. 8 - 11 (1975).
In Russian.

For six rich clusters the infrared flux of thermal radiation from intergalactic dust is reestimated within the framework of suggestions that optical absorption in the clusters is due to dust and that X-ray radiation from the central regions of the clusters is of thermal origin.

160.024 Properties of central regions of clusters of galaxies in the optical, radio and X-ray spectral regions.
B. V. Komberg.
In-t prikl. mat. AN SSSR. Preprint No. 85. Moskva, 1974. 65 pp. In Russian. − Abstr. in Referativ. Zhurn. 51. Astron., 3.51.821 (1975).

160.025 Core radii and central densities of 15 rich clusters of galaxies. N. A. Bahcall.
Astrophys. Journ., Vol. 198, 249 - 254 (1975).

The projected galaxy distributions and centers are determined for 15 rich regular clusters of galaxies with redshifts up to about 0.1. The galaxy distributions in all the clusters are similar and resemble the distribution of a projected, bounded, Emden isothermal gas sphere. The core radii obtained for the distributions exhibit a relatively small dispersion ($\pm 15\%$) around a constant linear value of $R_c = 0.25$ Mpc ($H_0 = 50$ km

s^{-1} Mpc^{-1}). This result suggests that the extension of the present study to high-redshift clusters, comparing their core radii with the low redshift values, will enable one to place some constraints on the cosmological deceleration parameter, q_0.

160.026 **Southern galaxy clusters identified with 3U X-ray sources.** J. Melnick, H. Quintana.
Astrophys. Journ., (*Letters*), Vol. 198, L97 - L99 (1975).

Plate inspection of fields within the error boxes of three high-latitude sources have been made. For two fields, more than one cluster of galaxies might be associated with the source. A morphology-luminosity criterion recently proposed is not sufficient to discriminate between the candidates.

160.027 **Sur un modèle décomposable d'univers hiérarchisé: déduction des corrélations galactiques sur la sphère céleste.** B. Mandelbrot.
Comptes Rendus Acad. Sci. Paris, Sér. A, Vol. 280, 1551 - 1554 (1975).

On étudie un modèle qui a été construit de telle façon que les corrélations entre deux ou trois points de la voûte céleste peuvent se déduire explicitement. La comparaison avec les données du réel montre un accord satisfaisant, et même surprenant.

160.028 **Collisions and relaxation in galaxy clusters.** P. D. Noerdlinger.
Publ. Astron. Soc. Pacific, Vol. 87, 333 - 334 (1975).

The rate of close collisions among galaxies in a cluster, and the rates of both gravitational two-body and violent relaxation are shown, in most cases, to be proportional to the Hubble constant. Therefore, the number of collisions or the degree of relaxation attained in one Hubble time are invariant to the Hubble constant.

160.029 **Possible systematic magnitude error in the Zwicky Catalogue of Galaxies.**
M. Burakowska, K. Rudnicki.
Acta Cosmologica, Fasc. 2, p. 7 - 12 (1974/75).

A discussion of the function N (m) in two regions of Zwicky's Catalogue indicates a possibility of a systematic magnitude error for bright galaxies. The correction for magnitude 13^m5 seems to be equal to $+0^m22$ when the largest magnitudes in the catalogue are accepted as correct. Such difference is of the same order as the mean random error of the catalogue.

160.030 **Distances of 31 clusters of galaxies (Zwicky-Kwast method).** P. Flin.
Acta Cosmologica, Fasc. 2, p. 33 - 35 (1974/75).

In this paper the distances of clusters of galaxies were determined from observed angular distances of the supernovae from the centre of the galaxies belonging to the clusters.

160.031 **Some remarks on the clustering of galaxies. I.** A. Zięba, P. Flin.
Acta Cosmologica, Fasc. 2, p. 117 - 125 (1974/75).

In this paper the preliminary results are given of a statistical analyses of distribution of galaxies in the Jagellonian field. It was found that galaxies have an evident tendency to clustering. Cells having a great number of galaxies have a trend toward forming aggregations and they are in isolation from cells having a small number of galaxies. Medium dense cells show a tendency to clustering with dense regions rather than with other medium dense region. These effects are not strong and they are superimposed on a stronger background.

160.032 **The redshift-distance relationship derived from clusters of galaxies.**
J. Kollerstrom, G. C. McVittie.

Observatory, Vol. 95, 90 - 97 (1975).

160.033 **Emission of X-rays from the Coma and the Virgo clusters of galaxies.** R. K. Thakur.
Bull. Astron. Soc. India, Vol. 2, 30 - 31 (1974). – Abstract.

160.034 **Scheinbare Virialsatz-Verletzung bei Galaxienhaufen infolge extrem langwelliger kosmischer Gravitationsstrahlung.** G. Dautcourt.
Publ. Astrophys. Obs. Potsdam, No. 108, Vol. 32, Heft 1, p. 15 - 24 (1974).

The well-known mass discrepancy for groups and clusters of galaxies might be connected with the existence of low-frequency intergalactic gravitational radiation, as was recently suggested by M. Rees. The virial discrepancy may be explained in terms of the gravitational background radiation. Some quantitative estimates are given.

160.035 **Systematic properties of clusters of galaxies.** A. Oemler, Jr.
Thesis, California Inst. Techn., Pasadena (USA). 91 pp. University Microfilms Order No. 74-17,951 (1973).

160.036 **Clusters of galaxies and the cosmic light.** S. A. Shectman.
Thesis, California Inst. Techn., Pasadena (USA). 57 pp. University Microfilms Order No. 74-14,286 (1974).

Supernovae in clusters of galaxies.
See Abstr. 125.048.

The structures and properties of 4C radio sources in Abell clusters – I. See Abstr. 141.002.

The nature of quasars. See Abstr. 141.062.

B 2.1028 + 31: a possible association of a quasar with a group of galaxies. See Abstr. 141.095.

Extended X-ray sources and missing masses in clusters of galaxies. See Abstr. 142.134.

About the structure of spherical galaxies and of clusters of galaxies. See Abstr. 151.033.

The motion of our Galaxy and the local group of galaxies. See Abstr. 155.032.

On the possibility of radioastronomical investigation of the birth of galaxies. See Abstr. 158.047.

cD galaxies in poor clusters. See Abstr. 158.082.

Extragalactic observational astronomy. See Abstr. 158.134.

Redshifts as distance indicators. See Abstr. 158.135.

Intergalactic H I in the Sculptor Group. See Abstr. 161.001.

Existence and amount of intergalactic dust. See Abstr. 161.004.

Steady state distribution of an ideal cluster of galaxies for negative cosmological constants. See Abstr. 162.003.

Spectrophotometry of faint cluster galaxies and the Hubble diagram: an approach to cosmology.

See Abstr. 162.005.

The redshift-magnitude relation for galaxies.
See Abstr. 162.021.

Steps toward the Hubble constant. V. The Hubble constant from nearby galaxies and the regularity of the local velocity field. See Abstr. 162.023.

Evidence for a spatially homogeneous component of the universe: single galaxies. See Abstr. 162.037.

Development of the correlation of galaxies in an ex-panding universe. See Abstr. 162.073.

Errata

160.901 **Erratum: 'Counts of galaxies in seven distant clusters'** [Astron. Journ., Vol. 79, 358 - 362 (1974)].
T. W. Noonan.
Astron. Journ., Vol. 80, 252 (1975).

160.902 **Erratum: 'Cosmological implications of available counts in clusters of galaxies'** [Astron. Journ., Vol. 79, 775 - 776 (1974)]. T. W. Noonan.
Astron. Journ., Vol. 80, 252 (1975).

161 Intergalactic Matter

161.001 **Intergalactic H I in the Sculptor Group.**
D. S. Mathewson, M. N. Cleary, J. D. Murray.
Astrophys. Journ., (*Letters*), Vol. 195, L97 - L100 (1975).
H I clouds have been found up to $1\overset{\circ}{.}5$ and $3\overset{\circ}{.}5$ from the galaxies NGC 55 and NGC 300, respectively, with velocities similar to the systemic velocities of the two galaxies. They are probably intergalactic gas clouds in the Sculptor Group with linear projected distances of up to 80 kpc from NGC 55 and 180 kpc from NGC 300. A long tail of H I was also found extending 2° to the SE of NGC 300 along its major axis. This tail and similar H I tails on M83 and IC 10 lie approximately in the plane of the Local Supergalaxy; the Magellanic Stream also lies near the supergalactic plane.

161.002 **Magnetic field in the intergalactic region.**
J. P. Vallée.
Nature, Vol. 254, 23 - 26 (1975).
An analysis of the redshift dependence of the rotation measures (RM) and of the intrinsic position angles of the polarised radiation of radio galaxies and QSOs is made. The results place an upper limit of 10 rad m^{-2} to the observed RM, coming from the contribution of the magnetic field in the intergalactic region.

161.003 **Search of an intergalactic component of Faraday rotation.** P. P. Kronberg, M. Normandin.
Bull. American Astron. Soc., Vol. 7, 339 (1975). — Abstr. AAS.

161.004 **Existence and amount of intergalactic dust.**
K.-H. Schmidt.
Astrophys. Space Sci., Vol. 34, 23 - 31 (1975). — Paper presented at the Symposium on Solid State Astrophysics, Cardiff, Wales, 9–12 July, 1974.
The existence of intergalactic dust has been proved by the following observational facts: the decrease of the numbers of distant galaxies and clusters of galaxies behind the central regions of near clusters of galaxies; the different distributions of RR Lyrae stars and galaxies near ι Microscopii (Hoffmeister's cloud); the dependence of colour excesses of galaxies on supergalactic coordinates as well as on the surface density of bright galaxies; the colour index vs redshift correlation of quasistellar objects. The densities of intergalactic dust are estimated to be between 5×10^{-30} g cm^{-3} (near the centers of clusters of galaxies) and 2×10^{-34} g cm^{-3} (in general intergalactic space). The grains may be formed either in the early phases of the Universe ($25 < z < 50$) or may be expelled from galaxies by the radiation pressure. The most effective destruction process seems to be the evaporation by soft cosmic rays.

161.005 **Statistical search for intergalactic matter.**
T. Kwast.
Acta Cosmologica, Fasc. 2, p. 65 - 85 (1974/75).
Two areas of especially low density of galaxies have been observed in the clusters of galaxies Zwicky 156-5 and 156-14. The question has arisen whether this phenomenon is caused by obscuring intergalactic clouds. The results of some investigations are discussed.

161.006 **Effects of the reheating of the intergalactic gas on the collapse of late density perturbations.**
L. Maraschi, G. C. Perola.
Astron. Astrophys., Vol. 40, 387 - 390 (1975).
The influence of the reheating of the intergalactic gas on the collapse of late density perturbations is studied in the approximation of spherical symmetry. It is found that if the IGG temperature is $> 5 \times 10^4$ °K out to $Z = 2$ at least, the collapse is adiabatic and is stopped by the rise of the internal pressure. These giant clouds of gas emit mostly X-rays with a power $10^{44}-10^{47}$ erg s^{-1}, for at least one tenth of the Hubble time. Over this time scale thermal instabilities could lead to the formation of galaxies in the inner regions of the cloud. It is suggested that high galactic latitude sources in the Uhuru catalogue, unidentified optically, could be interpreted as hot intergalactic clouds.

Energy loss of relativistic electrons and positrons traversing cosmic matter. See Abstr. 066.033.

Theoretical problems of high energy astrophysics. See Abstr. 141.101.

Interaction of protoclusters of galaxies with intergalactic matter. See Abstr. 158.017.

Ring galaxies and intergalactic gas clouds. See Abstr. 158.020.

The structure of the radio galaxy NGC 1265. See Abstr. 158.040.

Observational problems of high energy astrophysics. See Abstr. 158.132.

The soft X-ray spectrum of the Perseus cluster. See Abstr. 160.020.

162 Structure and Evolution of the Universe, Cosmology

162.001 **Distortions of the 3°K background radiation spectrum: observational constraints on the early thermal history of the universe.** K. L. Chan, B. J. T. Jones.
Astrophys. Journ., Vol. 195, 1 - 11 (1975).

The distortions of the cosmic microwave background radiation spectrum due to heat input prior to recombination is examined analytically and also by direct numerical solution of the time-dependent Kompaneets equation. The analytic approximations are found to be very good over a large range of thermal histories that are compatible with observation. The constraints on the cosmic thermal history imposed by observations of the Rayleigh-Jeans part of the spectrum are re-evaluated.

162.002 **The general metrical fundamental form of the de Sitter universes.** H.-J. Treder.
Astron. Nachr., Vol. 296, 15 - 18 (1975).

162.003 **Steady state distribution of an ideal cluster of galaxies for negative cosmological constants.**
M. C. Chen, S. K. Sachdev.
Astron. Nachr., Vol. 296, 19 - 24 (1975).

Space distribution of an ideal cluster of galaxies in steady state for various negative cosmological constants is solved numerically. The galaxy members in the ideal cluster are considered point objects with equal masses. This is an extension of the classical Lane-Emden isothermal sphere to negative cosmological constant.

162.004 **Nucleosynthesis and matter—antimatter cosmologies.**
F. Combes, O. Fassi-Fehri, B. Leroy.
Nature, Vol. 253, 25 - 26 (1975). – Letter.

162.005 **Spectrophotometry of faint cluster galaxies and the Hubble diagram: an approach to cosmology.**
J. E. Gunn, J. B. Oke.
Astrophys. Journ., Vol. 195, 255 - 268 (1975).

New spectrophotometric data are presented for large-red-shift galaxies in clusters. A new approach to aperture corrections and the analysis of the Hubble diagram is outlined, including the explicit incorporation of evolutionary effects. The importance of selection effects on the usual methods of analysis are in principle overcome in the analysis, although the heterogeneity of the sample still makes conclusions about cosmology slightly suspect. Formal values of the deceleration parameter are derived under several sets of assumptions, yielding results between q_0 = +0.33 and q_0 = −1.27 with formal standard deviations of order 0.7.

162.006 **The behaviour of point masses in an expanding cosmological substratum.** P. Mészáros.
Astron. Astrophys., Vol. 37, 225 - 228 (1974).

The author investigates the properties of point masses imbedded in a smooth, expanding cosmological substratum, which has an equation of state similar to that of radiation. He formulates a model in which the dynamics of the whole ensemble of point masses is taken into account, and studies its evolution beginning from small perturbations through a stability analysis. The results imply that there cannot be at present hidden mass in the form of neutrinos or gravitational waves of density significantly exceeding that of galaxies. The author also applies the results to models of the early universe in which the presence of primeval black holes is hypothesized, and concludes that over scales less than the horizon, fluctuations in the black hole number density cannot grow as long as the equation of state of the universe is of the radiation type.

162.007 **Die Hubble-Konstante und ihre Bestimmung.**
M. Reinhardt.
SuW, Vol. 14, 49 - 52 (1975).

162.008 **Primeval black holes and galaxy formation.**
P. Mészáros.
Astron. Astrophys., Vol. 38, 5 - 13 (1975).

The author presents a scheme of galaxy formation, based on the hypothesis that a certain fraction of the mass of the early universe is in the form of black holes. One is thus able to achieve galaxy and cluster formation at the right redshifts, and at the same time the black holes would account for the recently proposed massive halos of galaxies, and for the hidden mass in clusters required by virial theorem arguments.

162.009 **Opening up the universe.** P. C. W. Davies.
Nature, Vol. 253, 594 (1975).

162.010 **Deuterium of cosmological origin and the mean density of the universe.** Ya. B. Zel'dovich.
Pis'ma v Astron. Zhurn., Vol. 1, No. 1, p. 10 - 13 (1975).
In Russian.

There is a contradiction between the observed abundance of deuterium in the interstellar gas and that calculated for the nucleosynthesis in the big-bang with conservative estimate of the mean density of the universe $10^{-30} g/cm^3 < \rho < 2 \times 10^{-30} g/cm^3$. The way out proposed here is to assume strongly nonuniform distribution of baryons (protons and neutrons) on a uniform radiation background during reactions of light element production.

162.011 **Background radiation.** I. L. Rozental'.
New problems of astrophysics, Publ. Astrophys. Winter School, 1972, (see 012.001), p. 91 - 104 (1974).
In Russian.

162.012 **The "principle of minimality" and evolution of entropy perturbations at early stages of the expansion of the universe.** G. V. Chibisov.
New problems of astrophysics, Publ. Astrophys. Winter School, 1972, (see 012.001), p. 105 - 113 (1974). In Russian.

162.013 **The cosmological redshift.** B. M. Lewis.
Southern Stars, Vol. 25, 208 - 213 (1974).

162.014 **Cosmological absorption of gravitational waves.**
R. Burman.
Nature, Vol. 254, 205 - 206 (1975).

Several people have investigated the absorption of electromagnetic and neutrino waves in cosmological models and discussed the self-consistency or otherwise, in the Wheeler—Feynman absorber theory of radiation, of retarded and advanced fields; a solution of the field equations is self-consistent if the absorption tends to completeness as the field propagates. Cosmological absorption of gravitational waves has been treated briefly by Hawking. Here, the completeness or otherwise of the absorption of retarded and advanced gravitational waves are investigated using a technique introduced, in the case of electromagnetic absorption, by Davies; the technique applies to the Robertson—Walker cosmologies of arbitrary spatial curvature.

162.015 **Statistical analysis of catalogs of extragalactic objects. V. Three-point correlation function for the galaxy distribution in the Zwicky catalog.**
P. J. E. Peebles, E. J. Groth.
Astrophys. Journ., Vol. 196, 1 - 11 (1975).

A three-point correlation function for the distribution of galaxies is defined and is estimated for the Zwicky catalog. The function varies like a simple power of angular scale. The index of the power law is close to twice the index for the two-point function, in agreement with the prediction of a gravitational instability model, or, more generally, with the assumption that the clustering satisfies a simple scaling property over the range of lengths for which the authors have estimates of the correlation functions.

162.016 Relativistic hierarchical cosmology. I: Derivation of a metric and dynamical equations.
P. S. Wesson.
Astrophys. Space Sci., Vol. 32, 273 - 284 (1975).

An account of hierarchical cosmology is given that is based on general relativity, being split up into three pieces for ease of assimilation. Part II will treat of specific solutions relevant to the observed universe, while Part III compares these models with available empirical data in an effort to pick a model that agrees with observation. Part I derives a metric for a system based on the assumptions of (i) spherical symmetry about any local observer, (ii) a density distribution falling off as $\varphi = \varphi_0(t)r^{-2}$ from any local origin with the hierarchy delineated by step functions, (iii) the universe was denser and more compact at some epoch in the past.

162.017 Relativistic hierarchical cosmology. II: Some classes of model universes. P. S. Wesson.
Astrophys. Space Sci., Vol. 32, 305 - 314 (1975).

Following the re-expression of the metric for a hierarchical cosmology, the metric curvature is considered in showing that the hierarchy models are not, in general, related to the Robertson/Walker models even though the metrics can be made to appear superficially similar. The classes of models examined are quasi-Robertson/Walker models with $\Lambda=0$ quasi-Robertson/Walker models with $\Lambda\neq0$, constant curvature ($k(t)$= constant) models and zero-curvature ($k(t)$=0) models.

162.018 Relativistic hierarchical cosmology. III: Comparison with observational data. P. S. Wesson.
Astrophys. Space Sci., Vol. 32, 315 - 330 (1975).

The adoption of q_0 = +1 allows one to show that a Lemaitre-type hierarchical universe with a long coasting or waiting time can give agreement with observations of the numbers of QSO's etc. if the age of the universe is more than 10^{13} yr. The dependence of the effective Hubble parameter on $k(t)$, $\dot{k}(t)$ and R leads one to suggest that a k=0, $\dot{k}$=0 hierarchy with $\Lambda \neq 0$ might be the simplest acceptable form of model universe. The author points out that further data on source count anisotropies should allow the component levels of the hierarchy to be delineated.

162.019 Expanding universes with shear. M. Kubo.
Publ. Astron. Soc. Japan, Vol. 27, 111 - 117 (1975).

The classification and the evolution of expanding universes with shear but without rotation and containing dustlike matter are examined taking the cosmological constant into account. The author assumes the isotropy of the Gaussian curvature in the hypersurface orthogonal to the 4-velocity. Eight types of evolution-curves are found on the parameter diagram constructed by the density parameter and the anisotropy parameter.

162.020 The gravitational theories of Poincaré and Milne and the non-Riemannian kinematic models of the universe. I. W. Roxburgh, R. Tavakol.
Monthly Notices Roy. Astron. Soc., Vol. 170, 599 - 610 (1975).

Milne's general kinematic discussion on cosmological models is reviewed, revised and extended and it is shown that Riemannian geometry is insufficient for describing the general class of Milne's theories and that these theories if they are to be geometrized need the more general framework of Finsler spaces. Poincaré's early relativistic gravitational theory is then reviewed and applied to the one-body and cosmological problems. It is shown that this theory is not Riemannian geometrizable but is Finsler geometrizable. The cosmological solution is explicitly evaluated and shown to be an example of Milne's general kinematic models.

162.021 The redshift-magnitude relation for galaxies.
D. Nanni, A. Vignato.
Astrophys. Space Sci., Vol. 33, 11 - 21 (1975).

An analysis of the redshift-magnitude data for the 98 clusters of the list of Sandage and Hardy (1973) is repeated taking into account both the effect of richness and Bautz-Morgan classification on the absolute magnitude of the brightest members. The analysis of richness 1 and 2 clusters support an open model of the universe ($q_0 < 0.5$) while the uncertainties in the attribution of richness to the three most distant clusters do not permit to discard the steady state.

162.022 The influence of strong interactions on the early stages of the universe.
H. Dehnen, H. Hönl.
Astrophys. Space Sci., Vol. 33, 49 - 73 (1975).

It is shown that the singularity of space-time in Einstein-Friedmann's cosmology can be avoided, if one takes into account the strong interaction of the elementary particles in the earliest stage of the universe. The minimum radius of curvature of the universe becomes 1.4×10^{11} km; the density in its neighbourhood remains within reasonable limits of the magnitude of the nuclear density. The early evolution of the universe with time will be discussed in detail.

162.023 Steps toward the Hubble constant. V. The Hubble constant from nearby galaxies and the regularity of the local velocity field. A. Sandage, G. A. Tammann.
Astrophys. Journ., Vol. 196, 313 - 328 (1975).

Distances to very nearby bright spiral galaxies from paper IV are combined with individual redshifts to obtain the mean local Hubble expansion rate of $H_0 = 57 \pm 3$ km s^{-1} Mpc^{-1}. This is the same as $H_0 = 57 \pm 6$ km s^{-1} Mpc^{-1} from the Virgo cluster alone ($D \cong 20$ Mpc) found in paper IV, and is also closely the same as the global value of $H_0 = 55 \pm 6$ km s^{-1} Mpc^{-1} found from remote Sc I galaxies ($D \gtrsim 60$ Mpc, $v_0 \gtrsim 3000$ km s^{-1}) in paper VI. The agreement of the three rates shows that there is no measurable velocity anisotropy in the mean velocity field of nearby spiral galaxies, and further that there is no systematic variation of the Hubble constant with distance. This is the same result as obtained earlier by Sandage, Tammann, and Hardy in 1972, using the nearer clusters and groups of E galaxies. To investigate further, the sample is increased by using luminosity-class distances to Sc I, Sc II, and Sc III galaxies classified by van den Bergh in the north and by us in the south. A bias is identified, and the sample is cut to 71 objects to form a distance-limited list. The mean Hubble rate for these is found to be the same in the hemisphere toward and away from the Virgo cluster to within 1 σ of the combined errors, despite the large density contrast of the Coma-Virgo complex.

162.024 Large-scale random gravitational waves.
W. L. Burke.
Astrophys. Journ., Vol. 196, 329 - 334 = Lick Obs. Bull., No. 676 (1975).

Light propagating in a region with large-scale random (i.e., wide-band) gravitational waves suffers frequency shifts which lead to a small-scale anisotropy and also a large-scale anisotropy. Here detailed calculations of the statistical properties of both of these effects are given. It is suggested that the presence of megaparsec-scale waves sufficient to close the universe and which would explain part of the apparent velocity dispersion in cluster galaxies could be detected by this large-scale anisotropy.

162.025 **A matter-antimatter separation mechanism.**
A. H. Nelson, G. Rowlands.
Astrophys. Space Sci., Vol. 33, L1 - L3 (1975).
The plasma two-stream instability is proposed as a matter-antimatter separation mechanism for the Alfvén-Klein cosmology.

162.026 **On the theroetical evaluation of the Hubble red-shift constant.** V. A. Krat, I. L. Gerlovin.
Astrophys. Space Sci., Vol. 33, L5 - L8 (1975).
The coefficient of the cosmological red-shift (H_0) is calculated on the basis of the new theory of fundamental field (theory of the physical vacuum). Its value $H_0 = 2.01 \times 10^{-17}$ is in satisfactory agreement with its experimental values.

162.027 **Apparent magnitudes, redshifts, and inhomogeneities in the universe.** R. C. Roeder.
Astrophys. Journ., Vol. 196, 671 - 673 (1975).
The apparent magnitude-redshift data of Sandage and Hardy have been analyzed using inhomogeneous zero-shear models of the universe and treating both the acceleration factor and the amount of intergalactic material as parameters to be determined by the data. The value found for the acceleration factor depends on the homogeneity factor and on the evolutionary correction assumed for the absolute magnitudes of galaxies, but may possibly be greater than 0.5, in which case the universe is "closed".

162.028 **Propagation of electromagnetic polarization effects in anisotropic cosmologies.** C. H. Brans.
Astrophys. Journ., Vol. 197, 1 - 4 (1975).
The influence of a background cosmological metric on the propagation of electromagnetic radiation through the universe is studied, with particular emphasis on the effect of metric anisotropies on polarization properties. An approximation is used which regards the electromagnetic field as a flat-space plane wave (null field) with respect to the Minkowski frames parallel-transported along a null geodesic. These frames are Lorentz transforms of the local comoving cosmological frames at each point. Thus the effects of the metric on the electromagnetic field show up in terms of these Lorentz transformations. It is found that a general anisotropic metric will produce rotations of planes of polarization by amounts dependent on the length of the path, causing a scrambling of any polarization effects from different sources.

162.029 **Cosmology and the Higgs mass.** M. Veltman.
Phys. Rev. Letters, Vol. 34, 777 (1975).
It is demonstrated that a very small or zero Higgs mass is excluded by experiment.

162.030 **Dissipation in early universe. I. Bianchi type I and V models.** Z. Klimek.
Acta Astron., Vol. 25, 79 - 93 (1975).
The role of dissipative processes in the lepton era is discussed for Bianchi type I and type V cosmological models. In this era the two models give the same results. In an approximation of viscous fluid two physical processes are investigated: scattering of neutrinos on electrons and graviton absorption by viscous matter. In both cases there exists a maximum initial anisotropy, above which dissipative effects do not occur, because their characteristic time exceeds the characteristic time of model expansion.

162.031 **Steps toward the Hubble constant. VI. The Hubble constant determined from redshifts and magnitudes of remote Sc I galaxies: the value of q_0.**

A. Sandage, G. A. Tammann.
Astrophys. Journ., Vol. 197, 265 - 280 (1975).
Newly measured redshifts are given for a sample of remote Sc I galaxies in the apparent magnitude range $13.5 \lesssim m_{pg} < 15.8$ that were found by inspecting 33 Palomar Sky Survey plates in both galactic polar caps. The redshifts range from 2700 to 21,000 km s^{-1}. A bias is identified, and the material is suitably restricted to change the sample from a magnitude-limited set to one that is distance-limited. The statistical properties of the unbiased sample, combined with the absolute magnitude calibration in paper IV (Sandage, Tammann, 1974), give the Hubble constant to be $H_0 = 56.9 \pm 3.4$ km s^{-1} Mpc^{-1}. Upper limits for the deceleration parameter are calculated by comparing H_0^{-1} with the Friedmann time T_0, found by adding 10^9 yr to the age of the halo globular clusters, with the result that $q_0(H_0 = 55) = 0.10$ [$-0.08(\sigma)$, $+0.16(\sigma)$] and $q_0(H_0 = 50) = 0.20$ [$-0.12(\sigma)$, $+0.26(\sigma)$]. If one assumes instead that the Friedmann time T_0 is 4×10^9 yr earlier than the galactic globular clusters, based on the interpretation that the apparent redshift cutoff in quasar redshifts near $z = 4$ is real and is due to the matter-horizon of the formation of the first galaxies, then $q_0(H_0 = 50) = 0.03$ (-0.03, $+0.07$).

162.032 **The porthole effect and rings of fire in finite metagalaxies.** R. A. Breuer, M. P. Ryan.
Monthly Notices Roy. Astron. Soc., Vol. 171, 209 - 218 (1975).
If the metagalaxy is finite and surrounded by other metagalaxies within a larger universe than the authors show that radiation from outside would produce a bright ring in the direction closest to the 'edge' of our metagalaxy. The possible absence of any such effect is also discussed and it is argued that the 3 K blackbody radiation might not have originated within our metagalaxy. Separately the authors calculate the appearance of the outside world (radiation incoming from infinity) to an observer freely falling into a Schwarzschild black hole.

162.033 **A principle of impotence allowing for Newtonian cosmologies with a time-dependent gravitational constant.** P. T. Landsberg, N. T. Bishop.
Monthly Notices Roy. Astron. Soc., Vol. 171, 279 - 286 (1975).
A principle of impotence is given. Subject to some additional assumptions, it leads for Newtonian cosmologies to a gravitational interaction. This allows for the normal zero-pressure Friedmann cosmologies, the Dirac cosmology and also other models not previously considered.

162.034 **On conditions of fitting of the Friedmann universe with the empty space.** M. P. Korkina.
Astron. Zhurn. Akad. Nauk SSSR, Vol. 52, 299 - 305 (1975). In Russian. English translation in Soviet Astron., Vol. 19, No. 2.
The conditions of the fitting of a sphere of dust with the empty space are considered. The fulfilling of Lichnerowicz's conditions and the possibility of the identification of the interior and exterior coordinates in some known solutions have been analysed (Hoyle – Narlicar, Oppenheimer – Snyder, Novikov and others). The exterior solution, which satisfied all the necessary requirements, if the interior solution is Friedmannian, has been obtained.

162.035 **Are angular diameters a cosmological test?**
R. C. Roeder.
Journ. Roy. Astron. Soc. Canada, Vol. 69, 41 (1975). – Abstr. Canadian Astron. Soc.

162.036 **The homogeneity of the universe, the formation of galaxies, and fluctuations in the microwave background.** M. Clutton-Brock.
Journ. Roy. Astron. Soc. Canada, Vol. 69, 41 (1975). – Abstr. Canadian Astron. Soc.

162.037 **Evidence for a spatially homogeneous component of the universe: single galaxies.**
E. L. Turner, J. R. Gott III.
Astrophys. Journ., (*Letters*), Vol. 197, L89 - L93 (1975).

A study of the distribution in the sky of galaxies brighter than 14th magnitude reveals two populations. One strongly clustered population has a covariance function $\omega(\theta) \sim \theta^{-1}$ and contains roughly 60 percent of all galaxies. The remaining galaxies are distributed almost uniformly with $\omega(\theta) \approx 0$. The two populations are defined by the presence or absence (respectively) of nearby ($\lesssim 45'$) companions. The implications for the definition of clusters and the field, the growth of structure in the universe, and cosmology are discussed briefly.

162.038 **A new theory of the universe.** P. C. W. Davies.
Nature, Vol. 255, 191 - 192 (1975).

162.039 **Modelli cosmologici.** G. Romano.
Coelum, Vol. 43, 14 - 29 (1975).

162.040 **Energy conditions and inhomogeneities in the universe.** M. Heller.
Astrophys. Space Sci., Vol. 33, L33 - L35 (1975).

The so-called dominant energy condition and the strong energy condition (of singularity theorems) cannot be violated by the negative effective 'pressure' arising from the inhomogeneous distribution of matter in the universe.

162.041 **Viscous universes without initial singularity.**
M. Heller, Z. Klimek.
Astrophys. Space Sci., Vol. 33, L37 - L39 (1975).

It is shown that within a certain class of cosmological models (homogeneity and isotropy, two-component non-interacting fluid, density-dependent coefficient of bulk viscosity) the introduction of bulk viscosity effectively removes the initial singularity.

162.042 **On the theory of gravitational instability of an isotropic universe.**
G. M. Vereshkov, Yu. S. Grishkan, N. V. Pelikhov.
Izv. Sev.-Kavkaz. nauch. tsentra vyssh. shkoly. Seriya estestv. n., 1974, No. 2, p. 78 - 82. In Russian. — Abstr. in Referativ. Zhurn. 51. Astron., 3.51.856 (1975).

162.043 **The gravitational instability of a vacuum and the cosmological problem.** P. I. Fomin.
AN USSR. In-t teor. fiz. Preprint ITF-74-90R. Kiev. 10 pp. Price 3 Kop. (1974). In Russian. — Abstr. in Referativ. Zhurn. 51. Astron., 3.51.857 (1975).

162.044 **On some geometric properties of the universe as a whole.** V. V. Makarov.
Probl. yader. fiz. i kosmich. luchej. Resp. mezhved. temat. nauch.-tekhn. sb., 1974, vyp. (No.) 1, p. 39 - 43. In Russian. Abstr. in Referativ. Zhurn. 51. Astron., 3.51.868 (1975).

162.045 **Turbulence in cosmology. I. The effect of the strong turbulence on the early evolution of the universe.**
L. S. Marochnik, N. V. Pelikhov, G. M. Vereshkov.
Astrophys. Space Sci., Vol. 34, 233 - 247, 249 - 263 (1975). In Russian and English.

Principles of the theory of turbulence in relativistic cosmology are developed. By averaging Einstein's equations over stochastic fields a self-consistent system of equations is obtained which describes statistically: (1) the influence of the turbulence on the basic state of the universe (the background) on which the turbulence develops; (2) the behaviour of the turbulence on the background 'distorted' by it.

162.046 **Turbulence in cosmology. II. The asymptotic effect of collective excitations on the cosmological expan**sion of the homogeneous and isotropic, on the average, universe. L. S. Marochnik, N. V. Pelikhov, G. M. Vereshkov.
Astrophys. Space Sci., Vol. 34, 265 - 279, 281 - 295 (1975). In Russian and English.

At the stage of a weak turbulence the interactions between excitations are negligible, and potential, vortical and gravitational perturbations may be considered independent. In this approximation the analytical solutions to exact equations of the turbulence theory are found. A model of the universe from gravitational waves is constructed. The influence of the turbulence on the course of the expansion is essential till the beginning of, and probably during, the synthesis of light elements. The rate of cosmological expansion and gravitational instability decreases if the potential turbulence predominates over the vortical one, and increases in the opposite case.

162.047 **Entropie maximale et cosmologie de Friedmann.**
J.-P. Petit, G. Monnet.
Comptes Rendus Acad. Sci. Paris, Sér. A, Vol. 280, 1245 - 1248 (1975).

En 1934, Milne et McCrea montrèrent que l'on pouvait, dans une approche newtonnienne, en se basant sur les équations d'Euler, nanties d'hypothèses concernant le champ de vitesse et l'homogénéité de la solution, retrouver les aspects essentiels de la cosmologie relativiste de Friedmann. L'analyse suivante atteint le même but, mais avec une économie d'hypothèses, la seule étant que l'entropie est maximale en tout point. La constance de la densité de matière dans tout l'espace et le champ de vitesse de Hubble sont ici déduits et non introduits a priori.

162.048 **Proposed direct measurement of the cosmological deceleration parameter q_0.** M. M. Davis.
Bull. American Astron. Soc., Vol. 7, 236 (1975). — Abstr. AAS.

162.049 **Limits on Brans-Dicke cosmologies posed by "observational" constraints on H_0, t_0, q_0, and ω.**
R. C. Barnes, C. Eisenmann, R. Prondzinski.
Bull. American Astron. Soc., Vol. 7, 236 (1975). — Abstr. AAS.

162.050 **Newtonian cosmology with diminishing G.**
J. P. Vinti.
Bull. American Astron. Soc., Vol. 7, 341 (1975). — Abstr. AAS.

162.051 **Einige Probleme der Entwicklung des Kosmos.**
H.-J. Treder.
Astron. in der Schule, 12. Jahrgang, p. 28 - 30 (1975).

162.052 **Radioactivity and evolution of the universe.**
Ya. M. Kramarovskij, V. P. Chechev.
Priroda, No. 5.75, p. 54 - 62 (1975). In Russian.

162.053 **Neutrino in Friedmann's universe.**
N. A. Chernikov, N. S. Shavokhina.
Probl. teorii gravitatsii i ehlementarn. chastits. Vyp. (No.) 5. Moskva, Atomizdat, 1974, p. 154 - 162. In Russian. — Abstr. in Referativ. Zhurn. 51. Astron., 4.51.845 (1975).

162.054 **Connection between a central and an accompanying reference system and principles of general covariance in cosmological models of general relativity.**
K. P. Stanyukovich.
Probl. teorii gravitatsii i ehlementarn. chastits. Vyp. (No.) 5. Moskva, Atomizdat, 1974, p. 123 - 141. In Russian. — Abstr. in Referativ. Zhurn. 51. Astron., 4.51.846 (1975).

162.055 **Mach's principle in general relativity.**
D. J. Raine.
Monthly Notices Roy. Astron. Soc., Vol. 171, 507 - 528 (1975).

Mach's principle is taken as a criterion for selecting cosmological solutions of the Einstein field equations, in which, in a well-defined manner, the metric arises from material sources alone. In such model universes inertial forces are due to the gravitational interaction of matter, and there is a relativity of accelerated motion. Mach's principle is found to be satisfied in Robertson–Walker models and in a simple class of inhomogeneous solutions. These results lead us to suggest that Mach's principle may play a role in explaining the observed gross features of the universe.

162.056 **A note on the Zeldovich $p = \rho$ cold big bang.**
E. P. Liang.
Monthly Notices Roy. Astron. Soc., Vol. 171, 551 - 553 (1975).
Entropy production in $p = \rho$ cold baryon models is studied in the context of primordial shear damping. Special features of the dynamics of density inhomogeneities in such models are discussed.

162.057 **Cosmological model of space-time with torsion.**
V. N. Ponomarev.
Vestn. Mosk. un-ta. Fiz., astron., Vol. 15, 730 - 735 (1974). In Russian. – Abstr. in Referativ. Zhurn. 51. Astron., 5.51. 779 (1975).

162.058 **Violation of the CP-invariance and strangeness of the hot universe.** B. M. Pontekorvo.
Izv. vyssh. ucheb. zavedenij. Fizika, 1974, No. 12, p. 18 - 23. In Russian. – Abstr. in Referativ. Zhurn. 51. Astron., 5.51. 780 (1975).

162.059 **La métrique de de Sitter et sa duale.** I. Popovici.
Comptes Rendus Acad. Sci. Paris, Sér A, Vol. 280, 1329 - 1332 (1975).
On montre que la métrique de de Sitter possède une duale, en décrivant un champ gravitationnel tachyonique. On associe un univers mixte et on étudie la possibilité de communication entre ses deux zones opposées à l'aide de la lumière.

162.060 **Nonsingular Friedmann cosmology.**
Eh. B. Gliner, I. G. Dymnikova.
Pis'ma v Astron. Zhurn., Vol. 1, No. 5, p. 7 - 9 (1975). In Russian.
A cosmology based on the assumption that pressure is negative for high density is considered. There is no singularity in this model both in the past and in the future. Solution of Friedmann's equations results in a closed universe whose mass increases by many orders of magnitude from the beginning of expansion.

162.061 **The effect of a variable mass-luminosity ratio on the cosmic density.** T. W. Noonan.
Publ. Astron. Soc. Pacific, Vol. 87, 335 - 336 (1975).
The effect on the cosmic density of the possibility that galaxian mass varies as a power of luminosity is examined. The smaller the exponent, the greater the mean density. It is possible to close the universe if the brighter galaxies have a mass-luminosity ratio of 250 (Hubble constant 100 km sec^{-1} Mpc^{-1}) and mass varies as the one-fourth power of luminosity.

162.062 **Large-scale magnetic field and global structure of the universe.** D. D. Sokolov.
Pis'ma v Astron. Zhurn., Vol. 1, No. 6, p. 3 - 5 (1975). In Russian.
A class of locally "glued together" Friedmann universes is considered. A large-scale ($z \sim 0.0025 - 1.7$) magnetic field has a direction and thus destroys isotropy. Such a field is shown to decrease strongly the number of possible kinds of "glued together" universes. The importance of this restriction for "ghosts" finding is pointed out.

162.063 **Entropie maximale et univers tournants.**
J.-P. Petit, G. Monnet.

Comptes Rendus Acad. Sci. Paris, Sér. B, Vol. 280, 733 - 736 (1975).
Dans ce travail la solution maxwellienne instationnaire générale, satisfaisant aux équations de Vlasov et de Poisson est développée. Il apparait que cette solution est nécessairement homogène en température et en densité. Le champ de vitesse macroscopique est une superposition du champ de Hubble et d'un mouvement instationnaire de rotation en corps solide où la vitesse angulaire varie comme la température. La dimension caractéristique R de ce système newtonien obéit à une pseudo équation de Friedmann déjà obtenue par Heckmann et Schücking.

162.064 **Formation of galaxies in the expanding universe.**
A. G. Doroshkevich, Ya. B. Zel'dovich.
Astrophys. Space Sci., Vol. 35, 43 - 53, 55 - 65 (1975). In Russian and English.
The possibility of the formation of galaxies both during the galactic clustering of small structural units originated earlier and at fragmentation of larger formations is considered. A self-similar spectrum of long-wave perturbations is obtained. The results of numerical calculations and conclusions of Press and Schechter's (1974) work are discussed.

162.065 **Radar distance in Robertson-Walker space-times.**
R. C. Jennison, G. C. McVittie.
Commun. Roy. Soc. Edinburgh, (Phys. Sci.), No. 1, p. 1 - 17 (1975).
Radar distance is defined in terms of an observer's proper-time for target objects moving in his line-of-sight and is applied as a distance parameter in its own right without implying man-made transmissions. Robertson-Walker metrics (cosmological models) are employed corresponding to zero-pressure and zero cosmical constant models, except for the de Sitter model. Analytical methods apply in Milne's model and the de Sitter and Einstein-de Sitter models. Numerical computations are used for five other models. A table of radar distance versus the target's redshift is presented for all eight models.

162.066 **On the interpretative paradox in cosmology.**
M. Heller.
Acta Cosmologica, Fasc. 2, p. 37 - 41 (1974/75).
The question of defining a fundamental particle in relativistic cosmology is considered; this should change with cosmic time.

162.067 **On the formulation of the Copernican cosmological principle.** I. D. Karachentsev.
Acta Cosmologica, Fasc. 2, p. 43 - 48 (1974/75).

162.068 **Some thermodynamic features of cosmological models with viscosity.** Z. Klimek.
Acta. Cosmologica, Fasc. 2, p. 49 - 52 (1974/75).
Assuming isotropy and homogeneity of the universe which contains viscous matter the rate of entropy generation is calculated. It depends on bulk viscosity only. For extreme cases of incoherent matter and pure radiation the rate of entropy increase is obtained and its course is discussed near the singular point of cosmological models ("radius" $R = 0$).

162.069 **Uniform relativistic models of the universe with pressure.** B. Krygier, J. Krempeć.
Acta Cosmologica, Fasc. 2, p. 53 - 64 (1974/75).

162.070 **On the zone-universe.** A. Zięba.
Acta Cosmologica, Fasc. 2, p. 113 - 116 (1974/75).
In this paper the author considers a model of the universe consisting of concentric spherical zones alternately filled with matter and empty.

162.071 **Kinematical description of quasi-stellar objects.**

E. Alvarez, L. Bel, J. Gracia-Bondía.
Astron. Astrophys., Vol. 40, 381 - 386 (1975).

The authors use the general relativistic kinetic theory to describe the cosmological fluid which they consider as being a mixture of two gases: the gas of galaxies and the gas of QSO's. Assuming a generalized equilibrium distribution function for the latter, together with rather natural supplementary simplifying assumptions, they derive the statistical observables of the model. Comparing the theoretical results to the observational data the authors establish that it is in principle plausible to describe the gas of QSO's by a rather broad distribution function; this would mean that an important number of these are kinematically peculiar objects, i.e., objects having high velocities with respect to the mean motion of the cosmological fluid.

162.072 **On the time of galaxy formation in a turbulent universe.** S. A. Bonometto, L. Danese, F. Lucchin. Astron. Astrophys., Vol. 41, 55 - 59 (1975).

In the framework of a turbulent theory of galaxy formation, it is shown that the red-shift at which non-linear effects may become relevant on galactic scales should be lowered from ~1500 down to ~300, the main reason being that the residual interaction of matter with radiation is significant even when hydrogen ionization reaches very low values. As a consequence, galaxies would be formed later, and a long period of practically neutral hydrogen would be present between a red-shift somewhat lower than 1500 and the redshift ~300 after which matter can no longer be treated as incompressible.

162.073 **Development of the correlation of galaxies in an expanding universe.** K. Miyoshi, T. Kihara. Publ. Astron. Soc. Japan, Vol. 27, 333 - 346 (1975).

Development of the correlation of galaxies in an expanding universe is represented by computer simulations. Periodic structure with cubic unit cells, each containing 400 galaxies, is assumed. The spatial correlation $g(r) = (r_0/r)^s$ with $s \cong 2$ is generated from motionless galaxies distributed at random and also from a weak initial correlation. The characteristic length r_0 increases in such a way that the number of correlating galaxies increases; this gives a model for the formation of clusters of galaxies.

162.074 **Big bang cosmology.** K. Brecher.
High energy astrophysics and its relation to elementary particle physics, (see 012.018), p. 77 - 108 (1974).

162.075 **Microphysics, cosmology, and high energy astrophysics.** F. Hoyle, with notes by S. Fulling, C. Dyer.
High energy astrophysics and its relation to elementary particle physics, (see 012.018), p. 297 - 344 (1974).

162.076 **Electrodynamics and cosmology.**
J. V. Narlikar, with notes by J. Kiskis, H. Tesser.
High energy astrophysics and its relation to elementary particle physics, (see 012.018), p. 345 - 385 (1974).

162.077 **Weak interactions and cosmology.** Y. Ne'eman.
High energy astrophysics and its relation to elementary particle physics, (see 012.018), p. 387 - 414 (1974).

162.078 **Modern riddles of cosmology.**
G. Burbidge, M. Burbidge.
The heritage of Copernicus, (see 003.020), p. 116 - 139 (1974).

162.079 **Particle creation from vacuum in a nonstationary isotropic universe.**
A. A. Grib, S. G. Mamaev, V. M. Mostepanenko.

Izv. vyssh. ucheb. zavedenij. Fizika, 1974, No. 12, p. 79 - 84. In Russian. − Abstr. in Referativ. Zhurn. 51. Astron., 6.51. 888 (1975).

162.080 **Closed-form solutions for the evolution of density perturbations in some cosmological models.**
E. J. Groth, P. J. E. Peebles.
Astron. Astrophys., Vol. 41, 143 - 145 (1975).

The authors present closed-form solutions for the evolution of linear density perturbations (a) in a cosmological model where pressure can be neglected and $\Lambda = 0$; and (b) in a cosmologically flat model containing non-interacting matter and radiation, with $\Lambda = 0$ and the characteristic size of the irregularities much smaller than ct.

162.081 **Curves giving the age of the universe in Friedmann models.**
L. Campusano, J. Heidmann, J. L. Nieto.
Astron. Astrophys., Vol. 41, 229 - 231 (1975).

The authors give a set of curves $H_0 t_0$ = constant in the observationally relevant region of the $(\log \sigma_0, \lambda_0)$ plane, where H_0 is the Hubble constant, t_0 the age of the universe, σ_0 the density parameter and λ_0 the reduced cosmological constant. From the analysis of the observational data on H_0, t_0 and σ_0, they provide an upper limit to the cosmological constant.

162.082 **Einstein's principle of equivalence used for an alternative relativistic cosmology considering the system of galaxies as limited and not as the universe.** O. Klein.
Kon. Danske Vidensk. Selsk., Mat.-Fys. Medd., Vol. 39, No. 2, p. 3 - 18 (1974). In French.

162.083 **Neutrino cosmology.** S. A. Bludman.
AIP (American Inst. Phys.) Conference Proc., No. 22, p. 284 - 298 (1974).

162.084 **Radiation in cosmological backgrounds.**
S. C. Change.
Thesis, Pittsburgh Univ., Pittsburgh, Pa. (USA). 129 pp. University Mircrofilms Order No. 74-24,175 (1974).

162.085 **Structure and development of the universe.**
J. Bicak.
Cesk. Cas. Fys., Vol. 24, 425 - 445 (1974). In Czech. − Extended version of the lecture "Cosmology in the 5th century after Copernicus" read in Prague on April 19, 1973.

162.086 **Perspective of a finite and discrete universe.**
S. Thyssen-Bornemisza, H. W. Grayson.
Pure and Appl. Geophys., (Switzerland), Vol. 112, 257 - 264 (1974).

The anomalies in relativistic theory resulting from the infinities present in Riemannian geometry as used by Einstein are discussed and a better solution proposed by introducing a discrete and indivisible time unit.

162.087 **Friedmann universes containing wave fields.**
A. Das, P. Agrawal.
General Relativity and Gravitation, (GB), Vol. 5, 359 - 370 (1974).

The Friedmann universes containing (1) a massless real scalar field, (2) a massive real scalar field, (3) electromagnetic fields, (4) the combined massive complex scalar and electromagnetic fields are investigated.

162.088 **Cosmological singularities and higher-order gravitational Lagrangians.**
M. Giesswein, R. Sexl, E. Streeruwitz.
Phys. Letters B, (Netherlands), Vol. 52B, 442 - 444 (1974).

General relativity is modified by adding terms proportional to R^2 and $R^{\mu\nu}R_{\mu\nu}$ to the Lagrangian. One class of solutions of the modified field equations is free of singularities, but does not lead to asymptotic behaviour (for large time) of the Friedmann type. A second class, which shows the correct asymptotic behaviour, does contain the usual singularities of Friedmann universes, collapse being modulated by small oscillations only. The quantum effects considered here are thus unable to prevent the occurrence of cosmological singularities under physically reasonable conditions.

162.089 **Generalized shear-free singularities.** A. R. King.
General Relativity and Gravitation, (GB), Vol. 5, 371 - 377 (1974).

The possibility of matter singularities in a class of cosmological models is considered. The results are applied to shear-free models and suggest that in these the fluid cannot simultaneously expand and rotate.

162.090 **Tilting at cosmological singularities.**
C. B. Collins.
Commun. Math. Phys., (Germany), Vol. 39, 131 - 151 (1974).

A detailed investigation is made of the simplest type of general relativistic perfect fluid cosmological models that possess a singularity at which all physical quantities are well-behaved. These models are spatially homogeneous, axisymmetric generalisations of the open Robertson-Walker universes. A pictorial description of the evolution of the models is obtained by using the qualitative theory of differential equations.

162.091 **Preferred frames and oscillating universes.**
J. Katz.
International Journ. Theor. Phys., (GB), Vol. 10, 165 - 173 (1974).

A metric theory of gravitation is presented. It is based on the existence of a preferred 'cosmic' time. It agrees with all present experimental facts regarding gravitation and leads to singularity-free oscillating universes.

162.092 **On recent experiments to detect advanced radiation.**
P. C. W. Davies.
Journ. Phys. A, (Math., Nuclear, General), Vol. 8, 272 - 280 (1975).

Inconsistencies in the usual interpretation of the absorber theory of radiation are exposed which invalidated an experiment proposed recently by Heron and Pegg (1974). An earlier experiment by Partridge (1973) necessarily gave a null result owing to absorption on the far side of the earth of any advanced radiation which may have been present.

162.093 **Homothetic motions and perfect fluid cosmologies.**
C. B. G. McIntosh.
Phys. Letters A, (Netherlands), Vol. 50A, 429 - 430 (1975).

A theorem on homothetic (self-similar) motions in space-times with a perfect fluid is derived. The main result is that a perfect fluid cosmological model cannot have a non-trivial homothetic motion orthogonal to the fluid 4-velocity vector.

162.094 **Tachyon cosmology.** J. R. Ray.
Nuovo Cimento Lettere, Ser. 2, Vol. 12, 249 - 252 (1975).

Using Petrov's classification of gravitational fields (1969) by the symmetry of the space time, it has been shown that the gravitational field corresponding to the symmetry group $G_4 IV$ on V_3 allows only tachyon dust solutions. Another tachyon dust solution for this metric is presented which supports the view that the inverse could expand as a tachyon universe at each singularity.

162.095 **The cosmic microwave background and the hadron era of the early universe: a possible connection.**
M. Dersarkissian.
Phys. Letters B, (Netherlands), Vol. 55B, 84 - 86 (1975).

A cosmological model is constructed which is a Friedmann model, but with a finite ultimate temperature (T). A plausible argument is presented which suggests that the existence of T and the cosmic microwave background restricts the form of the hadronic level density.

162.096 **A perfect fluid distribution conformal to de Sitter universe.** K. P. Singh, S. Ram.
Indian Journ. Phys., Vol. 48, 1046 - 1048 (1974).

The authors construct a metric of space-time representing perfect fluid distribution conformal to the de Sitter universe which can be transformed into Schwarzschild interior solution with constant matter density. It is shown that the only static spherically symmetric solution conformal to the de Sitter universe representing a perfect fluid is given by Schwarzschild's interior solution.

162.097 **Origin of the kinetic energy of expansion of the universe in the Fermi energy of the Ylem.**
L. S. Levitt.
Nuovo Cimento Lettere, Ser. 2, Vol. 12, 537 - 541 (1975).

Gamow's theory of the origin of the universe by a catastrophic explosion of a small, hot and dense primordial cluster of material is considered. It is shown that the kinetic energy of expansion is equal to the potential energy of attraction. This means that the universe is periodic in both space and time — an oscillating pulsating universe.

162.098 **Closed gravitational-wave universes: analytic solutions with two-parameter symmetry.**
R. H. Gowdy.
Journ. Math. Phys., New York, Vol. 16, 224 - 226 (1975).

Explicit solutions are presented for Einstein's vacuum field equations for spacetimes which have two-parameter spacelike symmetry, a space-reflection symmetry and space sections homeomorphic to either $S^1 \times S^2$ or S^3.

162.099 **On characteristic turbulent quantities at the bounce epoch in our hadron-dominated model of the universe.** H. Nariai.
Progr. Theor. Phys. Japan, Vol. 53, 287 - 289 (1975).

A phenomenological modification of Einstein's field equations has been proposed in order to arrive at an isotropic model universe which may time-symmetrically bounce in the hadron era, tends rapidly to the Friedmann universe, and remains regular with respect to the excitation of gravitational and rotational response.

162.100 **Observational consequences of a 'domain' structure of the universe.** A. E. Everett.
Phys. Rev. D, Particles and Fields, Vol. 10, 3161 - 3166 (1974).

Studies observational consequences of Weinberg's suggested 'domain' structure in the universe — the preferred direction in charge space is not the same at all points in ordinary space. Shows that a discontinuous boundary between such domains would form a perfect mirror reflecting all incident electromagnetic radiation while a smooth transition of thickness that is large compared with a wavelength of the radiation would allow total transmission.

162.101 **Regge-calculus model for the Tolman universe.**
P. A. Collins, R. M. Williams.
Phys. Rev. D, Particles and Fields, Vol. 10, 3537 - 3538 (1974).

Approximates the Tolman universe with 600 equilateral tetrahedrons, each containing blackbody radiation, connected together to form a closed space. Using Regge calculus to study the time evolution of this universe.

162.102 **The universe is the unity of infinitude and finiteness.**

S.-Z. Bian.
Acta Phys. Sinica, Vol. 23, 83 - 94 (1974). In Chinese.

162.103 **Kosmische Evolution.** A. Unsöld.
Naturwiss. Rundschau [Wiss. Verlagsgesellschaft, Stuttgart], Vol. 28, 3 - 14 (1975).

162.104 **Contribution à l'étude d'une masse sphérique plongée dans un univers cosmologique.**
J. Eisenstaedt.
Thesis, Univ. Paris VI. AO–CNRS–10583, 115 pp. (1974).

On explicite la forme générale des équations du champ à symétrie sphérique pour un fluide parfait et dans certaines conditions permettant de construire des modèles d'univers au sens de Lichnerowicz et d'écrire sous une forme simple les équations du champ en utilisant un système de coordonnées spatialement isotopiques et entraînées. On étudie la structure des équations du champ et on adopte une classification des solutions.

162.105 **Some cosmological consequences of imaginary mass.**
P. C. W. Davies.
Nuovo Cimento B, Ser. 11, Vol. 25, 571 - 580 (1975).

162.106 **Flat anisotropic models of the universe with torsion and without singularity.** B. Kuchowicz.
Journ. Phys. A, (Math. Nuclear, General), Vol. 8, L29 - L31 (1975).

The intrinsic angular momentum of the cosmological substratum is able to prevent the appearance of the cosmological singularity in the Einstein-Cartan theory. Conditions for this to occur in Bianchi type I cosmologies are given, and several new cosmological models whose anisotropy diminishes with expansion are explicitly constructed.

162.107 **Cosmological models and non-denumerable singularities.** R. Schlegel.
International Journ. Theor. Phys., (GB), Vol. 12, 217 - 223 (1975).

A review of the standard cosmological models shows that the positive cosmological constant gives rise to exponential increase in spatial extension. Such an increase is contradictory if extended over infinite future time, since even in an infinite universe there can only be a denumerable infinity of finite spatial units.

Edwin Hubble and the universe outside our Galaxy.
See Abstr. 004.080.

Strong interactions, gravitation and cosmology.
See Abstr. 022.085.

Solar test of Dirac's large numbers hypothesis.
See Abstr. 043.001.

A determination of the rate of change of G.
See Abstr. 043.003.

Speculations on detection of the "neutrino sea".
See Abstr. 061.016.

Effects of primordial fluctuations on the abundances of light elements. See Abstr. 061.017.

On emission lines in the cosmic gamma-ray background. See Abstr. 061.047.

Time delays for multiply imaged quasars.
See Abstr. 066.003.

Cosmological effects of primordial black holes.
See Abstr. 066.006.

On the origin of the microwave background.
See Abstr. 066.032.

On the ability of current experiments to test π^0 - **decay gamma-ray background theories.**
See Abstr. 066.041.

Gravitational waves retaining the homogeneity of space. See Abstr. 066.051.

Unphysical solutions of Yang's gravitational-field equations. See Abstr. 066.094.

Cosmological singularities in a theory of gravitation with second order perturbations. See Abstr. 066.127.

A solar model with low neutrino emission.
See Abstr. 080.023.

The use of supernovae for determining the Hubble constant and estimating extragalactic distances.
See Abstr. 125.029.

The spatial distribution and cosmological evolution of scintillating radio sources. See Abstr. 141.025.

Significance of the angular diameter-redshift relation. See Abstr. 141.046.

The identification of absorption redshift systems in quasar spectra. See Abstr. 141.056.

On the pregalactic helium to hydrogen abundance ratio. See Abstr. 155.030.

Systematic errors in the velocities of galaxies.
See Abstr. 158.025.

Statistical analysis of catalogs of extragalactic objects. VI. The galaxy distribution in the Jagellonian field.
See Abstr. 158.029.

Spatial distribution of galaxies. See Abstr. 158.123.

The influence of the Scott effect on the determination of q_0. See Abstr. 160.012.

Does the Local Supercluster rotate?
See Abstr. 160.013.

Apparent galaxy clustering caused by low-frequency gravitational radiation. See Abstr. 160.014.

Studies of rich clusters of galaxies–III. Photometry of the cluster A1930 and the m^*-z **relation.** See Abstr. 160.015.

Core radii and central densities of 15 rich clusters of galaxies. See Abstr. 160.025.

Sur un modèle décomposable d'univers hiérarchisé: déduction des corrélations galactiques sur la sphère céleste.
See Abstr. 160.027.

Collisions and relaxation in galaxy clusters.
See Abstr. 160.028.

Search of an intergalactic component of Faraday rotation. See Abstr. 161.005.

Author Index

The authors are listed in alphabetical order
according to the initial letter following the first names.

Banfi, V.
047.020
Bangaru, B. R. P.
022.026
Bania, T. M.
155.035
Banos, G.
073.031
Bansal, B. M.
094.154 .491 .495 .496
.559
Bappu, M. K. V.
008.067
075.002
103.100
114.076 .358
141.099
158.129
Bar-Nun, A.
082.022
099.003
131.054
Barabashov, N. P.
094.106
Baranne, A.
133.032
Baranov, A. S.
151.003 .030
Baranov, A. V.
072.065
Baranov, V. B.
074.081
Baranovskij, I. V.
071.006
Barbanis, B.
008.096
Barbera, R.
003.031
Barbier, M.
112.013
114.017
121.064
Barbieri, C.
041.019
Barbon, R.
158.039
Barbour, J. B.
003.025
Bardas, D.
142.047
Bardeen, J. M.
066.010
Bardwell, C. M.
098.072
Barfield, J. N.
084.219
Barish, F. D.
062.060
Barkat, Z.
066.029
Barker, C.
094.514
Barker, E. S.
093.021 .024 .036 .037
097.053
Barkhatova, K. A.
013.014
Barkstrom, B. R.
091.026
Barlow, B. V.
003.032

Barlow, M. J.
064.002
114.070
Barnard, A. J.
022.007 .042
114.025
Barnes, J. V.
113.019
131.063
Barnes, R. C.
162.049
Barnes III, T. G.
122.003 .033
Barnhart, P. E.
073.114
Barnothy, J. M.
141.071 .072 .082 .344
Barnothy, M. F.
141.071
Barocas, V.
010.012
014.010
075.003
Baron, R. L.
094.179 .541
Barouch, E.
143.020 .095
Barrett, A. H.
131.124
Barricelli, N. A.
094.116 .238
Barros, S.
103.109 .122 .131 .132
.137
Barrow, C. H.
077.040
Barrowes, S.
143.026
Barsuhn, J.
022.071
Bartholdi, P.
096.008 .009
Bartlett, I. R.
034.011
Bartoe, J.-D. F.
034.015 .122
076.031
Bartolini, C.
121.048
123.017
Barua, J.
061.075
Baschek, B.
003.001
114.028
Basford, J. R.
094.502 .532
Bash, F. N.
141.074
Basko, M. M.
065.080
Bastin, J. A.
009.022
Basu, B.
151.040
155.057
158.001
Basu, D.
141.090
Basu, S.
083.032

Batchelor, R. A.
132.032
Bateson, F. M.
010.024
122.066
123.009 .010 .011
Bath, G. T.
117.015
Batson, R. M.
003.121
092.018
Battaglia, M. C.
099.233
Batten, A. H.
117.002
119.011
121.022 .060
Battistini, P.
121.029
141.095
Baturina, G. D.
041.055
Bauer, J.
122.117
Bauer, J. F.
094.447
Bauer, M.
094.534
Bauer, S. J.
031.203
Baum, P. J.
062.039
Baum, W. A.
091.024
Baur, T. G.
034.041 .043
Bazilevskaya, G. A.
078.010
143.008
Bazilevskij, A. T.
097.023
Beals, C. S.
094.110
Beard, D. B.
084.252
Beardsley, W. R.
114.364
Beaver, E. A.
141.009
Beavers, W. I.
008.003
096.002
Beck, F. B.
033.058
Beck, J. N.
105.094
Becker, G.
044.002
Becker, R. H.
125.004 .037
142.058
Becker, R. S.
131.064
Beckers, J. M.
021.005
071.031
073.002
080.001
Becklin, E.
132.027

Cartwright, D. C.
084.010
Carusi, A.
094.462
107.003
Carvalho Sr., G.
003.031
Carver, E. A.
104.009
Carver, J. H.
094.111
Casamassima, F.
075.010
Casasayas, F.
031.418
Casasent, D.
031.418
Casini, C.
158.055
Cassen, P. M.
094.010
Casserly Jr., R. T.
084.029
Cassinelli, J. P.
064.049
Castellani, V.
065.050 .077
154.011
Castelli, J. P.
077.055 .069
Casti, J. L.
091.010
Castillo, N. D.
099.030 .045
Castor, J. I.
064.004
Caswell, J. L.
142.053
Catalano, S.
085.001
Cataldo, J. C.
155.002
Catchpole, R. M.
113.054
114.361
Cato, B. T.
131.529
Catura, R. C.
034.082
074.005 .067
076.043
125.036
142.038
Cavallo, G.
158.133 .134
Cavani, C.
142.001
Cavarretta, G.
094.462
Cayrel, R.
064.036
Cayrel De Strobel, G.
064.036
Cazenave, A.
081.028
Cazenave, M.
004.060
Cazzola, P.
065.107
Cecchini, S.
143.094

Cecere, A.
084.214
Celis S., L.
114.088
Ceplecha, Z.
104.003
Cesarsky, C.
131.110
Cesarsky, C. J.
131.087
Cess, R. D.
091.011
093.015 .028
099.059
Chaffee, F. H.
064.017
Chaisson, E. J.
103.100
131.010 .057
Chajko, O. N.
014.011·
Chalikov, D. V.
003.041
Chalonge, D.
007.000
Chamberlain, J. W.
082.010
093.004
Chambers, R. M.
034.078
Champion, K. S. W.
082.065
Chan, K. L.
066.058 .071
162.001
Chan, Y.-W. T.
141.058 .111
Chandler II, P. P.
004.076
Chandra, S.
083.017 .066
Chandra, S. K.
066.128
Chandrasekhar, S.
066.075
Chang, S.
094.402 .515
Chang, T. C.
082.009 .021
Change, S. C.
162.084
Chanmugam, G.
141.362
Chao, E. C. T.
094.140
Chao, J. K.
084.271
106.048
Chao, N. C.
065.092
Chapelle, J.
022.079
Chapkynov, S. K.
034.067
Chapline, G. F.
066.006
Chapman, C. R.
098.001 .002 .004 .011
.032 .033 .035 .050
Chapman, G. A.
034.042

Chapman, G. A.
071.009
072.021
Chapman, R. D.
072.037
074.070
076.020
Chapman, S.
003.024
Chappell, C. R.
084.022
Chapront, J.
042.008 .031
Charakhch'yan, A. N.
143.075
Charakhch'yan, T. N.
078.010
143.008 .075
Chareton, M.
113.065
Charland, Y.
114.006
Charles, P. A.
125.011 .017 .035
132.012
142.030 .083
Charles, R.
010.041
Charman, W. N.
061.010
Charugin, V. M.
141.314
Chase, R. C.
072.036
074.054 .099
Chase, S. C.
099.031 .068 .069
Chashej, I. V.
074.076 .147
106.023
141.311 .321 .360
Chashey, I. V.
See Chashej, I. V.
Chasovitin, Yu. K.
083.057
Chatfield, C. N.
010.008
Chattopadhyay, T.
077.020
Chaturani, P.
061.077
Chaumont, J.
094.160
Chauville, M.-T.
119.014
Chavira, E.
122.097
Chebotarev, G. A.
098.027
Chechetkin, V. M.
065.076
Chechev, V. P.
162.052
Chejdo, G. P.
031.402
Chen, D.
099.209
Chen, J. C.
094.573
Chen, J. H.
105.025

Davis, J.
158.035
Davis, J. E.
158.081
Davis, J. H.
141.030
Davis, J. M.
072.036
076.029
Davis, M.
034.022
158.036
Davis, M. M.
162.048
Davis, P. R.
094.167
Davis, R.
105.021
Davis, R. J.
158.314
Davis, T. N.
084.013
Davis, W. D.
077.038
Davis Jr., L.
099.025 .046 .100
Davis Jr., R.
094.534
Davison, P.
142.126
Davison, P. J. N.
134.002
142.029 .033 .083
Davydov, V. D.
097.075
Davydovskij, V. Ya.
062.031
Day, C.
123.033
Day, G. A.
132.039
De, Bibhas R.
062.023 .050
073.058
De, M.
031.278
De Bary, E.
082.026
De Bergh, C.
100.027
De Boer, K. S.
114.304
122.032
142.116
De Callatay, V.
003.050
De Doncker, E.
071.010
De Feiter, L. D.
073.025 .034
076.055
De Freitas Mourao, R. R.
031.222
De Freitas Pacheco,
J. A.
142.002 .025
De Graaf, T.
061.065
De Graauw, T.
012.009
034.063 .105

De Groot, T.
007.000
010.019
De Hon, R. A.
094.433
De Jager, C.
064.038
071.044
073.003
099.234
De Korte, P. A. J.
061.073
De La Cotardiere, P.
092.002
095.002
097.039
103.101
De La Reza, R.
071.015
Le Loore, C.
065.104
114.305 .330
142.063
De Lorenzo, L. J.
033.050
De Meyer, F.
081.016
Le Moraes, R. V.
052.033
Le Mottoni Y Palacios,
G.
097.038 .091
Le Ruiter, H.
141.105
Le Sa, A.
031.254
De Sanctis, G.
098.053
De Santis, R.
121.047
Le Vaucouleurs, A.
158.034
Le Vaucouleurs, G.
097.035
158.034
De Vries, T.
004.045
De Zotti, G.
065.039
Dean, J. F.
158.024
Dean, W. A.
083.056
Dearborn, D. S.
065.003
Deasy, V.
014.010
Debarbat, S.
041.007 .008
Debehogne, H.
031.219 .220 .222
098.065
Dedkov, G. V.
063.006
Deerenberg, A. J. M.
061.004
Deforest, S. E.
084.232
DeGasparis, A. A.
105.057

Deharveng, J. M.
158.012 .022
Deharveng, L.
131.535
Dehmel, G.
034.129
Dehnen, H.
141.055
162.022
Deines, P.
105.127
Deitz, P. H.
031.210
Dejaiffe, R.
054.010
097.095
Dekker, W. L. B. J.
054.011
Del Wiseman Jr., J.
009.001
Delaboudiniere, J. P.
034.081
Delaney, T. J.
061.010
Delano, J. W.
094.474
Delcourt, J.
104.020
Delgado, R. F.
032.044
DeLoach, A. C.
072.039
074.051
Delplace, A. M.
114.010
Delsemme, A. H.
102.001 .020 .025
DeLuisi, J. J.
082.015
Delvoye, L.
034.034
Demaret, J.
065.022
Demarque, P.
065.046
080.007
Dembovskij, A. V.
097.061
Dement'eva, N. N.
097.019
Demianski, M.
066.113
Demin, V. G.
052.040
Demin, V. V.
131.113
Deming, D.
118.011
Deming, L. D.
115.009
Dempsey, M. J.
131.082
Den Boggende, A.
142.115
Den Boggende, A. G. F.
142.116
Dence, M. R.
094.436 .443
Denis, J.
065.078
117.037

Dumortier, B.
158.101
Duncan, A. R.
094.405 .407 .493
Duncan, B.
099.037
Duncan, B. J.
034.039
Duncombe, R. L.
007.000
021.007
Dungey, J. W.
084.034 .234
Dunham, D. W.
096.009
115.001
Dunham, T.
100.208
Dunkelman, L.
106.046
Dunlap, J. R.
125.060 .102 .103
Dunlap, Y.
125.060 .102 .103
Dunlop, S. R.
031.014
Dunn, J. R.
094.200 .201
Dunn, R. B.
034.118
071.042
Dunphy, P. P.
143.090
Duorah, H. L.
065.086
Duorah, K.
065.086
Dupal, I.
095.012
Dupree, A. K.
114.346
Dupuy, D. L.
122.033
Duquesne, M.
003.057
Durgapal, M. C.
141.110 .115
Durisen, R. H.
065.007
Durkovic, P. M.
047.021
Durney, B. R.
074.021 .025
080.038
Durrani, S. A.
094.186 .547
Duruy, M.
034.092
158.316
Duthie, J. G.
008.107
141.357
Duval, M.
114.312
Duvall, T. L.
009.015
Duxbury, T. C.
097.203
099.044 .214
Dvornikov, V. M.
143.032

Dworak, T. Z.
121.037
122.082
158.117
Dworetsky, M. M.
153.005
Dwornik, E. J.
094.485 .492
D'yachenko, V. A.
082.083
D'yachkov, A. V.
097.019
D'yachkov, V. E.
033.018 .020
Dyadichev, V. N.
083.044
Dyal, P.
094.196 .215
099.025 .100
Dyck, H. M.
064.009
114.302
141.601
Dyer, C.
162.075
Dyer, E. P.
076.032
Dyer, G. C.
005.009
Dyer, J. W.
053.016
Dymek, R. F.
094.442
Dymnikova, I. G.
162.060
Dynan, S. E.
022.073
Dynkin, S. D.
066.088
Dyson, P. J.
031.027
Dzhuman, B. M.
082.006
Dzhun', I. V.
031.228
Dzurisin, D.
092.020
094.226
Dzyamko, S. S.
097.074

Eachus, L. J.
141.106
Eardley, D. M.
066.031
Earl, J. A.
143.019
Easson, I.
065.052
Eastlund, B. J.
141.330
Eastwood, J. W.
084.204
Eaton, J. A.
121.020
Eberhardt, P.
074.122
094.509 .565
Eberhart, J.
099.102

Ebisawa, S.
097.065 .071 .072
Ecklund, W. L.
084.020
Eddy, J. A.
034.120
072.045
Eden, R. C.
034.008
Ederer, D. L.
031.230
Edgar, B. C.
082.059
083.069
099.002
Edmonds Jr., F. N.
071.036
Edmondson, F. K.
008.020
098.016
Edmunds, M. G.
022.014
061.079
065.041
066.086
155.018
Edwards, I. E. S.
012.004
Edwards, K. J.
083.021
Edwards, P. L.
158.305
Eelsalu, H.
155.011 .012
Efanov, V. A.
131.113
Efimov, A. I.
097.021
Efimov, Yu. E.
078.027
Efremov, A. I.
077.034
084.203
Efremov, Yu. N.
003.012 .037
122.057
Efurd, D. W.
105.094
Egan, W.
097.048
Egan, W. G.
097.092
100.010
Eggen, O. J.
113.017 .018
115.901 .902
122.009 .019 .045
155.025
Egger, F.
007.000
Eggleton, P. P.
065.003
Eggleton, R. E.
094.213
Eglinton, G.
094.167 .494
097.082
Egorov, I. V.
094.136 .228 .244
Egret, D.
113.064

Hart, M. H.
082.056
Hart Jr., H. R.
094.536
143.036
Harten, A.
151.047
Harten, R. H.
157.003
Hartle, J. B.
065.067
066.002 .030
Hartle, R. E.
092.005 .010
Hartman, R. C.
061.039
143.041
Hartmann, R.
074.114
Hartmann, W. K.
003.077
091.021
094.129
097.052
107.007
Hartner, W.
004.056
Hartoog, M. R.
114.046 .351
Hartung, H.
004.065
Hartung, J. B.
094.103 .120 .176 .182
. 185 .429
Hartwick, F. D. A.
131.007
154.008
Hartzman, M. J.
094.456
Harvey, C. C.
077.002
Harvey, J.
071.032
Harvey, J. W.
071.007 .008
072.001
076.029
Harvey, K. L.
072.001
Harvey, P. M.
141.605
Harwit, M.
133.016
Harwood, J. M.
096.001
Hasegawa, A.
003.078
084.239
Haser, L.
031.267
Haskin, L. A.
094.495
Haslam, C. G. T.
125.028
Hass, G.
031.030
Hassan, S. M.
153.022
Hatfield, B.
123.035

Hatfield, H.
008.053
Hatfield, H. R.
010.012
Hattori, A.
097.070
Hauck, B.
113.036 .037 .047
114.080
Hauge, Oe.
071.016 .048
Haupt, H.
009.020
Haupt, W.
122.005
Haviland, R. P.
015.004
Havlen, R. J.
131.130
155.038
Havnes, O.
114.011
Hawarden, T. G.
153.901
Hawke, B. R.
094.489
Hawkes, R. L.
082.013
104.005 .010
Hawkins, F. J.
142.014 .039
Hawkins, G. S.
004.036
083.001 .065
Hayakawa, S.
003.079
061.001 .004 .019
131.078
142.087
Hayatsu, R.
105.089
131.137
Hayden, E.
122.117
Hayes, D. P.
131.056 .109
Hayes, D. S.
012.008
082.044
113.038
114.326 .347
131.095
Hayes, J. M.
094.516
Hayes, R. W.
022.004
Hayes, S. H.
101.009
114.326
Haymes, R. C.
142.118
Hays, J. F.
094.453
Hays, P. B.
082.034 .072 .081 .082
Hayward, R. R.
008.009
Hazen, N. L.
032.041
Hazra, L. N.
031.278

Head, J. W.
094.144 .229
Healy, A. W.
080.901
Heap, S. R.
133.007
Heaps, M. G.
099.002
Heard, J. F.
007.000
119.005 .006
Hearn, A. G.
064.044 .069 .070
Hearn, D. R.
142.123
Hearnshaw, J. B.
114.014 .029
Heasley, J. N.
071.034
073.077 .112
Heaton, J. W.
054.004
Heaton, K. C.
065.032
Heck, A.
111.003
Heckmann, O.
041.037
Hedeman, E. R.
072.044 .061
Hedervari, P.
094.275
Hedgecock, P. C.
034.080
054.016
074.130
084.267
Heeschen, D. S.
008.033 .051 .130
033.013
158.027
Hegwer, F.
073.004
Heidmann, J.
032.020
158.055 .127
162.081
Heidmann, N.
064.068
158.002
Heier, K. S.
094.483
Heiken, G.
094.476 .481
Heiken, G. H.
094.147
105.063
Heiles, C.
131.085 .513
157.901
Heiles, C. E.
131.100
Heinemann, M. A.
117.041
Heintz, W. D.
008.121
118.002 .018
Heintze, J. R. W.
114.057
Heintzmann, H.
065.011 .101

Kilner, J. R.
142.091
Kilston, S.
114.034
Kilston, S. D.
065.093
Kimball, D. S.
084.001
Kimmer, E.
061.042
Kimpara, A.
083.023
Kinard, W. H.
106.026
King, A. R.
162.089
King, D. S.
122.004 .054
King, I. R.
154.016
King, M. W.
114.364
King III, C.
141.013
King Jr., E. A.
094.475
King-Hele, D. G.
004.013
009.023
012.004
081.004
082.074
Kingsley, S. P.
083.033
Kinman, T. D.
158.099 .304
Kinoshita, H.
043.011
Kintanar, R. L.
047.007 .008 .014
Kiosa, M. N.
042.078
Kiplinger, A. L.
124.101
Kippenhahn, R.
003.087
Kipper, T.
031.264
Kirakosyan, R. M.
033.005
Kiraly, P.
143.044
Kirichenko, A. G.
098.045
Kirichuk, V. V.
046.001 .027
Kirillov-Ugryumov, V. G.
142.017 .070
Kirpatovskij, V. M.
032.038
Kirshner, R. P.
125.016
Kirsten, T.
094.007 .506
Kisabeth, J. L.
084.232
Kiselev, A. A.
034.108
041.066
112.015

Kiseleva, T. P.
041.062 .063
Kishonkov, A. K.
074.116
Kiskis, J.
162.076
Kislyakov, A. G.
077.054
Kislyuk, V. S.
094.243 .255
Kissel, J.
104.032
106.007
Kissin, S. A.
105.070
Kivelson, M.
084.022
Kivelson, M. G.
084.220 .262 .410
Kiviniemi, A.
081.018
Kizilirmak, A.
007.000
Kizyun, L. N.
094.257
Klaasen, K.
092.014 .019
Klaasen, K. P.
092.015 .017
Klaine, B. I.
See Klajn, B. I.
Klajn, B. I.
084.208 .240
Klare, G.
155.047
Klebesadel, R. W.
142.091
Klein, H. P.
097.079
Klein, L.
094.173
Klein, L. A.
032.039
Klein, M. J.
099.005 .060 .061
Klein, O.
162.082
Klein, R. I.
064.004 .074
Kleinmann, D. E.
158.100
Klejmenova, N. G.
083.058
Klement, G.
098.009
Klement, G. T.
079.100
Klemperer, W. K.
141.021
Klepczynski, W. J.
012.005
099.036
103.105
Klikh, Yu. A.
042.081
Klimas, A. J.
062.012 .060
Klimek, Z.
121.017
162.030 .041 .068

Kliment, V.
094.411
Klimishin, I. A.
131.111
Klimuk, P. I.
082.048
Klinglesmith, D. A.
031.231
114.319
Kliore, A.
099.018 .032
Kliore, A. J.
099.043 .044 .227
Klobuchar, J. A.
083.001 .065
Klose, J. Z.
022.060
Klosko, S. M.
081.010
Klvana, M.
034.068
Kment, M.
095.012
Knacke, R. F.
008.119
100.006 .012 .214
131.129
152.002
Knapp, G. R.
131.027 .028
132.001 .011
155.016
Knapp, S. L.
115.010
131.027 .028
Kneubuehl, F. K.
071.012
Kniffen, D. A.
061.039
142.062
Knight, C. A.
141.021
Knight, J. W.
074.047
Knight, P. R.
123.034
Knoska, S.
072.051
Kobanov, N. I.
034.002 .003
Kobayashi, N.
035.010
Kobrick, M.
094.213
Kobzev, A. A.
034.048
Kobzev, V. A.
078.028
Koch, D.
134.005
142.036
Koch, G. F.
033.036
Koch, K.-R.
081.012
Koch, M.
009.016
Kocharov, G. E.
011.040
034.102
061.032 .064

MacLeod, J. M.
131.103
158.074
MacQueen, R. M.
034.120 .126
074.020 .057 .095 .103
MacRae, D. A.
008.105
114.051
Macris, C.
008.008
Macris, C. J.
073.084
Macris, C. L.
073.072
Macris, G.
042.036
Macy, W.
093.021
Macy Jr., W.
099.219
100.005
101.012
Madhyastha, V. L.
061.070
Madore, B. F.
121.076
122.021
153.009
Maeckle, R.
064.014
114.316
Maeda, K.
073.074
084.018
Maeder, A.
065.063 .074
113.035
Maehl, R. C.
143.088
Maeva, S. V.
094.002
Maffei, P.
002.008
122.073
123.019
Magee Jr., N. H.
061.015 .045
Magerramov, V. A.
071.043
Magnan, C.
063.015
114.350
Magnant-Crifo, F.
074.024
Magnenat, P.
113.037 .047
Magni, G.
107.003
Magri, W.
015.013
Maguire, W. C.
093.026
Mahajan, V. N.
082.030
Maheswaran, M.
141.362
Mahmood, A.
094.174
Mahra, H. S.
122.107

Maihara, T.
034.017 .018
Maillard, J. P.
100.027
Maillie, H. D.
099.074
Mainstone, J. S.
084.211
Maistrov, L. E.
004.040
Maitzen, H. M.
114.073
Major, S. P.
044.006
Makalkin, A. B.
091.016
Makarenko, N. L.
003.084
Makarikhin, S. I.
052.016
Makarov, G. V.
046.017
Makarov, I. N.
044.015
Makarov, O. F.
042.081
Makarov, V. V.
162.044
Makarov, Yu. V.
094.417
Makino, F.
061.001 .019
142.087
Maksimov, I. V.
085.014
Maksimova, I. I.
031.228
Maksumov, M. N.
151.043
Maksyukov, N.
061.049
Malacara, D.
031.024 .036
Malacara, Z.
122.011
Malanin, V. V.
052.017 .070
Malaroda, S.
153.010
Malcuit, R. J.
094.109
Malecek, B.
007.000
009.026
Malin, M. C.
092.020
094.226
097.045
Malin, S. R. C.
084.214
Malitson, H. H.
077.044 .045
Malkov, Ya. V.
097.016
Mallama, A. D.
114.341
121.023 .043 .113
Mallia, E. A.
034.071
154.003 .018

Mallik, D. C. V.
131.516 .537
Mallove, E. F.
002.002 .004
Mal'tsev, Yu. P.
084.258
Mal'tseva, O. A.
083.027 .028
Malville, J. M.
073.043
Mama, H. P.
013.004
Mamaev, S. G.
162.079
Mammano, A.
114.030 .333
Mamoshina, I. P.
097.029
Mamrukova, V. P.
143.007
Manabe, S.
155.019
Manara, A.
054.017
Manchanda, R. K.
142.064 .087
Manchester, R. N.
141.303 .318 .355
Mancuso, S.
121.014
Mandel, L.
031.225
Mandelbrot, B.
160.027
Mandel'shtam, S. L.
032.028
073.012
076.012
Manella, J.
122.037
Mankin, W. G.
034.120
Manning, P. G.
131.062
Manokhina, A. V.
106.002
Manoyan, V. M.
004.062
Mantas, G. P.
083.012
Mao, H. K.
094.439 .466 .471
Maran, S. P.
009.013
054.012
103.100
Marandino, G. E.
141.021
Maraschi, L.
161.006
Marchal, C.
052.034
Marchenko, O. A.
078.008
Marcus, H. L.
094.542
105.055
Marechal, A.
012.022
Margon, B.
142.039 .043 .050 .090

Maurer, P.
094.509 .565
Maurette, M.
074.040 .120 .123
094.160 .161 .409 .494
Maurice, E.
031.412
034.096 .098
Mauron, N.
121.088
Mavashev, Yu. Z.
011.002
Mavraganis, A.
042.002
Mavridis, L. N.
008.124
Max, C. E.
131.040
Maxwell, J. R.
094.167
Mayall, M. W.
010.001
Mayall, R. N.
010.001
Mayeda, T. K.
094.152 .162
Mayer, E.
123.039 .040 .042
124.104 .108
Mayer, E. H.
031.270
Mayer, W.
142.121
Mays, B. J.
094.494
Maza, J.
082.095
Mazets, E. P.
061.023
106.001 .015
Mazurek, T. J.
125.047
McAdoo, D. C.
097.041
McAllister, H. C.
071.041
076.009
McBreen, B.
134.005
142.036
McCall, H. F.
034.085
McCall, S. L.
125.049
McCallister, R. A.
103.016 .128
McCallister, R. H.
094.478
McCallum, I. S.
094.445 .558
McCartan, D. G.
031.254
McCarthy, C. C.
103.016 .107 .120 .124
.125 .126 .128
McCarthy, M. F.
031.232
113.046
McClintock, J.
142.128

McClintock, J. E.
142.089
McClure, R. D.
113.016 .033
McCluskey, G. E.
121.042
McClymont, A. N.
076.011
McCord, T. B.
034.053
092.009
094.227 .438
097.049
098.002 .011 .036
McCormac, B. M.
008.093
McCormick, G. R.
105.106 .114
McCoy, D. G.
094.111
McCoy, J. E.
094.209
McCray, R.
125.013
McCrea, W. H.
003.095
066.026
085.013
159.016
McCrosky, R. E.
103.010 .013 .014 .108
.109 .120 .124 .125
.126 .128
124.110
McCurnin, T. W.
034.013
McCutcheon, W. H.
131.067
McDiarmid, I. B.
084.006
McDonald, F. B.
078.006
099.028 .081
McDonald, L. H.
154.002
McDonnell, J. A. M.
094.180
106.006
McDonough, T. R.
042.048
099.014 .072
McEllin, M.
141.006
McElroy, M. B.
093.025
099.206 .221
McGee, R. X.
132.032
McGimsey, B. Q.
158.305
McGuire, J. P.
072.039
074.051
McHone, J.
105.028 .029
McIlwain, C. E.
099.029 .040
McIntosh, B. A.
082.062
McIntosh, C. B. G.
162.093

McIntosh, P. S.
073.042
080.036
McKay, D. S.
094.147 .184 .425 .463
.476
105.063
McKay, G. A.
094.153 .404
McKay, S. M.
094.405
McKee, C. F.
125.006
141.056
McKenzie, D. L.
076.003
McKibben, R. B.
099.026
143.096
McKibbin, D. D.
099.024
McKinney, W. R.
034.079
McLean, I. S.
082.037
McMillan, J. D.
052.073
McNally, D. M.
003.096
McNamara, B. J.
115.005
McNutt, D. P.
034.035
McPeters, R. D.
036.001
McPherron, R. L.
084.221
106.031
McVittie, G. C.
160.032
162.065
McWhirter, R. W. P.
074.087
076.032
Meaburn, J.
034.014
131.532 .533
132.006 .024
Mead, G. D.
084.228 .229
Meadows, A. J.
003.065
004.002 .050
082.020
091.012
102.007
Means, J. D.
074.130
106.042
Mebold, U.
131.504
Medenbach, O.
094.464
Medrano, R. A.
078.021
Medrano-Balboa, R. A.
074.112
Meegan, C. A.
143.019 .089
Meekins, J. F.
132.016

Scherrer, V.
074.002
Scherrer, V. E.
073.055
074.017
076.017
Schiefer, K.
036.005
Schiefer, U.
036.005
Schiffer III, F. H.
141.604
Schild, R. E.
064.017
Schilizzi, R. T.
158.109
Schilz, W.
033.051
Schindler, K.
084.237
Schjaer-Jacobsen, H.
033.043
Schlegel, R.
162.107
Schlickeiser, R.
142.135
Schlosser, W.
034.087
103.100
122.005
155.009 .049
Schmadebeck, R.
094.148
Schmahl, E. J.
073.040
076.038
Schmid-Burgk, J.
061.034
063.025
064.071 .072
Schmidt, E. G.
034.093
114.040
122.016
Schmidt, K.-H.
061.080
161.004
Schmidt, P. J.
143.041
Schmidt-Kaler, T.
012.013
034.087
113.005
151.036
159.021
Schmidt-Rohr, U.
009.017
Schmitt, H. H.
094.441
Schmitt, R. A.
094.125 .422 .488
Schmoys, J.
031.025
Schmugge, T. J.
082.009
Schneider, A.
014.002
Schneider, E.
094.103 .178
Schneider, M. V.
033.059 .063

Schneider, R.
031.211
Schnopper, H.
142.115
Schnopper, H. W.
034.032
142.112
Schober, H. J.
098.073
Schoembs, R.
114.315
Schoen, J.
009.019
Schoenfelder, V.
061.053
084.414
Scholefield, A. J.
082.017
Scholer, M.
084.202 .401
Scholl, H.
002.010
099.035
Scholz, D.
075.007
Scholz, G.
116.002
Scholz, M.
064.071 .072
Schonfeld, E.
094.123 .151 .428 .499
Schoolman, S.
073.118
Schoolman, S. A.
073.052 .099
Schorn, R. A. J.
093.024 .027
Schramm, D. N.
065.038
066.084
125.015 .057
Schreiber, R.
013.005
066.092
077.030
Schreier, E.
142.115
Schrijver, J.
142.115 .116
Schroeder, D. J.
034.031
Schroeder, H.
031.009
Schroeder, M.
099.031
Schroeder, R.
103.100
Schroeder, W.
084.003
Schroeter, E. H.
073.098
Schroll, A.
072.056
Schubart, J.
098.010
Schubert, G.
094.195 .214 .216 .260
097.004
Schuch, N. J.
141.017

Schuchardt, K. G. H.
082.063
Schuessler, M.
065.029
Schuhmann, S.
094.426 .498
Schukowski, M.
014.005 .009
Schuller, F.
022.047
Schultz, L.
094.494 .524
105.069
Schultz, P. H.
094.114 .236
Schultz, R. B.
072.064
Schultze, W.
051.013
Schulz, M.
084.404
Schupler, B.
099.031 .068 .069
Schurbohm, C.
003.128
Schuster, H. E.
041.011
131.130
132.004
Schutz, B. E.
046.005
052.029 .073
Schwan, H.
041.027
Schwarcz, H. P.
105.070
Schwartz, A. A.
031.208
Schwartz, G.
103.010 .013 .101 .108
.109 .125 .126 .128
124.110
Schwartz, K.
094.131 .216
Schwartz, R. D.
064.020 .054
132.005 .901
Schwarzschild, M.
064.003
071.008
158.090
Schweitzer, E.
010.043
034.091
098.063
120.003
Schwerer, F. C.
094.197 .200 .546
Scialom, G.
083.034
Sciama, D. W.
066.042
Sclar, C. B.
094.447 .567
Scoon, J. H.
094.494
Scott, D. H.
094.212 .434
Scott, E.
105.071

Sweetnam, D.
099.018
Sweetnam, D. N.
099.043 .044 .227
Sweetnam, D. W.
099.032
Sweigart, A. V.
080.007
Swenson Jr., G. W.
033.025
Swider, A.
104.036
Swider, W.
083.056
Swift, D. W.
084.024
Swihart, T. L.
141.022
Swindell, W.
003.063
099.030 .042
Swings, J.-P.
103.127
Swits, G.
094.203
Swope, H. H.
121.085
Sykes, D. E.
103.126
Sykora, J.
074.089
079.100
Synek, I.
103.100
Syrovatskij, S. I.
061.061
073.108
Sysoev, A. G.
105.133
Syunyaev, R. A.
117.032
142.071 .079
158.047 .064
Syzdykov, A. S.
042.020
Szafraniec, R.
121.051
Szalkowski, F. J.
094.542
Szczodrowska-Kozar, B.
042.084
Sze, N. D.
093.025
Szebehely, V.
042.046 .061
081.029
Szecsenyi-Nagy, G.
122.077
Szymanski, W.
075.005

Taam, R. E.
117.031
Tabak, R. G.
131.038
Tabor, J. E.
074.050
Tackett, C. D.
105.138

Taddeucci, A.
094.462
Tademaru, E.
141.335
Tadzhidinov, Kh. G.
022.076
Taff, L. G.
151.009 .016
Taieb, C.
083.034
Tajbo, R.
041.056
Takacs, S.
065.102
Takada, M.
022.902
Takagi, K.
131.120
Takagi, S.
044.012
045.007
Takahara, F.
151.048
Takahashi, H.
143.084
Takahashi, K.
085.023
125.026
Takakura, T.
076.052 .056
077.022 .024
Takeda, H.
094.470
105.075
Talbert, F. D.
153.021
Talbot Jr., R. J.
066.060
151.015
Tallant, P. E.
031.234
Talon, R.
076.057 .064
Talwani, P.
094.550
Talwar, S. P.
062.022
Tamazawa, S.
066.020
Tammann, G. A.
155.044
158.009
162.023 .031
Tanaka, H.
074.003
075.013
Tanaka, T.
094.497
105.060
Tanaka, Y.
061.001 .004 .019
142.087
Tandberg-Hanssen, E.
073.038 .077
074.049
Tandon, S. N.
011.032
Tang, F.
072.025
Tang, J. T.
062.026

Tannenwald, P. E.
033.041
Tanner, R. W.
094.110
Tapia, S.
113.020
141.048
158.311
Tapley, B. D.
046.005
052.029 .073
Tarafdar, S. P.
064.016
080.047
131.029 .139
Taranova, O. G.
099.016
Tarashchuk, V. P.
103.110
Tarasov, V. B.
033.016
Tarbell, T. D.
031.233
080.027
Tarenghi, M.
158.039
160.021
Tarrant, D.
033.048
Tarter, C. B.
008.075
142.089
Tartois, L.
007.000
Taton, R.
004.060
Tatsumoto, M.
094.508
Tavakol, R.
162.020
Tavastsherna, K. N.
031.273
Tayler, R. J.
065.056
154.014
Taylor, D. J.
034.010 .093
Taylor, F. W.
093.020
Taylor, G. E.
097.090
Taylor, G. J.
094.187 .484
Taylor, G. N.
083.016
Taylor, G. R.
034.085,
Taylor, H. C.
094.572
Taylor, H. C. J.
094.480
Taylor, J. H.
141.303 .304 .318 .355
Taylor, L. A.
094.460 .471
Taylor, L. S.
082.029
Taylor, M.
031.218
073.111

Subject Index

Molecules
  Interstellar Matter
    131.003 .004 .005
      .006 .009 .010
      .013 .022 .024
      .027 .028 .030
      .032 .034 .036
      .037 .039 .041
      .042 .045 .047
      .048 .053 .054
      .057 .058 .059
      .060 .061 .065
      .066 .068 .069
      .070 .077 .079
      .084 .090 .112
      .117 .119 .120
      .121 .122 .123
      .124 .125 .128
      .137 .143 .146
      .540
    132.025
    141.610
    155.041
    158.074
Molecules
  Jupiter Atmosphere
    099.080
Molecules
  Meteorites
    105.136
    131.137
Molecules
  Meteors
    104.029
Molecules
  Planetary Atmospheres
    091.030
Molecules
  Spectra
    022.010 .020 .035
      .036 .055 .071
Molecules
  Stellar Atmospheres
    064.065
Molecules
  Stellar Spectra
    114.360 .361
Molecules
  Sunspots
    072.003
Molecules
  Venus Atmosphere
    093.032
Moon
  Albedo
    094.105 .111
Moon
  Chemical Composition
    061.031
    094.113 .132 .406
    107.009
Moon
  Convection
    094.274
Moon
  Craters
    092.020
    094.117 .225 .229
      .239 .243 .258

Moon
  Density
    094.112 .222
Moon
  Dynamics
    012.003
Moon
  Electric Conductivity
    094.131 .228 .244
Moon
  Element Abundances
    094.235 .249 .402
Moon
  Evolution
    094.011 .129
Moon
  Global Properties
    012.003
Moon
  Gravity
    094.112 .223 .242
      .267
Moon
  Impacts
    094.116
Moon
  Infrared Radiation
    114.329
Moon
  Interior
    094.010 .102 .115
      .136 .137 .244
Moon
  Laser Observations
    094.001
Moon
  Laser Ranging Stations
    094.221
Moon
  Local Properties
    012.003
Moon
  Magnetic Field
    074.078
    094.104 .107 .228
      .234
Moon
  Mare Origin
    094.109 .238
Moon
  Mascons
    094.223 .242
Moon
  Meteorite Impact
    094.103 .408
Moon
  Models
    094.012
Moon
  Orbit
    094.004
Moon
  Origin
    094.008 .132
    107.002 .007 .012
Moon
  Potential
    094.012
Moon
  Radio Radiation
    094.130

Moon
  Regolith
    094.101 .248 .569
Moon
  Rocks
    094.401 .406 .413
      .554 .555 .557
      .559 .561 .574
Moon
  Samples
    015.012
    094.118 .240 .404
      .405 .407 .411
      .412 .419 .553
      .554 .556 .558
Moon
  Seismicity
    094.114 .115 .236
Moon
  Soil
    094.222 .402 .408
      .410 .572
Moon
  Solar Wind
    074.040 .078 .111
    094.260 .431
Moon
  Surface Structures
    094.108 .225 .275
Moon
  Thermal History
    094.011 .110 .131
Moon
  Viscosity
    094.237
Moon
  Volcanism
    094.230
Moon Dynamics
    094.000
Moon Global Properties
    094.100
Moon Local Properties
    094.400
Moon Surface
    094.135 .224 .233
      .259
Multiple Galaxies
    158.000
Multiple Stars
    117.000
Multiple Stars
  Kinematics
    117.006

N-Body Problem
    042.056
    066.046
Navigation
    046.000
Navigation
  Space Vehicles
    052.000
Nearby Dwarfs
  Parallaxes
    111.002
    126.010
Nearby Dwarfs
  Photometry
    111.002

ASTRONOMY AND ASTROPHYSICS ABSTRACTS

A Publication of the
Astronomisches Rechen-Institut Heidelberg
Member of the Abstracting Board
of the International Council of Scientific Unions

Editors:
S. Böhme, U. Esser, W. Fricke, U. Güntzel-Lingner. I. Heinrich,
F. Henn, D. Krahn, L. Schmadel, H. Scholl, G. Zech

Vol. 1   Literature 1969, Part 1, X + 435 pp. (1969)
Vol. 2   Literature 1969, Part 2, X + 516 pp. (1970)
Vol. 3   Literature 1970, Part 1, X + 490 pp. (1970)
Vol. 4   Literature 1970, Part 2, X + 562 pp. (1971)
Vol. 5   Literature 1971, Part 1, X + 505 pp. (1971)
Vol. 6   Literature 1971, Part 2, X + 560 pp. (1972)
Vol. 7   Literature 1972, Part 1, X + 526 pp. (1972)
Vol. 8   Literature 1972, Part 2, X + 594 pp. (1973)
Vol. 9   Literature 1973, Part 1, X + 610 pp. (1973)
Vol. 10   Literature 1973, Part 2, X + 661 pp. (1974)
Vol. 11   Literature 1974, Part 1, X + 579 pp. (1974)
Vol. 12   Literature 1974, Part 2, X + 699 pp. (1975)
Vol. 13   Literature 1975, Part 1, X + 632 pp. (1975)